ORGANIC CHEMISTRY

2nd Edition

Michael B. Smith
University of Connecticut
Storrs, CT

Contributing Edito
James R. Hermanson
Indiana University/Pur
Indianap

An Imp

An American BookWorks Corporation Production

HarperCollins books may be purchased for educations, business, or sales promotional use. For information please write: Special Markets Department, HarperCollins Publishers, 10 East 53rd Street, New York, NY 10022.

Library of Congress Catalog-in-Publication has been applied for.

BN: 13: 978-0-06-088154-2
: 10: 0-06-088154-2

09 10 CW 10 9 8 7 6 5 4 3 2 1

Finally, I must thank my wife Sarah and my son Steven for their patience, understanding and support during the preparation of this manuscript.

Michael B. Smith
Storrs, Connecticut
September, 2005
michael.smith@uconn.edu

Periodic Table of Elements

Key:

element name **atomic number** element symbol 1997 atomic weight (mean relative mass)

1	2		3	4	5	6	7	8	9	10	11	12	13	14	15	16	17	18
hydrogen 1 H 1.00794(7)																		helium 2 He 4.002602(2)
lithium 3 Li 6.941(2)	beryllium 4 Be 9.012182(3)												boron 5 B 10.811(7)	carbon 6 C 12.0107(8)	nitrogen 7 N 14.00674(7)	oxygen 8 O 15.9994(3)	fluorine 9 F 18.9984032(5)	neon 10 Ne 20.1797(6)
sodium 11 Na 22.989770(2)	magnesium 12 Mg 24.3050(6)												aluminium 13 Al 26.981538(2)	silicon 14 Si 28.0855(3)	phosphorus 15 P 30.973761(2)	sulfur 16 S 32.066(6)	chlorine 17 Cl 35.4527(9)	argon 18 Ar 39.948(1)
potassium 19 K 39.0983(1)	calcium 20 Ca 40.078(4)		scandium 21 Sc 44.955910(8)	titanium 22 Ti 47.867(1)	vanadium 23 V 50.9415(1)	chromium 24 Cr 51.9961(6)	manganese 25 Mn 54.938049(9)	iron 26 Fe 55.845(2)	cobalt 27 Co 58.933200(9)	nickel 28 Ni 58.6934(2)	copper 29 Cu 63.546(3)	zinc 30 Zn 65.39(2)	gallium 31 Ga 69.723(1)	germanium 32 Ge 72.61(2)	arsenic 33 As 74.92160(2)	selenium 34 Se 78.96(3)	bromine 35 Br 79.904(1)	krypton 36 Kr 83.80(1)
rubidium 37 Rb 85.4678(3)	strontium 38 Sr 87.62(1)		yttrium 39 Y 88.90585(2)	zirconium 40 Zr 91.224(2)	niobium 41 Nb 92.90638(2)	molybdenum 42 Mo 95.94(1)	technetium 43 Tc [98.9063]	ruthenium 44 Ru 101.07(2)	rhodium 45 Rh 102.90550(2)	palladium 46 Pd 106.42(1)	silver 47 Ag 107.8682(2)	cadmium 48 Cd 112.411(8)	indium 49 In 114.818(3)	tin 50 Sn 118.710(7)	antimony 51 Sb 121.760(1)	tellurium 52 Te 127.60(3)	iodine 53 I 126.90447(3)	xenon 54 Xe 131.29(2)
caesium 55 Cs 132.90545(2)	barium 56 Ba 137.327(7)	57-70 *	lutetium 71 Lu 174.967(1)	hafnium 72 Hf 178.49(2)	tantalum 73 Ta 180.9479(1)	tungsten 74 W 183.84(1)	rhenium 75 Re 186.207(1)	osmium 76 Os 190.23(3)	iridium 77 Ir 192.217(3)	platinum 78 Pt 195.078(2)	gold 79 Au 196.96655(2)	mercury 80 Hg 200.59(2)	thallium 81 Tl 204.3833(2)	lead 82 Pb 207.2(1)	bismuth 83 Bi 208.98038(2)	polonium 84 Po [208.9824]	astatine 85 At [209.9871]	radon 86 Rn [222.0176]
francium 87 Fr [223.0197]	radium 88 Ra [226.0254]	89-102 **	lawrencium 103 Lr [262.110]	rutherfordium 104 Rf [261.1089]	dubnium 105 Db [262.1144]	seaborgium 106 Sg [263.1186]	bohrium 107 Bh [264.12]	hassium 108 Hs [265.1306]	meitnerium 109 Mt [268]	ununnilium 110 Uun [269]	unununium 111 Uuu [272]	ununbium 112 Uub [277]		ununquadium 114 Uuq [289]				

*lanthanides	lanthanum 57 La 138.9055(2)	cerium 58 Ce 140.116(1)	praseodymium 59 Pr 140.90765(2)	neodymium 60 Nd 144.24(3)	promethium 61 Pm [144.9127]	samarium 62 Sm 150.36(3)	europium 63 Eu 151.964(1)	gadolinium 64 Gd 157.25(3)	terbium 65 Tb 158.92534(2)	dysprosium 66 Dy 162.50(3)	holmium 67 Ho 164.93032(2)	erbium 68 Er 167.26(3)	thulium 69 Tm 168.93421(2)	ytterbium 70 Yb 173.04(3)
**actinides	actinium 89 Ac [227.0277]	thorium 90 Th 232.0381(1)	protactinium 91 Pa 231.03588(2)	uranium 92 U 238.0289(1)	neptunium 93 Np [237.0482]	plutonium 94 Pu [244.0642]	americium 95 Am [243.0614]	curium 96 Cm [247.0703]	berkelium 97 Bk [247.0703]	californium 98 Cf [251.0796]	einsteinium 99 Es [252.0830]	fermium 100 Fm [257.0951]	mendelevium 101 Md [258.0984]	nobelium 102 No [259.1011]

Contents

Preface

Although Organic Chemistry is defined as the chemistry of carbon, a typical two semester course focuses on the bonding of carbon compounds, manipulation of functional groups and formation of carbon-carbon bonds for these molecules. Perhaps the most difficult part of an Organic Chemistry course is discovering what one does not know, *before* the exam. The only way to do this is to use study problems (homework) to test yourself. For any given textbook, there are only so many problems of a given type for a given topic. The purpose of this book is to provide a 'book of questions' which can be used as an adjunct to most currently used Organic Chemistry textbooks. There are many examples of directly worded 'reaction questions' where the student must give a specific structure as the product of a specific reaction. There are also several synthesis type problems.

This book is organized into 29 chapters that roughly parallel the most popular books currently in use. Rather than using a lengthy discussion followed by some questions, as in other books of this type, the material is taught completely in the form of questions. Several leading, discussion type questions are asked in order to present the necessary background information for a given topic. Other questions build upon this base and short answer questions punctuate the information. In many cases, the questions are short and obvious extensions of preceding questions. This tactic is used to reinforce key concepts. Where possible, complete mechanisms are included for pertinent and important reactions, presented in a question-and-answer format.

It is hoped that this book of questions will prove useful to students taking Organic Chemistry for the first time, who need extra practice. It should also be useful for those students who must 'refresh their knowledge of Organic Chemistry for pre-medical, pre-veterinary or pre-pharmacy exams. Graduate students who are studying for preliminary or orientation exams in Organic Chemistry, for entrance into a graduate program, should also find this book useful. The 'all question' format of this book should facilitate a 'self-guided' tour to review Organic Chemistry, or assist in studying for exams when taking the course for the first time. It should be a useful adjunct for many of the textbooks now used to teach Organic Chemistry. It is offered with these goals in mind.

I thank Mr. Fred Grayson of American BookWorks, who coordinated this project.

All manuscript structures and the proton NMR structures in Chapter 18 were drawn using CSC ChemDraw Ultra™ (Serial Number 551713). The infrared diagrams were taken from the Sadtler Indices. The picture of the Dean-Stark apparatus in the end-of chapter problems was drawn with Canvas 3.0™ (User Number 4032160730). The 3D drawings were rendered with MacSpartan v2.0.

CHAPTER 1

Atomic Orbitals and Bonding

Organic chemistry is the science that studies molecules containing the element **carbon**. Carbon can form bonds to other carbon atoms or to a variety of atoms in the Periodic table. The most common bonds observed in the first organic chemistry course are C–C, C–H, C–O, C–N, C-halogen (Cl, Br, I), C–Mg, C–B, C–Li, C–S and C–P.

This chapter will introduce the carbon atom and the covalent bonds that join carbon atoms together in organic molecules. The most fundamental properties of atoms and of covalent bonds will be introduced, including hybridization, electronic structure, and a brief introduction to using molecular orbital theory for bonding.

1.1. ATOMIC ORBITALS

What Is the Model Structure of an Atom?

Each atom of a given element possesses a fixed number of protons, neutrons and electrons. The protons and neutrons comprise the nucleus and the electrons are located in discrete energy levels (quanta) surrounding the nucleus. The nucleus is electrically positive while electrons are electrically negative. When carbon forms a covalent bond (two electrons are in each bond, represented by C–X, where X is any atom; see Section 2.3), it uses electrons from the outermost (valence) shell. These electrons are conveniently described by their 'shape' and distance relative to the nucleus.

What Are Atomic Orbitals and What Are Molecular Orbitals?

The space occupied by electrons is described by the term orbital. Different orbitals are described by their distance from the nucleus (the energy required to hold the electron) as well as the three-dimensional configuration of their electrons. Electrons associated with the atom of a free element are said to be in **atomic orbitals**. Once bonds have been formed, the atomic elements become part of molecules, and the electronic positions are described by **molecular orbitals**.

What Is the Source of the Orbital Pictures We Use?

Using the Schrödinger equation and making certain assumptions about the number of particles and constraints on the location of those particles in space, solutions are generated for each atom as increasing numbers of protons, neutrons, and electrons are added. The position of each electron is defined at a particular energy level relative to the nucleus. When this data is converted to a three-dimensional expression of the relative position of the electron(s), the familiar pictures of s, p, d, and f orbitals emerge.

S Orbitals and P Orbitals

What Is An s Orbital?

For the elements hydrogen (H) and helium (He), electrons reside in a **spherically symmetrical orbital** at a discrete distance from the nucleus. This corresponds to the first quantum level. All spherically symmetrical orbitals are referred to as **s orbitals** and have the general shape of ***1.1***. The **1s orbital** represents the first energetically favorable level where electrons can be held by the nucleus This results from the electrostatic attraction of the positive nucleus and the negatively charged electron(s).

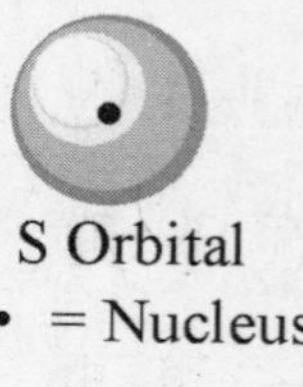

S Orbital
• = Nucleus
1.1

What Is A p Orbital?

Beginning in the second row, there is a second energy level for elements consisting of a single 2s level followed by a 2p level composed of three identical p orbitals. A p orbital is 'dumb-bell' shaped (as in ***1.2***) with electron density on either side of the nucleus having a point of zero electron density between the electron lobes.

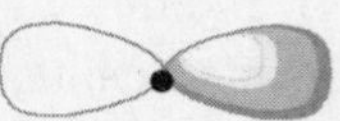

P Orbital
• = Nucleus

1.2

What Is a Node?

A node is a point of zero electron density.

Do the Electrons in a P Orbital Migrate from One Lobe to the Other?

No! The picture of the p orbital represents the probability of finding the electrons at a specific location in space. The diagram shows an equal probability of finding the electrons above and below the node. Therefore, the electrons are found in the **entire p orbital** (*both* lobes), and the diagram simply indicates the probability of finding electrons at any specific location.

How Many Orbitals Are There in Each Valence Shell?

Each orbital can hold two electrons. For the first valence shell containing H and He, there is one s orbital. For the next valence shell (including, for example, B, C, N, O, F), there is one 2s orbital, and also three 2p orbitals. The 2p orbitals have different spatial orientations, coincident to the x, y and z axes of a three-dimensional coordinate system. The three p orbitals are therefore labeled p_x, p_y and p_z.

Electron Configuration

What Is Electronic Configuration?

As each new orbital (energy level) moves further out from the nucleus, the electrons are held less tightly. Each orbital can hold a maximum of two electrons (as in ***1.3***). In addition, each set of orbitals will contain different numbers of electrons (two electrons for s orbitals, six electrons for three p orbitals and ten electrons for five d orbitals). Orbitals will fill with electrons from lowest-energy to highest-energy orbital, according to the order shown in ***1.4***.

H $\underline{\uparrow}$ $1s^1$ He $\underline{\uparrow\downarrow}$ $1s^2$

1.3

FILLING ORDER →				
1s	2s	2p	3s	3p
4s	3d	4p	5s	4d
5p	6s	4f	5d	6p

1.4

What Is the AUFBAU Principle?

Orbitals 'fill' according to the Aufbau Principle. Orbitals in a sublevel s, p or d will contain one electron before any contain two. Orbitals containing two paired electrons will have opposite spins and are said to be spin-paired, $\underline{\uparrow\downarrow}$.

Describe the Order in Which Orbitals Will Be Filled with Electrons through the 2P Level. Ignore the 1S and 2S Levels.

The order for the 2p orbitals will be $2p_x \rightarrow 2p_y \rightarrow 2p_z \rightarrow 2p_x \rightarrow 2p_y \rightarrow 2p_z$:

$\underline{\uparrow}\ \_\ \_ \rightarrow \underline{\uparrow}\ \underline{\uparrow}\ \_ \rightarrow \underline{\uparrow}\ \underline{\uparrow}\ \underline{\uparrow} \rightarrow \underline{\uparrow\downarrow}\ \underline{\uparrow}\ \underline{\uparrow} \rightarrow \underline{\uparrow\downarrow}\ \underline{\uparrow\downarrow}\ \underline{\uparrow} \rightarrow \underline{\uparrow\downarrow}\ \underline{\uparrow\downarrow}\ \underline{\uparrow\downarrow}$

Molecular Orbitals

What Is the Difference between a Molecular Orbital and an Atomic Orbital?

Once two atoms are joined in a covalent bond, their electrons are in a different position relative to the nuclei of the two atoms. This energy level is different from that in the atom, and the pertinent orbitals are referred to as **molecular orbitals** rather than atomic orbitals.

Linear Combination of Atomic Orbitals

How Are Molecular Orbitals Formed from Atomic Orbitals?

To form a molecular orbital, a mathematical device known as LCAO (Linear Combination of Atomic Orbitals) is used. Taking diatomic hydrogen (H_2) as an example, the atomic orbitals of two hydrogen atoms are 'mixed' to form the molecular orbitals of H_2 (***1.5***). Each hydrogen atomic orbital (H A.O.) contains one electron in the 1s orbital. Each of these orbitals is at the same energy level. When they mix to form the molecular orbital, the electrons are in a 'different' position relative to the nuclei, and their energy is different. If two atomic orbitals mix, two molecular orbitals are generated: one higher and one lower in energy than the original atomic orbitals.

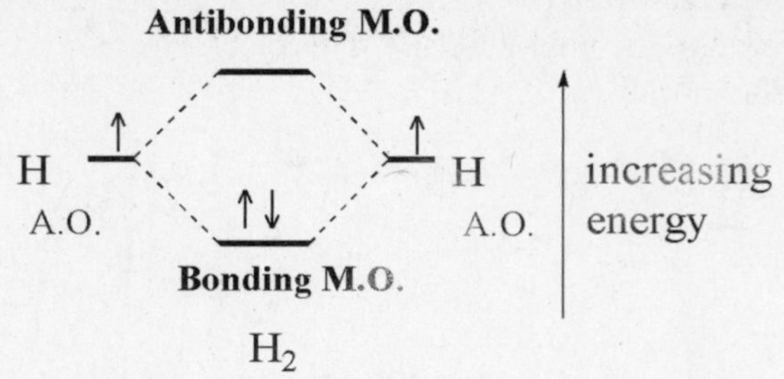

Antibonding M.O.

H A.O. ↑ — ↑↓ — ↑ H A.O. increasing energy

Bonding M.O.

H_2

1.5

Describe Bonding and Antibonding Molecular Orbitals.

When two molecular orbitals (M.O.) are formed by mixing the atomic orbitals, one will be formed at a higher and one at a lower energy level, as in ***1.5***. Two electrons will go into the lowest available energy level before the higher-level orbital is filled. These two electrons will again be spin-paired to minimize energy. A covalent bond requires two electrons, and the electrons in the lowest molecular orbital are used to form this bond. This molecular orbital is called the **bonding molecular orbital**. The higher orbital does not contain electrons, but becomes the next available quantum level if extra electrons are added to the system or if a lower-energy electron is energetically 'promoted' to the higher level. This potential site of electron density is called the **antibonding molecular orbital**. In the ground state (normal energies), the bonding M.O. is filled while the antibonding M.O. is empty.

1.2. BONDING

Ionic Bonding

What Is a Lewis Dot Structure?

A Lewis electron dot formula simply uses dots for electrons, where each bond is represented by two dots between the appropriate atoms, and unshared electrons are indicated by dots (1 or 2) on the appropriate atom.

Draw Lithium Fluoride As a Lewis Dot Structure.

The Lewis dot structure is shown in ***1.6***.

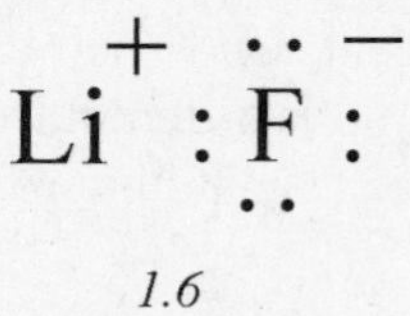

$$\mathrm{Li}^{+}\ \ :\ddot{\underset{\cdot\cdot}{\mathrm{F}}}:^{-}$$

1.6

What Is an Ionic Bond?

An ionic bond occurs when two atoms are held together by electrostatic forces. Sodium chloride (NaCl), for example, exists in the solid state as Na^+Cl^-.

Why Does Na in NaCl Assume a Positive Charge?

If the valence electrons for each atom are represented as dots (one dot for each electron), the structure for NaCl will be ***1.7***. All of the electrons are on chlorine (Cl), and none are on sodium (Na). There is special stability associated with a filled shell (i.e., the noble gases: He, Ne, Ar, etc.). In order for sodium (Na) with the electronic configuration $1s^2 2s^2 2p^6 3s^1$ to achieve a 'filled' shell, it can either lose one electron (**Ionization Potential**) to mimic the Ne atom or gain seven electrons to mimic the Ar atom. (The ability to gain one electron is called **Electron Affinity**.) The loss of one electron requires much less energy than gaining seven. Loss of the electron leads to a positive charge on the remaining atom (see Formal Charge in Section 2.4).

$$\overset{+}{Na}\;:\ddot{\underset{\cdot\cdot}{Cl}}:^{-}$$

1.7

Why Does Chlorine Assume a Negative Charge in NaCl?

Chlorine (Cl) with the electronic configuration $1s^2 2s^2 2p^6 3s^2 3p^5$ can either gain one electron to mimic the Ar atom or lose seven electrons to mimic the Ne atom. Clearly, the loss of seven electrons will require a great deal of energy. Energetically, it is far easier for Cl to gain an electron, which leads to the formation of a negatively charged atom. The strong electrostatic attraction between the positively charged sodium and the negatively charged chlorine binds the two atoms together in an **ionic bond**. The chlorine will be the negative 'pole,' since it is the more electronegative atom.

Covalent Bonding

Define a Covalent Bond.

A covalent bond is usually composed of two electrons that are shared by two atoms. In the case of hydrogen (H_2), this is represented as H:H or H–H, where the (:) or the (−) indicates the presence of two electrons with the bulk of the electron density localized between the hydrogen nuclei. This type of bond usually occurs when the atom cannot easily gain or lose electrons. (Another way to view this is that there is a small electronegativity difference between atoms.)

Draw Diatomic Hydrogen As a Lewis Dot Structure.

The Lewis dot structure of diatomic hydrogen is ***1.8***.

$$H : H$$

1.8

Draw a Carbon-Carbon Single Bond As a Lewis Dot Structure, Ignoring All Other Electrons.

This particular bond is shown in ***1.9***.

$$C : C$$

1.9

Describe the US System and the IUPAC System for Group Number in the Periodic Table.

The US system is the older system in which H and Li are in Group 1A, Be is 2A, B is 3A, C is 4A, out to He and Ne which are 8A. The transition metals, Sc through Ni, for example, are in Groups 3B → 8B. In this system, Cu is in Group1B and Zn in 2B.

In the IUPAC (International Union of Pure and Applied Chemistry), group numbers begin with 1 for H, Li, etc., and end with 18 for the noble gases (He, Ne, etc.). In this system, the transition metals Sc → Zn are in Groups 3 → 12. Boron is in Group 13, C in Group 14, N in Group 15 and O in Group 16.

What Is Valence?

Valence is usually defined as the number of bonds an atom can form to satisfy the octet rule and remain electrically neutral. This is not to be confused with valence electrons, which involve the number of electrons in the outermost shell. For convenience, we will use the US system for Group number, since this correlated directly with the number of valence electrons for the second row. In the second row, the valence is the same as [8- the group number] (4 for carbon, 3 for nitrogen, 2 for oxygen and 1 for fluorine). Boron, however, is an exception. There are only three valence electrons and, therefore has the possibility of forming only 3 covalent bonds and still remaining neutral. The valence of boron is thus 3, but it is electron deficient, making it a Lewis acid.

Why Does Carbon Have a Valence of Four?

With carbon ($C1s^22s^22p^2$), there are four electrons in the outermost or valence shell (n=2). The gain or loss of four electrons would require a prohibitively high amount of energy. However, carbon readily forms bonds with many different elements, including other carbon atoms, by sharing electrons. In other words, carbon will form covalent bonds to other carbon atoms, hydrogen and oxygen, as well as many other elements in the Periodic Table. For example, each carbon in structure ***1.10*** shares eight electrons. Four electrons from one carbon are shared with one electron from each of three hydrogens and one electron from another carbon. The other carbon similarly has eight shared electrons. In a covalent bond, the electrons are mutually shared between the nuclei, so each nucleus has a filled outer shell (eight in the case of carbon and two in the case of hydrogen). The most common way to show mutual sharing of electrons for two carbon atoms is to draw a single line between the two atoms (C–C) rather than using the Lewis dot structure shown for ***1.10***. The two electrons are equally distributed between the two carbon atoms, as shown in ***1.11***. Covalent bonds can occur between many atoms other than carbon or hydrogen. This concept is shown for two atomic nuclei with the electrons distributed between each nucleus (***1.12***).

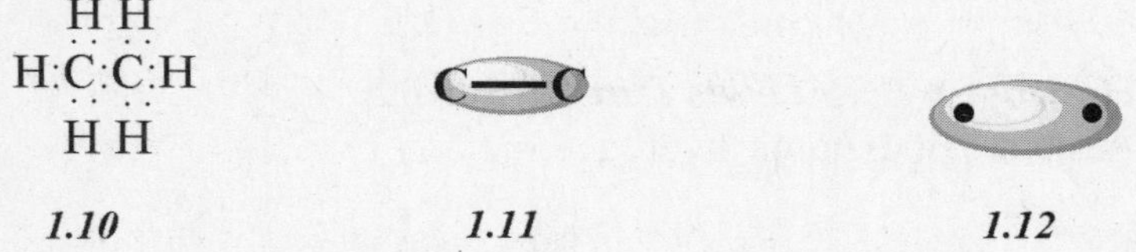

1.10 ***1.11*** ***1.12***

Polar Covalent Bonds

What Is Electronegativity?

Electronegativity is a measure of the attraction that an atom has for the bonding pair of electrons in a covalent bond.

How Is the Electron Density between the Nuclei of Two Atoms Affected If the Two Atoms Are Part of a Covalent Bond, and One Atom Is More Electronegative Than the Other?

The shared electron density is distorted towards the more electronegative atom, rather than being symmetrically distributed represented by ***1.11*** and ***1.12***.

Is the C–H Bond Considered to Be Polarized?

No! Although C and H have different electronegativities (H = 2.1 and C = 2.5 on the Pauling electronegativity scale), the C–H bond is not considered to be polarized. This assumption is based on the polarity of molecules containing only C–H bonds, as well as the chemical reactivity of molecules containing only C–H bonds.

What Is a Polarized Covalent Bond?

When a bond is formed between two atoms that are not identical, the electrons do not have to be equally shared. Electronegativity is the ability of an atom to attract electrons to itself. If one atom is more electronegative than another, then the higher the electronegativity and the greater the share of electrons from the bond that will be pulled towards it. The higher the propensity of an atom to attract electrons, the greater the distortion of electron density in the covalent bond towards that atom. The result is a 'polarized' covalent bond (***1.13***), in which polarization is represented by distortion of electron density towards the more electronegative atom. Since an electronegative atom has more electron density relative to the other, that atom will be 'more negative.' This distortion of electron density leads to a 'partial charge,' represented by δ^+ at the atom with the least electron density and δ^- at the atom with the most electron density.

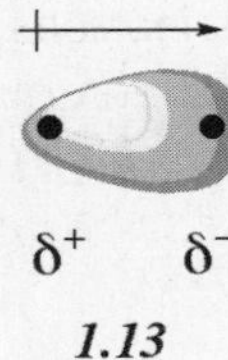

1.13

What Symbols Are Used to Represent the Dipole of a Polarized Covalent Bond?

One way to represent this distortion of electrons is with the symbol $+\!\!\!\longrightarrow$, with the + representing the positive atom and $\rightarrow$ the negative. It is also common to use a δ^+ at the atom with the least electron density and δ^- at the atom with the most electron density. The covalent bond is polarized, leading to a dipole moment. Any covalent bond between two atoms where electrons are unequally shared is called a polar covalent bond.

Which of the following Are Polar Covalent Bonds? For the Polarized Bonds, Identify the Negative and Positive Poles.

(A) C–O (B) C–C (C) O–H (D) H–H (E) Br–Br (F) C–N (G) *N*–N (H) C–Li (I) O–O (J) H–Br (K) Na–Cl

Only those bonds between dissimilar atoms will be polarized; therefore (a), (c), (f), (h) and (j) are polarized covalent. NaCl (k) is an ionic bond because the electronegativity difference is very large. The negative poles will be oxygen in (a) and (c), nitrogen in (f), carbon in (h) (carbon is more electronegative than lithium) and bromine in (j). Note that the positive pole can be a variety of atoms (C, H, Li).

Dipole-Dipole Interactions

What Is Van der Waal's Attraction?

When there are no polarizing atoms in the molecules, the only attraction between molecules results from the approach of the two clouds of electrons; one molecule approaches the other. When two atoms or nonpolar molecules approach, there is a temporary change in dipole moment due to a brief shift of orbital electrons to one side of one atom or molecule, which creates a similar shift in the adjacent atom or molecule (an induced dipole). This instantaneous dipole leads to a weak attractive force that is known as **Van der Waal's attraction**, also sometimes called **London forces**.

What Causes Two Molecules with C–O Or C=O Bonds to Associate Together in the Liquid Phase?

When two molecules, each with a polarized covalent bond, come into close proximity, the charges for one bond will be influenced by the charge on the adjacent molecule. In the example shown (***1.10***), the negative oxygen of one molecule is attracted to the positive carbon of the second molecule. Likewise, the positive carbon of that molecule is attracted to the negative oxygen of the other. This intermolecular electrostatic interaction is referred to as a **dipole-dipole interaction**. The net result of this interaction is that these molecules will be associated together, and some energy will be required to disrupt this association. The greater the dipole moment, the stronger the interaction and the greater the energy required to disrupt the molecules. The model used for ***1.14*** applies, more or less, to any molecule that contains a bond between carbon and an atom other than C or H.

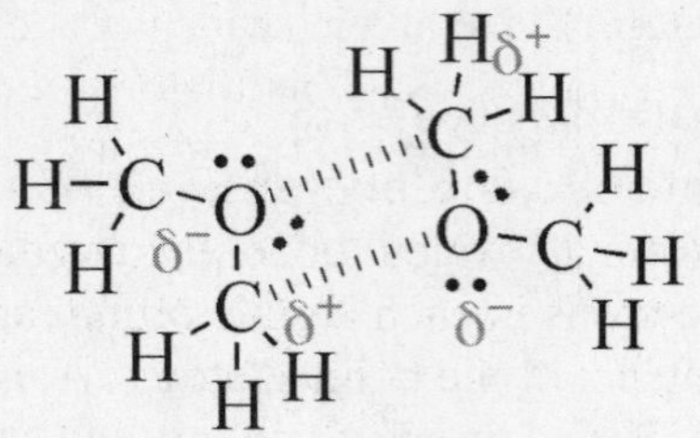

1.14

What Is a Heteroatom?

A heteroatom is defined as any atom other than carbon or hydrogen. Examples are O, N, S, P, Cl, Br. F. Mg, Na, etc.

How Does the Physical Size of the Groups Attached to the Heteroatom Influence Dipole-Dipole Interactions?

The electrostatic interaction of the two dipolar molecules is diminished by the physical size imposed by the carbon groups. The groups compete for the same space in what is called **steric hindrance**, and they repel each other, counteracting the electrostatic attraction to some extent.

Hydrogen Bonds

Why Is the Attraction between Two Groups Bearing an O–H Group Stronger Than between Two Groups Bearing AC=O Group?

When hydrogen forms a polar covalent bond with heteroatoms, which are atoms other than carbon or hydrogen (the most common are O–H, *N*–H, S–H), the hydrogen takes the δ^+ charge of the dipole. Since the classical Brønsted acid is H^+, a polarized hydrogen in O–H is somewhat acidic because it has some positive character. (This will be discussed in Sections 6.3 and 11.1.) The increased acidity and bond polarity lead to a stronger interaction with a negative heteroatom or unshared pair of electrons when brought into close proximity with the polarized hydrogen. This interaction is significantly stronger than a normal dipole-dipole interaction and is referred to as a **hydrogen bond** (***1.15***). The space occupied by an oxygen atom and a hydrogen atom is significantly less than that observed with the carbon groups in a dipole-dipole interaction (see structure ***1.14***). Since steric hindrance is less, the electrostatic interaction is much stronger and dominates, so the hydrogen bond is stronger.

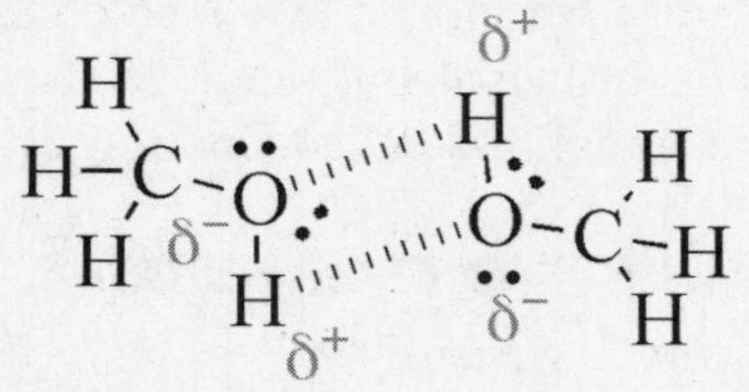

1.15

Which Is Stronger, a Dipole-Dipole Interaction or a Hydrogen Bond?

The attraction between a polarized hydrogen atom and a heteroatom in a hydrogen bond is generally significantly stronger than the dipole-dipole attraction between the two atoms in a polarized covalent bond (not involving H) on two different molecules.

1.3. HYBRIDIZATION

Molecular Orbital Theory

Give a Brief Working Definition of Molecular Orbital Theory As It Is Applied to Covalent Bonds in Simple Diatomic Molecules Such As Hydrogen.

When two atoms combine to form a covalent bond, the atomic orbitals of each atom are combined to form a molecular orbital. The electrons in each atomic orbital are transferred from energy levels near the atom to the space between the nuclei of the bonded atoms. A useful device for tracing this process is called **Molecular Orbital Theory**. In the simplest version, the atomic orbitals of each 'free' atom are mixed to form molecular orbitals. An example was shown above in ***1.5*** for diatomic hydrogen (H_2) using the Linear Combination of Molecular Orbitals. For molecules containing more electrons than hydrogen, and for those containing electrons in other than s orbitals, the diagram is more complex.

What Are 'Core' Electrons?

Core electrons are electrons found in closed shells within the valence shell and *are not involved in covalent bonding.*

What Are Valence Electrons?

Only those electrons in the outermost electronic shells (**the valence electrons**) are involved in covalent bonds. In the case of carbon-carbon single bonds, the two 2s orbitals combine to give two molecular orbitals. The three identical p-orbitals – orbitals with the same energy are referred to as **degenerate orbitals** – combine to give six molecular orbitals. The total number of orbitals remains constant, but the molecular orbitals are split into high-energy and low-energy components. The electrons cannot occupy the same energy in the molecule as they did in the atom. Since the energies must be different and since each orbital set cannot be of the same energy, one is higher and one is lower in energy relative to the original atomic orbitals. The orbitals are symmetrically split, as shown in ***1.12***.

Give the Molecular Orbital Diagram for Diatomic Carbon (C–C).

Atomic carbon has an electronic configuration $1s^2 2s^2 2p^2$. The molecular orbital diagram is shown in ***1.16***.

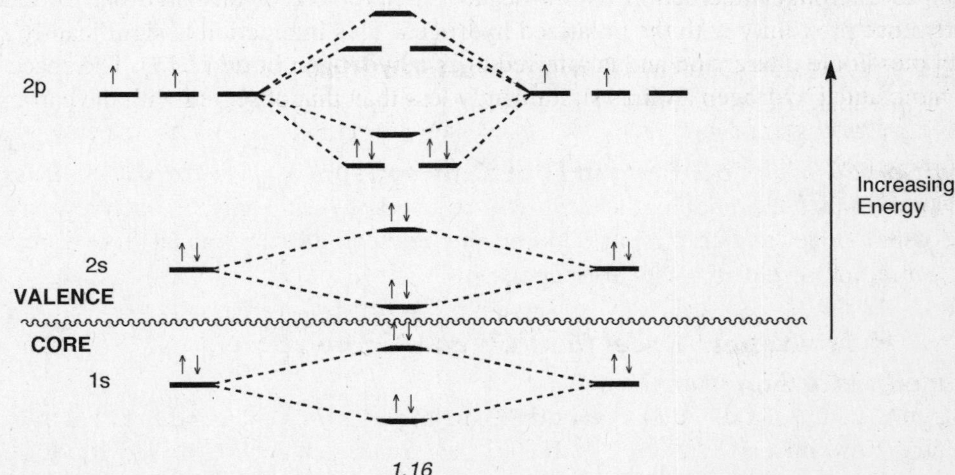

1.16

Explain Why the Molecular Orbital Diagram for Diatomic Carbon Is Incorrect.

The molecular orbital diagram ***1.16*** for carbon (C–C bond) suggests that there is more than one type of bond. The 2s molecular orbitals combine to form one type of bond; the p-orbitals combine to form another type of bond. **This is not correct**. In fact, it is known from many years of experiments that each carbon atom forms four identical bonds to other carbon atoms or to hydrogen atoms. Based on what we know to be the correct bonding in carbon compounds, the molecular orbital diagram does **not** predict the correct bonding. A new model is required. In this new model, it is recognized that all the bonds are equal in the final molecule. This requires four identical bonds in the valence shell of carbon.

sp^3 Hybridization

What Is Hybridization As Applied to a Covalently Bound Atom?

The atomic orbitals of an atom can be rearranged prior to bond formation using the Linear Combination of Molecular Orbitals model. When the atomic orbitals are rearranged, they mix and form hybrids, which are used for bonding. For carbon and other elements of the second row, the hybridization is limited to mixing one 2s orbital and one or more of the three 2p orbitals.

How Are Hybrid Orbitals Different When Using Different Numbers of 2p-Orbitals?

There are three basic types of hybridization: sp^3, sp^2 and sp^1 (or just sp). In each case, the sp refers to the hybridization of the atom. The superscript indicates the number of p orbitals used to form the hybrid combination with the 2s orbital.

Define an sp^3 Hybrid Orbital.

In sp^3 hybridization, all three p orbitals are mixed with the s orbital to generate four new hybrids that can form four identical covalent bonds.

Define an sp^2 Hybrid Orbital.

In sp^2 hybridization, two p orbitals are mixed with the s orbital to generate three new hybrids that can form three identical covalent bonds. The 'unused' p orbital will participate in π type bonding (see Chapter 8.1).

Define an sp Hybrid Orbital.

In sp hybridization, one p orbital is mixed with the s orbital to generate two new hybrids that can form two covalent bonds. The two 'unused' p orbitals will participate in two, mutually perpendicular π type bonds (see Chapter 8).

Can One Gain or Lose Orbitals During Hybridization?

No! The number of orbitals must be preserved in the mixing process.

Is the Hybridization Model Rigid in That Only sp^3, sp^2 or sp Hybrids Are Possible?

No! If d orbitals are available, the atom will adjust its hybridization in such a way as to form the strongest possible bonds and keep all its bonding and lone-pair electrons as far from each other as possible, thereby minimizing electron-electron repulsion.

What Is the Hybridization Model for a Carbon-Carbon Bond in a Tetrahedral Carbon Atom?

The device used to 'fix' the molecular orbital diagram for the C–C bond is called **hybridization**. In this model, the 2s orbital and all three 2p orbitals of each carbon atom are 'mixed' to form four, identical

sp^3 hybrid atomic orbitals. When these identical orbitals – they are the same energy and, therefore, degenerate – from two 'hybridized' carbon atoms are mixed to form a molecular orbital, the core molecular orbitals remain the same; however, the covalent orbitals now show four identical bonds, with two electrons per bond, as in ***1.17***. Since the correct answer was known in advance, the modified model must give the correct answer. This hybridization model is used extensively to correlate bonding in organic molecules and to make predictions concerning reactions, bonding and geometry.

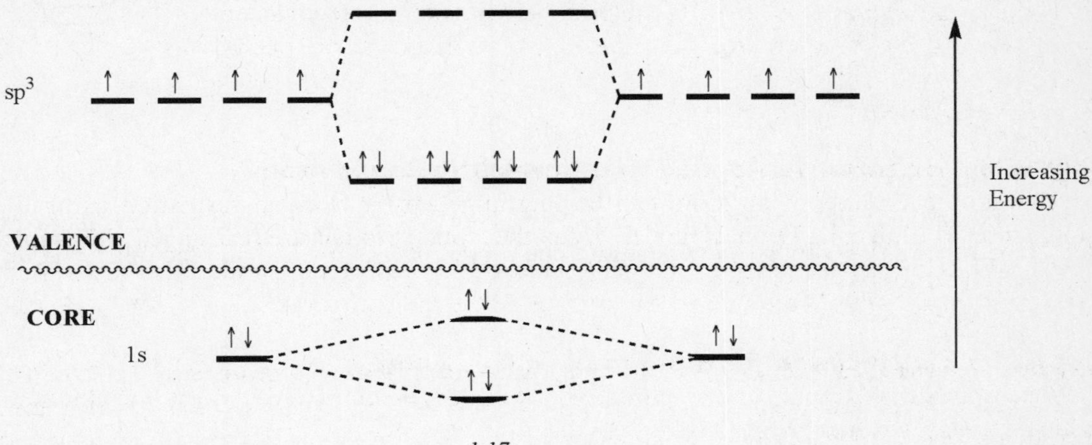

1.17

What Is a Sigma Bond?

A sigma (σ) bond is a normal covalent bond between two atoms in which the electron density is concentrated between the two nuclei, essentially on a 'line' between the two nuclei. The σ-bond is associated with sp^3 hybridization.

What Is the Classification for the Bond in a C–C Unit?

When drawn with a single line to represent the covalent bond between two carbon atoms, that bond is assumed to be a covalent σ-bond.

What Is a π-Bond?

A pi (π) bond occurs when two sp^2 hybridized are connected by a covalent σ bond, and each atom has an 'unused' p orbital as described above. When the p orbitals are parallel and on adjacent atoms, they can share electron density via 'sideways' overlap to form a new bond that is much weaker than the σ-bond. Effectively, there are two bonds between the atoms: a strong σ-bond and a weak π-bond, as shown in ***1.18***.

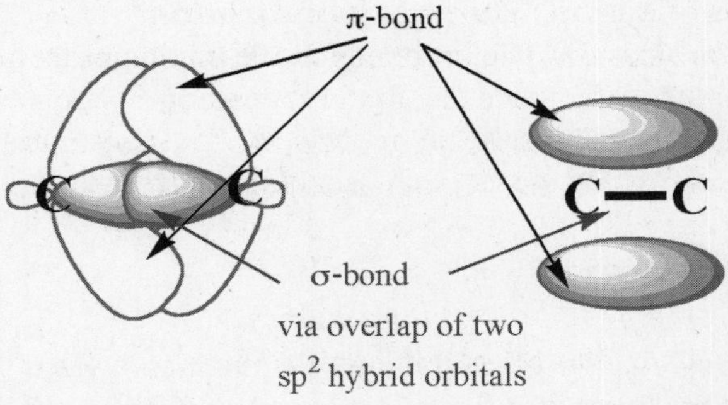

1.18

Show the Double Bond between Two Carbon Atoms, Ignoring the Other Valences on Carbon.

This representation is shown in ***1.19***.

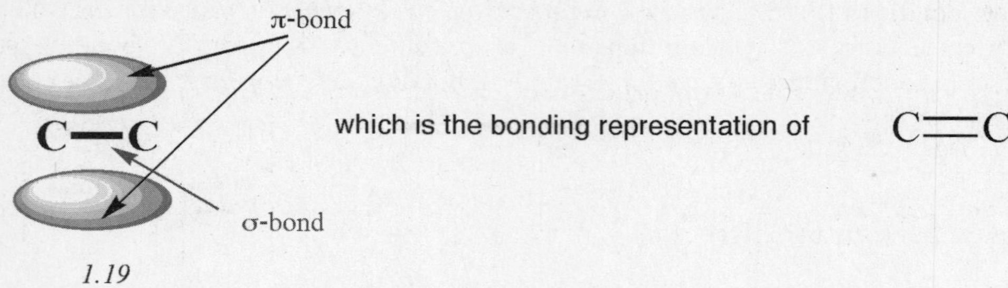

1.19

Note that in the C=C representation in ***1.19***, it is not possible to indicate which is the σ-bond and which is the π-bond. We simply know that one line represents the σ-bond and the other line represents the π-bond.

What Is the Classification for the Bond in a C–C Unit?

When drawn with a single line to represent the covalent bond between two carbon atoms, that bond is assumed to be a covalent σ-bond.

1.4. RESONANCE

What Is a Carbon-Carbon Single Bond?

This is the normal covalent σ-bond between two carbon atoms, involving mutual sharing of two electrons.

What Is a Carbon-Carbon Double Bond?

This is the C=C unit described above, containing one σ-bond and one π-bond (see ***1.19***)

Can Bonds Be Formed That Are 'in between' Single and Double Bonds?

In some cases, it is possible for electron density to be delocalized in such a way that the actual bond is in between a single and a double bond. This phenomenon occurs most commonly when a charge is present in an atom; it is also commonly associated with the presence of conjugated π-bonds.

What Is Resonance?

Bonds between carbon atoms, as well as bonds within and between functional groups, can be intermediate between single and double bonds due to the sharing of electrons in complex ways. These intermediate values make simplistic chemical bond representations inaccurate. In effect, the electron density is 'smeared' over several atoms, and we call this phenomenon 'resonance,' meaning to resonate between more than one state.

What Is a Point Charge?

When a charge is completely localized on a single atom, as in the chloride ion (Cl^-), for example. The two electrons that comprise the negative charge are described as being in an orbital localized on the chlorine atom.

Which Is More Stable, a Point Charge or a Delocalized Charge?

If a charge is delocalized over several atoms rather than localized on a single atom, the charge density on each atom is diminished. Lower charge density is usually associated with lower energy, so we can say that a delocalized charge should be more stable (less reactive) than a localized charge.

What Structural Features Are Required for Resonance?

Typically, a molecule of three or more atoms containing a π-bond with a third atom attached with a (+) charge as in ***1.20***, a (−) charge as in ***1.21***, or a single electron (a radical) as in *1.22*. In addition, when there are two atoms with a (+) charge on one atom and an attached heteroatom contains unshared electrons (as in ***1.23***), resonance can occur.

1.20 *1.21* *1.22* *1.23*

What Is a Resonance Contributor?

Taking ***1.20*** as an example, electron delocalization is indicated by moving the A=B π-bond and the charge in ***1.20***. Moving the π-bond towards the (+) charge results in formation of a new π-bond, with the new (+) charge left on the other terminal atom, as shown in ***1.24***. Neither of these two structures is correct by itself, and both are drawn, with a double-headed arrow, to indicate delocalization of an electron pair and resonance. Indeed, the actual bonding picture is in between the two structures of ***1.24***. Taking ***1.23*** as an example, moving the electron pair towards the (+) charge as shown results in a new structure with a new π-bond and a (+) charge on the adjacent atom (see ***1.25***). Again, the actual bonding picture is in between the two resonance structures shown for ***1.25***. Note the use of a curved arrow to indicate transfer of two electrons from one resonance contributor to the other.

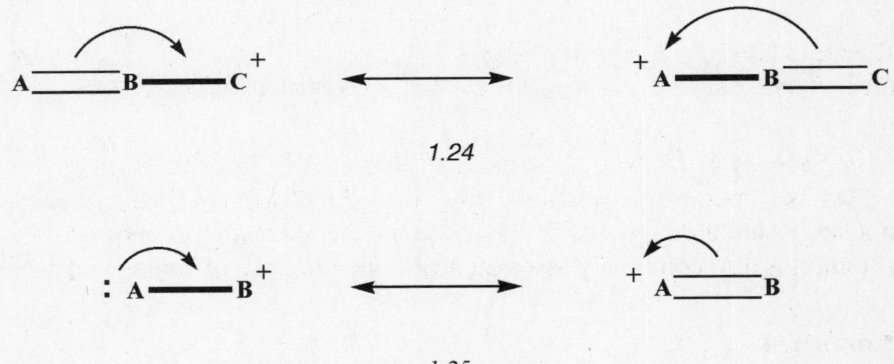

1.24

1.25

Are the Two Structures Represented by 1.24 in Equilibrium with Each Other?

No! They are not equilibrating structures, but rather two structures that, taken together, represent the bonding in a molecule that has bonds in between single and double bonds due to delocalization of the charge and electrons.

Draw Two Types of Real Molecules That Exhibit Resonance.

Two common resonance stabilized units are the carboxylate anion ***1.26*** with two resonance contributors and the oxygen-stabilized cation ***1.27*** with two resonance contributors.

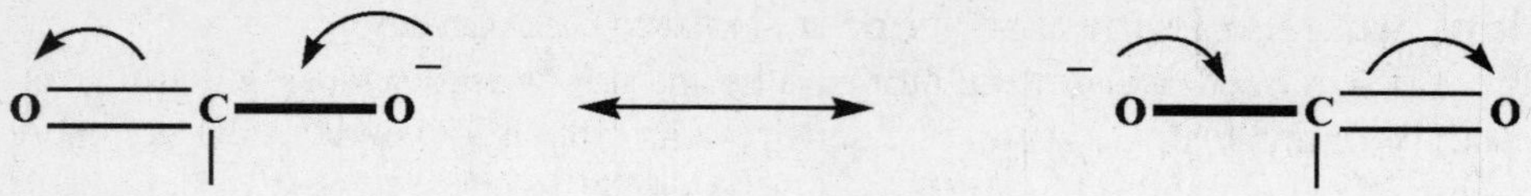

1.26

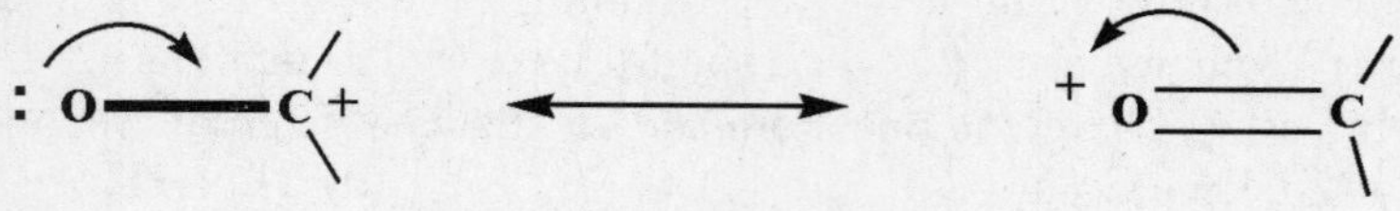

1.27

This chapter has provided the most fundamental tools of organic chemistry. In the following chapters, these tools will be applied to the many functional groups that comprise organic molecules and their reactions.

Test Yourself

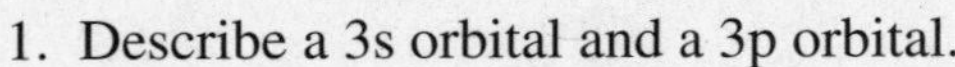

1. Describe a 3s orbital and a 3p orbital.
2. What is the difference between a $2p_x$ and a $2p_y$ orbital?
3. Give the electronic configuration for each of the following:
   (a) O (b) F (c) Cl (d) S (e) Si
4. Identify each of the following as having an ionic bond or a covalent bond:
   (a) K–Cl (b) Na–C≡N (c) H_3C–Br (d) H_2N–H
   (e) H_2N–Na (f) HO–Na (g) HO–H.
5. Give the number of covalent bonds each atom can form and still remain electrically neutral:
   (a) C (b) N (c) F (d) B (e) O.
6. Why is BF_3 considered to be a Lewis acid?
7. Which atom in each bond has the greatest δ- charge?
   (a) C–C, (b) C–N, (c) C–O, (d) C–F.
8. Identify the valence electrons in each of the following and identify the atom involved:
   (a) $1s^22s^22p^3$ (b) $1s^22s^22p^63s^1$ (c) $1s^2$ (d) $1s^22s^22p^63s^23p^3$
9. Indicate which bonds will generate hydrogen bonding, which will generate dipole-dipole interactions, and which will generate only Van der Waals interactions:
   (a) C–O–H (b) C–F (c) C–C (d) N–H
10. Indicate which structures are likely to have resonance contributors. In those cases where there is resonance, draw the resonance contributors.

(a) $C=C-C^+$ (b) $^+Cl=C$ (c) $C-C=C-C^-$ (d) H–O–C(–C)=O

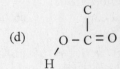

CHAPTER 2

Structure and Molecules

This chapter introduces a shorthand method for drawing organic molecules and discusses the VSEPR model, which is used in the preliminary analysis of simple molecules. The VSEPR model is a useful tool for beginning to look at the three-dimensional shapes of organic molecules. The chapter also discusses the dipole moment, isomers, functional groups and physical properties.

2.1. BASIC STRUCTURE OF ORGANIC MOLECULES

What Is the Valence of Each of the Following: C, H, N, O, F?

The valence of each atom is determined by subtracting the group number (we will use the US system throughout) from 8 for C, N, O, F since these atoms are in the 2nd row. For H, subtract the group number from 2. Therefore, C = 4, H = 1, N = 3, O = 2, and F = 1. The valence is the number of bonds that can be formed to each of these atoms and the resulting atom will be neutral (no charge).

Draw the Molecule with One Carbon Connected to Four Hydrogen Atoms by Sigma-Covalent Bonds.

```
    H
    |
H - C - H
    |
    H
```

2.1

What Is the Experimentally Determined Geometry of Methane (CH_4)?

Experiments have determined that CH_4 (see *2.1*) is tetrahedral about the carbon atom, with H–C–H bond angles of 109°28'. All the bond angles are identical, leading to the tetrahedral shape.

Using the Geometry of Methane, What Can You Infer About the Geometry of Organic Molecules Based on Carbon?

With a valence of four, each carbon atom, which has 4 groups attached in an organic molecule, should be tetrahedral (see *2.2*). The bond angles will vary slightly depending on the attached atom or group, but each carbon should have a tetrahedral-type geometry.

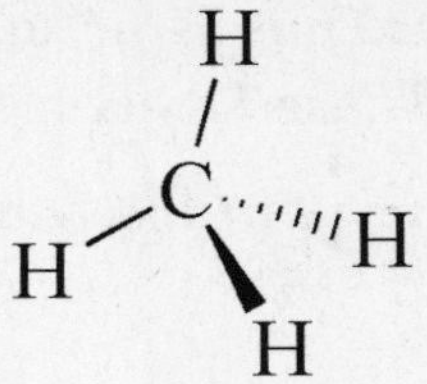

2.2

In *2.2*, note the use of a solid wedge and a dashed line. The solid wedge indicates the hydrogen atom is projected out of the page towards you, and the dashed line indicates that hydrogen atom is projected behind the page, away from you. The normal lines are used to indicate that the bonds and atoms are in the plane of the paper. This convention is used throughout the book to indicate the three-dimensional shape of molecules around a specific atom. Inspection of *2.3* shows how *2.2* represents a tetrahedron.

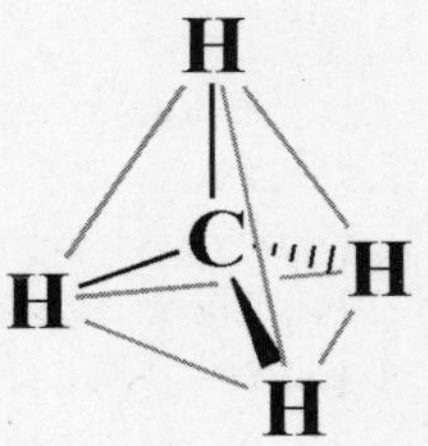

2.3

Can Carbon Form Bonds to Itself?

Yes! The property of carbon, which allows it to form covalent bonds to other carbon atoms, leads to a huge number of different molecules. Carbon can also form covalent bonds to many other atoms in the Periodic Table, leading to a great variety of molecules.

Connect Six Carbons Together in As many Different Ways As Possible and Complete the Remaining Valences on Each Carbon with a Hydrogen Atom.

There are five different possibilities, *2.4-2.8*.

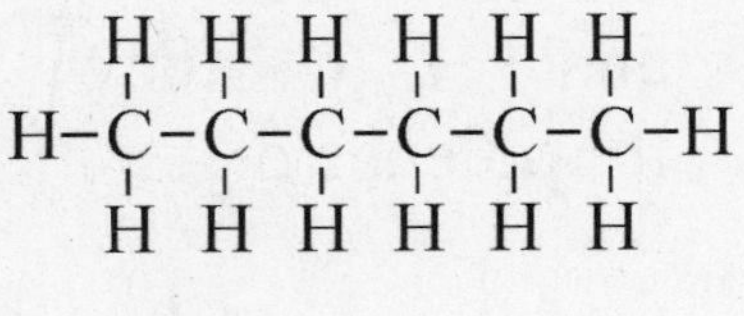

2.4

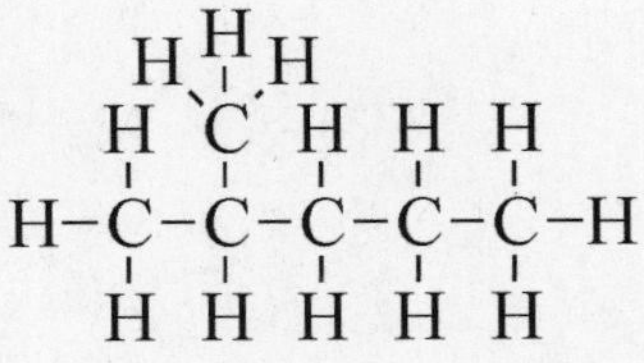

2.5

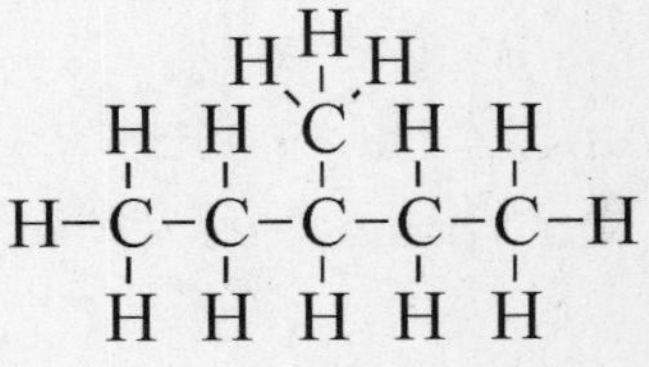

2.6

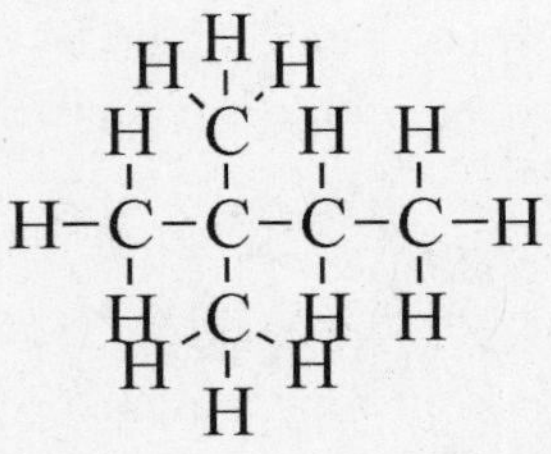

2.7

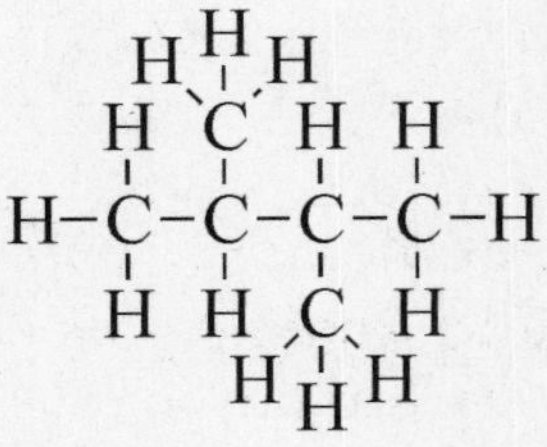

2.8

A Closer Look at 2.5 and 2.6 Suggests That Structures 2.9 and 2.10 Can Also Be Drawn. Briefly Explain Why These Structures Were Not Included.

Determine the Molecular Formula for All of the Molecules You Drew in the Previous Question.

```
 H H H                                      H H H
  \|/                                        \|/
   C  H  H  H  H                        H  H  H  C  H
   |  |  |  |  |                        |  |  |  |  |
H- C -C -C -C -C -H                  H- C -C -C -C -C -H
   |  |  |  |  |                        |  |  |  |  |
   H  H  H  H  H                        H  H  H  H  H
```

2.9 *2.10*

A close look at ***2.9*** shows that there are six carbons connected in a linear fashion and that structure is identical to ***2.4***. Since ***2.4*** and ***2.9*** are the same, only one of the structures is included. It is not important whether the atoms are up, down, etc. The way in which they are connected is the important consideration. Likewise, ***2.5*** and ***2.10*** are identical; they have five linear carbons connected together, with one carbon connected one carbon from the end. The formula is C_6H_{14}.

Define Line Notation for Drawing Organic Molecules.

Line notation is a shorthand method of drawing organic molecules. Each line represents a single covalent bond. The intersection of two lines is understood to be a carbon atom, and the remaining valences are understood to be hydrogen atoms. If the bond is between carbon and hydrogen, the hydrogen does not have to be drawn. If the atom is not carbon and hydrogen, it is inserted at the end of a line or at the intersection of two or more lines. For the moment, we will use line notation only for molecules that contain only C and H. We can draw ***2.4*** in line notation (see ***2.11***) to illustrate.

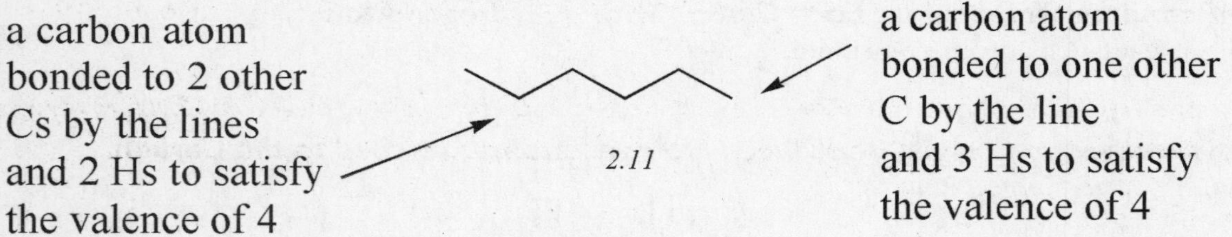
a carbon atom bonded to 2 other Cs by the lines and 2 Hs to satisfy the valence of 4
a carbon atom bonded to one other C by the line and 3 Hs to satisfy the valence of 4

2.11

Draw ***2.5-2.8*** using Line Notation.

2.12 *2.13* *2.14* *2.15*

Structure ***2.12*** corresponds to ***2.5***, ***2.13*** to ***2.6***, ***2.14*** to ***2.7***, and ***2.15*** to ***2.8***.

What Is the Term for Molecules That Have the Same Molecular Formula but Different Structures?

Isomers. Specifically, constitutional isomers.

Draw the Molecule That Has Three Carbon Atoms Attached to a Nitrogen Atom, and Each Carbon Atom Has Three Hydrogen Atoms.

```
         H  H
     H    \ /
     |     C-H
   H-C—N: /
     |   \   H
     H    C'
          | \
          H  H
```

2.16

Note that the electron pair not used to form covalent bonds is included in ***2.16*** as a lone pair.

Draw the Molecule That Has Two Carbon Atoms Attached to a Nitrogen Atom, with a Hydrogen on the Nitrogen and Three Hydrogen Atoms on Each of the Carbon Atoms.

```
         H  H
     H    \ /
     |     C-H
   H-C—N: /
     |   \
     H    H
```

2.17

In ***2.17***, the lone electron pair is again included. Clearly, nitrogen can form covalent bonds to C and H.

Draw the Molecule That Has One Carbon Attached to an Oxygen Atom, One Hydrogen Atom Attached to Oxygen and Three Hydrogen Atoms Attached to the Carbon.

```
     H    H
     |   /
   H-C—O:
     |  ..
     H
```

2.18

In ***2.18***, there are two lone electron pairs on the oxygen, which has a valence of 2, and forms two covalent bonds.

Draw the Molecule That Has Two Carbon Atoms Connected to an Oxygen Atom and Six Hydrogen Atoms Attached to the Carbons.

In ***2.19***, there are also two lone electron pairs on the oxygen, which has a valence of 2, and forms two covalent bonds. Clearly, carbon can form bonds to both H and C.

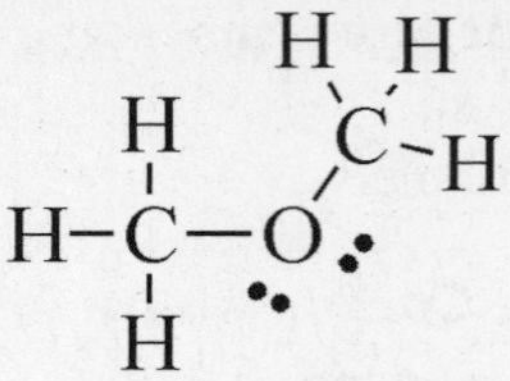

2.19

Another Way to Draw 2.19 is CH_3OCH_3. Explain!

The three hydrogen atoms on each carbon are 'condensed' to CH_3 (or H_3C). Drawing CH_3O means that the carbon atom is bonded to the three H atoms to its right, as well as the oxygen to complete the fourth valence. Likewise, the oxygen is understood to be bonded to C, not to H. In CH_3OCH_3, the oxygen is bonded to the carbon on its left, and the one on its right. Remember, the valence of O is 2. The carbon to the right of O is bonded to the O, as well as the three hydrogen atoms to its right. The valence of carbon is 4. This is an introduction to the shorthand method of drawing organic molecules that will be used extensively in this book.

2.2. VSEPR MODEL AND MOLECULAR GEOMETRY

The VSEPR Model

Organic chemists use this model for a preliminary analysis of simple molecules in an attempt to gain important information about the structure and properties of that molecule. The model is overly simplistic and gives poor results in many cases. Nonetheless, it is a useful tool to begin looking at the three-dimensional shapes of organic molecules.

What Do We Mean by the Shape of a Molecule?

The shape refers to the three-dimensional structure of a molecule, which is determined by the bond angles and bond distances. Shapes can be measured in some cases by using various experimental techniques; they can also be inferred in other cases by indirect methods. If we use carbon as an example, the tetrahedral array of atoms attached to a given carbon leads to the tetrahedral shape, or tetrahedral geometry, that we associate with organic molecules.

How Can We Determine the Shape of a Molecule?

Most organic molecules are three-dimensional, of course, and a useful model that will predict the general shape of a given molecule ***formed from atoms in the 2nd row of the Periodic Table***, is the **V**alence **S**hell **E**lectron **P**air **R**epulsion (**VSEPR**) model. There are two key components in this model: (1) the geometry of groups surrounding sp^3 carbon, oxygen and nitrogen atoms are assumed to be tetrahedral (sp^3 carbon forms a tetrahedral array of atoms around each carbon), and (2) electron pairs occupy space and are counted as 'groups.'

How Does the VSEPR Model Predict the Shape of Covalent Molecules?

The electrons in each bond around carbon will repel one another (like charges repel), but the bonds are connected to a central locus (the carbon atom), and the four atoms or groups attached to carbon cannot dissociate from the carbon atom. Therefore, the most efficient spatial arrangement minimizing electronic repulsion is the placing of each atom or group at the corner of a regular tetrahedron (bond angles are 109.5°). Using this observation, the model assumes a tetrahedral array of atoms and electrons around the other atoms in the 2nd row, specifically oxygen and nitrogen.

Can We Use the VSEPR Model for Boron Compounds?

Boron has only three groups around it to use the three valence electrons, and molecules of boron tend to be planar.

Use the VSEPR Model to Predict the Three-Dimensional Shape of CH_4, CH_2Cl_2, H_2O and NH_3.

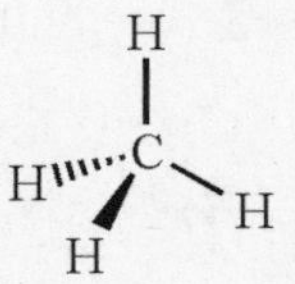

2.20 methane

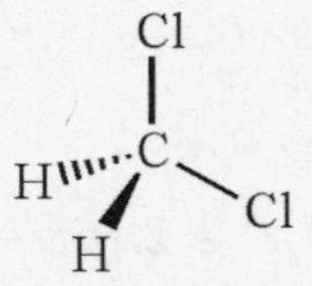

2.21 dichloromethane

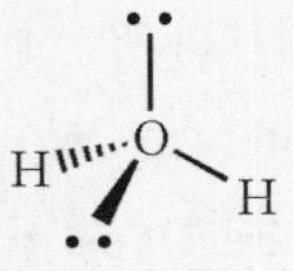

2.22 water

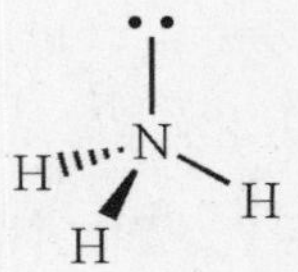

2.23 ammonia

For ***2.20***, a molecule named methane (see Section 3.3), the four hydrogen atoms are distributed to the corners of a regular tetrahedron with carbon as the central locus. There are no unshared electron pairs, and the 'shape' of the entire molecule, which is dictated by the covalent bonds, is **tetrahedral**. When two of the hydrogens are replaced with chlorine (molecule ***2.21***, called dichloromethane - see Section 3.3), again there are no unshared electron pairs on carbon, and the overall shape remains tetrahedral. When water (molecule ***2.22***) is analyzed, there are two hydrogens and two lone electron pairs, distributed to the corners of a tetrahedron. When viewing the molecule, however, only the *atoms* are observed (H–O–H), and these atoms assume an angular or **bent** shape for the water molecule. When ammonia (molecule ***2.23***) is analyzed, the three hydrogens and one lone pair distribute to the corners of the tetrahedron. On viewing the *atoms*, however, the N–H–H–H atoms assume a **trigonal pyramidal** shape, with nitrogen at the apex of the pyramid.

What Are the Main Shortcomings of the VSEPR Model?

This VSEPR model is flawed because it sometimes does not predict differences in shape resulting from the size of the various atoms. It also ignores attractive and repulsive forces that are present in some molecules. It does a reasonable job for simple molecules, however, and is usually invoked as the 'first guess' of the shape of a molecule.

Dipole Moment

The VSEPR model can be used to predict the three-dimensional shape of molecules. If the individual dipole moment of each bond in that molecule is superimposed on the VSEPR model, the dipole moment for the molecule can be estimated. The relative polarity (polar vs. nonpolar) of the molecule can also be estimated.

What Is Dipole Moment?

The dipole moment for a bond is defined as the product of the total amount of positive or negative charge and the distance between the atoms. If there is a difference in electronegativity of the elements involved in the bond ($3.0-2.1 = 0.9$, for example) there is a shift in electron density towards the more electronegative atom. This leads to polarization of the bonding electrons and of the whole molecule. The dipole moment can therefore be associated with an individual bond and also with the entire molecule. Charge polarization (or, in general, molecular polarity) can be quantified by a dipole moment.

Draw the Dipole Moment for the C–F bond.

$$\overset{\longmapsto}{\text{C–F}}$$
δ+ δ–

2.24

The dipole moment in ***2.24*** is indicated by the ⟼, although, in many cases, δ+ and δ− are used to indicate the presence of a dipole.

How Can the VSEPR Model be Used to Predict the Dipole Moment?

In a molecule such as methane (***2.20***), the C–H bond is not considered to be polarized; the electronegativity of C and H are assumed to be the same, although in reality they are not the same. Since there is no dipole moment for any bond on methane, the dipole moment for the molecule (the sum of all bond dipole moments equals Σ μ) where μ is the bond moment. For dibromomethane (***2.25***), there are two dipole moments, one each along the C–Br bonds. The negative pole of each bond is the bromine. Since dipole moments are directional, the dipole moment for the *molecule* is the vector sum of all the individual bond moments. The direction of the dipole moment for dibromomethane is, therefore, along a line that bisects the Br–C–Br angle and will be equal in magnitude to the vector sum of the two individual bond moments. The dipole moment is shown in ***2.26***, a model of dibromomethane, where the dipole moment was calculated to be 1.32 Debyes. Note the unit for dipole moment is the Debye.

2.20
methane

2.25
dibromomethane

2.26

Draw the Dipole Moment for the Molecule $CHCl_3$ (known as chloroform).

2.27

2.28

In *2.27*, the dipole moment for each C–Cl bond is shown. Since the three dipole moments for the bonds (bond moments) are in the same general direction due to the tetrahedral shape of the molecule, the dipole moment for the molecule is the vector sum of the three C–Cl bond moments, as well as any bond moment resulting from the C–H bond. This vector sum placed the direction of the dipole moment for the molecule is that shown, bisecting the base of the tetrahedron. The bond moment (calculated to be 1.16 Debyes) is represented by the arrow in *2.28*. Note that the unit for dipole moment is the Debye, and it has both magnitude and direction.

2.3. FUNCTIONAL GROUPS

Collections of atoms can have certain chemical and physical properties that are unique. When these collections of atoms appear in different organic molecules, those molecules often have similar chemical and physical properties. For this reason, it is convenient to categorize organic molecules by such collections of atoms. These are called functional groups. The functional group is also the basis upon which we name most organic molecules, as we will see starting in Section 3.3.

What Is a Functional Group?

When atoms other than carbon or hydrogen are incorporated into an organic molecule, certain collections of heteroatoms taken as a unit have unique chemical and physical properties and are called **functional groups**. Functional groups include collections of atoms that contain π-bonds, including C=C and C≡C. Double bonds are discussed in Section 8.1 and triple bonds are discussed in Section 8.3 .

What Functional Groups Involve Only Carbon and Hydrogen?

The C=C and C≡C units are functional groups for alkenes and alkynes, respectively.

What Functional Groups Involve Oxygen, Carbon and/or Hydrogen?

These include OH, C–O–C, C=O, CO_2H, CO_2C.

What Functional Groups Involve Nitrogen, Carbon and/or Hydrogen?

Groups with 1, 2, or 3 carbons on a nitrogen atom, with hydrogen attached to nitrogen when there are only 1 or 2 carbon groups. These are the **amines**. In addition, C≡N is a functional group.

What Functional Groups Involve Sulfur, Carbon, and/or Hydrogen?

These include SH and C–SC.

Show the Functional Group That Correlates with the Following Names: Hydroxyl, Thiol, Carbonyl, Carboxyl, Amino, Cyano, Alkene, Alkyne and Ether.

When these functional groups are incorporated into a molecule they usually form a **class** of molecules that tend to have unique chemical and physical properties. Examples of each type of functional group are obtained by attaching one or more **R** groups to each functional group, where **R** is a generic carbon group. The key functional groups are:

O–H hydroxyl	S–H thiol	C=O carbonyl
H–O–C=O carboxyl	C–N amino	C≡N cyano (nitrile)
C=C alkene	C≡C alkyne	C–O–C ether

Show a Generic Formula for Each of the Following: Alcohol, Thiol, Aldehyde, Ketone, Carboxylic Acid, Amine, Nitrile, Alkene, Alkyne and Ether.

When the functional group OH (hydroxyl) is in a molecule, the class is called **alcohols**. When the carbonyl (C=O) is incorporated, there are two structural variations. The carbonyl can be connected to a hydrogen (H) and a carbon group (R) to give an **aldehyde** or the carbonyl can be connected to two carbon groups (R) to give a **ketone**. If the carbonyl also contains a hydroxyl (-COOH), the molecule is called a **carboxylic acid**. If a trisubstituted nitrogen is in the molecule, the class is called an **amine**. If there is one R group and two hydrogens, it is a primary amine (1° = primary). If there are two R groups and one H, it is a secondary (2°) amine, and three R groups give a tertiary (3°) amine. The class of compounds that contains a nitrile group (C≡N) is called a **nitrile**. If the molecule contains a carbon-carbon double bond (C=C), it is an **alkene,** and if it contains a carbon-carbon triple bond (C≡C), it is an **alkyne**. The class of compounds characterized by a C–O–C bond is the **ether**. These are the major functional groups that will be discussed in this book. A few others will be added later as the body of information increases, and the applications of these groups arise.

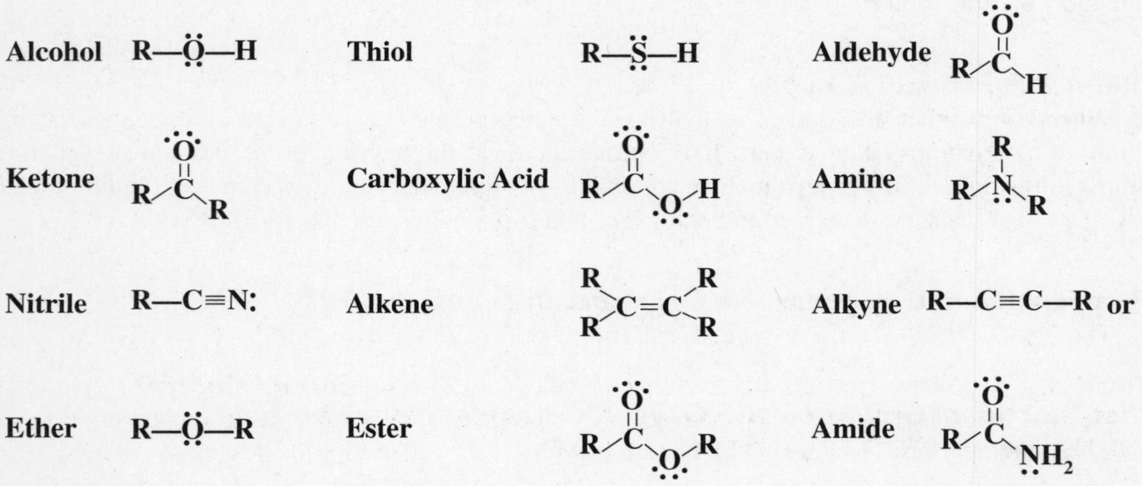

Alcohol	R—Ö—H	**Thiol**	R—S̈—H	**Aldehyde**	R–C(=O)–H
Ketone	R–C(=O)–R	**Carboxylic Acid**	R–C(=O)–O–H	**Amine**	R–N(R)–R
Nitrile	R—C≡N:	**Alkene**	$R_2C=CR_2$	**Alkyne**	R—C≡C—R or
Ether	R—Ö—R	**Ester**	R–C(=O)–O–R	**Amide**	R–C(=O)–NH_2

2.4. FORMAL CHARGE

How does one know if a structure is real? We begin by examining the molecule for charge, which is a function of the number of bonds to each atom, the number of unshared electrons, if any, and whether the valence for each atom is satisfied. This information will allow us to ask the following questions. Does the structure have positive or negative charges? Is it neutral? These two questions can be answered by calculating the **formal charge**. It is based on the number of valence electrons, the number of electrons used to form covalent bonds, and the number of unshared electrons.

What Is Formal Charge?

In a given structure, an atom that has one less bond than its valence but has two electrons will have a negative (−) charge, whereas an atom that has one less bond than its valence, but is missing two electrons will have a positive (+) charge. If an atom uses an unshared electron pair to form one more bond than its valence, it usually will have a + charge. If an atom forms one more bond than its valence, but the new atom or group provides both electrons for the bond, it usually will have a − charge. Formal charge is the charge assigned to each atom in a Lewis structure, and the sum of all the charges of the atoms gives the formal charge of the molecule.

What Is the Formula Used to Determine Formal Charge?

The formula is:

$$\text{Formal Charge} = \Omega = (\text{Valence Number}) - (\text{Number Unshared Electrons}) - \frac{1}{2}\,(\text{Number Shared Electrons})$$

Determine the Formal Charge of Each Atom and Also for the Entire Molecule of Dimethylamine ([CH$_3$]$_2$NH), 2.29

Before determining the formal charge for the entire molecule, we must determine the formal charge for each atom. The sum of all atomic formal charges will be the formal charge for the molecule. For C^1, $\Omega = 4 - 0 - \frac{1}{2}(8) = 0$. For C^2, Ω also $= 0$. In both cases, the valence number is the group number (4 for C). There are no unshared electrons and eight covalently shared electrons. For the nitrogen, the valence (group) number is 5, and Ω is $5-(0)-(1,2)(8) = +1$. The formal charge for each hydrogen is 0 $[1 - 0 - \frac{1}{2}(2)]$. To determine the formal charge for the molecule Ω_{mol}, the sum of the formal charge for all atoms is used: $\Omega_{mol} = \Sigma\,(\Omega_{atom})$. For this example, $\Omega_{mol} = 0+0+[8\times 0]+(+1) = +1$. Therefore, ***2.29*** exists as a cationic species.

```
   H  H  H
  1|  |  |2
 H-C-N-C-H
   |  |  |
   H  H  H
```

2.29

What Is the Criterion for a Molecule to Be Real, Based on Formal Charge?

If a structure is drawn and has a charge of greater than $+2$ or less than -2, that structure is likely to be very high in energy and its existence should be questioned. For the molecules encountered in the first organic course, it is very unusual for a molecule to exist as anything but a neutral compound (formal charge $= 0$), a mono-cation (formal charge of $+1$), or a mono-anion (formal charge of -1).

2.5. PHYSICAL PROPERTIES

The presence or absence of a functional group in an organic molecule has a profound effect on the boiling point, melting point, adsorption characteristics, etc. These are physical properties, commonly measured to help identify a unique molecule. This section will briefly introduce the physical properties commonly associated with organic molecules. A physical property is a parameter associated with a pure molecule that is unique to that molecule and can be used to help identify it. Examples are dipole moment, polarity, boiling point, melting point, adsorptivity, refractive index and solubility. While some of these physical properties may overlap with similar or related molecules, it is highly unusual for *all* of the physical properties to overlap. This set of physical properties allows each molecule to be identified as an individual.

Boiling Point

The molecules in the gas phase over a liquid phase and the molecules in the liquid phase are at equilibrium as long as the temperature is constant. When the vapor pressure of the liquid just equals ambient atmospheric pressure, boiling occurs. This is the boiling point.

What Are the Factors That Influence Boiling Point?

To get molecules into the gas phase, the intermolecular forces holding them together in the liquid phase must be disrupted. This requires heat. In general, the greater the mass of the molecule, the higher the boiling point, but the intermolecular interactions of the molecule are very important. Three examples will illustrate this statement: (a) $CH_3CH_2CH_3$ (propane), (b) $H_3C–O–CH_3$ (dimethyl ether) and (c) CH_3OH (methanol). In case (a), there are no dipole interactions, and the only intermolecular forces are Van der Waal's forces (sometimes called London forces). These are quite weak, and the boiling point of propane is −42.1°C for a molecular weight of 44.10. For case (b), the polarized C–O bond leads to a stronger intermolecular interaction relative to London forces, and the dipole-dipole interactions help keep the molecules together. It takes more energy to disrupt these interactions, and the boiling point is −24.8°C for a molecular weight of 46.07. The modest increase in mass cannot account for the 17.3°C increase in temperature required to boil dimethyl ether. When methanol in case (c) is examined, the boiling point of 64.7°C for a molecular weight of 32.04 is far higher than either propane or dimethyl ether. This is due to hydrogen bonding interactions between the oxygen and the hydrogen atom attached to oxygen, forces which are much stronger than the dipole interactions found in dimethyl ether.

Melting Point

Melting point is defined as the temperature at which the molecules in the solid phase are in equilibrium with those in the liquid phase.

What Factors Influence Melting Point?

An increase in molecular weight often leads to an increase in melting point. The presence or absence of dipoles, hydrogen-bonds, and symmetry in a molecule will influence the melting point. The more symmetrical a molecule, the higher the melting point. Conversely, the more amorphous a molecule, the lower the melting point. If we compare pentane ($CH_3CH_2CH_2CH_2CH_3$, ***2.30***) with 2,2-dimethylpropane ($(CH_3)_4C$, ***2.31***), the melting points are −129.7°C for pentane and −16.6°C for dimethylpropane. Both molecules have the same molecular formula (same number and kind of atoms), so they clearly have the same mass. Note that ***2.31*** is more compact, however, and will 'fit' into a crystal structure better than the 'floppy' linear molecule ***2.30***. This enhanced 'packing' leads to a higher melting point for ***2.31***.

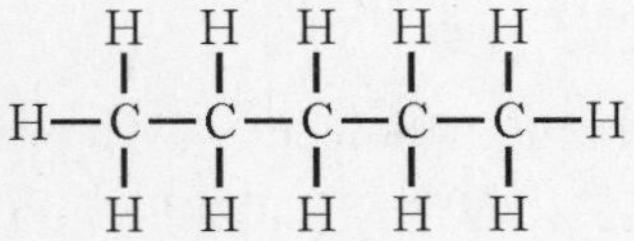

```
  H H H H H
  | | | | |
H-C-C-C-C-C-H
  | | | | |
  H H H H H
```

2.30

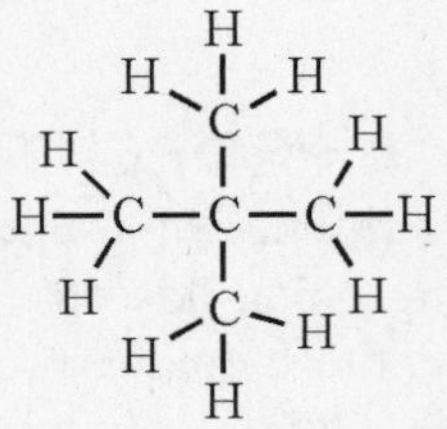

2.31

Draw 2.30 and 2.31 in Line Notation.

The line notation for ***2.30*** is ***2.32*** and ***2.31*** is represented by ***2.33***.

2.32

2.33

Solubility

Solubility is the ability of one molecule (a solid, a liquid or a gas) to dissolve into another molecule that is in the liquid state.

How Does Solubility Give Information about the Structure of a Molecule?

The old axiom 'like dissolves like' can be used with remarkable accuracy. In general, nonpolar molecules will dissolve well in nonpolar liquids such as hydrocarbons, but not in polar liquids such as water. Conversely, a polar molecule will not dissolve in a nonpolar liquid, but will dissolve in a polar liquid. If a molecule contains one or more C–heteroatom bonds, and the net dipole moment is not zero, it is considered to be polar. If it contains no polarized bonds, it is considered to be nonpolar. There are exceptions to this latter statement that use the qualifying statement for the heteroatom case, a molecule such as CCl_4, which contains polar bonds but has a net dipole moment of zero for the molecule and is still nonpolar. The actual test of polarity (and usually solubility) is, therefore, the presence of a dipole moment for the molecule. There are degrees of polarity and degrees of solubility. Molecules that undergo extensive hydrogen bonding are very polar and very soluble in polar liquids such as water. A molecule with a small dipole (CH_3Cl) will be less polar than one with a larger dipole ($CHCl_3$). This will be reflected in partial solubilities in polar solvents but some solubility in nonpolar solvents. Molecules that contain the polar O–H or *N*–H bonds can form hydrogen bonds to the O or H in water, suggesting solubility in water. If there are more than 5–8 carbon atoms, however, the one polar hydrogen bond is counteracted by the nonpolar carbon atoms (with the attached hydrogen atoms). Clearly, this can be a difficult property to quantify.

Test Yourself

Line notation will be used for these and subsequent problems. When atoms other than C and H are present, the atom or group will be shown, as in problem 1. A line is drawn to the Br, OH, etc. Where there is a multiple bond, two or three lines are used to indicated the double or triple bond, respectively. Examples are in problem 1b. In some cases, the functional group is condensed, as in the COOH unit shown in problem 6c when compared to problem 1b.

1. For each series of compounds, identify the one with the highest boiling point. Explain.

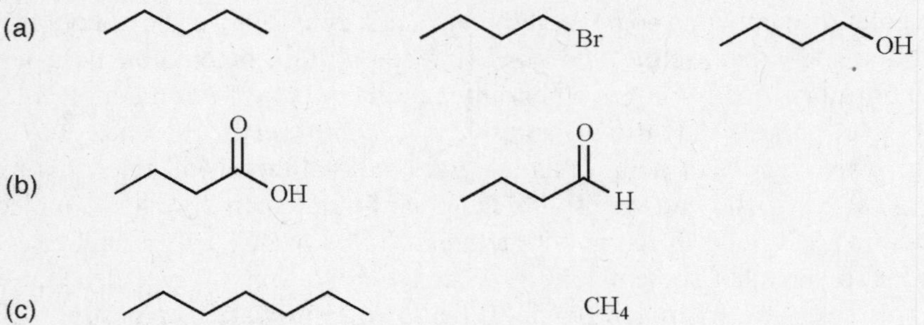

2. Predict the shape of each of the following:
   (a) CH_3OCH_3 (b) Cl_3CH (c) CH_3OH (d) $NH(CH_3)_2$
3. Indicate the direction of the dipole moment on the VSEPR model for each of the following:
   (a) CH_3OCH_3 (b) Cl_3CH (c) CH_3OH (d) NH_3 (e) CH_2BrCl
4. What is the functional group identified with each of the following?
   (a) alcohol (b) ketone (c) alkyne (d) aldehyde (e) thiol (f) nitrile
5. Determine the formal charge for all atoms in each molecule. Calculate the final charge for each molecule, assuming all have octets except hydrogen.

   (a) H^1, H^2, C^2, C^1, O^2, O^1, N—H^3, H^5, H^4

   (b) $H^1—C^1(O)(H^2)—C^2(H^3)(H^5)—C^3(H^4)(H^6)—N{\equiv}C^4$

6. Indicate which molecule has the higher boiling point in each of the following pairs. In each case explain your answer.

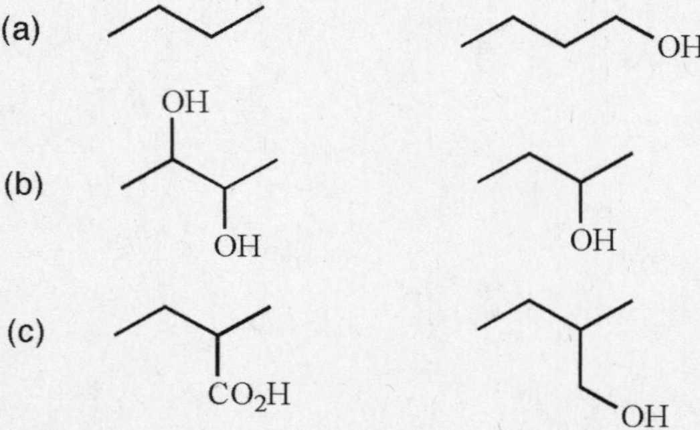

7. Which molecule has the higher melting point? Explain.

(a)

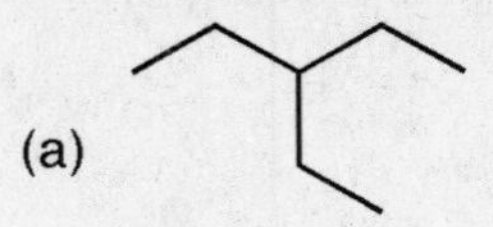

(b)

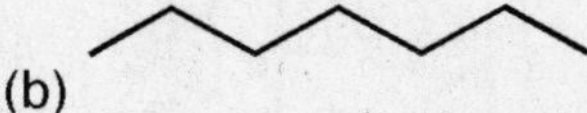

(c)

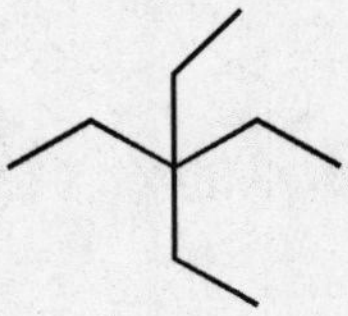

CHAPTER 3

Alkanes, Isomers, and Nomenclature

In this chapter the most fundamental of organic molecules, the alkanes, will be discussed. Alkanes are molecules that contain only carbon and hydrogen connected by single covalent bonds. The method for naming alkanes will be introduced as well as their chemical and physical properties.

3.1. DEFINITION AND BASIC NOMENCLATURE

Organic molecules are defined by the atoms in the molecule and the presence or absence of functional groups. These fundamental parameters are also the basis for naming organic molecules.

What Is a Hydrocarbon?

A hydrocarbon is a molecule whose structure contains only carbon and hydrogen atoms.

What Is the Definition of an Alkane?

An alkane is a hydrocarbon composed of only carbon and hydrogen that possesses the simplest empirical formula among organic molecules that do not contain a functional group. All alkanes have the general formula C_nH_{2n+2}. This formula is important because it represents the maximum number of hydrogens that can be attached to a given number of carbon atoms. There can be fewer hydrogen atoms than $2n+2$ when functional groups are introduced, but never more.

What Is the General Approach to Naming Alkanes?

Alkanes are named according to a set of rules established by the International Union of Pure and Applied Chemistry (IUPAC). The rules define a prefix, which shows the number of carbon atoms, and a suffix, which defines the functional group. In this case, there is no functional group, and the suffix is -ane.

What Are the Prefixes Used to Identify the Number of Carbon Atoms in an Alkane?

The most common prefixes used for linear alkanes that contain 1–20 are shown in the table. In each case, the hydrogens are added to satisfy the valence of four and the C_nH_{2n+2} formula.

C_1CH_4 **meth**ane
$C_2CH_3CH_3$ **eth**ane
$C_3CH_3CH_2CH_3$ **prop**ane
$C_4CH_3(CH_2)_2CH_3$ **but**ane
$C_5CH_3(CH_2)_3CH_3$ **pent**ane
$C_6CH_3(CH_2)_4CH_3$ **hex**ane

C_7 $CH_3(CH_2)_5CH_3$ **hept**ane	C_{14} $CH_3(CH_2)_{12}CH_3$ **tetradec**ane
C_8 $CH_3(CH_2)_6CH_3$ **oct**ane	C_{15} $CH_3(CH_2)_{13}CH_3$ **pentadec**ane
C_9 $CH_3(CH_2)_7CH_3$ **non**ane	C_{16} $CH_3(CH_2)_{14}CH_3$ **hexadec**ane
C_{10} $CH_3(CH_2)_8CH_3$ **dec**ane	C_{17} $CH_3(CH_2)_{15}CH_3$ **heptadec**ane
C_{11} $CH_3(CH_2)_9CH_3$ **undec**ane	C_{18} $CH_3(CH_2)_{16}CH_3$ **octadec**ane
C_{12} $CH_3(CH_2)_{10}CH_3$ **dodec**ane	C_{19} $CH_3(CH_2)_{17}CH_3$ **nonadec**ane
C_{13} $CH_3(CH_2)_{11}CH_3$ **tridec**ane	C_{20} $CH_3(CH_2)_{18}CH_3$ **icos**ane

What Is a Linear Alkane?

Examination of the table shows that all carbon atoms are connected in a linear manner with no branches. As a practical matter, this definition means that for all alkanes of three or more carbon atoms, there is a methyl group connected to some number of methylene units, ending in another methyl group, so the generic formula for a linear alkane is $CH_3(CH_2)_nCH_3$ (n is simply an integer, 0,1,2,3….). Ethane (n = 0) has two methyl groups connected together, with no intervening methylene unit.

What Are the General Physical Properties of Alkanes?

Alkanes are very nonpolar molecules and generally insoluble in water or other highly polarized liquids. They have no functional group and, therefore, associate in the liquid phase only by Van der Waal's forces, leading to low boiling points. Once the mass of the alkane is sufficiently large, however, the boiling point also becomes rather high, and very large alkanes are solids. In the following list, the first number is the boiling point and the number in brackets [ ] is the melting point: propane = −42.1°C [−187.7°C]; pentane = +36.1°C [−129.7°C]; decane = +174.1°C [−29.7°C]; icosane = +343.8°C [+36.4°C].

What Is the IUPAC Name for the Eleven Carbon Linear Alkane?

Undecane.

Draw Undecane Using Line Notation.

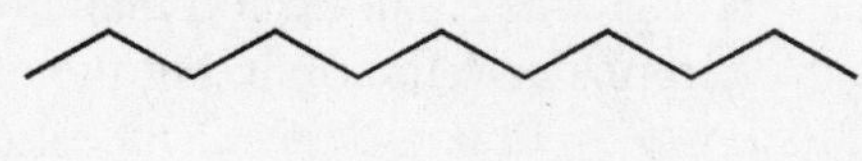

3.1

3.2. STRUCTURAL ISOMERS

With a valence of four (4), each carbon in a molecule will form four covalent bonds. One of the important features of carbon that distinguishes it from most other atoms in the Periodic Table is its ability to form covalent bonds to another carbon, as well as to other atoms. This leads to a very large number of molecules that contain carbon, all with different structures. It also leads to an interesting phenomenon called **isomerism**. Specifically, molecules that have the same molecular formula can have different structures, with the atoms connected together in different ways. Such compounds are called isomers..

Define Isomer.

Isomer refers to two or more molecules with the same molecular formula, but with the individual atoms connected in different ways, leading to *different molecules*.

Are There Isomers for the Formula C_3H_8?

No! The formula $CH_3CH_2CH_3$ for propane is the only way to attach these atoms and satisfy all valences to keep the molecule neutral. Therefore, propane has no isomers.

Are There Isomers for the Formula C_4H_{10}? If Yes, Draw Them Using Line Notation.

Yes! The linear connection of carbon atoms is butane, ***3.2***. It is also possible to connect three carbons in a chain, and then branch the fourth carbon from the middle carbon, as in ***3.3***.

Note the emergence of a pattern. First, make the longest continuous chain. Next, make a structure with one fewer carbon atoms in the longest continuous chain and attach the extra carbon at different branch points along the longest chain. In this case, there is only one place to attach the extra carbon (***3.3***).

3.2 *3.3*

As Drawn, Is 3.4 an Isomer of 3.2?

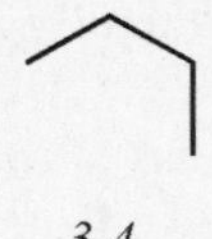

3.4

No! The longest continuous chain is four, so ***3.4*** and ***3.2*** are identical (butane).

How Can We Draw Hexane in the So-called Condensed Structure?

In ***3.5*** (hexane), the six carbons are connected in a linear manner. Adding the hydrogens leads to a molecule that satisfies the molecular formula. A more expedient way to represent this formula, the condensed structure ($CH_3\underline{C}H_2CH_2CH_2CH_2CH_3$), is shown. In this notation, the carbon indicated is connected to two other carbons (one to the right and one to the left), and the two hydrogen atoms to the *right* of that carbon are connected directly to it. All atoms other than carbon that are connected to a given carbon are shown to the right. (For ***3.6***, one of the carbons is to the right, but the other is to the left.)

```
  H H H H H H
H-C-C-C-C-C-C-H
  | | | | | |
  H H H H H H
```

≡ $CH_3CH_2CH_2CH_2CH_2CH_3$

3.5 *3.6*

What Are the Isomers for the Formula C_6H_{14}? Draw Them Using Condensed Structures.

Structural isomerism can be illustrated by a molecule with the molecular formula C_6H_{14}. Each carbon will have four bonds, and each hydrogen will have one bond. There are, however, five (5) different ways to assemble six carbon atoms and 14 hydrogen atoms such that these criteria are satisfied. In ***3.6*** (hexane), the six carbons are connected in a linear manner; adding the hydrogens leads to a molecule that satisfies the molecular formula. A different structure can also be drawn (***3.7***) where there are five carbons in a continuous chain, with a one-carbon fragment attached at the next-to-last carbon. That single carbon can be connected to the 'right' or to the 'left,' but the two structures are identical, since they can be completely superimposed, one on the other. A third structure appears when the single carbon fragment is placed on the 'middle' carbon (in ***3.8***). The same structure appears when four carbons are placed in a line, and a two-carbon fragment is attached. Examining the five encircled carbons shows the structures to be the same. A fourth molecule (***3.9***) can be drawn where four carbons are placed in a continuous chain and two one-carbon fragments are added. The fifth molecule (***3.10***) also has a longest continuous chain of four, but the two extra carbon atoms are on the same carbon (C2). These five different molecules are called isomers, molecules with the same molecular formula, but different points of attachment for the individual atoms.

$CH_3CH_2CH_2CH_2CH_2CH_3$

3.6

CH_3 (on C2)
$CH_3CHCH_2CH_2CH_3$
1 2 3 4 5
≡
CH_3 (on C2)
$CH_3CH_2CH_2CHCH_3$
5 4 3 2 1

3.7

CH_3 (on C3)
$CH_3CH_2CHCH_2CH_3$
1 2 3 4 5
≡
CH_3CH_2
$CH_3CHCH_2CH_3$

3.8

CH_3 (on C2)
$CH_3CHCHCH_3$
1 2 3 4
CH_3 (on C3)

3.9

CH_3 (on C2)
$CH_3CCH_2CH_3$
1 2 3 4
CH_3 (on C2)

3.10

Why Is the Concept of Isomerism Important for Alkanes?

Due to the ability to form isomers, there are literally millions of possible alkanes. The five different isomers of hexane were shown in the previous question. As the number of carbons increase, the number of constitutional isomers increases dramatically: 9 for C_7H_{16}, 4347 for $C_{15}H_{32}$, and more than 62×10^9 for $C_{40}H_{82}$.

Since isomers will all be different molecules (4347 for $C_{15}H_{32}$) with different physical properties (boiling points, melting points, solubility, etc.), a system is required to identify each individual molecule and distinguish it from all other isomers. The two rules listed above, which are based on the number of carbons and the suffix defining the alkane group, are insufficient. Additional rules must be added.

3.3. IUPAC NOMENCLATURE

Basic Nomenclature

In order to name millions of molecules, there must be a set of systematic rules based on structure to guide us. These nomenclature rules were established and accepted by IUPAC. They use the prefix-suffix idea to describe structural variations. The rules become more complicated as the complexity of the structure increases, but the fundamental rule can be presented for alkanes and then expanded as new functional groups are discussed in later chapters.

What Are the IUPAC Rules Pertinent to Alkane Nomenclature?

The following section describes the basic rules of nomenclature for organic molecules. In this chapter, the main focus will be on alkanes. Additional rules will be introduced as each new functional group is discussed.

The first rule must identify the length of the carbon chain, but a rule must be added to locate the number and type of carbon groups attached to that initial carbon chain. To do this, the carbon chain must be numbered. If we take the five isomers of hexane from Section 3.2, the necessity for these rules becomes apparent. In one of these cases, there was a five-carbon linear chain with a one-carbon group attached to it. The rules required to distinguish this molecule are as follows:

(1) Determine the longest continuous chain of carbon atoms and assign the alkane base name (methane → icosane).
(2) Number the longest chain such that the substituent (atom or groups attached to the longest chain) closest to the end of the chain receives the lowest possible number.
(3) Determine the number of carbon atoms in the substituent and assign a name based on the alkane names, but drop the *-ane* ending and add *-yl.* A one-carbon substituent is meth*yl*, a two-carbon substituent is eth*yl*, etc.

What Is a Substituent?

A substituent is an atom or group attached to the longest continuous carbon chain.

How Are Substituents Named?

Use the prefix for the number of carbon atoms and add the suffix –yl. A one-carbon substituent becomes methyl, a two-carbon substituent, ethyl, etc.

What Is the Substituent in 3.7?

In *3.7*, the longest continuous chain is five, so this compound is named as a pentane. There is a one-carbon fragment (substituent) attached to the pentane chain, which is a methyl group.

What Is the Rule for Using Substituents in a Name?

Always place the substituent in front of the base name (alkane *3.7* is methylpentane, for example) and indicate the position of the substituent with a number. The longest chain is five, which is a pentane. The substituent is a one-carbon fragment (circled), therefore = methyl. Number the pentane chain to give the methyl group the lowest number (2 rather than 4), and this is named 2-methylpentane. With this numbering system, it is now obvious why we labeled these two structures as identical. They not only superimpose, they receive the same name.

Why Is 3.7 Not Named 4-Methylpentane?

The nomenclature rule clearly states that the longest continuous chain is to be numbered to give the substituent the lowest possible number. Therefore, *3.7* is 2-methylpentane and *not* 4-methylpentane.

Why Is an Isomer of Hexane (3.7) Named As a Pentane?

Although there are six carbons, the longest continuous carbon chain is only five carbons, and it is, therefore, named as a pentane. Another isomer of hexane (*3.8*) illustrates the versatility of this nomenclature system. This is also a pentane (5 carbons), and the substituent is also a methyl group. The position of the methyl substituent in *3.8* is now different than in *3.7* and it is named 3-methylpentane - a different isomer. The next isomer of hexane raises a slightly different problem. There are two methyl substituents in *3.9* with a four-carbon chain, but here are also two methyl substituents in *3.10*. They differ by the points to which the methyl groups are attached.

What Is an Alkyl Group?

Any carbon substituent is generically called an alkyl group. Methyl, ethyl butyl, hexyl are all alkyl groups.

What Nomenclature Rule Applies When There is More Than One of the Same Group in the Same Molecule?

A new rule must be added to deal with this situation:

(4) **When there are two or more identical substituents, assign each one a number according to its position on the carbon chain and use the prefix di- (two), tri- (three), tetra- (four), penta- (five), etc. Remember to number the chain in a manner to give the lowest numbers to the substituents.** If there are two methyl groups the name will be dimethyl. Four ethyl groups will be tetraethyl.

Give the Names for 3.9 and 3.10.

The longest continuous chain is four, so ***3.9*** is a butane. Giving the substituents the lowest number, there are methyl groups on C2 and C3 in ***3.9***, but both methyl groups are on C2 in ***3.10***. Using the fourth rule, ***3.9*** is named 2,3-dimethylbutane and ***3.10*** is named 2,2-dimethylbutane. Note that each methyl group has a number, despite the fact that ***3.10*** has two methyl groups on the same carbon.

How Are Substituents Named When They Are Not the Same As in 3.11?

```
            CH3
  7         |6   5      4    3    2    1
  CH3CHCH2CHCH2CH2CH3
  1         2    3   |4    5    6    7
                   CH2CH3
```

3.11

A corollary is necessary for this rule when the substituents are not the same. **(Corollary to Rule 4.) When the substituents are different, assign each a number according to its position on the carbon chain and arrange the substituents alphabetically in the final name.** If a methyl and an ethyl group are substituents, the name will be ethylmethyl, as in 6-ethyl-3-methyldecane. (Draw this structure in line notation as a useful exercise.) As noted in this example, the *lowest combination of numbers* is used.

Give the Name for 3.11.

The longest continuous chain is seven (a heptane). There is a one-carbon substituent (methyl) and a two-carbon substituent (ethyl), making this an ethylmethylheptane. It could be named as 4-ethyl-2-methylheptane or as 6-ethyl-4-methylheptane. The 4,2 combination is smaller than the 6,4 combination, and ***3.11*** is named 4-ethyl-2-methylheptane.

How Are Halogens Treated in the Nomenclature System?

There is one case where the heteroatom is treated as a substituent when it is attached to an alkane 'backbone.' That case is when the heteroatom is a halogen.

(5) **If a halogen is present in the molecule, treat it as a substituent, dropping its -ine ending and adding -o. Fluorine becomes fluoro, chlorine becomes chloro, bromine becomes bromo, and iodine becomes iodo.**

Give the Name for 3.12.

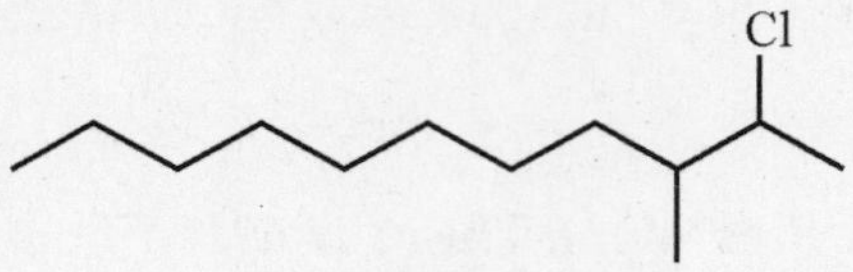

3.12

The longest continuous chain is 11, so this is an undecane. Giving the substituents the lowest numbers, Cl is at C2, and the methyl group is at C3. The chlorine is treated as a substituent, and is named chloro. Therefore, ***3.12*** is named 2-chloro-3-methylundecane.

Give the Name for 3.13.

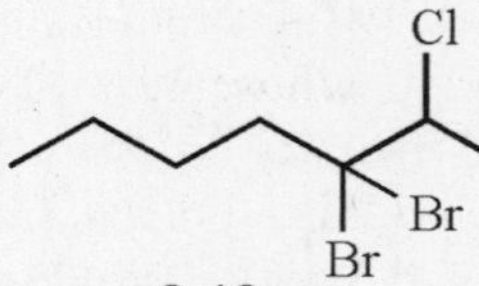

3.13

The longest continuous chain is 7, so this is a heptane. Giving the substituents the lowest numbers, Cl is at C2, and the two bromines are on C3. The chlorine is treated as a substituent, and is named chloro. Likewise, the bromines are bromo, and there are two of them, so dibromo is used. Therefore, ***3.13*** is named 2-chloro-3,3-dibromoheptane.

How Are Alkanes Named When the Substituent is More Complex Than the Simple Linear Alkyl Groups?

In many isomeric alkanes, the substituent is much more complex than a simple linear fragment such as methyl, ethyl, etc. The substituent can also be a group or atom other than a carbon group. In general, if a functional group is present, that group dominates the name (*see nomenclature sections in later chapters*).

For carbon substituents containing complex structures, the rule is more complex.

(6) If the carbon substituent is complex (not linear), identify the longest continuous chain of the substituent and use rules 1 and 3 from above to determine if it is an ethyl, butyl, hexyl group, etc. Determine the substituents on the substituent chain, number the substituent chain such that the point of attachment to the main chain is always 1, and assign a number to the secondary substituent. Set apart the complex substituent in parentheses.

Name the Fragment 3.14.

Let R represent an undefined alkyl group that constitutes the longest continuous chain. (We will use R for generic alkyl groups throughout the book.)

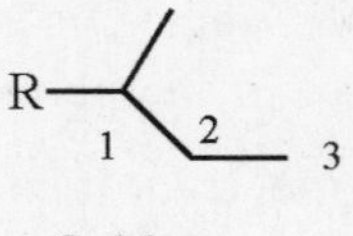

3.14

The point of attachment to R (the longest chain) is the second carbon of a four-carbon fragment. Number the longest carbon chain of the fragment from point of attachment, and we see that this is a propyl group with a methyl group is attached to C1 of the fragment. Therefore, this fragment is named 2-methylpropyl.

Give the Name for 3.15.

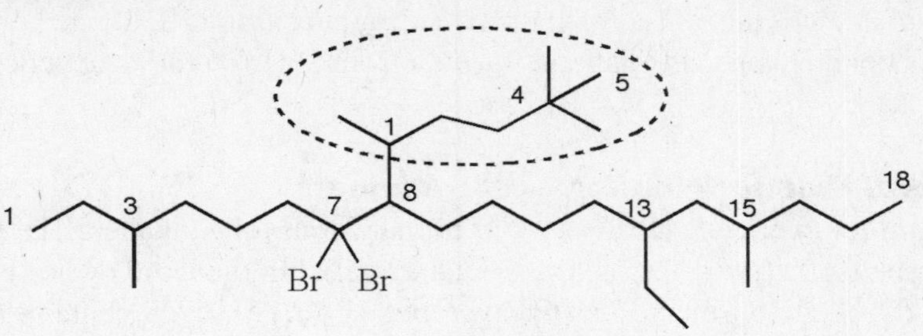

3.15

This specific example illustrates several rules. The longest chain is 18, and this is an octadecane. There are two methyl groups at C3 and C15, two bromines at C7 and an ethyl (two-carbon fragment) at C13. At C8, there is a complex substituent whose longest chain is five carbons from the point of attachment to the octadecane chain (1 $\rightarrow$ 5), and is therefore a pentyl side chain. Substituents are on the side chain –three methyls are attached to C1 and at C4. This side chain is, therefore (1,4,4-trimethylpentyl). The name of ***3.15*** is, therefore, 7,7-dibromo-13-ethyl-3,15-dimethyl-8-(1,4,4-trimethylpentyl)octadecane.

Common Names

A system of naming organic molecules developed over many years and persists today for some molecular fragments and small molecules. This system of common names is based on the number of methyl groups on a fragment and their relative position. The IUPAC rules are relatively recent.

Discuss the Structural Basis for Common Names.

The iso- group is used when one carbon at the end of the chain bears a CH and two methyl groups. The total number of carbon atoms dictates the name (isopropyl for three carbons, isobutyl for four carbons, etc.). The *secondary* (or *sec*-) group has a CH, one methyl and an alkyl group (not methyl). This leads to *sec*-butyl for four carbons and *sec*-pentyl for five carbons. The tertiary (*tert*- or just *t*-) group is a carbon with two methyls and one alkyl (which may be methyl). This leads to *t*-butyl for four carbons, *t*-pentyl for five carbons, etc. The -$CH_2C(CH_3)_3$ group is called neo-, as in neopentyl for the five carbon fragment. Common names are used primarily with simple molecules, although they are used extensively to denote small groups (an isopropyl group, a *t*-butyl group, etc.).

Give the Structure for Isopropyl, Isobutyl, Isopentyl, Secondary Butyl, Secondary Pentyl, Tertiary Butyl, Tertiary Pentyl and Neopentyl.

-$CH(CH_3)_2$ isopropyl
-$CH_2CH(CH_3)_2$ isobutyl
-$CH_2CH_2CH(CH_3)_2$ isopentyl
-$CH(CH_3)CH_2CH_3$ *secondary* (*sec*)-butyl
-$CH(CH_3)CH_2CH_2CH_3$ *sec*-pentyl
-$C(CH_3)_3$ *tertiary* (*tert*-, *t*-) butyl
-$C(CH_3)_2CH_2CH_3$ *tert*-pentyl
-$CH_2C(CH_3)_3$ neopentyl

3.4. Cyclic Alkanes

Alkanes are found where the carbons form a ring. In other words, the continuous chain of carbon atoms forms a ring, or cyclic structure. These compounds are known as cyclic alkanes, or cycloalkanes. There are important differences between cyclic alkanes and the linear alkanes just discussed.

What Is the General Formula for a Cyclic Alkane?

There is a class of alkanes that exists with the two terminal carbons tied together into a ring (a cyclic alkane). Forming this bond changes the general formula from C_nH_{2n+2} for an acyclic (no rings) alkane to C_nH_{2n} for a cyclic alkane.

What Is the General Rule for Naming Cyclic Alkanes?

The nomenclature for cyclic alkanes is similar to that for acyclic alkanes. The total number of carbons dictates the alkane name (propane, butane, pentane, etc.), but the word cyclo- is added to show that it is a ring. This leads to cyclopropane (***3.16***), cyclobutane (***3.17***), cyclopentane (***3.18***), cyclohexane (***3.19***) and cycloheptane (***3.20***), etc.

Draw the Structures of Cyclopropane, Cyclobutane, Cyclopentane, Cyclohexane and Cycloheptane, Using Line Notation.

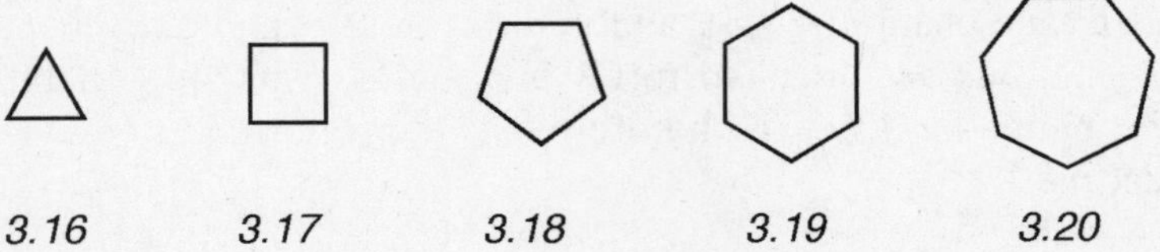

3.16 3.17 3.18 3.19 3.20

What Is the Rule for Cyclic Alkanes That Have One Substituent?

The rules are the same. Place the substituent name with a number in front of the cycloalkane name.

If There Is Only One Substituent, As in Methylcyclopentane 3.21, Is the Number Required?

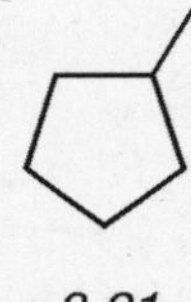

3.21

No! Giving the methyl group the smallest number (1) is obvious, and it is not necessary to use it in the name. Compound ***3.21*** is simply methylcyclopentane.

Give the Name for 3.22.

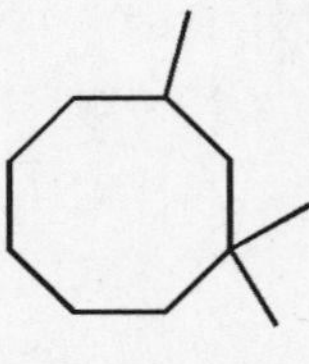

3.22

Compound ***3.22*** is an eight-membered ring and a cyclooctane. There are three methyl groups. The rules change a bit. Begin at the point of attachment (C1) and number the substituents on the ring so that the *second* substituent has as low a number as possible. If there is still ambiguity, number so that the *third* or *fourth* substituent has as low a number as possible until a point of difference is found. In this case, we use 1,1,3 rather than 1,3,3 or any other combination. Compound ***3.22*** is, therefore, 1,1,3-trimethylcyclooctane.

Give the Name for 3.23.

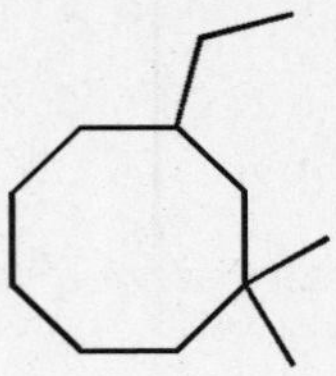

3.23

Compound ***3.23*** is also an eight-membered ring and a cyclooctane. There are now two methyl groups and an ethyl group. Before assigning number, we must decide which substituent is on C1. Remember that we number to give the smallest number to the substituent, so it is 1,1,3 rather than 1,3,3. Therefore, ethyl is on C3, and the methyl groups are one C1.The ethyl group is placed before methyl (e before m), and *3.23* is named 3-ethyl-1,1-dimethylcyclooctane.

Test Yourself

1. Indicate which of the following formulae correspond to an alkane.
   (a) C_6H_{12} (b) C_8H_{18} (c) C_5H_8 (d) C_6H_6 (e) $C_{40}H_{80}$ (f) $C_{40}H_{82}$
2. Briefly explain why 3-butyl-5-methylundecane is an improper name for the molecule shown.
3. Give the IUPAC name for each of the following.

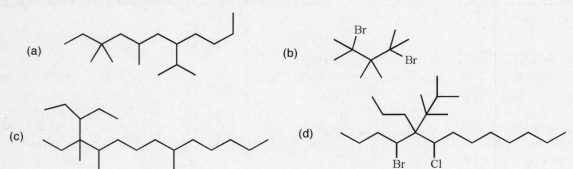

(a) 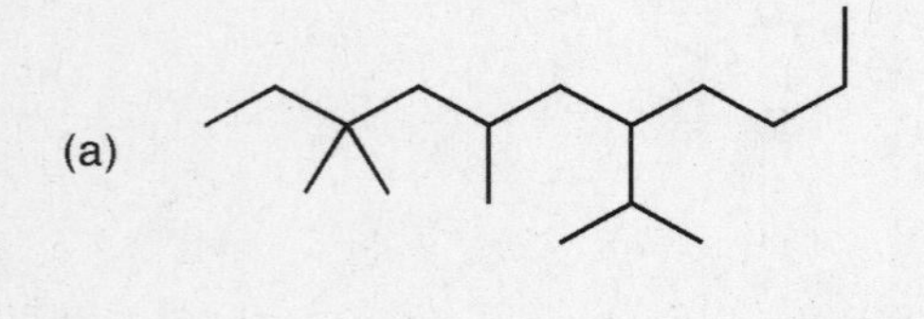

(b) 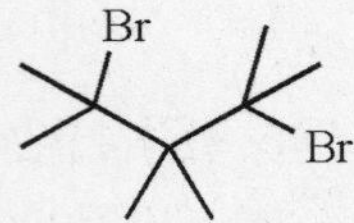

(c) 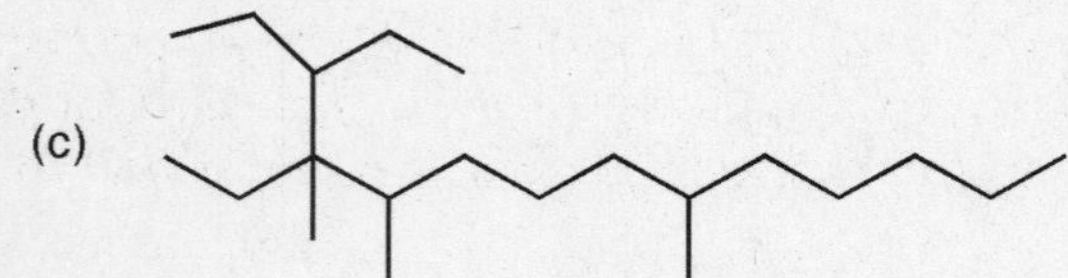

(d)

Br Cl

4. Give the structures in line formulae for each of the following IUPAC names:
   (a) 3,3,6,6-tetramethylpentadecane (b) 1-bromo-4-(2,2-dimethylbutyl)dodecane
   (c) 5,7,8-tetramethylicosane (d) 3,3,4,4-tetraethyloctane
   (e) 7-(3,3-dimethylpentyl)hexadecane.
5. Draw the structures of
   (a) neopentyl bromide (b) isobutane (c) *tert*-pentyl chloride
   (d) isopentane (e) *tert*-butyl iodide.
6. Draw eight different isomers for each of the following, using condensed structures:
   (a) C_8H_{18} (b) C_7H_{16}
7. Give the IUPAC name for each of the following.

(a) 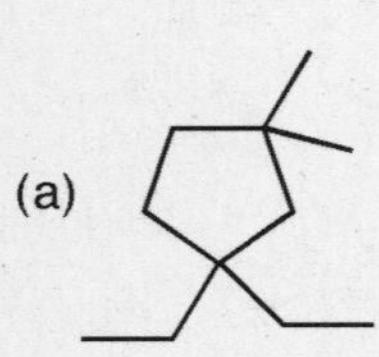

(b) 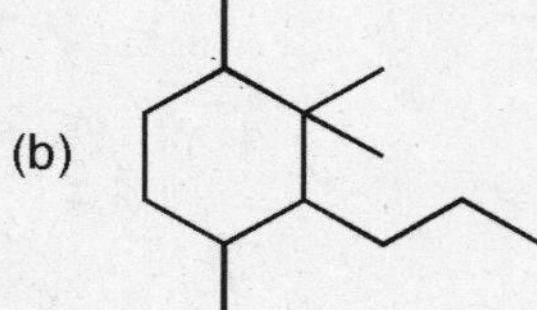

(c) 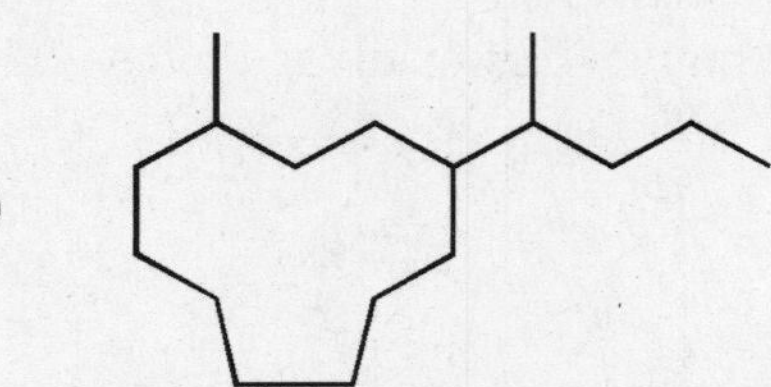

CHAPTER 4

Conformations

4.1. ACYCLIC CONFORMATIONS

Rotation of 360° can occur about covalent carbon-carbon single bonds. This rotation leads to different arrangements of the atoms, although it does not change the connectivity. This change in arrangement of the atoms is a dynamic process, and there is literally an infinite number of possibilities. The specific arrangements are called rotamers. The lowest energy arrangement is effectively the actual shape of the molecule and is called its conformation. Due to steric and electronic interactions, some rotamers and conformations are lower in energy than others.

What Does the Term Acyclic Mean?

A molecule that does not have a ring.

What Is the Result of a Molecule Dissipating Excess Energy by Rotation about a Single Covalent Bond?

The atoms in a covalent bond can rotate around that bond. This occurs because the molecule must dissipate excess energy (heat), and molecular motion (vibration, rotation, etc.) is one way the molecule can do this. When the atoms or groups on a carbon-carbon single bond rotate, the spatial arrangement of the atoms or groups changes with the rotations. These are called **rotamers**.

Rotation Around Carbon Bonds (Ethane)

Draw Ethane.

The two-carbon alkane (ethane) is shown as ***4.1***.

$$H_3C{-}CH_3$$

4.1

Discuss What Occurs If There Is Rotation about the Carbon-Carbon Bond of Ethane.

If the carbon-carbon bond in ethane is examined for rotamers, it is clear there is virtually an infinite number of possibilities. We will focus attention on two rotamers, the highest energy and the lowest energy, since these will determine the barriers to overall rotational motion of the molecule. In one of these (***4.2***), it appears that one hydrogen on the 'back' carbon bisects the angle formed by H–C–H on the 'front' carbon. The other (***4.3***) is shown at the right, and the two hydrogen atoms 'eclipse' each other.

Why Focus on Rotation about the C–C Bond Rather Than One of the C–H Bonds?

A hydrogen atom is small and symmetrical, so there is little change in the interaction between atoms during the rotation. However, in ethane each carbon has three hydrogen atoms attached in a tetrahedral array, so rotation about the C–C bond will lead to different interactions and different rotamers.

What Is a Sawhorse Diagram?

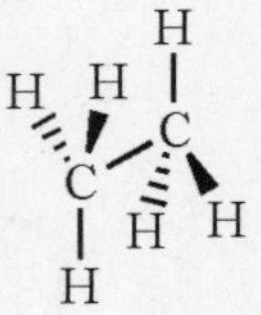

4.2 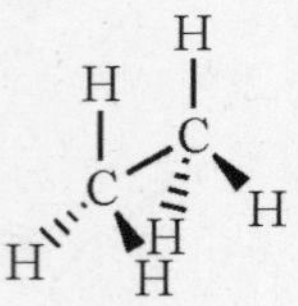 *4.3*

A sawhorse diagram is a way to represent a molecule by focusing on a specific bond. In effect, the molecule is turned so that one carbon atom is offset from the other, and the tetrahedral geometry of the remaining bonds is represented by a line, a solid wedge (projected out), and a dashed line (projected back). Two rotamers for ethane are shown (***4.2*** and ***4.3***) to illustrate.

What Is a Newman Projection?

A Newman projection is a way to represent a molecule by focusing on a specific bond. The molecule is turned so we sight down the bond, with one carbon atom in front and the other carbon atom immediately behind it. For convenience, the front carbon atom is the intersection of the three bonds, and the rear carbon is drawn as a circle. The tetrahedral geometry of the remaining bonds is shown by having three bonds projected from the dot and three lines projected from the circle. One such rotamer of ethane is shown in Newman projection (***4.4***) to illustrate.

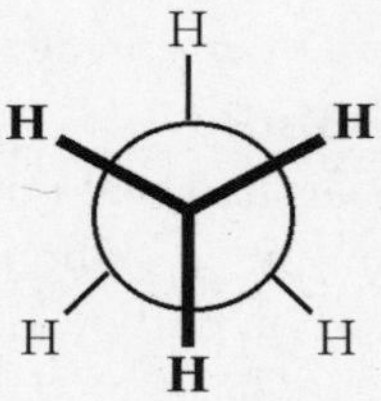

4.4

Why Is 4.2 the Lowest Energy Rotamer of Ethane?

In principle, rotation about the C–C of ethane can lead to an infinite number of rotamers. We can focus attention on only two, the lowest energy and the highest energy rotamers. The low-energy structure (***4.2***) is called the **anti** rotamer and the high-energy structure (***4.3***) is called the **eclipsed** or **syn** rotamer. Structure ***4.2*** is the lowest energy since the hydrogens are further apart than in any other rotamer. As the molecule rotates about the carbon-carbon bond, the hydrogen atoms come into close proximity in some rotamers (such as ***4.3***) and compete for the same space. The electrons in the bonds repel as they get close, and the atoms themselves repel if they get too close. This repulsion is called a **steric interaction**, or **steric hindrance**. Rotamer ***4.2*** is the lowest energy rotamer because steric interactions are minimized.

Draw the Anti- and Eclipsed Conformations of Ethane in Newman Projection.

If these two rotamers are turned so that one 'sights' down the carbon-carbon bond, one carbon atom is in front and the other is behind it, and we use a **Newman Projection** to represent the rotamer. This view allows the steric interactions and especially the high/low energy rotamers to be seen more clearly. The anti rotamer is ***4.5*** and the syn rotamer is ***4.6*** (Note that the front carbon is offset slightly to make it easier to see the atoms and bonds.) If the rear carbon of the anti rotamer is imagined to be immobilized, and the front carbon is rotated 60° clockwise, the syn rotamer results. In other words, keeping the rear carbon fixed and rotating the front carbon by 60° converts ***4.5*** to ***4.6***. In these Newman projections, the anti rotamer clearly shows the hydrogens to be as far apart as possible. Conversely, the hydrogens are very close together in the syn rotamer, illustrating the higher energy of this rotamer. To get a better sense of the size of the atoms and what information the Newman projection is conveying, rotamer ***4.5*** is shown as a space-filling model (***4.7***). Similarly, ***4.6*** is shown as space-filling model ***4.8***. Finally, ***4.6*** is also shown as ***4.9***, in what is essentially a sawhorse projection (see ***4.3***). In the space-filling models, the size of the atoms is conveyed more accurately, and it is clear that the hydrogen atoms are closer together in ***4.8***.

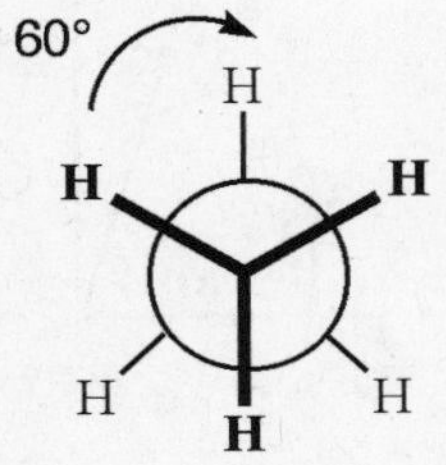

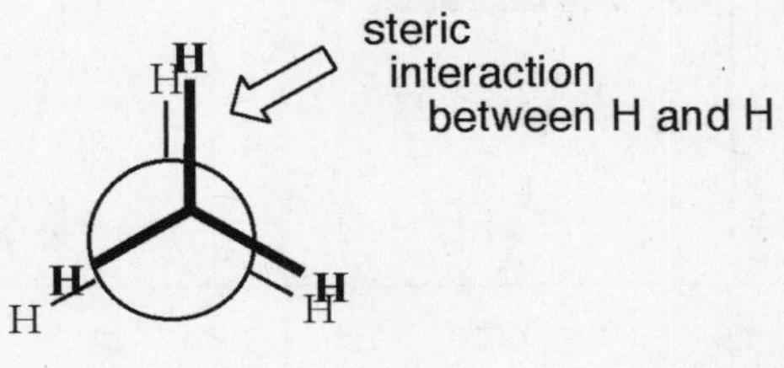

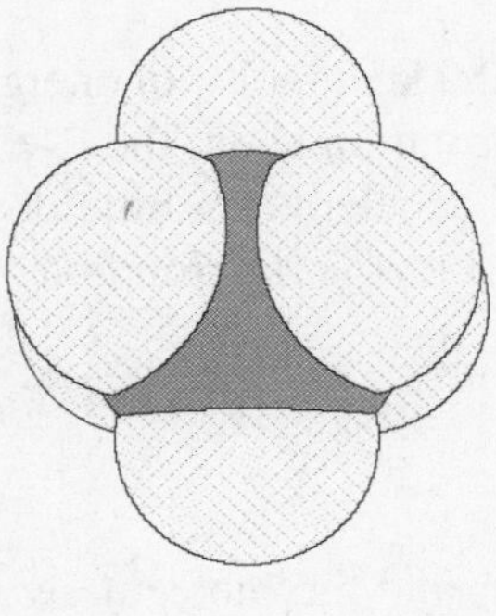

What Is an Energy Diagram for Rotamers in a Molecule?

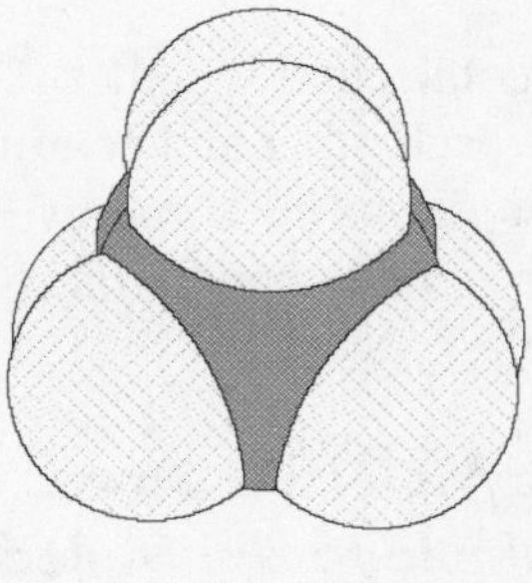

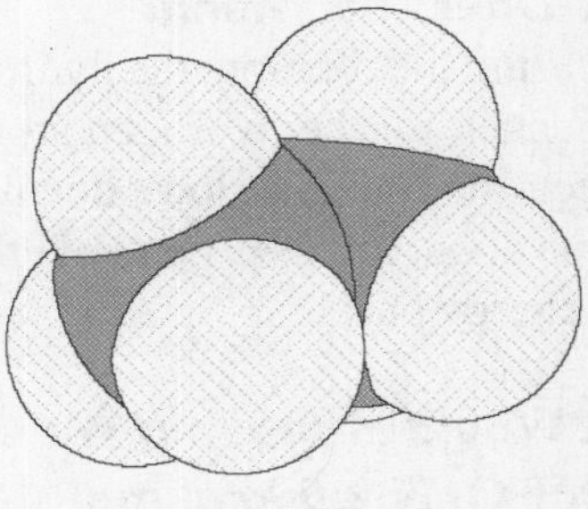

A rotation about the C–C bond occurs through 360°. The energy of the resulting rotamer will increase with increasing steric interactions and decrease as the steric interactions diminish. There will be energy minima and maxima. By plotting the energy of key rotamers at certain angles, a picture emerges of the energy demands for rotation.

Draw the Energy Diagram for Rotation of the Carbon-Carbon Bond of Ethane Through 360°.

Diagram *4.10* shows that the energy rises with a rotation of 60° to give the syn rotamer and falls to an energy minimum for the anti rotamer. There are three hydrogen atoms attached to each carbon, so there are three maxima and three minima.

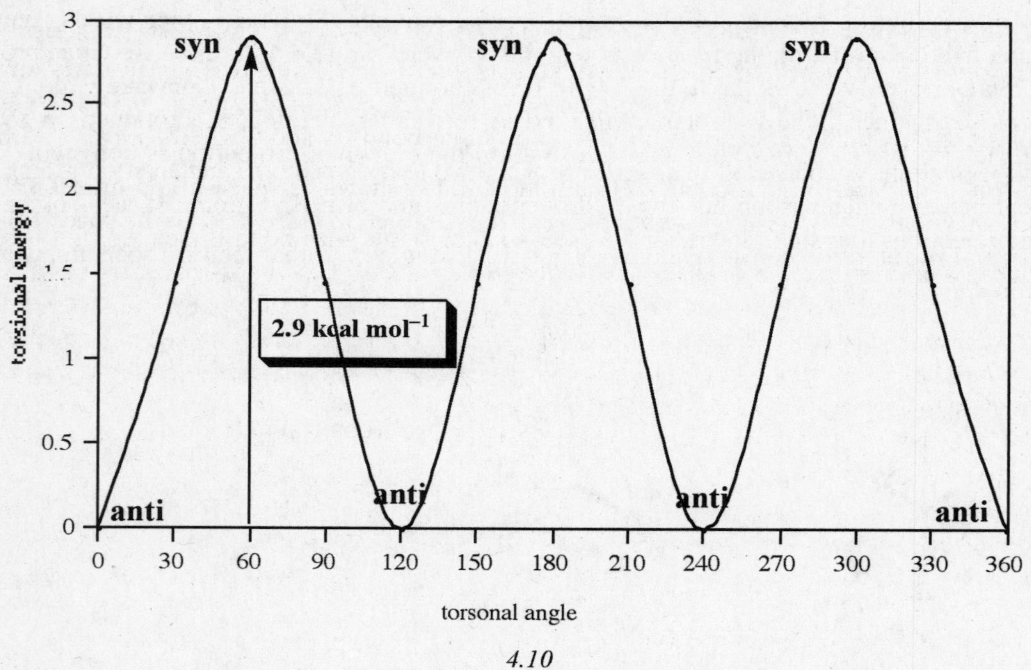

4.10

In 4.10, the Energy Maximum Was Measured to Be 2.9 kcal mol^{-1}. What Does This Mean?

During the rotation, the hydrogen atoms come close together. At least 2.9 kcal mol^{-1} of energy must be expended for the hydrogen atoms to 'pass' each other and continue the rotation about the C–C bond. In effect, the rotation slows down a little as it climbs each energy hill and accelerates a little downhill. Rotation about the C–C bond of ethane is not a smooth, continuous rotation, but a 'jerky' rotation because of the energy barriers.

What Would Happen If We Cooled the Molecule Down So That Only 2.0 kcal mol^{-1} Were Available to the Molecule?

There would not be enough energy to pass the 2.9 kcal mol^{-1} barrier, and the molecule would be essentially 'frozen' in that rotation would not be possible and would exist primarily in anti and near-anti rotamers.

What Does the Energy Diagram 4.10 Mean?

When ethane is rotated by 60° about the carbon-carbon bond, a series of three high-energy (syn) and three low-energy (anti) conformations appear as the bond rotates through 360°. This defines the 'free' rotation of ethane. Each time the rotation brings the hydrogens together, an *energy barrier* hinders the rotation (slows it down). This barrier is measured at 2.9 kcal mol^{-1}. Since at least 23 kcal/mol of energy are usually available at ambient temperatures, ethane rotates as shown in the graph, and one *cannot* isolate or 'freeze' the different rotational isomers.

Conformations

When Changing from Ethane to Butane, There Are Three C–C Bonds. Which Bond Is Most Important from the Standpoint of the Highest and Lowest Energy Rotamer?

There are three C–C bonds in butane, bonds A (C1–C2), B (C2–C3) and C (C3–C4) (see ***4.11***). Inspection of ***4.11***, which sights down bond A, shows that bonds A and C are identical: a carbon with three hydrogen atoms is connected to a carbon bearing two hydrogen atoms and an ethyl group. Bond B is different because each carbon has two hydrogen atoms and a methyl group. If we look at the anti Newman projections that sight down bond A (see ***4.12***) or sight down bond B (see ***4.13***), it is clear that there is more steric interaction in ***4.13***. The space-filling model ***4.14*** that focuses on bond A in ***4.11*** shows the steric interaction of the eclipsing H and CH_2CH_3 in ***4.12*** is less than the interaction of the two eclipsing methyl groups in ***4.13***. Space-filling model ***4.15*** (based on ***4.13***) shows that the two methyl groups are quite close together, leading to a significantly higher steric interaction. Therefore, the maximum energy demands will be for bond B.

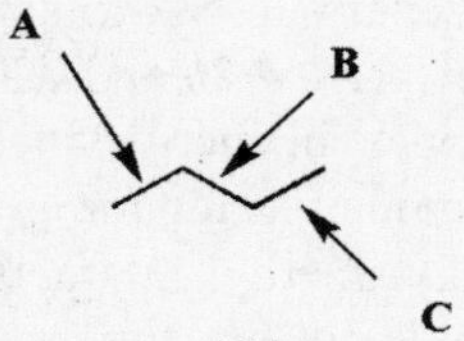

A B C

4.11

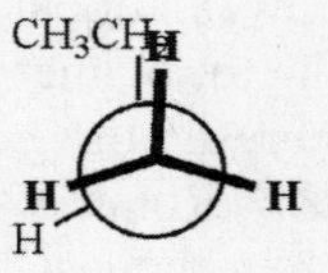

CH_3CH_2 H H H H H

4.12

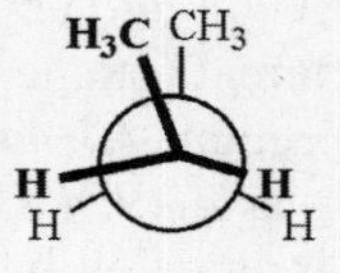

H_3C CH_3 H H H H

4.13

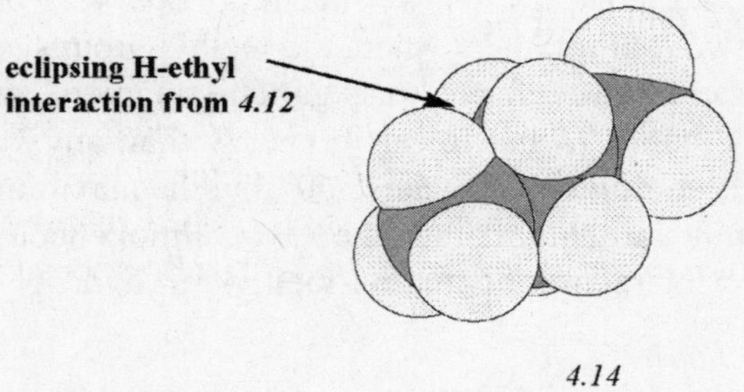

eclipsing H-ethyl interaction from *4.12*

4.14

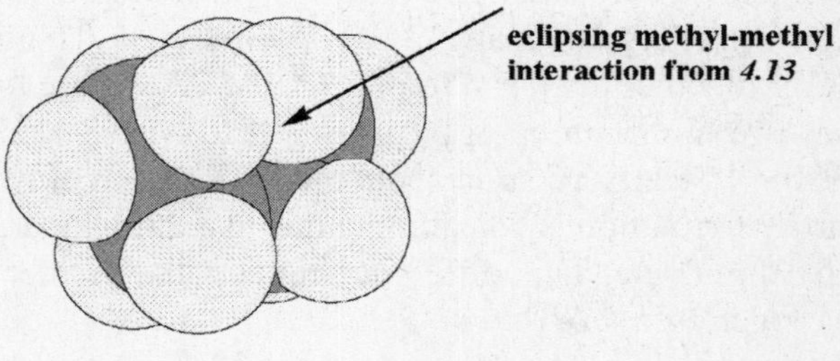

eclipsing methyl-methyl interaction from *4.13*

4.15

What Is the Effect on Butane When All Three C–C Bonds Are Considered Rather Than Just One?

Look at ***4.14*** and ***4.15***. It is clear that the molecules have different shapes. In ***4.14***, all three bonds are drawn as their anti rotamer (see ***4.16***), whereas in ***4.15***, bonds A and C are anti, but bond B is syn. When the rotamer for each bond is considered, the effect is to give the molecule as a whole a certain shape. This is known as its conformation. The conformations ***4.14*** and ***4.15*** are different, despite the fact that they are both butane. Therefore, molecules will exist as many conformations, since there are many possible rotamers for each bond, but a few conformations often dominate.

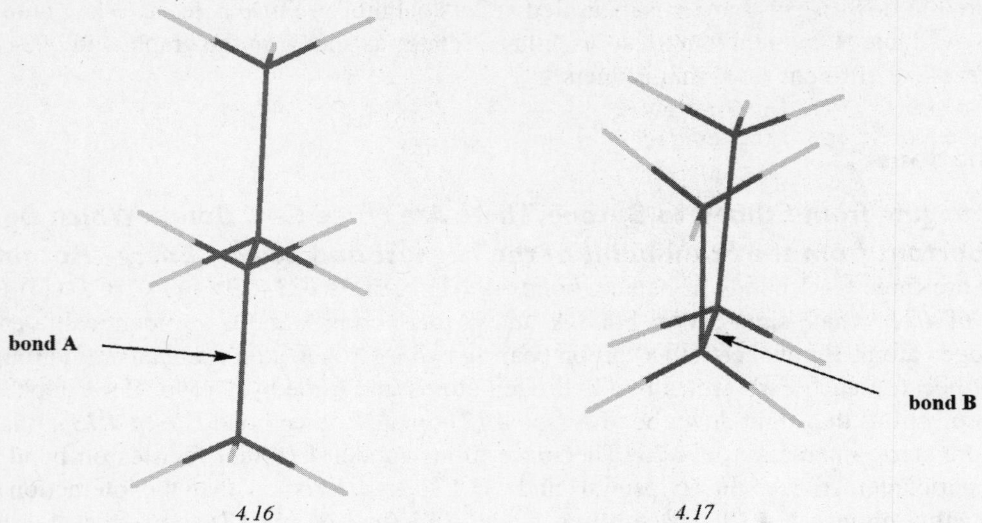

4.16 4.17

What Are the Important Conformations for the C2–C3 Bond of Butane?

With longer chain alkanes such as butane, we must focus on one bond at a time. Looking at bond B for butane (see ***4.11***), there are *anti* and *syn* conformations, but the relative positions of the two methyl groups are different in different rotamers, leading to rotamers of different energies. Six rotamers are considered to be the most important (***4.18-4.23***). There are three *anti* rotamers (***4.18, 4.20*** and ***4.22***) and three *syn* (eclipsed) conformations (***4.19***, ***4.21*** and ***4.23***). Rotamer ***4.18*** is the lowest in energy since the methyl groups are *anti*, and the interaction between these largest groups is responsible for the greatest steric interaction. Similarly, ***4.21*** must be the highest in energy since the methyl-methyl interaction is maximized. (This energy barrier has been measured to be 6.0 kcal/mole.) There are two additional *syn* rotamers that are equal in energy (3.4 kcal/mole), ***4.19*** and ***4.23***. These are lower in energy than ***4.21*** but higher in energy than any of the *anti* rotamers. There are two *anti* rotamers where the methyl groups are closer together than in ***4.18*** (***4.20*** and ***4.22***). These two rotamers are referred to as **gauche** rotamers, and they are higher in energy (measured to be 0.8 kcal/mol) than ***4.18*** but lower in energy than any *syn* rotamer. Clearly, rotation about the C_2–C_3 bond of butane is complex as it rotates 360°, but the maximum barrier to rotation is 6 kcal/mole due to ***4.21***, which is significantly higher than the 2.9 kcal/mol encountered in ethane. This difference reflects the greater steric interaction of methyl-methyl vs. hydrogen-hydrogen (see ***4.14*** and ***4.15***).

4.18 4.19 4.20

4.21 4.22 4.23

4.2. CONFORMATIONS OF CYCLIC MOLECULES

When a C–C bond is confined to a ring, rotation about 360° is impossible without breaking a bond and disrupting the ring. A twisting motion about the bond, called pseudo-rotation, occurs in order to dissipate excess energy. Pseudo-rotation of the bonds in rings leads to changes in the shape of the molecule—different conformations.

Is 'Free Rotation' Possible Around Carbon-Carbon Bonds in Cyclic Molecules? Why or Why Not?

No! Each of the carbons in the cyclic alkanes is, of course, three-dimensional with the usual tetrahedral geometry. Cyclic molecules *cannot* undergo the 'free' rotation observed with acyclic alkanes since the bonds are connected in a ring. In order to dissipate excess energy, however, the molecules distort by partial-rotation (pseudo-rotation) of the carbon-carbon bonds, leading to unique conformations for cyclic alkanes.

Baeyer and Torsion Strain

What Is Torsion Strain?

The Newman projections for planar cyclopentane (*4.24*) and planar cyclohexane (*4.25*) show a significant interaction of the eclipsing hydrogen atoms, which destabilizes both planar structures. This interaction is referred to as **torsion strain** or **Pitzer strain**.

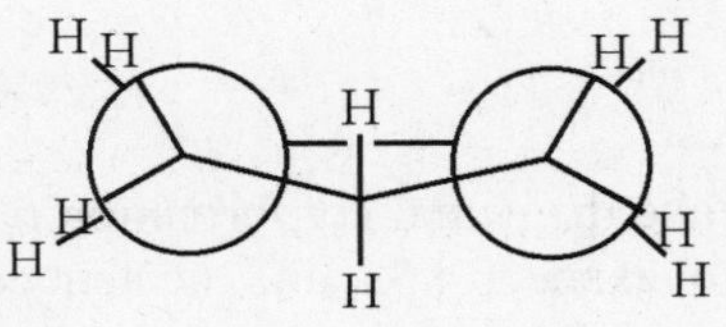

4.24 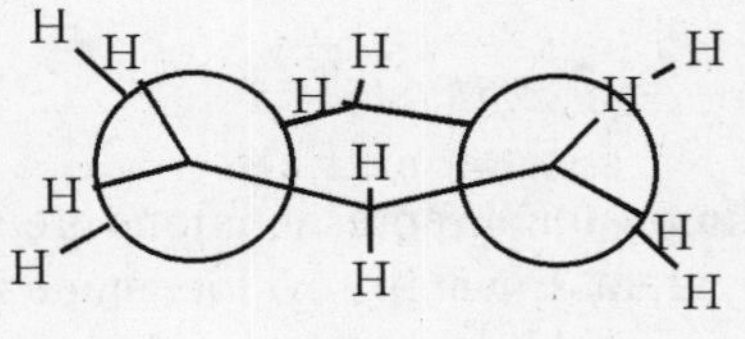 4.25

What Is Angle Strain?

When the bond angles in a ring are distorted from the 'ideal' tetrahedron of 109°28' (bond angles for methane), strain (higher energy) is introduced in the molecule. Such strain is called **angle strain** or **Baeyer strain**.

How Do Cyclic Molecules Relieve Torsion Strain and Angle Strain in C_3, C_4 and C_5 Cyclic Alkanes?

In cyclopropane, the angle strain is severe (bond angles of 60°). Angle strain is also high in cyclobutane (bond angles of 90°), although less than in cyclopropane. In planar cyclopentane, the bond angles of 108° show little angle strain, but significant torsion strain (see ***4.24***). In planar cyclohexane, the bond angles are 120°, and there is also great torsion strain (see ***4.25***). The steric and electronic interactions in these molecules are relieved by pseudo-rotation, and the lowest energy conformation is known to be ***4.26*** for cyclopentane, ***4.27*** for cyclohexane **and** ***4.28*** for cyclobutane. In other words, with the exception of cyclopropane, **cyclic alkanes are not planar**. Conformation ***4.26*** (cyclopentane) is referred to as the **envelope** conformation; conformation ***4.27*** (cyclohexane) is referred to as the **chair** conformation; and ***4.28*** is a **bent** conformation sometimes called the **butterfly** conformation. The Newman projections for envelope cyclopentane (***4.29***) and chair cyclohexane (***4.30***) show that the torsion strain has been relieved, and in ***4.30,*** each H–C–C–H bond resembles the gauche conformation of butane.

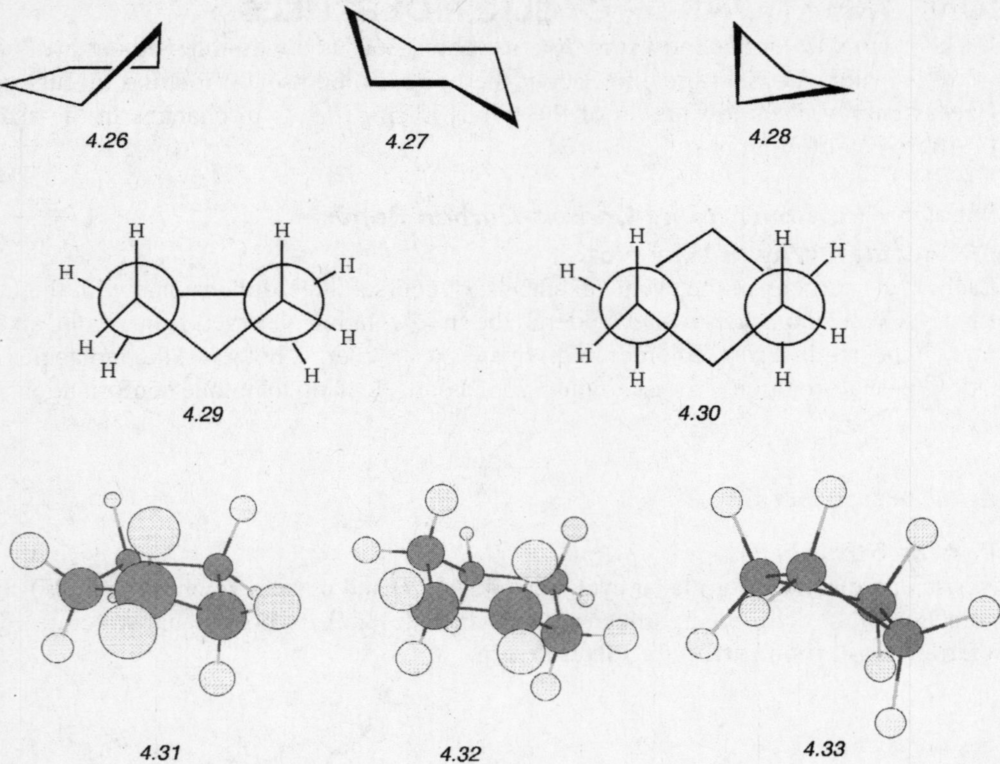

4.26 4.27 4.28

4.29 4.30

4.31 4.32 4.33

Three-dimensional representations are shown so as to make the actual conformation of each molecule more apparent. Envelope cyclopentane is ***4.31***, chair cyclohexane is ***4.32***, and the bent conformation of cyclobutane is ***4.33***. In each case, there is distortion from planarity, and the diminished steric interactions of the hydrogen atoms leads to less torsion strain.

What Are the Bond Angles in Chair Cyclohexane?

In cyclohexane, the bond angles are 111°. There is no angle strain, and there is minimal torsion strain.

In the Envelope Form of Cyclopentane the Angle Strain Is Increased Relative to the Planar Form. Why Then Is the Envelope Form of Cyclopentane the Lowest Energy Form?

In cyclopentane, the angle strain is increased (bond angles of 105°), but the significant reduction in torsion strain makes the envelope much lower in energy than planar cyclopentane.

Chairs, Boats, Twist-Boats and Half-Chairs

There Are Two Identical Chair Conformations of Cyclohexane. Discuss How One Is Converted into the Other by Partial Rotation About Carbon-Carbon Bonds.

Cyclohexane is particularly flexible, leading to several different conformations. The most important conformation is the **chair**, but there are two energetically identical chair conformations for cyclohexane

(*4.34* and *4.38*). They appear to be the same thing, but if a substituent is placed on the ring, the two chairs become different. This observation will be discussed later. Pseudo-rotation of chair ***4.34*** leads to a **twist-boat** conformation (or just a **twist conformation**, ***4.35***), and further pseudo-rotation leads to a **boat** conformation (***4.36***). Twisting the other side of the boat leads to another twist-boat (***4.37***), which then leads to the other chair conformation (***4.38***). There are also two significantly higher energy conformations, the two **half-chair** conformations, (***4.39***) and (***4.40***). Twist boats ***4.35*** and ***4.37*** are about 5.5 kcal mol^{-1} higher in energy than the chairs (***4.34*** and ***4.38***), but pseudo-rotation passes through the half chairs (***4.39*** or ***4.40***). The activation energy for this transition is about 10.8 kcal mol^{-1}. The boat conformation (***4.36***) is about 5.7 kcal mol^{-1} higher in energy than chairs (***4.34*** and ***4.38***).

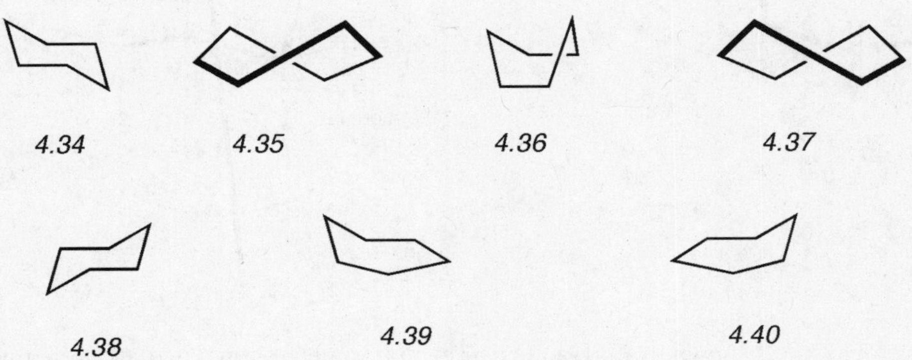

What Is a Transannular Steric Interaction?

When atoms or groups that are not connected by a bond are held close together in space due to the conformation of the molecule, there is a steric interaction. In cyclic molecules, such interactions are literally 'across the ring,' or transannular.

What Is a Flagpole Interaction in the Boat Conformation?

In the boat conformation, two of the hydrogen atoms, at C1 and C4 (across the ring), are held in relatively close proximity (see ***4.41***). These hydrogens, termed flagpole hydrogens, are close together because of the peculiar conformation of the ring, and there is a steric interaction called a flagpole interaction. The flagpole steric interaction contributes to the higher energy of this conformation. The interaction is a little clearer in the three-dimensional representation, ***4.42***.

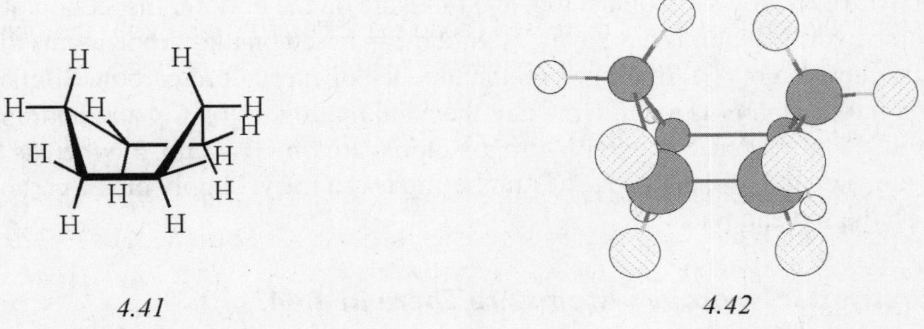

Axial and Equatorial Positions in Chair Conformations

Draw Chair Cyclohexane As a Planar Molecule and Also As a Chair, Being Careful to Include All Hydrogen Atoms.

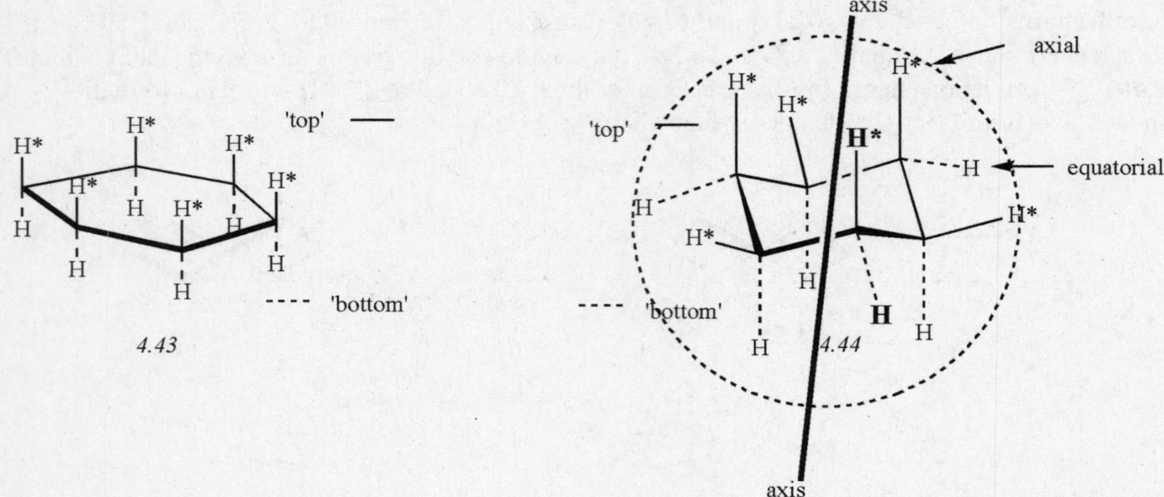

Planar cyclohexane is shown in ***4.43***, and chair cyclohexane is shown in ***4.44***. Note that these structures have a 'top' and a 'bottom' (see hydrogens [H*] in ***4.43*** and ***4.44***). When planar cyclohexane is distorted into a chair (see ***4.44***), the H* are in different relative positions. A 'globe' is superimposed on cyclohexane showing the imaginary axis. In both structures, all six H* are on 'top' of the molecule, but in ***4.44*** three of them are in the same direction as the imaginary axis, whereas three are aligned with the imaginary equator of the molecule. The 'vertical' H* are called **axial** hydrogens, and those bonds connected to H* are called **axial** bonds. The 'horizontal' H* are called **equatorial** hydrogens, and the bonds connected to those hydrogens are called **equatorial** bonds. The 'top' of the molecule has three axial positions and three equatorial positions. Likewise, the 'bottom' of the molecule has three axial and three equatorial bonds. Note that if a carbon atom has an axial bond on the 'top,' it will have an equatorial bond on the 'bottom'. Note that the terms 'top' and 'bottom' are arbitrary and simply a device to keep track of the 'sidedness' of the cyclohexane ring.

In Chair Cyclohexane, What Is the Relationship of the Axial Hydrogen Atoms on the Top and Bottom of the Ring?

Three axial hydrogen atoms are on the top, and three are on the bottom. Inspection of the top of ***4.44*** shows that the three hydrogen atoms have a '1,3' relationship based on the carbon atoms they are attached to. In other words, there is an axial hydrogen on the top face of every third carbon. Similarly, every third hydrogen on the bottom of ***4.44*** is axial. Note that the axial hydrogens on top and bottom appear on different carbon atoms. On the top, the axial hydrogen atoms are on C1, C3, C5, whereas the bottom has axial hydrogen atoms on C2, C4, and C6. The numbering is arbitrary; simply pick a carbon with an axial hydrogen atom on the top and make it C1.

How Many Equatorial Hydrogen Atoms Are There in 4.44?

Six.

Substituted Cyclic Alkanes

When a Hydrogen of Cyclohexane Is Replaced with a Methyl group the Two Chair Conformations Become Nonequivalent. Why?

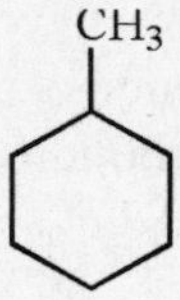

CH_3 H CH_3 CH_3 H

4.45 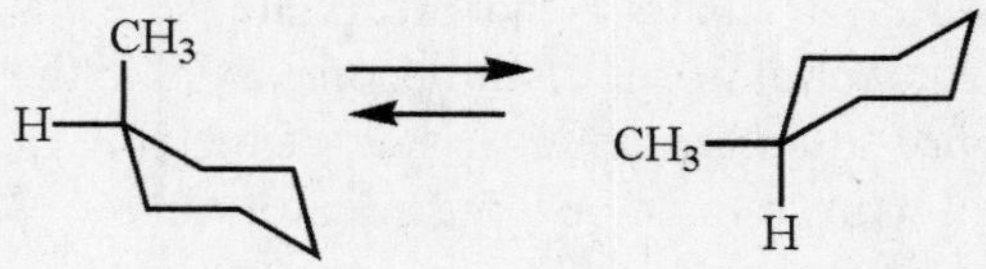 *4.46* *4.47*

1,3-diaxial steric interaction of methyl and Hs "A-Strain"

CH_3 H H H H H

normal 1,3-diaxial interaction of Hs

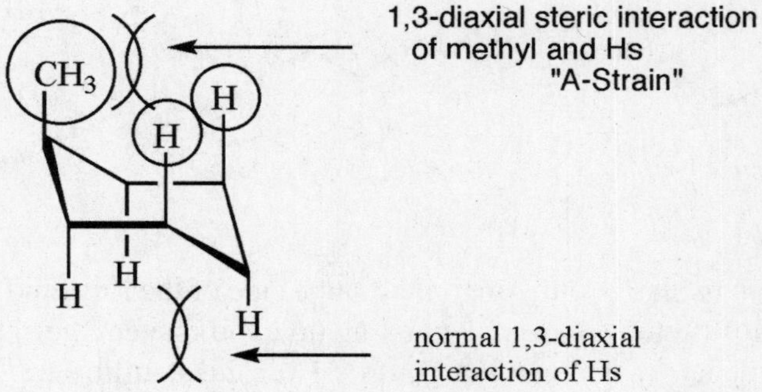

4.48

When the hydrogen atoms of cyclohexane are replaced with alkyl groups (or another group), that group can be in an axial position or an equatorial position. When methylcyclohexane (***4.45***) is converted to the two possible chair conformations, it is apparent that the methyl is in an axial position in one chair conformation (***4.46***) and in the equatorial position in the other (***4.47***). These two chairs are in equilibrium. As shown in structure ***4.48***, the presence of a methyl group in the axial position leads to a significant steric interaction with the axial hydrogens on that side of the ring (a 1,3-diaxial interaction called **A-strain**). The interaction of the three axial hydrogens on the 'bottom' of structure ***4.48*** is taken as the normal interaction and is not referred to as A-strain. The greater size of the methyl group, however, dictates that the steric interaction of the axial methyl with the axial hydrogens on the 'top' is greater than the interaction of the three axial hydrogens on the 'bottom.' The larger the group in the axial position (and the larger number of groups in the axial position), the greater the A-strain, and that conformation will have a higher energy. Examination of ***4.46*** and ***4.47***, therefore, suggests that ***4.46*** is higher in energy than ***4.47***, creating an energy difference (ΔE) for the two chair conformations.

Which Is Present in Greatest Amount at Equilibrium, 4.46 or 4.47?

Since ***4.46*** is at a higher energy level, the equilibrium will shift to favor the lower energy conformation, ***4.47***.

As Drawn, What Is the Expression for the Equilibrium Constant for 4.46 and 4.47?

The equilibrium constant expression is $K = \frac{[4.47]}{[4.46]}$

As Drawn, Is K Larger Than 1 or Smaller Than 1?

Since the equilibrium favors ***4.47***, that concentration term must be larger, and K is greater than 1.

Comparing 4.46 and 4.47, Why Is One Methyl Axial and the Other Equatorial?

When the chair flips, all atoms that are axial in one chair become equatorial in the other. Conversely, all atoms that are equatorial in one chair become axial in the other. This phenomenon occurs because axial and equatorial bonds alternate on each side of the ring (see ***4.44***). If an axial bond on the top pseudo rotates to the other chair, it must become equatorial.

For Each Molecule Drawn (A–C), Determine Which Chair Conformation Has the Greatest Amount of a-Strain?

(a) 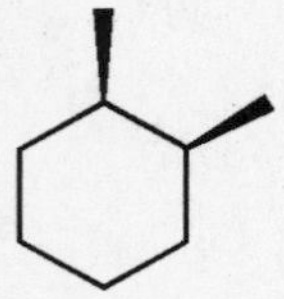 *4.49* (b) 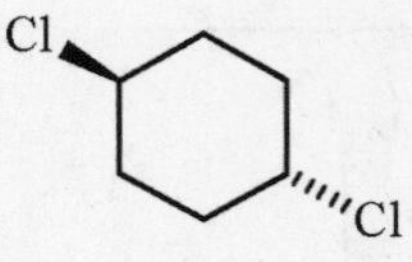 Cl, Cl *4.50* (c) 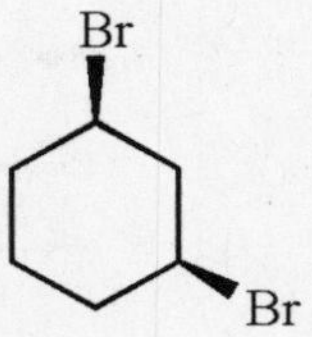 Br, Br *4.51*

For ***4.49*** in (a), both methyl groups are on the same side of the ring, and the two chair conformations are ***4.52*** and ***4.53***. Both structures have one axial methyl and one equatorial methyl group, and they are expected to be equal in energy. In other words, K = 1 for this equilibrium. For ***4.50*** in (b), the chlorine units are on opposite sides of the ring, and the two chair conformations are ***4.54*** and ***4.55***. Structure ***4.54*** has two axial chorine atoms whereas ***4.55*** has two equatorial chlorine atoms. Clearly, the A-strain in ***4.54*** is much greater than in ***4.55***, and the equilibrium will favor ***4.55***. In other words, K > 1 for this equilibrium. For ***4.51*** in (c), the bromine units are on opposite sides of the ring, and the two chair conformations are ***4.56*** and ***4.57***. Structure ***4.56*** has two axial bromine atoms, whereas ***4.57*** has two equatorial bromine atoms. Clearly, the A-strain in ***4.56*** is much greater than in ***4.57***, and the equilibrium will favor ***4.57***. In other words, K > 1 for this equilibrium. Note that the axial-equatorial relationship of the groups changes when the groups are on the same or opposite sides of the ring, as well as when there is a 1,2-, 1,3-, or 1,4-relationship between the groups. It is possible to memorize this pattern, but it can be confusing. Instead, draw out both chairs in all questions such as this in order to determine the relationship.

Why Does the Equilibrium Shift in Such a Way That There Is Only One Low-Energy Conformation for Tert-Butylcyclohexane, for all Practical Purposes?

The bulky *tert*-butyl group leads to a large amount of A-strain in ***4.58*** and virtually none in ***4.59***. Therefore, the equilibrium constant for the equilibrium between ***4.58*** and ***4.59*** is large, favoring ***4.59***. At equilibrium, there will be a higher percentage of ***4.59***, since it is significantly lower in energy than ***4.58***. In general, the lowest energy conformation will predominate for any given cyclic alkane. In this case, the energy difference is so large that there is much less than 1% of ***4.58*** present at equilibrium.

(a) 4.49 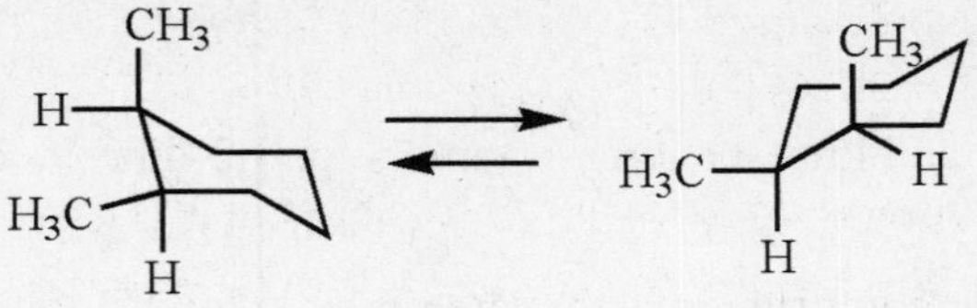 — 4.52 ⇌ 4.53

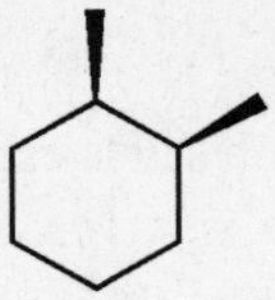

CH3 H H3C H CH3 H3C H H

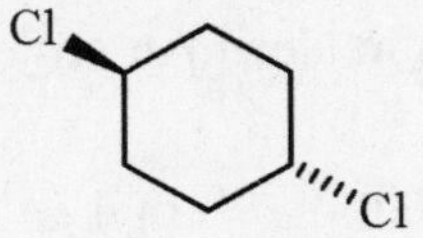

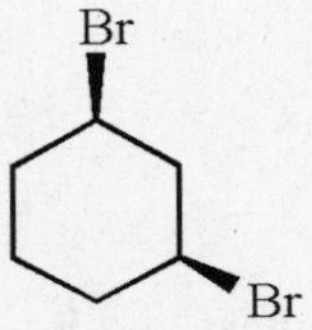

(b) 4.50 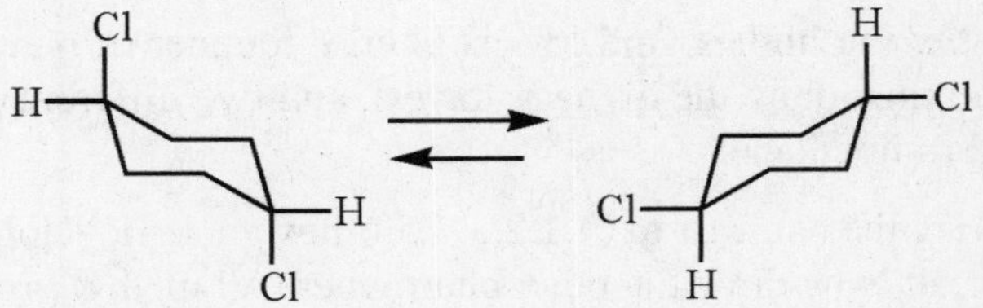 — 4.54 ⇌ 4.55

Cl Cl Cl H H Cl H Cl Cl H

(c) 4.51 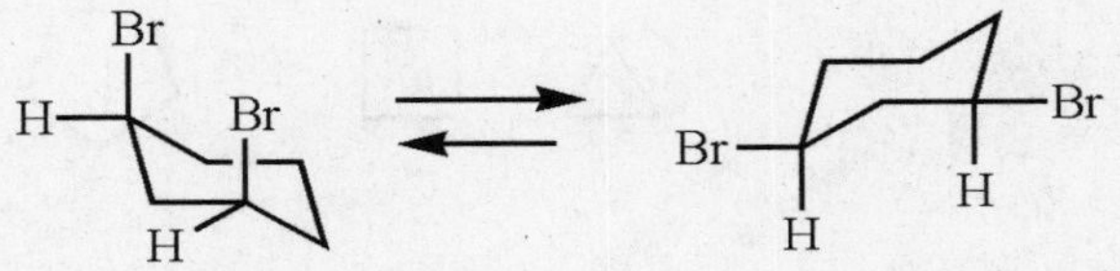 — 4.56 ⇌ 4.57

Br Br Br H Br H Br Br H H

Why Do Substituted Cyclohexane Derivatives Exist Largely in the Chair Conformation Rather Than the Other Conformations Available to Cyclohexane Derivatives?

Chair cyclohexanes predominate as the major conformation of cyclohexane because the boat, half-chair, and twist boat conformations are much higher in energy.

Similarly, the envelope form of cyclopentane predominates over the planar form because it is lower in energy. Likewise, methylcyclohexane exists largely in the chair form with the methyl group equatorial.

Test Yourself

1. Draw the highest and lowest energy rotamers for 1,2-dibromoethane, both in sawhorse diagrams and in Newman projection.
2. Draw the two boat conformations that are possible for 1,4-dichlorocyclohexane, where both chlorine atoms are on the same side of the molecules.
3. Of the two boat conformations drawn for question 2, which is present in higher percentage in the equilibrium mixture? Why?
4. Draw the highest and lowest energy rotamer in Newman projection for the C_3–C_4 bond of hexane, and then draw the highest, lowest, and two different gauche rotamers in Newman projection for 1,2-dibromoethane.
5. Draw the chair form of 1,2,3,4,5,6-hexamethylcyclohexane such that all the methyl groups are equatorial. Now draw the other chair where all methyl groups are axial.
6. Which of the following molecules has the smallest angle strain? Which has the greatest torsion strain? Which has the angle strain greater than the torsion strain?

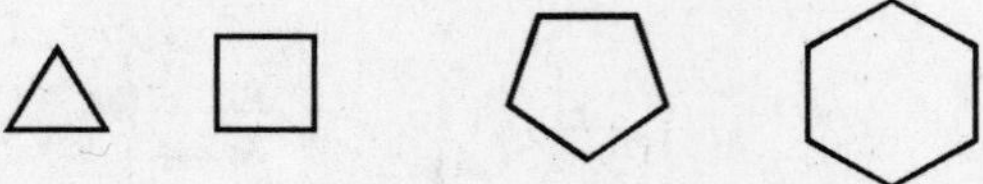

7. Draw both chair conformations for each of the following. Label the highest and lowest energy chair conformation in each case.

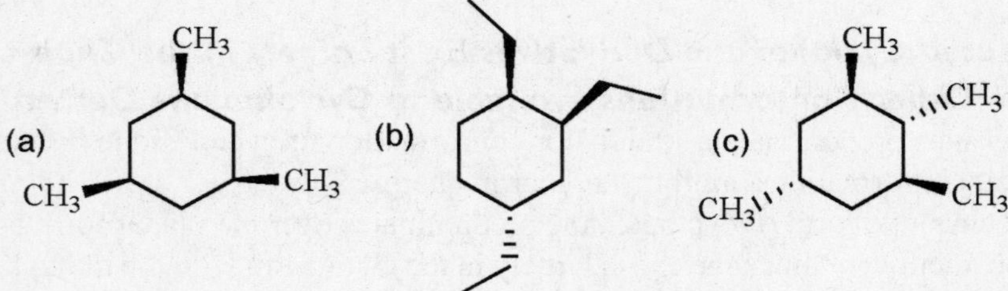

CHAPTER 5

Stereochemistry

Organic molecules are three-dimensional, and it is possible to form isomers that differ only in their spatial arrangement. Such molecules are called stereoisomers, and special properties are associated with stereoisomers that possess stereogenic atoms (an atom that possesses four different groups or atoms). This chapter will introduce the concept of chirality, including enantiomers, diastereomers, meso compounds, and absolute configuration. The Cahn-Prelog-Ingold selection rules will be introduced in order to determine the absolute configuration of a stereogenic center. The optical properties of stereogenic molecules will also be discussed, including the highly important physical property of specific rotation.

5.1. CHIRALITY

When a molecule is asymmetric about a given atom (there is no symmetry in the molecule), it exhibits a property known as **chirality** at that atom. Formally, the mirror image of that molecule is a different molecule. The property of chirality is important in chemical reactions and in biological systems.

Define a Stereoisomer

When two different molecules have the same molecular formula and the same connectivity (all atoms are attached in an identical manner), but differ in their spatial arrangement about a given atom or point in the molecule, they are called **stereoisomers**.

What Is a Stereogenic Center?

A stereogenic center, also called an asymmetric center or a chiral center, is defined as an atom (usually carbon, but it does not have to be carbon) that has four *different* atoms or groups attached to it and possesses no symmetry; it is asymmetric. An example is 2-butanol ($CH_3CH_2\underline{C}H(OH)CH_3$) where $\underline{C}$ is the stereogenic center. The stereogenic center has a H, an OH, a CH_3 and a CH_2CH_3 attached, so there are four different atoms or groups.

Which of the Following Molecules Have a Stereogenic Center?

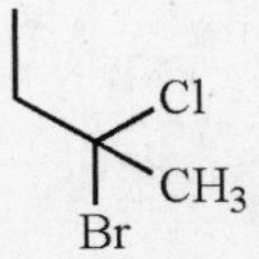

Cl CH3 Br

5.1

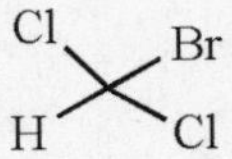

Cl Br H Cl

5.2

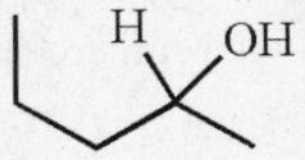

H OH

5.3

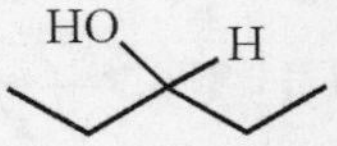

HO H

5.4

Of these four molecules, *5.1* and *5.3* have four different groups or atoms attached to a central carbon atom. Therefore, *5.1* and *5.3* have stereogenic centers. Compound *5.2* has two chlorine atoms, and compound *5.4* has two ethyl groups. Neither has a stereogenic center.

Give a Working Definition of Chirality.

Chirality is a property of molecules by which the spatial arrangement of atoms leads to asymmetry, and the mirror image of that molecule will be a different molecule. This difference in three-dimensional structure leads to a new type of isomer called a **stereoisomer**. Stereoisomers differ only in the spatial position of their groups. A molecule with this property is said to be **stereogenic**. Many natural substances are stereogenic, and this property is a key factor in their reactivity, particularly with amino acids and enzymes, saccharides, DNA, and RNA. Stereogenic molecules are also produced by plants and bacteria, and these substances have various properties, including defensive or attractive substances. Stereogenic molecules are also responsible for the chemicals that trigger odor responses in humans when they are released into the air from plants or animals. There is an endless list of important chiral molecules.

Enantiomers

When a mirror is held up to a molecule possessing a stereogenic center, its *mirror image* is observed. If one tries to superimpose, or match every atom in both molecules by laying one on the other, it becomes clear that the stereogenic molecule and its mirror image are *not superimposable*. They are, therefore, *different molecules*. **Enantiomers** are defined as molecules that have non-superimposable mirror images. In a more precise definition: two stereoisomers that are non-superimposable mirror images are enantiomers. It is important to understand that recognizing enantiomers as stereoisomers is an important part of the definition.

Draw the Enantiomers for 2-Bromobutane.

In 2-bromobutane, C_2 is stereogenic (a bromine, a methyl, an ethyl, and a hydrogen are attached to it). One stereoisomer is *5.5*, and its mirror image is *5.6*. If *5.6* is rotated 180°, the structure shown below *5.5* is the result (also labeled *5.6*). If one tries to superimpose these molecules (see *5.7*), the methyl and hydrogen do not match, and they are different molecules. Two non-superimposable mirror images of molecules that contain a stereogenic center are called **enantiomers**. Therefore, *5.5* is the enantiomer of *5.6,* and *5.6* is the enantiomer of *5.5*.

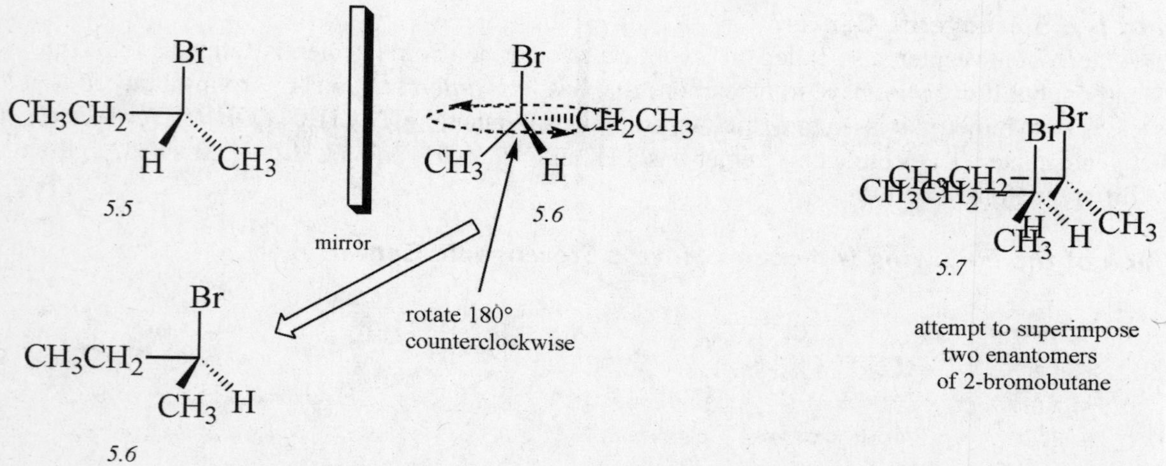

5.5

mirror

5.6

rotate 180° counterclockwise

5.6

5.7

attempt to superimpose two enantomers of 2-bromobutane

Are 5.8 and 5.9 Enantiomers?

H_3CH_2C—C(Br)(H)—CH_2CH_3 | (Br)C(H)(CH_2CH_3)—CH_2CH_3

5.8   *5.9*

No!. Two ethyl groups are connected to the carbon of interest, so there is no stereogenic center; therefore, ***5.8*** and ***5.9*** are the same molecule, not enantiomers. In other words, two molecules that *are* superimposable are the same.

Fischer Projections

A convenient notation for molecules containing a stereogenic center involves crossed lines. The horizontal line represents the bonds and the attached atoms projected out of the plane of the paper *towards you.* The vertical line represents bonds projected behind the plane of the paper *away from you.* This representation is called a **Fisher projection**. Structure ***5.10*** is the Fisher projection of one of the 2-bromobutane enantiomers from above (***5.5***). If ***5.10*** is the Fisher projection of enantiomer ***5.5*** from above, then its mirror image is ***5.6***, and the Fisher projection is ***5.11***. Enantiomer ***5.10*** is drawn again as ***5.12***, and ***5.11*** is drawn as ***5.13*** as a reminder of the actual stereochemistry represented by the Fischer projection.

Br / CH_3—C—CH_2CH_3 / H ≡ Br / CH_3—+—CH_2CH_3 / H | Br / CH_3CH_2—+—CH_3 / H ≡ Br / CH_3CH_2—C—CH_3 / H

5.12   *5.10*   mirror   *5.11*   *5.13*

Draw the Fisher Projections for Both Enantiomers of 3-Methylheptane.

The two Fischer projections are ***5.14*** for one enantiomer and ***5.15*** for the other.

CH_3 / $CH_3CH_2CH_2CH_2$—+—CH_2CH_3 / H | CH_3 / CH_3CH_2—+—$CH_2CH_2CH_2CH_3$ / H

5.14   mirror   *5.15*

5.2. OPTICAL ROTATION

Enantiomers have identical physical properties except for one. Each enantiomer rotates plane-polarized light in a different direction. This physical property is known as **optical rotation** and can be used to differentiate enantiomers.

What Is a Chiral Molecule?

A molecule that does not possess symmetry and rotates plane-polarized light is called a **chiral molecule**. Generally, a chiral molecule will possess one or more stereogenic centers.

Is It Possible to Distinguish Enantiomers by Their Physical Properties?

If a molecule contains one stereogenic center, it exists as two enantiomers that are different molecules. They have absolutely identical physical properties (melting point, boiling point, density, solubility, etc.) except for their ability to interact with plane-polarized light. One enantiomer will rotate plane-polarized light to the left (counterclockwise), and the other enantiomer will rotate it to the right (clockwise).

How Is the Rotation of Plane-Polarized Light Measured for a Given Enantiomer?

The instrument used to measure this property is called a **polarimeter**. A polarimeter consists of a light source with a polarizing filter. The resulting polarized light is confined to a single plane and is directed *through* a chamber containing a *solution* of the stereogenic molecule in an appropriate solvent. An appropriate solvent is one that dissolves the stereogenic molecule and does not itself have a stereogenic center. One sights through the tube containing the sample —in modern instruments this is done electronically —and when compared to a blank (a tube containing only solvent), the angle of the plane-polarized light, which changes as it passes through the sample, is measured. This angle is measured in degrees (°) and is defined as the **observed rotation, α**. If α for a molecule is measured to be $+60°$ for one enantiomer, α for the other enantiomer under identical conditions will be $-60°$. The rotation is of the same magnitude but opposite in sign, as (+) denotes clockwise rotation of the light and (−) denotes counterclockwise rotation of the light. The temperature of the experiment (25°C in this case) and the type of light used (sodium D line) are usually included for observed rotation data: i.e. α_D^{25}.

What Is Optical Rotation?

Since the length of the polarimeter tube and the concentration and solvent used may vary, some standardization is required. If the tube is longer, there are more molecules present to interact with plane-polarized light for a given concentration, and the observed rotation will be larger. As the concentration increases, there are also more molecules interacting with the light, and α increases. A parameter called **specific rotation** is the optical rotation for a molecule under a specific set of conditions. It is given the symbol $[\alpha]_D^{25}$ =, measured in degrees (°). The parameters that determine this physical property are path length (l) recorded in decimeters (dm), concentration (c) in g mL^{-1} and the observed rotation, α. Specific rotation is then given by the expression: $[\alpha]_D^{25} = \dfrac{\alpha}{(l)(c)}$.

If $\alpha = +60°$, $l = 5$ dm and $c = 1.25$ g mL^{-1}, What Is $[\alpha]_D^{25} = ?$

Using the formula given above, $[\alpha]_D^{25} = +9.6°$.

If the Optical Rotation of One Enantiomer of 2-Butanol Is +13.5°, What Is the Optical Rotation of the Other Enantiomer?

The magnitude of the specific rotation is the same for both enantiomers. This parameter differs only in sign. If one enantiomer of 2-butanol has an optical rotation of +13.5°, the other enantiomer must have an optical rotation of −13.5°.

What Is a Racemic Mixture?

A special case arises when there is an equal mixture of the two enantiomers (50:50 mixture of the + and − enantiomer). The specific rotation of this mixture is zero (0), and the mixture is called a **racemic mixture**. It may sometimes be called a **racemic modification** or simply a **racemate**. This implies that the specific rotation of enantiomeric mixtures are additive.

If One Enantiomer Has an Optical Rotation of +60°, and Its Enantiomer is −60°, What Is the Optical Rotation of a 70:30 Mixture (+ : −)?

Given that the optical rotations are additive, the optical rotation of this mixture can be calculated: $[\alpha]_D^{25} = (+60°)(0.7) + (-60°)(0.3) = +42° + (-18°) = +24°$.

5.3. SEQUENCE RULES

If enantiomers are different compounds, they must have different names. A set of rules has been devised to allow different names to be assigned to enantiomers.

Specific Rotation vs. Absolute Configuration

Give the IUPAC Name for 5.5A and for 5.6A.

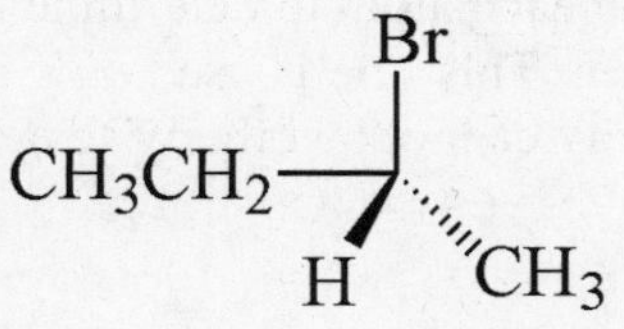

Br / CH_3CH_2 / H / CH_3

5.5A

Br / CH_3CH_2 / CH_3 / H

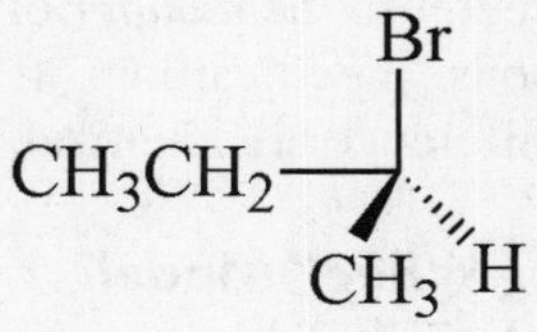

5.6A

The IUPAC name for *both compounds* is 2-bromobutane. However, another term must be added to the name in order to distinguish their stereochemical relationship.

For Enantiomer 5.5 of 2-Bromobutane, Is the Optical Rotation Positive (+) or Negative (−)?

Examining the optical rotation data for 2-bromobutane raises an interesting question: Is the optical rotation of enantiomer **5.5** positive (+) or negative (−)? There is *no way* to tell from the structure, and specific rotation tells nothing about the relative positions of the groups on the stereogenic carbon. There is an experiment in which the optical rotation is determined over a range of concentrations and temperatures that does give information about the relative positions of groups on a stereogenic center. This phenomenon is called circular dichroism, but it will not be discussed.

What Is Absolute Configuration?

Absolute configuration is the natural spatial arrangement of atoms around a stereogenic center. It is an interpretive property and not a physical property.

What Is the Distinction Between Optical Rotation and Absolute Configuration?

Optical rotation is a physical property of a molecule that is determined from the interaction of the enantiomers with plane-polarized light. Absolute configuration is the natural spatial arrangement of atoms around a stereogenic center and is an interpretive property, not a physical property. An enantiomer with a specified absolute configuration could have either a (+) or a (−) optical rotation.

Cahn-Ingold-Prelog Sequencing Rules

The set of rules used to determine absolute configuration is called the **Cahn-Ingold-Prelog Selection Rules**. The rules are used to assign a designator (**R** or **S**) to each enantiomer. The rules assign a priority (a,b,c,d, where a = highest and d = lowest priority) to each **group** or **atom** on the stereogenic center. The priority for a group is based on the **atom** attached to the stereogenic center. These sequencing rules are as follows:

(1) Working from the point of attachment to the stereogenic center, the first atom encountered is prioritized according to its **atomic number**. Therefore, F > O > N > C > H. [Corollary: If the atoms are the same, higher mass isotopes take the higher priority. Therefore, $^{3}H > {}^{2}H > {}^{1}H$ and $^{18}O > {}^{16}O$].

(2) If the atoms are identical (two carbons, for example), proceed outward from the stereogenic center ***to the first point of difference*** based on the atoms, not the entire group. Then use rule (1) to determine the priority.

(3) If the atoms at the first point of difference are identical, but the number of substituents on those atoms is different, use the *number* of groups on each atom to determine the priority. For example, three carbons > two carbons > one carbon. This rule is used *only* if the two atoms at the first point of difference are identical, and priority cannot be otherwise determined.

What Is the 'Steering Wheel' Model?

The 'Steering Wheel' Model essentially sights down the base of a tetrahedron that 'surrounds' the stereogenic carbon. The molecule must be turned so that the lowest priority group (d) is *always* to the rear in the model *before* determining absolute configuration. With the (d) group to the rear, a line is drawn from a→b→c, and if the direction of this line is clockwise, the molecule is assigned the R configuration (***5.16***). If that line is counterclockwise, the molecule is assigned the S configuration (***5.17***).

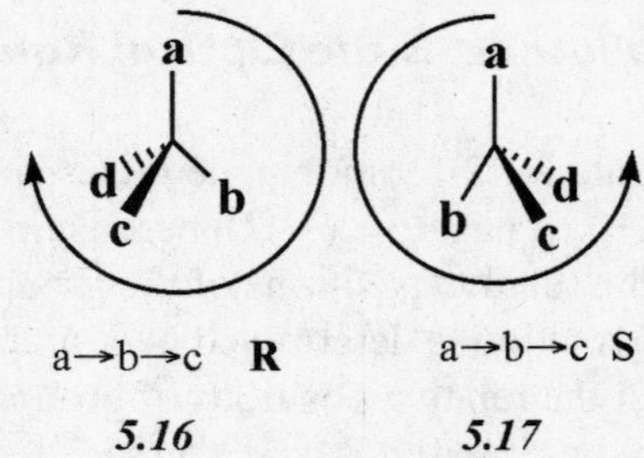

5.16 ***5.17***

How Do We Use the Priority from the Sequence Rules to Assign a Name?

With these three rules, many molecules can be assigned as having R or S configuration using the so-called 'steering wheel' model (***5.16*** and ***5.17***). In this model, the molecule is rotated to place the lowest priority group away from the viewer. If a line drawn from the highest group (a) → (b) and then to (c) rotates clockwise, the molecule is assigned the **R configuration**. Therefore, an R- is placed before the name of that enantiomer. If that line rotates from (a) → (b) → (c), but is counterclockwise, the molecule is assigned the **S configuration**. Therefore, an S- is placed before the name of that enantiomer.

Assign the Absolute Configuration (R or S) to 5.18 and Then Give the Proper IUPAC Name.

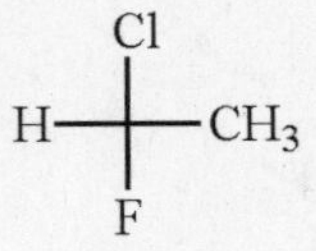

Cl
H—+—CH$_3$
F

5.18

In this example, the atom with the highest atomic mass is chlorine (a), followed by fluorine (b), the carbon of the methyl (c) and, finally, hydrogen (d). This assignment gives a model with the low priority group (d) pointed towards the viewer (see ***5.19***). To make the (d) group point to the rear in the steering wheel model, the (c) group is rotated to the left by about 120° (see ***5.19***). This motion *does not break any bonds*, and moving (c) moves (a) to the right, and (b) rotates to the right, giving ***5.20***. The a→b→c priority is then clockwise, and A has the R configuration. This illustrates Rule 1. The IUPAC name for ***5.18*** is (*R*)-1-chloro-1-fluoroethane.

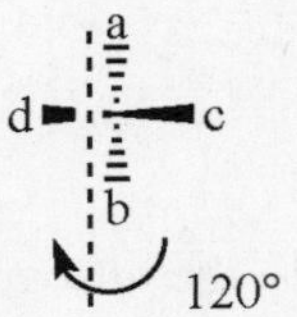

a, d, c, b, 120°

5.19

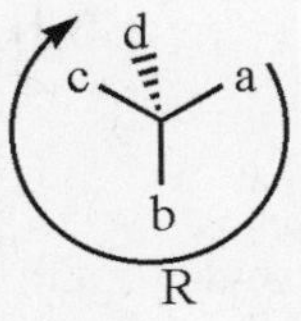

d, c, a, b, R

5.20

Determine the Absolute Configuration of 5.21, an Enantiomer of 1,3-Hexanediol, and Assign the Proper IUPAC Name.

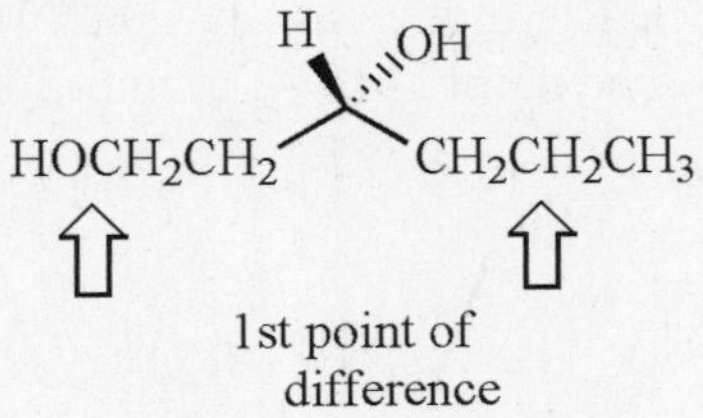

H OH
HOCH$_2$CH$_2$ CH$_2$CH$_2$CH$_3$
1st point of difference

5.21

In this example, the highest priority atom is the oxygen of the hydroxyl group (a) attached to the stereogenic carbon, and the lowest priority group (d) is the hydrogen. The next atoms are both carbon, and **Rule 1** does not allow the priority to be assigned. By **Rule 2**, the first point of difference is the second carbon from the stereogenic center. The first carbons both have one carbon and two hydrogens attached [indicated using special symbolism in which the carbon of interest is assigned superscript atoms that indicate the attached atom; here a carbon of interest attached to a carbon and two hydrogen atoms is C^{CHH}], but the next carbons are different (the first point of difference) with one having a C^{CHH} (the propyl) and the other having C^{OHH} (the hydroxyethyl) – see ***5.22***. Since O takes priority over C, the hydroxyethyl group takes priority (b), and the propyl group takes priority (c). This gives a model with the (d) group forward (***5.23***). The model must be rotated by 180° to the left, along the axis shown in ***5.23***, to give a model with the (d) group to the rear (see ***5.24***), showing ***5.21*** to have the R configuration. The name is (*R*)-1,3-hexanediol.

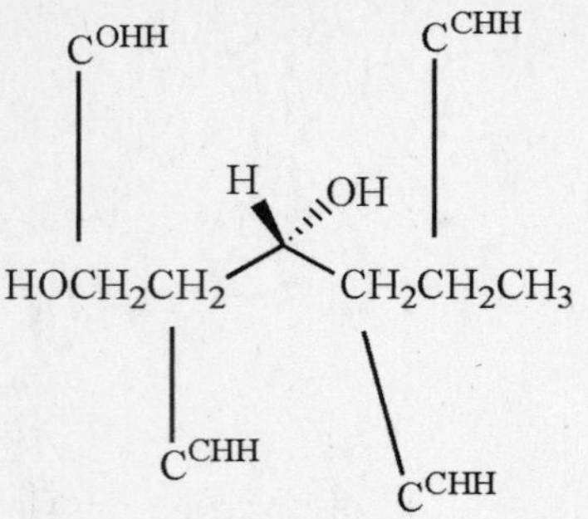

C^{OHH} C^{CHH} H OH $HOCH_2CH_2$ $CH_2CH_2CH_3$ C^{CHH} C^{CHH}

5.22

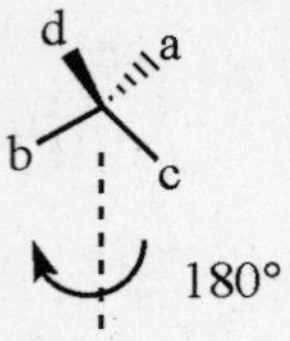

d a b c 180°

5.23

a d c b R

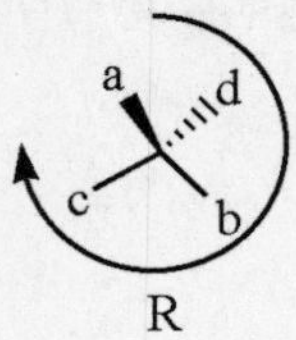

5.24

Determine the Absolute Configuration of 5.25 and Assign the Name.

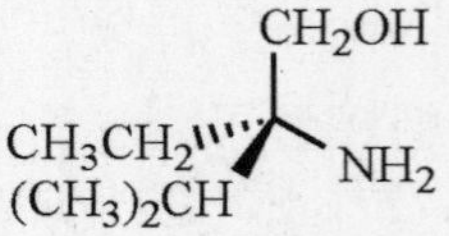

CH_2OH CH_3CH_2 NH_2 $(CH_3)_2CH$

5.25

In this example (2-amino-2-ethyl-3-methyl-1-butanol), the nitrogen attached to the stereogenic carbon is the highest priority (a), but the next three atoms are all carbon. At the first point of difference (see ***5.26***), the 'hydroxymethyl' carbon has two hydrogens and an oxygen [C^{OHH}], the isopropyl group is C^{CCH} and the ethyl group is C^{CHH} (see ***5.27***). The C^{OHH} is clearly higher in priority and takes (b). Using Rules 1 and 2, however, the only atoms available for the (c) and (d) groups are carbon and hydrogen, which are indistinguishable. This requires **Rule 3**, where the number of similar atoms on the carbon at the first point of difference are counted. The C^{CCH} has two carbons to one for the C^{CHH} and the isopropyl group becomes (c) and the ethyl group is (d) (see ***5.27***). The model now has the (d) group directed to the rear (***5.28***) and this molecule has the S configuration. The name is (*S*)-2-amino-2-ethyl-3-methyl-1-butanol.

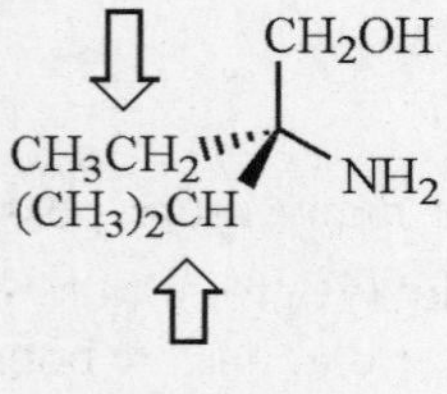

CH_2OH CH_3CH_2 NH_2 $(CH_3)_2CH$

5.26

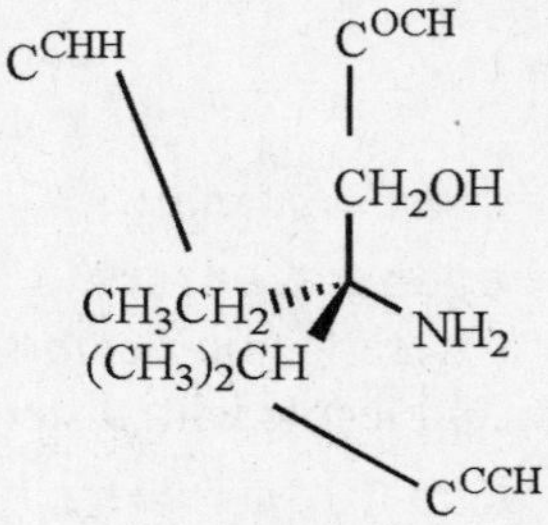

C^{CHH} C^{OCH} CH_2OH CH_3CH_2 NH_2 $(CH_3)_2CH$ C^{CCH}

5.27

b d c a S

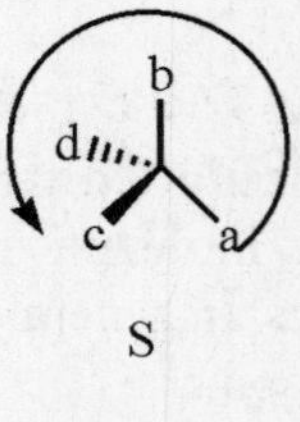

5.28

What Is the Rule When Groups on the Stereogenic Center Contain Multiple Bonds (Double or Triple)?

Rule (4): **If a group on the stereogenic center contains a double or a triple bond, the number of bonds to that atom is taken as the total number of atoms.** In other words, a C–<u>C</u>=O is taken to be $\underline{C}^{COO}$, where the underlined carbon is attached to one carbon and *two* oxygens (one O for each bond of the double bond).

What Is the Absolute Configuration of 5.29, an Enantiomer of 1-Penten-3-ol? Assign the Name.

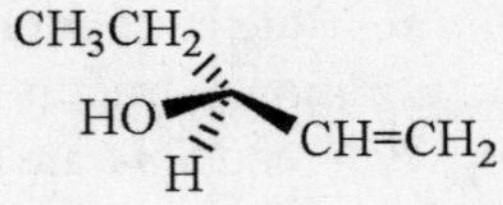

CH_3CH_2 HO H CH=CH_2

5.29

The multiple bond rule is illustrated by molecule ***5.29*** where the O attached to the stereogenic carbon is the highest priority (a), and the hydrogen is the lowest priority (d). At the first point of difference (see ***5.30***), the first carbon of the C=C group is assigned to be C^{HCC} due to the double bond. The indicated carbon has one bond to H and two bonds to C, so the rule assumes the carbon is attached to two carbon atoms and a hydrogen atom. The ethyl group is C^{CHH}. Rule 3 must be invoked since only carbon and hydrogen are present. In this case, C^{CCH} takes priority (b) over the C^{CHH}, which takes priority (c) —see ***5.31***. Tilting the model slightly forward from ***5.31*** puts the (d) group to the rear (***5.32***) and molecule ***5.29*** has the S configuration. The name is *S*-1-penten-3-ol.

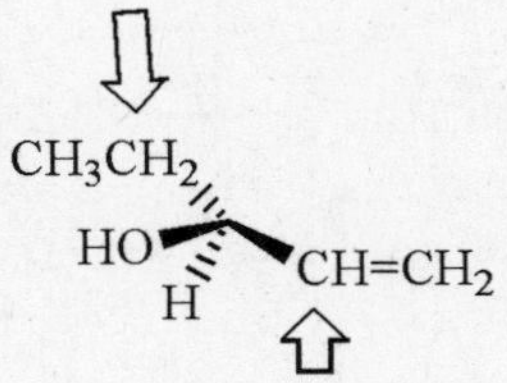

CH_3CH_2 HO H CH=CH_2 c a b d   d c a b

5.30 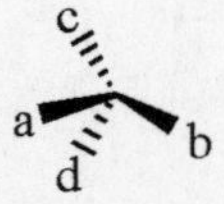   *5.31* 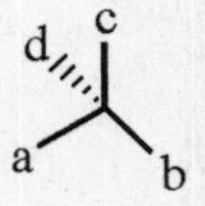   *5.32*

5.4. DIASTEREOMERS

When a molecule contains two or more stereogenic centers, it is possible to generate stereoisomers that are different molecules (non-superimposable) and are not mirror images. Such stereoisomers are given the term diastereomer.

What Is the Maximum Number of Stereoisomers for a Molecule Containing N Stereogenic Centers?

When a molecule has two or more stereogenic centers there are many more possibilities for stereoisomers. In general, for a stereogenic molecule with **n** stereogenic centers, there will be a *maximum* of 2^n stereoisomers.

If a Molecule Has Four Stereogenic Centers, What Is the Maximum Number of Stereoisomers That Are Possible?

For a molecule with four stereogenic centers, the maximum number of stereoisomers is $2^4 = 16$.

Draw All Stereoisomers for 2-Bromo-3-Pentanol.

When a molecule has two stereogenic centers, the 2^n rule predicts 2^2 or four stereoisomers. One enantiomer of 2-bromo-3-pentanol (***5.33***) is drawn in Fisher projection, but two other representations

(***5.34*** and ***5.35***) are also given to show the stereochemical relationship. Molecule ***5.33*** has an enantiomer, ***5.36***. However, this only constitutes two stereoisomers. Molecule ***5.33*** has the 2R,3S configuration, and its enantiomer (***5.36***) has the 2S,3R configuration. If one of the stereogenic centers is now *inverted* from 2R to 2S, which results in molecule ***5.37***, the absolute configuration is now 2S,3S. Molecule ***5.37*** also has an enantiomer (***5.38***) with the 2R,3R configuration. Structures ***5.33***, ***5.36***, ***5.37*** and ***5.38*** are the four stereoisomers that were predicted. Structures ***5.33*** and ***5.34*** are enantiomers, and structures ***5.37*** and ***5.38*** are also enantiomers.

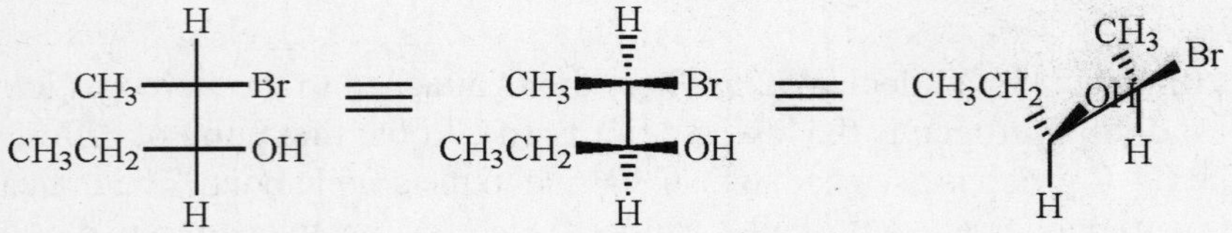

5.33 ≡ *5.34* ≡ *5.35*

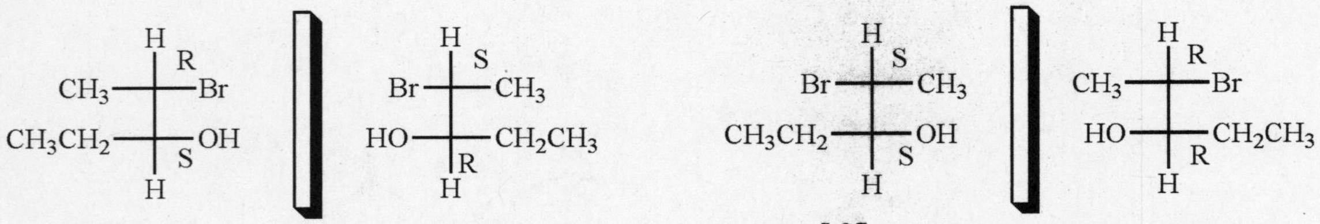

5.33 | mirror | *5.36* | *5.37* | mirror | *5.38*

What Is the Relationship of 5.33 and 5.37 or 5.36 and 5.38?

The indicated molecules are clearly stereoisomers, since they have the same molecular formula and the same connectivity. They are, however, non-superimposable, non-mirror image stereoisomers. In other words, they are *different compounds*. The term for stereoisomers that are not superimposable and not mirror images is **diastereomer**. Compound ***5.33*** is a diastereomer of ***5.37*** and ***5.38***. Likewise, ***5.36*** is a diastereomer of ***5.37*** and ***5.38***. Compound ***5.37*** is a diastereomer of ***5.33*** and ***5.36***, and ***5.38*** is a diastereomer of ***5.33*** and ***5.36***. Remember that ***5.33*** and ***5.36*** are enantiomers, and that ***5.37*** and ***5.38*** are enantiomers.

Meso Compounds

Is It Possible to Have Fewer Than the Number of Stereoisomers Predicted by the 2^N Rule?

Yes. In some cases, a molecule with two or more stereogenic centers will give two stereoisomers that are enantiomers, but the other stereoisomer will have a *superimposable* mirror image, that is, the two structures are the same molecule. Such a molecule is termed a **meso compound**. This gives a total of three different stereoisomers, not the predicted four, for two stereogenic centers.

Discuss Why 2,3-Dibromobutane Has Only Three Stereoisomers.

A simple example of a **meso compound** is 2,3-dibromobutane. Structure ***5.39*** has a non-superimposable mirror image (***5.40***), thus ***5.39*** and ***5.40*** are enantiomers. The diastereomer of ***5.39*** and ***5.40*** is ***5.41***, which also has a mirror image ***5.42***. A close look at ***5.41*** and ***5.42***, however, shows that they are identical.. Simple rotation of ***5.42*** by 180° gives ***5.41***. *Therefore, they are superimposable mirror images, and one compound.* Since ***5.41*/*5.42*** is one compound, there are only *three stereoisomers*, not four. The presence of a meso compound leads to a diminished number of stereoisomers. Closer inspection of ***5.41*/*5.42*** (see ***5.44***) shows that there is a plane of symmetry that bisects the molecule. In other words, the 'top' half of the molecule is identical with the 'bottom' half *when drawn as the eclipsed rotamer*. This **symmetry** is characteristic of meso compounds. If an eclipsed rotamer can be found where every atom matches, as in ***5.44*** (Br matches Br, methyl matches methyl and H matches H), this will be a meso compound. Using the same model for ***5.39*** (the enantiomer of ***5.40*** and the diastereomer of ***5.41*/*5.42***), ***5.43*** shows that there is no symmetry in this plane; the Br and H do not match. This lack of symmetry leads to the presence of enantiomers.

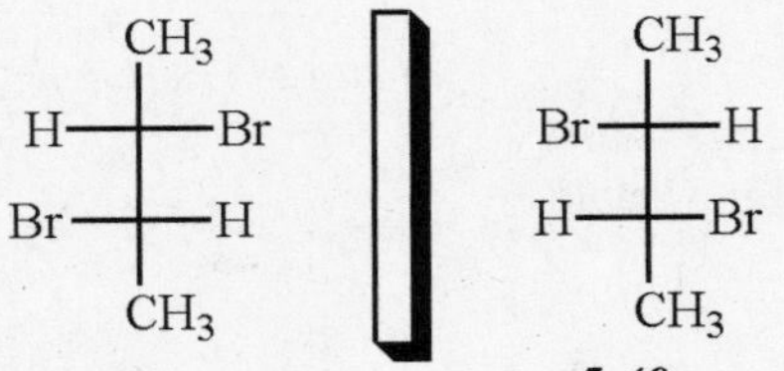

CH_3 H—Br Br—H CH_3 ***5.39*** | CH_3 Br—H H—Br CH_3 ***5.40***

CH_3 Br—H Br—H CH_3 ***5.41*** | CH_3 H—Br H—Br CH_3 ***5.42*** rotate by 180° superimposable

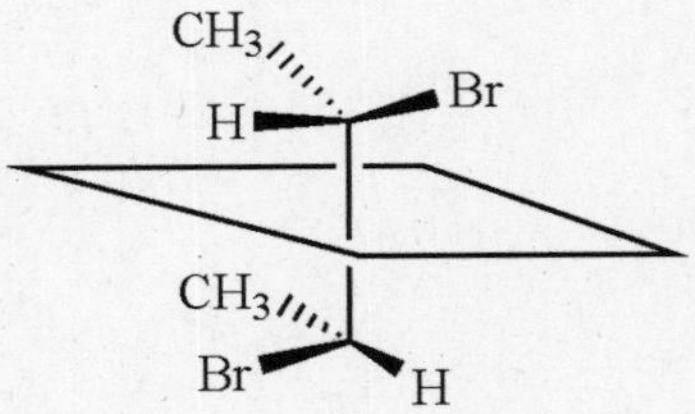

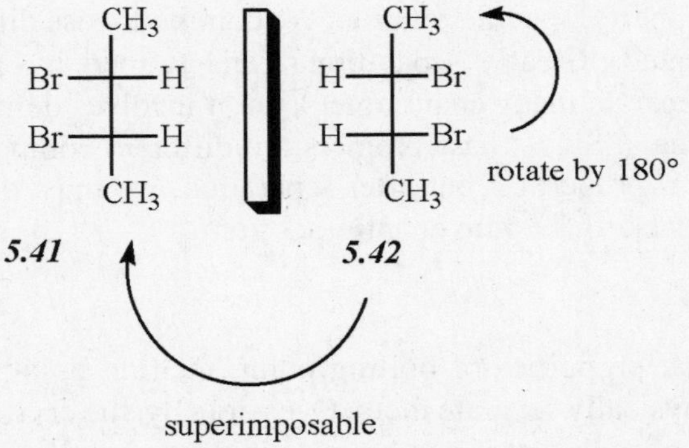

5.43 ***5.44***

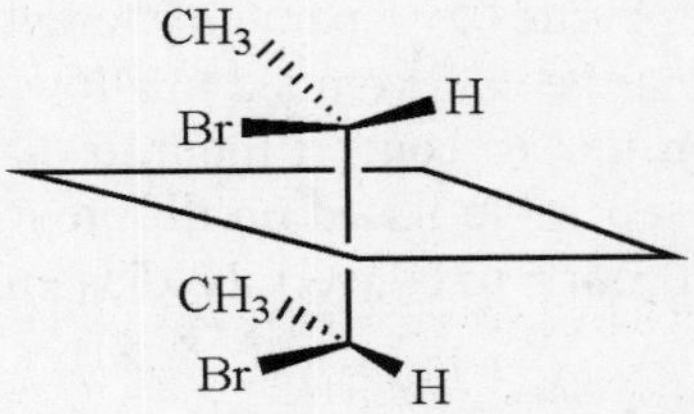

Does 1,2-Cyclopentanediol (5.45) have a Meso Compound?

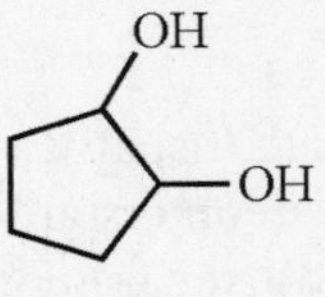

OH OH

5.45

Yes! To answer this question, however, requires a recognition that ***5.45*** is drawn without specifying stereochemistry, so we must assume it is a mixture of the cis and trans diols, ***5.46*** and ***5.48*** respectively. Structure ***5.45*** is deliberately drawn this way to emphasize the need to show stereochemistry. When drawn in this way, no stereochemistry is indicated, and this representation indicates a mixture as mentioned. We can draw ***5.46*** and ***5.47*** as enantiomers since they are non-superimposable mirror images. There is a plane of symmetry in ***5.48***, however, so the mirror image is superimposable, and this is a meso compound.

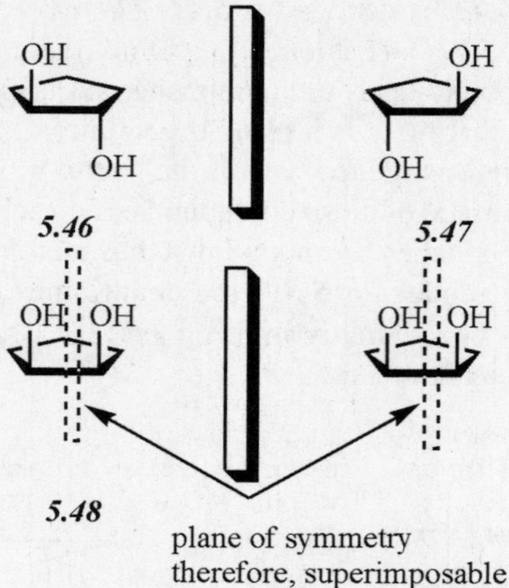

5.5. OPTICAL RESOLUTION

Enantiomers differ only in one physical property: specific rotation. We commonly use differences in physical properties to separate different compounds. Clearly, separation of enantiomers is a problem. A technique has been developed to allow us to separate many enantiomers, but it involves doing a chemical reaction first to convert them to diastereomers. Since diastereomers are different compounds, they should be separable based on different physical properties, but after separation, we must do a second chemical reaction to convert the diastereomer back to the pure enantiomer first.

Is It Possible to Separate Enantiomers?

Since enantiomers have the same physical properties of boiling point, melting point, solubility, adsorptivity, etc., it is virtually impossible to physically separate them. Occasionally, the crystal structure of one solid enantiomer is noticeably different enough that is can be selectivity removed as in Pasteur's separation of tartaric acid by physically picking out the different crystals under a microscope. More commonly, the only way to separate enantiomers is by a method called **optical resolution**.

What Is Optical Resolution?

In this technique, the enantiomeric mixture (usually a racemic mixture) is reacted with another chiral molecule to produce diastereomers. Diastereomers are different compounds, with different physical properties. They can, therefore, be physically separated. Once separated, another chemical reaction cleaves the bond between the enantiomer of interest and the second stereogenic molecule, *resolving* the individual enantiomers.

Since We Have Not Yet Discussed Chemical Reactions, Draw a Diagram Using A, B, C, etc., to Illustrate Optical Resolution.

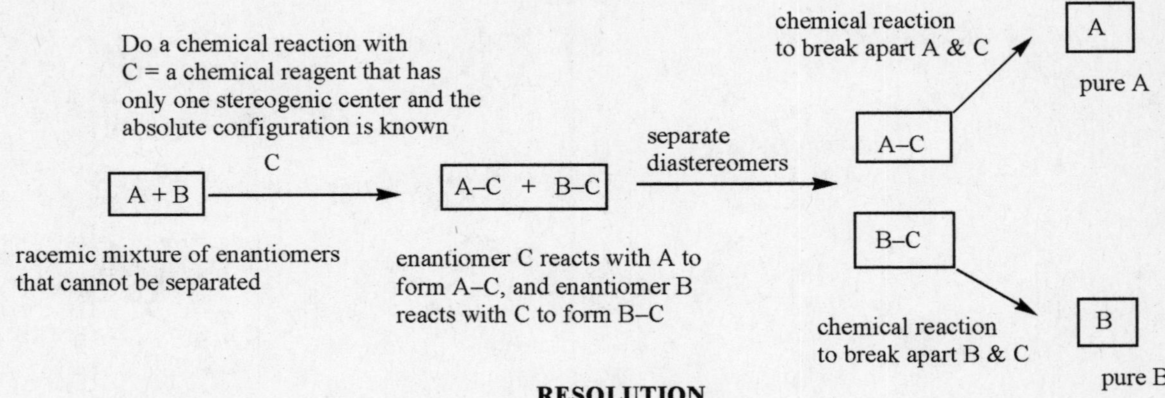

RESOLUTION

Assume that enantiomer A has the (*R*)-configuration, and B has the (*S*)-configuration. Likewise, assume that C has the (*R*)-configuration. When C reacts with A, the product A–C will have the *R–R* configuration, whereas when B reacts, B–C will have the *S–R* configuration. These two compounds are diastereomers, and because they have different physical properties, they can be separated. Once A–C is obtained in pure form, a chemical reaction will break apart A and C so that pure A (with the (*R*)-configuration) can be isolated. A similar process applied to B–C will result in pure B, with the (*S*)-configuration. It is assumed that C can be separated from A and from B, and hopefully recovered, purified, and used again.

In the following chapters, many reactions will be discussed that involve stereogenic centers. The principles introduced in this chapter will be used throughout.

Test Yourself

1. Determine the R or S configuration for each stereogenic center in the following molecules.

(a) 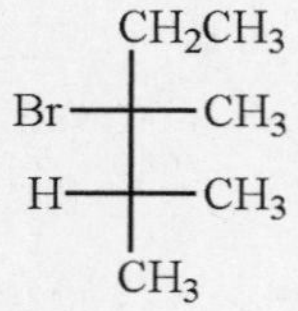
(b) 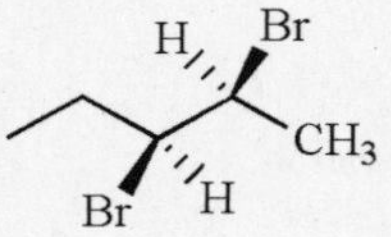
(c) 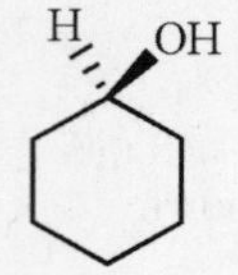
(d) 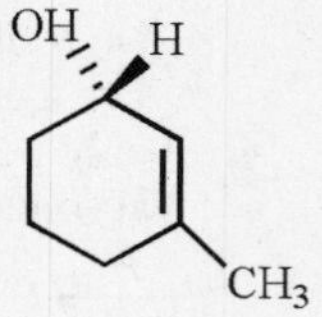

(e) 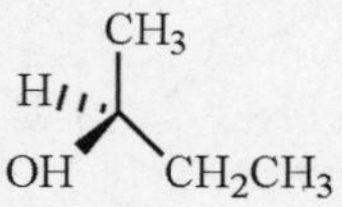
(f) 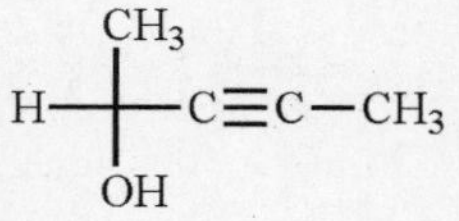
(g) 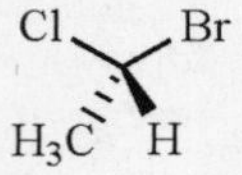
(h) 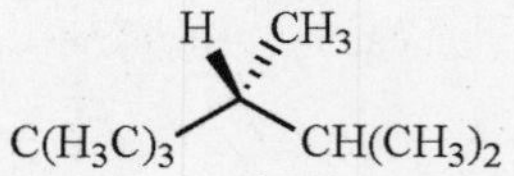

2. Which of the following are not suitable for use as a solvent in determining the specific rotation of a stereogenic unknown? Explain.

CH_3OH H_2O 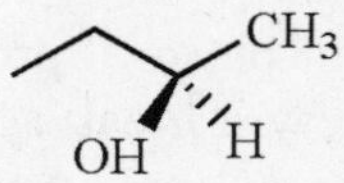 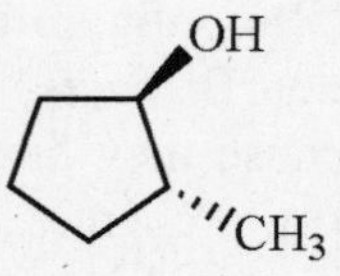 CH_2Cl_2

3. For a polarimeter with a path length of 10 dm, determine the specific rotation for each of the following. The concentration is given in brackets with each observed rotation value.
   (a) −24.6° (c = 0.47 g/mL) (b) +143.4° (c = 1.31 g/mL)
   (c) +0.8° (c = 0.65 g/mL) (d) −83.5° (c = 5.0 g/mL)
4. Calculate the % of R and S enantiomers present in the following *mixtures* of R + S enantiomers. In each case, the specific rotation value for the R enantiomer is +120°.
   (a) [α] = −14.8° (b) [α] = +109.2° (c) [α] = +4.6° (d) [α] = −18.3°
5. Draw all diastereomers for (a) 2-bromo-3-heptanol and (b) 4-methyl-3-octanol using Sawhorse diagrams. Which of the stereoisomers are diastereomers?
6. Draw all different stereoisomers of 2-bromo-3-chloropentane in Fischer projection, and assign the absolute configuration to each stereogenic center. Name each different compound.
7. Discuss the number of stereoisomers possible for 2,3-butanediol and for 2,3-cyclopentanediol.

CHAPTER 6

Acids and Bases

Acid-base reactions are perhaps the most important category of chemical reactions in all of organic chemistry. It will be seen that many, if not most organic reactions have an acid-base component. Seeing this relationship usually requires modifying the concepts of what is an acid and what is a base. This chapter will introduce the most fundamental principles of acids and bases.

6.1. ACIDS AND BASES

There are two fundamental definitions of acids and bases, the Lewis definition and the Brønsted-Lowry definition. We will view both definitions through the lens of electron donation and electron acceptor ability.

What Is a Brønsted-Lowry Acid?

A Brønsted-Lowry acid is defined as a hydrogen atom donor. Therefore, a Brønsted-Lowry acid must have an acidic hydrogen atom.

What Structural Feature Makes a Molecule Able to Donate a Hydrogen Atom?

The hydrogen atom must be polarized on H^+ or $H^{\delta+}$. In HCl, the H has a positive dipole and can be donated to a base by the Brønsted-Lowry acid definition. Likewise, a molecule such as C–OH might be a Brønsted-Lowry acid since the H has a positive dipole.

Are There Differences in Brønsted-Lowry Acid Strength?

Yes! In a simple case, compare HCl and HOH. The H in HCl is much more polarized than the H in HOH. Therefore, HCl is expected to be a stronger Brønsted-Lowry acid. This analysis is overly simplistic, since it ignores the products formed after the hydrogen is donated and, of course, does not mention the base to be used. Nonetheless, it is a simple illustration of the fact that different molecules can be stronger or weaker Brønsted-Lowry acids.

What Is a Lewis Acid?

A Lewis acid is an electron pair acceptor.

Give Two Examples of a Lewis Acid.

Both BCl_3 and $AlCl_3$ are Lewis acids. Both boron and aluminum are in Group III of the Periodic Table. Therefore, they can form three covalent bonds, using the three valence electrons, to generate a neutral molecule. In both cases, however, the B and the Al are electron deficient. These atoms can attain an octet by accepting an electron pair from another molecule, a Lewis base.

What Is a Brønsted-Lowry Base?

A Brønsted-Lowry base is defined as a hydrogen atom acceptor.

What Is a Lewis Base?

A Lewis base is an electron pair donor.

What Is the Product When the Lewis Acid $AlCl_3$ Reacts with Ammonia (NH_3)?

The product is the Lewis Acid–Lewis base adduct, called an 'ate' complex, ***6.1***.

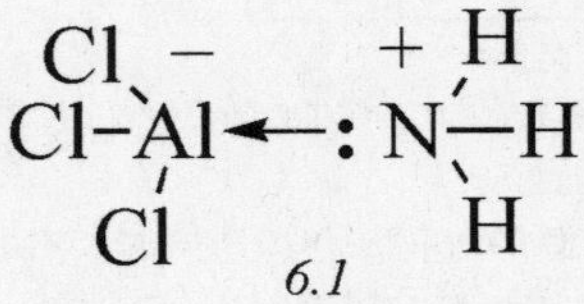

$$Cl_3Al^{-} \leftarrow :N^{+}H_3$$

6.1

Is It Possible to View a Brønsted-Lowry Acid in Terms of the Lewis Acid Definition?

Although the definitions are different, to donate a hydrogen atom, another molecule must donate an electron pair to the hydrogen atom. Therefore, we can view a Brønsted-Lowry acid and a hydrogen atom that accepts an electron pair from a suitable base.

Is It Possible to View a Brønsted-Lowry Base in Terms of the Lewis Base Definition?

Although the definitions are different, to accept a hydrogen atom, a Brønsted-Lowry base must donate an electron pair to the hydrogen atom. Therefore, we can view a Brønsted-Lowry base as a molecule that donates an electron pair to a hydrogen atom.

What Structural Features Are Required for a Molecule to be Considered an Acid?

For a Brønsted-Lowry acid, the molecule must have a hydrogen atom, which is generally attached to a heteroatom, that has a positive dipole. For a Lewis acid, the atom must be electron deficient and able to accept a pair of electrons from another molecule.

What Structural Features Are Required for a Molecule to be Considered a Base?

For a Brønsted-Lowry base, the molecule must be able to form a bond to a hydrogen atom and have an unused pair of electrons that can be donated to a hydrogen atom with a positive dipole. For a Lewis base, the atom must be electron-rich and able to donate a pair of electrons to another molecule.

6.2. THE ACIDITY CONSTANT, Ka

What Is a Conjugate Acid?

A conjugate acid is the product of the reaction between a base and an acid. Specifically, when a Brønsted-Lowry base accepts a hydrogen atom, the product will have that hydrogen incorporated in the molecule, and that product is called the conjugate acid of the initial base. An example is the reaction of ammonia with the acid H+, where NH_3 is the base, and the product NH_4+ (the ammonium ion, ***6.2***) is the conjugate acid of ammonia.

$$H_3N: \curvearrowright H^+ \longrightarrow H_3\overset{+}{N}\text{–}H$$

6.2

What Is a Conjugate Base?

A conjugate base is the product of the reaction between a base and an acid. Specifically, when a Brønsted-Lowry base donates a hydrogen atom, the product will be missing that hydrogen, and that product is called the conjugate base of the initial acid. In the reaction of the base hydroxide (the positive counterion is omitted for clarity), the base donates an electron pair to the acid hydrogen of HCl. In this process, the hydroxide accepts the hydrogen, and the HCl loses the hydrogen to form chloride ion. In this reaction, chloride ion is the conjugate base of HCl.

$$Cl\text{–}H \;\; :\overset{-}{O}H \longrightarrow HO\text{–}H + Cl^-$$

6.3

Which Is More Important, the Acid or the Conjugate Acid?

This is a trick question. Both are important. In the reaction of hydroxide with HCl (an acid -base reaction), the products are water and chloride ion. One can also view chloride as a base, and water as an acid. Only an understanding of the fact that HCl is a stronger acid than water and that hydroxide is a stronger base than chloride ion allows for a full understanding of the reaction. As we will see, one must examine the acid strength of the acid and conjugate acid, as well as the base strength of the base and the conjugate base in every acid-base reaction to really understand acidity and basicity.

Define the Equilibrium Constant for an Acid-Base Reaction.

$$H\text{–}A + B^- \underset{}{\overset{K}{\rightleftharpoons}} A^- + HB$$

6.4

Acid-base reactions are equilibrium reactions. For a typical reaction shown in ***6.4***, the acid (HA) reacts with the base (B^-) to form the conjugate base (A^-), and the conjugate acid (HB). HA and B^- is an acid/base pair, but A^- and HB is another acid-base pair. Therefore, this is an equilibrium, and the equilibrium constant for the reaction is K. Since it is an acid-base reaction, K is given the symbol Ka (the acidity constant). In the usual definition of an equilibrium constant, the products are written on the right of the equation (in this case, the conjugate acid and conjugate base), and the starting materials are written on the left of the equation (in this case, the acid and the base). This arrangement is taken as the standard definition of an acid-base reaction, with the acid and base written on the left side of the equation. With this definition,

$$K_a = \frac{[A][HB]}{[HA][B]}$$

What Is the Definition of a Strong Acid?

With the definition of Ka just given, and with the acid/base pair written on the left of the equation, if Ka is large ($\gg 1$, so that HA and B^- reacted to form A^- and HB), then HA is considered to be a strong acid. If Ka is small (<1), then HA and B^- did not react to form A^- and HB, but rather A^- and HB reacted to push the equilibrium back towards HA and B^-. In other words, a small value of Ka means there is more starting material than product, and the acid/base pair did not react very well. Note that the standard definition of a strong acid or strong base is their complete dissociation in water. A large number of reactions in organic chemistry are acid-base reactions, but water is not the solvent. To get away from this limited standard definition, the terms are defined as they will be commonly used in organic reactions.

What Is pK_a?

The definition of pK_a is $pK_a = -\log_{10} Ka$. Conversely, $Ka = 10^{-pK_a}$.

If the Ka for an Acid Is 1.5×10^{-5}, What Is the pK_a?

$pK_a = -\log_{10}(1.5 \times 10^{-5}) = 4.8$

If the pK_a for an Acid Is 4.35, What Is the Ka?

$Ka = 10^{-4.35} = 4.47 \times 10^{-5}$.

How Is Ka Different from K in a Non-Acid-Base Equilibrium?

There is no fundamental difference other than the equilibrium involves a forward acid-base reaction and a reverse acid-base reaction.

What Factors Contribute to Making the Acid More Acidic?

Four major factors influence acidity, at least from the standpoint of the molecules we are going to see. Focusing on Brønsted-Lowry acids, the bond polarization of the X–H bond in the acid is important. In this case, X = C, O, N, S, halogen, etc. The more polarized the X–H bond, the more positive the dipole on the hydrogen atom, making it more susceptible to reaction with a base. In one sense, this means that the X–H bond is weaker as it is more polarized, which makes it easier to break. Therefore, C–H is less acidic than N–H, which is less acidic than O–H. The stability of the conjugate base and conjugate acid are important. If those products are more stable, the equilibrium is pushed to the right, and the starting acid is considered to be more acidic. Conversely, if the conjugate acid and/or base are rather unstable (more reactive), the equilibrium is pushed to the left, and the starting acid is considered to be less acidic. Closely related to the second point, the relative acidity of the acid and conjugate acid, as well as the base and conjugate base, is important. If the conjugate acid is weaker than the starting acid, the equilibrium is pushed to the right. If the conjugate acid is stronger than the staring acid, the equilibrium is pushed to the left. Similar comments apply to the base and conjugate base. Finally, the solvent plays an important role, both in assisting loss of the acidic hydrogen and in stabilizing the conjugate acid and conjugate base. If the solvent stabilizes the products, the equilibrium is shifted to the right.

Which Is Expected to be More Acidic, HOH or H_2NH, Using Only Bond Polarization of the Acid As the Determining Criterion?

Since the O–H unit is more polarized than the N–H unit, one expects that HOH is more acidic than NH_3 using this single criterion.

What Factors Contribute to Making the Conjugate Acid More Stable?

More stable means less reactive. Therefore, the X–H bond in the conjugate acid should be stronger and less polarized than the X–H bond in the starting acid.

How Does This Contribute to the Equilibrium?

If the conjugate acid is more stable (less reactive), the reaction with the conjugate base is less likely, and the equilibrium is likely to lie to the right rather than the left.

What Factors Contribute to Making the Conjugate Base More Stable?

Large ions tend to be more stable than small ions because they are more easily solvated, and the charge is dispersed over a larger area. Therefore, iodide ion is more stable than chloride ion. Resonance also makes the conjugate base more stable by dispersing the charge over several atoms. More polarized conjugate bases tend to be solvated to a greater extent in polar media, making them somewhat more stable. In general, anything that makes it more difficult for the conjugate base to donate electrons makes it a weaker base, and therefore less reactive.

How Does This Contribute to the Equilibrium?

If the conjugate base is more stable (less reactive), the reaction with the conjugate acid is less likely, and the equilibrium is likely to lie to the right rather than the left.

If the Acid Is Stronger Than the Conjugate Acid, How Does This Influence the Equilibrium and pK_a?

The equilibrium is pushed to the right, and Ka is large, making pK_a smaller.

If the Conjugate Acid Is Stronger Than the Acid, How Does This Influence the Equilibrium and pK_a?

The equilibrium is pushed to the left, and Ka is small, making pK_a larger.

If One Compares Two Reactions, and Acid X Has the Structure R–OH, and Acid Y Has the Structure R_2NH, Which Is More Acidic and Why?

Since the OH bond in ROH is more polarized than the *N*–H bond in R_2NH, ROH is expected to be more acidic, since the H is more easily lost. In addition, the conjugate base RO^- is more stable (less reactive) than the conjugate base R_2N^-, shifting the equilibrium further to the right for the reaction of ROH, making it more acidic. Since oxygen is more electronegative than nitrogen, the RO^- unit is less likely to donate electrons, making it a weaker base. Oxygen is also larger than nitrogen, so the charge dispersal on oxygen is greater, again making it less able to donate electrons.

If One Compares Two Reactions, and Both Acid X and Acid Y Have the Structure R–OH, but Acid X Gives a More Stable Conjugate Base, How Does This Influence the Equilibrium and Ka?

If the conjugate base is more stable in one reaction, that equilibrium will be pushed further to the right, making Ka larger. This is equated with that acid being more acidic.

Draw the Conjugate Base Formed When Formic Acid (HCOOH) Reacts with NaOH.

$$\underset{\textit{6.5}}{\mathrm{HC(=O)O^{\delta-}H^{\delta+}}} + \mathrm{\overset{+}{Na}\,\overset{-}{O}H} \underset{}{\overset{K_a}{\rightleftharpoons}} \underset{\textit{6.6}}{\mathrm{HC(=O)O^{-}\,Na^{+}}} + \mathrm{H{-}OH}$$

When formic acid (***6.5***) reacts with the base hydroxide, the conjugate base is the formate anion, ***6.6***. Note that the sodium counterion from NaOH is transferred to the negatively charged oxygen in ***6.6***.

Draw the Resonance Contributors for the Formate Anion.

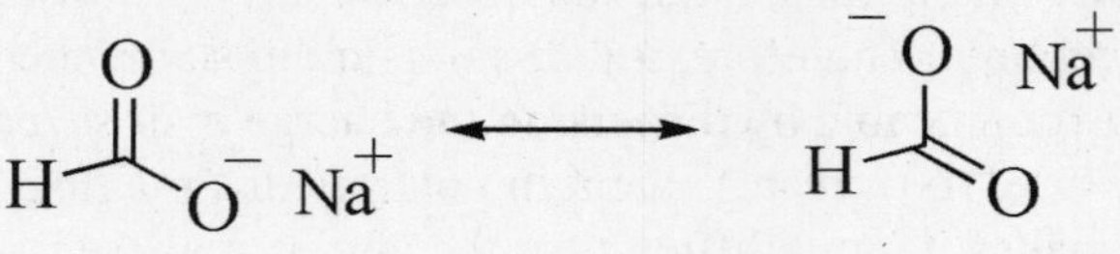

6.7

The two resonance forms of the formate anion are shown in *6.7*.

Which Is the Stronger Acid, Formic Acid (6.5) or Methanol (CH_3OH)? Explain Your Answer.

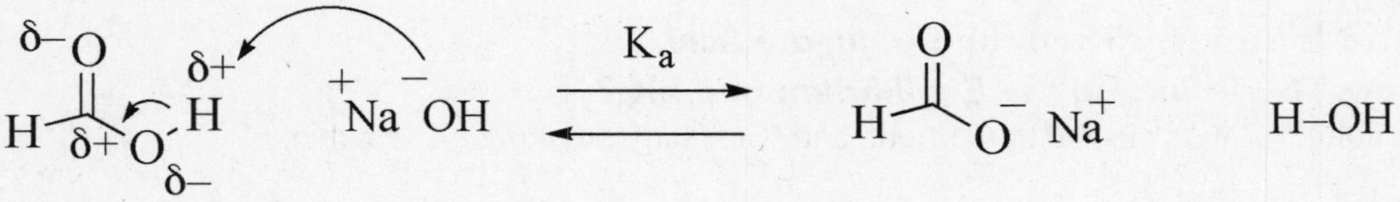

6.5 6.6

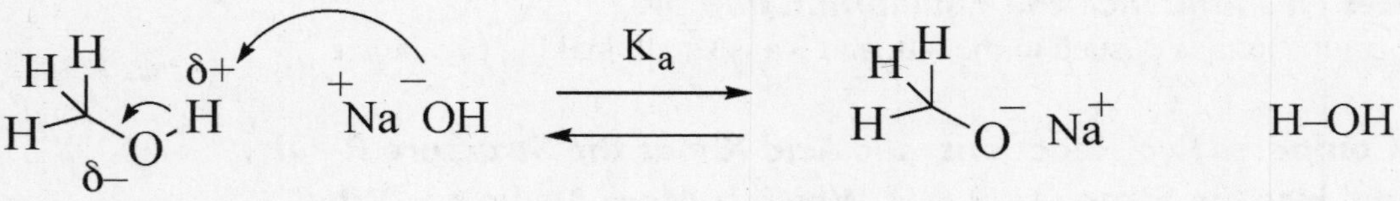

6.8 6.9

Formic acid is the stronger acid. When methanol (*6.8*) reacts with hydroxide, the product is the methoxide anion, *6.9*. In *6.9*, the charge is localized on the oxygen, making the electrons very available for donation (i.e., it is a good base). The formate anion (*6.6*), however, is resonance stabilized, meaning the electron pair is delocalized over three atoms. Therefore, the oxygen atom is less basic because it is less able to donate electrons. If *6.6* is more stable than *6.9*, the equilibrium for formic acid is shifted to the right, making it more acidic. In addition, the C=O unit in *6.5* is polarized as shown, making the O–H unit in *6.5* more polarized, and more reactive (i.e., more acidic) relative to the OH unit in methanol *6.8*. Note that this simple analysis has ignored any solvent effects, which can play a significant role.

6.3. STRUCTURAL FEATURES THAT INFLUENCE ACIDITY

Many classes of organic compounds can be viewed as acids, but they have vastly different acid strengths. Perhaps the most distinguishing feature of carboxylic acids (RCOOH) is that they are protic acids and react with a variety of suitable bases. Sulfonic acids (RSO_3H) are also strong organic acids. Alcohols (ROH) are much weaker acids than the carboxylic acids and slightly less acidic than water. For the time being, we will focus on carboxylic acids as the main organic acid.

What Does the Structure RCOOH Look Like in Terms of Bonding?

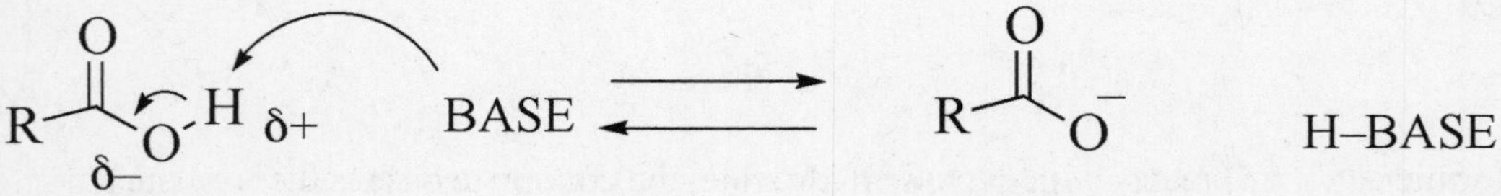

6.10 6.11

Carboxylic acids have the generic formula of ***6.10***, where R is any carbon group or a hydrogen atom. Carboxylic acids have a carboxyl group characterized by the C=O unit (a carbonyl), with an OH unit directly attached to the carbonyl carbon. After reaction with a base, the carboxylate anion ***6.11*** is generated as the conjugate base. Carboxylate anions are resonance-stabilized, just as we saw with the formate anion, ***6.7***.

What Is the pK_a of Acetic Acid (Ethanoic Acid = CH_3COOH)?

The pK_a of acetic acid (ethanoic acid) is 4.76.

Give the pK_a Values of the Following Carboxylic Acids: A) Propanoic Acid (CH_3CH_2COOH) B) Butanoic Acid ($CH_3CH_2CH_2COOH$) C) Formic Acid (HCOOH).

(a) propanoic = 4.89; (b) butanoic acid = 4.82; (c) formic acid = 3.75.

Why Is Acetic Acid a Weaker Acid Than Formic Acid?

Acetic acid has a methyl group attached to the electropositive carbonyl. Relative to hydrogen (in formic acid), methyl is an electron-releasing group. This 'pushes' electrons towards the O–H bond, strengthening it, and making that hydrogen less acidic (less positive = less like H^+).

What Is the pK_a of 2-Chloroethanoic Acid (Chloroacetic Acid = $ClCH_2COOH$, 6.12)?

O
OH
Cl

6.12

The pK_a of chloroacetic acid is 2.85.

Why Is 6.12 a Stronger Acid Than Acetic Acid?

The chlorine has an electron withdrawing effect. Cl polarizes the α-carbon $\delta+$, making the carbonyl carbon withdraw electrons from the O–H bond, thereby weakening it (more positive H, more like H^+), and making it a stronger acid.

'Through Bond' Inductive Effects

Certain electronic effects are transmitted from nucleus to nucleus through the adjacent covalent bonds, and these are collectively called **'through-bond' effects**, a type of **inductive effect**.

When an Electron Withdrawing Group Is Attached to the Carbonyl of a Carboxyl, What Is the Effect on the O–H Bond?

The electron withdrawing group induces a $\delta+$ charge on the α-carbon, which is adjacent to the $\delta+$ charge on the carbonyl carbon. This is destabilizing and, to compensate, the carbonyl carbon withdraws electrons from neighboring atoms to diminish the electron withdrawing effects of the α-carbon. This makes the O–H bond more polarized, and the H more positive and more acidic.

Will an Electron Withdrawing Substituent Increase or Decrease Acidity?

In general, electron withdrawing groups increase acidity (larger Ka, smaller pK_a).

Explain Why 3-Chlorobutanoic Acid (6.13) Is a Weaker Acid Than 2-Chlorobutanoic Acid (6.14).

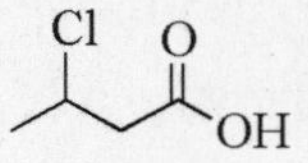

Cl O OH
6.13

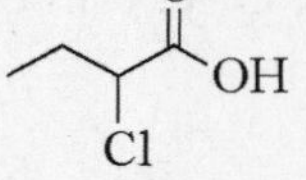

O OH Cl
6.14

The electron withdrawing chlorine atom is further away from the carbonyl carbon in 3-chlorobutanoic acid (***6.13***) than in 2-chlorobutanoic acid (***6.14***). The electron withdrawing ability of an atom or group diminishes as the distance from the O–H group and the carbonyl increases.

Explain Why 2,2,2-Trichloroethanoic Acid (6.15) Is a Stronger Acid Than 2-Chloroethanoic Acid (6.12).

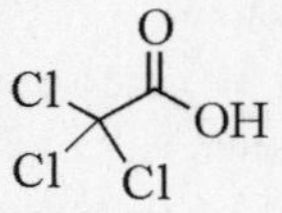

O Cl OH Cl Cl
6.15

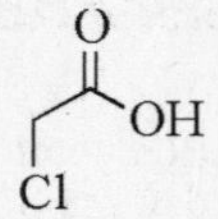

O OH Cl
6.12

If one chlorine atom withdraws electrons from the α-carbon, and this makes the O–H bond weaker, the presence of three electron withdrawing chlorines will withdraw even more electron density. This will make the α-carbon, the carbon next to the carbonyl, more positive and the O–H bond weaker and, therefore, more acidic. If the pK_a of 2-chloroethanoic acid (***6.12***) is 2.89, the pK_a of 2,2,2-trichloroethanoic acid (***6.15***) is 0.64, making the latter a significantly stronger acid.

Will an Electron-Releasing Substituent Increase or Decrease Acidity?

An electron-releasing substituent 'feeds' electrons towards the electropositive carbonyl, and that carbonyl carbon will withdraw *less* electron density from the O–H bond. The O–H bond is stronger, and the acid is a weaker acid (more difficult to remove the H from O–H).

In Each of the Following Series, Indicate the Strongest Acid.

(A) 3-CHLOROPROPANOIC ACID OR 3-CHLOROPENTANOIC ACID
(B) 2-METHOXYETHANOIC ACID OR PROPANOIC ACID

In (a), both acids have an electron withdrawing chlorine at C_3, but 3-chloropentanoic acid also has an ethyl group attached to C_3. The presence of the electron-releasing alkyl group, which is missing in 3-chloropropanoic acid, will make 3-chloropentanoic acid a weaker acid. Therefore, 3-chloro-propanoic acid is the stronger acid. In (b), methoxy is electron withdrawing, primarily by through-space effects (see below), whereas the methyl (attached to the α- carbon of acetic acid) is electron-releasing. Since methoxy is electron-withdrawing, 2-methoxyethanoic acid is expected to be the stronger acid [pK_a of methoxyacetic acid = 3.6; pK_a of propionic acid = 4.9].

'Through Space' Field Effects

When a heteroatom group is sufficiently close in space to the acidic hydrogen of the carboxyl, 'through-space' electronic effects, often called field effects, will influence the acidity of the acid. These through-space effects are usually a stronger influence on the acidity than the through-bond effects, depending on where they are.

Explain Why 3-Nitropropanoic Acid (6.16) Is a Stronger Acid Than 4-Nitrobutanoic Acid (6.17).

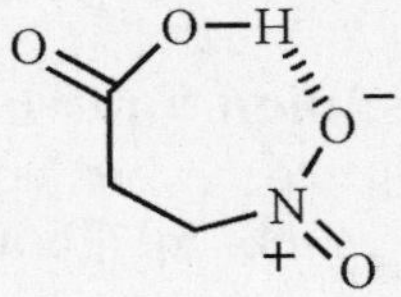

6.16 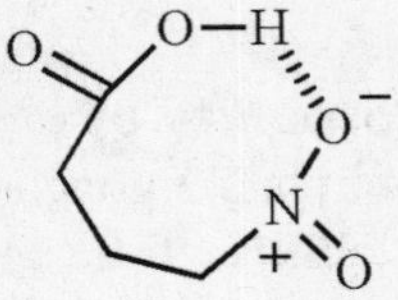 *6.17*

In 3-nitropropanoic acid (*6.16*), the electron-withdrawing group is closer to the carboxyl group. As shown in *6.16*, the H of the O–H bond is physically close to the electronegative oxygens of the nitro group. A 'through-space' hydrogen bonding effect, effectively through a seven-membered ring unit, will weaken the O–H bond, making the OH bond weaker, and *6.16* more acidic. In *6.17*, the extra carbon atom positions the nitro group further from the COOH unit, requiring a larger eight-membered ring unit for the internal hydrogen bonding. It is more difficult to attain this hydrogen-bonded unit, making *6.17* less acidic.

Describe the Relative Acidity of Ethanoic Acid (Acetic Acid) and 2-Chloroethanoic Acid in Terms of a Through-Space Effect.

Examination of *6.18* for 2-chloroethanoic acid shows that the chlorine and the H of the OH can hydrogen bond (through space) in what is effectively a five-membered ring. This is energetically accessible, and the electron withdrawing effect of this through space hydrogen bonding is very strong, making chloroacetic acid a stronger acid. This internal hydrogen bonding effect cannot occur in CH_3COOH (ethanoic acid), so the OH bond in *6.18* is weaker than in ethanoic acid, and *6.18* is the stronger acid.

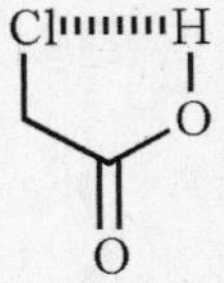

6.18

Nature of Carboxylate Anions

What Is the Structure of the Carboxylate Anion Formed by Removal of the Acidic Hydrogen from the Acid?

Treatment of a carboxylic acid with a base ($NaHCO_3$ is a sufficiently strong base for this reaction) generates the carboxylate anion, *6.19* (see *6.7* above). This ion is resonance stabilized with the charge dispersed equally on the two oxygen atoms. The carboxylate anion is a relatively weak base because it is more difficult to donate electrons as a base if the electrons are delocalized by resonance.

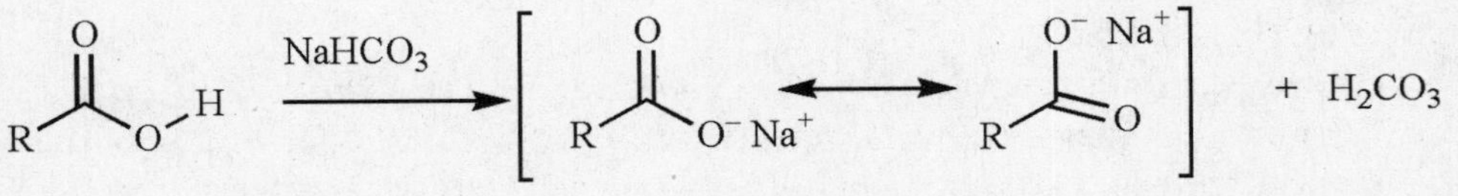

$NaHCO_3$ + H_2CO_3

6.19

Why Is the Carboxylate Anion Considered to be Very Stable?

As shown in ***6.19***, the carboxylate anion is resonance stabilized. The negative charge is delocalized over three atoms (O–C–O).

How Does the Special Stability of the Carboxylate Anion Influence Acidity?

The product (the carboxylate anion) is resonance stabilized, and an equilibrium reaction will be influenced by the relative stability of the species that make up that equilibrium. A more stable product tends to shift the equilibrium towards that product. In this case, if the equilibrium is shifted towards the carboxylate, Ka is larger. A large Ka is indicative of a stronger acid.

What Is the Product When Potassium Acetate (Draw It) Is Dissolved in Aqueous Solution at pH 4?

The relatively weak base (potassium acetate, ***6.20***) is protonated in the acidic solution (pH 4) to give acetic acid (ethanoic acid).

$$\left[CH_3C(=O)O^-K^+ \longleftrightarrow CH_3C(O^-K^+)=O \right] \xrightarrow{H^+\ Cl^-} CH_3C(=O)OH + KCl$$

6.20

Test Yourself

1. Which is more acidic, HCOOH or CH_3COOH?
2. Which is more basic, CH_3OCH_3 or CH_3NHCH_3?
3. Convert each of the following Ka values into pK_a.
   (a) 3.4×10^4 (b) 2.33×10^{-9} (c) 5.66×10^{-2} (d) 8.9×10^7
4. Convert each of the following pK_a values into Ka.
   (a) 2.33 (b) 23.55 (c) 17.05 (d) 4.78 (e) 10.15
5. Draw all resonance forms of Na_2CO_3.
6. Give a rationale for the observation that A is a stronger acid than B.

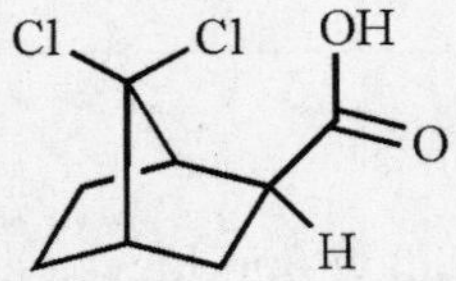

Cl Cl OH O H **A** Cl Cl H OH O **B**

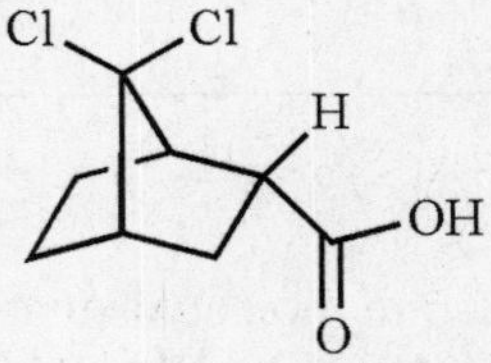

7. Which is the stronger acid, A or B?

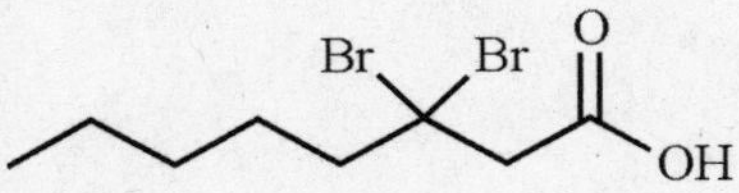

Br Br O OH **A** O OH Br Br **B**

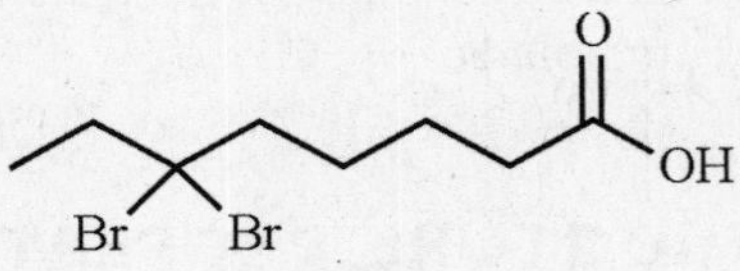

8. Draw a diagram to illustrate through-space inductive effects in 4-chloropropanoic acid.
9. Why is acetic acid a stronger acid in water than in ethanol?

CHAPTER 7

Alkyl Halides and Substitution Reactions

Alkyl halides are molecules that have one or more fluorine (alkyl fluorides), chlorine (alkyl chlorides), bromine (alkyl bromides) or iodine (alkyl iodides) atoms attached to a carbon. Each C–halogen bond (C–X) is polarized $C^{\delta+}-X^{\delta-}$, and a variety of reagents, usually called nucleophiles, can react with $C^{\delta+}$ to generate other types of functional groups. The reaction just described is known as a substitution reaction, where one atom or group attaches to carbon and replaces the X group of the C–X bond. Reactions that generate halides will be discussed, as well as the reactions of halides to form other compounds.

7.1. STRUCTURE, PROPERTIES AND NOMENCLATURE OF ALKYL HALIDES

Alkyl halides are a class of molecules that contain one or more halogen atoms. They are commonly used as reactive substrates in reactions where the halogen is replaced (substitution reactions) or eliminated from the molecule to form an alkene or an alkyne (elimination reactions). Some alkyl halides, such as dichloromethane (CH_2Cl_2), carbon tetrachloride (CCl_4), or chloroform ($CHCl_3$) are used as solvents in some reactions.

What Is an Alkyl Halide?

An alkyl halide is any molecule that contains one or more halogen atoms attached to a carbon atom. The carbon atom will be sp^3 hybridized.

Why Is the C–X Bond of an Alkyl Halide Polarized?

Alkyl halides are characterized by a C–X bond, where X = F, Cl, Br or I. The C–X bond is highly polarized with a $\delta+$ carbon and a $\delta-$ halogen. Fluorine is the most electronegative element, and the C–F bond is the most polarized. Fluorine is a small atom, however, and that bond is relatively strong. Although less polarized, the C–I bond is relatively weak due to the long bond length dictated by the large iodine atom when bonded to carbon.

Describe, in a General Way, the Physical Properties of Alkyl Halides.

Despite the polarized C–X bond, alkyl halides are only weak polar compounds,when compared to water. Alkyl halides are more polar than alkanes, but tend to be insoluble in water. The C–halogen bond length increases from F to I, while the electronegativity trend is 1.51 (F), 1.56 (Cl), 1.48 (Br), 1.29 (I) Debye. The boiling point trend is R–F < R–Cl < R–Br < R–I, and the density follows the same trend.

This chapter will describe the chemical properties of alkyl halides, but one can state that alkyl halides are generally reactive when mixed with certain molecules containing atoms that can donate electrons.

Is the Halogen in an Alkyl Halide considered to Be a Functional Group?

No! In an alkyl halide, the halogen is considered to be a substituent.

What Is the Nomenclature Rule for Naming Alkyl Halides?

Determine the longest continuous chain of carbon atoms and assign the appropriate prefix and suffix. Each halogen is treated as a substituent, and assigned a number to designate its position on the carbon chain. Drop the *–ine* from fluorine, chlorine, bromine, or iodine and replace that suffix ix –o. Fluorine becomes *fluoro*, chlorine becomes *chloro*, bromine becomes *bromo*, and iodine becomes *iodo*.

What Is the IUPAC Name of Compound 7.1?

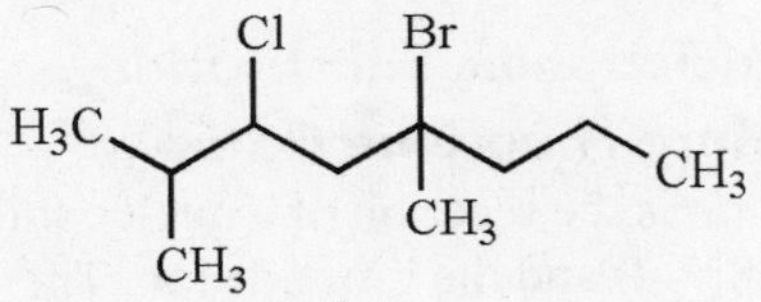

Cl Br
H_3C CH_3
CH_3
CH_3

7.1

Compound *7.1* is 5-bromo-3-chloro-2,5-dimethyloctane.

What Is the Nomenclature Rule If the Halogen Atom is Part of a Substituent Rather Than Attached to the Longest Continuous Chain?

Name the halogen as a substituent on the substituent. Determine the longest chain of the substituent from the point of attachment to the longest continuous chain and then assign the halogen the appropriate number on the substituent chain. For example, $-CH_2Cl$ is chloromethyl, and $-CH(Br)CH_2CH_3$ is 1-bromopropyl.

Name 7.2.

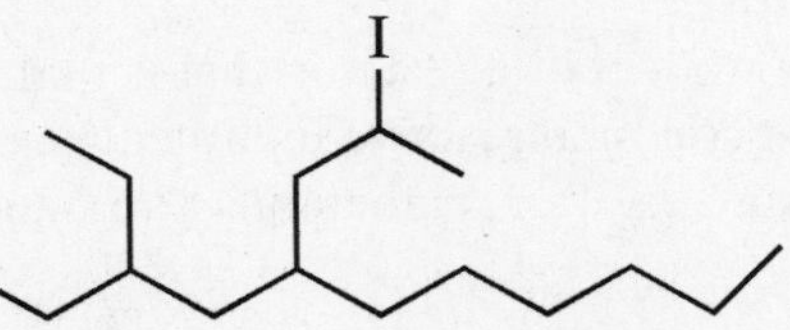

I

7.2

Compound *7.2* is 3-ethyl-5-(2-iodopropyl)undecane.

7.2. SECOND ORDER NUCLEOPHILIC SUBSTITUTION (S_N2) REACTIONS

When an atom or group replaces another atom or group, at carbon, the process is known as a substitution reaction. When the substitution proceeds without an intermediate reaction, but rather via a concerted process (single step), it is known as nucleophilic bimolecular substitution and given the designator S_N2.

Define a Substitution Reaction.

A substitution reaction is characterized by one atom or group replacing another atom or group at an sp^3 atom (usually carbon). This transformation is made possible by the presence of a polarized bond in an alkyl halide (C–Cl, C–Br or C–I), where the carbon is electron-deficient and has a polarity of δ^+. That carbon is most likely to react with an electron-rich species.

Define a Nucleophile.

If an electron-rich species donates its electrons to a carbon, it is called a **nucleophile**. This definition is used to distinguish reactions at carbon from other atoms. When an electron-rich species donates electrons to an atom other than C or H, it is known as a Lewis base.

What Is an Electrophilic Carbon?

An electrophilic carbon is the $\delta+$ polarized carbon that reacts with a nucleophile, forming a new bond to the nucleophile.

Give a Simple Example of a Nucleophilic Substitution Involving Molecules with No More Than One Carbon.

A simple example of a substitution is the reaction of a nucleophile (iodide) with bromomethane (*7.3*, the electrophile) to give iodomethane (*7.4*) and the bromide ion. The positive counterion is omitted so we can focus on the substitution reaction. Note the use of the curved arrows, which indicate that the iodide ion donates two electrons to the electrophilic carbon to form a new C–I bond. The C–Br bond is broken, and the electrons in that bond are transferred to bromine to form the bromide ion.

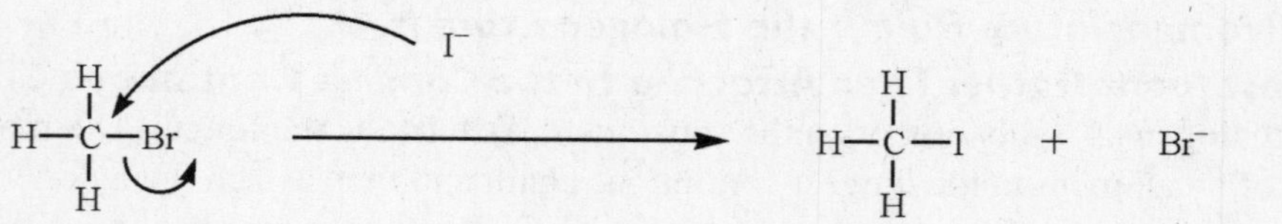

$$I^- + H_3C{-}Br \;(7.3) \longrightarrow H_3C{-}I \;(7.4) + Br^-$$

Walden Inversion

What Is a Reactive Intermediate?

A reactive intermediate is the product of a reaction that is high in energy, and reacts further to give a stable, isolable product. The most common reactive intermediates involving carbon atoms that we will see include carbocations (carbenium ions, *7.5*), carbanions (*7.6*), and radicals (*7.7*).

$$R_3C^+ \;(7.5) \qquad R_3C{:}^- \;(7.6) \qquad R_3C^{\bullet} \;(7.7)$$

Given the Bond Polarity of the C–Br Bond in 7.3. What Type of Charged Intermediate Might Be Expected If Bromine Leaves?

The bond polarity of the C–Br bond is such that bromine has a negative dipole and carbon has a positive dipole. Therefore, cleavage of the C–Br bond might occur in some cases to give a positive carbon as in the carbon cation *7.5*.

Has a Reactive Intermediate Ever Been Observed in Substitution Reactions Like the Conversion of 7.3 to 7.4?

No! This transformation is believed to be a concerted one-step process.

What Is a Concerted Reaction?

A concerted reaction is one that occurs without a reactive intermediate. In other words, bond breaking and bond making occur simultaneously once the activation energy for bond breaking/making has been attained, and the result is the product. There are no reactive intermediate products along the way.

What Is a Transition State?

A transition state is the logical mid-point of a reaction in which bonds are beginning to break and other bonds are beginning to form. The transition state for the conversion of *7.3* to *7.4* should be *7.8* if this is a concerted process. In *7.8*, the C–I bond is beginning to form as the C–Br bond is beginning to break. As the reaction proceeds, the C–I bond is made, with expulsion of bromide ion after complete breakage of the C–Br bond. Note the use of dashed lines in the transition state to indicated bonds being broken or formed, and the use of the bracket to indicate a transient entity called the **transition state**.

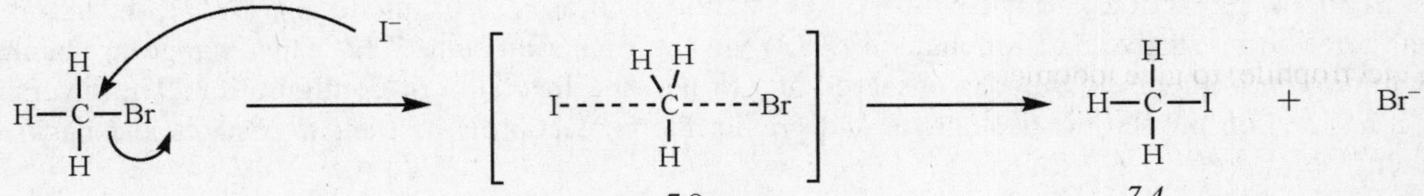

I^- + H–C(H)(H)–Br (*7.3*) → [I- - - - -C(H)(H)(H)- - - - -Br] (*7.8*) → H–C(H)(H)–I (*7.4*) + Br^-

Can You Isolate a Transition State?

No! A transition state is the midpoint of bond breaking/bond making, but is not a product and cannot be isolated.

Given That 7.3 Is Three-Dimensional (Tetrahedral), from What Angle Will the Bromide Ion Approach the Carbon Atom to Minimize Steric and Electronic Repulsions?

There will be great electronic repulsion, as well as steric repulsion, as the electron-rich iodide approaches over the electron-rich bromine atom. If iodide approaches any of the hydrogen atoms, there will be steric repulsion. If the iodide approaches from 180° relative to the bromine (see *7.9*), electronic and steric repulsion will be minimized. This is the preferred trajectory for this reaction.

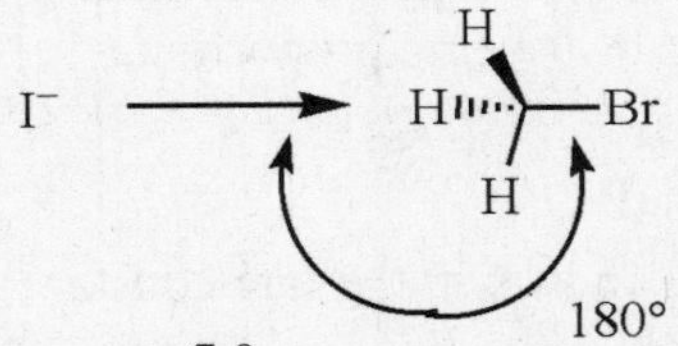

I^- → H₃C–Br (180°)

7.9

What Is the Transition State for Nucleophilic Bimolecular Substitution at Carbon?

The incoming nucleophile,such as the iodide ion, must *collide* with the electrophile (the δ^+ carbon of CH_3Br) for the substitution reaction to occur. The iodide is negatively charged and will be repelled by the δ^-bromine in *7.3*. If iodide approaches the carbon over one of the hydrogens (remember that *7.3* is three-dimensional and tetrahedral), the iodide and hydrogen repel (*steric repulsion*) as they attempt to occupy the same space. The path of least resistance is, therefore, **backside attack**, where iodide approaches the carbon 180° away from the bromine. As iodide approaches carbon, the hydrogen atoms are pushed away until a high energy five-coordinate **transition state** is formed (***7.10***) where the three hydrogens and the carbon are coplanar, the C–I bond is beginning to form and the C–Br bond is beginning to break. This transition state is the logical mid-point of the reaction and is *not* a product that can be isolated or even observed. As the C–I bond is formed, the hydrogens are pushed further away and the C–Br bond is broken to form *7.4*. The net result of this collision via backside attack is **inversion of configuration at the electrophilic carbon.** This inversion of configuration is known as **Walden inversion,** proceeding through a pentacoordinate transition state such as ***7.10***.

I^- + H, H, H–C(δ+)–Br(δ−) (*7.3*) ⟶ [I(δ−)- - - C(δ+)(H)(H)(H) - - -Br(δ−)] (*7.10*) ⟶ I–C(H)(H)(H) (*7.4*)

How Do We Know That Inversion of Configuration Occurs? Iodomethane 7.3 Does Not Have a Stereogenic Center.

If the S_N2 reaction is done with an alkyl halide such as *R*–2-bromobutane (***7.11***), inversion of configuration must give *S*–2-iodobutane (***7.13***) via the transition state ***7.12***. This change in absolute configuration is precisely what is observed and proves the inversion of configuration. The inversion of configuration is used to substantiate and predict the pentacoordinate transition states and backside attack.

I^- + CH_3, H, CH_3CH_2–C(δ+)–Br(δ−) (*7.11*) ⟶ [I(δ−)- - - C(δ+)(CH_3)(H)(CH_2CH_3) - - -Br(δ−)] (*7.12*) ⟶ I–C(CH_3)(H)(CH_2CH_3) (*7.13*)

Kinetics

What Is the Rate of a Reaction?

The rate of a reaction is, in effect, how rapidly the starting materials react to form the isolated product. The rate of a reaction is a dynamic property where the concentration of the starting materials changes as a function of time during the course of the reaction. The rate is therefore obtained from a differential equation. In general, we can describe rate as being proportional to the molar concentration of the starting materials and the time:

rate ∝ [starting materials] [time]

The proportionality constant is taken as k = the rate constant for the reaction.

rate = k [starting materials] [time]

What Is a First Order Reaction?

A first order reaction is one that obeys first order kinetics. In effect, the rate of the reaction depends only on one molecule, A. The first order rate equation is obtained from a differential equation that describes the reaction,

$$\text{rate} = \frac{-d[A]}{dt} = \text{k}\,[A]$$

where the rate is proportional to both concentration of A and to time. The integral equation is

$$\int_{[A]_o}^{[A]_t} \frac{\mathbf{d[A]}}{\mathbf{[A]}} = \int_{o}^{t} \mathbf{k}dt = \mathbf{k}\int_{o}^{t} dt$$

where $[A]_o$ is the concentration of A at time = 0, and $[A]_t$ is the concentration of A at any specified time. When this equation is integrated such that [A] is $[A]_o$ at t=0 and is $[A]_t$ at t= 'end time,' the expression obtained is:

$$\mathbf{In}\frac{[A]_o}{[A]_t} = \mathbf{k}\,(t_{\text{end time}} - t_o) \text{ and if } t_o = 0 \text{ (as defined),}$$

$$\text{then ln}\,\frac{[A]_o}{[A]_t} = \mathbf{k}\,t$$

What Is the Rate Constant for a First Order Reaction?

If we plot the concentration of the starting material A as a function of time (1 mole of A at time = 0; 0.9 molecules of A at t = 40 min; 0.8 moles of A at t = 0.68 min, etc.), a curve is obtained. If the time is plotted against the ln [A], however, a straight line is obtained, and the slope of that line is minus the rate constant (−k).

What Is a Second Order Reaction?

A second order reaction is one that obeys second order kinetics. The rate of the reaction depends on two molecules (A + B) reacting to form a different molecule. To calculate the rate constant we first recognize that *both* reactants are important for reaction to occur, and the rate expression is

$$-\frac{d[A]}{dt} = \text{k}\,[A]\,[B] \text{ or, rate} = \text{k}\,[A]\,[B]$$

The differential equation for this reaction is solved to give a complex expression:

$$\frac{1}{[A]_o[B]_o}\ln\frac{[B]_o[A]_t}{[A]_o[B]_t} = \text{k t}$$

where the $[\ ]_o$ terms denote the initial concentrations of A or B (at time = 0) and the $[\ ]_t$ terms indicate the concentrations of A and B at a specified time. For convenience, we will assume that the rate for a second order reaction is rate = $k[A]^2$ rather than rate = k[A][B]. This assumption simplifies things quite a bit, and the expression changes to: $\frac{1}{[A]_t[A]_o} = \text{k t}$

What Is the Rate Constant for a Second Order Reaction?

Normally, a plot similar to $\ln \frac{[A]_t}{[B]_t}$ versus time gives a straight-line plot, and the slope of this line is the rate constant, $-k$.

How Can We Determine If a Reaction is First Order or Second Order?

We must obtain the rate data (concentration vs. time) and plot the data. If the data gives a straight-line plot with $\ln \frac{[A]_O}{[A]_t} = k\,t$ it is first order. If the data gives a straight-line plot with

$$\frac{1}{[A]_o[B]_o} \ln \frac{[B]_o[A]_t}{[A]_o,[B]_t} = k\,t$$

then the reaction is second order.

What Is the Half-Life of a Reaction?

The half-life for a reaction (see Section 6.5.B) is given the symbol $t_{\frac{1}{2}}$ and is calculated from the simple formula,

$$\text{half-life} = t_{\frac{1}{2}} = \frac{K}{ln2} = \frac{K}{0.693}$$

The half-life of a reaction is the time required for half of the starting material to be consumed.

What Is the Half-Life If the Rate Constant (k) for a First Order Reaction is 12 M min^{-1} (Moles Per Liter Per Minute)

In this case, $t_{\frac{1}{2}}$ is $\frac{12}{0.693} = 17.3$ minutes.

What Is the Half-Life if the Rate Constant (k) for a First Order Reaction is 1.4×10^{-5} M Sec^{-1} (Moles Per Liter Per Second)

In this case, $t_{\frac{1}{2}} = \frac{1.4\times10^{-5}}{0.693} = 2.02\times10^{-5}$ seconds, and the reaction will rapidly be completed.

How Many Half-Lives Are Usually Required for a Reaction to Be Deemed Complete?

If we begin with 1 mole of starting material, 0.5 mole will remain after one half life - i.e., 50% of the starting material has reacted. After another half life, 0.5 (0.5) will remain, = 0.25. In other words, 75% of the starting material has reacted. If we take five half lives, we obtain $1\times0.5\times0.5\times0.5\times0.5\times05\times0.5 = 0.031$ moles remaining - i.e., 99.97% of the starting material has reacted. Therefore, it takes about five half-lives for a reaction to be deemed complete.

How Long Will It Take for Five Half-Lives for the Reaction with a Rate Constant (k) of 12 M min^{-1}?

In this case, $t_{\frac{1}{2}}$ is $\frac{12}{0.693} = 17.3$ minutes. Therefore, five half lives $= 17.3\times5 = 86.5$ minutes.

How Long Will It Take for Five Half-Lives for the Reaction with a Rate Constant (k) of 1.4×10^{-5} M Sec^{-1}?

In this case, $t_{\frac{1}{2}} = \dfrac{1.4 \times 10^{-5}}{0.693} = 2.02 \times 10^{-5}$ seconds. Therefore, five half-lives $= 2.02 \times 10^{-5}$ seconds $\times 5 = 10.1 \times 10^{-5}$ seconds $= 0.000101$ seconds.

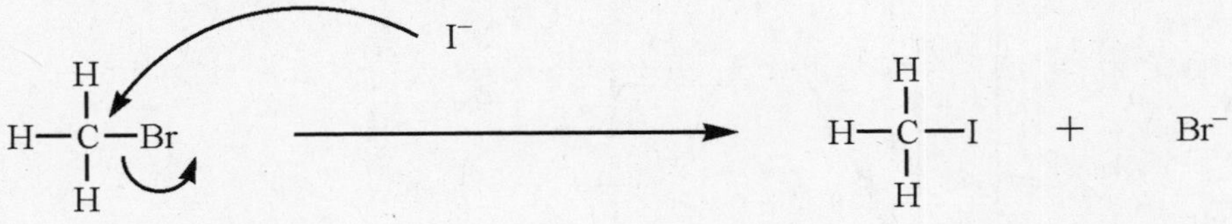

CH_3Br + I^- → CH_3I + Br^-

7.3 *7.4*

How Can One Determine How Fast a Reaction Occurs Relative to Another One?

The conversion of ***7.3*** into ***7.4*** is a collision process involving two molecules. It follows second order kinetics, which makes it a *bimolecular reaction.* It is a second order bimolecular substitution and is termed a S_N2 reaction. The rate of the reaction (rate = disappearance of the starting material in moles/sec or appearance of product in moles/sec, usually measured as disappearance of starting material) is given by the expression: rate = k [nucleophile] [halide], where [ ] represents the concentration in moles/liter of each molecule. For a S_N2 reaction, increasing the concentration of either nucleophile or halide will increase the rate of the reaction. The parameter (k) is called the rate constant. Larger rate constants are associated with faster reactions.

Influence of Steric Hindrance

Explain Why Bromomethane Reacts Much Faster Than 2-Bromopropane with Sodium Iodide.

For S_N2 reactions, primary halides react much faster than secondary halides with the same nucleophile via the pentacoordinate transition state ***7.14***. Tertiary halides are essentially unreactive under the same conditions. The **relative rate constants** are: $CH_3X = 1.0$; 1° (such as CH_3CH_2X) $= 0.033$; 2° (such as Me_2CHX) $= 8.3 \times 10^{-4}$; and 3° (such as Me_3CX) $= 5.5 \times 10^{-5}$. The S_N2 reaction with methyl, primary and secondary halides is useful with a variety of nucleophiles, but S_N2 reactions with 3° halides give no substitution. The reason is steric hindrance in the transition state. As seen in ***7.14*** when the alkyl group (R) is small, the four co-planar atoms (C,R,R,R) are easily accommodated in the five-center transition state. When the R groups are large (R=R=R=methyl), the steric crowding is too great, leading to a very high activation energy for formation of that transition state (see ***7.15***). For primary and secondary systems, the activation energy is low enough that a reasonable rate is observed.

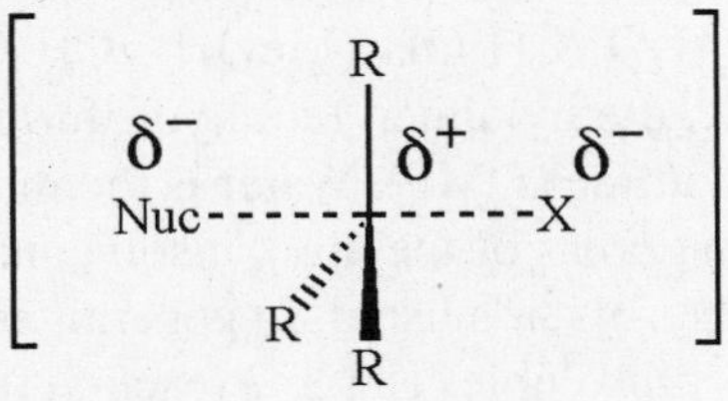

$[\overset{\delta^-}{Nuc} \cdots \overset{\delta^+}{C}R_3 \cdots \overset{\delta^-}{X}]$

7.14

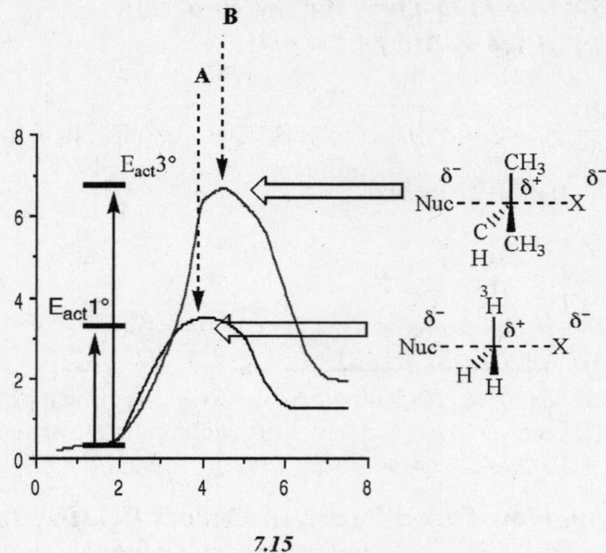

7.15

What Is 'A' in Diagram 7.15?

This point is the transition state for the reaction with the primary halide.

What Is the Term Used for the Energy Difference Between the Starting Material and the Transition State?

This corresponds to the activation energy (E_{act}) for that reaction. The higher the transition state energy, the more energy is required to begin bond making/breaking, and the reaction will be slower.

7.3. SOLVENT EFFECTS

For substitution reactions, the solvent plays a large role, both in stabilizing or destabilizing the transition state, and also in assisting the initial reaction with the starting materials. The solvent can also help stabilize the products and help to drive the reaction to completion.

Polar vs. Nonpolar Solvents

There are two major classifications for solvents: polar, which contain highly polarizable atoms and bonds and usually have a high dipole moment, and nonpolar, which have a low or zero dipole moment and generally contain no heteroatoms or heteroatoms whose individual bond moments cancel.

Distinguish Protic and Aprotic Solvents.

Each solvent can also be characterized as protic, or containing an acidic hydrogen -(X–H where X = O, S or N), or aprotic, which contain no acidic hydrogensor very weakly acidic hydrogens such as C–H. Common protic solvents are H_2O, CH_3OH (MeOH), CH_3CH_2OH (EtOH), NH_3, CH_3COOH. Common aprotic solvents are pentane, diethyl ether, tetrahydrofuran (THF), dichloromethane, dimethyl sulfoxide (DMSO) and dimethylformamide (DMF). Water is the most polar of the protic solvents listed. DMSO followed by DMF are the most polar of the aprotic solvents listed. Pentane, a hydrocarbon with no heteroatoms, is the least polar of the solvents listed. In general, polar protic solvents are water soluble and nonpolar aprotic solvents are water insoluble. The S_N2 reaction proceeds best in polar aprotic solvents and is slowest in polar protic solvents.

How Does a Protic Solvent Influence the Pentacoordinate Transition State (7.14) for a S_N2 Reaction?

If the protic solvent can solvate both cations and anions, then it may solvate the incoming nucleophile when that species bears a negative charge. The electrophilic carbon atom will be somewhat solvated as well, requiring that the solvent be 'moved aside' before collision can occur. This solvation, or formation of a solvation 'shell,' will slow down the S_N2 reaction. If an aprotic solvent is used, no solvation shell is formed, and the nucleophile will be poorly solvated. Collision should be more efficient with fewer solvent molecules to move aside, and the S_N2 reaction should be faster.

How Does a Polar Solvent Influence the Pentacoordinate Transition State (7.14) for a S_N2 Reaction?

In a polar solvent, solvation is more efficient, whereas solvation is much poorer in an aprotic solvent. Typically, the S_N2 reaction requires sufficient polarity in the solvent to ensure solubility and stabilization of the transition state. Given the solvation effects noted, the best solvent for a S_N2 reaction is a polar, aprotic solvent.

7.4. FIRST ORDER SUBSTITUTION (S_N1) REACTIONS

Substitution reactions occur with tertiary halides in highly polar, protic media that follow first order kinetics and proceed by a carbocation intermediate. Such first order nucleophilic substitution reactions are termed S_N1.

Explain Why 3-Bromo-3-Methylpentane Reacts with Potassium Iodide in Aqueous Ethanol to Give 3-Methyl-3-Iodopentane, When the Same Reaction in Anhydrous DMF Gave No Reaction.

In aqueous media, the carbon-bromine bond of 3-methyl-3-bromopentane (***7.16***) can be broken, and loss of bromine (as bromide ion) gives a carbocation ***7.17***. Carbocation ***7.17*** is the product of the first reaction shown, but this product is so high in energy and so reactive that it is not isolated; rather, it reacts further to give another product. Carbocation ***7.17*** is therefore properly viewed as a reaction intermediate.

The water 'pulls' the bromine atom from the carbon in ***7.16*** by hydrogen bonding (H–O---H----Br). As the bond begins to weaken, the charge increases and water begins to solvate the developing charges until the ionic intermediates (cation ***7.17*** and bromide ion) is separated and completely solvated by the water. Therefore, water not only assists in pulling off the bromine but also solvates the charges as they form. This carbocation is quickly attacked by the nucleophilic iodide to give the product, 3-iodo-3-methylpentane, ***7.18***. This process constitutes a completely different mechanism than the process discussed for the S_N2 reaction.

Br KI aq. EtOH I^- I

7.16 *7.17* *7.18*

Does the Transformation of 7.16 to 7.18 Follow First Order or Second Order Kinetics?

The reaction follows first order kinetics.

Should You Know That This Reaction Is First Order by Simply Looking at It Before Studying the Reaction?

No! This fact is determined by plotting the concentration of ***7.16*** as a function of time, and the outcome of that plot tells us it is first order.

If the Reaction Is First Order, That Means the Rate Depends Only on 7.16, but 7.18 Can Only Be Formed If There Is a Reaction with Iodide. How Can This Be?

This is a two-step process. The first is ionization of ***7.16*** to the cation, ***7.17***. The second reaction is that of ***7.17*** with iodide to give the product, ***7.18***. For the first order kinetic data to be consistent, the first reaction must be very slow and the second very fast. If this is the case, rapid collision of ***7.17*** with iodide leads to product, but does not have a significant influence on the overall rate of reaction. In other words, ionization of ***7.16*** to ***7.17*** may have a half-life of 4.5 seconds, whereas the reaction of *7.17* with iodide to give ***7.18*** may have a half-life of 0.00005 seconds. If we look at the overall rate of the transformation of ***7.16*** to ***7.18***, the rate is essentially dictated by the slower ionization.

Which Halide Reacts Faster via S_N1 with KI in Aqueous Ethanol, 2-Bromo-2-Methylpropane or 1-Bromoethane?

The tertiary halide reacts much faster under S_N1 conditions. The relative rate of reaction for tertiary, secondary and primary halides that react with KI in aqueous acetone is:

$CH_3Br = 1.0$; $CH_3CH_2Br = 1.0$; $(CH_3)_2CHBr = 11.6$; $(CH_3)_3CBr = 1.2 \times 10^6$

This data reflects the energy required to form the intermediate cation (high energy for the slow reactions of bromomethane and bromoethane and low energy for the faster reactions of 2-bromo-propane and 2-bromo-2-methylpropane). The general reaction rate for halides in the S_N1 reaction is: $3° > 2° >> 1° >>> X–CH_3$.

What Is the Nature of a Carbocation Intermediate Such As 7.17?

Inspection of ***7.17*** shows that the carbocation is tricoordinate (three substituents) and is electron deficient (see Section **7.2**). The cation is **planar** around the electron deficient carbon. The two ethyl groups and the hydrogen are **coplanar**. The cation is thought of as an sp^2 hybridized carbon bearing an **empty p orbital**.

Which Is the More Stable Cation, That Derived from Ionization of 2-Bromo-2-Methylpropane or That from Ionization of Bromoethane?

The rate data shown in the question above clearly suggests that the tertiary halide ionizes (reacts) faster than the primary. This is correlated with the relative stability of the intermediate cations. The tertiary cation is more stable than the secondary, which is more stable than the primary, i.e., $3° > 2° > 1°$ $\oplus CH_3$.

Why Is a Tertiary Carbocation More Stable Than a Secondary, etc.?

The reason for this order of stability is the presence of carbon groups on the cationic carbon. Carbon groups are **electron releasing**. If electron density is pushed towards the empty p orbital (the cationic center), the net formal charge of the cation is diminished. Lower charge is associated with greater stability, and the more carbon groups attached to the cationic center, the more stable that cation will be The tertiary cation has three carbon groups, the secondary has two, and the primary has one. Methyl has no carbon groups.

Why Does 2-Bromo-2-Methylpropane Ionize to a Cation When It Is a Perfectly Stable Molecule?

The bromine, called a **leaving group**, does not 'fly off' the molecule. Two things must happen. First, there must be something to help 'pull' it off. This is the protic solvent water ($^{\delta+}H{-}O^{\delta-}{-}H^{\delta+}$) where the electropositive hydrogen coordinates to the electronegative bromine and 'pulls' it. This pulling lengthens the C–Br bond making it weaker, which increases the $^{\delta\oplus}$ charge on the C_2 carbon of 2-bromo-2-methylpropane. The water therefore assists removal of the leaving group and this is essential to a S_N1 reaction. The second important consideration is stability of the ion being formed. 3° cations are relatively stable and relatively easy to form; 1° cations are relatively unstable and difficult to form. The cation product is stabilized by the aqueous solvent where the $^{\delta-}$O of water donates electron density to the positive center of the cation, further stabilizing it by **solvation**. A solvent that can solvate the cation product is essential. Solvation of the cation is important for another reason. Water solvates both cations and anions, with the solvent separating them (as with NaCl). Coordination of water to the $\delta\oplus$ carbon of the halide and to the $Br^{\delta}s-$ will help separate them by solvation. This accelerates the rate of ionization.

Why Does the Reaction of 7.19 with KI in Aqueous Ethanol Lead to 7.20, where the Iodine Atom Is on a Different Carbon?

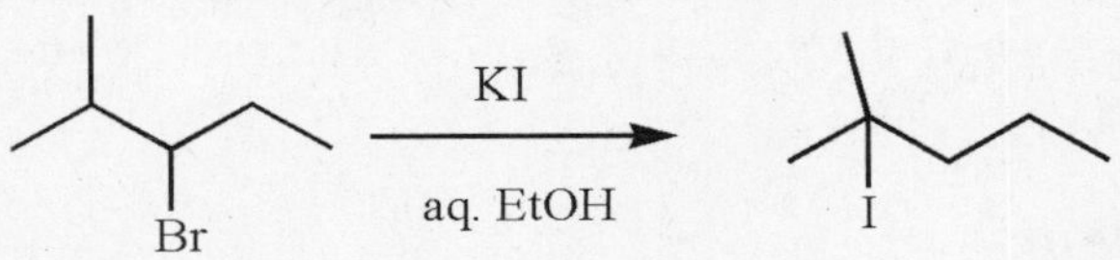

Br KI aq. EtOH I

7.19 *7.20*

Comparing the starting material ***7.19*** and the product ***7.20*** clearly reveals that there has been a **skeletal rearrangement**, since the bromine in ***7.19*** is on C3 and the iodide of ***7.20*** is on C2. This rearrangement occurs in the intermediate cation, a common occurrence in carbocations. Ionization of ***7.19*** generates a secondary cation (***7.21***). There is rotation about the C–C single bonds, and in one rotamer, the adjacent C–H bond on the tertiary carbon can be parallel to the p orbital on the cationic center, and the electron density of that bond can migrate towards the electron deficient center. If the bond carrying the hydrogen moves to that carbon (C3 from C2), a more stable tertiary cation will be formed (***7.23***). Since ***7.23*** is lower in energy than ***7.21***, this process is **exothermic** (by about 12-15 kcal/mol). The mid-point of this rearrangement is represented as ***7.22***, where the hydrogen migrates from C_2 to C_3. This hydrogen migration is called a **1,2-hydride shift**.

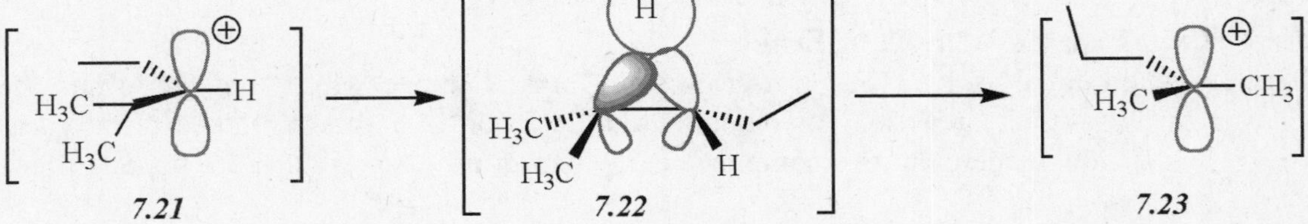

⊕ H_3C H H_3C H H_3C H_3C H ⊕ H_3C CH_3

7.21 *7.22* *7.23*

Can A Tertiary Carbocation Undergo a 1,2-Hydride Shift to Give a Secondary Carbocation?

No! The tertiary cation is more stable than the secondary, so a shift would be endothermic (require energy) and would not occur. Rearrangement always occurs from a less stable cation into a more stable cation.

Can Groups Other Than Hydrogen Migrate in This Type of Cationic Rearrangement?

Yes! In the reaction of *7.24* with KI in aqueous ethanol, cation *7.25* is formed, and migration of than adjacent methyl group (a 1,3-methyl shift) occurs to give the more stable *7.26* and subsequent reaction with iodide gives the final product, *7.27*.

aq. EtOH; KI

7.24 7.25 7.26 7.27

In 7.21, Why Did the Hydrogen Atom Migrate and Not One of the Methyl Groups When There Are Two Methyl Groups and Only One Hydrogen Atom?

The hydrogen is smaller than a methyl group and requires less energy to migrate via a transition state such as *7.22*. In general, given a choice, the smaller group will migrate unless there are special electronic or steric effects.

7.5. COMPETITION BETWEEN S_N2 VS. S_N1 REACTIONS

There are many reactions where the S_N2 and S_N1 process compete with each other, resulting in mixtures of products.

Protic vs. Aprotic Solvents

Is the Reaction of Bromomethane with KI Faster in Aqueous Ethanol or in Tetrahydrofuran?

In general, protic solvents such as water and ethanol solvate both cations and anions, separating those charges. The more polar solvent (water) efficiently separates and solvates ions. Aprotic solvents solvate cations, but anions are poorly solvated, making separation of ions difficult if not impossible. The unimolecular process (S_N1) requires separation of both cations and anions and, therefore, is efficient only in protic solvents. Bimolecular processes demand a collision. If the nucleophile and/or halide is solvated, the solvent molecules 'get in the way' of the collision, slowing it down. Aprotic solvents only solvate cations. The nucleophile, which is usually anionic, is not solvated and can approach the electrophilic center of the halide much easier, facilitating collision and reaction. For this reason, bromoethane reacts with KI (by an S_N2 process) faster in tetrahydrofuran (an aprotic solvent) than in the protic solvent, aqueous ethanol.

Explain Why 2-Bromo-2-Methylpentane Does Not Give the S_N2 Product When Heated with KI in Ether.

The activation energy for the S_N2 transition state *7.14* (R = Me, Me, Et) of the tertiary halide is too high to allow facile reaction. In other words, collision of iodide with the tertiary carbon does not provide enough energy to overcome the energy barrier necessary to achieve the S_N2 transition state.

Why Does 2-Bromo-2-Methypentane Give 2-Iodo-2-Methylpentane When the Reaction with KI Is Done in a Mixture of Ether and Water?

In the presence of water, and given that the S_N2 reaction is so slow that for all practical purposes does not occur, slow ionization of the bromide gives an intermediate carbocation, which reacts with the nucleophile (iodide in) to give the product. In other words, the water allows the S_N1 reaction to proceed.

Structural Features of The Halide

Explain Why Bromomethane Reacts with KI to Give Iodomethane via an S_N2 Mechanism in Aqueous Dimethylformamide.

Tertiary halides give essentially no reaction under S_N2 conditions, but react efficiently under S_N1 conditions. Primary halides react rapidly under S_N2 conditions and generally give S_N2 displacement or no reaction under S_N1 conditions. Secondary halides such as 2-iodobutane are midway in reactivity, and secondary cations are midway in stability. Primary carbocations are very difficult to form due to their instability and high activation energy for formation. The conditions described in this question are S_N1 conditions, but the S_N2 process can compete. The S_N2 reaction is just slower in water, not impossible. Given the difficulty in ionization to form a primary carbocation, the S_N2 process 'wins' and the reaction proceeds via S_N2. In general, primary halides give exclusively S_N2 reactions in protic and aprotic solvents, including water. Tertiary halides give no reaction under S_N2 conditions and ionization follows substitution under S_N1 conditions. The product formed from secondary halides depends upon the nucleophile and the reaction conditions. In polar, aprotic solvents, S_N2 reactions dominate. In water, S_N1 competes with S_N2. In polar, protic solvents, elimination dominates if the nucleophile is also a base (see Section **9.1**).

Does 2-Bromobutane React with KI via an S_N1 or S_N2 Pathway in Aqueous Dimethylformamide?

Secondary halides such as 2-iodobutane are midway in reactivity, and secondary cations are midway in stability. The conditions described in this question are S_N1 conditions, but the S_N2 process can compete, as described in the preceding question. Direct collision of iodide with the secondary carbon leads to product by the S_N2 pathway, whereas ionization to a carbocation followed by reaction with iodide is required for the S_N1 pathway. In general, the S_N2 pathway requires fewer bond making/bond breaking occurrences and should be faster. The presence of water in the solvent will slow down the S_N2 process, and we expect a product to be formed by both pathways. Which pathway is major depends on several factors, and additional information is required to prove or disprove which pathway dominates. An educated guess would say S_N2.

How Can You Tell If a Reaction Proceeded by S_N2 or S_N1 in the Absence of Kinetic Data?

If we use an alkyl halide as a substrate that has a configurationally pure stereogenic center (R or S), then the S_N2 process will show inversion of configuration, and the specific rotation of the product can be measured and compared with the theoretical amount. As we will see in the next few questions, ionization to a carbocation leads to a racemic product. Again, measuring the extent of asymmetric induction will allow us to decide how much S_N2 occurred and how much S_N1.

Stereochemical Features

If a Pure Enantiomer Is Subject to an S_N1 Reaction, Will the Product be Enantiopure or Racemic?

Carbocations are planar, and an incoming nucleophile can react from either face, giving both enantiomers in equal amounts (a racemic mixture). Therefore, in the reaction described in the question, we anticipate that the pure enantiomer starting material will be converted to a racemic product.

Predict the Product When 7.28 Reacts with KCN in DMF and Explain the Stereochemistry.

The product of the reaction of (2*S*)-iodoheptane (*7.28*) and potassium cyanide is (2*R*)-cyano-heptane, *7.29*. This S_N2 reaction proceeds with complete inversion of configuration at the stereogenic center.

7.28 —KCN, DMF→ 7.29

Predict the Product When 7.28 Reacts with KCN in Aqueous Ethanol and Explain the Stereochemistry.

The presence of water makes the S_N1 reaction competitive, and it proceeds by a *planar* carbocation. Formation of a carbocation from ***7.28*** via loss of the iodide results in a planar, achiral cationic carbon. This cationic carbon can be attacked by cyanide from either face. Attack from one face generates the *S*-enantiomer (***7.29***) with net inversion of configuration. Attack from the other face generates the *R*-enantiomer (***7.30***) with net retention of configuration. The planar nature of the cation therefore leads to a racemic mixture. It is noted that the S_N2 reaction can also occur in this medium, leading to more ***7.29***.

7.28 —KCN, EtOH/H_2O→ 7.29 + 7.30

A Reaction of (2S)-Bromopentane and Potassium Iodide Gave Only (2R)-Iodopentane. Did This Reaction Proceed by S_N2 or by a Mixture of S_N2 and S_N1?

The fact that 100% inversion of configuration occurred (note that the solvent was omitted from the question) indicates that the reaction was a clean S_N2 process.

7.6. RADICAL HALOGENATION OF ALKANES

Under certain conditions, chlorine and bromine can break apart to form chlorine or bromine radicals. Such radicals react with alkanes by removing a hydrogen atom and form HCl or HBr and a carbon radical. The carbon radical then reacts with more chlorine or bromine to give an alkyl chloride or an alkyl bromide.

What Is the Term for Breaking a Bond Such That Each Atom Receives One Electron? What Is the Name of the Resulting Intermediate?

Certain molecules are characterized by breaking a bond in an homolytic manner, or **homolytic cleavage**, where each atom of the covalent bond receives one electron, and the resulting products have a single electron. These intermediates are called **free radicals** and are capable of reacting with alkanes and other molecules to remove a hydrogen, generating a carbon radical.

What Is the Normal Reactivity of the Radicals Once Formed?

In one common reaction, radicals react with other atoms, removing the atom and generating a new radical. In another common reaction, radicals react with other radicals to give a **coupling reaction**, where a new bond is formed between the radical atoms (e.g. X• + Y• give X–Y), and each radical donates one electron to the new two-electron covalent bond. When a radical removes a hydrogen atom from a carbon, the resulting carbon radical reacts with another molecule of halogen, and the overall process replaces a hydrogen of the alkane precursor with a halogen atom. This is often called a **radical substitution**

reaction. In general, this is a poor reaction unless the radical reacts faster with one type of hydrogen rather than another, meaning that the rate of reaction of the radical with a primary hydrogen may be slower than the rate of reaction for a tertiary hydrogen.

What Is the Product When Diatomic Chlorine or Bromine Is Heated to 300°C?

When diatomic chlorine (Cl_2) or bromine (Br_2) are heated to 300°C or greater, they fragment homolytically to give chlorine radicals (Cl •) or bromine radicals (Br •). Exposing chlorine or bromine to light, usually ultraviolet(UV), also generates the radical. Radicals are highly reactive molecules and are intermediates (not isolated) that initiate reactions with alkanes.

What Is the Product When Methane Reacts with Chlorine Gas at 300°C?

Formation of chlorine radicals, removal of a hydrogen atom from methane, and reaction of the methyl radical with more chlorine leads to chloromethane and HCl as the products.

Draw the Mechanism for the Following Reaction:

$$Me_3C\text{–}H + Cl_2 \xrightarrow{hv} Me_3C-Cl+HCl.$$

A radical chain process involves formation of a radical in what is called a **chain initiation step**. That radical reacts with an alkane (*7.31*), usually producing a neutral molecule and a new radical, the chain carrier or propagator, *7.32*. The chain carrier radical reacts with a molecule to produce the 2-chloro-2-methylpropane product (*7.33*) and to regenerate the chain carrying radical. These are called **chain propagation steps**. If two of the radicals collide (a coupling reaction), they form a neutral molecule, but do not produce a new radical chain carrier. This stops the radical process and is called a **chain termination step**. To begin the process again, a new chain initiation step is necessary.

$$Cl\text{-}Cl \xrightarrow{\text{heat or hv}} 2\ Cl\bullet \qquad \textbf{Initiation}$$

$$(CH_3)_3C\text{–}H\ (7.31) + Cl\bullet \longrightarrow (CH_3)_3C\bullet\ (7.32) + H{:}Cl \qquad \textbf{Propagation}$$

$$(CH_3)_3C\bullet \xrightarrow{Cl\text{-}Cl} (CH_3)_3C\text{–}Cl\ (7.33) + Cl\bullet \qquad \textbf{Propagation}$$

$$Cl\bullet + Cl\bullet \longrightarrow Cl\text{-}Cl \qquad \textbf{Termination}$$

$$(CH_3)_3C\bullet \xrightarrow{Cl\bullet} (CH_3)_3C\text{–}Cl \qquad \textbf{Termination}$$

Radical Chlorination of Methane

Discuss Why 2,4-Dimethylpentane Reacts with Chlorine to Give Three Different Chlorinated Products. Draw Them.

In reactions with alkanes, halogen radicals abstract one of the hydrogens from the alkane to produce H–X and a carbon radical such as ***7.32*** from 2-methylpropane (***7.31***). That radical then reacts with additional halogen to produce an alkyl halide (2-chloro-2-methylpropane, ***7.33***, and the chain carrying radical such as Cl•). *Every* hydrogen atom in the alkane can be removed and replaced with a halogen. 2,4-Dimethylpentane (***7.34***), for example, can react with chlorine gas at 300°C to give three different products, ***7.35***, ***7.36*** and ***7.37***. These three products arise by removal of three different kinds of hydrogen, H_a, H_b and H_c. Replacement of H_a (there are 12 H_a since all four methyl groups are chemically identical) gives ***4.3.5***; replacement of H_b (there are two H_b) gives ***7.36*** and replacement of H_c (there are two H_c) gives ***7.37***. It is important to note that *every* hydrogen atom in *7.34* is replaced by a chlorine atom. Replacement of some hydrogens leads to the same product, and in *7.34* there are only three different kinds of hydrogen atoms, so there are three different products. In other words, if all 16 hydrogen atoms are replaced with chlorine, and those 16 structures are compared, some will be identical and only 3 different isomers can be found; the ones shown.

H_a H_b H_c *7.34* —(Cl_2 , 300°C)→ Cl *7.35* + Cl *7.36* + Cl *7.37*

Can We Predict the Relative Ratio of Products 7.35-7.37?

If each hydrogen is replaced at a given rate (see below), then the relative amounts of ***7.35***, ***7.36*** and ***7.37*** can be predicted based on the number of each kind of hydrogen. If the rates of reaction are different, this must be factored in.

Predict the Relative Percentages of 7.35, 7.36 and 7.37 from the Chlorination of 2,4-Dimethyl-Pentane.

For alkanes, a chlorine radical removes primary hydrogens with a relative rate of 1, secondary hydrogens with a relative rate of 3.9 and tertiary hydrogens with a relative rate of 5.2 [3°:2°:1° = 5.2:3.9:1]. For ***7.35***, there are 12 primary H_a's (all identical hydrogens), 2 tertiary H_b's and 2 secondary H_c's. The relative amount is then:

$$\text{Relative \% } \mathbf{7.35} = \frac{(12H_a x1)}{(12H_a \times 1) + (2H_a \times 1) + (2H_b \times 5.2) + (2H_a \times 3.9)} = \frac{12}{30.2} \times 100$$

Relative % ***7.35*** = 0.397 × 100 = 39.7%.

In this equation, the number of hydrogen atom types is adjusted for the different reaction rates. The 12 primary H's have a relative rate of 1, the 2 secondary H's a relative rate of 3.9 and the two tertiary H's a relative rate of 5.2.

Similarly, the relative % ***7.36*** = $\frac{10.4}{30.2} \times 100$ = 34.5% and the relative % ***7.37*** = $\frac{7.8}{30.2} \times 100$ = 25.8 %.

The generic formula to calculate the percentage of each different type of hydrogen is

$$\text{Relative \%} = \frac{\textit{Number of hydrogen atoms of a particular type}}{\textit{Total number of hydrogen atoms in the molecule}} \times 100$$

Bromination of 2,4-Dimethylpentane Leads to Nearly 95% of a Single Isomer, In Contrast to the Chlorination Reaction. Identify This Product and Discuss the Selectivity.

Bromine and chlorine react at different rates since the midpoint of the reaction for chlorine comes earlier than the midpoint of bromine. The factors that influence the stability of the radical intermediate are, therefore, more important for the bromine reaction, and the relative rates for bromination of a hydrogen are: 1°:2°:3° = 1:82:1640. Therefore, in a reaction with ***7.34*** with bromine, there will be a preponderance of one product. Using the same calculations as above, replacement of Ha will give 1-bromo-2,4,4-trimethylpentane ***(7.38),*** $\frac{(12\times1)}{(12\times1)+(2\times82)+(2\times1640)} = \frac{12}{3456} = 0.35\%$. Likewise there will be 4.75% of 2-bromo-2,4,4-trimethylpentane (***7.39***) via replacement of H_b but 94.91% of 3-bromo-2,4-dimethylpentane (***7.40***) via replacement of H_c.

α-Halogenation

Why Does 1-Propene React So Efficiently with Chlorine Radicals?

When the alkane hydrogen is on a carbon adjacent to a π-bond (see Section **8.1**), that hydrogen is removed by radicals at a much faster rate than other alkane hydrogens. This is because the resulting radical is **resonance stabilized**. This means that the radical is **delocalized** over several atoms rather than localized on a single atom. If 1-propene (***7.41***) is reacted with chlorine, the chlorine radical (Cl•) reacts with ***7.41*** to produce the so-called allyl radical, ***7.42***. The radical is in a p orbital, as shown in ***7.44***, which is parallel to the p orbitals of the π-bond. These orbitals can overlap in a manner that delocalizes the radical over all the carbons, and *both* structures shown for ***7.44*** are necessary to represent the actual structure. Therefore, ***7.42*** is represented as ***7.44***. This phenomenon is called **resonance**. A resonance stabilized structure is more stable and lower in energy. In the radical reaction, this extra stability means the radical is formed *faster*, allowing reaction with more chlorine to give the allyl chloride (3-chloro-1-propene, ***7.43***).

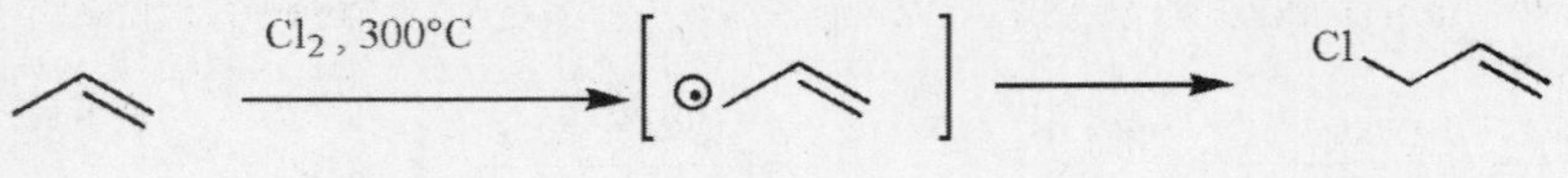

Cl_2, 300°C

Cl

7.41 7.42 7.43

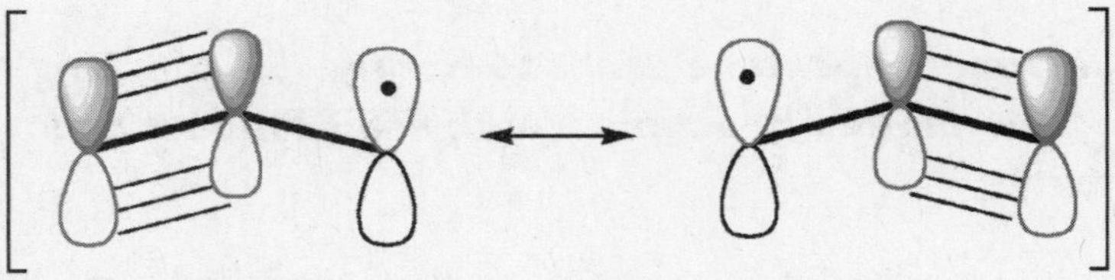

7.44

Draw the Products from the Reaction of 2-Butene (7.45) with Chlorine and Show all Resonance Structures for the Radical Intermediates. Do the Same for the Reaction of Toluene (7.48) with Chlorine.

If 2-butene (***7.45***) is reacted with chlorine, the implications of the resonance intermediates can be seen in the two products, ***7.46*** and ***7.47***. A chlorine radical attaches to both of the radical sites in the allylic intermediate. Similarly, if toluene (***7.48***) reacts with chlorine, a resonance stabilized radical is formed (***7.49***), leading to chloromethylbenzene, ***7.50***. For reasons to be discussed in Chapter 16, the 'intact' benzene ring in ***7.50*** is more stable than the product generated by attaching a chlorine to any of the resonance contributors where that ring does not contain the full compliment of six pi-electrons, i.e., where the radical is on the six-membered ring. The chlorine in the final product will therefore be attached *only* to the benzylic carbon as shown in ***7.50***. In both of these reactions, photochemical energy, provided by a sunlamp, was used to break diatomic chlorine into chlorine radicals.

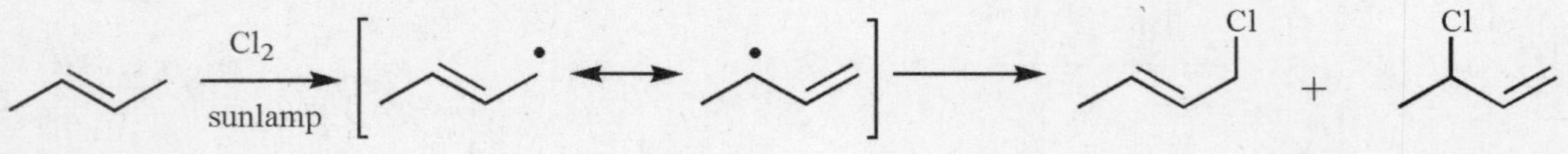

Cl_2 sunlamp Cl Cl

7.45 7.46 7.47

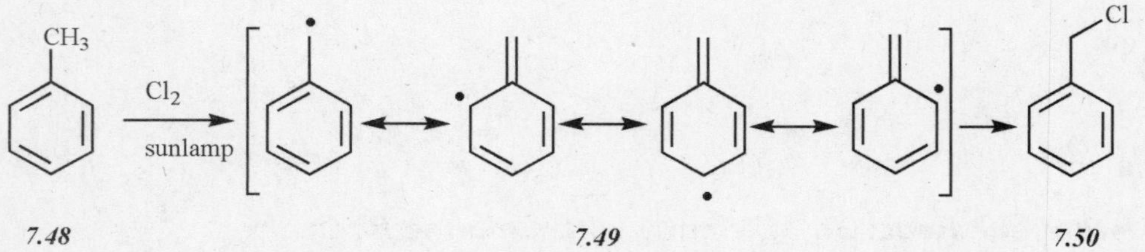

CH_3 Cl_2 sunlamp Cl

7.48 7.49 7.50

N-Bromosuccinimide and N=Chlorosuccinimide

What Is the Structure of N-Chlorosuccinimide (Abbreviated NCS)?

N-Chlorosuccinimide (***7.51***, called NCS) is a cyclic derivative of succinic acid (see Section 24.6).

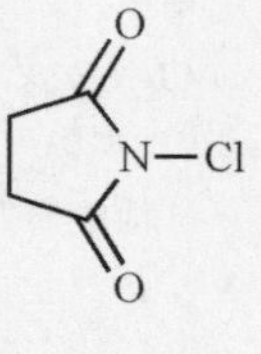

O N—Cl O

7.51 (NCS)

What Is the Structure of N-Bromosuccinimide (Abbreviated NBS)?

N-Bromosuccinimide (*7.52*, called NBS) is a cyclic derivative of succinic acid (see Section 24.6).

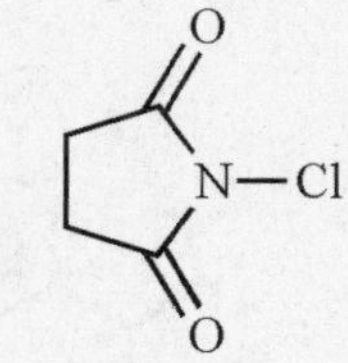

7.51 (NCS) 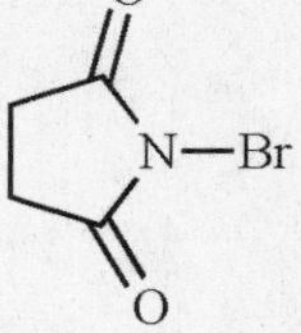 *7.52* (NBS)

What Is the Product When NCS Reacts with 2,2-Dimethylpropane (7.53)? When NBS Reacts?

When heated in a solvent such as carbon tetrachloride in the presence of photochemical energy (a sunlamp), ***7.53*** reacts with the chlorine radicals generated by the light reaction with NCS (***7.51***) to give ***7.54***. Since ***7.53*** has only one kind of hydrogen atom, there is only one product. Similarly, ***7.52*** is a source of bromine and bromine radicals. Therefore, when NBS is reacted with ***7.53***, ***7.55*** is formed. Again, there is only one product.

NBS , sunlamp NCS , sunlamp

7.55 *7.53* *7.54*

Test Yourself

1. Give the IUPAC name for each of the following.

(a) 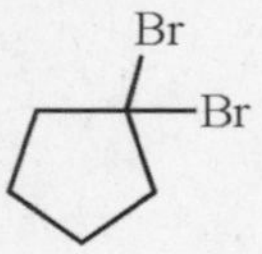 Br, Br (b) 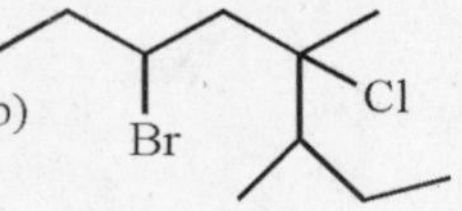 Br, Cl (c) 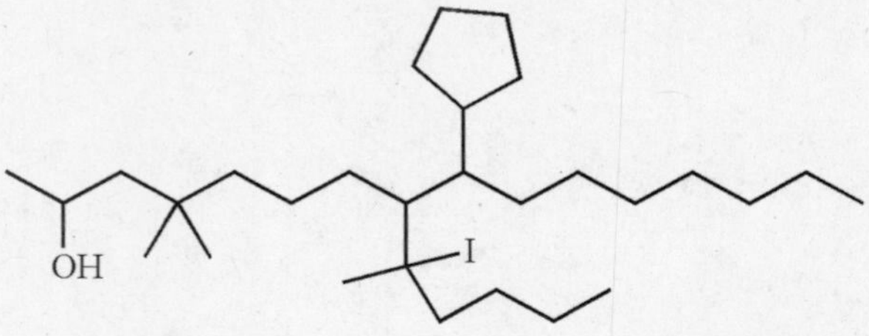 OH, I

2. Give the structure for each of the following, using line drawings.
   (a) 4,4-dimethyl-5-(2-fluoropropyl)dodecane
   (b) 2*R*-cyclopropyl-3*S*-iodohexane
   (c) 1-bromo-8*S*-chloro-2*S*-methyl-4,4-diethyltetradecane

3. Explain why cation A is less stable than cation B.

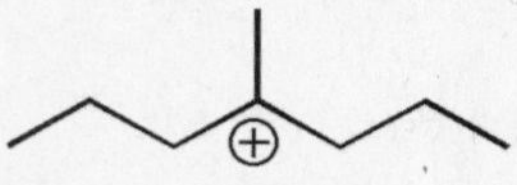

A 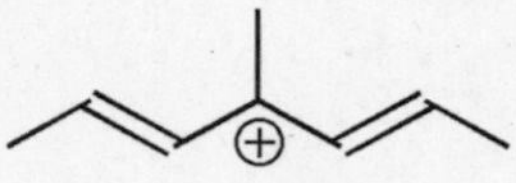 B

4. Show all products and the relative % of each product for the following reaction.

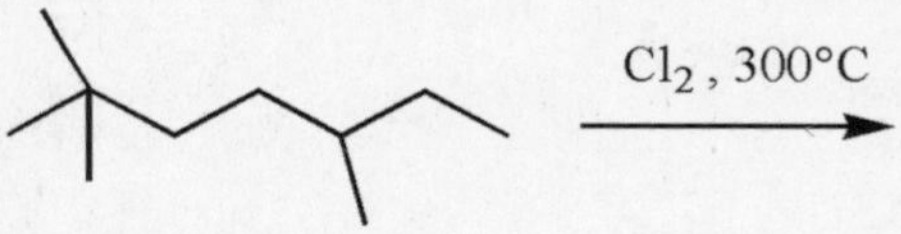

Cl_2, 300°C

5. If the concentration of KI in the reaction of KI and 1-bromopentane is increased to 10 equivalents, what is the effect on the rate of that reaction? Explain.

6. Explain why the rate of reaction of 1-bromo-2,2-dimethylpropane in a S_N2 reaction is 3.3×10^{-7} when compared to the rate of reaction of bromomethane under the same conditions, although the former is a primary halide.

7. Explain why the rate of reaction of allyl bromide in an S_N2 reaction is 1.3 times faster than the rate of reaction of bromomethane under the same conditions.

8. Draw the structures of THF, DMF, DMSO and dichloromethane.

9. Write the rate expression for an S_N1 reaction. What is the effect on the rate if 10 equivalents of 2-bromo-2-methylpropane are added to 1 equivalent of KI when compared to the reaction when one equivalent of both reagents is used?

10. Explain why the hydrogen migrates in this reaction rather than the methyl and provide the complete mechanism.

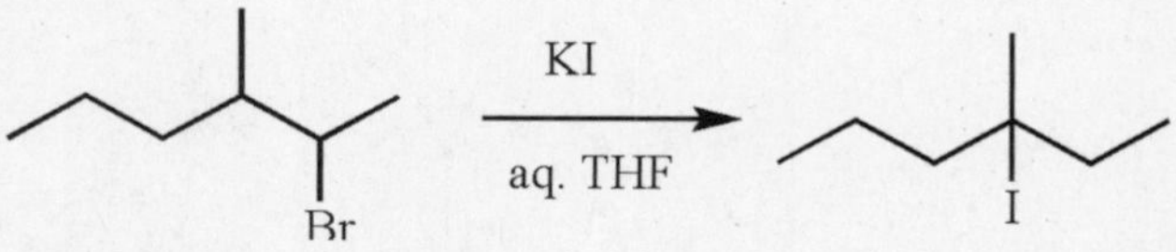

Br KI aq. THF I

11. Explain why this reaction gives a mixture of enantiomeric chlorides, but there is a slight preponderance of the inversion product.

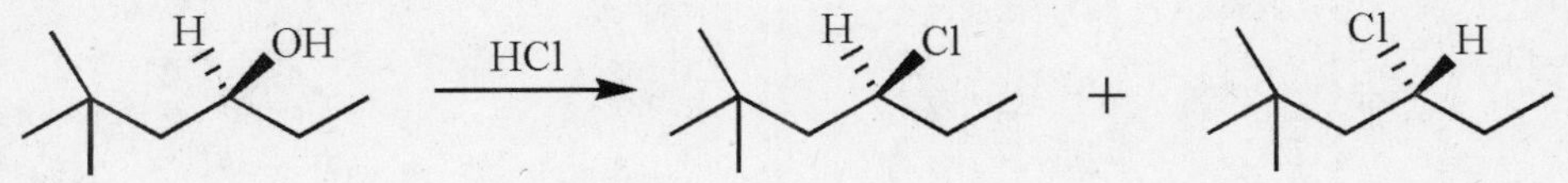

H OH — HCl → H Cl + Cl H

12. In each case, give the major product of the reaction. Remember stereochemistry, and if there is no reaction, indicate this by N.R.

(a) Cl — KI, THF, 25°C →

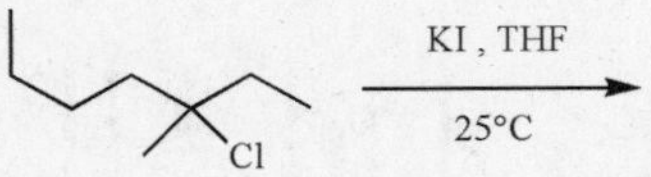

(b) Br — H_2O, THF, heat, 7 days →

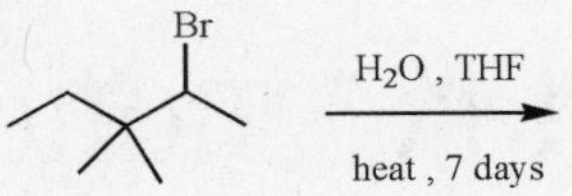

(c) I — NaN_3, THF →

(d) H Br — KI, THF, H_2O →

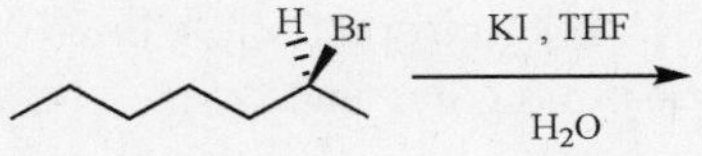

(e) — Br_2, hν →

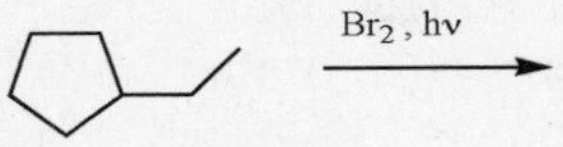

(f) — Cl_2, 300°C →

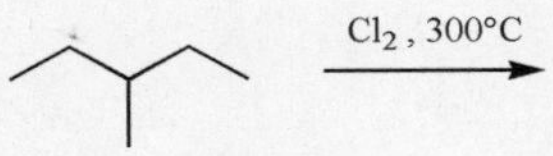

(g) — Br_2, hν →

(h) — NBS, hν →

(i) Me H I — KCN, DMF →

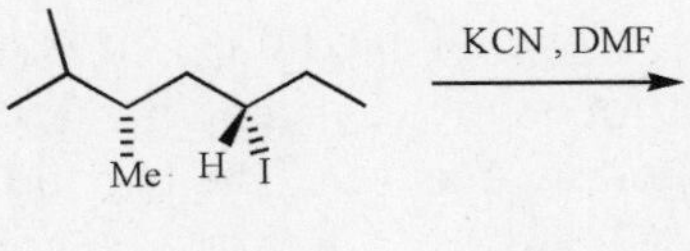

(j) Me Br — KI, THF →

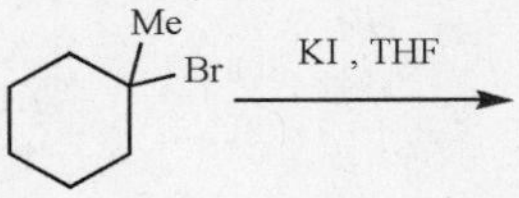

(k) H Br — KI, THF →

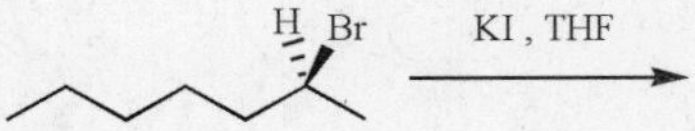

(l) $C_5H_{11}-C\equiv C{:}^{-}\ Na^{+}$ — 1-iodobutane, DMF →

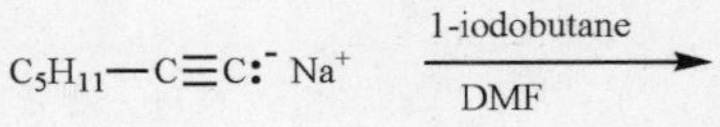

(m) I — KCN, DMF, 50°C →

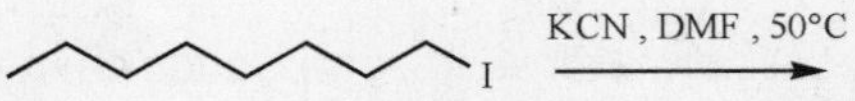

(n) Br — KCN, aq. EtOH →

(o) Br — KCN, aq. EtOH →

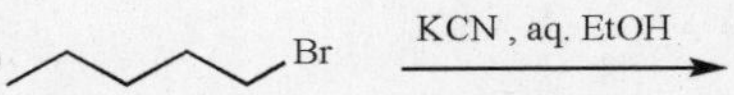

(p) H — 1. Br_2, hν; 2. KCN, aq EtOH, heaat →

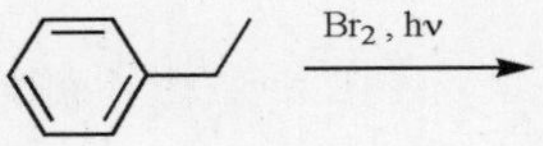

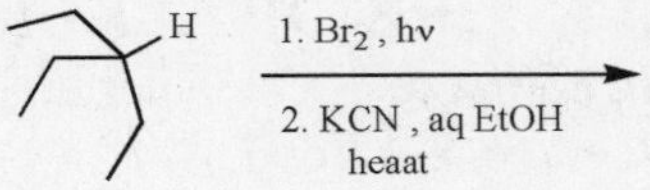

CHAPTER 8

Alkenes and Alkynes: Structure and Nomenclature

Alkenes are a class of compounds characterized by a carbon-carbon double bond (a π-bond), which is highly reactive with a variety of reagents. This π-bond can function as a base in the presence of a suitable acid. It is also a nucleophile in the presence of a sufficiently strong electrophile. Reaction of the π-bond with acid generates cations, which can then react with nucleophiles. Alkenes are among the most important of all functional groups due to the versatility of their chemistry. This chapter will also introduce the important class of reactions called addition reactions, which are characteristic of alkenes.

8.1. STRUCTURE OF ALKENES

Alkenes are compounds characterized by one or more C=C units, each comprised on one C–C covalent sigma bond and one π-bond. Alkynes are compounds characterized by one or more C≡C units, each comprised on one C–C covalent sigma bond and two π-bonds.

Describe the Bonding in Ethene.

Ethene (C_2H_4 or $CH_2{=}CH_2$) is the simplest member of the alkene family. It is a planar molecule and has two different kinds of bonds between the carbon atoms: a C–C single bond, called a sigma [σ] bond, and a weaker bond called a π-bond. This is usually drawn as ***8.1***. The σ-bond is a strong bond directed along the line between the two carbon nuclei (see ***8.2*** and C–C in ***8.3***). The π-bond is formed by overlap of the two parallel p orbitals on the sp^2 carbons. This bond is formed by sharing a portion of the electron density of the p orbitals (see ***8.2*** and ***8.3***) with electron density above and below the plane of the carbons and hydrogens. This 'sideways overlap' is characteristic of a π-bond. It is important to emphasize that structures ***8.2*** and ***8.3*** have only one π-bond with lobes above and below the plane of the carbon atoms.

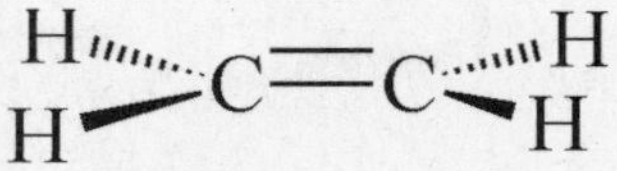

H H C=C H H

8.1

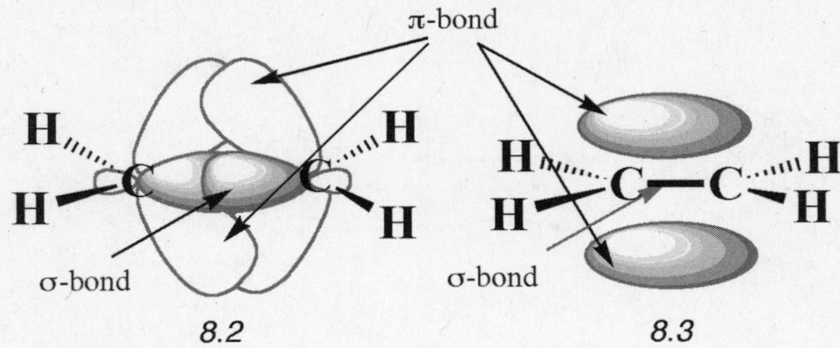

8.2 *8.3*

Is Rotation about the C=C Unit Possible in 8.1 or in Any Alkene at Room Temperature?

No! The π-bond effectively locks the molecule and prevents rotation. Therefore, in ***8.1***, two hydrogen atoms are on one side of the molecule, and two are on the other side.

In 8.4, Are the Bromine and Chlorine on the Same Side or Opposite Sides of the Molecule?

H, Br–C=C–Cl, H

8.4

The bromine and chlorine atoms are on opposite sides. The 'sides' are front and back as the molecule is drawn. The Br and H on the front side (solid wedges) are projected towards you, whereas the H and Cl on the back side (dashed lines) are projected behind the plane of the paper. Therefore, Br and Cl are on opposite sides. We will use this 'sidedness' convention throughout. Note that because there is no rotation about the C=C unit, the Br and Cl are locked on the respective sides.

Which Is the Stronger Bond in Ethene, the σ-Bond or the π-Bond?

Since the σ-bond has electron density concentrated in a line between the two carbon nuclei, it is very strong. The π-bond has much less electron density concentrated between the carbon nuclei due to the 'sideways overlap' of the p orbitals. Since it has less electron density between the carbon atoms, the π-bond is a weaker bond and more easily broken.

What Is the Hybridization of the Carbon Atoms in Ethene?

Each carbon of ethene is sp^2 hybridized. Each is trigonal planar, attached to three other atoms and possesses a p orbital (as part of the π-bond). This sp^2 hybridization is characteristic of the carbon-carbon double bonds of all alkenes.

Draw the Molecular Orbital Diagram for the C=C Bond of Ethene Using sp^2 Hybrid Orbitals.

The molecular orbital diagram for ethene (***8.5***) uses three degenerate sp^2 hybridized orbitals from each carbon to form three identical molecular orbitals. The extra p electron resides in a higher energy orbital and mixes with the other p orbital to form the bonding molecular orbital, which represents the π-bond. The molecular orbital for the π-bond is lower in energy than the unfilled antibonding orbital from the sp^2 hybrids.

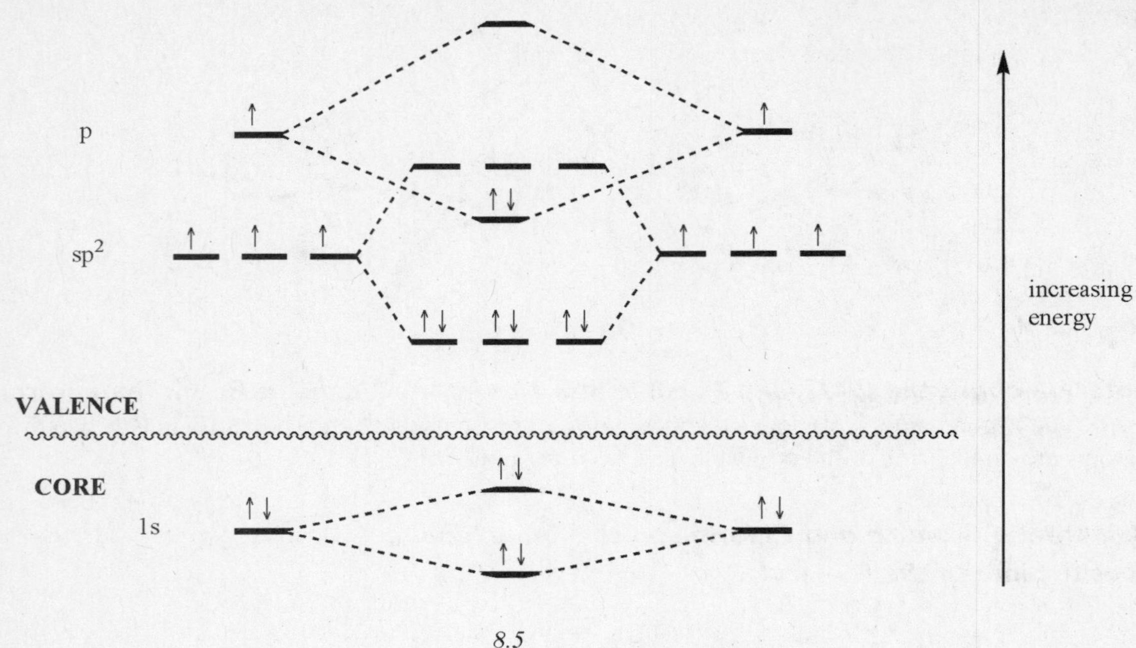

8.5

Which Alkene Is the Most Thermodynamically Stable, 2-Methyl-2-Butene or 3-Methyl-1-Butene?

In general, alkyl groups release electron density to the π-bond of the alkene. This extra electron density strengthens the π-bond, making it more stable. A tetrasubstituted alkene is, therefore, more stable than a trisubstituted alkene, which is more stable than a disubstituted alkene. In this example, 2-methyl-2-butene (a trisubstituted alkene) is more stable than 3-methyl-1-butene (a mono-substituted alkene).

What Is the IUPAC Ending Used of Alkenes?

The *-ane* ending is dropped from the alkane name and replaced with *-ene*. The only rule change that is important involves the C=C unit. The double bond *must* be part of the longest continuous chain and the first carbon of C=C that is encountered must receive the lowest possible number, regardless of the substitution pattern (assuming that C=C is the highest priority functional group).

Give the Correct IUPAC Name for 8.6–8.9.

Alkene ***8.6*** has an 8-carbon chain containing the double bond with an ethyl substituent. The name is, therefore, 3-ethyl-3-octene. Alkene ***8.7*** is a cyclic 6-carbon alkene and is called cyclohexene. Numbering to give the geminal dimethyls (both methyl groups on the same carbon) the lowest number, this becomes 4,4-dimethylcyclohexene (remembering that C_1 and C_2 of cyclohexene must contain C=C). Alkene ***8.8*** is a 7-carbon chain containing one Br, two methyls and one ethyl. The name is 2-bromo-3-ethyl-4,6-dimethyl-1-heptene. Alkene ***8.9*** is a cyclooctene with a methyl and an ethyl group. Since the groups are named alphabetically, this is 1-ethyl-2-methylcyclooctene.

Br

8.6 8.7 8.8 8.9

What Is a Diene?

A diene is simply a molecule that has two C=C units.

What Is the Name of $CH_3CH{=}CH{=}CH_2$?

This is a five-carbon molecule with two alkene (*-ene*) units, so it is a diene. Number the first carbon of each C=C unit to give the lowest number, making this molecule 1,3-pentadiene.

What Is a Conjugated Diene?

A conjugated diene has the two sp^2 carbons of the C=C units directly attached with no intervening sp^3 hybridized atoms. 1,3-Pentadiene is a conjugated diene.

Draw a Nonconjugated Hexadiene.

An example would be 1,5-hexadiene, $CH_2{=}CH{-}CH_2CH_2CH{=}CH_2$.

What Is an Allene?

An allene is a diene that has a central sp hybridized carbon that is part of two π-bonds (allene is $CH_2{=}C{=}CH_2$). The two π-bonds are mutually perpendicular.

What Is the Systematic Name of Allene?

Using the same nomenclature used for other dienes, allene is 1,2-propadiene.

What Is the Structure of 3-Chloro-1,2-Butadiene?

This molecule would be ***8.10***.

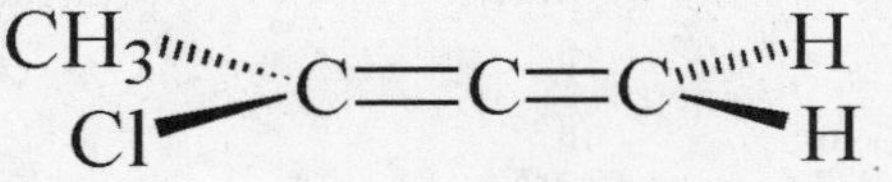

$$(CH_3)(Cl)C{=}C{=}C(H)(H)$$

8.10

8.2. ALKENE STEREOISOMERS

The fact that rotation about a C=C is not possible leads to sidedness for substituted alkenes, as we saw with ***8.4***. This property leads to the possibility of a new type of isomer for alkenes, a stereoisomer.

Alkene 8.11 Is Named 3-Hexene, but There Are Two Different 3-Hexenes. Explain.

$$CH_3CH_2CH{=}CHCH_2CH_3$$

8.11

The two different isomers of 3-hexene are ***8.12*** and ***8.13*** The π-bond prevents rotation about the C=C double bond. The two ethyl substituents are, therefore, *locked* in place and cannot be interchanged without breaking and making bonds. These two compounds differ only in the relative spatial position of the groups and are called **stereoisomers**. Since they are different compounds, they require different names to properly identify them.

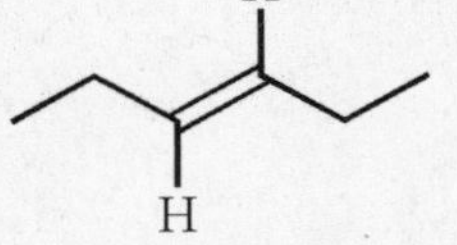

8.12

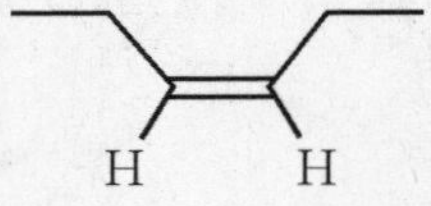

8.13

Are the Two Structures 8.14 and 8.15 Stereoisomers?

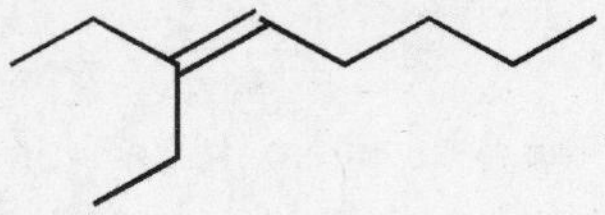

H

8.14 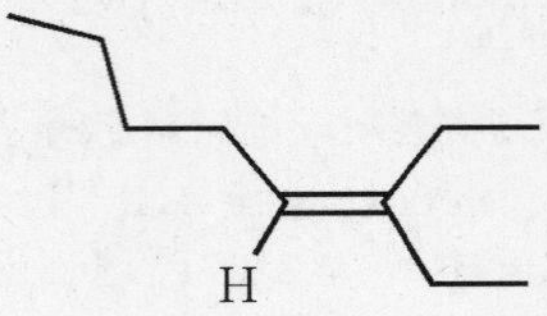 8.15

No! Note that one carbon of the C=C unit has two attached ethyl groups. This means there will be an ethyl group on each side of the molecule, making the two 'sides' identical. Therefore, ***8.14*** and ***8.15*** are the same molecule (one compound) and not stereoisomers.

Definition of Cis and Trans Isomers

What Is a Cis- Isomer?

A cis- stereoisomer has two identical groups attached to the C=C unit on the same side of the molecule. Examination of ***8.13*** shows that there are *identical substituents* (ethyl) *on the same side of the molecule*. When two identical groups are on the same side of an alkene, the term *-cis* is incorporated into the name. Alkene ***8.13*** is, therefore, *cis*-3-hexene.

What Is a Trans-Isomer?

A trans-stereoisomer has two identical groups attached to the C=C unit on opposite sides of the molecule. Alkene ***8.12*** has *identical substituents* (ethyl) *on opposite sides of the molecule*. When two like groups are on opposite sides of an alkene, the term *-trans* is incorporated into the name. Alkene ***8.12*** is, therefore, *trans*-3-hexene.

Give the IUPAC Names for 8.16 and 8.17.

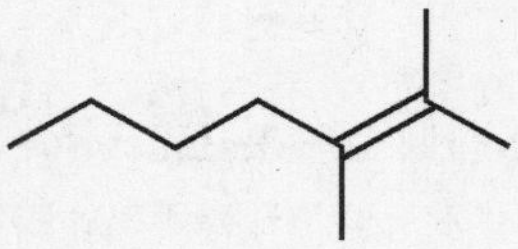

8.16 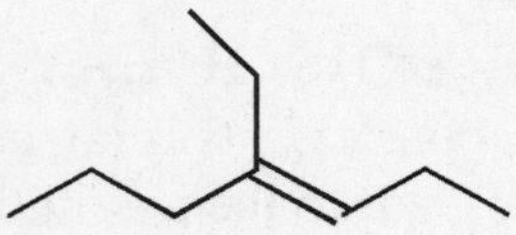 8.17

In ***8.16***, the two identical groups are methyl. Since the methyl on C_1 and that on C_2 are on opposite sides of the double bond, this is a *trans* alkene (*trans*-2,3-dimethyl-2-heptene). In ***8.17***, the two identical groups are the ethyl groups, and they are on the same side of the double bond. This is a *cis-* alkene (*cis*-4-ethyl-3-heptene).

When the Name of 2-Methyl-2-Octene (8.18) Is Given, There Is No Cis- or Trans- Designator. Explain.

The two identical groups are methyl, but they are both on C_1. The *cis-* and *trans-* nomenclature are used only when one of the like groups is on C_1 and the other is on C_2 of the C=C unit. Since the like groups are on the same carbon in ***8.18***, there are no *cis- / trans-* isomers and the name is simply 2-methyl-2-octene.

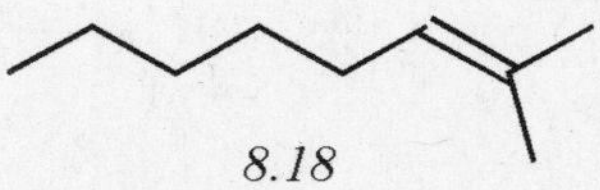

8.18

E/Z Nomenclature

Explain Why the Cis- and Trans- Names Do Not Apply to 2-Chloro-3-Ethyl-5-Methyl-2-Hexene (8.19).

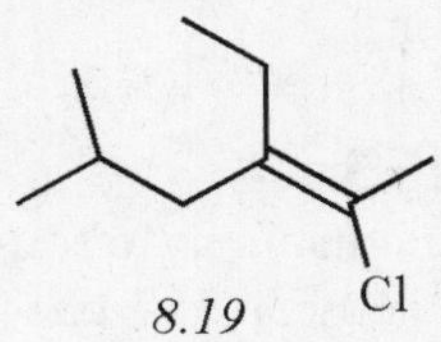

Cl

8.19

There are no like groups in ***8.19***, and the *cis- / trans-* nomenclature does not apply. Is the Cl *cis-* to the ethyl or the 2-methylpropyl group? A different nomenclature system is required to properly name compounds such as this.

In 8.19, a Methyl and a Chlorine Atom Are on C2. Using the Cahn-Ingold-Prelog Selection Rules, Which Is Higher in Priority?

The chlorine atom is higher in priority relative to the methyl group.

In 8.19, an Ethyl and a 2-Methylpropyl are on C3. Using the Cahn-Ingold-Prelog Selection Rules, Which Is Higher in Priority?

The 2-methylpropyl group is higher in priority relative to the ethyl group.

In 8.19, Are the High Priority Groups on C2 and C3 on the Same Side or the Opposite Side?

In ***8.19***, the Cl and the 2-methylpropyl groups are on the same side.

Describe the E-/Z- Nomenclature System.

This system is based on a comparison of the substituents on C_1 and C_2 of the double bond. The substituents are assigned a priority based on the Cahn-Prelog-Ingold selection rules introduced in Section 3.4. Using ***5.3.6*** as an example, one carbon of the double bond has a chlorine and a methyl substituent. Since Cl > C, the chlorine is the highest priority group. The other carbon of the double bond has an ethyl group and a 2-methylpropyl group. Using the sequence rules, 2-methyl-propyl is the higher priority group. Comparing the Cl and the 2-methylpropyl shows these higher priority groups to be on the same side of the molecule. They are given the notation **Z** (for **zusammen** = together). This protocol involves determining the priority of the two groups on one carbon of the double bond, determining the priority of the two groups on the other carbon of the double bond and then determining if the higher priority groups are on the same side (*Z*-) or the opposite side (*E*-, for **entgegen** = apart or opposite).

Using the E-/Z- Nomenclature, Name 8.19.

The name of ***8.19*** is *Z*-2-chloro-3-ethyl-5-methyl-2-hexene.

Give the Name for 8.20.

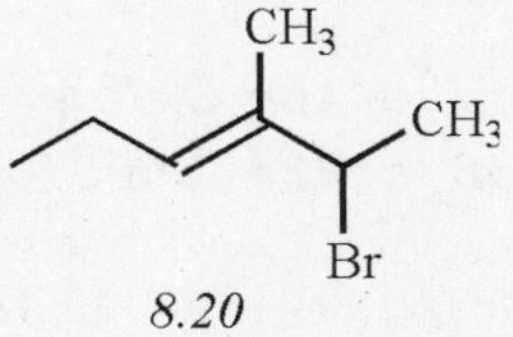

CH_3

CH_3

Br

8.20

Alkene *8.20* is an example of an *E*- alkene. The first carbon has a methyl group and a 1-bromoethyl group, with the latter having the higher priority. The other carbon has a methyl and a hydrogen, with the ethyl being the higher priority. The ethyl group and the 1-bromoethyl group are on opposite sides of the double bond, making this an *E*- alkene. The name is *E*-2-bromo-3-methyl-3-hexene.

8.3. STRUCTURE OF ALKYNES

Alkynes are organic hydrocarbons that contain a carbon-carbon triple bond, which is composed of two π-bonds that are perpendicular to one another. The chemistry of alkynes is very similar to that of alkenes due to the presence of π-bonds. The proximity of the π-bonds, however, leads to some interesting differences between alkenes and alkynes.

What Is the Characteristic Feature of an Alkyne?

Alkynes possesses a carbon-carbon triple bond:–C≡C–

What Is the Structure of Acetylene?

Acetylene is a two-carbon molecule with the formula C_2H_2. Its structure is H–C≡C–H, where both hydrogens and both carbons are linear. There is one σ-bond and two π-bonds. The two π-bonds are perpendicular to one another. This is represented in ***8.21***.

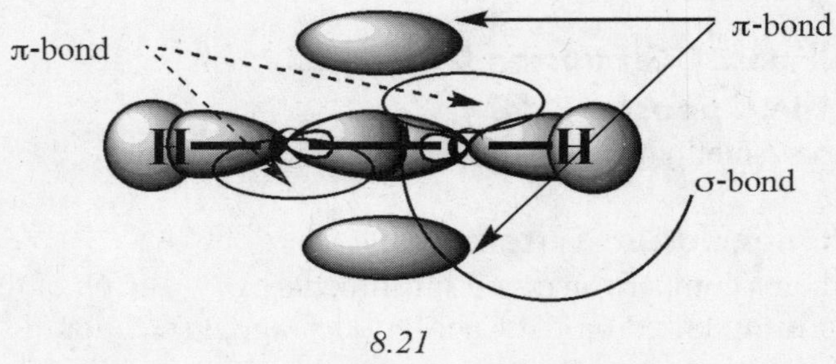

8.21

What Is the Hybridization of each Carbon in Acetylene?

The carbons are sp hybridized.

What Is the Relationship of C1 and C4 in 8.22?

$$\overset{4}{H_3C}-\overset{3}{C}\equiv\overset{2}{C}-\overset{1}{CH_3}$$

8.22

Since the triple bond is a linear unit, C1 and C4 must be linear (180° apart).

Give the Molecular Orbital Diagram for the Carbon-Carbon Triple Bond of Ethyne Assuming sp Hybridization for the Two Carbon Atoms.

The molecular orbital diagram is shown in ***8.23***. There are two electrons in sp hybrid orbitals for the sp σ-bonds, and two electrons in p orbitals for the two π-bonds for each carbon.

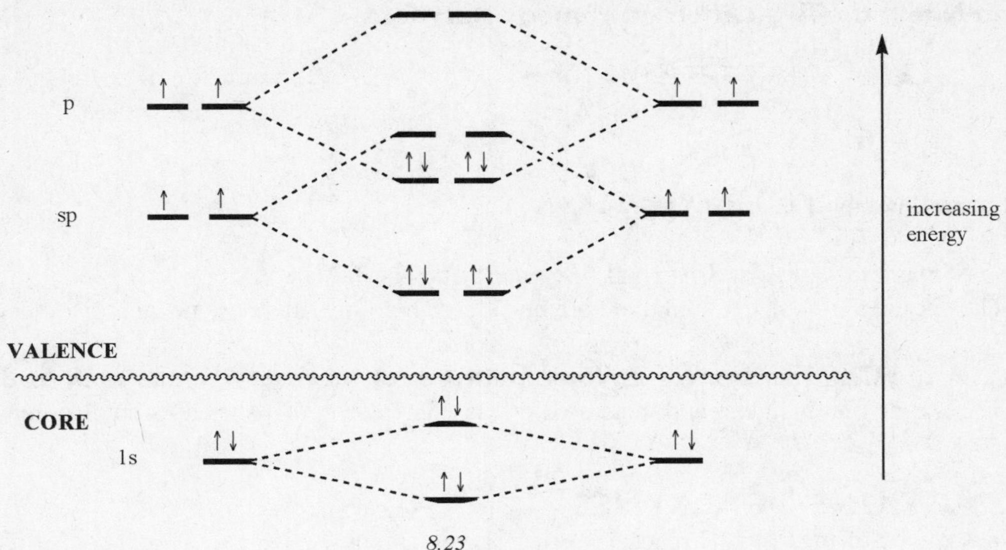

8.23

Nomenclature

What Is the IUPAC Ending for Alkynes?

The *-ane* ending of an alkane is dropped and replaced with the alkyne ending, *-yne*. A four-carbon alkyne is, therefore, butyne. The position of the first carbon of the triple bond is given the lowest number, and the C≡C unit must be in the longest continuous chain. The IUPAC name for $CH_3CH_2CH_2C{\equiv}CCH_3$ is 2-hexyne, and $CH_3CH_2C{\equiv}CH$ is 1-butyne.

Give the Correct Structure for Each of the Following.

(A) 5-CHLORO-4-PHENYL-2-HEPTYNE (B) 5,5-DIBROMO-1-HEXYNE
(C) 3-CYCLOHEXYL-1-BUTYNE (D) 2,5-OCTADIYNE (E) HEPTA-5E-EN-2-YNE
(F) 3,3-DIMETHYLOCTA*N*-5-YN-2-OL

The answer to (a) is ***8.24***, to (b) is ***8.25***, to (c) is ***8.26***, to (d) is ***8.27***, to (e) is ***8.28*** and to (f) is ***8.29***.

Ph Cl *8.24* Br Br H *8.25* H *8.26*

8.27 *8.28* OH *8.29*

Are Isomers Possible for Alkynes That Do Not Have Substituents Attached to the Longest Continuous Chain?

Yes! Monosubstituted (terminal) alkynes have the generic formula R–C≡C–H. Disubstituted (internal) alkynes have the generic formula R–C≡C–R.

Name the Two Different Structural Types of Alkynes.

Monosubstituted alkynes and disubstituted alkynes (terminal and internal, respectively).

Draw and Name the Two Different Isomeric Butynes.

8.30 8.31

Alkyne *8.30* is 1-butyne and *8.31* is 2-butyne.

Are There Stereoisomers for Internal Alkynes Such As 8.31?

No! The alkyne unit is linear, and there are no sides. Therefore, there are no stereoisomers.

Can We Use the Cis / Trans or E / Z Nomenclature for Internal Alkynes such as 8.31?

No! The alkyne unit is linear, and there are no sides. Therefore, there are no stereoisomers.

8.4. REACTIONS OF ALKYNES

Alkynes can be formed by S_N2 reactions using alkyne anions.

Are Alkynes Acids?

The hydrogen of a terminal alkyne is a weak acid. The terminal hydrogen atom is attached to the sp hybridized carbon of the triple bond, and this leads to acidity for that hydrogen. Internal alkynes do not have a hydrogen atom attached to the sp hybridized carbon and do not have an acidic hydrogen.

What Is the pK_a of the C–H Bond of Acetylene (Ethyne)?

The pK_a of the hydrogen in acetylene (H–C≡C–**H**) is about 25. Compared to water (pK_a of 15.8) and acetic acid (pK_a of 4.7), acetylene is a *very* weak acid and requires a strong base to remove the acidic hydrogen. The product of the acid/base reaction is called an **acetylide** (also known as an **alkyne anion**), with a negative charge residing on carbon: H–C≡C–H + BASE → H–C≡C:⁻

What Bases Can Be Used to Remove the Hydrogen of an Alkyne?

As just mentioned, a powerful base is required to remove the weakly acidic alkyne hydrogen. The most common bases used are sodium amide ($NaNH_2$ in liquid ammonia), *n*-butyllithium ($CH_3CH_2CH_2CH_2Li$, see Chapter 10.3) in ether solvents or sodium hydride (NaH) in anhydrous solvents. Sodium hydride is very attractive, since the by-product of the acid/base reaction is hydrogen gas $\left(\frac{1}{2}H_2\right)$ and the sodium salt of the alkyne (a sodium acetylide, $R–C≡C:^- Na^{\oplus}$).

Why Is the Hydrogen of an Alkyne More Acidic Than the Hydrogen of an Alkene or an Alkane?

This is related to the hybridization of the C–H bond. In general, acidity increases with increased 's character' of the bond: $sp > sp^2 > sp^3$. An alkyne is, therefore, more acidic than an alkene, which is more acidic than an alkane. An s orbital is generally better able to stabilize a negative charge than a p orbital and an sp orbital has 50% s character. An sp^3 orbital has only 25% s-character and an sp^2 orbital has 33% s character.

What Is the Reactivity of an Acetylide?

Since the negative charge resides on carbon, acetylides are powerful **nucleophiles** and acetylides behave as **carbanions**.

Can an Acetylide Anion Participate As a Nucleophile in S_N2 Reactions?

Yes!

Alkylation with Alkyl Halides

Define the Reaction Type When the Acetylide of 1-Butyne Reacts with Iodomethane.

A nucleophilic acetylide reacts with the δ⊕ carbon of an alkyl halide in an S_N2 reaction. In this case, the product is 2-pentyne ($CH_3CH_2C{\equiv}C{:}^-Na^+ + MeI \rightarrow CH_3CH_2C{\equiv}CCH_3$).

What Solvents Are Appropriate for the Reaction of Acetylides and Alkyl Halides?

Since this is an S_N2 reaction, a polar aprotic solvent should be used such as THF, DMF or DMSO.

Give the Major Product of Each of the Reactions of 8.32–8.35.

(a) 8.32 — 1. *n*-BuLi, ether; 2. CH_3OH

(b) 8.33 — 1. THF, $NaNH_2$; 2. CH_3I

(c) 8.34 — 1. NaH, THF; 2. (benzyl bromide, Br)

(d) H—≡—H 8.35 — 1. $NaNH_2$, THF; 2. CH_3Br; 3. *n*-BuLi, ether; 4. (1-iodopentane, I)

In reaction (a), alkyne ***6.3.4*** is treated with base to generate the acetylide. Subsequent treatment with methanol, which is an acid in the presence of the basic acetylide, protonates the acetylide to give back the starting material, and the product of the reaction is ***8.32***. In reaction (b) alkyne ***8.33*** is first converted to the acetylide with $NaNH_2$, and subsequent reaction with iodomethane generates ***8.36***. In reaction (c), alkyne ***8.34*** is converted to the acetylide with base, and alkylation with benzyl bromide gives the disubstituted alkyne product, ***8.37***. In reaction (d), acetylene (***8.35***) is deprotonated and reacted with iodomethane to give 1-propyne. A second deprotonation-alkylation sequence (with iodopentane) gives the final product, 2-octyne, ***8.38***.

8.36 *8.37* *8.38*

Test Yourself

1. What is the geometry of the atoms in 2,3-dimethyl-2-butene?
2. Name $CH_3CH_2CH_2C{=}C(Br)Cl$.
3. Discuss why alkene A is more stable than alkene B.

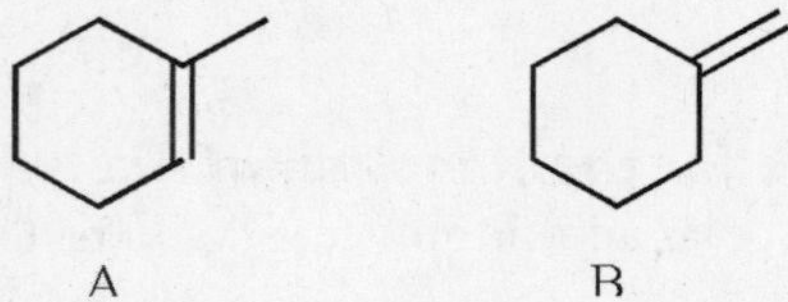

4. Give the IUPAC name for each of the following.
5. Explain how *Z*-3-chloro-3-hexene is also properly labeled as a *trans*-alkene (i.e., why does *Z* *not* translate directly to *cis*?).
6. Give the correct IUPAC names for each of the following.
7. Which hydrogen is removed first when $HC{\equiv}C\text{–}CO_2H$ is treated with one equivalent of *n*-butyllithium? Explain.

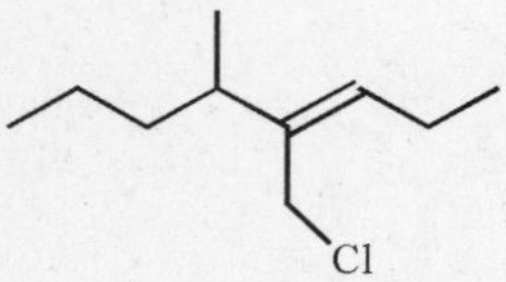

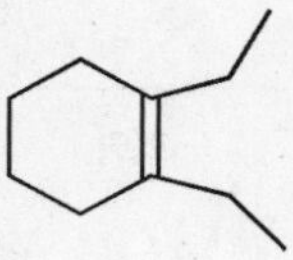

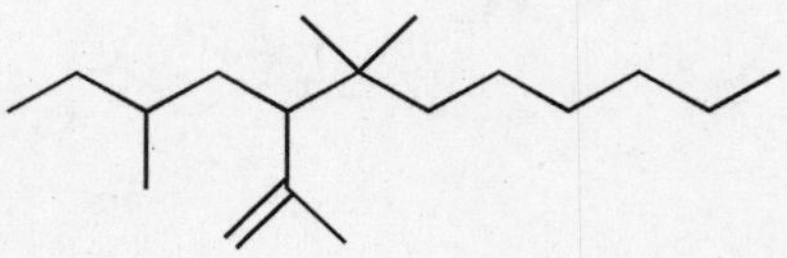

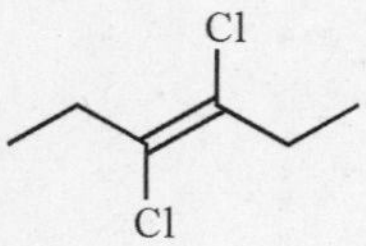

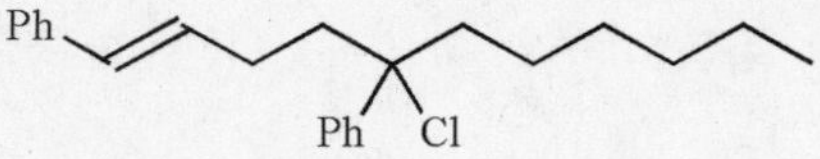

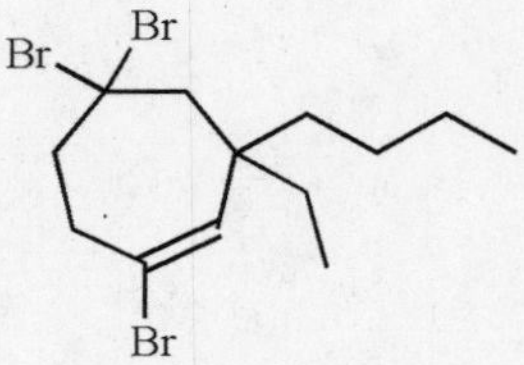

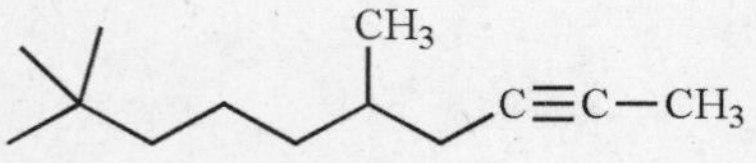

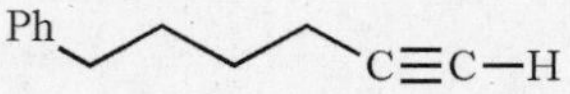

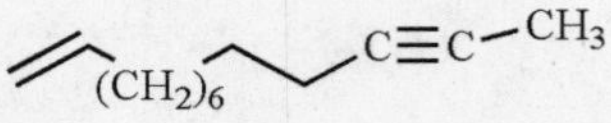

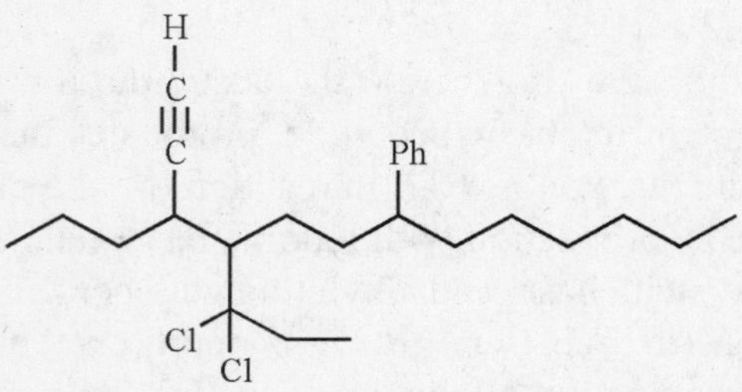

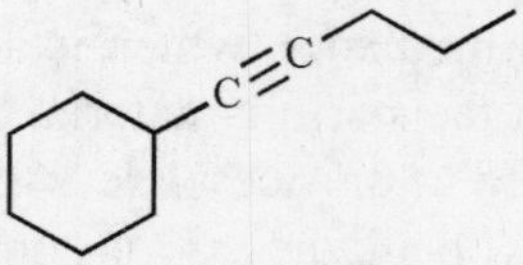

8. In each case, give the major product of the reaction. Remember stereochemistry, and if there is no reaction, indicate this by N.R.

(a) 1. THF , $NaNH_2$ 2. 2*R*-bromobutane

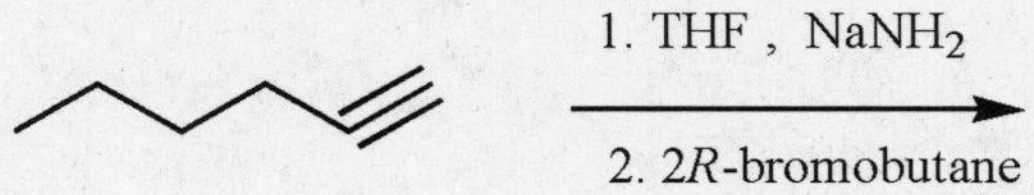

(b) 1. *n*-BuLi , ether 2. 1-iodobutane

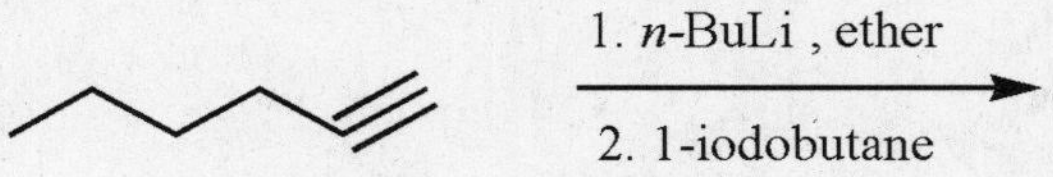

(c) n-BuLi ether

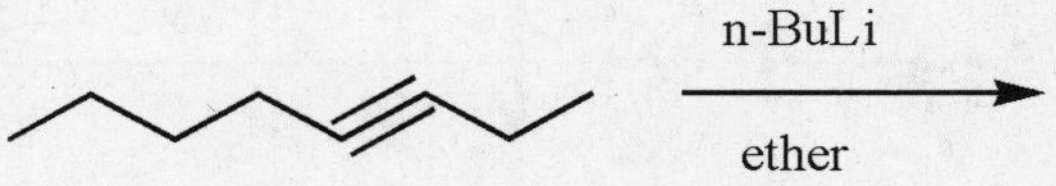

(d) Br H $PhC{\equiv}C{:}^- \ Na^+$ DMF

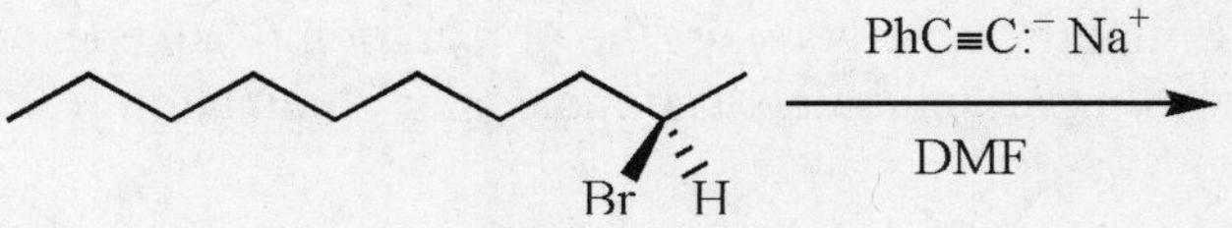

CHAPTER 9

Elimination Reactions

9.1. THE E2 REACTION

There are several important methods for the preparation of alkenes and alkynes. This section will focus on one of the most important methods, bimolecular elimination from alkyl halides.

If the Carbon Bearing the Bromine Atom in 9.1 Is Called the α-Carbon, What Is the β-Carbon?

Br

9.1

The carbon atoms of the three identical methyl groups could be considered β-carbons, i.e., they are attached to the α-carbon.

Where Is a β-Hydrogen in 9.1?

A β-hydrogen must be attached to one of the β-carbon atoms. Therefore, any of the hydrogen atoms of the three equivalent methyl groups in ***9.1*** is a β-hydrogen.

When 2-Bromo-2-Methylpropane (9.1) Is Mixed with a Nucleophile, an S_N2 Reaction Is Not Possible. Why Not?

2-Bromo-2-methylpropane (*9.1*) is a tertiary halide, and the activation energy for an S_N2 reaction at a tertiary carbon is so high that for all practical purposes, the reaction cannot proceed.

When 2-Bromo-2-Methylpropane (9.1) Is Mixed with Koh in Ethanol, an S_N2 Reaction Is Not Possible, but a Rapid Reaction Takes Place Giving 2-Methylpropene (9.2) As the Product. Explain This Process.

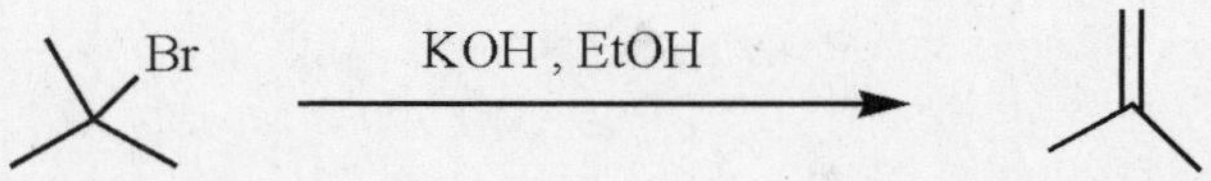

Br — KOH , EtOH →

9.1 *9.2*

The hydroxide (^{-}OH) reagent is nucleophilic and will be attracted to the carbon bearing the bromine. The energy required for this S_N2 reaction is too high (see Section 7.2), and the substitution reaction is so slow that for all practical purposes it will not occur. Bond polarization makes the β-hydrogen (the hydrogen attached to the second carbon away from the bromine, the β-carbon) electropositive. The hydroxide attacks the β-hydrogen in an acid-base reaction, as shown in ***9.3***. Removal of this hydrogen leads to a transition state such as ***9.4*** where water (H–OH) is being formed, the bromine is being displaced, and a carbon-carbon double bond is beginning to form. The final product is ***9.2***.

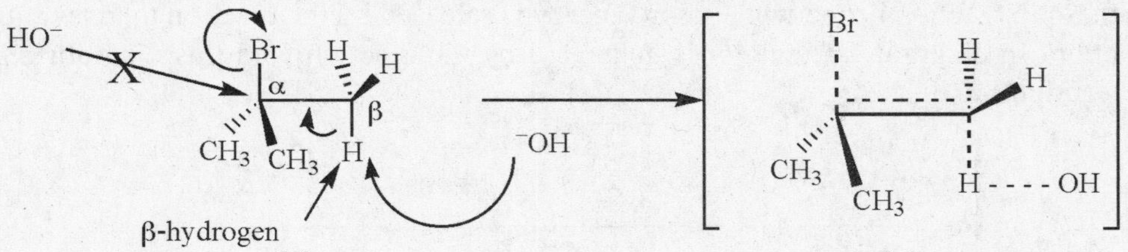

9.3

9.4

Is the Reaction That Converts 9.1 to 9.2 Concerted or Does It Have an Ionic Intermediate?

There is no intermediate, so it is a concerted reaction proceeding via transition state ***9.4***.

Does the Reaction That Converts 9.1 To 9.2 Follow First Order or Second Order Kinetics?

The reaction is bimolecular. It requires a collision of OH and the β-hydrogen. It is, therefore, second order in its kinetics (rate = k [RX] [^{-}OH]).

What Is the Orientation of the Bromine and the β-Hydrogen in 9.3 When the Reaction with Hydroxide Ion Occurs to Generate Transition State 9.4?

The β-hydrogen atom in ***9.3*** is removed by hydroxide in an acid-base reaction. Since the hydrogen atom is removed as a proton, the electrons in that C–H bond stay with the substrate (the two electrons in the new H–OH bond come from hydroxide). In effect, the two electrons from the C–H bond expel the bromide to generate the new π-bond in ***9.2*** Since the bromine in ***9.3*** is displaced by the electrons in the C–H_β moiety, those electrons should displace Br from the rear (backside attack). This requires that the β-hydrogen be oriented at an angle of close to 180° relative to the leaving group (Br). This reaction, therefore, requires an *anti-* orientation of the β-hydrogen and the bromine. In general, attack of the base (^{-}OH) on the β-hydrogen to initiate this reaction occurs *only* in the rotamer where the leaving group and β-hydrogen have an *anti-* relationship. This is an **elimination reaction,** since the elements of Br and H are *eliminated* from ***9.3***. This bimolecular elimination is termed **E2**.

We Call This an Elimination Reaction, but Can We Use Another Reaction Type Description to Describe it?

The conversion of ***9.1*** to ***9.2*** is an acid-base reaction, involving attack of a base on the β-hydrogen, with concomitant displacement of the leaving group.

Why Doesn't the E2 Reaction Occur When 9.1 Is Reacted with Potassium Iodide in Ethanol?

Potassium iodide is not very basic. The conjugate acid is HI, strong acid → weak base. So it cannot remove the β-hydrogen in ***9.1***. If there is no base to remove this hydrogen, the E2 reaction is not possible.

What Is Standard Free Energy?

Standard free energy ($\Delta G°$), *assumes that the reaction is done in a standard state*, in solution with a concentration of 1 Molar and at one atmosphere of pressure for gases and 25°C. The standard free energy can be calculated from enthalpy (H°) and entropy (S°) by the equation:

$\Delta G° = \Delta H° - T\Delta S°$

What Constitutes an Exothermic Reaction?

An exothermic reaction is one where more energy is released by the reaction than is required to initiate the reaction. In diagram ***9.5***, energy is generated by the reaction, and this diagram represents an exothermic reaction.

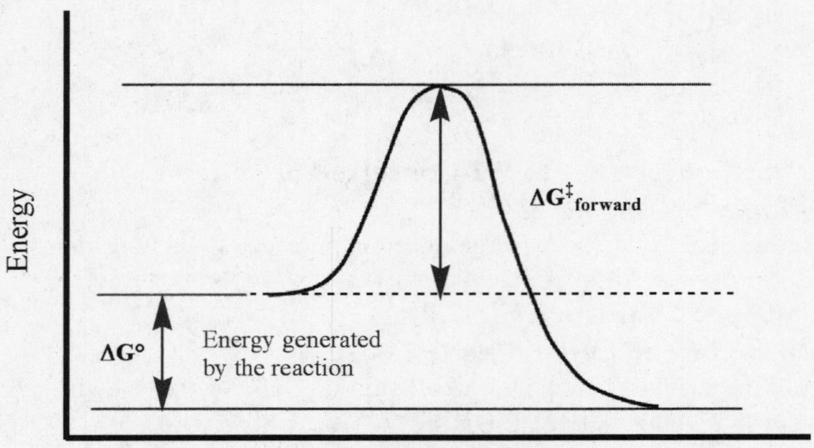

9.5

What Do We Call the Amount of Energy Required to Initiate Bond Making/Bond Breaking?

This amount of energy is the activation energy, and it is marked on the diagram ***9.6***.

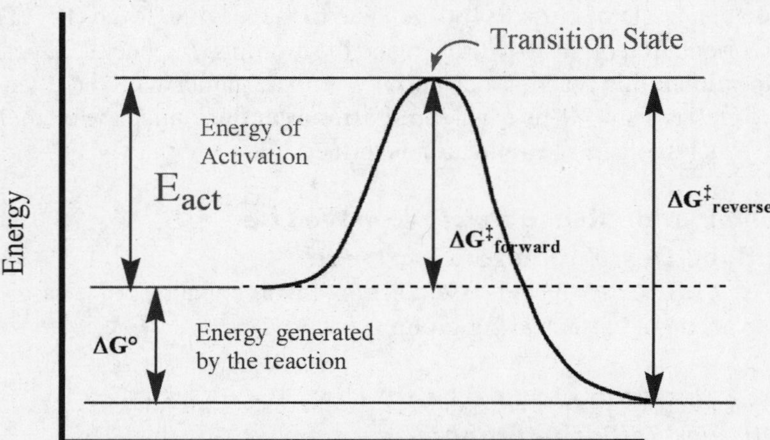

9.6

Where Is the Transition State Energy in a Reaction?

In marked diagram ***9.6***, the transition state is the midpoint of the reaction at the top of the energy barrier. The amount of energy required to achieve the transition state is the activation energy.

Draw a Reaction Diagram for the Conversion of 9.1 to 9.2 and Assume the Reaction is Exothermic.

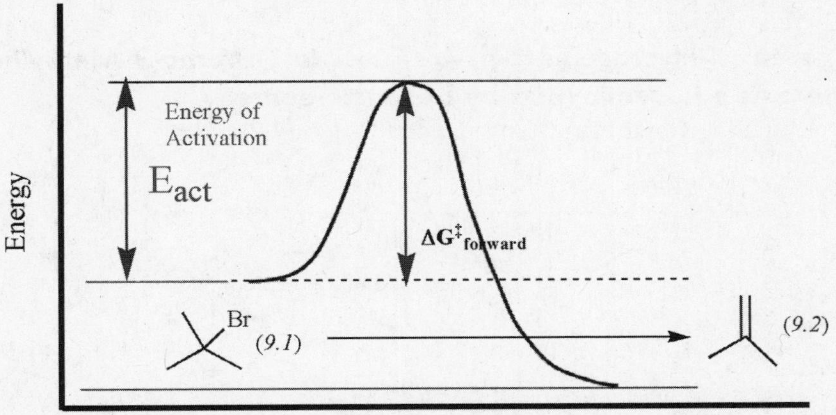

9.7

Diagram ***9.7*** shows the reaction coordinate with 2-bromo-2-methylpropane marked as the starting material and 2-methylpropene marked as the product.

On the Diagram from the Previous Question, Indicate the Position of the Transition State 9.4, and Draw the Transition State.

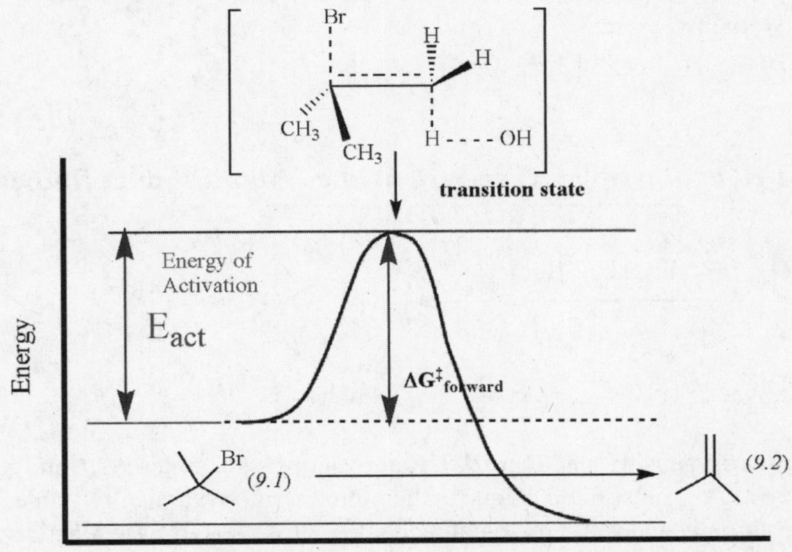

9.8

The transition state position and structure is indicated in ***9.8***.

What Is the Hammond Postulate?

The Hammond postulate states that for an endergonic reaction step, the transition state will structurally resemble the product of that step.

What Is a Late Transition State?

A late transition state is one that is closer in free energy to the products than the starting material.

How Many Different β-Hydrogen Atoms Are There in 2-Bromo-3-Methylhexane, Represented Here As a Racemic Mixture of Diastereomers?

There are two different β-hydrogen atoms, H_a and H_b (see ***9.9***).

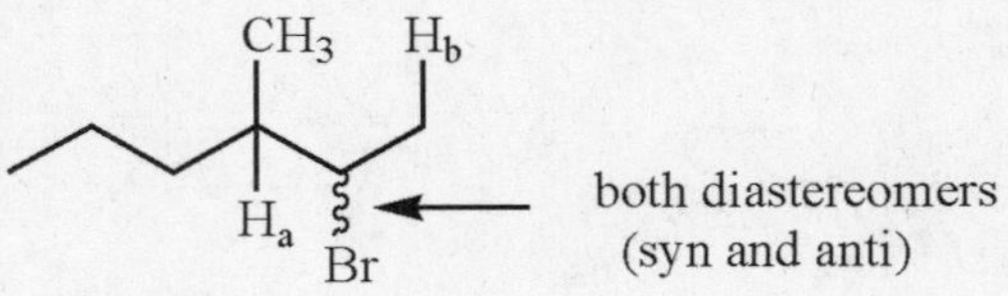

CH_3 H_b H_a Br

both diastereomers (syn and anti)

9.9

Removal of Two Different β-Hydrogen Atoms from 9.9 Will Lead to Different Alkenes. Draw Them!

The alkene products are ***9.10*** and ***9.11***. However, ***9.10*** is actually a mixture of E- and Z-isomers, so ***9.10*** actually represents two compounds. Therefore, the elimination reaction of ***9.9***, written as a racemic mixture of diastereomers, leads to three products.

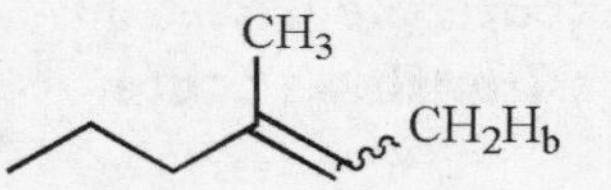

CH_3 CH_2H_b

9.10 E + Z

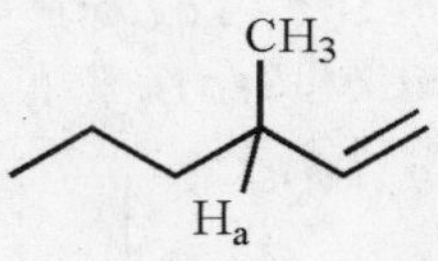

CH_3 H_a

9.11

Why Does the E2 Reaction of 9.9 Give 9.10 As the Major Product Rather Than 9.11?

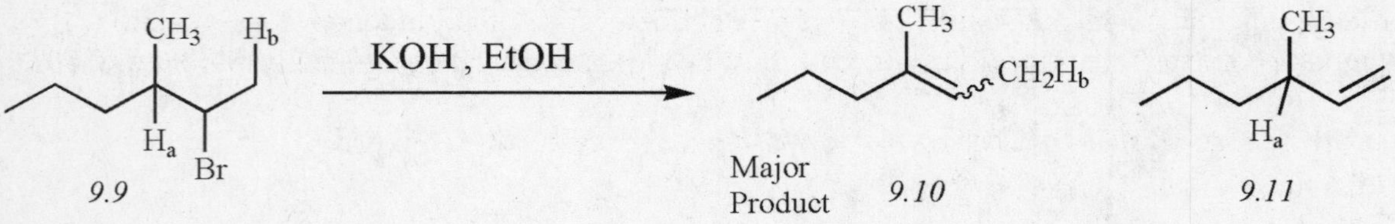

CH_3 H_b H_a Br *9.9* — KOH , EtOH → CH_3 CH_2H_b Major Product *9.10* ; CH_3 H_a *9.11*

There are two β-hydrogens, H_a and H_b in ***9.9***. Removal of H_a generates ***9.10***, and removal of H_b generates ***9.11***. The more highly substituted alkene is the more thermodynamically stable. It has been determined that the E2 reaction is under thermodynamic control. It is an acid-base reaction and, therefore, an equilibrium reaction. The major product will always be the more thermodynamically stable alkene, in this case ***9.10***. The more thermodynamically stable alkene, and the major product of the E2 reaction, will *always be the most highly substituted alkene.*

Draw the Transition State for the Conversion of One Enantiomer of 9.9 (2s,3s) to 9.10, and Also Draw the Transition State for the Conversion to 9.11.

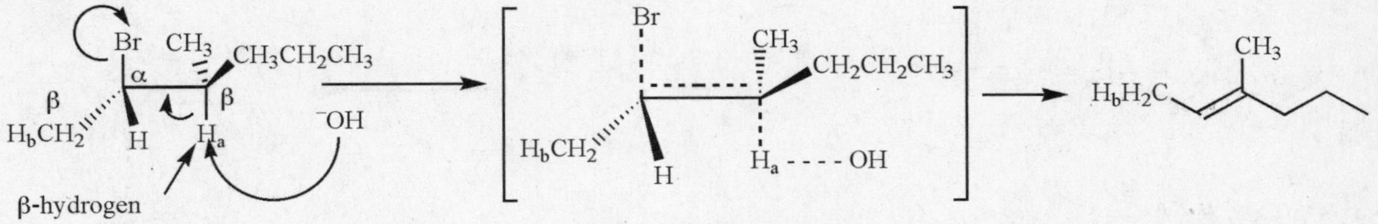

9.12　*9.10*

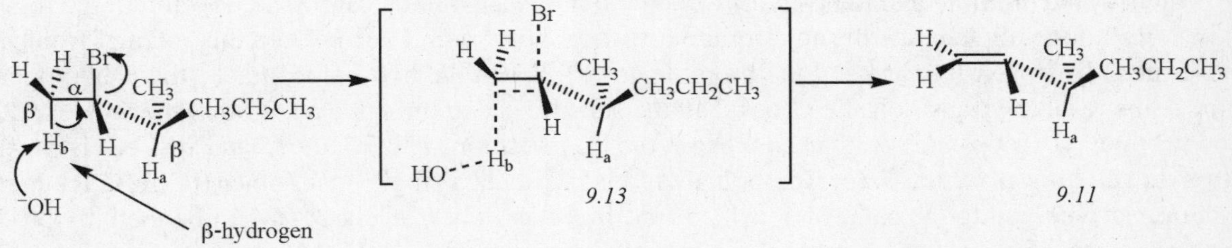

9.13　*9.11*

The transition state for the formation of ***9.10*** is ***9.12***. The transition state for formation of ***9.11*** is ***9.13***. Note that the E2 reaction of the (2*S*,3*S*)-enantiomer of ***9.9*** leads exclusively to the cis-alkene, ***9.10***. The (2*R*,3*R*) enantiomer will lead to the cis-isomer, but the (2*R*,3*S*) and (2*S*,3*R*) enantiomers will lead to the trans-alkene. Check this out for yourself by drawing the appropriate transition state. Therefore, *the E2 reaction is* ***stereospecific***.

What Is Zaitsev (Also Called Saytzeff) Elimination?

A rule formulated in 1875 by Alexander Zaitsev. He observed that the alkene formed in greatest amount is the one that corresponds to removal of the hydrogen from the β-carbon having the fewest hydrogens. Formation of the more substituted, more stable alkene via an E2 reaction is now termed **Zaitsev elimination.**

If 9.12 and 9.13 Represent Late Transition States, and 9.10 Is the Observed Major Product, Which Is More Important: the Acidity of the β-Hydrogen Atom in the Starting Material, or the Relative Stability of the Alkene Products, 9.10 and 9.11?

If the more stable product (***9.10***) is the major product, transition state ***9.12*** is more important, since in a late transition state, the stability of the product determines the outcome of the reaction. By the Hammond postulate, late transition state ***9.12*** must be structurally and energetically similar to ***9.10***.

Determine the Absolute Configuration of the Stereogenic Centers in 9.14 and Name That Compound.

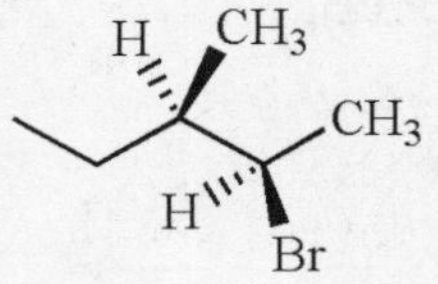

9.14

The absolute configuration is 2R and 3R, so the name of ***9.14*** is 2R-bromo-3R-methylpentane.

The E2 Reaction of 9.14 Gives Exclusively One Isomer, 9.16. Explain Why the Other Stereoisomer Is Not Formed.

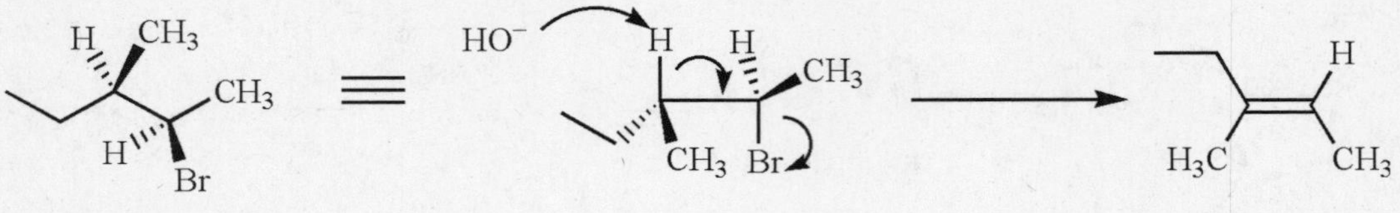

9.14 *9.15* *9.16*

In ***9.14***, the bromine and the β-hydrogen on the more substituted carbon are *not* anti in the rotamer drawn. Rotation will generate the appropriate rotamer (shown as ***9.15*** ; to obtain this rotamer from ***9.14***, move the β-hydrogen forward and the bromine down). When the hydroxide attacks the β-hydrogen as shown, the resulting transition state 'locks' all the substituents in the same position as they were in ***9.15***. The two methyl groups are *cis*- to each other in the E2 transition state from ***9.15*** and **will be cis- to each other in the final product, *9.16***. Compound ***9.14*** is drawn as a single enantiomer ([C_2R, C_3R], but the enantiomer will lead to the same stereochemistry. In other words, alkene ***9.16*** will be the only product from either enantiomer. Only rotamer ***9.15*** will lead to the E^2 product, ***9.16***. The E^2 reaction is termed **stereospecific**.

What Does Stereospecific Mean?

The term stereospecific means that, of two or more possible products, one stereoisomer gives one and only one product, whereas the other stereoisomer will give a different product. In this case, stereoisomer ***9.14*** gives the *E*-alkene ***9.16***, and the diastereomer of ***9.14*** will give the *Z*-alkene.

Draw the Diastereomer of 9.14, Name It, and Give the Product Formed from Reaction with Koh in Ethanol.

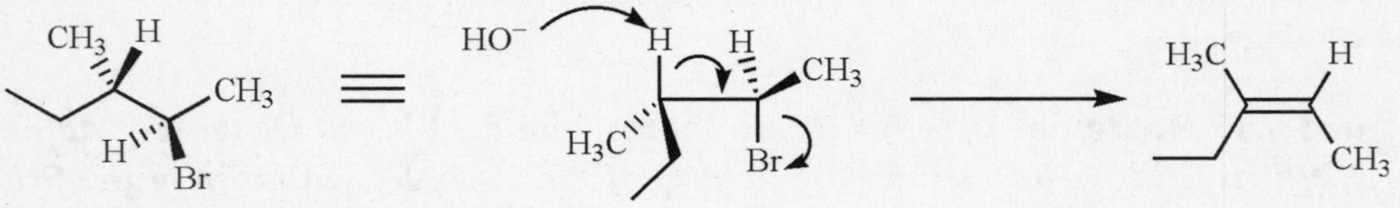

9.17 *9.18* *9.19*

The diastereomer of ***9.14*** is ***9.17*** (2R-bromo-3S-methylpentane). Upon reaction with KOH in ethanol, the proper rotamer of ***9.17*** required for the E2 reaction is ***9.18***, where the two methyl groups are on opposite sides, leading to the Z-alkene, ***9.19***.

Cyclohexane Derivative 9/20 Gives 9.21 As the Alkene Product of the E2 Reaction. Explain This Observation.

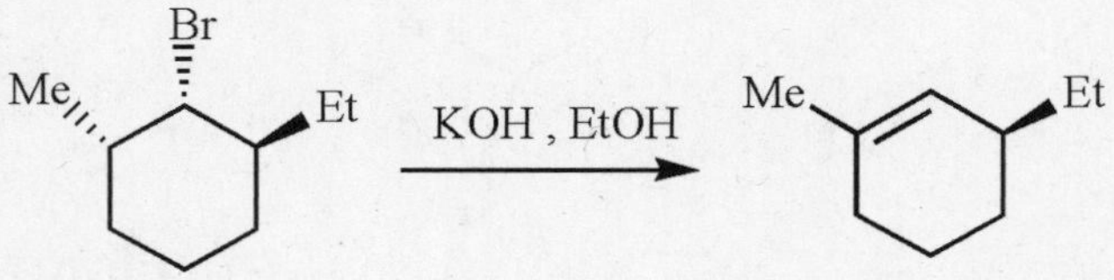

KOH , EtOH

9.20 *9.21*

This question cannot be answered using the two-dimensional drawing shown. Cyclohexane derivatives exist primarily as chair conformations (Section 4.2), and ***9.22*** and ***9.23*** are the two equilibrating chair conformations of ***9.20***. There are two β-hydrogens, but only ***9.22*** has H_a and the Br *anti-* . They have a *trans*-diaxial relationship. In ***9.23***, the β-hydrogen H_a is equatorial and essentially at right angles to Br. In other words, it cannot be removed via an E2 reaction. The β-hydrogen H_b is axial, but the bromine atom is equatorial, and again, there is no possibility of an E2 reaction. Only removal of H_a in ***9.22*** can give the E[2] product. Because of this phenomenon, the alkene is formed *only* towards the methyl (Me) and not towards the ethyl (Et) (see ***9.21***). The bromine-bearing carbon is probably too sterically hindered to give a S_N2 reaction.

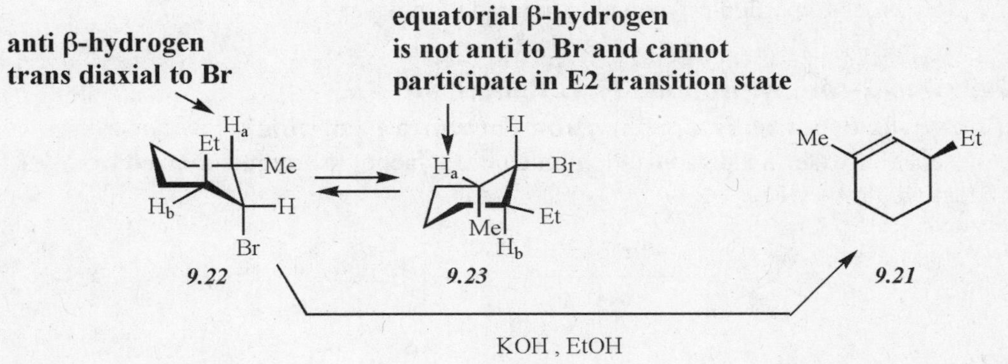

Cyclohexane Derivative 9.24 Gives No Elimination Products At All under E2 Reaction Conditions. Explain This Observation.

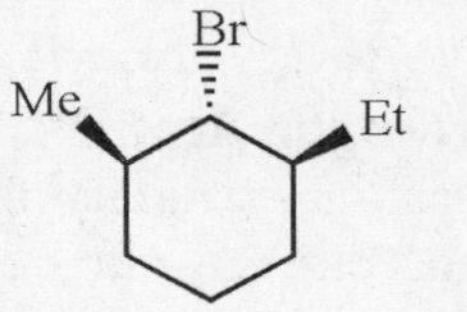

9.24

The equilibrating chair conformations for ***9.24*** are ***9.25*** and ***9.26***. For an E2 reaction, there must be a β-hydrogen atom with a trans-diaxial relationship to the bromine. The Br is axial *only* in ***9.25*** (it is equatorial in ***9.26***), but both β-hydrogens are equatorial. In ***9.26***, both β-hydrogens are axial, but the Br is equatorial. An E2 reaction is *not* possible from either ***9.25*** or from ***9.26***, and treatment with hydroxide gives no reaction (N.R.). The bromine-bearing carbon is probably too sterically hindered to give an S_N2 reaction.

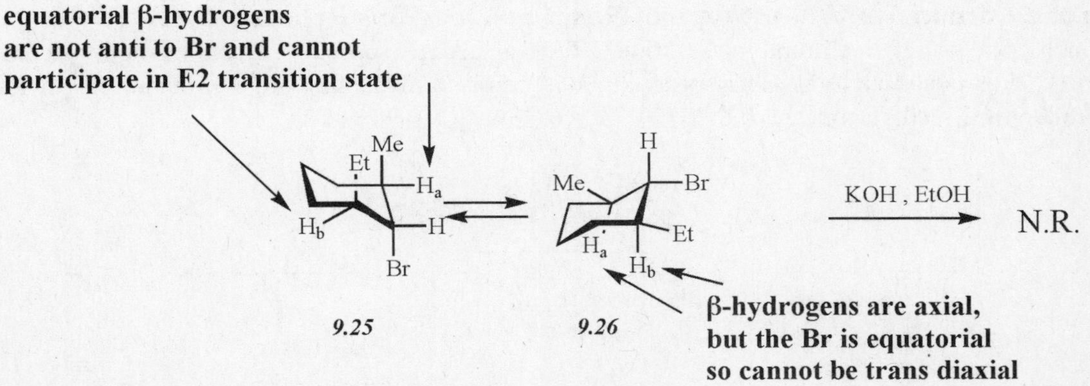

9.2. THE E1 REACTION

When a highly ionizing medium such as water is present, and substitution is slow, elimination can occur via a carbocation intermediate. Typically, E2 reactions are faster, but this first order elimination reaction, termed E1, can occur and under certain conditions, dominate the reaction.

When 2-Bromo-2-Methylpropane (9.1) Reacts with KOH in Dry Ethanol, Is Carbocation Formation Possible?

Since ethanol is a polar protic solvent, ionization is possible. However, ionization in pure ethanol is very slow. For example, ask yourself if NaCl is very soluble in pure ethanol. The answer is no, and ionization is a slow process, particularly when compared with water.

When 2-Bromo-2-Methylpropane (9.1) Reacts with KOH in Aqueous Ethanol, Is Carbocation Formation Possible?

The presence of water makes ionization much more facile. If bromine ionized from ***9.1***, the carbocation product would be ***9.27***.

Br – Br⁻ [⊕] [Br⊖]

9.1 *9,27*

Draw the Carbocation Produced by Ionization of the Bromide In 9.1.

The carbocation is ***9.27***.

Which Site in 9.27 Is More Attractive to Hydroxide, the Carbocation Carbon or a β-Hydrogen Atom?

The carbon has a formal positive charge, whereas the β-hydrogen atoms have a δ+ dipole. Therefore, the hydroxide ion should be most strongly attracted to the positive carbon.

If Hydroxide Reacts with the Carbon in 9.27 Rather Than a β-Hydrogen, What Type of Reaction Does This Represent?

Collision with carbon makes hydroxide a nucleophile, and such a reaction is essentially like an S_N1 reaction.

If Hydroxide Reacts with a β-Hydrogen in 9.27 Rather Than the Carbon Atom, What Type of Reaction Does This Represent?

Such a reaction is an elimination reaction to form an alkene (see ***9.28***). The reaction proceeds via an intermediate carbocation, follows first order kinetics, and is termed a first order elimination reaction, E1. The product is 2-methylpropene, ***9.2***.

⊖OH

Br – Br⁻ [H ⊕] [Br⊖]

9.1 *9.28* *9.2*

What Is the Product When Cyclohexanol (9.29) Reacts with an Acid, H^+?

When cyclohexanol reacts with H^+, the oxygen atom behaves as a base, and the product is the oxonium ion, ***9.30***. Note that this acid-base reaction is reversible.

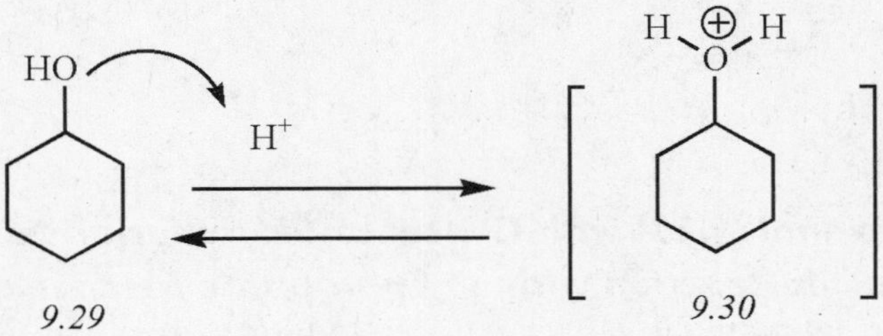

Describe Possible Reactions of Intermediate 9.30.

In oxonium ion ***9.30***, the HOH unit (water) is a good leaving group. Three possible reactions are loss of water by removal of a β-hydrogen to give the alkene directly (E2), ionization to carbocation ***9.32*** by loss of water, or reaction with a nucleophile with direct displacement of water (S_N2) to give ***9.33***. Note that loss of a β-hydrogen from ***9.32*** will give ***9.31*** by an E1 mechanism and that collision of ***9.32*** with a nucleophile will give ***9.33*** by an S_N1 mechanism.

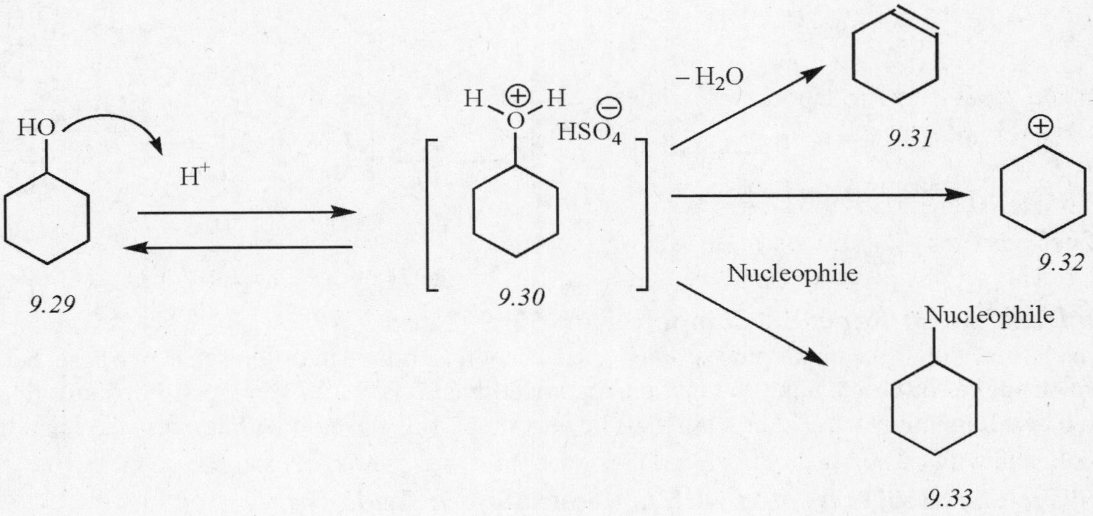

In the Reaction of Cyclohexanol (9.29) with Concentrated Sulfuric Acid (Little Water), Is There a Nucleophilic Species Present?

In concentrated sulfuric acid, water is not considered to be present in sufficient quantity to be a nucleophilic species. The only other nucleophile is the hydrogen sulfate anion (see ***9.30***), which is resonance stabilized and a very weak nucleophile.

Draw All Resonance Forms of The Hydrogen Sulfate Anion.

The three resonance forms are shown for ***9.34***, where the negative charge is delocalized, and upon reaction with a carbon atom, somewhat sterically hindered. Therefore, electrons cannot be readily donated, and ***9.34*** is a weak nucleophile. Interestingly, ***9.34*** is a weak base because it is easier to react with a hydrogen atom than with a carbon.

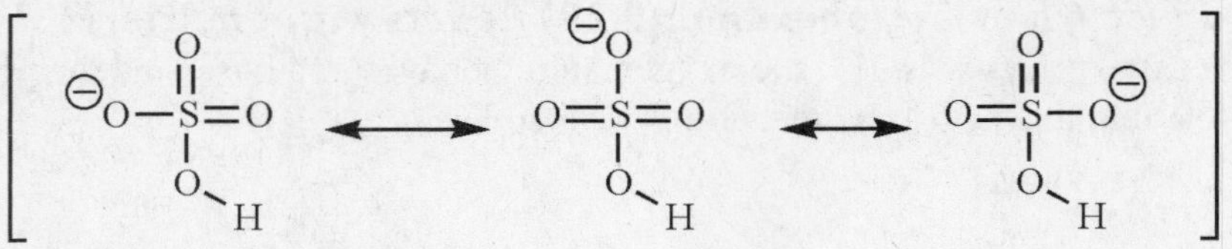

9.34

Treatment of Cyclohexanol (9.29) with Concentrated Sulfuric Acid Generates Cyclohexene. Explain This Reaction with an Appropriate Mechanism.

The oxygen of ***9.29*** reacts with the acidic hydrogen to form an oxonium salt, ***9.30***. The by-product of losing H^+ from H_2SO_4 is the hydrogen sulfate anion, HSO_4^-, which is a very poor nucleophile. This fact makes both an S_N2 displacement or an S_N1 reaction very slow. In concentrated sulfuric acid, which is a good dehydrating agent, ***9.30*** readily loses water to form carbocation ***9.32***. Since the hydrogen sulfate anion is a weak base, the β-hydrogen in ***9.32***, which is acidic due to its proximity to the cation center, can be removed via an acid-base reaction to generate the alkene, cyclohexene (***9.31***). This is a unimolecular process (ionization of ***9.29*** to the cation ***9.32*** via *9.30*) and is an elimination. Therefore, it is called an **E1 reaction**.

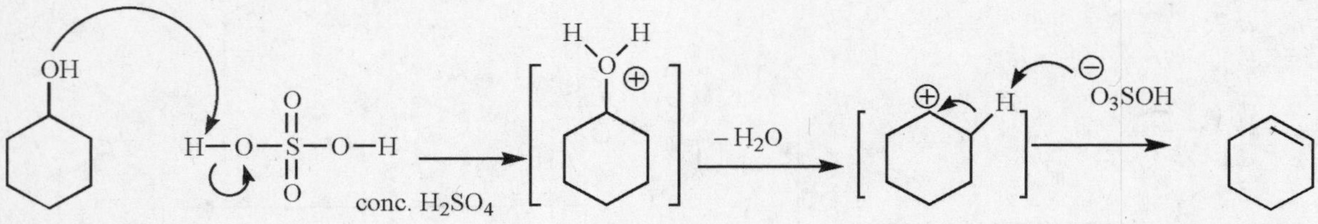

conc. H_2SO_4 — $-H_2O$ — O_3SOH

9.29 9.30 9.32 9.31

When Does the E1 Reaction Compete with E2, S_N2 and S_N1?

The E1 reaction is the major process *only* when a cation is formed in the presence of a base, but there is no nucleophile in the medium that can react competitively via the S_N1 process. For ionization to a cation to be competitive, the S_N2 process must be very slow, as is the case with tertiary alkyl halides.

When Cyclohexanol Is Treated with Aqueous Sulfuric Acid, Cyclohexanol Is Recovered. Explain This Reaction.

Oxonium ion ***9.30*** is formed, and ionization to the cation (***9.32***) also occurs. In aqueous media, however, water is present and can function as a nucleophile. If water attacks ***9.32***, ***9.30*** is formed, and in the presence of water, loss of a proton can regenerate the starting alcohol. The E1 reaction occurs as the major process only when no good nucleophile is present. If water is added to this reaction, it will function as a nucleophile, making the S_N1 process more likely than the E1 reaction.

What Are the Requirements (Structure of the Halide or Alcohol and Reaction Conditions) for an E1 Reaction?

In general, a tertiary halide or alcohol precursor must be used, since they will generate a relatively stable tertiary cation. Primary substrates do not give the cation, and secondary substrates give relatively stable secondary cations. The secondary cation may give some E1 product, but other reactions often compete. With alkyl halides, water is usually necessary to help form and stabilize the cation, but water will be a nucleophile and can give the S_N1 reaction as the major process. In such cases, E1 is usually a minor process. With alcohols, anhydrous acids can be used to generate the cation, but the conjugate base of that acid must not be nucleophilic. This restricts the acid to molecules such as sulfuric acid (H_2SO_4), perchloric acid ($HClO_4$), and tetrafluoroboric acid (HBF_4).

Give the Major Product(s) for the Reactions of 9.35–9.38, and Offer an Explanation for Your Choice.

(a) OH — aq. EtOH, cat. H^+ / KI →

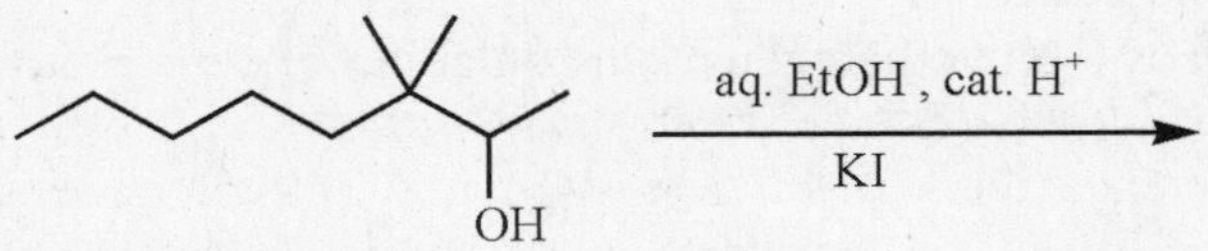

9.35

(b) CH_3 H I — KCN, DMF, heat →

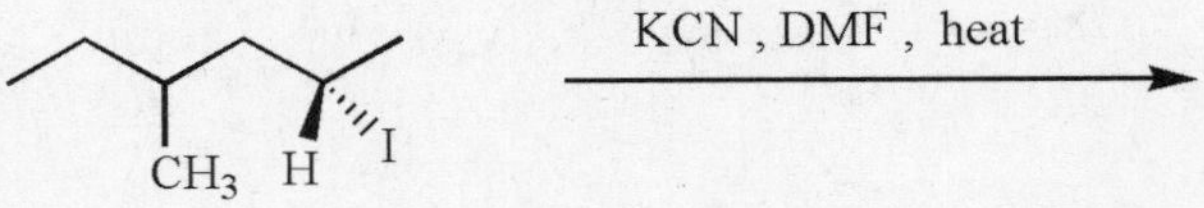

9.36

(c) H_3C H Br CH_3 — KOH, EtOH →

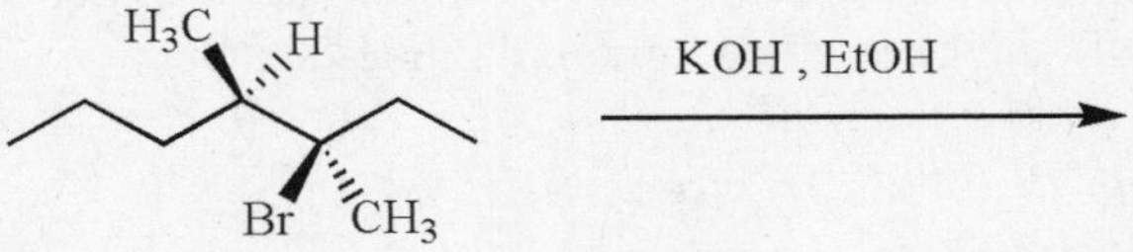

9.37

(d) Br CH_3 — KI, THF / 25°C →

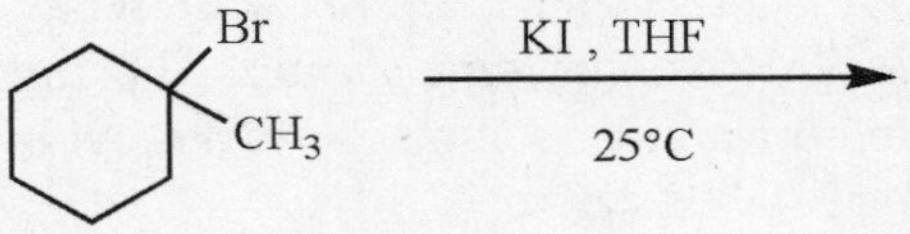

9.38

In reaction (a), the aqueous solvent and acid catalyst suggest an S_N1 reaction. Ionization to a cation is followed by rearrangement (the methyl group migrates to give a more stable tertiary cation), and iodide attacks the cation to give the product, ***9.39***. Reaction (b) is an S_N2 reaction and proceeds with 100% inversion of configuration at the stereogenic center. The nucleophile ($^-C{\equiv}N$) displaces the leaving group (I) at the *S*- stereogenic center to give the *R*-nitrile, ***9.40***. Reaction (c) suggests an E2 reaction (tertiary halide – KOH), and elimination of the β-hydrogen will give a tetrasubstituted alkene as the more stable product. Since ***9.37*** represents one diastereomer, a single alkene product will be formed, ***9.41***. In reaction (d), the tertiary halide cannot react with KI under these S_N2 conditions, and the correct answer is no reaction.

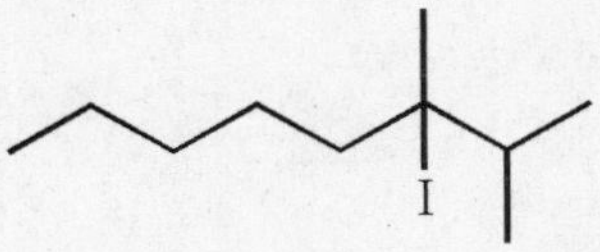

I

9.39

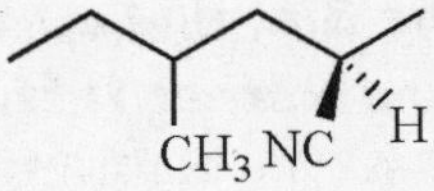

CH_3 NC H

9.40

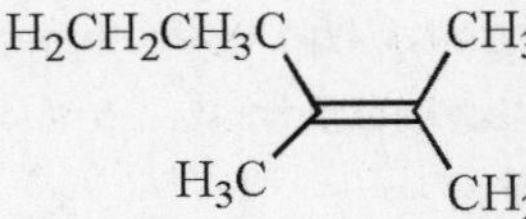

$H_2CH_2CH_3C$ CH_3 H_3C CH_2CH_3

9.41

9.3. PREPARATION OF ALKYNES

Most alkynes are prepared in one of two ways: elimination of geminal dihalides with strong base or by alkylation of acetylene or monosubstituted alkynes, as seen in Chapter 8. This section will focus on the preparation of alkynes, as well as vinyl halides, by elimination reactions.

What Is a Geminal Dihalide?

A geminal dihalide has both halogen atoms on the same carbon, as in 2,2-dibromobutane, ***9.42***.

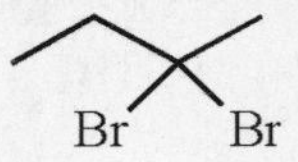

Br Br

9.42

What Is a Vicinal Dihalide?

A vicinal dihalide has the two halogen atoms on adjacent carbon atoms, as in 2,3-dibromobutane, ***9.43***.

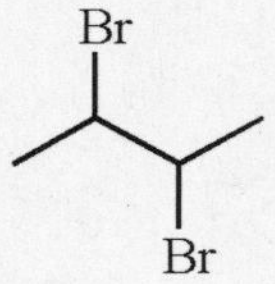

Br

Br

9.43

How Are Geminal Dihalides Prepared?

Geminal dihalides (both halogens on the same carbon) are prepared from alkynes. Sequential addition of HBr or HCl to an alkyne gives the geminal dihalide (see Sections 8.4 and 13.1).

If 1,2-Dibromopentane (9.44) Is Treated with One Equivalent of $NaNH_2$, What Is the Product?

Sodium amide ($NaNH_2$) is a very strong base. In the presence of the bromine atoms, the most likely reaction is an E2, but there are two bromine atoms. The initial product of the reaction of 1,2-dibromopentane (***9.44***) and sodium amide in ammonia is a **vinyl halide** (***9.45***), which is derived from an E^2 reaction to produce the more stable alkene.

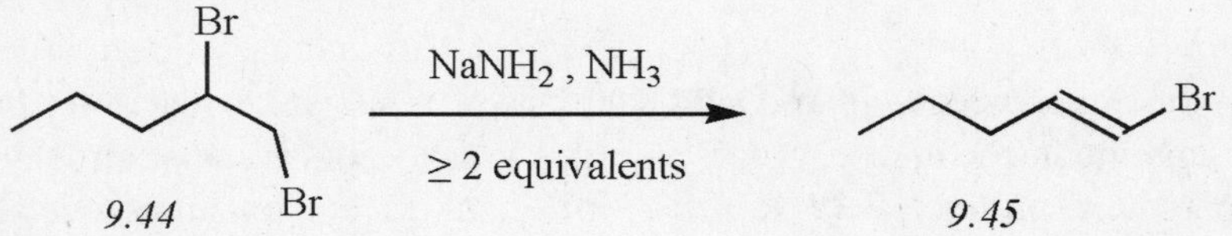

Br

$NaNH_2$, NH_3

≥ 2 equivalents

9.44 Br

Br

9.45

What Is a Vinyl Halide?

A vinyl halide is an alkene where a halogen atom is directly attached to one of the sp^2 hybridized carbon atoms of the C=C unit.

Another Product, 9.46, Is Possible in the Elimination Reaction of 9.44, via Loss of the Hydrogen Atom β- to the Bromine on C2. Offer an Explanation As to Why 9.45 Is Formed Rather Than 9.46.

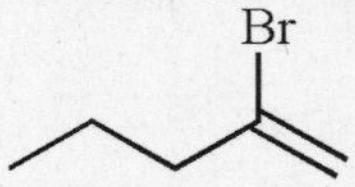

Br

9.46

Alkene ***9.46*** is a disubstituted alkene where the bromine atom and the propyl group are on the same carbon. In ***9.45***, the bromine atom and propyl group are on opposite carbon atoms, so there is less steric hindrance; ***9.45*** is an *E*-alkene, so the groups are as far apart as possible. The E2 reaction is under thermodynamic control and gives the more stable alkene, which in this case is ***9.45***.

What Is the Product When 1,2-Dibromopentane Is Treated with an Excess of Sodium Amide in Liquid Ammonia?

In the presence of at least two equivalents of sodium amide ($NaNH_2$), ***9.44*** is first converted to ***9.45***. Sodium amide is a sufficiently strong base to remove the hydrogen on the carbon-carbon double bond, and in the presence of an excess of sodium amide, ***9.45*** reacts to give the alkyne, ***9.47*** (1-pentyne).

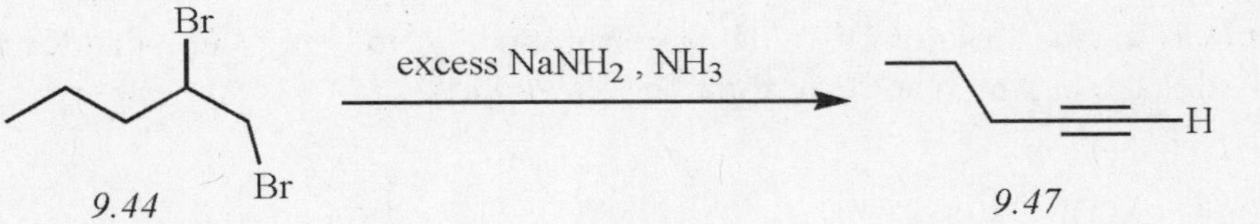

Br Br excess $NaNH_2$, NH_3 → H

9.44 *9.47*

Why Is Sodium Amide Commonly Used for the 'Second' Elimination Reaction Rather Than an Alkoxide Base?

Removal of an alkene hydrogen (C=C–H) in ***9.45*** is more difficult, since it is much less acidic than the methyl hydrogen (on an sp^3 carbon) in ***9.44***. For this reason a strong base is required, and $NaNH_2$ is a very powerful base, i.e., pK_a 38 vs. pK_a 18.

What Is the Product When Geminal Dibromide 9.48 Reacts with One Equivalent of Sodium Amide?

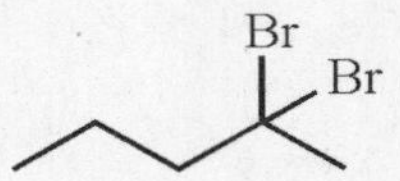

Br Br

9.48

Reaction of ***9.48*** with the strong base will again give an E2 reaction. Removal of a hydrogen atom from the methyl group leads to a disubstituted alkene, whereas removal of a hydrogen atom from C3 leads to the more stable trisubstituted alkene, 2-bromo-2-pentene. However, it is possible to form both the *E*- and *Z*-isomers because there is no stereochemical bias in ***9.48***, so this reaction produces two products, ***9.49*** and ***9.50***.

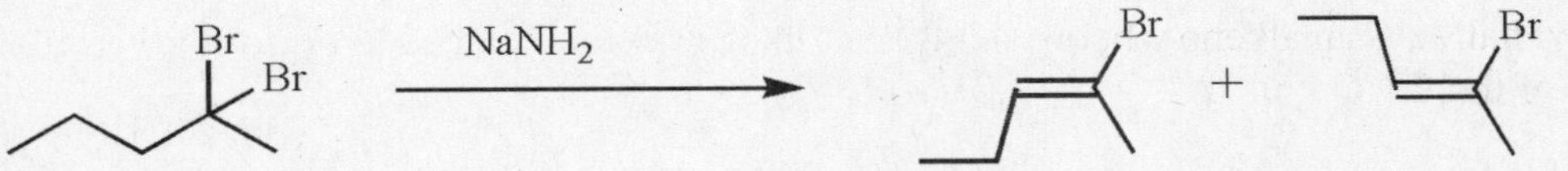

Br Br $NaNH_2$ → Br + Br

9.48 *9.49* *9.50*

What Is the Product When Geminal Dibromide 9.48 Reacts with an Excess of Sodium Amide?

The reaction of either ***9.49*** or ***9.50*** with a sodium amide will remove a vinyl hydrogen atom from the C=C unit to give 2-pentyne, ***9.51***.

Br + Br $NaNH_2$ →

9.49 *9.50* *9.51*

Test Yourself

1. Draw the transition state leading to the major product for the reaction of 2-bromo-3-methylbutane with KOH in ethanol and draw that product.

2. Briefly explain why 3-bromo-2,2,4,4-tetramethylpentane does not give an E2 product when heated with KOH in ethanol.

3. In each of the following reactions give the major product. Remember stereochemistry, and if there is no reaction, indicate this by N.R.

(a) 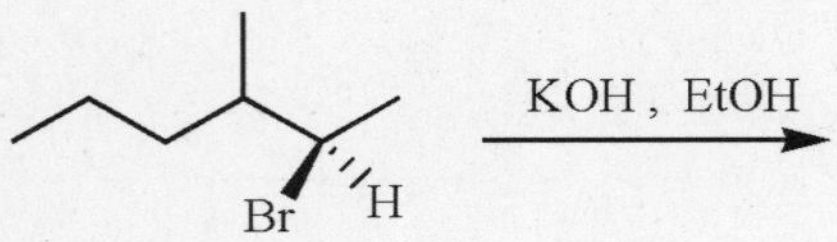 Br, H; KOH , EtOH →

(b) 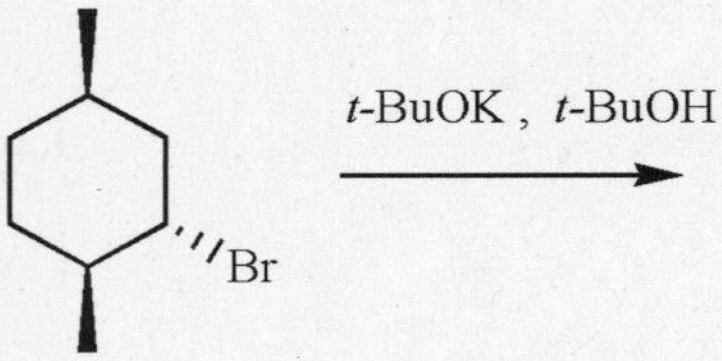 Br; t-BuOK , t-BuOH →

(c) 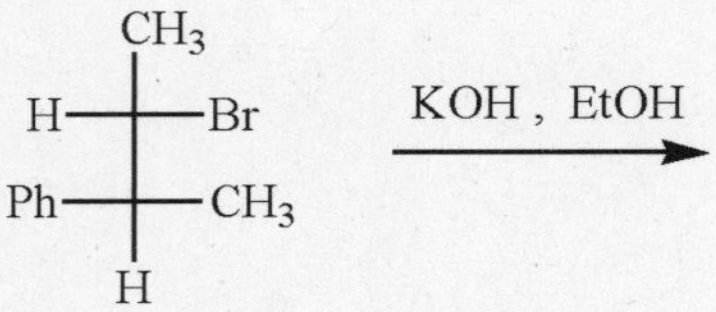 CH_3; H—Br; Ph—CH_3; H; KOH , EtOH →

(d) 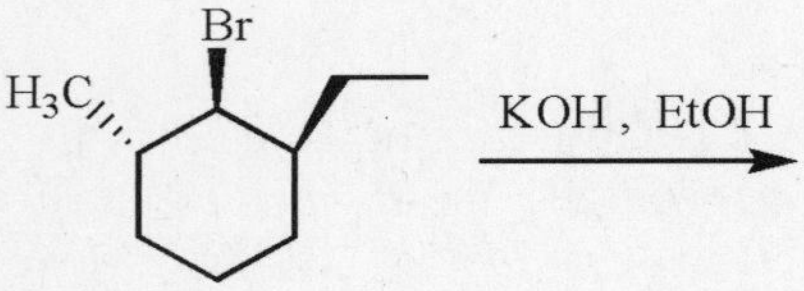 Br; H_3C; KOH , EtOH →

(e) 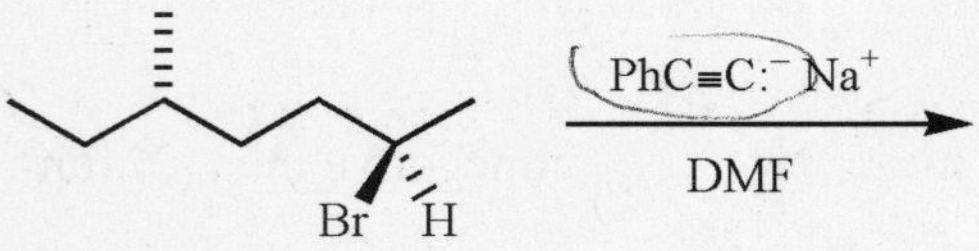 Br H; $PhC{\equiv}C{:}^{-}\ Na^{+}$ / DMF →

(f) 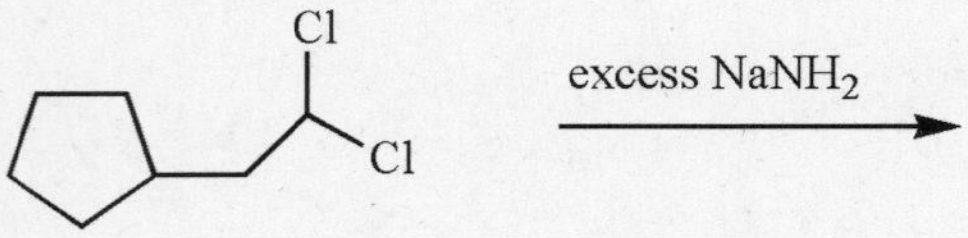 Cl; Cl; excess $NaNH_2$ →

(g) 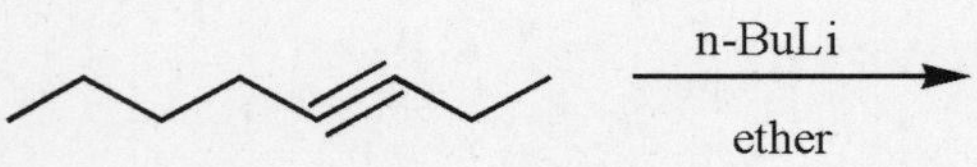 n-BuLi / ether →

(h) 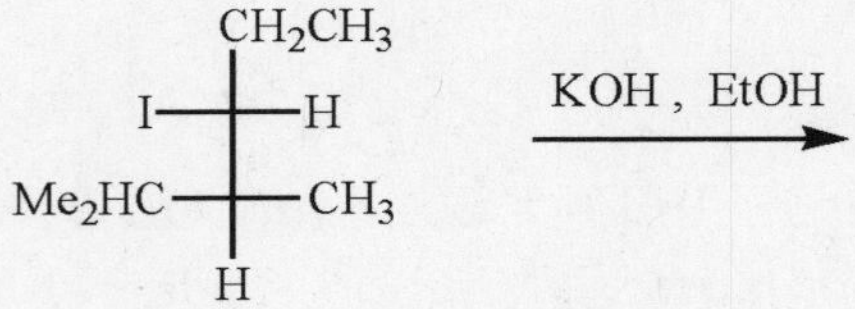 CH_2CH_3; I—H; Me_2HC—CH_3; H; KOH , EtOH →

(i) 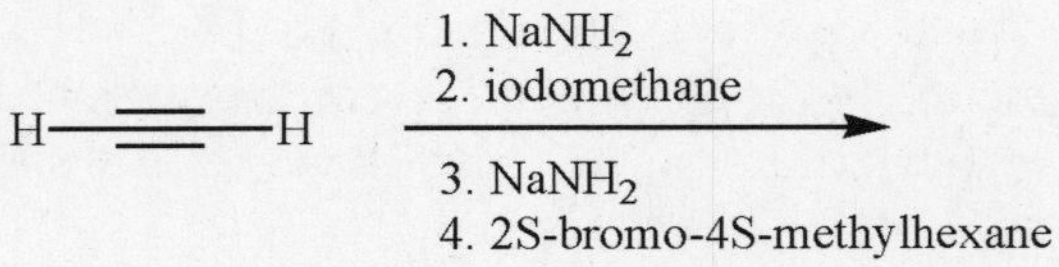 H—≡—H; 1. $NaNH_2$ 2. iodomethane 3. $NaNH_2$ 4. 2S-bromo-4S-methylhexane →

(j) 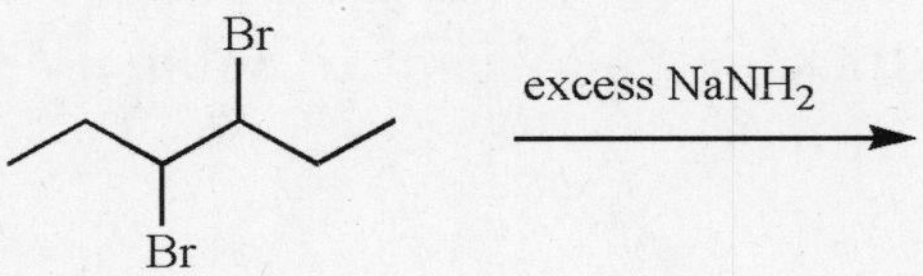 Br; Br; excess $NaNH_2$ →

(k) Br; H_3C; KOH , EtOH →

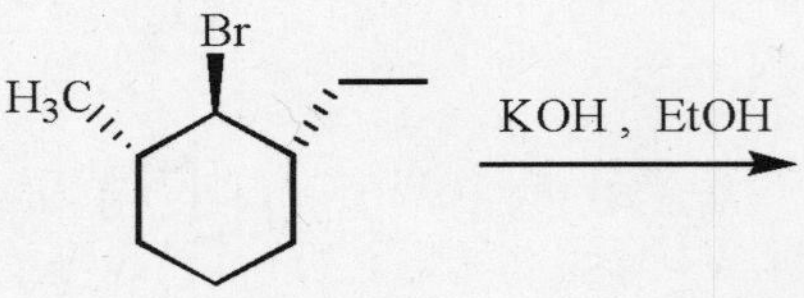

(l) 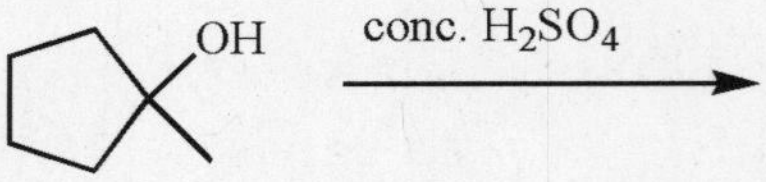 OH; conc. H_2SO_4 →

(m) 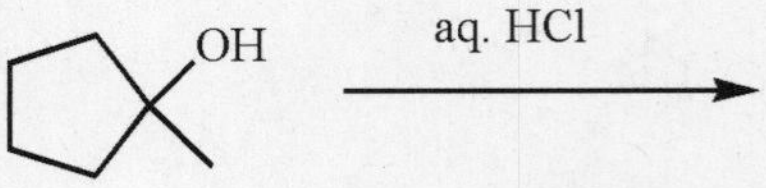 OH; aq. HCl →

(n) Cl; NaCl , EtOH →

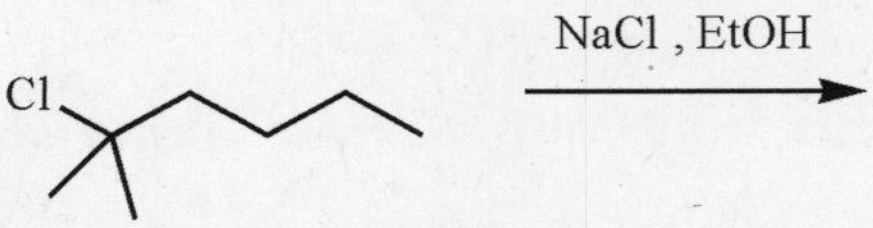

4. Draw the mechanism for the E1 conversion of cyclopentanol to cyclopentene in concentrated sulfuric acid.

CHAPTER 10

Organometallic Compounds

10.1. ORGANOMETALLICS

Carbon can form covalent bonds to many elements found in the Periodic Table. When carbon forms covalent bonds to certain metals, an interesting and highly useful class of compounds is formed that are known as **organometallic compounds**.

What Is an Organometallic?

An organometallic is a compound that contains at least one carbon–metal bond.

What Is the Valence of Group I Metals Such As Lithium or Sodium?

Both lithium and sodium have a valence of 1. Therefore, such metals should form a C–Li or C–Na bond.

What Is the Valence of Group II Metals Such As Magnesium?

Magnesium has a valence of 2. Therefore, such metals should form a C–Mg –X bond, where X is another group or atom. In most cases, the X in C-Mg-X will be a halogen atom such as chlorine or bromine.

Is the C–LI Bond Polarized? If So, Indicate the Bond Polarization.

Yes! The C–Li bond is polarized because carbons are more electronegative than lithium. Therefore, the bond polarization is $^{\delta-}$C—Li$^{\delta+}$.

Is the C–Br Bond Polarized? If So, Indicate the Bond Polarization.

Yes! The C–Br bond is polarized because carbon is less electronegative than bromine. Therefore, the bond polarization is $C^{\delta+}-Br^{\delta-}$. This bond polarization is the normal dipole we have seen in previous chapters.

Is the C–Mg Bond Polarized? If So, Indicate the Bond Polarization.

Yes! The C–Mg bond is polarized because carbons are more electronegative than magnesium. Therefore, the bond polarization is $C^{\delta-}-Mg^{\delta+}$.

Focusing on the Carbon Atom, Would a C–LI Unit React As a Nucleophile or an Electrophile with a Positively Polarized Carbon Atom?

Since the carbon has a negative dipole, it should react as a nucleophile in the presence of a positively polarized carbon atom.

10.2. ORGANOMAGNESIUM COMPOUNDS

Magnesium metal reacts with alkyl halides to form an organic molecule that contains magnesium (an organometallic compound). First exploited by Victor Grignard in the early 1900s, this type of compound is known as a **Grignard reagent**, and its reactions with ketones and aldehydes are known as **Grignard reactions** (see Section 21.4).

What Is the Product When Iodomethane Reacts with Magnesium Metal in Diethyl Ether?

The product is H_3C–Mg–I (methylmagnesium iodide). In ether, magnesium inserts between the C–I bond. This is a relatively unstable molecule, and the ether is necessary to stabilize it by coordination of ether (via the oxygen) with the magnesium.

What Is the Nomenclature System for Compounds Such As RMgX?

If we take CH_3MgI as an example, the name is methylmagnesium iodide. Use the nomenclature term for the alkyl group, followed by magnesium as one word, and the name of the halide. Likewise, $(CH_3)_2CH_2CH_2MgCl$ is 3-methylpropylmagnesium chloride.

What Is the Role of the Ether Solvent?

The oxygen of the ether is a Lewis base, and Mg is a Lewis acid. The ether donates electrons to (a) assist in the insertion of the magnesium between the C–X bond and (b) to stabilize the Grignard reagent via coordination with the magnesium.

What Is a Grignard Reagent?

A **Grignard reagent** is the product of the reaction between an alkyl halide and magnesium metal. A Grignard reagent has the generic structure R–Mg–X, where R is an alkyl group and X = Cl, Br, I.

What Is the Bond Polarization of the C–Mg Bond?

Since Mg lies to the left of carbon in the Periodic Table, carbon is the more electronegative atom. The bond polarity is: $^{\delta-}$C–Mg$^{\delta+}$ and the carbon of the Grignard reagent behaves as a **carbanion** (nucleophilic carbon) in most of its reactions.

Is It Possible to Form the Grignard Reagent from Vinyl Halides?

Yes! Vinyl halide (C=C–X) reacts with magnesium to form C=C–MgX, but the reaction is sluggish compared to alkyl halides.

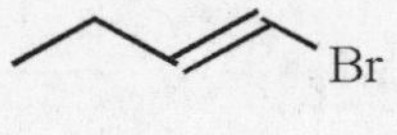

Br

10.1

What Is the Product Formed When Vinyl Halide 10.1 Reacts with Magnesium in THF?

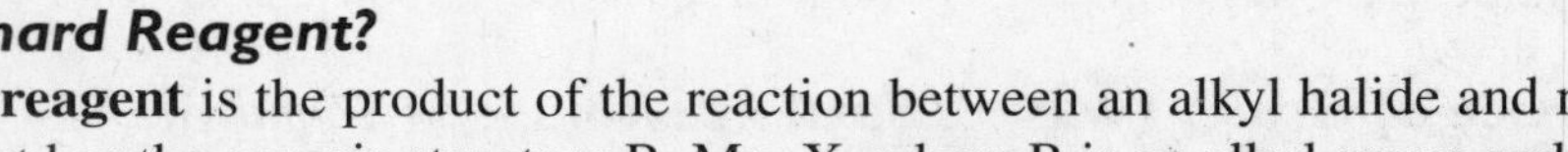

Br —— Mg° , THF ——> MgBr

10.1 *10.2*

Name 10.2.

Compound ***10.2*** is named 1-butenylmagnesium bromide.

Why Was THF Used Rather Than Ether?

Formation of the vinyl Grignard reagent ***10.2*** is slow, requiring the use of THF. The solvent THF is more basic than diethyl ether, providing assistance for the formation of ***10.2*** and also extra stabilization for ***10.2*** after it is formed.

Is It Possible to Form the Grignard Reagent from Aryl Halides Such As PhBr (10.3)?

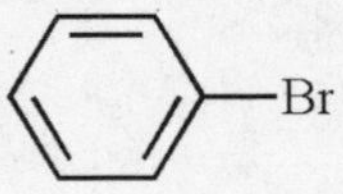

Br

10.3

Yes! Aryl halides (derivatives of benzene - see chapter 16 - Ar–X) react with magnesium to form Ar–MgX, but the reaction is sluggish compared to alkyl halides.

What Does the Symbol Ph Mean?

The symbol Ph indicates a benzene ring as a substituent. Compound ***10.3*** is abbreviated PhBr.

What Is the Product When 10.3 Reacts with Magnesium in THF?

In the example shown, ***10.3*** reacts with magnesium to form ***10.4***. Note that THF is used as the solvent. Formation of aryl Grignard reagents is slow, again requiring the assistance of the better Lewis base, THF, relative to diethyl ether.

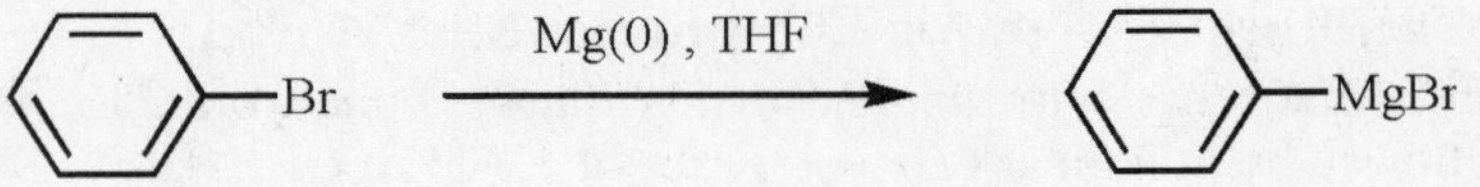

Br —Mg(0) , THF→ MgBr

10.3 *10.4*

Name 10.4.

Compound ***10.4*** is named phenylmagnesium bromide.

Give the Major Products of the Reactions of 10.5–10.7.

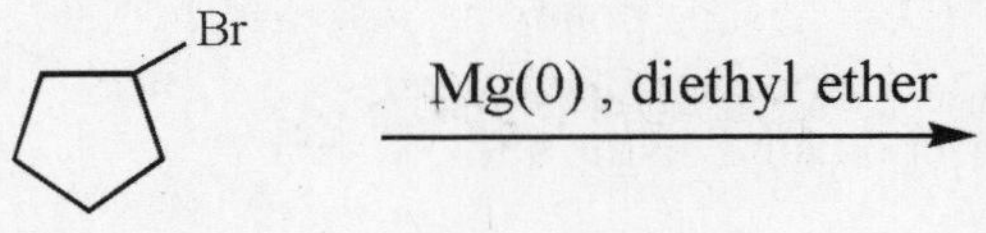

Br —Mg(0) , diethyl ether→

10.5

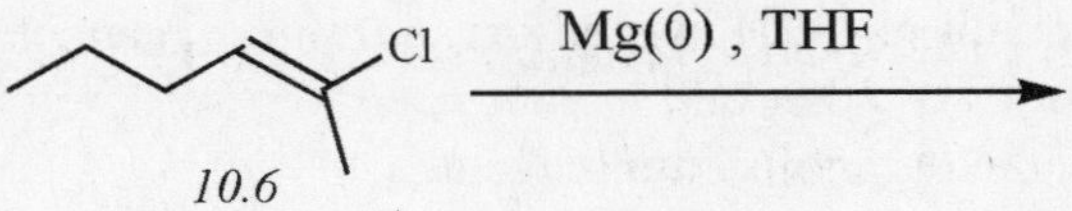

Cl —Mg(0) , THF→

10.6

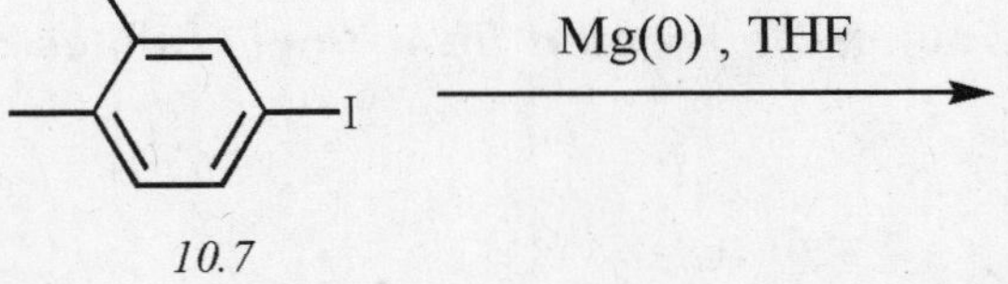

I —Mg(0) , THF→

10.7

In all three cases, the product is the corresponding Grignard reagent. 2-Bromocyclopentane (***10.5***) gives ***10.8***, *E*-2-chloro-2-hexene (***10.6***) gives ***10.9***, and 4-iodo-1,2-dimethylbenzene (***10.7***) gives ***10.10***. In the latter two cases (the vinyl halide and the benzene derivative), THF is used rather than diethyl ether. THF is a stronger Lewis base than diethyl ether, and the extra coordination with Mg is required to stabilize these Grignard reagents.

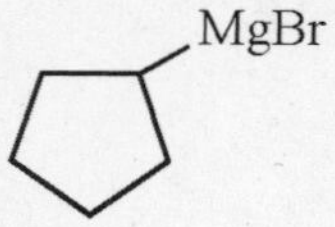

MgBr — *10.8*

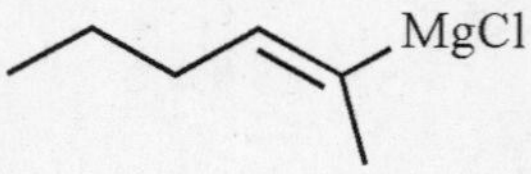

MgCl — *10.9*

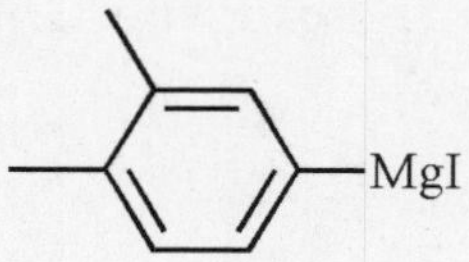

MgI — *10.10*

10.3. ORGANOLITHIUM COMPOUNDS

Just as magnesium reacts with alkyl halides to form Grignard reagents, lithium reacts with alkyl halides to form organolithium reagents. This is another example of an organometallic compound.

What Is the Reaction Product When Lithium Metal Reacts with 1-Bromobutane?

This reaction produces *n*-butyllithium ($CH_3CH_2CH_2CH_2Li$). Unlike Grignard reagents, lithium is monovalent, forming C–Li. The generic reaction is:

R–X + Li(0) → R–Li + Li–X, where Li° actually exists as Li–Li.

What Is the Nomenclature System for Compounds Such As RLi?

If we take CH_3Li as an example, the name is methyllithium - one word. Use the nomenclature term for the alkyl group followed by lithium as one word. Likewise, $(CH_3)_2CHCH_2Li$ is 3-methylpropyllithium and $CH_3CH_2CH_2CH_2Li$ is butyllithium or *n*-butyllithium, where *n*-indicates *normal* or straight-chain butane. These compounds can also be named where lithium is a substituent, indicated by lithio and the number of the carbon to which it is attached. In other words, CH_3Li is lithiomethane, $(CH_3)_2CHCH_2Li$ is 1-lithio-3-methylpropane, and $CH_3CH_2CH_2CH_2Li$ is 1-lithiobutane.

What Is the Predicted Reactivity of an Organolithium Reagent Such As Butyllithium?

Organolithium reagents are characterized by a C–Li bond, which is polarized similarly to the C–Mg bond of a Grignard reagent (the C–Li has the polarization $^{\delta-}C–Li^{\delta+}$). Organolithium reagents will, therefore, function as nucleophiles in the presence of an electrophilic species, such as acetone or other ketones and aldehydes and other electrophilic and acidic sites.

What Solvent Is Used to Form Organolithium Reagents?

Typically, the reaction done is ether. However, organolithium reagents decompose in ether solvent, and it is common to see mixtures of ether and hexane used, or even pure hexane (and sometimes other alkanes) to prolong the life of the organolithium reagent.

Can Organolithium Compounds Be Formed from Vinyl Halides and Aryl Halides?

Yes! Both are known.

Give the Major Products of the Reactions of 10.11–10.13.

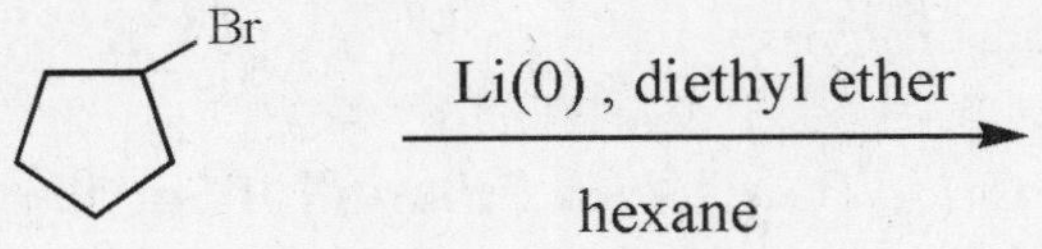

Br

Li(0) , diethyl ether

hexane

10.5

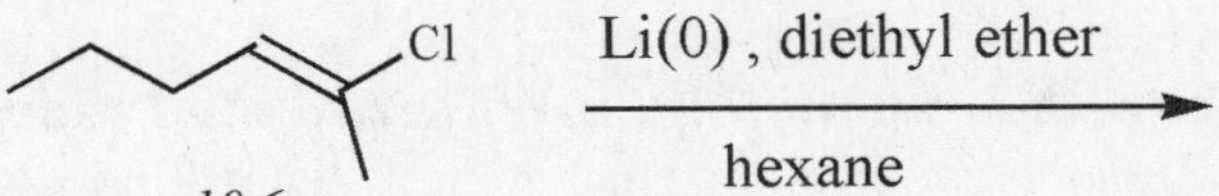

Cl

Li(0) , diethyl ether

hexane

10.6

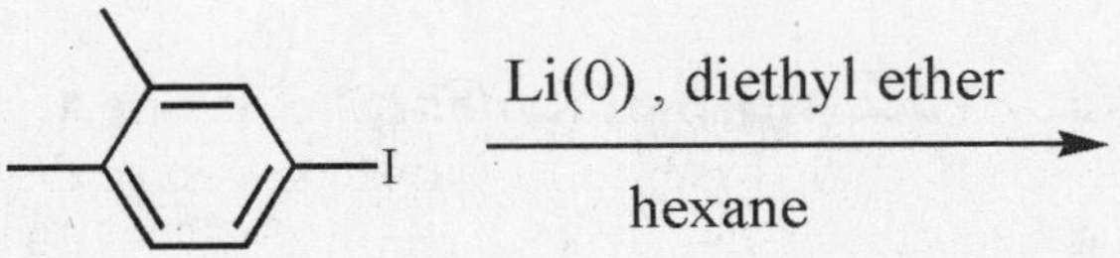

I

Li(0) , diethyl ether

hexane

10.7

In all three cases, the product is the corresponding organolithium reagent. 2-Bromocyclopentane (***10.5***) gives ***10.11***, *E*-2-chloro-2-hexene (***10.6***) gives ***10.12*** and 4-iodo-1,2-dimethylbenzene (***10.7***) gives ***10.12***. It is noted that formation of vinyllithium reagents and aryllithium reagents can be sluggish or proceed to give poor yields. In such cases, a change in solvent is undertaken or nitrogen-containing compounds are added as additives to facilitate formation of the organolithium reagent. For the moment, we will not discuss the use of such additives and simply assume that the organolithium reagent can be formed without problem.

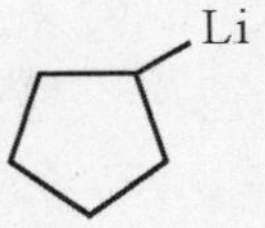

Li

Li

Li

10.11 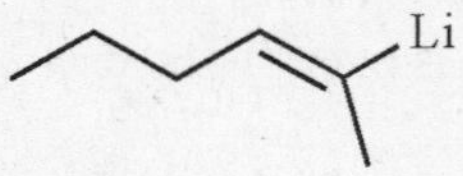 *10.12* 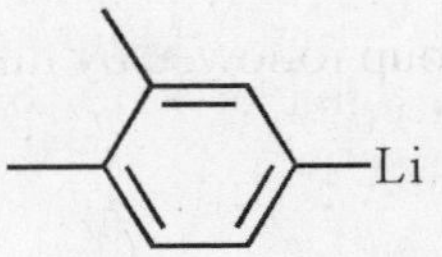 *10.13*

10.4. BASICITY

The negatively polarized carbon atom of a Grignard reagent or an organolithium reagent can function as a nucleophile in the presence of a positively polarized carbon atom. Such negatively polarized carbon atoms are also powerful bases in the presence of compounds having an acidic hydrogen atom. Such bases are so powerful that they react with some compounds that have not heretofore been considered acids.

Why Is Methylithium Considered to Be a Base?

The carbon bearing the lithium atom is polarized with a negative dipole and is attracted to an acidic hydrogen which has a strong positive dipole. In an acid-base reaction, methylithium reacts with the acid (H^+, to give a conjugate acid CH_3–H (methane), which is an extremely weak acid. Using the old axiom, strong acids give weak conjugate bases, and weak acids give strong conjugate bases; if methane is the conjugate acid, methylithium must be a very strong base.

What Is the Conjugate Base of the Reaction Between Methylithium and Water?

When methylithium reacts with water, the products are methane and lithium hydroxide. In this reaction, lithium hydroxide is the conjugate base.

Can Methylmagnesium Bromide React with Ethanol? If So, What Are the Products?

Yes! Methylmagnesium bromide is a strong base, and ethanol has an acidic hydrogen, with a pK_a of about 17. The reaction gives methane as the conjugate acid and $EtO^- \ ^+MgBr$ as the conjugate base.

Can Butyllithium React with Ammonia? If So, What Are the Products?

Yes! Butyllithium is a strong base, and ammonia has an acidic hydrogen, with a pK_a of about 38. The reaction gives butane as the conjugate acid and $Li^+ \ ^-NH_2$ as the conjugate base.

Can Propylmagnesium Iodide React with 1-Butyne? If So, What Are the Products?

Yes! propylmagnesium iodide is a strong base, and 1-butyne has an acidic hydrogen, with a pK_a of about 25. The reaction gives propane as the conjugate acid and $IMg^+ \ ^-C{\equiv}CEt$ as the conjugate base.

Which Is More Polarized, C–Li or C–Mg?

The C–Li bond is more polarized.

Explain Why Butyllithium Is a Stronger Base Than Butylmagnesium Bromide.

The C–Li bond is more highly polarized than the C–Mg bond. Therefore, the $\delta-$ charge on the carbon of the organolithium reagent is greater (more carbanionic). One must also focus on the conjugate acid by-product (acid + base → conjugate acid + conjugate base). The conjugate acid of the organolithium, and also of the Grignard reagents, is an alkane, a very weak acid (pK_a >40): $R–Li + H^+X^- \rightarrow R–H + LiX$. When comparing butyllithium and butylmagnesium bromide, the acids (R–H) are the same for both (butane), and the greater bond polarity of C–Li is usually the dominant factor, which makes butyllithium the stronger base.

What Is the Major Product of the Reactions of 10.14, 10.11, and 10.15?

Li —— H-O-H ——→

10.14

Li —— $(CH_3)_2N$-H ——→

10.11

–Li —— CH_3CH_2O-H ——→

10.15

Each of these reactions is an acid-base reaction where the organolithium is the base. In the first reaction, pentyllithium (also named 1-lithiopentane, ***10.14***) reacts with water (the acid) to give pentane (***10.15***) and LiOH. In the second case, cyclopentyllithium (also lithiocyclopentane, ***10.11***) reacts with dimethylamine to produce cyclopentane and lithium dimethylamide (***10.16***). Amines are usually basic due to the lone electron pair on nitrogen. In this case, the organolithium is a much more powerful base than the aliphatic secondary amine, making the N–H bond an acid (pK_a 23–25). Phenyllithium (***10.14***) reacts with ethanol to form benzene and lithium ethoxide (***10.17***). Alcohols are amphoteric, and in the presence of the powerful base (***10.15***), ethanol is an acid.

+ LiOH
10.15

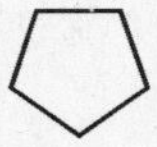

+ $(CH_3)_2N$-Li
10.16

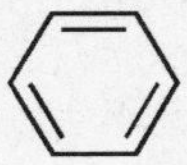

+ CH_3CH_2OLi
10.17

10.5. ORGANOCUPRATES

Carbon forms bonds to copper in several ways. One of the more useful is to form a cuprate. In various reactions, the carbon atom in an organocuprate functions as a nucleophile.

What Is the Product When a Grignard Reagent or an Organolithium Reagent Reacts with 1-Bromopentane?

Grignard reagents and organolithium reagents give very poor yields of substitution products when they react with alkyl halides. *n*-Butyllithium, for example, reacts with 1-bromobutane, but very little octane (Bu–Bu, the substitution product) is produced or none at all. Elimination (see Chapter 9) and **disproportionation** (self oxidation-reduction) products predominate. In this case, butene and butane are the major products. In general, alkyl halides do not give good yields of substitution products with Grignard reagents or organolithium reagents, *unless a transition metal is added to the reaction.*

What Is an Organocuprate?

An organocuprate has the generic structure R_2CuLi.

What Is a Gilman Reagent?

To honor the work of Henry Gilman (1893–1986), an organic chemist at the University of Iowa, the organocuprate reagent R_2CuLi is commonly referred to as a **Gilman reagent**.

How Are Organocuprates Named?

Such reagents are named by the metal (lithium), followed by the alkyl group (dialkyl), and finally, cuprate, to indicate the oxidation state of the copper. Therefore, Me_2CuLi is lithium dimethyl cuprate, and $(CH_3CH_2CH_2)_2CuLi$ is lithium dipropyl cuprate.

What Is the Structure of Lithium DI-n-Butylcuprate?

When two equivalents of 1-lithiobutane (butyllithium) react with cuprous iodide (CuI), which is Cu(I), the copper forms an **'ate' complex**, which occurs when a metal expands its valence and assumes a negative charge. In this case, R_2Cu^- forms with Li^+ as the counterion. Dibutylcuprate is, therefore, $[Bu–Cu–Bu]^-$, and lithium di-*n*-butylcuprate is $LiBu_2Cu$ or Bu_2CuLi.

How Does an Organocuprate React with Alkyl Halides?

Dialkylcuprates are carbanion-like reagents that react with carbonyls to a limited extent (see Chapter 21 for acyl addition reactions), but react rapidly and in high yield with alkyl halides. In this reaction, one of the alkyl groups displaces the halogen of the alkyl halide to give the substitution product (a coupling product). 1-iodoethane, for example, reacts with lithium dibutylcuprate to give hexane:

$$LiBu_2Cu + CH_3CH_2\text{–}I \rightarrow Bu\text{–}CH_2CH_3$$

Test Yourself

1. Explain why ethanol cannot be used as a solvent for the reaction of ethylmagnesium bromide and 2-butanone.
2. Explain why butyllithium is a good reagent to convert 1-pentyne to the alkyne anion, 1-lithiopentyne.
3. Draw the structure of the Grignard reagent formed for each of the following, and name each product.

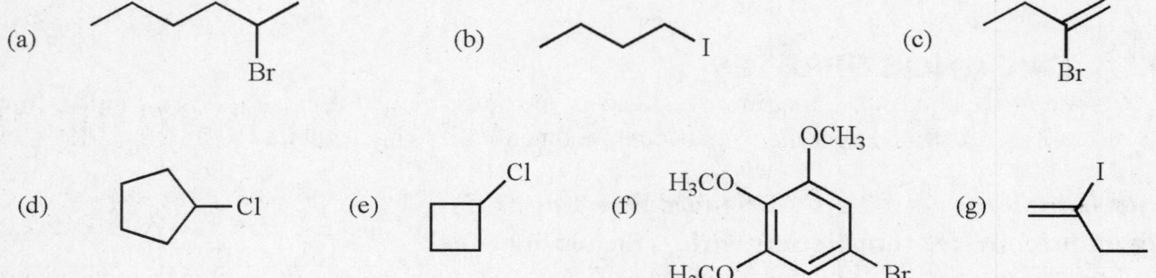

4. Draw the Gilman reagent formed for each of the following reactions.

(a) Li (b) Li (c) Li (d) Li

5. Amines (R_2NH) are usually viewed as bases due to the presence of the lone electron pair. Explain why Et_2NH (diethylamine) behaves as an acid in the presence of butyllithium.
6. In each case, give the major product of the reaction. Note that in reaction (c), the reaction with an aldehyde, this acyl addition reaction is not covered until Chapter 21. It is included here as a thought question.

(a) Br — 1. Mg(0), ether; 2. CH_3NH_2 →

(b) I — Mg(0), THF →

(c) I — 1. Li(0), ether; 2. propanal; 3. H_3O^+ →

(d) OH — Li →

(e) I — 1. Li(0), ether; 2. CuI, -10°C; 3. $PhCH_2Br$; 4. H_2O →

(f) I — 1. Bu_2CuLi, ether; 2. H_3O^+ →

(g) Li — 1. 0.5 CuI, ether, −10°C; 2. 1-iodopentane →

CHAPTER 11

Alcohols. Preparation, Substitution and Elimination Reactions

Alcohols are one of the most important functional groups in organic chemistry. Alcohols can be prepared from alkenes, halides, ketones, aldehydes and acid derivatives, as well as other functional groups. Alcohols can be chemically transformed into a wide range of other functional groups, including halides, alkenes, ethers, aldehydes, ketones and acid derivatives. This chapter will describe their physical and chemical properties, including reactions that form alcohols and some that transform alcohols into other functionality. Since alcohols are pivotal in the formation of many other functional groups, they will appear in reactions throughout the remainder of this book.

11.1. STRUCTURE AND NOMENCLATURE OF ALCOHOLS

Alcohols are organic molecules that are characterized by the hydroxyl unit, -OH. Alcohols are relatively polar compounds, and the proton on the oxygen is slightly less acidic than the proton in water. Alcohols are considered to be functional groups, with the nomenclature suffix *-ol*.

Structure

What Is the Distinguishing Feature of an Alcohol?

An alcohol is characterized by a C–O–H bond.

What Is the Distinguishing Feature of a Thiol?

A thiol is the sulfur analog of an alcohol and is characterized by a C–S–H bond.

What Is the Bond Polarization for an O–H Group?

Since the oxygen is more electronegative than hydrogen, the oxygen is the negative pole, and the hydrogen is the positive pole ($^{\delta-}$O–H$^{\delta\oplus}$). Therefore, the hydrogen is slightly acidic ($pK_a \approx 16–18$).

Name Some Common Bases That Will Deprotonate an Alcohol in an Acid-Base Reaction.

Relatively strong bases are required for this reaction. Typical bases include sodium hydride, organolithium reagents such as methyllithium or butyllithium sodium amide, or even sodium metal.

What Is the Conjugate Base When an Alcohol (Roh) Loses Its Acidic Proton?

The conjugate base of an alcohol (ROH) is the alkoxide, RO^-. Clearly, the positive counterion will be that associated with the base used to deprotonate the alcohol, typically sodium, lithium, or potassium.

Draw the Products Formed When 1-Butanol Reacts with Methyllithium.

The reaction of 1-butanal with methyllithium gives lithium butoxide (***11.1***) as the conjugate base and methane (HCH_3) as the conjugate acid.

$$\text{Bu}-\overset{\delta-}{O}-\overset{\delta+}{H} \quad \overset{\delta+}{Li}-\overset{\delta-}{CH_3} \longrightarrow \text{Bu}-O^{\ominus}\ Li^{\oplus}\ (\mathit{11.1}) \quad H-CH_3$$

Why Is the Intermolecular Interaction of Two C–O–H Groups Greater Than the Interaction of Two C–O–C Groups?

The O–H bond is more polarized than the C–O bond, leading to a greater attraction of the oxygen of one OH for the hydrogen of another. Since the intermolecular O-----H bond (a hydrogen bond) is quite strong, the OH interactions are stronger than the C–O interaction in the ether unit.

What Is the Associative Interaction Between Two Alcohol Groups Called?

Hydrogen bonding.

Compared to an Alkane, How Much Is the Boiling Point Increased by Introducing an OH Group?

A comparison of the boiling points of several alkanes with their homologous alcohol derivatives leads to the conclusion that replacing a hydrogen with an OH does not increase the boiling point by a set amount. There is a clear and significant upward trend, but the difference in boiling point diminishes as the size of the carbon chain increases.

Ethane, bp = −88.5°C and ethanol, +78.3°C (Δ = 166.8°C); propane, bp = −42°C and propanol, bp = +92.2°C (Δ = 134.2°C); butane, bp = 0°C and butanol, bp = +117.7°C (Δ = 117.7°C); octane, bp = +126°C and octanol, bp = +195°C (Δ = 69°C).

Why Is the Boiling Point of an Alcohol Higher Than That of an Ether (R–O–R) with Approximately the Same Molecular Weight?

A comparison of diethyl ether (MW = 74.12, bp = 34.5°C) and 1-butanol (MW 74.12, bp = 117.7°C) clearly shows the trend. The strong hydrogen bonding possible in the alcohol far outweighs the dipole interactions possible in the C–O bond of the ether.

What Is the Geometry of an Alcohol If Attention Is Focused upon the Oxygen?

One cannot 'see' the lone electron pairs. If the methyl group and hydrogen of methanol are included with the lone pairs in a VSEPR model, however, the geometry around the oxygen is 'tetrahedral.' If one only focuses upon the atoms one can see, the geometry is angular (bent) around the oxygen (see ***11.2***).

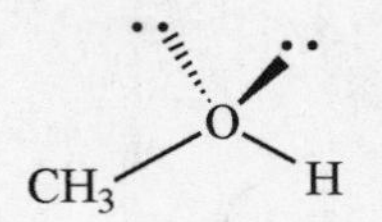

11.2

What Is the Approximate Number of Carbons Required for an Alcohol to Be Soluble in Water?

Low molecular weight alcohols such as methanol and ethanol are completely miscible in water. High molecular weight alcohols such as 2-decanol or 1-pentadecanol are generally insoluble. There is a competition between the 'water-soluble' (hydrophilic) OH group and the 'water-insoluble' (hydrophobic) hydrocarbon backbone. The 'break point' is about five carbons (pentanol), where 1-butanol has a solubility of 7.4 g/100 mL of water, 1-pentanol has a solubility of 2.7 g/100 mL, and 1-hexanol has a solubility of 8 g/100 mL. Branched chain alcohols are more soluble than linear alcohols [isobutyl alcohol (10g/100 mL) vs. *tert*-butanol (infinitely soluble in water)].

Nomenclature

What Is the Suffix for an Alcohol Using the IUPAC Rules of Nomenclature?

The IUPAC ending for an alcohol is *-ol*, where the *-e* of the alk**ane** is dropped (or *-e* in alk**ene** or *-e* in alk**yne**) and replaced with *-ol*, as in ethane → ethanol.

Why Is the C6 Alcohol with the Formula $C_6H_{14}O$ Called Hexanol Rather Than Hexol?

A six-carbon alcohol named hexol could have an alkane backbone ($CH_3CH_2CH_2CH_2CH_2CH_2OH$, ***11.3***), an alkene backbone ($CH_3CH{=}CHCH_2CH_2CH_2OH$ with a formula of $C_6H_{12}O$, ***11.4***) or an alkyne backbone ($CH_3C{\equiv}CCH_2CH_2CH_2OH$ with a formula of $C_6H_{10}O$, ***11.5***). The formula C_6H_{14} clearly indicates a saturated hydrocarbon, so alkene and alkyne units are not a possibility. *However, if you simply see the name hexol, you don't know that.* In order to specify which one is being considered, either the alkan-, alken- or alkyn- prefix is retained, as in 1-hexanol for ***11.3***, 4-hexen-1-ol for ***11.4***, and 4-hexyn-1-ol for ***11.5***.

What Is the Nomenclature Protocol for Cyclic Alcohols?

The usual rules apply, with the *-ol* suffix to indicate an alcohol. Therefore, ***11.6*** is cyclohexanol, ***11.7*** is 3-bromo-4-methylcycloheptanol, and ***11.9*** is cyclopropanol. Note that in cyclic alcohols, the carbon bearing the OH unit is always C1, and the 1- is omitted from the name, since it is obvious.

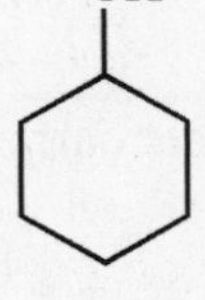

11.6 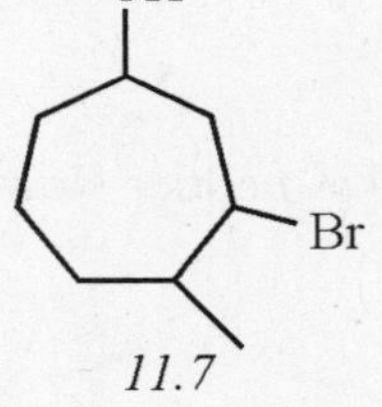 11.7 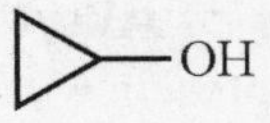 11.8

Give the IUPAC Names for 11.9–11.14.

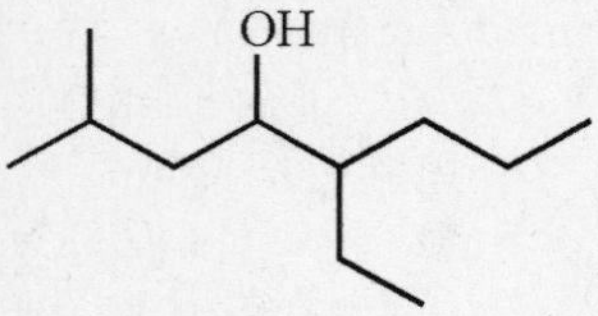

OH Cl OH Ph Ph OH

11.9 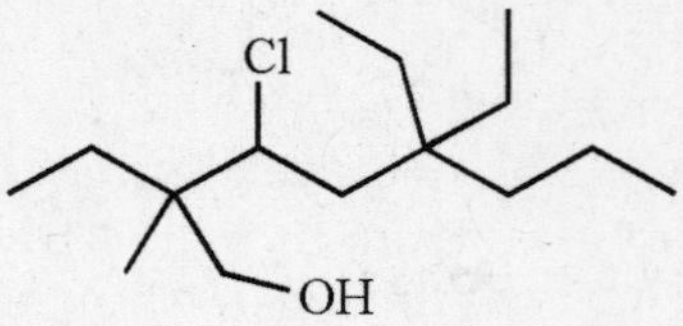 11.10 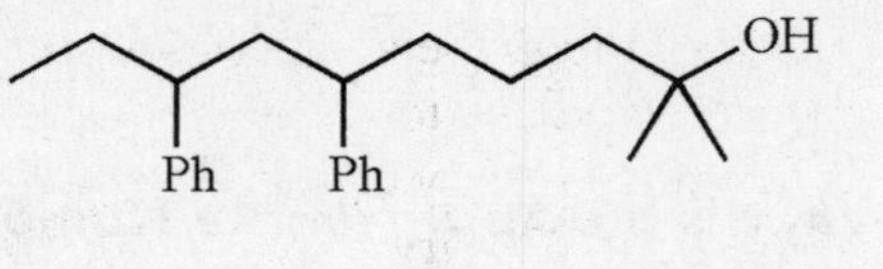 11.11

OH OH OH

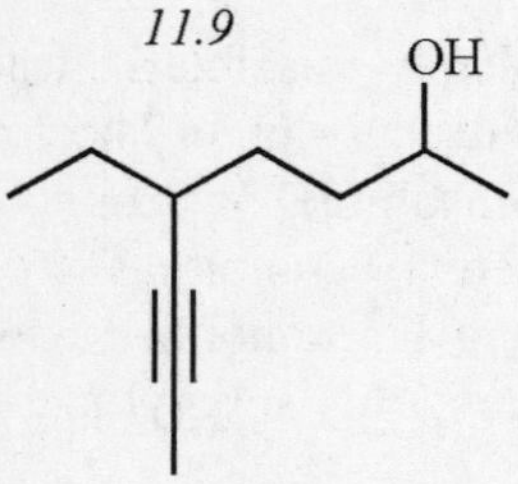

11.12 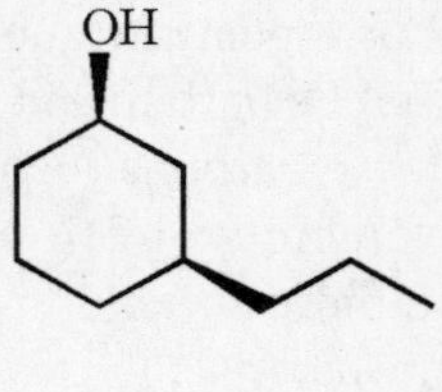 11.13 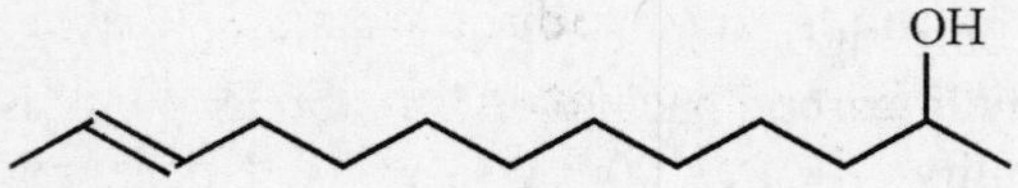 11.14

The names are: 5-ethyl-2-methyl-4-octanol for ***11.9***; 3-chloro-2,5,5-triethyl-2-methyl-1-octanol for ***11.10***; 2-methyl-6,8-diphenyl-2-decanol for ***11.11***; 5-ethyl-oct-6-yn-2-ol for ***11.12***; *cis*-3-propylcyclohexanol for ***11.13***; *E*-tridec-11-en-2-ol for ***11.14***.

Give the Common Names for 11.15–11.18.

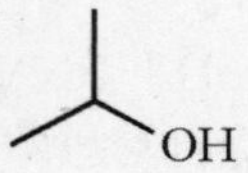

OH OH OH OH

11.15  11.16 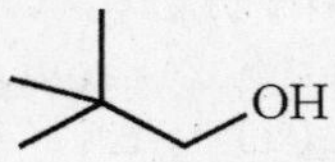 11.17 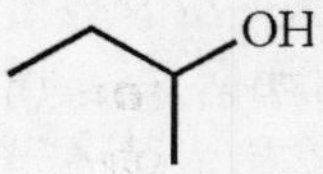 11.18

There are several alcohols, usually with lower molecular weights, that have common names. The common name of ***11.15*** is isopropyl alcohol (rubbing alcohol); ***11.16*** is *tert*-butyl alcohol; ***11.17*** is neopentyl alcohol and ***11.18*** is *sec*-butyl alcohol.

11.2. PREPARATION OF ALCOHOLS

There are several different reactions that produce alcohols, and they involve several different functional groups. The two most common functional groups used as precursors to produce alcohols are alkenes and carbonyl compounds (ketones and aldehydes or acid derivatives).

Solvolysis of Alkyl Halides

In the presence of water, tertiary alkyl halides slowly ionize to an intermediate carbocation, which reacts with water to give an alcohol.

What Is Solvolysis?

Solvolysis is most commonly defined as the nucleophilic displacement of a leaving group by the nucleophilic atom of a solvent such as the oxygen atom in water or an alcohol.

Why Is Aqueous Solvolysis of a Tertiary Halide Rather Easy, If Slow, but a Similar Reaction with a Primary Halide Is So Slow That It Effectively Does Not Occur?

This reaction relies on ionization of the leaving group to give an intermediate carbocation, which then reacts with the nucleophilic solvent. Ionization is only facile for tertiary halides, relative to primary or secondary, due to the higher activation barrier for ionization in the latter compounds and greater stability of the tertiary cation. Ionization of the tertiary halide leads to the more stable tertiary carbocation, whereas the primary halide would have to give a primary carbocation. The great difficulty in ionization of the latter makes the reaction rather specific for the tertiary halide.

What Is the Product When 2-Methyl-2-Bromopentane (11.19) Is Heated in Water for a Week?

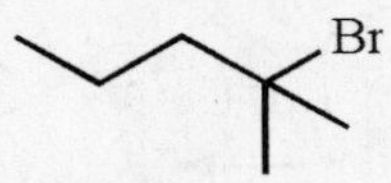

Br

11.19

The product is 2-methyl-2-pentanol, ***11.20***.

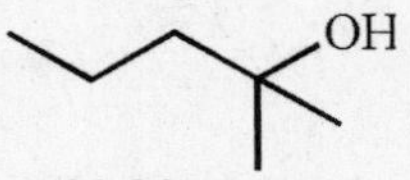

OH

11.20

Draw the Mechanism for the Conversion of 11.19 to 11.20.

Slow ionization of the bromine atom in ***11.19*** leads to the tertiary carbocation ***11.21***. Water (the solvent) donates electrons from oxygen to the positive carbon atom to give oxonium ion ***11.22***. Loss of the proton from ***11.22*** generates the neutral alcohol product *11.20*. Oxonium ions such as this are acidic and will react with water - the base - to form the neutral alcohol; note the arrow indicating attack of the base on the acidic hydrogen atom in ***11.22***.

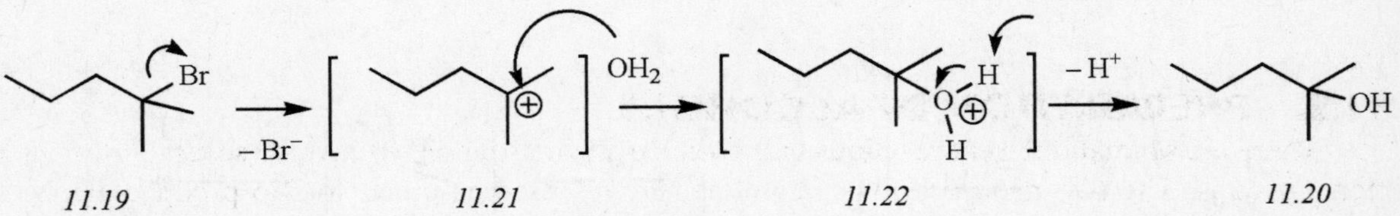

Br → $-\,Br^-$ [⊕] OH_2 → [H O⊕ H] $-\,H^+$ → OH

11.19 *11.21* *11.22* *11.20*

Does the Bromine Spontaneously Leave 11.19, or Is It Pulled Off? If It Is Pulled Off, What Does the Pulling?

2-Bromo-2-methylpentane is a perfectly stable compound, and the bromine most certainly does not spontaneously fly off. The bromine has a negative dipole and is attracted to a hydrogen atom in water, which has a strong positive dipole. This hydrogen-bonding effect pulls off the bromine, weakening the C–Br bond, and eventually causing it to break. Once broken, water will solvate both the cation and anion, which assists in the overall ionization process.

What Is the Product When an Alkyl Halide Is Heated with an Alcohol Such As Ethanol?

Solvolysis can occur in alcohol solvents, giving an ether.

Give the Product and the Mechanism of Its Formation When 11.19 Is Heated in Ethanol for Several Days.

Ionization of the bromide gives carbocation ***11.21*** as before. The nucleophilic species in ethanol, however, is the oxygen of the alcohol, and reaction with the carbocation gives oxonium ion ***11.23.*** Loss of a proton gives the ether, ***11.25***.

Br
$-Br^-$
HOEt
H
O
Et
$-H^+$
OEt

11.19 *11.21* *11.23* *11.24*

Can Rearrangement Occur During Solvolysis Reactions Such As These?

Yes! Since the intermediate in these solvolysis reactions is a carbocation, rearrangement to a more stable ion is always a possibility.

Give the Product and the Mechanism of Its Formation When 2-Bromo-1-Methylcyclopentane (11.25) Is Heated in Water for Several Days.

Initial ionization of the bromide from ***11.25*** generates the secondary carbocation ***11.26***. A 1,2-hydrogen shift (the rearrangement is indicated by the 'curly' arrow) leads to the more stable tertiary carbocation ***11.27***. Subsequent reaction with water gives oxonium ion ***11.28***, which loses a proton to give the final product, 1-methylcyclopentanol, ***11.29***.

Br
$-Br^-$
1,2- H shift
OH_2

11.25 *11.26* *11.27*

H
O
H
$-H^+$
OH

11.28 *11.29*

Give the Major Product for the Reactions 11.30–11.32.

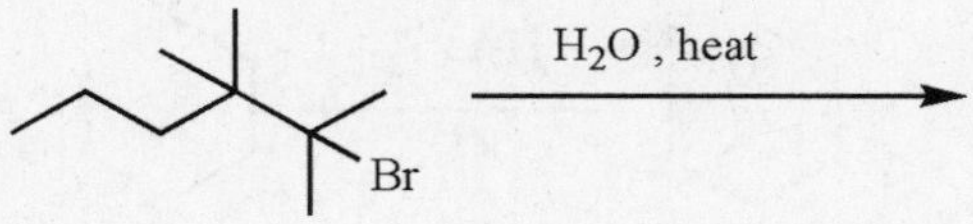

Br

H_2O , heat

11.30

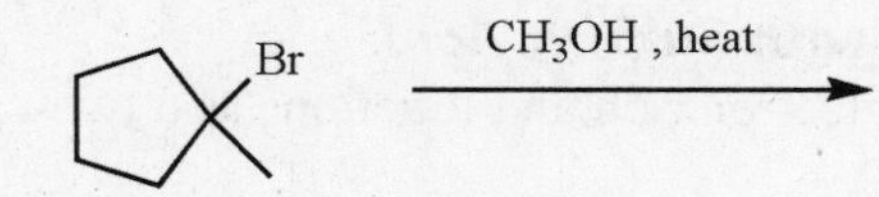

Br

CH_3OH , heat

11.31

Br

H_2O , heat

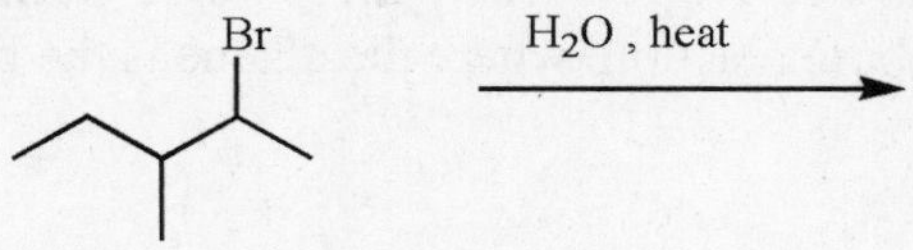

11.32

The solvolysis product of ***11.30*** is alcohol ***11.33***. Since ***11.30*** is a tertiary halide, no rearrangement is observed. Solvolysis of ***11.31*** in methanol gives the ether, ***11.34***. Solvolysis of the secondary halide ***11.32*** proceeds by rearrangement to a more stable tertiary carbocation and formation of alcohol ***11.35*** as the final product.

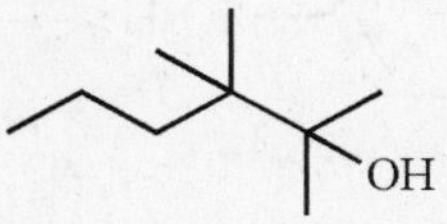

OH

11.33

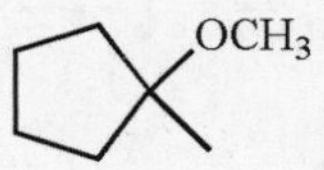

OCH_3

11.34

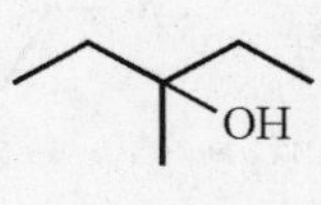

OH

11.35

Hydration of Alkenes

When an alkene reacts with an acid such as HCl, a carbocation intermediate is generated. That cation reacts with the nucleophilic counterion of the acid to give the final addition product. In order for this reaction to work, the acid must be strong enough to react with the weakly basic alkene. Weak acids give no reaction unless an acid catalyst is provided.

What Is the Product When 2-Methylcyclohexene Is Heated in Water?

There is no reaction. Water is not a sufficiently strong acid to react with the weak base (the π-bond of the alkene).

If a Catalytic Amount of H_2SO_4 Is Added to the Reaction of 1-Methylcyclohexene (11.36) and Water, What Is the Result?

The strong acid catalyst (H_2SO_4) reacts with the alkene moiety of ***11.36*** to form the more stable cation, ***11.37***. Once this cation is formed, water is a sufficiently strong nucleophile to attack the positive center, generating oxonium ion ***11.38***. Loss of a proton completes the reaction, giving 1-methyl-1-cyclohexanol (***11.39***) as the final product.

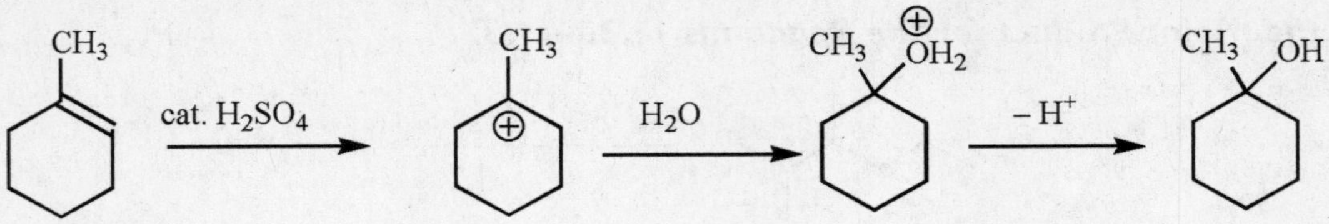

CH$_3$ (11.36) —cat. H_2SO_4→ CH$_3$ ⊕ (11.37) —H_2O→ CH$_3$ $\overset{\oplus}{O}H_2$ (11.38) —$-H^+$→ CH$_3$ OH (11.39)

11.36 *11.37* *11.38* *11.39*

Why Does the Alkene React with Sulfuric Acid?

The π-bond of the alkene is electron rich and functions as a two-electron donor (a base) to the proton of sulfuric acid.

What Type of Reaction Is the Reaction of an Alkene with Sulfuric Acid?

This reaction is an acid-base reaction, where the alkene is the base and the proton of sulfuric acid is the acid.

What Is Para-Toluenesulfonic Acid?

This is *para*-toluenesulfonic acid, RSO_3H, where R = *p*-methylphenyl. See ***11.40***, where the sulfonic acid is drawn out to indicate the bonding about sulfur and also drawn using the abbreviated form of the sulfonic acid, $-SO_3H$. This organic acid is soluble in most organic solvents and is commonly used as an acid catalyst.

O, S—OH, O ≡ SO_3H

11.40

What Is p-TsOH?

This is an abbreviation for *para*-toluenesulfonic acid, RSO_3H, where R = *p*-methylphenyl.

Give the Major Product for the Reactions of 11.41 and 11.42.

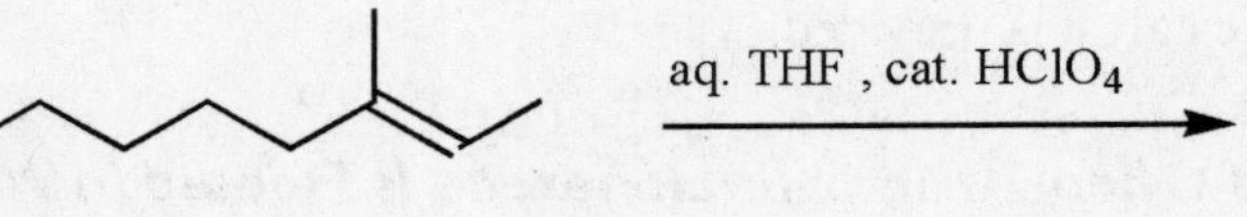

aq. THF , cat. $HClO_4$

11.41

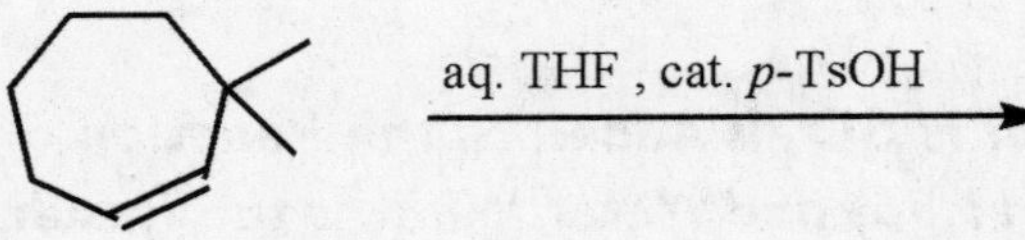

aq. THF , cat. *p*-TsOH

11.42

The reaction of water with the first alkene (***11.41***) generates 3-methyl-3-hexanol (***11.43***). Note the use of perchloric acid as the catalyst. Clearly, acids other than sulfuric can be used. The second reaction with ***11.42*** involves a 1,2-methyl shift from the initially formed secondary cation, generated by reaction of the alkene with toluenesulfonic acid, to the more stable tertiary cation. Trapping water leads to the final product, 1,2-dimethyl-1-cycloheptanol (***11.44***).

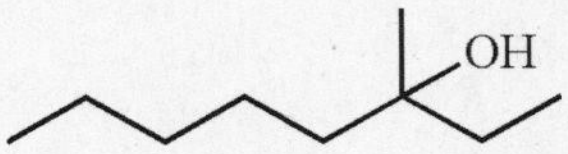

OH OH

11.43 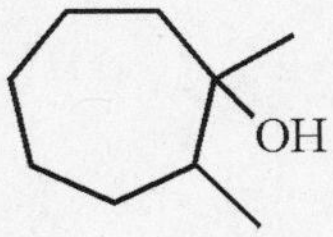 *11.44*

Oxymercuration

Alkenes can react with Lewis acids as well as Brønsted-Lowry acids to form a carbocation. Once formed, reaction with water leads to an alcohol. When mercury is employed as the Lewis acid, a C–Hg bond is formed that stabilizes the carbocation, but requires a second chemical step to remove the mercury from the final product.

What Is the Structure of Mercuric Chloride?

The structure of mercuric chloride is $HgCl_2$.

What Does Mercuric Mean?

This is the term used for divalent mercury, Hg^{+2}.

What Is the Structure of Mercuric Acetate?

The structure of mercuric acetate ($Hg(OAc)_2$) is ***11.45***. The term acetate refers to the unit CH_3–C(=O)–O, derived from acetic acid.

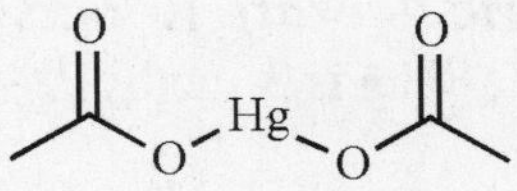

O O Hg O O

11.45

Classify Mercuric Acetate As a Reagent.

The mercury is electron deficient and behaves as a Lewis acid in its reactions with alkenes. An alkene will donate electrons to mercury, displacing one of the acetate groups.

What Is the Product of a Reaction Between 2-Methyl-1-Butene and Mercuric Acetate?

The initial product of the Lewis acid–Lewis base reaction is the organomercury cation, ***11.46***. The mercury stabilizes the cationic center by 'back donation' of electrons, as indicated by the dashed line. Subsequent reaction of this carbocation with water gives ***11.47*** via the usual oxonium ion.

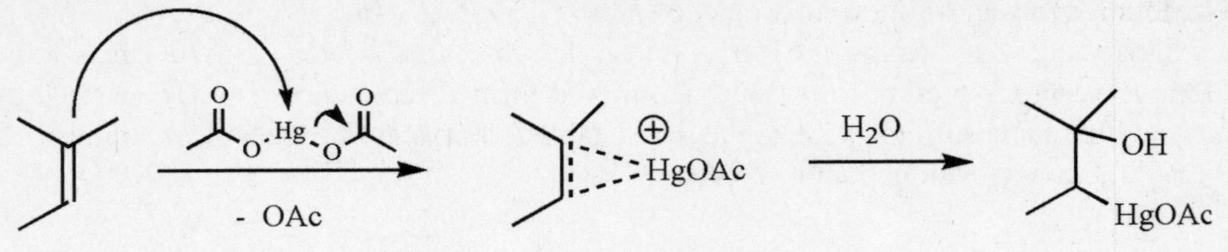

11.46

11.47

If the Reaction Is Done in an Aqueous Medium, What Is the Product?

Organomercury cation ***11.46*** behaves, more or less, as any other cation, and in the presence of a nucleophile, such as water, will be converted to an oxonium ion, and loss of a proton gives a hydroxymercury compound, ***11.47***.

How Can the Mercury Be Removed from the Product?

Removal of mercury requires cleavage of the C–Hg bond. This can be done very efficiently by addition of sodium borohydride ($NaBH_4$) to ***11.47***, followed by aqueous hydrolysis. The final product is the alcohol that is formed by addition of water to the more substituted carbon (***11.48***).

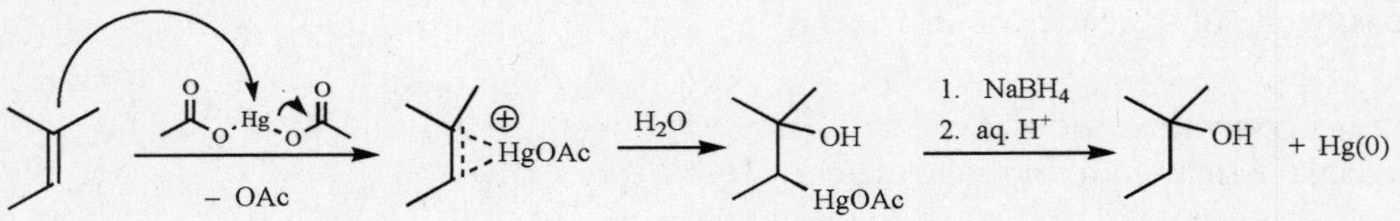

11.46 11.47 11.48

This overall sequence is called **oxymercuration** or, sometimes, **oxymercuration-demercuration**.

What Is the Product When Mercuric Acetate Reacts with 3-Methyl-1-Pentene?

The product (after treatment with $NaBH_4$) is 3-methyl-2-pentanol, ***11.49***.

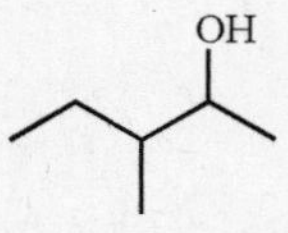

11.49

Is That Product the Result of a Rearrangement?

No! The mercury-stabilized cation does not result in rearrangement under these conditions prior to reaction with water. Therefore, the final alcohol product is the one obtained without rearrangement.

Why Is the Alcohol Formed Without Rearrangement of the Intermediate Carbocation?

There is no rearrangement of the intermediate mercuric cation. The mercury stabilizes the secondary cationic center by **back donation** of its d orbitals. This increased stability diminishes the possibility of rearrangement, and trapping with water is usually faster than the rearrangement.

Give the Products of the Reactions of 11.50–11.52.

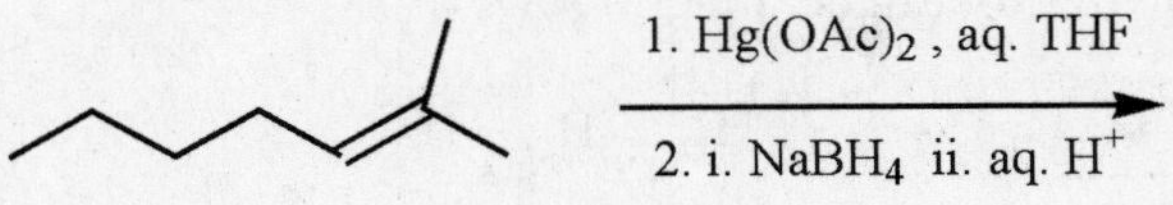

1. $Hg(OAc)_2$, aq. THF
2. i. $NaBH_4$ ii. aq. H^+

11.50

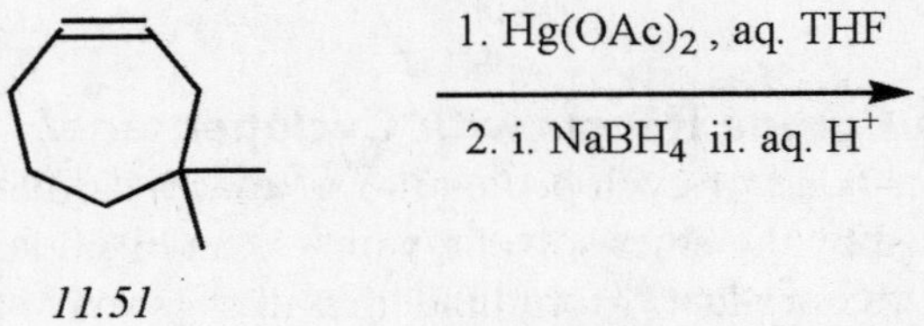

1. $Hg(OAc)_2$, aq. THF
2. i. $NaBH_4$ ii. aq. H^+

11.51

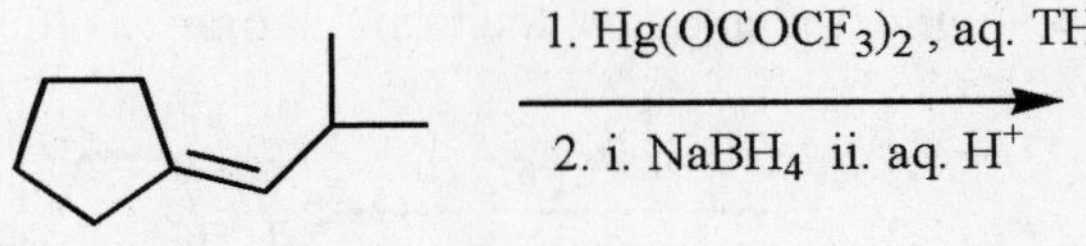

1. $Hg(OCOCF_3)_2$, aq. THF
2. i. $NaBH_4$ ii. aq. H^+

11.52

The first reaction converts ***11.50*** into the tertiary alcohol, 2-methyl-2-heptanol (***11.53***). This hydration reaction proceeds as if it were a normal cation intermediate in an addition reaction. The second reaction of ***11.51*** poses a problem. There is no regiochemical preference for addition of the mercury. The overall sequence will, therefore, produce a roughly 1:1 mixture of two alcohols, ***11.54*** and ***11.55***. In the final reaction, mercury trifluoroacetate is used rather than mercuric acetate. The reaction of ***11.52*** proceeds normally, however, producing the tertiary alcohol, ***11.53***.

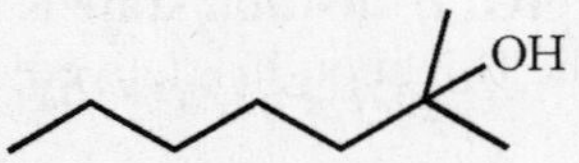

OH OH HO OH

11.53 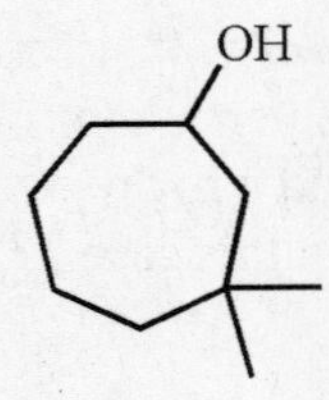 *11.54* 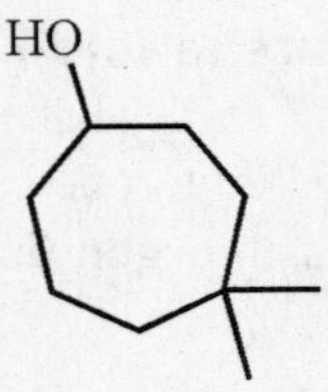 *11.55* 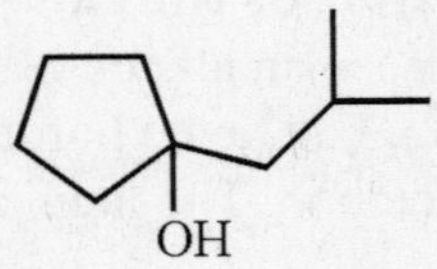 *11.56*

What Is the Advantage of Using Mercuric Trifluoroacetate Rather Than Mercuric Acetate?

Trifluoroacetate is a better leaving group than acetate. In addition, trifluoroacetate withdraws more electron density from mercury, making it more electrophilic. Both of these things combine to make the trifluoroacetate derivative more reactive.

Hydroboration

Alkenes react with borane and its derivatives to give alkylboranes. Treatment of these alkylboranes with various reagents leads to a variety of new functional groups. These reactions are known collectively as **hydroboration**. The hydroboration of alkenes to produce alcohols will be the focus of this section.

What Is the Structure of Diborane?

Diborane (B_2H_6) is a hydrido-bridged dimer, ***11.57***.

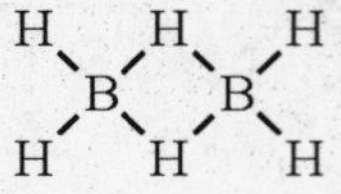

11.57

What Is the Product When Borane Reacts with Cyclopentene?

Diborane reacts with the π-bond of cyclopentene to give a trialkylborane (tricyclopentylborane, *11.58*). This reaction is catalyzed by the ether solvent, which is an essential component of the reaction, although it does not appear in the product. More hindered alkenes may stop at the monoalkyl or the dialkylborane. With relatively unhindered alkenes, the trialkylborane is the most common product.

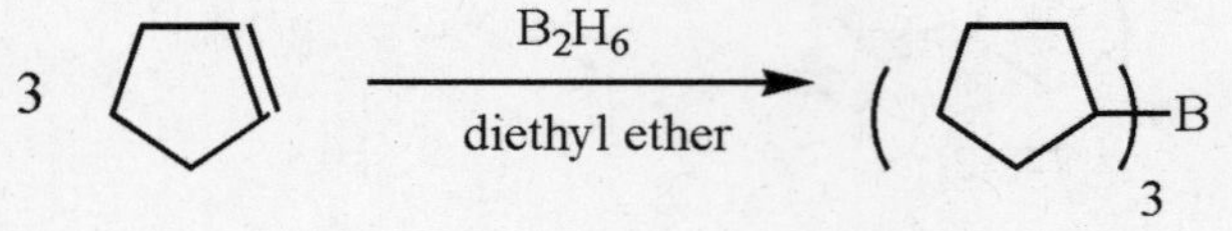

B_2H_6
diethyl ether

11.58

Why Is an Ether Solvent Such As THF Used When Diborane Reacts with Alkenes?

Ether solvents catalyze the reaction between borane and alkenes. Without the ether solvent, the reaction can require temperatures up to 200–250°C.

What Is the Nature of the Transition State for the Reaction of the Alkene 1-Butene and Diborane?

When an alkene reacts with an alkene such as 1-butene, a **four-centered transition state** is formed (***11.59***), where the boron is delivered to one carbon and a hydrogen of the B–H bond is delivered to the other carbon. The final product is an **organoborane**, in this case ***11.60***.

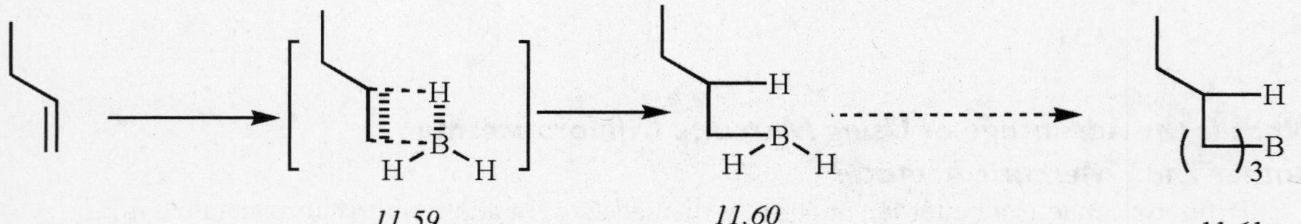

11.59 *11.60* *11.61*

Why Did We Draw the Product of Cyclopentene As a Triborane and the Product of 1-Butene As a Monoborane?

Monoborane ***11.60*** is simply the initial product. Note the presence of two more B–H units in ***11.60***. each of them can react with the alkene (here, 1-butene), and the final product will be the triborane (*11.61*), where all three B–H units have reacted with an alkene. In other words, one molar equivalent of borane can react with three molar equivalents of simple alkenes. Note that ***11.61*** is also drawn as $(CH_3CH_2CH_2CH_2)_3B$.

Why Does the Reaction of 1,2-Dimethylcyclohexene (11.62) Lead to a Product (11.63) Where the Stereochemistry of the Methyl Groups Is Cis–?

When 1,2-dimethylcyclohexene (***11.62***) reacts with borane, the B and the H of the B–H bond MUST add in a *cis-* manner. Therefore, B and H in *11.63* are *cis-* (BR_2 represents the formation of a dialkyl or trialkylborane where R = dimethylcyclohexyl and/or hydrogen). The two methyl groups in ***11.62*** will be pushed away from the boron in the four-center transition state and will be *cis-* to each other. The boron and hydrogen are opposite to the methyl groups in ***11.63***.

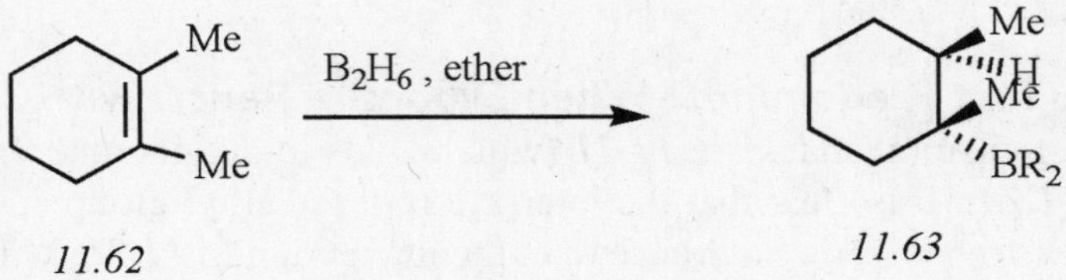

Me, Me — B_2H_6, ether → Me, H, Me, BR_2

11.62 *11.63*

Draw the Four-Center Transition State for the Conversion of 11.62 to 11.63.

This stereochemistry in ***11.63*** arises because the reaction proceeds by a **four-center transition state** such as ***11.64*** for addition of the borane to the π-bond.

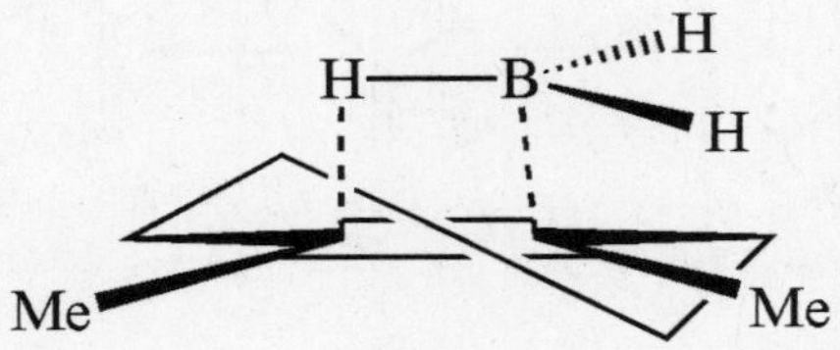

H—B, H, H, Me, Me

11.64

The Reaction of Borane and 1-Pentene (11.65) Can Give Two Possible Products.

The boron can attach to either the first carbon (path a) or the second (path b). The two products are ***11.66*** (via path a) and ***11.67*** (via path b). The major product is ***11.66*** (80–90% yield), and monosubstituted alkenes such as ***11.65*** *always* give attachment at the less sterically hindered carbon as the major product.

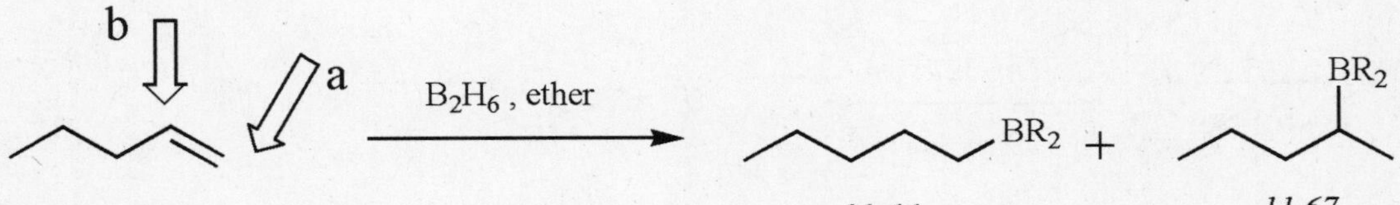

b, a — B_2H_6, ether → BR_2 + BR_2

11.65 *11.66* *11.67*

(R = 1-pentyl R = 2-pentyl)

Draw the Transition States for the Formation of 11.66 from 11.65 and Also That for 11.67 from 11.65. Use These Drawings to Rationalize Which Is the Major Product.

Comparison of transition state ***11.68*** (which leads to ***11.66***) with ***11.69*** (which leads to ***11.67***) reveals that the BH_2 moiety interacts sterically with the propyl group of ***11.65*** in transition state ***11.69***. In ***11.68***, however, the bulky BH_2 group interacts only with hydrogens. Since transition state ***11.68*** is less sterically hindered than ***11.69***, it is lower in energy and leads to the major product.

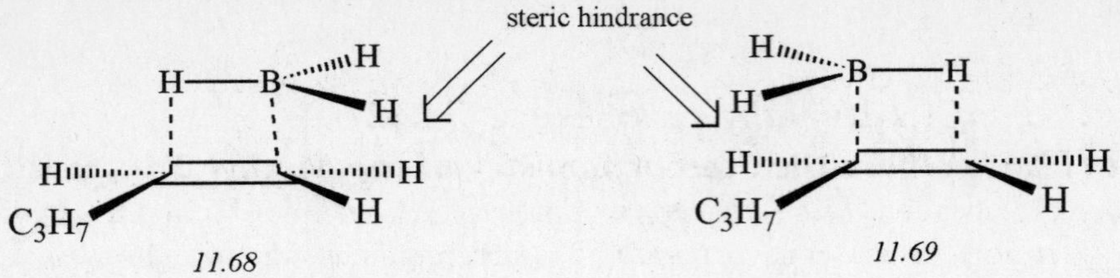

steric hindrance

11.68 11.69

What Is the Regiochemical Preference of the Reaction of 1-Butene and Diborane?

The major product is the one where boron is attached to the *less* substituted carbon (see ***11.66***), although about 15% of the other isomer (boron attached to the secondary carbon) is also produced.

Why Does One Product Predominate When Diborane Reacts with 2-Methyl-1-Butene?

The two possible transition states are ***11.71*** (which leads to alkylborane ***11.70***) and ***11.72*** (which leads to alkylborane ***11.73***). It is clear that the interaction of the ethyl group and methyl group and the BH_2 unit in ***11.72*** causes more steric hindrance than the interaction in ***11.71***. Minimization of this steric interaction drives the reaction to give the major product where boron is attached to the less substituted carbon, ***11.70***.

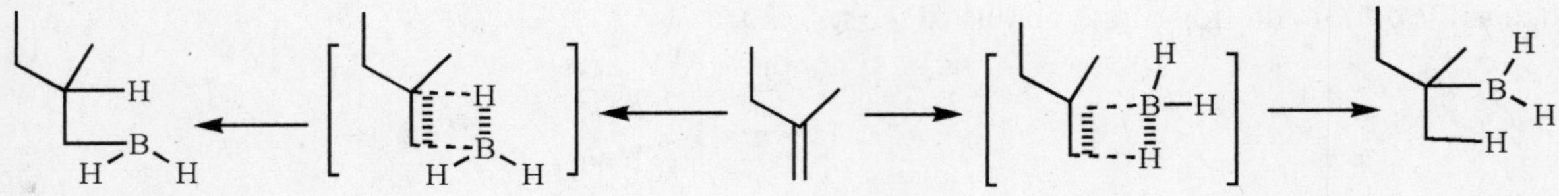

11.70 11.71 11.72 11.73

Give the Structure and Show a Reaction for the Preparation of Disiamylborane, Thexylborane, and 9-BBN.

When 2-methyl-2-butene is treated with diborane, a dialkylborane (***11.74***) is formed. This product is given the common name disiamylborane (siamyl = *sec*-isoamyl). When 2,3-dimethyl-2-butene is treated with diborane, a monoalkylborane (***11.75***) is formed. The common name of this product is thexylborane (thexyl = *tert*-hexyl). When 1,5-cyclooctadiene is treated with diborane, boron adds to one alkene to form an alkylborane, but then adds intramolecularly to the other alkene to give ***11.76***, 9-borabicyclo[3.3.1]nonane (9-BBN).

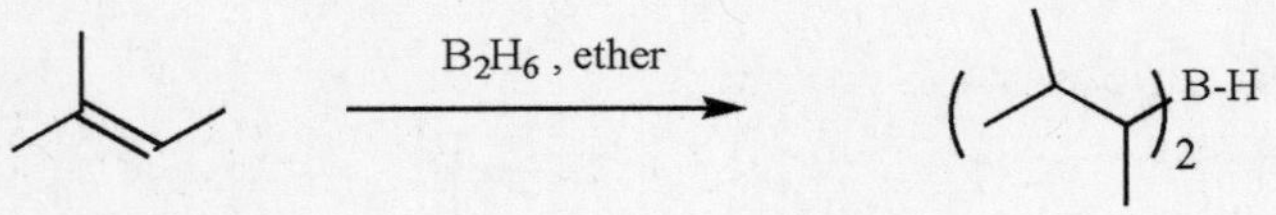

B_2H_6 , ether → 11.74

B_2H_6 , ether → 11.75

B_2H_6 , ether → 11.76

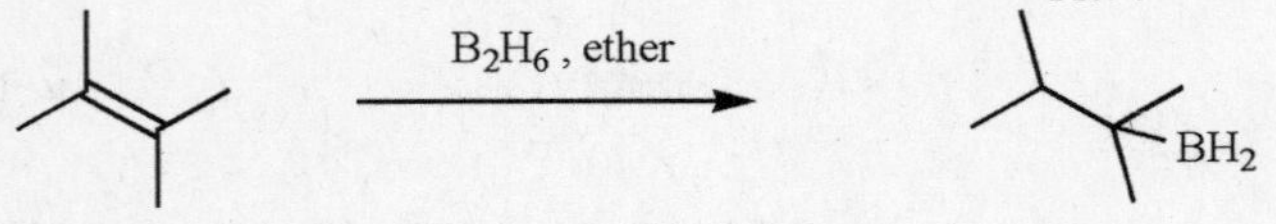

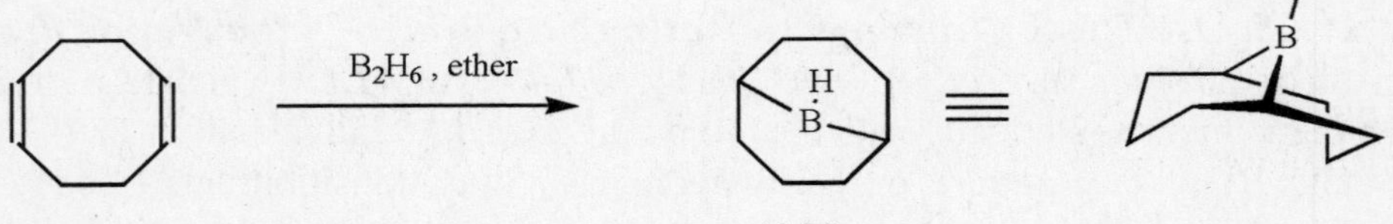

Explain Why 9-BBN Gives a Higher Percentage of the 1-Substituted Product When Compared to Diborane in a Reaction with 1-Butene.

Note the use of the abbreviated cartoon for ***11.76***, where the curved lines represent the bicyclic ring structure. Examination of the two possible transition states (***11.77*** and ***11.79***) for the reaction of 9-BBN and 1-butene shows that ***11.79*** is much more sterically hindered due to the bulky nature of the bridged ring system of 9-BBN. This leads to a much greater preference for ***11.77*** and formation of ***11.78*** as the major product rather than ***11.80*** (usually >99:1).

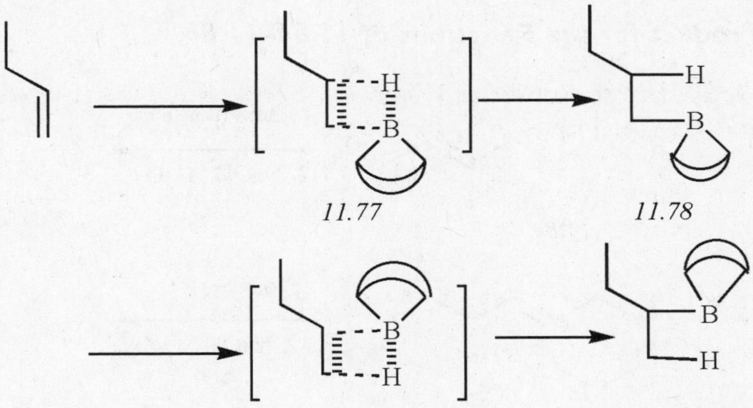

11.77 *11.78*

11.79 *11.80*

Oxidation of an Alkylborane to an Alcohol

The main application of hydroboration is the ability to convert the monoalkyl, dialkyl, and trialkylborane products into alcohols by treatment with basic hydrogen peroxide in an oxidation step. Many other transformations of organoboranes are also possible, but this section will focus only on the conversion to alcohols.

When the Borane Product Resulting from the Reaction of Cyclopentene and Diborane Is Treated with Sodium Hydroxide and Hydrogen Peroxide, What Is the Product?

The product is cyclopentanol. Boric acid [$B(OH)_3$] is also formed as a by-product.

What Is the Active Reagent When Sodium Hydroxide Is Mixed with Hydrogen Peroxide?

The initial reaction is between hydrogen peroxide and hydroxide to produce the hydroperoxide anion (HOO^-).

What Is the Mechanism of the Reaction of Cyclopentene and Diborane?

The hydroperoxide anion (HOO^-) attacks the boron of the tricyclopentylborane (***11.81***) generated from cyclopentene to form ***11.82***. A B→O carbon shift is accompanied by loss of hydroxide (^-OH) and leads to a so-called borinate (***11.83***). Two additional reactions of HOO*o* followed by the B→O alkyl shift produces boric acid [$B(OH)_3$] and three equivalents of the alcohol (cyclopentanol). In general, a trialkylborane leads to three equivalents of the alcohol, and a dialkylborane leads to two equivalents.

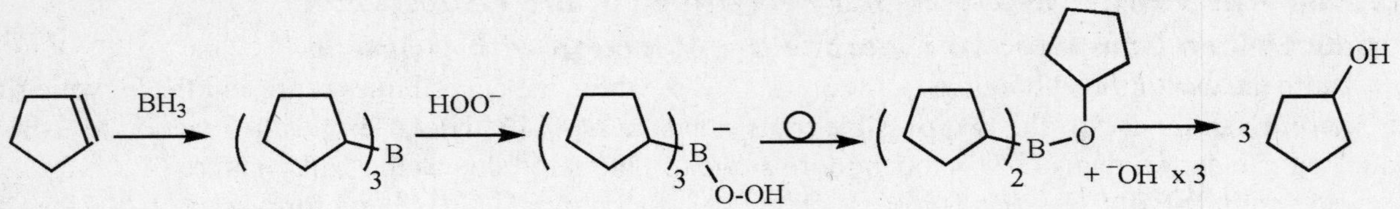

BHз HOO⁻ B 3 B 3 O-OH 2 B–O + ⁻OH x 3 3 OH

11.81 11.82 11.83

Give the Major Product for the Reactions of 11.84–11.86.

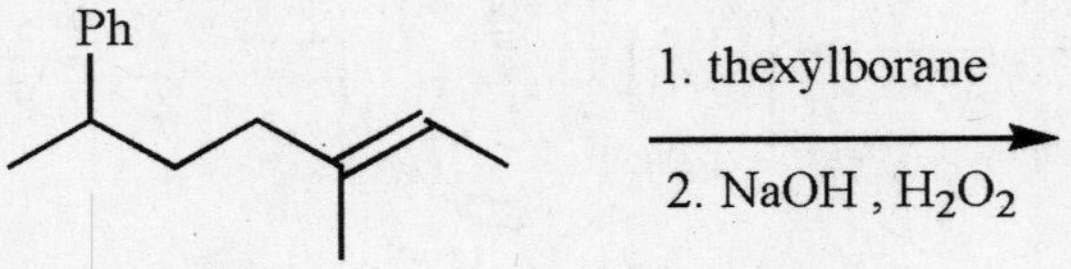

Ph 1. thexylborane 2. NaOH , H_2O_2

11.84

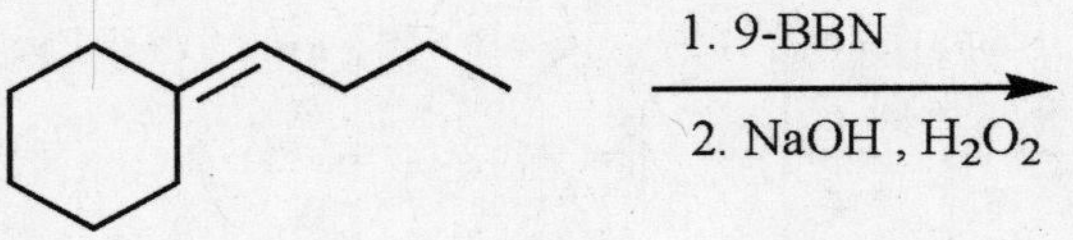

1. 9-BBN 2. NaOH , H_2O_2

11.85

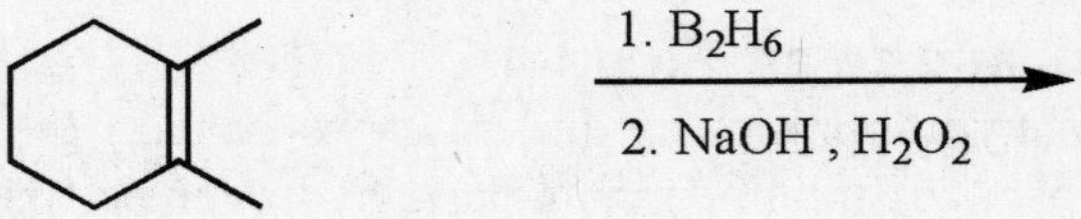

1. B_2H_6 2. NaOH , H_2O_2

11.86

In the first case, the product is the alcohol formed by 'delivery' of boron (and thereby OH after the oxidation step) to the less substituted carbon of ***11.84***. The product is 3-methyl-6-phenyl-2-heptanol (***11.87***). In the second reaction, ***11.85*** is converted to the less hindered carbon of the C=C unit to give secondary alcohol, ***11.88***. In the last case, alkene ***11.86*** (1,2-dimethylcyclohexene) is converted to ***11.89*** where the *cis* addition leads to the two methyl groups, *cis* to each other but *trans* to the OH.

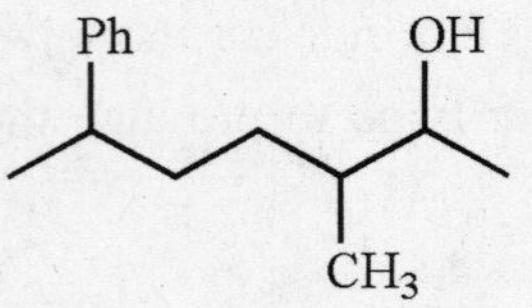

Ph OH CH_3 OH CH_3 CH_3 OH

11.87 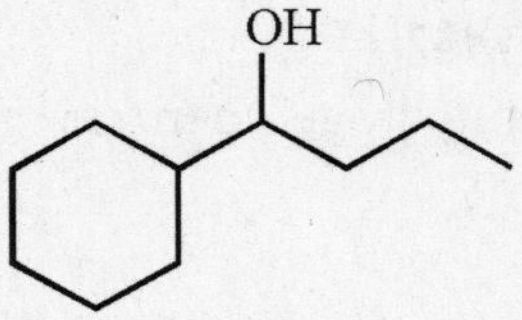 11.88 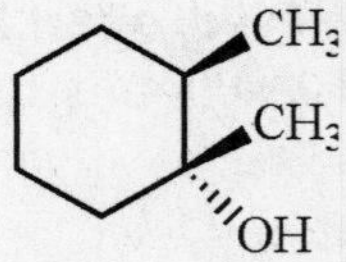 11.89

Radical Halogenation of Alkanes

This reaction is discussed in Section 7.6.

11.3. REACTIONS OF ALCOHOLS

Once an alcohol is formed, there are a number of important reactions that are possible. Most of these reactions involve the O-H bond.

List Several Reagents That Transform Alcohols into Alkyl Halides.

A standard method for the preparation of alkyl halides involves treating alcohols (R–OH, see Section 11.1) with a reagent that contains excess halogen. Typical chlorination reagents are thionyl chloride ($SOCl_2$), phosphorous trichloride (PCl_3), phosphorus oxychloride ($POCl_3$), phosgene ($Cl_2C{=}O$) and phosphorus pentachloride (PCl_5). The usual brominating agents are phosphorous tribromide (PBr_3) and thionyl bromide ($SOBr_2$). Reagents to make the iodide are often unstable and are prepared as needed by reaction of red phosphorus and iodine. Both HCl and HBr can be used, in many cases, to prepare the chloride or bromide, respectively.

The Reaction of Alcohols and Acids (HX)

What Is the Product of the Reaction Between Ethanol and HBr?

The products are bromoethane and water.

What Is the Product of the Reaction Between 3-Methyl-3-Pentanol and HCl?

The products are 3-chloro-3-methylpentane and water.

What Is the Mechanism for the Conversion of 2-Methyl-2-Pentanol (11.90) into 2-Chloro-2-Methylpentane (11.91)?

Alcohols function as a *base* in the presence of strong acids such as HCl or HBr. The alcohol (2-methyl-2-pentanol, ***11.90***, for example) is protonated by HCl to form an oxonium salt (***11.92***). This cation loses water (H_2O) to form carbocation ***11.93***. A carbocation is a tricoordinated carbon that is *planar* and obviously electron deficient. The cation will react with any species that can donate an electron pair. In this solution, both water and chloride ion (Cl^-) can provide electrons to the carbon. Remember, a species that *donates an electron pair to carbon* is called a **nucleophile**. A nucleophile literally means 'nucleus loving,' where the nucleus is a positive species. Cation ***11.93*** therefore reacts with the chloride counterion (produced during the reaction of the alcohol and HCl) to give 2-chloro-2-methylpentane (***11.91***). As we have seen in other chapters, the molecules in brackets are called **intermediates** and are *transient products* that are not isolated. The alcohol and the alkyl chloride are the *starting material* and the *product*, respectively.

What Is a Carbocation?

The cation intermediate (***11.93***) is reproduced as ***11.94***, where its planar nature is seen. As we discussed in previous chapters, a carbocation can be viewed as an *empty p orbital* (a cation, therefore no electrons). The nucleophilic chloride can attack from either the bottom or top of the planar cation to give ***11.91***. Carbocations are high energy species, making them unstable to isolation. The electropositive carbon of the cation will draw electrons from any available source to diminish the net charge and thereby increase its stability (less charge = more stable = lower energy). Carbon groups are capable of releasing electrons to an electropositive center. *Alkyl groups are electron-releasing when attached to the electron deficient C+.*

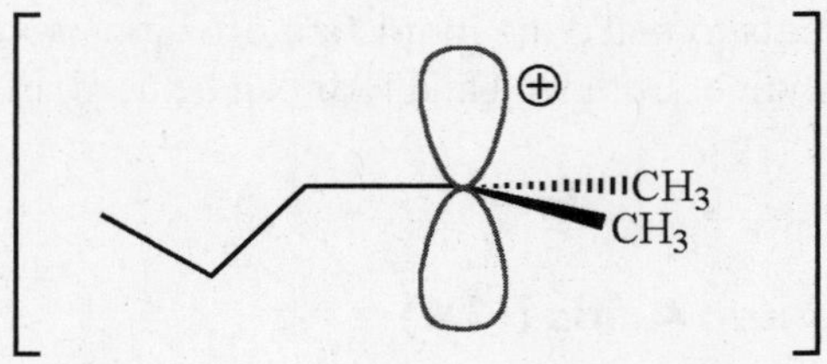

CH_3 CH_3

11.94

Why Does the Cation Produced from a Stereogenic Alcohol Lead to a Racemic Product?

If stereogenic alcohol ***11.95*** is treated with HCl, the cation that is produced (***11.96***) is planar, and the stereogenic center has been lost. The nucleophilic chloride ion can attack the cation from either the 'top' or the 'bottom,' as shown, to produce a mixture of two chlorides, ***11.97*** and ***11.98*** (a racemic mixture).

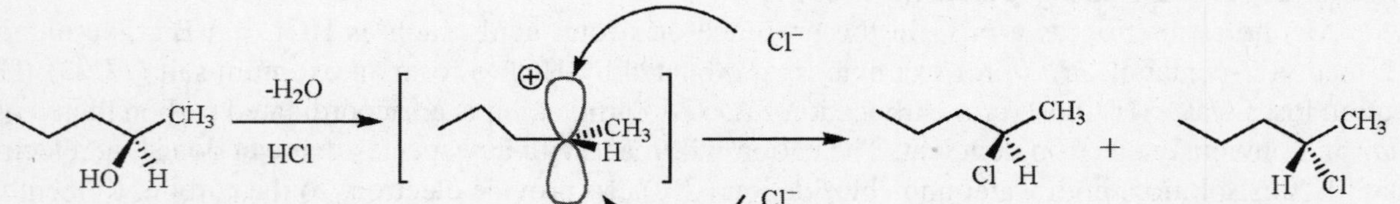

CH_3 HO H —$-H_2O$, HCl→ [CH_3 H] Cl^- Cl^- → CH_3 Cl H + CH_3 H Cl

11.95 *11.96* *11.97* *11.98*

Determine the Most Stable Carbocation (Methyl, 1°, 2° or 3°).

There are four types of simple alkyl cations, tertiary (three carbon groups [R = alkyl carbon group] on the positive carbon as in ***11.99***; secondary (two carbon groups on the carbon as in ***11.100***); primary (one carbon group on the carbon as in ***11.101***) and methyl (three hydrogens on the carbon, as in ***11.102***). Three carbons will release more electron density to the electropositive center, providing more stabilization than the electron density released by one carbon group in the primary cation. A tertiary cation will, therefore, *be more stable than either a secondary or a primary cation.* The order of cation stability is: tertiary > secondary > primary > methyl.

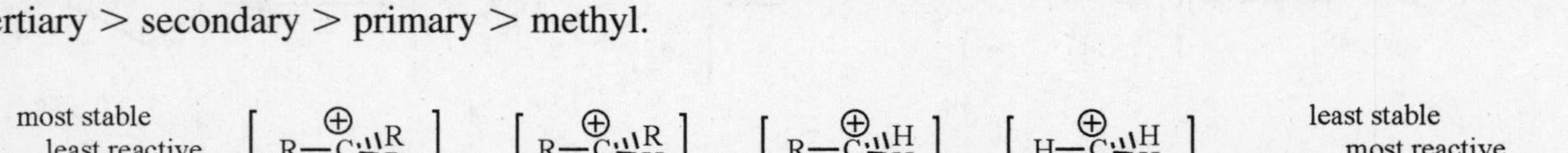

most stable
least reactive
easiest to form

[R—C⊕RR] [R—C⊕RH] [R—C⊕HH] [H—C⊕HH]

11.99 *11.100* *11.101* *11.102*

least stable
most reactive
most difficutl to form

What Is the Least Reactive Carbocation: Methyl, 1°, 2° or 3°?

Since the tertiary cation is the most stable, it will be the least reactive with the nucleophile. Conversely, the high energy (less stable) methyl cation will be the most reactive. This is misleading, however, in terms of the actual reactivity, since formation of a methyl cation requires significantly more energy than the tertiary (higher activation barrier) and is very difficult. The tertiary reacts better because it is relatively easy to form, whereas the primary cation does not react because it does not form. *If* a primary cation could form, however, it would be extremely reactive.

Primary Alcohols Are Converted to Halides upon Treatment with HX. Is the Mechanism of This Reaction the Same As Observed with Tertiary Alcohols?

No! the difference in energy between primary and tertiary cations is evident in the reaction of the primary alcohol 1-butanol with HCl. The initial reaction to form the oxonium salt (***11.103***) is the same as above. Loss of water to form a primary cation, however, requires too much in energy and does not occur. The chloride will instead attack the *carbon* connected to the $^{\oplus}OH_2$ species to form a new C–Cl bond and breaking the C–O bond. This is an S_N2 reaction (see Section 7.2). Polarization of the C–O bond shows the carbon to be electropositive, *but not cationic*. Such a carbon is termed *electrophilic* (literally 'electron loving') and will be attacked by nucleophiles. Displacement by chloride leads to the product, 1-chlorobutane, ***11.104***. The difficulty in forming a primary cation leads to a different mechanism for this reaction when compared to the tertiary alcohol. It is noted that methanol (CH_3OH) reacts with HCl to give chloromethane ($ClCH_3$) by an identical mechanism and that secondary alcohols will give products resulting from ***both*** mechanisms (cationic and direct displacement at carbon).

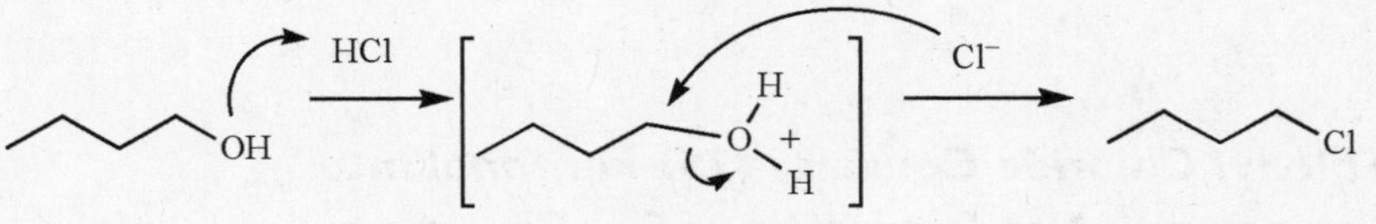

OH → (HCl) → [11.103] → (Cl⁻) → Cl

11.103 *11.104*

Both HBr and HI react by a similar mechanism to give alkyl bromides and alkyl iodides. The reaction with HI is not as efficient, however. Substitution of OH with chloride ion via a cation (***11.90*** → ***11.91***) is called **Unimolecular Nucleophilic Substitution** (S_N1), and direct substitution by Cl (butanol → *11.104*) is called **Bimolecular Nucleophilic Substitution** (S_N2). These reactions are introduced in Chapter 7.

The Reaction of Alcohols with Sulfur and Phosphorus Halides

Give the Structure of Thionyl Chloride, Thionyl Bromide, Phosphorus Trichloride, Phosphorus Pentachloride, and Phosphorus Tribromide.

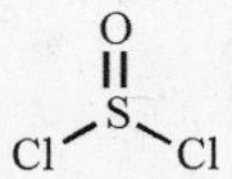

= $SOCl_2$ *11.105*

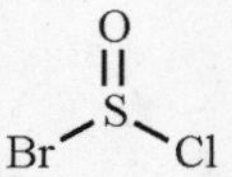

= $SOBr_2$ *11.106*

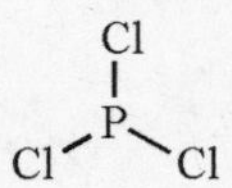

= PCl_3 *11.107*

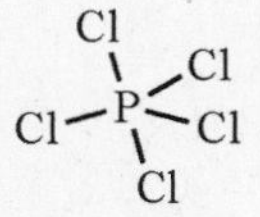

= PCl_5 *11.108*

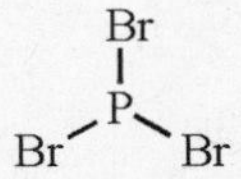

= PBr_3 *11.109*

What Is the Product of a Reaction Between Cyclopentanol and Thionyl Chloride?

Thionyl chloride reacts with cyclopentanol to give chlorocyclopentane, ***11.110***.

What Is the Product of a Reaction Between 2-Methyl-2-Pentanol and Phosphorus Tribromide?

Phosphorus tribromide reacts with 2-methyl-2-pentanol to give 2-bromo-2-methylpentane, ***11.111***.

What Is the Product of a Reaction Between 1-Hexanol and Thionyl Bromide?

Thionyl bromide reacts with 1-hexanol to give 1-bromohexane, ***11.112***.

Explain How Thionyl Chloride Converts (2S)-Pentanol into (2S)-Chloropentane, with Net Retention of Configuration.

Thionyl chloride ($SOCl_2$) reacts with primary, secondary and tertiary alcohols to give the corresponding chloride. Thionyl bromide ($SOBr_2$) reacts similarly to give the primary, secondary or tertiary bromide. There are two versions of the reaction with thionyl chloride, but *not* with thionyl bromide, with and without base. When an amine such as triethylamine is added, it functions as the base (see Sections 6.1 and 27.4), and the mechanism is different from the reaction without a base. In the first version, (2S)-pentanol (***11.113***) reacts with thionyl chloride, as shown, to give chloride ***11.115***. The alcohol oxygen attacks the sulfur, displacing a chloride, forming a sulfinate ester (***11.114***) and HCl. This molecule decomposes by losing SO_2 (sulfur dioxide), and the Cl is delivered to the carbon bearing the oxygen, *intramolecularly*, to give (2S)-chloropentane (***11.115***). This decomposition is known as an **S_Ni reaction (nucleophilic substitution, internal)**. Note that the Cl in ***11.115*** is on the same side as the OH in ***11.113***.

When Triethylamine Is Added to the Reaction of (2S)-Pentanol (11.113) and Thionyl Chloride, the Product Is (2R)-Chloropentane (11.117). Why?

When an amine base (triethylamine, NEt_3) is added to this reaction, ***11.113*** reacts with thionyl chloride as before to give ***11.114*** and HCl. The amine, which is a base, reacts with HCl to form triethylammonium hydrochloride, ***11.116***. The formation of ***11.116*** means there is a nucleophilic chloride in the reaction medium. Therefore, a simple explanation is that an intramolecular decomposition leading to ***11.115*** is slower that the reaction of ***11.114*** with the chloride ion *intermolecularly* to give (2R)-chloropentane (***11.117***), where the chloride is on the opposite side of the molecule from the OH in ***11.113***.

O
S
Cl Cl
OH
H
- HCl
H
O
Cl
S
O
Cl
H
11.113
11.114
11.117
Et_3N-H+ Cl^-
Et_3N:
Cl delivered on side opposite O
11.116

The Williamson Ether Synthesis

Why Is the Hydrogen of the O–H Bond Considered to Be Acidic?

The bond polarization of the O–H bond makes the hydrogen electrophilic ($\delta\oplus$). Since an 'acid' is considered to be a proton, an electropositive hydrogen is 'close to being an acid.'

What Is the pK_a of an Alcohol Proton?

Most alcohols have a pK_a of about 16–18. The acidic hydrogen of the O-H group in phenol has a pK_a of about 10. Water has a pK_a of 15.8.

What Type of Base Is Required to Remove the Hydrogen of an O–H Bond?

Most of the common bases will remove the acidic hydrogen of an alcohol. All organolithium reagents and Grignard reagents will remove that hydrogen. Sodium hydroxide and alkoxide bases such as sodium ethoxide will remove the proton. Sodium hydride (NaH) and amide bases (NR_2^-) are also suitable.

What Is the Conjugate Base of an Alcohol Called? Is It a Strong Base or a Weak Base?

The conjugate base of an alcohol is an alkoxide. Since alcohols are relatively weak acids, alkoxides are relatively strong bases. Alkoxides are somewhat stronger bases than hydroxide, but much weaker than carbanion bases such as organolithium reagents.

Classify Sodium Methoxide As a Reagent.

The oxygen in sodium methoxide is classified both as a base and a nucleophile.

What Is the Product When Sodium Methoxide Reacts with 1-Iodoethane?

In reactions with alkyl halides such as iodoethane, methoxide is a nucleophile (Na is the positive counterion) and will react at the electropositive carbon that bears the iodine. The product of this reaction is an ether, ethyl methyl ether ***11.118***.

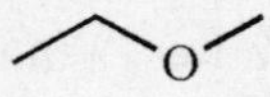

11.118

What Type of Reaction Is This?

This is a bimolecular substitution reaction, S_N2.

What Is the Common Name of This Reaction?

The **Williamson Ether Synthesis**.

What Is the Product When Potassium tert-Butoxide Reacts with 1-Iodoethane?

The product of the nucleophilic *tert*-butoxide and iodoethane is ethyl *tert*-butyl ether, ***11.119***.

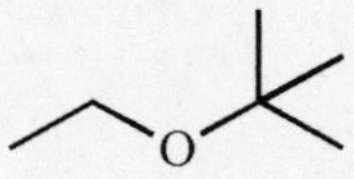

11.119

What Is the Product When Sodium Ethoxide Reacts with 2-Bromo-2-Methylbutane?

The nucleophile is ethoxide, but the electrophilic center is a tertiary halide. The S_N2 reaction will not occur at a tertiary center. Since ethoxide is also a base, removal of a β-hydrogen from 2-bromo-2-methylbutane will lead to an E2 reaction, and the final product will be 2-methyl-2-butene.

Give the Major Products of the Reactions of 11.120–11.122.

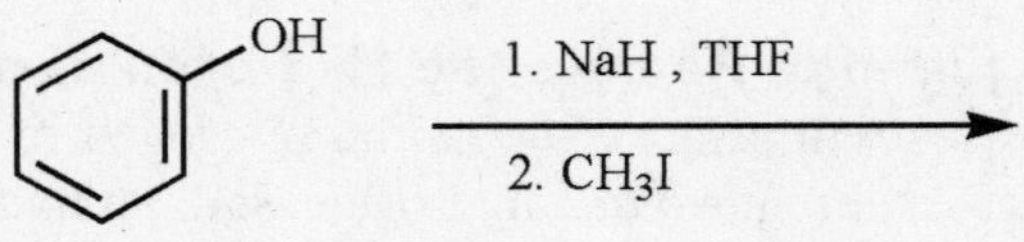

OH

1. NaH , THF

2. CH_3I

11.120

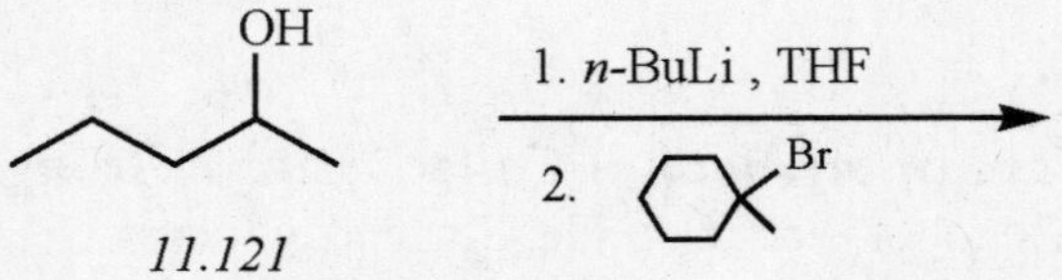

OH

1. *n*-BuLi , THF

2. Br

11.121

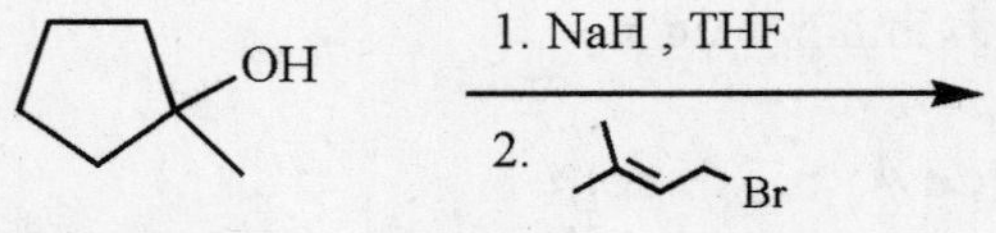

OH

1. NaH , THF

2. Br

11.122

In the first reaction, phenol reacts with NaH to form the resonance stabilized phenolic anion, a weak nucleophile. This nucleophile can react with iodomethane, however, to produce anisole (***11.123***). In the second reaction, *n*-butyllithium deprotonates 2-pentanol to form the alkoxide, ***11.124***. While the tertiary halide cannot undergo an S_N2 reaction, an E2 reaction is possible, and the products after addition of water to quench the strong base are the starting alcohol (2-pentanol) and methylcyclohexene (***11.125***). In the last reaction, deprotonation of 1-methylcyclopentanol generates an alkoxide, which reacts with crotyl bromide to give the ether, ***11.126***.

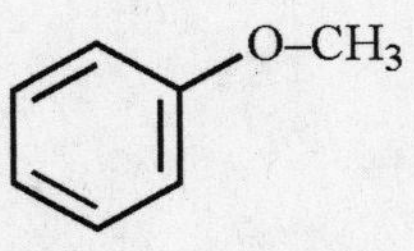

O–CH_3 O$^-$ O

11.123 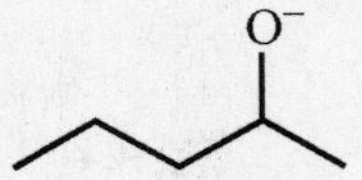 *11.124* 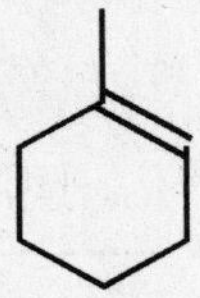 *11.125* 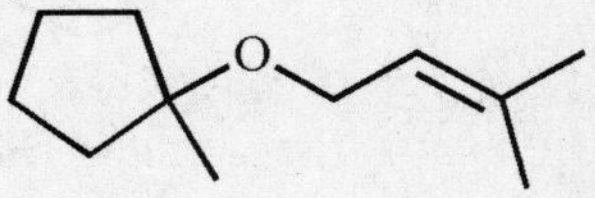 *11.126*

Test Yourself

1. Draw Lewis dot formulas for (a) phosgene, (b) thionyl bromide.
2. Give the complete mechanism for each of the following.

(a) OH → HBr → Br

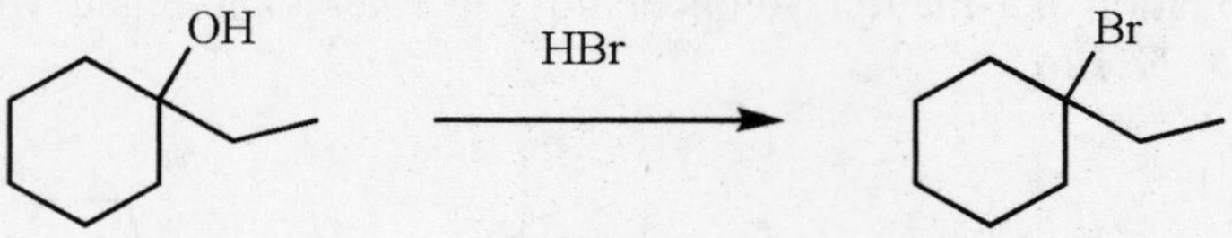

(b) OH → HCl → Cl (racemic)

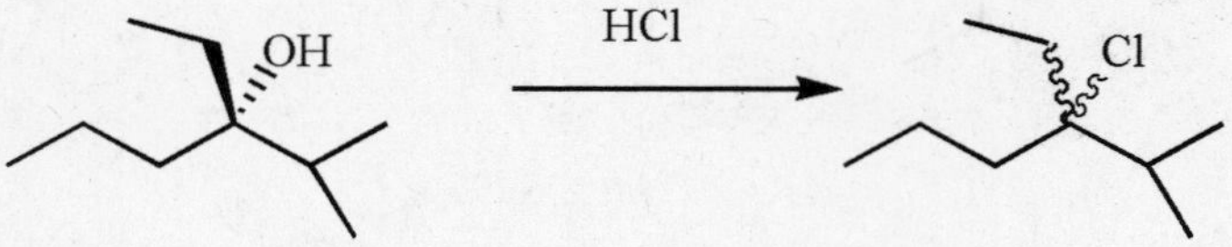

3. What is the intermediate, if any, when 1-pentanol is treated with HCl under anhydrous conditions? What is the product, if any?
4. What is the intermediate, if any, when 3-pentanol is treated with HCl under anhydrous conditions? What is the product, if any?
5. Give the major product for the reactions of *A-D*.

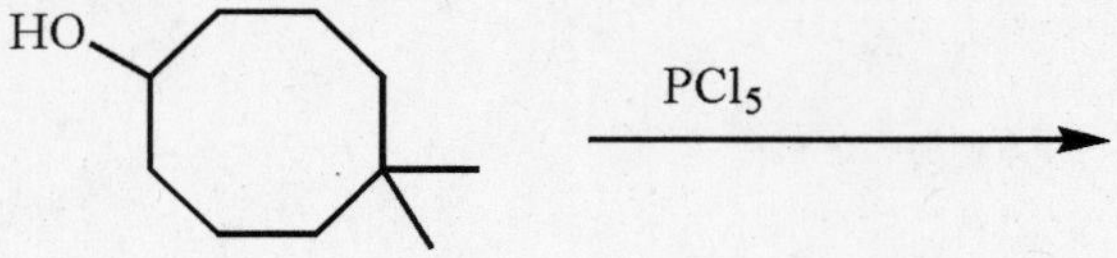

HO → PCl_5 →

A

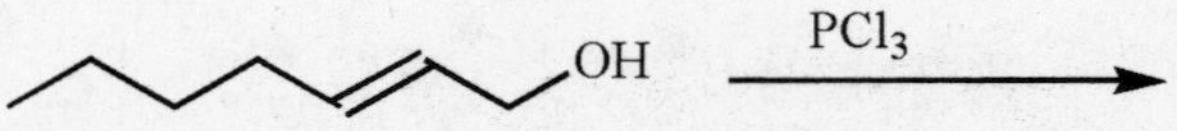

OH → PCl_3 →

B

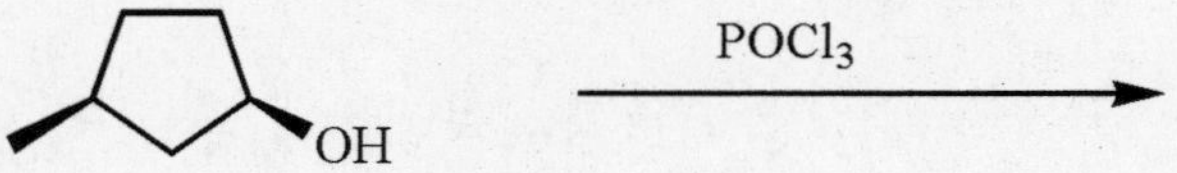

OH → $POCl_3$ →

C

OH → PBr_3 →

D

6. Explain why this reaction gives a mixture of enantiomeric chlorides, but there is a slight preponderance of the inversion product.

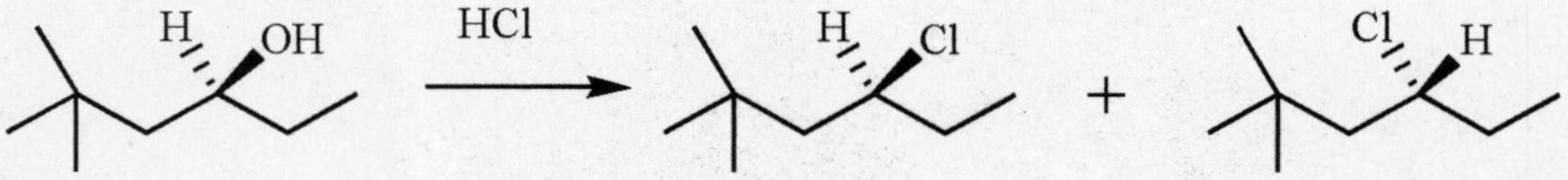

H OH → HCl → H Cl + Cl H

7. What is the major product in the reactions of *A-C*?

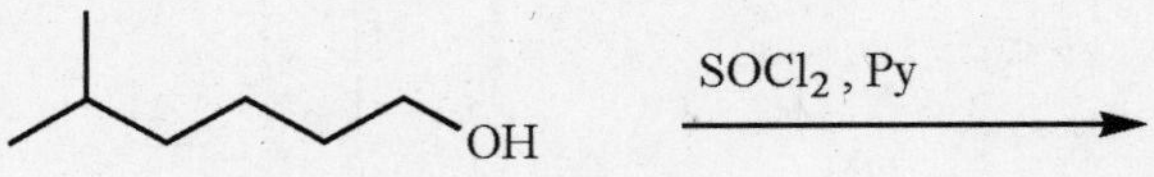

OH $\xrightarrow{SOCl_2, Py}$

A

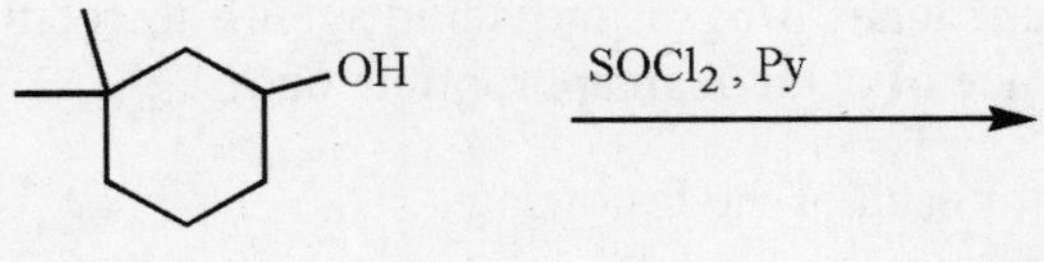

OH $\xrightarrow{SOCl_2, Py}$

B

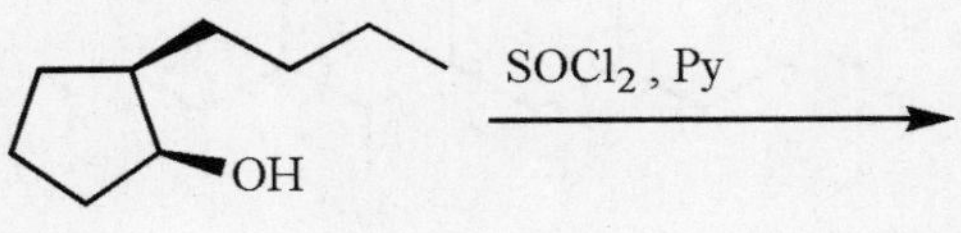

OH $\xrightarrow{SOCl_2, Py}$

C

8. Explain why ethanol cannot be used as a solvent for the reaction of ethylmagnesium bromide and 2-butanone.

9. In each case, give the major product of the reaction. Remember stereochemistry, and if there is no reaction, indicate this by N.R.

(a) OH $\xrightarrow{HBr}$

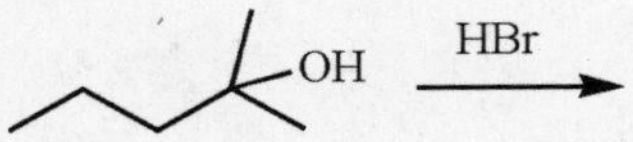

(b) OH $\xrightarrow{HCl}$

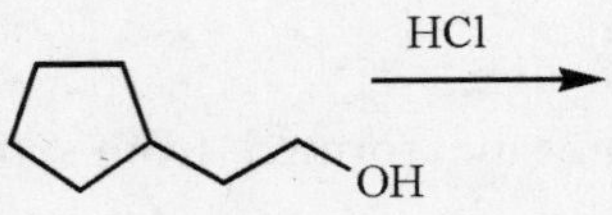

(c) H, H, OH $\xrightarrow[\text{pyridine}]{SOCl_2}$

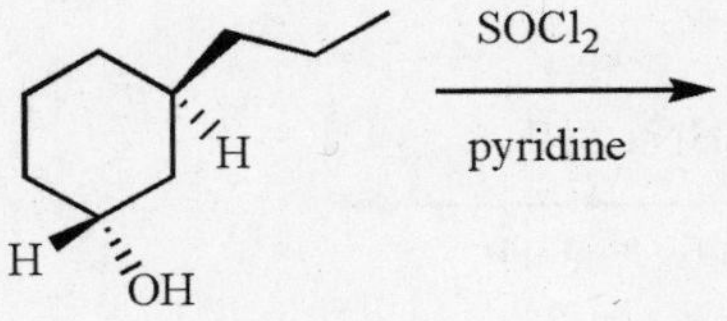

(d) OH $\xrightarrow{PBr_3}$

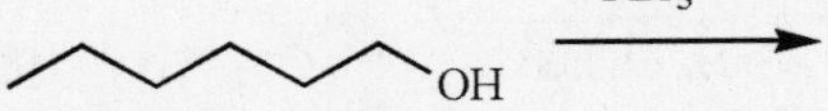

(e) OH $\xrightarrow{HBr}$

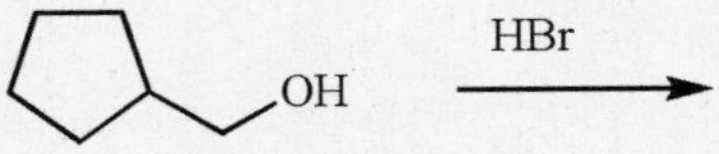

(f) Cl — 1. Mg(0), ether; 2. 2-butanone; 3. H_3O^+

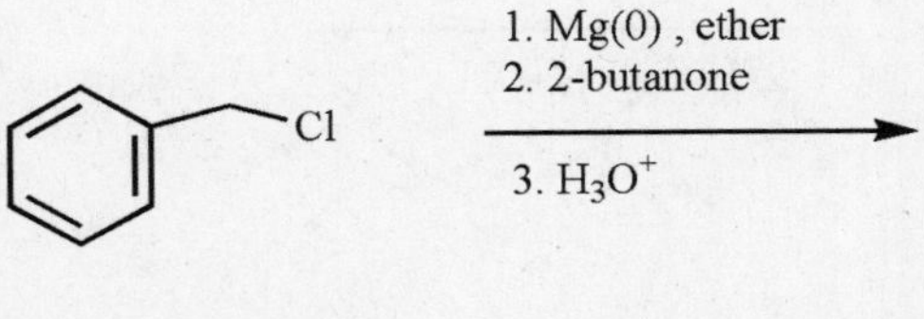

(g) O — 1. PhMgBr, THF; 2. H_3O^+

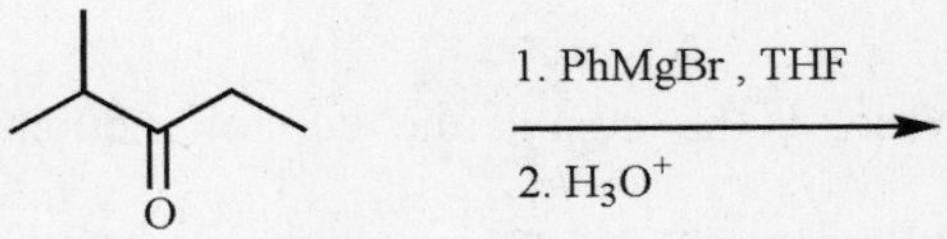

(h) O, H — 1. MgCl, ether; 2. H_3O^+

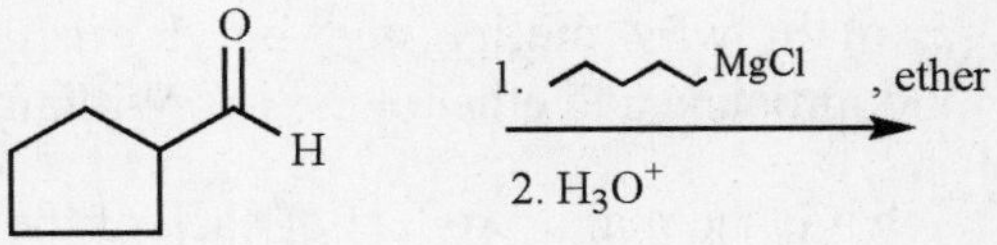

(i) I — 1. Li(0), ether; 2. propanal; 3. H_3O^+

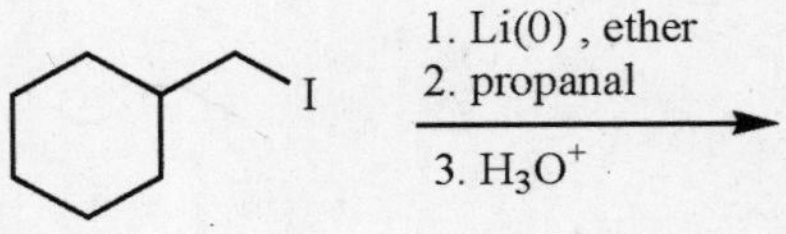

(j) OH $\xrightarrow{Li}$

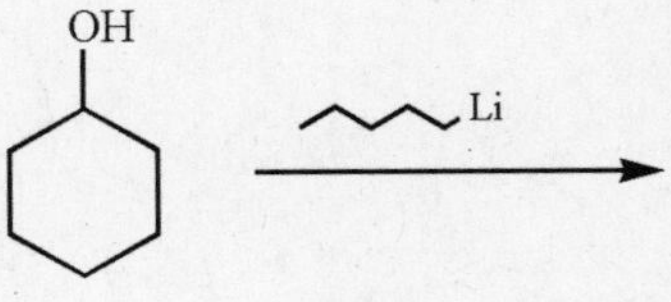

10. Which of the following is likely to have the higher boiling point? Explain.
    (a) pentene
    (b) 1-pentanol
    (c) dodecane
    (d) dodecanol

11. The lipophilic character of many drugs is measured against its relative solubility in an octanol/water mixture. Why is the choice of octanol an appropriate one?

12. Give the IUPAC name for each of the following.

(a)
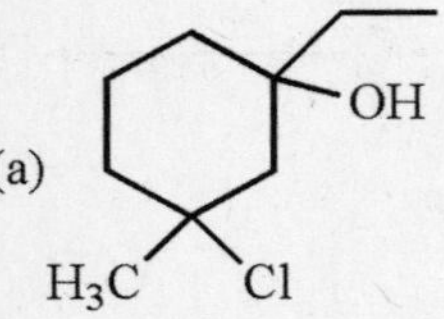

(b)
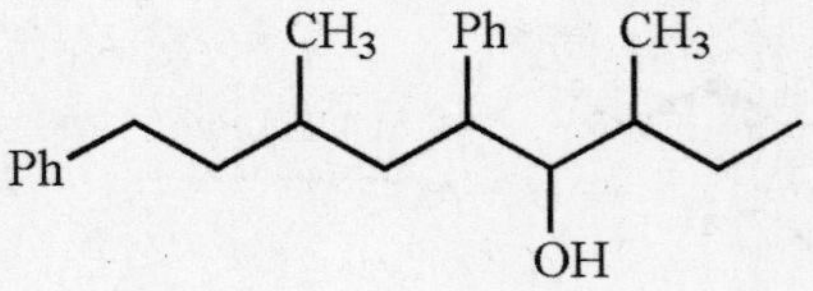

(c)
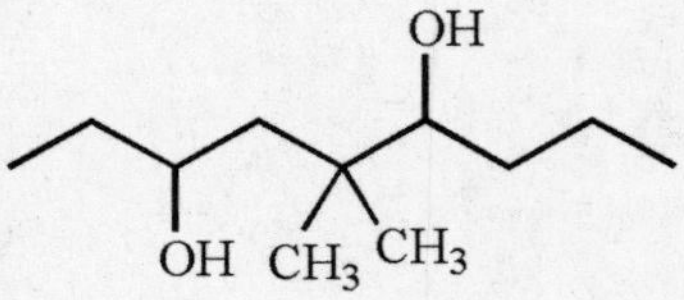

(d)
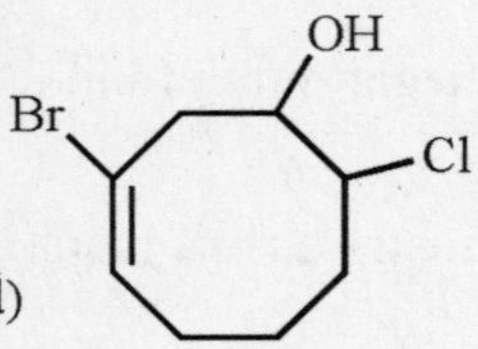

(e)
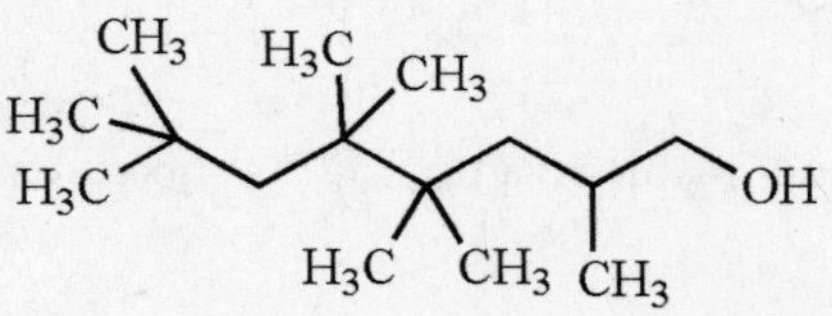

(f)
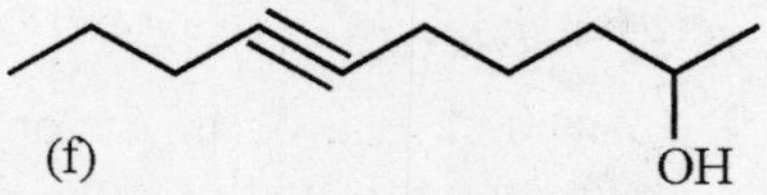

13. Which is the most acidic alcohol: methanol of *tert*-butanol? Explain.

14. Is the sodium salt of 2-propen-1-ol resonance stabilized? Explain.

15. Explain why methyl *tert*-butyl ether [$CH_3OC(CH_3)_3$] cannot be formed from 2-methyl-2-iodopropane and ethanol using a Williamson ether synthesis.

16. What is the major product of each of the reactions of *A*-*C*?

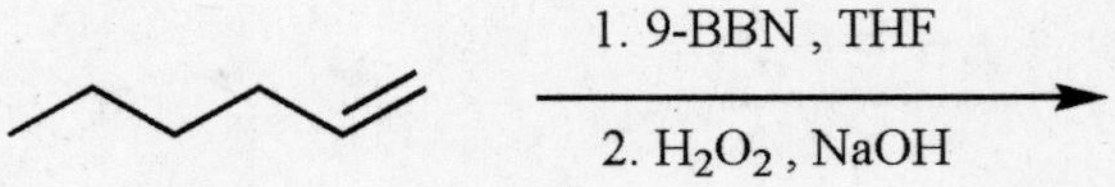

A

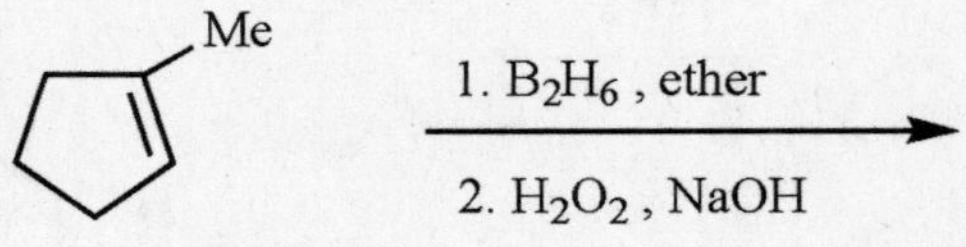

B

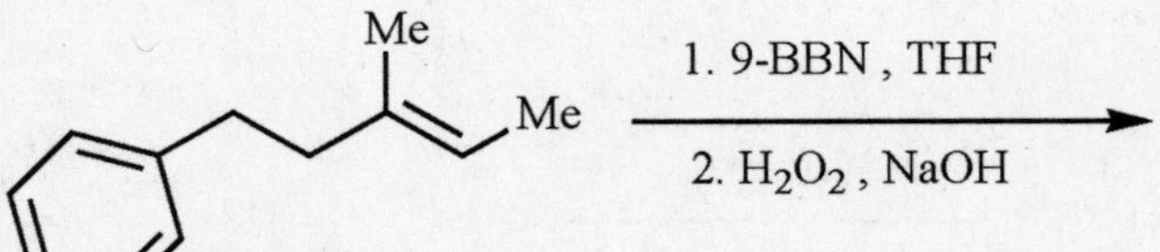

C

17. For each of the following reactions, give the major product. Remember stereochemistry where appropriate, and if there is no reaction, indicate this by N.R.

(a) 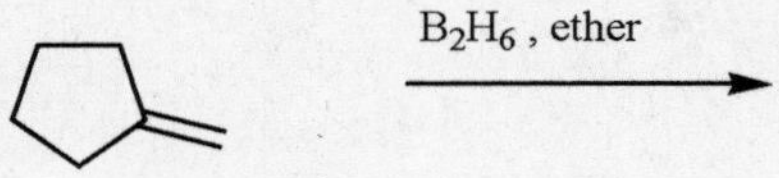 B_2H_6, ether

(b) Me; 1. 9-BBN, ether; 2. NaOH, H_2O_2

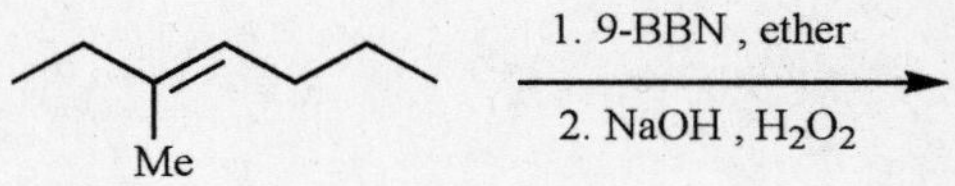

(c) 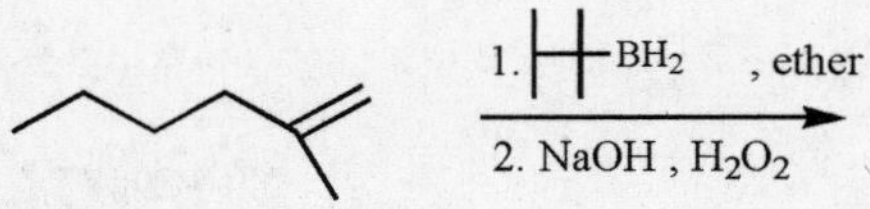 1. BH_2, ether; 2. NaOH, H_2O_2

(d) 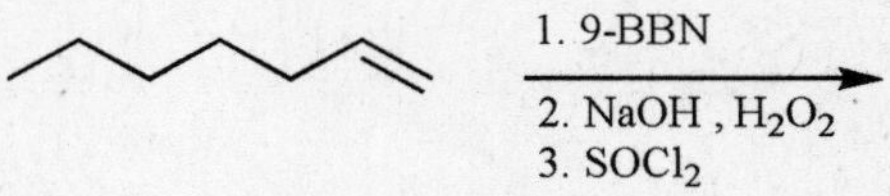 1. 9-BBN; 2. NaOH, H_2O_2; 3. $SOCl_2$

(e) 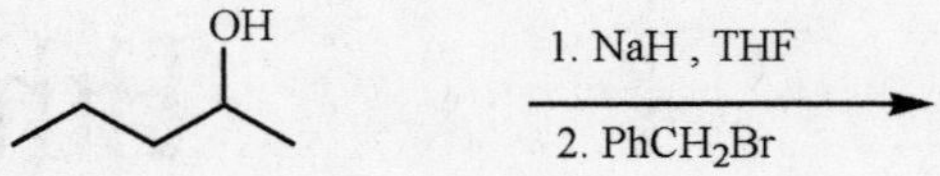 OH; 1. NaH, THF; 2. $PhCH_2Br$

(f) Me Me; OH; aq. HBr

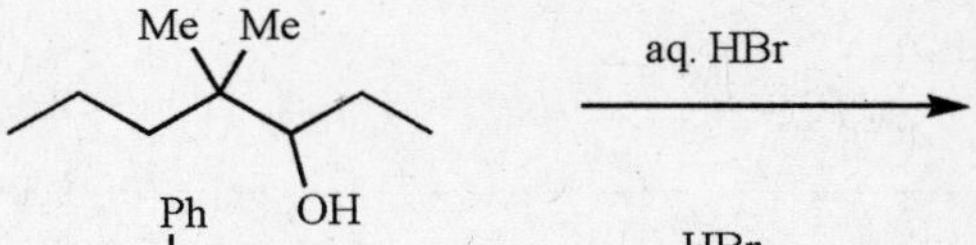

(g) 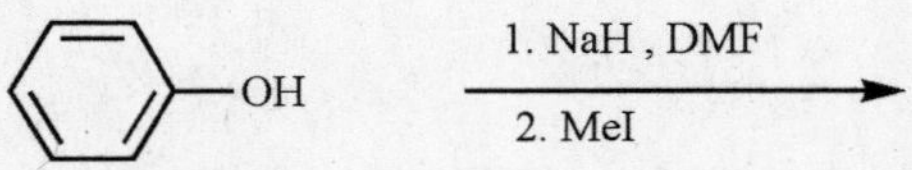 OH; 1. NaH, DMF; 2. MeI

(h) Ph; OH; Me Me; HBr

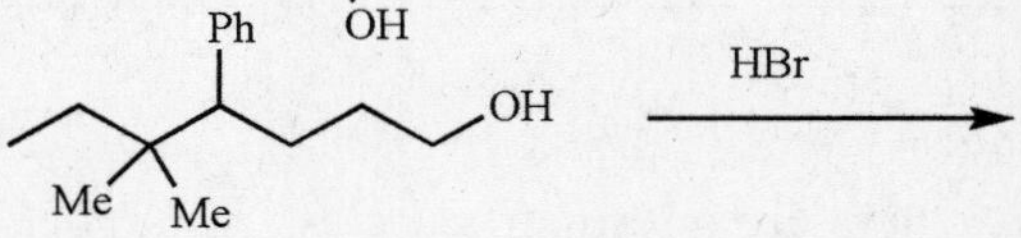

(i) 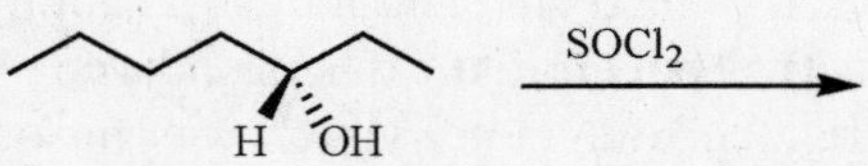 H OH; $SOCl_2$

(j) 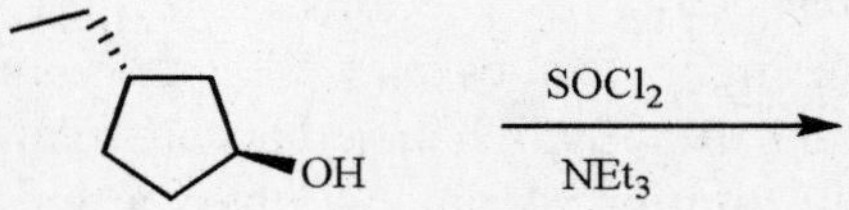 OH; $SOCl_2$; NEt_3

(k) 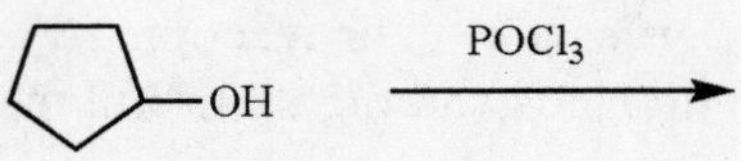 OH; $POCl_3$

(l) 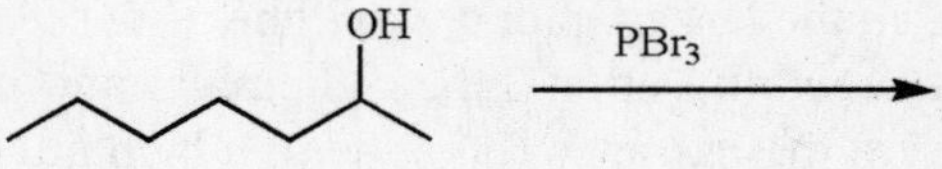 OH; PBr_3

(m) 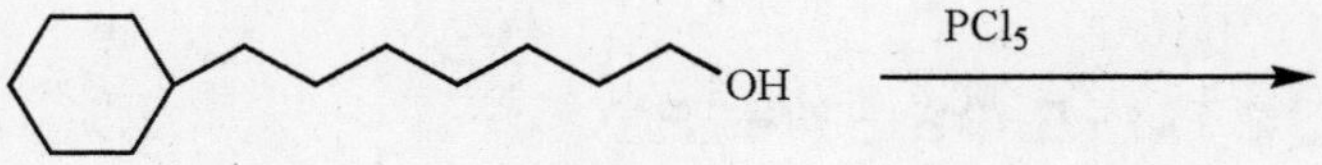 OH; PCl_5

18. In each case, give the major product of the reaction. Remember stereochemistry, and if there is no reaction, indicate this by N.R.

(a) 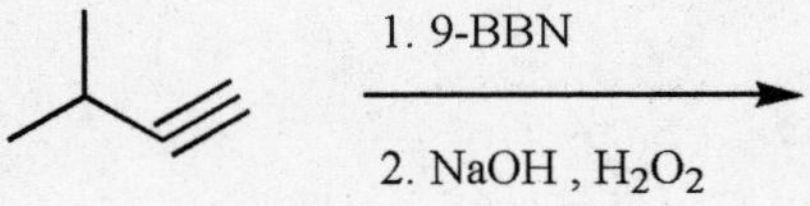 1. 9-BBN; 2. NaOH, H_2O_2

(b) 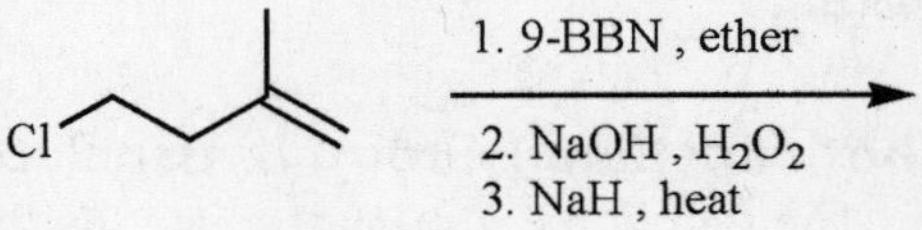 Cl; 1. 9-BBN, ether; 2. NaOH, H_2O_2; 3. NaH, heat

19. Give the mechanism for the conversion of an alkylborane *A* to the alcohol *B*.

H; B; H; HOOH/NaOH; OH

A *B*

CHAPTER 12

Ethers

Ethers are an important class of molecules for two reasons. Simple ethers are used as solvents in many organic reactions. Ethers are also found as structural components of many naturally occurring molecules, as well as those molecules used in medicine and industrial applications. As a class of compounds, aliphatic ethers are relatively unreactive, but aromatic ethers undergo electrophilic aromatic substitution. Aromatic ethers are chemically related to anisole, which was introduced with ***11.123*** in the last chapter and will be discussed in Chapter 16. This chapter will discuss the various methods of preparing ethers, their chemical and physical properties and their reactions. The most reactive ethers are the oxiranes (epoxides), which are common compounds and highly useful in chemical transformations.

12.1. STRUCTURE OF ETHERS

Ethers are organic compounds characterized by a C–O–C bond.

What Is an Ether?

An ether is a molecule containing an oxygen, but it does not have the OH for C=O functional groups.

What Functional Group Is Associated with an Ether?

The functional group associated with an ether is the C–O–C moiety.

What Is an Alkyl Ether?

An alkyl ether is an ether with two alkyl groups on either side of the oxygen ($R–O–R^1$).

Give Two Different Examples of Alkyl Ethers.

Two examples are ethylmethyl ether ($CH_3CH_2OCH_3$) and diethyl ether ($CH_3CH_2OCH_2CH_3$).

What Is the Structure of Tetrahydrofuran?

Tetrahydrofuran is a cyclic ether where the oxygen is part of a five-membered ring (***12.1***). This is also called oxolane.

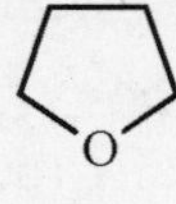

12.1

Give Examples of Cyclic Ethers in a Four-Membered Ring, Six-Membered Ring and Seven-Membered Ring.

The four-membered ring ether is oxetane (*12.2*), the six-membered ring ether is oxane (also called tetrahydropyran, *12.3*) The seven-membered ring ether is *12.4*.

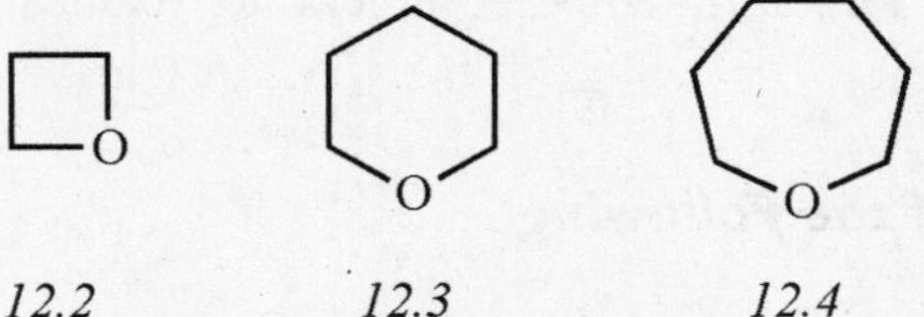

12.2 *12.3* *12.4*

What Is a Three-Membered Ring Ether Called?

An oxirane, but it is also called an epoxide.

What Is an Aryl Ether?

An aryl ether has an aryl group as at least one of the carbon groups attached to oxygen.

Give Two Different Examples of Aryl Ethers.

Two examples are anisole ($PhOCH_3$) and diphenyl ether (PhOPh).

Give Two Different Examples of Arylalkyl (Aralkyl) Ethers.

The term aralkyl ether is used to describe an ether with one aryl group and one alkyl group, as in anisole. Another example is phenylpropyl ether ($PhOCH_2CH_2CH_3$).

12.2. NOMENCLATURE

Ethers are named in two ways. In the first, the two alkyl units connected to the oxygen are named, followed by the word *ether*. In the second method, one RO unit is named as an alkoxy group, attached to the longest chain alkane, alkene, or alkyne unit (the 'other' R group).

What Is the Nomenclature Suffix Associated with an Ether?

Ethers are usually named as 'ethers' or as alkoxy derivatives.

Name the Following Compounds Using the 'Ether' Nomenclature System.

a) $CH_3CH_2CH_2CH_2OCH_2CH(CH_3)_2$ ***b)*** $PhOC(CH_3)_3$ ***c)*** $(CH_3)_2CHCHOCH_2CH_3$

Ether (a) is named *n*-butyl isobutyl ether as a common name or 1-butyl-3-methylpropyl ether. Ether (b) is phenyl *tert*-butyl ether or phenyl 1,1,-dimethylethyl ether. Ether (c) is ethyl isopropyl ether or ethyl 1-methylethyl ether.

Describe the 'Alkoxy' Nomenclature System for Ethers.

The RO group is an alkoxy group. $ROCH_2CH_2CH_3$ could, therefore, be called a 1-alkoxypropane.

Name the Following Compounds Using the 'Alkoxy' Nomenclature System.

a) $CH_3CH_2CH_2CH_2OCH_2CH(CH_3)_2$ b) $PhOC(CH_3)_3$ c) $(CH_3)_2CHCHOCH_2CH_3$

Ether (a) is 1-butoxy-3-methylpropane. Ether (b) is 1-phenoxy-1,1,-dimethylethane, and (c) is 1-ethoxy-1-methylethane. Note that the IUPAC names for the cyclic ethers (oxetane, oxolane and oxane) were given above.

What Is the Nomenclature System for Cyclic Ethers?

In general, cyclic ethers are named using a base name indicated by the size of the ring and the presence of the oxygen. Three-membered ethers are oxiranes, four-membered ethers are oxetanes, five-membered ethers are oxiranes, and six-membered ethers are oxanes. Substituents are given numbers based on oxygen = 1.

Give the Name for Each of the Following.

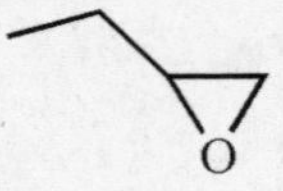

O O O Ph Cl O

12.5 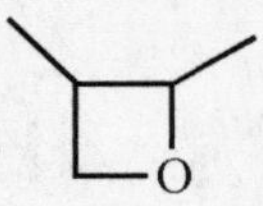 *12.6* 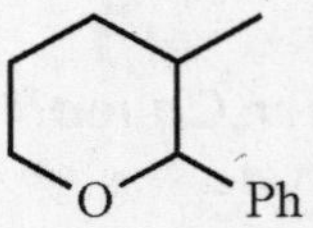 *12.7* 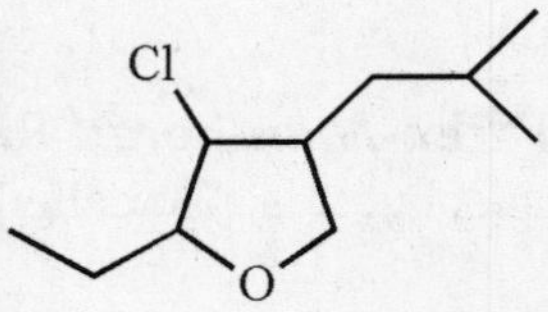 *12.8*

Compound ***12.5*** is 2-ethyloxirane; ***12.6*** is 2,3-dimethyloxetane; ***12.7*** is 3-methyl-2-phenyloxane; ***12.8*** is 3-chloro-2-ethyl-4-(2-methylpropyl)oxetane.

12.3. PREPARATION OF ETHERS

There are two main preparation of ethers, the Williamson ether synthesis and alkoxymercuration.

What Is the Name Given to the Reaction That Forms Ethers by S_N2 Reaction of Alkoxides with Alkyl Halides?

This reaction is called the Williamson ether synthesis.

What Is the Product When 2-Butanol is Treated with Sodium Hydride? With Methyllithium?

The reaction of 2-butanol and sodium hydride gives the sodium salt of 2-butoxide (***12.9***), and the reaction with methyllithium gives the lithium salt of 2-butoxide (***12.10***).

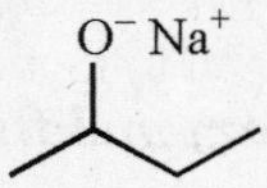

$O^- Na^+$ $O^- Li^+$

12.9 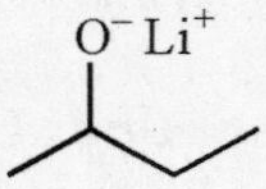 *12.10*

What Is the Major Product When 1-Butanol Is Treated First with NaH and Then with Iodomethane?

The product after reaction with NaH is sodium 1-butoxide ($CH_3CH_2CH_2CH_2O^- Na^+$). The oxygen of this product is a good nucleophile, reacting with the carbon of iodomethane by an S_N2 reaction to give the final product, an ether - methyl butyl ether.

Describe a Synthetic Route to Ethyl tert-Butyl Ether from Molecules Containing Four Carbons or Less.

The synthesis requires the use of *tert*-butanol and its conversion to an alkoxide (with NaH), followed by reaction with bromoethane (iodoethane is more reactive) to give the product, ***12.11***. It is not possible to begin with ethanol and make sodium ethoxide, since the subsequent reaction with the tertiary halide will give elimination rather than substitution.

1. NaH , THF
2. EtBr

12.11

Is It Possible to Prepare Ethers via an S_N1 Mechanism. If So, Give an Example.

Yes, it is possible. If *tert*-butanol is treated with a catalytic amount of acid in ethanol solvent, the initially formed tertiary cation (***12.12***) can react with ethanol to give ***12.13***. Subsequent loss of a proton from this oxonium intermediate gives ethyl *tert*-butyl ether, ***12.11***.

H^+ + EtOH $- H^+$

12.12   ***12.13***   ***12.11***

What Is the Product When 4-Bromobutanol Is Treated with NaH in Ether?

Reaction of bromohydrin ***12.14*** with NaH will form the alkoxide, ***12.15***. This initially formed alkoxide will react with the bromide at the end of the molecule in an intramolecular Williamson ether synthesis to give tetrahydrofuran, ***12.16***.

NaH

12.14   ***12.15***   ***12.16***

Alkoxymercuration

Alkoxymercuration is the process by which an alkene is treated with a mercuric salt and an alcohol to give an hydroxy-mercuric compound. Subsequent treatment with sodium borohydride gives the ether.

What Is the Nature of the Intermediate Formed by Initial Reaction of 1-Pentene with Mercuric Acetate, Before Reaction with Ethanol?

This intermediate is mercury-stabilized carbocation **12.17**. Note that the mercuric is on the primary carbon in ***12.18***, suggesting that ***12.17*** reacts as if it were the more stable secondary carbocation. Therefore, the use of mercury generates the more stable carbocation in terms of its reaction with a nucleophile.

HgOAc

12.17

Note also that ***12.17*** is stabilized by coordination with the mercury (called back-donation), as with oxymercuration in Section **11.2**.

What Is the Product Formed from 1-Pentene, Mercuric Acetate and Ethanol?

The product is *12.18*.

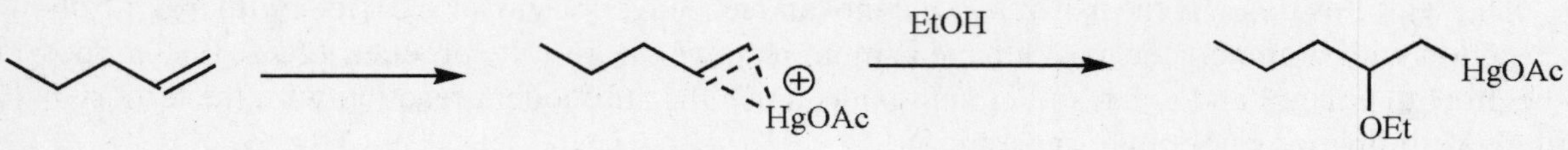

EtOH

$\oplus$ HgOAc

HgOAc

OEt

12.17

12.18

What Is the Purpose of Treating the Hydroxy-Mercuric Compound 12.18 with Sodium Borohydride?

To remove the mercury (cleave the C–Hg bond) and convert it to a C–H unit.

What Is the Final Product When 1-Pentene Is Reacted with 1. $Hg(OAc)_2$, EtOH 2. $NaBH_4$ 3. Aq. NH_4Cl?

The product is ethyl 1-methylbutyl ether, *12.19*.

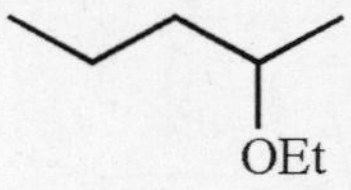

OEt

12.19

What Is the Purpose of the Ammonium Chloride?

The reaction with sodium borohydride forms a boron complex, and the aqueous ammonium chloride functions as an acid, in the presence of the water, to decompose that complex. In other words, reaction with the C–Hg unit forms an intermediate boron complex, and the aqueous ammonium chloride is required to liberate the C–H unit.

Why Does This Process Lead to an Ether Rather Than an Alcohol (As with Oxymercuration)?

The only nucleophilic species provided to the cation intermediate is the alcohol solvent, leading to the ether.

What Is the Product of Each of the Following Reactions?

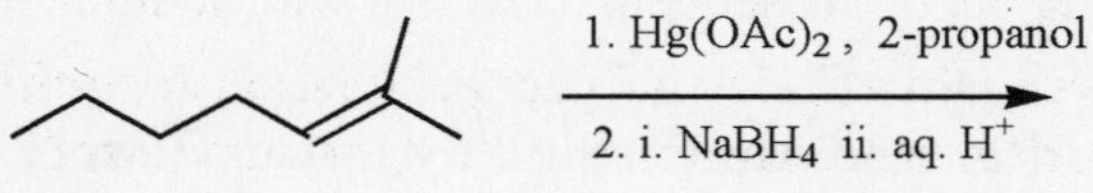

1. $Hg(OAc)_2$, 2-propanol

2. i. $NaBH_4$ ii. aq. H^+

12.20

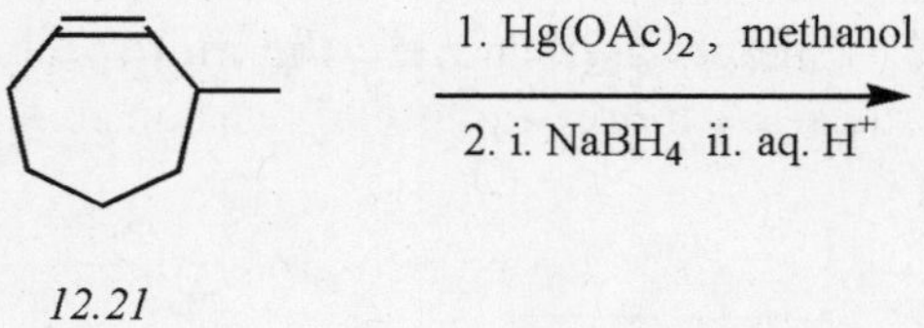

1. $Hg(OAc)_2$, methanol

2. i. $NaBH_4$ ii. aq. H^+

12.21

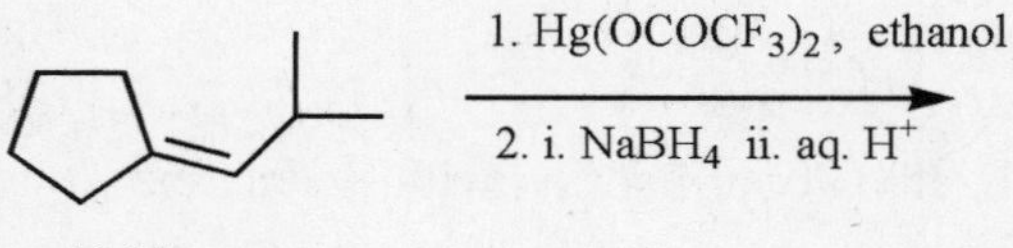

1. $Hg(OCOCF_3)_2$, ethanol

2. i. $NaBH_4$ ii. aq. H^+

12.22

The more stable cation from alkene ***12.20*** is at the tertiary position, and the final product is ***12.23***. Reaction of alkene ***12.21*** poses two problems. First, both possible cation intermediates are secondary, so there will be a mixture of two products (regioisomers). In other words, the OCH_3 unit will be incorporated at each carbon of the C=C unit to give two different products. Second, one of the intermediate carbocations is a secondary carbocation, but is next to a potential tertiary carbocation site. Remember that in oxymercuration and alkoxymercuration reactions, the mercury stabilizes the cation, and rearrangement does not occur. Therefore, the two products are ***12.24*** and ***12.25***. In the last reaction of ***12.22***, the carbocation forms at the more stable tertiary position, and the final product is ***12.26***.

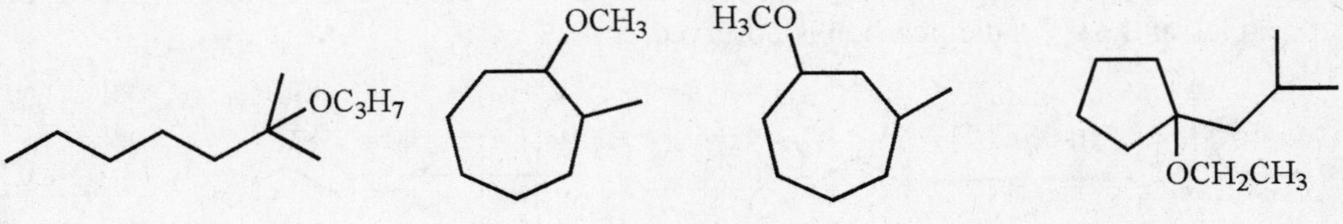

OC_3H_7 OCH_3 H_3CO OCH_2CH_3

12.23 *12.24* *12.25* *12.26*

12.4. REACTIONS OF ETHERS

Ethers give only a limited number of reactions, as described below.

If an Alkyl Ether Is a Base, Is It a Strong Base or a Weak Base?

Ethers are relatively weak Lewis bases, requiring strong Lewis acids for a reasonable reaction.

What Acids Would Be Strong Enough to Protonate an Ether and Lead to Further Reaction?

HI is the most common, and any acid stronger than HI would also protonate an ether.

What Is the Reaction Product When Diethyl Ether Is Treated with Hydroiodic Acid (HI)?

The powerful acid HI cleaves diethyl ether into ethanol and iodoethane.

What Is the Product When Diethyl Ether Is Treated with Hydrochloric Acid (HCl)?

In general, HCl is too weak to induce cleavage of the ether, and we should recover the ether without any reaction.

Give a Mechanistic Rationale for How an Alkyl Ether Is Cleaved by HI.

The reaction occurs by protonation of the ether oxygen to give an oxonium ion such as ***12.27*** (the ether oxygen is basic in the presence of the powerful acid). The nucleophilic iodide counterion attacks the less sterically hindered carbon, displacing ethanol and forming iodoethane.

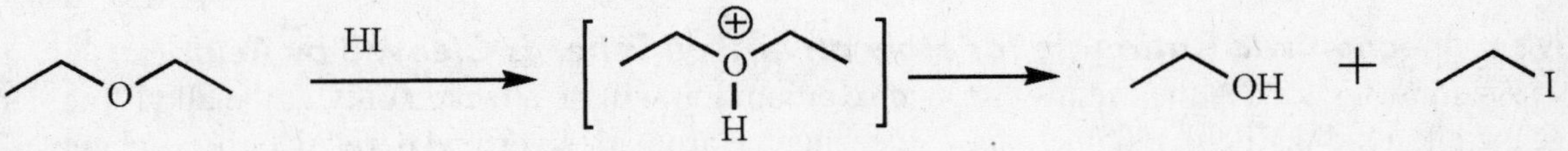

O HI ⊕ O H OH + I

12.27

What Is the Product When Methyl Isopropyl Ether (12.28) Is Treated with HI? Show a Mechanism.

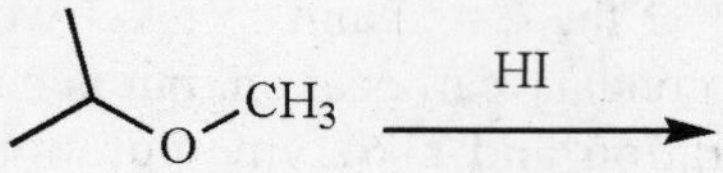

HI

12.28

The products are isopropanol and iodomethane. The oxygen of ***12.28*** is protonated to give ***12.29***. Iodide attacks the less sterically hindered carbon in an S_N2 displacement, as shown, to give the products. Very little attack at the secondary carbon is observed.

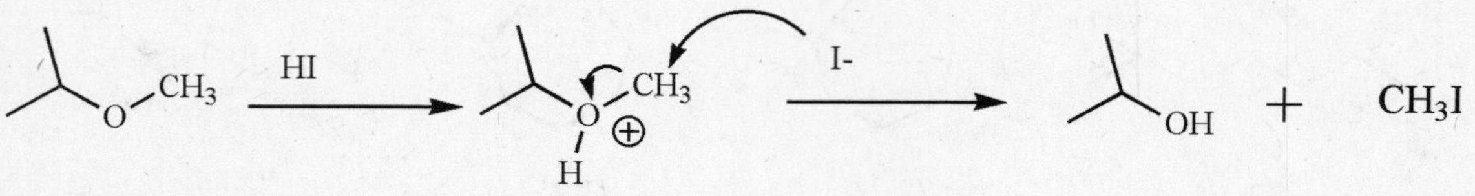

HI I- OH + CH_3I

12.28 12.29

What Acid or Acids Are Capable of Cleaving Alkyl Ethers?

Very few protonic acids are strong enough to cleave ethers. HI is the most common, but HBr also works. Relatively strong Lewis acids will also cleave ethers. The most commonly used reagent is boron tribromide, BBr_3.

Is Diphenyl Ether (12.30) Cleaved with HI?

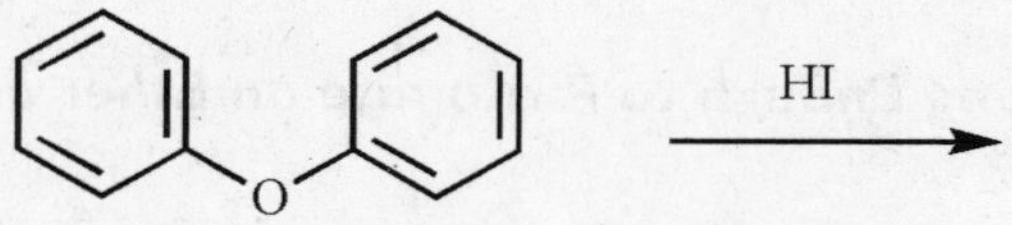

HI

12.30

No! Although protonation of the ether will occur, S_N2 displacement at the aromatic carbon is not possible and cleavage does not occur.

Which Is Preferred, Liberation of the Alkyl Alcohol or the Phenol When Aralkyl Ethers Are Cleaved with Acid?

An alkyl aryl ether such as anisole, ***12.31***, will be protonated (to give ***12.32***), but attack by iodide is possible only at the alkyl fragment (in this case the methyl group). This cleavage will always liberate the phenol (***12.33***) and an alkyl iodide (in this case iodomethane).

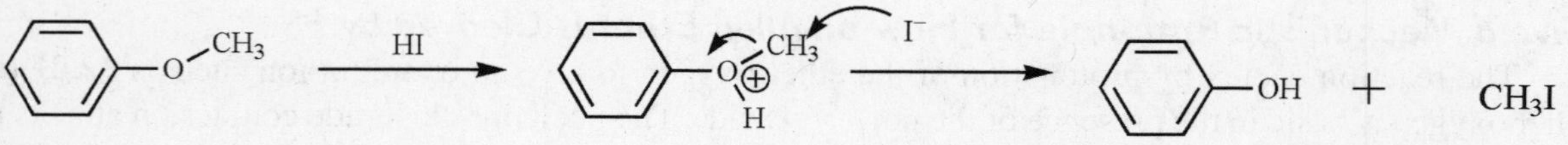

HI I^- OH + CH_3I

12.31 12.32 12.33

Give a Mechanistic Rationale for How an Aralkyl Ether Is Cleaved by Acid.

As shown in ***12.32***, the initially formed oxonium ion will be attacked only at the alkyl fragment to give the alkyl iodide (here, iodomethane), releasing the aromatic hydroxyl derivative (here, phenol). This reaction is an S_N2 process and will not occur on the phenyl ring.

Give the Major Product or Products of the Reactions of 12.34-12.36.

HI

12.34

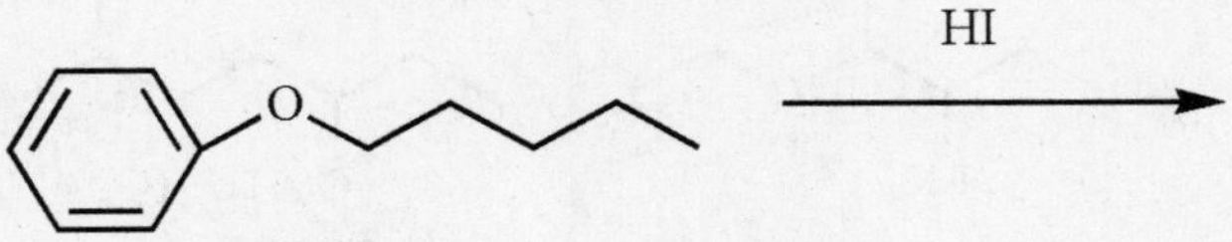

HI

12.35

HI

12.36

Cleavage of ether ***12.34*** will give neopentyl alcohol (2,2-dimethyl-1-propanol, ***12.38***) and 1-iodobutane, ***12.37***. Nucleophilic attack at the neopentyl carbon (adjacent to the tertiary center) is too sterically hindered for the S_N2 reaction with iodide ion. In the second reaction, ether ***12.35*** is cleaved to give phenol (***12.33***) and iodopentane, ***12.39***. In the last reaction, dicyclopentyl ether (***12.36***) is a symmetrical ether and is cleaved to give one molecule of cyclopentanol (***12.40***) and one molecule of iodocyclopentane (***12.41***).

I HO

12.37 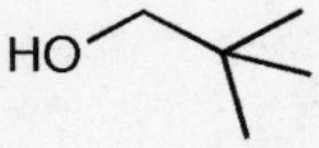 *12.38*

OH I OH I

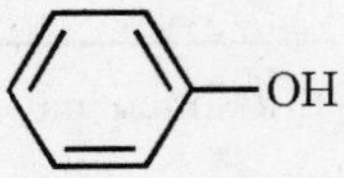

12.33 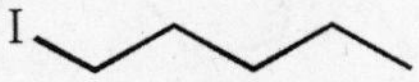 *12.39* 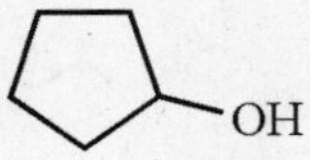 *12.40* 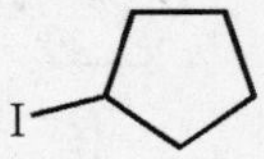 *12.41*

Test Yourself

1. Give the correct IUPAC names for each of the following.

(a) 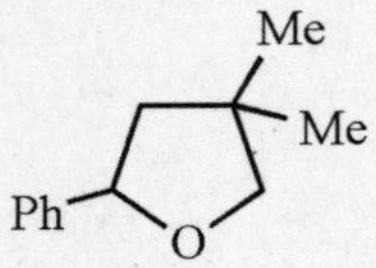 Ph, O, Me, Me (b) 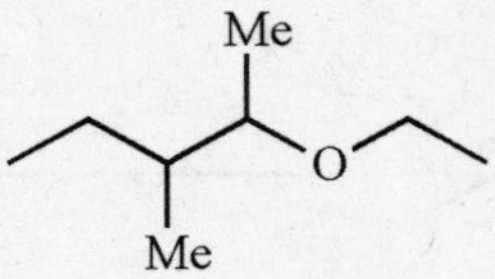 Me, O, Me (c) 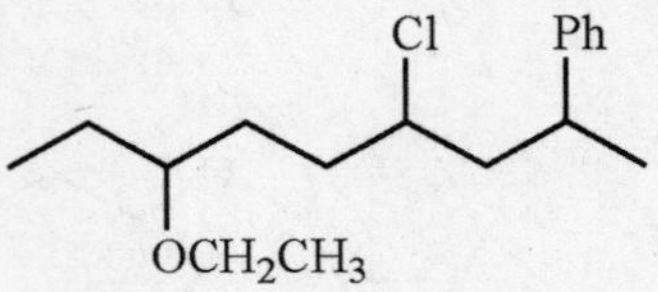 Cl, Ph, OCH_2CH_3 (d) 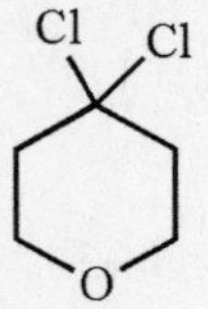 Cl, Cl, O

(e) 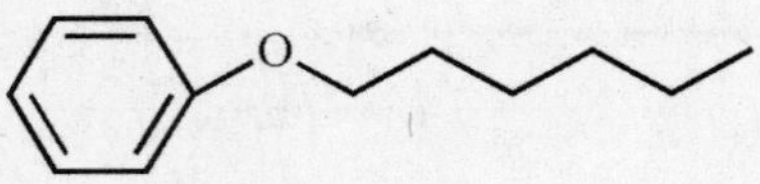 O (f) 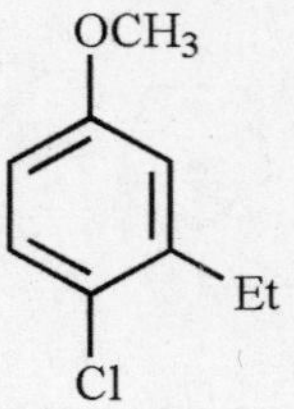 OCH_3, Et, Cl

2. In each case, give the structure of the major product. Remember stereochemistry where appropriate, and if there is no reaction, indicate this by N.R.

(a) 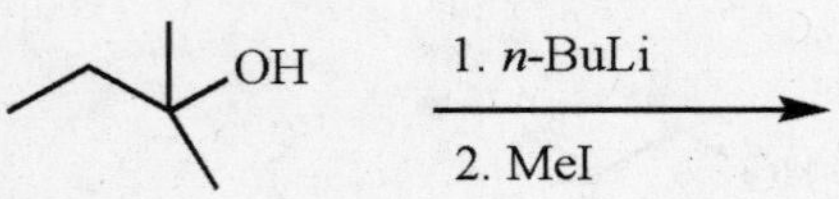 OH → 1. *n*-BuLi 2. MeI

(b) 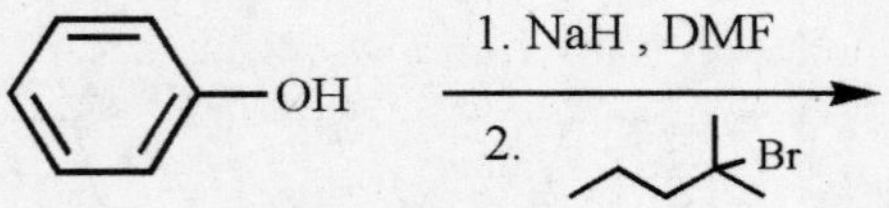 OH → 1. NaH, DMF 2. Br

(c) 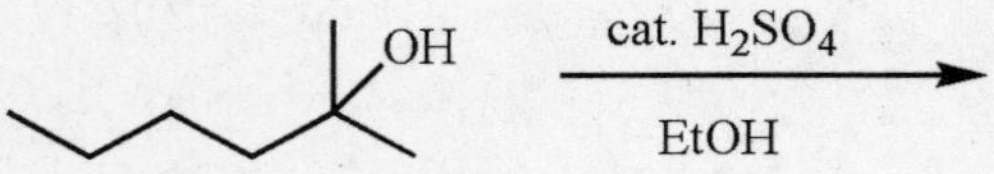 OH → cat. H_2SO_4, EtOH

(d) 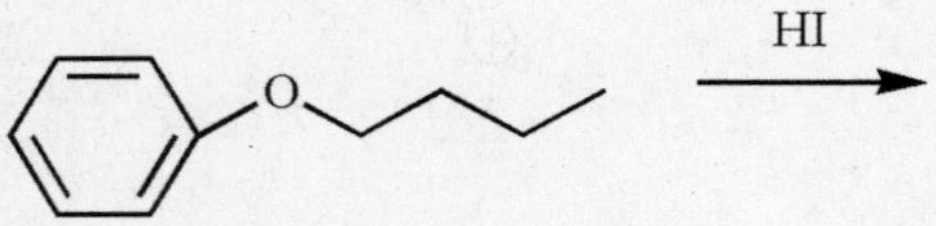 O → HI

(e) 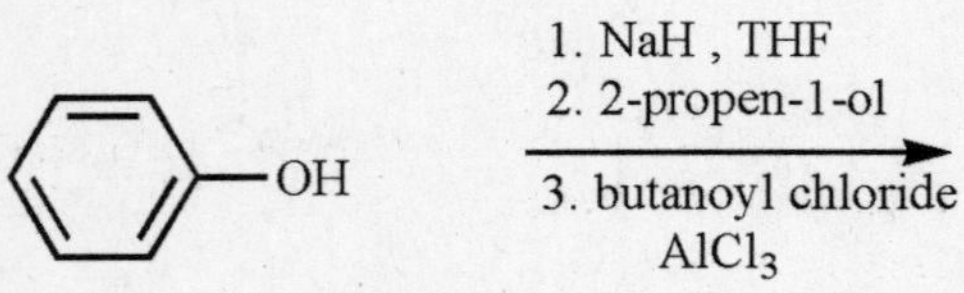 OH → 1. NaH, THF 2. 2-propen-1-ol 3. butanoyl chloride $AlCl_3$

(f) 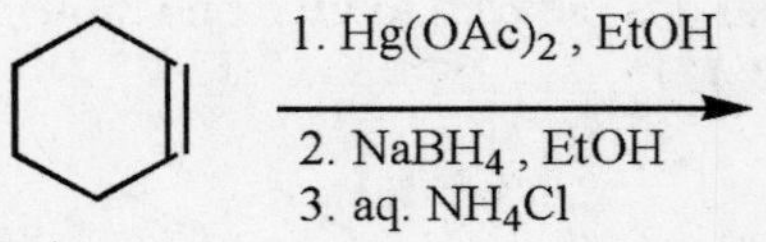 → 1. $Hg(OAc)_2$, EtOH 2. $NaBH_4$, EtOH 3. aq. NH_4Cl

(g) 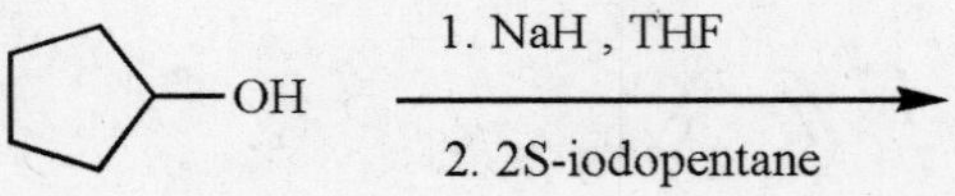 OH → 1. NaH, THF 2. 2S-iodopentane

(h) O → HCl

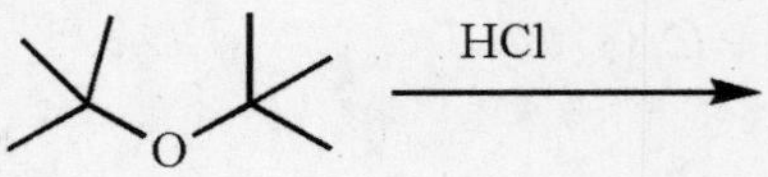

3. Give the mechanism for the conversion of 2-methyl-2-propanol to di-*tert*-butyl ether upon treatment with a catalytic amount of sulfuric acid.

CHAPTER 13

Addition Reactions

13.1. REACTIONS OF ALKENES WITH ACIDS, HX

In the presence of an acid such as H–X, the π-bond of an alkene will function as a base, donating two electrons to H^+. This will generate a new C–H bond, a carbocation and a nucleophilic counterion, X^-. If X^- reacts with the cation, the transformation is C=C → H–C–C–X, where the resulting product contains both H and X. The elements of HX are said to have *added* to the alkene, and this is an **addition reaction**.

Is the π-Bond of an Alkene Electron Rich or Electron Poor?

The π-bond of an alkene is electron rich. Since the π-bond is much weaker than a covalent σ-bond, the electrons of a π-bond can be donated to an electron deficient atom in many cases.

If an Alkene Were Mixed with an Acid, H^+, What Reaction Is Possible?

The π-bond of the alkene can donate two electrons to H^+. This reaction breaks the π-bond, and one carbon of the C=C unit is used to form a new C–H bond, leaving behind a carbocation on the remaining carbon. Using ethene as an example, reaction with H^+ generates carbocation *13.1*.

$$CH_2{=}CH_2 + H^{\oplus} \longrightarrow H_2C{-}H\,|\,CH_2^{\oplus}$$

13.1

The Reaction of Ethene and H^+ Is What Type of Reaction?

The reaction of the π-bond of an alkene with an acid (H^+) is an acid-base reaction; here the alkene is the base.

When 1-Propene Reacts with H^+, the Proton Can Attach to One of Two Carbon Atoms of the C=C Unit. Is One Preferred?

Yes! The two possible products are ***13.2*** and ***13.3***. However, ***13.2*** is a primary carbocation, and ***13.3*** is a secondary carbocation. The secondary carbocation is much more stable and is formed preferentially. In other words, the transition state leading to ***13.3*** is lower in energy than that leading to ***13.2***, and the product of this reaction is ***13.3***.

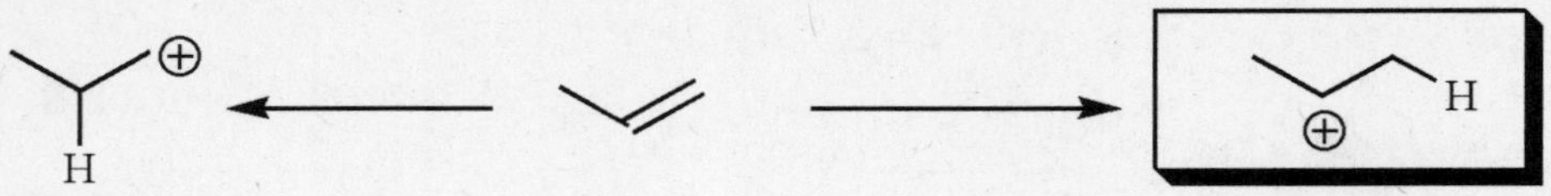

13.2 ⟵ ⟶ 13.3 major product

Which Is the More Stable Cation, That Derived from 1-Propene or That from 2-Methyl-1-Propene? Explain.

The carbocation formed by the reaction of 1-propene and H^+ is ***13.3***, but of the primary and secondary carbocations possible from 2-methyl-1-propene, tertiary carbocation ***13.4*** is formed preferentially. Comparing the secondary carbocation ***13.3*** and the tertiary carbocation ***13.4***, ***13.4*** is more stable.

13.4

When a Carbocation Such As 13.4 Is Formed, What Can It React With?

The carbocation can react with any reasonable nucleophile. In most cases, the counterion of the acid (H^+) is X^- and will react with the carbocation to form a new C–X bond.

What Is the Major Product When 2-Methyl-1-Propene Reacts with HCl?

The major product of this reaction is 2-chloro-2-methylpentane (***13.5***), derived from the cationic intermediate ***13.4***.

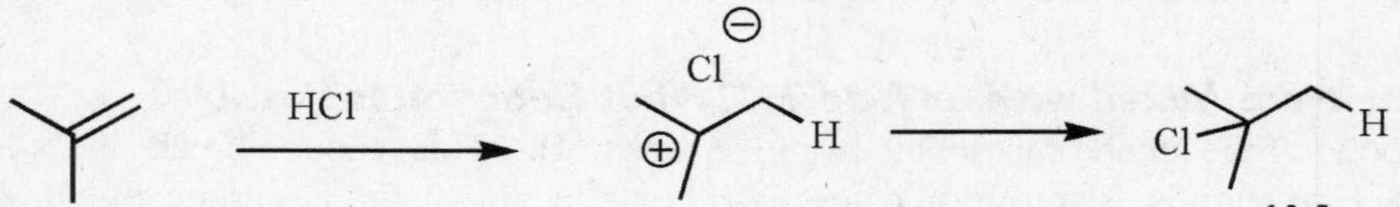

HCl ⟶ 13.4 ⟶ 13.5

Explain Why Tertiary Cations Are More Stable Than Secondary Cations.

As noted in Section 7.4, tertiary cations have more electron releasing carbon groups attached to the electropositive carbon than the secondary cations. Electron release diminishes the formal charge on the cation, and greater stability is associated with less change.

Give a Mechanistic Rationale for the Reaction of 1-Propene with HCl.

The π-bond of the alkene behaves as a base, donating two electrons to the acidic H^+ of HCl. This reaction forms a new C–H bond and generates a carbocation (***13.3***). The chloride ion attacks the cationic center to give the final product, 2-chloropropane (***13.6***).

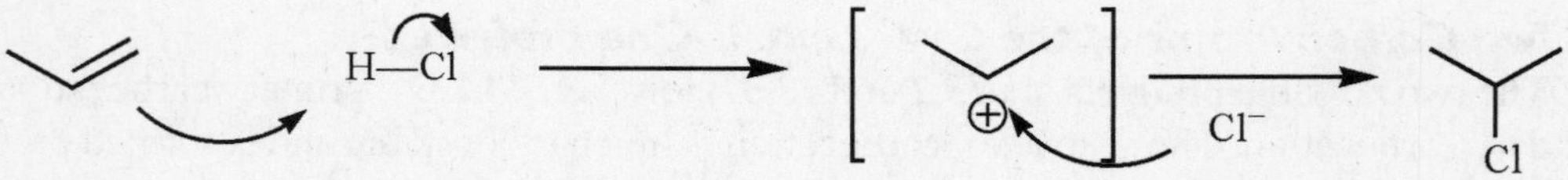

H—Cl ⟶ [13.3] ⟶ Cl^- 13.6

Give the Mechanism for the Transformation of 13.7 into 13.8.

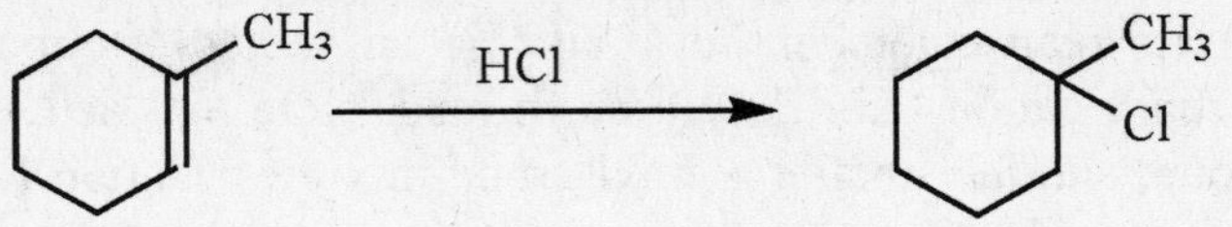

CH3 HCl CH3 Cl

13.7 *13.8*

The π-bond in ***13.7*** behaves as a base and attacks H^+ of HCl, giving a carbocation intermediate, ***13.9***. The nucleophilic by-product (Cl^-) attacks the electrophilic carbon of ***13.9*** to give the final product, 1-chloro-1-methylcyclohexane, ***13.8***.

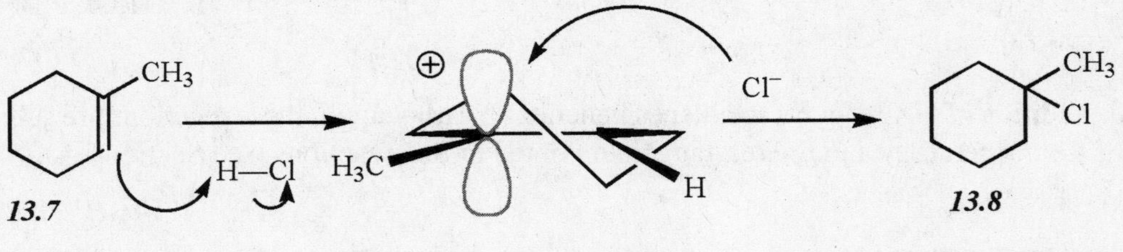

CH3 H—Cl H3C H Cl⁻ CH3 Cl

13.7 *13.9* *13.8*

When 2-Methyl-1-Pentene (13.10) Reacts with HCL, the H^+ Can Be Transferred to Two Different Carbons. Draw Both Possible Cationic Intermediates and Then the Major Product.

The two carbocations are ***13.11*** and ***13.12***, and the tertiary carbocation ***13.11*** is more stable and constitutes the major product. Therefore, subsequent reaction with chloride ion will give ***13.13*** as the major isolated product.

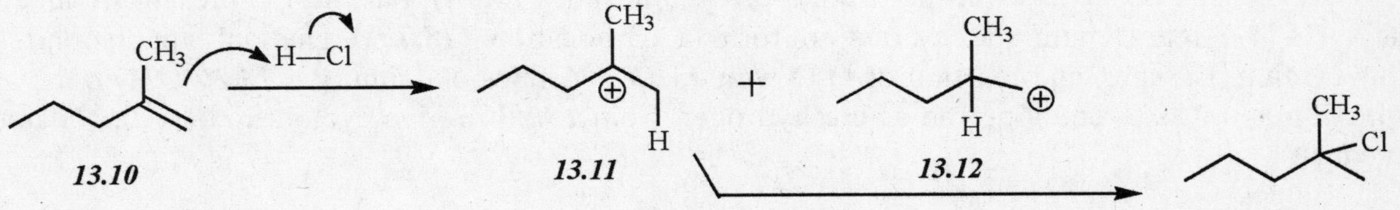

CH3 H—Cl CH3 H + CH3 H CH3 Cl

13.10 *13.11* *13.12* *13.13*

What Is the Name Given to the Reaction of Alkene with Acids of the Type HX, Where the X Group in the Major Product Is Always on the More Substituted Carbon?

Since the reaction proceeds by formation of the more stable carbocation, the X group will always appear on that carbon (the more substituted). Prior to determination of the mechanism, a Russian chemist observed that the X group always appeared on the more substituted carbon, and the reaction is named in his honor. It is called the Markovnikov addition.

Give the Mechanism of the Transformation of 13.14 into 13.15.

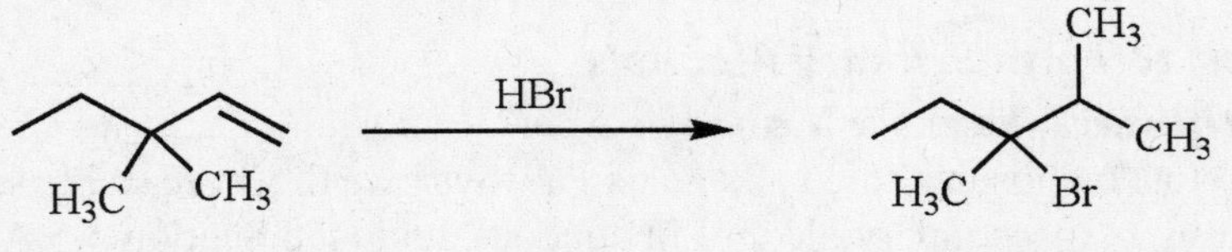

H3C CH3 HBr CH3 CH3 H3C Br

13.14 *13.15*

3,3-Dimethyl-1-pentene (***13.14***) initially reacts with the H^+ of HBr to give a secondary cation, ***13.16***. No hydrogen is available for rearrangement, but if an adjacent methyl group migrates via a 1,2-methyl shift, the more stable tertiary cation (***13.17***) will be formed. Note that in ***13.16***, there are two methyl groups and one ethyl group, and the smaller methyl group migrates. Cation ***13.17*** then reacts with bromide to give the major product, ***13.15***.

HBr; 1,2-methyl shift; Br-

13.14 *13.16* *13.17* *13.15*

Since addition of HX reagents to alkenes generates a carbocation, these reactions are subject to both the loss of stereochemistry and the rearrangement noted in S_N1 reactions (see Section 7.4).

13.2. HYDRATION

It is possible to add the elements of water (H–O–H) to an alkene, but only if another reagent is added to initiate the reaction. Water is inert to alkenes under neutral conditions.

Is There a Product When Water Is Mixed with an Alkene Such As Cyclopentene?

No! Water is not a strong enough acid (i.e., the alkene is too weak a base) for a reaction to occur.

Explain the Fact That Cyclohexene Reacts with Aqueous Acid to Give Cyclohexanol.

Water does not react with the π-bond of cyclohexene (***13.18***), but the H^+ present in aqueous acid (H_3O^+) reacts with the π-bond to form a carbocation (***13.19***). The only nucleophile in this medium is water, and reaction of H_2O with ***13.19*** generates oxonium ion ***13.20***. Water can also function as a base, removing the acidic hydrogen from ***13.20*** to give cyclohexanol (***13.21***) as the product.

H+; + H_2O; - H^+

13.18 *13.19* *13.20* *13.21*

Is the Acid in the Aqueous Acid Catalytic or Stoichiometric?

Since H^+ is regenerated when ***13.20*** is converted to ***13.21***, the reaction is catalytic in H^+. Therefore, when cyclohexene is reacted with a catalytic amount of acid in the presence of water, the product is cyclohexanol.

Why Is It Difficult to Form Primary Alcohols from Monosubstituted Alkenes in Aqueous Acid?

Addition of H^+ to a monosubstituted alkene will always give the more stable secondary cation as the major product, leading to a secondary alcohol in aqueous acid (see Section 5.5.A). Formation of a primary cation in order to form a primary alcohol requires too much energy to be formed preferentially under these conditions.

Give the Major Product for the Reaction of 1-Pentene under the Following Conditions.

H_2O , THF

cat H_2SO_4

13.22

Under these conditions, the alkene reacts with the acid catalyst to form the secondary carbocation ***13.23***. Subsequent reaction with water leads to ***13.24,*** and loss of the proton leads to the final product, 2-pentanol (***13.25***).

H+ + H_2O - H^+ OH H H

13.22 *13.23* *13.24* *13.25*

If an Alkene Reacted with HX and the X Group Appeared on the Less Substituted Carbon, What Name Would Be Applied to That Reaction?

It would be called an anti-Markovnikov addition.

If the Reaction between an Alkene and HX Always Gives a Carbocation, Is Anti-Markovnikov Addition Possible?

No! The reaction will always form the more stable carbocation, which is always at the more substituted position in the absence of resonance effects.

How can Anti-Markovnikov Addition Be Accomplished?

The mechanism between the alkene and HX must be modified. Specifically, a radical mechanism must be operative.

What Is the Major Product from the Reaction of 13.26 and HBr in the Presence of Di-tert-Butyl Peroxide?

HBr

t-Bu-O-O-t-Bu

13.26

The reaction of alkene ***13.26*** does not give the 2-bromo-hexane expected from the reaction with HBr via a cation. The major product is 1-bromohexane (***13.27***), where the presence of the peroxide changes the mechanistic course of the reaction from a cationic reaction to one involving a radical intermediate.

Br

13.27

What Is the Product When HBr Reacts with Di-tert-Butylperoxide?

Di-*tert*-butylperoxide has the structure $Me_3C–O–O–CMe_3$ (***13.28***) and homolytically cleaves upon heating to give two $Me_3CO\bullet$ radicals (***13.29***). When this radical reacts with HBr, $Me_3C–O–H$ (2-methyl-2-butanol = *tert*-butanol) is formed along with a bromine radical, Br•.

Explain Why Bromine Resides on the Less Substituted Carbon When HBr Reacts with 1-Hexene (13.30) in the Presence of t-BuOOt-Bu (13.28).

In the reaction of ***13.30***, the π-bond of the alkene reacts with Br• (produced by cleavage of hydroperoxide ***13.28*** to give *tert*-butanol and Br•) rather than with H^+, and the initial product is a carbon radical, ***13.31***. The bromine ends up on the less substituted carbon because Br• reacts with the alkene to form the more stable radical, in this case a secondary radical rather than a primary radical. This radical intermediate reacts with additional HBr to produce the product (1-bromohexane, ***13.32***) and a bromine radical (Br•), which can continue the radical chain reaction.

Give the Relative Order of Stability of Alkyl Radicals.

Carbon radicals can be considered as electron deficient species for this purpose. Just as with cations, carbon groups are electron releasing and stabilize the radical. Therefore, a tertiary radical is more stable than a secondary, which is more stable than a primary. 3° > 2° > 1° > $\bullet CH_3$. An important difference in radical chemistry is the observation that carbon radicals do *not* rearrange as carbocations do. Another important consideration is that the reaction shown for HBr is not general. It works with HBr and a peroxide, but not with HCl or HI.

13.3. ADDITION OF HALOGENS

Alkenes react with halogens such as diatomic chlorine, bromine or iodine to give **vicinal** dihalides. A vicinal halide has two halogen atoms on *adjacent* atoms (X–C–C–X). The mechanism of this addition is different from the reactions with HX and leads to *trans*-dihalides.

What Is Polarizability?

Polarizability is the property of an atom to take on an induced dipole when in proximity to a polarized atom. In general, proximity to a positive dipole will induce a negative dipole, and vice versa.

What Are Common Molecules That Are Polarizable?

The halogens, diatomic chlorine, bromine and iodine are highly polarizable molecules.

Why Does an Alkene React with Diatomic Bromine (Br_2)?

The Br–Br bond is polarizable, and in the presence of the alkene (the alkene comes in close proximity to one of the bromines), the closest Br becomes δ⊕ by *induced polarization*. In the presence of the alkene, therefore, Br–Br becomes $^{\delta\oplus}Br–Br^{\delta-}$. With this induced dipole, the π-bond can transfer two electrons to $Br^{\delta\oplus}$ and generate Br^- as a leaving group.

Is the Reaction Between Cyclohexene and Diatomic Bromine Concerted or Ionic in Nature?

The reaction has a cationic intermediate, so it is ionic in nature.

What Is the Intermediate When Cyclohexene Reacts with Diatomic Bromine?

When the π-bond of cyclohexene (***13.33***) reacts with Br–Br, a **bromonium ion** intermediate ***13.34*** is formed. When the double bond transfers electron density to Br, positive charge 'builds up' on the adjacent carbon. The lone electron pairs on the Br form a covalent bond with this positive center, forming the three-membered ring shown. The bromonium ion is re-drawn as ***13.35*** to show the proper stereochemical relationship of the ion and the bromide ion it reacts with. The bromide counterion acts as a nucleophile, opening the three-membered ring to give the *trans*-dibromide product, ***13.36***.

Br-Br

Br ⊕

Br ⊕

Br^-

Br

Br

13.33 *13.34* *13.35* *13.36*

When Intermediate 13.34 Reacts with Bromide Ion, from Which Face Will It Approach, from the Same Side As the Bromine or from the Opposite Side? Justify Your Answer.

This is best answered by inspection of ***13.35***. If Br^- approached from the 'top,' there is both steric and electronic repulsion. This repulsion is minimized by 'backside attack,' where Br^- approaches the three-membered ring *anti-* to the Br, as shown. This backside attack dictates an *anti-* relationship for the two bromines (a *trans*-dibromide, ***13.36***). This *trans-* geometry is characteristic of the intermediacy of a bromonium ion (halonium ion is the generic term: chloronium, bromonium and iodonium for Cl, Br, I, respectively).

When a Cyclic Alkene Reacts with Chlorine, Bromine or Iodine, Is the Cis-Dihalide Ever Observed?

No! The intermediacy of the halonium ion (chloronium, bromonium, iodonium) always leads to the trans product.

What Term Is Used for a Reaction That Gives One and Only One Diastereomer of Two or More Possible Diasteromeric Products?

The term is **diastereospecific**. Since the reaction of halogens and alkenes gives a single diastereomer, the reaction is diastereospecific.

What Does Diastereoselective Mean?

Diastereoselective means that of two or more possible products, both are formed, but one is major.

What Is the Product When E-4-Ethyl-3-Octene (13.37) Reacts with Br_2 IN CCl_4?

This reaction of bromine and ***13.37*** also proceeds by a bromonium ion intermediate (***13.38***). *Anti-*attack is also required, which dictates an *'anti-'* relationship for the bromines in the product, ***13.39***. Since this is an acyclic molecule, *cis- / trans-* isomers are not possible. Diastereomers are possible, however. *Anti-* attack dictates that a single diastereomer is formed (***13.39***). Since the bromine can form the bromonium ion on either the 'top' or the 'bottom' of the alkene, diastereomer ***13.39*** is *racemic*. Formation of a racemic diastereomer is the norm for this reaction, which is diastereospecific, but shows no enantioselectivity at all. Note that Br^- attacks ***13.38*** at the less sterically hindered carbon. Attack at the other carbon of the bromonium ion would generate the enantiomer.

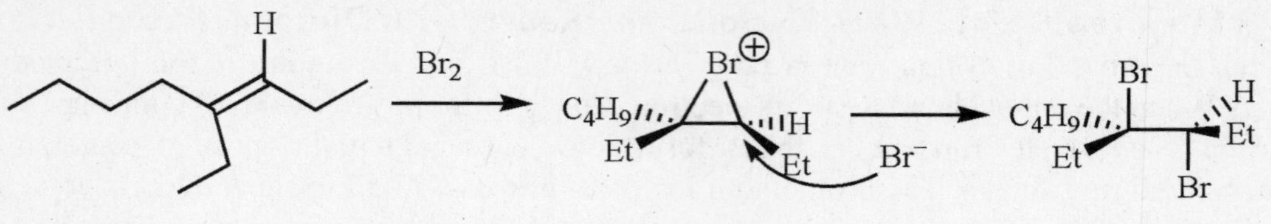

13.37 13.38 13.39

How Many Stereoisomers Are Produced When cis-2-Butene Reacts with Br_2?

cis-2-Butene (**13.40**) reacts with Br_2 to give bromonium ion **13.41**. *Anti-* attack by Br^- leads to **13.42** as the only diastereomer, but it is racemic. If the 'bottom' carbon is rotated by 180°, the other rotamer shown for **13.42** results. The mirror image of **13.42** is not superimposable, and since this diastereomer is racemic, two stereoisomers result from this reaction.

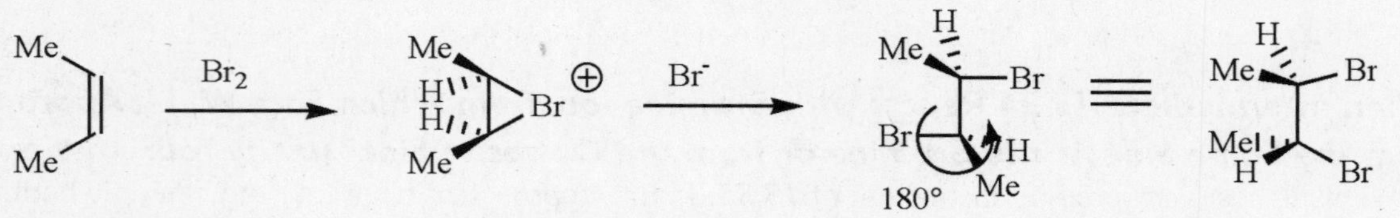

13.40 13.41 13.42

If *trans*-2-butene were reacted with Br_2, only one diastereomer would result, meso 2,3-dibromobutane.

Draw Meso-2,3-Dibromobutane.

This diastereomer is **13.43**. Note the symmetry in the molecule that leads to its identification as a meso compound.

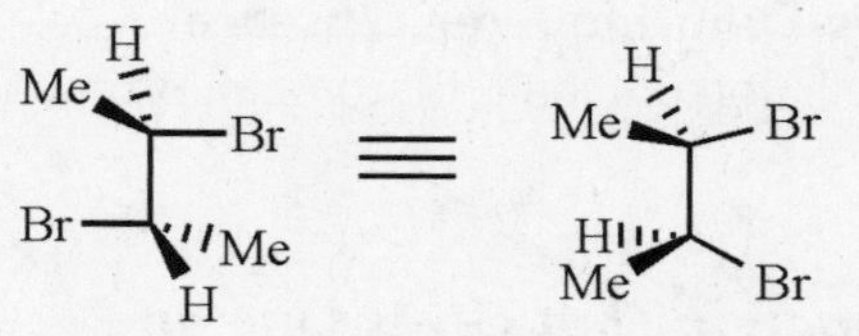

13.43

What Is the Product When Chlorine Is Dissolved in Water?

When chlorine(Cl_2) is dissolved in water, hypochlorous acid (HOCl) is formed. This molecule is polarized as $Cl^{\oplus}$ OH^-. Similarly, HOBr is formed when bromine (Br_2) is dissolved in water.

What Is the Product of the Reaction between Cyclopentene (13.44) and HOCl?

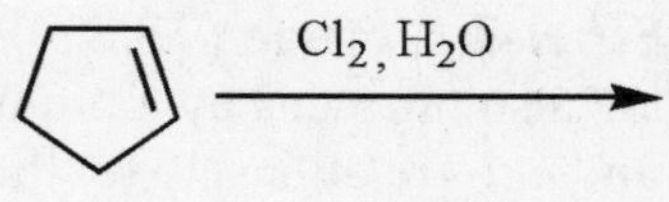

Cl_2, H_2O

13.44

The product is a *vicinal* (on adjacent carbons) chlorohydrin (chlorine and OH in the same molecule), **13.46**. Initial reaction of **13.44** with the electrophilic chlorine atom of the HOCl generates a chloronium ion, **13.45**. Although water is present, the three-membered ring remains largely intact, and the nucleophile ($^-$OH) attacks on one of the electrophilic carbon atoms (from the face opposite the Cl in **13.45**) of **13.45** to give the chlorohydrin, **13.46**.

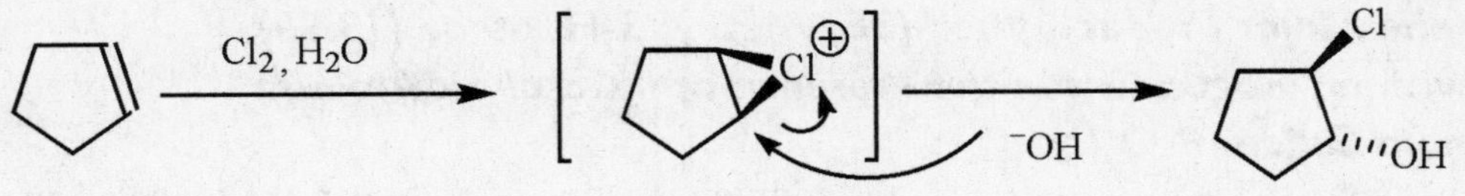

13.44 13.45 13.46

What Is the Product of the Reaction between Methylcyclopentene (13.47) and HOCl?

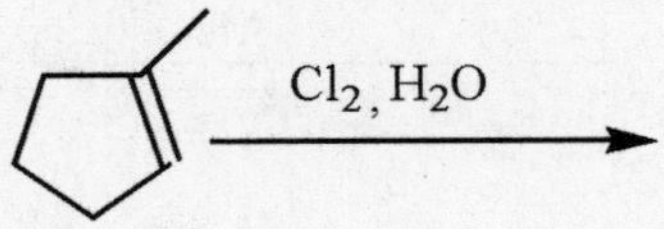

13.47

Initial reaction of *13.47* with the electrophilic chlorine atom of the HOCl generate a chloronium ion, *13.48*. The reaction follows Markovnikov's rule (Cl will end up on the carbon with the most hydrogen atoms), so the nucleophile ($^{-}$OH) attacks the more substituted carbon to give the racemic diastereomer, *13.49*.

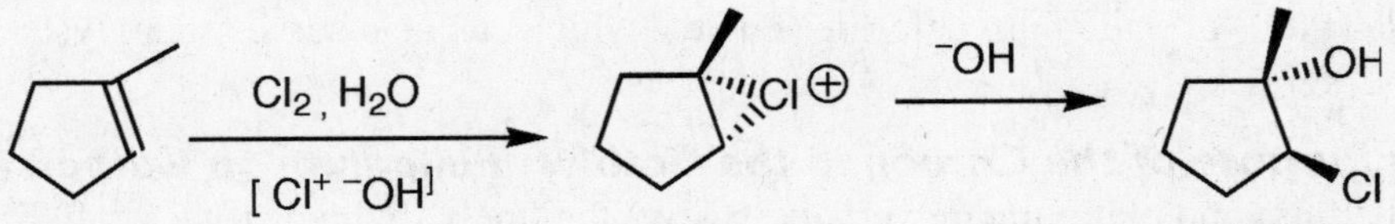

13.47 13.48 13.49

13.4. CATALYTIC HYDROGENATION

Although hydrogen does not react directly with the C=C unit of an alkene, in the presence of a transition metal catalyst, hydrogen can add to the double bond. This reaction is known as **catalytic hydrogenation**.

Explain Why 1-Hexene Does Not React with Hydrogen Gas When No Other Reagents Are Present.

Diatomic hydrogen gas (H–H) is not polarized and is not very polarizable. When the alkene comes into close proximity to H_2, there is no 'H$^+$' or 'H•' for the alkene to react with. The only way H_2 can react with an alkene is when another reagent is present that can react with H_2 to break the H–H bond.

Can Diatomic Hydrogen React with an Alkene If the HH Unit Is Cleaved by Some Other Molecule Added to the Reaction?

Yes! Transition metals such as nickel, palladium, platinum, or metal salts such as platinum oxide react with hydrogen gas, breaking the H–H bond and forming metal–H bonds. Once the hydrogen bond is broken by the metal, the alkene can react with the hydrogen atoms on the surface of the metal.

In the Reaction with Hydrogen Gas, Is a Catalytic or a Stoichiometric Amount of the Metal Required?

The reaction occurs at the surface, and after reaction with the alkene and transfer of the hydrogen atoms from the metal to carbon, the metal is free to react with more hydrogen gas. Therefore, only a catalytic amount of the metal is required.

What Is the Major Product When (3E)-Methyl-3-Heptene (13.50) Reacts with Hydrogen Gas in the Presence of a Catalytic Amount of Platinum Oxide (PtO_2)?

The platinum oxide is a *catalyst* for the addition reaction of hydrogen gas to the alkene, converting ***13.50*** to the alkane (3-methylheptane, ***13.51***). Only a catalytic amount of PtO_2 is required to promote this reaction, which is known as **catalytic hydrogenation**.

H_2 , PtO_2 , MeOH

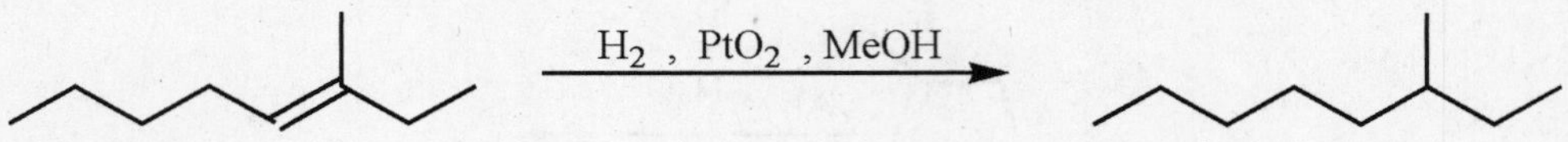

13.50 *13.51*

List Several Common Catalysts Used for Catalytic Hydrogenation.

The most common catalysts are the transition metals platinum (Pt), palladium (Pd), nickel (Ni), rhodium (Rh) and ruthenium (Ru). These expensive metals must be finely divided, since catalytic hydrogenation occurs in heterogeneous systems as a surface reaction (the higher the surface area of the catalyst, the faster the rate of hydrogenation). The metals are often converted to metal compounds such as platinum oxide (PtO_2) or palladium chloride ($PdCl_2$), which can also serve as catalysts.

What Is the Purpose of the Carbon in the Catalyst Palladium on Carbon (Pd/C)?

Since palladium and platinum are very expensive, the finely divided metal is often mixed with inert materials that have a high surface area, such as carbon black. The carbon is called a **solid support** and 'dilutes' the catalyst, as well as increases the relative surface area. Metals can also be adsorbed or mixed with calcium carbonate ($CaCO_3$), barium carbonate ($BaCO_3$), alumina (Al_2O_3) or Kieselguhr (a form of diatomaceous earth).

Give a Mechanistic Rationale for the Catalytic Hydrogenation of 2-Butene with Palladium on Carbon.

If the catalyst is treated as a surface (as in ***13.52***), hydrogen approaches the metal. The metal transfers electrons to hydrogen, and a homolytic cleavage leads to hydrogen atoms, which bind to the metal (as in ***13.53***). When the alkene approaches the metal, the π-bond binds to the surface of the metal, and a hydrogen atom is transferred to the alkene, generating a carbon radical, which is also bound to the metal, ***13.54***. Transfer of a second hydrogen liberates the alkane product (here, butane) and regenerates the catalyst, which can react with more hydrogen.

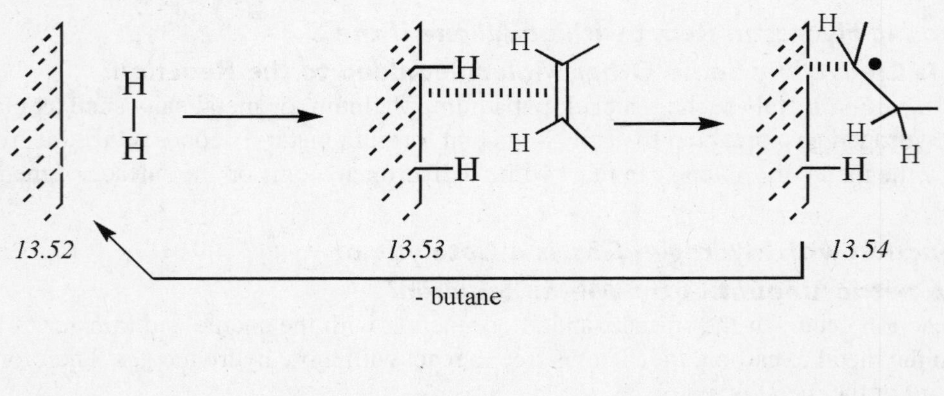

13.52 *13.53* *13.54*

- butane

Give the Major Product of the Reactions of 13.55–13.58.

H_2 , Pd/C , EtOH

13.55

H_2, Pt, $BaSO_4$

EtOH

13.56

2 H_2, Pd/$BaSO_4$

MeOH

13.57

H_2, carbon black

MeOH

13.58

Reaction of ***13.55*** generates cyclopentane. Reaction of ***13.56*** gives 2,5-dimethylhexane. Diene ***13.57*** has two double bonds, and with two equivalents of hydrogen, both are hydrogenated to give 2-methylheptane. In the reaction of ***13.58***, there is no metal catalyst, and this gives no reaction.

What Is Heat of Hydrogenation?

Each covalent bond requires that a certain amount of energy be added to break it. That same amount of energy is inherent to the bond and is called the **bond dissociation energy**. When a carbon-carbon double bond is broken during catalytic hydrogenation, the bond dissociation energy for that bond is released and is called **heat of hydrogenation** (alternatively, the amount of heat that must be added to disrupt the C=C bond).

What Information Does Heat of Hydrogenation Provide About Alkenes?

The stronger a C=C bond, the more energy is required to break it. Conversely, for a weaker bond, less energy is required. If the C=C bond of an alkene is stronger than another C=C bond (tetrasubstituted vs. monosubstituted, for example), the heat of hydrogenation will be higher for the stronger alkene bond. Heat of hydrogenation, therefore, gives information concerning the inherent strength of different alkenes.

13.5. ADDITION REACTIONS WITH ALKYNES

Just as alkenes react with HX reagents and X_2 reagents to give the 1,2-addition products, alkynes react with these reagents. The final products usually contain a double bond, since only one of the two π-bonds react with the reagent. In some cases, the initially formed vinyl compound reacts further to give a product that does not contain a carbon-carbon π-bond.

Addition of Acids (HX)

When 1-Pentyne (13.59) Reacts with HBr, to Which Carbon of the Triple Bond Is the Bromine Attached?

When HBr reacts with ***13.59***, the hydrogen will add to the less substituted carbon to give the positive charge on the more substituted carbon, and the bromine will be attached to the more highly substituted carbon in the final product, ***13.61***. One of the π-bonds of the alkyne donates an electron pair to H^+ generating a 'vinyl cation,' in this case ***13.60***. The addition of HBr can generate a less stable primary vinyl cation or a more stable secondary vinyl cation. The more stable cation forms, and reaction with Br^- leads to the final product, ***13.61*** (2-bromo-1-pentene).

H–Br
Br-

13.59 *13.60* *13.61*

What Is a Vinyl Cation?

A vinyl cation is an intermediate with a positive charge on one of the sp^2-hybridized carbons (C=C⊕).

What Is a Vinyl Halide?

A vinyl halide is an alkene with a halogen attached to one of the sp^2-hybridized carbons (C=C–X).

What Is the Relative Stability of a 'Vinyl Cation'?

A secondary vinyl cation is much less stable than a normal secondary cation, but once formed, it is more reactive with a nucleophile. Vinyl cations do not rearrange and react very quickly. In general, alkyl cations are more stable than vinyl cations, and a 2° vinyl cation is more stable than a 1° vinyl cation.

What Are the Products When HCl Reacts with 2-Hexyne (13.62)?

The H^+ can add to either carbon of the triple bond of ***13.62*** to give two vinyl cations, ***13.63*** or ***13.64***. Both are secondary vinyl cations, and since there is no difference in relative stability, both will form. There is no energy difference to drive one cation to be formed preferentially over the other. When chloride reacts with ***13.63*** and ***13.64***, two vinyl chlorides are formed, ***13.65*** and ***13.66***. Since the internal alkyne generates a vinyl cation with the possibility of *Z*- and *E*-isomers, and there is no reason to form one in preference to the other, the final alkene products are a mixture of *E*- and *Z*-isomers. A total of four products are formed in this reaction: *E*- and *Z*-***13.65*** and *E*- and *Z*-***13.66***.

H–Cl
Cl^-
Cl^-

13.62 *13.63* *13.65* *13.64* *13.66*

Give the Major Products of Each of the Reactions of 13.67–13.69.

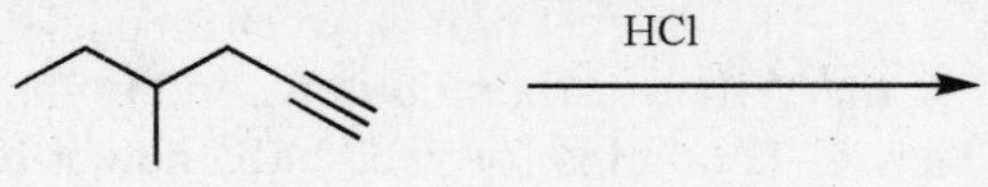

HCl

13.67

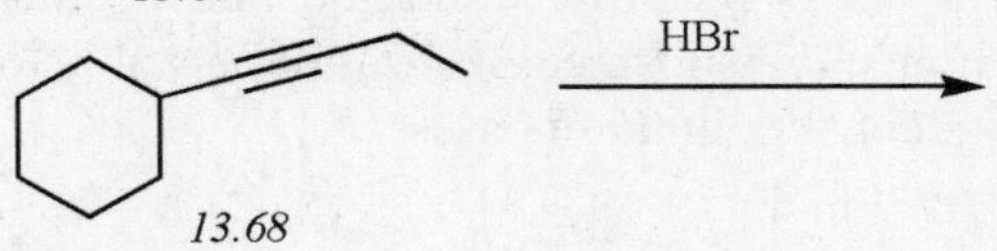

HBr

13.68

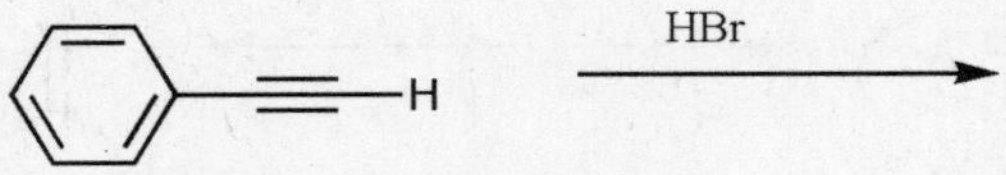

HBr

13.69

In the reaction of ***13.67***, the more stable 2° vinyl cation is formed, and the final product is ***13.70***. In the reaction of ***13.68***, two 2° vinyl cations of essentially equal stability are formed, leading to two products, *E*- and *Z*-***13.71*** and *E*- and *Z*-***13.72***. Alkyne ***13.69*** is a terminal alkyne, and formation of the secondary vinyl cation leads to the major product, ***13.73***.

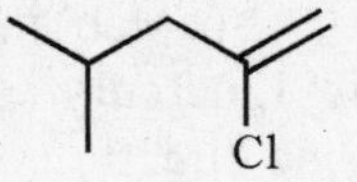

13.70 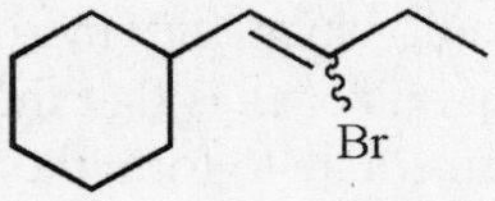 *13.71* 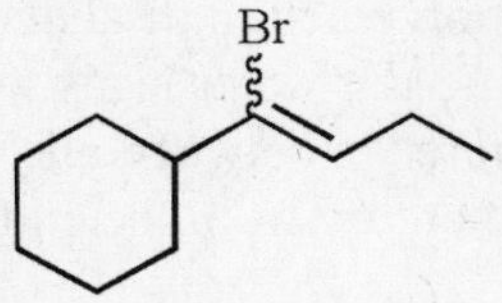 *13.72* 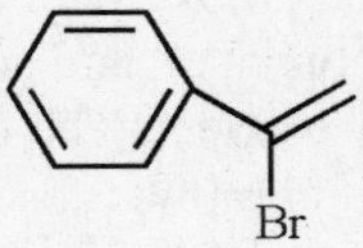 *13.73*

Hydration

Does 1-Pentyne React with Water When No Other Reagent Is Present?

No. Water is too weak an acid to react with the very weakly basic π-bonds of an alkyne. A reaction can only occur if a strong acid is added to the alkyne to generate a vinyl cation, allowing water to behave as a nucleophile.

What Is the Initial Product When 2-Pentyne Reacts with Aqueous Acid? What Is the Fate of That Initial Product?

The initial reaction of 2-pentyne and H^+ generates vinyl cation ***13.74***. Water is a nucleophile in this system, and it attacks the cation to form an oxonium ion (***13.75***), which loses a proton to form an *enol* (***13.76***). Enols are relatively unstable and transfer a hydrogen (from the O–H) to the adjacent carbon of the carbon-carbon double bond to generate a carbonyl, in this case ketone ***13.77*** (2-pentanone). The *initial product* is, however, enol ***13.76***.

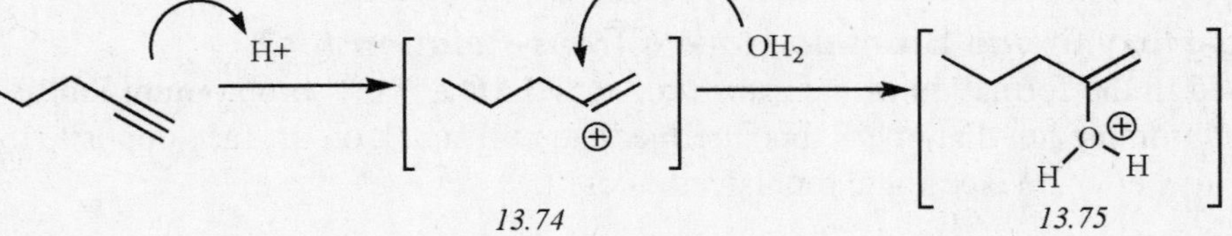

H+ OH_2

13.74 *13.75*

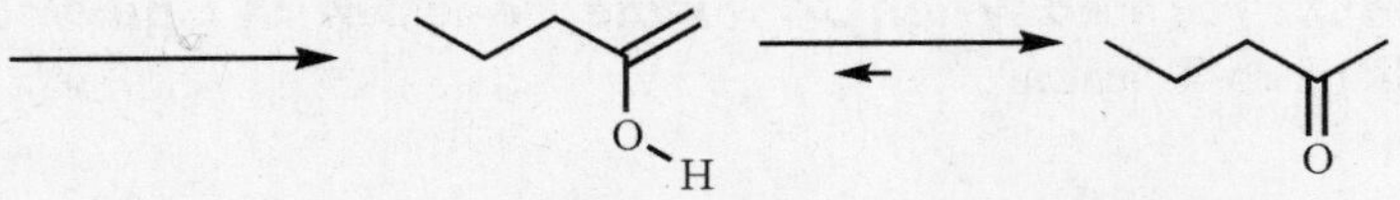

13.76 *13.77*

What Is Keto-Enol Tautomerism?

Keto-Enol tautomerism is the equilibrium reaction between an enol and its carbonyl partner (as with ***13.76*** and ***13.77***). If the π-bond of the carbon-carbon double bond in the enol (see ***13.78***) donates electrons to the acidic O-H bond, a new C–H bond is formed, and a new π-bond is formed between carbon and oxygen (a carbonyl - see ***13.79***). This equilibrium almost always favors the keto form over the enol form. One should assume that if an enol is formed, it will not be stable to isolation, and the final product will be the carbonyl (ketone or aldehyde, in these cases).

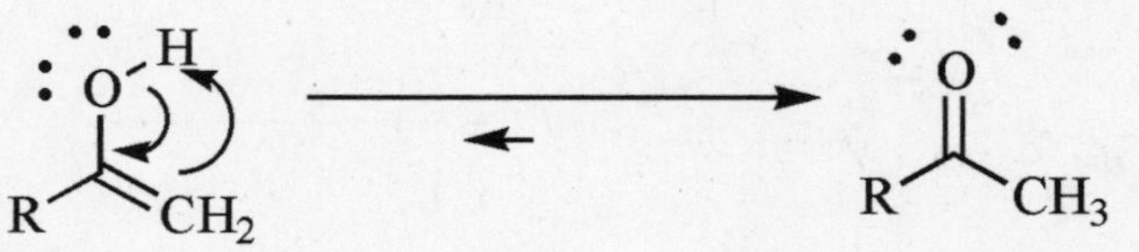

13.78 (enol) *13.79* (keto)

What Is the Product When 1-Pentyne Is Treated with a Mixture of Mercuric Acetate and Mercuric Sulfate in Aqueous Media?

Just as in oxymercuration reaction of alkenes, the mercuric salts stabilize a vinyl cation (the more stable secondary), and addition of water leads to an enol, which tautomerizes to 2-pentanone (***13.77***) as the major product. The difference between alkynes and alkenes is that the vinyl mercury compound is unstable and rapidly loses mercury from the enol intermediate to form the ketone directly.

Addition of Halogens

What Is the Transient Intermediate in the Reaction of Bromine to 2-Butyne and What Is the Product?

Just as bromine adds to an alkene to form a bromonium ion (see ***13.41***), bromine adds to one π-bond of 2-butyne to form ***13.80***. The bromide ion formed during reaction of diatomic bromine with the alkyne attacks ***13.80*** as a nucleophile, opening it to form the vicinal vinyl dibromide, in this case 2,3-dibromo-2-butene, ***13.81***. In general, the *trans*-isomer predominates, since the ring opening requires *anti*-attack, as with reactions of alkenes.

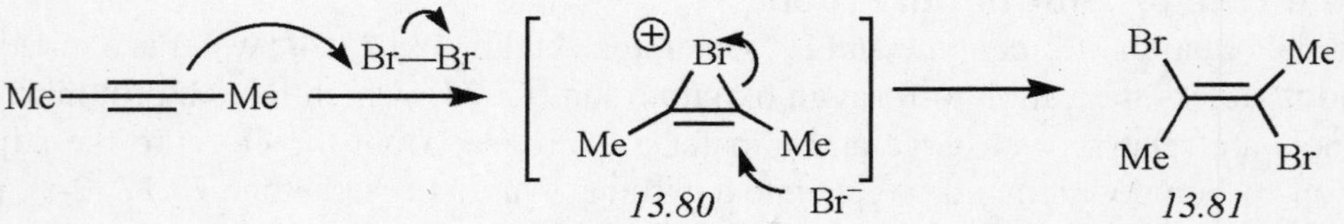

13.80 *13.81*

Why Do the Bromine Atoms in the Product of a Reaction between 3-Hexyne and Bromine Have a Trans-Relationship?

As noted in the formation of the trans product ***13.81***, a 'vinyl bromonium ion' is formed, and the nucleophilic bromide ion that opens this intermediate will attack on the face opposite the bromine (*anti*-attack), due to steric and electronic repulsive forces.

Give the Name of the Product Formed When 3-Hexyne Reacts with Chlorine.

The product is trans-3,4-dichloro-3-hexene.

If Two Equivalents of Bromine Are Added to 3-Hexyne, What Is the Final Product? Draw it.

The product is 3,3,4,4-tetrachlorohexane. The C=C unit of the initially formed 3,4-dibromo-3-hexene reacts with additional bromine to give ***13.82***.

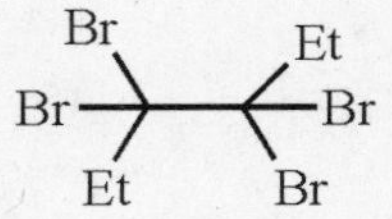

13.82

What Is the Major Product When 5,6-Dimethyl-1-Phenyl-3-Octyne (13.83) Is Treated with Bromine in CCl_4?

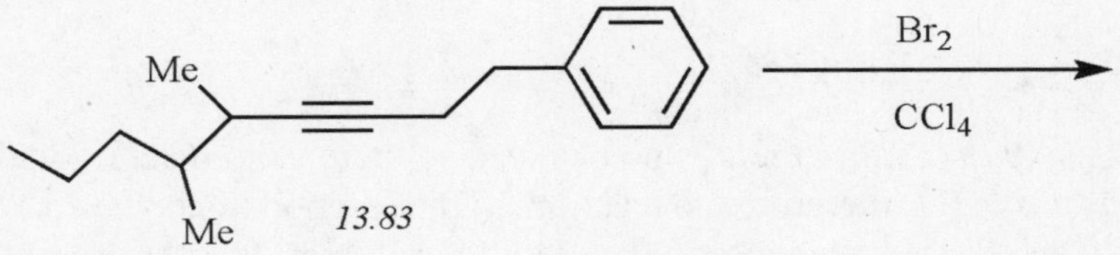

13.83 — Br_2 / CCl_4 →

The major product is *E*-3,4-dibromo-5,6-dimethyl-1-phenyl-3-octene, ***13.84***.

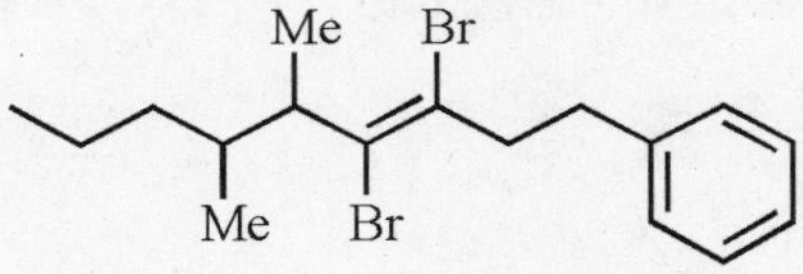

13.84

Explain Why the Product 13.84 Does Not Have a Chlorine Atom, Despite the Fact That Carbon Tetrachloride Was Used.

Carbon tetrachloride is the solvent in this reaction and does not react with the alkyne.

Hydrogenation

When Only One Equivalent of Hydrogen Is Used, What Is the Product of Catalytic Hydrogenation of an Alkyne?

A single equivalent of hydrogen, with an appropriate catalyst, will give an alkene as the final product. Some alkane may be formed if the alkene product is more reactive to the catalyst than is the alkyne. The reaction is: $R–C\equiv C–R + H_2 + Pd/C \rightarrow RCH=CHR$.

How Many Equivalents of Hydrogen Are Required to Convert an Alkyne to an Alkane?

The conversion of an alkyne to an alkane is $R–C\equiv C–R \rightarrow RCH_2CH_2R$ and requires a minimum of two equivalents of hydrogen gas, along with an appropriate catalyst.

What Catalysts Can Be Used for Hydrogenation of an Alkyne?

In general, the same catalysts used for hydrogenation of an alkene, nickel, platinum and palladium.

What Is the Major Product in the Reactions of 13.85–13.87?

H_2 , EtOH →

13.85

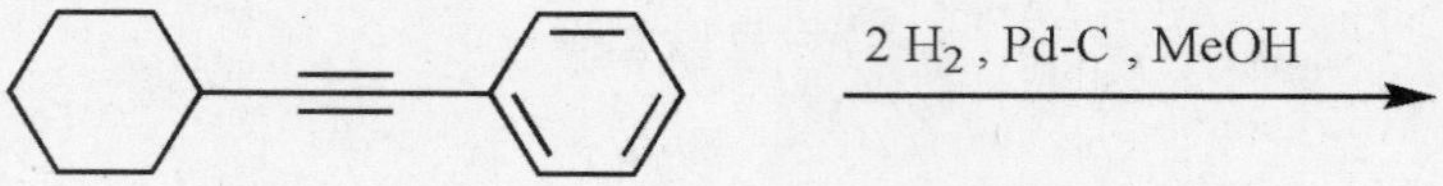

2 H_2 , Pd-C , MeOH →

13.86

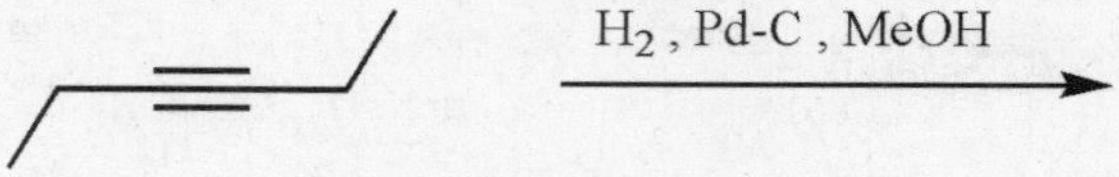

H_2 , Pd-C , MeOH →

13.87

In the reaction 1-hexyne (***13.85***), there is no catalyst, so there is no reaction with hydrogen, and no product is formed. The answer is, therefore, no reaction. In the reaction of alkyne ***13.86***, reaction with two equivalents of hydrogen, in the presence of a palladium catalyst, leads to the alkane ***13.88.*** When 3-hexyne (***13.87***) reacts with one equivalent of hydrogen in the presence of the palladium catalyst, the product is an alkene. However, there is little selectivity, and ***13.89*** is a mixture of *cis* and *trans* isomers.

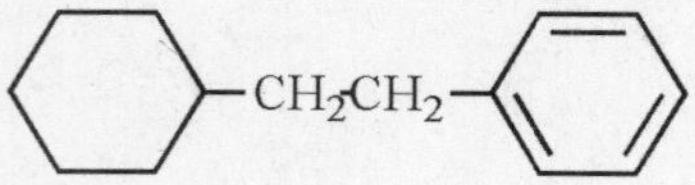

CH_2CH_2

13.88

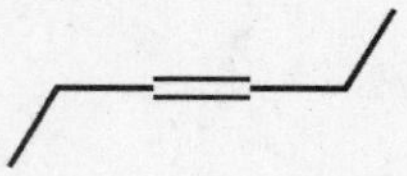

cis + *trans*

13.89

How Many Products Are Possible When 3-Hexyne Is Treated with One Equivalent of Hydrogen in the Presence of a Platinum Catalyst?

There are three possible products, two major and one minor. The two major products are *cis*- and *trans*-3-hexene (***13.89***), and the minor product is hexane. In this case, reduction of the alkyne group in 3-hexyne to the alkene gives either a *cis*-alkene or a *trans*-alkene. In this case, there is no way to control the reaction and *both* stereoisomers are formed. In many cases, the thermodynamically more stable *trans*-isomer is formed in greater amount than the *cis*-, but both are formed.

Why Is It Important to Carefully Control the Number of Equivalents of Hydrogen Added to an Alkyne When the Alkene Is the Desired Product?

If too much hydrogen gas is used, the alkene product of the reduction can be further reduced to give the alkane, as seen above. Another problem is that the alkene product may be more reactive to the catalyst (or to hydrogen after being bound to the catalyst) and react faster than unreacted alkyne. For this reason, the alkane often accompanies the alkene as a minor product.

If One Equivalent of Hydrogen Gas Is Used, but the Reaction Conditions Are Vigorous (Heat, Pressure), What Types of Products Are Possible?

Since the alkene product may be more reactive than the alkyne starting material, vigorous conditions (heat and pressure) may cause over-reaction. To minimize this, both the temperature and the pressure of the reaction are controlled, along with the number of equivalents of hydrogen gas. Another important factor in this reaction is the nature of the catalyst. In general, palladium catalysts are used to reduce alkynes to alkenes, although rhodium and ruthenium catalysts are also very useful.

What Is the Lindlar Catalyst?

The original Lindlar catalyst was a mixture of palladium chloride ($PdCl_2$), which was precipitated on calcium carbonate ($CaCO_3$) in acidic media and deactivated with lead tetraacetate [$Pb(OAc)_4$] to give the named Pd–$CaCO_3$–PbO catalyst. Later, palladium on barium carbonate ($BaCO_3$) or calcium carbonate ($CaCO_3$) was deactivated with quinoline (***13.90***) and found to give similar reactivity. This latter catalyst is now known as the Lindlar catalyst.

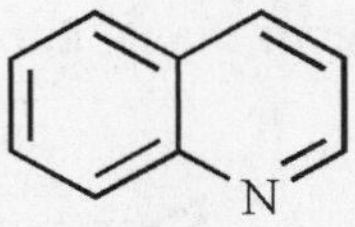

N

13.90

Why Is the Lindlar Catalyst Important?

The importance of the Lindlar catalyst is shown by the conversion of alkyne ***13.86*** to ***13.91***. The Lindlar catalyst gives almost exclusively the *cis*-alkene from the alkyne, with little or no contamination by the *trans*-alkene or from further reduction of the alkene product (in this case, further reduction would give 2-phenylethyl cyclohexane). *The preferred use of the Lindlar catalyst is, therefore, formation of cis-alkenes from alkynes.*

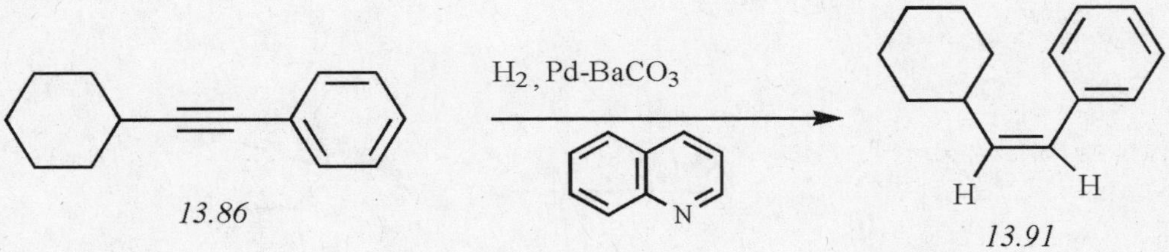

13.86 — H_2, Pd-$BaCO_3$, N (*13.90*) → H H *13.91*

What Is the Major Product of the Reaction of 5-Methyl-2-Hexyne and One Equivalent of Hydrogen Gas in the Presence of the Lindlar Catalyst?

The reaction of 5-methyl-2-hexyne (***13.92***) under these conditions gives the *cis*-alkene, ***13.93*** (*cis*-5-methyl-2-hexene) as the major product.

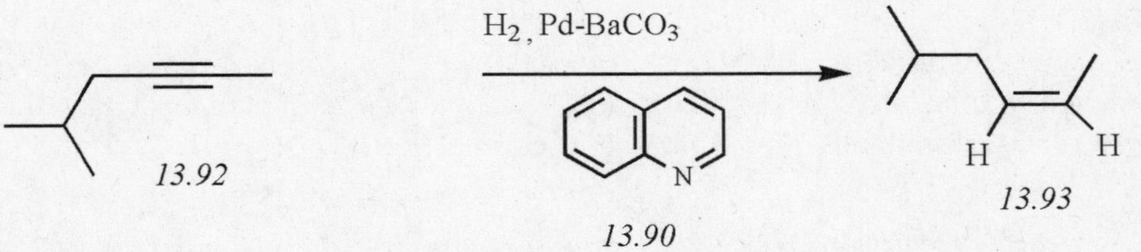

13.92 — H_2, Pd-$BaCO_3$, N (*13.90*) → H H *13.93*

Test Yourself

1. Give the mechanism for the following reaction.

HCl

Cl

2. Why does 2-methyl-2-butene react faster with HCl than does 1-butene?
3. Explain why the H^+ of HBr adds to C1 rather than C2 to give the major product.

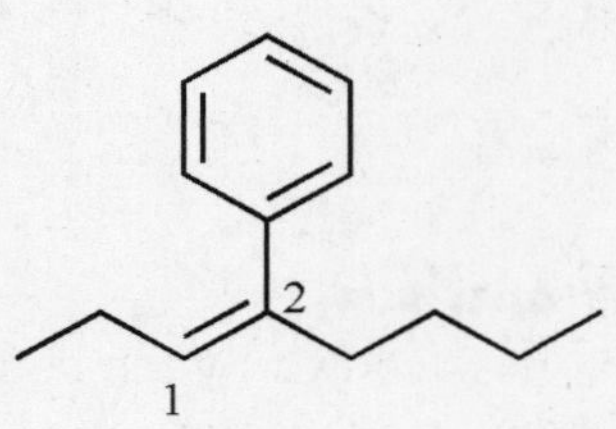

2

1

4. Give at least four *different* transition metals that can be used in hydrogenation reactions.
5. Give the mechanism for the following reaction. Why was perchloric acid used rather than HCl?

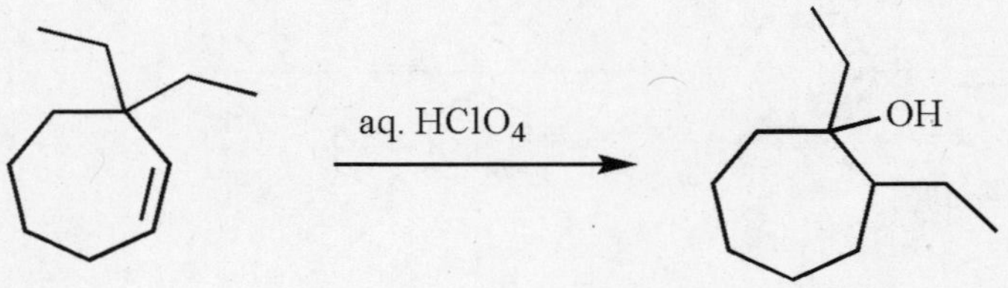

aq. $HClO_4$

OH

6. What is an induced dipole?
7. Does *trans*-2-butene react with iodine to give a meso compound or the d,l pair?
8. In the reaction of 2-methyl-1-pentene and bromine, which carbon is attacked by bromide ion? Explain.

9. In each of the following reactions, give the major product. Remember stereochemistry, and if there is no reaction, indicate this by N.R.

(a) 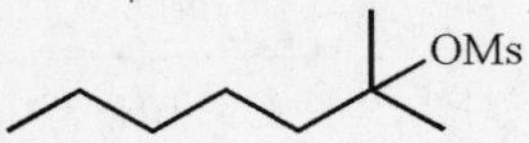 OMs; KCN , EtOH →

(b) 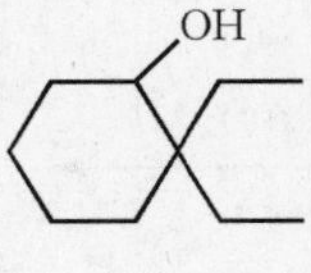 OH; aq. $HClO_4$ →

(c) 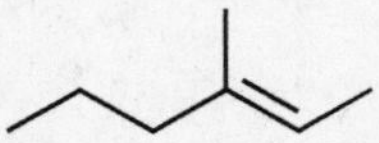 HBr →

(d) 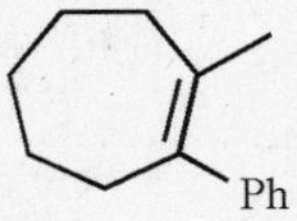 Ph; HCl →

(e) 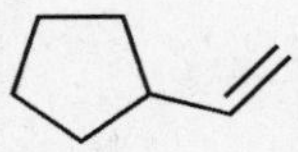 HBr , *t*-BuOO*t*-Bu →

(f) 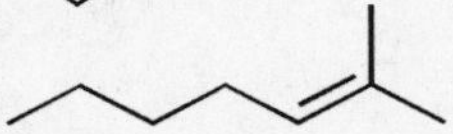 H_2 , Pd-$BaSO_4$ →

(g) 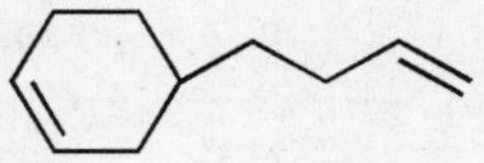 2 H_2 , PtO_2 →

(h) 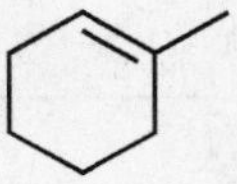 1. $Hg(OAc)_2$, H_2O 2. $NaBH_4$ →

(i)  1. $Hg(OAc)_2$, H_2O 2. $NaBH_4$ →

(j)  I_2 , CCl_4 →

(k) 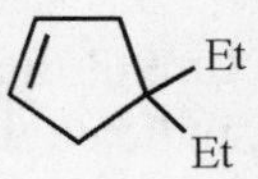 Et, Et; Br_2 , CCl_4 →

(l) 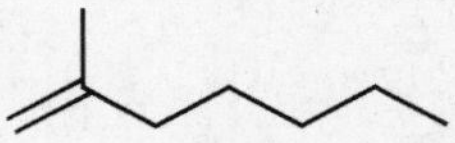 Cl_2 , H_2O →

10. In each case, give the major product of the reaction. Remember stereochemistry, and if there is no reaction, indicate this by N.R.

(a) 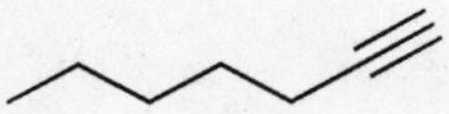 1. HBr 2. HBr →

(b) 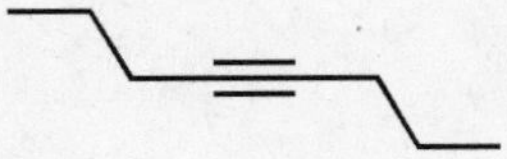 H_2 , Pd-$BaSO_4$ quinoline →

(c) 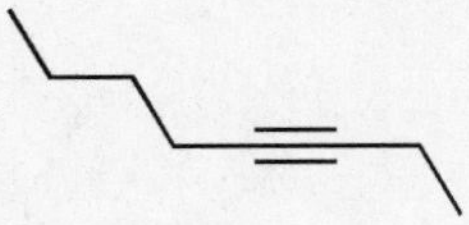 n-BuLi ether →

(d) 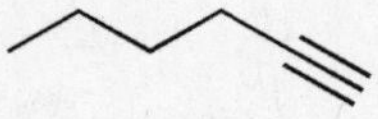 3 H_2 →

(e) 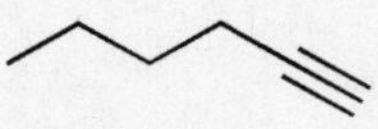 1. Br_2 2. excess $NaNH_2$ →

(f) 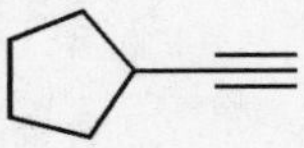 HI →

(g) 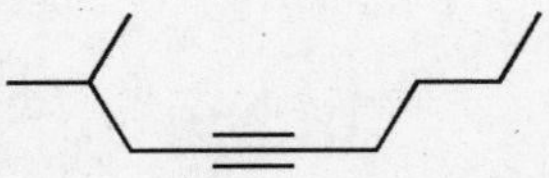 HCl →

(h) 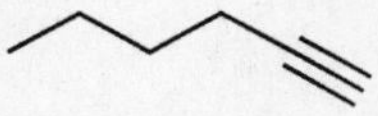 $Hg(OAc)_2$ $HgSO_4$ →

(i) 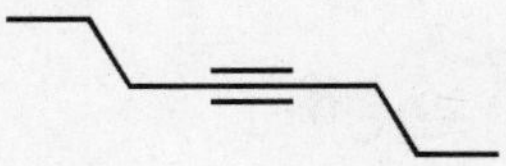 Br_2 , CCl_4 →

(j) 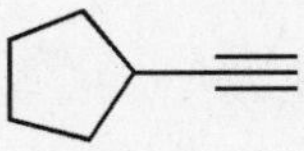 2 Cl_2 →

(k) 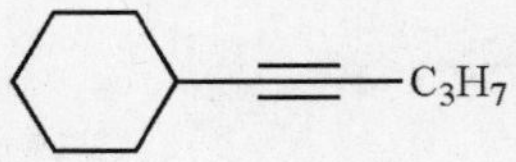 C_3H_7; H_2 , Pd-$BaSO_4$ quinoline →

11. For each of the following reactions, give the major product. Remember stereochemistry where appropriate, and if there is no reaction, indicate this by N.R.

(a) aq. THF , cat. *p*-TsOH

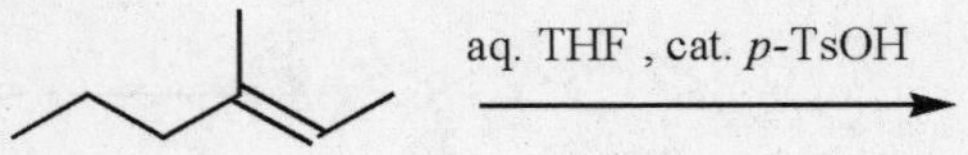

(c) 1. $Hg(OAc)_2$, H_2O 2. $NaBH_4$

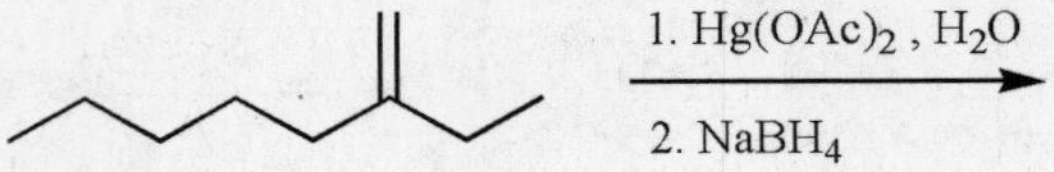

(b) cat. H_2SO_4 aq. THF

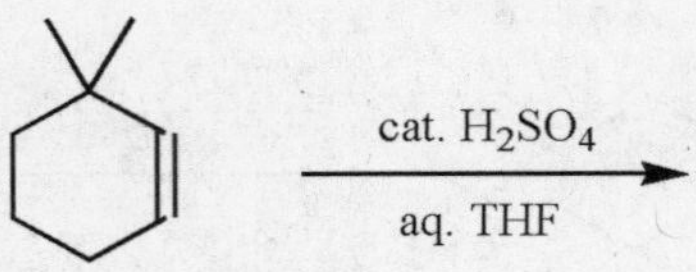

(d) 1. $Hg(OAc)_2$, H_2O 2. $NaBH_4$

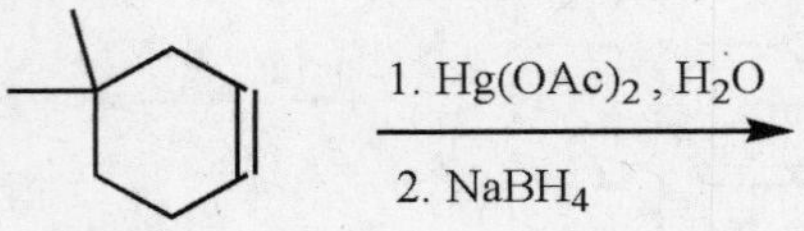

12. Give the complete mechanism for the following reaction.

$Hg(OAc)_2$, EtOH

OEt

HgOAc

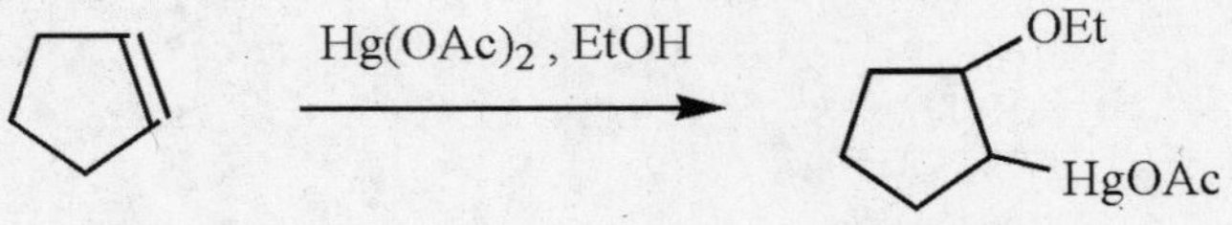

13. In each case, give the structure of the major product. Remember stereochemistry where appropriate, and if there is no reaction, indicate this by N.R.

Me

HOBr

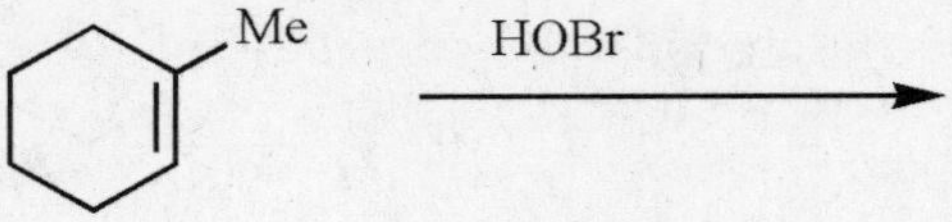

CHAPTER 14

Oxidation Reactions of Alkenes and Alkynes

The primary reduction reaction of alkenes involves catalytic hydrogenation, which was introduced in Section 13.4. For that reason, this chapter will focus on oxidation reactions of alkenes and alkynes. Although the reactions to be discussed can be considered addition-type reactions, there are sufficient differences that they are discussed separately.

14.1. HYDROXYLATION

It is possible to add two hydroxyl groups across an alkene, giving a vicinal diol. There are two major reagents that give this product, $KMnO_4$ (potassium permanganate) and OsO_4 (osmium tetroxide).

Potassium Permanganate

Potassium permanganate has the structure *14.1*, where the manganate anion is resonance stabilized.

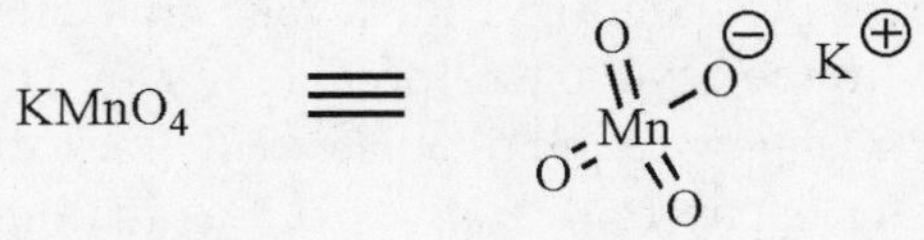

$KMnO_4$ ≡ $O{=}Mn({=}O)_2{-}O^{\ominus}\ K^{\oplus}$

14.1

What Is the Major Product When Dilute $KMnO_4$ Reacts with Cyclopentene in Aqueous KOH?

The major product of this reaction is *cis*-1,2-cyclopentanediol, ***14.3***. The reaction proceeds by initial formation of a cyclic manganese compound (***14.2***), which is decomposed by the hydroxide in solution to give the diol.

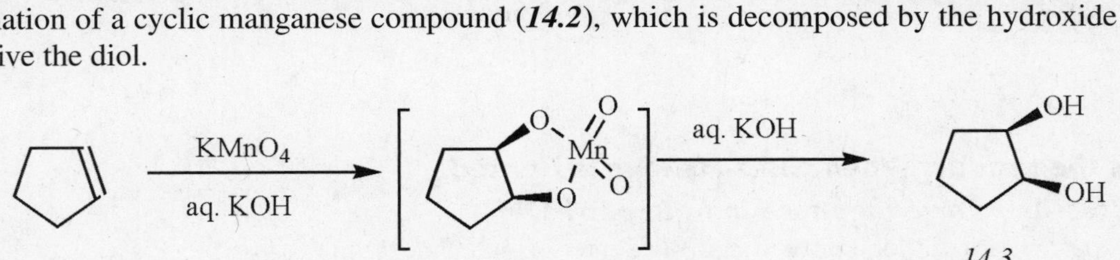

cyclopentene —($KMnO_4$, aq. KOH)→ [cyclic manganate ester] *14.2* —(aq. KOH)→ *cis*-1,2-cyclopentanediol (OH, OH) *14.3*

What Is a Manganate Ester?

A manganate ester is a compound containing two OR units attached to manganese, with the generic structure $O_2Mn(OR)_2$. The intermediate in the reaction just presented (*14.2*) is a **manganate ester**.

What Is the Purpose of the KOH in the Reaction That Formed 14.3?

Intermediate ***14.2*** is the initial product of the reaction of cyclopentene and permanganate (***14.1***), and a second chemical reaction (with hydroxide) is required to convert ***14.2*** into the diol product (***14.3***). Hydroxide attacks the Mn in ***14.2*** and breaks the Mn–O–C bond between Mn and O to form a species like ***14.4***. Two attacks by hydroxide on ***14.2*** and proton transfer to the O^- species gives the diol.

KOH
HOH

14.1 *14.2* *14.4*

Why Does This Reaction Produce the Cis- Diol?

As seen in ***14.2***, the alkene reacts with two of the oxygens of MnO_4^-. The two bonds are formed close together in time, probably in a concerted manner, generating the *cis*- stereochemistry in ***14.2***. The *trans*- compound could form only if there were a bond rotation prior to formation of the second C–O bond. The bond-making process is too fast to allow this, and the *cis*-stereo-chemistry is characteristic of this reaction. Since hydroxide attacks Mn rather than carbon, the stereochemistry at the carbon atom is retained, and the final diol product (***14.3***) retains the *cis*- relationship of the oxygen atoms in ***14.2***.

What Is the Major Product When trans-3-Hexene Reacts with $KMnO_4$ and Aqueous KOH?

Since the *cis*-diol is formed without formation of the *trans*-diastereomer, the reaction is diastereospecific. It will form the racemic *cis*-diol. Starting with acyclic molecules such as *trans*-3-hexene (***14.5***), the terms *cis-/trans-* do not apply to the acyclic diol product (***14.7***). During the course of the reaction, however, formation of the cyclic manganese intermediate will 'lock' the *trans*-stereochemistry of the ethyl groups into place. Hydroxide opens ***14.6*** to give ***14.7*** as the only diastereomeric product, as a racemic mixture.

$KMnO_4$
aq KOH

14.5 *14.6* *14.7*

What Is the Product When Cis-3-Hexene Is Treated with Potassium Permanganate in Aqueous KOH?

This particular reaction converts cis-3-hexene (***14.8***) to ***14.10*** via ***14.9***.

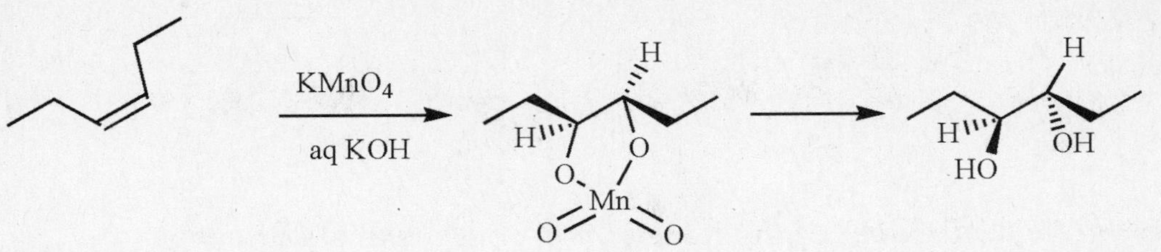

KMnO$_4$
aq KOH

14.8 14.9 14.10

How Many Stereoisomers Are Represented by Structure 14.10?

Diol ***14.10*** is a meso compound, so it represents a single compound, not a mixture of enantiomers.

What Is the Term for Conversion of an Alkene to a Vicinal Diol?

This particular transformation is called **dihydroxylation**.

Is the Reaction between an Alkene and Hydroxide/KMnO$_4$ Selective or Specific?

Specific! The *cis*-dihydroxylation reaction is diastereospecific, since only one diastereomer is formed.

Osmium Tetroxide

Osmium tetroxide has the structure ***14.11***.

OsO_4 ≡

14.11

How Does OsO$_4$ React with Cyclohexene?

Osmium tetroxide (OsO_4, ***14.11***) reacts very similarly to $KMnO_4$. The alkene reacts with the oxygens of OsO_4 to form what is known as an osmate ester (***14.12***). A reagent other than KOH is required to decompose this osmate and generate the diol product (***14.13***). The usual reagent is sodium thiosulfite ($NaHSO_3$), which can be added in an aqueous solution in a second chemical step, although it is common to mix the alkene with OsO_4 and $NaHSO_3$ in the same solution. The combination of OsO_4 and aqueous $NaHSO_3$ converts alkenes to vicinal *cis*- diols.

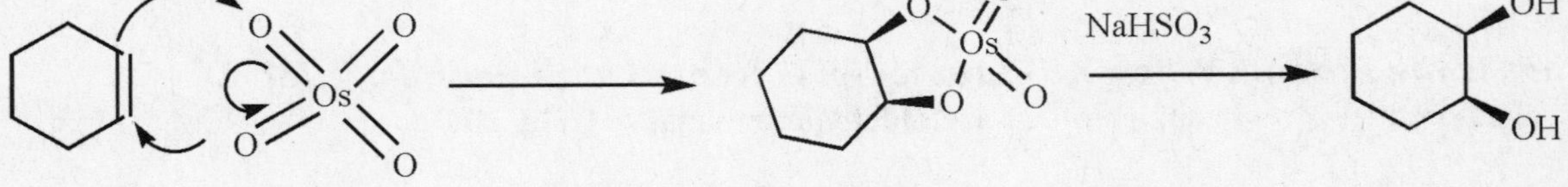

$NaHSO_3$

14.11 14.12 14.13

Are Other Reagents Available to Convert the Osmate Ester to the Diol?

Yes! Two common reagents are *tert*-butylhydroperoxide (*14.14*) and *N*-methylmorpholine *N*-oxide (***14.15***). In both cases, the alkene is mixed with OsO_4 and one of these two reagents in aqueous solutions. These reagents allow a catalytic amount of OsO_4 to be used, and the product is the diol.

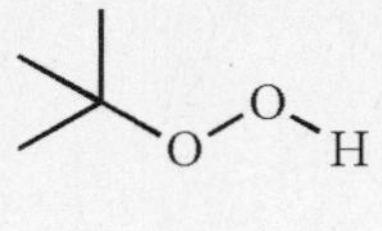

14.14 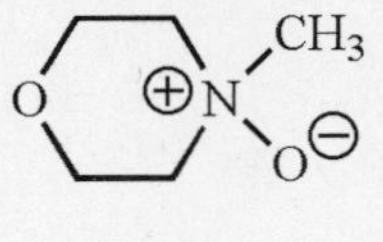   14.15

What Is the Major Product of the Reaction of E-Hexene (14.5) and Osmium Tetroxide, in the Presence of Aqueous Sodium Thiosulfate?

Hydroxylation with OsO_4 is also diastereospecific, as with the permanganate hydroxylation. In this case, 2*E*-hexene is converted to the racemic diastereomer ***14.16*** because of the *cis*- addition of osmium tetroxide.

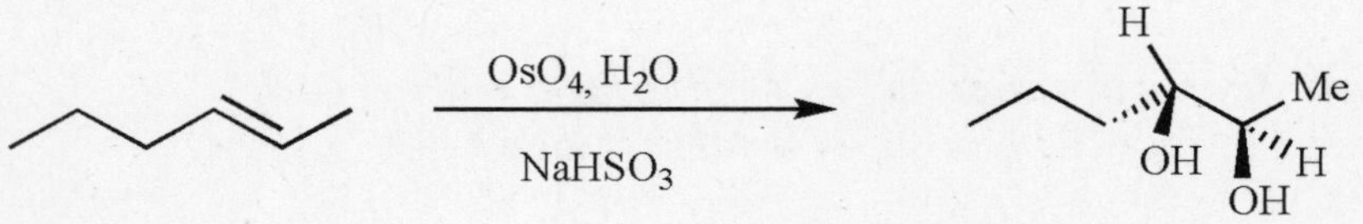

OsO_4, H_2O / $NaHSO_3$

14.5   14.16

If $KMnO_4$ and OsO_4 Lead to the Same Products, Why Use OsO_4?

The yields of *cis*-diol tend to be higher with OsO_4 and there are fewer side reactions. If permanganate reactions become too hot or too concentrated, oxidative cleavage of the alkene can result. This is rarely a problem with OsO_4. There are drawbacks, however, since OsO_4 is expensive and toxic. For these two reasons, OsO_4 is used for small scale applications in reactions with very expensive alkenes or when the $KMnO_4$ reaction cannot be controlled. Otherwise, potassium permanganate is used.

14.2. OXIDATIVE CLEAVAGE

Alkenes can be oxidatively cleaved to ketones, aldehydes or carboxylic acids with a variety of reagents. The most common are ozone and potassium permanganate.

What Is the Structure of Periodic Acid?

Periodic acid is usually written as HIO_4 but exists as the dihydrate, written as H_5IO_6. The two structures are used interchangeably.

What Is the Product of the Reaction of 1,2-Cyclohexanediol and Periodic Acid?

Periodic acid cleaves 1,2-diols into aldehydes. In this case, the product is a dialdehyde, 1,6-hexanedial [OHC–$(CH_2)_4$–CHO].

What Is the Product When 2,3-Butanediol Is Treated with Periodic Acid?

The product is two equivalents of acetaldehyde (ethanal = CH_3CHO).

What Is the General Mechanism of This Reaction?

Periodic acid is H_5IO_6 (see above), but is usually drawn as HIO_4 ***14.17*** and reacts with the 1,2-butanediol (***14.18***) to form a cyclic periodate ester (***14.19***). A concerted electronic rearrangement allows cleavage of the carbon-carbon bond, liberating two equivalents of the carbonyl compound (here, ethanal, ***14.20*** and HIO_3, ***14.21***).

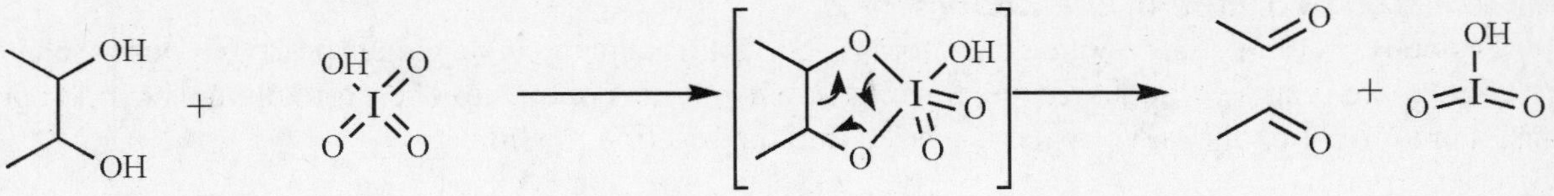

14.18 14.17 14.19 14.20 14.21

How Are 1,2-Diols Usually Formed?

As discussed in Section 14.1, 1,2-diols are usually formed by treatment of an alkene with dilute $KMnO_4$ and hydroxide or with OsO_4 and sodium thiosulfite.

What Is the Product When 2-Heptene (14.22) Is Treated with a Mixture of Osmium Tetroxide and HIO_4?

In this case the OsO_4 converts 2-heptene (***14.22***) to the corresponding 1,2-diol *in situ* (in other words, the diol is generated in the reaction medium during the course of the reaction, and then it reacts with the periodic acid), in the presence of HIO_4. The HIO_4 then cleaves the diol into carbonyl compounds, in this case ethanal (***14.20***) and pentanal (***14.23***).

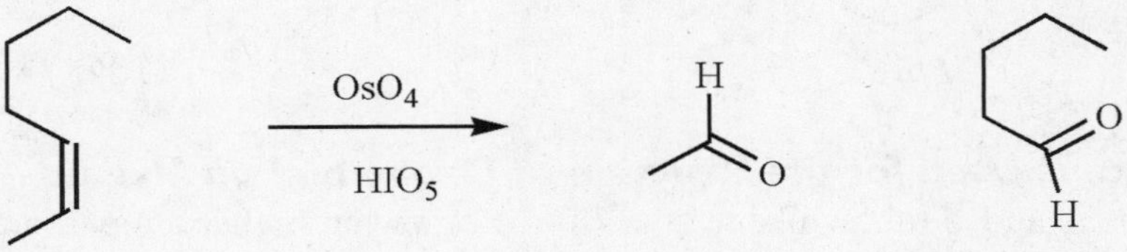

OsO_4, HIO_5

14.22 14.20 14.23

What Is the Product When 2,3-Butanediol is Treated with Hot and Concentrated Potassium Permanganate?

Hot and concentrated permanganate is capable of cleaving 1,2-diols to the corresponding carboxylic acids via oxidative cleavage. In this case, cleavage of the symmetrical diol leads to two equivalents of ethanoic acid (acetic acid = CH_3COOH).

14.3. OZONOLYSIS

Ozone (O_3) adds to alkenes, but the initially formed product rearranges to a more stable product called an **ozonide**. Subsequent treatment with an oxidizing or reducing agent leads to net cleavage of the carbon-carbon double bond and formation of aldehydes, ketones or carboxylic acids.

Draw All the Resonance Forms of Ozone.

There are four resonance forms that bear positive and negative charges, ***14.24-14.27***.

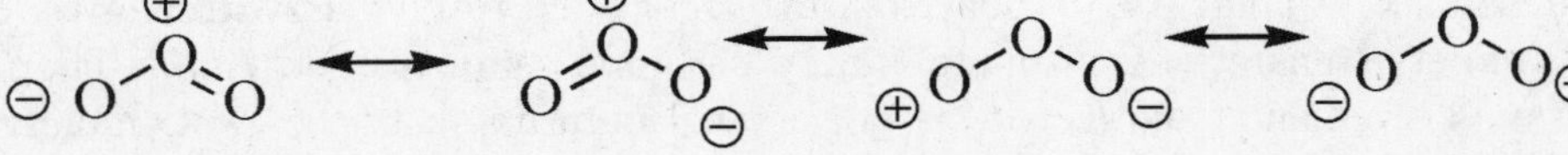

14.24 14.25 14.26 14.27

How Is Ozone Formed in the Laboratory?

Typically, a commercial ozone generator is used. In this apparatus, oxygen is passed through a chamber where there is an electric discharge, forming small amounts of ozone. Therefore, in most ozone reactions, a mixture of ozone and oxygen is passed into the reaction medium.

Which Resonance Form(s) is(are) Most Likely to Predict the Reactivity of Ozone with the π-Bond of an Alkene?

In a process that is similar to the one observed with OsO_4 and MnO_4^-, ozone can be thought to react with alkenes such as cyclohexene via ***14.26*** or ***14.27***, where the π-bond of an alkene attacks the positive oxygen, and the negatively charged oxygen of ozone in turn attacks the developing positive charge on the carbon (see ***14.28***) to form a cyclic species, ***14.2.*** The five-membered ring that contains three oxygen atoms is known as a 1,2,3-trioxolane—also known as a **molozonide**. Clearly, these are simply resonance forms and not discrete structures, but using ***14.26*** and/or ***14.27*** makes the analysis easier to see.

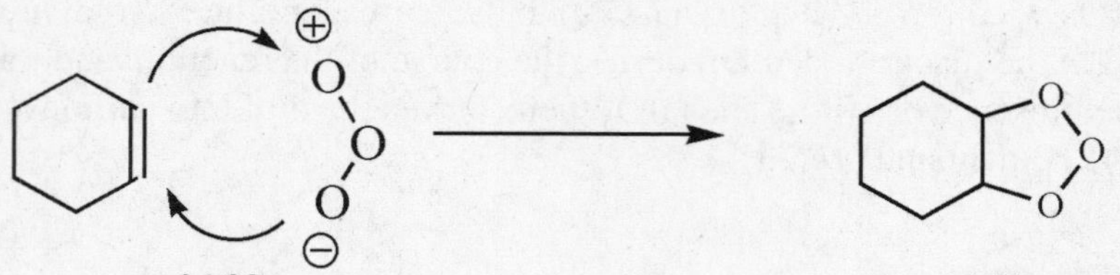

14.28 *14.29*

What Temperature Is Used for the Reaction of Ozone and an Alkene?

Typically, ozone is added to the alkene at −78°C. Somewhat higher temperatures can be used, but the highly reactive nature of ozone and the lability of the products formed usually require the lower temperature.

When Ozone Reacts with 2-Butene, the Observed Product Is Not the Initial 1,2,3-Trioxolane. Explain.

The initially formed 1,2,3-trioxolane (***14.30***) is unstable and rearranges to the more stable 1,2,4-trioxolane, ***14.31***. This final product is called an ozonide.

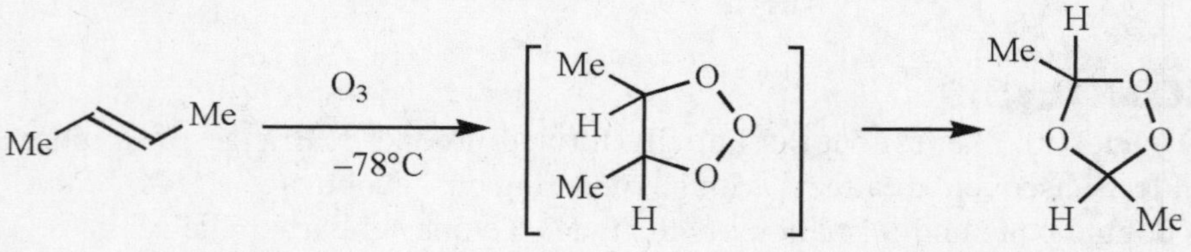

O_3, −78°C

14.30 *14.31*

What is the Mechanism for Transformation of 1,2,3-Trioxolane 14.30 to 1,2,4-Trioxolane 14.31?

In this mechanism, ozonide ***14.30*** cleaves at the O–O bond to give a **zwitterion** (a dipolar ion), ***14.32***. The O^- moiety transfers electrons to form a carbonyl (ethanal) and ***14.33***. The O^- of ***14.33*** attacks the carbonyl of ethanal, and the oxygen of ethanal in turn attacks the C=O moiety of ***14.33*** to give the ozonide, ***14.31***. This latter reaction occurs before the two molecules (shown in ***14.33***) can 'drift apart.'

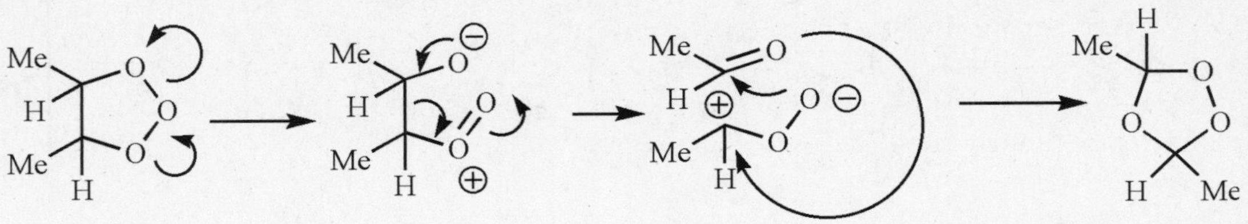

14.30 14.32 14.33 14.31

When the Ozonide Derived from 1-Hexene Is Treated with Hydrogen Peroxide, What Are the Products?

The ozonide formed from 1-hexene is ***14.32***. Both C1 and C2 in ***14.32*** have at least one hydrogen attached. These carbons are derived from the C=C bond. With an oxidizing agent such as hydrogen peroxide, both C1 and C2 are converted to a carboxylic acid (initial oxidation to an aldehyde is followed by rapid oxidation, *in situ* —in the reaction medium without isolation— to the acid; for a discussion of carboxylic acids, see Chapter 24). In this case, the two acid products (two carbonyl products from cleavage of the C=C bond) are formic acid (***14.33***) and pentanoic acid (***14.34***). If a hydrogen is not present, the product is a ketone. These reactions can be generalized:

$$RCH{=}CHR^1 \rightarrow RCO_2H + R^1CO_2H$$
$$R_2C{=}CHR^1 \rightarrow R_2C{=}O + R^1CO_2H$$
and $$R_2C{=}CR^1_2 \rightarrow R_2C{=}O + R^1_2C{=}O.$$

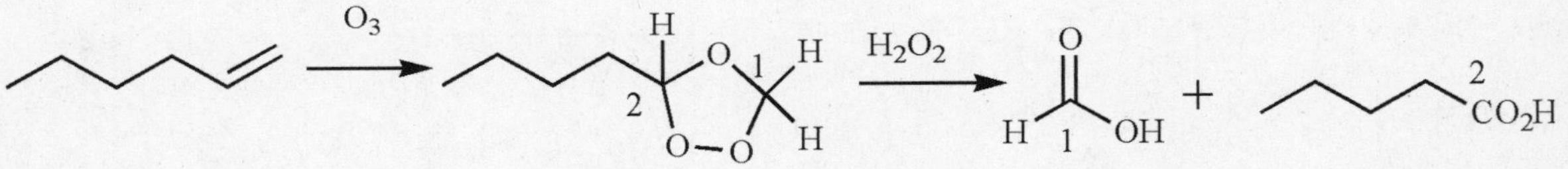

O_3 H_2O_2 14.32 14.33 14.34

When the Ozonide Derived from 2,3-Dimethyl-2-Pentene (14.35) Is Treated with Hydrogen Peroxide, What Are the Products?

Cleavage of the C=C bond in ***14.35*** leads to ozonide ***14.36***. In this case, both C1 and C2 have two alkyl groups attached and no hydrogen atoms attached to those carbon atoms. Oxidation of ***14.36*** therefore leads to two ketone products, acetone (***14.37***) and 2-pentanone (***14.38***).

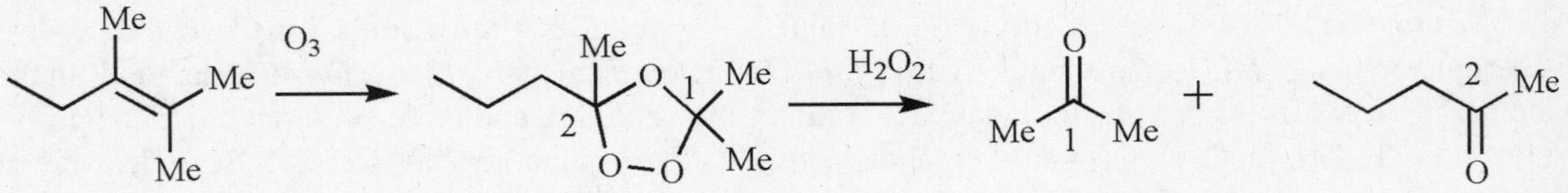

O_3 H_2O_2 14.35 14.36 14.37 14.38

When the Ozonide Derived from 1-Hexene Is Treated with Zinc and Acetic Acid, What Are the Products?

The ozonide (***14.32***) derived from 1-hexene can be reduced rather than oxidized by using zinc and acetic acid or dimethyl sulfide. When ***14.32*** is reduced with zinc metal in acetic acid (AcOH; CH_3CO_2H), the products are aldehydes rather than acids ***14.37*** and ***14.38***. In this case, formaldehyde (***14.39***) and pentanal (***14.40***) are formed by ozonolysis of 1-hexene followed by reduction of ***14.32***.

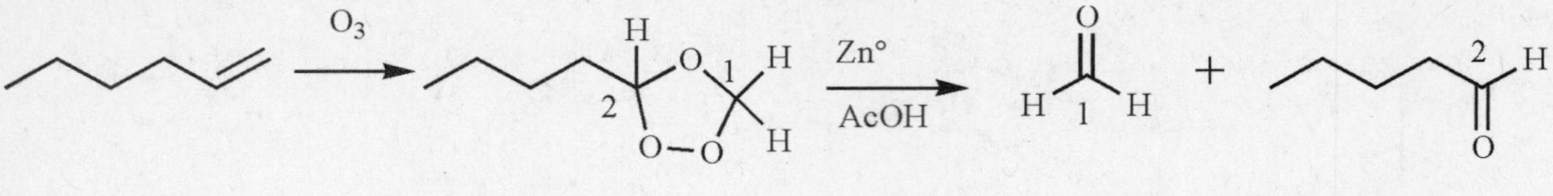

O_3 · Zn° / AcOH

14.32 · 14.39 · 14.40

What Is Dimethyl Sulfide?

Dimethyl sulfide is a **thioether** (the sulfur analog of an ether), with the structure CH_3–S–CH_3.

When the Ozonide Derived from 1-Hexene Is Treated with Dimethyl Sulfide, What Are the Products?

The products are ***14.39*** and ***14.40***, the same as when ***14.32*** is treated with zinc and acetic acid.

Give the Product(s) of the Reactions of 14.41–14.42.

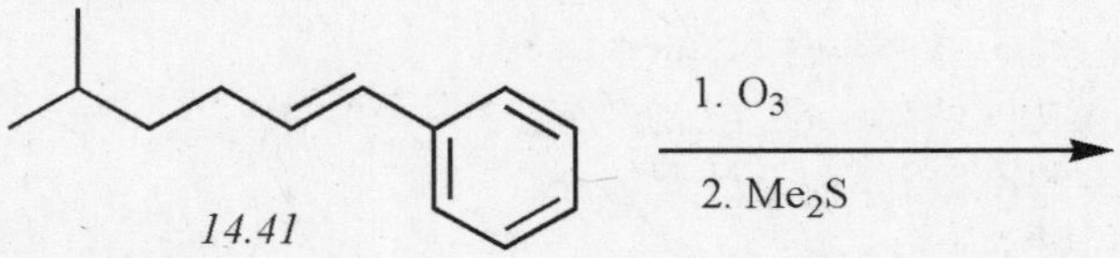

14.41: 1. O_3; 2. Me_2S

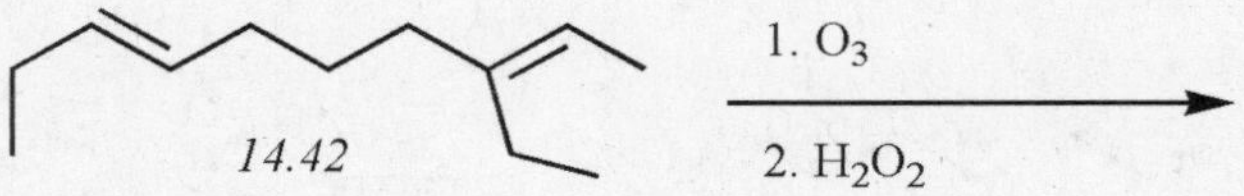

14.42: 1. O_3; 2. H_2O_2

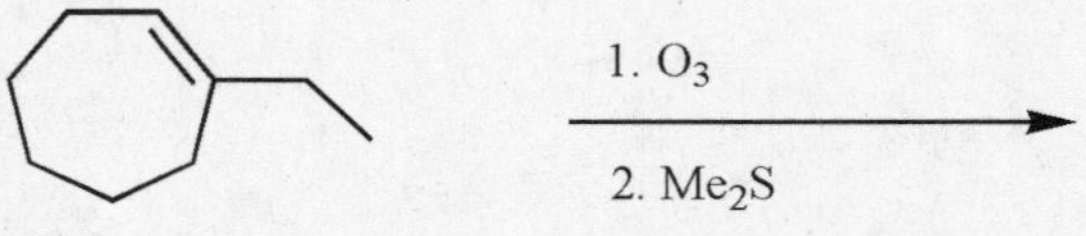

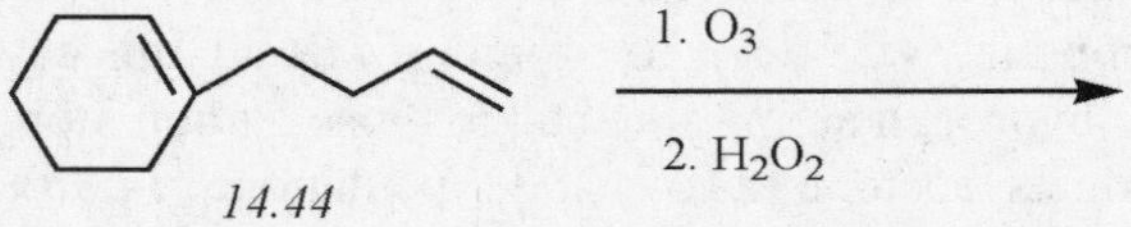

14.43: 1. O_3; 2. Me_2S

14.44: 1. O_3; 2. H_2O_2

Ozonolysis of ***14.41*** gives the ozonide, and Me_2S reduces the ozonide into two aldehydes: 4-methylpentanal (***14.45***) and benzaldehyde (***14.46***). In the reaction of ***14.42***, the two double bonds in the diene are cleaved (followed by an oxidative workup of the ozonide), and there are three products: propionic acid (CH_3CH_2COOH) + 5-keto-heptanoic acid (***14.47***) + ethanoic acid (CH_3COOH). Cleavage of the double bond in cyclic alkene ***14.43*** gives a single molecule with two functional groups: keto-aldehyde ***14.48***. Finally, diene ***14.44*** is cleaved to two products, methanoic acid, also called formic acid, HCOOH, and keto-diacid ***14.49***. Note that the cyclohexene unit in ***14.44*** is cleaved to a keto-acid, and the terminal alkene unit of ***14.44*** is cleaved to formic acid and the second COOH unit in ***14.49***.

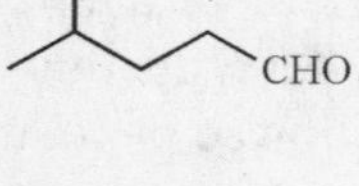

CHO · OHC · HO · HO · OH

14.45 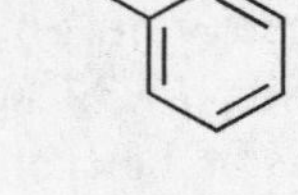 · 14.46  · 14.47 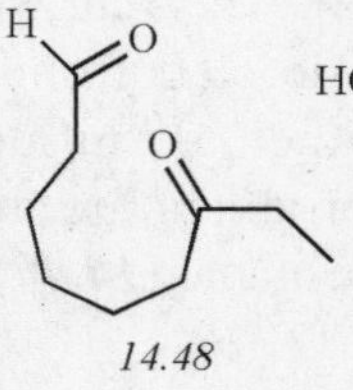 · 14.48 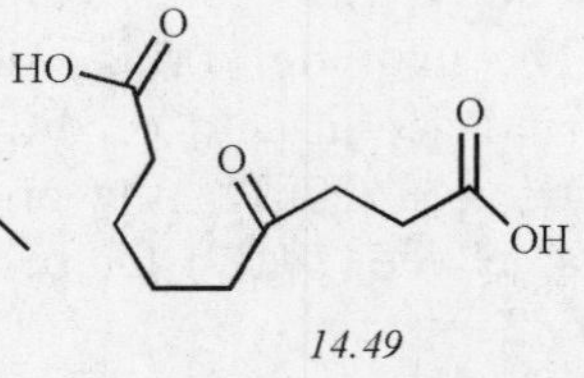 · 14.49

14.4. HYDROBORATION AND OXIDATION

This reaction was presented in Section 11.2. The products are aldehydes and ketones, a net oxidation, and the reactions are shown again in that context.

Give the Initial Product When 2-Pentyne Is Treated with Diborane.

The product is a vinylborane, ***14.50***. Oxidation generates an enol (***14.51***) which tautomerizes (see Section 13.5 and Chapter 21) to pentanal, ***14.52***.

B_2H_6 → BR_2 (*14.50*) → NaOH , H_2O_2 → OH (*14.51*) ⇌ H, O (*14.52*)

What Is the Final Product of the Hydroboration/Oxidation (1. Diborane 2. NaOH, H_2O_2) of 2-Butyne?

The final product is 2-butanone (***14.54***), via tautomerization of the initially formed enol (***14.53***).

Me—≡—Me → 1. B_2H_6 , ether 2. NaOH , H_2O_2 → OH, Me, Me (*14.53*) ⇌ O, Me, Me (*14.54*)

What Is the Final Product of the Hydroboration/Oxidation (1. Diborane 2. NaOH, H_2O_2) of 2-Hexyne?

Since this is not a symmetrical alkyne, the reaction with diborane generates two vinylboranes (***14.55*** and ***14.56***). Oxidation converts them to the corresponding enol, and tautomerization leads to two products, 2-hexanone (***14.57***) and 3-hexanone (***14.58***), respectively.

B_2H_6 → BR_2 (*14.55*) + R_2B (*14.56*) → NaOH , H_2O_2 → O (*14.57*) + O (*14.58*)

Test Yourself

1. In each of the following reactions, give the major product. Remember stereochemistry and if there is no reaction, indicate this by N.R.

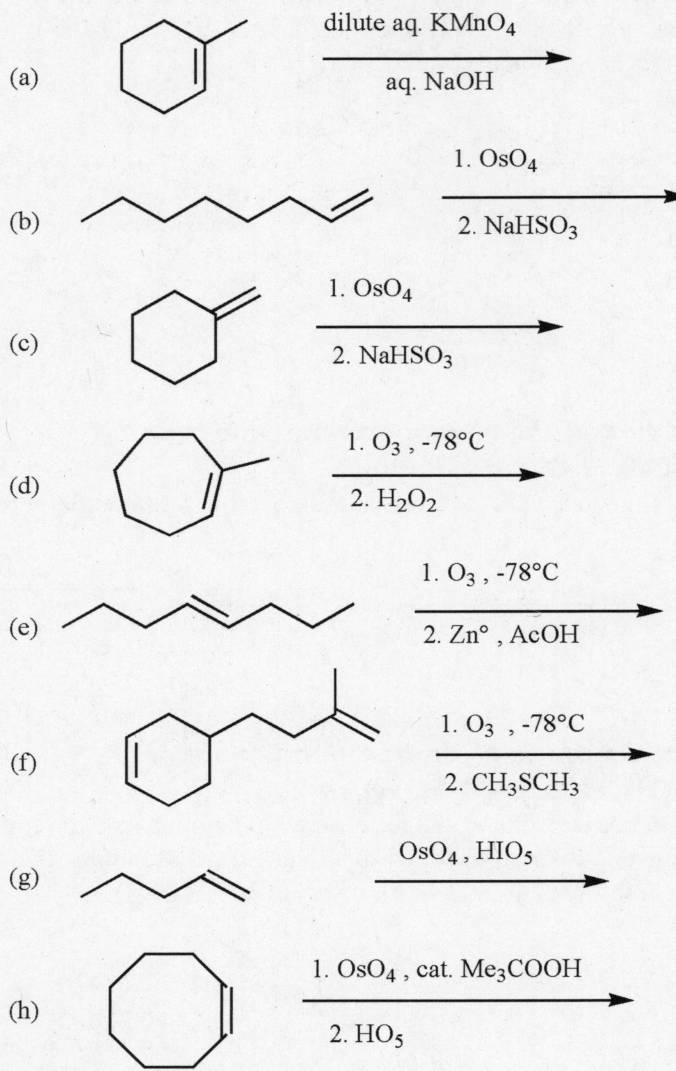

2. Give the products of ozonolysis and hydrogen peroxide workup of 2,3-dimethyl-2-hexene.
3. Give the products of ozonolysis and dimethyl sulfide workup of 2-methyl-2-pentene.

4. What is the major product or products when 2-octene is reacted with ozone and then with hydrogen peroxide?

5. Give the major products from each reaction.

(a) [1-methylcyclohexene] $\xrightarrow[\text{2. } H_2O_2]{\text{1. } O_3\text{, -78°C}}$

(b) [3-ethyl-3-octene] $\xrightarrow[\text{2. } H_2O_2]{\text{1. } O_3\text{, -78°C}}$

6. In each case, give the major product of the reaction. Remember stereochemistry, and if there is no reaction, indicate this by N.R.

(a) [4-octyne] $\xrightarrow[\text{quinoline}]{H_2\text{, Pd-BaSO}_4}$

(b) [1-hexyne] $\xrightarrow{3\ H_2}$

(c) [cyclohexyl]$-C\equiv C-C_3H_7$ $\xrightarrow[\text{quinoline}]{H_2\text{, Pd-BaSO}_4}$

(d) [ethynylcyclopentane] $\xrightarrow[\text{2. NaOH , } H_2O_2]{\text{1. } BH_3\text{ , ether}}$

(e) [3-hexyne] $\xrightarrow[\text{2. NaOH , } H_2O_2]{\text{1. } BH_3\text{ , ether}}$

7. What is the major product when 3-hexene is treated with hot concentrated potassium permanganate?

8. What is the product when 3-hexene is treated with cold and dilute potassium permanganate?

CHAPTER 15

Epoxides

Epoxides, or oxiranes, are three-membered ring ethers. The chemistry of ethers was presented in Chapter 12, but the strain inherent to the three-membered ring of epoxides gives them significant reactivity relative to other ethers. For this reason, epoxides are discussed in a separate chapter.

15.1. STRUCTURE AND BONDING

Epoxides are three-membered ring ethers, and the ring is highly strained.

What Is an Epoxide?

An **epoxide** is a molecule that contains a three-membered ring ether. The most simple example of this type of ether is ethylene oxide, ***15.1***. Nomenclature will be described in Section 15.2.

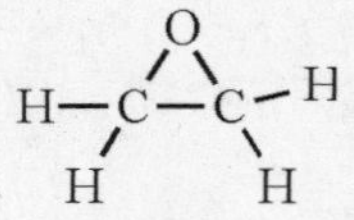

15.1

Describe the Hybridization in a Three-Membered Ring Ether.

The hybridization of an epoxide is similar to that of cyclopropane. The C–C and C–O sigma bonds of an epoxide are weaker than sp^3 hybrids and stronger than sp^2 hybrids. These bonds are usually said to have $sp^{2.33}$ hybridization.

Why Is a Three-Membered Ring Ether Expected to Be More Reactive Than Other Cyclic Ethers?

The strain inherent to a three-membered ring makes the electron density between the nuclei (the σ-bond) distort to a position away from the line between the nuclei. There is less electron density between the nuclei, and the bond is weaker.

Why Are the Sigma Bonds of a Three-Membered Ring Considered to Be 'Bent'?

As described above, the strain of the three-membered ring makes the σ-bonds 'bend' away from the line between the nuclei. This leads to 'bent' bonds (sometimes called 'banana bonds'), as shown in ***15.2***. This leads to the idea of $sp^{2.33}$ hybridization.

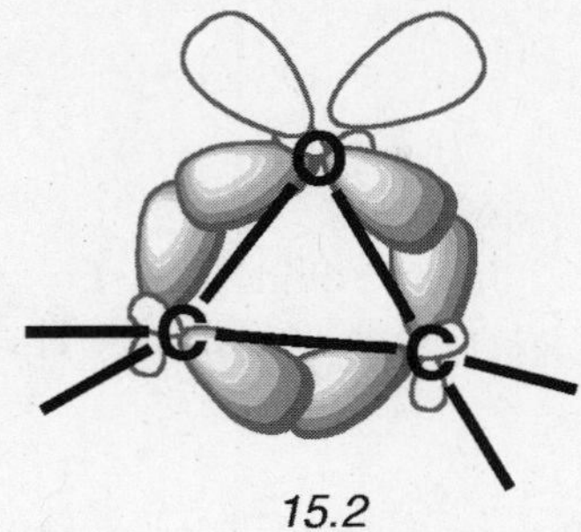

15.2

What Effect Does the Strain Have on the Relative Strength of the Sigma Bond?

Since there is less electron density between the nuclei, the bonds are significantly weaker.

Describe the Dipole of the Carbon Atoms in 15.1.

Since oxygen is more electronegative than carbon, each carbon in ***15.1*** will have a δ^+ dipole.

15.2. NOMENCLATURE

The generic name of three-membered ring ethers is 'epoxides.' An epoxide can be named by three nomenclature systems.

'Epoxy' Nomenclature

Epoxides can be named as an 'epoxy' alkane. Numbers are assigned to the two carbons connected to the oxygen atom, using the lowest numbers possible. The epoxide formed from 1-pentene, for example, is known as 1,2-epoxypentane.

Give the Name of Each of 15.3–15.5.

O, Cl — *15.3*; O — *15.4*; Ph, Ph, O — *15.5*

The name of ***15.3*** is 4-chloro-1-cyclopentyl-1,2-epoxyhexane. Epoxide ***15.4*** is 1,2-epoxycyclohexane. Epoxide ***15.5*** is called 1,4-diphenyl-1,2-epoxyhexane.

'Oxide' Nomenclature

Epoxides are named using the alkene as the base, with the term *epoxide* as a separate word.

Can Epoxides Be Named Using Common Names?

Yes! Epoxides can be named as oxides when the alkene is named with one of the common names: ethylene, styrene, propylene, etc. The epoxide derived from ethylene is then called ethylene oxide. This nomenclature system is considered to be a 'common name.' The simplest alkene (ethylene) is a precursor for the simplest epoxide (oxirane), called ethylene oxide.

Give the Names for 15.3–15.5 Using This Common Nomenclature.

Epoxide ***15.3*** is 4-chloro-1-cyclopentyl-1-hexene oxide, ***15.4*** is cyclohexene oxide, and ***15.5*** is 1,4-diphenyl-1-hexene oxide.

'Oxirane' Nomenclature

Oxirane is the base name for the three-membered ring ether, ***15.1***.

How Is Oxirane Used to Name Epoxides?

The epoxide derived from ethene (ethylene oxide, ***15.1***) is given the IUPAC name *oxirane*. It is, therefore, reasonable to name epoxides as oxirane derivatives. The epoxide derived from 1-propene, for example, is called 1-methyloxirane.

Give the Name of 15.3–15.5 Using the IUPAC Nomenclature.

Epoxide ***15.3*** is 1-cyclopentyl-2-(2-chlorobutyl)oxirane, ***15.4*** is 1,2-tetramethyleneoxirane, and ***15.5*** is 1-phenyl-2-(2-phenylbutyl)oxirane.

15.3. PREPARATION

Epoxides are generally prepared by two synthetic routes from halo-alcohols and from alkenes.

S_N2 From Chlorohydrins

What Is a Chlorohydrin?

A chlorohydrin is a molecule containing both a chlorine and an alcohol in the same molecule.

In General, How Are Chlorohydrins Formed?

Chlorohydrins are formed by the reaction of an alkene with hypochlorous acid (HOCl), which is formed by adding chlorine gas to water. See Section 13.3.

How Is 2-Chloro-1-Butanol Formed From 1-Butene?

Reaction of 1-butene with hypochlorous acid (HOCl - see Section 13.3) leads to this chlorohydrin, ***15.6***.

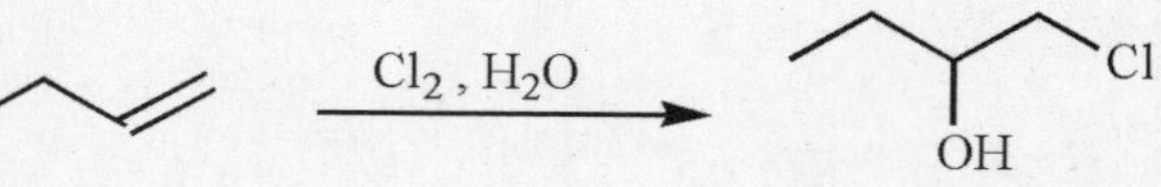

Cl_2, H_2O → Cl, OH

15.6

If 15.6 Were Treated with a Base Strong Enough to Form the Alkoxide from the OH Unit, What Is the Final Product?

The product will be an epoxide. Treatment of ***15.6*** with a suitable base leads to ***15.7***.

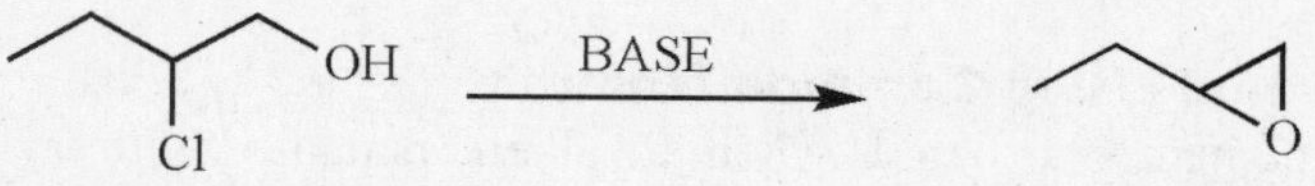

OH, Cl → BASE → O

15.6 *15.7*

Give a Mechanistic Explanation of How Treatment of 2-Chloro-1-Butanol (15.6) with NaH Leads to 1-Ethyloxirane (15.7).

When ***15.6*** is treated with the base sodium hydride (NaH), alkoxide ***15.8*** is formed. An intramolecular S_N2 displacement of the chloride leads to the epoxide, ***15.7***.

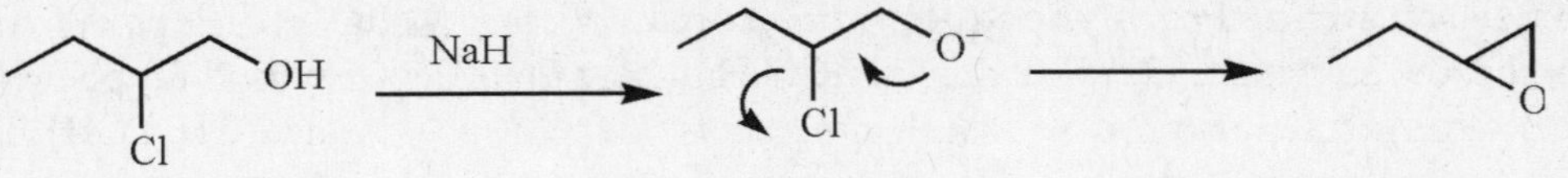

NaH

15.6 15.8 15.7

Is There Another Name for the Reaction That Produced 15.7?

This is another example of an intramolecular Williamson ether synthesis (see Section 11.3).

Give the Major Product of the Following Reaction.

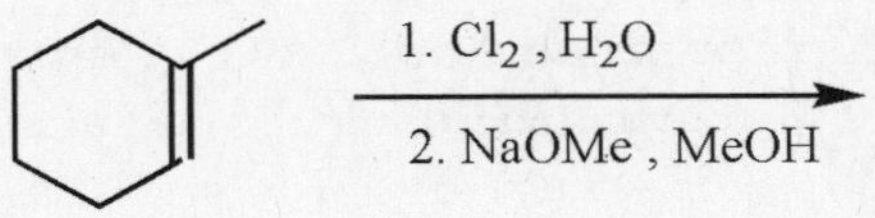

1. Cl_2 , H_2O

2. NaOMe , MeOH

An initial reaction of methylcyclohexene with HOCl, formed from chlorine and water, gives chlorohydrin ***15.9***. Subsequent treatment with the base sodium methoxide gives alkoxide ***15.10***, and the intramolecular S_N2 reaction gives epoxide ***15.11***.

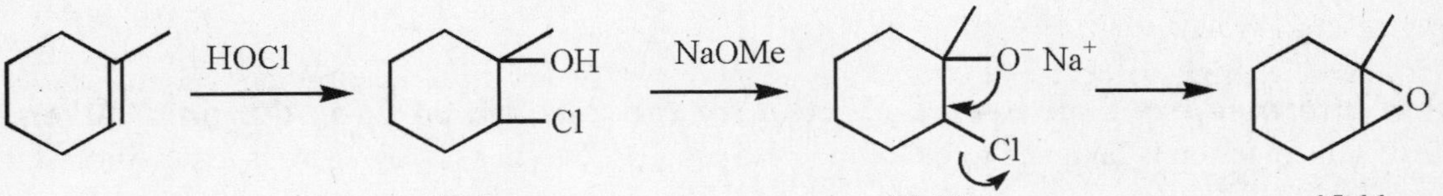

HOCl NaOMe

15.9 15.10 15.11

What Bases Can be Used to Form Epoxides in this Manner?

Typical bases include sodium hydride (NaH), as well as sodium and potassium alkoxides (NaOR and KOR) from lower molecular weight alcohols such as methanol, ethanol, 2-propanol, and *tert*-butyl alcohol. Sodium amide ($NaNH_2$) can be used. Organolithium bases such as butyllithium or methylithium have been used, and sodium hydroxide can be used in some cases, although it is usually not the best choice.

Can Bromohydrins Be Used to Form Epoxides?

Yes! Treatment of a bromohydrin with a base leads to the epoxide, just as with chlorohydrins.

How Are Bromohydrins Formed?

Bromohydrins are formed by treating an alkene with bromine and water, which forms HOBr.

From Alkenes and Peroxyacids

What Is the Generic Structure of a Peroxyacid?

A peroxyacid has the generic formula RCO_3H and is characterized by an R–(C=O)–OOH unit.

How Are Peroxyacids Named?

The word peroxy is added to the front of the carboxylic acid name (either IUPAC or common). Formic acid is the parent of peroxyformic acid. Benzoic acid is the parent of peroxybenzoic acid. Butanoic acid is the parent of peroxybutanoic acid, and acetic acid is the parent of peroxyacetic acid.

Draw the Structures of Peroxybenzoic Acid, Peroxy Acetic Acid, and Peroxyformic Acid.

Peroxybenzoic acid is ***15.12***, abbreviated $PhCO_3H$, where Ph is the term used for phenyl (a benzene substituent). Peroxyacetic acid (or peroxyethanoic acid) is ***15.13***, abbreviated CH_3CO_3H or $MeCO_3H$, where Me = methyl. Peroxyformic acid (or peroxymethanoic acid) is ***15.14***, abbreviated HCO_3H.

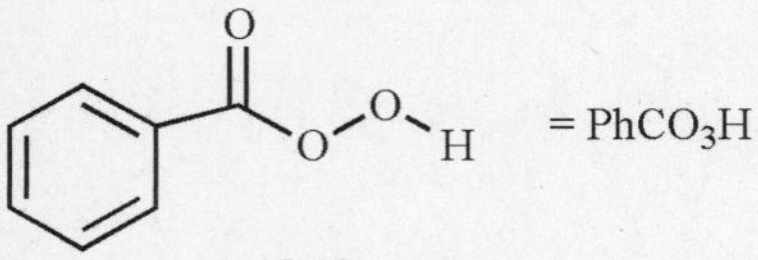

= $PhCO_3H$ = CH_3CO_3H = HCO_3H

15.12 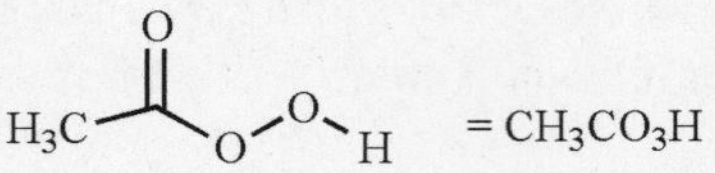   *15.13* 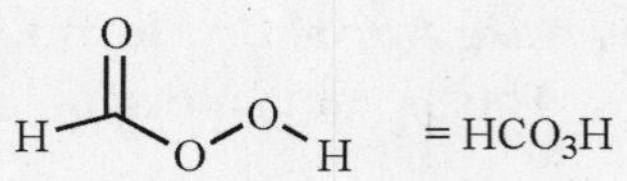   *15.14*

What Is the Product When Cyclopentene Reacts with Peroxyacetic Acid?

The products are the epoxide (cyclopentene oxide, ***15.15***) and the carboxylic acid, acetic acid (***15.16***).

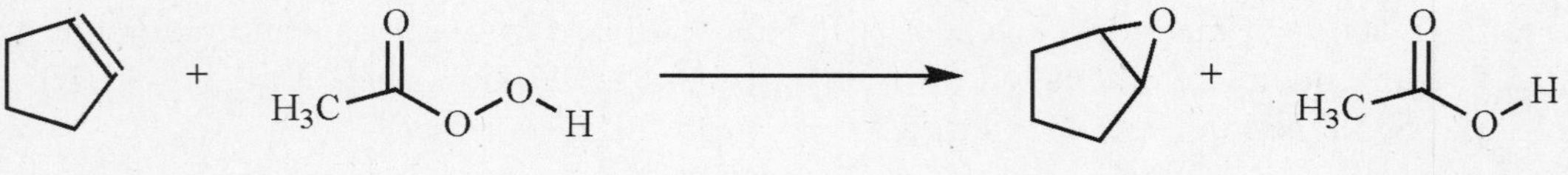

15.13   *15.15*   *15.16*

Has an Intermediate Ever Been Detected for the Peroxyacid Epoxidation of Alkenes?

No! This reaction is believed to be a concerted process. In other words, there is a transition state, but no formal intermediate.

Give a Mechanistic Rationale for Conversion of Cyclopentene to the Epoxide with Peroxybenzoic Acid.

The bond polarization of the peroxyacid is such that the 'second' oxygen from the carbonyl is electrophilic ($\delta\oplus$), and it is attacked by the electrons in the π-bond of cyclopentene. There is no intermediate, and this reaction is thought to proceed by a concerted mechanism involving transition state ***15.17***. In this transition state, the alkene attacks the electropositive oxygen and transfers the hydrogen to the oxygen of the peroxyacid. Synchronous bond breaking leads to formation of the second bond of the epoxide (***15.15***) and expulsion of benzoic acid (***15.18***), the by-product. In this reaction, part of the driving force for the reaction is loss of the leaving group, benzoic acid. This is a generic mechanism in that alkenes attack the electropositive oxygen of a peroxyacid, form the epoxide, and expel the corresponding carboxylic acid.

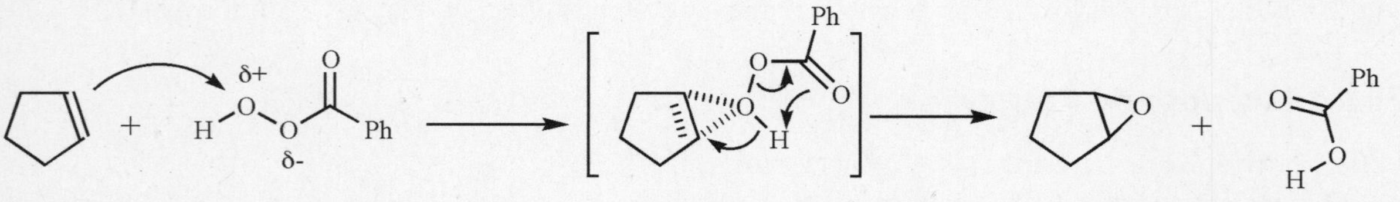

15.17   *15.15*   *15.18*

Explain Why Peroxyacids React Faster with More Highly Substituted Alkenes Than with Less Substituted Alkenes.

The initial reaction involves attack of the alkene π-bond on the electropositive oxygen of the peroxyacid. The more electron density in that π-bond, the faster the reaction. Since carbon substituents are electron releasing, more highly substituted alkenes are richer in electron density and react faster.

Are Peroxy Acids Strong Acids As Are Carboxylic Acids (RCO_2H)?

No! The bond polarization is such that the OH bond is not as polarized in the peroxyacid, and the hydrogen atom is not as acidic.

What Is the Structure of meta-Chloroperoxybenzoic Acid, and Why Is It a Popular Choice for Epoxidations?

The structure is ***15.19***. It is popular because it is commercially available, a crystalline solid that is readily purified, stable in storage and gives *meta*-chlorobenzoic acid (***15.21***) which is usually easily removed from the product. In a typical reaction, 2-butene reacts with ***15.19*** to give the epoxide, ***15.20***, and ***15.21***.

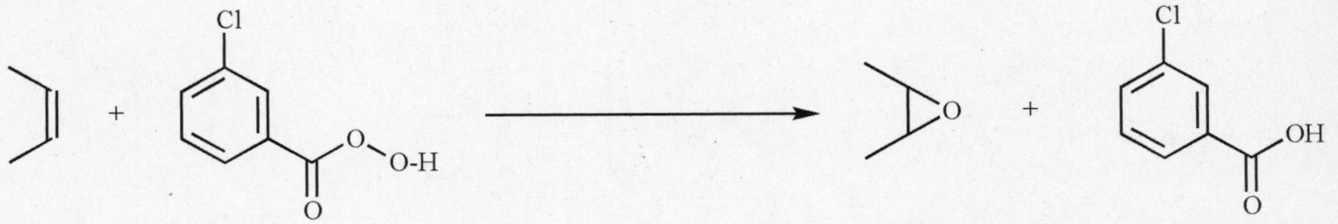

Cl O O-H O OH

15.19 *15.20* *15.21*

Give the Major Products for the Reactions of 15.22–15.24.

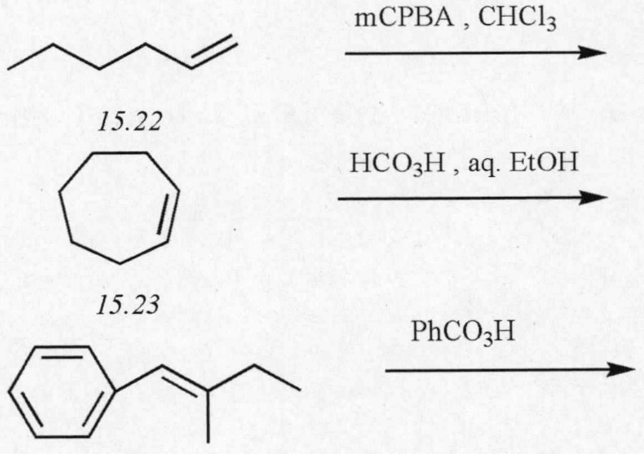

mCPBA , $CHCl_3$

15.22

HCO_3H , aq. EtOH

15.23

$PhCO_3H$

15.24

Epoxidation of 1-hexene (***15.22***) with *m*-chloroperoxybenzoic acid gives 1-hexene oxide (***15.25***). Epoxidation of cycloheptene (***15.23***) gives cycloheptene oxide (***15.26***), and epoxidation of ***15.24*** leads to the epoxide ***15.27***.

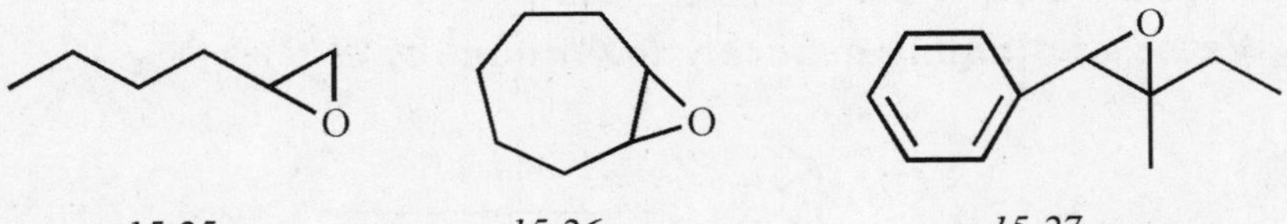

15.25 *15.26* *15.27*

15.4. REACTIONS

Epoxides are highly strained molecules and react with a variety of reagents. The most common reactions involve opening the three-membered ring with acid or with a nucleophile and reduction with various reducing agents.

Acid Catalyzed Ring Opening

What Is the Initial Product When 1-Methyloxirane Reacts with HCl?

As with any other ether, an oxirane reacts with HCl to form an oxonium ion, in this case ***15.28***.

What Is the Final Product of the Reaction of 1-Methyloxirane and HCl?

Once the oxonium ion (***15.28***) is formed, the three-membered ring will be opened by a nucleophile. The only available nucleophile in this reaction is the chloride ion, leading to ring opening and formation of chlorohydrin ***15.29***. Depending on the solvent, cation ***15.28*** has considerable ionic character at the more stable secondary site, leading to incorporation of the chlorine atom at the more substituted position, ***15.29***.

H—Cl

H_3C ; H_3C ; H ; O ⊕ ; Cl^- ; Cl ; H_3C ; OH

15.28 *15.29*

Why Does This Reaction Favor Opening the Three-Membered Ring?

The highly strained three-membered ring is converted to an oxonium ion, making the 'OH' unit an excellent leaving group. The combination of these two facts makes opening the epoxide by even weak nucleophiles very favorable.

What Is the Product Formed When Epoxide 15.30 Is Treated with HBr?

O ; H–Br

15.30

The product is bromohydrin ***15.31***.

O ; H–Br ; Br ; OH

15.30 *15.31*

What Is the Product Formed when Epoxide 15.30 Is Treated with Water Containing a Catalytic Amount of H^+?

O ; H_2O , cat H^+

15.30

The product is the diol, ***15.34***, formed by protonation of the epoxide to form ***15.32***, followed by opening of the three-membered ring by water to form ***15.33*** and loss of a proton to give ***15.34***.

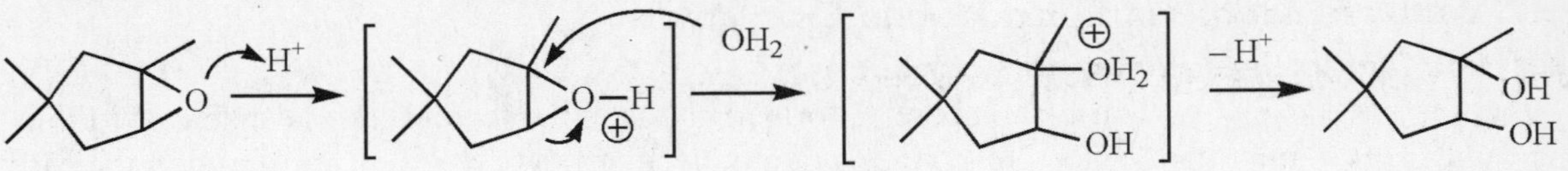

15.30 15.32 15.33 15.34

Nucleophilic Ring Opening

What Is the Product When 1-Methyloxirane Reacts with Sodium Hydroxide?

It is not necessary to first protonate the ether oxygen to induce ring opening. Epoxides are sufficiently strained that reaction with good nucleophiles gives the ring-opened product. In this case, 1-methyloxirane is opened by hydroxide to give the anion of diol, 1,2-propanediol, ***15.35***. Treatment of ***15.35*** with an acid is necessary to protonate the alkoxide unit and generate the final product, diol ***15.36***. Note that the nucleophile attacks the least sterically hindered carbon in what is essentially an S_N2 process.

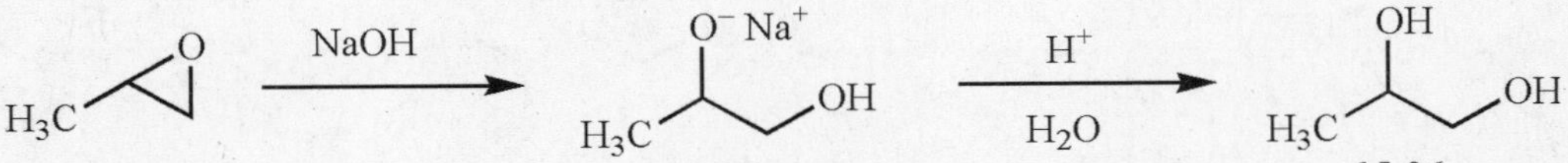

15.35 15.36

Explain Why the Reaction Is Highly Regioselective for the Less Substituted Carbon of the Epoxide.

If this reaction is an S_N2-like process, the transition state for the less sterically hindered carbon is lower in energy than that for the more highly substituted carbon.

What Is the Product When Cyclohexene Epoxide Reacts with: A) NaN$_3$ B) NaCN C) NaI D) Sodium Acetylide E) Na$^+$ $^-$C≡C-ET?

In each case, there is a second step where the alkoxide product is treated with dilute aqueous acid to generate the alcohol. In every case, the epoxide is opened by the nucleophile (Y^-) to give a *trans-* product. Since both bonds to the oxygen of an epoxide are on the same side of the ring, the nucleophile must approach from the opposite face (see ***15.37***) to give trans alkoxide ***15.38***. The first reaction generates the azido alcohol (***15.39***). Reaction with cyanide gives the cyanohydrin, ***15.40***. Iodide opens the epoxide ring to give the iodohydrin, ***15.41***. Sodium acetylide is a powerful nucleophile and the anion derived from acetylide gives ***15.42***, whereas that derived from 1-butyne gives ***15.43***.

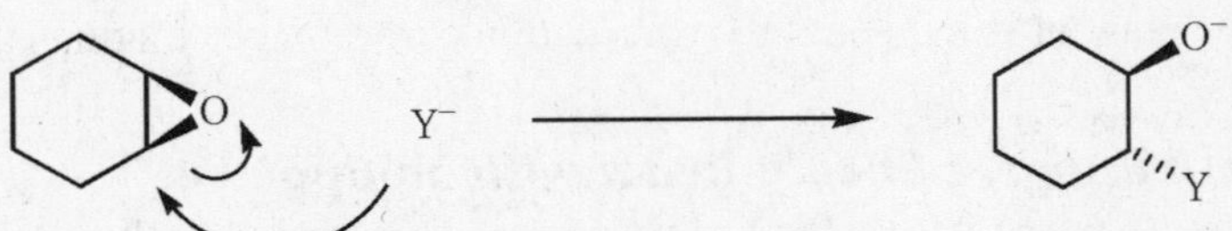

15.37 15.38

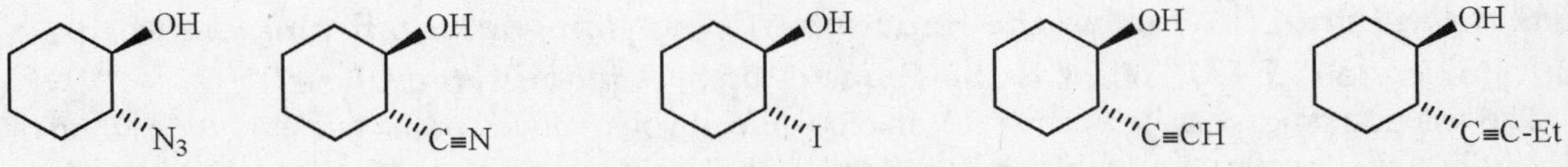

15.39 15.40 15.41 15.42 15.43

Why Are the Reactions of Epoxides and Cyanide or Acetylide Derivatives Particularly Important?

In both cases, a new carbon-carbon bond is formed in one of the few reactions capable of forming a C–C bond. This is important, since the synthesis of a molecular target relies on constructing the carbon skeleton by a series of carbon-carbon, bond-forming reactions.

What Is the Product or Products When Sodium Cyanide Reacts with 2,3-Epoxyhexane (15.44)? Explain.

The cyanide will attack the less sterically hindered carbon of the epoxide if given a 'choice.' In this case, ***15.44*** has a similar substitution pattern at each carbon of the epoxide unit, and there is little steric difference between methyl and propyl. In such cases, a mixture of two products is expected from attack at *both* carbons of the epoxide. In this reaction the two products are cyanohydrins ***15.45*** and ***15.46***. There may be a slight preference for ***15.45***, since the methyl group is slightly smaller than the propyl.

NaCN, DMF; H_3O^+

OH, C≡N, C≡N, OH

15.44 *15.45* *15.46*

What Solvents Are Most Favored for Nucleophilic Ring Opening Reactions with Epoxides?

Since this is considered to be an S_N2 process, polar aprotic solvents such as ether, THF, DMSO or DMF are the most commonly used.

Reaction with Grignard Reagents

What Is the Generic Structure of a Grignard Reagent?

A Grignard reagent is an alkylmagnesium halide, RMgX, where X = Cl, Br, I.

How Are Grignard Reagents Formed?

Grignard reagents are formed by the reaction of an alkyl, aryl, or vinyl halide with magnesium metal in diethyl ether or THF.

What Do Grignard Reagents React with?

Since the carbon of RMgX has a negative dipole, it is considered a nucleophile in reactions at carbon. Therefore, a Grignard reagent is expected to react with a carbon atom that has a positive dipole.

Explain Why a Grignard Reagent Should React with an Epoxide.

A Grignard reagent is a source of nucleophilic carbon. This nucleophilic carbon will attack and open the electropositive carbon of an epoxide, generally at the less sterically hindered site.

What Is the Initial Product of the Reaction of Phenylmagnesium Bromide and 1-Butyloxirane (15.47)? What Is the Product after Aqueous Hydrolysis?

The reaction process, as with any other nucleophile, involves attack of the nucleophilic carbon of the Grignard reagent at the less sterically hindered carbon of the epoxide (***15.47***) to give an alkoxide. Hydrolysis gives 1-phenyl-3-hexanol, ***15.48***.

1. PhMgBr , THF
2. H_3O^+

15.47 → 15.48

What Is the Product (or Products) When 3-Heptene Oxide (15.49) Reacts with Ethylmagnesium Iodide?

As with other nucleophiles, ethylmagnesium iodide will react at both electropositive carbons of ***15.49***. There is no steric bias to favor attack at one carbon over the other. Both carbons are equally substituted, and the result will be a roughly equal mixture of two alcohols, 3-ethyl-4-heptanol (***15.50***) and 4-ethyl-3-heptanol (***15.51***).

1. EtMgI , ether
2. H_3O^+

15.49 → 15.50 + 15.51

Hydride Reduction

What Is the Molecular Formula of Lithium Aluminum Hydride?

Lithium aluminum hydride has the formula $LiAlH_4$.

What Is the Molecular Formula of Sodium Borohydride?

Sodium borohydride has the molecular formula $NaBH_4$.

What Chemical Reactivity with Epoxides Is Exhibited by the Two Hydrides Mentioned?

Both of these reagents are reducing gents, and reduce an epoxide to an alcohol.

What Is the Major Product When 1,2-Epoxycyclohexane (15.37) Is Treated First with $LiAlH_4$ and Then with Dilute Aqueous Acid?

In this reaction, the hydride reagent behaves as a 'nucleophile' delivering the 'H^-' to the less substituted carbon of the epoxide. This symmetrical epoxide (***15.37***) is reduced, and the product is cyclohexanol, ***15.52***.

1. $LiAlH_4$, ether
2. H_3O^+

15.37 → 15.52

What Is the Purpose of the Second Step (Treatment with Dilute Acid)?

The initial product of the reaction with lithium aluminum hydride is the alkoxide, ***15.53***, and the second step is required to protonate the alkoxide to give the alcohol (***15.52***).

$\xrightarrow{LiAlH_4,\ ether}$ $\xrightarrow{H_3O^+}$

O⁻ Li⁺ OH

15.37 15.53 15.52

Explain Why 'Hydride' Is Generally Delivered to the Less Substituted Carbon of the Epoxide.

In the four-centered transition state required for reaction of $LiAlH_4$ with the electropositive carbon of the epoxide, the hydride will be delivered to the less sterically hindered carbon so that all steric interactions are minimized.

What Is the Product When 15.37 Reacts with 1. $NaBH_4$ 2. AQ NH_4Cl?

The product is ***15.52***. Sodium borohydride gives effectively the same reaction, but it is much less reactive than $LiAlH_4$. This means that the reaction with epoxides is slower, and may require longer reaction times and the use of solvents with higher boiling points so the reaction can be heated. In fact, the common solvent for reductions using $NaBH_4$ is ethanol.

What Is the Purpose of the Ammonium Chloride Solution?

The ammonium chloride solution acts as an acid to protonate the boron alkoxide intermediate (***15.54***) formed by the initial reaction with sodium borohydride.

$\xrightarrow{NaBH_4,\ EtOH}$ $\xrightarrow{aq\ NH_4Cl}$

O⁻ Li⁺ OH

15.37 15.54 15.52

Give the Major Product from the Reactions of 15.55–15.57.

15.55: 1. $LiAlH_4$, ether; 2. H_3O^+

15.56: 1. $LiAlH_4$, ether; 2. H_3O^+

15.57: 1. $NaBH_4$, EtOH; 2.aq NH_4Cl

In the first reaction, epoxide ***15.55*** is reduced to a mixture of 2-heptanol (***15.58***) and 3-heptanol (***15.59***) due to the lack of a less substituted carbon on the epoxide. Reduction of ***15.56*** shows the expected delivery of hydride to the less substituted carbon, forming 1-cyclopentyl-1-ethanol, ***15.60***. In the last reaction, reduction of ***15.57*** at the less substituted carbon leads to ***15.61*** as the major product.

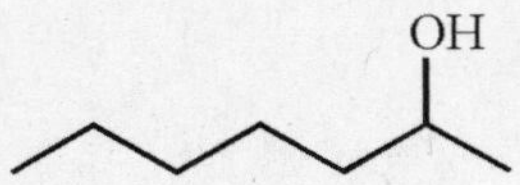

OH OH OH OH

15.58 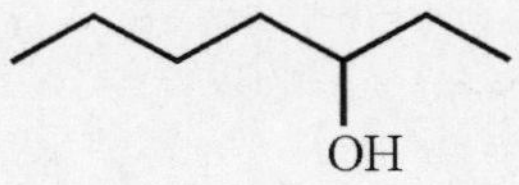 *15.59* 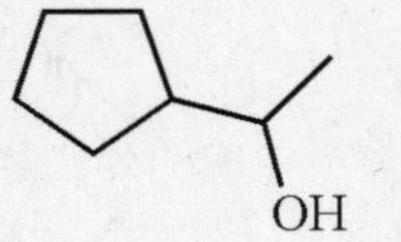 *15.60* 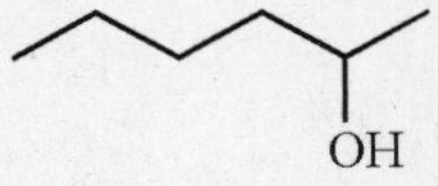 *15.61*

Test Yourself

1. Give the IUPAC name for each of the following.

(a) CH_3 O

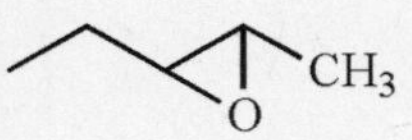

(b) CH_3 O CH_3

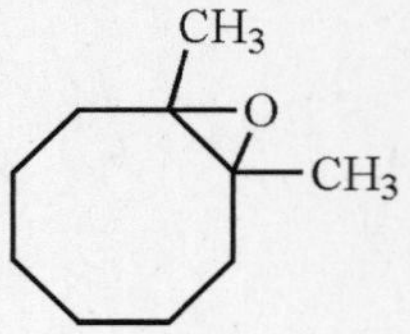

(c) O Ph CH_3 CH_3 Cl

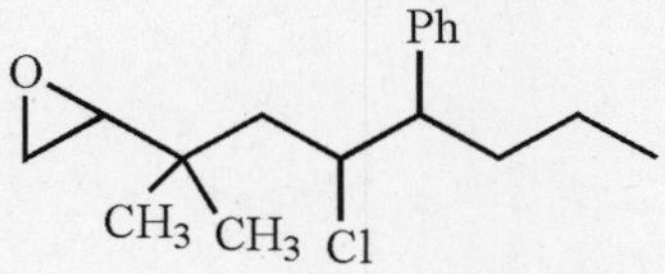

(d) O CH_3

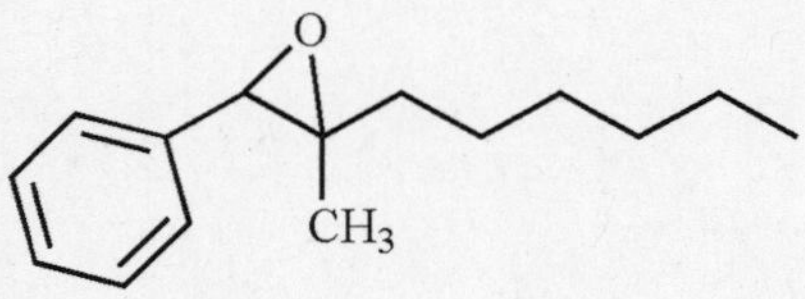

(e) OMe Cl Cl O Cl

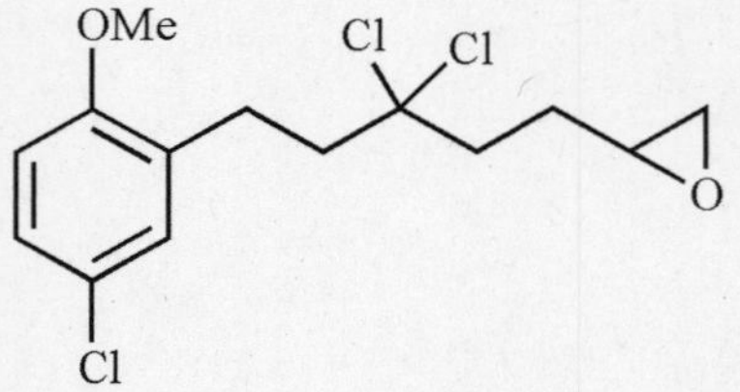

2. Give the final product, and the mechanism of its formation for the following transformation.

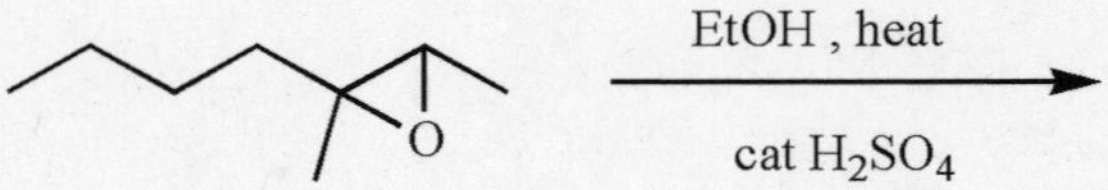

O; EtOH , heat; cat H_2SO_4

3. In each case give the structure of the major product. Remember stereochemistry where appropriate, and if there is no reaction, indicate this by N.R.

(a) OH Br; NaH , THF

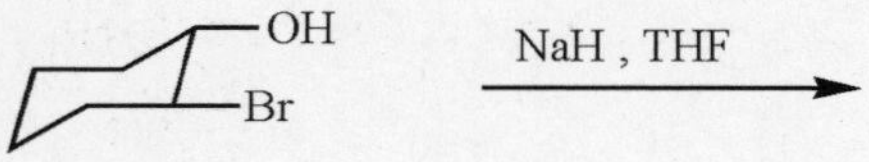

(b) Cl CO_3H

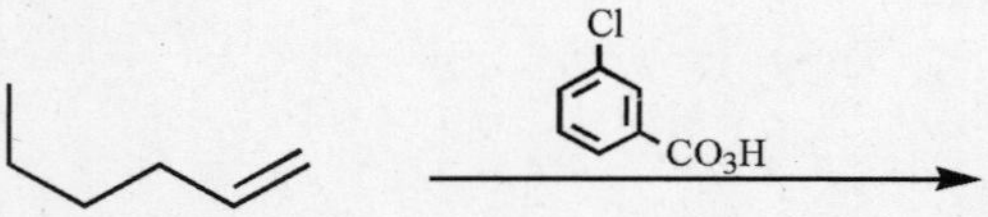

(c) CF_3CO_3H; NaOAc

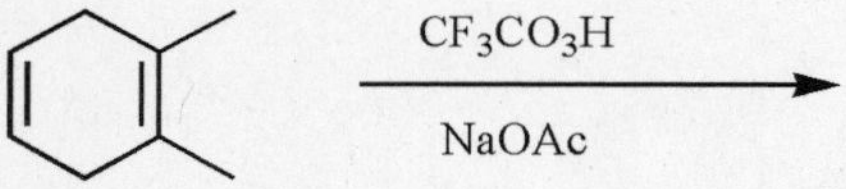

(d) O; HBr

(e) 1. mCPBA; 2. NaH/1-butyne; 3. H_2O

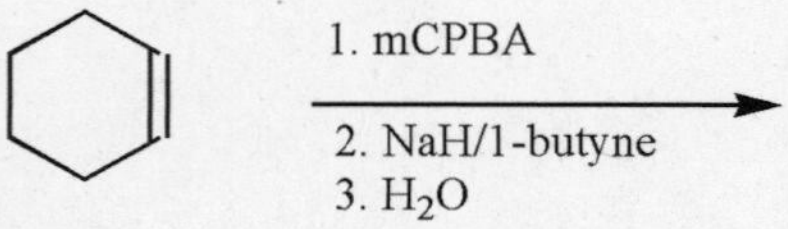

(f) O; KCN , DMF

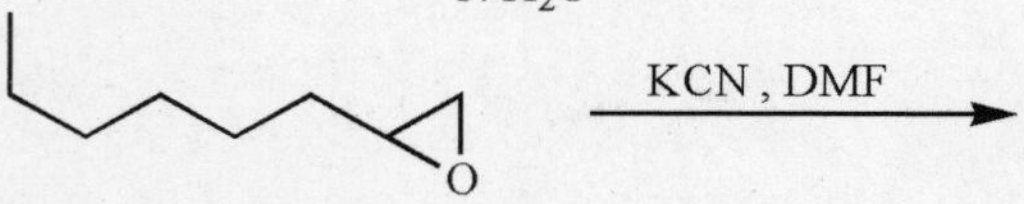

(g) O; 1. NaN_3 , DMF; 2. H_3O^+

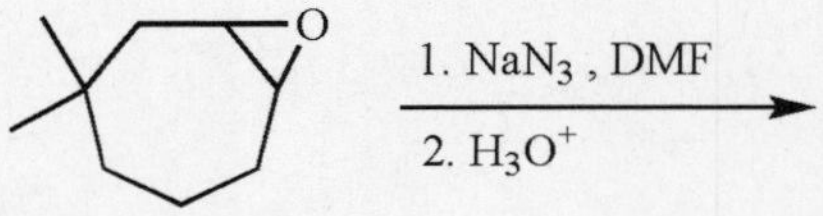

(h) O; 1. EtMgBr , ether; 2. H_3O^+

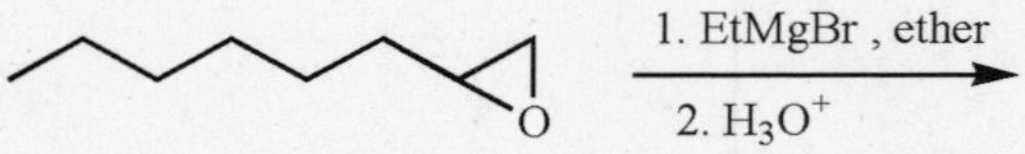

(i) Me Br; 1. KOH , EtOH; 2. mCPBA; 3. BuMgBr; 4. H_3O^+

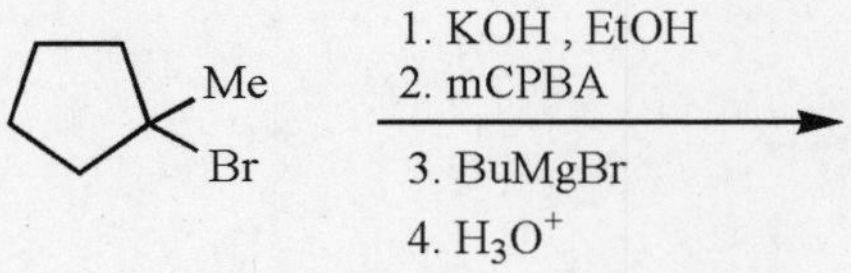

(j) O Me; 1. $LiAlH_4$, ether; 2. aq. NaOH

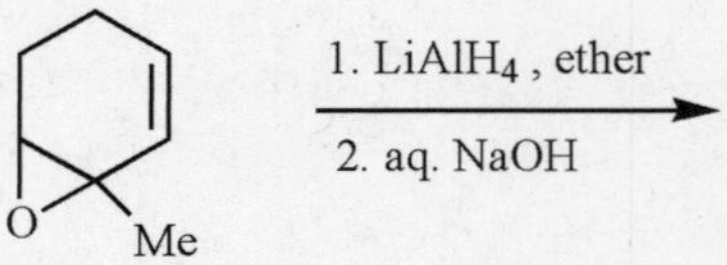

(k) 1. Cl_2 , H_2O; 2. NaH , DMF

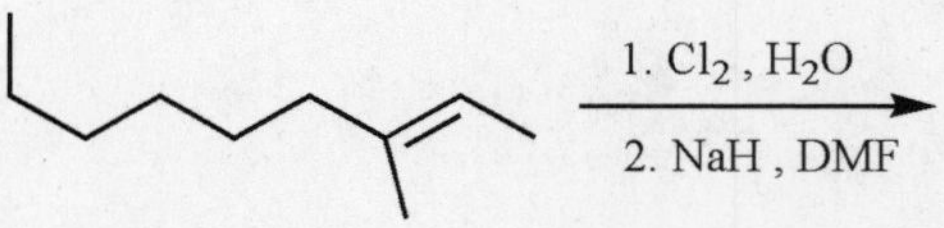

CHAPTER 16

Benzene, Aromaticity, and Benzene Derivatives

Benzene is one of the most important organic molecules known because it is representative of a wide variety of molecules known as aromatic compounds. The special stability associated with its structure leads to special types of reactions that warrant close scrutiny.

16.1. STRUCTURE

The structure of benzene is unique in that it defines the physical and chemical properties for an entire class of organic molecules.

What Is the Molecular Formula of Benzene?

The formula is C_6H_6.

Draw the Structure Commonly Used for Benzene.

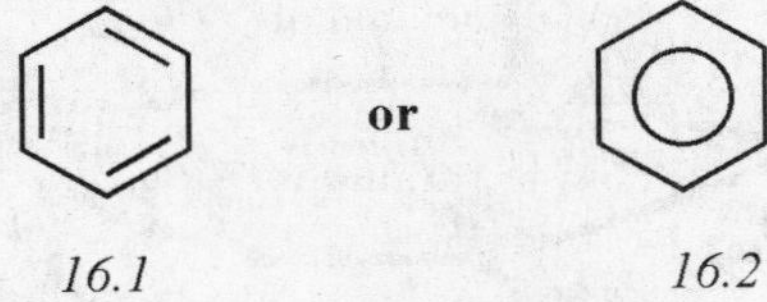

16.1 or 16.2

Structure 16.1 Suggests the Name Cyclohexatriene. Why Is This Molecule Called Benzene?

The name cyclohexatriene suggests alternating single and double bonds, with the electrons localized between carbon atoms for each type of bond. This structure also suggests alternating long-short bond distances for the C–C vs. C=C units. In fact, all of the bond lengths and bond strengths are identical, and a different name was required to establish that the molecule is ***not*** cyclohexatriene, but something else. The name settled upon was benzene. Note that structure ***16.1*** does not really represent the true structure of benzene.

What Does Structure 16.2 Represent?

The circle is drawn to indicate that all bond lengths are identical and that the electrons are delocalized over all six carbon carbons, and not localized between C—C and C=C units.

How Do We Know That 16.1 Is Not the Correct Structure?

We can measure the bond distances between carbon atoms. When these measurements are done, all bond distances are identical (1.40 Å = 140 pm). The normal bond distance for a sp^2–sp^2 single bond is 146 pm and 134 pm for a normal sp^2–sp^2-double bond. The measured bond distance for benzene is 'in between.'

When Electron Density Is Delocalized (As Represented by 16.2), Does This Represent a More Stable or Less Stable Situation?

If the C=C units in benzene were the same as in cyclohexene and cyclohexadiene, a bond dissociation energy (obtained by thermally decomposing benzene and measuring the heat obtained) should be 3×28.6 kcal. In other words, we would expect a bond dissociation energy that is consistent with cyclohexatriene. The fact is that less energy is liberated, and this observation is taken to mean that benzene is inherently more stable than either the alkene or the diene. This 'extra stability' is about 36 kcal/mol and is due to the special lowering of energy that occurs when the p orbitals are parallel, contiguous (every p-orbital has a p orbital neighbor on every adjacent carbon) and able to share electron density with their neighboring p orbitals.

What Is the Term Commonly Applied to Delocalization of Electron Density in P orbitals over Several Atoms?

Benzene is stabilized by resonance.

Draw a Picture That Represents the Electron Delocalization in Benzene Due to Resonance.

Benzene is a planar molecule consisting of six sp^2 hybridized carbons connected in a ring, with six p orbitals perpendicular to the plane of the carbons (see ***16.3***). There are six π-electrons in the six contiguous orbitals that are delocalized over all six orbitals.

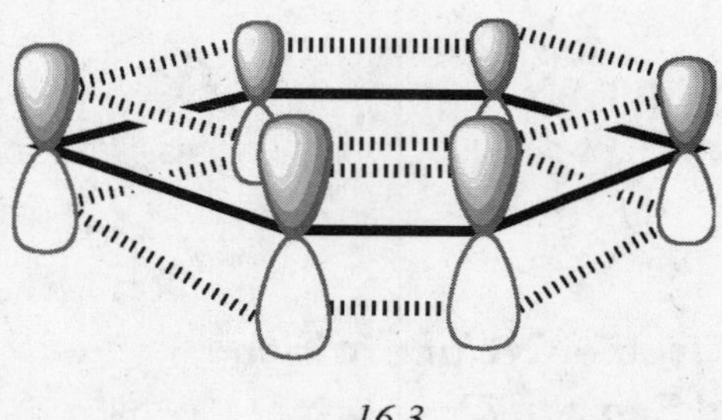

16.3

How Can We Represent 16.3 Using More Conventional Structures?

We use Kekulé structures, ***16.1*** and ***16.4***. **Kekulé structures** are representations of benzene where the bonds are localized, and different structures are used to show that the bond positions 'change.' *Both structures are required* in order to represent delocalization of the electrons and to represent the fact that the structure does not have true double or single bonds, but rather a bond that is 'in between,' shorter than a normal single bond, but longer than a normal double bond.

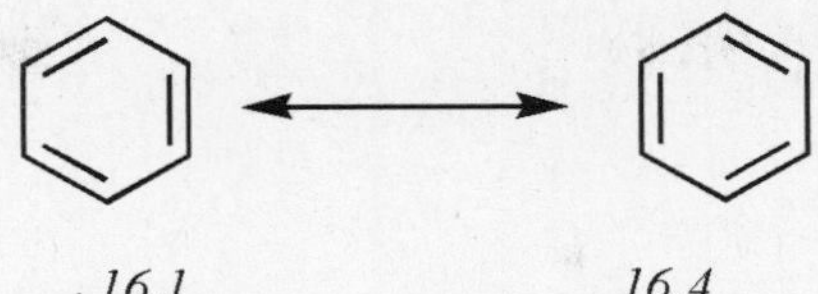

16.1 *16.4*

What Is the Double-Headed Arrow Used with 16.1 and 16.4?

The double-headed arrow is used to indicate resonance. In other words, ***16.1*** and ***16.4*** are two resonance contributors that, taken together, represent the resonance delocalized benzene structure. Note that the resonance arrow does ***not*** indicate an equilibrium. Benzene is ***not*** equilibrating between ***16.1*** and ***16.4***, but rather both structures are required to represent the electron delocalization.

What Are the Necessary Structural Features Required for Resonance?

A continuous and contiguous array of p orbitals are required for resonance.

16.2. AROMATICITY

When a molecule has a continuous array of sp^2 atoms (no intervening sp^3 atoms) that are contiguous (no breaks) and are confined to a ring (a cyclic molecule), each atom has a p orbital, and the p orbitals will be parallel. In such a system, electron delocalization leads to enhanced stability called aromaticity when there are 2, 6, 10, 14, 18, 22 π-electrons (or, electrons = 4n+2, where n = a series 0,1,2,3,4......).

What Are the Structural Features Required for Aromaticity?

Several criteria are required for aromaticity: a *continuous* and contiguous *array of p-orbitals* in a *cyclic* system are required (every carbon must be sp^2 hybridized) for a total of *4n+2 p-electrons* (where n is a series 0,1,2,3,.... generating a new series 4n+2 = 2,6,10,14,18,22,.......). If the number of π-electrons is equal to one of the numbers in this series, the compound may be aromatic. All of the emphasized criteria are required for aromaticity. If one of them is missing, the compound is not aromatic.

What Is the Name of the 4N+2 Rule?

This rule is sometimes called the Hückel rule for aromaticity.

Which of the Following Molecules Are Aromatic?

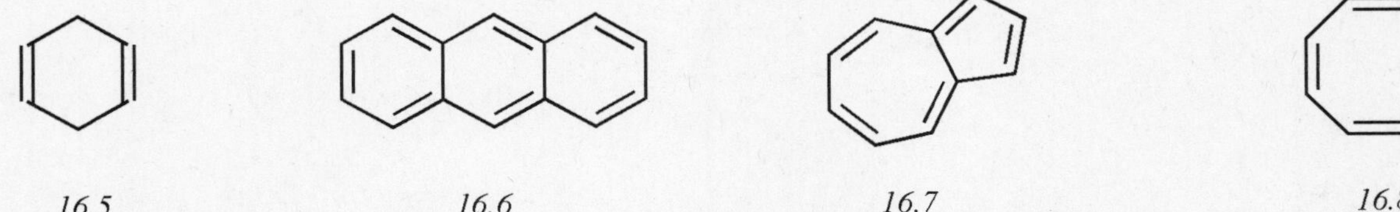

16.5 *16.6* *16.7* *16.8*

1,4-Cyclohexadiene (***16.5***) is not aromatic. Although there is a ring, there are sp^3-hybridized atoms between the C=C units, so there is no continuous or contiguous array of π-bonds. Compound ***16.6*** is aromatic. It is cyclic (there are multiple rings fused together, and this counts), with every carbon being sp^2 hybridized, and there is a total of 14 π-electrons (2 π-electrons for each C=C unit). Since 14 π-electrons are part of the 4n+2 series, this molecule is aromatic. Compound ***16.7*** is aromatic. It is cyclic, with every carbon being sp^2 hybridized, and there is a total of 10 π-electrons (2 π-electrons for each C=C unit). Since 10 π-electrons are part of the 4n+2 series, this molecule is aromatic.

Cyclooctatetraene (***16.8***) is not aromatic. Although there is a ring and a continuous and contiguous array of π-electrons, there are 8 π-electrons, which does not fit the 4n+2 series.

Can a Cation or Anion Be Aromatic?

Yes. As long as the criteria given above are met, a cation or anion can be aromatic.

Classify 16.9-16.18 As Aromatic or Non-Aromatic.

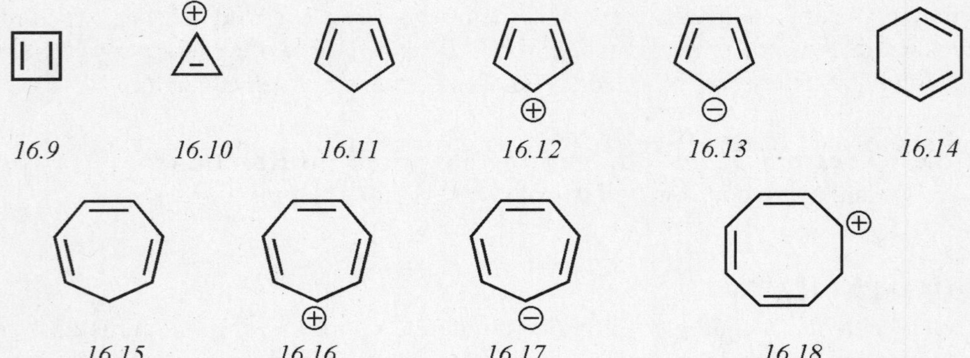

Cyclobutadiene (***16.9***) has only 4 π-electrons and is not aromatic. The cyclopropenyl cation (***16.10***) has 2 π-electrons and is aromatic. Cyclopentadiene (***16.11***) does not have a continuous array of p orbitals and is not aromatic. The cyclopentadienyl cation (***16.12***) has only 4 π-electrons and is not aromatic. The cyclopentadienyl anion (***16.13***) has 6 π-electrons and is aromatic. 1,2-Cyclohexadiene (***16.14***) does not have a continuous array of p orbitals and is not aromatic. Cycloheptatriene (***16.15***) does not have a continuous array of p orbitals and is not aromatic. The cycloheptatrienyl cation (***16.16***) has 6 π-electrons and is aromatic, whereas the cycloheptatrienyl anion (***16.17***) has 8 π-electrons and is not aromatic. The cyclooctatetraene cation (***16.18***) has 6 π-electrons, but there is an intervening sp^3 carbon, and it is not aromatic.

Is 16.18 Resonance Stabilized?

Yes! The positive charge can be delocalized over the C=C network, but the molecule is ***not*** aromatic.

If Treated with a Suitable Base, Can H_a in Cyclopentadiene (See 16.19) Be Considered an Acid?

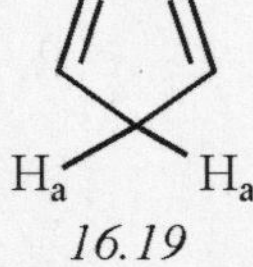

Yes! Removal of H_a leads to the aromatic cyclopentadienyl anion (***16.13***). Formation of a very stable product shifts the equilibrium of an acid-base reaction, and cyclopentadiene is a weak acid, with a pK_a of about 14-15. In other words, cyclopentadiene is more acidic than water (pK_a = 15.8).

Are Molecules Other Than Benzene Considered to Be Aromatic?

Yes. As illustrated by the previous questions, there are many examples of polycyclic molecules that are aromatic. As long as the criteria listed above are met, a molecule may be aromatic. Typical examples are naphthalene (***16.20***), anthracene (***16.21***) and phenanthrene (***16.22***).

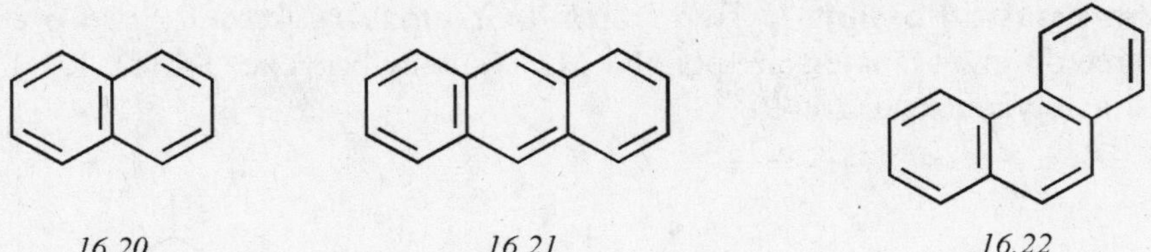

16.20 16.21 16.22

Is It Possible to Have Aromatic Compounds with a Heteroatom in the Ring?

Yes! As long as all of the criteria for aromaticity are met, such compounds are well known. The major difference is that one of the lone electron pairs on a heteroatom can be used as part of the aromatic system, while other electron pairs are not.

Give the Structure of Furan, Pyrrole, Thiophene, Pyridine, Pyrimidine, Quinoline, Isoquinoline, Indole, and Purine.

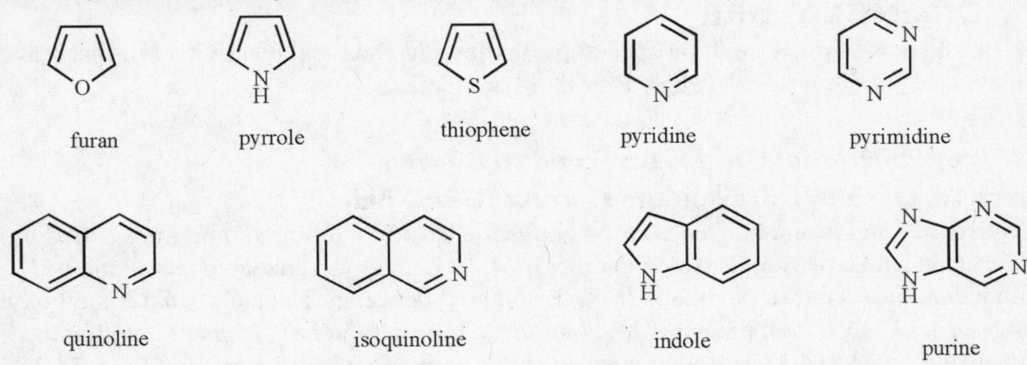

furan pyrrole thiophene pyridine pyrimidine

quinoline isoquinoline indole purine

16.3. AROMATIC DERIVATIVES

There is a multitude of aromatic compounds, some benzene derivatives, some not. In this particular section, only derivatives of benzene will be considered.

Give a List of Functional Groups That Can Be Attached to a Benzene Ring.

Several of the functional groups are halogen (as in ***16.23***), hydroxyl (***16.24***), amino (***16.25***), cyano (***16.26***), carboxylic acid (***16.27***), ester (***16.28***), aldehyde (***16.29***), ketone (***16.30***), methoxy (***16.31***), ethoxy (***16.32***), alkyl groups such as methyl (***16.33***) or ethyl (***16.34***), sulfonic acid (***16.35***), alkene (***16.36***) and nitro (***16.37***).

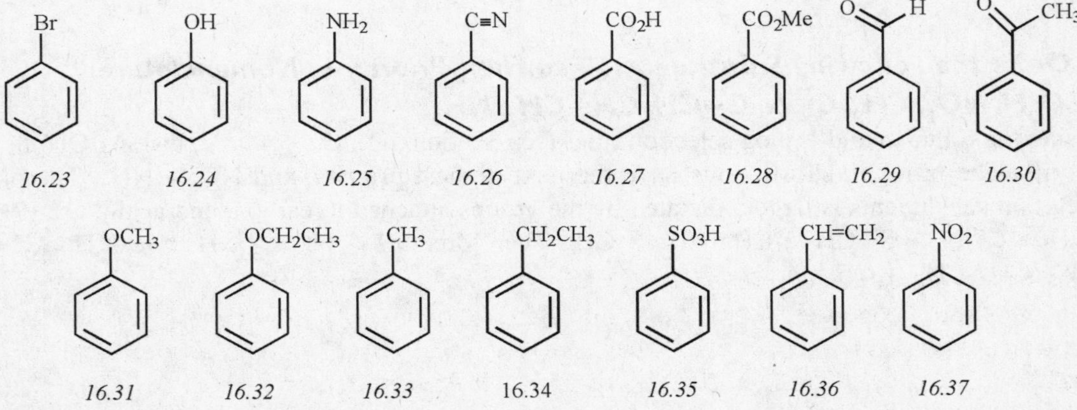

16.23 16.24 16.25 16.26 16.27 16.28 16.29 16.30

16.31 16.32 16.33 16.34 16.35 16.36 16.37

How Many Isomers Are Possible If Two Methyl Groups Are Attached to a Benzene Ring?

A total of three different isomers are possible, 1,2-dimethylbenzene (***16.38***), 1,3-dimethylbenzene (***16.39***) and 1,4-dimethylbenzene (***16.40***).

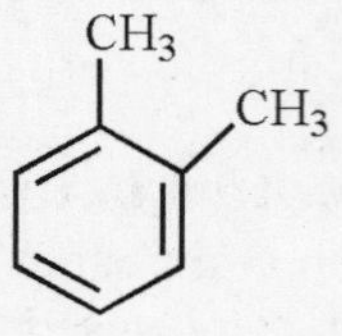

CH_3 CH_3 CH_3 CH_3 CH_3 CH_3

16.38 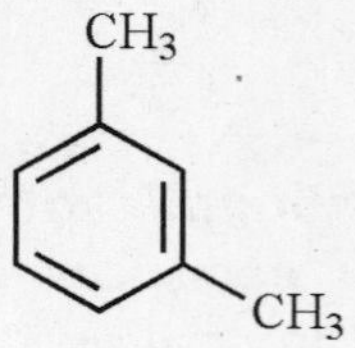 *16.39* 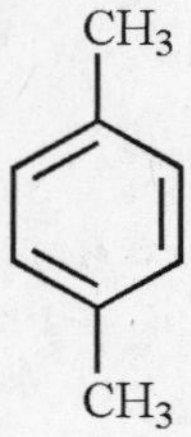 *16.40*

16.4. NOMENCLATURE

Benzene and its derivatives exhibit special properties and chemical reactivity. For that reason, a different nomenclature system was developed for these compounds.

For the List of Functional Groups Attached to Benzene (Structures 16.23-16.37), Provide the Correct IUPAC Name.

Derivative ***16.23*** is bromobenzene, ***16.24*** is phenol, ***16.25*** is aniline (sometimes called aminobenzene), ***16.26*** is benzonitrile, ***16.27*** is benzoic acid, ***16.28*** is methyl benzoate (see Section 25.1), ***16.29*** is benzaldehyde, ***16.30*** is acetophenone, ***16.31*** is methoxybenzene or anisole, ***16.32*** is ethoxybenzene or phenetole, ***16.33*** is methylbenzene (the common name is toluene), ***16.34*** is ethylbenzene, ***16.35*** is benzenesulfonic acid, ***16.36*** is ethenylbenzene (the common name is styrene), and ***16.37*** is nitrobenzene.

What Is the Protocol When There Is More Than One Substituent?

The base name of the benzene derivative is derived from the group or substituent having the highest priority. The carbon bearing that group or substituent is labeled C1, and the remaining groups or substituents are given the lowest combination of numbers.

What Is the Protocol When There Are Identical Substituents?

If there are identical substituents, the base name is used, giving the substituents the lowest combination of numbers.

Rank Order the Following Substituents As to Their Priority in Nomenclature: OH, CO_2H, NO_2, CH_3, C≡N, C≡CH, CH=CH_2, NH_2.

Using the Cahn-Ingold-Prelog selection rules (see Section 5.3), O > N > C, making OH the highest priority. The nitrogen substituents have the next highest priority, and $NO_2 > NH_2$. The priority of the carbon substituents is largely dictated by the groups attached to carbon, and again, O > N > C, so CO_2H > C≡N > C≡CH > CH=CH_2 > CH_3. Therefore, OH > NO_2 > NH_2 > CO_2H > C≡N > C≡CH > CH=CH_2 > CH_3.

Name the Following Molecules: 16.38, 16.41–16.44.

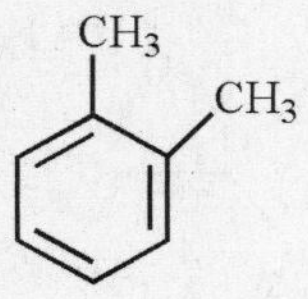

16.38 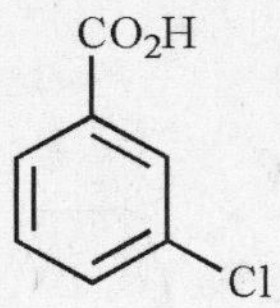 *16.41* 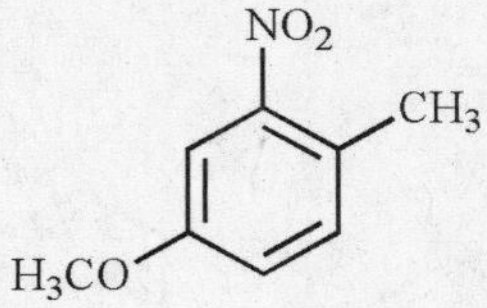 *16.42* 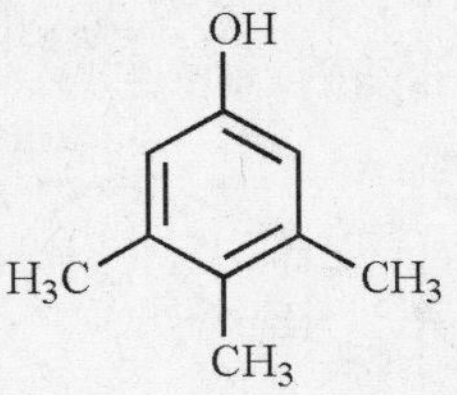 *16.43* 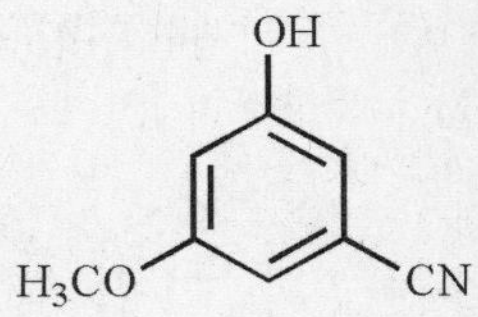 *16.44*

Compound ***16.38*** is 1,2-dimethylbenzene. Compound ***16.41*** is 3-chlorobenzoic acid. Compound ***16.42*** is 4-methyl-3-nitroanisole. Compound ***16.43*** is 3,4,5-trimethylphenol, and ***16.44*** is 3-hydroxy-5-cyanoanisole.

When Benzene Is Treated As a Substituent, What Is the Proper Name?

A benzene ring that is treated as a substituent is given the name ***phenyl***.

If a Benzene Ring Is Attached to a Long Chain Alkane, Is It an Alkylbenzene or Phenyl Alkane?

The general 'rule' is that if the alkyl chain is greater than six carbons, the benzene ring is treated as a substituent, and it is a phenylalkane. If the chain is less than or equal to six carbons, it is named as a benzene derivative (alkyl benzenes such as ethylbenzene or hexylbenzene).

What Is the Name of 16.45 and of 16.46?

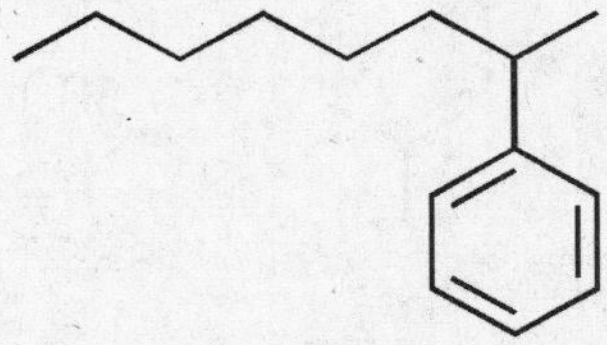

16.45 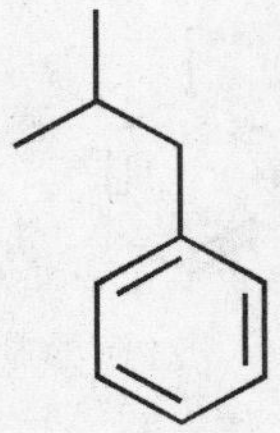 *16.46*

Benzene derivative ***16.45*** is 2-phenyloctane, and ***16.46*** is (2-methylpropyl)benzene.

Test Yourself

1. Which of the following are aromatic?

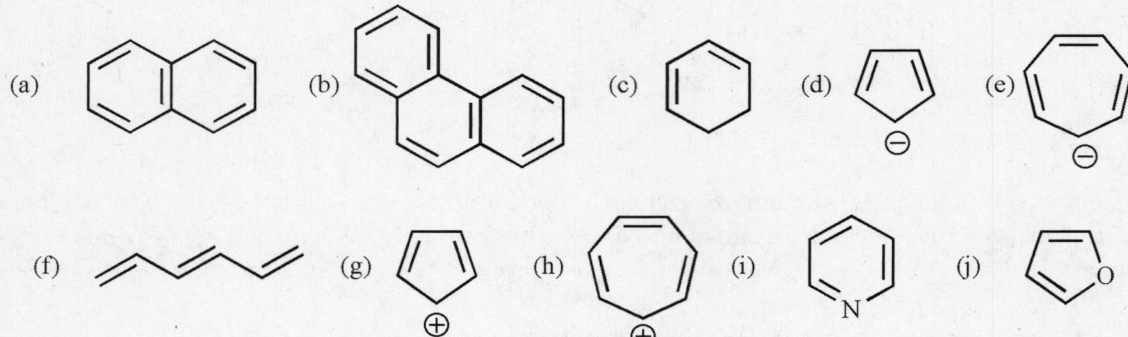

2. Draw all Kekulé structures for the following molecules.

(a) 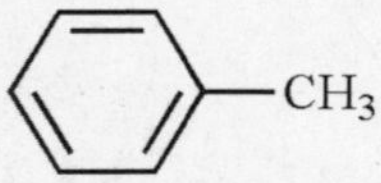 —CH_3 (b) 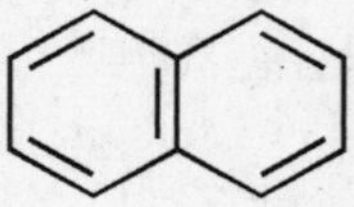 (c)

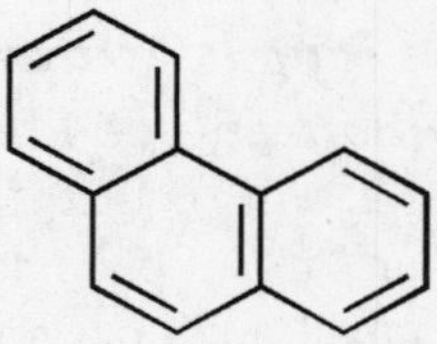

3. Give the correct IUPAC name for each of the following.

(a) 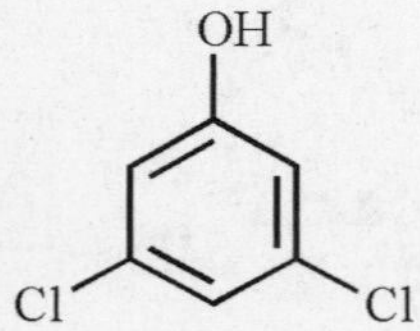 OH, Cl, Cl

(b) 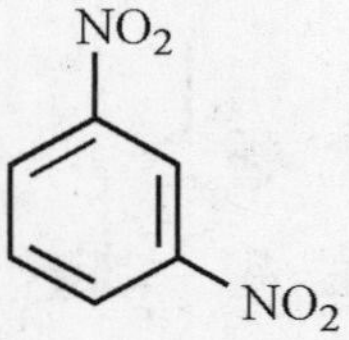 NO_2, NO_2

(c) 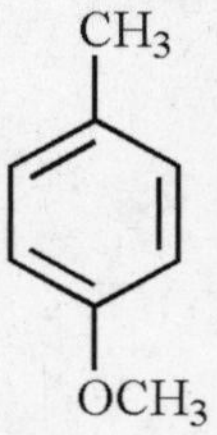 CH_3, OCH_3

(d) 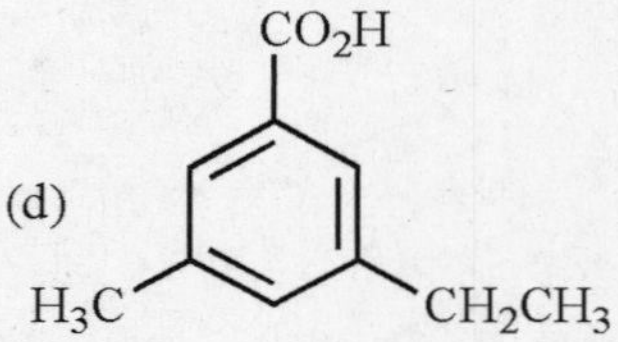 CO_2H, H_3C, CH_2CH_3

(e) Cl, Cl, Cl, Cl, Cl, Cl

(f) OH

(g) CH_3, H_3C, $NHCH_3$

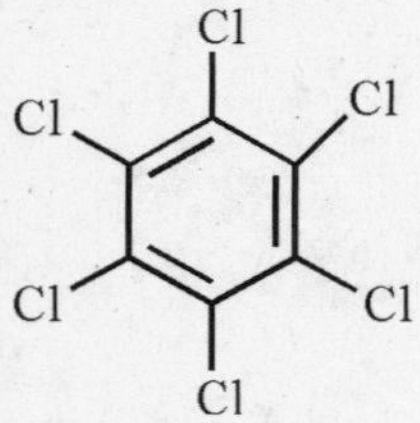

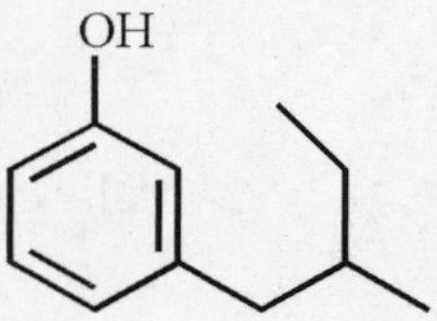

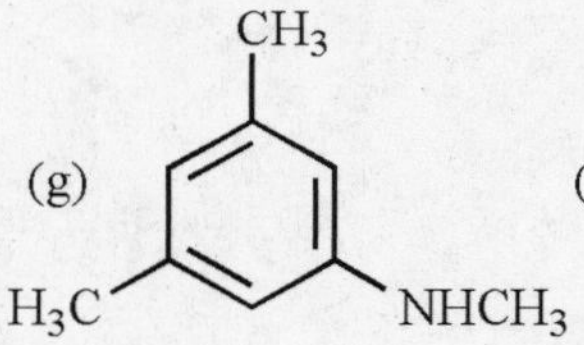

(h) SO_3H, Br

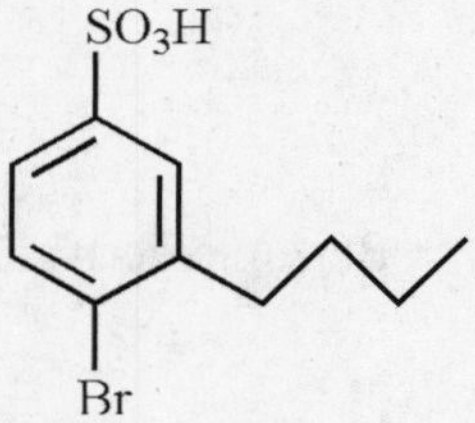

4. Is the following molecule 2-phenyloctane or 1-methylheptylbenzene?

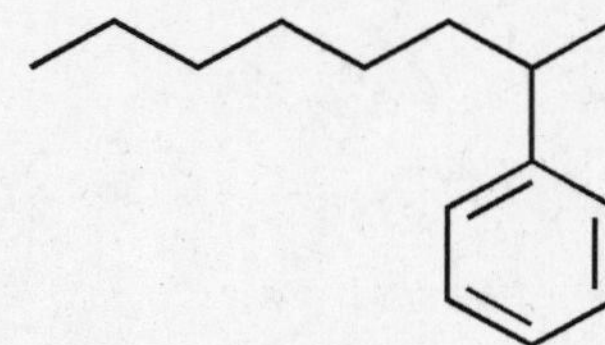

CHAPTER 17

Electrophilic Aromatic Substitution

WHAT IS ELECTROPHILIC AROMATIC SUBSTITUTION?

When benzene reacts with a suitable electrophilic species (E^+) that group replaces a hydrogen on the benzene ring (Ph–H → Ph–X). In this chapter, the E^+ group will be cationic, replacement of H by E^+ is a substitution, and benzene is aromatic. This process is called **electrophilic aromatic substitution**. This chapter will introduce many derivatives of benzene, along with a new nomenclature system. The concept of resonance stabilized intermediates will be much more important for predicting stability of intermediates and distribution of products in this chapter. Several new types of molecules will also be introduced, along with their special chemistry.

How Are Aryl Halides Named by the IUPAC System?

They are named as benzene derivatives using the chloro, bromo, iodo, fluoro prefixes (see Section 3.3). The names are, therefore, fluorobenzene, chlorobenzene, bromobenzene, and iodobenzene.

Give the Structure of 1,4-Dibromo-3-Chlorobenzene.

The structure is ***17.1***, *but this is an incorrect name*. The substituents are numbered to give the smallest numbers. The correct name of ***17.1*** is, therefore, 1,4-dibromo-2-chlorobenzene.

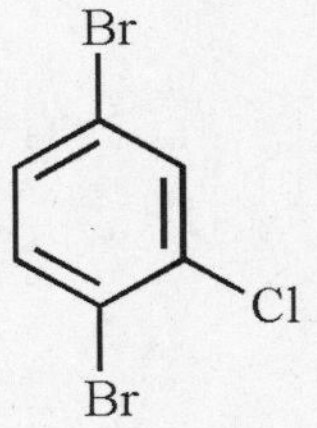

Br

Cl

Br

17.1

17.1. REACTION OF BENZENE WITH CATIONS

Although benzene does not react directly with chlorine or bromine, it can react with cations. In order for this reaction to occur, there must be a chemical step in which cations are produced in the presence of benzene or other aromatic compounds.

Explain Why Benzene Does Not React with HBr or Br_2 Unless Other Reagents Are Added.

Benzene is a very weak Lewis base. To function as a Lewis base, benzene must donate the π-electrons, but these are 'tied up' in the aromatic cloud that gives benzene its special stability. Both HBr and Br–Br are insufficiently strong Lewis acids to react. The loss of resonance energy associated with donation of electrons by benzene precludes its reaction.

What Is the Product When Benzene Reacts with a Cation, E^+?

The product is a cation, ***17.2***, obtained by benzene donating two electrons to E^+ to form a new C–E bond, leaving a positive charge on the ring.

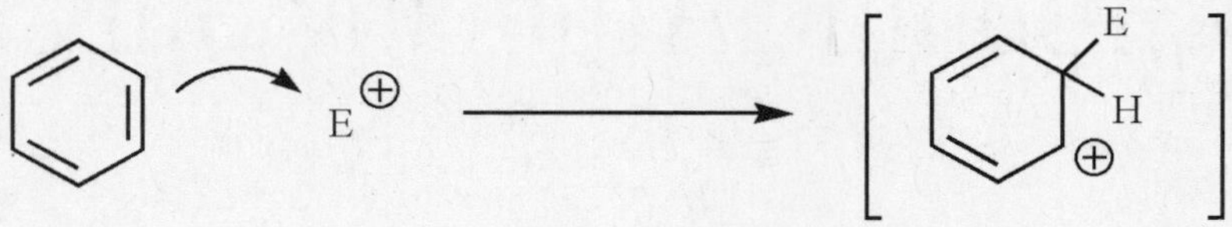

17.2

Is 17.2 Resonance Stabilized? If Yes, Draw the Resonance Contributors.

Yes! The positive charge is adjacent to a π-bond and can be delocalized over both π-bonds. The resonance contributors are ***17.2–17.4***.

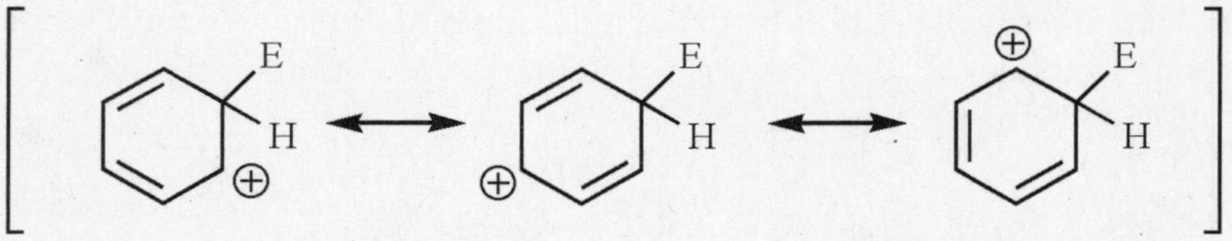

17.2 *17.3* *17.4*

Since 17.2 Is a High Energy, Unstable Intermediate, What Is the Final Product of This Reaction?

The final product is ***17.5*** formed by loss of the proton and formation of the aromatic ring. This step is fundamentally analogous to an E1 reaction.

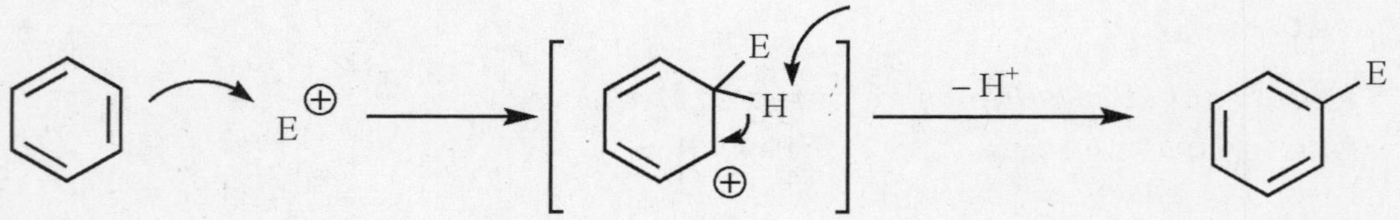

$- H^+$

17.2 *17.5*

Give a Mechanistic Rationale for the Reaction of Benzene with a Cation, E⊕.

We can consolidate the two previous questions, which define the mechanism of the reaction. Benzene donates a pair of electrons to E⊕, forming a C–X bond and leaving a positive charge on the ring, as in ***17.2–17.4***. This carbocation intermediate is resonance stabilized, since the positive charge can be delocalized by the adjacent π-bonds. A total of three resonance forms can be drawn, representing the relative stability of the cationic intermediate. A proton is lost from ***17.2*** to regenerate the aromatic ring in the final product, ***17.5***. This overall process involves cationic intermediates and leads to substitution of the hydrogen by E. It is therefore called **electrophilic aromatic substitution**.

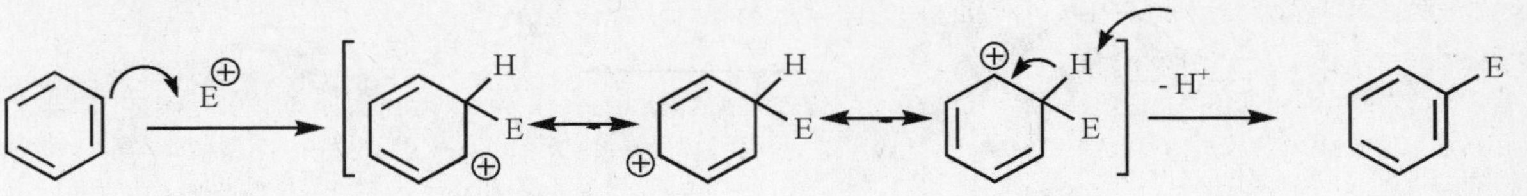

17.2 17.3 17.4 17.5

What Is the Name Given to the Resonance Stabilized Intermediate (17.2–17.4) in This Reaction?

This intermediate is sometimes referred to as a **Wheland intermediate** or a **Meisenheimer adduct**.

The Reaction That Generates 17.5 Shows an Arrow Pointed to a Hydrogen Atom, Which Implies Something Attacks That Hydrogen Atom to Remove It. What Could Attack That Proton?

This is an acid-base reaction, and a base must attack the proton. The E^+ species must have a negatively charged counterion, and this is often the base that attacks the proton.

Why Does the Resonance Stabilized Intermediate Lose a Proton to Give the Substitution Product?

To form ***17.2–17.4***, the aromatic character of benzene must be disrupted. Loss of the hydrogen adjacent to the E group is very facile because reformation of the aromatic ring is an exothermic process, and product ***17.5*** is aromatic.

What Is the Name Given to This Entire Process?

As mentioned earlier, this process is called **electrophilic aromatic substitution**.

17.2. AROMATIC SUBSTITUTION: FORMATION OF CATIONS

The key to electrophilic aromatic substitution is formation of the cation (E^+) in the presence of benzene. There are various methods of generating cations, leading to formation of several benzene derivatives.

Halogenation

What Is the Product of a Reaction between Benzene and a Halogen Cation Such As Cl^+ or Br^+?

Benzene reacts with halogens in the presence of a suitable Lewis acid and dihalide to give a halobenzene, in this case chlorobenzene or bromobenzene, respectively.

Why Is There No Reaction When Benzene Is mixed with Diatomic Bromine?

Diatomic bromine (or chlorine) is polarizable, but benzene is not a sufficiently strong Lewis base to induce a large enough positive dipole to initiate the reaction.

What Is the Product When Diatomic Bromine Reacts with Ferric Chloride ($FeCl_3$)?

The bromine reacts as a Lewis base, with iron as the Lewis acid. The product is a Lewis acid-Lewis base '-ate' complex (*17.6*), which has a bromine cation. Note that this reaction forms a bromine cation, despite the use of ferric chloride.

$$Br{-}Br \quad Cl{-}FeCl_2 \longrightarrow \overset{\oplus}{Br} \quad \overset{\ominus}{FeBrCl_3}$$

17.6

What Is the Product When Diatomic Bromine Reacts with Ferric Bromide ($FeBr_3$)? With Aluminum Chloride ($AlCl_3$)?

When $FeBr_3$ comes in close proximity to diatomic bromine, the closest bromine takes on a negative charge (an induced dipole), inducing a positive charge on the other bromine (see ***17.7***). Bromine is very polarizable, and induced polarity is relatively facile in the presence of a suitable Lewis acid. The bromine (behaving as a Lewis base) attacks the iron, which is a Lewis acid, generating the usual complex. This Lewis acid-Lewis base complex, ***17.8***, contains an electrophilic bromine, Br^+.

$$FeBr_3 \quad \overset{\delta^-}{Br}{-}\overset{\delta^+}{Br} \longrightarrow \overset{\oplus}{Br} \quad Br{-}\overset{\ominus}{Fe}Br_3$$

17.7 17.8

Similar reaction with $AlCl_3$ generates the aluminum complex, ***17.9***, also containing $Br^{\oplus}$.

$$Br_2 + AlCl_3 \longrightarrow Br^+ \; BrAlCl_3^-$$

17.9

Give the Mechanism for the Conversion of Benzene to Bromobenzene by Reaction with Bromine and $FeBr_3$.

The intermediate ***17.10*** leads to ***17.11*** as the product (bromobenzene), and this is the mechanism for all such reactions.

$$\text{Benzene} \xrightarrow{\overset{+}{Br}\ \overset{-}{Br{-}FeBr_3}} [\text{resonance forms of arenium ion (H, Br, } \oplus)] \longrightarrow \text{Bromobenzene}$$

17.10 17.11

Why Does the Reaction of Bromine and $AlCl_3$ with Benzene give Bromobenzene and Not Chlorobenzene?

The initial product of the reaction between bromine and aluminum chloride is ***17.9***, where the electrophilic species is $Br^{\oplus}$. Since $Br^{\oplus}$ is the species that reacts with benzene to form ***17.10***, the final product is bromobenzene (***17.11***), not chlorobenzene.

Is This Reaction Restricted to Bromine?

No! Chlorine is converted to $Cl^{\oplus}$ in the presence of a Lewis acid and generates chlorobenzene. Formation of fluorobenzene and iodobenzene is possible by this method with a suitable Lewis acid, but there are several problems with these reactions. Therefore, fluorobenzene and iodobenzene are commonly made by procedures that will be discussed later, and electrophilic aromatic substitution is largely restricted to formation of chlorobenzene and bromobenzene.

Nitration

What Is the Electrophilic Species Formed When Nitric Acid Reacts with Sulfuric Acid?

The electrophilic species is the nitronium ion, $NO_2^{\oplus}$ (***17.13***). Nitric acid reacts with sulfuric acid to form ***17.12***, with the hydrogen sulfate anion as the counterion. In a second step, ***17.12*** reacts with additional sulfuric acid to give an intermediate that fragments to form the nitronium ion (***17.13***), as well as the hydronium ion with a hydrogen sulfate counterion.

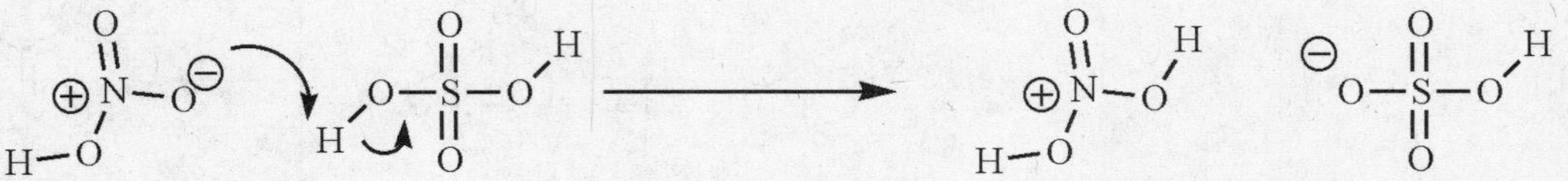

17.12

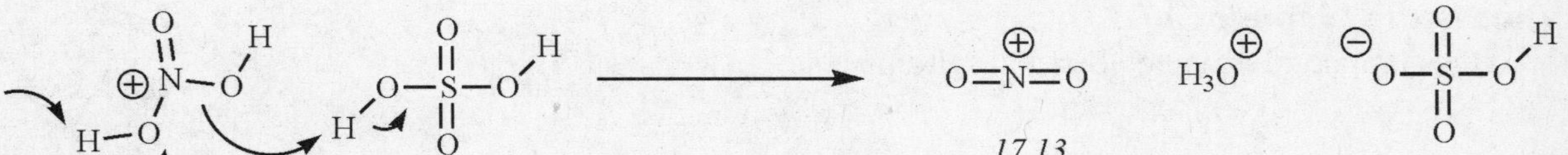

17.13

In the Reaction of Nitric Acid and Sulfuric Acid, Which Is the Acid and Which Is the Base?

Since nitric acid accepts the proton from sulfuric acid, an oxygen of nitric acid functions as a base, and sulfuric acid is the acid. Sulfuric acid is the stronger acid.

Draw the Resonance Intermediates and the Final Product for the Reaction of Benzene with HNO_3/H_2SO_4.

Initial reaction with $NO_2^{\oplus}$ generates the Wheland intermediate, ***17.14***. Loss of a proton generates the final substitution product (***17.15***).

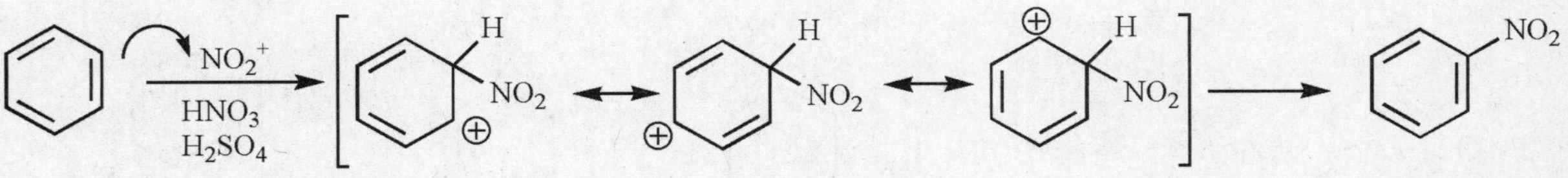

17.14

17.15

What Is the Name of 17.15?

This compound is named nitrobenzene.

Sulfonation

What Is Fuming Sulfuric Acid?

Fuming sulfuric acid is sulfuric acid that is saturated with sulfur trioxide, SO_3.

What Is the Electrophilic Species in Fuming Sulfuric Acid?

Both SO_3 and $HSO_3^{\oplus}$ are electrophilic species in this medium, but SO_3 is taken to be the reactive entity.

Draw the Resonance Intermediates in the Reaction of Benzene with Fuming Sulfuric Acid.

The benzene ring attacks the sulfur of SO_3 to generate a Wheland intermediate such as ***17.16***. Loss of hydrogen in the usual manner regenerates the aromatic ring with transfer of a hydrogen to the oxygen of the SO_3^- moiety, forming ***17.17***.

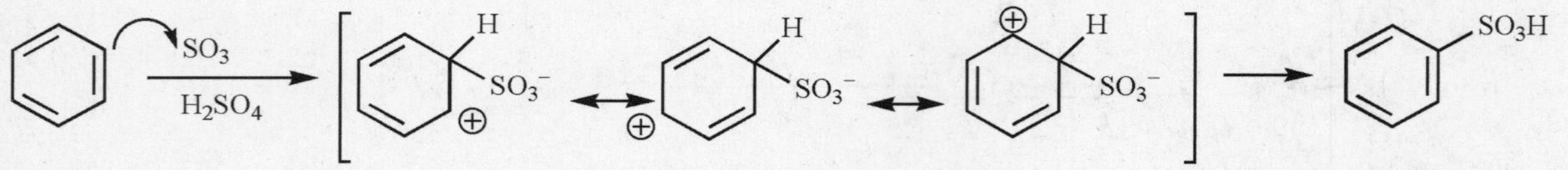

SO_3 / H_2SO_4 → [H, SO_3^-, ⊕ ↔ H, SO_3^-, ⊕ ↔ ⊕, H, SO_3^-] → SO_3H

17.16 *17.17*

What Is the Name of 17.17?

This compound is named benzenesulfonic acid.

17.3. SUBSTITUTED BENZENE DERIVATIVES (DISUBSTITUTION)

Electrophilic aromatic substitution occurs on substituted benzene compounds. However, with one substituent in place, the second substituent can be attached at C2, C3, or C4 relative to the first substituent. This means that the reaction can produce regioisomers.

What Is the Common Name of Two Substituents That Are Situated on a Benzene Ring in a: (A) 1,2 Relationship (B) 1,3-Relationship (C) 1,4-Relationship?

The common name for (a) 1,2-substitution is **ortho**, (b) for 1,3-substitution is **meta**, and (c) for 1,4-substitution is **para**.

What Is the IUPAC Method for Naming 17.18–17.22?

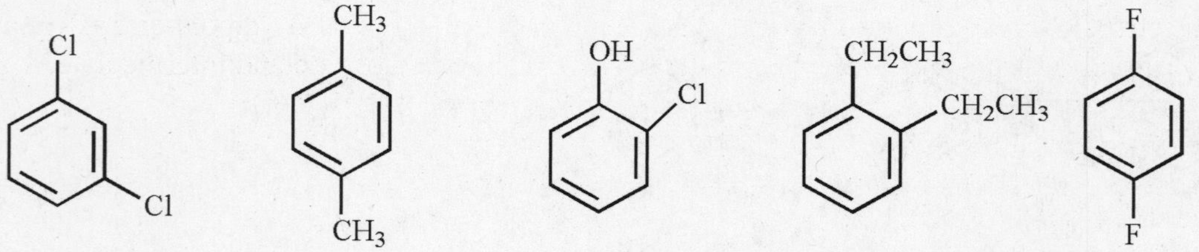

Cl, Cl — *17.18*; CH_3, CH_3 — *17.19*; OH, Cl — *17.20*; CH_2CH_3, CH_2CH_3 — *17.21*; F, F — *17.22*

The numbers are used with the highest priority functional group being assigned to C1. The ring is numbered to give the substituents the lowest possible number. Therefore, ***17.18*** is 1,3-dichlorobenzene; ***17.19*** is 1,4-dimethylbenzene; ***17.20*** is 2-chlorophenol; ***17.21*** is 1,2-diethylbenzene; and ***17.22*** is 1,4-difluorobenzene.

Using the Ortho, Meta and Para Nomenclature, Give the Name of 17.18–17.22.

Compound ***17.18*** is *ortho*-dichlorobenzene; ***17.19*** is *para*-xylene; ***17.20*** could be called *ortho*-chlorophenol, if the two groups are not the same. The *ortho/meta/para* nomenclature is not used very often. In this case, 2-chlorophenol is the better name. Compound *17.21* is *ortho*-diethylbenzene, and *17.22* is *para*-difluorobenzene.

Using the IUPAC Nomenclature, Give the Name of Each of the Following.

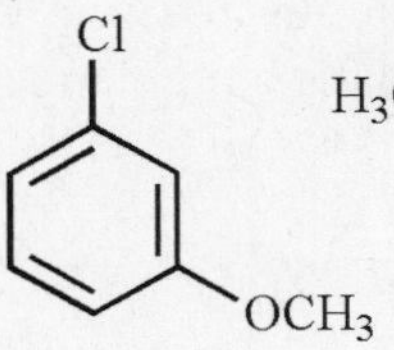

Cl, OCH_3 — *17.23*; H_3C, CO_2H — *17.24*; NH_2, Cl — *17.25*; H_3C, Br — *17.26*; CH_2CH_3, $CH_2(CH_2)_3CH_3$ — *17.27*

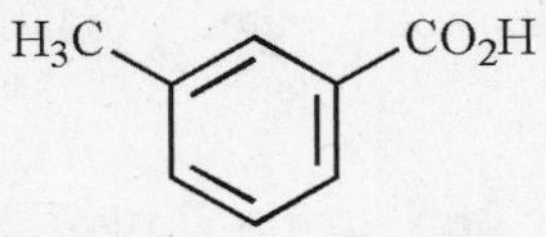

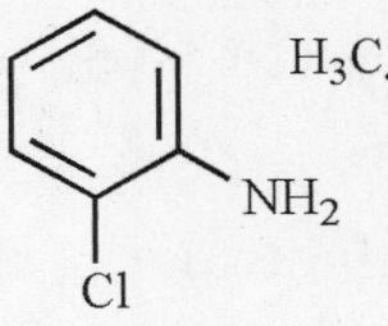

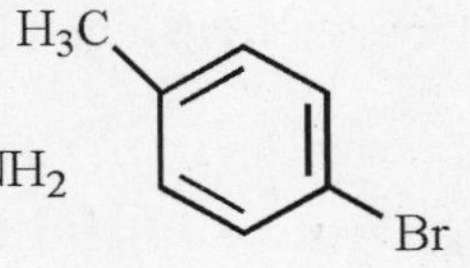

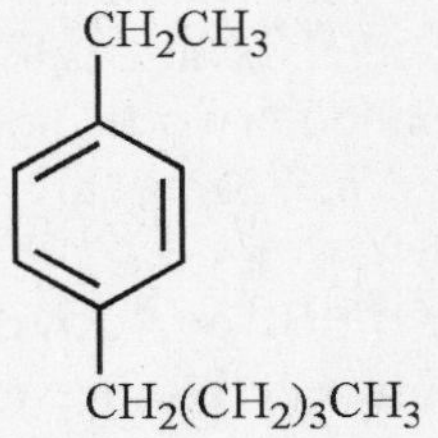

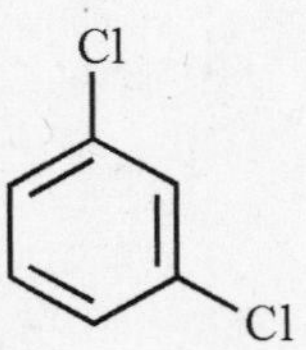

Cl, Cl — *17.28*; CH_3, CH_3 — *17.29*; CH_3, OH — *17.30*; C≡N, F — *17.31*; I, Br — *17.32*

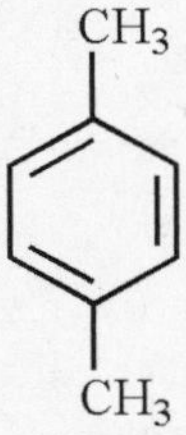

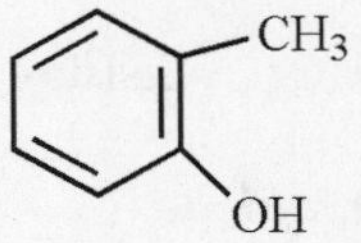

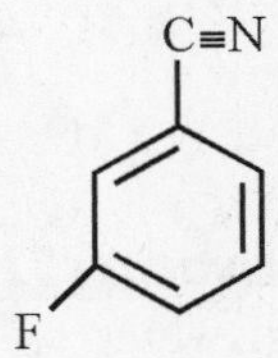

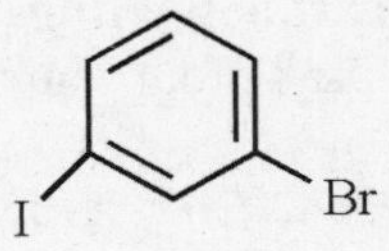

Compound ***17.23*** is 3-chloroanisole; ***17.24*** is 3-methylbenzoic acid; ***17.25*** is 2-chloroaniline; ***17.26*** is 4-bromotoluene (or 4-bromomethylbenzene); ***17.27*** is 4-ethylpentylbenzene (1-[4-ethyl-phenyl]pentane could also be used); ***17.28*** is 1,3-dichlorobenzene; ***17.29*** is 1,4-dimethylbenzene; ***17.30*** is 2-methylphenol; ***17.31*** is 3-fluorobenzonitrile; and ***17.32*** is 3-bromoiodobenzene.

Electron Releasing Groups

What Is an Electron Releasing Group in This Context?

An electron releasing group is one that has a negative charge or a δ (*o*,

What Is an Electron Withdrawing Group in This Context?

An electron withdrawing group is one that has a positive charge or a δ^+ dipole directly attached to a benzene ring. It is electron withdrawing to the positive charge in the Wheland intermediate.

What Is an Ipso Carbon?

In Wheland intermediates ***17.33***, the ipso carbon is the carbon bearing the substituent, which in this case is bromine.

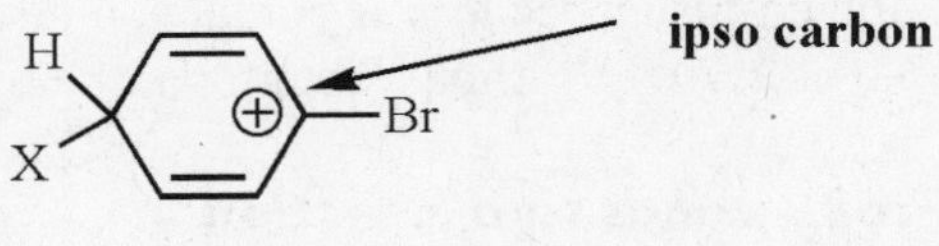

H, X, ⊕, Br — **ipso carbon**

17.33

What Is an Activating Group?

An activating group is one attached to benzene that releases electrons to the Wheland intermediate and causes electrophilic aromatic substituent to proceed ***faster*** than a similar reaction with a compound that does not have that group attached. In other words, an activated aromatic compound reacts ***faster*** than benzene.

What Is a Deactivating Group?

A deactivating group is one attached to benzene that withdraws electrons from the Wheland intermediate and causes electrophilic aromatic substituents to proceed ***slower*** than a similar reaction with a compound that does not have that group attached. In other words, an activated aromatic compound reacts ***slower*** than benzene.

Give a List of Activating Groups in the Order of their Activating Power.

most activating	OH, NH_2, NR_2
	OR
	N(C=O) [an amide], R, O(C=O)R
[an ester]	
least activating	R, Ar (R = alkyl, Ar = aromatic)

Alkyl groups are considered to be electron releasing .

Why Is OCH_3 More Activating Than CH_3?

This question can be answered by inspection of a Wheland intermediate from attack at C4 of anisole (***17.34***) and that from attack at C4 of toluene (***17.35***). Note that the oxygen of anisole has lone electron pairs and that the oxygen will have a relatively large δ^- dipole. By contrast, the carbon atom of the methyl group is electron releasing, but does not have a formal dipole. The proximity of the electron pairs (the oxygen is a δ^- dipole) to the positive charge makes that atom more electron releasing and more stabilizing. If the intermediate is more stabilized, it forms faster, and the overall reaction is faster. Therefore, OMe is a more powerful activating group relative to Me.

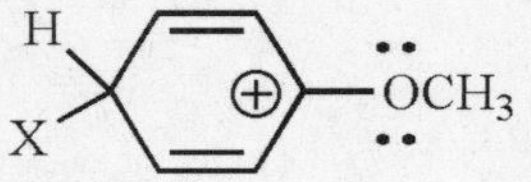

17.34 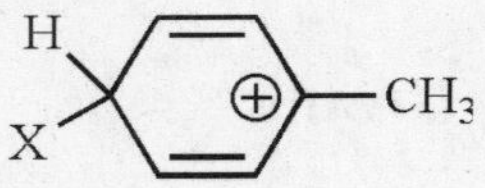 *17.35*

Why Are Electron Releasing Groups Referred to As Activating Groups?

When an electron releasing group is attached to a benzene ring, the rate of electrophilic aromatic substitution is greatly increased. Phenol, for example, undergoes bromination about 6×10^{11} times faster than benzene, and toluene undergoes chlorination about 340 times faster than benzene. The fact that the methyl group in toluene increases the rate of the reaction relative to benzene causes it to be placed among the activating substituents.

Which Molecule Reacts Faster with Bromine/$FeBr_3$, Anisole or Benzene?

As mentioned, anisole reacts much faster because it contains a strongly electron releasing group. Bromination of anisole, for example, is 1.79×10^5 times faster than the bromination of benzene.

Explain Why an Activating Group Leads to a Reaction That Is Faster Than a Similar Reaction with Benzene.

The faster rate of reaction can be explained by examining the ipso carbon in the resonance stabilized intermediate for the bromination of anisole. In this case, there are four resonance forms (numbered

17.35–17.38). Structures *17.36*, *17.37* and *17.35* are the usual structures associated with a Wheland intermediate. Structure *17.38*, however, initially places the positive charge on the ipso carbon, immediately adjacent to the oxygen of Ome, and is formed by donation of the unshared electrons on oxygen to the electrophilic center, generating a positive charge on oxygen. This 'extra' resonance form represents the increased stability realized by delocalization of the positive charge out to the oxygen. Activating groups release electrons to a positive charge on the ipso carbon in resonance forms such as *17.35*, providing this extra stability. If the intermediate is more stable, it is formed more easily and accelerates the overall rate of reaction. It is noted that *17.35–17.38* represent *one* intermediate. The different numbers are simply used to illustrate how the extra stability arises.

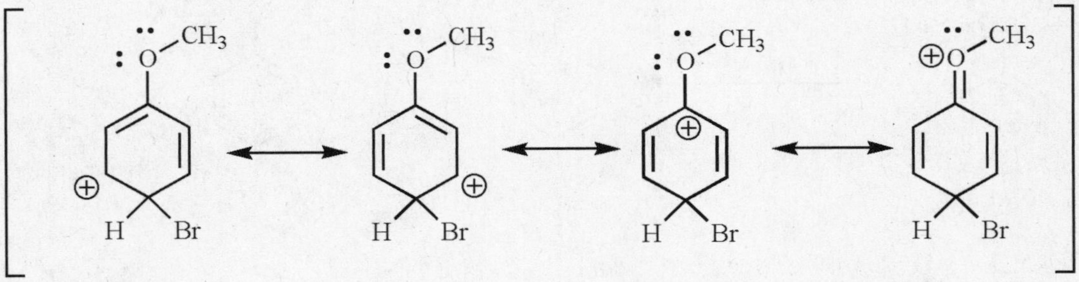

17.36 17.37 17.35 17.38

Although NH_2 Is an Activating Group, the Reaction of Bromine/$AlCl_3$ and Aniline (17.39) Does Not Give a Bromoaniline Derivative. Explain.

The NH_2 group in *17.39* functions as a strong Lewis base and reacts with the Lewis acid $AlCl_3$ to form the 'ate' complex *17.40* *before* $AlCl_3$ can react with bromine, which is a weaker Lewis base than aniline. Therefore, no Br^+ is available for electrophilic aromatic substitution.

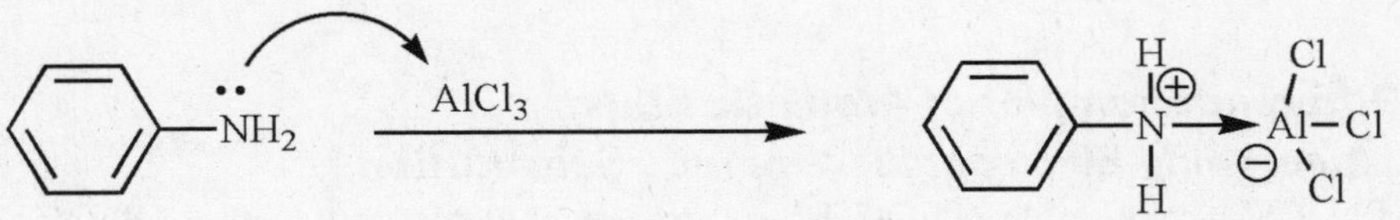

17.39 17.40

How Can Aniline Be Chemically Changed to Allow Reaction with Br^+?

If the lone electron pair on nitrogen is somehow 'delocalized,' the basicity is diminished. The most common method for doing this is to **protect** the nitrogen as an amide (see Section 25.5). If aniline is reacted with acetyl chloride (*17.41*), the acetamide derivative (*N*-acetyl aniline or acetanilide, *17.42*) is formed. See Section 25.5 for the reaction of amines and acid chlorides.

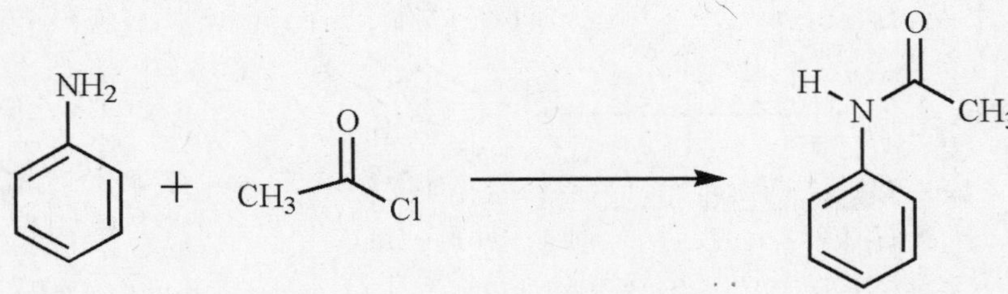

17.41 17.42

What Is the Product Formed When N-Acetyl Aniline (17.42) Reacts with Bromine and Aluminum Chloride?

If ***17.42*** is reacted with bromine and $AlCl_3$, a mixture of *ortho*- and *para*-bromo derivatives (***17.43*** and ***17.44***, respectively) is formed. Treatment with aqueous acid (or i. NaOH ii. H_3O^+ – see Section 25.3 for hydrolysis of amides) will remove the acetyl group, regenerating the NH_2 group, giving ***17.45*** and ***17.46***.

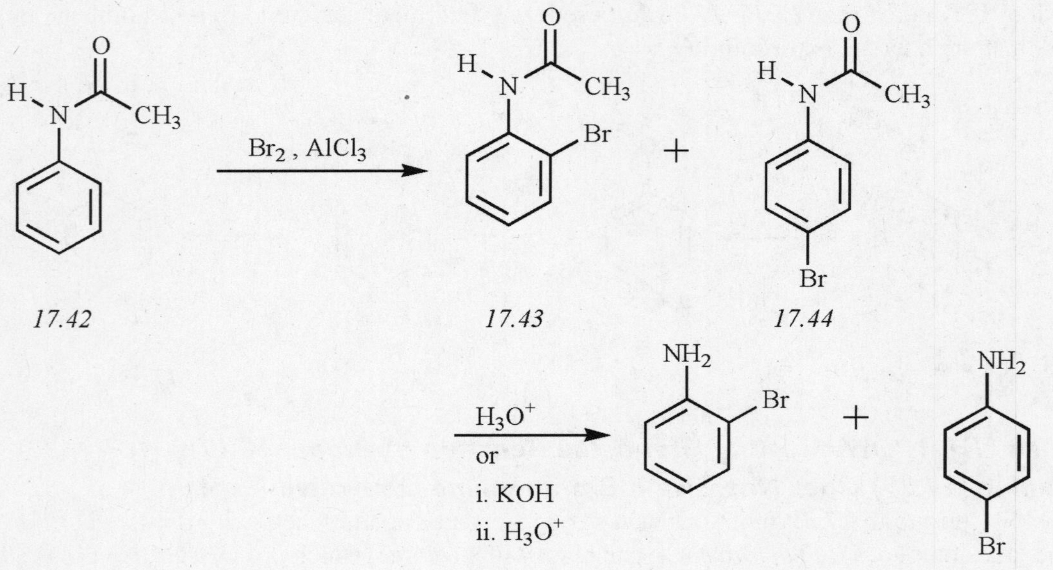

Br_2, $AlCl_3$

17.42 *17.43* *17.44*

H_3O^+ or i. KOH ii. H_3O^+

17.45 *17.46*

Give Three Different Examples of Aromatic Ethers Capable of Undergoing Electrophilic Aromatic Substitution.

Anisole ($PhOCH_3$), phenetole ($PhOCH_2CH_3$) and phenyl *tert*-butyl ether [$PhOC(CH_3)_3$] will all undergo electrophilic aromatic substitution with a variety of reagents.

What Is the Product (or Products) When Phenyl Ethyl Ether (Phenetole) Reacts with Bromine and Ferric Bromide? Draw Them!

Bromination of phenyl ethyl ether (phenetole) leads to a mixture of 2-bromophenetole (***17.47***) and 4-bromophenetole (***17.48***).

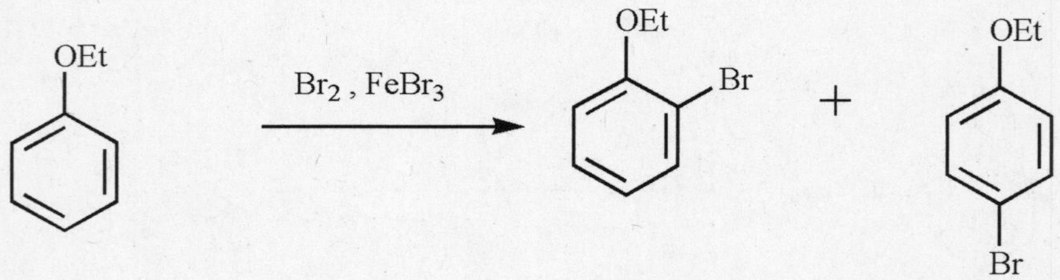

Br_2, $FeBr_3$

17.47 *17.48*

Draw all Resonance Intermediates When Br^+ Reacts with Anisole at the Ortho Position and Then at the Para Position.

Attack at the *ortho* position generates a resonance stabilized intermediate (***17.49***) with four contributors. Attack at the *para* position also generates an intermediate with four resonance contributors (see ***17.35–17.38***, above).

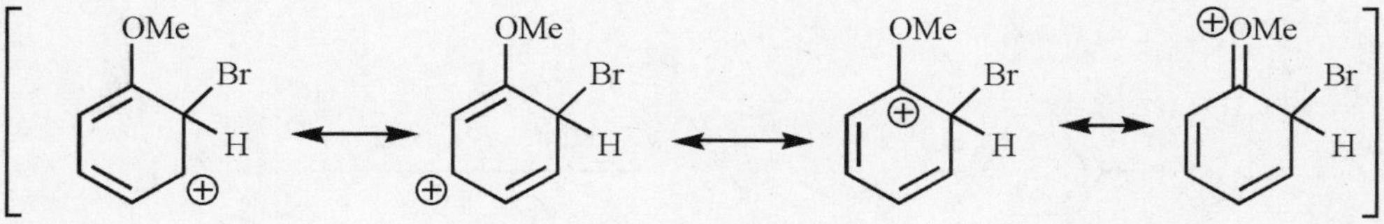

17.49

Is There Any Difference in the Relative Stability of 17.35–17.38 vs. 17.49 for These Two Modes of Attack?

No. Both have an equal number of resonance contributors. There may be some dipole interactions in ***17.49*** that stabilize this intermediate relative to ***17.35–17.38***, sometimes called the ortho effect, and with heteroatom substituents, the *ortho* product often predominates. Sometimes there is a steric factor for *ortho* attack that diminishes the amount of *ortho* product, however.

Draw All Resonance Intermediates for Attack of Br^+ at the Meta Position of Anisole.

Resonance intermediate ***17.50*** is formed, with three resonance contributors.

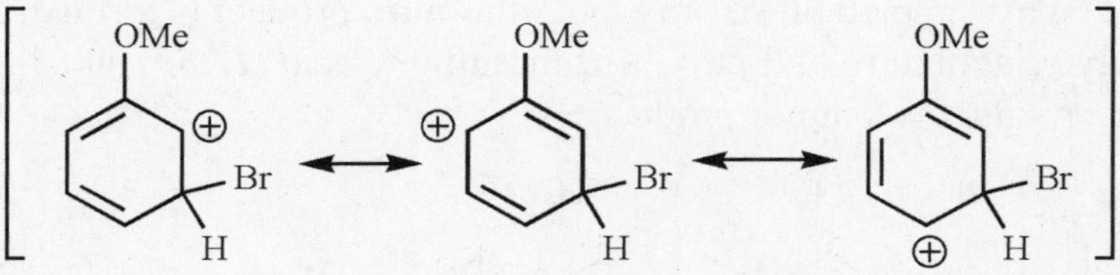

17.50

Does 'Meta Attack' Lead to a More Stable or less Stable Intermediate? Explain.

It leads to a less stable intermediate. The positive charge on the ring in ***17.50*** is never positioned on the carbon bearing the OMe (the ipso carbon), and delocalization of electron on the oxygen is not possible. With fewer resonance contributors, the intermediate is less stable, and the reaction is slower.

What Are the Major Products When Anisole Reacts with $Br_2/FeBr_3$? Explain.

The two major products are 2-bromoanisole (***17.51***) and 4-bromoanisole (***17.52***). Loss of H^+ from the more stable intermediates ***17.49*** and ***17.35–17.38*** (which is one intermediate; the four numbers are arbitrary) leads to the two major products. Since there are only three resonance contributors for ***17.50***, its formation is much slower than the others, and very little meta-bromoanisole is formed.

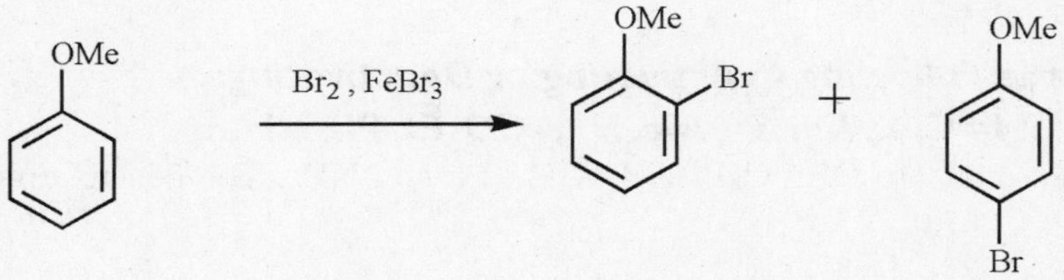

Br_2, $FeBr_3$

17.51 *17.52*

Give the Major Product(s) for the Reactions of 17.53–17.55.

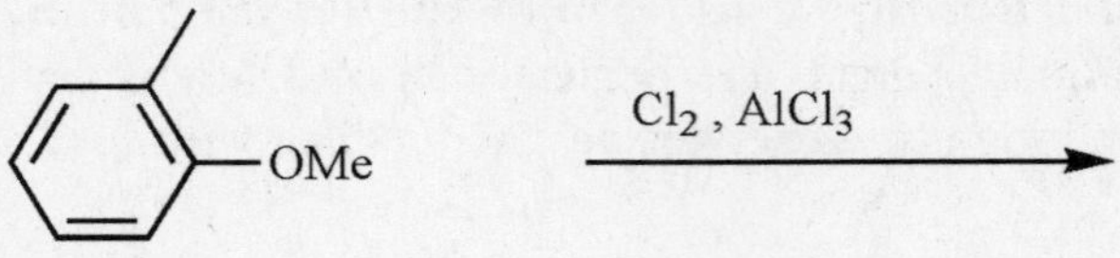

OMe

Cl_2 , $AlCl_3$

17.53

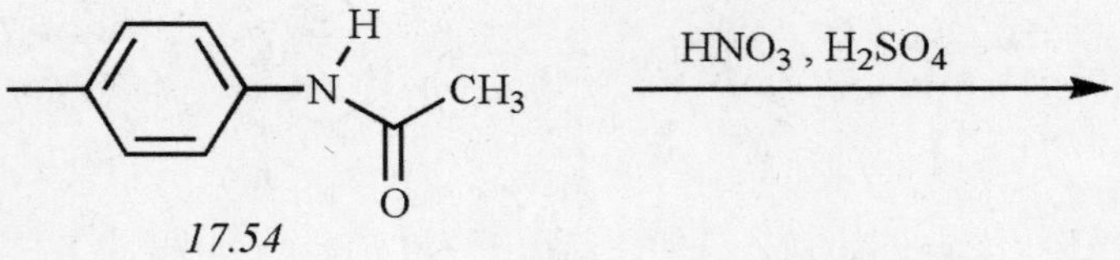

H N CH_3 O

HNO_3 , H_2SO_4

17.54

CH_2CH_3

SO_3 , H_2SO_4

17.55

The reaction of 2-methylanisole (***17.53***) and $Cl_2/AlCl_3$ generates a mixture of 2-chloro-5-methylanisole (***17.56***) and 4-chloro-2-methylanisole (***17.57***). Note that the bromines are incorporated ortho and para to the more powerful activating OMe group, rather than to the methyl. *N*-Acetyl 4-methylaniline (***17.54***) undergoes nitration to give *N*-acetyl-2-nitro-4-methylaniline (***17.58***). Note that the para- position is blocked by the methyl, so only the ortho-nitro product is formed. Sulfonation of ethylbenzene (***17.55***) generates a mixture of 2-ethylbenzenesulfonic acid (***17.59***) and 4-ethylbenzenesulfonic acid (***17.60***). The *meta* product is a minor product.

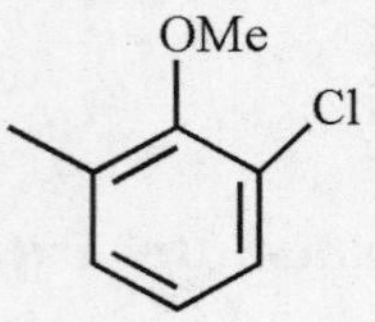

OMe Cl | OMe Cl | NHAc NO_2 | Et SO_3H | Et SO_3H

17.56 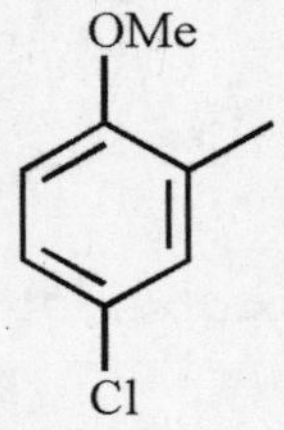 *17.57* 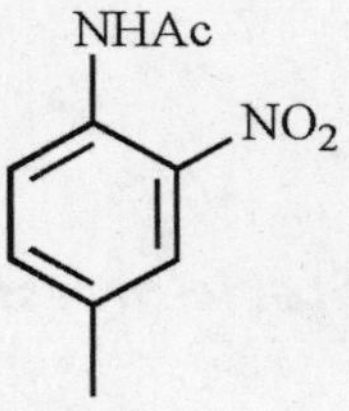 *17.58* 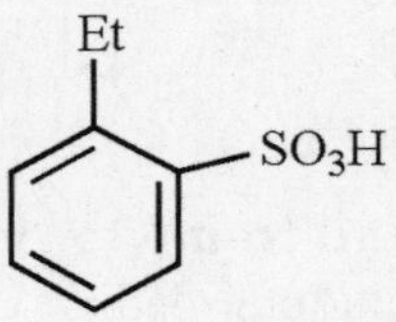 *17.59* 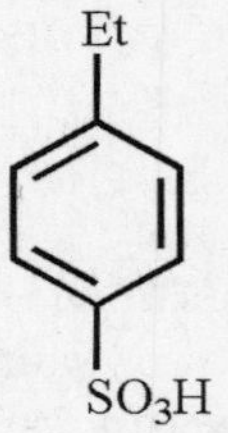 *17.60*

Electron Withdrawing Groups

Why Are Electron Withdrawing Groups Referred to As Deactivating Groups?

When the benzene ring attacks an electrophilic species (E^+), the resonance stabilized intermediate has a positive charge in the ring. If this positive charge is adjacent to a ⊕ or δ⊕ charge on an electron withdrawing group substituent, the electrostatic repulsion of like charges destabilizes the intermediate.

Classify Each of the Following As Activating or Deactivating: -NO_2, OEt, CH_3, CH=CH_2, Me_2S⊕, Me_3N⊕, -CO_2Et, Ph, NH_2.

The activating groups are OEt, CH_3, CH=CH_2, Ph and NH_2. The deactivating groups are -NO_2, Me_2S⊕, Me_3N⊕ and -CO_2Et.

When Compared with Nitrobenzene, Does Benzene React Faster or Slower with Bromine and Aluminum Chloride?

Benzene reacts faster than nitrobenzene in electrophilic aromatic substitution reactions.

Why Does an Electron Withdrawing Group Slow Electrophilic Aromatic Substitution?

The nitro group is an example of an electron withdrawing group, and it has a positive charge on the nitrogen. Repulsion of this charge with the positive charge of the intermediate derived from attack of an electrophile raises the energy of that intermediate and slows the overall reaction.

Give a List of Deactivating Group in Relative Order of their Ability to Deactivate a Benzene Ring.

$-CN$, $-^{\oplus}NR_3$, $-^{\oplus}SR_2$, CF_3 > $-NO_2$, $-SO_3H$, $RC{=}O$ > X (Cl, Br, F, I), $-CH_2X$

Why Is Nitro (NO_2) a More Powerful Deactivating Group Than the Carbonyl of a Ketone (As in Acetophenone)?

The nitrogen of nitro has a formal charge of +1 [$O{=}N^{\oplus}{-}O$ *o*], whereas the carbon of a carbonyl is part of polarized bond and has a $\delta^{\oplus}$ charge. The greater the electrophilic character of the atom (the larger the positive charge), the greater will be its deactivating ability.

Draw All Resonance Intermediates When $Br^{\oplus}$ Reacts at the Ortho Position of Nitrobenzene; at the Para Position.

There is one intermediate for *ortho* attack, with three resonance contributors, ***17.61–17.63***. There is also one intermediate for *para* attack, with three resonance contributors, ***17.64–17.65***.

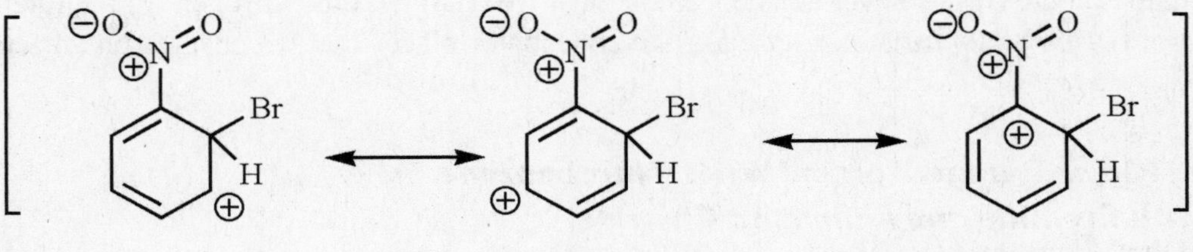

17.61 *17.62* *17.63*

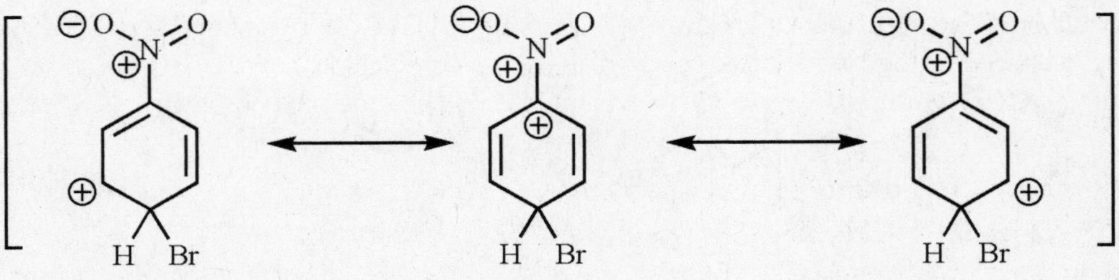

17.64 *17.65* *17.66*

Which Is the Highest Energy Resonance form for Para Attack? For Ortho Attack?

For *para* attack, the highest energy contributor is ***17.65*** where the two positive charges are on adjacent atoms, one on the ipso carbon and the other on the nitrogen. In this contributor, repulsion by the like charges is maximized. Similarly, contributor ***17.63*** has two like charges on adjacent atoms and is the highest energy contributor for *ortho* attack. These two resonance contributors greatly destabilize the intermediate, making it more difficult to form, slowing the overall rate of reaction.

Does Nitrobenzene React Faster or Slower Than Benzene?

Nitrobenzene reacts about 10^6 times *slower*. The destabilized intermediates just discussed make the reaction much more difficult and slower.

Draw All Resonance Intermediates When $Br^{\oplus}$ Reacts at the Meta Position of Nitrobenzene.

Attack at the *meta* position generates one intermediate with three resonance contributors (***17.67–17.69***).

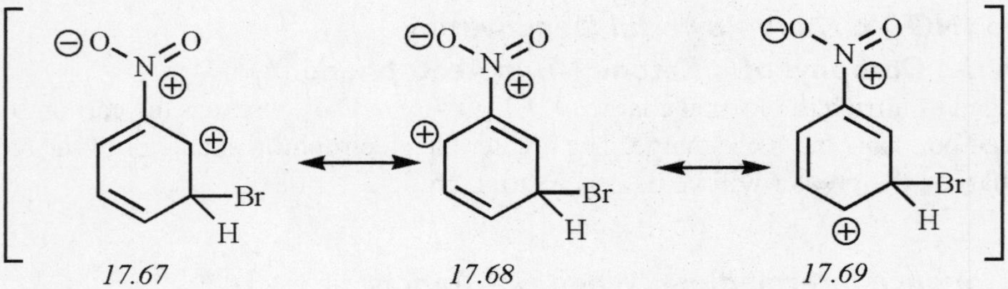

17.67 *17.68* *17.69*

In Reactions of Nitrobenzene, What Is the Lowest Energy Intermediate: Ortho, Meta or Para Attack? Explain.

The lowest energy intermediate arises from *meta* attack. Examination of ***17.67–17.69*** reveals that the positive charge on the ring is never adjacent to the positive charge on the nitrogen. In the intermediates resulting from *ortho* and *para* attack, at least one contributor places the two charges on adjacent atoms (***17.63*** and ***17.65***).

Draw the Major Product Formed When Nitrobenzene Reacts with Bromine and Aluminum Chloride.

The major product will be 3-bromonitrobenzene (***17.70***), the *meta* product.

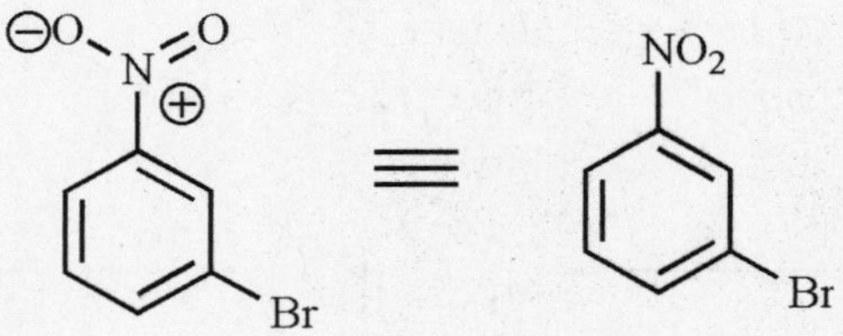

17.70

Give the Major Product(s) for the Reactions of 17.71–17.73.

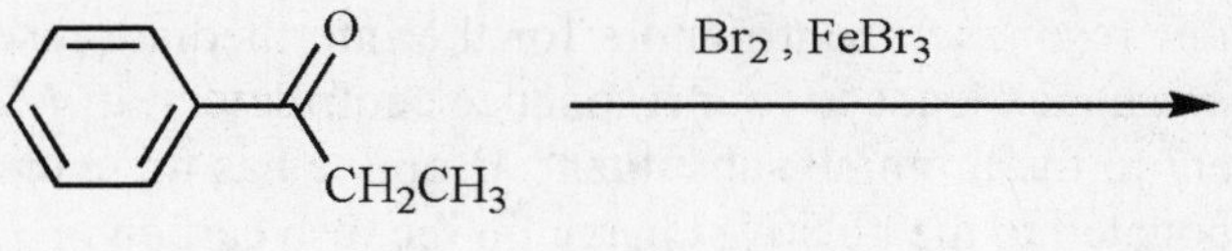

O, CH_2CH_3 — Br_2 , $FeBr_3$ →

17.71

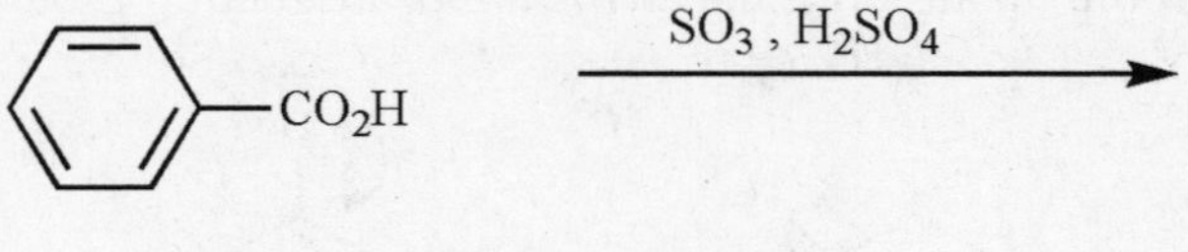

CO_2H — SO_3 , H_2SO_4 →

17.72

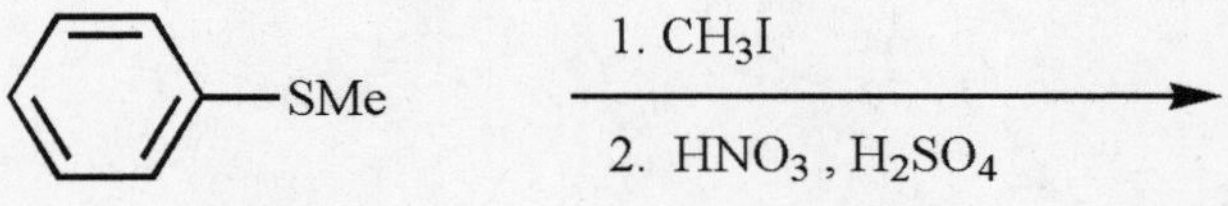

SMe — 1. CH_3I 2. HNO_3 , H_2SO_4 →

17.73

The carbonyl of the ketone in ***17.71*** is deactivating, since the carbon has a positive dipole, and it is a *meta* director. The product is, therefore, ,1-(3-bromophenyl)-1-propanone. The carboxyl group in ***17.72*** also has a carbon with a positive dipole and is also deactivating. The major product is benzenesulfonic acid-3-carboxylic acid (***17.75***). Initial reaction of ***17.73*** with iodomethane generates a dimethylsulfonium ion via an S_N2 reaction with sulfur as the nucleophile, forming ***17.76*** containing the $-SMe_2^+$ group, a deactivating group, whereas –SMe is activating. Sulfonium salt ***17.76*** undergoes nitration, slowly if at all, and the major product of nitration would be 3-(dimethylsulfonium)nitrobenzene, ***17.77***.

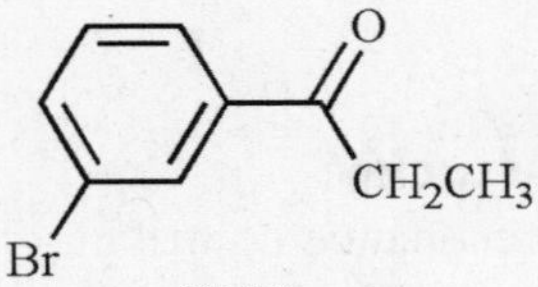

O, CH_2CH_3, Br — *17.74*

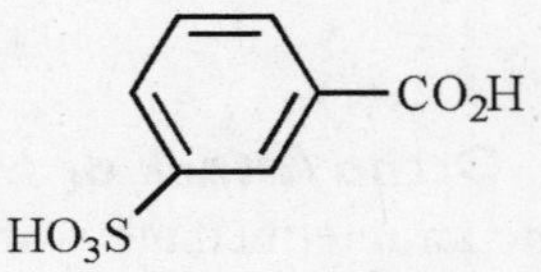

CO_2H, HO_3S — *17.75*

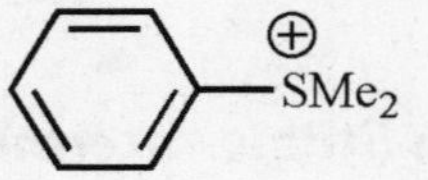

$\oplus$ SMe_2 — *17.76*

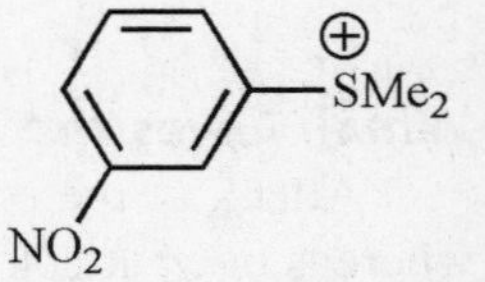

$\oplus$ SMe_2, NO_2 — *17.77*

Halobenzenes

Does Bromobenzene React Faster or Slower Than Benzene in Electrophilic Aromatic Substitution Reactions?

Bromobenzene reacts *slower* than benzene in electrophilic aromatic substitution. Other aryl halides (fluorobenzene, chlorobenzene and iodobenzene) also react slower than benzene.

Is Br an Activating or a Deactivating group?

Bromine is a deactivating group. The other halogens (F, Cl, I) are also deactivating.

Draw all Resonance Intermediates When Br^+ Reacts at the Ortho Position of Bromobenzene.

There is a total of four resonance contributors for the intermediate resulting from *ortho* attack, ***17.78–17.81***. Similarly, *para* attack leads to four resonance contributors. In ***17.80***, the positive charge is on the ipso carbon, adjacent to the bromine substituent. Bromine has three unshared pairs of electrons, and two electrons can be donated to the positive charge on the ipso carbon of ***17.80***, as shown, to form a new resonance contributor, ***17.81***. The presence of the unshared electrons on bromine, and the proximity of the charge on the ipso carbon, makes bromine an *ortho/para* directing group.

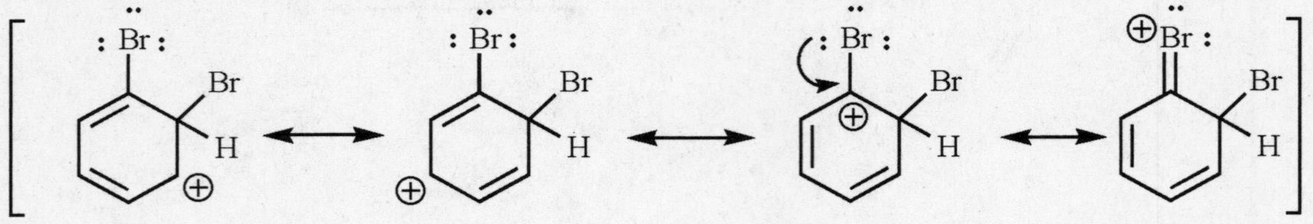

17.78 *17.79* *17.80* *17.81*

Draw All Resonance Intermediates When $Br^{\oplus}$ Reacts at the Meta Position of Bromobenzene?

Reaction at the *meta* position generates one intermediate (***17.82***) with only three resonance contributors.

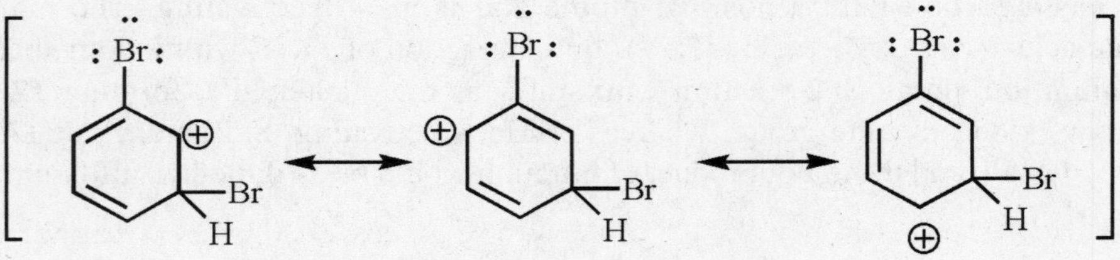

17.82

Which Gives the More Stable Intermediate, Ortho Attack or Meta Attack?

Attack at the *ortho* and *para* positions generate an intermediate with four resonance contributors, whereas *meta* attack generates an intermediate with only three resonance contributors. The intermediates arising from *ortho* and *para* attack are, therefore, more stable. Halogens on a benzene ring are *ortho/para* directors, although they are deactivating.

What Is the Major Product(s) When Bromobenzene Reacts with Cl_2 and $AlCl_3$?

When bromobenzene reacts with chlorine and $AlCl_3$, the major products are a roughly equal mixture of 2-chlorobromobenzene and 4-chlorobromobenzene.

17.4. FRIEDEL-CRAFTS REACTIONS

Arenes are benzene derivatives that have alkyl substituents. They can be formed by electrophilic aromatic substitution.

What Is an Arene?

An **arene** is an allyl benzene (a benzene with an alkyl substituent).

What Is the Common Name of 1,2-Dimethylbenzene? Of 1,3-Dimethylbenzene? of 1,4-Dimethylbenzene?

Common names are often used for these compounds. A dimethylbenzene is referred to as a **xylene**. 1,2-Dimethylbenzene is called *ortho*-xylene. 1,3-Dimethylbenzene is called *meta*-xylene and 1,4-dimethylbenzene is called *para*-xylene.

What Is the IUPAC Name of Each of the Following Compounds?

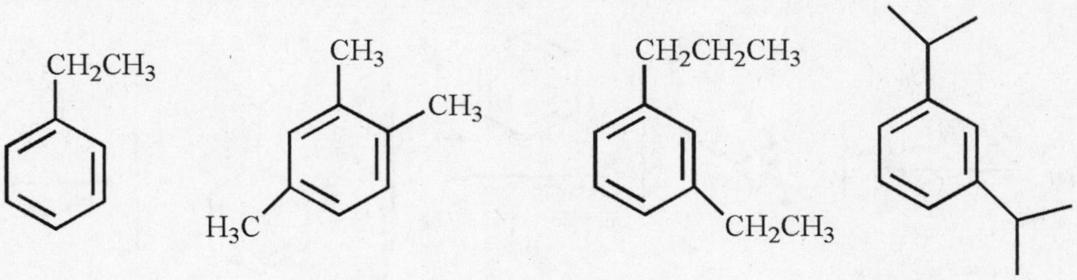

CH_2CH_3 | CH_3 CH_3 H_3C | $CH_2CH_2CH_3$ CH_2CH_3

17.83 *17.84* *17.85* *17.86*

The IUPAC name of ***17.83*** is ethylbenzene. Arene ***17.84*** is 1,2,5-trimethylbenzene; *17.85* is 3-ethylpropylbenzene; and ***17.86*** is 1,2-(di-1-methylethyl)benzene, although it is most commonly referred to as 1,2-diisopropylbenzene.

What Is the Intermediate Product When an Alkyl Halide Such As 2-Bromopropane Reacts with $AlCl_3$?

The intermediate of this reaction is a carbocation. When 2-bromopropane reacts with $AlCl_3$, the product is $Me_2HC^{\oplus}$ $AlCl_4^-$.

What Is the Product When 1-Chloro-2,2-Dimethylpropane Reacts with $AlCl_3$?

The product is the tertiary cation, ***17.87***.

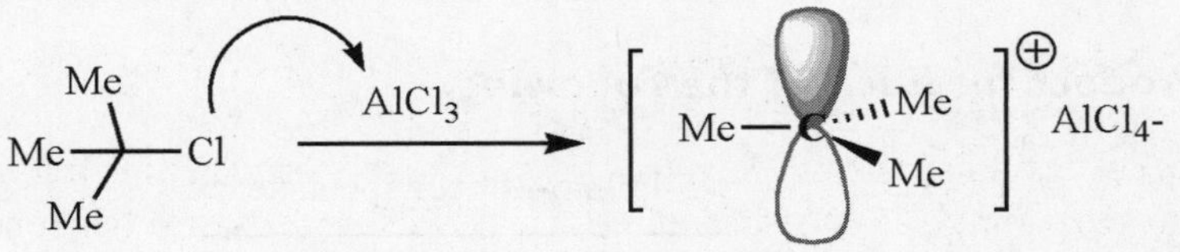

Me, Me, Me—Cl $\xrightarrow{AlCl_3}$ [Me—C(Me)(Me)]$^{\oplus}$ $AlCl_4$-

17.87

What Is the Initial Product When 1-Chlorobutane Reacts with $AlCl_3$?

The initial product of this reaction is a primary cation (***17.88***), but this rearranges to the more stable secondary cation (***17.89***), as expected with any carbocation.

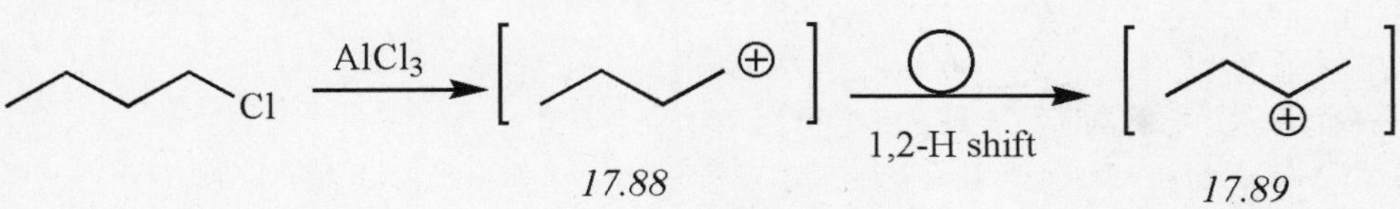

Cl $\xrightarrow{AlCl_3}$ [⊕] $\xrightarrow[\text{1,2-H shift}]{}$ [⊕]

17.88 *17.89*

Explain Why Benzene Reacts Faster with Tertiary Halides Than with Primary Halides.

The energy required to form a tertiary cation is relatively low, and $AlCl_3$ (or another Lewis acid) easily removes the halide to generate the cation. With primary halides, the activation energy for removing the halide to generate a relatively unstable primary cation is high, and the reaction is slow. The rate of the reaction is strongly influenced, however, by the strength of the Lewis acid and the solvent and reaction temperature that are used.

What Is the Major Product When Benzene Reacts with 1-Chloropropane and $AlCl_3$?

The initially formed primary cation rearranges to ***17.90***. Subsequently, benzene reacts with ***17.90*** to give (1-methylethyl)benzene [isopropylbenzene, ***17.92***] via the usual cationic intermediate, ***17.91***.

Cl $\xrightarrow{AlCl_3}$ [⊕] ⟶ [⊕] ⟶

17.90 *17.91* *17.92*

Explain Why Formation of Linear 1-Phenylalkanes $[Ph(CH_2)_nCH_3]$ Is Virtually Impossible via Friedel-Crafts Alkylation.

Virtually all primary halides will react with the Lewis acid to give a primary cation, but this unstable intermediate almost always rearranges to a more stable secondary or tertiary cation. Since the relative population of primary cation is very low, very few substitution products arise from that species (no linear arenes).

What Is the Name Given to Alkylation of Benzene Derivatives by Electrophilic Aromatic Substitution?

This reaction is called **Friedel-Crafts alkylation**, after the two chemists who discovered the process in the 19th century.

Give the Major Product for Each of the Following.

$\xrightarrow[FeCl_3]{\text{2-bromo-2-methylpentane}}$

$\xrightarrow{Ph_3CH\text{ , }AlCl_3}$

$\xrightarrow[AlCl_3]{\text{1-iodo-2-methylbutane}}$

In the first reaction, benzene reacts with the tertiary cation formed from 2-bromo-2-methylpentane to give ***17.93***. In the second reaction, benzene reacts with the so-called trityl cation (Ph_3C^+), formed from triphenylmethane, to give ***17.94***. In the last reaction, initial reaction of 1-iodo-2-methylbutane to give a primary cation is followed by rapid rearrangement to a tertiary cation, and reaction with benzene gives ***17.95*** as the final product.

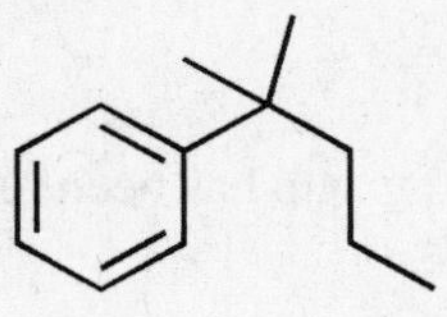

CPh$_3$

17.93 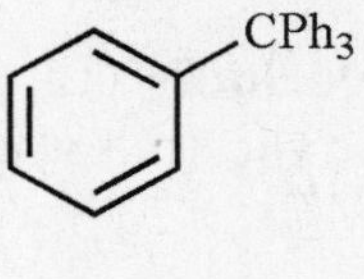 *17.94* 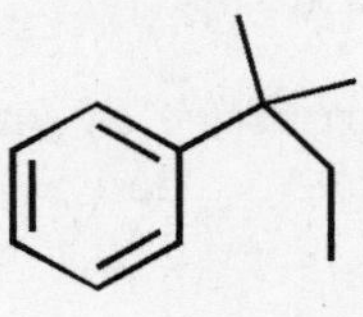 *17.95*

Polyalkylation

If Para-Xylene and Ortho-Xylene Are Formed by the Reaction of Toluene and Iodomethane/AlCl$_3$, Explain How 1,2,4-Trimethylbenzene Is also Formed in the Reaction.

Toluene reacts with iodomethane to give a new methyl group at the *ortho* and *para* positions, one of which is *para*-xylene (***17.96***). This reaction is not instantaneous, so in the early stages (see Section 7.2) of the reaction, the product is formed in the presence of the reactants, iodomethane and aluminum chloride. Since xylene has two alkyl groups, it is more activated than in toluene, i.e., xylene reacts faster with the MeI, $AlCl_3$ reagent than does toluene. Further reaction gives a new product, where a third methyl group will be incorporated *ortho* to the methyls (as in ***17.97***), which are both *ortho/para* directors.

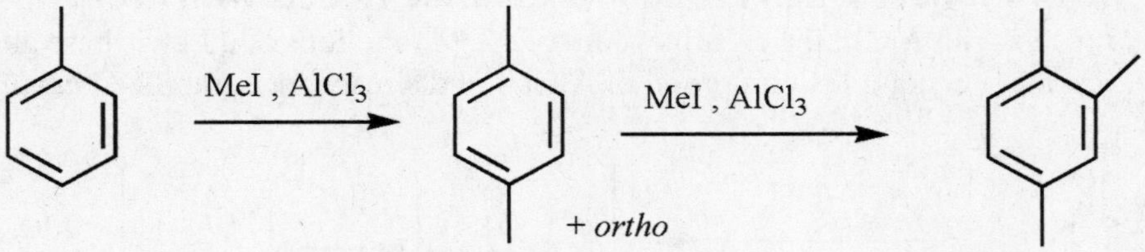

MeI , AlCl$_3$ MeI , AlCl$_3$

+ *ortho*

17.96 *17.97*

In the Reaction of Iodomethane with Toluene in the Presence of Aluminum Chloride, What Is the Major Product?

This question is difficult to answer. If an excess of iodomethane is used, ***17.97*** is likely be the major product. If a large excess of toluene is used, ***17.96*** (both *ortho* and *para*) will be the major product. If one equivalent of toluene and one equivalent of iodomethane are used, in a different solvent, a mixture of products will be formed.

What Is This Process Called?

Formation of arenes with more than one alkyl group is called **polyalkylation**, and it is a major problem with Friedel-Crafts alkylation.

Which Is the More Reactive Benzene Derivative, 1,2-Diethylbenzene or Toluene?

As in the formation of ***17.97***, 1,2-diethylbenzene is more activated than toluene and will react faster in electrophilic aromatic substitution reactions.

How Can Polyalkylation Be Suppressed in Electrophilic Aromatic Substitution?

If a large excess of the initial aromatic substrate is used (a large excess of toluene in the examples cited), the mono-alkylation product (xylene) will usually predominate. Alternatively, although it is usually impractical, the initial substitution product can be removed from the reaction medium as it is formed.

Friedel-Crafts Acylation

What Is the Structure of an Acid Chloride?

A acid chloride is a derivative of carboxylic acids where the –OH group has been replaced with –Cl, as in ***17.98*** (see Section 25.1).

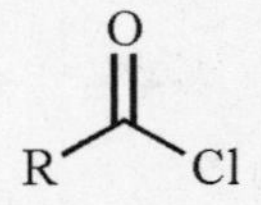

17.98

How Are Acid Chlorides Prepared?

Acid chlorides are prepared by treating a carboxylic acid with a halogenating reagent, such as $SOCl_2$, PCl_3, PCl_5 or $POCl_3$ (see Section 11.3).

How Are Acid Chlorides Named?

The *-oic* acid ending for carboxylic acids is replaced with *–oyl*, chloride. The acid chloride of propionic acid is, therefore, propanoyl chloride (see Section 25.1).

What Is the Major Product When Propanoyl Chloride Reacts with $AlCl_3$?

When ***17.98*** reacts with $AlCl_3$, the chlorine atom of ***17.98*** functions as a Lewis base and attacks Al (a Lewis acid) to form an 'ate' complex composed of $AlCl_4$– and a resonance stabilized *acylium ion*, ***17.99***.

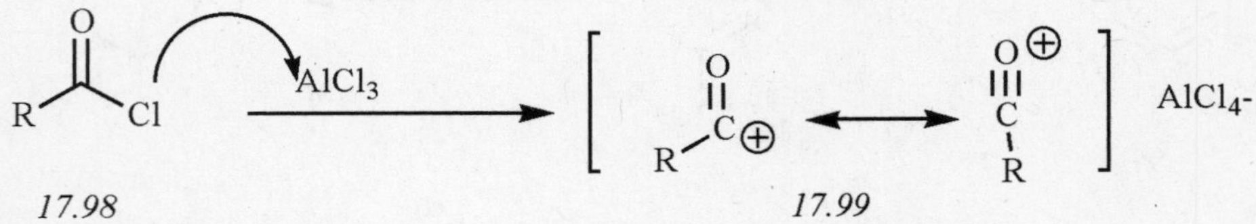

$AlCl_3$ $AlCl_4$-

17.98 *17.99*

If an Aliphatic Carbocation Undergoes Rearrangement, Does an Acylium Ion Rearrange?

No! An acylium ion is stabilized by the oxygen atom to give the resonance contributors shown in ***17.99***. This additional stability essentially stops cationic rearrangement.

Assume That Benzene Reacts with an Acylium Ion. Draw the Wheland Intermediate and the Final Product.

The initial reaction produces the usual Wheland intermediate, resonance stabilized cation ***17.100***, with the carbonyl attached to the ring. Loss of a proton generates the phenyl ketone product, ***17.101***.

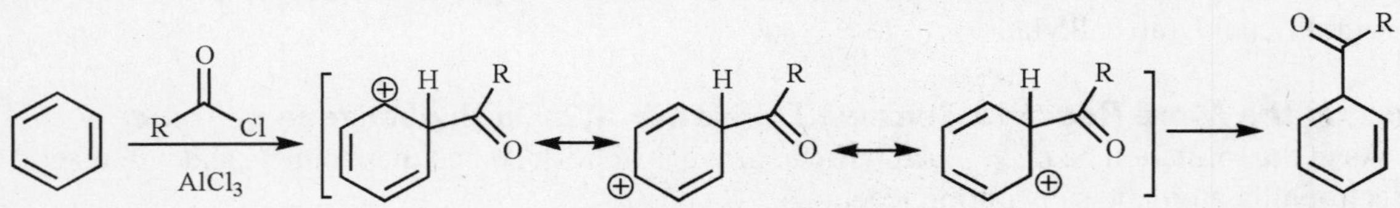

$AlCl_3$

17.100 *17.101*

What Is the Product When Propanoyl Chloride Reacts with $AlCl_3$ and Benzene?

When propanoyl chloride (***17.102***) reacts with benzene and $AlCl_3$, the product is ***17.103***.

17.102

$AlCl_3$

17.103

Give the IUPAC Name for 17.103.

This ketone is named 1-phenyl-1-propanone.

Why Does the Benzene Ring in 17.103 Not Undergo Further Reaction to Give a Disubstituted Product?

The carbonyl group in ***17.103*** is deactivating. Therefore, the product is less reactive than benzene.

What Is the Product When Anisole Reacts with Butanoyl Chloride and $AlCl_3$?

The OMe group of anisole is strongly activating and an *ortho/para* director. The reaction with butanoyl chloride (***17.104***) therefore gives a mixture of (2-methoxyphenyl)-1-propanone (***17.105***) and (4-methoxyphenyl)-1-propanone (***17.106***).

OMe; C_3H_7; Cl; 17.104; $AlCl_3$; OMe; C_3H_7; + OMe; C_3H_7

17.105 17.106

What Is the Product When Nitrobenzene Reacts with Propanoyl Chloride and $AlCl_3$?

In this case, the NO_2 group strongly deactivates the ring, and Friedel-Crafts acylation does ***not*** occur.

What Is the Product When Aniline ($PhNH_2$) Reacts with Propanoyl Chloride and $AlCl_3$?

Since the nitrogen of aniline can function as a Lewis base, the first reaction with $AlCl_3$, generates the 'ate' complex ***17.107***. No further reaction will occur, since the ring is now deactivated.

NH_2; $AlCl_3$; $H_2\overset{+}{N}\rightarrow\overset{-}{Al}Cl_3$

17.107

Is There a Method Available That Will Allow Friedel-Crafts Reactions with Aniline, or a Derivative?

Yes! This problem can be solved by *protecting* the nitrogen as an amide (see Section 25.5 for formation of amides). Initial reaction of aniline with acetic anhydride (***17.108***) gives the acetamide derivative (***17.109***, see Section 25.1). The amide is less basic than the amine and does not form an 'ate' complex with the Lewis acid. Therefore, ***17.19*** reacts with an acid chloride, in this case propanoyl chloride, *17.102*, to give two ketone products (the *para* product ***17.110*** and the *ortho* product, which is not shown). The acetamide is removed (*deprotected*) by acid hydrolysis followed by neutralization with acid (see Section 25.3) to give the amine, ***17.111***.

NH_2 H N O CH_3 O Et Cl $AlCl_3$ 1. H_3O^+ 2. OH^- + ortho O Et

17.108 *17.109* *17.110* *17.111*

Is the Product of a Friedel-Crafts Acylation More Reactive or Less Reactive Than Benzene?

The product is a ketone and the C=O group is deactivating, due to the δ⊕ charge on the carbonyl carbon. This makes the ketone product *less* reactive than benzene.

Is Rearrangement a Problem in Friedel-Crafts Acylation?

An acylium ion (***17.99***) is stabilized by the oxygen, leading to the resonance forms shown previously. This stability makes rearrangement very unlikely.

Is Polyacylation a Problem in Friedel-Crafts Acylation?

No! Since the carbonyl is a deactivating group, the product is generally less reactive than the starting material. Further acylation is not, therefore, a problem.

Primary Arenes Cannot Be Formed by Friedel-Crafts Alkylation, but Can They Be Formed by Friedel-Crafts Acylation?

Yes, if the C=O (carbonyl) group can be removed from the molecule and changed into a methylene ($-CH_2-$) group. In fact, there are several reagents that reduce the C=O group to $-CH_2-$.

Wolff-Kishner Reduction

What Is the Product When Acetophenone (17.112) Is Treated with Hydrazine and KOH?

The reduction of a ketone C=O to a $-CH_2-$ with hydrazine (NH_2NH_2) under basic conditions is called the **Wolff-Kishner Reduction**. When acetophenone (***17.112***) is treated with KOH and NH_2NH_2, the product is ethylbenzene, ***17.113***.

$H_2N\text{-}NH_2$ / KOH

17.112 → 17.113

Give the Mechanism of the Wolff-Kishner Reduction of Acetophenone.

Initial reaction of the NH_2 group with the carbon of the carbonyl in ***17.112***, which has a positive dipole, leads to a hydrazone (***17.114***). (See Chapter 21 for the chemistry of ketones and aldehydes and the mechanism for formation of hydrazones from ketones.) The hydroxide removes a proton from the NH_2 group to give the resonance stabilized anion, ***17.115***. The 'carbanion resonance contributor' of ***17.115*** reacts with water, which is an acid in this system, to form ***17.116***. Reaction with additional hydroxide removes the hydrogen from ***17.116*** to give the anion, ***17.117***. Loss of diatomic nitrogen (N≡N), which is an excellent leaving group, forms the carbanion intermediate, ***17.118***, which reacts with water complete the reduction, producing ***17.113***.

17.112 —($H_2N\text{-}NH_2$, KOH)→ 17.114 ⇌(^-OH) 17.115 ⇌(H-O-H) 17.116 ⇌(^-OH) 17.117 →($-N_2$) 17.118 →(H-O-H) 17.113

Clemmenson Reduction

What Is Zinc Amalgam?

An **amalgam** is an element compounded with mercury. **Zinc amalgam** is, therefore, written as Zn/Hg or Zn(Hg).

What Is the Product When Acetophenone (17.112) Is Treated with Zinc Amalgam and HCl?

When acetophenone is treated with HCl and zinc amalgam, the product is ethylbenzene (***9.7.13***). This reduction process is called **Clemmenson Reduction** and is the acid medium complement of the basic medium Wolff-Kishner Reduction. Ethanol is often used as a solvent in this reaction.

Zn-Hg , EtOH / HCl

17.112 → 17.113

Suggest a Set of Reactions That Will Convert Benzene into 1-Phenylhexane.

Using Friedel-Crafts acylation, the reaction of hexanoyl chloride (***17.119***) and benzene will produce 1-phenyl-1-hexanone, ***17.120***. Subsequent reduction (here Clemmenson reduction) leads to the target, 1-phenylhexane, ***17.121***.

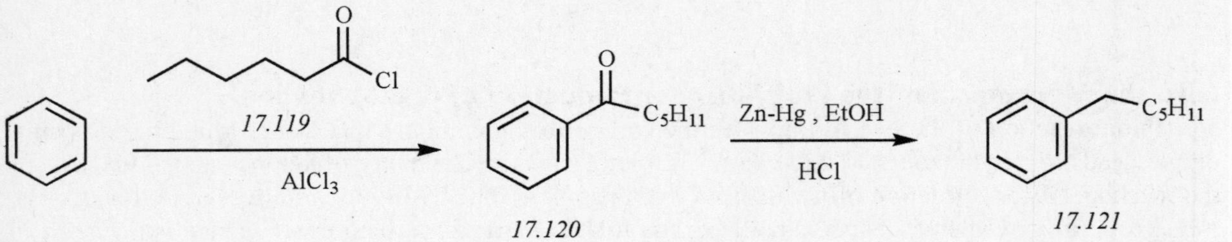

O Cl 17.119 $AlCl_3$ O C_5H_{11} Zn-Hg, EtOH HCl C_5H_{11}

17.120 *17.121*

17.5. TRISUBSTITUTION

When two substituents are attached to a benzene ring, electrophilic aromatic substitution will give a trisubstituted derivative. All of the 'rules' described for mono-substitution apply to this reaction.

Draw the Resonance Intermediate Formed When Br⊕ Reacts at C2 of 4-Nitroanisole. AT C3.

Reaction of 4-nitroanisole (***17.122***) with bromonium ion generates a resonance stabilized intermediate (***17.123***) with four resonance contributors. In one resonance contributor, positive charge is on the ipso carbon bearing the OMe, where the charge is dispersed to the oxygen, further stabilizing this intermediate. When Br^+ attaches to C3, resonance intermediate ***17.124*** is formed with three resonance contributors. Since the charge in one resonance contributor resides on the carbon bearing the positive nitrogen of the nitro group, this intermediate is very destabilized.

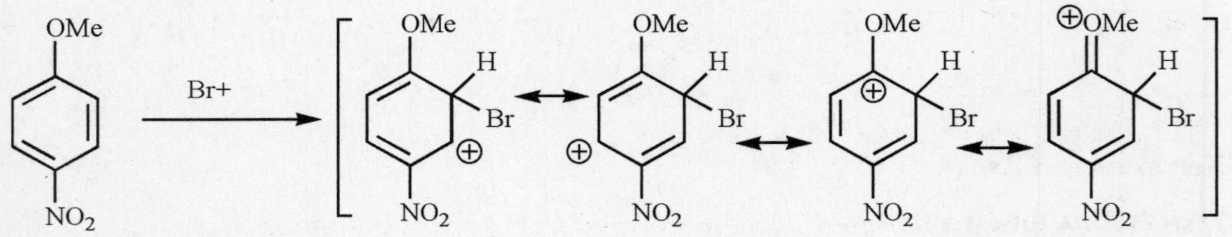

OMe NO_2 Br+ OMe H Br ⊕ NO_2 OMe H Br ⊕ NO_2 OMe ⊕ H Br NO_2 ⊕OMe H Br NO_2

17.122 *17.123*

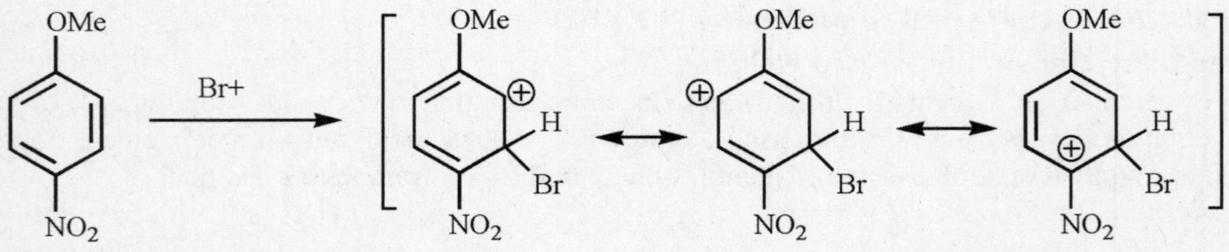

OMe NO_2 Br+ OMe ⊕ H Br NO_2 OMe ⊕ H Br NO_2 OMe H ⊕ Br NO_2

17.122 *17.124*

Which Is Preferred, Attack at C2 or C3? Explain.

The comparison of ***17.123*** with ***17.124*** clearly shows that attack at C2 generates a more stable intermediate. Attack at C3 places the positive charge adjacent to the positive nitrogen and leads to a very unstable intermediate compared with the highly stabilized ***17.123*** resulting from attachment of Br to C2. Therefore, attack at C2 is preferred.

What Is the Major Product of This Reaction?

The major product of this reaction is 2-bromo-4-nitroanisole.

If an Activating and a Deactivating Group Are in a Molecule, Which Dominates the Electrophilic Substitution Reaction? Explain.

By definition, an activating group increases the rate of electrophilic aromatic substitution, whereas a deactivating group diminishes the rate of the reaction. If both are in the same molecule, electrophilic aromatic substitution will be directed by the reaction with the faster rate dictated by the activating group.

The Reaction of 3-Nitroanisole and Bromine/$AlCl_3$ Can Generate Three Isomeric Products, Assuming Methoxy Is the Directing Group. Draw Their Structures.

When 3-nitroanisole (***17.125***) reacts with $Br_2/AlCl_3$, the bromine will attach at the *ortho* and *para* positions, relative to OMe. This leads to three different products, 2-bromo-3-nitroanisole (***17.126***), 4-bromo-3-nitroanisole (***17.127***), and 5-bromo-3-nitroanisole (***17.128***).

OMe NO2 —(Br_2, $AlCl_3$)→ Br OMe NO2 + OMe NO2 Br + OMe Br NO2

17.125 *17.126* *17.127* *17.128*

Of the Three Products Drawn, Two Are Formed in Near Equal Amounts, but 2-Bromo-3-Nitroanisole Is a Minor Product in This Mixture. Explain.

Both ***17.126*** and ***17.127*** are formed without problem. In order to form ***17.128***, the bromine, which is relatively large, must be inserted between the OMe and NO_2 groups. There is some steric hindrance as the bromine approaches this position (and in the cation intermediate above, that carbon is sp^3 hybridized and tetrasubstituted), raising the energy required for its formation and leading to diminished yields of this isomer.

In This Reaction, 6-Bromo-3-Nitroanisole Is the Major Product. Explain.

As the bromine approaches the aromatic ring, there is some coordination of $Br^{\oplus}$ with the electron pairs on the methoxy oxygen. This coordination leads to a greater preference for attack at the *ortho* position and a greater percentage of that isomer. This is known as the **ortho effect**.

What Is the Product When 4-Methoxyanisole Reacts with Nitric and Sulfuric Acid?

4-Methoxyanisole (***17.129***) has two identical *ortho/para* directing groups, and the only available position is *ortho*. The sole product is, therefore, 2-nitro-4-methoxyanisole, ***17.130***.

OMe, OMe — HNO_3, H_2SO_4 → OMe, NO_2, OMe

17.129 *17.130*

What Is the Product When 4-Methylanisole Reacts with Fuming Sulfuric Acid?

There are two *ortho/para* directors in 4-methylanisole (***17.131***), but the OMe group is more powerful than the methyl group (Me). The OMe dominates the reaction, directing the SO_3 group *ortho* to OMe, not to Me. The final product is 2-methoxy-5-methylbenzenesulfonic acid, ***17.132***.

OMe, Me — SO_3, H_2SO_4 → OMe, SO_3H, Me

17.131 *17.132*

What Is the Product (or Products) When 2-Methylanisole Reacts with Chlorine/$AlCl_3$?

2-Methylanisole (***17.133***) reacts with chlorine to form 4-chloro-2-methylanisole (*11.134*) and 6-chloro-2-methylanisole (***17.135***), where the more powerful activator OMe directs the reaction. Note that the chlorine atom in the products is *meta*- to the methyl group.

OMe, Me — Cl_2, $AlCl_3$ → OMe, Me, Cl + Cl, OMe, Me

17.133 *17.134* *17.135*

What Is the Product When 4-Nitroacetophenone Reacts with Cl_2/$AlCl_3$?

4-Nitroacetophenone (***17.136***) reacts with chlorine to give 3-chloro-4-nitroacetophenone, ***17.137***. Both the carbonyl and the nitro groups are *meta* directors and deactivating groups. The nitro group is *more deactivating*, however, and a nitro directed reaction will be *slower* than a carbonyl directed reaction. For this reason, the product will have the chlorine *meta* to the carbonyl, which is the *less deactivating* group.

Cl$_2$, AlCl$_3$

NO$_2$ NO$_2$ Cl

17.136 *17.137*

What Is the Product When Nitric Acid and Sulfuric Acid React with 1,3-Dinitrobenzene?

Both of the nitro groups in 1,3-dinitrobenzene (***17.138***) are deactivating and *meta* directors. The product will form with difficulty, if at all, and the yield may be low, but it will be 1,3,5-trinitrobenzene (***17.139***).

NO$_2$ NO$_2$

HNO$_3$, H$_2$SO$_4$

NO$_2$ NO$_2$ NO$_2$

17.138 *17.139*

17.6. SYNTHESIS VIA AROMATIC SUBSTITUTION

The conversion of benzene to ***17.121*** above illustrated the synthesis of ***17.121*** from benzene. In other words, we used a series of chemical reactions to convert one molecule into another in more than one chemical step. This approach can be used for the construction of many molecules.

What Is Synthesis?

Synthesis is a progression of chemical steps that begins with a molecule (the starting material) and adds functional groups or carbon-carbon bonds via chemical reactions until a new molecule (the target or product) is constructed.

Provide a Suitable Synthesis for the Conversion of Benzene to 1-(4-Aminophenyl)-Butanone (17.144).

The first step is to choose whether the NH$_2$ group is incorporated first or the carbonyl. Since the carbonyl is a *meta* director and deactivating, ***17.144*** is most likely formed by first incorporating the amine unit, an *ortho/para* director. To incorporate nitrogen, the first reaction is a nitration to give nitrobenzene, ***17.140***. The nitro group is *reduced* with hydrogen, and a palladium catalyst (see Section 27.5) to give aniline, ***17.141***.

Before a Friedel-Crafts reaction can be done, the basic amino group must be protected as the amide. Treatment of aniline with acetic anhydride leads to *N*-acetyl aniline (***17.142***), and subsequent reaction with butanoyl chloride and AlCl$_3$ leads to the ketone, ***17.143***, along with the expected *ortho* substituted ketone. To complete the synthesis of ***17.144***, the *ortho* and *para* ketones are separated, and the amide group in ***17.143*** is removed with aqueous acid, followed by neutralization to give. In this sequence, not only the reactions are important, but also the *order* in which the reactions are performed. This requires planning of the synthesis before the first chemical step is performed.

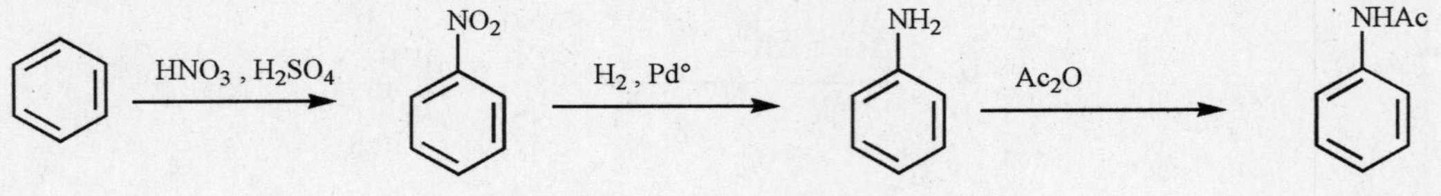

NO_2 HNO_3, H_2SO_4 H_2, Pd° NH_2 Ac_2O NHAc

17.140 17.141 17.142

NHAc NH_2 Cl O $AlCl_3$ 1. H_3O^+ 2. OH^- O O

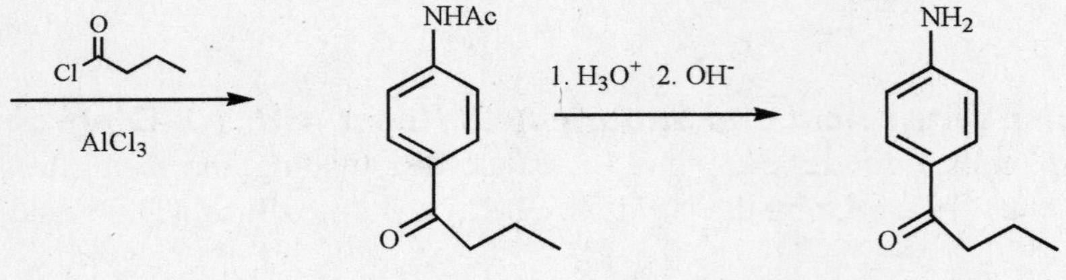

17.143 17.144

Is There a Shorthand Notation That Can Be Used to Indicate We Want to Make 17.144 from Benzene?

Yes! The following notation is used, with the reverse arrow.

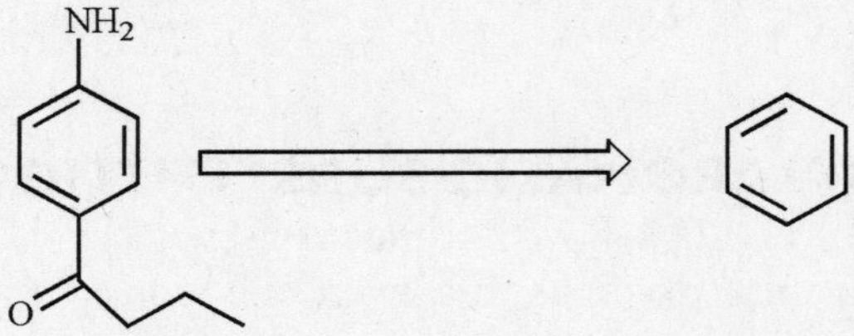

NH_2 O

17.144

What Does the Reverse Arrow Shown Above Mean?

This reverse arrow is used to indicate a specific problem. In the case of ***17.144***, the arrow means that ***17.144*** is to be prepared by synthesis using benzene as a starting material.

Give a Complete Synthesis for 17.145–17.147 from the Indicated Starting Material.

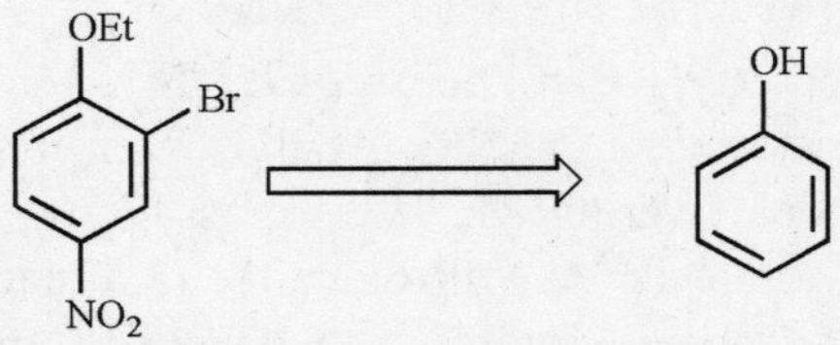

OEt Br NO_2 OH

17.145

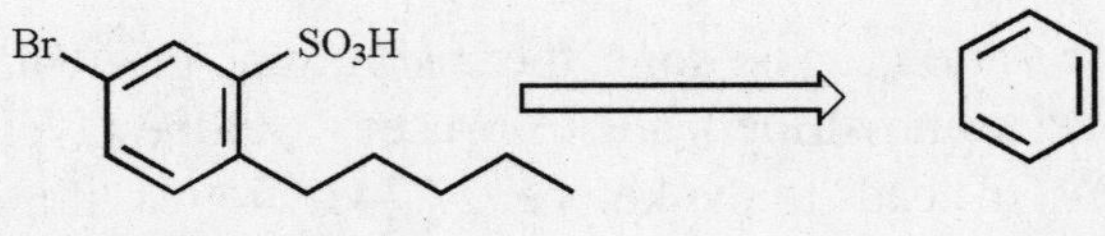

Br SO_3H

17.146

O_2N

17.147

The following are reasonable syntheses of the indicated targets from the designated starting material.

(a) i. NaH ii. EtBr (b) HNO_3 , H_2SO_4 (c) Br_2, $AlCl_3$

(a) pentanoyl chloride , $AlCl_3$ (b) NH_2NH_2 , KOH (c) Br_2 , $FeBr_3$ (d) H_2SO_4 , SO_3

(a) Me_3CCl , $AlCl_3$ (b) HNO_3 , H_2SO_4

Test Yourself

1. Name each of the following molecules using the *ortho*, *meta* or *para* nomenclature.

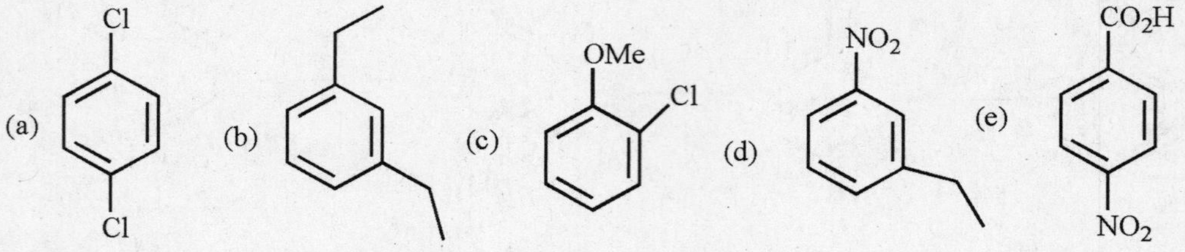

(a) Cl, Cl (b) (c) OMe, Cl (d) NO_2 (e) CO_2H, NO_2

2. Give the IUPAC name for each compound in Question 1.
3. Why is methyl considered to be an activating group in electrophilic aromatic substitution?
4. Draw all resonance contributors for the reaction product of $Br^{\oplus}$ and *N*-acetylaniline when Br reacts at the *ortho* position.
5. How can a pure sample of *ortho*-chlorotoluene be obtained from the reaction of chlorine, toluene and aluminum chloride?
6. Give the IUPAC name for each of the following.

(a) CO_2H, C_4H_9, C_4H_9 (b) Cl, O (c) O, Cl (d) O, Cl, Br, CH_3

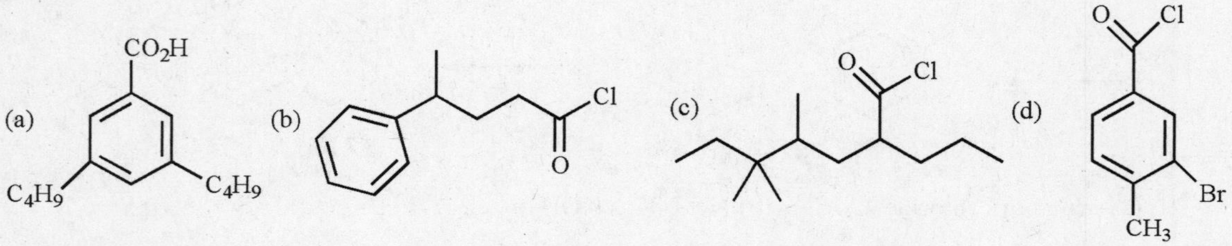

7. Why does the nitration of anisole give more *ortho* product than does the nitration of toluene under similar reaction conditions?
8. Provide a suitable synthesis for each of the following. Give all intermediate products and show all reagents. Do not give mechanistic intermediates.

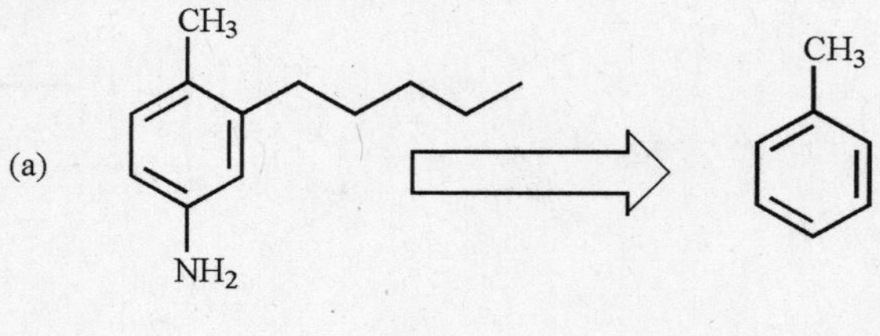

(a) CH_3, NH_2 ⇒ CH_3

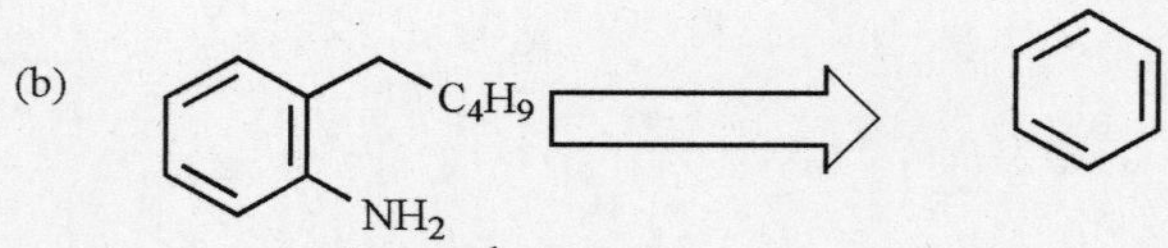

(b) C_4H_9, NH_2 ⇒

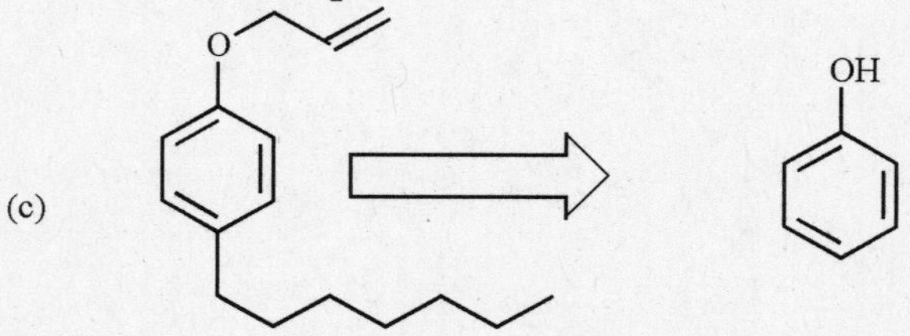

(c) O ⇒ OH

9. For each reaction give the major product. If there is no reaction, indicate this by N.R.

(a) benzene $\xrightarrow{HNO_3,\ H_2SO_4}$

(b) benzene $\xrightarrow{Br_2,\ AlBr_3}$

(c) anisole (OMe) $\xrightarrow{HNO_3,\ H_2SO_4}$

(d) aniline (NH_2) $\xrightarrow[2.\ Br_2,\ FeBr_3]{1.\ Ac_2O,\ pyridine}$

(e) propiophenone $\xrightarrow{Cl_2,\ FeCl_3}$

(f) fluorobenzene (F) $\xrightarrow{HNO_3,\ H_2SO_4}$

(g) ethyl 4-bromobenzoate (CO_2Et, Br) $\xrightarrow{AlCl_3,\ PhCOCl}$

(h) 3,4-dimethylanisole (OMe, CH_3, CH_3) $\xrightarrow{H_2SO_4,\ SO_3}$

(i) 1,3-dinitrobenzene (NO_2, NO_2) $\xrightarrow[AlCl_3]{CH_3CH_2CH_2COCl}$

(j) ethylbenzene (CH_2CH_3) $\xrightarrow[AlCl_3]{\text{(alkyl chloride)}}$

(k) benzene $\xrightarrow[AlCl_3]{\text{(alkyl bromide)}}$

(l) 4-propylanisole (OMe) $\xrightarrow{H_2SO_4,\ SO_3}$

(m) 1,4-diphenylbenzene (Ph, Ph) $\xrightarrow[AlCl_3]{\text{(acyl chloride)}}$

(n) hexanoic acid (CO_2H) $\xrightarrow{SOCl_2}$

(o) benzophenone (Ph, Ph) $\xrightarrow{aq.\ KOH,\ N_2H_4}$

(p) toluene (CH_3) $\xrightarrow[2.\ Zn^{\circ}(Hg),\ EtOH]{1.\ (CH_3)_2CHCH_2COCl,\ AlCl_3}$

(q) 4-cyanophenyl methanesulfonate ($C{\equiv}N$, SO_3CH_3) $\xrightarrow{Br_2,\ AlBr_3}$

10. What is the product of this reaction? Explain your answer.

$PhOCH_2CH_2Ph \xrightarrow{Br_2,\ FeBr_3}$

CHAPTER 18

Spectroscopy

The answer involves the manipulation of electromagnetic radiation and its interaction with organic functional groups. Chapter 7 introduced the use of ultraviolet light to assist in the identification of conjugated molecules such as dienes and α,β-unsaturated carbonyl derivatives. This chapter will introduce three techniques that can be used to give structural information for most organic molecules. The first technique is mass spectrometry, which bombards organic molecules with 70 electron volts of energy and monitors the fragmentation of the resulting ionic articles. When infrared energy is absorbed by an organic molecule, the bonds in the molecule dissipate that energy by vibrational motion. The frequency of vibration is different for different bonds within functional groups, and this is the basis for identifying functional groups. When an organic molecule is placed in a strong magnetic field, each hydrogen in the molecule behaves as a tiny electromagnet. When irradiated with electromagnetic energy in the radio frequency range, the hydrogens absorb the energy and change their nuclear spin state (with or against the large external magnetic field). The frequency of this absorption depends on the magnetic and chemical environment of each hydrogen and can be used to identify different types of hydrogens. These three techniques can be used to identify the structure of an organic molecule.

18.1. THE ELECTROMAGNETIC SPECTRUM

Chemists use energy to probe the structure of molecules, mostly based on the electromagnetic spectrum.

What Is the Electromagnetic Spectrum?

The **electromagnetic spectrum** is simply the range of energies associated with various forms of this energy. This includes not only ultraviolet, visible, and infrared energy, but also low energy microwaves and high energy x-rays and cosmic rays. A simplified version of the electromagnetic spectrum is shown in ***18.1*** in both wavelength (in meters, m) and frequency (in hertz, ν).

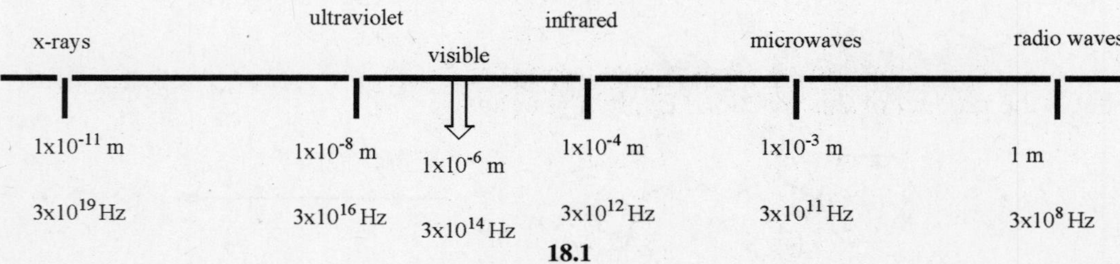

18.1

What Is the Relationship between Wavelength (λ) and Frequency (ν)?

They are inversely proportional by the equation $c = \nu\lambda$, where c is the speed of light (3×10^8 m/sec). In general, a high frequency is related to a small wavelength.

Which Is Higher in Energy, Infrared or Ultraviolet Light?

High energy is associated with a high frequency (low wavelength), and low energy is associated with low frequency (long wavelength). Since ultraviolet light has a higher frequency than infrared light, UV light is higher in energy. Similarly, x-rays are very high energy radiation, and radio waves are very low energy radiation.

At What Wavelengths Do Molecules Absorb Infrared Light?

The general absorption frequency range for organic molecules is 2.5×10^{-6} m to 16×10^{-6} m. Since an angstrom (Å) is 1×10^{-8} meters, infrared light absorbs in the range 2500-1600 Å.

Are Radio Waves More or Less Energetic Than Infrared Light?

Radio waves are much lower in energy than infrared light.

18.2. MASS SPECTROMETRY

Mass spectrometry can be used to determine the molecular formula of molecules once the molecular mass is known, as well as provide structural information based on fragmentation patterns of high energy ions derived from the molecule.

Ionization of Molecules

What Is an Electron Impact Mass Spectrometer?

An electron impact mass spectrometer is an instrument that bombards an organic molecule, or any other molecule, with high energy electrons to ionize the molecule, usually to a radical cation, although other ions can also be produced under the proper conditions. Usually, the initially formed ion fragments into other, smaller mass ions. An electric field accelerates and focuses the ions toward a large magnet, which separates the ions that are produced according to their mass and charge. The ions then pass into a detector where they are recorded.

What Is the Function of an Electron Gun?

An electron gun focuses a high energy beam of electrons towards the organic molecule. A typical electron beam will bombard the molecule with 70 electron volts (eV) of energy, although this is adjustable. This is usually enough energy to cause ionization and fragmentation.

How Many Kilocalories Correlate with One Electron Volt (ev)?

One electron volt is equivalent to 22.85 (about 23) kcal/mole of energy.

What Is the Purpose of the Magnetic Field Portion of the Mass Spectrometer?

The ion produced by electron bombardment is accelerated by the applied voltage and passes into a strong magnetic field. The magnetic field deflects the flight path of the ion. Ions that are too heavy or too light will not traverse the instrument, but will 'crash' into its sides and be lost (not detected). By adjusting both the magnetic field and accelerating voltage, this separation of 'heavy' and 'light' can be refined to separate ions differing by fractions of a mass unit.

How Are Molecular Ions Accelerated in the Mass Spectrometer?

A positively charged ion is forced into a chamber where it is accelerated by a strong electrostatic field.

What Is the General Result of a Molecule Being Bombarded with High Energy Electrons?

The high energy electron will expel an electron from the molecule, forming a radical cation (***18.2***). In this example, 2-pentanone is bombarded with an electron beam, expelling one electron, probably from oxygen, and leaving a positively charged oxygen atom that also has an unshared electron. This ion is a radical cation.

e^- → + e^-

18.2

What Is the Molecular Weight of 18.2?

The molecular weight of ***18.2*** is identical to that of 2-pentanone, minus the mass of one electron. For all practical purposes, the mass of the ion formed from the molecule introduced into the mass spectrometer is the same as that of the initial molecule.

What Is the Molecular Ion?

When a molecule is bombarded with electrons, an electron is ejected form the molecule to form a radical cation. This initial product has the same mass as the parent molecule that was introduced into the mass spectrometer and has been called the **molecular ion**. The correct term is actually **molecular ion**.

How Is a molecular Ion Formed?

Ion ***18.2*** is an example of a molecular ion. It is formed by ejection of a single electron from the molecule to form the radical cation.

What Is the Significance of the Molecular Ion?

The molecular ion has the same mass as the neutral molecule, *if* the charge of the ion is +1. The unit of measure used for the mass is actually m/*z* (m = mass, *z* = charge). Mass spectrometry and identification of the molecular ion therefore allows determination of the molecular weight of the molecule.

What Unit Is Used to Identify the Molecular Ion in Mass Spectrometry?

The mass to charge ratio, m/*z*.

What Are the Natural Abundance Isotopic Ratios of ^{2}H, ^{13}C, ^{34}S, ^{18}O, ^{37}Cl, ^{15}N and ^{81}Br?

For deuterium (^{2}H) the natural abundance is 0.015%; ^{13}C is 1.11%; ^{34}S is 4.22%; ^{18}O is 0.204%; ^{37}Cl is 24.47% and ^{81}Br is 49.46%.

For a Given Organic Molecule, What Is the Natural % of Deuterium and ^{13}C in the Molecule?

In organic molecules composed of carbon and hydrogen, the 'hydrogen' atoms in that molecule are actually a mixture of 99.985% ^{1}H and 0.015% of ^{2}H. Similarly, the carbon atoms are a mixture of 98.89% ^{12}C and 1.11% of ^{13}C.

If the Molecular Ion Includes a Deuterium, What Is the M/Z Value for That Fragment?

The m/z value will be the parent m/z + 1. If the parent m/z is 90 without ^2H it is 91 with ^2H.

Does the Molecular Ion Remain Intact or Does It Fragment into Smaller Fragments? Explain.

In general, the 70 eV ion beam contains enough energy to cause the molecular ion (molecular ion) to fragment into smaller radical cations. Each of these fragmentation ions can be accelerated and separated by the mass spectrometer and detected.

What Are the Smaller Fragment Ions Called?

Daughter ions.

The M+1 and M+2 Fragments and Determination of Molecular Formula

If the Molecular Ion Represents the Molecular Weight, Why Are There Higher Mass Fragments That Show Peaks of +1 and +2 Mass Units Greater Than the Parent?

These are the isotope peaks that arise from small amounts of ^{13}C, ^{15}N and ^2H. All three of these isotopes contribute to the molecular ion + 1. The isotope of oxygen (^{18}O) will contribute to the molecular ion + 2.

What Is the Significance of the M+1 Fragment?

Since the isotopic ratios of $^{12}C/^{13}C$, $^2H/^1H$ and $^{15}N/^{14}N$ are fixed, the ratio of the M+1 to the M (molecular ion or P, parent) ion allows calculation of the number of carbons and nitrogens in the molecular ion. The isotopic ratio of deuterium is so small that calculation of the number of hydrogens is not useful using this method.

What Is the Significance of the M+2 Fragment?

Since the isotopic ratios of $^{18}O/^{16}O$, $^{34}S/^{32}S$, $^{37}Cl/^{35}Cl$ and $^{81}Br/^{79}Br$ are fixed, the M+2 ion can be used to calculate the number of oxygens, sulfur, chorine and bromine atoms in the molecular ion.

Based on the Isotopic Ratios of Atoms, How Can the M+1 Fragment Be Used to Calculate the Number of Carbons and Nitrogen Atoms in the Molecule?

The ratio %M+1 to %P can be used to determine the formula: M+1 = [(#C(× (1.11)]+ 0.36 (#N).

If the ratio of $\frac{M}{M+1}$ is 6.66%, this implies a total of 6 carbons. Important in this calculation is an estimation of the number of nitrogens. This calculation requires that we make a working ***assumption***. If the molecular weight is *odd*, the molecule must contain an *odd* number of nitrogens (1,3,5,7,.....), and the first assumption is the presence of 1 nitrogen rather than 3 or more. If the molecular weight is *even*, the molecule must contain an *even* number of nitrogens (0,2,4,6,8,.....). The first assumption is that there are zero nitrogens rather than 2. If the molecular weight is 100, one assumes #N = 0, and if the M+1 is 6.66% of P, there are 6 carbons in the molecule: 6.66% × (1 carbon/1.11%) = 6C. In these calculations, the parent is assumed to be 100% and the M+1 is some % of P.

If the Parent Is Assumed to Be 100%, Does This Mean It Is the Highest Peak in the Entire Spectrum?

No! Not necessarily. We are simply taking the ratio of M+1 and M+2 relative to the parent, P. For that calculation, P must be 100% relative to the other ions.

How Can the M+2 Fragment Be Used to Calculate the Number of Oxygens in the Molecule?

The isotopic ratios lead to the formula $M+2=\left[\frac{\{[\#C][1.11]\}^2}{200}+\frac{\{[\#^2H][0.016]\}^2}{200}\right]+0.20\,(\#O)$.

The first term is the contribution of naturally occurring ^{14}C. The second term is the contribution of naturally occurring deuterium (^2H), and the third term is due to isotopic oxygen (^{18}O). Since the deuterium term is quite small, the formula usually incudes only the ^{14}C and the ^{18}O terms. This formula could be extended to include ....+ 4.22 (#S) + 24.47 (#Cl) + 100 (#Br), but these terms are usually omitted, since the calculation is relatively obvious, and all of these terms are sufficiently large that they will 'swamp out' the smaller number for oxygen. For a molecule with a molecular weight of 100, the M+1 term indicates 6 carbons, and the M+2 term is measured to be 0.42. For 6 carbons, the first term is 0.22, and solving for #O leads to one (1) oxygen. The partial formula will therefore be C_6O.

How Can We Estimate the Number of Hydrogen Atoms in the Formula?

The hydrogens must be calculated by their difference from the molecular ion. The mass of C_6O from the preceding problem is 88, and the mass of the molecular ion is 100. Therefore, 100−88 = 12. This mass is assumed to be due to the hydrogen atoms, and the formula will be $C_6H_{12}O$.

If an Organic Molecule Contains an Odd Number of Nitrogen Atoms, Will the Molecular Weight Be Even or Odd? What Is the Significance of This Observation?

With an odd number of nitrogens, the molecular weight will also be odd. This allows one to estimate the number of nitrogens in the M+1 term. Molecules with 3 or more nitrogens are somewhat rare for the type of molecule encountered in introductory organic chemistry, and one assumes that a molecule with an odd mass will have one (1) nitrogen. Likewise, a molecule with an even number of nitrogens will have an even mass, since molecules with four or more nitrogens are somewhat unusual for the type of molecule encountered in introductory organic chemistry. The presence of 2 nitrogens is common, however, so one assumes either zero (0) or two (2) nitrogens. For simple molecular weights, always assume zero nitrogens in that first calculation. It should *always* be checked, however.

If the M+1 Peak Is 13.69% of the Parent, and the M+2 Peak Is 1.09% of the Parent, What Is the Molecular Formula of a Molecule with a Molecular Ion M/Z of 191?

The odd mass suggests one (1) nitrogen. Using the M+1 formula, one nitrogen leads to M+1 = (13.69−0.37)/1.11 = 12 C. Using the M+2 formula for C_{12}, M+2 = (1.09–0.89)/0.2 = 1 O. The partial formula is, therefore, $C_{12}NO$ (mass = 174). The number of hydrogens is calculated by 191−174 = 17H, and the final formula is $C_{12}H_{17}NO$.

Given the Appropriate M+1 and M+2 Ratios, Calculate the Molecular Formula for Each of the Following.

(a) [M = m/z 128] M+1 = 9.99, M+2 = 0.50; *(b) [M = m/z 114] M+1 = 7.77, M+2 = 0.50;*
(c) [M = m/z 129] M+1 = 9.25, M+2 = 0.39; *(d) [M = m/z 149] M+1 = 10.36, M+2 = 0.70;*
(e) [M = m/z 150] M+1 = 5.55, M+2 = 98% of P.

The formula for (a) is C_9H_{20}; (b) is $C_7H_{14}O$; (c) is $C_8H_{19}N$; (d) is $C_9H_{11}NO$; (e) $C_5H_{11}Br$.

What Is the Significance of A M+2 Peak That Is 25–33% of the Parent?

This is a clear indication that there is one chlorine in the molecule. The ^{37}Cl isotope is about 25−33% of the ^{35}Cl peak.

A Molecule with the Formula C_4H_9Cl Has a Molecular Weight of 92.45, Using the Weights from the Periodic Table. What Is the Mass of the Molecular (Parent) Ion Derived from This Molecule?

There are two isotopes of chlorine, 35 and 37. The mass from the Periodic Table reflects the near 3:1 ratio of these isotopes, or 35.45. The molecular ion will be formed from the lower mass isotope, ^{35}Cl. Therefore, the mass of the molecular ion is 48 (12×4) for C, 9 for H, and 35 for Cl = m/z 92.

What Is the Significance of an M+2 Peak That Is 100% of the Parent?

This is a clear indication that there is one bromine in the molecule. The ^{79}Br and ^{81}Br isotopes have about the same abundance.

A Molecule with the Formula $C_6H_{11}Br$ Has a Molecular Weight of 162.9, Using the Weights from the Periodic table. What Is the Mass of the Molecular (Parent) Ion Derived from This Molecule?

There are two isotopes of bromine, 79 and 81. The mass from the Periodic Table reflects the near 1:1 ratio of these isotopes, or 79.9. The molecular ion will be formed from the lower mass isotope, ^{79}Br. Therefore, the mass of the molecular ion is 72 (12×6) for C, 11 for H, and 79 for Cl = m/z 161.

This Is an Odd Mass. Does the Nitrogen Assumption Apply?

No! The mass of the parent is odd, but there is no nitrogen. The assumption used previously about nitrogen will not apply here; hence, the caution to check the assumption carefully before using it. In light of a large M+2, nitrogen can usually be eliminated as a possibility.

What Is the Significance of an M+2 Peak That Is About 4–5% of the Parent?

This is consistent with one sulfur in the molecule where the ^{34}S is about 4% of the ^{32}S.

The Mass Spectrum

A mass spectrum is the plot of m/*z* ions as a function of the abundance of those ions. The abundance is usually taken as the ratio of each ion to the most abundant ion, but it does not have to be done this way. A typical mass spectrum is shown in ***18.3***.

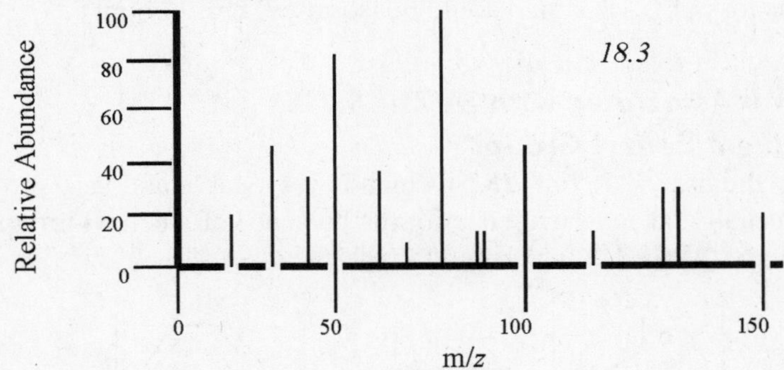

What Peaks in the Mass Spectrum Are Taken to Be the Most Important?

The molecular ion (along with the M+1 and M+2 peaks, if they are strong enough) and the most abundant ion, which is called the base peak. In spectrum ***18.3***, the base peak (B) appears at about 75–80 (obviously this will be determined exactly), and the molecular ion will be the highest mass ion, in this case m/*z* = 150. Other important peaks will be those ion fragments that give structural information, but this will vary with the molecule.

What Is the Base Peak?

The base peak is that daughter ion (sometimes it is also the parent) that has the highest abundance, meaning that more ions of this m/z migrate through the instrument and are recorded by the detector. Often this represents the most stable ion or the lowest energy fragmentation mode.

What Is the Most Convenient Way to Show the Relationship of a Fragment Ion to the Molecular Ion?

The easiest method is to report both the m/z value and also the *difference in m/z* between the daughter ion of interest and the parent. This is reported as P–? (M–15, P–28, P–56, etc.).

What Is the Expected Mass Spectrum for a Long Chain Alkane?

The mass spectrum of a long chain alkane such as decane will look something like the spectrum ***18.4***. In general, it will show a weak molecular ion, since there is extensive fragmentation. There is a consistent pattern of P–CH_2 (M–14) for cleavage of each C–C bond. This leads to a characteristic pattern where the higher mass fragments show very low abundance, and the lower mass fragments are prominent (m/z 85, 71, 57, 43 and 29 usually show up in the mass spectrum of a linear alkane).

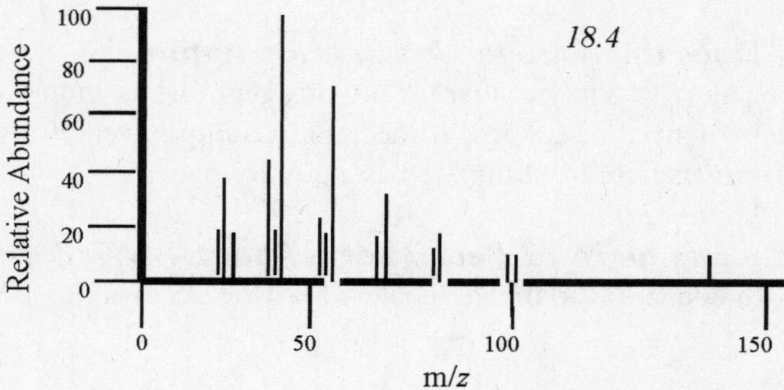

18.4

What Fragment Is Associated with an M–15 Peak? P–29? P–43?

An M–15 fragment is usually associated with loss of methyl (–CH_3), P–29 with loss of ethyl (–CH_2CH_3) and P–43 is loss of C_3H_7 which could be isopropyl or *n*–propyl.

What Daughter Ion Is Associated with M/Z = 91 in Molecules Containing a Benzyl Group?

This ion is usually the tropylium ion (***18.5***) which is very stable and very prominent. The presence of this daughter ion is usually taken as evidence for the presence of a benzyl group in the molecule.

$$[C_6H_5CH_2]^{\bullet +} \longrightarrow [C_7H_7]^{\bullet +}$$

18.5

What Is a Benzyl Group?

The benzyl group is a substituent, –CH_2Ph.

What Is an M–1 Fragment Associated with?

An M–1 peak is usually loss of a single hydrogen (loss of H from *N*–H in an amine, for example).

Suggest Possible Structures for M/Z = 57.

A daughter ion with m/*z* of 57 is probably C_4H_9. This could be *n*-butyl, isobutyl, *sec*-butyl or *tert*-butyl, pointing up the ambiguity of assigning specific structures via mass spectrometry based only on fragmentation of the molecular ion.

What is α-Cleavage?

The fragmentation of an ion at the bond between a carbon and a carbon bearing a heteroatom (C×C–O, C×C–S, C×C–N) or between a carbon and a functional group (C×C=O, C×C≡N, C×CO_2R, etc.) is called **α-cleavage**.

What Functional Groups Undergo α-Cleavage?

Ethers, amines, alcohols, acid derivatives, ketones and aldehydes and nitriles all undergo α-cleavage. An example is the cleavage of the molecular ion of diethyl ether (***18.6***) to give ***18.8*** and ***18.9***. A second example is the cleavage of the molecular ion of butanal (***18.9***) to give two daughter ions, ***18.10*** and ***18.11***.

18.6 α-cleavage → *18.7* + *18.8* (CH_3)

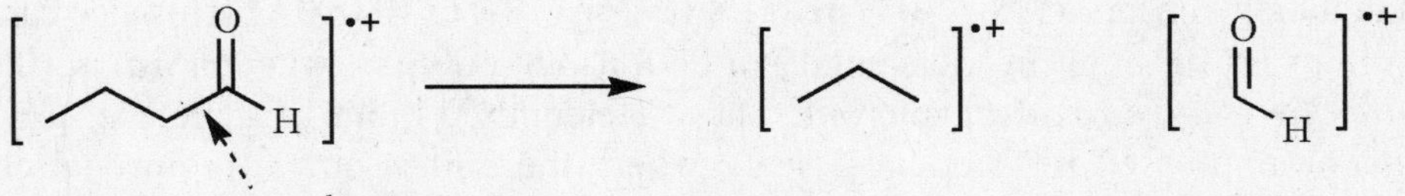

18.9 α-cleavage → *18.10* + *18.11*

Note that ***18.7*** corresponds to M–15 using ***18.6*** as the parent, or loss of a methyl group.

Why Does an Alcohol Usually Exhibit an M–18 Fragment?

A mass of 18 corresponds to water, and M–18 is characteristic of a molecule that loses a molecule of water (dehydration). Under ion bombardment, most alcohols readily lose water from the molecular ion. Usually, the molecular ion is not observed at all, and the first ion actually observed in significant concentration is the M–18 daughter ion.

What Is a McLafferty Rearrangement?

A McLafferty rearrangement occurs in aldehydes and ketones that have a hydrogen attached to a γ-carbon. The γ-carbon is the third carbon away from the carbonyl, **C**–C–C–C=O. In a typical example, the hydrogen is transferred to oxygen via a 6-centered transition state (see ***18.12***) to form a neutral alkene and an enol daughter ion (***18.13***).

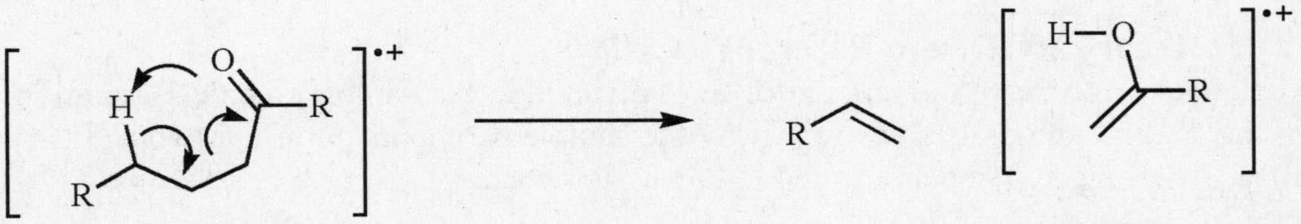

18.12 → R–CH=CH_2 + *18.13*

What Is the Significance of Losing a Neutral Fragment?

Loss of most fragments means loss of an odd mass (H, CH_3, etc.). Loss of a neutral fragment usually indicates loss of an even mass [$- H_2O = 18$; $- CH_3CH_2CH{=}CH_2$ (R = Et) = 56].

Show a McLafferty Rearrangement for 2-Hexanone.

The McLafferty rearrangement for this ion is shown in the conversion of ***18.14*** to ***18.15***.

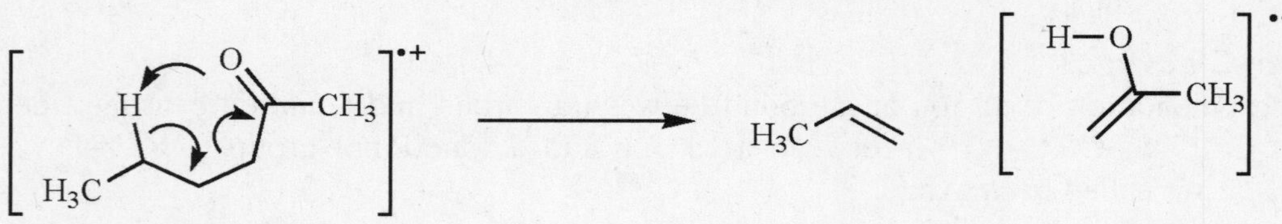

18.14 *18.15*

18.3. INFRARED SPECTROSCOPY

When a molecule is irradiated with infrared radiation, there is insufficient energy to make or break bonds. The molecule absorbs the energy, and the bonds vibrate. We can measure these vibrational frequencies and use that information to obtain structural information, particularly for functional groups.

Infrared Radiation and the Infrared Spectrometer

What Is the Wavelength Range for Infrared Energy in the Electromagnetic Spectrum?

Infrared radiation appears at 7.8×10^{-7} m to 1×10^{-4} m (78Å to 10,000Å). This also can be converted to 7.8×10^{-5} cm to 1×10^{-2} cm (in wavelength) or 0.78 μm to 100 μm, where a μm is 10^{-4} cm or 10^{-6} meters, *usually abbreviated as* μ, or microns. The wavelength (λ) unit is, therefore, μm. This energy range can also be expressed in frequency (ν ~), where the unit is cm^{-1} (reciprocal centimeters or *wavenumbers*). The frequency range for infrared radiation is 12821 cm^{-1} to 100 cm^{-1} (1.282–0.01 μ).

What Wavelength Range Is Used in Infrared Spectroscopy?

In most infrared spectrophotometers, the infrared radiation is measured between 4000–625 cm^{-1} or 2.5–16 μ. Most organic molecules absorb infrared radiation in this energy range.

Describe How an Infrared Spectrophotometer Works.

Infrared light is directed towards a prism and split into two equal parts. These two beams are focused by a mirror system, one passing through the sample, and the other through a reference cell. The beams are redirected with mirrors and another prism, and the two are compared. If an organic molecule in the sample beam absorbed infrared light, the sample beam will be less intense than the reference beam. The spectrophotometer scans through a range of wavelengths (2.5–16 μ), and those regions of the spectrum that were absorbed by the organic molecule appear as peaks. This series of peaks is the infrared spectrum.

Why Is It Necessary to Have a Reference Cell?

It is essential to compare the amount of light directed at the sample (incident radiation) with the amount of infrared radiation after absorption by the sample. Comparison of the two signals allows the instrument to determine how much infrared light was absorbed.

What Materials Are Used to Make the Infrared Cells That Contain the Organic Sample?

The material must be transparent to infrared light (not absorb infrared light). Pressed plates of NaCl or KBr are the most common materials.

When an Organic Molecule Absorbs Infrared Light, What Happens to the Molecule?

If a molecule absorbs infrared light, this excess energy is dissipated by molecular vibrations. Bending and stretching vibrations are the most common. There are symmetric and asymmetric stretching vibrations as well as symmetric and asymmetric bending vibrations, both 'in plane' and 'out of plane.' These vibrations occur at different infrared frequencies, also measured in wavelength, and lead to the various 'peaks' in the infrared spectrum.

How Does a Change in Dipole Moment for a Bond Influence Infrared Absorption?

If the dipole moment of a bond changes during a bending or stretching vibration, that absorption is particularly strong. The larger the change in dipole moment, the stronger the absorption. If the vibration is not accompanied by a change in dipole moment, it is usually a very weak absorption. When a symmetrical bond such as C–C) vibrates, the infrared absorption is very weak. The C–C signal in the infrared is very weak (essentially nonexistent).

What Types of Bonds Are Expected to Give Strong Infrared Signals? Weak Infrared Signals?

Bonds with significant dipole moments will give strong signals in the infrared. Examples are O–H, C–O, C=O, N–H, Cl–C=O and N–C=O, which will usually give strong signals. Bonds that give weak signals are C–C, O–O and N–N. Interestingly, the C–H signal is relatively strong.

How Is an Organic Compound Introduced into the Infrared Spectrophotometer?

If the compound is an oil, a drop is placed on pressed KBr or pressed NaCl plates, which are clear and allow light to pass through them. A second KBr or NaCl plate is placed on top to make a 'sandwich' of the sample. The plates are placed in a holder and inserted into the instrument to allow the spectrum to be determined.

Why Can't We Use Water to Wash the Salt Plates?

Both NaCl and KBr are soluble in water and will dissolve.

Why Can't We Use Acetone to Wash the Salt Plates?

Acetone usually has water in it.

Hook's Law

Hook's Law is an equation that describes the vibrational frequency of two masses connected by a spring. The frequency of the vibration is a function of both the mass of the two objects connected to the spring and the strength of the spring. This is an excellent model for predicting vibrational frequencies of two atoms connected by a bond capable of vibrational motion. Hook's Law is:

$$\nu \sim = \left(\frac{1}{2\pi c}\right)\sqrt{\frac{\mathrm{k}}{\mathrm{m}}}$$

where c = speed of light = 3×10^{10} cm sec^{-1}; k = force constant (a parameter to describe the tightness of the spring); and m = mass. For organic molecules, the *reduced mass* is used $\left(= \dfrac{mM}{m + M}\right)$, since the small mass of both atoms are close in magnitude.

If Hook's Law Is Used to Describe Two Masses Connected by a Spring, describe the Relevance of This Law to a Covalent Bond.

Two atoms function as masses, and the covalent bond functions as a spring. As the bond strength (force constant) increases, the vibrational frequency increases, whereas a weaker bond leads to a lower vibrational frequency. As the two atoms change, of course, there will also be differences in the frequency.

As a Bond Gets Stronger, Is the Wavelength of Infrared Absorption Expected to Increase or Decrease?

A stronger bond absorbs infrared light at a higher frequency (smaller wavelength). A comparison of the C≡C bond with the C=C bond reveals that C≡C absorbs at 2100–2260 cm^{-1} (4.76–4.42 μ) at higher frequency (higher energy) than the absorption for C=C at 1650–1670 cm^{-1} (6.06–5.99 μ).

Which Is Associated with Higher Energy, Longer Infrared Wavelength or Shorter Infrared Wavelength?

Since short wavelengths are generally of higher energy, a low infrared wavelength (2.0–4.0 μ, for example, as compared to 11–14 μ) will be of higher energy. Analysis of the triple bond vs. the double bond shows this to be true. It takes more energy to make the stronger C≡C bond vibrate than it does for the weaker C=C bond.

What Is the Reduced Mass for a C–O Bond?

Reduced mass = $\dfrac{mM}{m + M}$ where m is the mass of the smaller atom (C), and M is the mass of the larger atom (O). It is important to remember that this is the mass of the individual atom, *not* the atomic mass for a mole of the atom. This mass is the atomic weight divided by Avagadro's number. For C, $m = \dfrac{12.00}{6.023 \times 10^{23}} = 1.992 \times 10^{-23}$ and for O, $M = \dfrac{15.9994}{6.023 \times 10^{23}} = 2.656 \times 10^{-23}$. With these masses, the reduced mass is $\dfrac{[1.992 \times 10^{-23}]\,[2.656 \times 10^{-23}]}{1.992 \times 10^{-23} + 2.656 \times 10^{-23}} = \dfrac{5.29 \times 10^{-46}}{4.648 \times 10^{-23}} = 1.14 \times 10^{-23}$. In general, one can simply use the atomic masses, where for C=C, $M = \dfrac{(12)(12)}{12 + 12} = 6$.

The Infrared Spectrum

An infrared spectrum is the plot of either **absorbance** or **% transmittance** as a function of wavelength and/or frequency. When an organic molecule absorbs infrared energy of a particular wavelength, it will appear as a 'peak.' If the signal absorbs a large amount of infrared radiation, it will show as a strong absorption (weak % transmittance), and if little energy is absorbed, the peak will show a small absorption (large % transmittance).

What Labels Are Applied to the Axes of an Infrared Spectrum?

Most infrared spectrophotometers are calibrated in % transmittance with the scale linear in frequency (cm^{-1}). The wavelength scale (μ) is also shown on the spectrum, *18.16*.

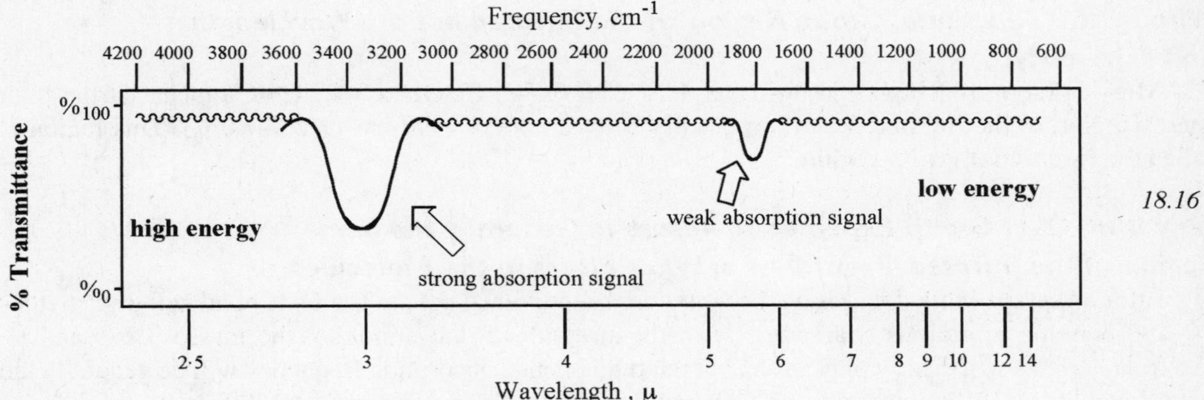

18.16

What Is Beers Law?

The most common form of Beer's Law (actually the **Beer-Lambert Law**) is: $T = \frac{I}{I_o}$ where T = % transmittance. Absorbance (A) is related to T by: $A = -\log_{10}(T)$ and $T = 10^{-A}$. In this equation, I = intensity of the sample beam, and I_o is the intensity of the incident (reference) beam.

Is 90% T Associated with a Strong Signal or a Weak Signal?

If 90% of the light is transmitted, only 10% was absorbed, so this is a weak signal.

Is 70% A Associated with a Strong Signal or a Weak Signal?

If 70% of the light was absorbed, this is a strong signal.

What Is the Relationship Between Wavelength and Wave Numbers?

They are inversely proportional: $\nu\ (cm^{-1}) \propto \frac{1}{\lambda(\mu)}$.

What Units Are Used for Wavelength in Infrared Spectroscopy?

Wavelength (λ) uses microns (μ) as the basic unit ($1\ \mu = 1 \times 10^{-4}$ cm).

What Units Are Used for Frequency in Infrared Spectroscopy?

Frequency (ν) uses reciprocal centimeters as the unit (cm^{-1}).

Is 700 cm^{-1} Associated with High Energy or Lower Energy?

An infrared absorption at 700 cm^{-1} is at the low energy end of the spectrum and would be associated with a weaker bond.

Is 3.45 μ Associated with High Energy or Low Energy?

This is a short wavelength absorption and is, therefore, higher in energy.

Which Is the Lower Energy Absorption, 10.57 μ or 2335 cm⁻¹?

An absorption at 10.57 μ is equivalent to 946 cm^{-1}. This signal is much lower in energy than the signal at 2335 cm^{-1}.

What Is the 'Functional Group Region' of the Infrared in Both Wavelength and Frequency?

Most common functional groups (OH, NH, CH, C=O, COOH, C≡C, C≡C) appear at the high energy region of the infrared spectrum, usually between 4000–1400 cm^{-1} (2.5–7.0 μ). This region is called the **functional group region**.

Why Is an O–H Group Expected to Absorb in Generally the Same Region of the Infrared, Regardless of What Else is in the Molecule?

Infrared spectroscopy focuses on the vibrations of individual bonds. The O–H bond will give stretching and bending vibrations characteristic of the strength of that bond and the masses of O and H. Electronic effects will clearly play a role, but the fundamental absorption frequency will be generally the same for all molecules containing an OH group. This absorption appears at 3200–3640 cm^{-1} (or at 2.94–2.75 μ as a strong, broad signal).

Which Is Expected to Absorb at Higher Energy, an O–H group or a C–H group? Explain.

The bond dissociation energy for O–H (from methanol) is 102 kcal/mole, and the bond dissociation energy for C–H (from methane) is 104 kcal/mole. They are obviously rather close, and bond dissociation energy calculations do not provide an answer. Using Hook's Law, the greater mass of oxygen, relative to carbon, might be expected to push the OH absorption to higher energy. In fact, OH absorbs at 3400–3640 cm^{-1} and C–H generally absorbs at 2850–2960 cm^{-1} (3.51–3.38 μ), making the OH signal the higher energy.

The O–H bond is significantly more polarized than the C–H bond and should, therefore, give a much stronger absorption. In fact, the OH absorption is very strong (usually 0–10%T) and quite broad. Hydrogen bonding will increase the relative bond polarity of the O–H, and the more extensive the hydrogen bonding, the broader the signal (broad = larger range of infrared absorption frequencies).

Which Group Is Expected to Give the Lowest Energy, Infrared Absorption, C–C, C=C or C≡C? Explain.

The weakest bond is the C–C bond, and it is expected to have the lowest energy absorption. For the C–C bond, this is so weak it is usually not reported, perhaps appearing at 1000–1100 cm^{-1} [10–9 μ]. The C=C bond absorbs at 1650–1670 cm^{-1} [6.06–5.99 μ] and the C≡C bond absorbs at 2100–2260 cm^{-1} [4.76–4.42 μ].

What Is the Fingerprint Region of the Infrared?

The low energy portion of the infrared spectrum between 1400–625 cm^{-1} (7.0–16 μ) contains a few functional group absorptions (C=C, C–O, C-halogen, C-aromatic, and aromatic carbons), but is usually the region where low energy bending and stretching vibrations absorb. Each individual molecule will have its own peculiar set of bending, stretching, rocking, twisting and wagging vibrations due to the carbon 'backbone' of the molecule as well as the functional group bonds. The combination of these vibrations leads to a 'fingerprint' of the molecule that can often be used to identify a specific molecule by 'matching fingerprints' if a library of known compounds is available. This region of the infrared spectrum is generally called the **fingerprint region**.

What Is the Use of the Fingerprint Region?

If there is an unknown organic molecule, and you have a 'library' of prerecorded infrared spectra, the fingerprint region can be used to find the proper 'match.' Fingerprinting the molecule in the library allows its identification. Obviously, if your molecule is not in the library, you cannot use this technique to identify it.

What Functional Groups Appear in the Fingerprint Region?

As mentioned above, C=C, C–O, C-halogen and aromatic carbons usually absorb in the fingerprint region of the infrared.

Functional Group Correlation Chart

Give Generic Infrared Absorptions for the Following Major Functional Groups:

A) Alcohols; B) Ketones; C) Aldehydes; D) Carboxylic Acids; E) Amines; F) Esters; G) Amides; H) Nitriles; I) Alkenes; J) Alkynes; K) Benzene Derivatives; L) Ethers; M) Acid Chlorides; N) Acid Anhydrides; O) Alkyl Halides.

(a) **alcohols**: <u>O–H</u> [3400–3610 cm^{-1}, 2.94–2.77 μ] and <u>C–O</u> [1050–1150 cm^{-1}, 9.52–8.70 μ]

(b) **ketones**: <u>C=O</u> [1725–1680 cm^{-1}, 5.80–5.95 μ]

(c) **aldehydes**: <u>C=O</u> [1740–1695 cm^{-1}, 5.75–5.90 μ] and O=<u>C–H</u> [2816 cm^{-1}, 3.55 μ]

(d) **carboxylic acids**: O=C–<u>O–H</u> [3300–2500 cm^{-1}, 3.03–4.0 μ], <u>O=C</u>–OH [1725–1680 cm^{-1}, 5.80–5.95 μ]

(e) **amines**: 1° amines *N*–<u>H</u> [3550–3300 cm^{-1} - a doublet of peaks for asymmetric and symmetric modes of *N*–H, 2.82–3.03 μ] 2° amines *N*–<u>H</u> [3550–3400 cm^{-1} a single peak for the only *N*–H, 2.82–2.94 μ]

(f) **esters**: <u>O=C</u>–O–C [1780–1715 cm^{-1}, 5.61–5.83 μ], O=C–<u>O–C</u> [1050–1100 cm^{-1}, 9.52–9.09 μ]

(g) **amides**: <u>O=C</u>–*N*- Amide I - 1° [1690 cm^{-1}, 5.92 μ] 2° [1700–1670 cm^{-1}, 5.88–6.00 μ]. The Amide II band is - 1° [1600 cm^{-1}, 6.25 μ] 2° [1550–1510 cm^{-1}, 6.45–6.62 μ]. amide *N*–<u>H</u> - 1° [3500 cm^{-1}, 2.86 μ], 2° [3460–3400 cm^{-1}, 2.89–2.94 μ]

(h) **nitriles**: <u>C≡N</u> [2260–2200 cm^{-1}, 4.42–4.56 μ]

(i) **alkenes**: C=<u>C–H</u> [3040–3010 cm^{-1}, 3.29–3.32 μ]; <u>C=C</u> [this is a relatively weak to moderate signal at 1680–1620 cm^{-1} [5.95–6.17 μ];

(j) **alkynes**: C=<u>C–H</u> [2250–2100 cm^{-1}, 3.03 μ], <u>C≡C</u> [terminal, 2140–2100 cm^{-1}, 4.67–4.76 μ; non-terminal, 2260–2150 cm^{-1}, 4.42–4.65 μ];

(k) **benzene derivatives**: Ar–<u>H</u> [3040–3010 cm^{-1}, 3.29–3.32 μ], Ar <u>C=C</u> [about 1600 and 1510 cm^{-1}, 6.25 and 6.63 μ]

(l) **ethers**: <u>C–O</u> [1150–1060 and 1140–900 cm^{-1}, 8.70–9.43 and 8.77–11.11 μ];

(m) **acid chlorides**: <u>O=C</u>–Cl [1815–1750 cm^{-1}, .51–5.71 μ], C–Cl [730–580 cm^{-1}, 13.70–17.24 μ];

(n) **acid anhydrides**: $(\underline{O=C})_2O$ [1850–1780 and 1790–1710 cm^{-1}, 5.41–5.62 and 5.59–5.85 μ];

(o) **alkyl halides**: <u>C–Cl</u> [730–605 cm^{-1}, 13.70–16.53], <u>C–Br</u> [645–605 cm^{-1}, 15.50–16.53 μ], <u>C–I</u> [600–560 cm^{-1}, 16.67–17.86 μ]. .

What Is the Effect of Conjugation on the Absorption of a Carbonyl? Compare Cyclohexanone with Cyclohexenone.

In general, conjugation shifts the absorption to longer wavelengths (lower frequencies-lower energy). The C=O group of cyclohexanone, for example, absorbs at 1725–1705 cm^{-1} [5.80–5.87 μ], but the conjugated C=O group of cyclohexenone will absorb at 1685–1665 cm^{-1} [5.93–6.01 μ]. The absorption was shifted to lower energy (longer wavelength) by conjugation.

Identify Each of the Following Functional Groups Based Entirely on the Provided Molecular Formula and the Infrared Spectrum.

(a) C_4H_6O

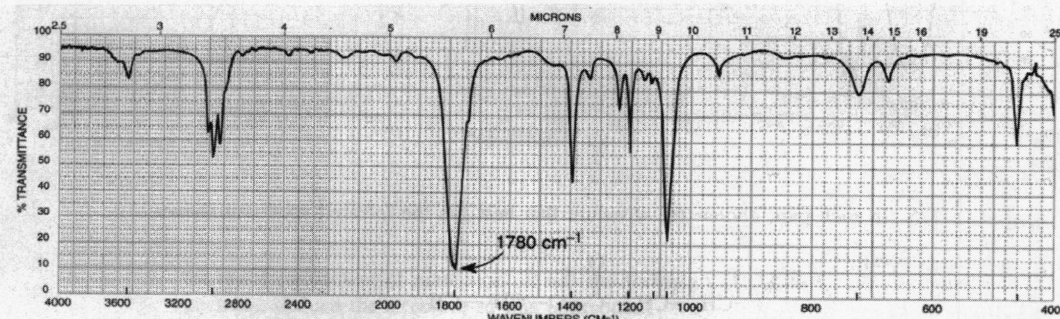

(b) $C_5H_{12}O$

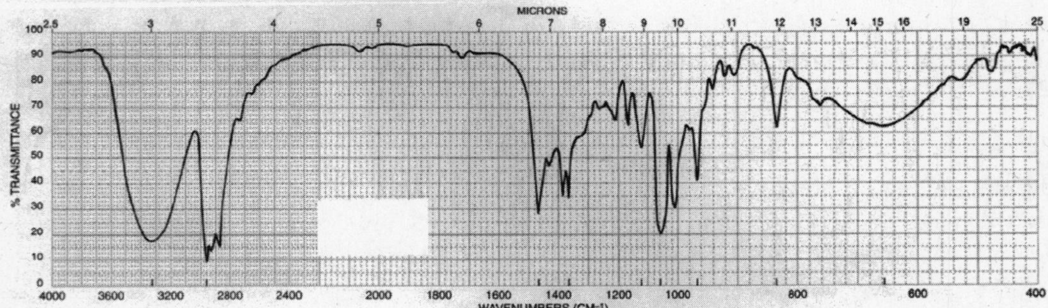

(c) $C_6H_{12}O_2$

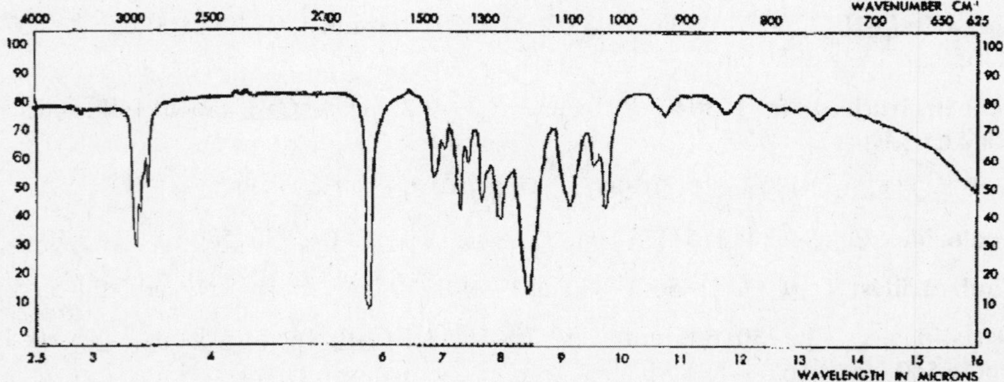

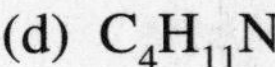
(d) $C_4H_{11}N$

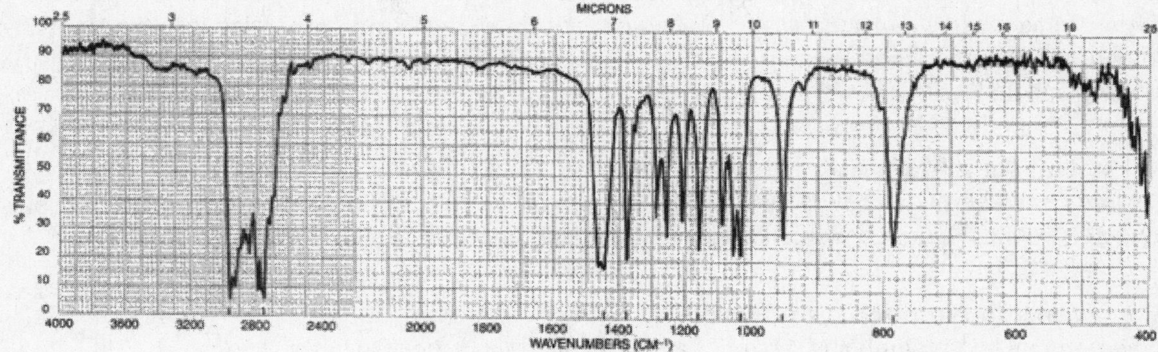

(e) C_5H_7

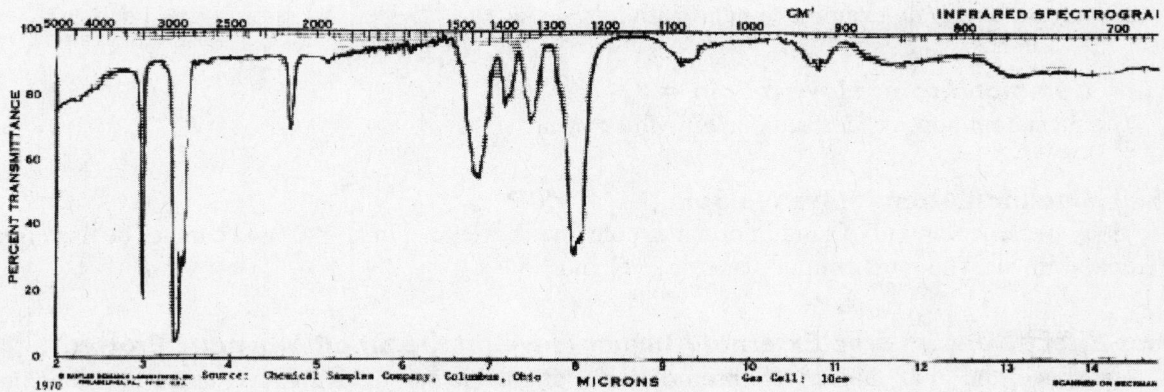

In all cases, note the strong absorption at 2850 cm^{-1} (3.50 μ). This absorption almost always appears, but is not due to a functional group. It is the C–H absorption and does not give useful information about structure, unless it is absent. In infrared (a), the molecule has only one oxygen, limiting the functional group choices to acetone or aldehyde, an ether or an alcohol. There is no O–H absorption, but there is a C=O absorption. There is no absorption at 2816 cm^{-1} characteristic of an aldehyde. This molecule is, therefore, likely to be a ketone, such as 2-butanone. Infrared (b) has one oxygen, so it could be an ether, an alcohol, a ketone or an aldehyde There is a strong band in the 1050–1150 cm^{-1} region suggesting a C–O bond and is an O–H at 3400–3610 cm^{-1}. This molecule is probably an alcohol, such as 1-pentanol. In (c), there are two oxygen atoms, meaning that we must add acid and ester units as possibilities, as well as diketones, diols, dialdehydes, etc. There is no strong O–H absorption but there is a C=O and there appears to be a C–O at about 8.5 microns. It cannot be an acid or an alcohol, and there is no band for an aldehyde, so all combinations of those functional groups are ruled out. The simplest functional group with two oxygen atoms that fits this spectra is an ester, such as ethyl butyrate (EtOC(=O)$CH_2CH_2CH_3$. In (d), the molecule contains one nitrogen and could be an amine or a nitrile. There is no CN absorption and the molecule cannot be a nitrile. If it is an amine, there are three choices, 1°, 2° or 3°. If it were a 1° or 2° amine, there would be a signal in the NH region of 3550–3300 cm^{-1}. Since there is no signal here, this molecule is a 3° amine. This data suggests a 3° amine, such as *N,N*-dimethylethylamine ($CH_3CH_2NMe_2$). In (e), there is clearly a triple bond at about 2120 cm^{-1}. There is no nitrogen atom, and this is consistent with an alkyne, probably a terminal alkyne, such as 1-pentyne.

18.4. NUCLEAR MAGNETIC RESONANCE SPECTROSCOPY (NMR)

Protons function as small magnets. Since the proton possesses the property of spin, it is possible for all hydrogen atoms in a molecule to absorb energy and change their spin states relative to the large magnetic field. We can monitor the energy of these signals and use this information to determine the structure of organic molecules.

The Proton As a Magnet

A proton is a positive nucleus surrounded by negative electrons. Since it has the property of spin, a spinning charge will generate a small magnetic field and, therefore, act as an electromagnet.

What Quantum Number Is Important in NMR?

The spin quantum number (I) is important to nuclear magnetic resonance.

Which Common Atoms Have a Spin $= \frac{1}{2}$?

The most commonly examined nuclei with $I = \frac{1}{2}$ are ^{1}H, ^{13}C, ^{19}F, ^{15}N $\left(\text{spin} = \pm\frac{1}{2}\right)$ and ^{31}P.

Which Common Atoms Have a Spin = 1?

The most commonly examined nuclei with I = 1 are ^{2}H and ^{6}Li.

Which Common Atoms Have No Spin (i.e. I = 0)?

Many nuclei commonly found in organic molecules have no spin (I = 0) and can *not* be used in an NMR experiment. The most common are ^{12}C, ^{16}O and ^{32}S.

What Effect Does a Large External Magnet Have on the Small Magnetic Proton?

The spin of the small magnet (the proton) can be aligned in parallel with the external magnetic field or can be opposed to it (see ***18.17***). When the proton field is aligned opposite the external field, it is at a higher energy than when it is aligned with the field. This creates an energy gap (ΔE). This energy difference is usually of about the same frequency as radio waves ($\nu = 3\times10^{6} - 3\times10^{8}$ Hz).

Compare the Influence of a 14,100 Gauss Magnet vs. a 63,450 Gauss Magnet on a Proton.

When the magnetic field is 14,100 gauss, the ΔE for the proton will be about 60 MHz (5.7×10^{-6} kcal/mole). When the magnet field strength is increased to 63,450 gauss, ΔE for the proton will increase to 270 MHz (25.7×10^{-6} kcal/mol).

Irradiation and Absorption of a Radio Signal

What Is the Energy Range of the ΔE Imposed by Introduction of a Proton into a Magnetic Field?

This ΔE is in the range of radio waves (1000×10^{-4} – 100000×10^{-4} cm [1000–100000 λ)= 10–0.1 cm^{-1}.

What Energy Source Can Cause the Proton to Absorb Energy = ΔE?

A radio signal of controlled frequency is directed towards the sample inside the strong magnetic field.

What Happens When a Proton in a Magnetic Field is Bombarded with Energy Equal to ΔE?

A proton with spin will *absorb* ΔE, and the nuclei will change its spin state (flips its spin state) as outlined in ***18.17***.

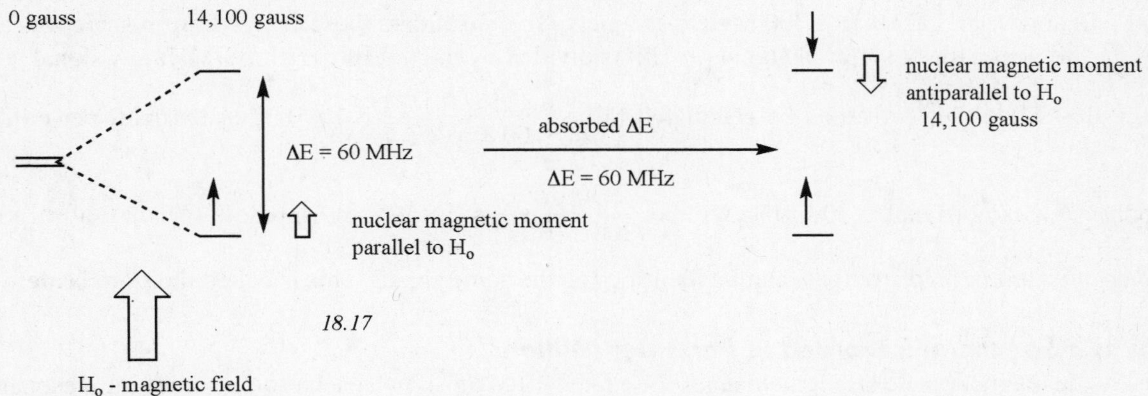

18.17

Is This Phenomenon an Absorption or Emission Process?

This is an absorption process.

Is ΔE Larger or Smaller As the Magnetic Field Strength Increases?

As the magnetic field strength increases, ΔE will become larger for the given nuclei. Important: If the ΔE is kept constant, say at 60 MHz, then the magnetic field strength needed to flip nuclei spins due to different nuclei (protons in different magnetic environments) will change slightly and these differences can be measured.

How the NMR Spectrometer Works

A sample is dissolved in an appropriate solvent and placed in a thin glass tube. This tube is lowered into a magnetic field and spun to average out in homogeneous areas in the magnetic field. A radio signal (linked to the magnetic field strength - a 14,100 gauss magnet requires a 60 MHz radio signal for protons to absorb the energy) is applied to the sample. The holder into which the sample tube is lowered is surrounded by a radio generating coil. The radio signal is kept constant, and the magnetic field is varied slightly. Those protons that absorb the radio signal at the various magnetic field strengths will be recorded as absorption 'peaks.' The collection of these absorption peaks is the NMR spectrum.

What Process Is Examined in the NMR?

The absorption of energy by a nucleus with spin in a magnetic field. The most commonly examined nuclei are the proton (^1H) and carbon-13 (^{13}C). Other nuclei can also be examined, however, including ^2H, ^6Li, ^{15}N and ^{19}F.

Why Do Different Nuclei Require Different Magnetic Field Strengths?

Each nucleus will absorb different amounts of energy for a given magnetic field strength (different ΔE). For a ^1H in a 14,100 gauss magnetic field, ΔE is 60 MHz but ΔE for ^{13}C is 15 MHz at 14,100 gauss.

What Is an NMR Spectrum?

The NMR spectrum are the signals produced as different protons in a molecule absorb energy when the sample is in a strong magnetic field. The signals are recorded in Hz as an absorption spectrum.

What Is the PPM Scale?

Different protons absorb at different magnetic field strengths, which represent energies of between <1 to 1200 Hz on a 60 MHz instrument. Since 1 MHz is 1×10^6 Hz, the changes in field strength represent millionths of the total field strength at 60 MHz. A convenient method for measuring the position of an absorption signal is in Hz. There is a problem, however, since a signal at 60 Hz in a 60 MHz instrument will appear at 300 Hz in a 300 MHz instrument. To consolidate these data, the **ppm scale** was created. In this scale, the absorption signal (in Hz) is divided by the field strength (in MHz). A signal at 60 Hz in a 60 MHz field will then be calculated to be $\left(\frac{60Hz}{60\times10^6Hz}\right) = 1\times10^{-6}$ or one part-per-million (1 ppm). The same signal at 300 MHz will be $\frac{300Hz}{300\times10^6Hz} = 1\times10^{-6} = 1$ ppm. In this way, absorption signals at different field strengths can be identified as the same signal. This is called the **ppm scale**.

Why Is a Spectrum Recorded in Parts per Million?

As the magnetic field strength changes for a fixed 60 MHz signal, each proton comes into resonance at a slightly different field strength. If the field is changed with a larger magnet, the absorption signals will change proportionally. The ppm scale allows the absorption process to be normalized and compared from one instrument to another.

If an NMR Is Recorded at 60 MHz and a Signal Appears at 345 Hz, What Is the Absorption in PPM?

This signal will appear at $\frac{345Hz}{60\times10^6\ Hz} = 5.45$ ppm.

If a Signal Appears at 4.50 PPM at 60 MHz Where Is the Signal, in PPM, at 500 MHz?

The position of the signal in ppm will not change with field strength. This signal will, therefore, also appear at 4.50 ppm at both 500 MHz and at 60 MHz.

If a Signal Appears at 3.25 PPM at 60 Hz, Where Is the Signal, in Hz, at 400 MHz?

A signal at 3.25 ppm (3.25×10^{-6}) appears at 3.25×10^{-6} $(60\times10^6) = 195$ Hz. at 400 MHz, the 3.25 ppm signal appears at 3.25×10^{-6} $(400\times10^6) = 1300$ Hz.

Where Is the 'Zero Point' in NMR?

The 'zero point' in NMR is the absorption signal for tetramethylsilane (TMS), which is added to the sample. The instrument is adjusted so that the TMS signal is set on zero.

What Solvents Are Used to Prepare Samples in the NMR?

Typical solvents include $CDCl_3$, d_6-acetone, D_2O and d_6-DMSO. Deuterated solvents are used so they will not contribute signals to the proton NMR that can obscure peaks, which are due to the sample of interest.

Can Ethanol Be Used As a Solvent in Proton NMR?

We cannot use CH_3CH_2OH because of the protons in these molecules. If we replaced every hydrogen atom with a deuterium atom, we could use it as a solvent. The NMR solvent should not have hydrogen atoms that can interfere with the molecule being examined.

Can Carbon Tetrachloride Be Used As an NMR Solvent?

Yes! since CCl_4 has no protons, it is a suitable solvent.

Why Is an Internal Standard Used in NMR?

There has to be a 'zero point.' Since the absorption peaks are measured in ppm, the scale makes no sense unless there is a point at zero ppm to be used as a reference standard. This must be taken as 0 ppm in all samples under all conditions. All absorption peaks are, therefore, reported in ppm with reference to the standard. The most common standard in NMR is the chemical tetramethylsilane (TMS), which is added to the sample.

What Is the Structure of Tetramethylsilane?

Tetramethylsilane has the structure $(Me_3)_4Si$.

Why Is Tetramethylsilane an Ideal Internal Standard?

Tetramethylsilane absorbs at a position in most magnetic fields at higher energy (higher external field strength) than most protons in most organic molecules. In a typical NMR spectrum (shown in ***18.18***), absorptions at higher magnetic fields appear on the right- hand side of the spectrum, and signals of lower field strength appear to the left. The TMS reference signal (assigned 0 ppm) will appear to the far right (high field, or upfield), and most signals due to the organic sample will appear at lower energy to its left (low field, or downfield). As the absorption position in ppm increase, the field strength decreases. This means that a signal at 7 ppm is of lower field strength than a signal at 2 ppm. Another way to say this is that 7 ppm is downfield of 1 ppm, or 1 ppm is upfield of 5 ppm.

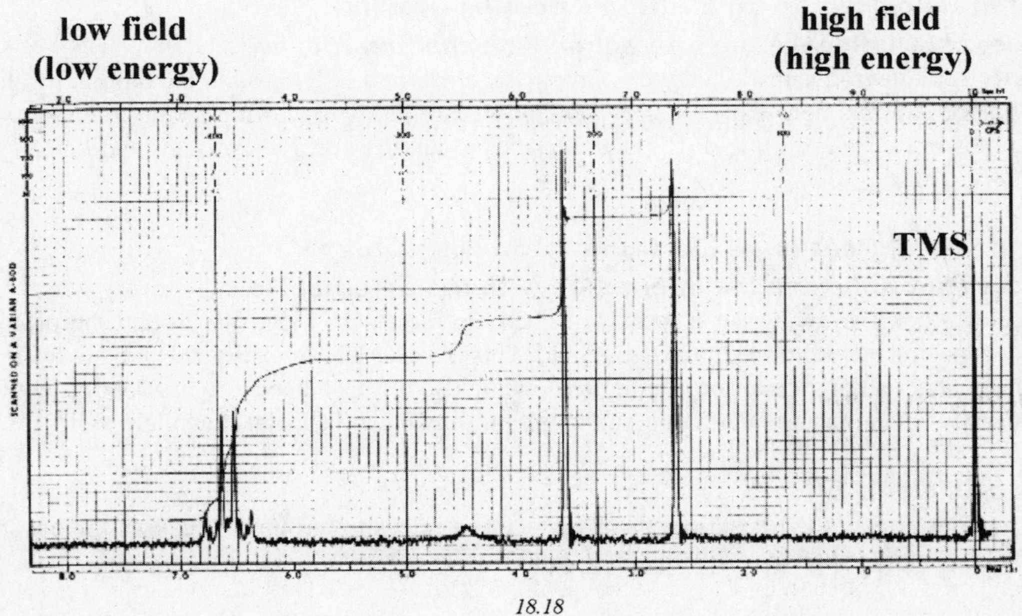

18.18

The Signal for TMS Appears at What PPM in the NMR at 100 MHZ?

At all field strengths and all radio frequencies, TMS appears at zero (0) ppm. The instrument is adjusted to *make* the signal for TMS appear at 0 ppm.

Chemical Shift

The position of an absorption relative to TMS is referred to as the **chemical shift** for that signal.

What Is an Upfield Signal? A Downfield Signal?

An upfield signal appears at low ppm relative to TMS (close to TMS). A signal at 0.5 ppm is considered to be 'upfield.' A downfield signal appears at larger ppm relative to TMS. The signal at 7.0 ppm in the above spectrum is 'downfield.' These terms can be used in a relative sense. The signal at 5.0 ppm is 'upfield' of the signal at 7.0 ppm. The signal at 5.0 ppm is 'downfield' of the signal at 2.45 ppm.

Is High Field Associated with Absorptions Closer to 8 PPM or Closer to 1 PPM?

High field (up field) is associated with signals closer to 1 ppm.

Is the Chemical Shift for a Given Functional Group As Constant As in Infrared Spectroscopy?

No! the variations in local structure of a molecule can greatly influence the absorption in ppm. This is because the change in signal relative to the applied field is very small (parts-per-million). Small local variations can lead to changes of 0.5–2.0 ppm (these are arbitrary numbers), which is a significant change in a 10 or 20 ppm scale.

Why Are PPM for a Given Group Reported As a Range of Numbers?

For a given group, local variation can lead to a range of absorption frequencies. If the hydrogen attached to the α carbon of a bromide (Br–C–H) appears at about 3.5 ppm, the signal could vary between about 3.2–3.9 ppm, depending on what other groups or atoms are in proximity to this proton. Usually, the variation is less than 0.5 ppm for a given functional group.

If a Proton Is Adjacent to an Electron Releasing Group, How Does That Influence the Interaction with the External Field?

An electron releasing substituent will increase the electron density around the proton, insulating the proton from the external field. This requires more of an external field to bring the proton into resonance, taking the absorption more upfield. Protons adjacent to an electron releasing group generally appear between 2–3 ppm.

If a Proton Is Adjacent to an Electron Withdrawing Group, How Does That Influence the Interaction with the External Field?

The electron withdrawing substituent will decrease the electron density around the proton, more effectively exposing the proton to the external field. This requires less of an external field to bring the proton into resonance, taking the absorption more downfield. Protons adjacent to an electron withdrawing group generally appear between 2–6 ppm, although the signal can be as far downfield as 15–18 ppm.

What Is a Shielding Effect?

A shielding effect is the description for an electron releasing group that causes the absorption to occur upfield. A shielded proton will absorb upfield.

Is Shielding Associated with an Upfield or Downfield Chemical Shift?

Shielding is associated with an upfield shift.

What Are Common Functional Groups That Cause Shielding Effects?

The silane (R_4Si) group induces a large shielding effect. This is why TMS appears far upfield relative to most other organic molecules. Alkyl groups are shielding relative to C–O, C–N, C-halogen, C–C=O, etc.

What Is a Deshielding Effect?

A deshielding effect is associated with an electron withdrawing group, pushing the absorption downfield. A deshielded proton absorbs downfield.

Is Deshielding Associated with an Upfield or Downfield Chemical Shift?

Deshielding is associated with a downfield shift.

What Are Common Functional Groups That Cause Deshielding Effects?

Most electron withdrawing groups induce downfield shifts via deshielding. These include C=O, C≡N, OR, NR_2, Ph, C=C, etc.

Which Atom Will Cause a Greater Downfield Shift, Oxygen or Bromine?

The Br induces greater bond polarization and, thereby, greater deshielding. A proton adjacent to a bromine (H–C–Br) absorbs at about 2.70–4.10 ppm, whereas bonding an oxygen to carbon leads to chemical shifts of 3.2–3.6 ppm for H–C–O.

Diamagnetic Anisotropy

Why Is the Signal for a Proton Attached to a C=C Group Further Downfield Than That Attached to a Cl?

The π-electrons in the C=C double bond induce a secondary magnetic field that is opposed to the external field (H_o, see *18.19*). If a proton is held in the center part of this secondary field, it will be shielded and moved upfield. If it is held in the outer portion of the secondary field, however, it will be deshielded and moved downfield. Since the proton attached to the C=C is held in the deshielding portion of the secondary field, it absorbs further downfield than is usual. This effect is in addition to the usual electron withdrawing effects, and H–C=C absorbs between 4.5–6.5 ppm, whereas H–C–Cl absorbs between 3.1–4.1 ppm.

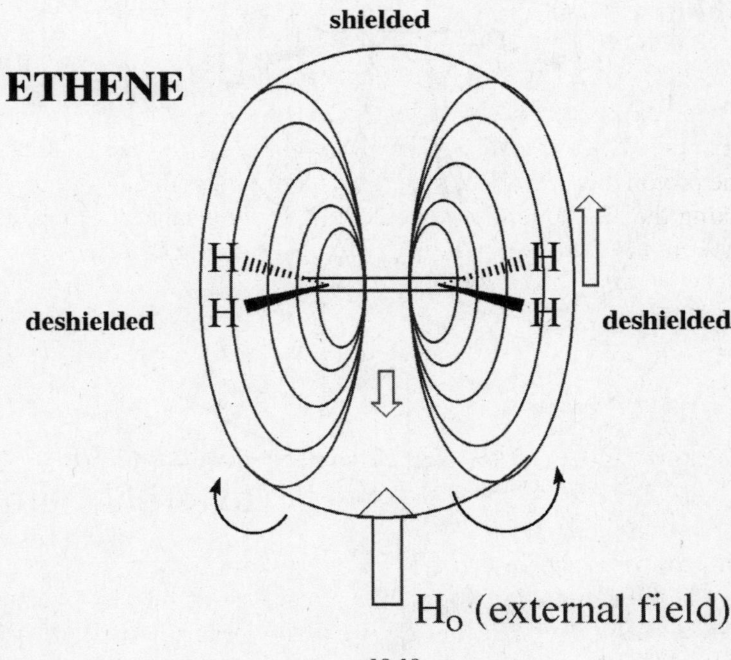

18.19

What Is the Name of the Effect Observed with Ethene in 18.19?

Diamagnetic anisotropy.

Define Diamagnetic Anisotropy.

When a molecule with π-electrons is placed in a magnetic field, the π-electrons are induced to circulate around the ring and called ring current. The moving electrons generate a magnetic field, and depending on the orientation of the molecule, its protons can be shielded or deshielded by this ring current. This phenomenon is called **diamagnetic anisotropy**.

Why Does the Hydrogen on a Benzene Ring Appear at 6.8–7.2 PPM, Further Downfield Than the Protons Found on Ethene?

There are significantly more π-electrons in benzene than in ethene (6 vs. 2), and the diamagnetic anisotropy effect is much greater. As shown in ***18.20***, the protons on the benzene ring are in the deshielding portion of the secondary field, pushing the signal far downfield (to 6.5–8.5 ppm).

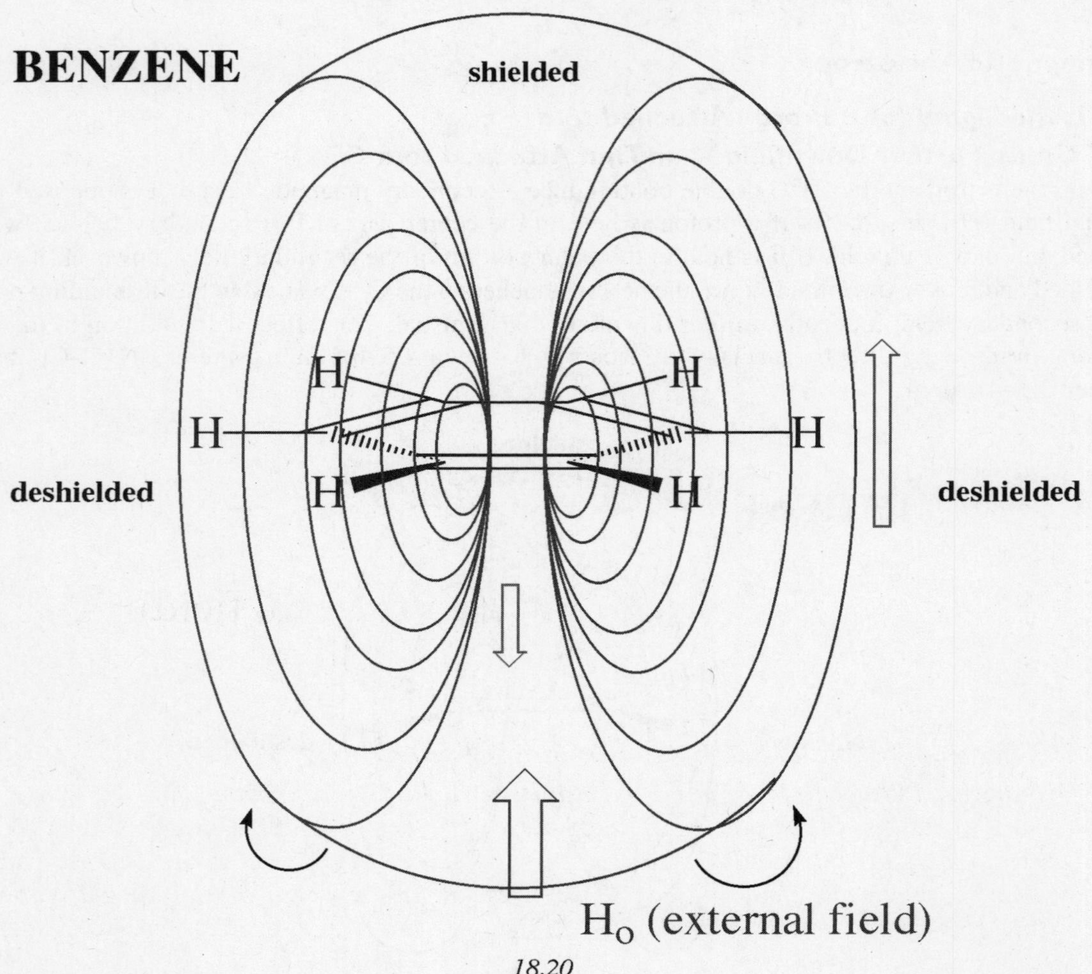

18.20

Explain Why the Proton of an Aldehyde Appears at 9–10 PPM, but the Proton of an Alkene Appears at 5–6 PPM.

In both cases the H–C= proton is deshielded by diamagnetic anisotropy. The aldehyde proton, however, is also highly polarized, further pushing that signal downfield of the polarized C=C bond.

Explain Why the Proton Attached to an Alkyne Is Upfield of the Proton Absorption for an Alkene.

The presence of two π-bonds changes the orientation of the molecule with respect to the external field (H_o). As shown in ***18.21***, the proton attached to C≡C is held in the shielding portion of the secondary field and is shielded. For this reason, the signal is more upfield than the analogous H–C=C signal.

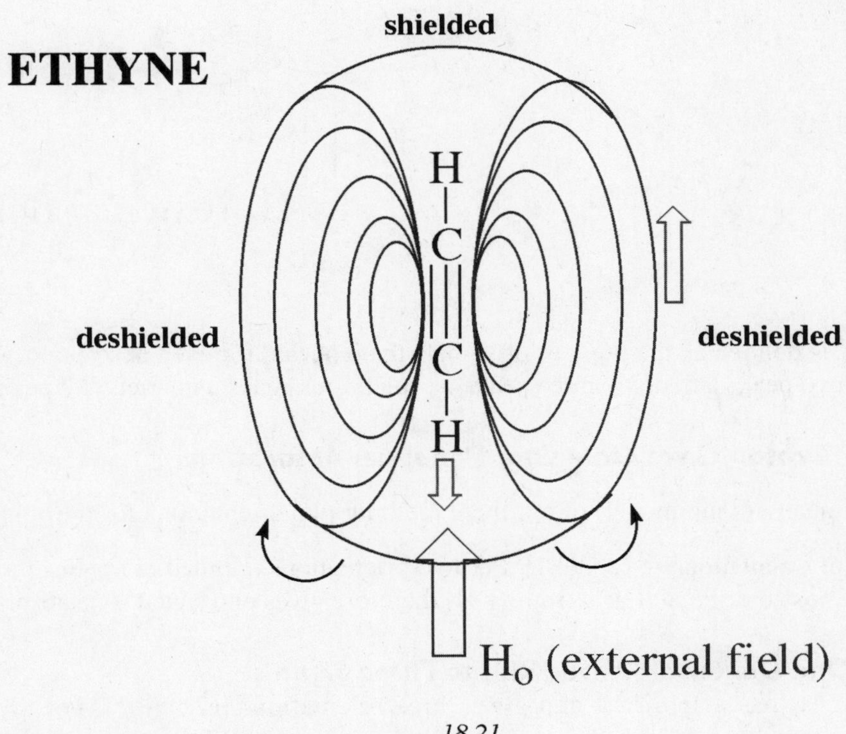

18.21

Explain Why the Proton Attached to the Carbonyl Carbon of an Aldehyde Is Downfield of the Proton Absorption for an Alkene.

If we use formaldehyde as an example, the ring current shown in ***18.22*** indicates a downfield shift for the proton attached to the carbonyl due to diamagnetic anisotropy. However, the carbonyl carbon is also polarized with a positive dipole due to the electronegative oxygen atom, so the proton is pushed further downfield. In other words, the deshielding effect of the bond polarization and the diamagnetic anisotropy effect are additive.

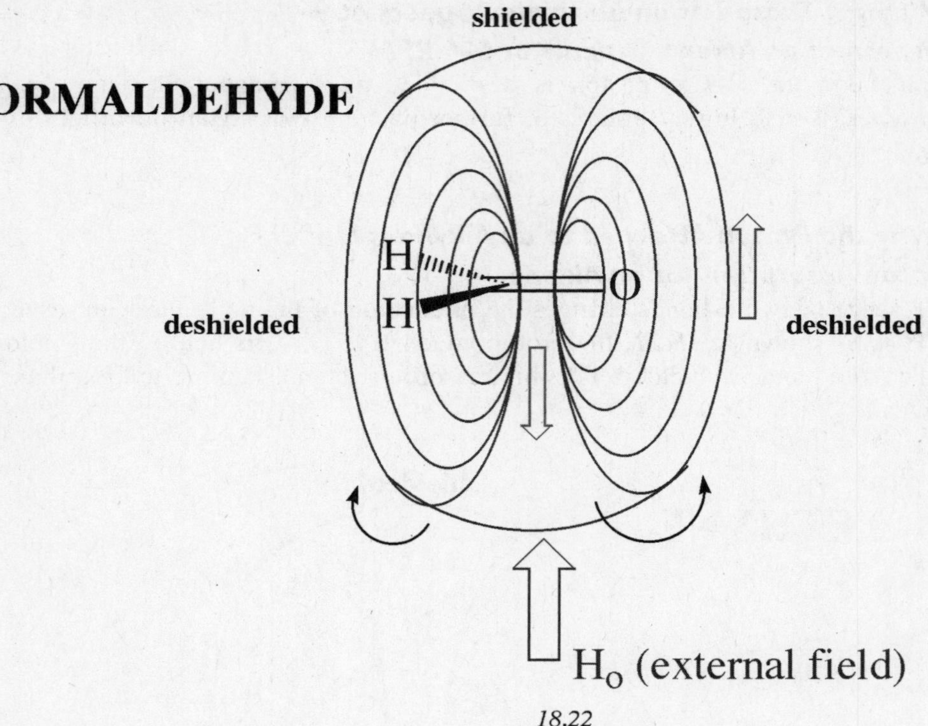

18.22

Multiplicity

Multiplicity is defined as the number of signals that appears for each absorption. A single peak is called a singlet, two peaks a doublet, three peaks a triplet, four peaks a quartet, five peaks a pentet, etc.

Explain Why a Proton Gives Only One Signal per Absorption.

With a spin quantum number (I) of $\frac{1}{2}$, there are only two orientations for the spin moment of the proton [number of orientations = (2*I)+1]. For two orientations (aligned or against the external field), there is only one possible signal. Each proton (^1H) therefore gives one signal (one absorption).

Explain Why One Deuterium Gives Rise to Three Signals.

Deuterium (^2H) has a spin of 1 and gives three orientations [(2*1)+1]. For three orientations (A,B,C), there are three signals [A→B, A→C, B→C]. Each atom of ^2H, therefore, gives three signals per absorption.

How Many Signals per Absorption Are Associated with ^{13}C? with ^{15}N? with ^{16}O?

Both ^{13}C and ^{15}N have I = $\frac{1}{2}$ and a spin of $\pm\frac{1}{2}$ and give one absorption per nuclei. Since ^{16}O has a spin of zero (0), it does not exhibit a signal in the NMR.

If a Proton Has One Non-Equivalent Hydrogen Neighbor, What Is the Influence of That Neighbor?

The neighboring proton will generate a small local field and will split the signal for the proton of interest into two peaks (a doublet).

How Many Signals Appear for Each Non-Equivalent Neighboring Hydrogen?

Each neighboring hydrogen will split the proton signal of interest into two peaks (a doublet).

Why Does 2,2-Dimethylpropane (18.23) Give Only One Signal in the Proton NMR?

$$\begin{array}{c} CH_3 \\ | \\ H_3C-C-CH_3 \\ | \\ CH_3 \end{array}$$

18.23

The NMR of ***18.23*** shows only one signal because all three of the protons of each methyl group are magnetically equivalent, and all four methyl groups are magnetically equivalent. For this reason, the protons are not considered to be neighbors, but are identical and absorb as a single unit (see ***18.24***).

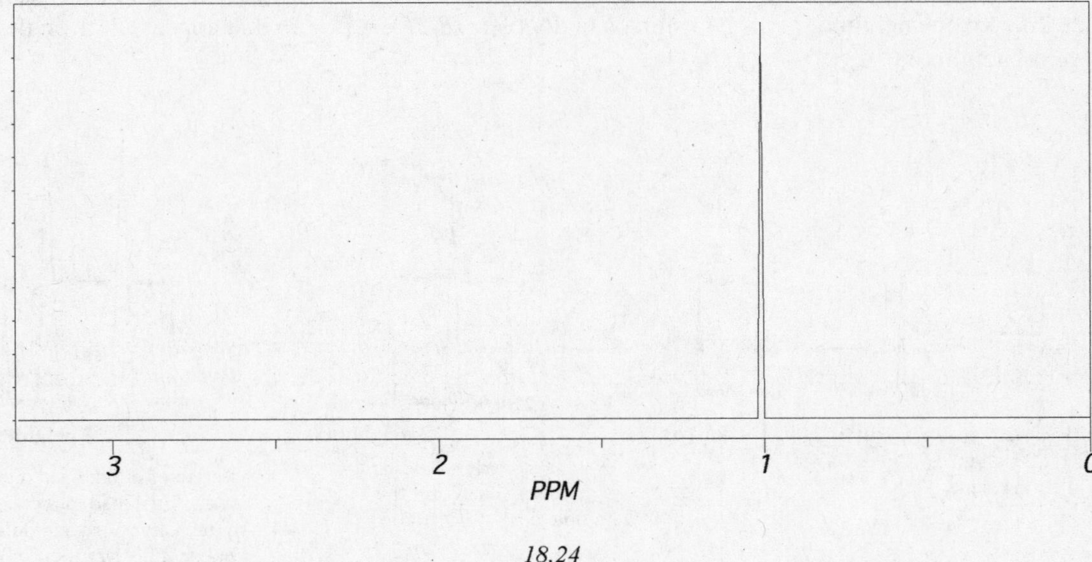

18.24

What Is the Coupling Constant?

When a signal is split into two peaks, the two peaks will be separated by some number of Hz. This distance (in Hz) is referred to as the **coupling constant** (J), as in ***18.25***.

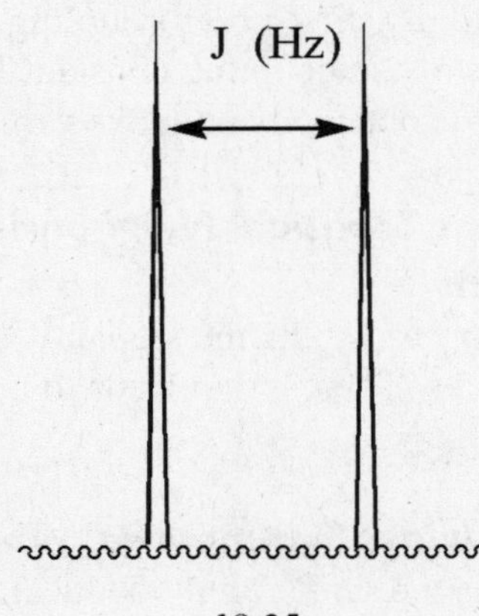

18.25

What Units Are Used to Describe the Coupling Constant?

The units of the coupling constant are hertz (Hz).

Of What Magnitude Are Coupling Constants?

Typical values of J are 0–15 Hz although much larger values are possible and are occasionally observed.

The N+1 Rule

For n non-equivalent neighbors, there will be n+1 peaks if the neighboring hydrogens all have the same coupling constant (J).

If a Proton Has One Neighbor, How Many Signals Will Be Generated?

For *n* neighbors, a proton for the methyl group in ***18.26*** will be split into *n+1* signals. For three neighbors, the proton will appear as four signals (a quartet, ***18.29***); two neighbors lead to three signals (a triplet, ***18.28***); one neighbor gives two signals (a doublet, ***18.27***). A proton that appears as a single peak will have no neighbors.

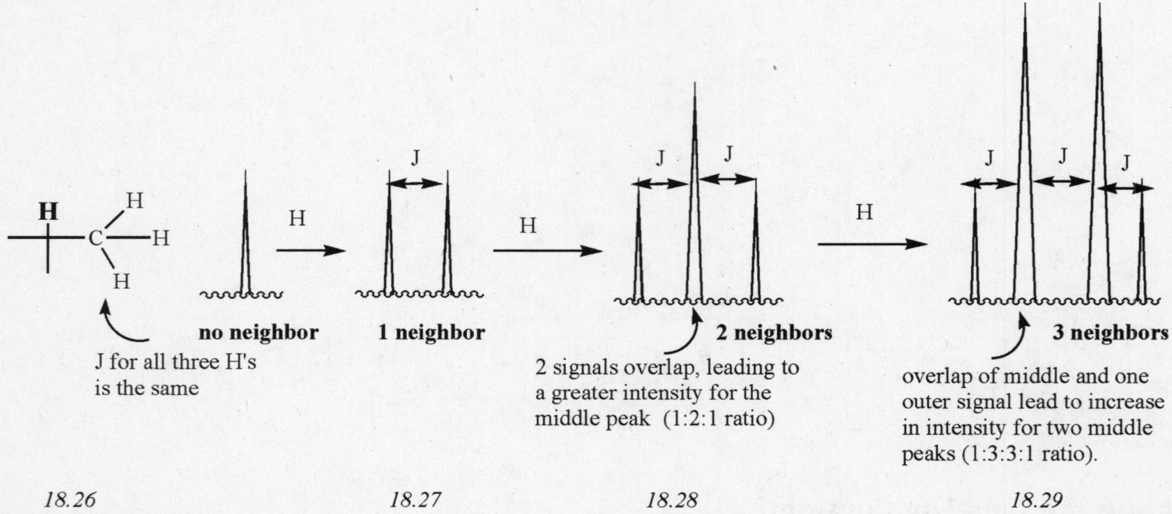

18.26 *18.27* *18.28* *18.29*

Why Does the N+1 Rule Work?

The n+1 rule only works when the coupling constants for all neighboring protons are the same, and it is called a first-order coupling. Formally, first-order coupling applies when the difference in ppm between two coupled proton ($\Delta\, \delta_{ppm}$) is $\gg$ the coupling constant J. If the coupling constants are different, a so-called non-first order coupling is obtained and looks very different.

Explain Why Three Non-Equivalent, Identical Neighboring Hydrogens Lead to Four Signals for a Given Proton.

When each identical neighboring proton splits the signal of interest, the coupling constant will be identical, leading to overlap of the signals. The overlap leads to a total four signals in an intensity ratio of 1:3:3:1.

Why Are These Multiple Signals Always Symmetrical When the N+1 Rule Is Applied?

Because the coupling constant (J) is the same for all identical neighboring hydrogens. If J is not the same, the resulting signal will be asymmetric.

For Each of the Signals 18.30–18.32, Give the Number of Identical Neighboring Hydrogens.

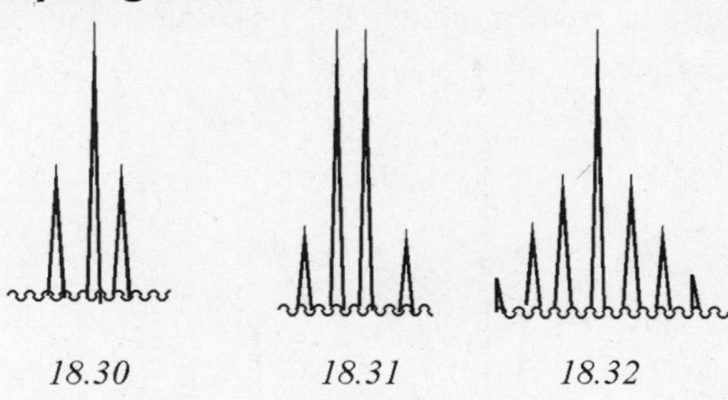

Signal ***18.30*** is a triplet and results from 2 neighbors. Signal ***18.31*** is a quartet and results from 3 neighbors. Signal ***18.32*** is a septet (7 peaks) and results from 6 identical neighbors.

What Is a Multiplet?

A multiplet is a cluster of several peaks, but it is not possible to count them. This may be due to very low intensity for the outermost (satellite) signals or due to unsymmetrical overlap of peaks.

Neighbors with Different Coupling Constants

If a Proton Has Two Different Kinds of Hydrogen Neighbors, What Is the Result?

The multiplet pattern will be asymmetric, reflecting one unique hydrogen splitting a proton into a doublet with one coupling constant. Each of those peaks will then be split often, but not always, into new doublets, but with a different coupling leading to little or no overlap of peaks.

What Is This Phenomenon Called?

It is called a non-first order NMR spectrum.

If the Coupling Constant for Neighbor A Is 2 Hz and the Coupling Constant for Neighbor B is 5 Hz, What Is the Signal for a Given Proton If There Are Two H_A and Two H_B?

The result of these asymmetric splitting patterns is the final 9-peak multiplet shown in ***18.33***. Both the 2 Hz and 5 Hz coupling constants can be discerned in this multiplet. Working 'backwards' from the multiplet, the number of neighbors and each coupling constant can be determined.

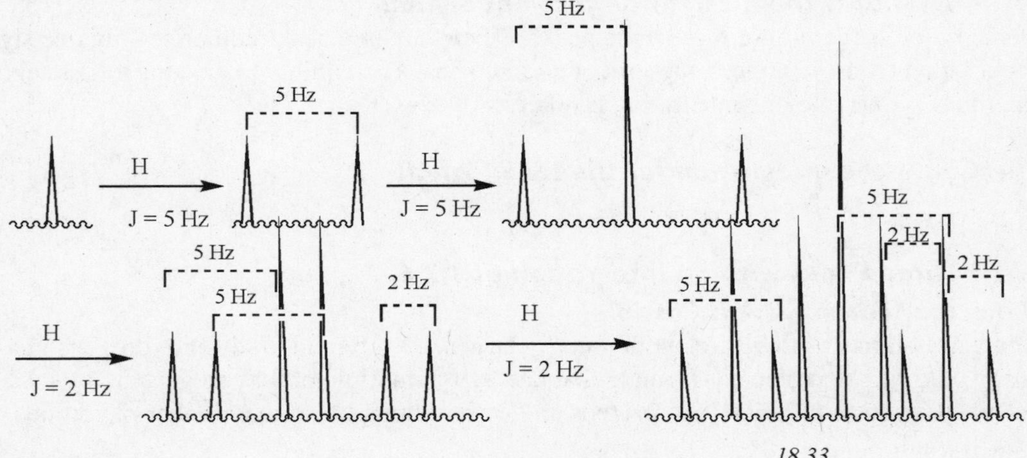

18.33

What Is an AB Quartet?

An AB quartet arises when two hydrogens have almost the same coupling constant, but those hydrogens have slightly different magnetic environments. The asymmetric coupling leads to 4 peaks with the distinctive pattern shown in ***18.34***.

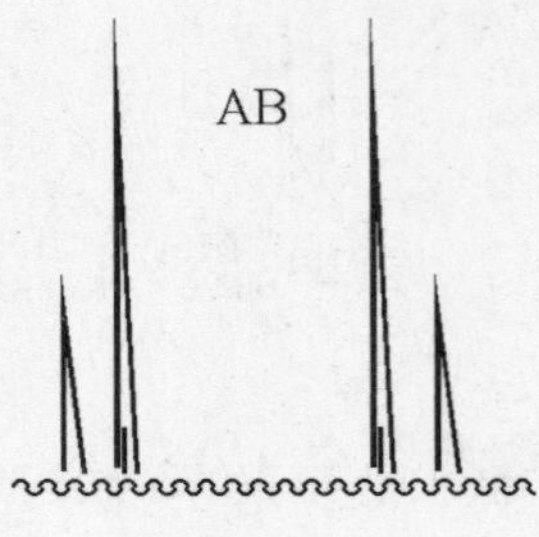

18.34

Integration

Each absorption signal appears as a 'peak' or a cluster of 'peaks.' If the area under these peaks is measured (integrated), comparing the peak area of one signal with another allows us to determine the relative ratio of hydrogens (1:1, 1:2, 2:5, etc.). This does not give the actual number of hydrogens unless the molecular formula is known.

If One Signal (Peak A) Has a Peak Area of 120 and Another (Peak B) Has a Peak Area of 40, What Is the Integration Ratio for the NMR?

The integration (ratio) of A:B $= \frac{120}{40}$ and $\frac{40}{40} = 3{:}1$. This ratio would be consistent with a molecule with 4 hydrogens, 8 hydrogens, 12 hydrogens, etc.

What Is the 'Width at Half Height' Method for Integration?

If the width of the absorption peak is measured at $\frac{1}{2}$ the highest point of the peak, multiplying the width at half-height times the height gives an excellent measure of the total area under the absorption curve.

If the Integration of a Triplet is 20, 40 and 20 for Each of the Three Signals, What Is the Integration for That Signal?

The integration is the sum of these three peaks. Since the three peaks comprise only one signal that is split into a triplet by its neighbors, the sum of all three peaks is required to give the total integration of that signal. For this example, the integration is 80 units.

Does One Count the Integration for the TMS Signal?

No!

If There Are Three Peaks with an Integration of 1:2:6, What Molecular Formulas Are Possible?

This integration means there are a total of 9 hydrogens (3 different kinds of hydrogens in a ratio of 1:2:6). This would fit any molecular formula that had 9 or a multiple of 9 hydrogens in it (9, 18, 27, 36, 45, etc.). This integration refers only to protons and says nothing about the number of carbons or other atoms in the formula.

What Information Is Required to Determine Exactly How Many Protons Are Associated with a Given NMR?

The molecular weight and the molecular formula, usually determined from the mass spectrum, are required. In the example above, if given a choice between $C_9H_{18}O$ or $C_9H_{20}O$, the only possibility is $C_9H_{18}O$.

Why Is the Integration Always Given As a Ratio of Signals?

The exact number of hydrogens is unknown without the specific molecular formula. The integration is based on the ratio of the peak, with the smallest area being divided into all other peaks.

If the Molecular Formula for an NMR is $C_6H_{14}O$, Identify the Number of Protons Associated with Each Signal in the NMR 18.35.

The integration in ***18.35*** is 3:2:2 (7 protons). First, note that the number of protons obtained by integration does not match the formula. The total number of protons by integration is seven. Since the molecular formula indicates there are 14 hydrogens, each integration signal must be multiplied by 2 to give the 14 protons. The signal at 0.9 ppm, therefore, represents 6 hydrogens; the signal at 1.5 ppm represents 4 hydrogens; and the signal at 3.4 ppm represents 4 hydrogens.

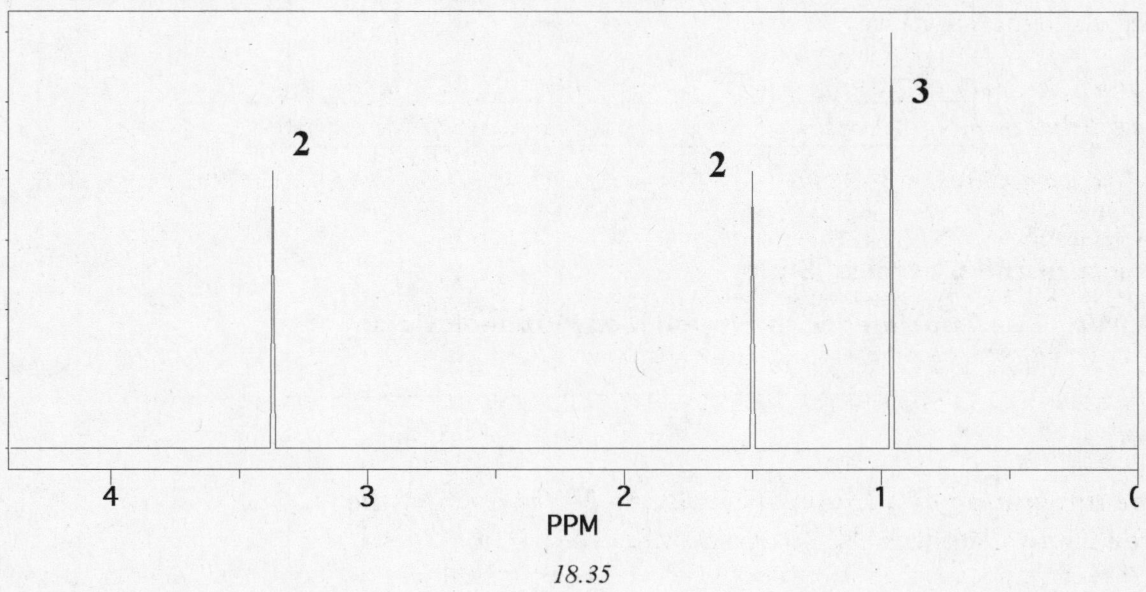

18.35

There Are Six Protons in Acetone, but Acetone Gives Only One Signal (a Singlet) in the NMR. Explain.

All six hydrogens in acetone are identical, magnetically and chemically. They have the same magnetic and chemical environments. They will, therefore, absorb the radio signal at exactly the same strength, leading to a single peak.

What Is the NMR for Benzophenone?

Benzophenone [$Ph_2C{=}O$] has a total of 10 aromatic protons. The two *ortho* protons are different from the two *meta* and the *para* protons, and the aromatic region (7.2–7.7 ppm) should reflect the presence of these three signals. Commonly, when the coupling constant is rather small, these protons will absorb as one broad signal at about 7.2–7.5 ppm.

What Is the NMR for 2,6-Dimethyl-4-Heptanone?

This molecule has a formula of $C_9H_{18}O$. Since it is symmetrical, it will exhibit only three different signals, two for the two isopropyl groups and one for the two $-CH_2-$ groups attached to the carbonyl. The integration is 1:0.5:3 or 4:2:12. Since there is a total of 18 hydrogens, this represents four hydrogens for the doublet at 2.3 ppm; 2 hydrogens for the nonet (marked as a multiplet because in this spectrum it is difficult to determine the precise number of peaks) centered at about 1.7 ppm; and 12 hydrogens for the doublet at about 1.3 ppm (see ***18.36***).

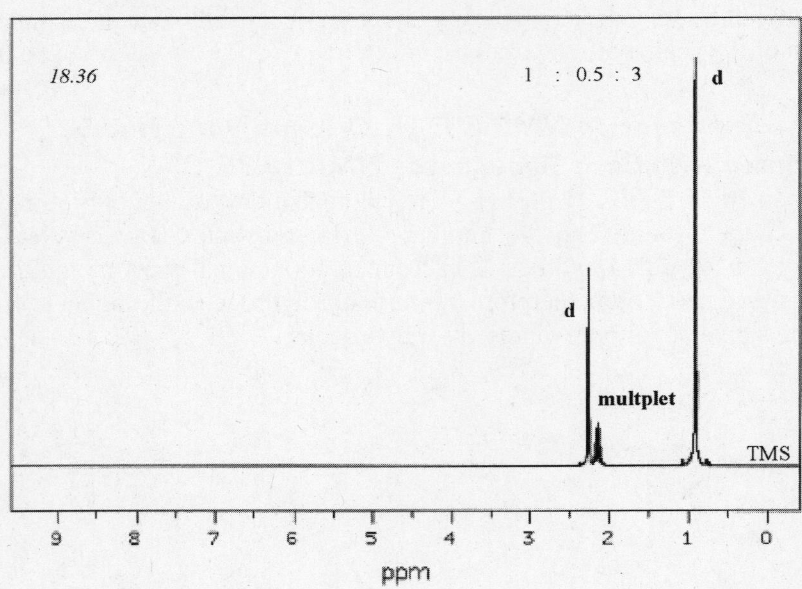

Structure and Chemical Shift

Is NMR 18.37 Associated with Propanol or Propanoic Acid?

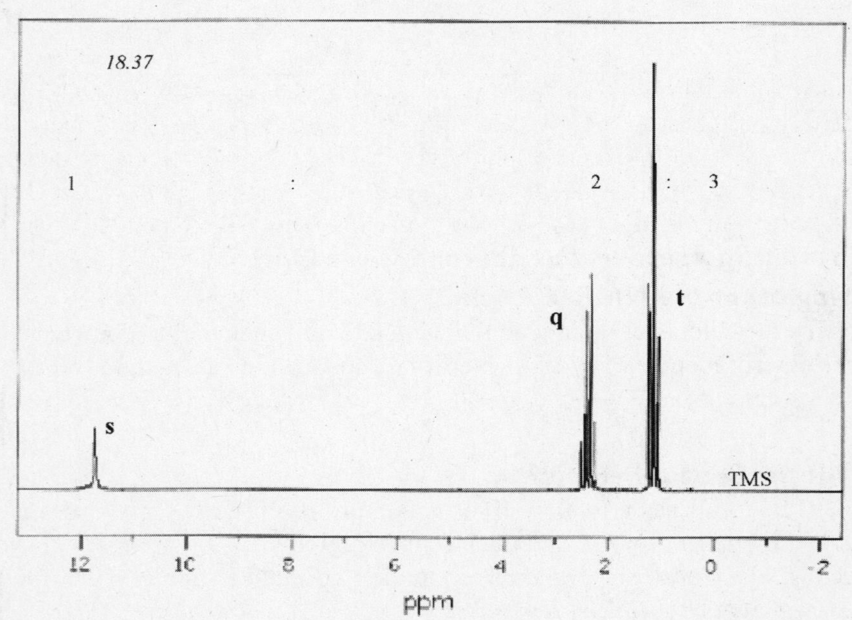

The triplet at 1.2 ppm and the quartet at 3.0 ppm are characteristic of an ethyl group (see ***18.37***). The chemical shift of 3.0 ppm for the $-CH_2-$ group is downfield of the signal normally observed for attachment of a CH_2 to a carbonyl [$CH_3CH_2-C{=}O$]. This signal usually absorbs at 2.2–2.6 ppm. The signal for one proton at 13.6 ppm is characteristic of a very deshielded proton on a carboxylic acid. This NMR spectrum is, therefore, due to propanoic acid rather than 1-propanol.

Is NMR 18.38 Associated with Butanal or 2-Butanone?

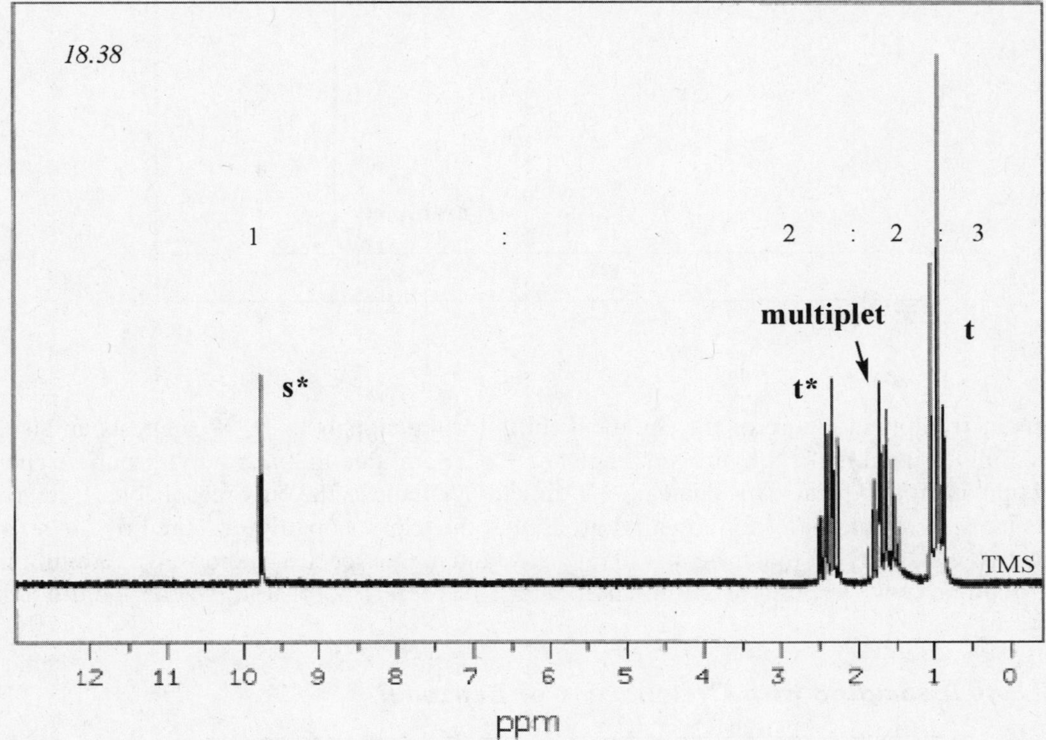

The triplet worth three hydrogens in ***18.38*** is a methyl group split by two neighbors ($-CH_2-$). The (*) is placed on this signal because it is not a clean triplet, but shows a small J value for coupling to another H, which is the aldehyde H. This doublet of triplets at 2.4 ppm, attached to a carbonyl, is a CH_2 group split by two neighbors ($-CH_2-$) and also an aldehyde H. The multiplet at about 1.8 ppm will be a sextet. The singlet, worth one hydrogen, at 9.2 is very characteristic of an aldehyde (O=C–H), but as the (*) indicates, it is not a clean singlet. In fact, this aldehyde H shows weak coupling to the CH_2 protons, and the signal at 9.2 is actually a triplet with a very small coupling constant. All of this information identifies this spectrum as coming from butanal rather than butanone.

Is NMR 18.39 Associated with 2,2,-Dimethylpentane or 2-Methylpentane?

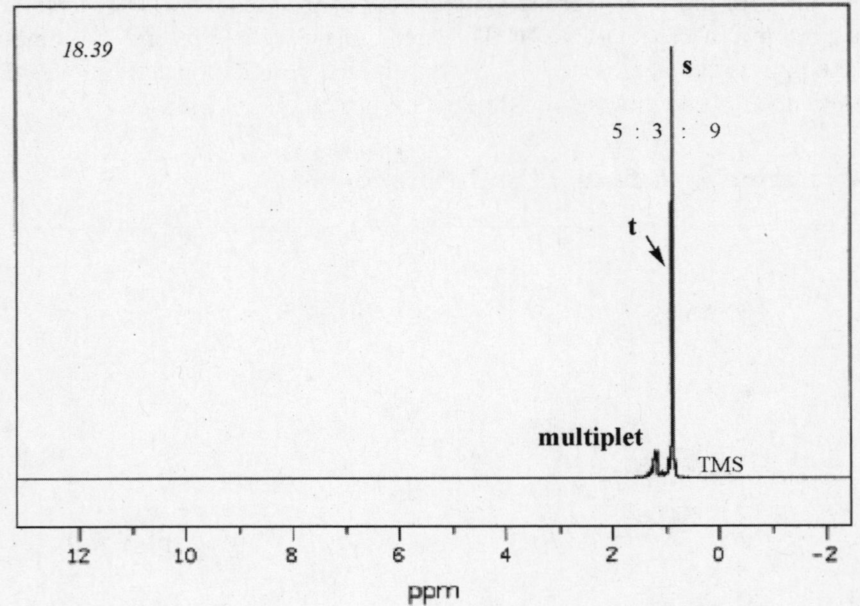

The distinctly upfield nature of the chemical shifts for the signals in ***18.39*** suggests an alkane. A clearly discernible signal is the 9 hydrogen singlet at −1.2 ppm, due to a *tert*-butyl group. A chemical shift of 0.9 ppm is more typical. This points to 2,2-dimethylpentane as the only reasonable structure. The spectrum is marked with a triplet for a methyl group, but the triplet is partly obscured by the *tert*-butyl singlet. The two CH_2 units are much weaker in intensity, and in this spectrum, show up as a multiplet. Of the two structures, however, only 2,2-dimethylpentane has a *tert*-butyl unit, so the identification is straightforward.

Is NMR 18.40 Associated with Cyclohexene or Benzene?

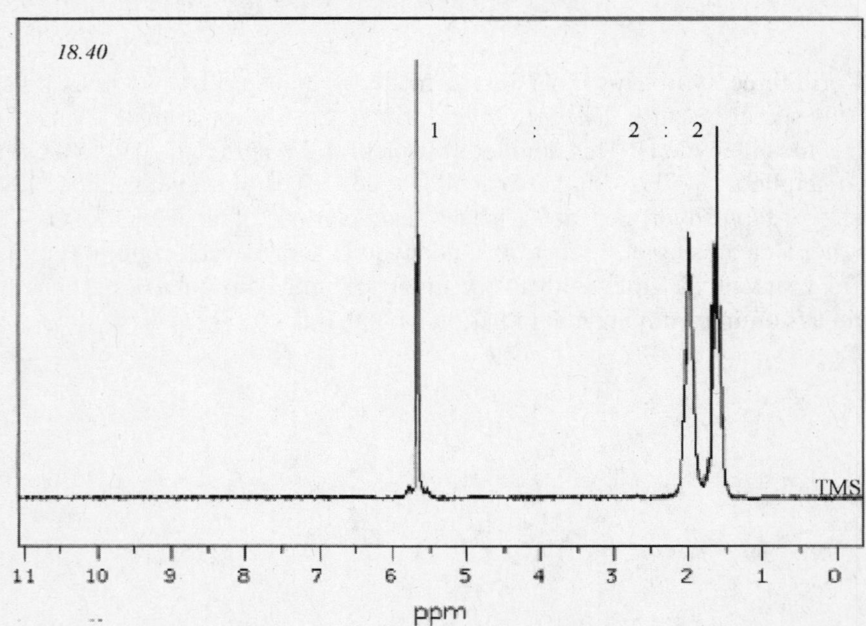

If this were benzene, it would show a singlet at about 7.1 ppm. NMR ***18.40*** shows two multiplets for two CH_2 units (not characteristic of benzene with no substituents) between 1.8–2.2 ppm. The signal at about 5.7 ppm, worth two hydrogens for cyclohexene, is characteristic of alkene hydrogens.

Is NMR 18.41 Associated with Diethyl Ether or Methyl Propyl Ether?

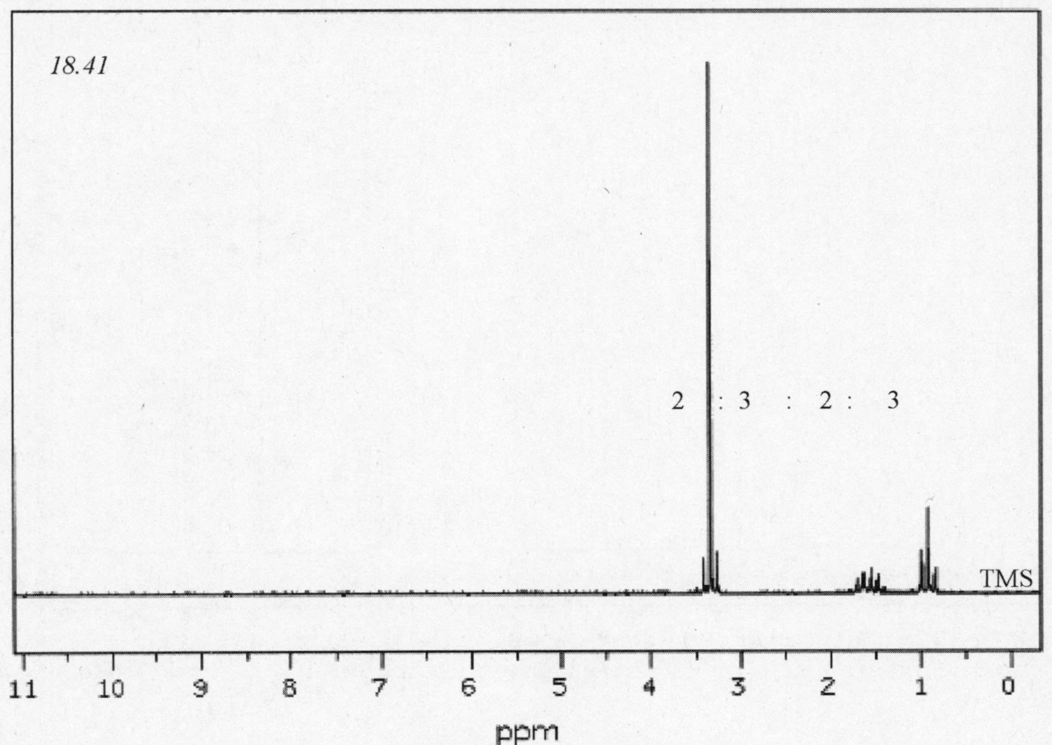

The singlet at about 3.4 ppm in ***18.41*** is due to a methyl group attached to an oxygen. The fact that the signal is a singlet indicates there are no neighboring protons, and this is clearly OCH_3. This fact alone makes this the NMR spectrum of methyl propyl ether rather than diethyl ether. The $–CH_2–$ group (triplet at 3.4 ppm) is partly obscured by the OMe group, but indicates the group $–CH_2–O$. The methyl (triplet at about 1.0 ppm) is attached to a $–CH_2–$ (at 1.6 ppm), and this evidence points to methyl propyl ether. Diethyl ether is symmetrical and would show only an ethyl group with the $–CH_2–$ signal at about 3.4 ppm and the methyl signal at about 1.5 ppm.

Describe the NMR Spectrum of Hexachlorobenzene.

The proton NMR will show *no* signals at all, since this molecule does not contain a hydrogen.

Is the Spectrum 18.42 Consistent with Ethyl 3-Ethoxypropionate ($CH_3CH_2OCH_2CH_2CO_2CH_2CH_3$) or with 2-Ethoxyethyl Acetate ($CH_3CH_2OCH_2CH_2O_2CCH_3$)? Since the Molecular Formulas are Different, the Integration Is Omitted.

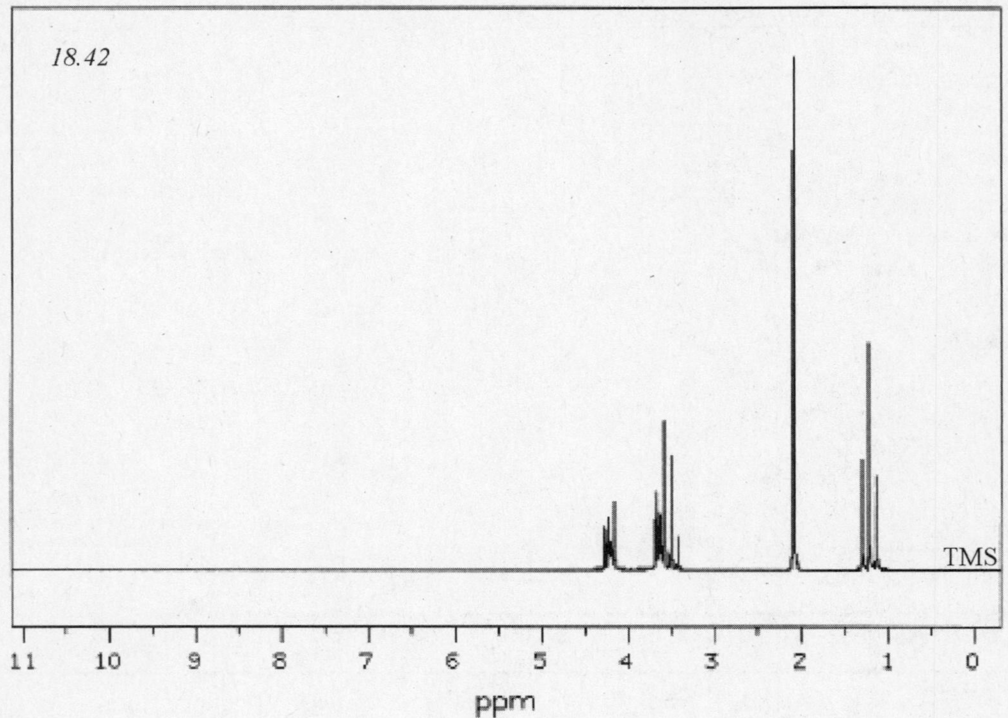

The signal at about 2.1 ppm appears to be a singlet and is consistent with a methyl group attached to a carbonyl. Only 2-ethoxyethyl acetate has such a group. The former compound would have two ethyl groups, which may or may not overlap. In fact, the methyl group of one ethyl is a triplet at 1.21 ppm, and the methyl from the other ethyl group is a triplet a 1.25 ppm. The latter compound would also show four CH_2 units downfield (>3.5 ppm) with a triplet at about 2.5 for the CH_2 next to the carbonyl. Spectrum ***18.42*** has one clean methyl triplet upfield at about 1.2, the singlet at 2.1 and three downfield CH_2 units. This is clearly 2-ethoxyethyl acetate.

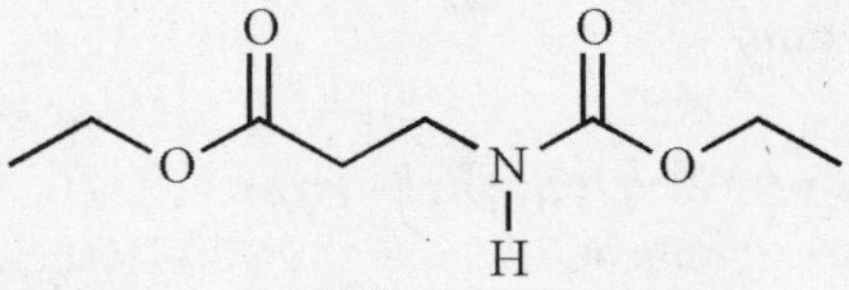

18.43

Explain How One Can Correlate Which CH_2 Is Attached to the Phenyl in 5-Phenyl-1-Pentanol.

In this molecule, one must distinguish between a $-CH_2Ph$, $HO-CH_2-$ and two $-CH_2-$ groups attached to a $-CH_2-$ group. The CH_2 connected to OH will appear at about 3.3–3.5 ppm, and the $-CH_2-$ groups connected to other $-CH_2$ units will appear below 2 ppm. The CH_2 group connected to Ph will have a chemical shift of 2.2–2.5 ppm and should be distinguishable from the other signals .

Test Yourself

1. Interchange each of the following:
   (a) 4.87×10^{-8} m = _____ Hz (b) 2350 Å = _____ cm (c) 6.91×10^{14} Hz = _____ nm
   (d) 3800 nm = _____ μ (e) 6.43 m = _____ Hz (f) 1641 cm^{-1} = _____ m
   (g) 875 cm^{-1} = _____ nm
2. Interchange each of the following:
   (a) 70 eV = _____ kcal = _____ kJ (b) 30 eV = _____ kcal = _____ kJ
   (c) 1200 kcal = _____ eV (d) 2145 kcal = _____ eV
3. Draw the molecular ion for each of the following:
   (a) 2-butanone; (b) *N,N*-dimethylbutanamine; (c) 1-propene; (d) ethanenitrile.
4. Calculate the P, M+1 and M+2 ratios for each of the following:
   (a) $C_5H_{10}O$; (b) $C_8H_{17}N$; (c) $C_5H_{12}O_2$; (d) $C_{10}H_{21}NO$.
5. Calculate the molecular formula for each of the following (molecular weight in brackets):
   (a) P (136) 100%, M+1 (137) 8.88%, M+2 (138) 0.794%
   (b) P (113) 14.5 mm, M+1 (114) 1.18 mm, M+2 (115) 0.044 mm
   (c) P (98) 79.8 mm, M+1 (99) 5.02 mm, M+2 (100) 0.123 mm
   (d) P (100) 100%, M+1 (101) 6.66%, M+2 (102) 0.22%
6. What can be gleaned from the following data?
   (a) P 100%, M+1 7.78%, M+2 100%
   (b) P 100%, M+1 11.1%, M+2 4.7%
   (c) P 100%, M+1 8.88%, M+2 31%
7. Suggest possible fragments for each of the following:
   (a) M–15; (b) M–18; (c) P–28; (d) P–29; (e) P–43; (f) P–44.
8. Which of the following can undergo a McLafferty rearrangement? Show the actual fragmentation in those cases.

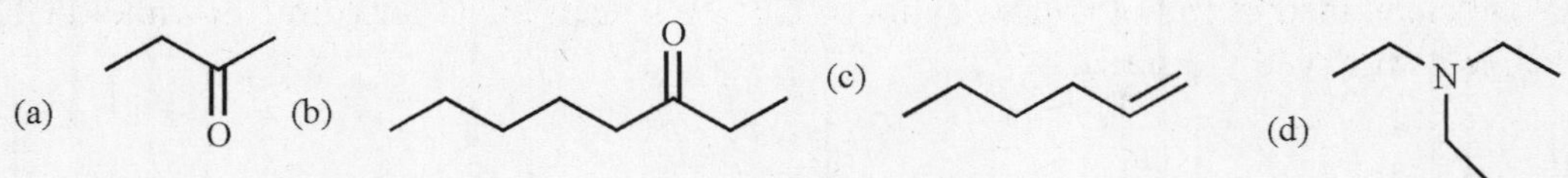

9. Why is it important not to take an infrared spectrum in an aqueous solution when using pressed KBr cells?
10. What is a 'bending' vibration? A stretching vibration?
11. Does the C≡C or the C–O bond give the most intense infrared absorption? Which gives the lowest energy absorption?
12. Calculate the reduced mass for (a) C–Cl (b) C–N (c) C–C
13. Describe the important infrared absorption bands for
    (a) 2-butanone; (b) 4-cyclohexyl-2-pentanol; (c) 6-bromohexanoic acid.

14. Identify how many peaks each of the following nuclei will generate in the NMR for each signal.
(a) ^{2}H; (b) ^{1}H; (c) ^{13}C; (d) ^{15}N; (e) ^{6}Li; (f) ^{12}C

15. If a 14000 gauss magnet required a ΔE of 60 MHz for resonance, and this is assumed to the standard, calculate ΔE for each of the following magnetic fields. (a) 28,574; (b) 125,672; (c) 297,113.

16. Calculate each of the following in ppm.
(a) 625 Hz at 120 MHz (b) 1475 Hz at 90 MHz
(c) 432 Hz at 60 MHz (d) 2122 Hz at 500 MHz

17. Draw the diamagnetic anisotropy field for the *N*-methyl imine of acetone.

18. For the following pairs of signals, identify which one will absorb further upfield.
(a) HC–BrHC–NR_2 (b) H–C=OH–C–C=O
(c) O=C–O–HC=C–C–H (d) C≡C–HC=C–H

19. Identify all magnetically equivalent hydrogens in each molecule.

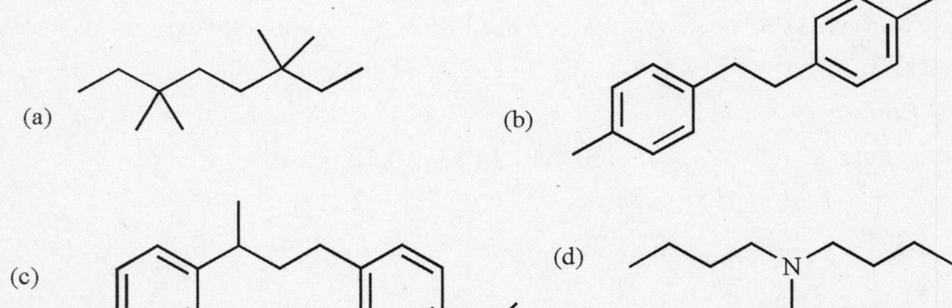

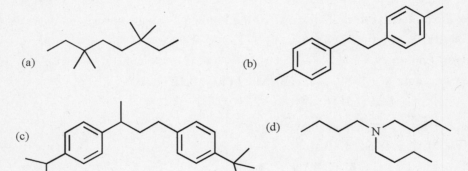

20. For each NMR signal, indicate how many neighboring hydrogens it will have.

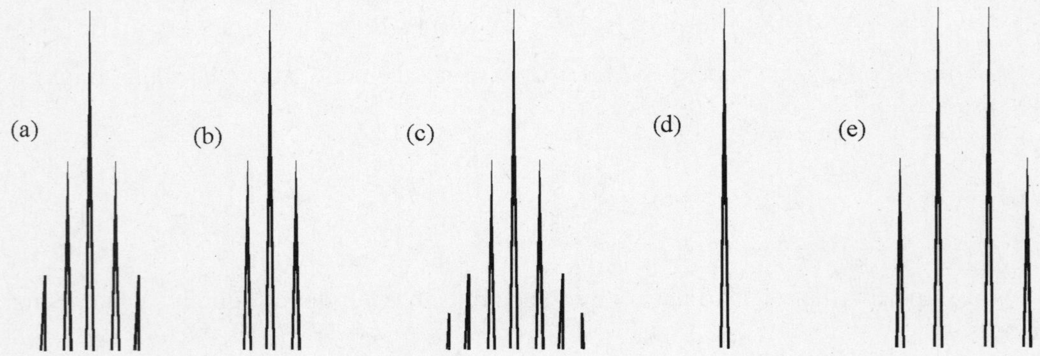

21. What is the actual signal for a hydrogen with 3 neighbors J = 12.0 Hz) and also 2 neighbors (coupling constant 8.0 Hz)?

22. Give the chemical structure for each of the following based on the spectra provided.
(a) Infrared stretching frequencies. 3350–2400 (bd, strong), 1675 (s), 1338–1220 (bd, strong), 996, 921, 683 cm^{-1}

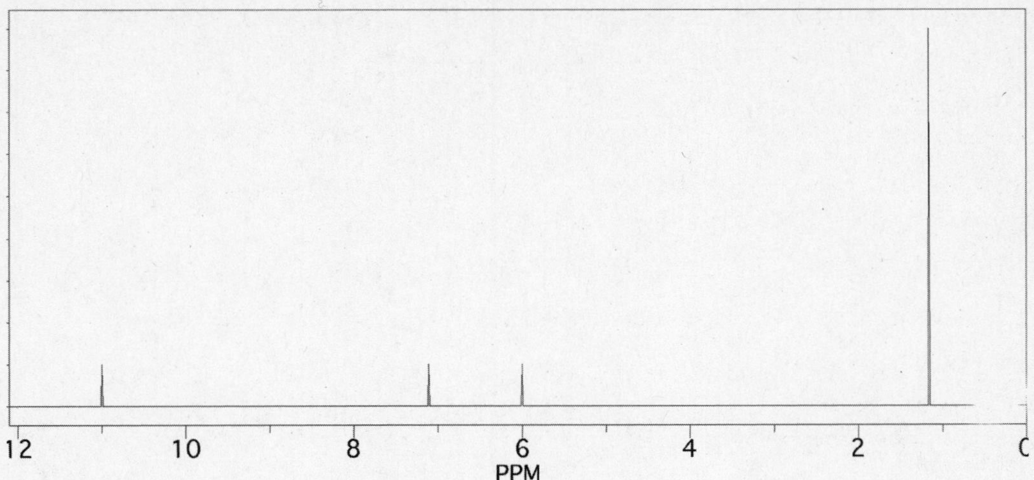

(b) Infrared stretching frequencies. 2933, 2770, 1613, 813 cm^{-1}

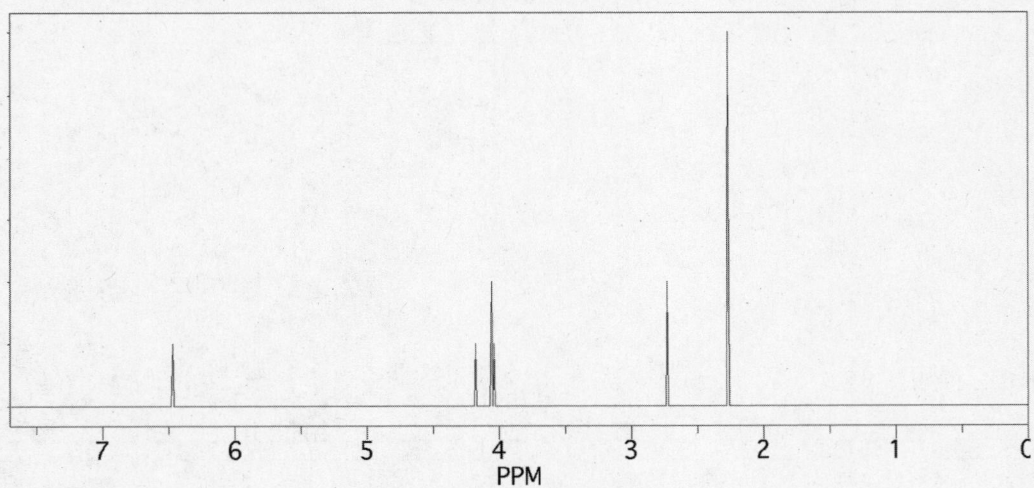

(c) Infrared stretching frequencies. 2950, 1721 (s), 1105, 735 695 cm^{-1}

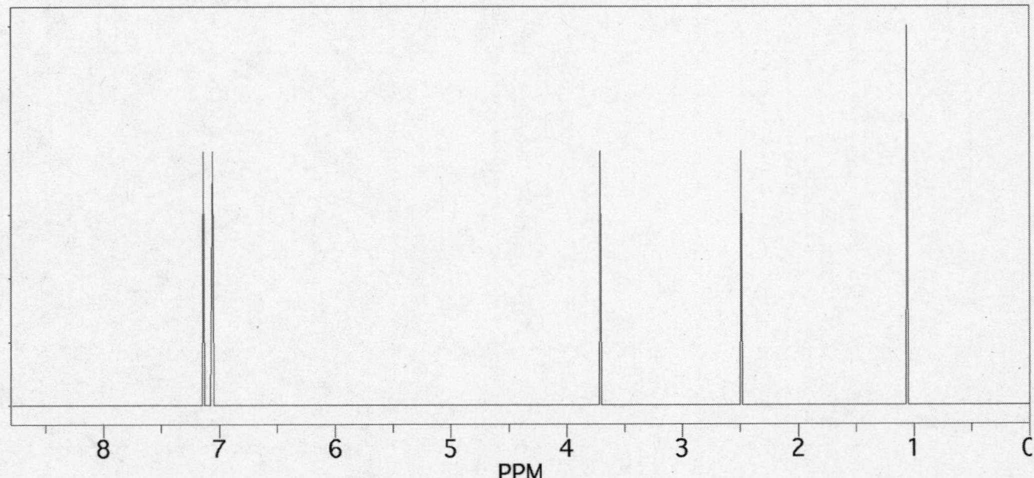

(d) Infrared stretching frequencies. 2941, 2833 (w), 1718 (s) 1366, 1220, 1079 cm^{-1}

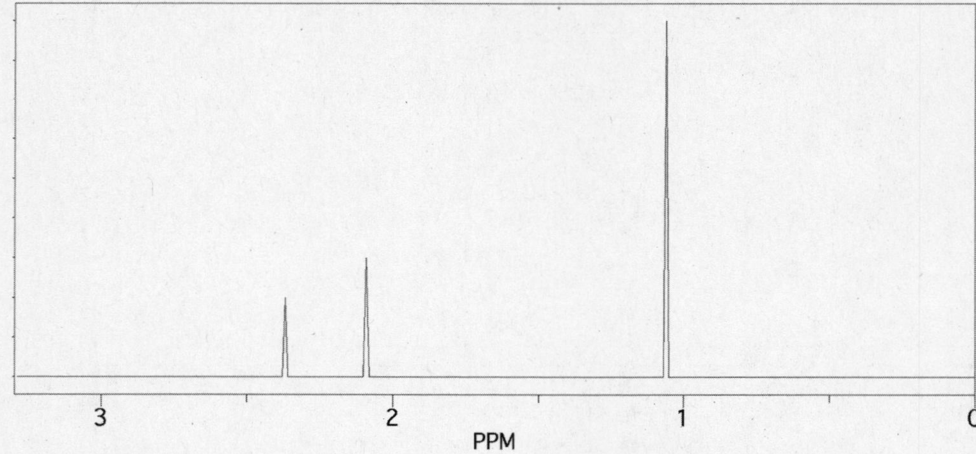

(e) Infrared stretching frequencies. 3448 (s), 2941, 1160, 1078, 833 cm^{-1}

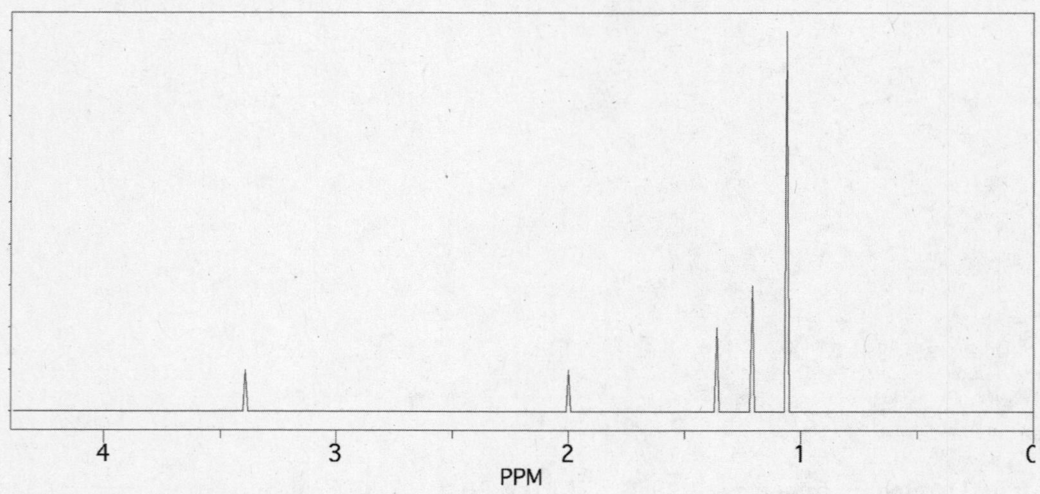

(f) Infrared stretching frequencies. 2262, 1493, 1412, 1071, 1015, 833, 794 cm^{-1}

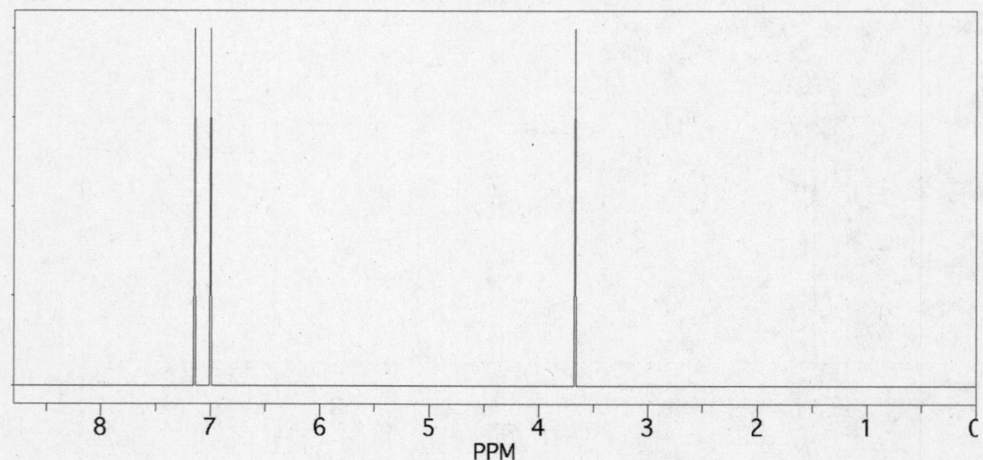

23. Given the following information, determine a reasonable structure for this molecule.

P (92) (100)   M+1 (93) (4.44)   M+2 (94) (33)

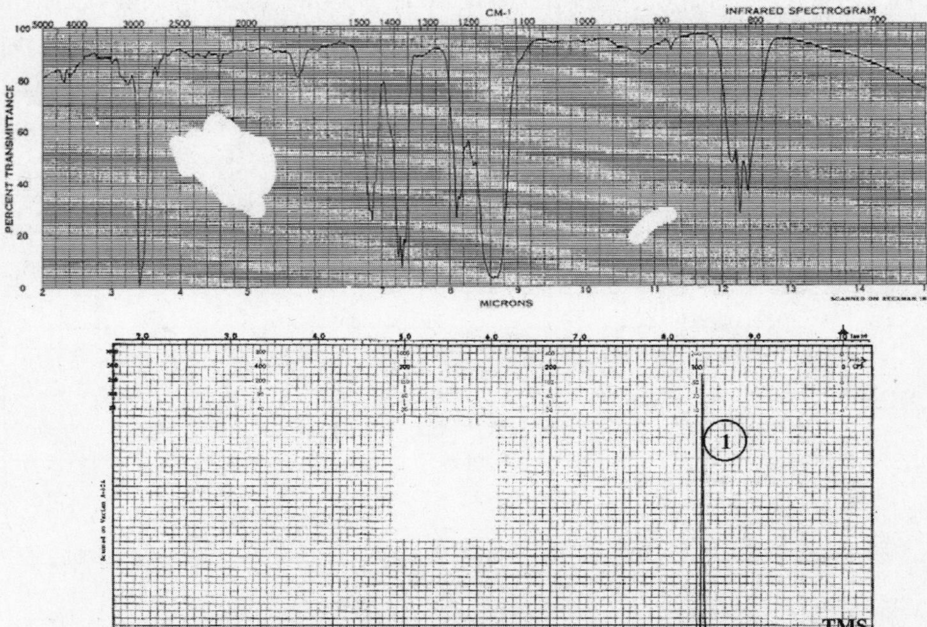

24. Given the following information, determine a reasonable structure for this molecule.

P (88) (100)   M+1 (88) (5.55)   M+2 (88) (0.35)

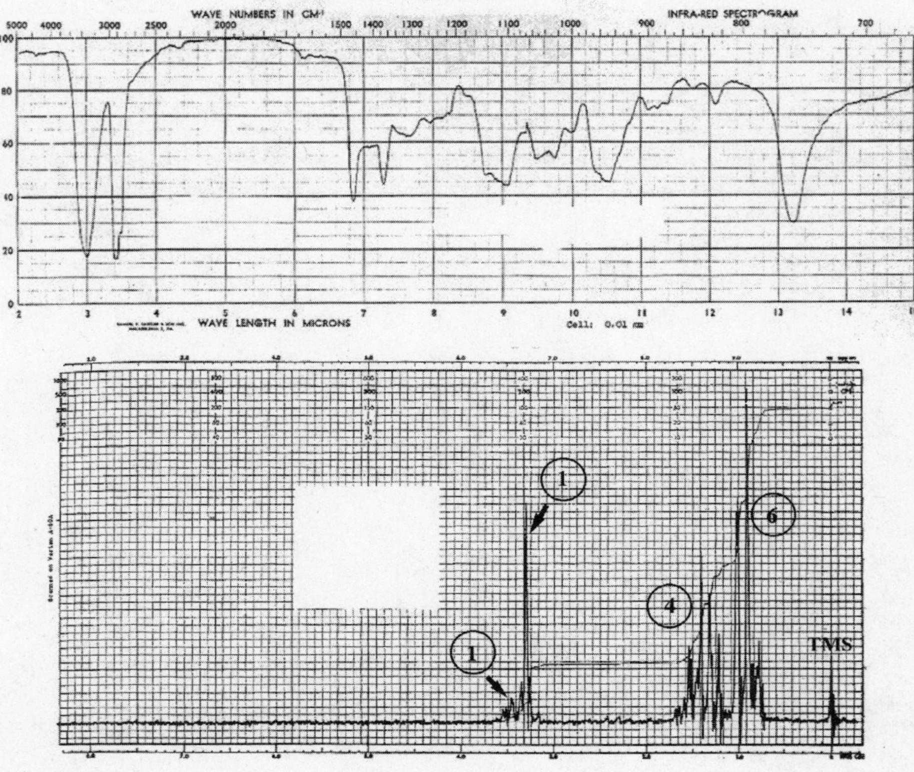

25. Given the following information, determine a reasonable structure for this molecule.

P (162) (100) M+1 (163) (12.21) M+2 (164) (0.95)

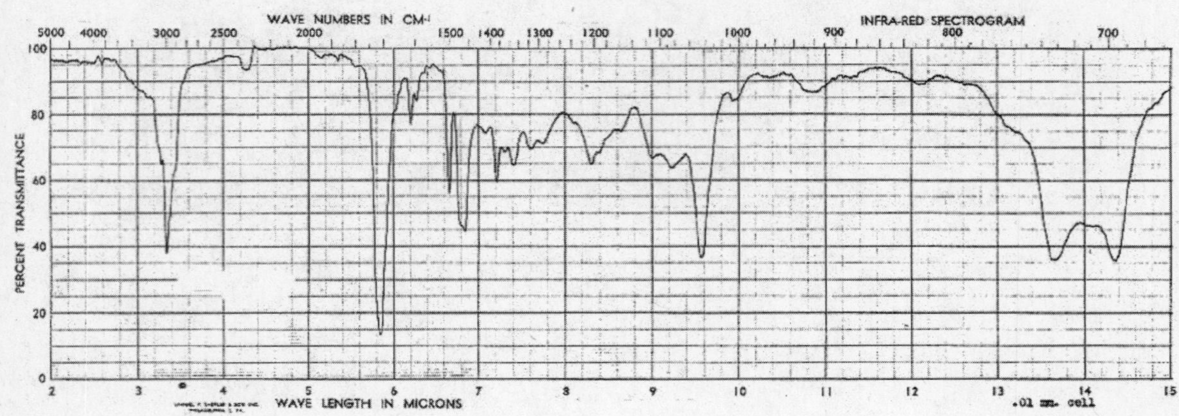

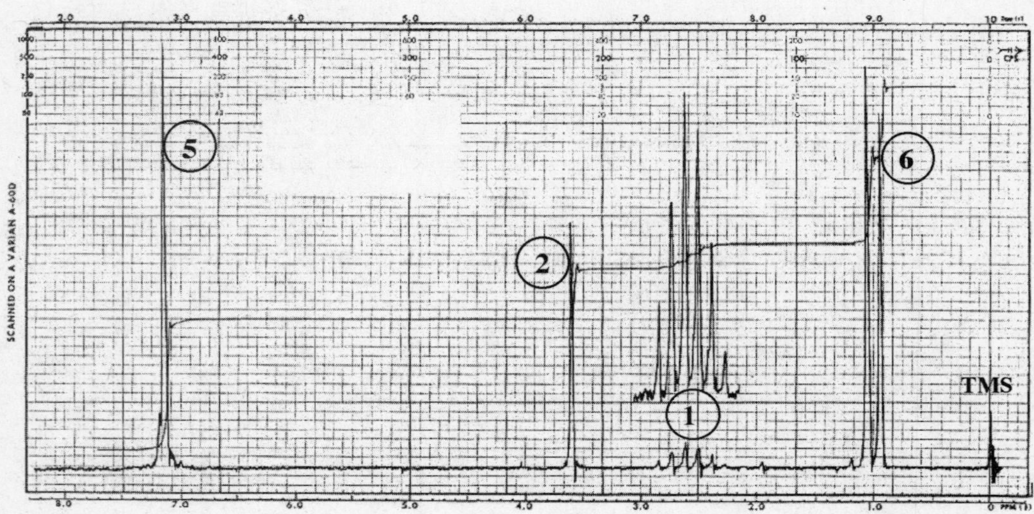

CHAPTER 19

Conjugation and Reactions of Conjugated Compounds

In a Diene or Triene Where Two π-Bonds Are Separated by at Least One σ-Bond, Does One C=C Unit Influence the Chemical Reactivity of Another C=C Unit?

When a π-bond is directly attached to another π-bond, one influences the reactivity and properties of the other. One π-bond can be part of an alkene, and the other can be part of an alkene (forming a diene), a carbonyl (giving α, β-unsaturated ketones, aldehydes and acid derivatives), or part of other functional groups such as imines, nitro compounds or nitriles. Note that a 1,2-diene, C=C=C, is known as an allene (Section 8.1).

What Is the Term Used to Define Molecules That Have a Continuous Array of sp^2 Atoms, with No Intervening sp^3 Atoms?

When there are no intervening sp^3 atoms, but rather a continuous array of sp^2 atoms such as C=C–C=C or C=C–C=C–C=C or C=C–C=O, this special arrangement of π-bonds is called **conjugation**. The important properties of conjugation include the interaction of these molecules with ultraviolet light (and other light such as infrared light), and changes in chemical reactivity that are found, relative to unconjugated molecules. The concept of resonance will be introduced and also the idea of competing reactions, 1,2- vs. 1,4-addition with conjugated π-bonds.

19.1. CONJUGATED MOLECULES

There are several types of conjugated systems, including dienes, trienes, diynes, enynes, alkenes and alkynes conjugated to carbonyl groups, and aromatic derivatives.

Nomenclature

When Two Double Bonds Are Present in a Diene, What Is the IUPAC Rule for Naming Such Compounds?

The prefix *di-* is used for two functional groups of the same kind (diene). *Tri-* is used for three groups (triene), and *tetra-* is used for four groups (tetraene). A molecule with two carbon-carbon double bonds is, therefore, a **diene,** and numbers are used to denote the first carbon of each double bond. The double bonds should be given the lowest possible numbers, and *both* double bonds must be part of the longest continuous chain.

What is the Structure of (A) Cyclopentadiene; (B) 1,5-Cyclooctadiene?

(a) The structure is ***19.1***; (b) The structure is ***19.2***.

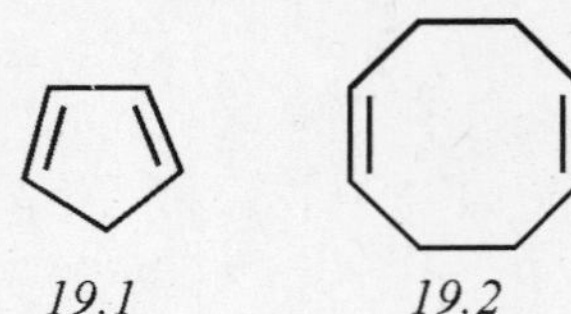

19.1 *19.2*

Give the Structure for (A) (E)-3-Bromo-4,5-Dimethyl-1,6-Octadiene; (B) 1,4-Diethyl-1,3-Cyclohexadiene; (C) 5,5-Dimethylcyclopentadiene; (D) (2e,4e)-Heptadiene; and (E) (2e, 4z)-Heptadiene.

Compound (a) is ***19.3***; compound (b) is ***19.4***; compound (c) is ***19.5***; compound (d) is ***19.6***; and compound (e) is ***19.7***.

CH_3 Br H_3C CH_3 H_3C CH_3

19.3 *19.4* *19.5* *19.6* *19.7*

How Are Enynes Named?

The triple bond takes priority over the double bond. Therefore, the suffix is *-yne*, and a number is assigned to the first carbon of the C=C unit with the first carbon of the triple bond receiving the lower number. Both the double bond and the triple bond must be part of the longest continuous chain.

How Are Diynes Named?

The prefix *di-* is used for two functional groups of the same kind (diyne). *Tri-* is used for three groups (triyne), and *tetra-* is used for four groups (tetrayne). A molecule with two carbon-carbon triple bonds is, therefore, a **diyne,** and numbers are used to denote the first carbon of each triple bond. The triple bonds should be given the lowest possible numbers, and *both* triple bonds must be part of the longest continuous chain.

Draw the Structure of (A) Hex-5-En-1-Yne; (B) HEX-(3z)-EN-1-YNE; (C) Non-(4e)-EN-1,8-Diyne; (D) 1,3,5-Heptatriyne.

Compound (a) is ***19.8***; compound (b) is ***19.9***; compound (c) is ***19.10***; and compound (d) is ***19.11***.

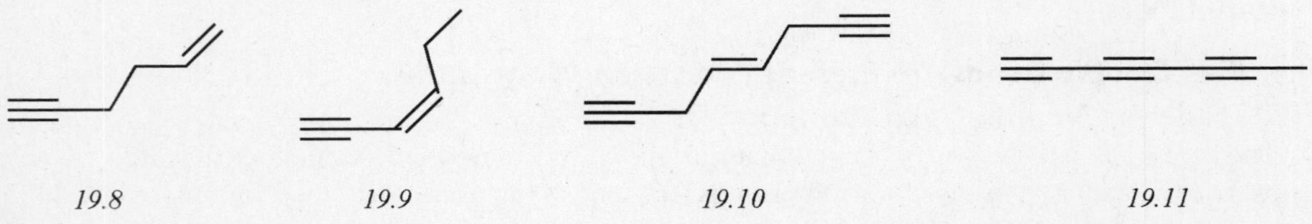

19.8 *19.9* *19.10* *19.11*

How Are Compounds Containing a C=C Unit and a Carbonyl Named?

The carbonyl group of an aldehyde, ketone, or carboxylic acid derivative has a higher priority than a carbon-carbon double or triple bond. Therefore, the suffix is derived from the carbonyl unit and given the smallest number, and the double and triple bonds are numbered accordingly. The longest chain should contain the C=O, the C=C and the C≡C units.

Draw the Structure of (A) Hex-(3z)-ENAL; (B) HEX-(3e)-EN-2-ONE; (C) Cyclohex-2-EN-1-ONE; (D) HEPT-4-YNE-3-ONE.

Compound (a) is ***19.12***; compound (b) is ***19.13***; compound (c) is ***19.14***; and compound (d) is ***19.15***.

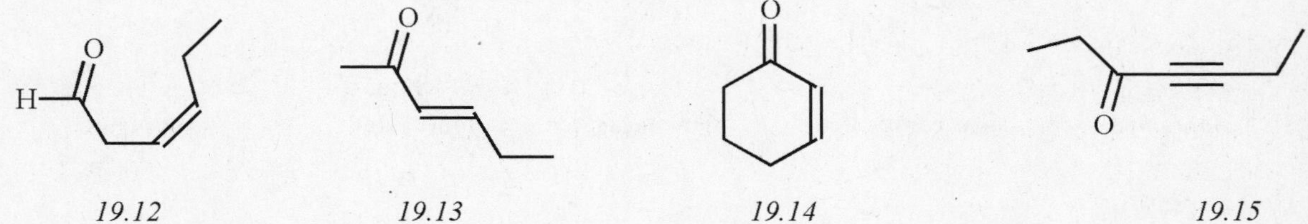

19.12 *19.13* *19.14* *19.15*

Conjugation

Is 1,5-Hexadiene Considered to Have a Conjugated Double Bond?

No. The double bonds are separated by two $-CH_2$ groups, and they are not conjugated.

What Is the Structure of 1,3-Butadiene?

The structure of 1,3-butadiene is CH_2=CH–CH=CH_2.

Is 1,3-Butadiene a Conjugated Diene?

Yes! The two C=C units are connected, with no intervening sp^3 atoms.

What Is the Structure of 1,5-Hexadiene?

The structure of 1,3-butadiene is CH_2=CH–CH_2CH_2CH=CH_2.

Is 1,5-Hexadiene Conjugated?

No! There are two sp^3 hybridized carbon atoms (the two methylene units) between the C=C units.

Is 1,3-Butadiene Resonance Stabilized?

1,3-Butadiene is *not* resonance stabilized. Each double bond has the π-bond localized between the two carbons.

Is the C2-C3 Bond Length of 1,3-Butadiene Shorter or Longer Than the C2-C3 Bond of 1-Butene? Explain.

There is some overlap of the p orbitals on the C_2 and C_3 carbons of 1,3-butadiene. For this reason, the bond length is somewhat shorter than the normal C–C single bond found between C_2 and C_3 in 1-butene. This overlap does not constitute resonance, however.

Draw each of the following and Indicate Whether or Not They Are Conjugated. (A) HEXA-(3E)-EN-2-YNE; (B) 1,4-Cycloheptadiene; (C) 1,5-Octadiyne; (D) 3,5-Octadiyne; (E) Hepta-(2E,4Z)-Diene.

Enyne (a) is ***19.16*** and is conjugated. Diene (b) is ***19.17*** and is not conjugated. Diyne (c) is ***19.18*** and is not conjugated. Diyne (d) is ***19.19*** and is conjugated. Diene (e) is ***19.20*** and is conjugated.

(a)  (b) 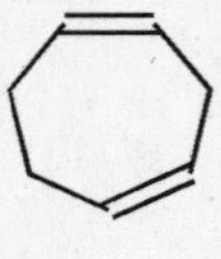 (c) 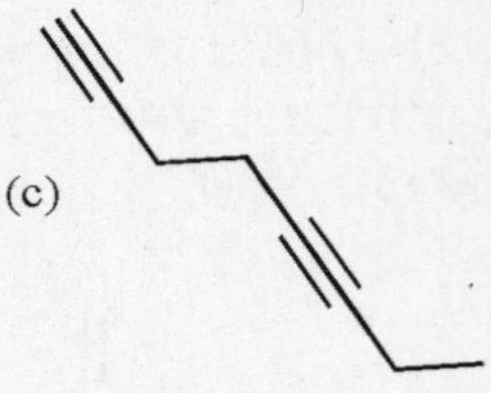 (d) 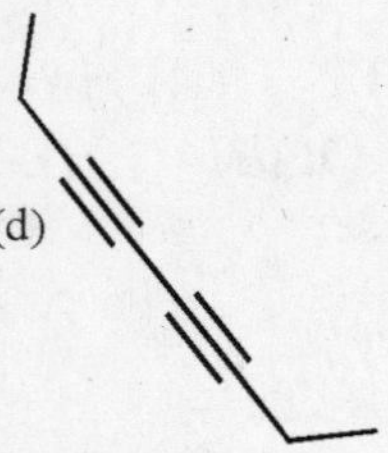 (e)

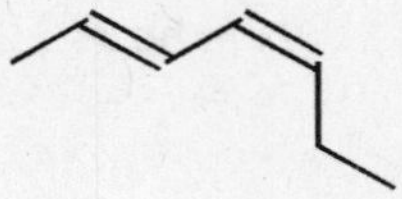

conjugated **non-conjugated** **non-conjugated** **conjugated** **conjugated**

19.16 *19.17* *19.18* *19.19* *19.20*

Draw Hex-5-En-2-One and Hex-(3z)-En-2-One and Indicate Whether or Not They Are Conjugated.

Hex-5-en-2-one is ***19.21***, and the C=C and C=O units have sp^3 atoms intervening so it is not conjugated. Hex-3Z-en-2-one is ***19.22***, and the C=C and C=O units are directly connected so it is conjugated.

O O

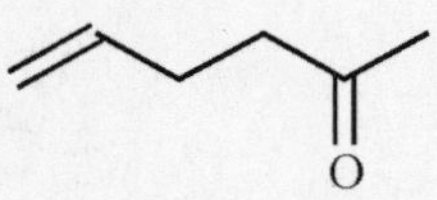

19.21 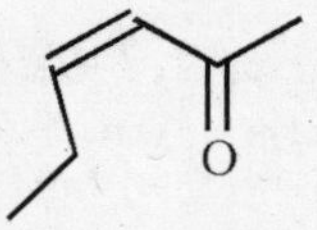 *19.22*

Is 19.23 Conjugated or Non-Conjugated? Name It.

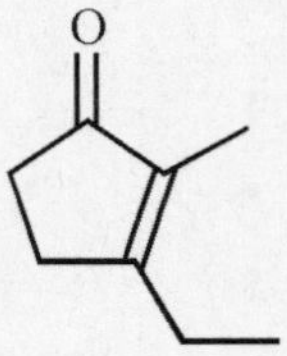

O

19.23

Compound ***19.23*** has the C=C and C=O units directly connected so it is conjugated. The name is 3-ethyl-2-methylcyclopenten-1-one.

Draw 1-Phenylethene, Otherwise Known As Styrene. Is It Conjugated?

The structure of styrene is ***19.24***, and since the C=C unit of the alkene and the C=C units of the benzene ring are directly attached, it is conjugated.

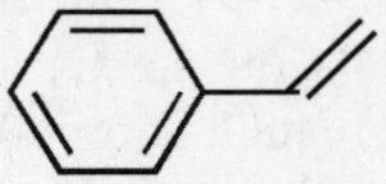

19.24

Draw Biphenyl. Is It Conjugated?

Biphenyl is ***19.25*** with two benzene rings directly attached. Since the C=C units are directly attached, this is a conjugated molecule.

19.25

Draw Benzaldehyde, Benzophenone, Acetophenone, and 1,5-diphenyl-3-pentanone. Indicate Which of These Compounds Are Conjugated.

Benzaldehyde is ***19.26***; benzophenone is ***19.27***; acetophenone is ***19.28***; and 1,5-diphenyl-3-pentanone is ***19.29***. Compound ***19.29*** is not conjugated, but ***19.26–19.27*** are conjugated.

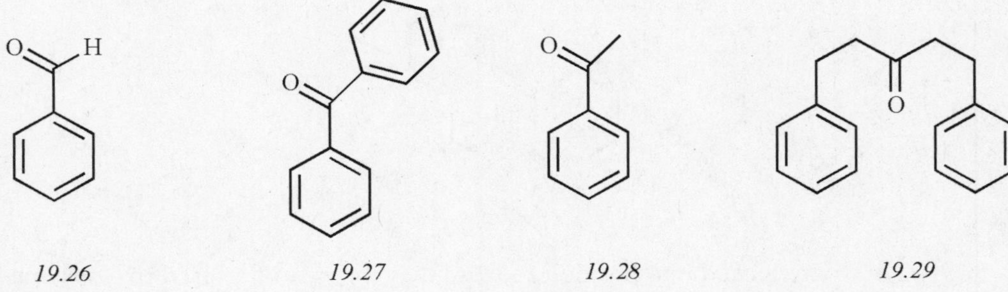

19.26 *19.27* *19.28* *19.29*

Conformation

A C–C Unit Connects the C=C Units of 1,3-Butadiene. Is Rotation Possible about This Bond?

Yes! There is normal rotation about this single bond.

Will There Be Syn and Anti Rotamers for 1,,3-Butadiene?

Yes! Rotation about the single bond will produce rotamers as with any other single bond.

Draw the Syn and Anti Rotamers for 1,3-Butadiene.

The syn rotamer is ***19.30***, and the anti rotamer is ***19.31***.

19.30 *19.31*

Draw the S-Cis and S-Trans Conformations of 1,3-Butadiene.

The *s-cis* (the old term *s-sym*, also referred to as *cisoid*) conformation has the two C=C groups on the same side of the molecule (as in ***19.30***) and is another name for the syn-rotamer. The *s-trans*, also referred to as *transoid*, conformation has the C=C groups on opposite sides (as in ***19.31***) and is another name for the anti rotamer. These are rotamers generated by rotation around the C2–C3 bond of the diene.

Which Is More Stable, S-Cis 1,3-Butadiene or S-Trans 1,3-Butadiene?

The s-trans form (***19.31***) is more stable because the hydrogens on the C=CH_2 groups sterically interact in ***19.30***. This steric interaction is missing in ***19.31***.

What Is the More Stable Rotamer of (2z,4z)-Hexadiene?

The *s-trans* rotamer (***19.33***) is more stable than the *s-cis* rotamer (***19.32***). In ***19.32***, the two methyl groups on the C=C groups come close together (steric hindrance), destabilizing that rotamer relative to the *s-trans* rotamer.

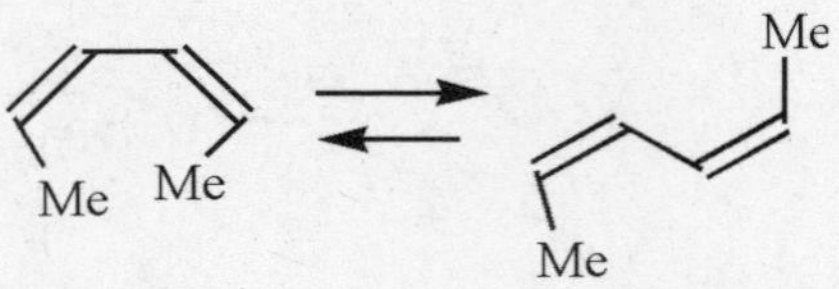

Me Me Me Me

19.32 *19.33*

Compound 19.34 Is Drawn in an S-Trans Conformation. Can It Rotate to Achieve the S-Cis Conformation?

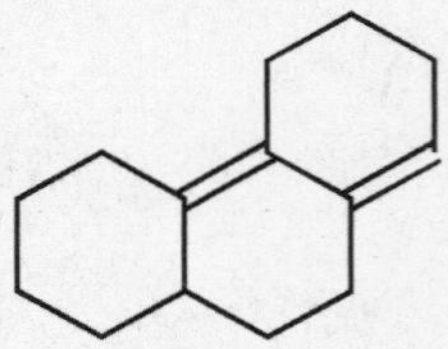

19.34

No! The C=C units are locked into position because they are part of a fused ring system. There is no change of rotation around the C–C bond that connects the C=C units.

19.3. ADDITION REACTIONS

Just as C=C and C≡C units of alkenes and alkynes undergo addition reactions, those units also react when they are a part of a conjugated system. However, the reaction of one C=C unit in the presence of a conjugated π-bond leads to interesting differences in reactivity and product distribution.

What Is the Intermediate When 1-Butene Reacts with HCl?

As discussed in Section 13.1, the π-bond of 1-butene (***19.35***) reacts as a base in the presence of the acidic HCl to form the more stable secondary carbocation ***19.36***, which then reacts with the nucleophilic chloride ion to give 2-chlorobutane, ***19.37***.

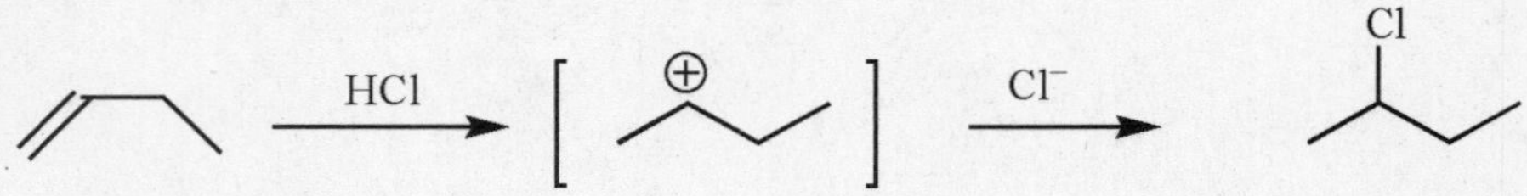

HCl ⊕ Cl^- Cl

19.35 *19.36* *19.37*

What Is the Intermediate When HCl Reacts with 1,3-Butadiene?

When 1,3-butadiene reacts with HCl, only one of the C=C units reacts to form a secondary carbocation. However, the unreacted C=C unit is adjacent to the positive charge, so the intermediate in this reaction is an allylic cation, ***19.38***. When chloride ion reacts with this cation, chlorine is delivered to *both* cationic carbons, generating two products (***19.39*** and ***19.40***).

H—Cl

19.38

Cl^-

Cl

Cl

19.39

19.40

Elucidate the Difference in Reactivity with HCl Between 1-Butene and 1,3-Butadiene.

A simple alkene reacts with HCl to give the more stable secondary carbocation (***19.36***), whereas 1,3-butadiene reacts to give an allylic carbocation, ***19.38***. A simple carbocation has one site of reactivity with chloride ion, whereas the allylic carbocation has two sites of reactivity.

Which Is more Stable, 19.36 or 19.38?

The resonance stabilized allylic cation ***19.38*** is more stable than ***19.36***, which is not capable of resonance delocalization of the positive charge.

Which Is more Stable, Alkene 19.39 or Alkene 19.40?

Since ***19.39*** is a disubstituted alkene, and ***19.40*** is a monosubstituted alkene, ***19.39*** is more stable.

Describe the Electron Distribution When a π-Bond Is Adjacent to a P Orbital.

When a π-bond is adjacent to a p orbital, as in an allylic cation, the three p orbitals, two from the π-bond and the other one (see ***19.41***) will share electron density. The electrons are, therefore, distributed over all three orbitals, and the positive charge will be dispersed to the two termini (shown by δ⊕). This is represented as two structures and is called **resonance**. The species represented by ***19.41*** is said to be **resonance stabilized**.

δ+ δ+ ≡ $H_2C{=}CH{-}\overset{\oplus}{C}H_2 \longleftrightarrow H_2\overset{\oplus}{C}{-}CH{=}CH_2$

19.41

Explain Why the Resonance Shown for $CH_2{=}CH{-}CH_2\oplus$ Does Not Occur in $CH_2{=}CHCH_2CH_2CH_2\oplus$.

In $CH_2{=}CHCH_2CH_2CH_2^{\oplus}$ the p orbitals of the π-bond are too far away from the p orbital containing the positive charge to share electron density. Therefore, it is impossible to disperse the charge, and it is localized on a single carbon.

When 1,3-Butadiene Reacts with HCl at −78°C, What Is the Major Product? Explain.

The major product is the '1,2-addition' product, ***19.40***. At low temperature, the electrophilic carbon of ***19.38*** with the secondary cationic position is more stable than the one with the positive charge on the primary carbon. Reaction with Cl *o* generates ***19.40*** as the major product. This is referred to as the **kinetic product**.

When 1,3-Butadiene Reacts with HCl at 30°C, What Is the Major Product? Explain.

At higher temperatures, both canonical forms of ***19.38*** are present, but the 'primary' cation has a disubstituted double bond, whereas the 'secondary' cation has a monosubstituted double bond. Under thermodynamic equilibration conditions (higher temperature), the more stable alkene product is favored, ***19.39***. This is referred to as the **thermodynamic product**.

What Are the Products Formed When HCl Reacts with (S-Cis) 1,3-Hexadiene?

There are four products: *E*- and *Z*-2-chloro-3-hexene (***19.42*** and ***19.43***, respectively) and *E*- and *Z*-4-chloro-2-hexene (***19.44*** and ***19.45***, respectively).

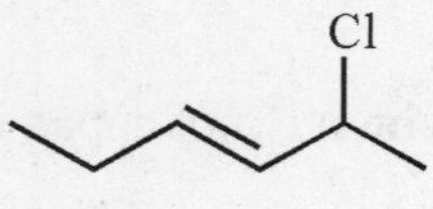

Cl Cl Cl Cl

19.42 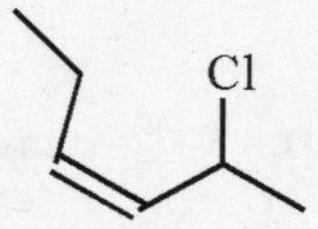 *19.43* 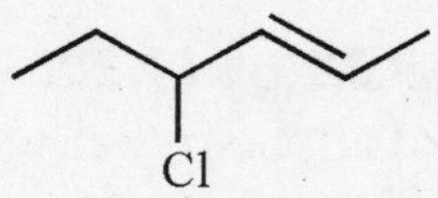 *19.44* 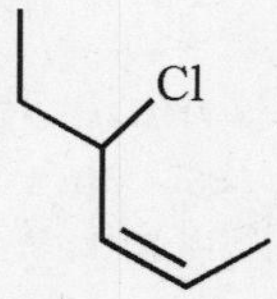 *19.45*

Explain Why a Mixture of Four Products Is Obtained.

When allylic cation ***19.46*** is generated by the reaction of 1,3-hexadiene and HCl, the charge is delocalized over all three atoms. Delocalization means that the C=C unit of each resonance contributor is not locked into place, but when the charge is localized in each contributor, the C=C unit can be formed as both the E and Z isomer. When reaction occurs with chloride ion, the π-bond in the product will be localized, but reaction can occur to give either the E or the Z. In other words, when the π-bond forms at the time of reaction of chloride ion with the cation, both isomers are possible. Since each localized π-bond exists as a mixture of E- and Z-isomers, four products are formed.

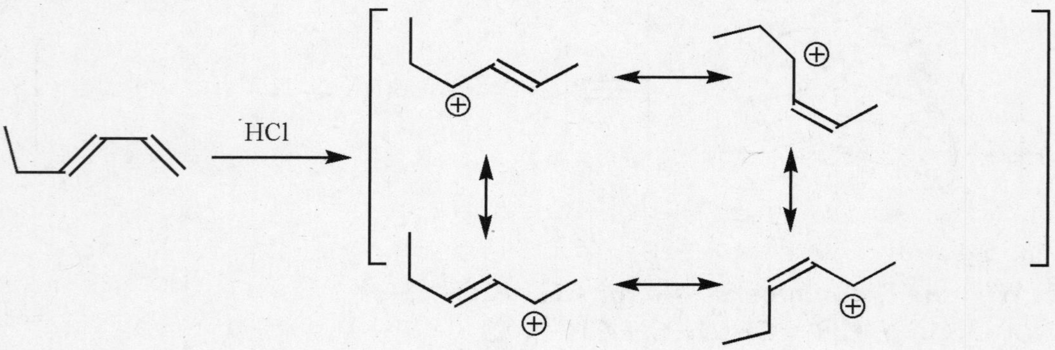

HCl

19.46

Hydration

Does 1,3-Butadiene React with Water?

No! Water is not a sufficiently strong acid to react with the π-bond of an allene or an alkyne.

What Are the Products When 1,3-Butadiene Reacts with a Catalytic Amount of Acid in Aqueous Media?

The initial product is the allylic cation ***19.38***. Subsequent reaction with water leads to two oxonium ions (***19.47*** and ***19.48***) via reaction with the two electrophilic carbon atoms of ***19.38***. Loss of a proton leads to the final products, 3-hexen-2-ol (***19.49***) and 3-hexen-1-ol as a mixture of E and Z isomers (***19.50***). Note the use of the 'squiggly' line to indicate the mixture of E and Z isomers.

H+ ; H_2O ; *19.38* ; $\oplus$ OH_2 ; – H^+ ; OH ; *19.47* ; *19.48* ; *19.49* ; *19.50*

What Are the Products When (2E,4E)-Hexadiene Reacts with a Catalytic Amount of Acid in Aqueous Media?

The products are a mixture of *E*- and (*Z*)-4-hexen-3-ol (***19.52***) and (*E*)- and (*Z*)-3-hexen-2-ol (***19.53***), both derived from the allylic cation, ***19.51***.

H+ ; *19.51* ; + H_2O ; - H^+ ; OH ; *19.52* ; *19.53*

Addition of Halogens

What Is the Mechanism and the Final Product of the Reaction between 1-Butene and Diatomic Bromine?

1-Butene (***19.54***) reacts with bromine to form a bromonium ion (***19.55***), and subsequent nucleophilic attack by the bromide counterion at the less substituted carbon leads to 1,2-dibromobutane, ***19.56***.

Br_2 ; Br^- ; *19.54* ; *19.55* ; *19.56*

What Is the Major Product Given by the Reaction of Bromine and 1,3-Butadiene at -40°C? Explain.

The major product is the '1,2-addition' product, 3,4-dibromo-1-butene, ***19.57***. The '1,4-addition' product (***19.58***) arises by Br^- attacking the C=C moiety and opening the bromonium ion (see below), but this is a higher energy process that is slow at this low temperature.

Br—Br

19.57 *19.58*

Does This Reaction Involve a Carbocation?

No! The initial reaction generates a bromonium ion, which is then opened by the nucleophilic bromide counterion.

What Is the Product When Bromine and 1,3-Butadiene React at +30°C?

At higher temperatures, the thermodynamic product (the '1,4-addition product', ***19.58***) predominates, but there is a mixture of ***19.58*** and ***19.57***.

If a Bromonium Ion Is the Intermediate in Reactions of Bromine and a Conjugated Diene, Explain How 1,2- and 1,4-Addition Can Occur.

A bromonium ion is formed by reaction of one C=C unit. There are two possible modes of attack when bromide reacts. The first involves simple S_N2-like attack at the carbon of the bromonium ion represented in ***19.59***, which leads to ***19.57***. The presence of the C=C unit leads to another mode of attack. As shown in ***19.60***, Br^- can attack the π-bond, transferring electron density towards Br^+, opening the three-membered ring. This is labeled an S_N2' reaction (nucleophilic substitution with allylic rearrangement) and generates ***19.58***.

S_N2 gives *19.57* — *19.59*

S_N2' - gives *19.58* — *19.60*

What Is Vinylogy?

Vinylogy is the extension of reactivity by adjacent π-bonds. The S_N2' reaction shown for ***19.60*** is an example. The susceptibility of the bromonium ion carbon atoms to nucleophilic attack is extended by the p-bond so that attack at the C=C unit allows cleavage of the C–Br bond by the π-electrons. The net result is substitution at the C=C unit and opening of the bromonium ion.

What Is a Working Definition of a S_N2' Reaction?

An S_N2' reaction is nucleophilic substitution at a π-bond connected to a carbon bearing a leaving group, with substitution occurring at the C=C unit and displacement of the leaving group. The formal definition is second order nucleophilic substitution with allylic rearrangement.

Explain Why Lower Reaction Temperatures Favor 1,2-Addition.

As mentioned above, attacking the C=C (***19.60***) requires more energy, since more bond making-bond breaking processes occur. When the reaction is kept cold, the higher energy process is slower than direct opening of the bromonium ion, and ***19.59*** is favored, leading to ***19.57*** as the major product.

Give the Major Product for the Reactions of 19.61–19.64.

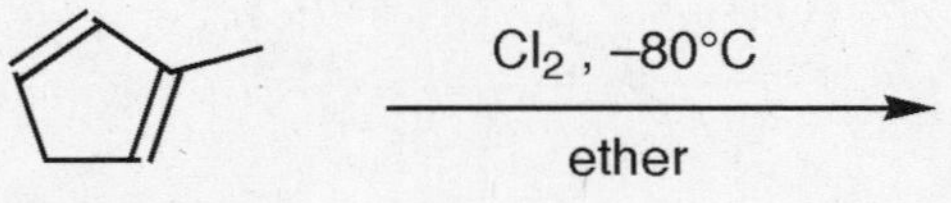

Cl_2 , –80°C
ether

19.61

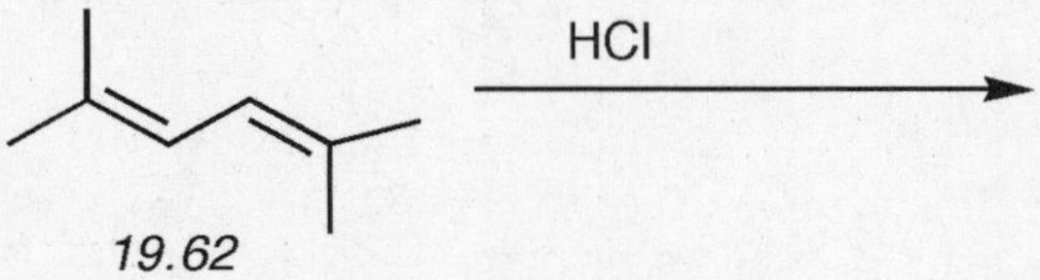

HCl

19.62

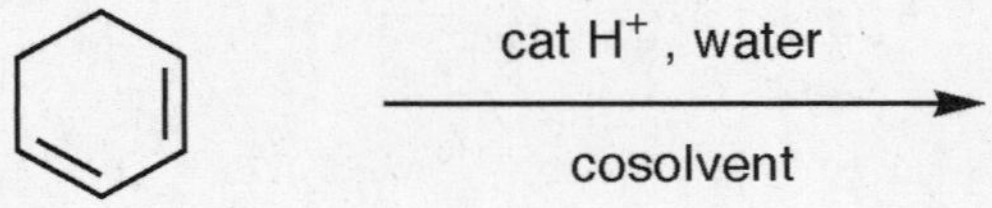

cat H^+ , water
cosolvent

19.63

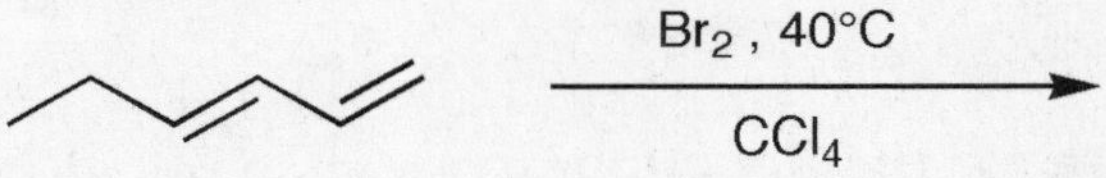

Br_2 , 40°C
CCl_4

19.64

The reaction of chlorine with diene ***19.61*** at low temperature leads to the 1,2-product ***19.65***. When diene ***19.62*** reacts with HCl, the resulting allylic cation leads to a mixture of ***19.66*** plus E/Z ***19.67***. The reaction of cyclohexadiene (***19.63***) and a catalytic amount of acid, in the presence of water, leads to allylic alcohol ***19.68***. Finally, 1,3-hexadiene (***19.64***) reacts with bromine at elevated temperatures to give the S_N2' product as the major product, *E/Z* ***19.69***.

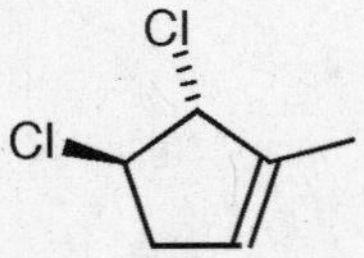

Cl Cl *19.65* | Cl *19.66* | Cl *19.67* | OH *19.68* | Br Br *19.69*

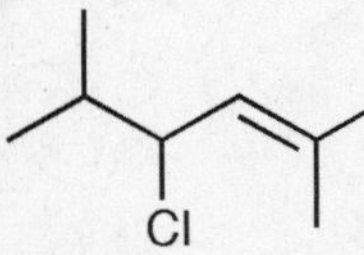

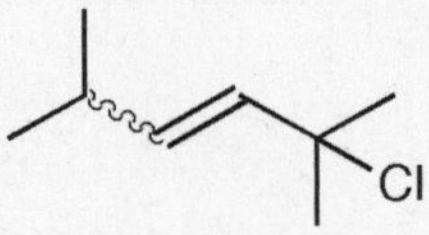

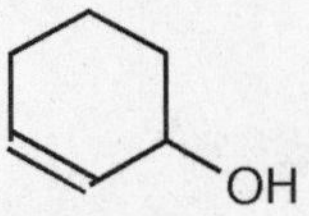

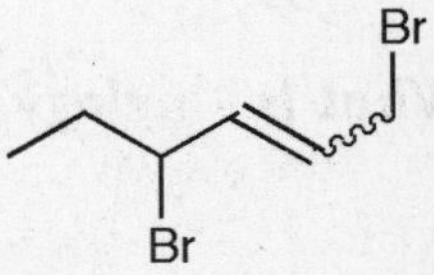

19.4. ALLYLIC CATIONS FROM ALLYLIC ALCOHOLS

The reaction of conjugated dienes with an acid generates an allylic carbocation. It is possible to generate allylic carbocations by other routes, however, and typical methodology will be introduced in this section.

What Are the Intermediates for the Reaction of 2-Propen-1-ol with H⁺?

When 2-propen-1-ol ($CH_2=CHCH_2OH$, *19.70*) reacts with an acid, an oxonium salt is formed ($CH_2=CHCH_2OH_2^+$, ***19.71***), which loses water to form the allyl cation, ***19.72***.

OH — H+ → [H–O⊕–H] — $-H_2O$ → [⊕ ↔ ⊕]

19.70 *19.71* *19.72*

What Is the Product Formed When 2-Propen-1-ol Reacts with HCl?

Allylic cation ***19.72*** reacts with the chloride counterion to form 3-chloro-1-propene (allyl chloride, ***19.73***).

OH — HCl → [⊕ ↔ Cl⊖ ⊕] → Cl

19.70 *19.72* *19.73*

What Is the Common Name for 2-Propene-1-ol?

Alcohol ***19.70*** is commonly known as allyl alcohol.

What Is the Intermediate for the Reaction of 3E-Penten-2-ol with HCl? Give the Major Product(s).

This alcohol (***19.74***) is an *allylic* alcohol, and after initial formation of oxonium ion ***19.75***, will lose water to form allylic cation ***19.76***. Note that the delocalization of the positive charge in the resonance stabilized allylic cation leads to loss of stereochemical integrity of the E-alkene unit. Since ***19.76*** is symmetrical, subsequent reaction with chloride ion leads to ***19.77*** as a mixture of E/Z isomers.

OH — HCl → [Cl⊖ H–O⊕–H] — $-H_2O$ → [⊕ ↔ Cl⊖ ⊕]

19.74 *19.75* *19.76*

+ Cl^- → Cl

19.77

What Is the Intermediate and Final Product(s) When 2-Phenyl-1-Ethanol (19.78) Is Treated with Aqueous Acid?

The initially formed cation (via loss of water from the oxonium ion, ***19.79***) is a primary cation, ***19.80***. A rearrangement (1,2-hydrogen shift) will generate the more stable, resonance stabilized cation, ***19.81***. This is known as a benzylic cation (the positive charge is on a carbon directly attached to the benzene ring), and the charge is delocalized into the benzene ring as shown. After reaction with water and loss of a proton, the final product is benzyl alcohol, ***19.82***.

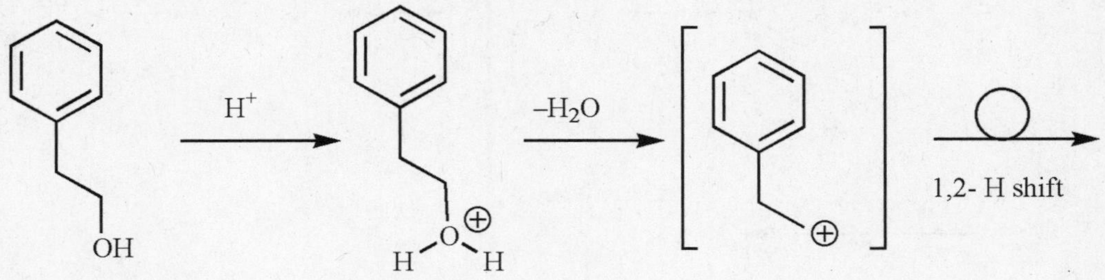

H^+ | $-H_2O$ | 1,2- H shift

OH | H O H |

19.78 | *19.79* | *19.80*

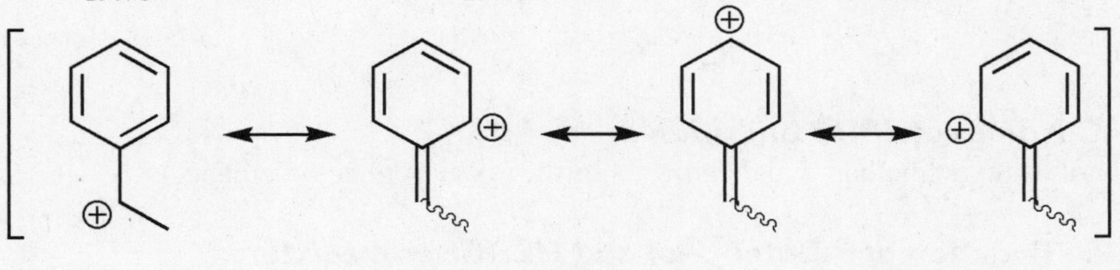

19.81

$+ H_2O$ | $- H^+$

H_2O | HO

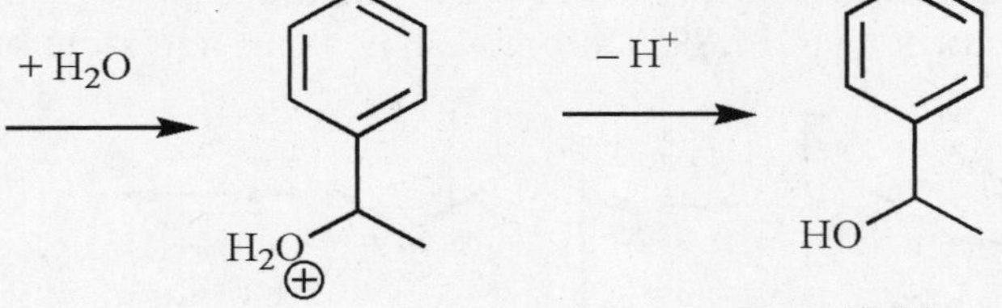

19.82

When Benzyl Alcohol (19.83) Is Treated with HCl, the Product Is Benzyl Chloride, 19.84. Explain Why No Products Are Isolated with a Chlorine Attached to the Benzene Ring.

HCl

OH | Cl

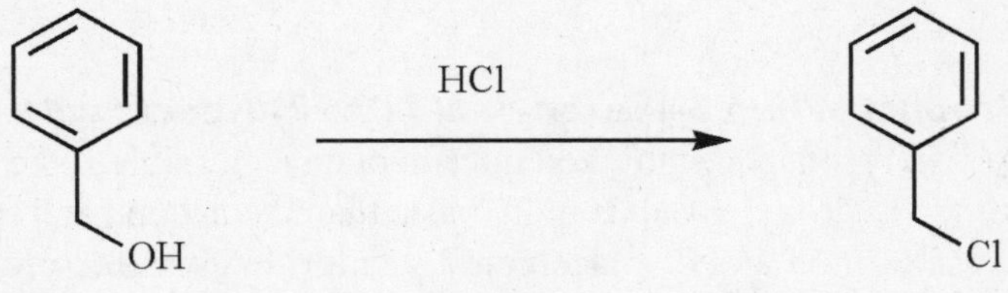

19.83 | *19.84*

The resonance stabilized intermediate for this reaction (***19.85***), formed after protonation of the alcohol and loss of water from the resulting oxonium ion, has the charge delocalized on the benzene ring. If chloride ion were to react with one of the resonance contributors such that chloride was attached to the benzene ring, the final product would not have an aromatic benzene ring (see Chapter 17). When chloride ion attaches to the benzylic carbon, the aromatic benzene ring is intact, making that product the lowest energy product that is possible of all that can be drawn from ***19.85***. Therefore, attachment at the benzylic position to form ***19.84*** is the preferred site of attack.

H–Cl

- H_2O

OH

19.83

19.85

+ Cl^-

Cl

19.84

Give the IUPAC Names of 19.83 and 19.84.

Benzyl alcohol is phenylmethanol and benzyl chloride is chlorophenylmethane.

Explain Why the Reaction of 3-Buten-1-ol and HBr Gives a Mixture of 3-Bromo-2-Butene and E and Z-1-Bromo-2-Butene.

The reaction of 3-buten-1-ol with HBr leads to oxonium ion ***19.86***, and loss of water gives allylic cation ***19.87***. The resonance delocalization leads to a mixture of *E/Z* isomers, and subsequent reaction with bromide ion gives ***19.88*** (3-bromo-1-butene) and ***19.89*** (1-bromo-2-butene) as a mixture of *E/Z* isomers.

H–Br

OH

$Br^{\ominus}$

H ⊕ H

-H_2O

$Br^{\ominus}$

19.86

19.87

+ Br^-

Br

+

Br

19.88

19.89

What Is the Major Product When 5-Hexen-1-ol (19.90) Reacts with HCl?

The product is ***19.91***. No resonance stabilized intermediate is possible if the alkene reacts with HCl. In this case, the oxygen of the hydroxyl group is more basic than the π-bond and reacts preferentially with HCl. Formation of the oxonium ion and displacement by chloride gives the product in an S_N2 reaction (see Section 7.2).

OH

H–Cl

- H_2O

Cl

19.90

19.91

What Is the Product When 2-Phenyl-1-Ethanol (19.92) Reacts with HBr?

Initial reaction of the alcohol and HBr gives an oxonium ion, and loss of water would lead to ***19.93***. A 1,2-hydrogen shift leads to the more stable benzylic cation ***19.94***, which reacts with bromide ion to give the final product, 1-bromo-1-phenylethane (***19.95***).

19.5. THE DIELS-ALDER REACTION

There is a class of chemical reactions that involve reactions of π-bonds that are known as **pericyclic reactions**. One of the most common, and most useful, of these reactions is the so-called [4+2]-cycloaddition reaction of a conjugated diene with an alkene. The product is a cyclohexene derivative, and the reaction is named after two German chemists who studied the reaction, Otto Diels and Kurt Alder. It is called the Diels-Alder reaction.

What Is the Definition of a Pericyclic Reaction?

A pericyclic reaction is one in which electrons are transferred within a π-system to form new bonds. This type of reaction generally involves transfer of double bonds from one position to another within a molecule as well as formation of sp^3 hybridized carbon-carbon bonds.

What Is a Cycloaddition?

A cycloaddition reaction is one in which two molecules react with each other, one adds to the other, to form a ring. This term is used primarily with pericyclic reactions.

What Does the Term [m+n] Refer to in a Pericyclic Reaction?

The m and n refer to the number of π-electrons that are transferred during the reaction. Typical examples are 2+2, 4+2, etc. The Diels-Alder reaction is an example of a 4+2 pericyclic reaction, where one molecule (a diene) with 4 π-electrons reacts with another molecule having 2 π-electrons (an alkene).

Frontier Molecular Orbital Theory

What Is a HOMO?

The term HOMO stands for *H*ighest *O*ccupied *M*olecular *O*rbital. It is the highest energy π-molecular orbital that contains valence electrons.

What Is the HOMO of 1,3-Butadiene?

1,3-Butadiene has four π-molecular orbitals that can react in pericyclic reactions (see ***19.96***). Note the different symmetry of the orbitals. The diagram in ***19.96*** represents four orbitals, and each orbital has four lobes, one for each sp^2 carbon. The lowest energy orbital has four lobes with the positive (shaded) portions of each lobe on the same side, and with the negative (unshaded) portions on the same side. **Collectively, this is one orbital.** The most symmetrical orbital is the lowest in energy, while the highest energy molecular orbital has the least amount of symmetry. Since 1,3-butadiene has a total of four π-electrons, the lowest energy orbital contains two electrons, and the next highest energy orbital has the next two electrons. This highest energy orbital that contains electrons is the HOMO. The HOMO is the highest energy molecular orbital that contains electrons.

Lowest Unoccupied Molecular Orbital (LUMO)

Highest Occupied Molecular Orbital (HOMO)

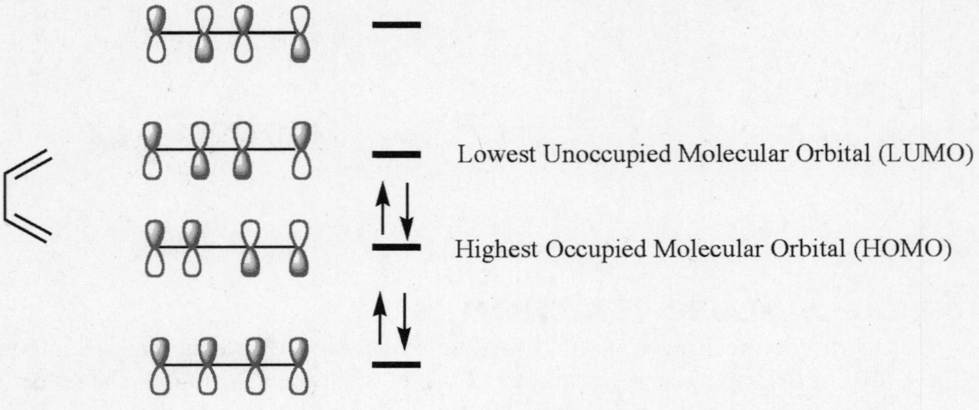

19.96

What Is the LUMO?

The LUMO is the *L*owest *U*noccupied *M*olecular *O*rbital, and the LUMO for 1,3-butadiene is marked in ***19.96***. Effectively, the LUMO is the lowest energy molecular orbital that does not contain electrons.

Does a Simple Alkene Such As Ethene Have a HOMO and a LUMO?

Yes! There are two π-electrons and two orbitals (see ***19.97***). The two electrons are in the lowest symmetrical molecular orbital, the HOMO. The highest energy orbital of the C=C unit does not contain electrons and is the LUMO.

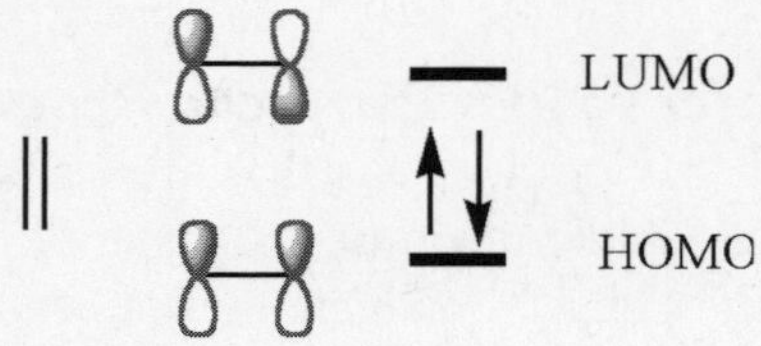

LUMO

HOMO

19.97

What Is the Homo of Ethene in Electron Volts? Of Methyl Acrylate? Of Methyl Vinyl Ether? Draw the Structure of these last two Compounds.

Alkenes have a HOMO that represents an energy level. Its relative energy is expressed in terms of electron volts (eV, where 1 eV ≈ 23 kcal/mol). The HOMO of ethene appears at −10.52 eV, that of methyl acrylate (CH_2=$CHCO_2Me$, ***19.98***) is at −10.72 eV, and that of methyl vinyl ether (CH_2=CHOMe, *19.99*) is at −9.05 eV. On this scale, −10.52 eV is lower in energy than −9.05 eV.

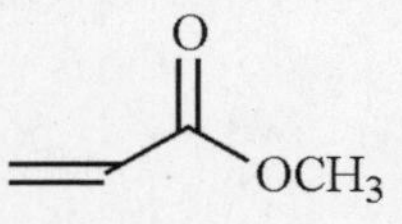

O

OCH_3

$O-CH_3$

19.98

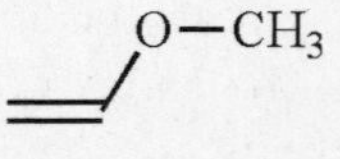

19.99

Experimentally, What Is the Energy of the HOMO?

The ionization potential for the π-electron.

What Is the LUMO of 1,3-Butadiene?

Inspection of the diagram ***19.96*** shows the LUMO is the third highest energy level for 1,3-butadiene and the highest energy level for ethene (see ***19.97***).

What Is the LUMO of Ethene? Of Ethyl Acrylate? Of Methyl Vinyl Ether?

The LUMO value for ethene is +1.5 eV, the value for methyl acrylate is 0 eV and for methyl vinyl ether it is +2.0 eV. The highest energy LUMO is +2.0 eV.

Experimentally, What Is the LUMO?

The electron affinity for putting a π-electron into the next available molecular orbital.

The Diels-Alder Reaction

A Diels-Alder reaction is the 4+2 cycloaddition of a 1,3-diene and an alkene to give a cyclohexene derivative. 1,3-Butadiene will react with ethene, for example, to give cyclohexene, but only at high reaction temperatures and at high pressure (250°C and 2500 psi are typical reaction conditions).

What Is the [m+n] Designator for a Diels-Alder Reaction?

A Diels-Alder reaction is a 4+2 cycloaddition (4 π-electrons from the diene and 2 π-electrons from the alkene).

What Reactants Are Required for a Diels-Alder Reaction?

A diene (called an enophile) and an alkene (an ene, called a dienophile).

What Reaction Conditions Are Common in Diels-Alder Reactions?

The Diels-Alder reaction is a thermal reaction, and temperatures in the range of 0°C to >300°C are common. Most common Diels-Alder reactions occur in the 60–180°C range at ambient pressure.

What Is the Product of a Diels-Alder Reaction?

The product is a cyclohexene derivative.

Using 1,3-Butadiene and Ethene, What Is the Product?

The product is cyclohexene, ***19.100***.

heat,pressure

19.100

Using 1,3-Butadiene and Ethene, Which Orbitals Can React?

Only those orbitals that have the proper symmetry can react. Specifically, the orbitals on C1 and C4 of butadiene can react with the two orbitals of ethene. In addition, the orbitals must have the correct symmetry to react. This means that only the HOMO of the diene can react with the LUMO of the alkene

(ΔE^1), or the LUMO of the diene can react with the HOMO of the alkene (ΔE^2), as in ***19.101***. In most cases, the important interaction is ΔE^1 ($HOMO_{diene}$ - $LUMO_{alkene}$). Such a reaction is said to proceed with normal electron demand.

ΔE^2

ΔE^1

19.101

How Do These Orbitals React?

The interaction of the $HOMO_{diene}$ and the $LUMO_{alkene}$ is drawn again in ***19.102***, focusing on the key orbitals and the lobes of each orbital that must react. This clearly shows the like symmetry of the C_1–C_4 orbitals of the diene HOMO and the alkene LUMO. The electron rich HOMO can be thought of as donating electrons to the electron poor LUMO, if the ΔE is sufficiently small. This initiates a reaction that transfers 6 π-electrons as shown, forming two new sigma bonds and moving the pi-bond. As drawn, this reaction will generate cyclohexene (***19.103***).

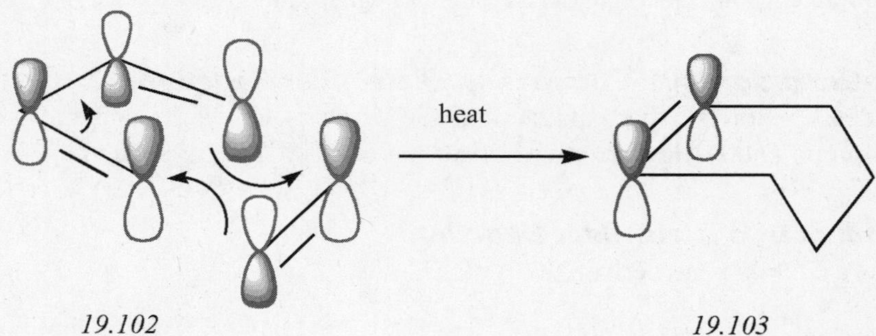

19.102 *19.103*

Why Is 19.102 Drawn As the S-CiS Rotamer?

The reaction occurs via the HOMO of the diene, but C1 and C4 react with the alkene to form new bonds in the product. The only way those atoms can react is to be close to the alkene carbons, and this can only occur via the *s-cis* rotamer.

Which Is More Reactive, 1,3-Butadiene and Ethene or 1,3-Butadiene and Ethyl Acrylate? Explain.

The HOMO of 1,3-butadiene is at -9.07 eV. This means the ΔE^1 for $HOMO_{butadiene}$-$LUMO_{ethene}$ –9.07 eV – 1.5 eV = 10.57 eV. Similarly, the ΔE^1 for $HOMO_{butadiene}$-$LUMO_{ethyl\ acrylate}$ is –9.07 eV - 0 eV = 9.07 eV. Since ΔE^1 for ethyl acrylate is lower than for ethene, ethyl acrylate is expected to react faster (or at a lower reaction temperature) with 1,3-butadiene than is ethene.

Which Is More Reactive, 1,3-Butadiene and Ethene or 1,3-Butadiene and Methyl Vinyl Ether? Explain.

Using the same rationale, if ΔE^1 for $HOMO_{butadiene}$-$LUMO_{ether}$ is 10.57 eV, ΔE^1 for $HOMO_{butadiene}$-$LUMO_{methyl\ vinyl\ ether}$ is –9.07 eV – 2.0 eV = 11.07 eV. Since the ΔE^1 for methyl vinyl ether is larger than for ethene, ethene will react faster (or at a lower temperature).

What Is a Dienophile?

A dienophile is a 'diene lover.' The species that reacts with diene in a Diels-Alder reaction is an alkene. The dienophile is that molecule that reacts with the diene, namely the alkene.

What Are Common Dienophiles?

Methyl acrylate, acrylonitrile, maleic acid and maleic anhydride, fumaric acid, acrolein, methyl vinyl ketone and cyclohexenone are typical alkenes that react with common dienes under relatively mild conditions.

What Is a S-CiS Diene? An S-TRANS Diene?

A *s-cis* diene is the rotamer of a 1,3-diene where the two C=C groups are syn (*s-cis*), as in ***19.104***. An *s-trans* diene is the rotamer of a 1,3-diene where the two C=C groups are anti (*s-trans*), as in ***19.105***. See ***19.30*** and ***19.31*** above. These two rotamers are in equilibrium, and both are present in an acyclic diene.

19.104 *19.105*

Which Is More Reactive in a Diels-Alder Reaction, a S-CiS Diene or A S-TRANS Diene?

Since it is the C1 and C4 carbons of the diene that undergo reaction, those carbons must *both* be in close proximity to the alkene orbitals. Only the *s-cis* conformation brings all the reactive carbons close enough together.

Of 19.106–19.111, Which Can Undergo a Diels-Alder Reaction with a Suitable Alkene?

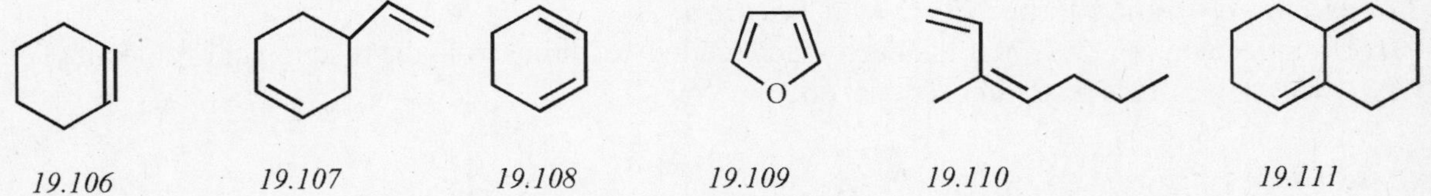

19.106 *19.107* *19.108* *19.109* *19.110* *19.111*

Compound ***19.106*** is a simple alkene, not a diene, so it cannot undergo a Diels-Alder reaction with another alkene. It *might* react with a suitable diene. The remaining compounds are all dienes, but ***19.107*** is *not* a conjugated diene and will not undergo the Diels-Alder reaction as a diene. Diene ***19.111*** is conjugated, but is 'locked' into an *s-trans* conformation. It is, therefore, impossible for that diene to undergo the Diels-Alder reaction. In all other cases, the diene reacts normally. Both cyclohexadiene (***19.108***) and furan (***19.109***) are 'locked' in the *s-cis* conformation and can react. Furan is relatively unreactive in the Diels-Alder reaction, requiring more vigorous reaction conditions, but it is prone to other competing reactions that can diminish the yield of the Diels-Alder product. Although diene ***19.110*** is drawn in the *s-trans* conformation, it is acyclic and is not locked in that rotamer. The *s-cis* conformation will be in equilibrium with ***19.110***, and the *s-cis* rotamer will react normally.

Which Is More Reactive, 1,3-Butadiene or Cyclopentadiene?

Since cyclopentadiene (***19.112***) is locked into the *s-cis* conformation, it is much more reactive than 1,3-butadiene.

What Is the Product Formed When Cyclopentadiene Reacts with Ethyl Acrylate?

When cyclopentadiene reacts with an appropriate alkene (such as ethyl acrylate) a mixture of two bicyclic products are formed, ***19.113*** and ***19.114***. In ***19.113***, the CO_2Et group is in the *exo* position, which means it is on the same side as the $–CH_2–$ bridge. This contrasts with ***19.114***, where the CO_2Et is in the *endo* position, on the opposite side of the $–CH_2–$ bridge. In general, the endo product (***19.114***) is preferred by about 3:1.

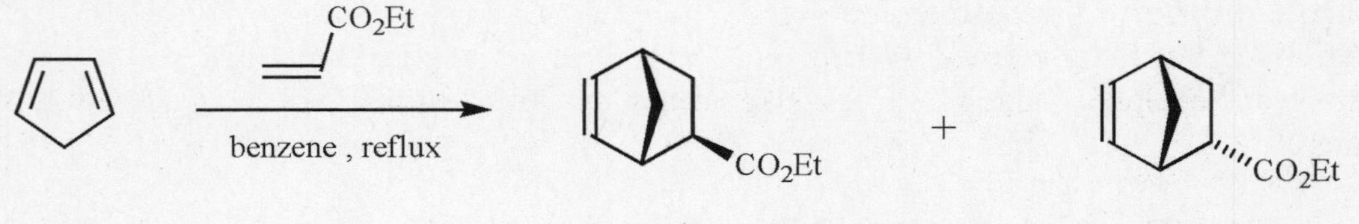

CO_2Et

benzene , reflux

CO_2Et + CO_2Et

19.112 *19.113* *19.114*

Draw 19.113 and 19.114 in a Manner That Illustrates the Endo/Exo Positions with Greater Clarity. In Other Words, Draw These Structures Such That the Six-Membered Ring is Tipped Over.

The typical way in which this strange request is answered is to draw ***19.113*** as ***19.115*** and ***19.114*** as ***19.116***. In this drawing, it is easy to see that the cyclohexene unit is in one plane, that the $–CH_2–$ bridge is on one side, and the CO_2Et unit is on the same side as the bridge or on the opposite side.

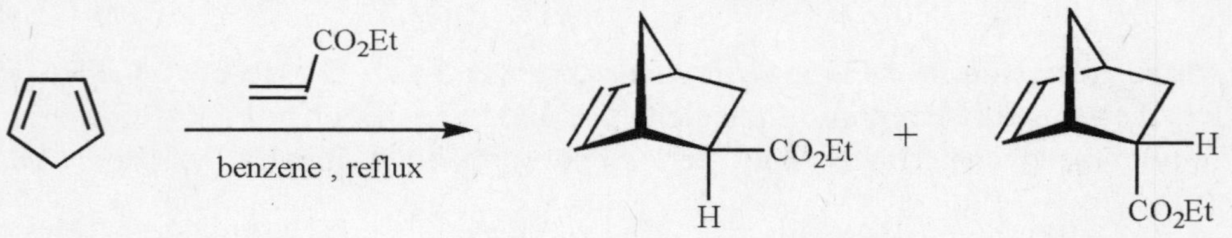

CO_2Et

benzene , reflux

CO_2Et H + H CO_2Et

19.112 *19.115* *19.116*

What Is Dicyclopentadiene? How Is It Formed?

Dicyclopentadiene (***19.117***) is the Diels-Alder adduct resulting from the reaction of cyclopentadiene (***19.112***) with itself. This reaction occurs at about 35°C.

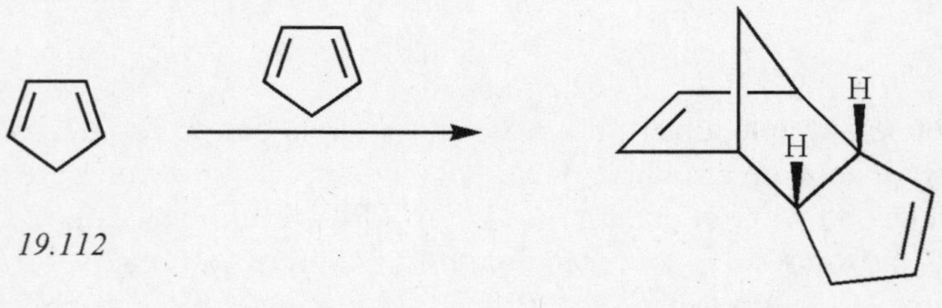

H H

19.112 *19.117*

It Is Known That a Bottle Labeled Cyclopentadiene Is Actually 19.117. How Can a Manufacturer Do This, and How Can We Obtain Cyclopentadiene?

When ***19.117*** is heated to >135°C, a reaction known as a **retro-Diels-Alder reaction** occurs to regenerate the two components, two molecules of cyclopentadiene. Effectively, ***19.117*** is heated and cyclopentadiene distills off and is trapped in a flask cooled in ice or dry ice. The retro-Diels-Alder reaction is the reverse of the Diels-Alder reaction, but occurs at much higher temperatures.

Rank the Order of the Following Alkenes As to Their Reactivity in a Diels-Alder Reaction: (a) CH_2=$CHCH_3$; (b) CH_2=$CHCO_2Et$; (c) CH_2=CHOMe.

If these alkenes react in a Diels-Alder reaction such that ΔE^1 dominates the HOMO-LUMO interaction (a normal electron demands Diels-Alder reaction), ethyl acrylate (b) will be the most reactive, since it has the lowest energy LUMO to react with the HOMO of the diene. Methyl vinyl ether (c) will be the least reactive. since it has the highest energy LUMO.

List Several Common Dienophiles.

Some of the more common dienophiles are acrylonitrile (***19.118***), acrolein (***19.119***), but-3-en-2-one or methyl vinyl ketone (***19.120***), methyl acrylate (***19.121***), diethyl fumarate (***19.122***), diethyl maleate (***19.123***), maleic anhydride (***19.124***), and N-phenylmaleimide (***19.125***).

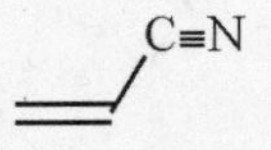

C≡N

19.118

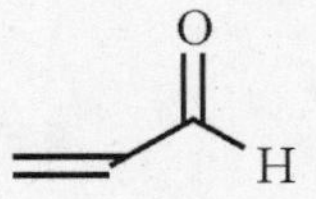

O H

19.119

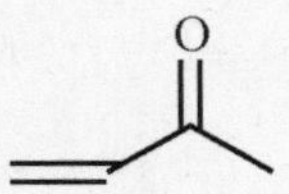

O

19.120

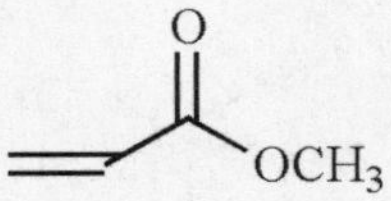

O OCH_3

19.121

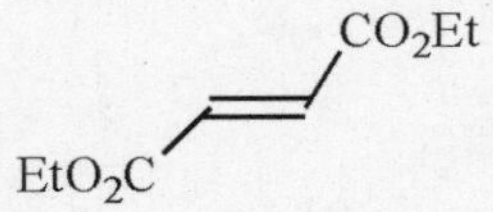

CO_2Et EtO_2C

19.122

EtO_2C CO_2Et

19.123

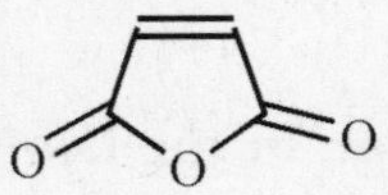

O O O

19.124

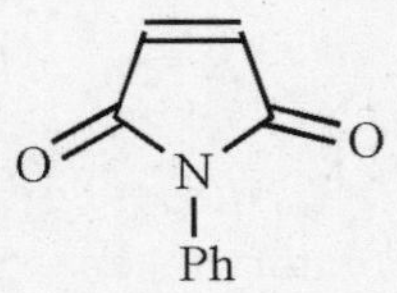

O N O Ph

19.125

Regiochemistry and Stereochemistry

What Is an 'Ortho' Product in the Diels-Alder Reaction? A 'Meta' Product? A Para Product?

An '*ortho*' Diels-Alder product has two substituents in a 1,2-position (***19.126***). Similarly, the '*meta*' adduct has a 1, 3 orientation of the substituents (***19.127***), and a '*para*' product has a 1, 4-relationship (***19.128***).

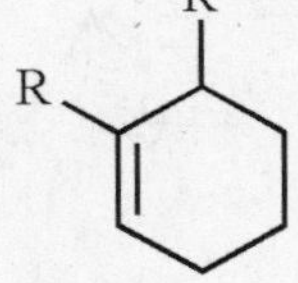

R R^1

19.126

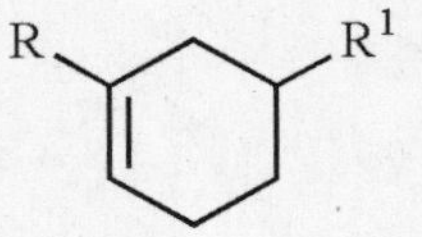

R R^1

19.127

R R^1

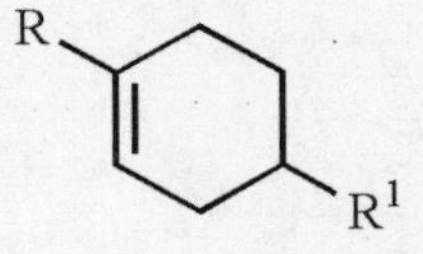

19.128

Is There a Preference for 19.126, 19.127 or 19.128 in the Diels-Alder Reaction?

Yes, but the preference depends on the nature of R and R,' and we will not discuss the factors that control regiochemistry here. Regiochemistry is predicted by knowing the magnitude of the orbital coefficients for both the alkene and the diene, and overlap of the largest magnitude coefficients will predict the major product. This topic is best suited for an advanced course.

What Is a Disrotatory Motion in the Diels-Alder Reaction?

Disrotatory refers to the motion of the groups at the termini of the diene during the Diels-Alder reaction. If we focus on the two 'outer' groups (methyls) in ***19.129***, they move towards each other, but in opposite directions (as shown). This motion is referred to as **disrotatory** (moving in opposite directions). The opposite directions refer to one methyl group moving clockwise and the other moving counterclockwise. This motion will lead to a *cis* relationship of the two methyl groups in the final cyclohexene product (***19.130***). Note that the two 'inner' groups (H atoms in ***19.129***) also move in opposite directions, but away from each other. If the outer groups move towards each other, the inner groups must move away from each other to maintain the opposite direction of rotation.

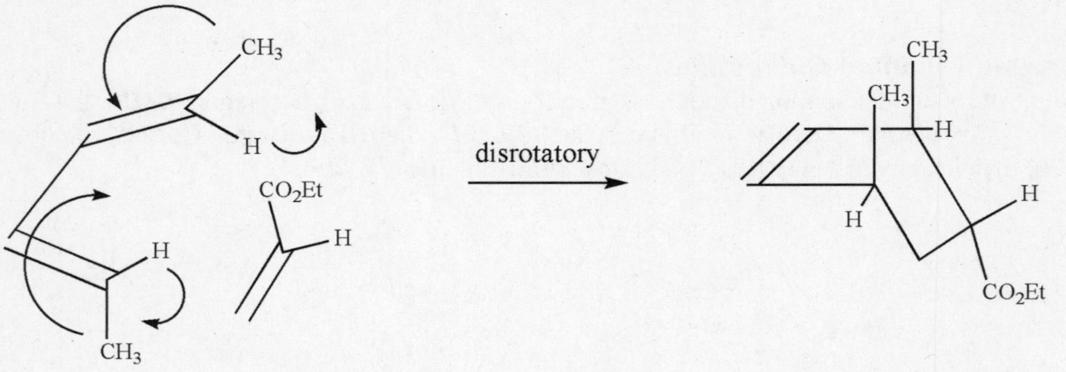

CH_3 H CO_2Et H H CH_3 disrotatory CH_3 CH_3 H H H CO_2Et

19.129 *19.130*

What Is a Conrotatory Motion in the Diels-Alder Reaction?

Conrotatory means the groups on the diene are moving in the same direction (the opposite of disrotatory). In ***19.131***, the two 'outer' groups (methyls) are moving in the same direction (counterclockwise), leading to a *trans* relationship in the cyclohexene product (***19.132***). Note that the two inner groups (H atoms) are moving in the same direction (clockwise — required if the outer groups move counterclockwise), and the two H atoms are *trans* in ***19.132***. Note also that moving the outer groups clockwise or counterclockwise is arbitrary as long as the groups move in the same direction.

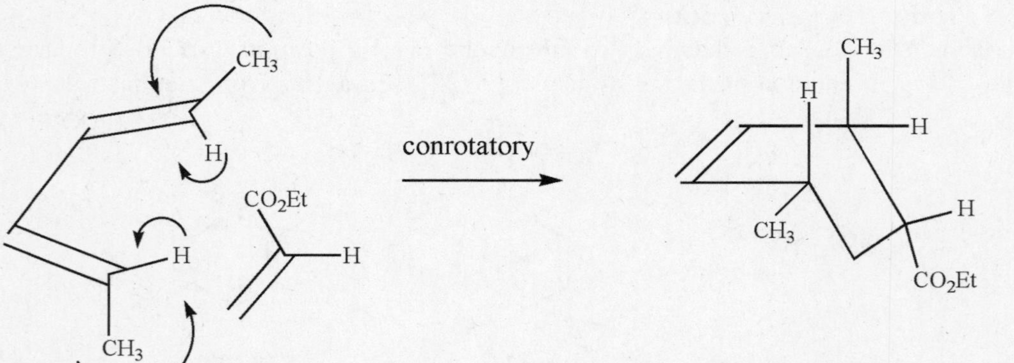

CH_3 H CO_2Et H H CH_3 conrotatory CH_3 H H CH_3 H CO_2Et

19.131 *19.132*

In Diels-Alder Reactions, Which Is Preferred, Disrotatory or Conrotatory?

The normal electron demand Diels-Alder reactions we will discuss all react with a disrotatory motion, and this fact allows us to make predictions as to stereochemistry.

What Is the Stereochemistry of the Groups in the Final Product When 1,3-Pentadiene Reacts with Maleic Anhydride?

The major product is ***19.116*** (the *endo* product). About 75% of ***19.116*** is produced, along with about 25% of the *exo* product (***19.115***). The alkene portion of ethyl acrylate reacts with cyclopentadiene, and in the transition state, the C=O of ethyl acrylate is 'tucked under' the cyclopentadiene ring. The approach of the carbonyl 'under' the diene is indicated in ***19.133***, although this is not the transition state. This approach will lead to the *endo* product (***19.116***).

O
H
OEt
H
CO_2Et
+
CO_2Et
H

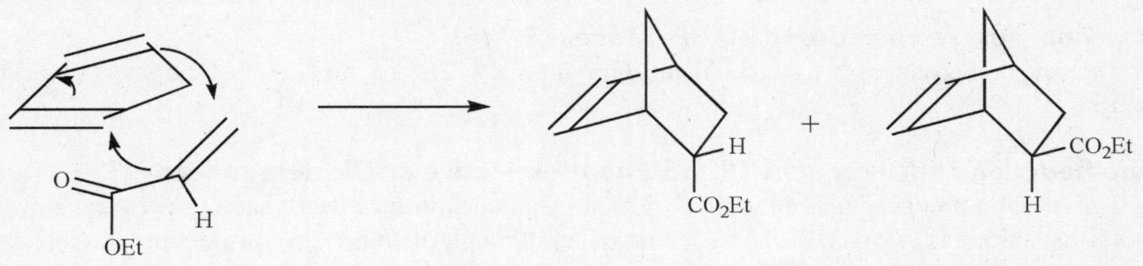

19.133 *19.116* *19.115*

Why Is the Endo Product Preferred in This Reaction?

Examination of ***19.133*** shows that the π-bond of the carbonyl can interact with the π-bonds of the diene (π-stacking) via what is known as **secondary orbital interactions**. These interactions slightly increase the stability of the *endo* transition state at the expense of the *exo* transition state, leading to the *endo* product (***19.116***) as the major product.

What Is the Stereochemistry of the Groups in the Final Product When (E,Z)-1 4-Diphenyl-1,3-Butadiene (19.134) Reacts with Ethyl Acrylate?

Reaction of (*E,Z*)-1,4-diphenyl-1,3-butadiene (***19.134***) with ethyl acrylate will give ***19.135***, assuming the *endo* transition state and disrotatory motion are the most favorable. The disrotatory motion of the groups on the terminal atoms leads to the *trans*- relationship of the two phenyl groups, and the *endo* transition state fixes the relative position of the CO_2Et group with respect to the two phenyl groups. This is indicated in ***19.136***, which leads to ***19.135*** drawn a second time as ***19.137*** to indicated the stereochemical outcome of ***19.136***. Only one enantiomer is shown, although this reaction will generate racemic product.

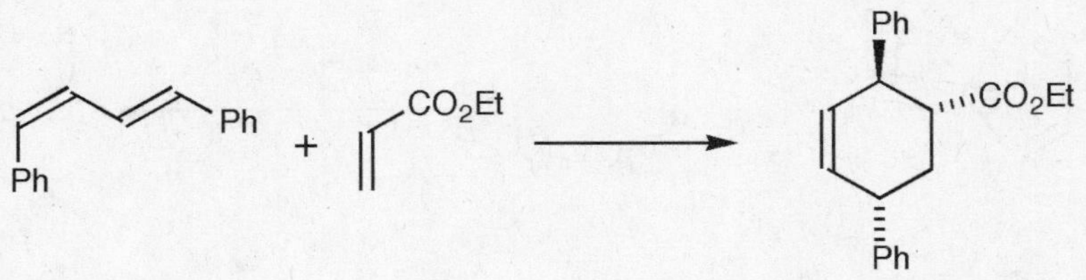

Ph
Ph
+
CO_2Et
Ph
CO_2Et
Ph

19.134 *19.135*

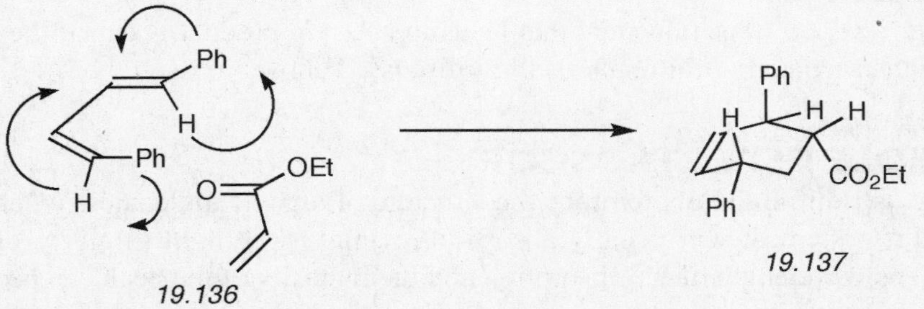

Ph
H
Ph
H
O
OEt
Ph
H
H
H
CO_2Et
Ph

19.136 *19.137*

How Many Stereogenic Centers Are There in 19.134? In 19.135?

Diene ***19.134*** has zero stereogenic centers, but product ***19.135*** has three.

What Is the Maximum Number of Stereochemical Products That Can Be Formed from the Diels-Alder Reaction of 19.134 and Ethyl Acrylate?

In ***19.135***, there are three stereogenic centers, and the 2^n rule predicts 2^3, or 8 stereochemical products.

How Many Are Formed Based on the Model 19.136?

This reaction produces two stereochemical products, ***19.135*** as a mixture of enantiomers (a racemic mixture).

Is the Reaction That Produces 19.135 Enantioselective or Diastereoselective?

The reaction produces a racemic mixture so it is not enantioselective or enantiospecific, but it produces a racemic diastereomer as the major product, although there are several possible diasteromeric products, so it is diastereoselective.

What Is the Stereochemistry of the Groups in the Final Product When Diethyl Fumarate Reacts with 1,3-Butadiene?

The reaction of diethyl fumarate (***19.122***, the diethyl ester of fumaric acid — see Section 25.1) and 1,3-butadiene gives ***19.138***. The *trans-* relationship of the two CO_2Et group of the dienophile is retained during the Diels-Alder reaction and in the final cyclohexene product.

CO_2Et + EtO_2C — benzene , reflux → CO_2Et, CO_2Et

19.122 *19.138*

What Is the Product When Cyclopentadiene Reacts with Diethyl Maleate?

The major product is ***19.139***, assuming an *endo* approach of both carbonyl groups in diethyl fumarate.

+ CO_2Et, CO_2Et — benzene , reflux → CO_2Et, CO_2Et

19.123 *19.139*

What Is the Alder Endo Rule?

The preference for an *endo* transition state that leads to an *endo* product is called the **Alder endo rule**. It is the result of the secondary orbital interactions discussed above.

19.6. ULTRAVIOLET SPECTROSCOPY

In Chapter 18, we used infrared light to probe the structure of organic molecules. When a molecule contains a conjugated π-system, it will usually absorb ultraviolet (UV) light to give a characteristic absorption spectrum. The wavelength of the absorption and the intensity of the peak can be used to gain structural information.

What Is the Relative Position (Relative Energy) of Ultraviolet Light in the Electromagnetic Spectrum?

From ***19.140*** it is apparent that ultraviolet (UV) light is *higher* in energy than visible light or infrared energy (≈ 71–143 kcal/mol for UV vs. 35–71.5 kcal/mol for visible).

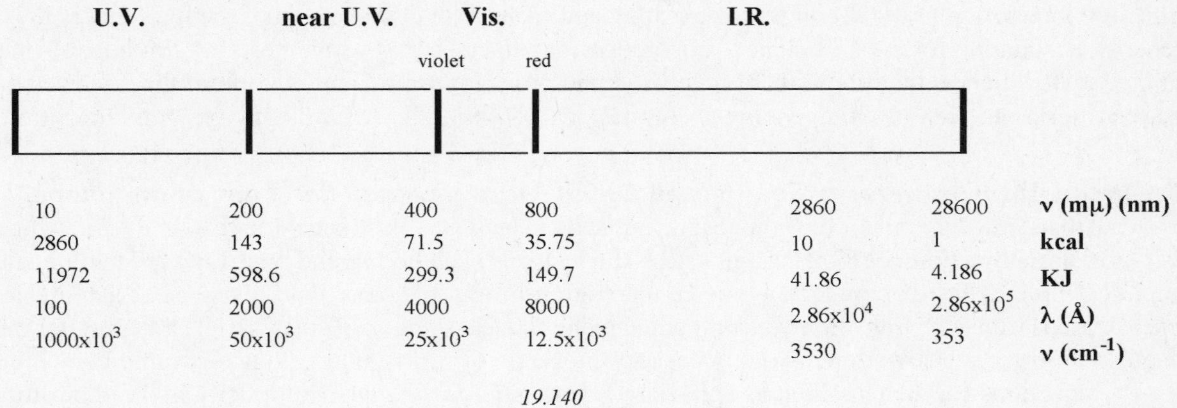

10	200	400	800	2860	28600	**ν (mμ) (nm)**
2860	143	71.5	35.75	10	1	**kcal**
11972	598.6	299.3	149.7	41.86	4.186	**KJ**
100	2000	4000	8000	2.86×10^4	2.86×10^5	**λ (Å)**
1000×10^3	50×10^3	25×10^3	12.5×10^3	3530	353	**ν (cm^{-1})**

19.140

What Functional Groups Absorb Ultraviolet Light?

The functional groups that best accept a photon of UV light are those that contain a conjugated π-bond or a polarized π-bond. Simple alkenes do not absorb UV light effectively, but conjugated dienes and conjugated carbonyls give strong absorption bands. The carbonyl group of ketones and aldehydes absorbs UV light, as does benzene.

Describe How a Photon of Ultraviolet Light Reacts with the π-Bond of an Alkene.

As shown by ***19.141***, a photon of light excites an electron in a valence, bonding molecular orbital and promotes it to an antibonding orbital, generating a high energy species. This high energy species can dissipate energy in several ways, one of which is for the electron to cascade back to the lower energy orbital, with emission of heat and/or light. In ***19.141***, the electron is promoted from one of the vibrational levels ($\nu_o \rightarrow \nu_3$) to a higher energy antibonding orbital. The energy difference between these energy levels (ΔE) is the energy quoted in the previous figure (in kcal/mol) for that absorption.

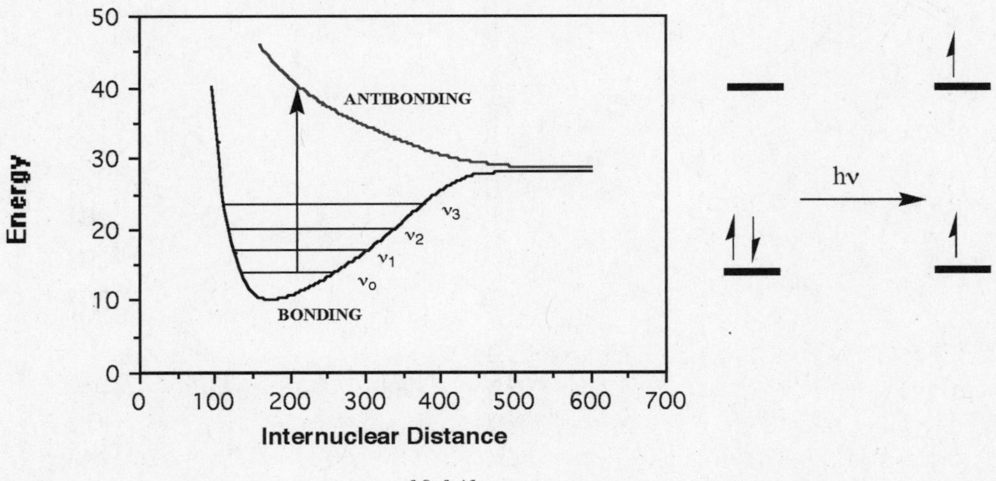

19.141

Why Is the UV Absorption Maximum for 1, 3-Butadiene Shifted Relative to That of an Alkene Such As 1-Butene?

For 1,3-butadiene (see following table), the most intense absorption in the UV appears at 217 nm (131.8 kcal/mol where 1 kcal = 28600 nm). For a simple alkene, this signal appears at about 165 nm (173.3 kcal/mol). It takes more energy for the simple alkene to absorb the light than it does the conjugated diene. The interaction of the π-bonds in the conjugated diene allows absorption of the light that is lower in energy, accounting for the difference in absorption maxima (see following question). For both alkenes and conjugated dienes, the orbital that accepts the photon is a π orbital and the orbital that receives the electron after promotion is a high energy π−orbital, a π* orbital. The ΔE is due to a π → π* transition.

Why Is the Absorption for a Conjugated Diene More Intense Than That of an Alkene?

Absorbance is the amount of light absorption and is calculated by **Beer's Law**: $A = \varepsilon \cdot l \cdot c$ where ε is the molar extinction coefficient, a physical constant for each compound, and l is the length of the sample cell in cm. The parameter c is the concentration in mol/liter. For both alkenes and conjugated dienes, there is a π → π* transition but conjugated dienes have a *smaller* ΔE than simple alkenes. For this reason, it absorbs at a lower energy. It is beyond the scope of most undergraduate organic books, but π → π* transitions can be classified as spin-*allowed* and spin-*forbidden*. Multiplicity is the sum of the spin quantum numbers of all the electrons in a species. If the multiplicity of the excited state is different from the multiplicity of the ground state, the transition is spin-forbidden. If they are the same, it is spin-allowed. In general, allowed transitions have larger extinction coefficients (more intense peaks) than forbidden processes. An overly simplistic way to give the answer to this question is, therefore, that conjugation leads to a lower energy spin-allowed transition that is more intense. This is also very dependent upon the solvent and the wavelength of light that is used.

What Are the Extinction Coefficients and Strongest Absorption Bands for Some Common Molecules?

All values from this table are taken from *Practical Handbook of Spectroscopy*, Robinson, J.W. (Ed.), CRC Press, Boca Raton, *1991*, pp. 566–630. They are arranged by increasing value of the strongest absorption band in nanometers (nm).

Molecule	Molar Extinction Coefficient	Strongest Band (nm)
1-butyne	4467	172
2-butanol	316	174
1-butene	15849	175
trans-2-butene	12589	177
2,3-dimethyl-2-butene	13813	198
2-methyl-1-butene	2754	200
cis-2-butenoic acid	11482	204
trans-2-butenoic acid	15849	205
acrolein	11221	207
cyclohexene	447	207
3-buten-2-one	6457	210
1,3-butadiene	20893	217

Molecule	Molar Extinction Coefficient	Strongest Band (nm)
2-methyl-1,3-butadiene	21379	224
1,3-pentadiene	22908	224
cyclohex-2-enone	8230	225
2,3-dimethyl-1,3-butadiene	21379	226
benzaldehyde	14125	241
2,5-dimethyl-2,4-hexadiene	13100	242
acetophenone	18600	252
2-nitrotoluene	4330	254
benzene	212	254
nitrobenzene	7180	259
1,3-cyclohexadiene	10000	259
toluene	238	261
3-nitrotoluene	7700	264
1,4-dinitrotoluene	14454	266
4-nitrotoluene	11700	274
1-phenyl-1,3-butadiene	22387	280
cyclohexanone	27	280
2-butanone	21	280
pentanal	13	283
1,4-diphenyl-1,3-butadiene	30199	313

Characteristics of an Ultraviolet Spectrum

What is λ_{max}?

The wavelength at which the maximum absorbance occurs is called $\lambda_{max.}$ It is used to help identify certain functional groups in a molecule.

What Is the Extinction Coefficient?

The extinction coefficient is the intensity of an absorption peak. A larger extinction coefficient indicates greater absorption. In other words, stronger absorption peaks have large extinction coefficients.

How Can UV Spectroscopy Distinguish between 1,3-Butadiene and 1,5-Hexadiene?

Since 1,3-butadiene contains conjugated π-bonds, it will have a λ_{max} at about 217 nm whereas 1,5-hexadiene (with only simple alkene units) will have a λ_{max} between 160–190 nm. In addition, the extinction coefficient for the conjugated diene will be much larger (a more intense signal at the same concentration) than the unconjugated diene.

What Is the λ_{max} for the Carbonyl of Acetone?

This can be answered by the table of absorption spectra presented here.

Molecule	Transition	λ_{max} (nm)	E (kcal) (KJ)	ε_{max}
C_4H_9–I	n → σ*	224	127.7 (534.6)	486
$CH_2=CH_2$	π → π*	165	173.3 (725.4)	10,000
HC≡CH	π → π*	173	165.3 (691.9)	6.000
acetone	π → π*	150	190.7 (798.3)	-
	n → σ*	188	152.1 (636.7)	900
	n → π*	279	102.5 (429.1)	15
$CH_2=CHCH=CH_2$	π → π*	217	131.8 (551.7)	21,000
$CH_2=CHCHO$	π → π*	210	136.2 (570.1)	11,500
	n → π*	315	90.8 (380.1)	14
benzene	π → π*	ca. 180	158.9 (665.2)	60,000
	π → π*	ca. 204	231.1 (853.9)	8,000
	π → π*	ca. 255	288.9 (1067.4)	215

Note that acetone has several absorption maxima and each different type of transition (π→ π*, n → σ*, or n → π*, where n is an unshared electron) will have its own λ_{max}.

What is ε_{max}?

This value is the molar absorbtivity at maximum absorption.

For Acrolein ($CH_2=CHCHO$), ε_{max} Is Correlated with Which Transition?

In this case, the π → π* transition is the strongest absorption.

What Is the λ_{max} for Methyl Vinyl Ketone (3-Buten-2-one)?

The π → π* maxima for this conjugated ketone is similar to that of the conjugated aldehyde (acrolein) shown: ≈ 210 nm. The n → π* maxima is ≈ 315 nm.

What Is the λ_{max} for Benzene?

There are three absorption signals for benzene at 180 nm, 204 nm, and 255 nm. The strongest transition is at 180 nm.

Is the UV Absorption at 279 nm for Acetone a Strong Signal or a Weak Signal?

The UV band at 279 nm is relatively weak (ε_{max} of only 15).

Test Yourself

1. Draw all resonance forms for each of the following.

(a) (b) (c) (d)

2. Which of the following contains a conjugated C=C bond?

a) (b) (c) (d)

(e) (f) (g) (h)

3. Give the IUPAC name for all molecules in Question 2. Give the structure of (3E,5E)-octadiene and (Z)-3-methyl-1,3-pentadiene.

4. Identify each of the following as having an *s-trans* conformation or an *s-cis* conformation and indicate for which molecules the *s-trans* and *s-cis* rotamers are in equilibrium.

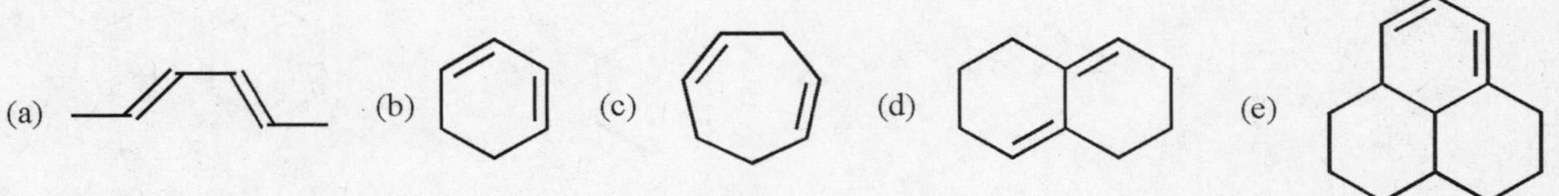

5. What are the products when HBr reacts with 1,3-pentadiene at –78°C?

6. Correlate each of the following:

83 kcal = ____ KJ = _____ cm^{-1} = _____ nm

132 KJ = _____ nm = _______ Å = _____ kcal

0.45 nm = _______ kcal = ________ cm^{-1}

2.4 kcal = _____ nm = _______ Å = _____ cm^{-1}

7. Which of the following are expected to give strong absorption bands in the UV?

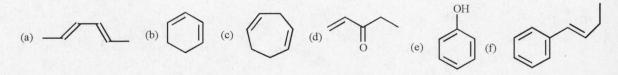

(a) (b) (c) (d) (e) (f)

8. If the extinction coefficient of 3-buten-2-one is 6457, and the extinction coefficient of 2-butanone is 17, what is the absorbance of each one if the concentration is 0.3 M in a 5 cm cell?

9. Explain why the extinction coefficient of 1,4-diphenyl-1,3-butadiene is larger than the extinction coefficient for 1,3-butadiene.

10. For each reaction, give the major product. If there is no reaction, indicate this by N.R.

(a) OH; aq. H^+, KCl →

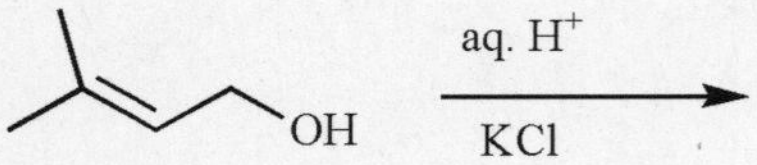

(b) Ph, Ph, OH; aq. $HClO_4$ →

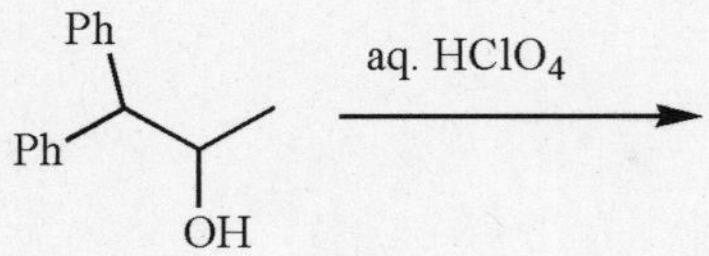

(c) Ph, Ph, OH; HBr →

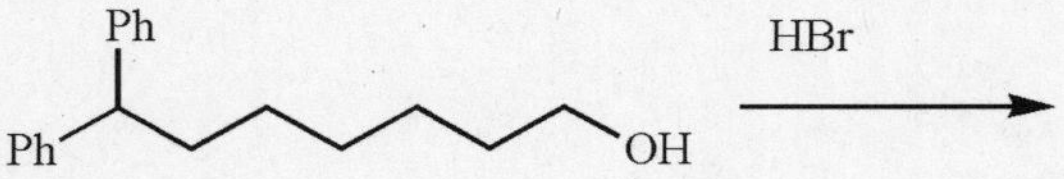

(d) HBr, +50°C →

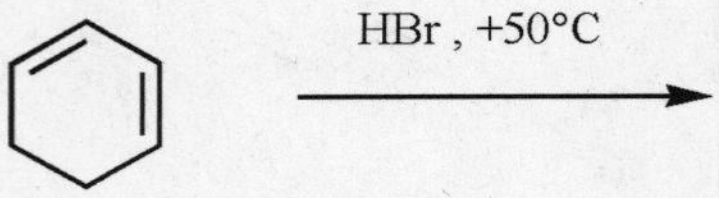

(e) HCl, -50°C →

(f) 2 Br_2, 30°C →

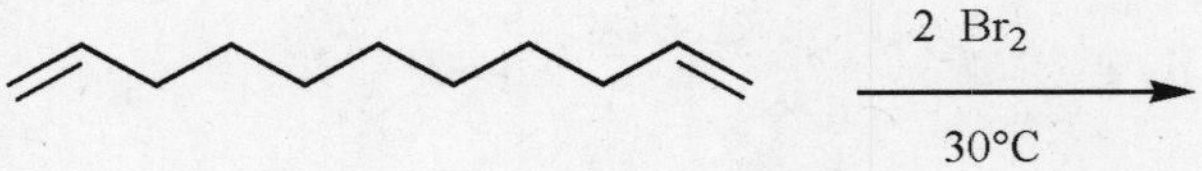

(g) Br_2, -40°C →

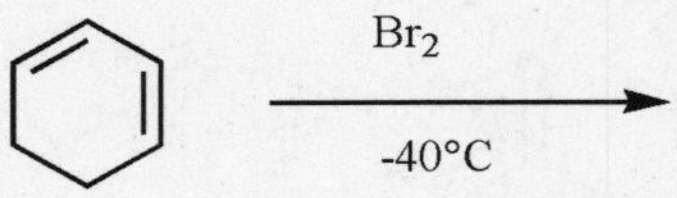

11. Draw the molecular orbital diagram of 1,3-butadiene, of ethene, of methyl acrylate and of methyl vinyl ether. Label all HOMO and LUMO orbitals. What is the ΔE for the $HOMO_{butadiene}$-$LUMO_{alkene}$ for all three alkenes? Which one reacts faster?

12. Which of the following can undergo a Diels-Alder reaction? Explain!

(a)  (b) 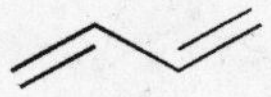 (c) 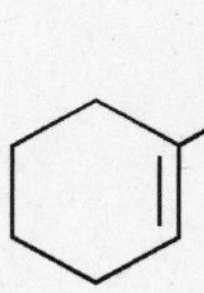 (d)

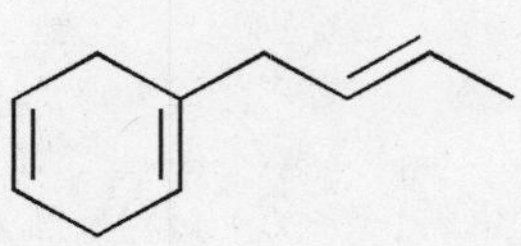

13. Predict the stereochemistry of each Diels-Alder product.

(a) + CO_2Et heat →

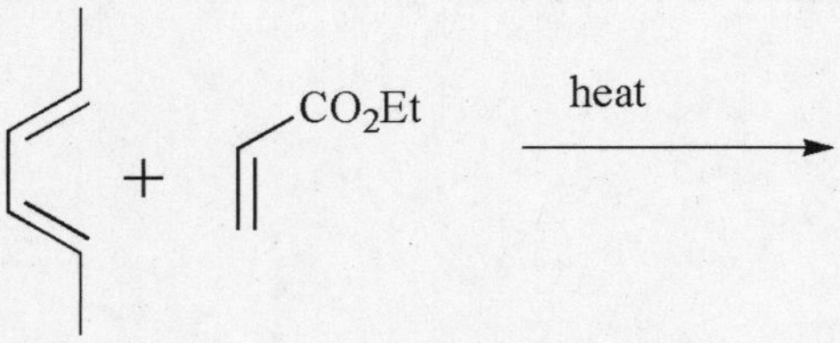

(b) + CO_2Et, EtO_2C heat →

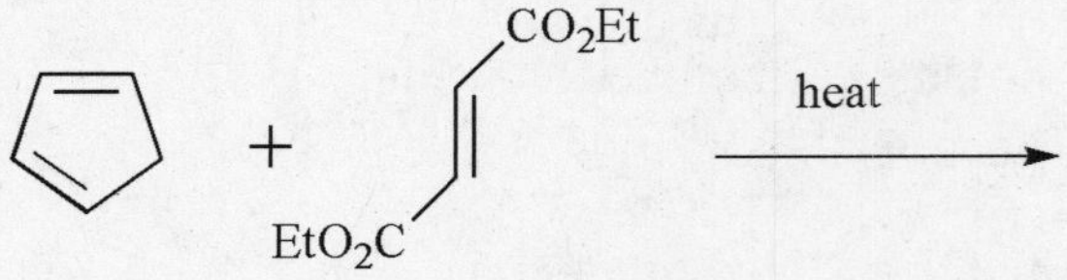

(c) Ph, Ph + C≡N, C≡N heat →

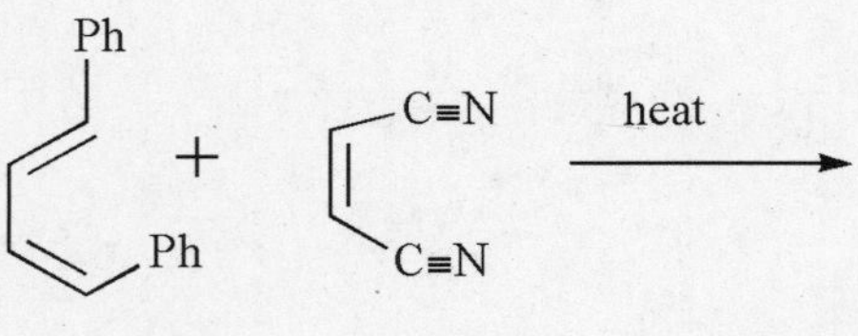

(d) + O, O, O heat →

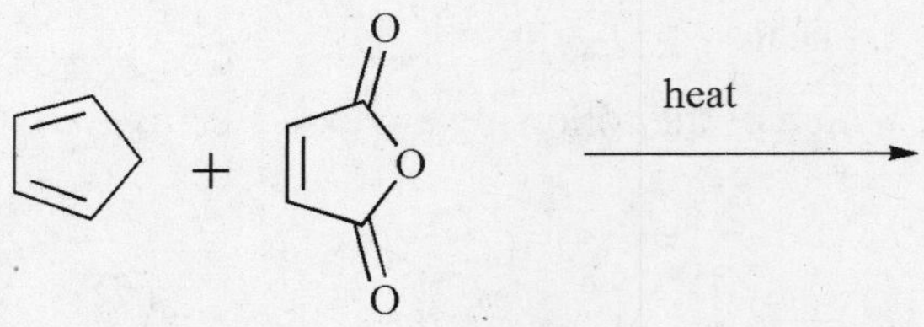

14. In each case, give the major product. Remember stereochemistry, and if there is no reaction, indicate this by N.R.

(a) 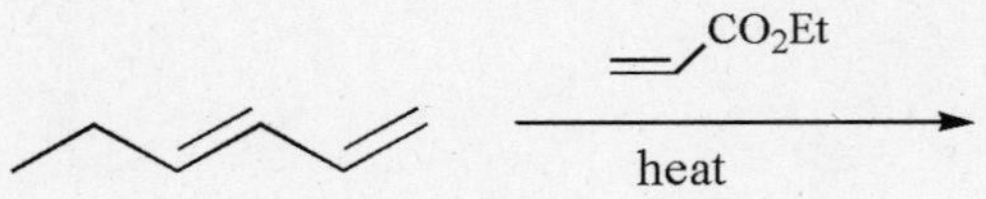 CO_2Et heat

(b) 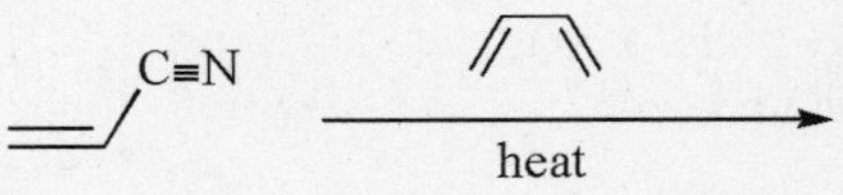 C≡N heat

(c) heat N-Ph O O

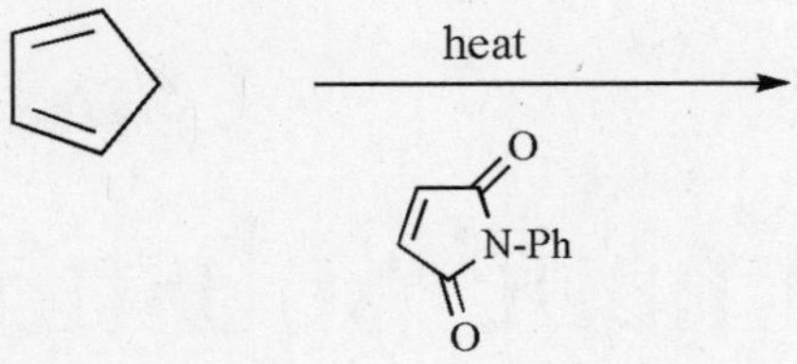

(d) 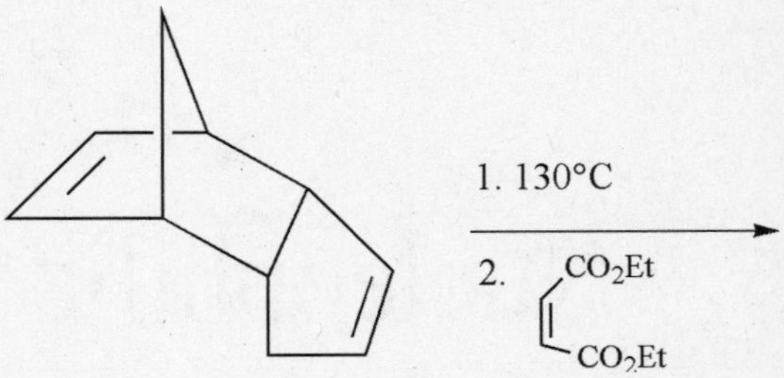 1. 130°C 2. CO_2Et CO_2Et

15. In each case, provide a suitable synthesis. Show all reagents and intermediate products.

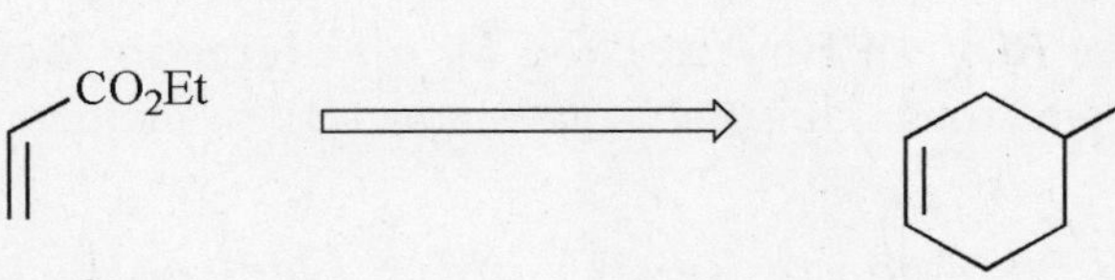

CO_2Et

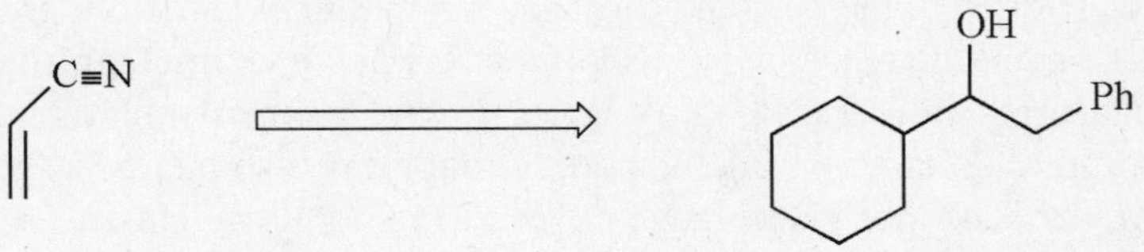

C≡N OH Ph

CHAPTER 20

Nucleophilic Aromatic Substitution

Is It Normal for a Benzene Ring to Be Attacked by an Electron Rich Reagent?

A benzene ring is rich in electrons and is **not** usually attacked by another nucleophilic species.

How Can a Benzene Ring Be Made to React with an Electron Rich Reagent?

If substituents are attached to the benzene ring that can withdraw electrons from it, however, nucleophiles can attack the ring and displace halogens in what is known as **nucleophilic aromatic substitution**. It is also possible to attach extraordinarily good leaving groups (such as nitrogen, N_2^+) that are readily displaced under mild conditions. Such leaving groups can also be lost to generate a highly reactive intermediate known as benzyne. These reagents lead to rather novel chemistry that is different from any discussed to this point.

What Is the Definition of Nucleophilic Aromatic Substitution?

When a leaving group on a benzene ring,such as a halogen, is displaced by a nucleophile, the reaction is known as nucleophilic aromatic substitution. In general, such a reaction is less facile than electrophilic aromatic substitution because an electron rich nucleophile must attack an electron rich benzene ring.

20.1. REACTIVITY OF ARYL HALIDES

Aryl halides are the most common partners in nucleophilic reactions of benzene derivatives.

What Is the Product When Chlorobenzene Reacts with KOH or NaOH in Aqueous Media at Ambient Temperatures?

There is no reaction. Neither KOH nor NaOH are sufficiently nucleophilic to attack a benzene ring and displace the chlorine. For this to occur, high temperatures and pressures are required.

What Is the Product When Chlorobenzene (20.1) Reacts with Aqueous KOH at 350°C and 2000–3000 PSI?

When chlorobenzene (*20.1*) reacts with KOH at elevated temperature and pressure the resulting product is phenol, *20.2*.

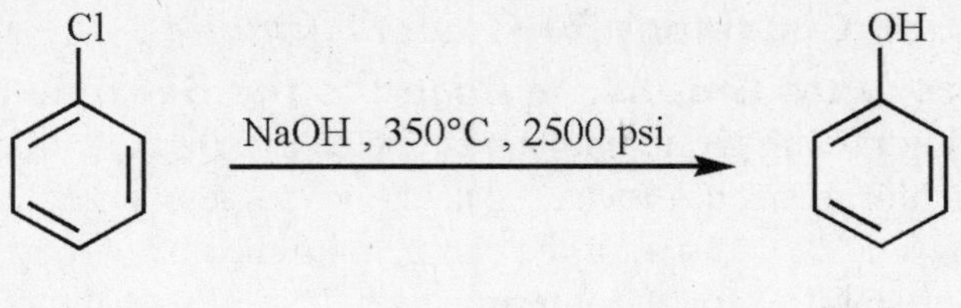

$$\text{20.1 (chlorobenzene)} \xrightarrow{\text{NaOH , 350°C , 2500 psi}} \text{20.2 (phenol)}$$

What Is the Mechanism of This Reaction?

The initial reaction involves collision of hydroxide with the chlorine-bearing carbon of chlorobenzene. The π-bond is broken, dumping the electrons into the ring as in ***20.3***. This is a resonance stabilized intermediate, and the electrons in the ring can displace the chlorine in the final step to generate phenol. This process has a high activation energy for the initial attack of hydroxide, since the negatively charged species must attack a benzene ring that is rich in electron density. This is the reason for the high reaction temperature and pressure. At much lower temperatures and pressures, no reaction occurs.

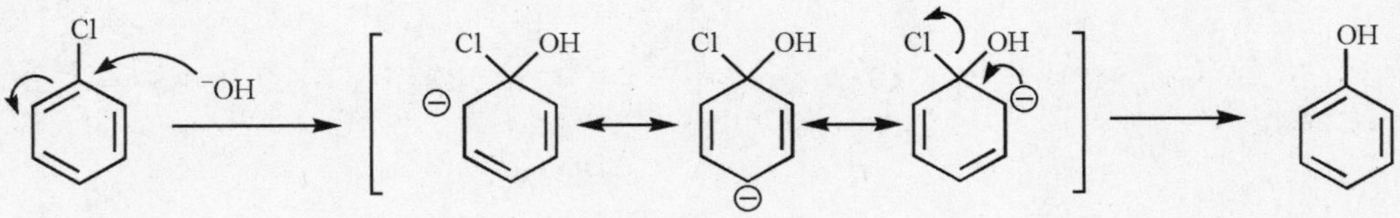

20.1 + $^-$OH → [20.3] → 20.2

What Is the General Name for This Reaction?

Nucleophilic aromatic substitution. This is an addition-elimination mechanism.

Can This Reaction Be Used for a General Synthesis for Phenols?

For phenol itself, this is a very useful method. The high reaction temperatures and pressures make it unattractive for the synthesis of all but the simplest phenol derivatives. It is very useful in industrial labs for the production of phenol, however.

Under What Conditions Will 4-Chloronitrobenzene React with Hydroxide?

The reaction involves treatment with aqueous NaOH at about 130°C, and the product is 4-nitrophenol, ***20.4***.

Under What Conditions Will 2-Chloro-1,5-Dinitrobenzene React with Aqueous Hydroxide?

The reaction involves the use of aqueous sodium hydroxide at about 100°C, and the product is 2,4-dinitrophenol, ***20.5***.

Under What Conditions Will 2-Chloro-1,3,5-Trinitrobenzene React with Hydroxide?

In this case, heating with warm water converts the chloride to 2,4,6-trinitrophenol, ***20.6***.

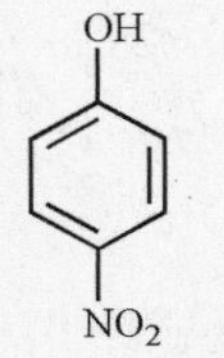

OH, NO_2 — *20.4*; OH, NO_2, NO_2 — *20.5*; OH, O_2N, NO_2, NO_2 — *20.6*

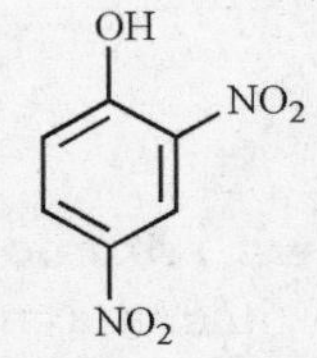

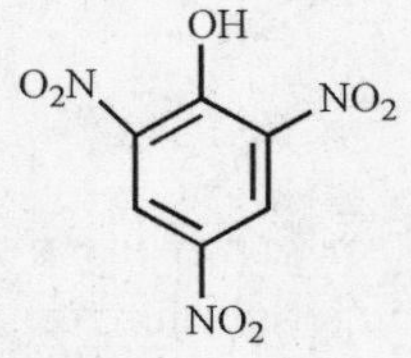

Explain Why the Reaction Conditions for Phenol Formation Become Milder As More Nitro Groups Are Added to the Benzene Ring.

Nitro is an electron withdrawing group with a positive charge on the nitrogen. Examination of intermediate *20.7*(from 4-nitrochlorobenzene) reveals that the negative charge is adjacent to the positively charged nitrogen. This allows delocalization of the charge onto the nitro group, formation of another resonance form, and increased stability for the intermediate. This allows the overall reaction to proceed under milder reaction conditions. As more nitro groups are added to positions that will bear a negative charge in the intermediate (C2, C4, C6 relative to the ipso carbon), more resonance stability is possible, the intermediate is more stable and easier to form, and the reactions occur under milder conditions.

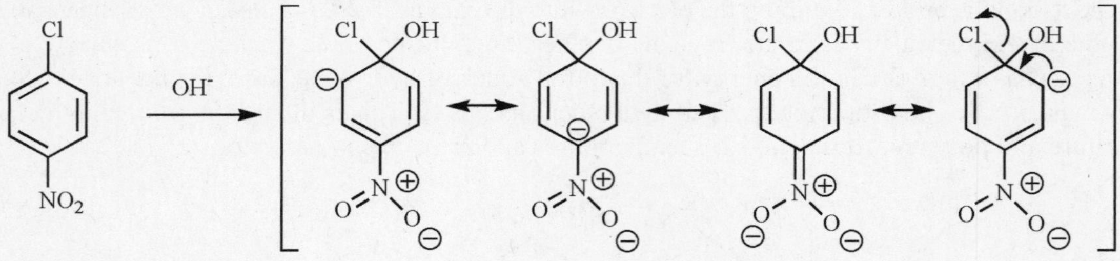

Cl NO$_2$ OH$^-$ Cl OH Cl OH Cl OH Cl OH

20.7

What Is the Product When Potassium Methoxide Reacts with 2-Chloro-1,3-Dinitrobenzene (20.8) at about 100°C ?

The product is 2,4-dinitroanisole, ***20.9***.

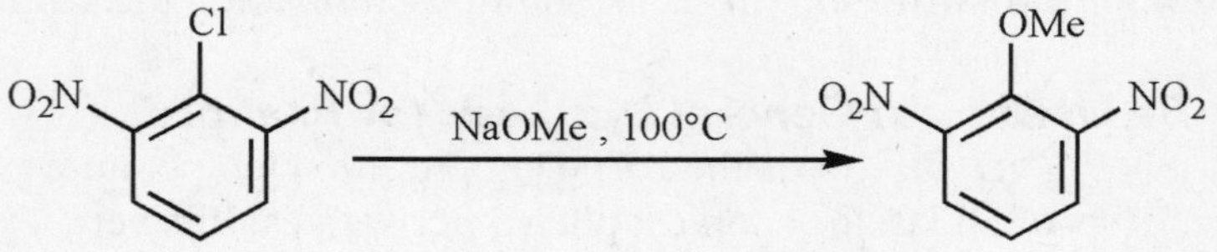

Cl O$_2$N NO$_2$ NaOMe , 100°C OMe O$_2$N NO$_2$

20.8 *20.9*

What Is the Product When 4-Chloroanisole Reacts with Hydroxide at 0°C?

There is no reaction. The electron releasing methoxy group makes the nucleophilic attack even more difficult than in chlorobenzene and, in general, no reaction occurs.

What Is the Product When Ammonia Reacts with Chlorobenzene at High Temperatures and Pressures?

The product is aniline, ***20.10***.

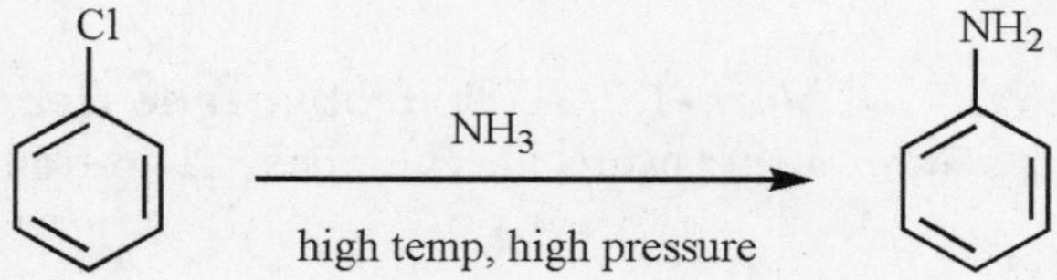

Cl NH$_3$ high temp, high pressure NH$_2$

20.1 *20.10*

What Is the Mechanism for the Reaction with Ammonia?

The mechanism is nucleophilic aromatic substitution, as noted with reactions of hydroxide and aryl halides. The intermediate for the reaction of chlorobenzene and ammonia is, therefore, that shown in ***20.11***.

Cl NH$_3$ - H$^+$ Cl NH$_2$ Cl NH$_2$ Cl NH$_2$ NH$_2$

20.1 *20.11* *20.10*

What Is the Product When Methylamine (CH_3NH_2) Reacts with 4-Chloronitrobenzene (20.12) in Refluxing Ethanol?

The reaction product is expected to be *N*-methylaniline (***20.13***), if the reaction is done at high temperatures and pressures.

Cl MeNH$_2$ EtOH, reflux NHMe NO$_2$ NO$_2$

20.12 *20.13*

Is This a General Synthesis for Aniline Derivatives?

It is a general synthesis only for simple compounds. The high temperatures and pressures required for unactivated aromatic rings make it unattractive for complex molecules. If there are electron withdrawing groups such as nitro attached to the benzene ring at the *ortho* and *para* positions relative to the halogen, the reaction conditions are much milder.

What Is the Product When Sodium Amide ($NaNH_2$) Reacts with Chlorobenzene at High Temperatures and Pressures?

The reaction produces aniline as the product, but the mechanism may not be the same as in nucleophilic aromatic substitution (see below).

What Is the Product When Sodium N,N-Diethylamide Reacts with 2-Chloro-1,3,5-Trinitrobenzene (20.14) at about 100°C?

The reaction is expected to give 2,4,6-trinitro-*N,N*-diethylaniline, ***20.15***. It is noted that this reaction probably occurs at a lower temperature.

Cl O$_2$N NO$_2$ NaNEt$_2$ 100°C NEt$_2$ O$_2$N NO$_2$ NO$_2$ NO$_2$

20.14 *20.15*

20.2. BENZYNE INTERMEDIATES

Some nucleophilic aromatic substitution reactions proceed by an elimination-addition mechanism to produce what is called a benzyne. This benzyne intermediate can react with a nucleophile to produce the substitution product.

What Is the Approximate pK_a of the Ortho-Hydrogen of Chlorobenzene?

The pK_a of the *ortho* hydrogen of chlorobenzene is approximately 40. A very powerful base is required for its removal as an acid.

What Bases Are Strong Enough to Remove This Ortho Hydrogen?

Suitable bases for removal of the *ortho* hydrogen are organolithium reagents, such as *n*-butyllithium and amide bases, such as sodium amide ($NaNH_2$).

If the Ortho Hydrogen Were Removed from Chlorobenzene, Draw the Product If a Carbanion Were Formed.

Deprotonation of chlorobenzene leads to a phenyl carbanion, ***20.16***. This is highly unstable and would likely react immediately to generate the benzyne. It is likely that ***20.16*** does not form at all, and deprotonation leads directly to benzyne via a concerted mechanism.

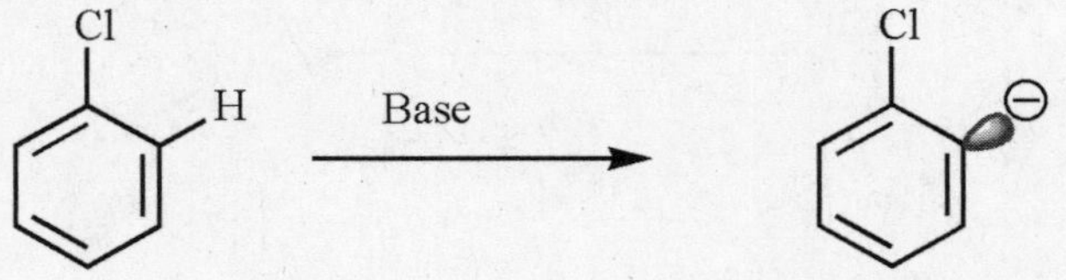

Cl H Base Cl ⊖

20.16

Assume 20.16 Is formed, What Is the Product If Chlorine Is a Leaving Group? In Other Words, What Is the Elimination Product (–H AND –Cl) from Chlorobenzene?

Loss of chlorine from ***20.16***, via displacement of Cl by the lone electron pair leads to a triple bond in the aromatic ring. This new π-bond is perpendicular to the plane of the aromatic π-cloud, as shown in ***20.17***. This is usually represented by structure ***20.18***.

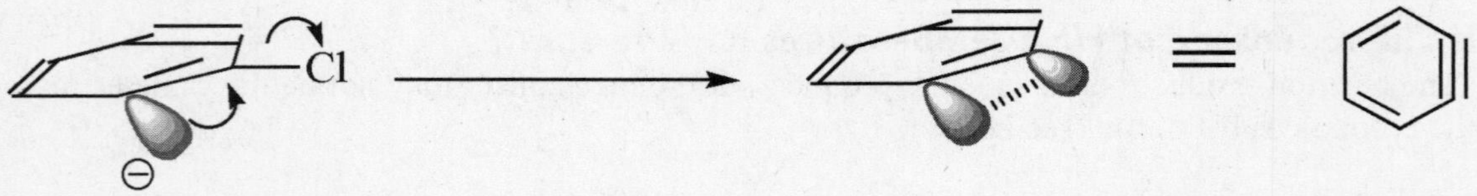

Cl ⊖

20.16 *20.17* *20.18*

What Is the Name of This Product (20.18)?

Structure ***20.18*** is known as **benzyne**.

What Amide Bases Are Commonly Used to Generate a Benzyne?

The sodium, lithium, or potassium salts of ammonia or simple primary and secondary amines such as methylamine, ethylamine, diethylamine or diisopropylamine.

What Is the Product When Potassium Amide Reacts with Chlorobenzene in Liquid Ammonia?

The product is aniline, ***20.10***.

What Is the Mechanism of This Reaction?

The reaction proceeds via a benzyne intermediate (***10.3.3***). Amide adds to the triple bond to form a new carbanion (***20.194***) which removes a proton from the ammonia solvent to give aniline.

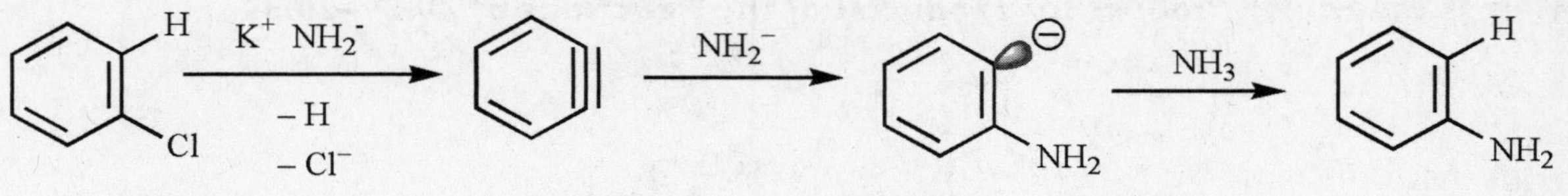

20.18 20.19 20.10

What Is the Product When Sodium Amide Reacts with 2-Bromo Ethylbenzene (20.20)? Discuss the Formation of Two Major Products.

The two products are 2-ethylaniline (***20.25***) and 3-ethylamine (***20.26***). The initially formed carbanion intermediate (***20.21***) from reaction of ***20.20*** with the amide base generates the benzyne intermediate (***20.22***). There are two sites capable of being attacked, the C2 position (leading to ***20.23***) and the C3 position (leading to ***20.24***). The electron releasing ethyl group is expected to destabilize the intermediate carbanion at C2 (*20.24*) relative to the carbanion at C3 (***20.23***), and ***20.25*** is expected to be the major product, but only slightly. This usually gives a 40–50:50–60 ratio of products favoring ***20.25***.

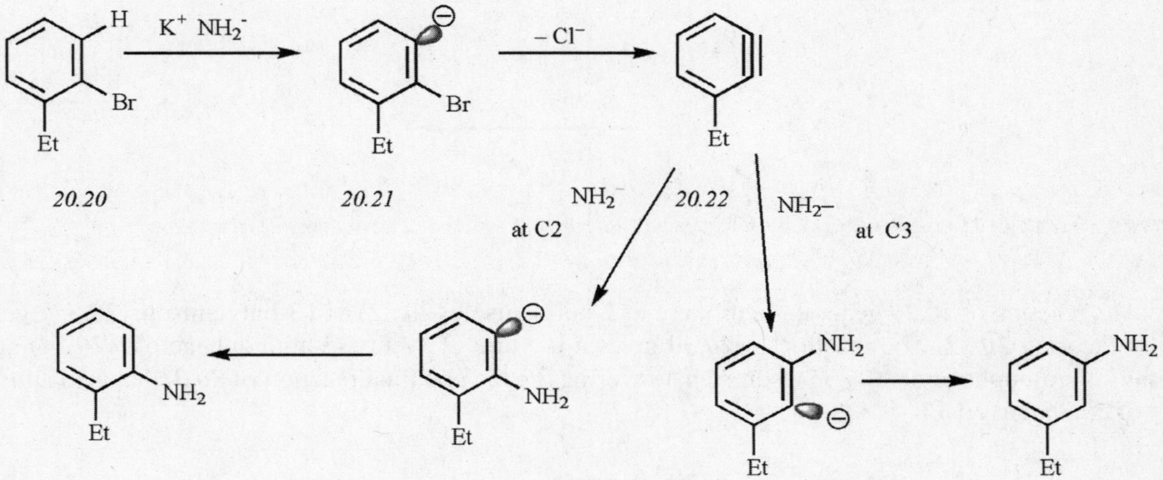

20.20 20.21 20.22 20.25 20.23 20.24 20.26

Why Is Phenyllithium a Suitable Base for the Generation of Benzyne?

Phenyllithium is a very powerful base. The pK_a of its conjugate acid, which is benzene, is greater than 40.

What Is the Major Product When Fluorobenzene Reacts with an Excess of n-Butyllithium?

The usual product is biphenyl, ***20.28***. Initial formation of benzyne (via loss of fluorine) is followed by addition of phenyllithium to give ***20.27***. Protonation then generates ***20.28***.

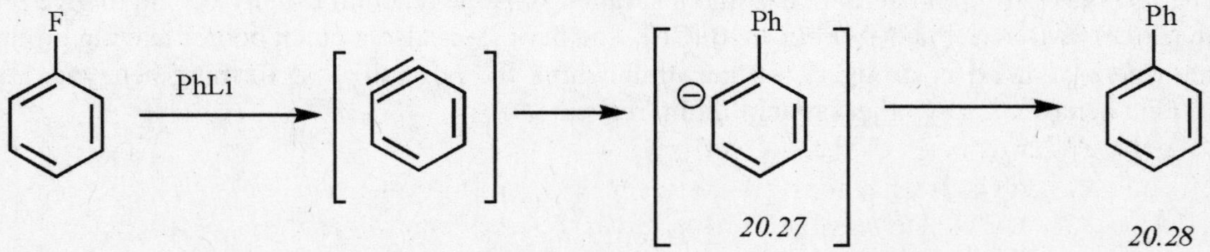

20.27 20.28

What Is the Major Product (or Products) of the Reactions of 20.29–20.31?

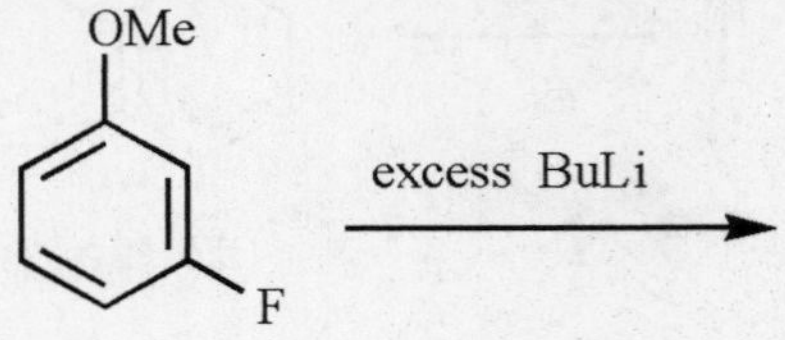

OMe
F
excess BuLi

20.29

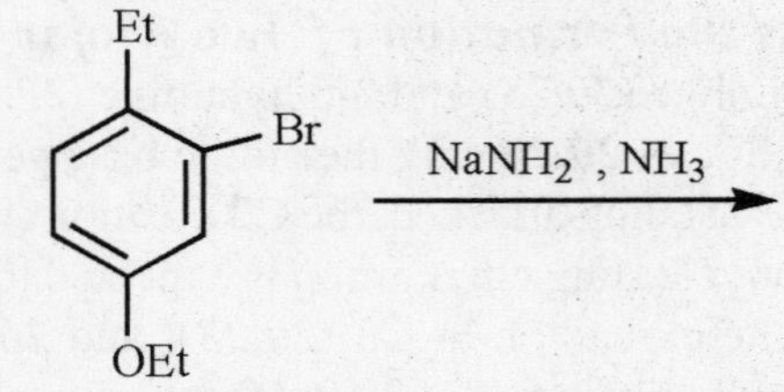

Et
Br
$NaNH_2$, NH_3
OEt

20.30

CF_3
$NaNH_2$, NH_3
Cl

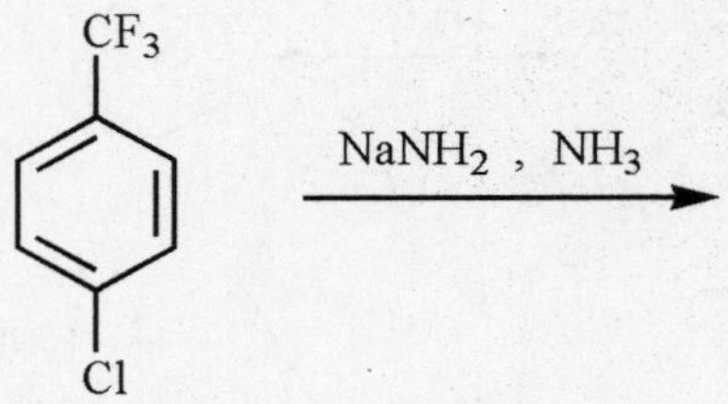

20.31

The reaction of ***20.29*** generates a mixture of 2-butylanisole (***20.32***) and 3-butylanisole (*20.33*), generally favoring ***20.32***. The reaction of ***20.30*** gives a mixture of 4-ethyl-3-aminophenetole (***20.34***) and 4-ethyl-2-aminophenetole (***20.35***), somewhat favoring ***20.35***. The final reaction of ***20.31*** leads to a mixture of ***20.36*** and ***20.37***.

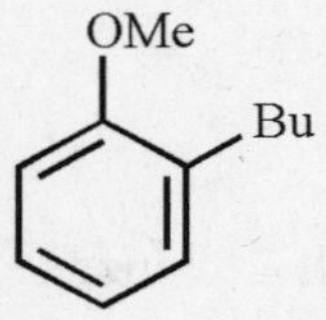

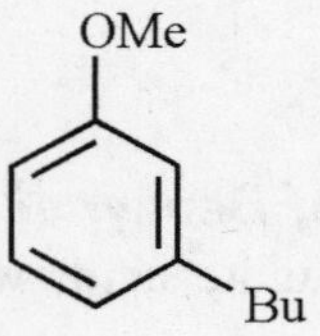

OMe Bu — *20.32*; OMe Bu — *20.33*; Et NH_2 OEt — *20.34*; Et NH_2 OEt — *20.35*; CF_3 NH_2 — *20.36*; CF_3 NH_2 — *20.37*

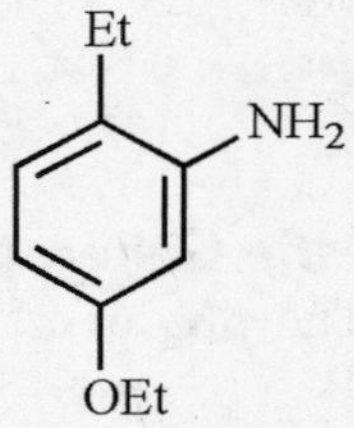

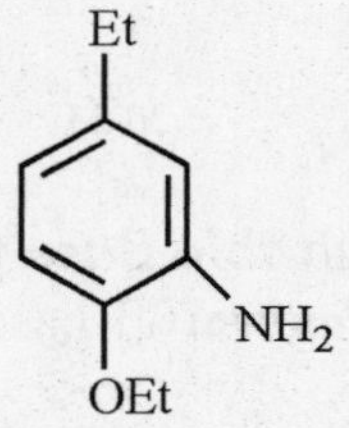

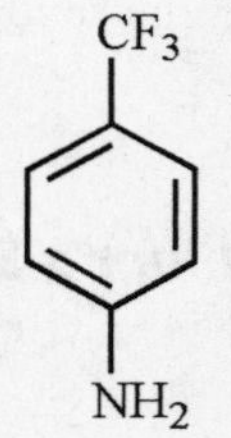

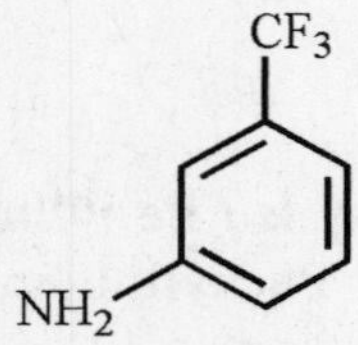

What Is the Product When Chlorobenzene Is Treated with tert-Butyllithium? Explain Why This Differs from the Reaction with Fluorobenzene.

The *tert*-butyl group is rather bulky, and a simple exchange reaction usually occurs to give phenyllithium (PhLi) [*t*-BuLi+PhCl → PhLi+*t*-BuCl]. The fluorine is also a much poorer leaving group than chlorine (the F–C bond is stronger), further diminishing the possibility of forming benzyne. Phenyllithium then behaves as any other organolithium reagent.

Test Yourself

1. Draw all resonance contributors to the intermediate of the reaction of KOH and 1-bromo-2,4,6-trinitrobenzene.

2. Describe the new π-bond formed in benzyne when an *ortho* hydrogen is eliminated from chlorobenzene.

3. Give a suitable synthesis for each of the following. In each case, give all intermediate products and show all reagents that are used.

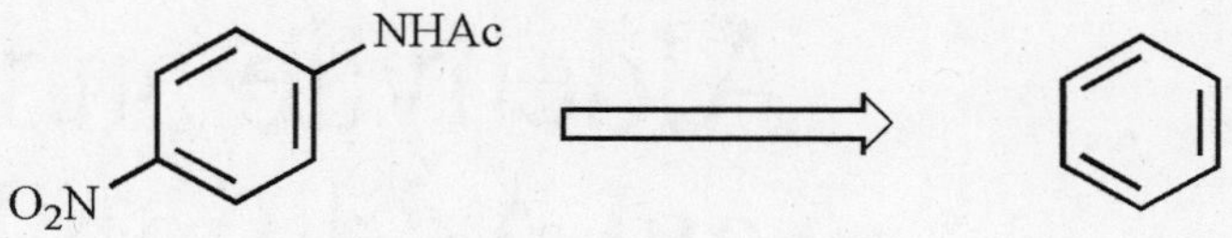

NHAc

O_2N

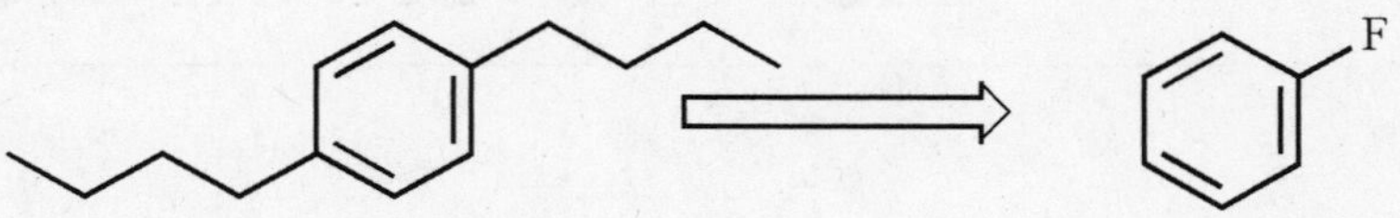

F

4. Give the major product of each of the following reactions. If there is no reaction, indicate this by N.R.

(a) Br, CH_3 — aq. KOH, 350°C / 2500 psi →

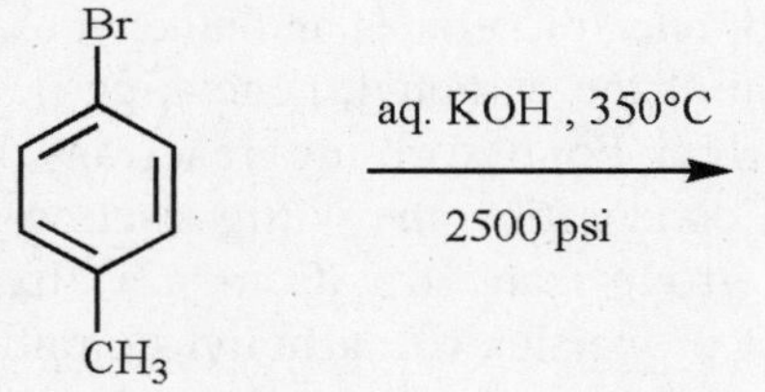

(e) Br — KNH_2, NH_3, 300°C / 3000 psi →

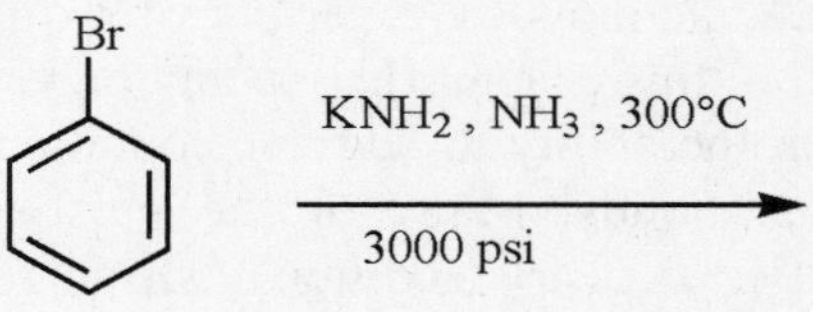

(b) Cl, NO_2, NO_2, NO_2 — H_2O, reflux →

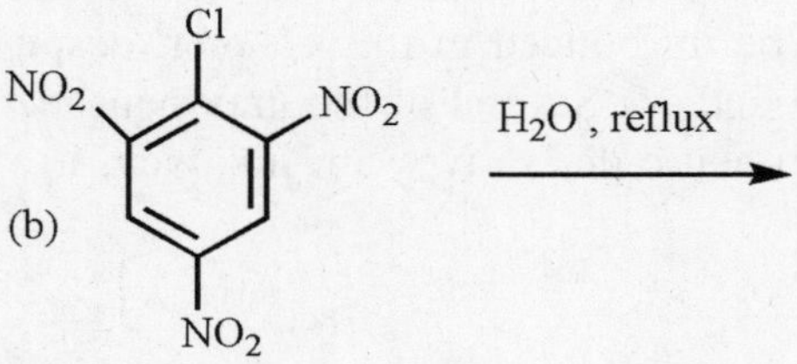

(f) Br, NO_2 — 1. aq. KOH, 200°C / 2. H_2, Pd-C →

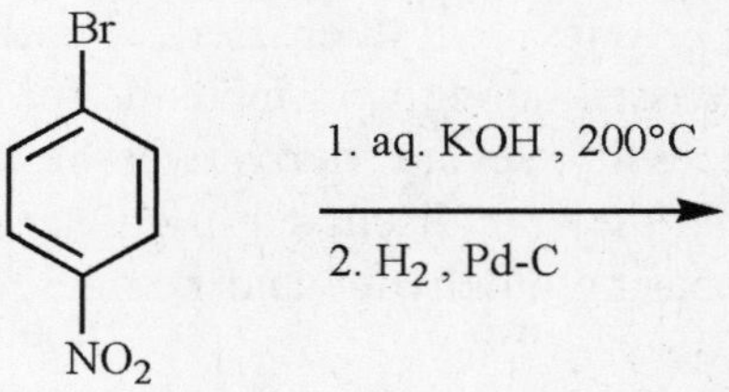

(c) Br, NO_2, NO_2, NO_2 — NaOMe, reflux / MeOH →

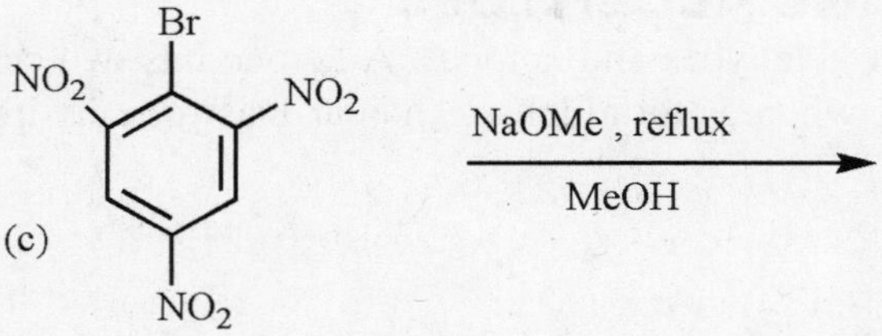

(g) F — excess butyllithium →

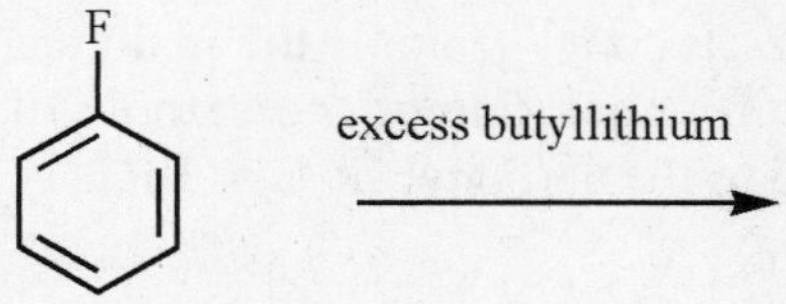

(d) Br, OMe, OMe — aq. KOH, reflux →

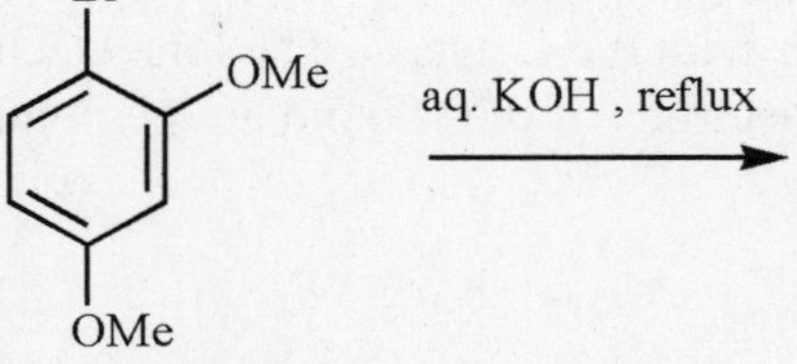

(h) Br — $NaNH_2$, NH_3 →

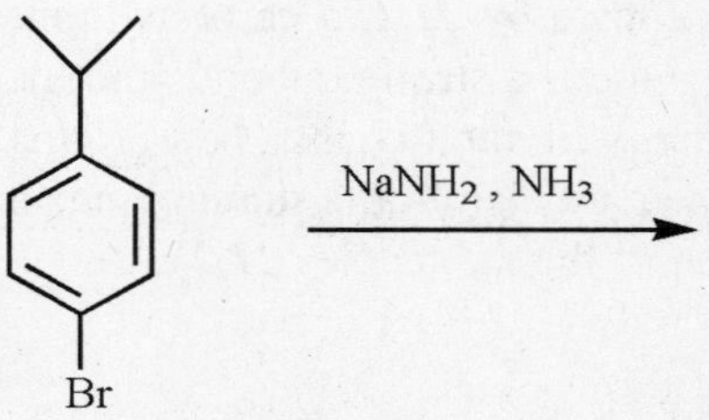

CHAPTER 21

Aldehyde and Ketones. Acyl Addition Reactions

Ketones and aldehydes, along with alkenes and alcohols, are among the most common and most used class of organic molecules. The carbonyl group (C=O) can be formed from alcohols by a variety of oxidative methods. Indeed, oxidation will be discussed in greater detail in this chapter than in any previous chapter. The importance of ketones and aldehydes, however, lies in the ability to add nucleophilic reagents to the carbon of the carbonyl, generating new carbon-carbon bonds. Many of the most important carbon-carbon bond forming reactions known in organic chemistry (addition of Grignard reagents, the Aldol Condensation, the Wittig reaction, for example) will be discussed in this chapter. Many functional group transformations are also possible using ketones and aldehydes. The chemical and physical properties of carbonyl compounds will be discussed, along with methods for their preparation and their transformation into other molecules. By necessity, several derivatives of carboxylic acids will be introduced in this chapter, despite the fact that they are discussed in greater detail in Chapters 14 and 15. Several of the important methods of preparing aldehydes and ketones, particularly when aromatic derivatives are involved, use acid derivatives.

21.1. STRUCTURES OF KETONES AND ALDEHYDES

The carbonyl group is the main characteristic of aldehydes and ketones. A ketone has two carbon groups on either side of the carbonyl [R(C=O)R], whereas an aldehyde has at least one hydrogen attached to the carbonyl [R(C=O)H].

Draw a Picture of a Carbonyl.

As shown by ***21.1***, a carbonyl consists of a C–O unit with two bonds between the carbon and the oxygen. One is a strong σ-bond, and the other is a weak π-bond. The geometry of the carbonyl is trigonal planar, with the two attached groups (R^1 and R^2) coplanar with the C and O. The lone electron pairs on oxygen are also in the same plane as the atoms, but the electrons of the π-bond are perpendicular to the plane of the atoms (see ***21.2***).

R^2 R^1 C=O ≡ R^2 R^1 C—O

21.1 *21.2*

Contrast and Compare a Carbon-Carbon Double Bond with a Carbon-Oxygen Double Bond.

The double bond of an alkene is very similar to that of a carbonyl in that there is a weak π-bond and a strong σ-bond. The carbonyl is polarized, however, due to the presence of the oxygen ($C^{\delta+}$ and $O^{\delta-}$). This leads to significant differences in reactivity, since nucleophiles will attack the electropositive carbon of the carbonyl. Such chemistry is not possible with an alkene.

What Is the Generic Structure of an Aldehyde?

An aldehyde has at least one hydrogen attached to the carbonyl, as in ***21.3***. This is usually abbreviated as RCHO.

What Is the Generic Structure of a Ketone?

A ketone has two carbon groups attached to the carbonyl, as in ***21.4***.

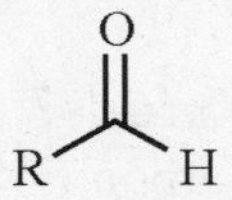

ALDEHYDE
(R = carbon: alkyl, aryl)
21.3

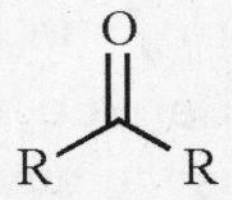

KETONE
(R = carbon: alkyl, aryl)
21.4

What Is the Only Aldehyde with Two Hydrogen Atoms Attached to the Carbonyl Group?

Formaldehyde (HCHO). Formaldehyde exists as monomeric gas, but also exists as a trimer [$(CH_2O)_3$] and as a solid polymer, paraformaldehyde [$(CH_2O)_n$].

Which Is More Reactive, an Aldehyde or a Ketone? Explain.

An aldehyde is usually more reactive. There is less steric hindrance to the approach of a reagent to the electropositive carbonyl, since there is at least one hydrogen and only one carbon group (H is less sterically hindered than any alkyl or aryl group).

What Is the Generic λ_{max} of a Carbonyl Group?

For unconjugated carbonyls, absorption occurs at 150 nm or lower, which is typically invisible to most ultraviolet spectrophotometers. A weak absorption appears at 260–290 nm which is typical of ketones and aldehydes. For conjugated carbonyls, a peak appears around 230 nm and around 300–310 nm.

What Transitions Are Possible for a Carbonyl?

The signal at 150 nm signal is due to a $\pi \rightarrow \pi^*$ transition. The 260–290 nm signal is weak and due to an $n \rightarrow \pi^*$ absorption. For conjugated carbonyls, the 230 nm absorption is the $\pi \rightarrow \pi^*$ absorption and the 300–310 signal is the weaker $n \rightarrow \pi^*$ transition.

How Can UV Spectroscopy Distinguish Between a Ketone or Aldehyde and an Alkene?

Ultraviolet spectroscopy cannot distinguish between an aldehyde and a ketone, but alkenes show a $\pi \rightarrow \pi^*$ transition at around 170–200 nm and no $n \rightarrow \pi^*$ transition.

Nomenclature

What Is the Base Name for an Aldehyde?

The IUPAC name for an aldehyde uses the alkane, alkene or alkyne prefix with the ending *-al*. Drop the *-ane* (or *-ene*, or *-yne*) ending and add *-al*, as in hexanal.

What Is the Base Name for a Ketone?

The IUPAC name for a ketone uses the alkane, alkene or alkyne prefix with the ending *-one*. Drop the *-ane* (or *-ene*, or *-yne*) ending and add *-one*, as in 2-hexanone, hex-3-en-2-one and hex-4-yn-2-one.

Why Is the Position Number of the Carbonyl Usually Omitted with an Aldehyde, but Not with a Ketone?

Since the carbonyl (C=O) is the functional group, the carbon of the carbonyl receives the lowest possible number when it is part of the longest continuous chain. Aldehydes always contain at least one hydrogen attached to the carbonyl carbon, which *must* be C1. For this reason the number is usually omitted.

Give the Correct IUPAC Names for 21.5–21.8.

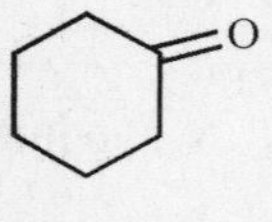

O

21.5

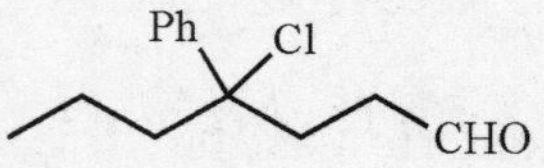

Ph Cl CHO

21.6

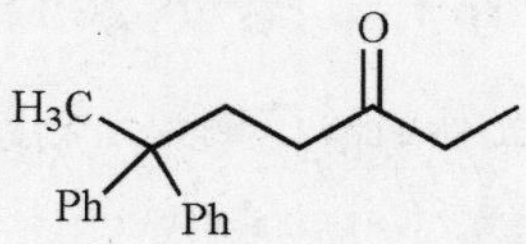

H_3C O Ph Ph

21.7

CHO

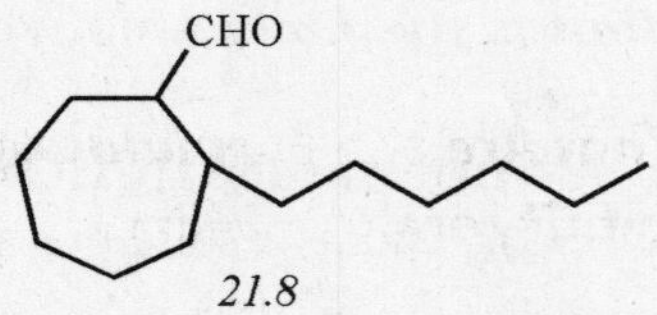

21.8

The name of ***21.5*** is cyclohexanone. For ***21.6***, this aldehyde is named 4-chloro-4-phenylheptanal. Ketone ***21.7*** is named 6,6-diphenyl-3-heptanone. When an aldehyde is attached to a ring, the naming system is changed somewhat. The parent aldehyde of ***21.8*** is cycloheptane carboxaldehyde. The name of ***21.8*** is, therefore, 2-hexylcylcoheptane carboxaldehyde.

How Are Phenyl Alkyl Ketones Named with the IUPAC System?

When the phenyl group is attached to C1, the ketone is named normally with the exception that the carbonyl carbon is C1. An example is ***21.9***, which is 1-phenyl-1-propanone. If the phenyl group is not attached to C1, the C_6H_5 unit is treated as a substituent and named as *phenyl*.

What Is the IUPAC Name of 'Diphenyl Ketone'?

Benzophenone, which has structure ***21.10***.

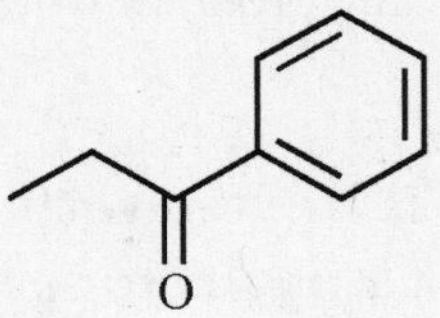

21.9 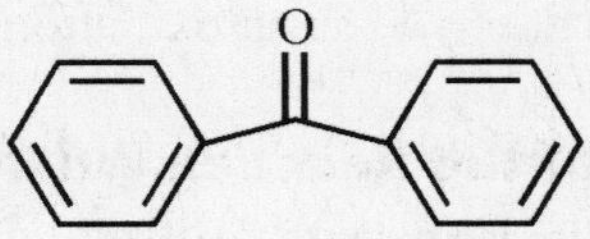 *21.10*

21.2. PREPARATION OF KETONES AND ALDEHYDES

There are several different methods for the preparation of ketones and aldehydes.

Oxidation of Alcohols

What Is the Most Common Method for Preparing Aldehydes or Ketones?

The most common method for the preparation of a ketone or aldehyde is the oxidation of an alcohol. In general, oxidation of a secondary alcohol is expected to give a ketone, and oxidation of a primary alcohol gives either an aldehyde or a carboxylic acid, depending on the oxidizing agent. This type of reaction will be discussed in more detail in Chapter 22.

Draw the Structure of Chromium Trioxide, CrO_3.

Chromium trioxide is a polymer, represented by $(CrO_3)_n$, where n = a large number. However, the monomeric structure of CrO_3 is shown as a trigonal planar species, ***21.11***.

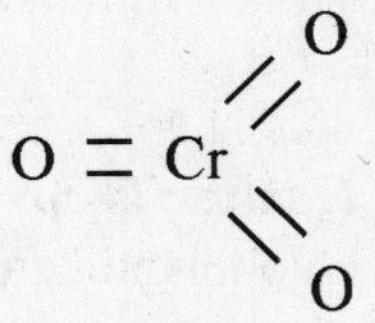

$$O = Cr(=O)=O$$

21.11

What Are the Formulas of Potassium Dichromate, Sodium Dichromate, and Chromic Acid?

Potassium dichromate is $K_2Cr_2O_7$. Sodium dichromate is $Na_2Cr_2O_7$. Chromic acid is $HCrO_4$.

If CrO_3 Is in Dilute Solution, What Is the Major Species?

The major species is CrO_3 or H_2CrO_4.

If CrO_3 Is in Concentrated Solution, What Is the Major Species?

The major species is dichromate, $H_2Cr_2O_7$.

What Is the Major Product When Cyclohexanol Is Treated with Chromium Trioxide in Aqueous Acid?

The major product is cyclohexanone.

Is This an Oxidation or a Reduction?

The alcohol → ketone transformation is an oxidation. The alcohol is oxidized to the ketone, but the chromium species is reduced. Since we are focused on the organic product, we view the transformation of alcohol to ketone or aldehyde as an oxidation.

What Is the Name of the Reaction When an Alcohol is Treated with Chromium Trioxide in Aqueous Sulfuric Acid, Usually Using Acetone As a Solvent?

This is called **Jones oxidation** or the **Jones reaction**.

What Is the Product When 1-Heptanol Is Treated with Chromium Trioxide in Aqueous Acid? Explain.

Since CrO_3 in aqueous acid is a powerful oxidizing agent, a primary alcohol is oxidized first to an aldehyde, heptanal. Aldehydes are very susceptible to oxidation, so the initially formed aldehyde is further oxidized under these conditions to a carboxylic acid. In this case, 1-heptanol is oxidized to heptanoic acid, which is the final product.

Give the Structures of Pyridinium Chlorochromate (PCC) and Pyridinium Dichromate (PDC).

PCC is *21.12* and PDC is *21.13*.

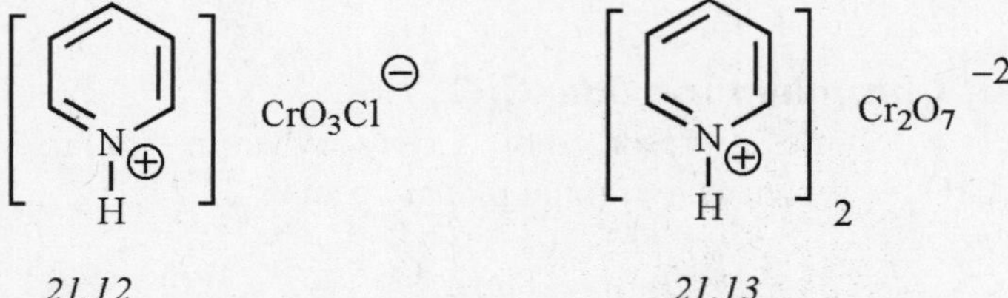

21.12 *21.13*

Why Would One Use These Reagents Rather Than Chromium Trioxide in Aqueous Acid?

Both PCC and PDC oxidize secondary alcohols to ketones, but both also oxidize primary alcohols to aldehydes without over-oxidation to the carboxylic acid.

What Is the Product When 1-Octanol Is Treated with PCC in Dichloromethane?

Since PCC is a much milder and more selective oxidizing agent, 1-octanol is oxidized to the aldehyde, octanal.

What Is the Product When 2-Methylcyclopentanol Is Treated with PDC in Dichloromethane?

Reaction of 2-methylcyclopentanol with PDC, therefore, gives 2-methylcyclopentanone as the product.

Friedel-Crafts Acylation

Friedel Crafts acylation was discussed in Section 17.4.

What Is the Product of a Reaction Between an Acid Chloride and an Aromatic Derivative in the Presence of a Lewis Acid?

The product of an acid chloride and an aromatic derivative in the presence of a Lewis acid is an aryl ketone.

What Is the Reactive Species When Propanoyl Chloride is Mixed with Aluminum Chloride?

An acylium ion $(RC{\equiv}O)^{\oplus}$ is the intermediate. In this case, R = CH_3CH_2.

What Is the Product When Anisole is treated with Propanoyl Chloride and Ferric Chloride?

An acylium reacts with aromatic rings to form an aralkyl ketone. Since the OMe group of anisole is activating and an *ortho/para* director, the products are 1-(2-methoxyphenyl)-1-propanone (***21.14***) and 1-(4-methoxyphenyl)-1-propanone (***21.15***).

OMe O Cl $FeCl_3$ OMe O + OMe O

21.14 *21.15*

What Is the Product When Nitrobenzene Reacts with Propanoyl Chloride and $AlCl_3$?

A strongly deactivating group such as nitro makes the benzene ring too unreactive for a Friedel-Crafts acylation reaction. In this case, the correct answer is no reaction.

Can Aldehydes Be Prepared by Friedel-Crafts Acylation with Acid Chlorides?

The required acid chloride would be the acid chloride of formic acid (methanoyl chloride), which is a rather unstable species. Formation of benzaldehyde derivatives by this method is, therefore, very difficult. The problem is not the Friedel-Crafts acylation, but rather the availability of the requisite acid chloride.

From Acid Derivatives

This chemistry is discussed in Chapter 25.

21.3. REACTIONS WITH WEAK NUCLEOPHILES

Aldehydes and ketones react primarily by the polarized carbonyl group. The most common reactions are nucleophilic acyl addition and acid-base reactions involving protonation of the carbonyl oxygen.

Define Nucleophilic Acyl Addition

This reaction involves addition of a nucleophile to the carbon of a carbonyl, accompanied by cleavage of the π-bond to form an alkoxide.

Water and Hydrates

What Is the Polarization of a Carbonyl?

The oxygen is more electronegative and takes on a negative dipole (electron rich), and the carbon takes on a positive dipole (electron deficient).

What Types of Reaction Are Possible for a C=O Unit?

The π-bond of the carbonyl can donate electrons to a suitable acid, just as the π-bond of an alkene reacted with an acid in Section 13.1. Therefore, the π-bond of a carbonyl can function as a base. Since the carbonyl is polarized and has a relatively weak π-bond, there is another type of reaction. A nucleophile can attack the carbonyl carbon, break the π-bond, and form an alkoxide.

What Is Acyl Addition?

Acyl is a term used for reactions at a carbonyl group, and addition implies that something reacts with the carbonyl. Acyl addition is formally the reaction of a nucleophilic species at the acyl carbon (the carbon of a C=O unit) to form a new bond, with the π-bond of the carbonyl group being broken and the electrons transferred to oxygen to form an alkoxide.

Can Alkenes Undergo This Type of Nucleophilic Reaction?

No! The C=C unit of a simple alkene is not polarized and cannot react with a nucleophile analogous to acyl addition.

Show a Generic Reaction for a Nucleophile Y That Reacts with Acetone.

The nucleophile (Y) donates two electrons to the acyl carbon of acetone, forming a new C–Y bond. The π-bond breaks, and the two electrons in the π-bond are transferred to the more electronegative oxygen atom, forming the alkoxide unit in the product, ***21.16***.

21.16

How Can 21.16 Be Converted to an Alcohol?

Since ***21.16*** is an alkoxide, it is a base. Specifically, it is the conjugate base of an alcohol, so simple treatment with aqueous acid will convert ***21.16*** to its conjugate acid, alcohol ***21.17***.

H_3O^+

21.16 *21.17*

What Is the Reaction Product When a Ketone Such As Acetone Reacts with an Acid, H^+?

When the π-bond of acetone functions as a base and donates two electrons to the acid (H^+), a new O–H bond is formed, but the oxygen takes on a positive change. This protonated carbonyl is resonance stabilized, as shown in ***21.18***.

21.18

Is the Formation of 21.18 Reversible or Irreversible?

It is an acid-base reaction and is reversible.

Does Acetone React with Water?

Water is not a sufficiently strong nucleophile to add to the carbonyl of a ketone. In this case, acetone is soluble in water, but does not react with it to any appreciable extent.

What Is the Mechanism of Reaction Between Acetone and Water with a Catalytic Amount of Acid Present? What Is the Product called?

The mechanism of reaction involves initial protonation of the carbonyl oxygen to give ***21.18***, which is resonance stabilized. This intermediate reacts with water to form ***21.19***. Loss of a proton gives the hydrate (***21.20***). The presumed product of this reaction is a hydrate (***21.20***), but this is a very unstable product that loses water and reverts back to acetone. In general, ketones and aldehydes are in equilibrium with a hydrate, and the equilibrium favors the carbonyl form.

O H+ ⊕OH OH $+H_2O$ OH $-H^+$ OH

OH_2 ⊕ OH

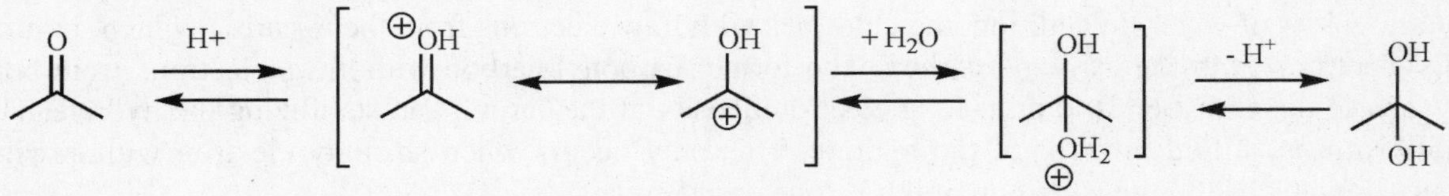

21.18 *21.19* *21.20*

Why Can't We Isolate Most Hydrates?

Hydrates are in equilibrium with the carbonyl precursor. The hydrate is more sterically hindered than the carbonyl, driving the equilibrium to the left (to the carbonyl compound). In addition, two OH groups withdraw a significant amount of electron density from the central carbon, further destabilizing it.

What Is an Enol?

An enol is a generally unstable compound that has an OH group directly attached to a C=C unit, as in ***21.21***. In ***21.21***, the R groups can be hydrogen atoms, alkyl groups, or aryl groups.

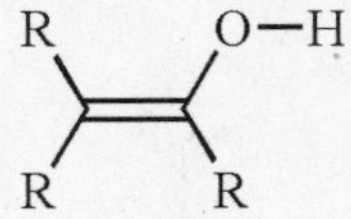

R O—H

R R

21.21

What Is Keto-Enol Tautomerism?

This is an internal acid-base reaction. The π-bond of the C=C unit in ***21.21*** can react with the proton on oxygen to generate the *keto* form, ***21.22***. Alternatively, the oxygen of ***21.22*** can react with the proton on the adjacent carbon to form ***21.21***. This equilibrium, called **keto-enol tautomerism**, greatly favors the carbonyl form (the *keto* form) as shown.

R O—H R O

H

R R R R

21.21 *21.22*

Why Does a Hydrate Convert Back to a Ketone or Aldehyde?

A hydrate loses a molecule of water to form an *enol*, which is a tautomer of the carbonyl. The keto-enol tautomerism favors the carbonyl.

Draw the Hydrate Product of Trichloroethanal (21.23).

Trichloroethanal (***21.23***) is commonly known as chloral. The predicted hydrate when chloral reacts with water is ***21.24***, known as chloral hydrate.

$$Cl_3C{-}CHO \xrightarrow{H_2O} Cl_3C{-}CH(OH)_2$$

21.23 *21.24*

Why Is Chloral Hydrate a Stable Product?

Chloral hydrate (***21.24***) is the hydrate of chloral (trichloroethanal, ***21.23***). There are no β-hydrogens that allow loss of water. In addition, the chlorines withdraw electrons from the α-carbon which, in turn, pulls electrons from the carbonyl carbon. The former carbonyl carbon withdraws electrons from both oxygens of the C–O bonds, diminishing electron density at the carbon and stabilizing the hydrate. The equilibrium is shifted in favor of the hydrate. This only occurs when strongly electron withdrawing groups are attached to the α-carbon of the ketone or aldehyde.

Can We Isolate the Hydrate Formed When Cyclopentanone Reacts with Water and an Acid Catalyst?

No! Although the hydrate may exist in equilibrium, in solution, attempts to isolate any hydrate of this type will lead to loss of water, formation of an enol, and regeneration of the ketone, in this case cyclopentanone.

Alcohols, Acetals and Ketals

What Is an Acetal?

An acetal is derived from an aldehyde and has two alkoxy units attached to the same carbon. The generic structure of an acetal is ***21.25***.

$$R{-}CH(OR^1)_2$$

21.25

What Is a Ketal?

A ketal is derived from a ketone and has two alkoxy units attached to the same carbon. The generic structure of a ketal is ***21.26***.

$$R_2C(OR^1)_2$$

21.26

What Is a Hemi-Acetal?

A hemi-acetal is a mechanistic intermediate between an aldehyde and an acetal, with an OH and an alkoxy group on the same carbon. The generic structure of a hemi-acetal is ***21.27***.

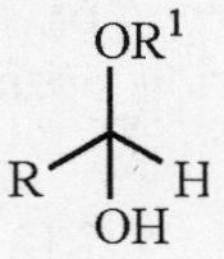

21.27

What Is a Hemi-Ketal?

A hemi-ketal is a mechanistic intermediate between a ketone and an acetal, with an OH and an alkoxy group on the same carbon. The generic structure of a hemi-ketal is ***21.28***.

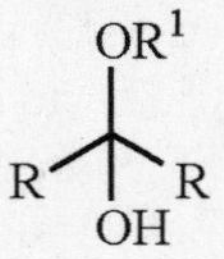

21.28

What Is the Product When Butanal (21.29) Is Reacted with an Excess of Ethanol in the Presence of a Catalytic Amount of P-Toluenesulfonic Acid?

This is analogous to the hydrate-forming reaction. Rather than water, the alcohol (ethanol) is the nucleophilic species, leading to a geminal diethoxy compound called an acetal (from an aldehyde) or a ketal (from a ketone). In this case, butanal generates 1,1-diethoxybutane (***21.36***).

Give the Complete Mechanism for This Reaction.

Initial protonation of butanal gives ***21.30***, and the acyl carbon is attacked by ethanol to give ***21.31***. Loss of a proton (an acid-base reaction, where the base can be ethanol or a molecule of butanal) generates ***21.32***, which is a **hemiacetal**. Addition of another proton gives intermediate ***21.33***. Either oxygen in ***21.32*** can be protonated, but protonation of the OEt leads back to ***21.31***, and only protonation of the OH can lead to the product, ***21.36***. If the OH group is protonated, the product is ***21.33***, and loss of water gives a new resonance stabilized cation, ***21.34***. Addition of a second molecule of ethanol to the acyl carbon gives ***21.35***, and loss of a proton from this intermediate completes the sequence to give the acetal, ***21.36***.

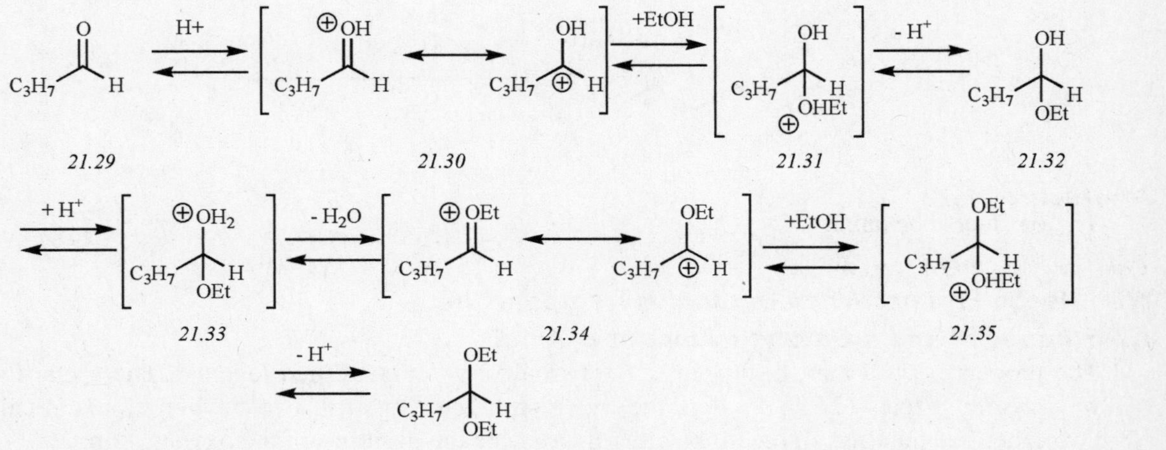

21.29 *21.30* *21.31* *21.32*

21.33 *21.34* *21.35*

21.36

Can We Isolate 21.36?

Yes! Acetals and ketals are generally stable, isolable compounds in neutral media and in pure form, in contrast to hydrates. This process is an equilibrium reaction. Removing water from the reaction drives the product towards the acetal. Adding a large excess of water drives it back to the carbonyl. Alternatively, the acetal product could be removed by distillation, again driving the equilibrium to the right.

If Water Is Removed from This Reaction As It Is Formed, What Is the Effect on the Reaction?

As mentioned, removal of one of the reaction products (water) will drive the equilibrium towards the right, to the acetal product.

What Are Some Methods That Can Be Used to Remove Water from a Reaction?

Drying agents such as calcium chloride ($CaCl_2$) or magnesium sulfate ($MgSO_4$) can be added, although these are not very efficient. A zeolite (molecular sieve) can be added that has a pore size sufficiently large to accommodate water, but not the larger organic molecules present in the mixture. The molecular sieves most commonly used are molecular sieves 3Å and 4Å. Another method for removing water is the use of a Dean-Stark trap (see End of Chapter Problem # 8). This glass apparatus relies on azeotropic distillation. Ethanol-water mixtures form an azeotrope, which is a constant boiling mixture of the two liquids boiling lower than each individual liquid. The azeotropic distillate is a fixed ratio of the two liquids. The Dean-Stark trap relies on a mixture of a solvent (such as benzene) in which water is mostly insoluble, but with which it forms an azeotrope. The azeotropic mixture distills off and is collected, the water sinks to the bottom, since it is more dense than benzene, and the benzene eventually overflows back into the reaction vessel. The water collected in this manner is removed from the reaction.

What Is the Product When an Acetal Is Reacted with Aqueous Acid?

As mentioned, if this equilibrium mixture of acetal (or ketal) and aldehyde (or ketone) is treated with aqueous acid, an excess of water will drive the reaction back to the carbonyl. In effect, this will remove the acetal group and replace it with the carbonyl group. In the mechanism shown above, ethanol replaced with water shifts the equilibrium to the left.

What Is the Product When Acetal 21.37 Is Reacted with Aqueous Acid?

OCH_3 OCH_3 aq. H^+ →

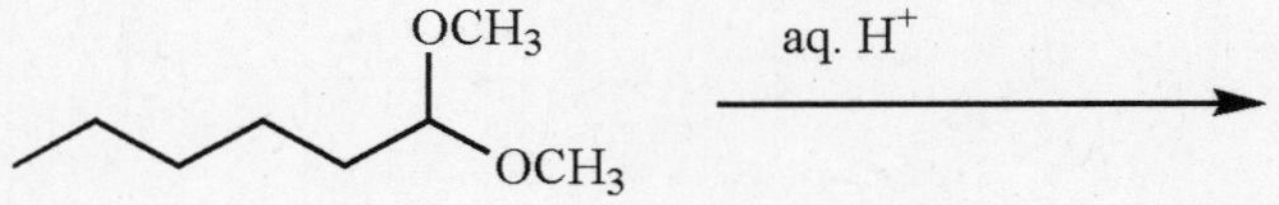

21.37

The product is hexanal.

What Is the Product When Butanal Is Treated with 1,2-Ethanediol and a Catalytic Amount of Acid?

The product is an acetal. In this case, the two alcohol units are tied together. The acetal formed will be a cyclic species (*21.38*), where the five-membered ring with two oxygen atoms is called a 1,3-dioxolane. The numbers of the dioxolane ring indicate the positions of the oxygen atoms.

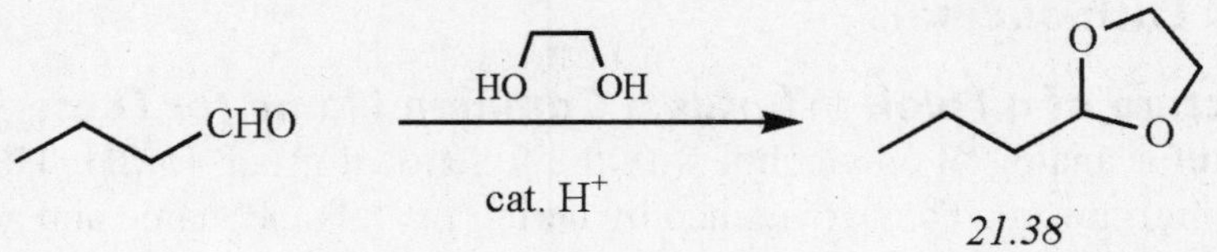

HO OH

CHO

cat. H^+

21.38

What Is the Product When 2-Pentanone Is Reacted with an Excess of Propanol in the Presence of a Catalytic Amount of p-Toluenesulfonic Acid?

The product is 2,2-dipropoxypentane, ***21.39***.

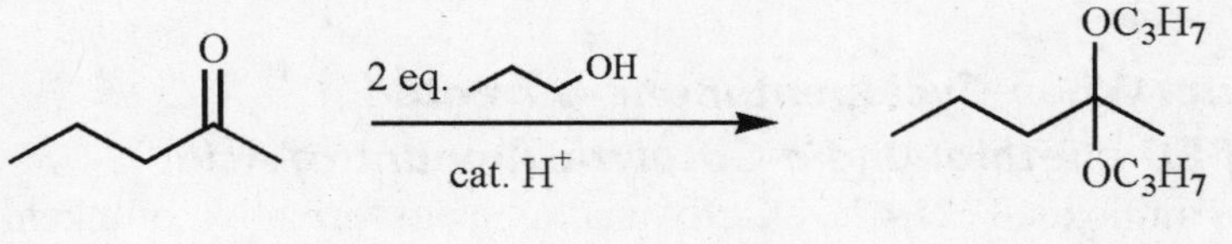

2 eq. OH

cat. H^+

OC_3H_7

OC_3H_7

21.39

What Is the Mechanism for this Reaction?

The mechanism is identical to that presented above. Substitute 2-pentanone for butanal and substitute 1-propanol for ethanol.

What Is the Product When 5 Equivalents of Cyclopentanol Are Reacted with Cyclohexanone in the Presence of a Catalytic Amount of P-Toluenesulfonic Acid? Assume There Are No Secondary Reactions of the Alcohol.

The product is 2,2-dipropoxypentane, ***21.40***.

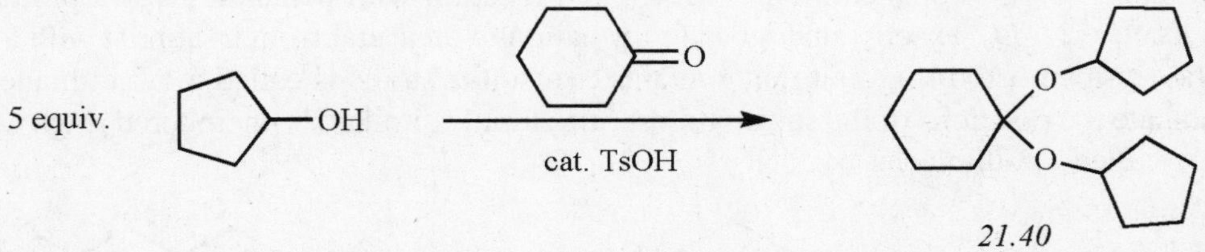

5 equiv. OH

cat. TsOH

21.40

What Is the Product When a Ketal Such As 21.39 Is Reacted with Aqueous Acid?

As with acetals, treatment of a ketone with aqueous acid shifts the equilibrium back to the carbonyl derivative. Treatment of ***21.39*** with aqueous acid will, for example, regenerate 2-pentanone and propanol.

What Is the Product When Cyclohexanone Is Treated with 1,2-Ethanediol and a Catalytic Amount of Acid?

The product is a ketal. As with dioxolane formation from aldehydes, ketones also form dioxolanes upon treatment with ethylene glycol. The product is ***21.41***.

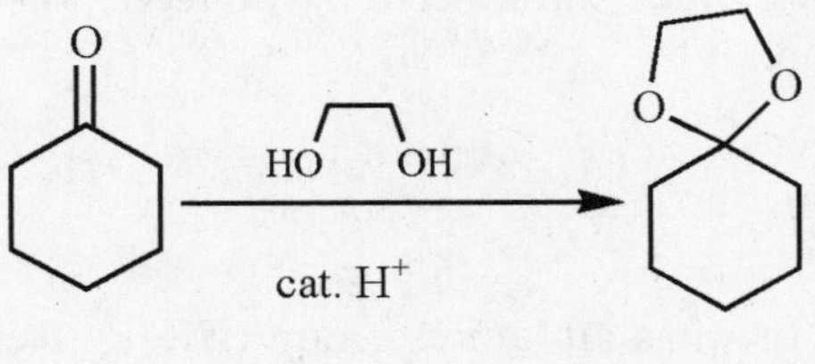

HO OH

cat. H^+

21.41

Dithioketals and Dithioacetals

What Is the Structure of a Thiol? What Is a Common Name for These Molecules?

A thiol is the sulfur analog of an alcohol with the functional group R–SH. The common name for these compounds is **mercaptan**. They are named by taking the 'alkane' name and adding the word *thiol* to it. The mercaptan $CH_3CH_2CH_2SH$ is, for example, propanethiol.

How Would You Predict a Thiol to React?

A thiol should react, more or less, in a manner similar to an alcohol, at least in reactions with carbonyl derivatives.

What Is the Product When Cyclopentanone Is Treated with an Excess of Ethanethiol and a Catalytic Amount of Acid?

The product is a dithioketal (*21.42*), exactly analogous to formation of a ketal from treatment with an alcohol.

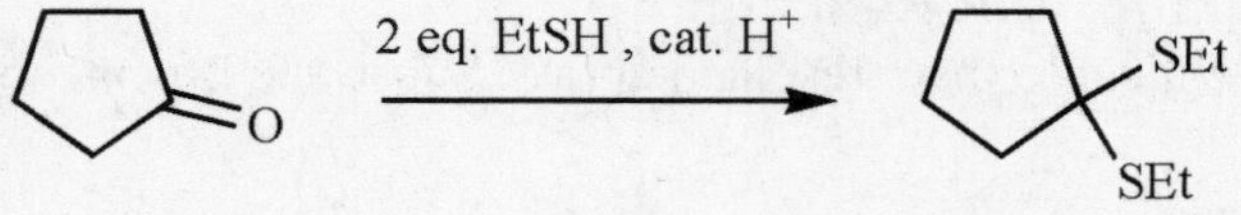

2 eq. EtSH , cat. H^+

SEt

SEt

21.42

What Is the Product When Pentanal Is Treated with 1,3-Propanedithiol in the Presence of a Catalytic Amount of Acid?

The structure of 1,3-propanedithiol is *21.43*, and reaction with pentanal gives a product that is a cyclic dithioacetal (*21.44*), exactly analogous to formation of an acetal from treatment with a cyclic diol. Note that the six-membered ring unit that contains two sulfur atoms is called a 1,3-dithiane, where the numbers indicate the positions of the sulfur atoms. Structurally similar six-membered rings with two sulfur atoms are called 1,3-dithianes.

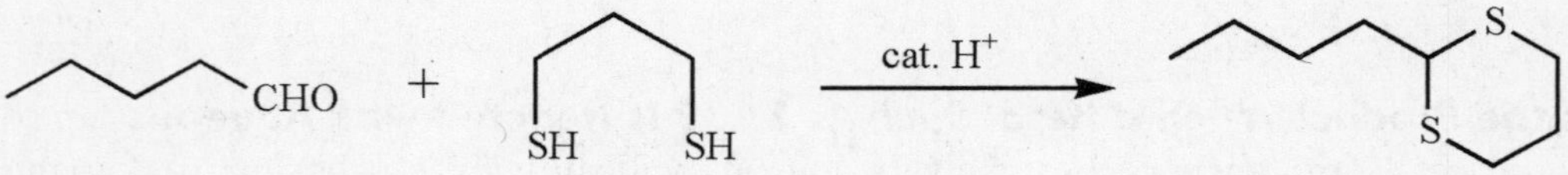

CHO + SH SH

cat. H^+

S S

21.43 *21.44*

What Is the Product When 2,2-Di(Methylthio)Hexane Is Reacted with Raney Nickel in Hot Acetone? Explain.

Nickel has a strong affinity for sulfur. Raney nickel is specially prepared and has a considerable amount of hydrogen adsorbed on the surface of the finely divided nickel. When a dithioketal is treated with this reagent, complete removal of the sulfur occurs, with reduction to a $-CH_2-$ group. In this case, the final product is the alkane, hexane.

Imines and Enamines

What Is an Enamine?

An enamine is a molecule that has an amino group (NR_2) attached directly to a carbon-carbon double bond, as in *21.45*. This particular example is the diethylamino enamine of 2-pentanone [named 2-(*N,N*-diethylamino)-1-pentene].

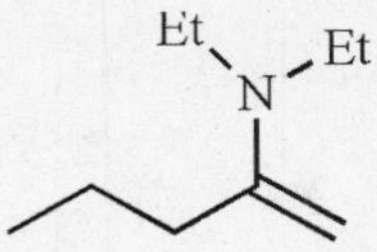

Et Et N

21.45

What Is an Imine?

An imine is a molecule characterized by the presence of a C=N–R unit, as in ***21.46***. This particular example is the *N*-ethyl imine of 2-pentanone.

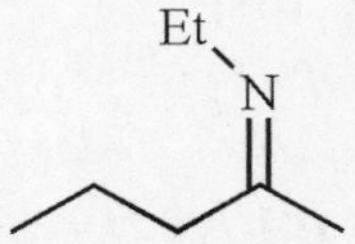

Et N

21.46

How Are Enamines Formed?

Enamines such as ***21.45*** are formed by reaction of a ketone (in this case 2-pentanone) with a ***secondary*** amine (HNR_2, in this case *N,N*-diethylamine), usually in the presence of an acid catalyst.

O $\xrightarrow{Et_2NH\text{ , cat }H^+}$ Et Et N

21.45

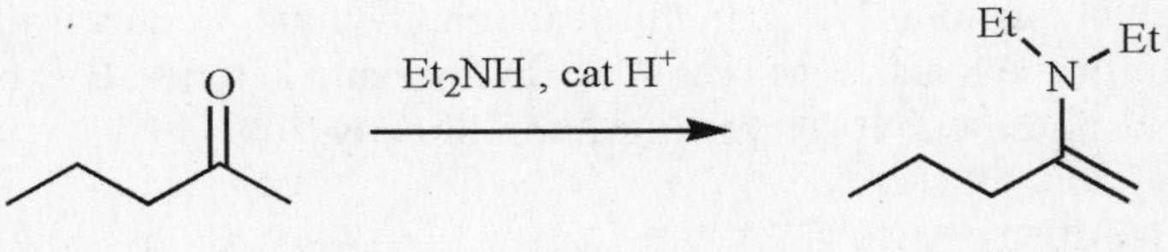

How Are Imines Formed?

Imines such as ***21.46*** are formed by reaction of a ketone (in this case 2-pentanone, but aldehyde can be used) with a ***primary*** amine (H_2NR, in this case *N*-ethylamine), usually in the presence of an acid catalyst.

O $\xrightarrow{EtNH_2\text{ , cat }H^+}$ Et N

21.46

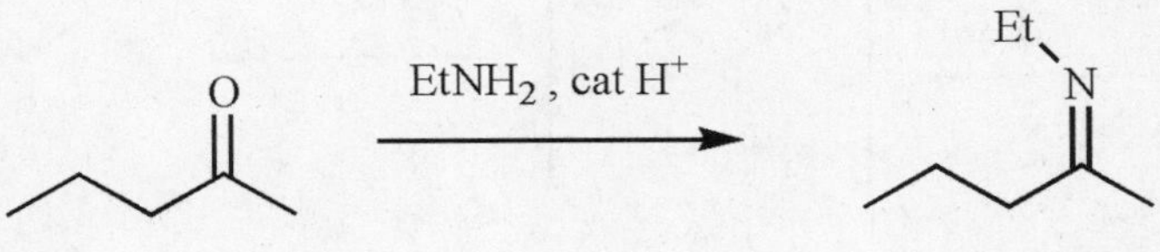

Give the Product and the Mechanism for Its Formation When Cyclopentanone Reacts with Ethylamine in the Presence of a Catalytic Amount of Acid.

The reaction begins by formation of the protonated carbonyl species (***21.47***) by reaction of the ketone and the acid catalyst. Addition of ethylamine to the acyl carbon leads to ***21.48***, and loss of the proton give the amino alcohol ***21.49***. Reprotonation of the amine would regenerate the carbonyl via ***21.48***, but protonation of the OH group (to form ***21.50***) allows expulsion of water by the amine to give iminium salt ***21.51***. Loss of the proton from nitrogen in ***21.51*** leads to the final product, imine ***21.52***. This is a general mechanism for the reaction of primary amines and ketones. It is noted that the proton on nitrogen in ***21.48*** can be shifted to oxygen, giving ***21.50*** directly.

H+ → [21.47] EtNH₂ → [21.48] - H⁺ → 21.49

[21.50] - H₂O⁺ → [21.51] - H⁺ → 21.52

Give the Product and the Mechanism for Its Formation When Cyclopentanone Reacts with Diethylamine in the Presence of a Catalytic Amount of Acid.

The reaction begins by formation of the protonated carbonyl species (***21.47***), as before, by reaction of the ketone and the acid catalyst. Addition of diethylamine to the acyl carbon leads to ***21.53***, and loss of the proton give the amino alcohol ***21.54***. Reprotonation of the amine would regenerate the carbonyl via ***21.53***, but protonation of the OH group to form ***21.55*** allows expulsion of water by the amine to give iminium salt ***21.56***. In contrast to ***21.51***, iminium salt ***21.56*** does **not** have a proton on nitrogen, so imine formation is not possible. There is a proton on an adjacent *carbon* that can be lost, but this generates a C=C unit and formation of enamine ***21.57***. In this reaction, the final product is the enamine, ***21.57***. This is a general mechanism for the reaction of secondary amines and ketones. It is noted that the proton on nitrogen in ***21.53*** can be shifted to oxygen, giving ***21.55*** directly.

H+ → [21.47] Et₂NH → [21.53] - H⁺ → 21.54

[21.55] - H₂O⁺ → [21.56] - H⁺ → 21.57

What Amines Are Commonly Used to Form Enamines?

The most common secondary amines used to form enamines are diethylamine (Et_2NH), dimethylamine (Me_2NH), pyrrolidine (***21.58***), piperidine (***21.59***) and morpholine (***21.60***). (See Section 27.1.)

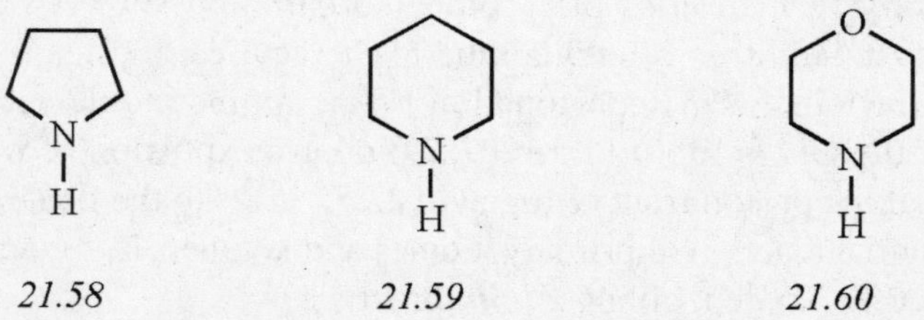

21.58 21.59 21.60

What Is the Product When Butanal Reacts with Diethylamine and a Catalytic Amount of Acid?

The initial product is 1,1-(*bis*-diethylamino) butane (***21.62***). The term *bis*- refers to the presence of two diethylamine species in the molecule. This product arises from intermediate ***21.61***, which reacts with additional diethylamine to form the disubstituted amino compound. In most cases, this geminal diamine will eliminate diethylamine to form iminium salt ***21.61*** as the major product. In many cases, however, ***21.62*** is the isolated product and must be reacted with base to give the enamine, ***21.63***.

Et_2NH , cat. H^+ ; + Et_2NH / - Et_2NH ; Base

21.61 *21.62* *21.63*

What Is the Product When 2-Pentanone Reacts with Butylamine?

In general, when a ketone or aldehyde reacts with a primary amine, the product is an imine. In this case, 2-butanone reacts to give ***21.64***.

n-$BuNH_2$, cat. H^+

21.64

What Is the Name Given to an Imine Derived from an Aromatic Aldehyde?

When an aldehyde such as benzaldehyde reacts with a primary amine, the final imine product is referred to as a **Schiff base** (***21.65***). The definition of a Schiff base is a molecule that contains a C=N group (an imine), with an aryl or alkyl group (not hydrogen) on the nitrogen atom.

21.65

21.4. REACTIONS WITH STRONG NUCLEOPHILES

Aldehydes and ketones react primarily by the polarized carbonyl group. The most common reactions are nucleophilic acyl addition to form alcohol products.

What Is the Product When a Ketone Such As Acetone (A) Reacts with a Nucleophile via Acyl Addition, and Then (B) is Treated with Aqueous Acid?

As seen above in Section 21.3, the nucleophile (Y) donates two electrons to the acyl carbon of acetone, forming a new C–Y bond. The π-bond breaks, and the two electrons in the π-bond are transferred to the more electronegative oxygen atom, forming the alkoxide unit in the product, ***21.16***. Since ***21.16*** is an alkoxide, it is a base. Specifically, it is the conjugate base of an alcohol, so simple treatment with aqueous acid will convert ***21.16*** to its conjugate acid, alcohol ***21.17***.

21.16 21.17

Cyanide

Is Cyanide a Nucleophile?

Yes! Cyanide ion is a good nucleophile.

What Is an Ambident Nucleophile?

An ambident nucleophile is one that has two reactive sites. In other words, there are two nucleophilic centers in the same molecule. Using cyanide as an example (see ***21.66***), both carbon and nitrogen have an electron pair that can be donated to an electrophilic carbon.

$$^{\ominus}:C{\equiv}N:$$

21.66

Which Is More Nucleophilic in the Cyanide Ion, Carbon or Nitrogen?

In general, the carbon of the cyanide ion (***21.66***) is the more nucleophilic in reactions with acyl carbons and in S_N2 reactions.

What Is the Product When Sodium Cyanide Reacts with Iodomethane?

The reaction of cyanide ion and iodomethane is an S_N2 reaction, and the carbon of cyanide displaces iodide to give acetonitrile, ***21.67***.

$$H_3C{-}I \xrightarrow{NaCN} H_3C{-}C{\equiv}N$$

21.67

What Is the Product When Sodium Cyanide Reacts with Acetone?

The product is ***21.68***.

21.68

Give the Mechanism for the Formation of 21.68.

Protonation of acetone gives the cation in the usual manner. Subsequent nucleophilic attack at the acyl carbon by cyanide gives ***21.68*** directly.

21.68

What Is the Common Name for a Product Such As 21.68?

A cyanohydrin.

What Is the Product When Cyclohexanone Reacts with HCN?

This reaction also generates a cyanohydrin ***21.69***.

HCN

HO C≡N

21.69

Acetylides (Alkyne Anions)

How Is an Alkyne Anion Formed from an Alkyne?

When an alkyne such as 1-butyne (***21.70***) is treated with a strong base such as sodium amide or butyllithium, the acidic proton of the terminal alkyne is removed, generating an alkyne anion such as ***21.71***.

$NaNH_2$ or BuLi

Na^+ or Li^+

21.70 *21.71*

What Is the Product When the Sodium Salt of 1-Butyne Reacts with 2-Pentanone?

When sodium butyne (Na^+ $^-C{\equiv}CCH_2CH_3$) reacts with 2-pentanone, the nucleophilic acetylide attacks the electropositive carbonyl carbon to produce an alkoxide via acyl addition. Hydrolysis liberates the alcohol product, 4-methyl-oct-3-yn-4-ol (***21.72***) in a second chemical step.

1. Et-C≡C:$^-$ Na^+, THF
2. H_3O^+

21.72

What Is the Product When the Sodium Salt of Acetylene Reacts with Benzaldehyde?

The product is 1-phenylprop-2-yn-1-ol, ***21.73***.

1. H-C≡C:$^-$ Na^+, THF
2. H_3O^+

21.73

Why Is This Type of Reaction Important in Organic Chemistry?

It is important because it is one of the few chemical reactions that form a new carbon-carbon bond. In addition, the product contains other functionality that will allow further chemical transformations. In this case, two functional groups, an alcohol and an alkyne, are introduced into the same molecule.

Grignard Reagents

Formation of Grignard reagents was discussed in Section 10.2.

What Is a Grignard Reagent?

A Grignard reagent has the structure R–Mg–X, where R is an alkyl, vinyl, or aryl group, and X is a halogen, Cl, Br, I.

What Is the Bond Polarization of a C-Mg Bond?

Since Mg is less electronegative than carbon the polarization is: $^{\delta+}$Mg—C$^{\delta-}$.

What Is the Product When 1-Bromobutane Reacts with Magnesium in Ether? What Is the Name of This Product?

The product is butylmagnesium bromide, $CH_3CH_2CH_2CH_2MgBr$.

What Is the Schlenk Equilibrium?

The Schlenk equilibrium describes several organometallic species that comprise the actual structure of a 'Grignard reagent' in solution. In its most simple form, the Grignard (RMgX) is described as: 2 RMgX $\rightleftarrows$ R_2Mg + MgX_2.

What Is the Role of the Ether in This Reaction?

The ether solvent acts as a Lewis base, donating electrons to the electron deficient magnesium, assisting formation of the Grignard reagent, and stabilizing it once it is formed.

What Is the Product Formed When 1-Bromopropane Is Mixed with Magnesium Metal in Diethyl Ether?

The product is ***21.74***.

Br —(Mg , Et_2O)→ MgBr

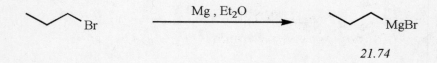

21.74

Give the Name of 21.74.

Grignard reagent ***21.74*** is propylmagnesium bromide.

What Is the Product Formed When 21.74 Is Treated with Water?

Since ***21.74*** is a strong base, water functions as an acid, and the MgBr is replaced with H. Therefore, the product is propane, the conjugate acid of ***21.74***.

What Is the Product of the Reaction Between Phenylmagnesium Bromide and Cyclohexanone?

The product is 1-phenylcyclohexanol, where the nucleophilic carbon of the Grignard reagent attacks the electropositive carbonyl carbon, forming a new carbon-carbon bond and producing an alkoxide. Hydrolysis gives the alcohol product.

What Is the Product of the Reaction Between Butylmagnesium Chloride and Hexanal?

The product, after hydrolysis, is 5-decanol.

What Aldehyde Reacts with Ethylmagnesium Bromide to Produce a Primary Alcohol?

Formaldehyde. When ethylmagnesium bromide reacts with formaldehyde (HCHO), hydrolysis gives the alcohol, 1-propanol.

Why Is the Reaction of a Ketone or Aldehyde and a Grignard Reagent Not Reversible?

A strong carbon-carbon bond is formed when the π-bond of the carbonyl is broken by $C^{\delta-}$ of the Grignard reagent. The strength of the carbon-carbon bond makes the reverse reaction highly endothermic, and it also has a high activation energy relative to the forward reaction. The reactions that form Grignard reagents from alkyl halides were described in Section 10.2.

What Is the Product When Butylmagnesium Bromide Reacts with Cyclohexanone?

The product is 1-butylcyclohexanol, resulting from nucleophilic attack of the carbon in the Grignard reagent at the electropositive carbonyl, followed by aqueous hydrolysis of the resulting alkoxide.

What Is the Product When Bromobenzene Reacts with Magnesium in THF?

The product is phenylmagnesium bromide, PhMgBr. The THF is required to stabilize the product, since it is less stable, and THF is a stronger Lewis base than diethyl ether.

Grignard Reaction with Carbonyls

When a Grignard Reagent Reacts with an Aldehyde, What Is the Major Product after Hydrolysis?

The major product is a secondary alcohol.

Which Aldehyde Leads to a Primary Alcohol upon Reaction with Butylmagnesium Bromide?

Formaldehyde.

When a Grignard Reagent Reacts with a Ketone and Is Followed by Hydrolysis, What Is the Major Product?

The major product is a tertiary alcohol.

What Product Results from the Reaction of Methylmagnesium Bromide and Acetone?

The reaction product is 2-methyl-2-propanol (***21.76***) in a two-step reaction. In the first step, the nucleophilic carbon of the Grignard reagent attacks the $\delta\oplus$ carbon of the ketone (acetone). When the $C^{\delta-}$ and $C^{\delta\oplus}$ collide, the π-bond of the carbonyl is broken, and those two electrons are transferred to the more electronegative oxygen, making the anion (***21.75***). In other words, the Grignard reagent reacts via acyl addition. A second step (*hydrolysis*) is required in which aqueous acid is added to ***21.75*** to protonate the alkoxide and generate the alcohol product (***21.76***).

O, MgBr, CH$_3$MgBr, H$_3$C, CH$_3$, H$_3$O+, OH

21.75 21.76

In general, the electrophilic carbon of a carbonyl (aldehyde or ketone) reacts with the nucleophilic carbon of a Grignard reagent to form an alkoxide. Hydrolysis then gives the alcohol.

Can We Correlate the Type of Alcohol Formed with the Structure of the Aldehyde or Ketone?

Yes! Ketones react with Grignard reagents to give tertiary alcohols, and aldehydes react to give secondary alcohols. Formaldehyde is the only aldehyde that reacts to give a primary alcohol.

What Is the Major Product When 21.77 Is Involved in the Three Reactions Indicated on the Arrow?

Initial reaction of 1-bromohexane (*21.77*) and lithium produces 1-lithiohexane (hexyllithium). This nucleophilic species reacts with cyclohexanone to give the alkoxide, and hydrolysis will generate the alcohol, ***21.78***. This is exactly analogous to the reaction of a Grignard reagent with a ketone or aldehyde.

Br; 1. Li°, ether; 2. (cyclohexanone) O; 3. H$_2$O; HO

21.77 21.78

What Is the Mechanism of Reaction When Methylmagnesium Bromide Reacts with 2-Butanone?

The nucleophilic carbon of the Grignard is attracted to the electropositive carbonyl of the carbonyl, and the electropositive magnesium is attracted to the electronegative oxygen of the carbonyl, leading to a four-centered transition state (***21.79***), allowing formation of a new carbon-carbon bond in the alkoxide product, ***21.80***. The alkoxide must be treated with aqueous acid, as shown, to generate the final alcohol product, ***21.81***.

O; CH$_3$MgBr, ether; Br, Mg, CH$_3$; Mg-Br, CH$_3$; H$_3$O+; OH, CH$_3$

21.79 21.80 21.81

What Reaction Is Required to Convert the Alkoxide Product of a Grignard Reaction into an Alcohol?

An acid-base reaction is required to convert ***21.80*** into the final alcohol product, 2-methyl-2-butanol, ***21.81***.

What Is the Product of the Reactions of 21.82–21.85?

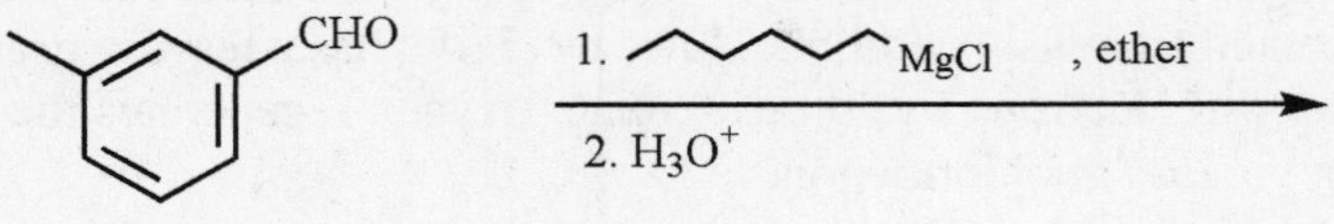

CHO

1. MgCl , ether

2. H_3O^+

21.82

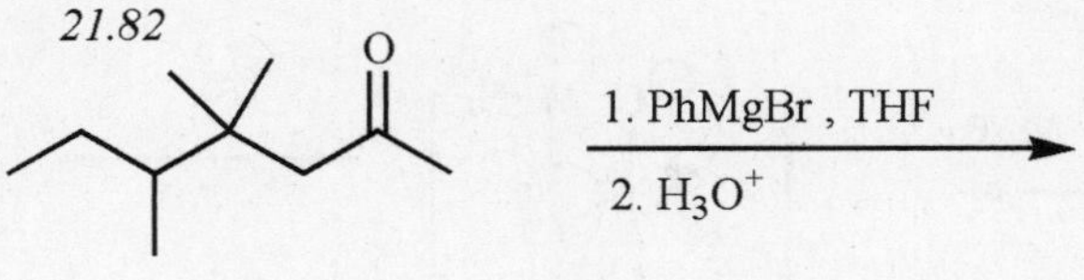

O

1. PhMgBr , THF

2. H_3O^+

21.83

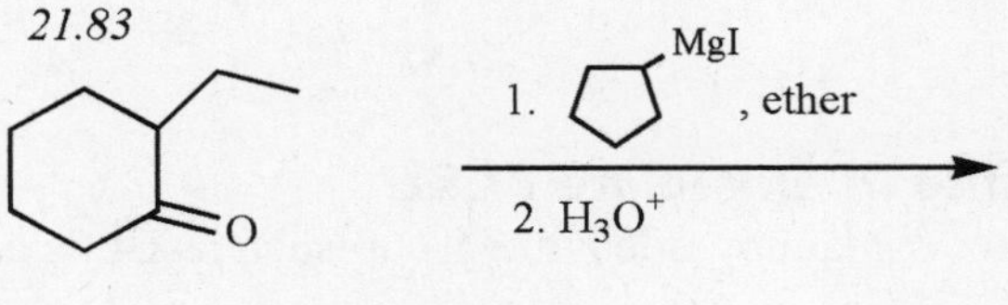

MgI

1. , ether

O

2. H_3O^+

21.84

1. PBr_3

2. Mg° , ether

OH

3. PhCHO

4. H_3O^+

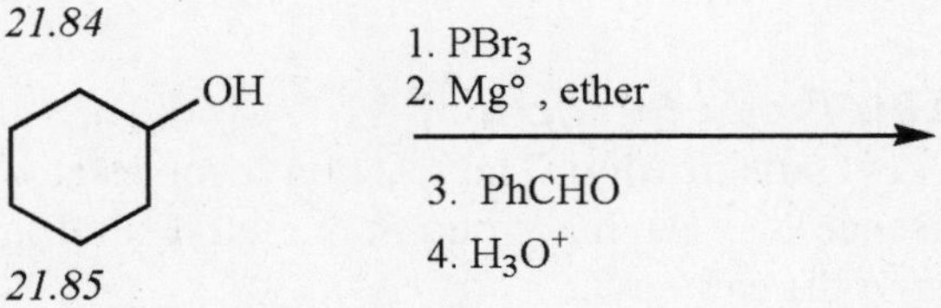

21.85

The reaction of 3-methylbenzaldehyde (***21.82***) and hexylmagnesium chloride is 1-(3-methylphenyl)-1-heptanol (***21.86***). When 4,4,5-trimethyl-2-heptanone (***21.83***) reacts with phenylmagnesium bromide, the product is 4,4,5-trimethyl-2-phenyl-2-heptanol, ***21.87***. Reaction of 2-ethylcyclohexanone (***21.84***) and cyclopentylmagnesium iodide leads to formation of 1-cyclopentyl-2-ethylcyclohexanol, ***21.88***. Finally, reaction of cyclohexanol (***21.85***) and PBr_3 forms the bromide (bromocyclohexane), and subsequent reaction with magnesium gives the Grignard reagent. When this Grignard reagent is condensed with benzaldehyde, the product is cyclohexylphenylmethanol, ***21.89***.

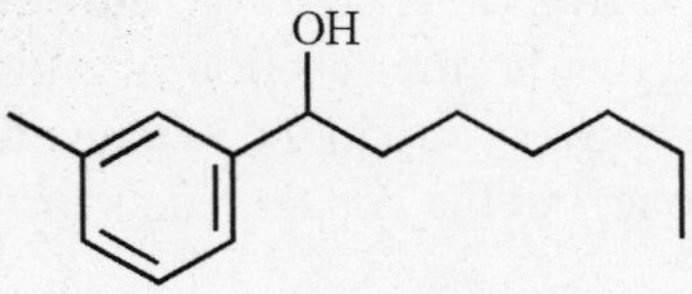

OH Ph OH OH OH Ph

21.86 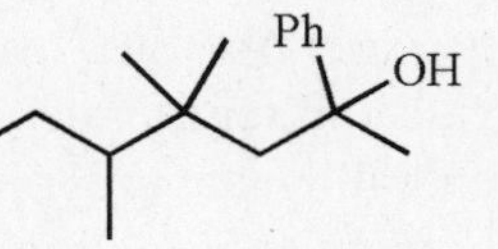 *21.87* 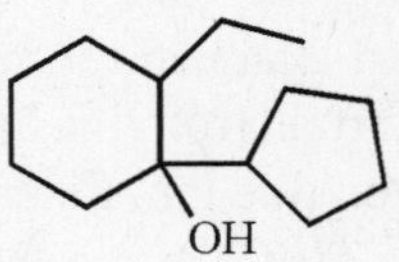 *21.88* 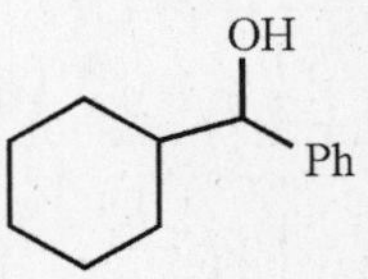 *21.89*

Grignard Reaction with Carboxylic Acid Derivatives

What Is the Usual Product When an Acid Chloride Is Treated with a Grignard Reagent? Explain Your Answer.

When a Grignard reagent reacts with an acid chloride, the initial product is a ketone (***21.90***), but this is also reactive with the Grignard reagent, and the final product, after hydrolysis, is the tertiary alcohol, ***21.91***.

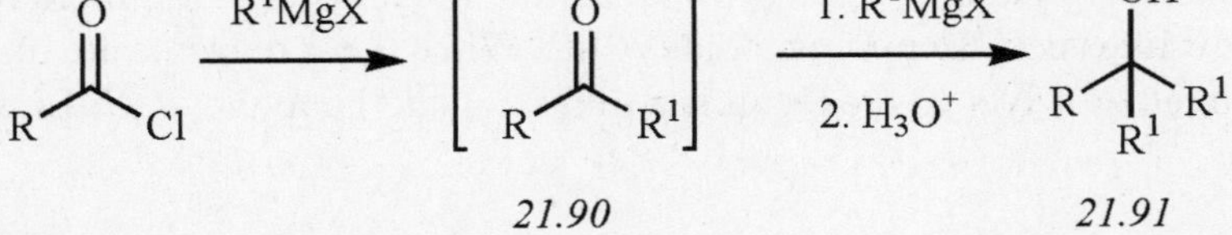

O R^1MgX O 1. R^1MgX OH

R Cl R R^1 2. H_3O^+ R R^1 R^1

21.90 *21.91*

Why Is the Product of a Grignard Reagent and an Acid Chloride a Ketone?

If we use methylmagneisum bromide as the Grignard reagent, initial reaction with the acid chloride generates ***21.92***, as expected from acyl addition. However, ***21.92*** has a leaving group attached to the acyl carbon, chlorine. Transfer of electrons from the alkoxide oxygen regenerates the carbonyl group in the ketone product (***21.93***), via loss of chloride ion.

$$RC(=O)Cl \xrightarrow{CH_3MgBr} \left[R\text{–}C(O^{\ominus}\ MgBr^{\oplus})(CH_3)\text{–}Cl \right] \xrightarrow{-\,Cl^-} RC(=O)CH_3$$

21.92 *21.93*

What Is the Name Given to Intermediate 21.92?

This type of intermediate, containing a leaving group such as Cl, is called a **tetrahedral intermediate**.

What Is the Name for This Type of Reaction?

This reaction is called **acyl substitution**. The reaction involves acyl addition of the nucleophile to the acyl carbon, but the presence of a leaving group in the tetrahedral intermediate leads to loss of the leaving group and formation of the carbonyl.

What Is the Expected Product When Benzoyl Chloride Is Reacted with Ethylmagnesium Bromide?

The product is the tertiary alcohol, 3-phenyl-3-pentanol.

What Is the Usual Product When an Ester Is Treated with a Grignard Reagent?

If we use methylmagneisum bromide as the Grignard reagent, initial reaction with the ester (***21.94***—for the chemistry of esters, see Chapter 25) generates ***21.95***, as expected from acyl addition. However, ***21.95*** has a leaving group attached to the acyl carbon, OEt. This is identical to the reaction of an acid chloride. Transfer of electrons from the alkoxide unit regenerates the carbonyl group in the ketone product (***21.96***), via loss of ethoxide ion. Esters generally react to form a ketone, but the ketone reacts further to give a tertiary alcohol (***21.97***) as the final product after hydrolysis.

$$RC(=O)OEt \xrightarrow{CH_3MgBr} \left[R\text{–}C(O^{\ominus}\ MgBr^{\oplus})(CH_3)\text{–}OEt \right] \xrightarrow{-\,Cl^-} RC(=O)CH_3 \xrightarrow[2.\ H_3O^+]{1.\ CH_3MgBr} RC(OH)(CH_3)CH_3$$

21.94 *21.95* *21.96* *21.97*

What Is an Organocuprate?

An organocuprate is a product formed by the reaction of excess organolithium reagents (RLi) and a cuprous [Cu (I)] derivative such as cuprous iodide (CuI). When these reagents are mixed at temperatures lower than $-10°C$, in ether solvents, the result is a lithium dialkyl cuprate, R_2CuLi, which is a source of nucleophilic carbon.

What Name Is Attached to the Lithium Dialkyl Cuprates?

These reagents are named after Henry Gilman, who discovered them. They are called **Gilman reagents**.

What Is the Product When Butanoyl Chloride Is Treated with Lithium Di-N-Butyl Cuprate?

Lithium di-*n*-butyl cuprate $(n\text{-Bu})_2CuLi$ is prepared by reaction of *n*-butyllithium and cuprous iodide. When this reagent reacts with butanoyl chloride, the product is 4-octanone. Under these conditions, the organocuprate does **not** react with the ketone product.

Why Does the Organocuprate Produce a Ketone Rather Than a Tertiary Alcohol?

An organocuprate is much less reactive than a Grignard reagent. It is strong enough to react with a highly reactive species such as an acid chloride, but not strong enough to react with the less reactive ketone.

What Is the Formula of a Dialkylcadmium Reagent? Show the Formula of Diethyl Cadmium.

A Dialkylcadmium reagent has the structure R_2Cd. An example is diethylcadmium, Et_2Cd.

How Is Diethyl Cadmium Prepared?

Diethylcadmium is prepared by the reaction of ethyllithium with cadmium chloride ($CdCl_2$). All organocadmium reagents are prepared in this way (from the appropriate organolithium), but the reaction demands anhydrous $CdCl_2$.

What Is the Product When Pentanoyl Chloride Reacts with Dibutylcadmium?

As with organocuprates, when a dialkyl cadmium reacts with an acid chloride, the product is a ketone. In this case, pentanoyl chloride reacts with dibutylcadmium to give 5-nonanone.

Why Is the Tertiary Alcohol Not Formed in Dialkyl Cadmium Reactions?

Dialkylcadmium reagents are relatively weak nucleophiles. They are strong enough to react with the highly reactive acid chloride but not strong enough to react with the less reactive ketone product of the reaction.

Grignard Reaction with Nitriles

What Is a Nitrile?

A nitrile is an organic compound that contains the cyano unit (CN).

What Is the Structure of Butanenitrile?

The structure is $CH_3CH_2CH_2C{\equiv}N$.

How Can Nitriles Be Prepared?

An S_N2 reaction of an alkyl halide and cyanide ion will give a nitrile. The reaction of 1-iodopropane and sodium cyanide in ether, for example, will give butanenitrile as the major product.

Why Does Ethyl Magnesium Bromide React with Butanenitrile? At What Atom Does the Grignard Reagent React?

The carbon of the nitrile group is polarized with a δ+ charge ($^{\delta+}C{\equiv}N^{\delta-}$), and the nucleophilic carbon of the Grignard reagent will be attracted to that electropositive atom. In this regard, the nitrile carbon shows many analogies to a carbonyl carbon.

What Is the Initial Product of the Reaction of Ethylmagnesium Bromide and Butanenitrile?

The initial product is an iminium salt, ***21.98***. When the Grignard reagent attacks the nitrile carbon, a new carbon-carbon bond is formed, and one of the π-bonds is broken, 'dumping' those electrons on the nitrogen, the most electronegative atom.

C≡N BrMg → $N^- MgBr^+$

21.98

What Is the Product When 21.98 Is Reacted with Aqueous Acid?

Aqueous hydrolysis of an iminium salt (with acid) leads to a ketone, in this case 3-hexanone, ***21.99***.

C≡N BrMg → $N^- MgBr^+$ —H_3O^+→ O

21.98 *21.99*

Give the Mechanism for This Latter Hydrolysis Reaction.

The initially formed iminium salt (***21.98***) is protonated to give an imine (***21.100***) and then protonated a second time to give an iminium salt, ***21.101***. The carbon of this iminium salt is electropositive and is attacked by water (the only nucleophile in the system) to give ***21.102***, which loses a proton. The proton is transferred to the more basic atom (nitrogen) in ***21.103***, and loss of ammonia (a good leaving group) generates the protonated carbonyl, ***21.104***. Loss of a proton from ***21.104*** completes the hydrolysis to generate the ketone, ***21.99***.

$N^- MgBr^+$ —$+H^+$→ NH —$+H^+$→ $\oplus NH_2$ —$+H_2O^+$→ NH_2, $\oplus OH_2$

21.98 *21.100* *21.101* *21.102*

—1. $-H^+$ 2. $+H^+$→ $\oplus NH_3$, OH —$-NH_3$→ $\oplus OH$ —$-H^+$→ O

21.103 *21.104* *21.99*

Give the Major Product for the Reactions of 21.105–21.107.

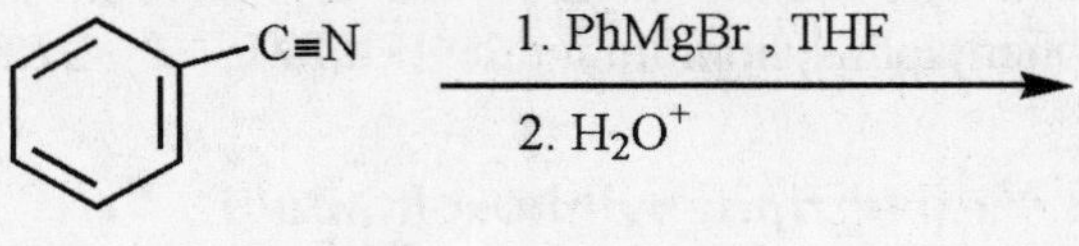

C≡N — 1. PhMgBr , THF; 2. H_2O^+ →

21.105

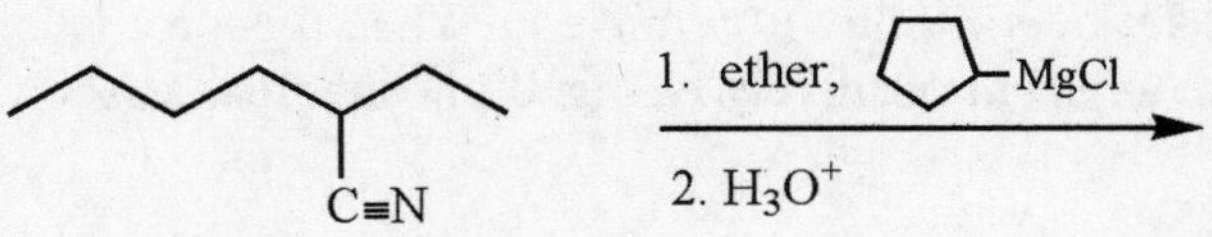

C≡N — 1. ether, cyclopentyl-MgCl; 2. H_3O^+ →

21.106

C≡N — 1. CH_3MgI , ether , reflux; 2. H_3O^+ , reflux →

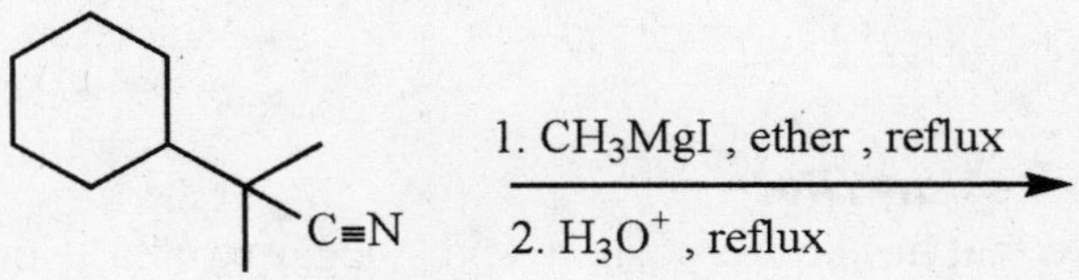

21.107

In the first reaction, benzonitrile (***21.105***) reacts with phenylmagnesium bromide to give benzophenone (***21.108***), after acid hydrolysis. In the second reaction, 2-ethylhexanenitrile (***21.109***) is treated with cyclopentylmagnesium chloride to give the corresponding ketone, ***21.109***. In the last example, ***21.107*** is treated with methylmagnesium iodide. In such sterically hindered nitriles, the reaction with a Grignard is sluggish and requires more vigorous conditions. The imine product is also sterically hindered and the hydrolysis can be slow, requiring vigorous conditions. The ketone product would be ***21.110***.

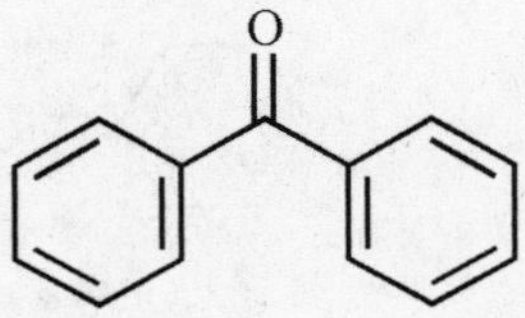

O O O

21.108 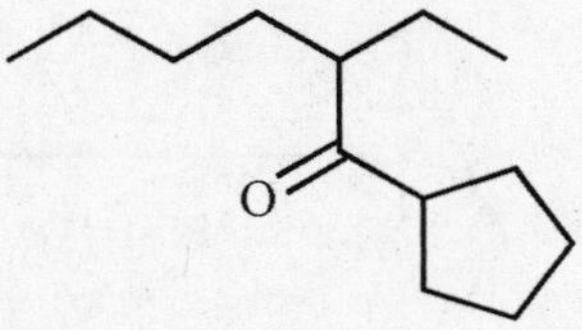 *21.109* 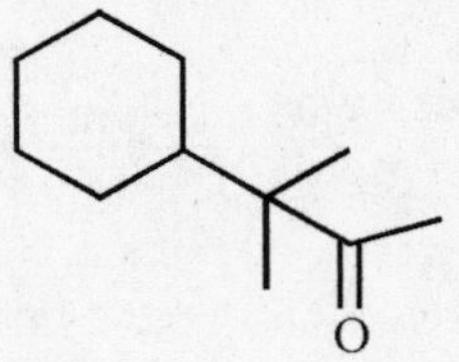 *21.110*

21.5. THE WITTIG REACTION

An ylid reacts with ketones or aldehydes to produce an alkene in what is known as the **Wittig reaction** or **Wittig olefination**.

What Is an Ylid?

An ylid is a molecule that has a positive and a negative charge on adjacent atoms.

What Is the Structure of Triphenylphosphine?

Phosphines are the phosphorus analogs of amines. Triphenylphosphine is Ph_3P.

What Is the Closest Analogy to the Reaction of a Phosphine?

Since phosphorus is beneath nitrogen in the Periodic Table, phosphines are expected to react similarly to amines, R_3N.

What Is the Product When Triphenylphosphine Reacts with Iodomethane?

The phosphorus of triphenylphosphine behaves as a nucleophile, displacing iodide from iodomethane to form methyltriphenylphosphonium iodide [$Ph_3PMe^+ I^-$, ***21.111***].

What Is the Product When Methyltriphenylphosphonium Iodide (21.111) Reacts with N-Butyllithium?

The proton (on the methyl group) that is α- to the phosphorus is acidic (a weak acid, pK_a 25–32) and requires a strong base such as *n*-butyllithium to remove it. When this proton is removed, the remaining carbanion is stabilized by the adjacent positively changed phosphorus, as shown in ***21.112***. Compound ***21.112*** is an ylid.

$$Ph_3P + CH_3I \longrightarrow \underset{21.111}{Ph_3\overset{\oplus}{P}\text{-}CH_3\ \ \overset{\ominus}{I}} \xrightarrow{n\text{-BuLi}} \underset{21.112}{\left[Ph_3\overset{\oplus}{P}\text{-}\overset{\ominus}{C}H_2 \longleftrightarrow Ph_3P{=}CH_2 \right]}$$

What Is the Definition of an Ylid?

An ylid is a molecule that has both a positive and a negative charge on adjacent atoms. As shown in ***21.112***, two resonance structures can be drawn. This product is named triphenylphosphonium methylid, or methylenetriphenylphosphorane. It is noted that *ylid* can also be spelled *ylide*.

When Triphenylphosphonium Methylid (Methylenetriphenylphosphorane) Reacts with Cyclohexanone, What Is the Major Product?

When a phosphonium ylid reacts with a ketone or aldehyde, an alkene is formed, along with triphenylphosphine oxide [$Ph_3P{=}O$]. In this case, ***21.112*** reacts with cyclohexanone to form methylene cyclohexane, ***21.113***.

$$\underset{21.112}{\left[Ph_3\overset{\oplus}{P}\text{-}\overset{\ominus}{C}H_2 \longleftrightarrow Ph_3P{=}CH_2 \right]} \xrightarrow[-\ Ph_3P=O]{\text{cyclohexanone } (C_6H_{10}{=}O)} \underset{21.113}{C_6H_{10}{=}CH_2}$$

What Name Is Associated with This Reaction?

The reaction is named after the person who discovered that phosphorus ylids reacted with aldehydes and ketones to give alkenes, Georg Wittig. It is known as the **Wittig reaction**.

How Does This Reaction Work?

The ylid (***21.112***) initially reacts with the carbonyl to form a dipolar ion, ***21.114***, often called a betaine. The phosphorus and oxygen react to form a four-membered ring compound, ***21.115***, known as an oxaphosphatane. The P–O bond is very strong, and its formation (exothermically) drives the cleavage of the four-membered ring differently from the way in which it was formed. The C–P and C–O bonds are cleaved, forming the alkene (***21.113***) and triphenylphosphine oxide.

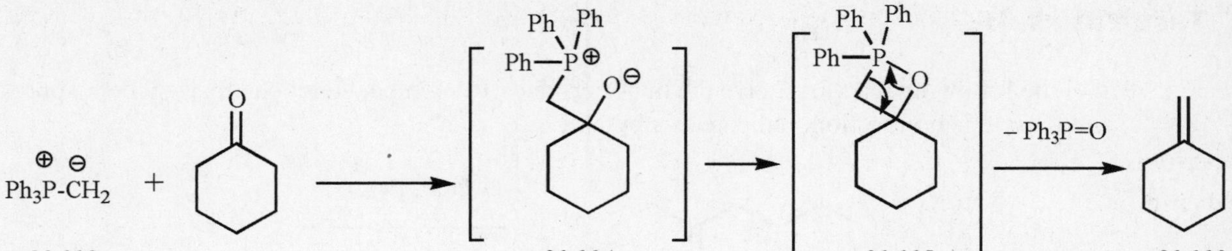

$Ph_3\overset{\oplus}{P}\text{-}\overset{\ominus}{C}H_2$ + cyclohexanone → [21.114] → [21.115] —($-Ph_3P{=}O$)→ 21.113

21.112 21.114 21.115 21.113

What Is the Distinction Between an Oxaphosphatane and a Betaine?

The betaine is the dipolar ion, ***21.114***. An oxaphosphatane is a four-membered ring containing both a phosphorus and an oxygen, such as ***21.115***.

What Is the 'Driving Force' for the Wittig Olefination Reaction? In Other Words, Why Does It Work?

Formation of the very strong P–O bond leads to cleavage of the oxaphosphatane to generate an alkene and triphenylphosphine oxide.

Give the Major Product of the Reaction of 21.116–21.118.

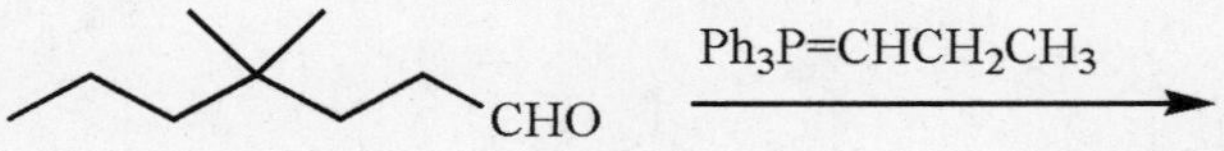

21.116 (CHO) —$Ph_3P{=}CHCH_2CH_3$→

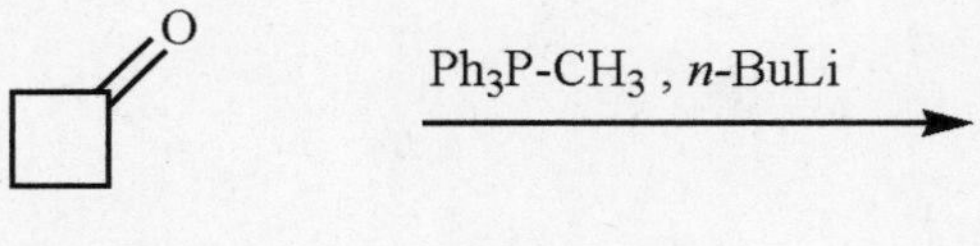

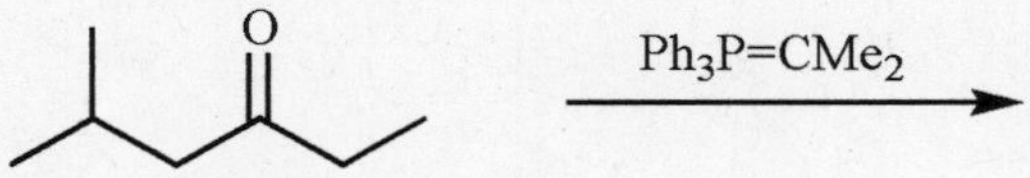

21.117 —$Ph_3\overset{+}{P}\text{-}CH_3$, *n*-BuLi→

21.118 —$Ph_3P{=}CMe_2$→

Aldehyde ***21.116*** reacts with the ylid to produce the linear alkene, ***21.119***. Cyclobutanone (***21.117***) reacts to form the exocyclic methylene compound, methylenecyclobutane, ***21.120***. Exocyclic means the C=C unit is outside the ring, or attached to it, rather than being part of the ring. In the last example, a methyl group is incorporated in the conversion of ketone ***21.118*** into the tetrasubstituted alkene, ***21.121***.

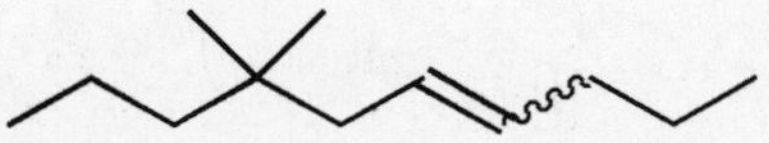

21.119  21.120 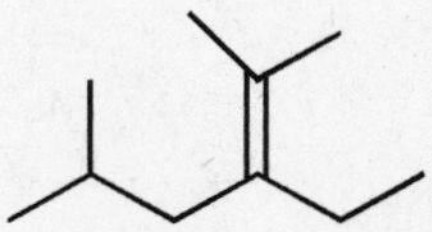 21.121

Test Yourself

1. For each of the following reactions, give the major product. Remember stereochemistry where appropriate, and if there is no reaction, indicate this by N.R.

(a) → 1. PhMgBr , THF; 2. H_3O^+

(b) CHO → 1. MgCl, ether; 2. H_3O^+

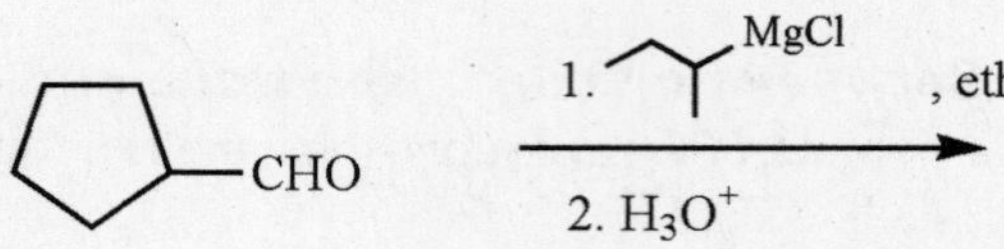

(c) Br → 1. Mg°, ether; 2. HCHO; 3. H_3O^+

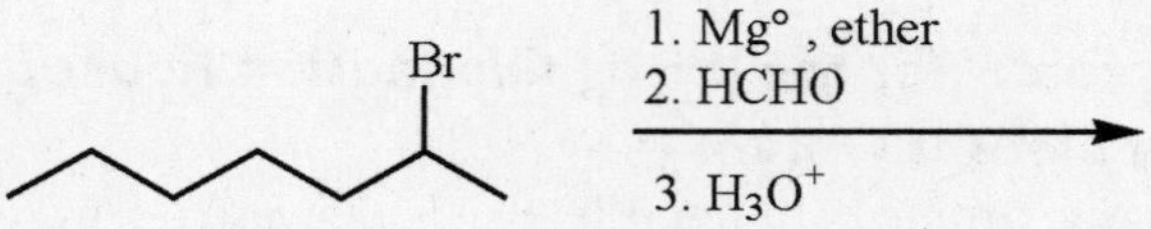

2. How can one distinguish between cyclohexanone and 2-cyclohexenone using spectroscopy?

3. Give the IUPAC name for each of the following.

(a) O

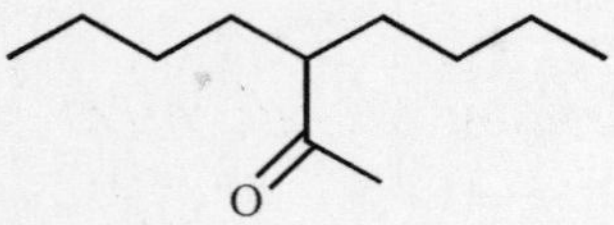

(b) CHO

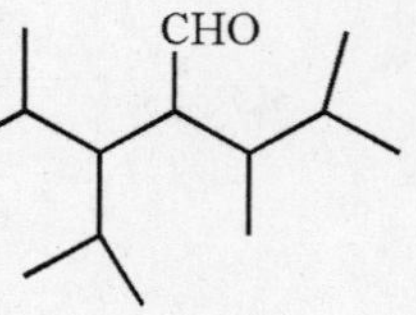

(c) CHO, Cl, Cl

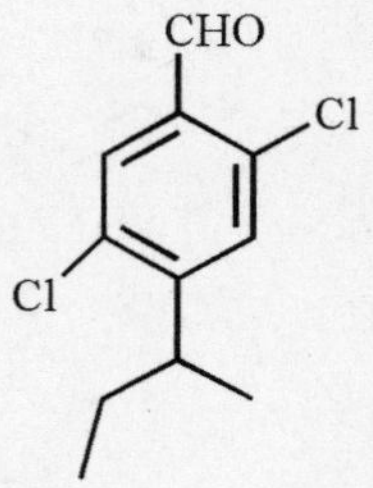

(d) Ph, Ph, O

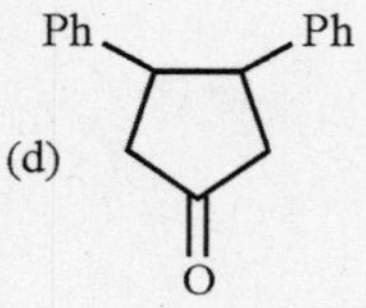

(e) O, O

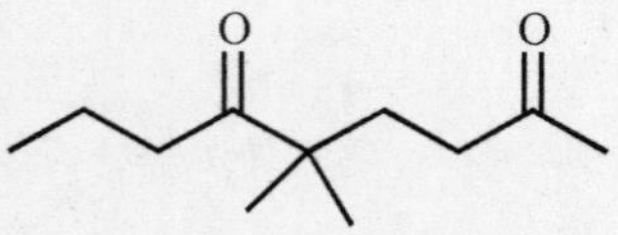

(f) CHO

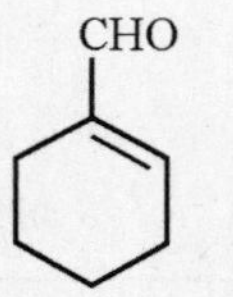

(g) O, Ph

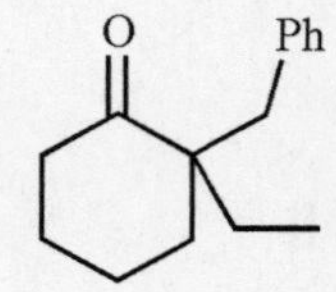

(h) O, H

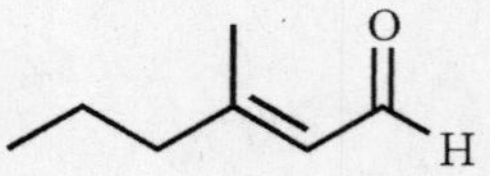

4. When butanoyl chloride is treated with one equivalent of butylmagnesium bromide, why is it difficult to obtain 4-octanone as the final major product?

5. Give the mechanism of the following reaction.

Me, Me, N^- $MgBr^+$ → H_3O+ → Me, Me, O

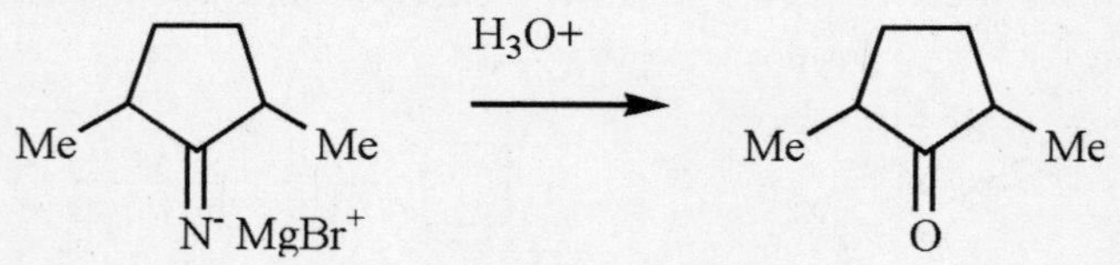

6. Give a complete mechanism for each of the following reactions.

(a) 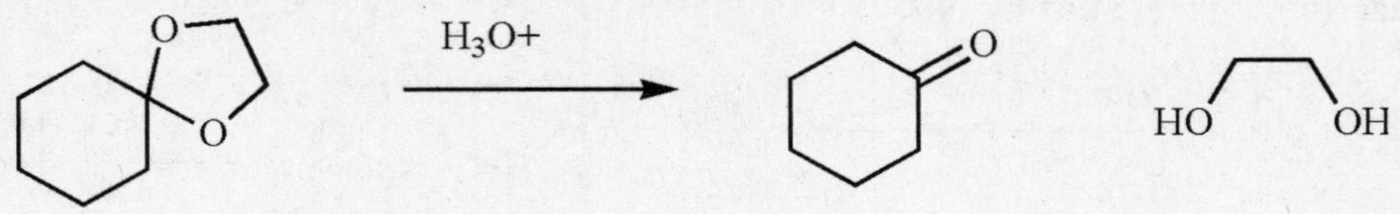 $\xrightarrow{H_3O+}$ HO OH

(b) 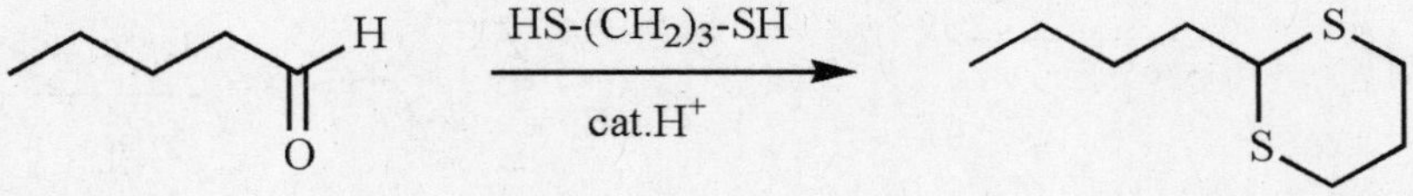 H O $\xrightarrow[\text{cat.}H^+]{HS\text{-}(CH_2)_3\text{-}SH}$ S S

7. What is the initial product when the hydrate of 2-butanone loses water? How is this converted back to 2-butanone?

8. Draw a picture of a Dean-Stark apparatus.

9. Which of the following might form stable hydrates?

(a) 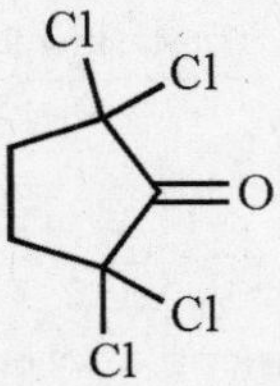 Cl Cl O Cl Cl (b) 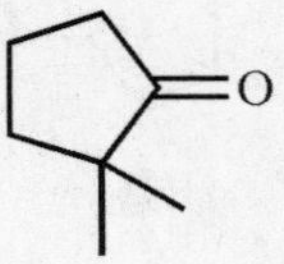 O (c) 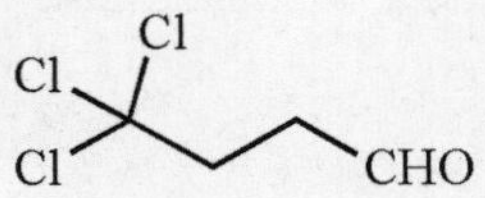 Cl Cl Cl CHO (d) 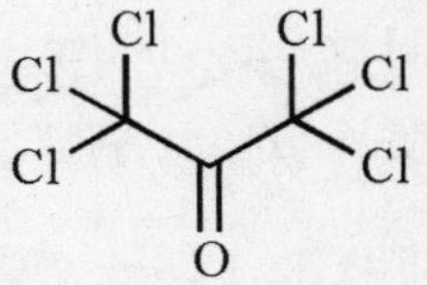 Cl Cl Cl Cl Cl Cl O

10. Give the structures of 1,3-dioxane, of 1,3-dioxolane, of 1,3-dithiane and of 1,3-dithiolane.

11. Give the IUPAC name of each of the following.

(a) 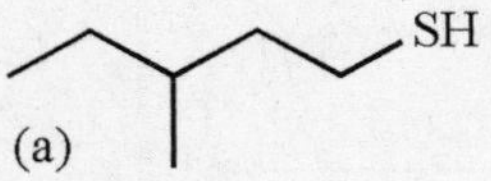 SH (b) 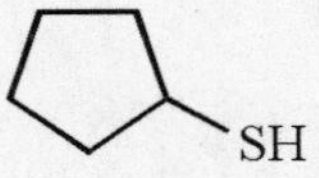 SH (c) 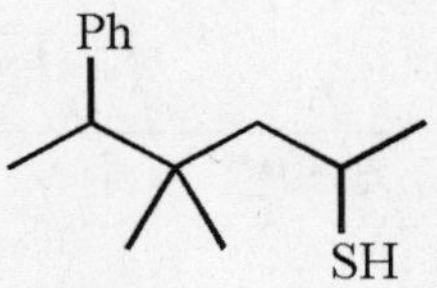 Ph SH (d) 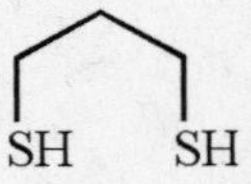 SH SH

12. In each case, give the major product Remember stereochemistry where appropriate, and if there is no reaction, indicate this by N.R

(a) 1. OsO_4 2. $NaHSO_3$ 3. H_5IO_6

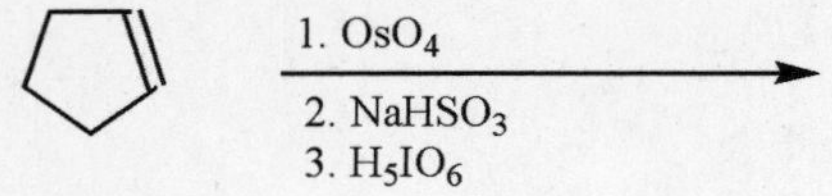

(b) 1. O_3 2. H_2O_2

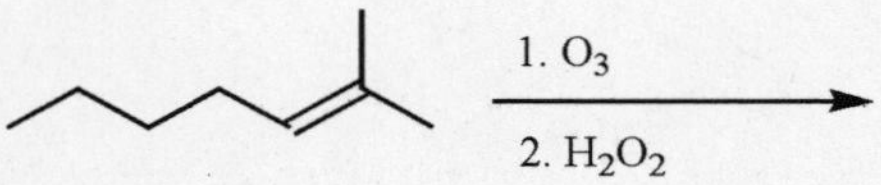

(c) Br; 1. KOH, EtOH 2. O_3 3. Me_2S

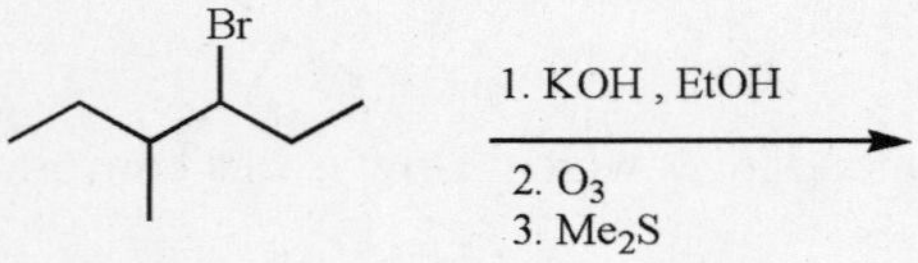

(d) OsO_4, H_5IO_6

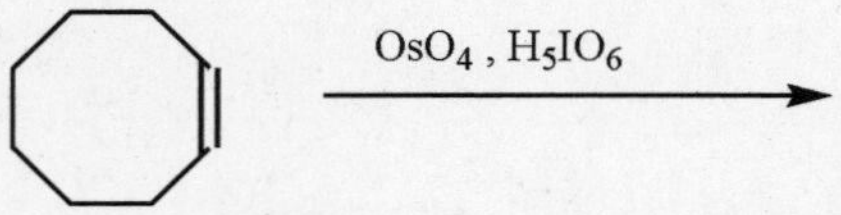

(e) O; cat. p-TsOH, EtOH

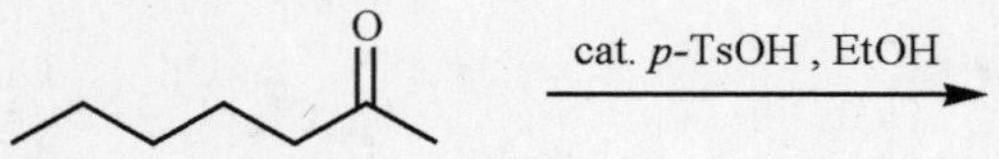

(f) CCl_3—CHO; H_2O

(g) O; 1. $HSCH_2CH_2SH$, cat. H^+ 2. Ni(R), acetone

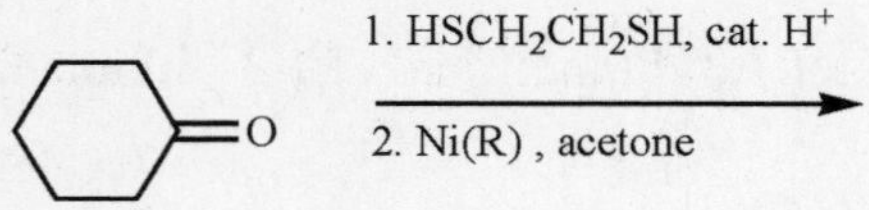

(h) CHO; 1. cat. H^+, OH OH 2. H_3O^+

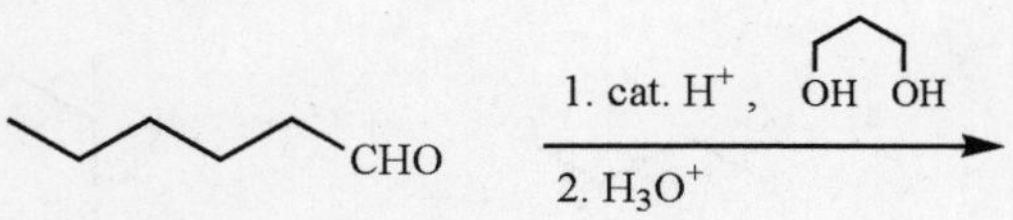

(i) O; HCN

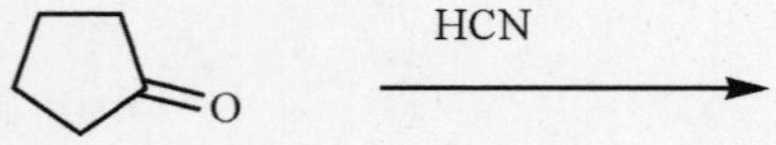

(j) O; 1. EtC≡C:$^-$, DMF 2. H_3O^+

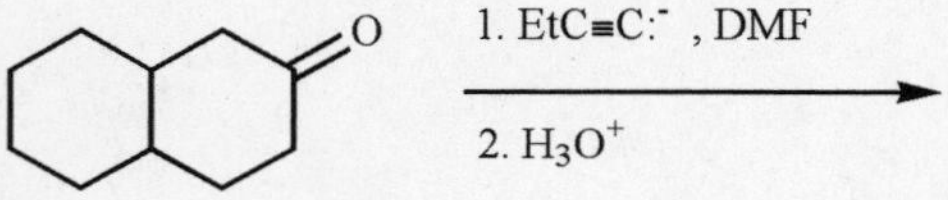

(k) 1. $NaNH_2$, THF 2. 2-butanone 3. H_3O^+

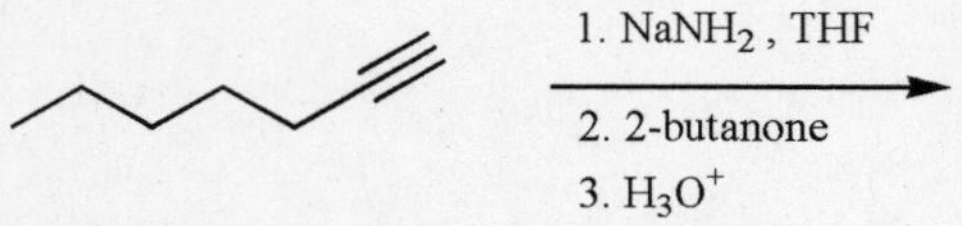

(l) C≡N; 1. C_3H_7MgBr 2. H_3O^+, heat

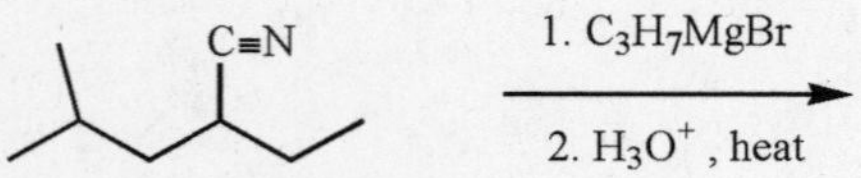

(m) Br; 1. KCN, DMF 2. PhMgBr 3. H_3O^+, heat

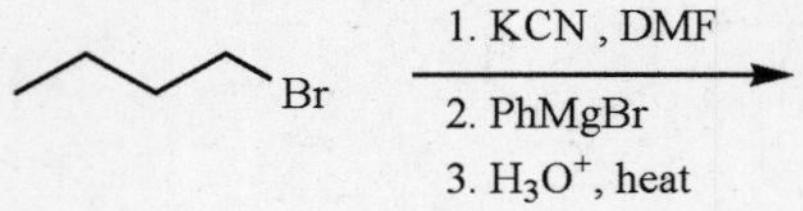

(n) Br; 1. Mg°, ether 2. HCHO 3. H_3O^+

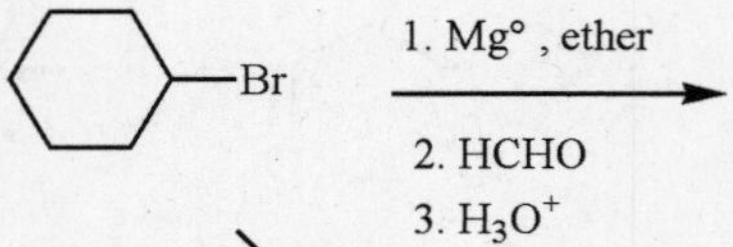

(o) O; 1. MgI, THF 2. H_3O^+

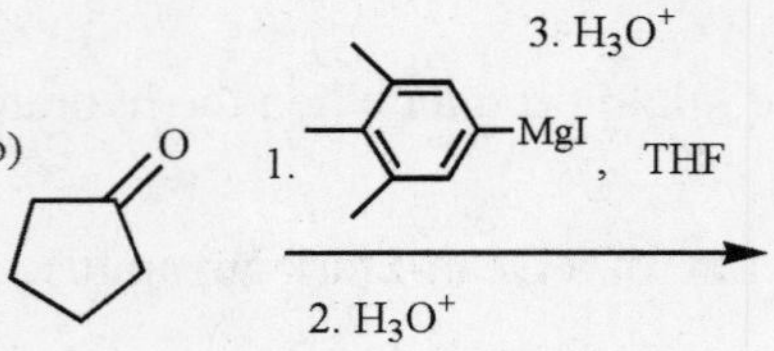

(p) O, OMe; 1. excess EtMgBr, ether 2. H_3O^+

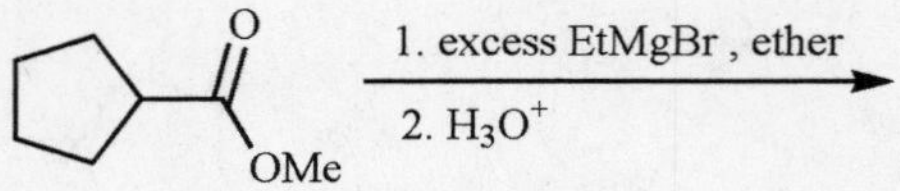

(q) I; 1. Mg, THF 2. 3,5-dimethylcyclohexanone 3. H_3O^+

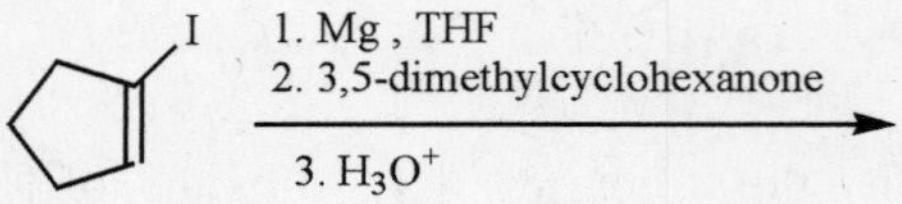

(r) O, Cl; Ph_2CuLi

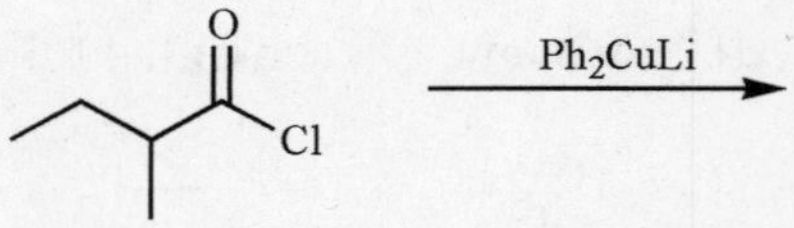

(s) OMe O, Cl; Et_2Cd

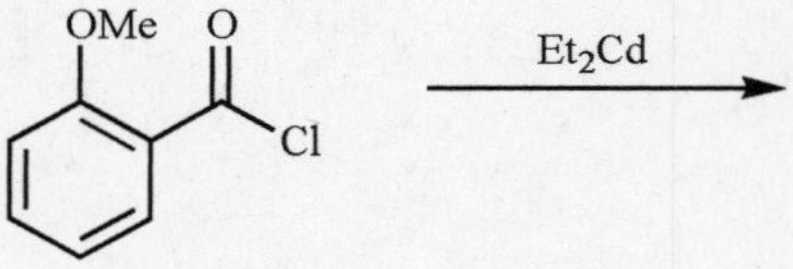

(t) CHO; Ph_3P=$CHCH_3$

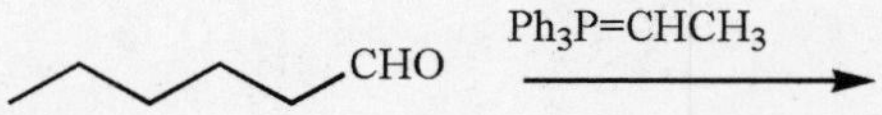

(u) PPh_3^+ I^-; 1. BuLi, THF 2. cyclohexanone

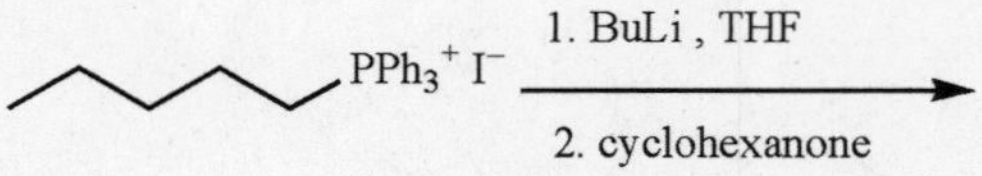

(v) I; Bu_3P, THF

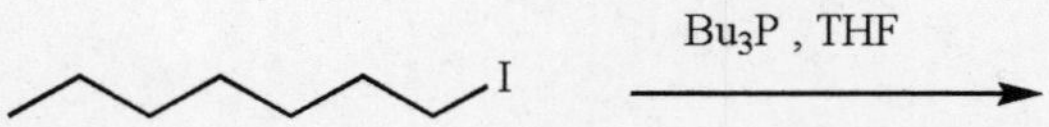

13. For each of the following provide a suitable synthesis. Show the structure of all intermediate products, and show all reagents.

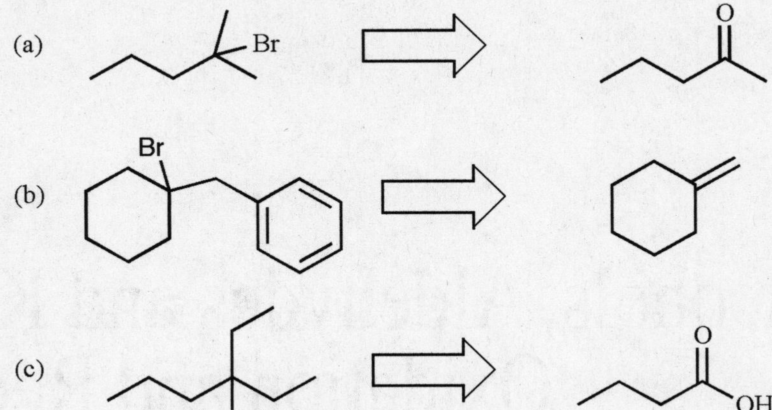

CHAPTER 22

Alcohols, Aldehydes and Ketones. Oxidation and Reduction

22.1. OXIDATION

Alcohols are oxidized to aldehydes or ketones. Oxidation is formally defined as loss of two electrons.

What Is the Definition of an Oxidation?

Oxidation is the reverse of reduction and is, therefore, a loss of two electrons in the course of a chemical reaction. Oxidation can also be defined as the loss of hydrogen or the gain of heteroatoms, such as in oxygen.

What Is the Oxidation Number for a Molecule?

The oxidation number for a molecule is the sum of the oxidation levels for each carbon in the molecule.

What Is an Oxidation Level?

Oxidation level is a number associated with gain or loss of electron density. The usual formalism involves assigning to carbon a -1 to each attached element that is less electronegative than carbon (such as H) and $+1$ to each attached element that is more electronegative than carbon (such as O, N, Br, etc.). Another attached carbon is assigned a value of zero (0). These numbers are added together to give the oxidation level for each carbon.

What Is the Change in Oxidation Number When Acetone Is Converted to Isopropanol?

In acetone (*22.1*), the two C–C bonds are each assigned a value of 0, and the two bonds to oxygen (C=O) give a value of $+2$ ($+1$ each). The oxidation level for the carbon of the carbonyl group is, therefore, $+2$. Examination of isopropanol (*22.2*) shows the C–O bond is valued at $+1$, the C–H bond at -1, and each C–C bond is 0. The carbon of isopropanol has an oxidation level of 0. The conversion from $+2 \rightarrow 0$ is a net *gain* of electrons. Electrons are negatively charged, and increasing electron density leads to a more negative number. Therefore, the conversion acetone to isopropanol *(22.1* $\rightarrow$ *22.2)* is a reduction.

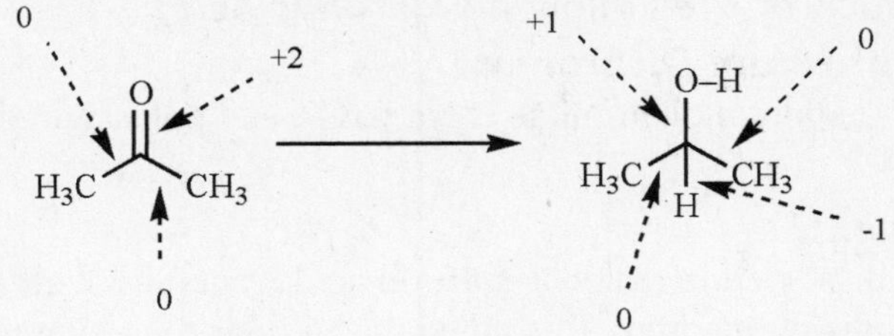

22.1 22.2

How Are Oxidation and Reduction Defined Using Oxidation Number?

In general, if the change in oxidation number is positive, it is an oxidation (see Section 11.4.D), and if it is negative, it is a reduction.

If the Alcohol Is Oxidized, What Happens to the Reagent That Causes the Oxidation to Occur?

Oxidation and reduction are linked. If the alcohol is oxidized, the reagent that caused that transformation must be reduced. Conversely, if a ketone is reduced, the reagent that caused that transformation must be oxidized.

How Are These Reagents Identified?

A reducing agent is a molecule that causes another molecule to be reduced. An oxidation agent is a molecule that causes another molecule to be oxidized..

What Is the Change in Oxidation Number When 2-Propanol Is Converted to 2-Propanone?

As noted above, the oxidation level of the oxygen-bearing carbon in 2-propanol is 0 and that of the carbonyl carbon unit in 2-propanone is +2. Conversion of the alcohol to the ketone involves a $0 \rightarrow +2$ transition. Going to the more positive number indicates a decrease in electron density (a loss of two electrons).

Why Is This Reaction Considered to Be an Oxidation?

The reaction involves a loss of two electrons and also involves loss of two hydrogens from the alcohol.

22.2. OXIDATION OF ALCOHOLS TO ALDEHYDES OR KETONES

Alcohols are oxidized to aldehydes or ketones. Several different reagents are available that accomplish this transformation.

Chromium Trioxide

What Is the Formula of Chromium Trioxide?

The basic formula of chromium trioxide is CrO_3, but this reagent is generally considered to be a polymer, $(CrO_3)_n$ where n is a large number. Chromium trioxide is a Cr (VI) reagent.

What Species Are Present in Solution When Chromium Trioxide Is Dissolved in Water?

There are several Cr(VI) species, including CrO_3, H_2CrO_4 (chromic acid) and dichromate (CrO_7^{-2}). In a large excess of water, dichromate is usually the major species, but this depends on what is added to the solution. In very dilute solutions, more CrO_3 and chromic acid are present.

Give the Formula for Each of the Following: Chromic Acid, Sodium Dichromate, Potassium Dichromate.

Chromic acid is $HCrO_4$; sodium dichromate is Na_2CrO_7; and potassium dichromate is K_2CrO_7.

What Is the Jones Reagent?

Jones reagent is CrO_3 dissolved in aqueous sulfuric acid, generally with an organic solvent such as acetone. It is a powerful oxidizing medium. It is noted that similar oxidation reactions can be performed using CrO_3 with other aqueous acids.

What Is the Product When Cyclohexanol Is Treated with Jones Reagent?

The Cr(VI) in Jones reagent converts cyclohexanol into cyclohexanone.

What Is the Mechanism for Oxidation of an Alcohol to a Ketone?

This reaction begins with reaction between the alcohol (*22.3*) and the chromium (VI) species to give a chromate ester (*22.4*). The hydrogen β- to the chromium is now acidic and can be removed by the water that is in the system (water behaves as a base here), expelling the chromium-leaving group and forming the ketone, *22.5*.

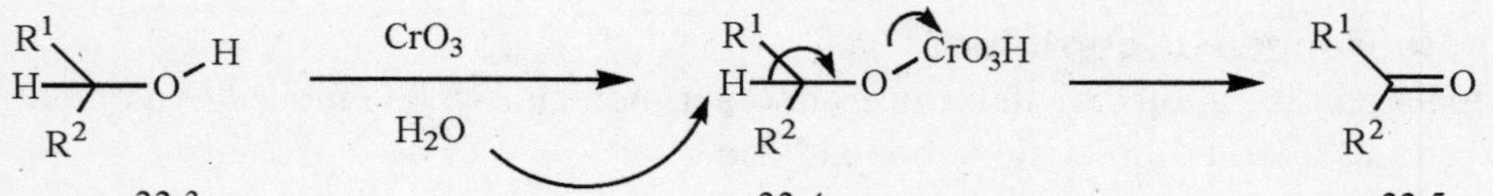

What Is the Nature of a Chromate Ester?

The presence of the chromium species converts CrO_3H into a good leaving group and the H–C–O–Cr moiety makes the α-hydrogen acidic.

How Is a Chromate Ester Converted to the Ketone?

The β-hydrogen (on the chromium atom) is removed by a base, usually water, to form a carbonyl π-bond and expel the chromium (III) species, which disproportionates, since the chromium is a good leaving group. Since acid (H_3O^+) is formed, the reaction proceeds best in acid media.

Why Is It Not Possible to Oxidize a Tertiary Alcohol to a Ketone?

There is no hydrogen β-to the chromate ester of this alcohol, and formation of a carbonyl would require breaking a C–C bond.

Why Is There a Difference in Reaction Rate for Chromate Esters 22.6 and 22.7?

The hydrogen attached to the OCr-bearing carbon in chromate ester in *22.6* (labeled H_a) is very difficult to approach due to steric hindrance imposed by the ring system. In *22.7*, however, the hydrogen (labeled H_b) is relatively accessible. Removal of the indicated hydrogen is the rate-determining step for the oxidation, and since H_a in *22.6* is hard to remove, the rate of oxidation is slower than that for *22.7*.

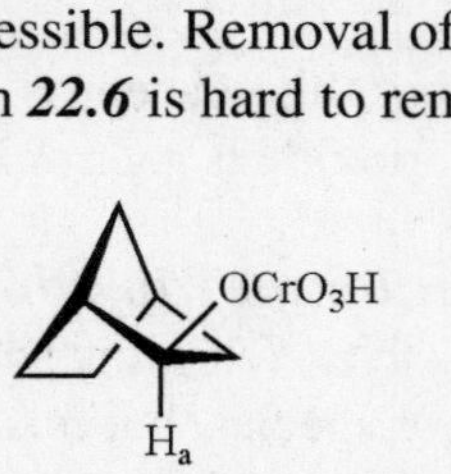

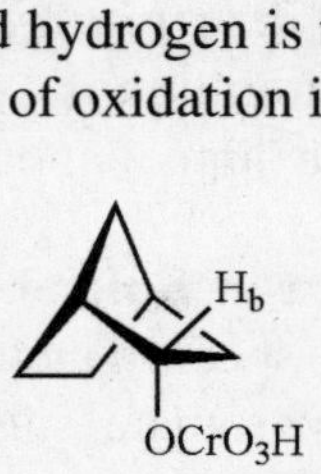

Give the Major Products for the Reactions of 22.8–22.10.

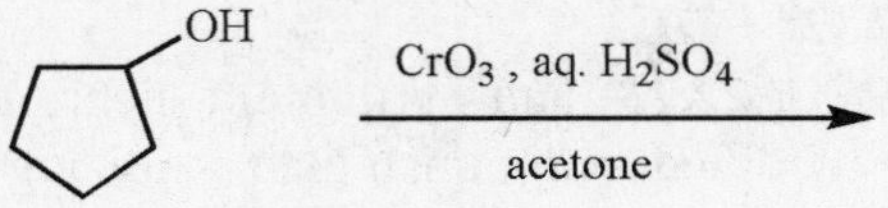

OH $\xrightarrow[\text{acetone}]{CrO_3\text{ , aq. }H_2SO_4}$

22.8

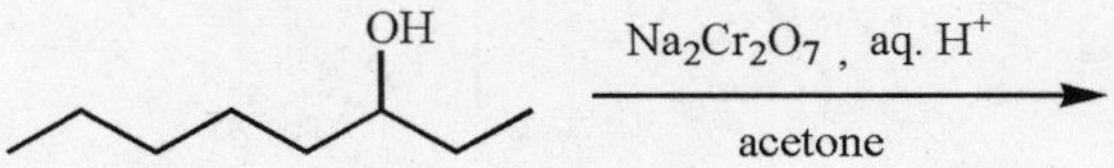

OH $\xrightarrow[\text{acetone}]{Na_2Cr_2O_7\text{ , aq. }H^+}$

22.9

OH $\xrightarrow[\text{acetone}]{CrO_3\text{ , aq. }H_2SO_4}$

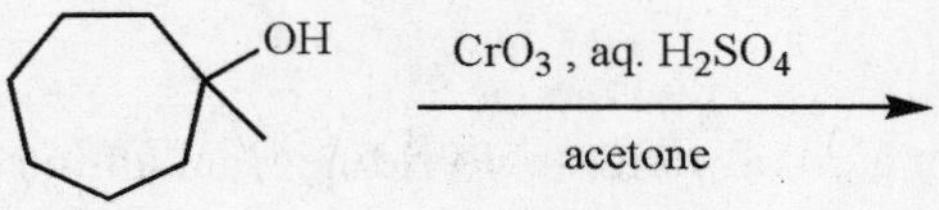

22.10

In the first reaction, cyclopentanol (***22.8***) is converted into cyclopentanone by the Jones reagent. In the second reaction, 3-octanol (***22.9***) is converted into 3-octanone. In the final example, methylcycloheptanol (***22.10***) was treated with Jones reagent. Since this is a tertiary alcohol, no oxidation takes place, and the correct answer is no reaction.

Primary Alcohols to Acids

What Is the Expected Product When a Primary Alcohol Is Oxidized with Cr (VI)?

A primary alcohol is expected to be oxidized into an aldehyde ($RCH_2OH \rightarrow RCHO$), but this is **not** the observed product.

What Is the Actual Product When a Primary Alcohol is Oxidized with CR (VI)? Explain.

The actual product isolated when a primary alcohol is oxidized with Jones reagent is usually a carboxylic acid ($RCH_2OH \rightarrow RCO_2H$). Chromium (VI) in acidic media is a powerful oxidizing medium, and aldehydes are very susceptible to oxidation to acids. Indeed, many low molecular weight aldehydes are oxidized to acids by exposure to oxygen in the air. The initially formed aldehyde from the oxidation will, therefore, be further oxidized to a carboxylic acid.

What Is the Major Product for the Reaction of 4-Phenyl-3, 5-Dimethyl-1-Hexanol (22.11) with Jones Reagent?

We assume that the primary alcohol is converted to the carboxylic acid by Jones oxidation. Therefore, ***22.11*** reacts with Jones reagent to give ***22.12***.

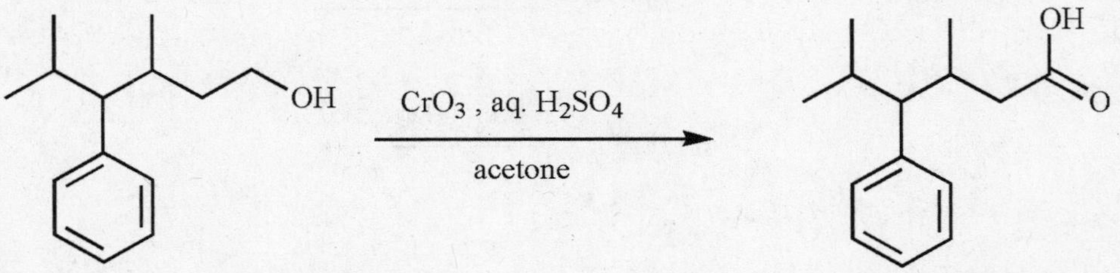

OH $\xrightarrow[\text{acetone}]{CrO_3\text{ , aq. }H_2SO_4}$ OH O

22.11 *22.12*

Pyridinium Chlorochromate and Pyridinium Dichromate

What Is the Structure of PCC? Of PDC?

Pyridinium chlorochromate (PCC, ***22.13***) and pyridinium dichromate (PDC, ***22.14***) are Cr (VI) reagents that have been structurally modified to diminish their oxidizing power and improve their solubility in organic solvents.

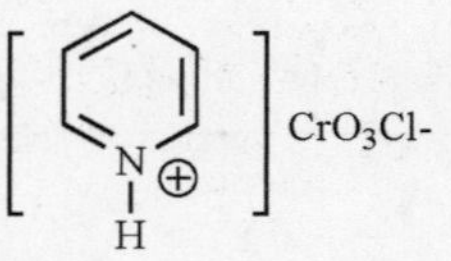

CrO_3Cl^- $Cr_2O_7^{=}$

22.13 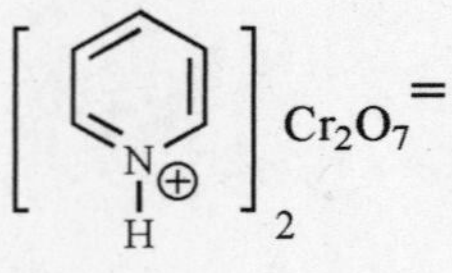 *22.14*

What Is Pyridinium?

Pyridinium is the conjugate acid of pyridine. Pyridine is an amine and has an unshared pair of electrons on nitrogen, which can function as a base. In other words, pyridine (***22.15***) reacts with an acid (HX) to give a pyridinium ion (***22.16***), where the counterion of this salt is X^-, from HX.

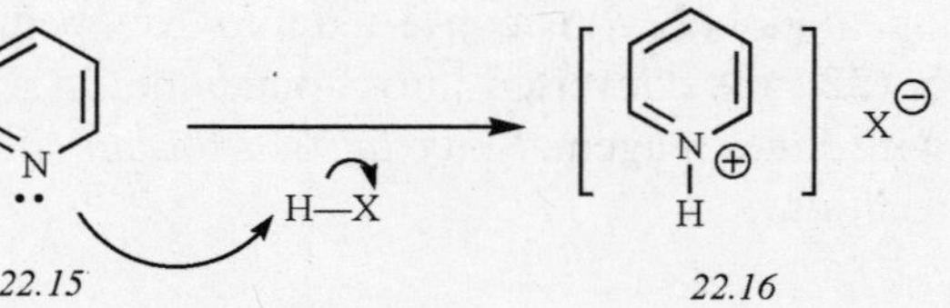

H—X $X^{\ominus}$

22.15 *22.16*

How Are the Reagents PCC and PDC Formed?

They are suspected carcinogens. PCC is formed by reaction of CrO_3 in aqueous HCl and pyridine. PDC is formed by reaction of CrO_3 with pyridine in aqueous solution, but without addition of HCl. In both cases, the reagent is isolated and purified as crystalline material.

Why Are PCC and PDC Used Rather Than Jones Reagent?

They are significantly milder oxidizing agents and will convert a primary alcohol to an aldehyde without further oxidation to the carboxylic acid. Both PCC and PDC will convert a secondary alcohol to a ketone.

What Is the Product of the Reactions of 11.4.51–11.4.54?

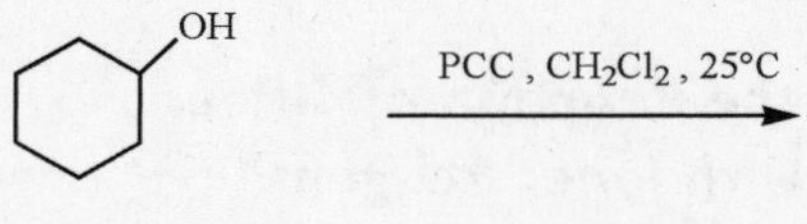

OH — PCC , CH_2Cl_2 , 25°C →

22.17

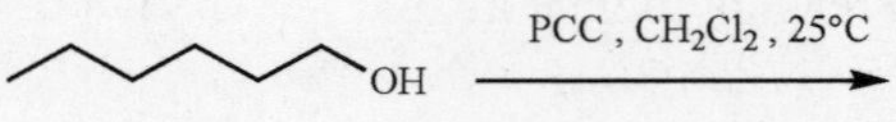

OH — PCC , CH_2Cl_2 , 25°C →

22.18

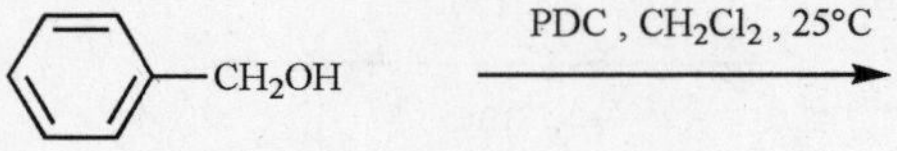

CH_2OH — PDC , CH_2Cl_2 , 25°C →

22.19

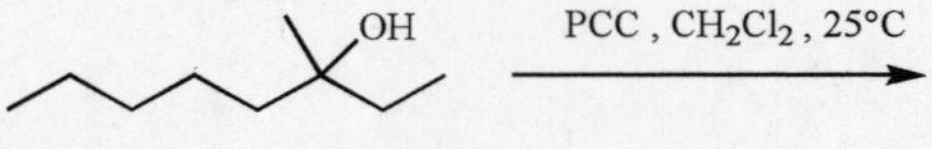

OH — PCC , CH_2Cl_2 , 25°C →

22.20

Cyclohexanol (*22.17*) is oxidized to cyclohexanone with PCC. 1-Hexanol (*22.18*) is oxidized to the aldehyde (hexanal) with PCC. PDC will oxide benzyl alcohol (*22.19*) to benzaldehyde. 3-Methyl-3-octanol (*22.20)* is a tertiary alcohol, and neither PCC nor PDC will oxidize this alcohol. The answer here is no reaction. In all cases, the reactions are performed at 25°C using dichloromethane as a solvent. These are typical reaction conditions.

22.3. REDUCTION OF CARBONYL COMPOUNDS

Aldehydes or ketones are reduced to alcohols. Several different reagents are available that accomplish this transformation.

What Is the Definition of a Reduction?

Reduction is defined as a reaction in which two electrons are gained in the final product. A more practical definition states a reduction is a reaction that adds hydrogen to the molecule or loses an electronegative element (such as oxygen) from the molecule.

Why Is the Conversion of a Ketone to an Alcohol Considered to Be a Reduction?

As seen above for *22.1* → *22.2*, the C=O → H–C–O–H conversion involves the gain of two electrons (formation of two σ bonds [C–H and O–H] with breaking of the π-bond). Clearly, the reaction adds two hydrogens to the carbonyl and, therefore, fits that definition.

Lithium Aluminum Hydride

What Is the Formula of Lithium Aluminum Hydride?

The formula is $LiAlH_4$, where the active species is tetrahydridoaluminate, AlH_4^-.

What Is the Bond Polarization for the Al–H Bond of Lithium Aluminum Hydride?

Aluminum is less electronegative than hydrogen, and the hydrogen takes the negative pole ($H^{\delta-}$). The polarization is $^{\delta+}Al–H^{\delta-}$.

What Is the Product When 3-Methylhexanal Is Treated with i. $LiAlH_4$ in Diethyl Ether ii. Aqueous Acid?

Under these conditions, 3-methylhexanal is converted to 3-methyl-1-hexanol. The aldehyde is reduced to the alcohol.

What Is the Mechanism of the Reaction When Lithium Aluminum Hydride Reacts with 2-Butanone?

The reaction proceeds by a four-centered transition state such as *22.21*. The electropositive aluminum is attracted to the electronegative oxygen. The electronegative hydrogen, a hydride, is attracted to the electropositive carbon. The product of this reaction is *22.22*, in which the π-bond has been broken, a new C–H bond has been formed, and an aluminate species (O–Al) is generated.

AlH_4^-, ether

22.21 *22.22*

To Reduce a Carbonyl, Hydrogen Must Be Delivered to the Acyl Carbon. What Is the Source of This Hydrogen?

As shown with *22.21* and *22.22*, the hydrogen from aluminum (AlH_4^-) attacks the acyl carbon.

What Kind of Reaction Is Necessary to Convert the Aluminum Alkoxide Product into an Alcohol? What Reagent Is Used for this Transformation?

In order to cleave the Al–H bond, an acid-base reaction is necessary. The usual reagent is dilute aqueous acid, although plain water or even aqueous hydroxide can be used. The treatment that leads to the isolation of the alcohol product with the greatest ease is the sequence: i. water; ii. 15% aq. NaOH; iii. 3n water, where n = the number of equivalents of $LiAlH_4$.

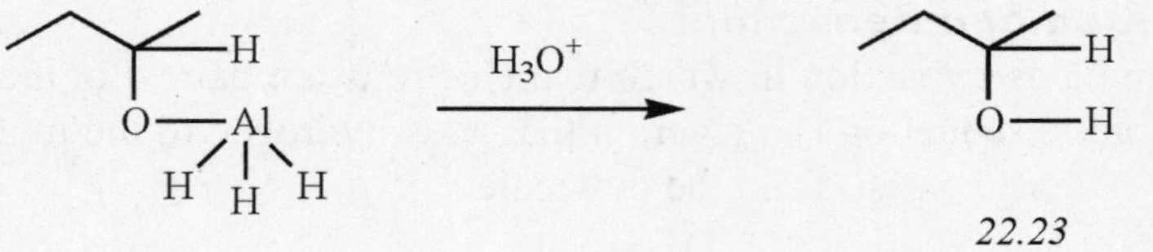

H
O—Al
H H H
H_3O^+
H
O—H

22.22

22.23

To Reduce a Carbonyl, Hydrogen Must Be Delivered to the Oxygen. What Is the Source of This Hydrogen?

As shown with *22.22* and *22.23*, the hydrogen from water replaces the aluminum. Therefore, the second hydrogen atom of the reduction is added during the hydrolysis step. Note that for reduction of a C=O unit, the hydrogen on carbon in the alcohol product is derived from $LiAlH_4$, whereas the hydrogen on oxygen (the OH) is derived from the water involved in the second step of the reaction (hydrolysis).

Can Water Be Used As a Solvent in Reduction Reactions Involving Lithium Aluminum Hydride?

No! $LiAlH_4$ reacts *violently* with water or alcohols (methanol or ethanol, for example). Therefore, protic solvents must be avoided.

What Solvents Are Used?

The most common solvents are diethyl ether and THF.

Give the major product of the reactions of *22.24–22.26*.

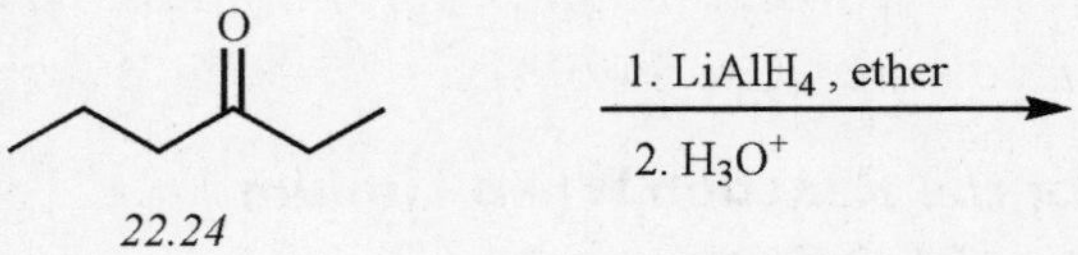

O
1. $LiAlH_4$, ether
2. H_3O^+

22.24

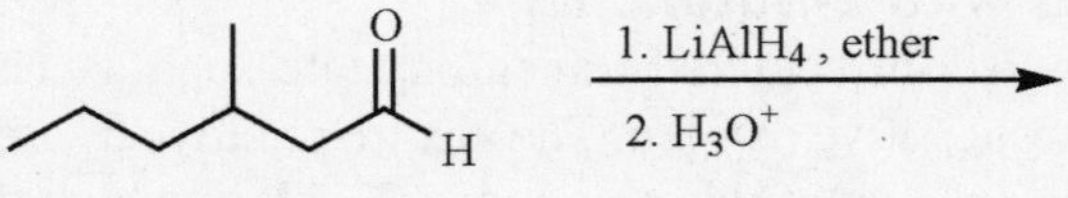

O
H
1. $LiAlH_4$, ether
2. H_3O^+

22.25

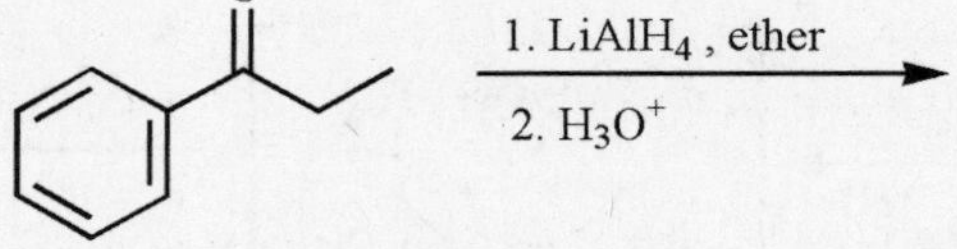

O
1. $LiAlH_4$, ether
2. H_3O^+

22.26

The product of the reduction of 3-hexanone (*22.24*) is 3-hexanol. Reduction of 3-methylhexanal (*22.25*) is 3-methyl-1-hexanol. Reduction of 1-phenyl-1-propanone (*22.26*) leads to 1-phenyl-1-propanol.

Sodium Borohydride

What Is the Formula of Sodium Borohydride?

Sodium borohydride is $NaBH_4$ and the active agent is tetrahydridoborate, BH_4^-.

Is Sodium Borohydride a Stronger or a Weaker Reducing Agent When Compared to Lithium Aluminum Hydride? Explain.

Sodium borohydride is a much weaker reducing agent than lithium aluminum hydride. It is capable of reducing only a few functional groups (aldehydes, ketones, acid chlorides). Esters can be reduced, but with difficulty, and the yields of alcohol products are often poor. In general, the B–H bond is less polarized than the Al–H bond, and the hydrogen of the B–H bond is a weaker hydride species.

What Is the Mechanism of the Reaction Between Sodium Borohydride and Cyclohexanone? What Is the Product after Hydrolysis?

The mechanism of reduction of $NaBH_4$ with ketones and aldehydes is identical to that shown for reduction with $LiAlH_4$. Reaction of cyclohexanone and borohydride proceeds by a four-centered transition state (*22.27*), which leads to an alkoxyborate product, *22.28*. Hydrolysis of *22.28* leads to cleavage of the B–O bond and formation of cyclohexanol, along with boric acid, $B(OH)_3$.

BH_4^- / EtOH → [22.27] → 22.28 —aq. NH_4Cl→ 22.29 (OH)

22.27 *22.28* *22.29*

Why Was Water Sufficient to Hydrolyze a Lithium Aluminum Hydride Reduction, but Aqueous Ammonium Chloride Was Used with the Sodium Borohydride Reduction?

Since $NaBH_4$ is a much weaker reducing agent, it does not react completely with water. Sodium borohydride dissolves in water and reacts to give a mono or dihydroxy borohydride species. An acid stronger than water is required to hydrolyze the B–O bond, and the slightly acidic ammonium chloride solution is a mild and excellent choice.

Why Was Ethanol Used As a Solvent?

Sodium borohydride is much less reactive than $LiAlH_4$. It reacts very slowly with alcohols, so slowly that there is little or no interference with the desired reaction with an aldehyde or a ketone. Therefore, it is less reactive with protic solvents, and thus methanol or ethanol can be used as the solvent.

Can Water Be Used As a Solvent with Sodium Borohydride?

Yes! Alcohol is more common, but water can be used in many cases.

What Is the Product for the Reactions of 22.30 and 22.31?

1. $NaBH_4$, EtOH
2. aq. NH_4Cl

22.30

1. $NaBH_4$, EtOH
2. aq. NH_4Cl

22.31

The $NaBH_4$ reduction of cycloheptanone (*22.30*) leads to cycloheptanol. Reduction of 2-ethylheptanal (*22.31*) gives 2-ethyl-1-heptanol.

Catalytic Hydrogenation

Catalytic hydrogenation of alkenes was discussed in Section 13.4. Reduction of ketones and aldehydes with hydrogen also requires a catalyst, usually platinum, and is very similar in mechanism to that shown for alkenes in Chapter 13.4.

What Is the Product When 2-Pentanone Is reacted with Hydrogen Gas?

As with the hydrogenation of alkenes, hydrogen gas does not react with ketones or aldehydes unless a catalyst is present. This attempted hydrogenation gave no reaction.

What Catalysts Are Used for the Catalytic Hydrogenation of Ketones and Aldehydes?

The most common catalyst is platinum, or platinum oxide. However, both palladium and nickel catalysts can be used, as well as others.

What Name Is Associated with Platinum Oxide?

Platinum oxide (PtO_2) is commonly known as Adams' catalysts, after Roger Adams (1889–1971, an organic chemist from the University of Illinois).

What Is the Product When 2-Pentanone Is reacted with Platinum Oxide and Hydrogen Gas?

In the presence of the platinum derivative catalyst, 2-pentanone is reduced to 2-pentanol.

Using Platinum Oxide and Hydrogen Gas, Is It possible to Reduce a Ketone or Aldehyde in the Presence of an Alkene Unit Elsewhere in the Molecule?

Yes! the reaction of hex-5-enal with H_2/PtO_2 leads to hex-5-en-1-ol.

Which Functional Group Is Easier to Reduce with Hydrogen, a Ketone or an Ester?

The carbonyl group of an aldehyde or a ketone is much easier to reduce by catalytic hydrogenation than is the carbonyl of an ester (see Chapter 25 for a discussion of esters). Esters are usually very difficult to reduce with hydrogen.

Give the Major Product of the Reactions of 22.32–33.35.

H_2, Pt°, MeOH

22.32

H_2, PtO_2, MeOH

22.33

H_2, $Pt°O_2$, EtOH

only 1 equiv of H_2

22.34

H_2, Pt°, MeOH

22.35

Catalytic hydrogenation of 3-octanone (***22.32***) leads to reduction of the carbonyl and formation of 3-octanol. Reduction of 4, 4-dimethylhexanal (***22.33***) with hydrogen gives 4, 4-dimethyl-1-hexanol. Reduction of 2-(2-propenyl)cyclohexanone (***22.34***) leads to 2-(2-propenyl)cyclohexanol. With only one equivalent of hydrogen gas and Adams' catalysts, the carbonyl is reduced first. Catalytic hydrogenation of ***22.35*** (3-methyl-2-phenyl-cyclopentanone) gives 3-methyl-2-phenylcyclopentanol as the major product.

Dissolving Metal Reduction

When Sodium Metal Is Mixed with Acetone, What Is the Initial Product?

Sodium transfers one electron to the carbonyl, forming a resonance stabilized **ketyl** (*22.36*). This resonance stabilized intermediate has characteristics of both a radical and a carbanion. It is an example of a radical anion.

Na (0)

22.36

Why Does Sodium Metal Transfer an Electron to the P-Bond?

Sodium metal lies in Group I of the Periodic Table. Energetically, it is easier to lose one electron (ionization potential) than to gain seven electrons (electron affinity) to achieve a full octet in the outer shell.

What Other Metals Are Capable of Transferring an Electron in This Manner?

Most metals in Group I and Group II of the Periodic Table are capable of single electron transfer. The most common metals used in this type of reaction are lithium, sodium, potassium, calcium and magnesium.

What Is the Complete Mechanism for the Reduction of 2-Butanone to 2-Butanol with Sodium Metal in Liquid Ammonia and Ethanol?

Na (0)

EtO-H

22.37

O–H Na (0) O–H EtO-H O–H

H

22.38 22.39

In this process, initial electron transfer from sodium metal to the carbonyl of 2-butanone generates a ketyl (*22.37*). The 'carbanion' portion of this ketyl reacts with an acid which, in this medium, generates a radical (*22.38*). Ammonia can function as an acid in this case, but the added ethanol is a more powerful acid (relative to ammonia), and the overall reduction is faster and more efficient in the presence of ethanol. Ketyl ***22.37***, therefore, reacts with ethanol to give ***22.38***. A second electron transfer from sodium gives anion ***22.39*** and proton transfer from ethanol completes the reduction to give 2-butanol. Note that in dissolving metal reductions both hydrogens come from the acid, in this case ethanol.

What Are the Major Products of the Reactions of 22.40 and 22.41?

O

Na°, NH_3, EtOH

22.40

K°, NH_3, EtOH

H

O

22.41

The dissolving metal reduction of cyclohexanone (***22.40***) with sodium leads to cyclohexanol. Similar reduction of 4,4-dimethylpentanal (***22.41***) with potassium leads to 4,4-dimethyl-1-pentanol.

Test Yourself

1. Which of the following reactions can be classified as reductions? Calculate the oxidation number of each reactive carbon and determine the net electron change.

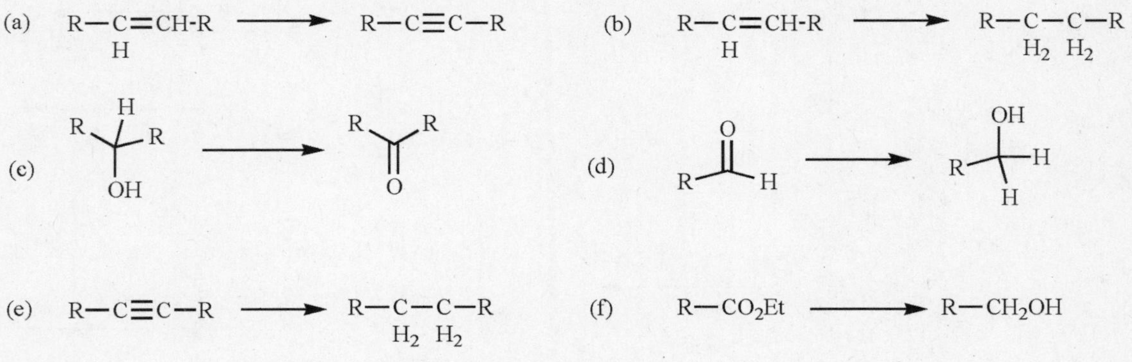

(a) R—CH=CH-R ⟶ R—C≡C—R

(b) R—CH=CH-R ⟶ $R—CH_2—CH_2—R$

(c) $R_2CH(OH)$ ⟶ $R_2C=O$

(d) R—CHO ⟶ $R—CH_2OH$

(e) R—C≡C—R ⟶ $R—CH_2—CH_2—R$

(f) $R—CO_2Et$ ⟶ $R—CH_2OH$

2. Reduction of 5-chloropentanal with sodium borohydride may produce a small amount of a by-product other than the expected 5-chloropentanol. What is this minor product and how is it formed?

3. Give the mechanism for the following reaction.

Cyclohexanone —(Na°, NH_3; no ethanol)⟶ cyclohexanol

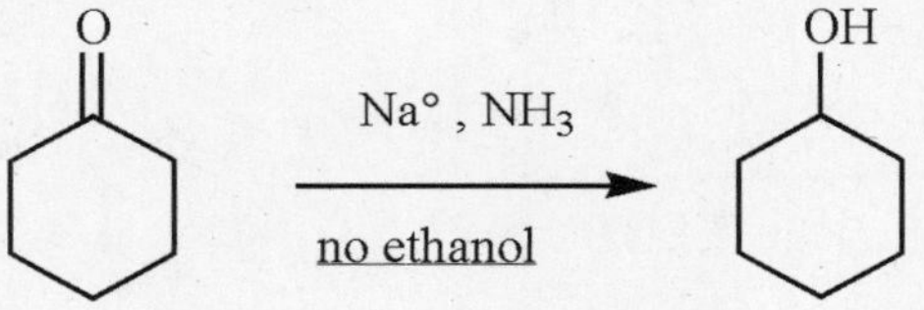

4. Explain why alcohol A is oxidized much slower than alcohol B.

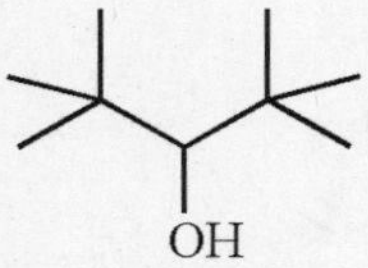

A: OH

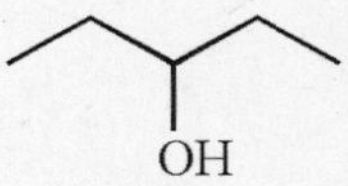

B: OH

5. For each of the following reactions, give the major product. Remember stereochemistry where appropriate, and if there is no reaction, indicate this by N.R.

(a) 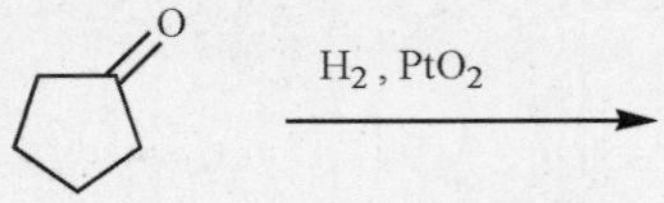 H_2, PtO_2

(b) 2 H_2, PtO_2

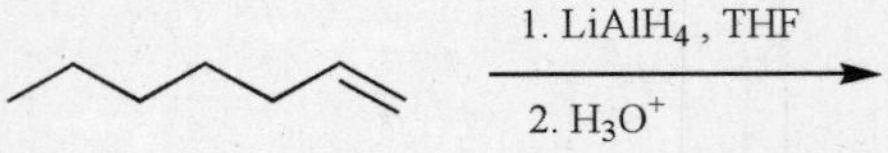

(c) 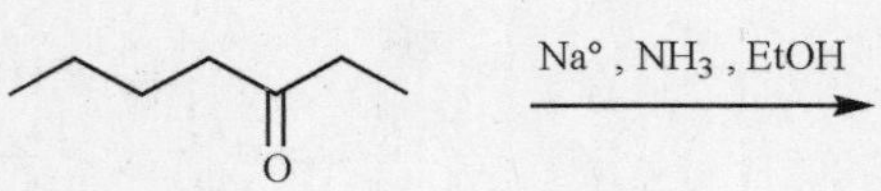 Na°, NH_3, EtOH

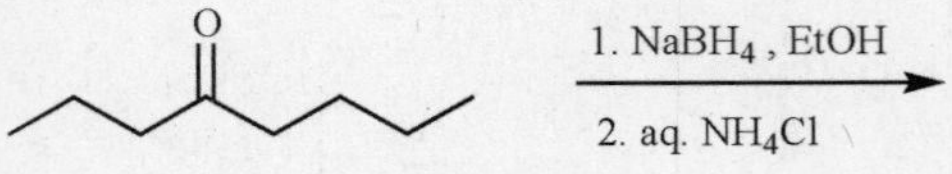

(d) 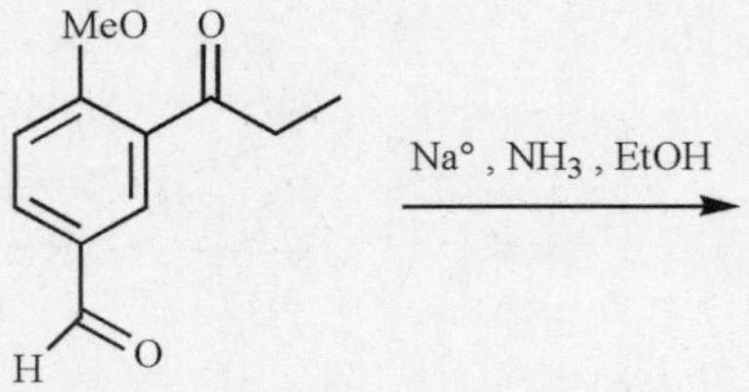 MeO O H O Na°, NH_3, EtOH

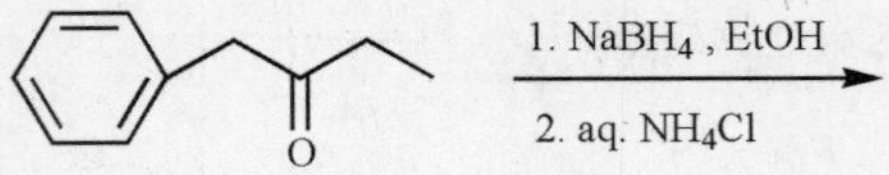

(e) 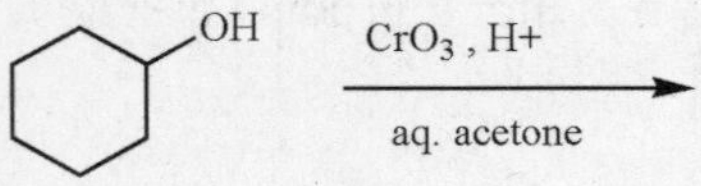 OH CrO_3, H+ aq. acetone

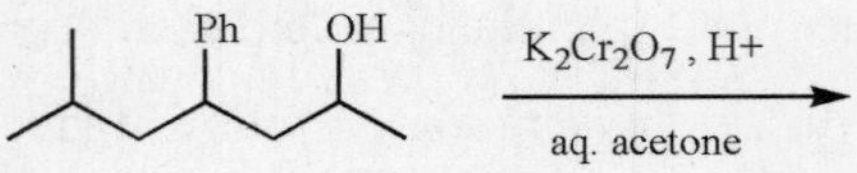

(f) 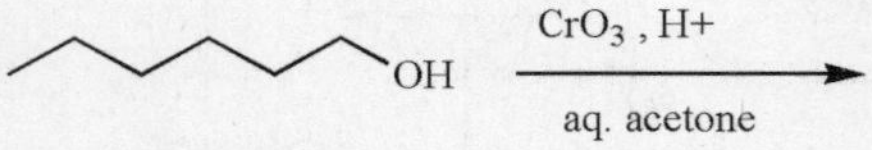 OH CrO_3, H+ aq. acetone

(g) 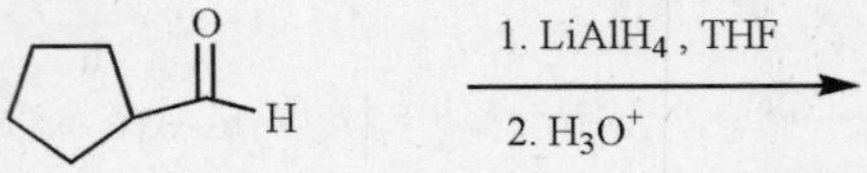 O H 1. $LiAlH_4$, THF 2. H_3O^+

(h) 1. $LiAlH_4$, THF 2. H_3O^+

(i) O 1. $NaBH_4$, EtOH 2. aq. NH_4Cl

(j) O 1. $NaBH_4$, EtOH 2. aq. NH_4Cl

(k) Ph OH $K_2Cr_2O_7$, H+ aq. acetone

(l) 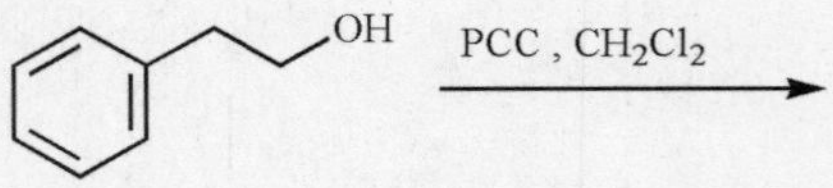 OH PCC, CH_2Cl_2

6. In each case, give the major product. Remember stereochemistry where appropriate, and if there is no reaction, indicate this by N.R.

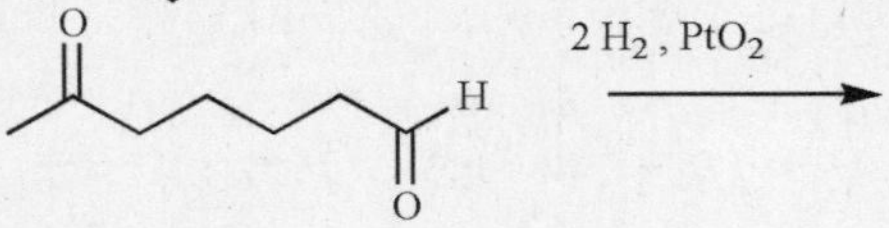

(a) 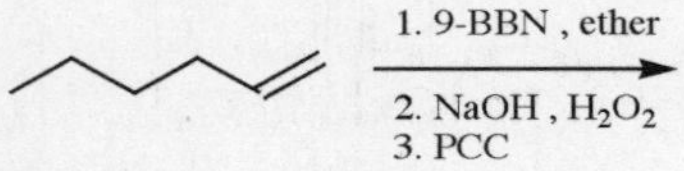 1. 9-BBN, ether 2. NaOH, H_2O_2 3. PCC

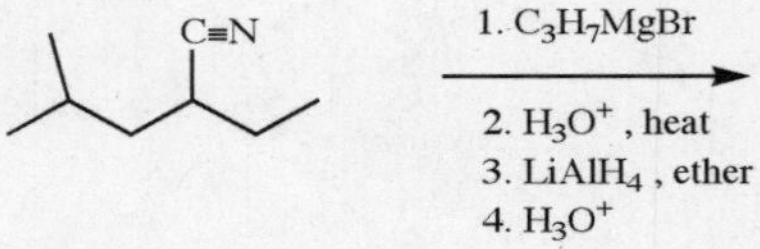

(b) 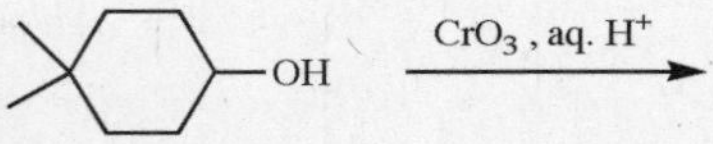 OH CrO_3, aq. H^+

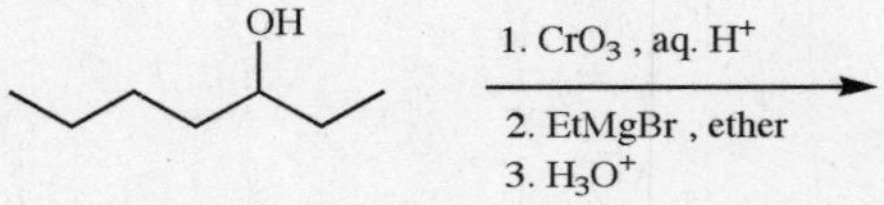

(c) 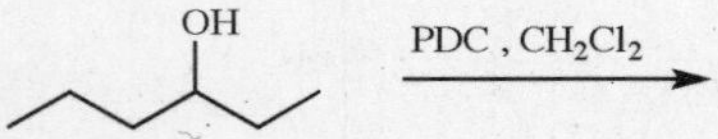 OH PDC, CH_2Cl_2

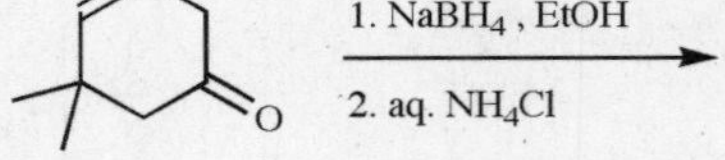

(d) 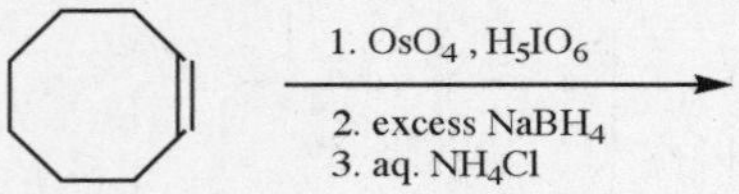 1. OsO_4, H_5IO_6 2. excess $NaBH_4$ 3. aq. NH_4Cl

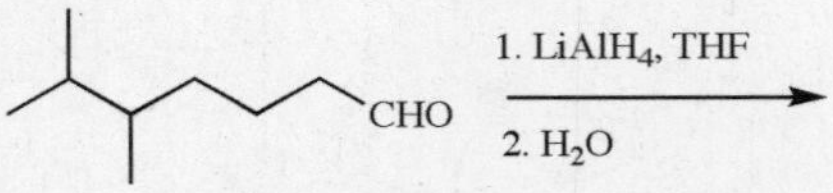

(e) 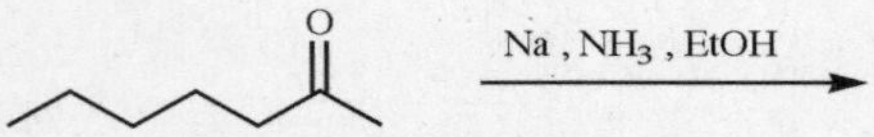 O Na, NH_3, EtOH

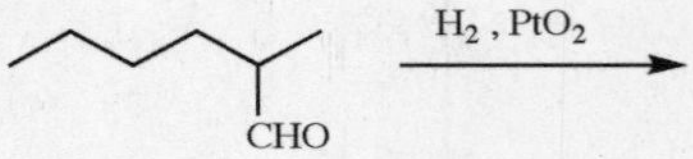

(f) 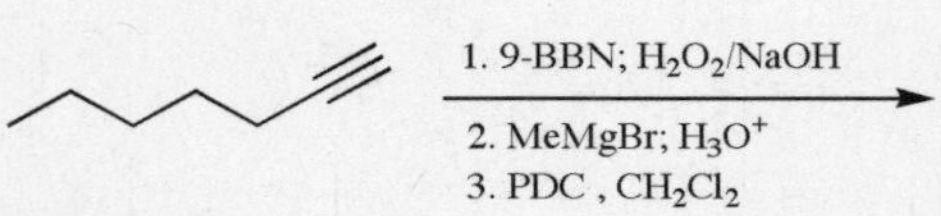 1. 9-BBN; H_2O_2/NaOH 2. MeMgBr; H_3O^+ 3. PDC, CH_2Cl_2

(g) C≡N 1. C_3H_7MgBr 2. H_3O^+, heat 3. $LiAlH_4$, ether 4. H_3O^+

(h) OH 1. CrO_3, aq. H^+ 2. EtMgBr, ether 3. H_3O^+

(i) O 1. $NaBH_4$, EtOH 2. aq. NH_4Cl

(j) CHO 1. $LiAlH_4$, THF 2. H_2O

(k) CHO H_2, PtO_2

(l) 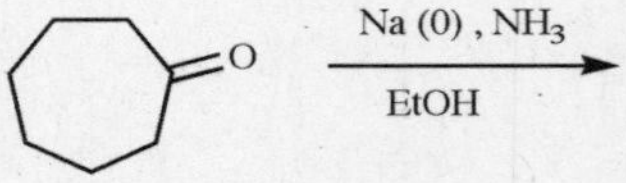 O Na (0), NH_3 EtOH

7. What is the major product when benzaldehyde is reacted with $NaBH_4$ and then aqueous ammonium chloride?

8. Give the major product from the reactions of **A** and **B**.

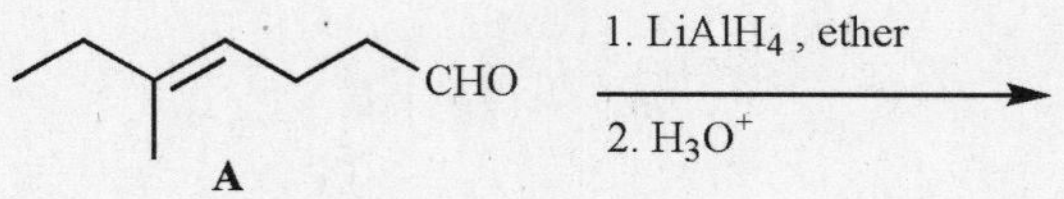

CHO 1. $LiAlH_4$, ether 2. H_3O^+

A

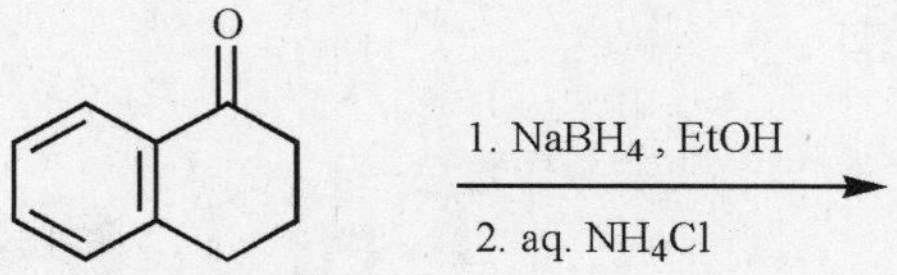

O 1. $NaBH_4$, EtOH 2. aq. NH_4Cl

B

9. Give the major product from the reaction of **A** and **B**.

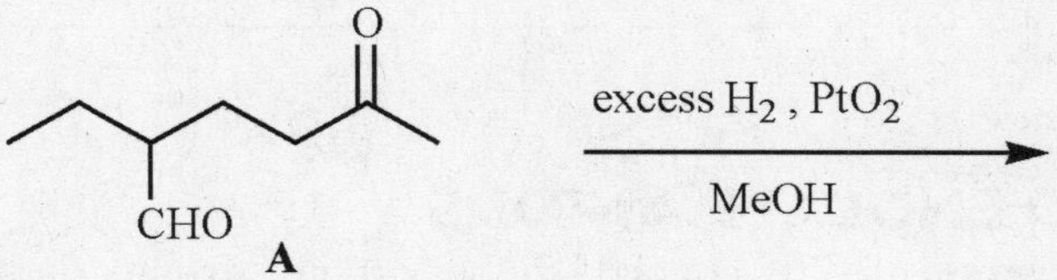

O excess H_2, PtO_2 MeOH

CHO

A

O excess H_2, PtO_2 MeOH

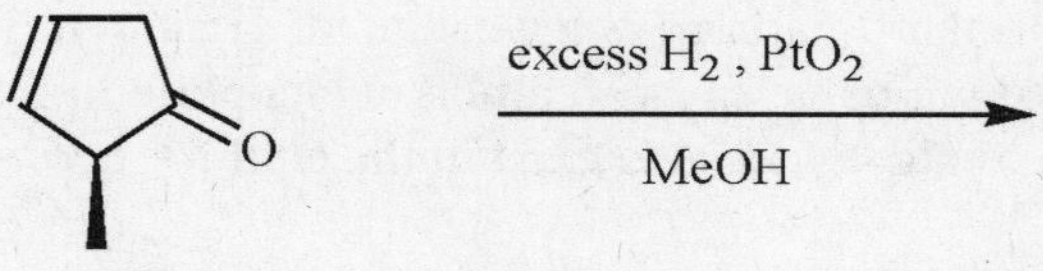

B

10. Give the major product from the reactions **A** and **B**.

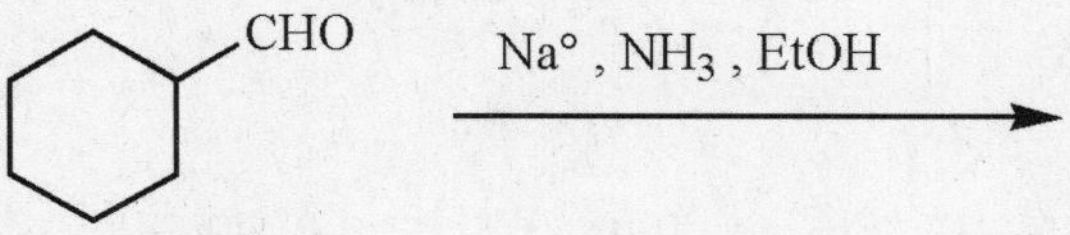

CHO Na°, NH_3, EtOH

A

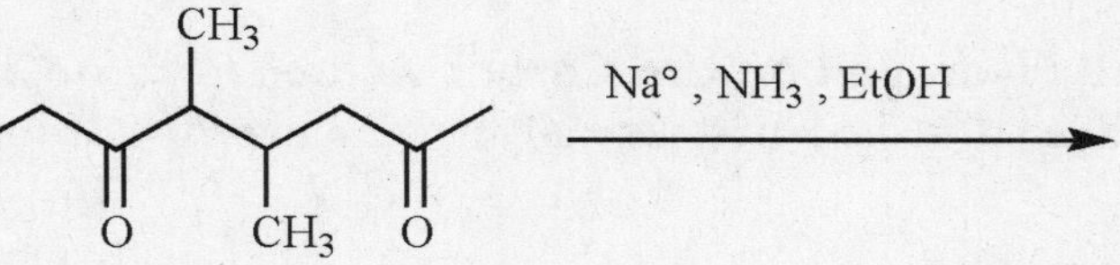

CH_3 Na°, NH_3, EtOH

O CH_3 O

B

CHAPTER 23

Enolate Anions and the Aldol Condensation

23.1. ENOLS AND ENOLATE ANIONS

The hydrogen on the carbon directly attached to the carbonyl group of a ketone or aldehyde (the α-proton, which has attached to the α-carbon) is a weak acid. Treatment with base removes this proton to generate an enolate anion, which behaves as a carbon nucleophile in reactions with alkyl halides, aldehydes or ketones. Ketones and aldehydes also exist in the enol form via keto-enol tautomerism.

Acidity of the α-Hydrogen

What Is the α-Carbon?

The α-carbon is the one directly attached to the carbonyl carbon of an aldehyde or a ketone.

What Is the α-Proton?

The α-proton is the one directly attached to the α-carbon of an aldehyde or a ketone.

What Is the pK_a of the Hydrogen on the Carbon Attached to a Carbonyl?

The pK_a of this so-called α-hydrogen is generally in the range 20–22.

Why Is This Hydrogen Acidic?

The electron-withdrawing carbon group of the carbonyl induces a δ^+ charge on that hydrogen, making it susceptible to attack by a base. In addition, most ketones and aldehydes are in equilibrium with an enol (see below), which has an acidic O–H moiety.

When the α-Carbon Is Substituted, What Is the Effect on the Acidity of That α-Proton?

Since a carbon group is electron releasing, its presence makes the α-hydrogen less acidic (larger pK_a) by about one pK_a unit for each alkyl substituent.

Keto-Enol Tautomerism

What Is an Enol?

An enol is a molecule that has an OH group directly attached to a carbon-carbon double bond, as in *23.2*.

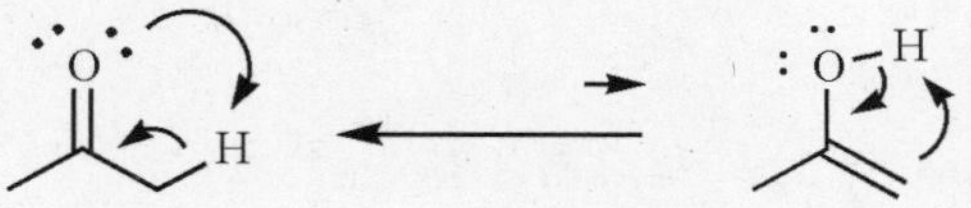

23.1 *23.2*

What Is Keto-Enol Tautomerism?

Most ketones and aldehydes (considered to be the keto form, ***23.1***) are in equilibrium with an enol form (***23.2***). This equilibrium usually favors the carbonyl partner to a very great extent (often >99%). In general, a ketone or aldehyde is expected to exist primarily in the carbonyl form. This equilibrium occurs when the acidic α-hydrogen is removed by the basic carbonyl oxygen, forming an alkene group. The reverse reaction has the less basic alkene π-bond removing the acidic hydrogen from the O–H, regenerating the carbonyl. Since O–H is more acidic than C–H, the equilibrium is generally shifted to the left, favoring the carbonyl.

Which Is Higher in Energy, the Keto Form of Acetone or the Enol Form?

The enol form is about 9.8 kcal mol^{-1} higher in energy than the keto form. Therefore, acetone exists primarily in the keto form as do most aldehydes and ketones.

What Is the Equilibrium Constant for the Keto/Enol Forms in Acetone?

The equilibrium constant $K = \frac{[enol]}{[keto]} = 6.3 \times 10^{-8}$ for acetone.

What Structural Features Can Lead to Greater Percentages of Enol?

If the enol form is stabilized by internal hydrogen bonding, a higher percentage of the enol form will result. This occurs primarily when an atom capable of hydrogen bonding (such as an oxygen) is located at the position β-to the carbonyl carbon, as in 1,3-dicarbonyl compounds such as 2,4-pentanedione (***23.3***).

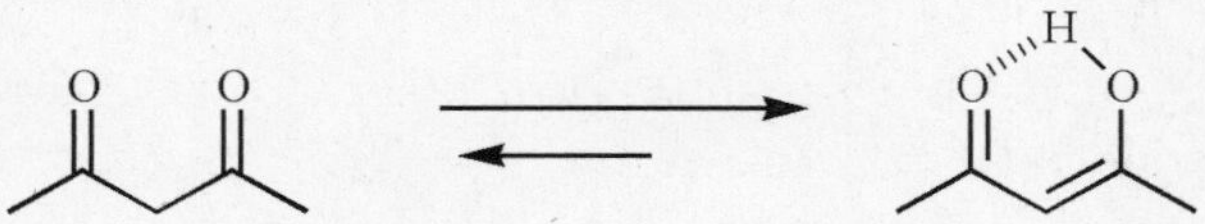

23.3 *23.4*

What Is the Equilibrium Constant (Enol/Keto) for the Pure Liquid 23.3?

The equilibrium constant $K = \frac{[\mathbf{23.4}]}{[\mathbf{23.3}]}$ is reported to be 6.2×10^{-2}.

What Does This Equilibrium Constant Mean?

In general terms, an equilibrium constant of 6.2×10^{-2} means there is approximately a 94:6 mixture of ***23.3*** and ***23.4***. This means that in the pure liquid the keto form predominates, but some enol is in equilibrium with the keto form.

What Is the Relative pK_a of 2,4-Pentanedione? Explain.

The pK_a of the hydrogens on the $-CH_2-$ group between the carbonyls is about 9. The presence of two C=O groups make these hydrogens the most acidic. This is explained primarily by the enol content of this compound (***23.3***), which is present in >90% in hydrocarbon solvents. Removal of this acidic hydrogen is much more facile than from the enol form of a simple ketone (see ***23.2***), which has a pK_a of about 19.

What Is the Common Name of 2,4-Pentanedione?

The common name of 2,4-pentanedione is acetylacetone.

Compare the Acidity of H_a in 23.5 and 23.6.

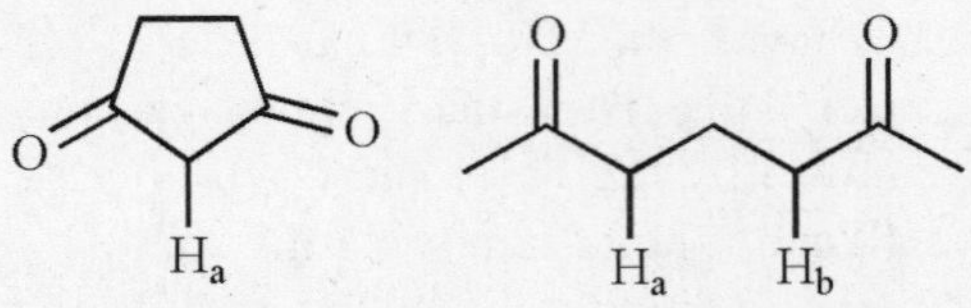

H_a H_a H_b

23.5 *23.6*

H_a in 1,3-cyclopentanedione (***23.5***) is very acidic ($pK_a \approx 9–10$) due to its proximity to two carbonyls and the high percentage of enol. Both H_a and H_b in *23.6* have the same pK_a (≈ 21). The enol content is about the same as in a simple ketone, as is the pK_a, due to the fact that the carbonyl groups are not close enough to provide internal hydrogen bonding, which would stabilize the enol form.

Formation of Enolate Anions

When the Acidic α-Proton of a Ketone Is Removed, What Is the Product?

If one assumes the hydrogen is removed from the enol form of ***23.7***, the product is an enolate anion (***23.8***), which is a resonance stabilized anion.

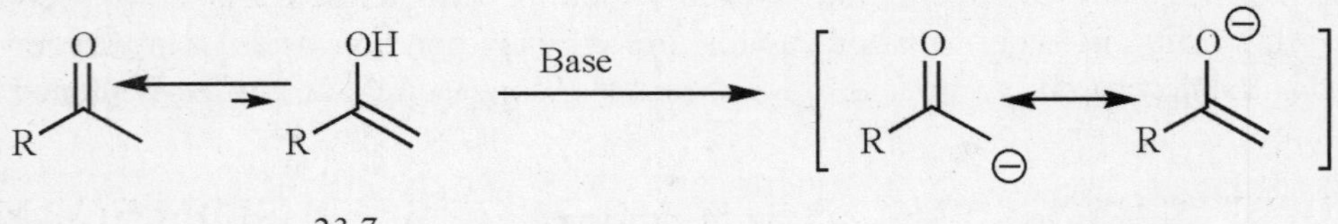

Base

23.7 *23.8*

Why Is an Enolate Anion Particularly Stable?

Enolate anions are resonance stabilized, as in ***23.8***.

What Bases Are Used to Remove This Acidic Proton?

A variety of bases can be used, including NaOEt, NaOH, NaOMe, K*t*-OBu and amide bases such as $NaNH_2$, $NaNR_2$ $LiNH_2$ and $LiNR_2$.

What Are 'Non-Nucleophilic' Bases?

Non-nucleophilic bases are usually amide bases (NR_2^-) where the R groups are bulky. This causes steric hindrance as the nitrogen approaches a carbon, behaving as a nucleophile, but does not hinder approach to a hydrogen, behaving as a base.

What Is LDA?

The most commonly used non-nucleophilic base is **lithium diisopropyl amide** [$LiN(iPr)_2$], often abbreviated LDA.

Is An Enolate Anion a Bidentate Anion ?

Yes! Both the oxygen and the carbon in ***23.8*** are electron rich and nucleophilic. This is a bidentate anion, which is a bidentate nucleophile.

Which Is Generally More Reactive As a Nucleophile, the Carbon or the Oxygen in 23.8?

In S_N2 reactions and acyl addition reactions, the carbon reacts faster than oxygen. However, in many cases (iodomethane and silyl halides such as R_3SiCl) the oxygen can react preferentially. **For the remainder of this book, we will *assume* that enolate anions react via carbon in all nucleophilic reactions.**

What Is the Product When 2,4-Pentanedione Is Reacted with LDA? With Sodium Ethoxide?

In both cases, the product is the same, enolate anion ***23.9*** (as the Li or Na salt, respectively). The presence of the second carbonyl group makes this enolate anion even more stable than ***23.8***.

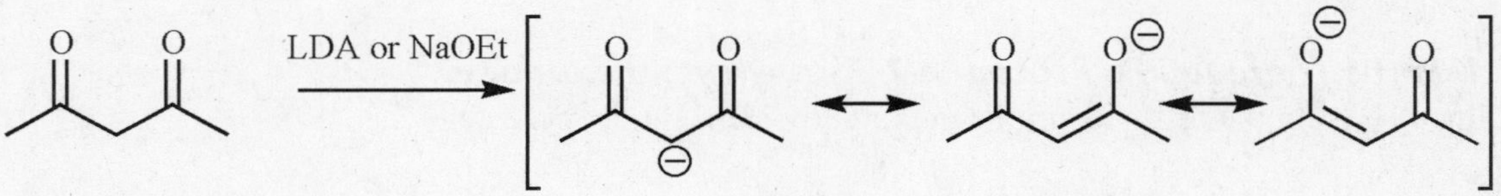

LDA or NaOEt

23.9

If 23.9 Is More Stable Than 23.8, Which Is More Reactive?

The greater stability of ***23.9*** means it is *less* reactive than the less stable ***23.8***.

What Type of Reagents Are Enolate Anions Expected to React with?

The carbon of the enolate anion is very nucleophilic and is expected to react with suitable electrophilic carbons. Typical reaction partners are alkyl halides and other carbonyl derivatives (ketones, aldehydes and acid derivatives).

23.2. KINETIC VS. THERMODYNAMIC ENOLATE ANION FORMATION

In unsymmetrical ketones, there are two α-protons. In general, the proton on the less substituted carbon is more acidic. The presence of two different α-protons can lead to regioisomeric products in nucleophilic reactions. The product formed is dictated by the reaction conditions used to form the enolate anion.

Which Is More Acidic, H_a in 23.10 or H_a in 23.11?

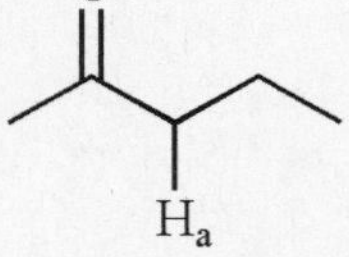

H_a H_a

23.10 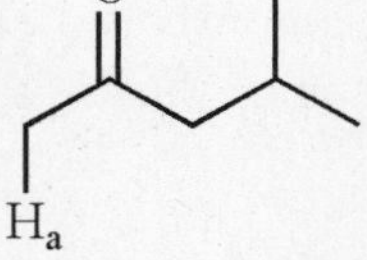 *23.11*

In general, the proton on the less substituted carbon atom is more acidic, since carbon groups are electron releasing. In ***23.10***, H_a is on an α-carbon that has one carbon substituent (remember that carbon is electron releasing), whereas H_a in ***23.11*** is on an α-carbon that has no carbon substituents. Therefore, H_a in ***23.11*** is more acidic.

In general, acidity follows the following order for the pK_a of H_a: $O{=}C{-}CH_2H_a$ (19), $O{=}CCHRH_a$ (20), and $O{=}CCR_2H_a$ (21). In other words, begin with a pK_a of about 19 and add one pK unit for each carbon substituent.

Label All Different Acidic Protons in 2-Methylcyclohexanone.

As shown in ***23.12***, there are two different acidic protons, H_a and H_b. There are two identical protons H_a, and then H_b.

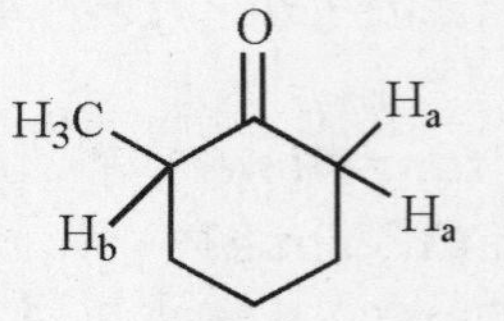

23.12

Which Is the More Acidic Proton in 2-Methylcyclohexanone?

Since H_a is on the less substituted position of ***23.12***, H_a is more acidic.

What Are the Two Possible Enolate Anions That Can Be Formed from the Reaction of 2-Butanone and a Suitable Base?

There are two acidic α-protons, H_a and H_b. Removal of H_a generates ***23.14***, and removal of H_b generates ***23.13***. When these isomeric enolate anions react, isomeric products can result by reaction of both ***23.13*** and ***23.14***.

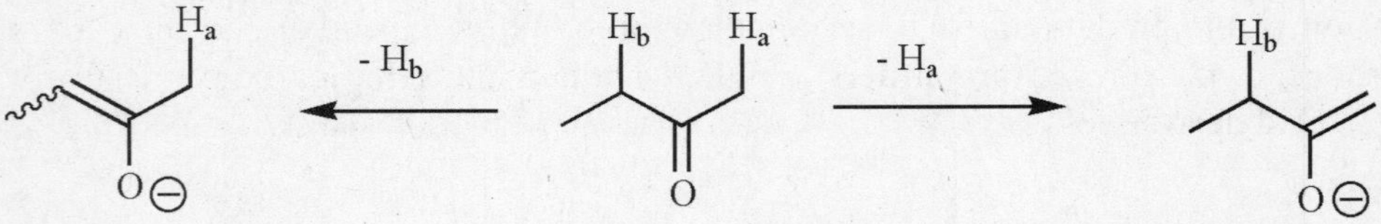

23.13 *23.14*

Which Enolate Anion Is Formed by Removing the Most Acidic Hydrogen?

Since an alkyl group is electron releasing, its presence will make the α-proton *less* acidic. The most acidic hydrogen will, therefore, be attached to the *less substituted carbon*. In the case of 2-butanone, H_a is the most acidic, leading to ***23.14***. This is called the **kinetic enolate**.

Which Enolate Anion Is the Thermodynamically More Stable?

The most stable enolate anion is the one with the most substituents. In the case of 2-butanone, removal of H_b leads to enolate anion ***23.13***, where the C=C unit has three substituents, whereas the C=C unit in ***23.14*** has only two substituents. In this case, ***23.13*** is the more stable enolate. This is called the **thermodynamic enolate**.

If the Most Acidic Hydrogen Is Always Removed First, How Can the Thermodynamic Enolate Anion Ever Form?

The only way to generate the thermodynamic enolate anion from the kinetic enolate anion is to establish an equilibrium between ***23.14***, 2-butanone, and ***23.13***. Since removal of H_a or H_b is an acid-base reaction, which is inherently an equilibrium process, reaction conditions that promote this equilibrium will lead to the thermodynamic enolate ***23.13*** as the major product, but not the exclusive product.

What Conditions Favor the Kinetic Enolate?

Since the kinetic enolate anion is formed faster (and first), and an equilibrium is required to convert the kinetic enolate anion into a thermodynamic enolate, conditions that disfavor an equilibrium will favor the kinetic enolate. In general, these conditions are: a polar, aprotic solvent (no acidic proton is available to reprotonate the enolate anion once it is formed); low temperatures (0 to −78°C are typical, which makes the proton transfer required for an equilibrium much slower); a strong base such as LDA, which removes the proton in a fast reaction and, most important, *does not produce a conjugate acid with an acidic proton*). If NaOEt is used, EtOH is the conjugate acid ($pK_a \approx 17$), which is strong enough to reprotonate the enolate). If LDA is used the conjugate acid is diisopropylamine ($PK_a \approx 23–25$), which is not strong enough to quickly reprotonate the enolate anion . Short reaction times also favor the kinetic enolate.

What Conditions Favor the Thermodynamic Enolate?

Any reaction conditions that promote an equilibrium will favor the thermodynamic enolate anion and, of course, those conditions will be the exact opposite of the kinetic conditions. In general, a protic (acidic) solvent such as water or an alcohol is used (also an amine such as ethylamine or ammonia), since the acidic proton can be transferred to the enolate anion, regenerating the ketone and promoting the equilibrium. Higher reaction temperatures (mild heating to reflux) will promote the equilibrium, as will long reaction times. A base that generates a stronger conjugate acid (NaOEt, NaOH, etc.) will also favor thermodynamic control.

Give the Major Enolate Anion from the Reactions of 23.15–23.17. Identify Each Reaction As Being Under Kinetic or Thermodynamic Control Conditions.

O
NaOEt , EtOH
reflux
23.15

O
H
LDA , THF
-78°C
23.16

LDA , THF
-78°C
O
23.17

In the first reaction, cyclohexanone (**23.15**) is deprotonated under thermodynamic control conditions. Since this is a symmetrical molecule, both kinetic and thermodynamic control give the same enolate anion (**23.18**). Aldehyde **23.16** has only one acidic α-proton and both kinetic and thermodynamic conditions will remove it. For reasons involving reactivity, kinetic control conditions (shown) are superior and will generate enolate anion **23.19**. In the last case, ketone **23.17** will give enolate anion **23.20** under the kinetic control conditions shown.

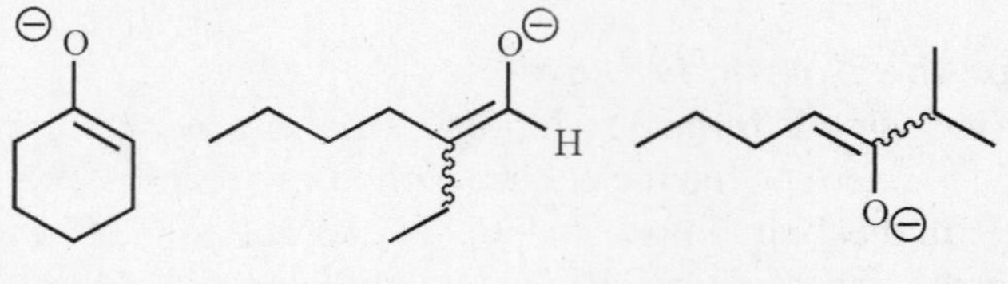

23.18 23.19 23.20

23.3. ENOLATE ANION ALKYLATION

Enolate anions are carbon nucleophiles and react with alkyl halides to produce substituted aldehydes or ketones. In addition, enolate anions react with epoxides to produce hydroxy ketones or aldehydes. These reactions are generally S_N2-like in their reactivity and stereochemistry.

Reaction with Alkyl Halides

What Is the Product When Cyclohexanone Reacts with i. LDA in THF ii. Iodomethane?

Initial reaction of cyclohexanone with LDA generates the enolate anion **23.21**. Subsequent reaction with iodomethane, where the carbon of the enolate anion is a nucleophile, gives the alkylation product, 2-methylcyclohexanone, **23.22**.

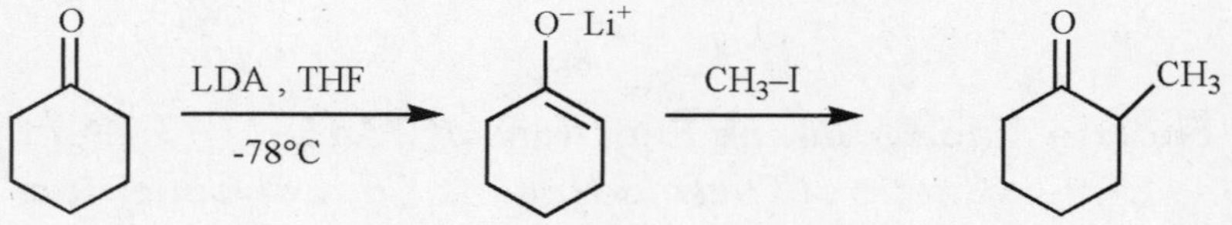

LDA , THF -78°C; $O^- Li^+$; CH_3–I; CH_3

23.21 23.22

When the Kinetic Enolate Anion Derived from 2-Pentanone Is Reacted with Benzyl Bromide, What Is the Major Product?

The kinetic enolate anion is **23.23**. After reaction with benzyl bromide, the final product is 1-phenyl-3-hexanone, **23.24**.

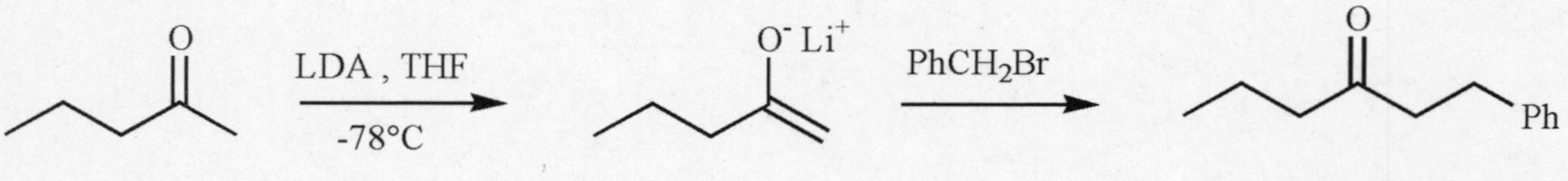

LDA , THF -78°C; $O^- Li^+$; $PhCH_2Br$; Ph

23.23 23.24

When 2-Methylcyclohexanone Is Reacted with Sodium Ethoxide in Ethanol and Then with Iodomethane, What Is the Expected Major Product?

Since these are thermodynamic control conditions, the most highly substituted enolate anion will form **23.25** and then react with iodomethane to give 2,2-dimethylcyclohexanone (**23.26**) as the major product.

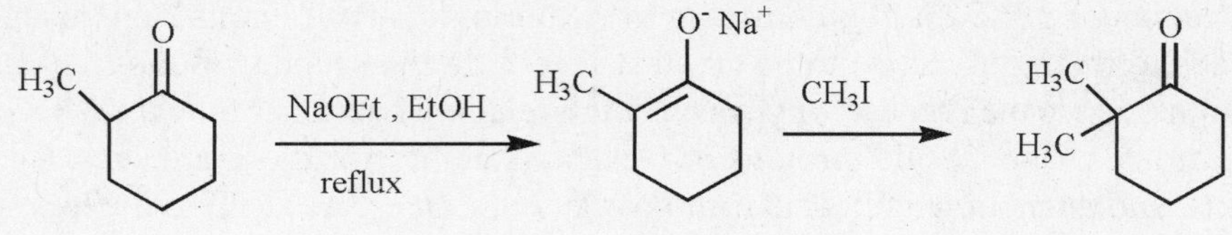

H_3C ... NaOEt, EtOH / reflux → $O^- Na^+$ ... CH_3I →

23.25 23.26

Will 23.26 Be the Only Product Formed in the Reaction of 2-Methylcyclohexanone Under These Conditions?

No! Under thermodynamic (equilibrium) conditions, both the kinetic and thermodynamic enolate anions are present at equilibrium, and both lead to an alkylation product. Since the thermodynamic enolate anion is present in greater amount, ***23.26*** will be the major product, but not the exclusive product.

Explain How the Reaction of Cyclohexanone with Sodium Amide in Ammonia and Treatment with a Large Excess of Iodomethane Can Lead to Small Amounts of 2,2,6,6-Tetramethylcyclohexanone.

The initially formed enolate anion of cyclohexanone will react with iodomethane to form 2-methylcyclohexanone. This can react with more $NaNH_2$ to form ***23.25***, and subsequent reaction with iodomethane will give ***23.26***. Since there are additional α-protons, deprotonation-methylation can occur sequentially on the other α-carbon of ***23.26*** to produce the tetramethyl compound (2,2,6,6-tetramethylcyclohexanone). This will occur only with excess iodomethane and under thermodynamic control conditions with a relatively strong base.

What Is the Product When 2,4-Cyclopentanedione Is Reacted with Sodium Carbonate and Then with Allyl Chloride? Why Can Na_2CO_3 Be Used Rather Than a More Powerful Base?

The initially formed enolate anion (***23.27***) reacts with allyl chloride to form ***23.28***. The presence of two carbonyls makes the proton on the 'middle' -CH_2- moiety very acidic ($pK_a \approx 9$). Therefore, a much weaker base such as Na_2CO_3 can be used for the deprotonation.

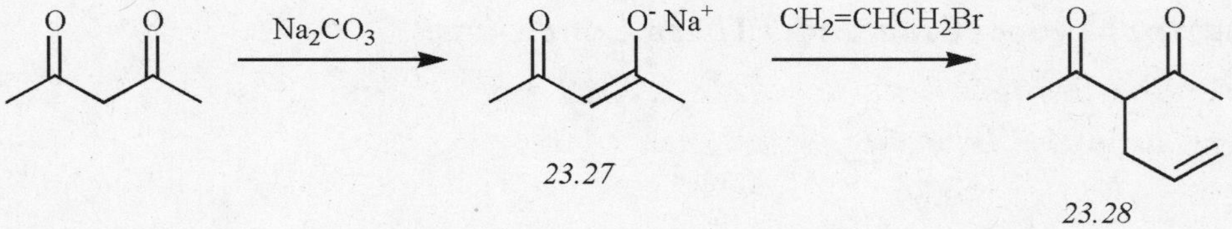

Na_2CO_3 → $O^- Na^+$ ... CH_2=$CHCH_2Br$ →

23.27 23.28

Give the Major Product from the Reactions of 23.29 and 23.30.

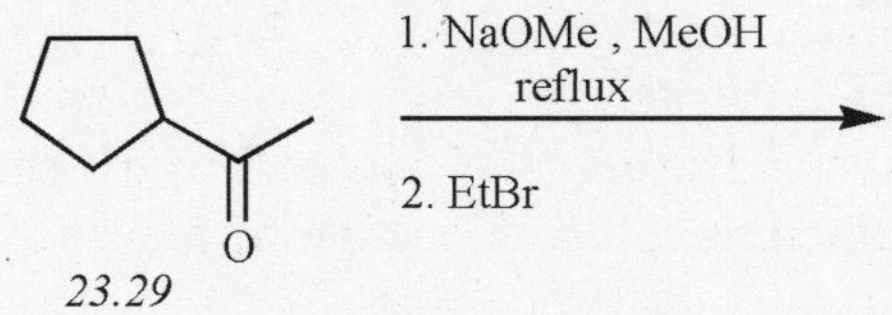

1. NaOMe, MeOH reflux
2. EtBr

23.29

1. LDA, THF, -78°C
2. D_2O

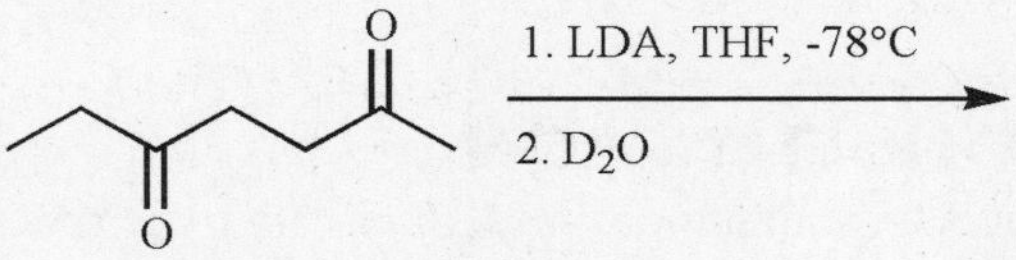

23.30

In the first reaction, ketone ***23.29*** is converted to the thermodynamic enolate anion, and reaction with bromoethane leads to the highly substituted product, ***23.31***. In the second reaction, diketone ***23.30*** has four acidic positions. Since the two carbonyls have a 1,4-relationship rather than a 1,3-relationship, there is no enhanced acidity of the 'inside' protons nor special stability for those enolates. These are kinetic control conditions, and quenching with deuterium oxide will transfer a D (^{2}H) to the less substituted carbon to give ***23.32***.

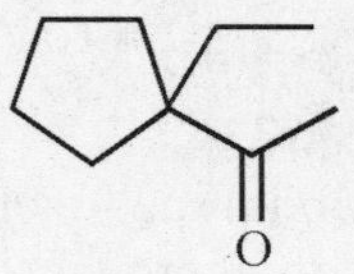

23.31 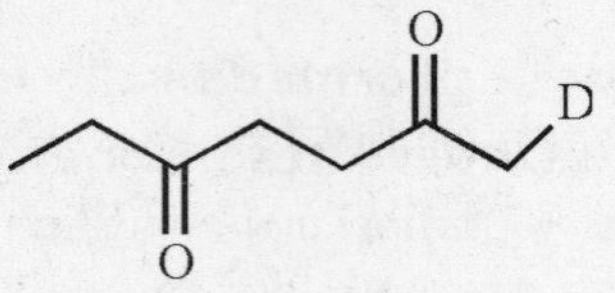 *23.32*

Reaction with Epoxides

What Is the Product When the Kinetic Enolate Anion of 2-Hexanone Reacts with 1-Ethyloxirane Followed by Hydrolysis?

The electropositive carbon of an epoxide is subject to attack by a nucleophile such as an enolate. The nucleophile generally attacks the less sterically hindered (less substituted) carbon of the epoxide. In this case, enolate anion ***23.33*** reacts at the less substituted position of the epoxide unit, and the product is ***23.34***.

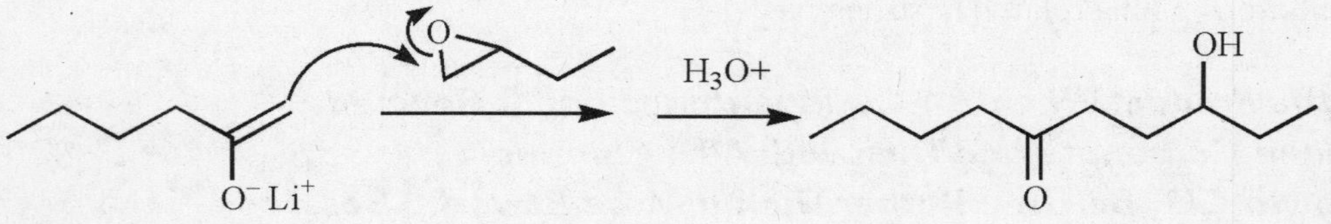

$O^- Li^+$ → H_3O+ →

23.33 *23.34*

Explain Why Epoxide 23.35 Does Not React with the Enolate Anion of Acetophenone, Even under These Conditions.

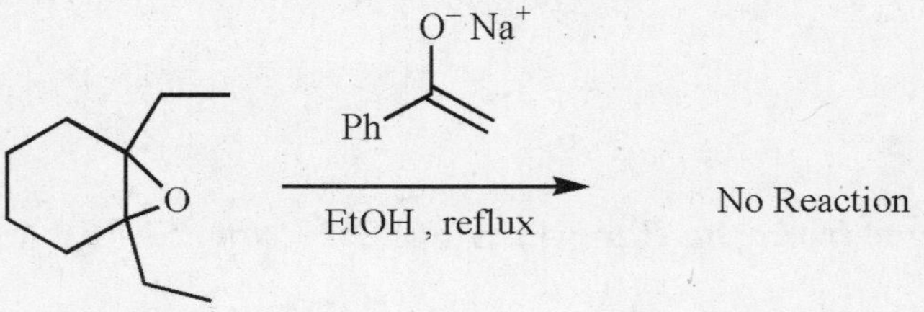

$O^- Na^+$, Ph; EtOH, reflux → No Reaction

23.35

The reaction of an enolate anion and an epoxide can be viewed essentially as a S_N2 reaction. The two tertiary centers of the epoxide are, therefore, too unreactive with the nucleophilic enolate anion to permit the reaction.

23.4. THE ALDOL CONDENSATION

When an enolate anion reacts as a nucleophilic with an aldehyde or a ketone, the acyl addition reaction leads to a β-hydroxy aldehyde or ketone, an aldol. This reaction is known as the **aldol condensation**.

What Is an ALDOL?

Aldol is the generic name for a β-hydroxy aldehyde or ketone (see ***23.36***).

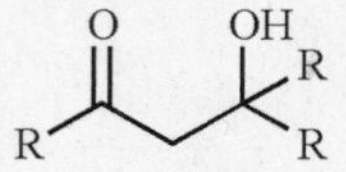

23.36

What Is the Product of the Reaction Between the Kinetic Enolate Anion of 2-Butanone and Cyclopentanone?

The nucleophilic carbon of the enolate anion will attack the electropositive carbon of the cyclopentanone carbonyl to form a new carbon-carbon bond, dump the electrons on oxygen, and form alkoxide ***23.37*** as the initial product. In other words, the enolate anion reacts as a normal nucleophile via acyl addition. Hydrolysis liberates the final alcohol product, ***23.38***. Such β-hydroxyketones (or β-ketoalcohol) are known generically as aldols.

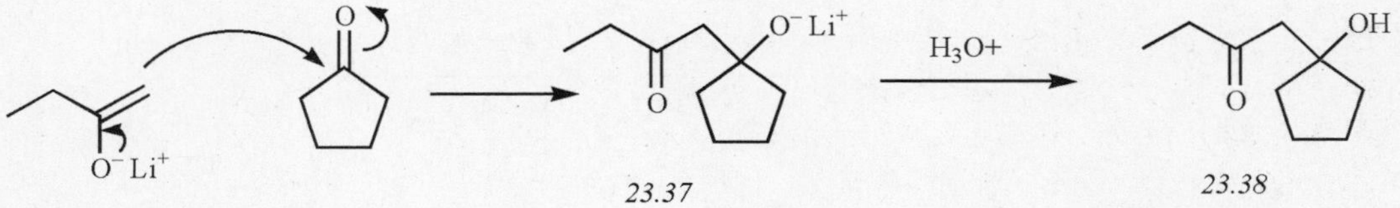

H_3O+

23.37 *23.38*

What Is the Reaction Product When 2-Butanone Is Treated with Sodium Ethoxide in Ethanol?

When 2-butanone is converted to the thermodynamic enolate, significant amounts of base and *unreacted* 2-butanone are present in the equilibrium mixture. It is, therefore, reasonable to conclude that the enolate anion of 2-butanone (***23.39***) will react with unchanged 2-butanone in a condensation reaction that produces the alkoxide as the product. Hydrolysis gives the aldol, ***23.40***.

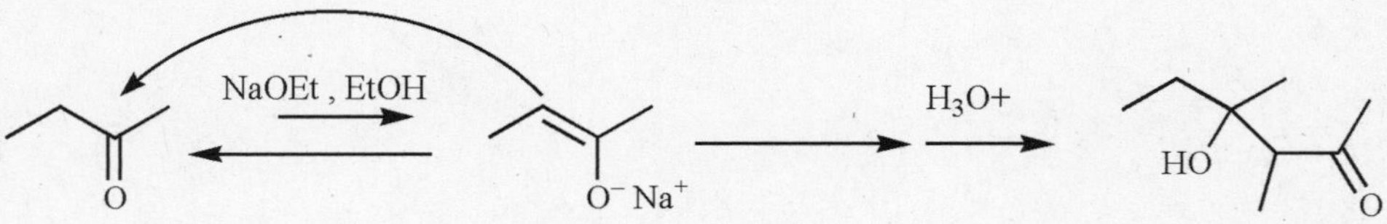

NaOEt , EtOH

H_3O+

23.39 *23.40*

Where Does the 'Second' Equivalent of 2-Butanone Come from to Form 23.40?

Remember that the chemical symbol used for 2-butanone indicates one mole, 6.023×10^{23} molecules. Therefore, if one molecule reacts to form an enolate anion, a lot of unreacted ketone is present. In an equilibrium reaction, perhaps 0.1–5% of the enolate anion is present, and again there is plenty of ketone for a reaction.

What Is the Name of This Reaction?

It is the **aldol condensation**, named after the β-hydroxy carbonyl product (an aldol).

Explain Why Conjugated Ketones or Aldehydes Are Often Isolated from This Reaction When the Product Is Treated with Aqueous Acid or Is Heated.

Loss of water (dehydration) is very facile since the final product (a conjugated carbonyl derivative such as ***23.42***) is more stable than an unconjugated carbonyl derivative, ***23.41***. Dehydration can be induced thermally by heating the aldol product or during the aqueous acid workup (H^+ will catalyze dehydration). It is important to note, however, that dehydration does **not** occur in all cases. With the acid workup, isolation of the aldol product is often straightforward.

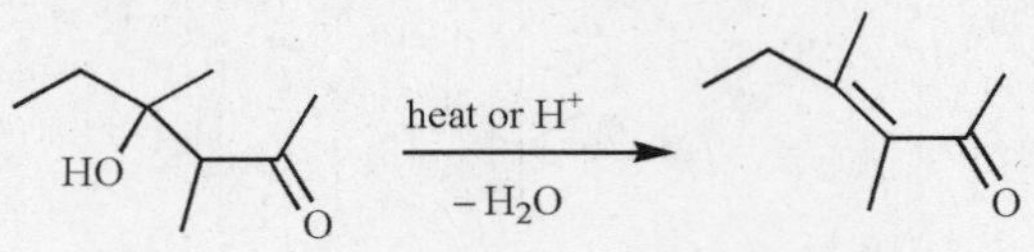

HO — heat or H^+ / $-H_2O$ → O

23.41 23.42

Give the Major Product of the Reactions Shown.

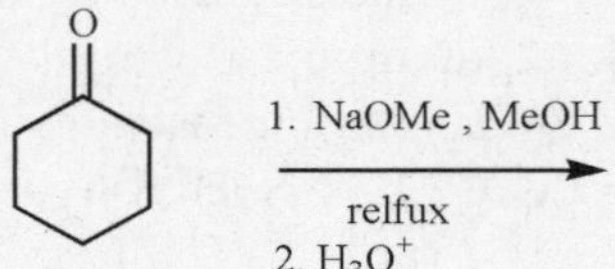

O — 1. NaOMe , MeOH / relfux / 2. H_3O^+

23.43

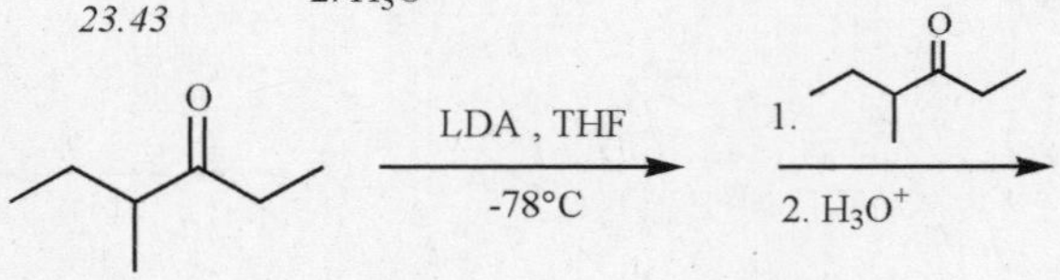

O — LDA , THF / -78°C → 1. O / 2. H_3O^+ →

23.44

Cyclohexanone (***23.43***) is expected to react with itself, under thermodynamic conditions, to give the aldol product (after hydrolysis), ***23.45***. In the second example, ***23.44*** is converted to the kinetic enolate anion under kinetic control conditions. No 'self-condensation' can occur, since all of the ketone was converted to the enolate. In order to get this condensation, a second equivalent of ***23.44*** must be added to the enolate, allowing condensation and formation of the aldol product, ***23.46*** (after hydrolysis).

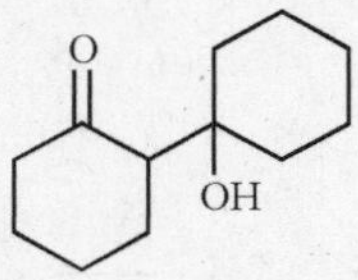

O OH O HO

23.45 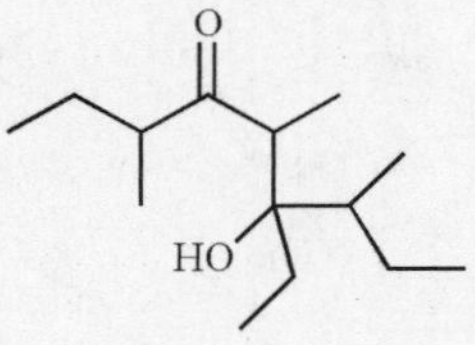 23.46

The 'Mixed' Aldol

What Is a 'Mixed' Aldol Condensation?

Formation of ***23.47*** by condensing 2-butanone and cyclopentanone under kinetic control conditions is an example of a mixed aldol. It simply means that the enolate anion of one ketone or aldehyde is reacted with a different ketone or aldehyde.

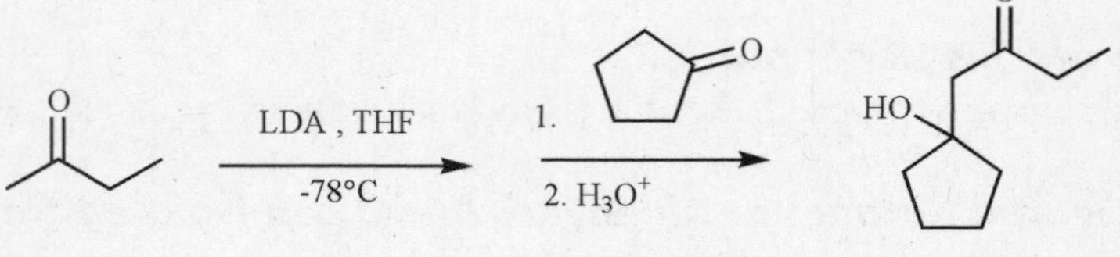

O — LDA , THF / -78°C → 1. O / 2. H_3O^+ → HO O

23.47

Why Are Thermodynamic Conditions Not Used Very Often to Produce Mixed Aldols?

If two different ketones are present in a reaction under thermodynamic control, two different enolate anions can be formed, and each enolate anion can react with either ketone, both of which remain in solution as part of the equilibrium, leading to four different aldol condensation products. As an example, the enolate anion of 2-butanone could react with both 2-butanone and cyclopentanone. Alternatively, the enolate anion of cyclopentanone could react with both 2-butanone and cyclopentanone.

What Is the Advantage of Using Kinetic Control in Forming Mixed Aldol Products?

Under kinetic control conditions, the enolate anion is formed essentially irreversibly. A second carbonyl partner is then added, giving a single 'mixed aldol' product. Under thermodynamic control conditions, both enolate anions must be present at the same time. This is not true for kinetic control conditions.

The Intramolecular Aldol

What Is the Product When 2,5-Hexanedione (23.48) Is Reacted with LDA and Then Hydrolyzed?

Aldol condensation reactions can occur intramolecularly as well as intermolecularly. Under kinetic control conditions, ***23.48*** reacts to form enolate anion ***23.49***. This enolate anion will attack the carbonyl at the other end of the molecule (as shown) to form the cyclic alkoxide, ***23.50***. Hydrolysis then gives the alcohol product, ***23.51***. In general, intramolecular attack is faster than intermolecular attack for formation of five-, six- and seven-membered rings. The enolate anion (***23.49***) is drawn as the 'carbanion' form and then re-drawn in a conformation that shows how the intramolecular reaction can occur.

LDA , THF
-78°C
O⁻ Li⁺
OH

23.48 *23.49* *23.50* *23.51*

What Is the Product When 2,5-Hexanedione Is Treated with Sodium Methoxide in Methanol and Then Hydrolyzed?

The major product is the cyclopentanone derivative, ***23.54***. Under thermodynamic conditions, enolate anions ***23.49*** and ***23.52*** are in equilibrium with the dione (***23.48***). The intramolecular aldol condensation of ***23.52*** would generate a three-membered ring (***23.53***), which is energetically unfavorable when compared to formation of the five-membered ring derivative (***23.49***) from enolate anion ***23.49***. The equilibrium will favor formation of the more stable product, which in this case is the alkoxide precursor of ***23.54***. Due to the relative ease of forming the five-membered ring product, the 'kinetic' enolate anion leads to the product, even under thermodynamic control conditions. Remember, thermodynamic control implies that the *most stable product* is formed.

NaOMe, MeOH
H_3O+
NaOMe
MeOH
H_3O+

23.48 *23.52* *23.53*

23.49 *23.54*

23.5. MICHAEL ADDITION

When a nucleophile adds in a 1,4-manner to a conjugated carbonyl compound, it is called Michael addition, after the English chemist Arthur Michael.

What Is a Michael Addition?

A Michael addition is the reaction of a nucleophile with a conjugated (α, β-unsaturated) ketone or aldehyde, usually a ketone, where the nucleophile adds to the π-bond of the alkene rather than to the carbonyl carbon.

Why Is the Terminal Carbon of a Conjugated Ketone System Subject to Attack by a Nucleophile?

Analysis of *23.55* shows that bond polarization extends from the electronegative oxygen through the bonds to the β-carbon, making it electropositive.

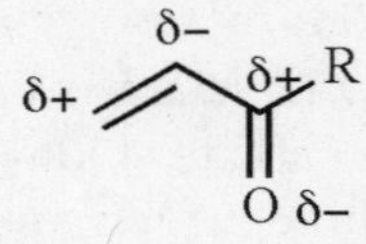

δ– δ+ δ+ R O δ–

23.55

What Is the Initial Product When a Nucleophile Reacts with Methyl Vinyl Ketone?

Reaction of methyl vinyl ketone (*23.56*), and a nucleophile will produce an enolate anion (*23.57*) via Michael addition.

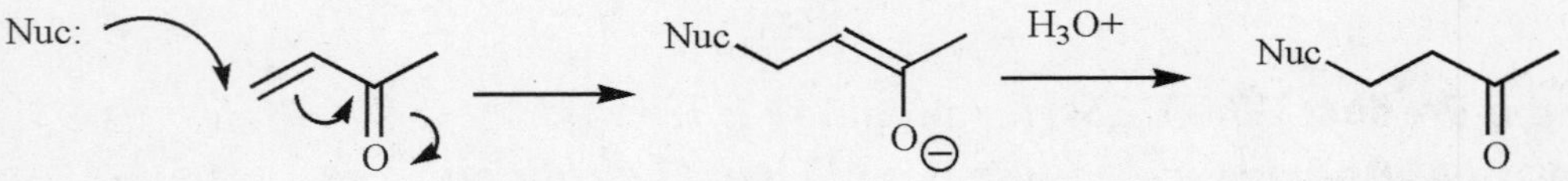

Nuc: → Nuc O⊖ → (H_3O+) Nuc O

23.56 *23.57* *23.58*

What Is the Final Product of the Reaction of 23.56 with a Nucleophile, after Hydrolysis?

When the initial product *23.57* is hydrolyzed, a ketone (*23.58*) will be formed.

What Are the Products of the Reactions of 23.59 and 23.60?

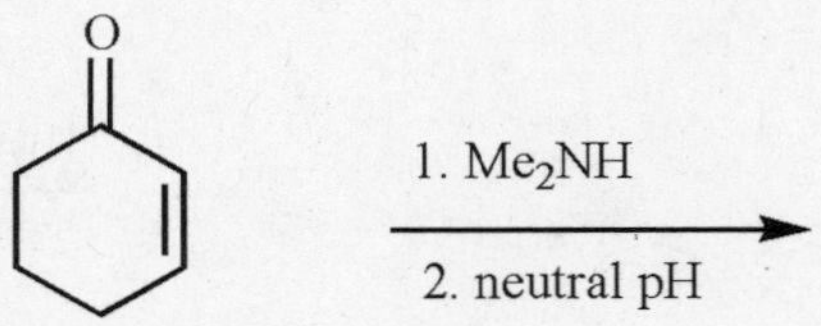

O

1. Me_2NH
2. neutral pH

23.59

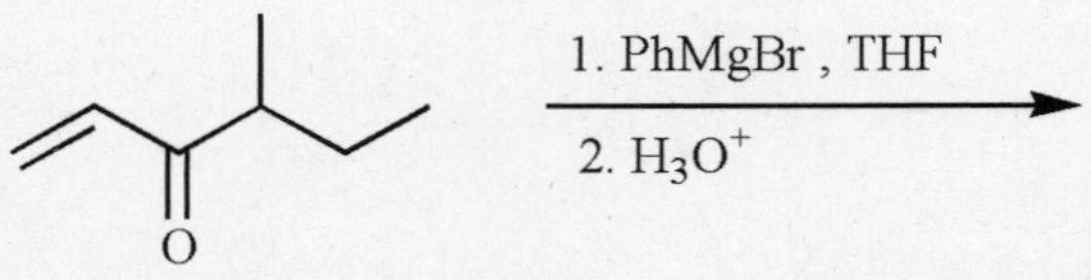

1. PhMgBr , THF
2. H_3O^+

O

23.60

If a Michael addition occurs with cyclohexenone (***23.59***), the 3-dimethylamino product ***23.61*** will be the result after careful neutralization of the amino ketone product. These **Michael adducts** are often unstable and are usually sensitive to heat and acid, losing the amine to reform the conjugated ketone. In the second case, the Grignard reagent adds to the double bond primarily because approach to the carbonyl is somewhat hindered by the adjacent alkyl substituents. A Michael addition of phenylmagnesium bromide to ***23.60*** leads to the phenyl derivative, ***23.62***.

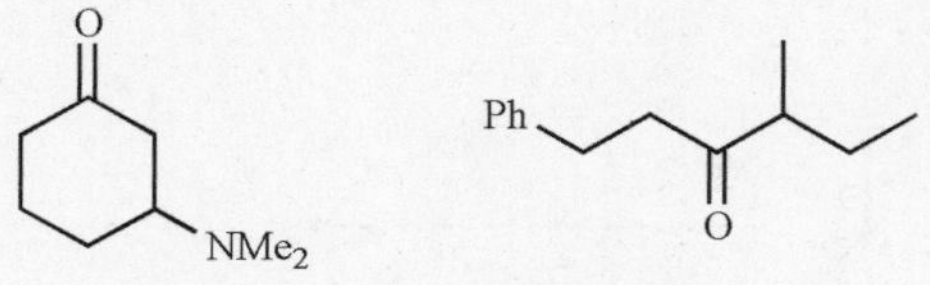

O NMe$_2$ Ph O

23.61 *23.62*

Robinson Annulation

The **Robinson annulation** is a cyclization reaction that combines an intramolecular aldol condensation after an initial Michael addition. When cyclopentanedione (***23.63***) is treated with methyl vinyl ketone under thermodynamic conditions, the product after hydrolysis is ***23.64***. This is a classical example of the Robinson annulation.

1. NaOEt , EtOH, reflux; 2. H_3O^+

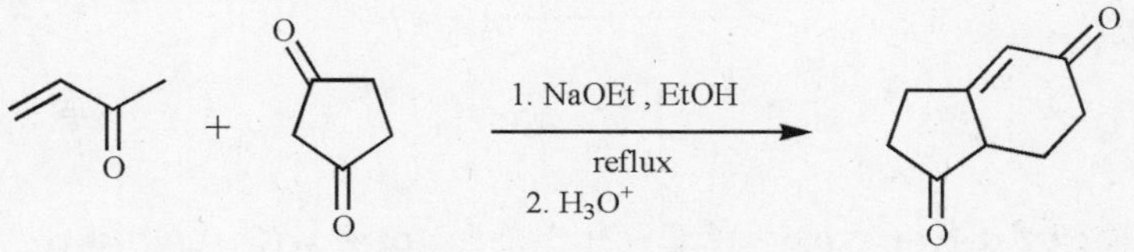

23.63 *23.64*

Give a Mechanistic Rationale for Why the Robinson Annulation Leads to a Six-Membered Ring.

The initial enolate anion (***23.65***) derived from the reaction of ***23.63*** with ethoxide adds to methyl vinyl ketone to produce enolate anion ***23.66*** via a Michael addition. An intramolecular aldol condensation from ***23.66*** would form a higher energy four-membered ring and is disfavored. Enolate anion ***23.66*** is formed under equilibrium conditions, which means that an equilibrium is established. Therefore, ***23.66***, the diketone ***23.67***, and the kinetic enolate anion ***23.68*** will be present at equilibrium. An intramolecular aldol condensation of enolate anion ***23.68*** will form the six-membered ring product ***23.69***. This alkoxide is protonated in a second chemical step by hydrolysis to give ***23.70***. In many cases, this is the isolated product. Heating or hydrolysis with strong aqueous acid, however, will induce dehydration to give the conjugated ketone, ***23.64***.

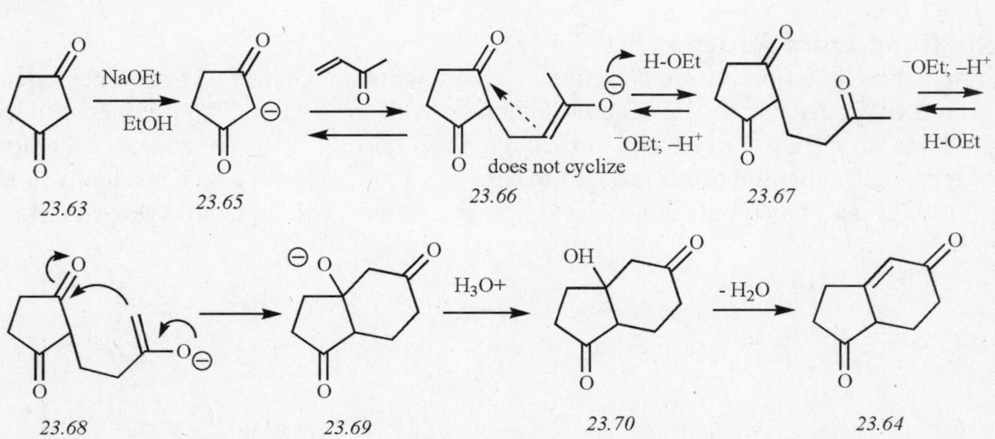

NaOEt, EtOH; H-OEt; $^-$OEt; –H$^+$; does not cyclize; $^-$OEt; –H$^+$; H-OEt

23.63 *23.65* *23.66* *23.67*

H_3O+; - H_2O

23.68 *23.69* *23.70* *23.64*

Conjugate Addition of Organocuprates

What Is the Structure of an Organocuprate?

An organocuprate has the structure R_2CuLi (see Section 10.5).

What is the Reaction Product when Lithium Dimethylcuprate Reacts with Hex-2-Enal?

Organocuprates react with conjugated ketones and aldehydes to give the Michael addition product, in this case enolate anion ***23.71***.

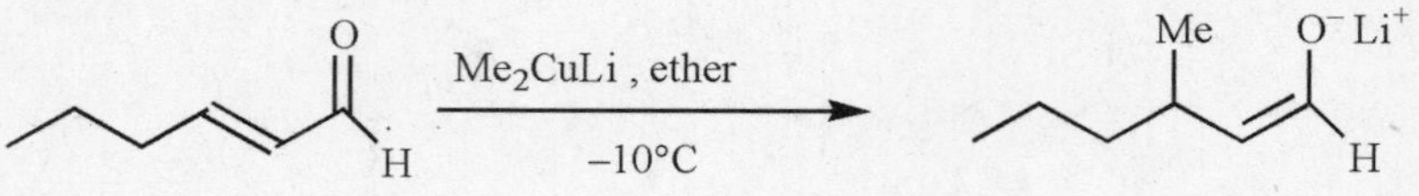

Me_2CuLi, ether, −10°C

23.71

What Chemical Step Is Required to Convert 23.71 to a Neutral Ketone?

The enolate anion is converted to its conjugate acid, aldehyde ***23.72***, by treatment with dilute aqueous acid.

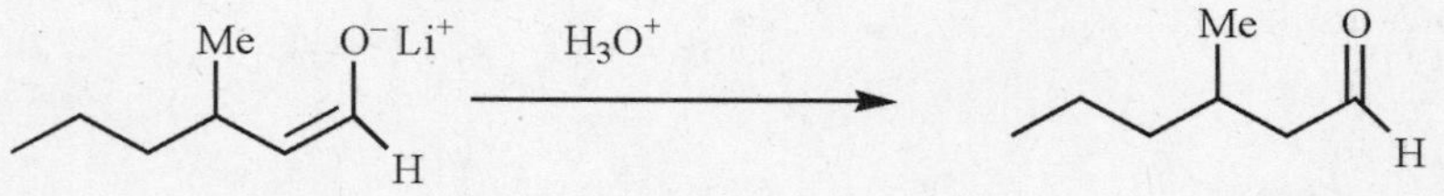

H_3O^+

23.71 *23.72*

What Is the Product of a Reaction of Cyclopentenone with Lithium Diphenylcuprate, Followed by Treatment with Aqueous Acid?

The product is 3-phenylcyclopentanone, ***23.73***.

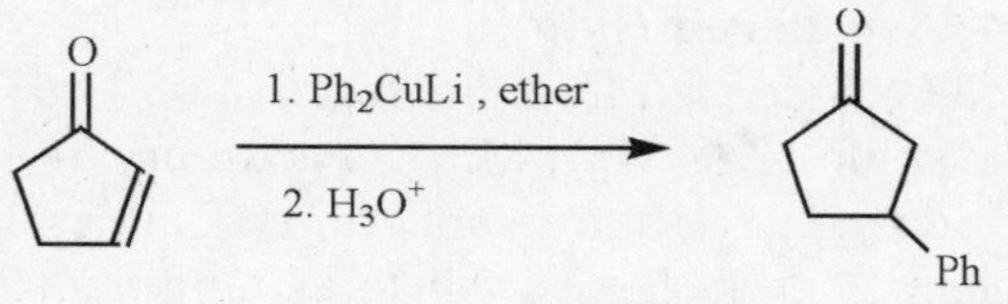

1. Ph_2CuLi, ether 2. H_3O^+

23.73

23.6. THE CANNIZZARO REACTION

The Cannizzaro reaction is a disproportionation reaction that converts some aldehydes into a mixture of a carboxylic acid and an alcohol.

What Is the Cannizzaro Reaction?

The **Cannizzaro Reaction** is a condensation reaction of aromatic aldehydes (or other aldehydes that do not have an α-hydrogen) that is induced by treatment with hydroxide. Two products are formed, an oxidation product (an acid), and a reduction product (an alcohol). This 'self oxidation-reduction' type of reaction is known as **disproportionation**. An example is the conversion of benzaldehyde (***23.74***) to benzoic acid (***23.75***) and benzyl alcohol (***23.76***) upon treatment with aqueous hydroxide, followed by neutralization.

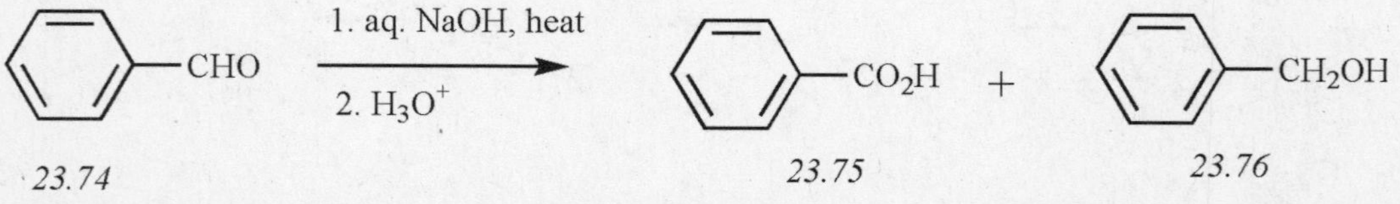

1. aq. NaOH, heat 2. H_3O^+

23.74 *23.75* *23.76*

Give a Mechanistic Rationale for this Reaction.

Hydroxide attacks the carbonyl of benzaldehyde to produce an alkoxide intermediate, ***23.77***. This is a reversible process, but given sufficient energy, ***23.77*** can react with another molecule of benzaldehyde, as shown. Transfer of hydrogen to the carbonyl (as *hydride, or* H^-) generates the acid and an alcohol. These react to form a carboxylate and an alcohol, and the acid is isolated upon hydrolysis.

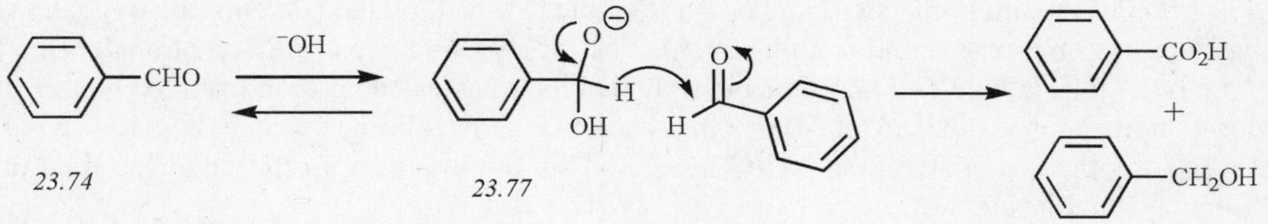

23.74 *23.77*

Explain the Limitations on the Carbonyl Partner of the Cannizzaro Reaction.

This only works if the aldehyde does not have an α-hydrogen. If it does, deprotonation leads to the enolate. Ketones do not give this reaction. Only aldehydes have the hydrogen that is transferred.

23.7. ENAMINES

Enamines react with aldehydes and ketones in a manner similar to enolate anions, but a hydrolysis step is required to convert the iminium salt product to an aldehyde or ketone

What Is the Structure of an Enamine?

An enamine has an amino group directly attached to a C=C unit, as in ***22.78***. See Section 21.3 for reactions that form enamines.

How Are Enamines Related to Enolate Anion S?

An enolate anion can donate electrons from oxygen to an electrophile (E^+), making the α-carbon nucleophilic (as in ***23.79***). Similarly, electrons can be donated from the nitrogen of an enamine, again making the α-carbon nucleophilic (as in ***23.78***). In a very practical sense, an enamine can be considered as a nitrogen analog of an enolate anion.

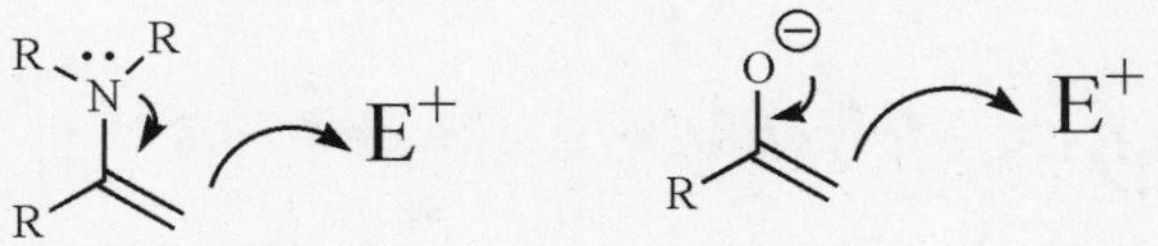

23.78 *23.79*

What Is the Initial Product When the Pyrrolidine Enamine of Cyclohexanone (23.80) Reacts with Benzyl Bromide?

When the α-carbon of the enamine (***23.80***) attacks the electropositive carbon of benzyl bromide, the product is an iminium salt, ***23.81***. The α-carbon has undergone alkylation.

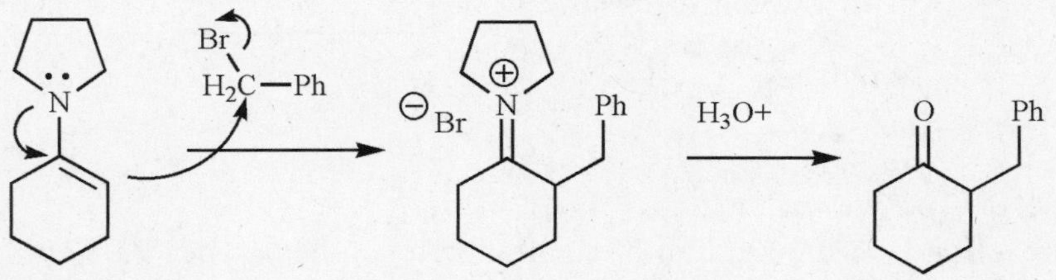

23.80 *23.81* *23.82*

What Is the Product When 23.81 Is Treated with Aqueous Acid?

When an iminium salt such as *23.81* is treated with aqueous acid, the final product is a ketone, *23.82*.

Give the Mechanism for This Latter Reaction.

The carbon of the iminium salt (*23.81*) is susceptible to attack by the nucleophilic oxygen of water, dumping electrons on nitrogen and forming ***23.83***. Loss of a proton gives ***23.84*** and transfer of a proton to the more basic nitrogen gives ***23.85***. It is possible that the transfer of a proton from oxygen to nitrogen can occur without the intermediacy of ***23.84***. Once protonated, pyrrolidine becomes a good leaving group and its loss gives the protonated carbonyl (***23.86***), and simple loss of a proton gives the final product, ***23.82*** (2-benzylcyclohexanone).

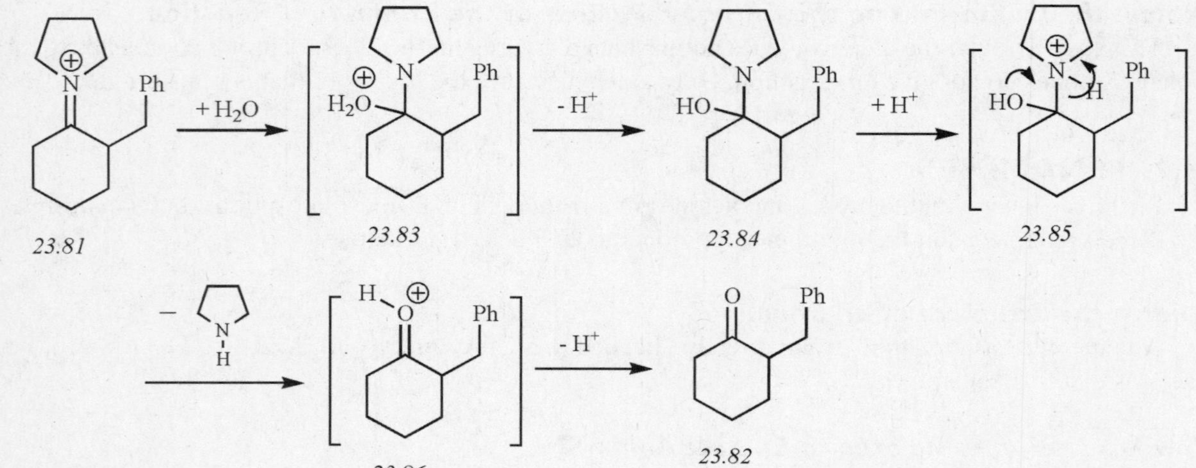

Ph; $+ H_2O$; H_2O; N; Ph; $- H^+$; HO; N; Ph; $+ H^+$; HO; N; H; Ph

23.81 *23.83* *23.84* *23.85*

– pyrrolidine (N–H); H, O; Ph; $- H^+$; O; Ph

23.86 *23.82*

Do Enamines React with Aldehydes and Ketones?

Since enamines behave as 'nitrogen enolates,' they will react with both aldehydes and ketones to produce iminium salts. Hydrolysis will then produce aldol-like products.

Give the Major Product for the Reaction of 23.87 and 23.88.

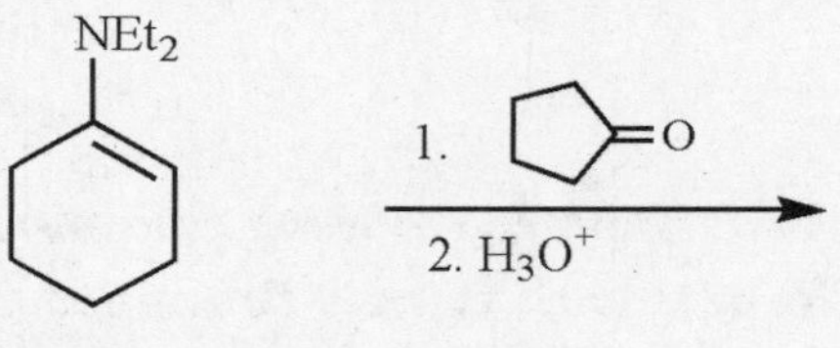

NEt_2; 1. cyclopentanone (=O); 2. H_3O^+

23.87

O; N; 1. CHO; 2. H_3O^+

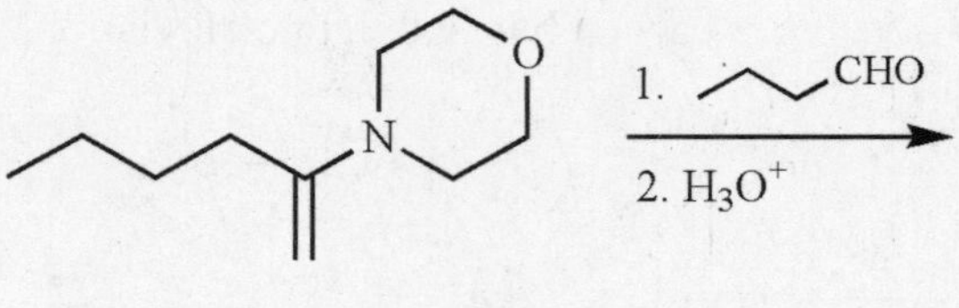

23.88

In the first reaction, the diethylamino enamine of cyclohexanone (***23.87***) is condensed with cyclopentanone to give, after hydrolysis, ***23.89*** (or possibly the conjugate ketone derived from dehydration). The second enamine (***23.88***) is condensed with butanal and gives, after hydrolysis, ***23.88*** (or possibly the conjugate ketone derived from dehydration). In both reactions, the aldol product is drawn. Under the hydrolysis conditions used to convert the iminium salt to the carbonyl, it is likely that elimination to the conjugated ketone will occur in both cases. The aldol products are shown to indicate the similarity to the aldol condensation products discussed above.

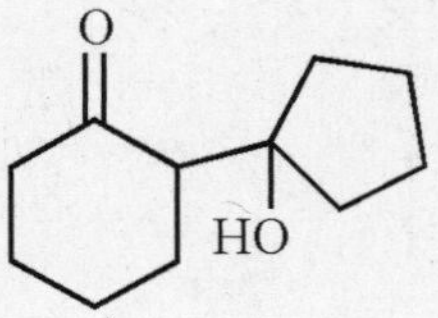

O HO

23.89

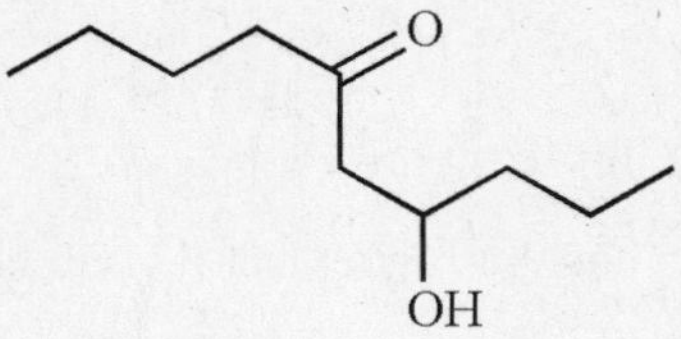

O OH

23.90

Test Yourself

1. What is the most acidic hydrogen in each of the following molecules?

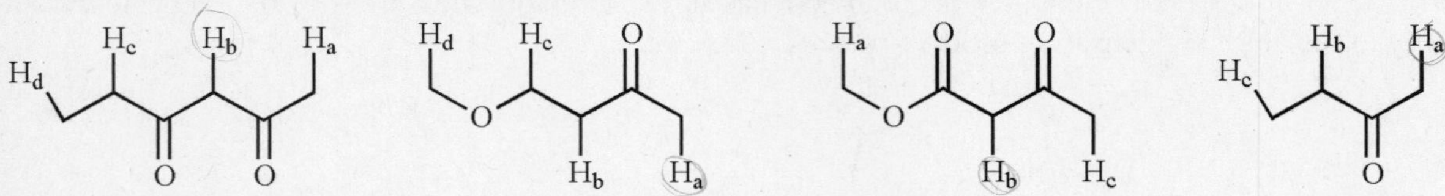

2. Which of the following has the highest enol content? Explain.

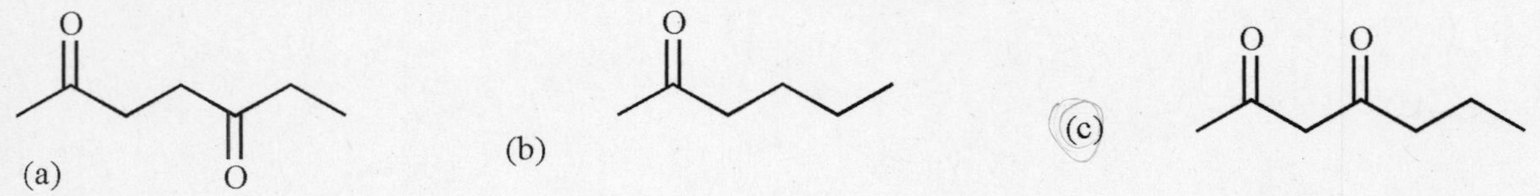

3. What reaction conditions favor kinetic control in forming an enolate anion from 2-butanone?

4. Give the mechanism of the following reaction.

, cat. H^+

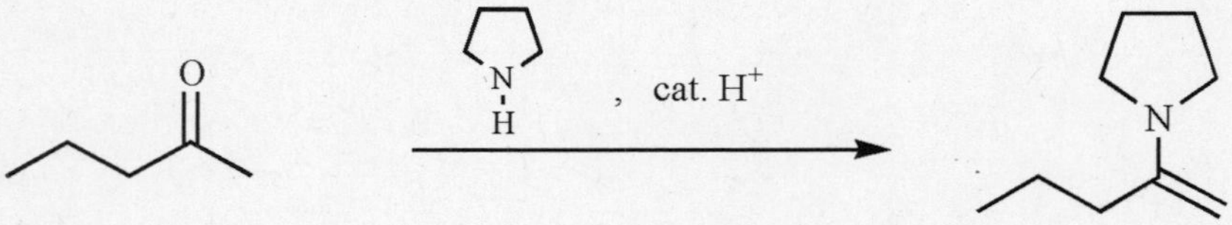

5. Give all aldol products that are possible from the following reaction.

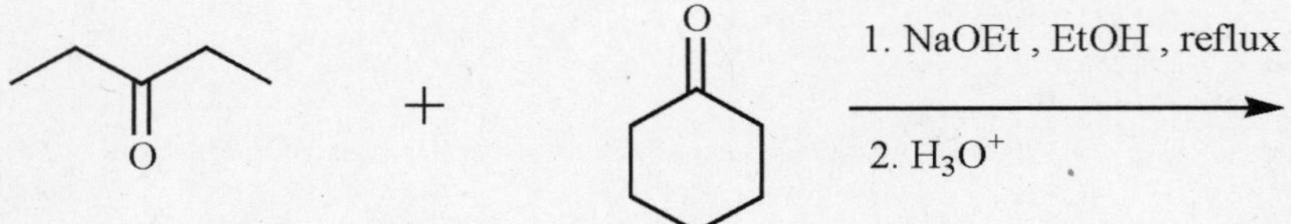

6. Predict the product from the reaction of but-3-en-2-one and lithium dibutylcuprate.

7. Why does the Cannizzaro reaction **not** occur with pentanal and aqueous sodium hydroxide?

8. In each case, give the major product. Remember stereochemistry where appropriate, and if there is no reaction, indicate this by N.R.

(a) H, O — t-B uOK , t-BuOH →

(b) CHO — 1. $LiN(iPr)_2$, THF, -78°C; 2. 4-nonanone; 3. H_3O^+ →

(c) O — O N-H, cat. H^+ →

(d) OH — 1. PCC , CH_2Cl_2; 2. NH_2 →

(e) H, N — BuLi , THF →

(f) N, H — O, cat. H^+ →

(g) O — 1. $LiNEt_2$, THF , -78°C; 2. O; 3. H_3O^+ →

(h) O, Me, O, Me — 1. $LiNEt_2$, THF, -78°C; 2. H_3O^+ →

(i) O — 1. LDA, THF, -78°C; 2. R-2-bromobutane; 3. H_2O →

(j) O — 1. MgBr THF; 2. H_3O^+ →

(k) O — 1. NaOEt , EtOH; 2. H_3O^+ →

(l) O O — 1. NaOEt , EtOH , reflux; O; 2. H_3O^+ →

(m) O — 1. LDA, THF, -78°C; 2. cycloheptanone; 3. H_3O^+ →

(n) CHO — 1. NaOH , H_2O, heat; 2. H_3O^+ →

(o) O, O — 1. NaOEt , EtOH; 2. H_3O^+ →

(p) N — 1-iodopentane →

(q) CHO — 1. LDA, THF, -78°C; 2. O; 3. H_3O^+ →

(r) O — 1. NaOEt , EtOH; 2. H_3O^+ →

(s) 2 Li — 1. CuI , ether; 2. cyclopentenone; 3.H_3O^+ →

(t) CHO — 1. $(PhCH_2)_2CuLi$ ether , –10°C; 2. H_3O^+ →

9. For each of the following provide a suitable synthesis. Show the structure of all intermediate products and show all reagents.

(a) O, CH_3 ⇒

(b) Ph, OH ⇒

CHAPTER 24

Carboxylic Acids

Carboxylic acids are the prototype organic acid. The name 'acid' describes much of their chemistry. These organic acids are important building blocks for the formation of other functional groups. Their acid properties are important in such diverse applications ranging from giving important physical and chemical properties to amino acids (see Chapter 28) to their use as acid catalysts in cationic reactions of alcohols and alkenes. This chapter will focus on the acid properties of carboxylic acids and the factors which influence acidity. This chapter will also show how carboxylic acids can be transformed into other 'acid derivatives,' such as acid chlorides, esters, anhydrides and amides.

24.1. STRUCTURE AND NOMENCLATURE OF CARBOXYLIC ACIDS

Carboxylic acids are characterized by a COOH unit, which is a carbonyl group with an OH unit attached to the carbon.

What Is a Carboxyl Group?

A carbonyl with an OH attached to it, HO–C=O = –COOH = –CO_2H. See ***24.1***.

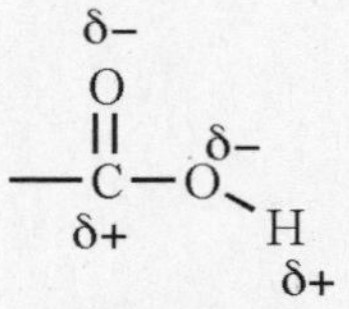

δ–
O
‖ δ–
—C—O
δ+ H
δ+

24.1

What Is the Distinguishing Feature of a Carboxyl Group?

A carboxyl group is characterized by the presence of a very acidic hydrogen (O–H) along with an electropositive carbonyl carbon of the C=O.

Describe the Bond Polarity in a Carboxyl Group.

The O–H group in ***24.1*** is polarized such that the H is acidic and the carbonyl carbon is electrophilic.

How Extensive Is the Hydrogen Bonding Capability of a Carboxyl Group?

The acidic hydrogen of the OH group is strongly hydrogen bonded to the carbonyl oxygen of a second carboxyl, as illustrated in ***24.2***.

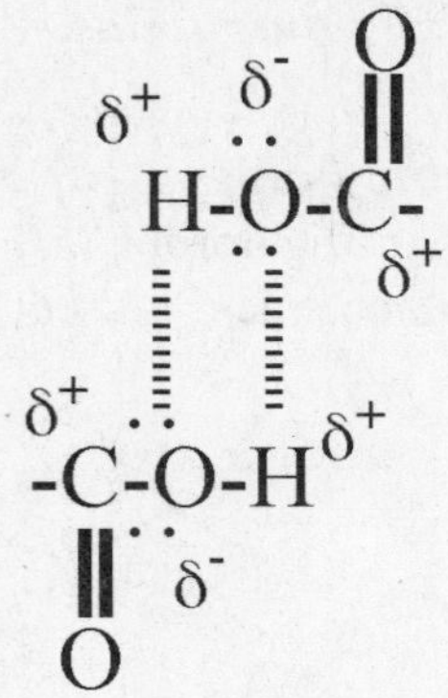

```
δ+   δ-   O
     ..   ||
   H-O-C-
     ..   δ+
   ⋮    ⋮
δ+ ..     δ+
-C-O-H
 || ..  δ-
 O
```

24.2

When Compared to an Alcohol of Similar Molecular Weight, a Carboxylic Acid Generally Has a Much Higher Boiling Point. Why?

As seen in Chapter 1, a carbonyl group is capable of extensive hydrogen bonding, much more extensive and stronger than observed with alcohols. This strong hydrogen bonding must be disrupted to bring the carboxylic acid into equilibrium with both gas and liquid (boiling point), requiring higher temperatures than with an alcohol.

Does a Carboxylic Acid Exhibit Important UV Absorption?

There are no important UV bands that can be used for unique identification of a carboxylic acid relative to another functional group.

What Is the Relative Polarity of a Carboxylic Acid?

A carboxyl group is very polar, with the dipole moment generally directed at an angle between the O–C=O bond.

Nomenclature

What Is the IUPAC Ending That Is Correlated with a Carboxylic Acid?

The IUPAC ending for a carboxylic acid is *-oic acid*. The C_6 straight-chain acid is hexanoic acid. Drop *-ane* from hexane and add *-oic acid*.

Why Is the Numerical Position of the Carboxyl Group Usually Omitted from the Name of a Carboxylic Acid?

Since the carbonyl carbon of the carbonyl *must* have the lowest number (it is the functional group) and that carbon contains the OH, it *must* be C1. For this reason, as with aldehydes, the number is usually omitted.

Why Is a C6 Acid Called Hexanoic Acid Rather Than Hexoic Acid?

A C6 acid could have an alkane, alkene or alkyne backbone. The *-an, -en* or *-yn* must be included to show which is present. This will lead to hexanoic acid, hexenoic acid and hexynoic acid for the three cases mentioned. In the last two cases, a number must be assigned to the C=C or C≡C group (as in hex-4-enoic acid or hex-5-ynoic acid, $CH_3CH{=}CHCH_2CH_2COOH$ and $HC{\equiv}CCH_2CH_2CH_2COOH$, respectively).

Give the Correct Structure for the Following Acids.
(a) 3,5-Diphenylheptanoic Acid; (b) 3,5-Dichlorobenzoic Acid; (c) 3-Ethyl-4-(3-Methylbutyl)Nonanoic Acid; (d) 3-Cyclopentylbutanoic Acid.

3,5-Diphenylheptanoic acid is *24.3*; 3,5-dichlorobenzoic acid is *24.4*; 3-ethyl-4-(3-methylbutyl) nonanoic acid is *24.5*; and 3-cyclopentylbutanoic acid is *24.6*.

(a) Ph Ph CO_2H *24.3*; (b) CO_2H Cl Cl *24.4*; (c) CO_2H *24.5*; (d) CO_2H *24.6*

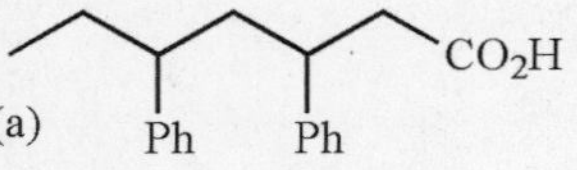

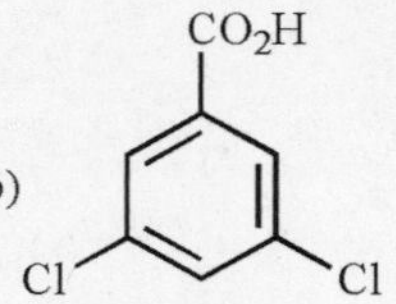

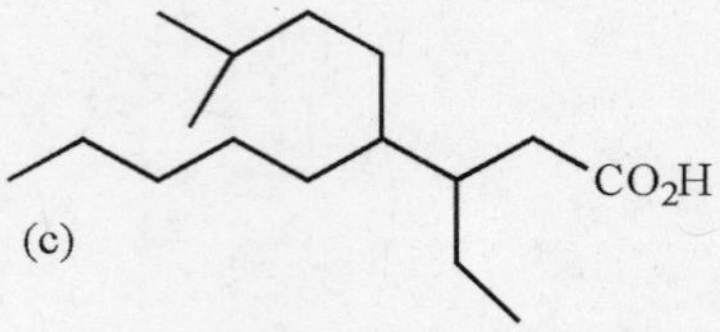

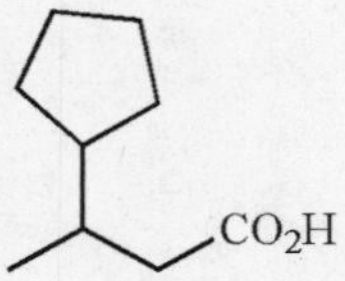

Give the Common Name of the Following Acids.
(a) CH_3CO_2H; (b) HCO_2H; (c) $PhCH_2CO_2H$; (d) $CH_3CH_2CH_2CH_2CH_2CH_2CH_2CO_2H$

The common name of (a) is acetic acid, and (b) is called formic acid. Using the acetic acid nomenclature, the common name of (c) is phenylacetic acid. Case (d) is an 8-carbon acid with the common name of caprylic acid. Many of the common lower molecular weight acids have common names, especially those with an even number of carbons:

C5 = valeric acid; C6 = caproic acid; C10 = capric; C12 = lauric; C14 = myristic; C16 = palmitic; C18 = stearic.

24.2. ACIDITY

Perhaps the most distinguishing feature of carboxylic acids is that they are protonic acids and react with a variety of suitable bases.

What Is the pK_a of Acetic Acid (Ethanoic Acid)?

The pK_a of acetic acid (ethanoic acid) is 4.76.

Give the pK_a Values of the Following Carboxylic Acids: A) Propanoic Acid; B) Butanoic Acid; C) Formic Acid.

(a) propanoic = 4.89; (b) butanoic acid = 4.82; (c) formic acid = 3.75.

Why Is Acetic Acid a Weaker Acid Than Formic Acid?

Acetic acid has a methyl group attached to the electropositive carbonyl. Relative to hydrogen (in formic acid), methyl is an electron releasing group. This 'pushes' electrons towards the O–H bond, strengthening it, and making that hydrogen less acidic (less positive = less like H^+).

What Is the Pk_a of 2-Chloroethanoic Acid (Chloroacetic Acid)?

The pK_a of chloroacetic acid is 2.85.

Why Is 2-Chloroethanoic Acid a Stronger Acid Than Acetic Acid?

The chlorine has an electron withdrawing effect. Cl polarizes the α-carbon $\delta+$, making the carbonyl carbon withdraw electrons from the O–H bond, thereby weakening it (more positive H, more like H^+) and making it a stronger acid. Certain electronic effects are transmitted from nucleus to nucleus through the adjacent covalent bonds, and these are collectively called **'through-bond' effects**, a type of **inductive effect**.

When an Electron Withdrawing Group Is Attached to the Carbonyl of a Carboxyl, What Is the Effect on the O–H Bond?

The electron withdrawing group induces a larger $\delta+$, charge on the carbonyl carbon. This is destabilizing, and, to compensate, the carbonyl carbon withdraws electrons from neighboring atoms to diminish the electron withdrawing effects of the α-carbon. This makes the O–H bond more polarized, the H more positive and more acidic.

Will an Electron Withdrawing Substituent Increase or Decrease Acidity?

In general, electron withdrawing groups increase acidity (larger Ka, smaller pK_a).

Explain Why 3-Chlorobutanoic Acid Is a Weaker Acid Than 2-Chlorobutanoic Acid.

The electron withdrawing chlorine atom is further away from the carbonyl carbon in 3-chloroacetic acid than in 2-chloroacetic acid. The electron withdrawing ability of an atom or group diminishes as the distance from the O–H group and the carbonyl increases.

Explain Why 2,2,2-Trichloroethanoic Acid Is a Stronger Acid Than 2-Chloroethanoic Acid.

If one chlorine atom withdrawing electrons from the α-carbon makes the O–H bond weaker, the presence of three electron withdrawing chlorines will withdraw even more electron density. This will make the α-carbon more positive and the O–H bond weaker and, therefore, more acidic. If the pK_a of 2-chloroethanoic acid (listed above) is 2.89, the pK_a of 2,2,2-trichloroethanoic acid is 0.64, making the latter a significantly stronger acid.

Will an Electron Releasing Substituent Increase or Decrease Acidity?

An electron releasing substituent 'feeds' electrons towards the electropositive carbonyl, and that carbonyl carbon will withdraw *less* electron density from the O–H bond. The O–H bond is stronger, and the acid is a weaker acid (more difficult to remove the H from O–H).

Explain Why 4-Nitrobenzoic Acid Is a Stronger Acid Than 4-Methoxybenzoic Acid.

The nitro group is strongly electron withdrawing. This effect is transmitted through the aromatic ring to the carbonyl carbon, making the acid stronger. When attached to benzene, the OMe group is electron releasing (relative the benzene ring), and when this effect is transmitted through the aromatic ring, the acid is weaker.

In Each of the Following Series, Indicate the Strongest Acid. (a) Benzoic Acid or 4-Methoxybenzoic Acid; (b) 3-Chloropropanoic Acid or 3-Chloropentanoic Acid; (c) 2-Methoxyethanoic Acid or Propanoic Acid; (d) 4-Nitrobenzoic Acid or P-Toluic Acid (4-Methyl-Benzoic Acid).

In (a), the electron releasing methoxy group makes 4-methoxybenzoic acid the weaker acid relative to benzoic acid [pK_a of benzoic acid = 4.19 and the pK_a of 4-methoxybenzoic acid = 4.46]. In (b), both acids have an electron withdrawing chlorine at C_3 but 3-chloropentanoic acid also has an ethyl group attached to C_3. The presence of the electron releasing alkyl group, which is missing in 3-chloropropanoic acid, will make 3-chloropentanoic acid a weaker acid. Therefore, 3-chloro-propanoic acid is the stronger acid. In (c), methoxy is electron withdrawing, primarily by through-space effects (see below), whereas the methyl (attached to the α- carbon of acetic acid, respectively) is electron releasing. Since methoxy is electron withdrawing, 2-methoxyethanoic acid is expected to be the stronger acid [pK_a of methoxyacetic acid = 3.6; pK_a of propionic acid = 4.9]. In (d), the nitro group is electron withdrawing, making it a stronger acid than toluic acid, which has an electron releasing methyl group at the *para-* position [the pK_a of 4-nitrobenzoic acid is 3.41 and the pK_a of toluic acid = 4.34].

What Is a Through-Space Inductive Effect?

When a heteroatom group is sufficiently close in space to the acidic hydrogen of the carboxyl, 'through-space' electronic effects, often called field effects, will influence the acidity of the acid. These through-space effects are usually a stronger influence on the acidity than the through-bond effects.

Explain Why 2-Nitrobenzoic Acid Is a Stronger Acid Than 4-Nitrobenzoic Acid.

In 2-nitrobenzoic acid, the electron withdrawing group is closer to the carboxyl group. As shown in *24.7*, the H of the O–H bond is physically close to the electronegative oxygens of the nitro group. A 'through-space' hydrogen bonding effect will weaken the O–H bond and strengthen the acid. Since the nitro group in the *para-* position *cannot* assume such an intramolecular hydrogen bonding array, it will be a weaker acid.

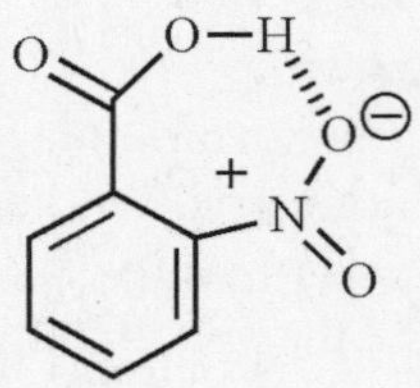

24.7

Describe the Relative Acidity of Ethanoic Acid and 2-Chloroethanoic Acid in Terms of a Through-Space Effect.

Examination of ***24.8*** shows that the chlorine and the H of the OH can hydrogen bond (through space) in what is effectively a five-membered ring. This is energetically accessible and the electron withdrawing effect of this through-space hydrogen bonding is very strong, making chloroacetic acid a stronger acid.

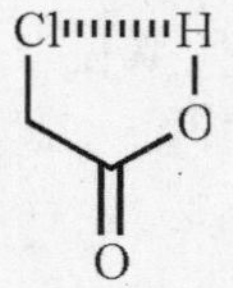

24.8

Why Is 2-Methoxybenzoic Acid (24.9) a Stronger Acid Than 4-Methoxybenzoic Acid (24.10) Although Both Are Electron Releasing and, in Principle, Weaken Acidity?

In ***24.9***, the oxygen of the methoxy is sufficiently close to the carbonyl that hydrogen bonding can occur between the lone electrons of oxygen and the acidic hydrogen of the acid. This makes the *ortho*-methoxy derivative more acidic than the *para*-methoxy derivative (***24.10***) where intramolecular hydrogen bonding is not possible. This enhancement in acidity due to the presence of a heteroatom in the *ortho*-position is often called the **ortho effect**.

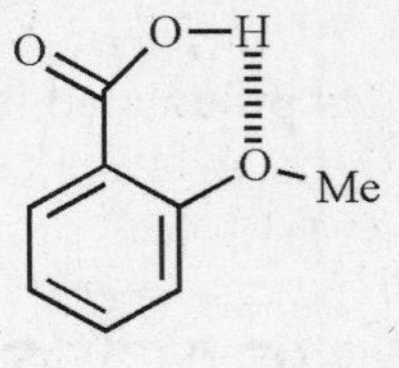

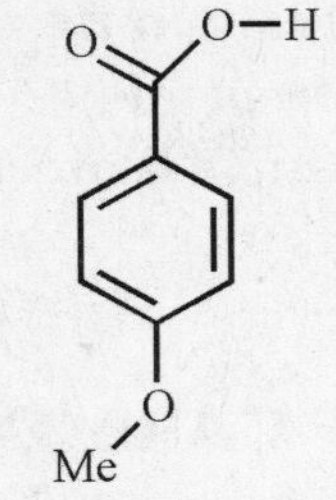

24.9 24.10

Explain why 2-Methoxybenzoic Acid Is a Stronger Acid Than 2-Methylbenzoic Acid.

This is another example of the *ortho* effect. The oxygen of the methoxy group can hydrogen bond with the acidic hydrogen of the carbonyl group when OMe is in the *ortho-* position. The methyl, of course, has no heteroatom for hydrogen bonding.

What Is the Conjugate Base of Propanoic Acid?

The conjugate base of propanoic acid is the propanoate anion, $CH_3CH_2CH_2CO_2^-$.

What Is the Structure of the Carboxylate Anion Formed by Removal of the Acidic Hydrogen from the Acid?

Treatment of a carboxylic acid with a base ($NaHCO_3$) is a sufficiently strong base for this reaction and generates the carboxylate anion, ***24.11***. This ion is resonance stabilized, with the charge dispersed to both oxygens. The carboxylate anion is a weak nucleophile and a relatively weak base.

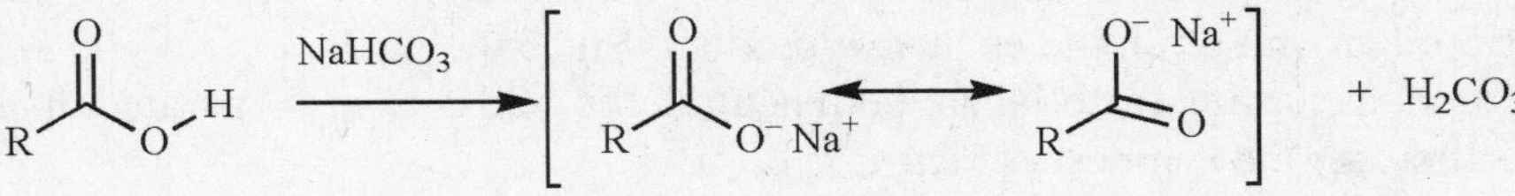

$NaHCO_3$ + H_2CO_3

24.11

Why Is the Carboxylate Anion Considered to Be very Stable?

As shown in ***24.11***, the carboxylate anion is resonance stabilized. The negative charge is delocalized over three atoms (O–C–O).

How Does the Special Stability of the Carboxylate Anion Influence Acidity?

The product, the carboxylate anion, is resonance stabilized, and an equilibrium reaction will be influenced by the relative stability of the species that make up that equilibrium. A more stable product tends to shift the equilibrium towards that product. In this case, if the equilibrium is shifted towards the carboxylate, Ka is larger. A large Ka is indicative of a stronger acid.

When Benzoic Acid Is Treated with NaOH, What Is the Product?

The product is sodium benzoate, ***24.12***, and water.

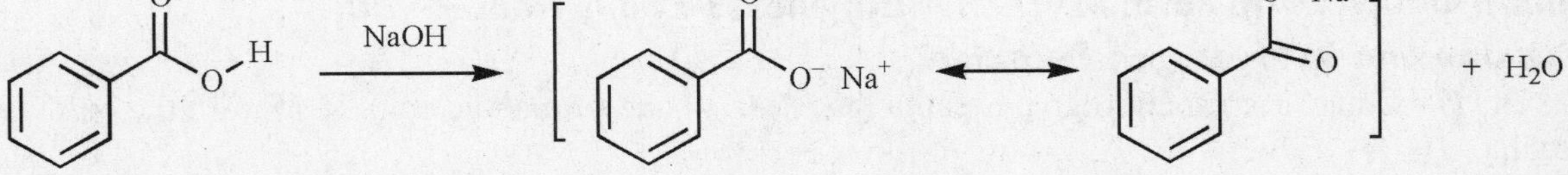

NaOH + H_2O

24.12

What Is the Product When Potassium Benzoate Is Dissolved in Aqueous Solution at pH 4?

The relatively weak base (potassium benzoate) is protonated in the acidic solution (pH 4) to give benzoic acid.

24.3. PREPARATION OF CARBOXYLIC ACIDS

Carboxylic acids can be prepared in several different ways, but oxidation of alcohols or alkenes is probably the most common method. Hydrolysis of cyanides derived from alkyl halides is another very common preparative method.

Oxidation of Aldehydes

The oxidation of alcohols to aldehydes and to carboxylic acids was discussed in Section 22.2.

What Is the Product When 1-Pentanol Is Reacted with Chromium Trioxide in Aqueous Acid?

Chromium trioxide (in aqueous sulfuric acid, it is usually called Jones reagent) is a powerful oxidizing agent. The initially formed aldehyde from the primary alcohol is further oxidized to a carboxylic acid. The product of oxidizing 1-pentanol is, therefore, pentanoic acid.

Is This a Useful Method for the Preparation of Carboxylic Acids or Does It Usually Give Mixtures of Products?

When Jones reagent is used for the oxidation of primary alcohols, the carboxylic acid is almost always the major product and often the exclusive product. Since the acidic carboxylic acid and the neutral aldehyde and alcohol starting material are both neutral, the acid is easily separated, making this a useful preparative method for the synthesis of carboxylic acid.

Oxidative Cleavage of Alkenes

This chemistry was described in Section 14.2. Alkenes are oxidatively cleaved to produce carboxylic acids if the carbon of the C=C group has at least one hydrogen attached to it.

What Is the Product When Cyclohexene Is Treated with a) Ozone and b) Hydrogen Peroxide?

The product is 1,5-hexanedioic acid, ***24.13***.

1. O_3, −78°C

2. H_2O_2

CO_2H

CO_2H

24.13

What Product(s) Are Formed When 4-Ethylhex-3-Ene Is Treated with a) Ozone and b) Hydrogen Peroxide?

The C=C unit is cleaved to give 3-pentanone (***24.14***) and propanoic acid ***24.15*** with the oxidative workup.

1. O_3, –78°C
2. H_2O_2

24.14 14.15

The Haloform Reaction

What Is a Haloform?

A haloform has the structure HCX_3 where X = F, Cl, Br, I.

What Is the Product When 2-Butanone Is Treated with Bromine and Aqueous Sodium Hydroxide?

This reaction leads to oxidative cleavage of the C–C bond in methyl ketones. In this case, there are two products, bromoform ($HCBr_3$) and propanoic acid.

What Is the Product When Acetone Is Treated with Iodine and Aqueous Sodium Hydroxide?

Iodine and base also gives oxidative cleavage to iodoform (HCI_3) and ethanoic acid.

What Is the Mechanism of This Reaction Using Acetone As a Substrate?

The reaction begins with formation of enolate anion ***24.17*** from reaction of acetone with hydroxide. Note that acetone also reacts with hydroxide via nucleophilic acyl addition to give ***24.16***, but this is reversible and favors the carbonyl starting material. Enolate anion ***24.17*** reacts with iodine (with loss of iodide) to form ***24.18*** (1-iodo-2-propanone). The presence of I on the α carbon makes the α-proton more acidic than the analogous proton in acetone, and formation of the enolate allows reaction with a second molecule of iodine to give ***24.19***. Similarly, ***24.19*** can form an enolate that reacts with a third equivalent of iodine to give ***24.20***. Just as hydroxide added to acetone gives ***24.16***, hydroxide adds to all ketone species in this reaction. In the case of acetone, ***24.18–24.20***, the reaction is reversible. When hydroxide adds to ***24.20*** to give ***24.21***, however, the $-CI_3$ is a good leaving group, since it is able to accommodate the negative charge. The charge is stabilized in ***24.22***. This allows cleavage to acetic acid and ***24.22***. Since ***24.22*** is a strong base, it deprotonates the acetic acid to give the carboxylate salt and iodoform, ***24.23***.

^-OH I_2 ^-OH, I_2 ^-OH, I_2 ^-OH

24.17 24.18 24.19 24.16 24.23 24.22 24.21 24.20

What Is the Iodoform Test? What Functional Group Does It Detect?

The iodoform test mixes a suspected methyl ketone (the functional group is a methyl group attached to a carbonyl [CH_3-C=O]) with iodine and aqueous hydroxide. If the methyl ketone functional group is present, a yellow precipitate of iodoform is observed. The reaction also works with methyl carbinols (HO–C–CH_3).

From Cyanides

How Can a Nitrile Be a Precursor to a Carboxylic Acid?

When a nitrile is reacted with aqueous acid under rather vigorous conditions (heat and relatively concentrated acid), a carboxylic acid is formed. Often, the nitrile can be converted to an amide and then further hydrolyzed to the acid.

What Is the Product When 1-Bromobutane Is Reacted with Sodium Cyanide in DMF?

Cyanide behaves as a carbon nucleophile, giving pentanenitrile ($CH_3CH_2CH_2CH_2C{\equiv}N$) via a S_N2 reaction (see Section 7.2).

What Is the Product When Hexanenitrile Is Treated with Cold and Dilute Aqueous HCl?

Nitriles can be hydrolyzed to carboxylic acids or amides, but this usually requires strong acid and vigorous reaction conditions. If any reaction occurs at all under these milder conditions, the product will be the amide, pentanamide. It is more likely that the nitrile will be recovered unchanged from this reaction medium.

What Is the Product When a 6 N Solution of HCl Is Added to Pentanenitrile and Heated to 80°C for Several Hours?

Under these vigorous reaction conditions, pentanenitrile is converted to the carboxylic acid, pentanoic acid. Since the reaction proceeds by initial hydrolysis to pentanamide, some pentanamide may also be isolated from this reaction. Given sufficient time, the acid will become the major product, however.

What Is a Cyanohydrin?

A cyanohydrin is a molecule that has both a hydroxyl (OH) and a nitrile (C≡N) group.

How Are Cyanohydrins Formed?

Treatment of an aldehyde or ketone with HCN will lead to a cyanohydrin, ***24.24***. Cyanohydrins are also formed by reaction of a ketone or aldehyde with sodium (or potassium) cyanide to give ***24.25***, followed by quenching with aqueous acid to give the cyanohydrin, ***24.26***.

OH, C≡N, R — dil. H_3O^+ ← O⁻, C≡N, R ← ⁻C≡N — O, R — H-C≡N → OH, C≡N, R

24.26 *24.25* *24.24*

What Is the Product When a Cyanohydrin Is Hydrolyzed?

Hydrolysis will convert the nitrile group to a carboxylic acid. Since the hydroxyl group will remain, the product will be an α-hydroxy acid. In many cases, however, treatment with acid induces loss of water via protonation of the OH, and the product is an α,β-unsaturated acid.

From Grignard Reagents and Carbon Dioxide

In What Way Is Carbon Dioxide Related to Formaldehyde?

Carbon dioxide is a carbonyl compound [O=C=O] just as formaldehyde [$H_2C=O$] contains a carbonyl. Both compounds can react with nucleophiles.

When Ethylmagnesium Bromide Reacts with Carbon Dioxide, What Is the Initial Product? What Is the Product after Hydrolysis?

The nucleophilic Grignard reagent will attack the carbonyl of CO_2 to produce a carboxylate salt (*24.27*). Hydrolysis will convert the salt to the acid, propanoic acid in this case.

BrMg

O=C=O ——ether——> O^- O ——H_3O+——> OH O

24.27

Give the Major Products from the Reactions of 24.28 and 24.29.

Br

1. Mg°, THF
2. CO_2
3. H_3O^+

24.28

OH

1. PBr_3
2. Mg°, ether
3. CO_2
4. H_3O^+

24.29

When bromobenzene (*24.28*) reacts with magnesium, phenylmagnesium bromide is formed. This Grignard reagent reacts with CO_2 to give benzoic acid after hydrolysis. When 2-heptanol (*24.29*) reacts with PBr_3, 2-bromoheptane is produced. Subsequent reaction with Mg leads to the Grignard reagent and condensation with CO_2 leads to 2-ethylhexanoic acid.

24.4. REACTIONS OF CARBOXYLIC ACIDS

Carboxylic acids can be converted into a variety of acid derivatives, but there are also many other reactions that can be useful.

Reduction

What Is the Product When a Carboxylic Acid Is Reduced?

In general, when a carboxylic acid is reduced, a primary alcohol is the product.

What Is the Product of the Reaction of Pentanoic Acid and $LiAlH_4$?

$LiAlH_4$ reduces acids to the corresponding primary alcohol, in this case 1-pentanol.

What Is the Product of the Reaction Between Butanoic Acid and $NaBH_4$?

$NaBH_4$ is not powerful enough to reduce the acid to the alcohol. $NaBH_4$ reacts with butanoic acid in an acid-base reaction to form the carboxylate salt of butanoic acid.

What Is the Product When Propanoic Acid Is Treated with Borane and then Hydrolyzed with Dilute Acid?

Borane has a great affinity for reaction with carboxylic acid, giving the corresponding alcohol. Borane will reduce propanoic acid to 1-propanol.

Explain Why Borane Is Often Called 'The Reagent of Choice' for the Reduction of Carboxylic Acids.

Borane will reduce an acid faster than an ester, a nitro compound, a nitrile or many other functional groups. If both an ester and a carboxylic acid are present in the same molecule (as in ***24.30***), reaction with borane will give selective reduction of the acid to give ***24.31***.

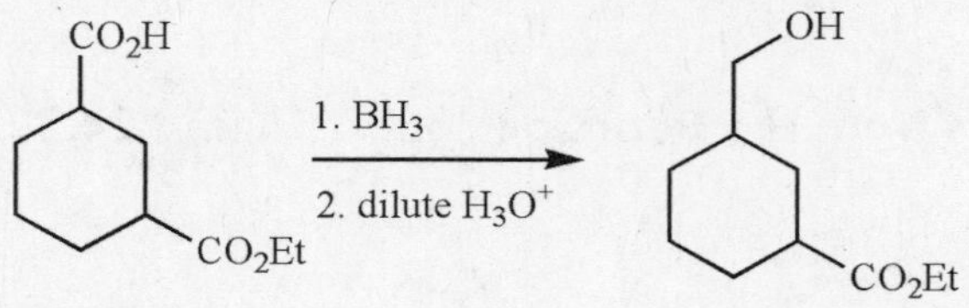

CO_2H CO_2Et — 1. BH_3 2. dilute H_3O^+ → OH CO_2Et

24.30 *24.31*

Preparation of Acid Chlorides

What Is the Structure of an Acid Chloride?

An acid chloride has a chlorine attached directly to a carbonyl, as in ***24.32***.

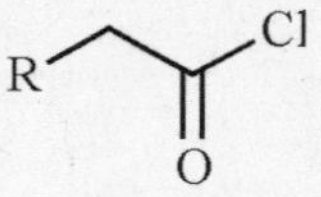

R Cl O

24.32

What Is the IUPAC Nomenclature System for Acid Chlorides?

Drop the *-oic acid* ending of the parent carboxylic acid and add *-oyl chloride*. For example, the acid chloride derived from pentanoic acid is pentanoyl chloride.

Give the IUPAC Name for 24.33–24.36.

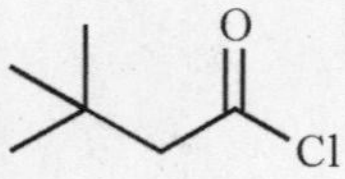

O Cl Cl O Cl O Cl O Cl NO_2 O Cl Ph

24.33 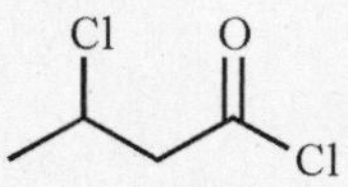 *24.34* 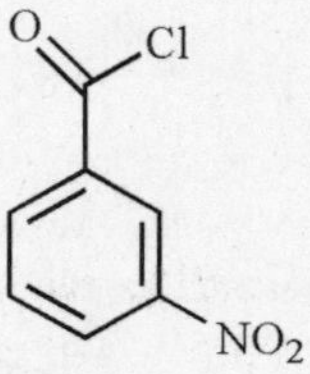 *24.35* 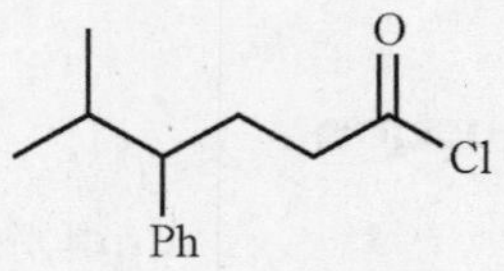 *24.36*

Acid chloride ***24.33*** is 3,3-dimethylbutanoyl chloride, and ***24.34*** is named 3-chlorobutanoyl chloride. Compound ***24.35*** is 3-nitrobenzoyl chloride, and ***24.36*** is 5-methyl-4-phenylhexanoyl chloride.

What Reagents Are Capable of Converting Benzoic Acid into Benzoyl Chloride?

Common halogenating reagents are thionyl chloride ($SOCl_2$), phosphorus oxychloride ($POCl_3$), phosphorus trichloride (PCl_3) and phosphorus pentachloride (PCl_5).

Can These Reagents Be Used to Convert Most Carboxylic Acids to the Corresponding Acid Chloride?

Yes!

What Is the Major Product When Decanoic Acid Is Treated With $SOCl_2$?

The product is the acid chloride, decanoyl chloride [$CH_3(CH_2)_7CH_2COCl$].

α-Halogenation

What Is the Product of the Reaction Between Bromine and Butanoic Acid?

The reaction of a carboxylic acid and bromine or chlorine is very slow. For all practical purposes, there is no reaction, since a halogen must react with the enol form of the acid, which is present in very small concentration. If the % of enol can be increased, halogenation is more viable.

What Is the Hell-Volhard-Zelinsky Reaction?

The conversion of a carboxylic acid to an α-halo acid by treatment with a halogen (usually bromine) and a halogenating agent such as PCl_3 or PBr_3 (P + Br_2) is the **Hell-Volhard-Zelinsky** reaction. In general, PBr_3 is the more commonly used reagent. An example is the conversion of butanoic acid to 2-bromobutanoic acid. Initial reaction with PBr_3 gives the acid chloride, which has an enol (***24.37***). The enol will then attack the bromine, as shown, displacing bromide and forming the α-bromo acid chloride (***24.38***). This acid chloride usually reacts with the starting acid to produce the final product, α-bromobutanoic acid (***24.39***) and the acid chloride, which can be recycled. For this reason, only a catalytic amount of PCl_3 is required.

O OH PBr3 → O Br → OH Br Br-Br *24.37* → O Br Br *24.38*

O OH → O OH Br *24.39* + O Br

Describe the Reaction of 2-Chloroethanoic Acid and Sodium Phenoxide. Explain the Product.

The chlorine at the 2-position is very susceptible to nucleophilic attack, usually reacting faster than nucleophilic attack at the acyl carbon. In this case, the basic phenoxide will form the carboxylate (***24.40***), and a second equivalent of phenoxide will give the final substitution product, the phenyl ether, ***24.41***.

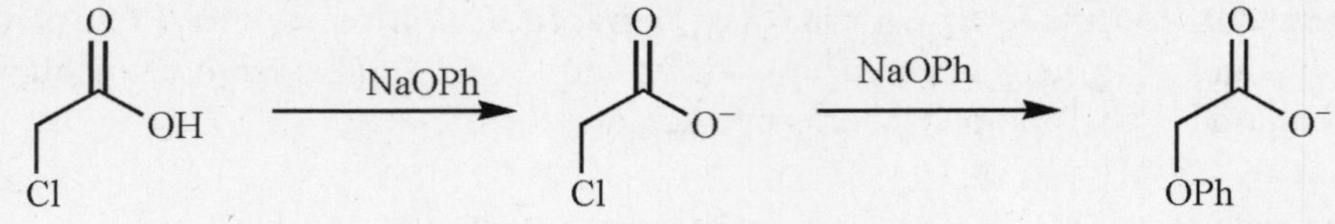

24.40

24.41

What Is the Product of the Reaction of 24.41 with Dilute Aqueous Acid?

Under these conditions, the carboxylate is converted to 2-hydroxypropanoic acid, ***24.42***.

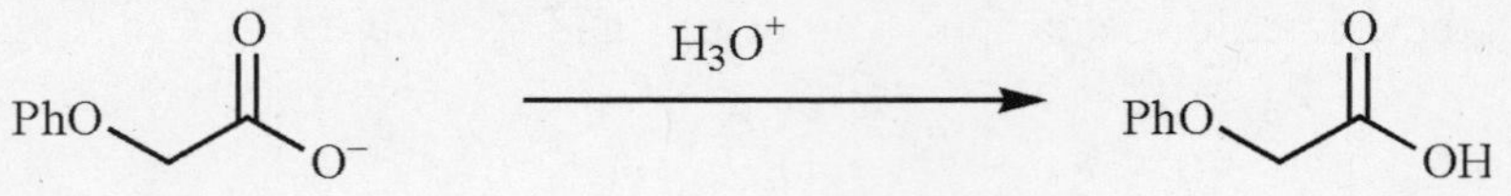

24.42

Reaction with Alcohols

What Is the Product When Butanoic Acid Is Refluxed in Ethanol That Contains a Catalytic Amount of P-Toluenesulfonic Acid?

The product is an ester [$CH_3CH_2CH_2CO_2Et$]. See Section 25.1.

Give the Complete Mechanism of This Reaction.

The reaction begins by protonation of the carbonyl oxygen, which requires an acid with a pK_a much lower than that of a carboxylic acid, to give ***24.43***; this can add a molecule of ethanol to give the oxonium ion, ***24.44***. When this ion loses a proton, usually to the ethanol solvent, the highly reactive *ortho* acid derivative ***24.45*** is formed. This is quickly protonated to give ***24.46*** and loses water to give the resonance stabilized ion ***24.47***. Loss of a proton gives the final product, ethyl butanoate. As shown, all steps in this reaction are reversible. An excess of ethanol or any alcohol and removal of water will drive the reaction to the ester. Conversely, the use of aqueous acid will drive the equilibrium back towards the acid.

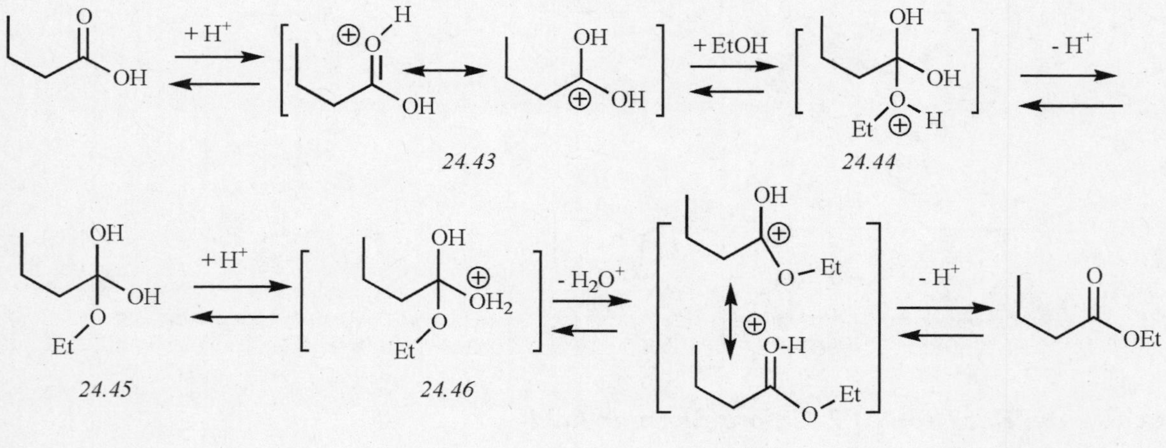

24.43

24.44

24.45

24.46

24.47

What Is the Generic Name of the Product of This Reaction?

This type of product is called an ester.

What Is the IUPAC Nomenclature System for Esters?

The *-oic acid* ending of the parent acid is dropped, and *-oate* is added. The alcohol used to form the ester is also part of the name. If ethanol is used, the name is ethyl; methanol leads to methyl, propanol leads to propyl, etc. The name of $CH_3CH_2CH_2CH_2CH_2CO_2CH_2CH_3$, for example, is ethyl hexanoate [ethyl from the ethanol precursor and hexanoate from the hexanoic acid precursor], as described in Section 15.1.C.

What Is the Major Product of the Reactions of 24.48–24.50?

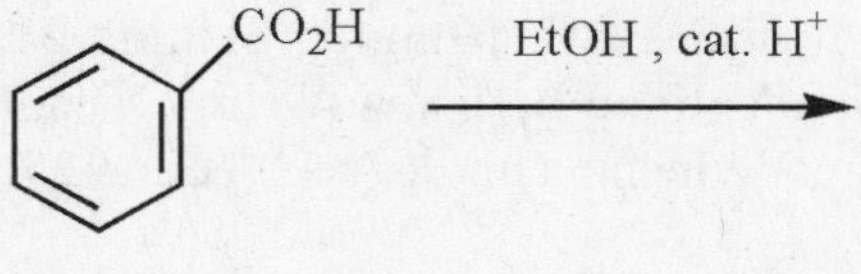

CO_2H EtOH , cat. H^+ →

24.48

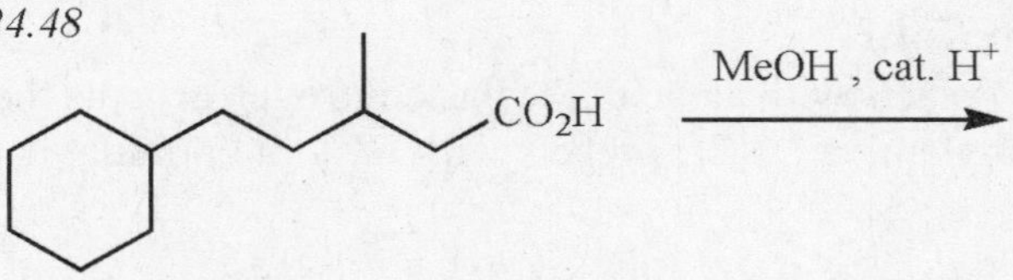

MeOH , cat. H^+ →

CO_2H

24.49

H_3O+ →

CO_2Et

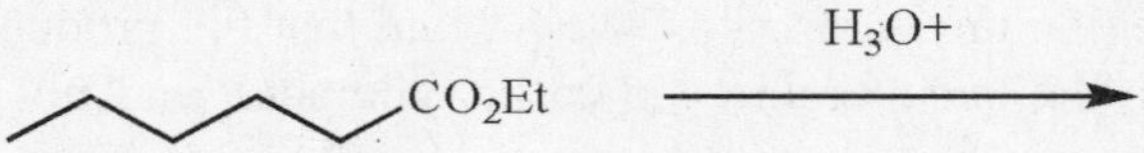

24.50

Benzoic acid (***24.48***) is converted to ethyl benzoate (***24.51***) upon treatment with ethanol and an acid catalyst. The second acid (***24.49***, 5-cyclohexyl-3-methylpentanoic acid) is converted to methyl 5-cyclohexyl-3-methylpentanoate (***24.52***) under these conditions. In the last example, ethyl hexanoate (***24.50***) is hydrolyzed with aqueous acid to ethanol and hexanoic acid.

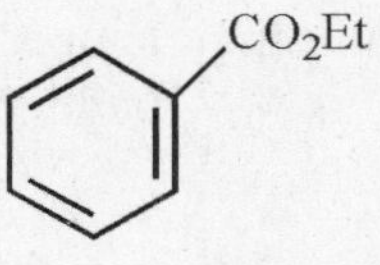

CO_2Et CO_2Me

24.51 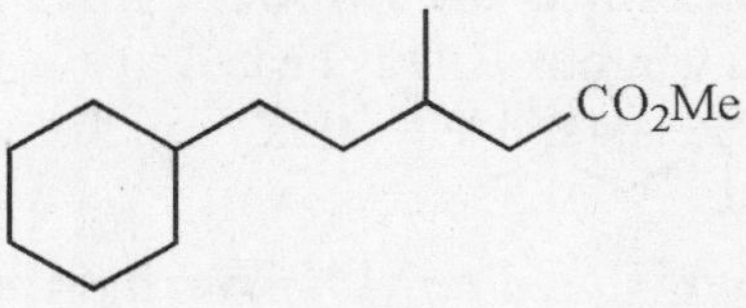 *24.52*

Reaction with Amines

What Is the Structure of an Amine?

An amine is a molecule containing nitrogen with the generic structure R_3N where R = alkyl, aryl or hydrogen. Amines can be classified as primary amines (RNH_2), secondary amines (R_2NH), or tertiary amines (R_3N). See Chapter 27.

What Is the Structure of an Amide?

An amide has a NR_2 group attached to a carbonyl. A primary amide has structure ***24.53***, an example of a secondary amide is ***24.54***, and an example of a tertiary amide is ***24.55***.

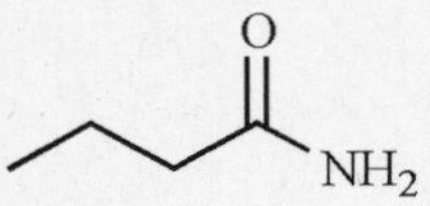

O NH$_2$ — 24.53

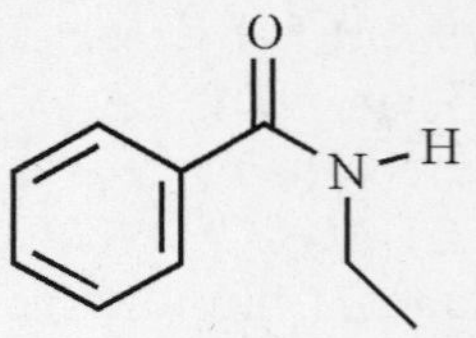

O N–H — 24.54

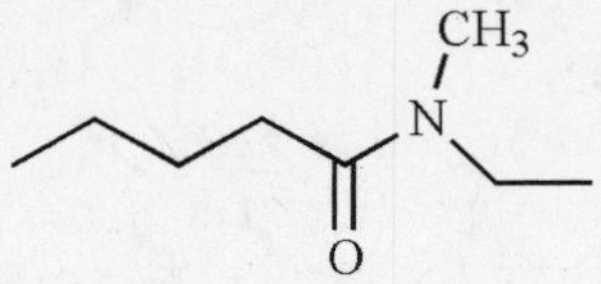

CH$_3$ N O — 24.55

How Are Amides Named by the IUPAC System?

The *-oic acid* is dropped from the pertinent acid, and the word *amide* is added. Pentanoic acid produces pentanamide, and benzoic acid produces benzamide. Amide ***24.53*** is butanamide. If a substituent is attached to the nitrogen, it is identified by using *N*- in the name. The name of ***24.54*** is *N*-ethylbenzamide and ***24.55*** is *N*-ethyl-*N*-methylpentanamide (see Section 25.1).

What Is an Ammonium Salt?

When an amine (NR_3) reacts with an acid, the basic nitrogen accepts the hydrogen to form R_3NH^+, an ammonium salt. If triethylamine [NEt_3] reacts with HCl, the product is triethylammonium chloride [Et_3NH^+ Cl^-].

What Is the Major Product When Butanoic Acid Is Mixed with Diethylamine?

The product is diethylammonium butanoate, ***24.56***. Note that the product is the salt, the result of an acid-base reaction between the basic amine and carboxylic acid, and **not** the amide, which would be ***24.57***.

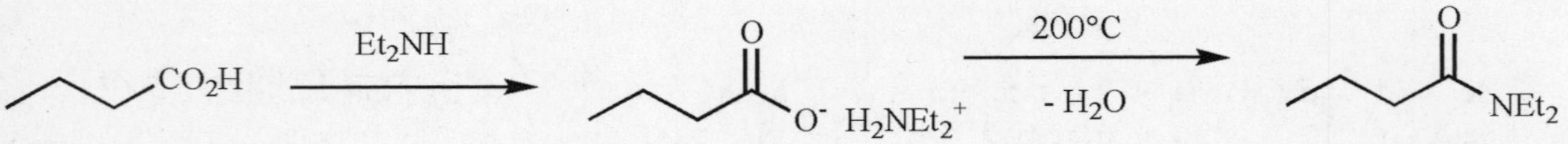

CO_2H —(Et_2NH)→ O, O^- $H_2NEt_2^+$ (24.56) —(200°C, - H_2O)→ O, NEt_2 (24.57)

Can 24.56 Be Converted into the Amide, 24.57?

Yes! If the ammonium carboxylate is heated to a sufficiently high temperature, in this case 200°C, water is expelled to generate an amide. In this case, the product is *N,N*-diethylbutanamide, ***24.57***.

What Is the Structure of Pyridine? Of Triethylamine?

Pyridine is an amine, ***24.58***. Pyridine is often abbreviated as *Py*. Triethylamine has the structure ***24.59*** (or NEt_3). Both of these amines are commonly used as bases.

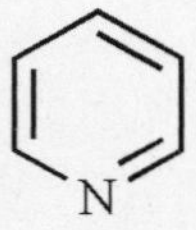

N — 24.58

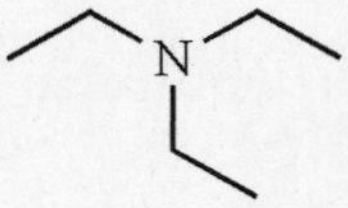

N — 24.59

What Is the Major Product When Pentanoyl Chloride (24.60) Is Mixed with Butylamine in the Presence of Pyridine?

The product is an amide, *N*-butylpentanamide, ***24.61***. The pyridine acts as a base to react with the HCl by-product, producing pyridinium chloride, ***24.62***.

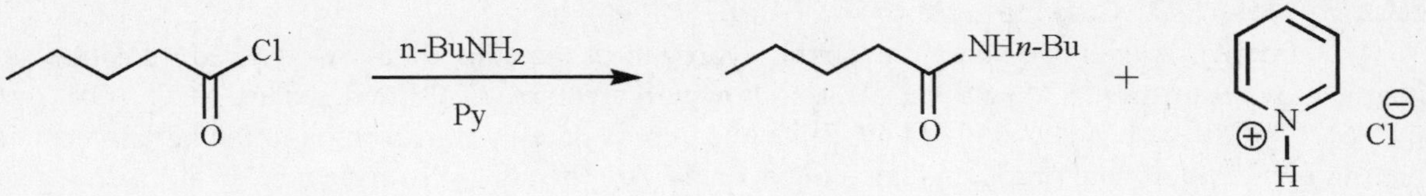

n-BuNH$_2$, Py

24.60 24.61 24.62

What Is the Product When Ethyl Pentanoate (24.63) Is Heated with Ammonia?

An amide is generally more stable than an ester. In other words, the ester is much more reactive than the amide. When ammonia is heated with an ester, ammonia behaves as a nucleophile, attacking the carbonyl and leading to loss of ethanol to form the amide. In this case, ester ***24.63*** reacts with ammonia to form a tetrahedral intermediate, ***24.64***. In this intermediate, a proton is lost from the NH_3 unit, and ethoxide is a better leaving group that NH_2, so loss of ethoxide leads to pentanamide, ***24.65***.

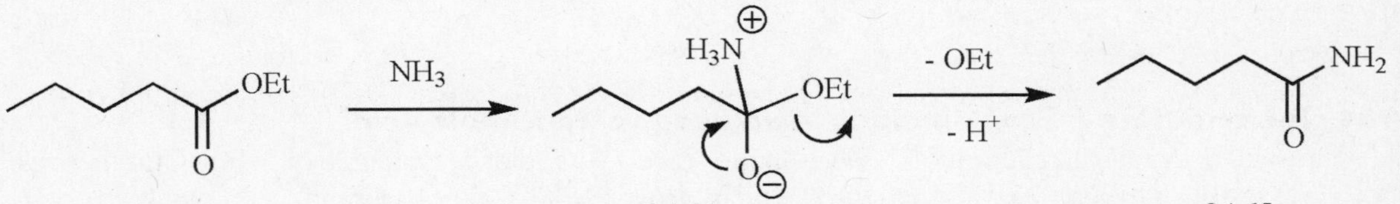

NH_3; - OEt, - H^+

24.63 24.64 24.65

What Is the Product When Methyl Benzoate (24.66) Is Heated with Diethylamine?

The product is *N,N*-diethyl-benzamide ***24.67***.

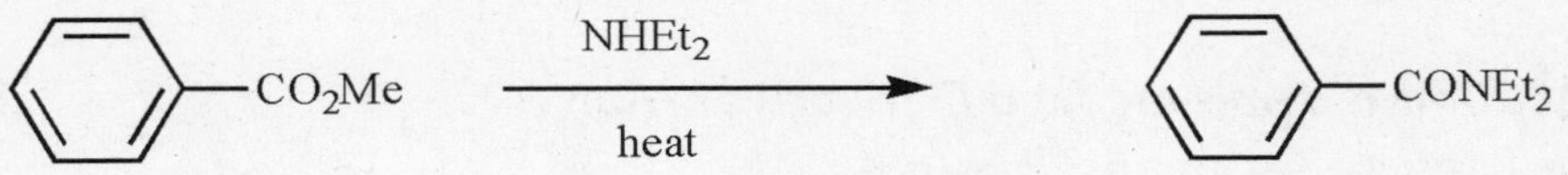

$NHEt_2$, heat

24.66 24.67

24.5. DECARBOXYLATION

When a carboxyl group is conjugated to a π-bond, heating often induces the loss of carbon dioxide in a reaction called decarboxylation.

What Is Decarboxylation?

Decarboxylation is the term used to describe loss of carbon dioxide from a carboxylic acid unit.

What Is the Structure of Malonic Acid?

Malonic acid is 1,3-propanedioic acid, $HO_2C–CH_2–CO_2H$.

What Is the Product Formed When 2-Propylmalonic Acid (24.68) Is Heated to 250°C?

Decarboxylation leads to butanoic acid.

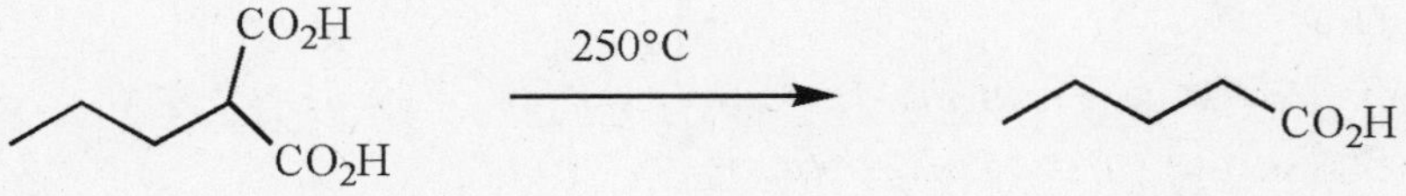

250°C

24.68

Describe How CO_2 Can Be Lost from 24.68.

The carbonyl oxygen can attack the acidic hydrogen of the acid via a six-centered transition state (illustrated by the arrows in ***24.68***). The electron flow will break the C–C bond, generate CO_2 (O=C=O) and the *enol* of the carboxylic acid, ***24.69***. Since the enol is quickly transformed to the ketone via keto-enol tautomerism, the final product is butanoic acid, ***24.70***. The acid *must* have a basic atom on a group that can form a six-membered or a five-membered transition state to facilitate loss of CO_2. The more basic that atom (O > C, for example), the lower the temperature for loss of CO_2. Loss of CO_2 in this manner is called **decarboxylation**.

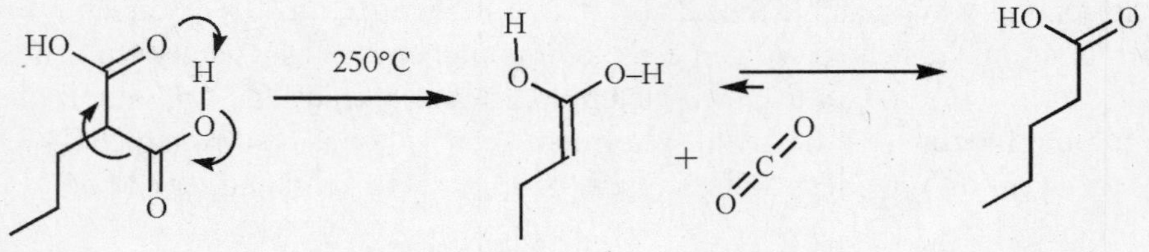

HO O H O O 250°C H O O–H + O C O HO O

24.68 *24.69* *24.70*

Why Does Heating 2-Propylmalonic Acid Lead to Pentanoic Acid?

The decarboxylation reaction of ***24.68*** is facile since 1,3-diacids are capable of losing CO_2 thermally. The C=O of one carboxyl can remove the acidic hydrogen of the other carbonyl.

Is Decarboxylation Restricted to 1,3-Dicarboxylic Acids?

No! β-keto acids, benzoic acid derivatives, and α,β-unsaturated acids can undergo decarboxylation. In general, β-keto acids require higher reaction temperatures than 1,3-dicarboxylic acid, and aromatic and α,β-unsaturated acids require even higher reaction temperatures.

Is Decarboxylation of Malonic Acid Derivatives More or Less Facile Than β-Ketoacids? Explain.

In general, the acidic proton of the COOH unit is transferred to the oxygen of a carbonyl or to the C=C unit of an alkene or aromatic ring. The presence of the OH unit on the carbonyl carbon of an acid unit leads to diminished reactivity relative to the alkyl groups attached to the carbonyl carbon of a ketone. For this reason, removal of the hydrogen of the COOH unit in the decarboxylation is slower in malonic acid, relative to keto acids. It therefore requires somewhat higher reaction temperatures and is slightly less facile.

Describe How CO_2 Can Be Lost from 3-Ketopentanoic Acid.

The carbonyl oxygen can attack the acidic hydrogen of the acid via a six-centered transition state (illustrated by the arrows in ***24.71***). The electron flow will break the C–C bond, generate CO_2 (O=C=O) and the *enol*, ***24.72***. Since the enol is quickly transformed to the ketone via keto-enol tautomerism, the final product is 2-butanone.

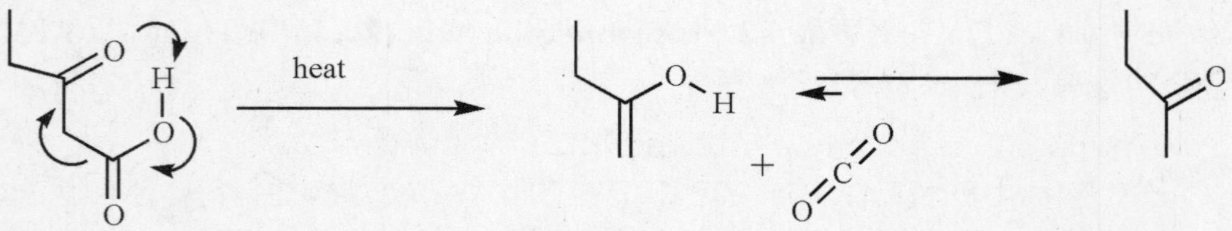

O H O O heat O H + O C O O

24.71 *24.72*

Why Is the Final Product of this Reaction a Ketone?

Since the initial product is an enol, tautomerism favors the carbonyl form, which is thermodynamically more stable, as introduced in Section 13.5.

What Are the Requirements for Thermal Loss of Carbon Dioxide from β-Ketoacids?

The acid must have a carbonyl, an alkene or an aromatic ring in the β-position relative to the carbonyl. That group must have a basic atom that can form a six-membered or a five-membered transition state with the acidic hydrogen.

Give the Major Products from the Reactions of 24.73–24.75.

CO_2H CO_2H *24.73* $\xrightarrow{200°C}$

CO_2H O *24.74* $\xrightarrow{200°C}$

CO_2H *24.75* $\xrightarrow{320°C}$

In the first case, the carboxyl carbonyl at the β-position can remove the hydrogen from the OH group. Such 1,3-diacids decarboxylate thermally to give the mono-acid. Therefore, heating ***24.73*** leads to octanoic acid. The carbonyl that is β-to the acid group in ***24.74*** is a ketone, and decarboxylation by heating leads to 2-methyl-3-pentanone. In the final case, there is no carbonyl β- to the acid, but there is an aromatic ring. Since the C=C group is much less basic than the C=O group of a carbonyl, higher temperatures are required for decarboxylation. Loss of CO_2 does occur, however, and ***24.75*** is converted into 1-phenylpentane.

24.6. DIBASIC CARBOXYLIC ACIDS

When a molecule contains two carboxyl groups, it is referred to as a dibasic acid. The IUPAC nomenclature system uses the ending dioic acid with each carboxyl identified by a number.

Is Malonic Acid a Dibasic Carboxylic Acid?

Yes!

What Is the IUPAC Nomenclature System for Dibasic Acids?

The carbon chain is indicted by the usual alkane, alkene or alkyne name, followed by *dioic acid*, where *di-* shows the presence of two carboxyl groups. The position of the carbonyl groups is given by numbers. An example is $HO_2C–(CH_2)_8–CO_2H$, which is named 1,10-decanedioic acid.

What Is the IUPAC Name for Compounds 24.76–24.77?

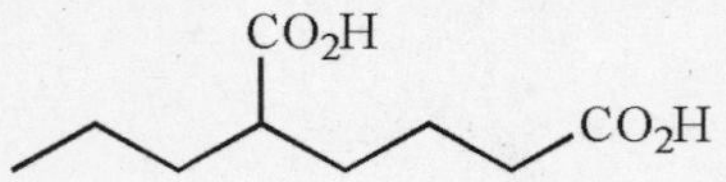

24.76 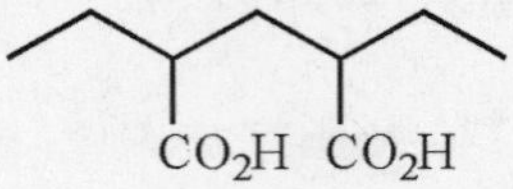 24.77 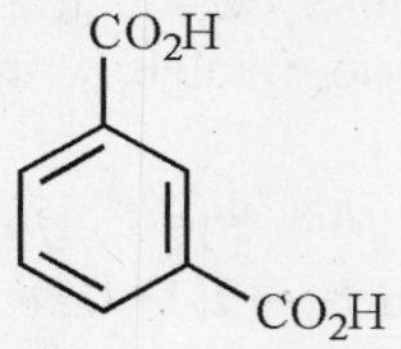 24.78

The name of the first diacid (*24.76*) is 2-ethyl-1,6-hexanedioic acid. The second example (*24.77*) is named 2,4-diethyl-1,4-butanedioic acid. The aromatic diacid (*24.78*) is named 1,3-benzenedicarboxylic acid. Note the name change for aromatic acids.

What Is the Structure of Oxalic Acid?

Oxalic acid is the common name of ethanedioic acid, $HO_2C–CO_2H$.

What Is the IUPAC Name of Oxalic Acid?

Ethanedioic acid.

There Are Two Different pK_a Values for Oxalic Acid. Explain.

There are two carboxyl groups and two acidic hydrogens. One of the hydrogens has a pK_a of 1.27, but the second pK_a is 4.27. Examination of ***24.79*** shows that internal hydrogen bonding via a five-centered interaction is expected to greatly enhance the acidity of one of the carboxyl hydrogens in the diacid. In addition, the carboxyl group is electron withdrawing, strengthening the first pK_a by an inductive effect. The carboxylate (***24.80***, formed by removal of the first hydrogen) is less electron withdrawing, and the internal hydrogen bonding is less extensive, weakening the acid. Removal of both acidic hydrogens generates the dicarboxylate salt, ***24.81***.

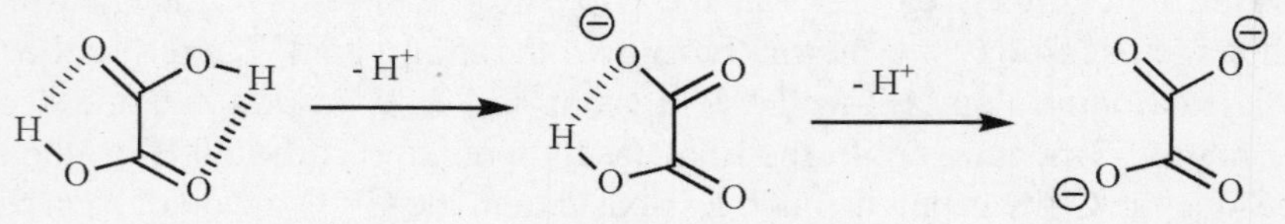

- H^+ - H^+

24.79 24.80 24.81

Compare the First pK_a of Oxalic Acid with That of Acetic Acid.

The pK_a of acetic acid is 4.76 The first pK_a of oxalic acid is 1.27, and the second is 4.27. These values show the strong electron withdrawing effect of a second carboxyl. Even the second pK_a of oxalic acid indicates that the second carboxyl is more acidic than acetic acid, again the result of the electron withdrawing carboxyl group.

What Is the IUPAC Name of Malonic acid?

The IUPAC name is 1,3-propanedioic acid.

Give the First and Second pK_a of Malonic Acid.

The first pK_a is 2.86, and the second pK_a is 5.70.

Compare the First and Second pK_a Value of Malonic Acid with Those of Oxalic Acid.

The first pK_a of malonic acid shows it to be a significantly weaker acid than oxalic. Likewise, the second pK_a is much higher, again reflecting a much weaker acid, weaker even than acetic acid. The reason for these changes is the presence of the -CH_2- group between the carboxyl groups. The increased distance between the acidic hydrogens and the carboxyl groups diminishes the inductive effects, and hydrogen bonding is weaker due to the increased distance between the carbonyl group and the acidic hydrogen.

Give the IUPAC Name and Structure of the C3–C6 Dibasic Acids Where the Carboxyl Groups Are at the Two Extreme Ends of Each Molecule.

The C_3 acid is $HO_2CCH_2CO_2H$ (1,3-propanedioic acid), the C_4 acid is $HO_2C(CH_2)_2CO_2H$ (1,4-butanedioic acid), the C_5 acid is $HO_2C(CH_2)_3CO_2H$ (1,5-pentanedioic acid), and the C_6 acid is $HO_2C(CH_2)_4CO_2H$ (1,6-hexanedioic acid).

What Is the Structure and IUPAC Name of Succinic Acid? Of Glutaric Acid?

Succinic acid is the common name of 1,4-butanedioic acid [$HO_2C(CH_2)_2CO_2H$], and glutaric acid is the common name of 1,5-pentanedioic acid [$HO_2C(CH_2)_3CO_2H$].

Explain Why Glutaric Acid Is a Significantly Weaker Acid Than Malonic Acid.

The first pK_a of glutaric acid is 4.34 and the second pK_a is 5.27. This decrease in acidity reflects the increased distance between the carboxyl groups and the diminished inductive effects.

What Is the Structure of Maleic Acid? Give the IUPAC Name.

Maleic acid is the common name of *cis*-2-buten-1,4-dioic acid (***24.82***).

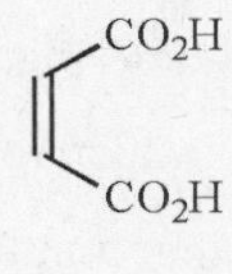

24.82

What Is the Structure of Fumaric Acid? Give the IUPAC Name.

Fumaric acid is the common name of *trans*-2-buten-1,4-dioic acid (***24.83***).

24.83

Explain Why Maleic Acid Is a Stronger Acid Than Fumaric Acid (Compare First pK_a).

In maleic acid (***24.82***), the carboxyl groups are on the same side of the molecule, and intramolecular hydrogen bonding, a through-space inductive effect, will make the acid much stronger (see ***24.84***). When the carboxyl groups are on opposite sides of the molecule as in fumaric acid (***24.83***), this interaction is not possible, and the result is a weaker acid. The first pK_a of maleic acid is 2.0, and the second pK_a is 6.3. The first pK_a of fumaric acid is 3.0 and the second pK_a is 4.4.

24.84

Test Yourself

1. Give the IUPAC name for each of the following.

(a) 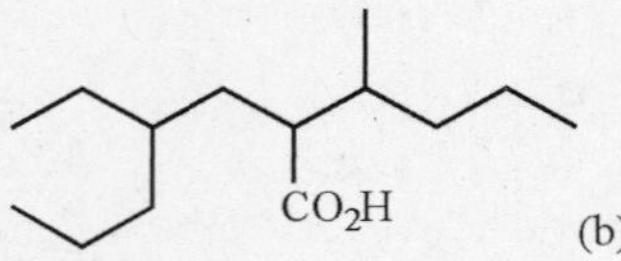 CO_2H (b) 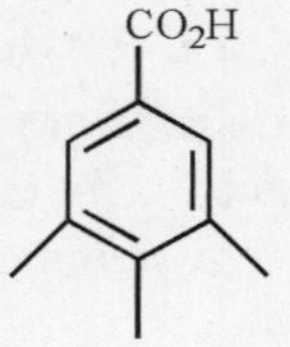 CO_2H (c) 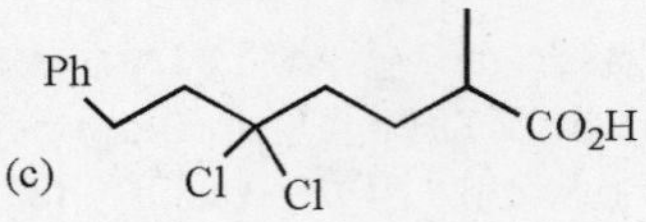 Ph Cl Cl CO_2H (d)  CO_2H

(e) 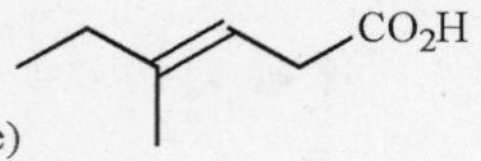 CO_2H (f) 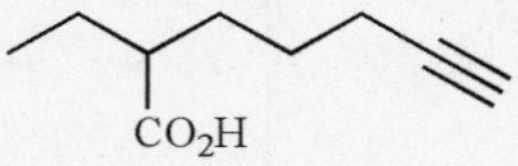 CO_2H (g) 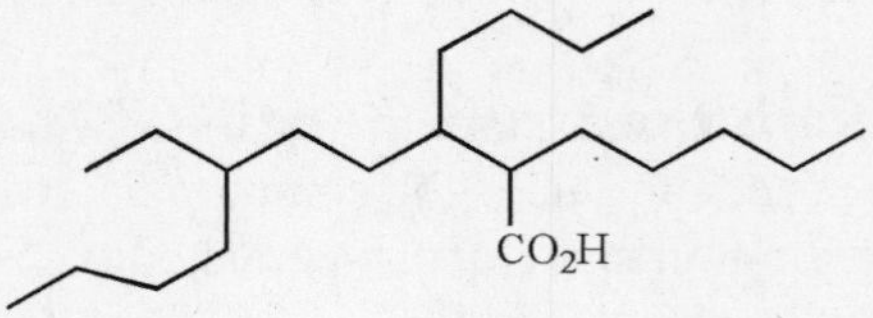 CO_2H

2. Which of the following is the strongest acid? Explain.

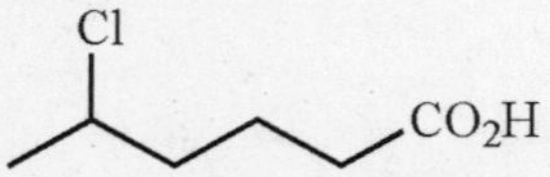

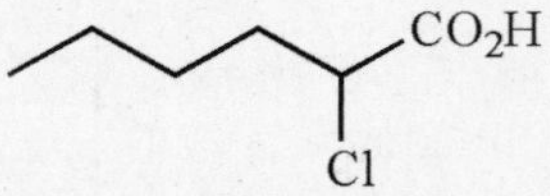

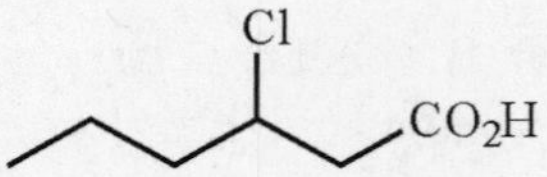

Cl CO_2H CO_2H Cl Cl CO_2H

3. Why is *ortho*-chlorobenzoic acid a stronger acid than *para*-chlorobenzoic acid?
4. Draw a diagram to illustrate through-space inductive effects in 4-chloropropanoic acid.
5. Why is acetic acid a stronger acid in water than in ethanol?
6. How can the 'iodoform test' be used to give information about the structure of ketones?
7. When 4-hydroxybutanoic acid was prepared and heated, the final product was not acidic as expected. Explain this observation.
8. Give the complete mechanism for the following reaction.

CO_2Et $\xrightarrow{H_3O+}$ CO_2H

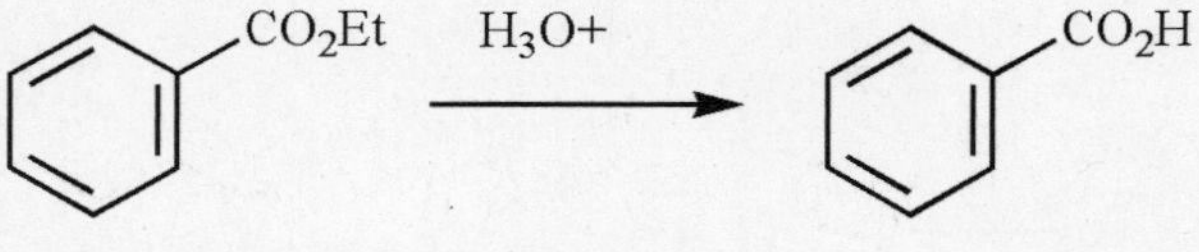

9. Give the complete mechanism for the following reaction.

CO_2Me $\xrightarrow{NH_2Me}$ O NHMe

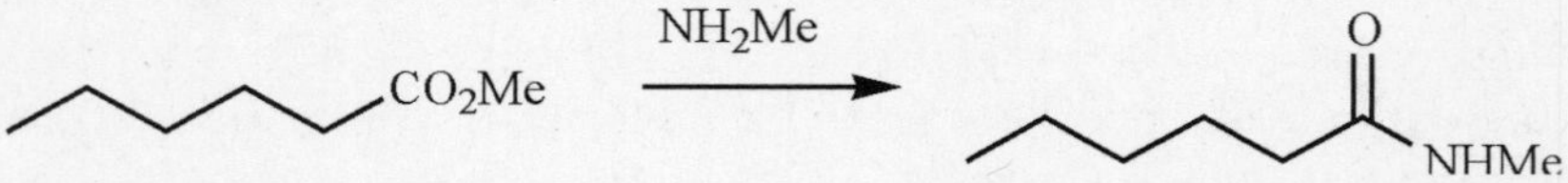

10. Give the correct IUPAC name for each of the following. Also give the common name.

(a) 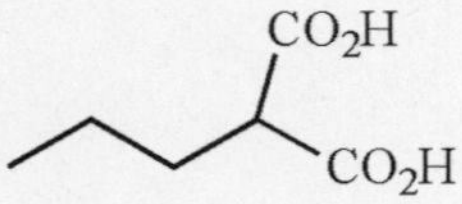 CO_2H CO_2H (b) 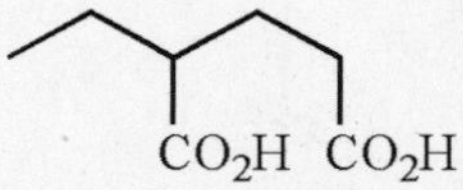 CO_2H CO_2H (c) 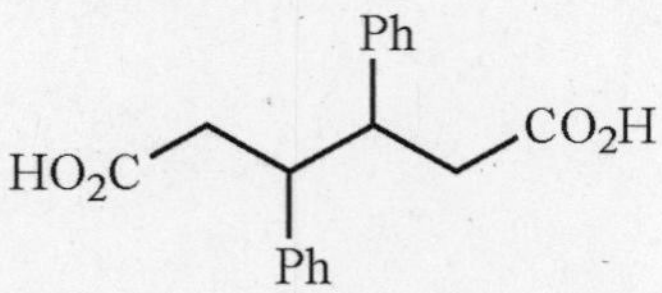 HO_2C Ph Ph CO_2H

11. Why is the first pK_a of malonic acid lower than the first pK_a of glutaric acid?
12. What is the common name of (a) 1,2-benzenedicarboxylic acid (b) 1,4-benzenedicarboxylic acid?

13. In each case, give the major product. If there is no reaction, indicate this by N.R.

(a) 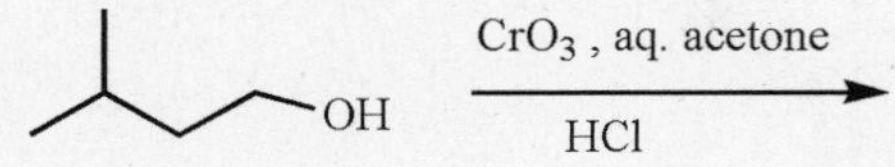 OH — CrO_3, aq. acetone / HCl →

(b) 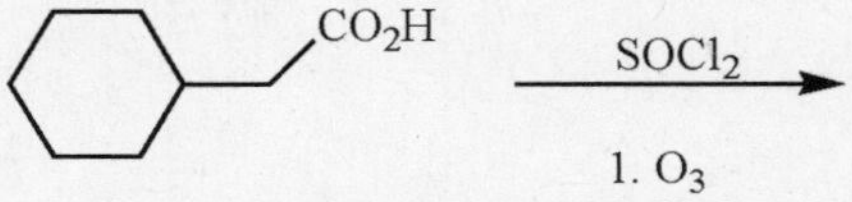 CO_2H — $SOCl_2$ →

(c) 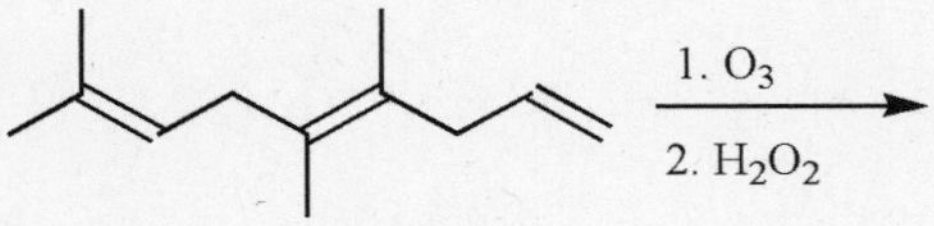 1. O_3 / 2. H_2O_2 →

(d) 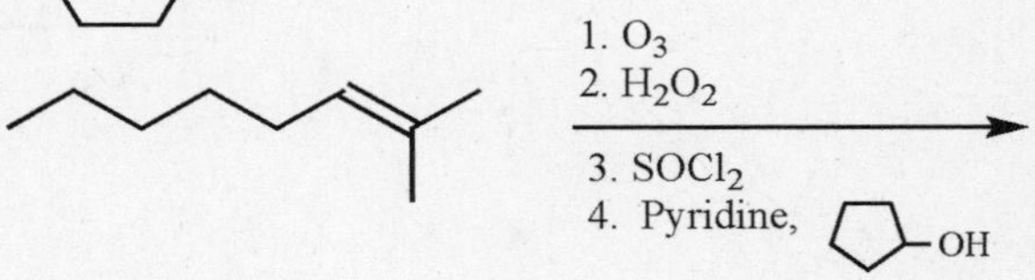 1. O_3 / 2. H_2O_2 / 3. $SOCl_2$ / 4. Pyridine, OH →

(e) 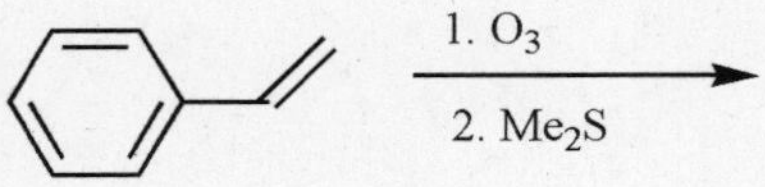 1. O_3 / 2. Me_2S →

(f) 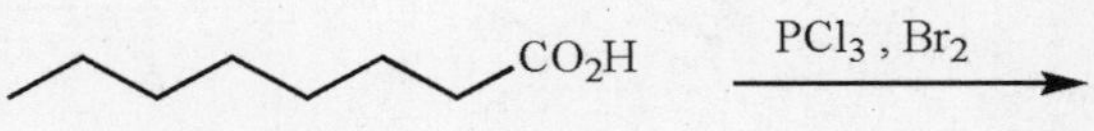 CO_2H — PCl_3, Br_2 →

(g) 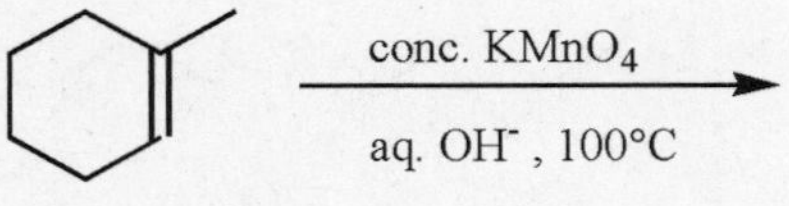 conc. $KMnO_4$ / aq. OH^-, 100°C →

(h) 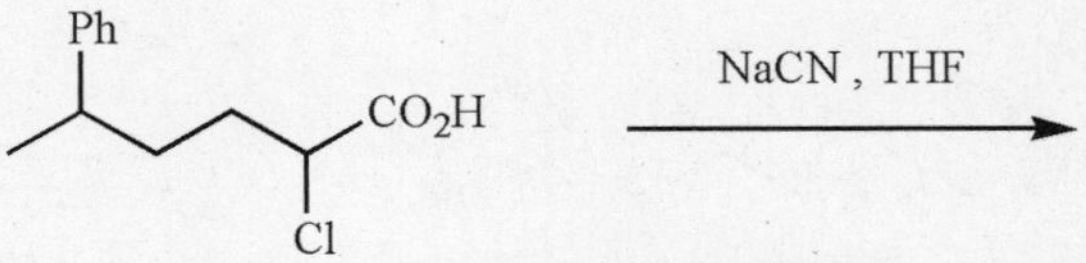 Ph, CO_2H, Cl — NaCN, THF →

(i) 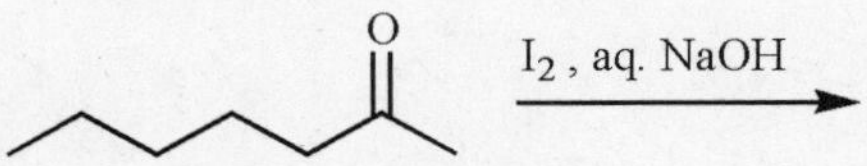 O — I_2, aq. NaOH →

(j) 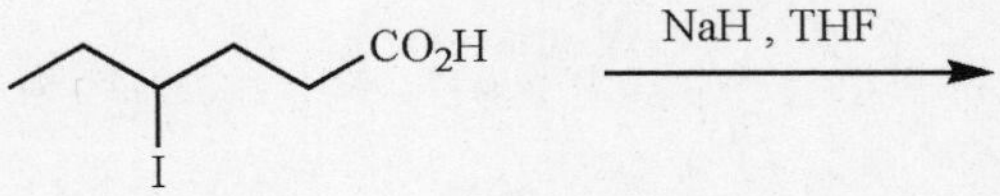 CO_2H, I — NaH, THF →

(k) 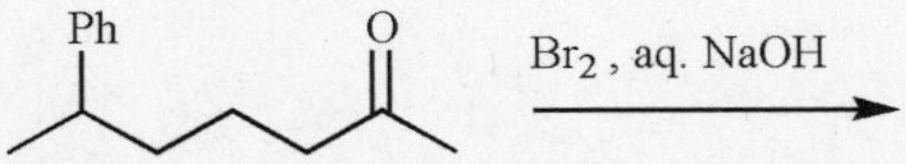 Ph, O — Br_2, aq. NaOH →

(l)  Cl — aq. EtOH, KCN →

(m) 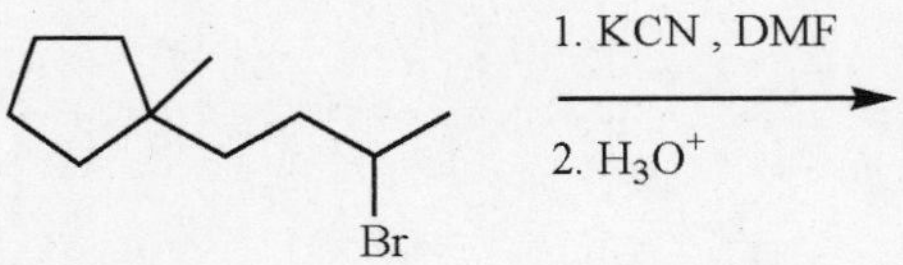 Br — 1. KCN, DMF / 2. H_3O^+ →

(n) 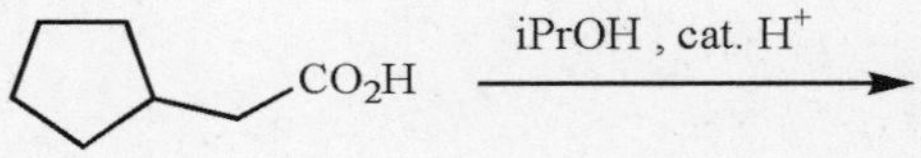 CO_2H — iPrOH, cat. H^+ →

(o) CHO — 1. HCN / 2. H_3O^+ →

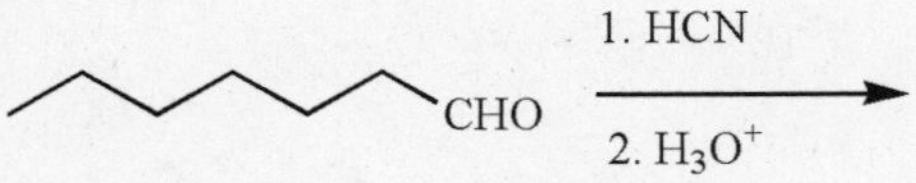

(p)  Ph CO_2H — $MeNH_2$, 200°C →

(q) 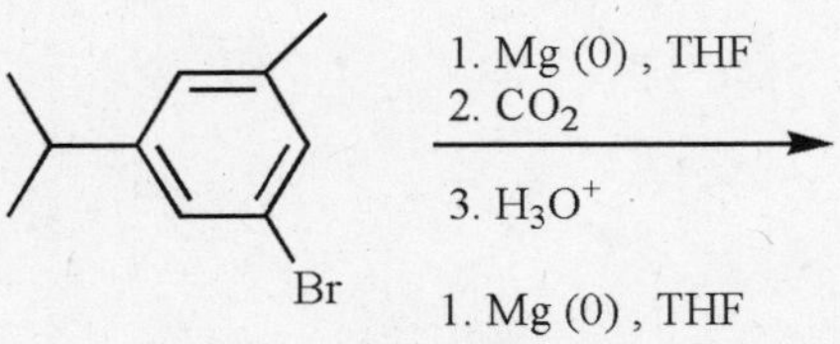 Br — 1. Mg (0), THF / 2. CO_2 / 3. H_3O^+ →

(r) 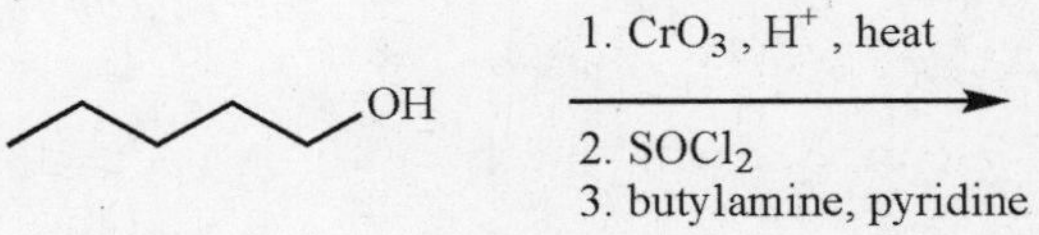 OH — 1. CrO_3, H^+, heat / 2. $SOCl_2$ / 3. butylamine, pyridine →

(s) 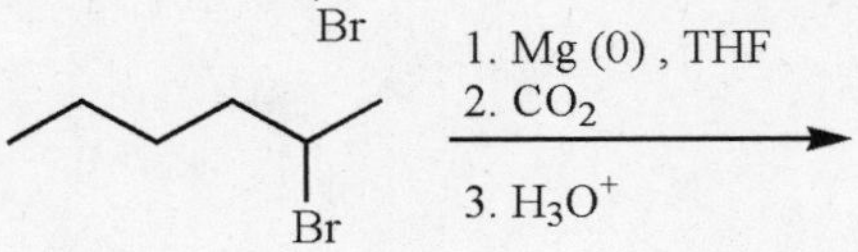 Br — 1. Mg (0), THF / 2. CO_2 / 3. H_3O^+ →

(t) 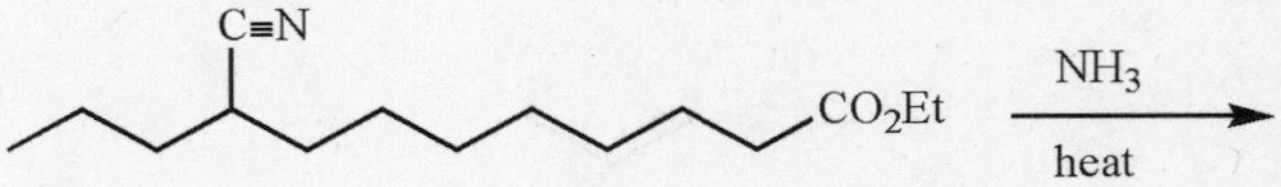 C≡N, CO_2Et — NH_3 / heat →

(u) 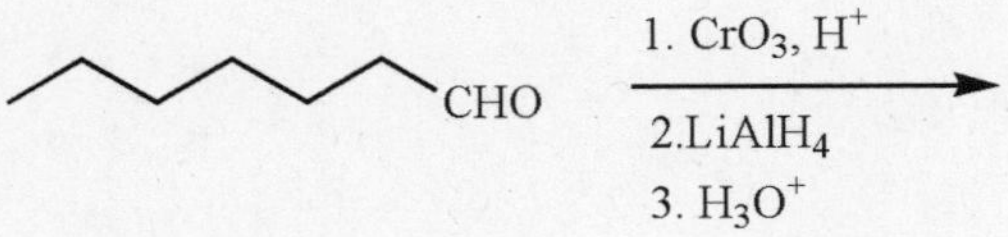 CHO — 1. CrO_3, H^+ / 2. $LiAlH_4$ / 3. H_3O^+ →

(v) 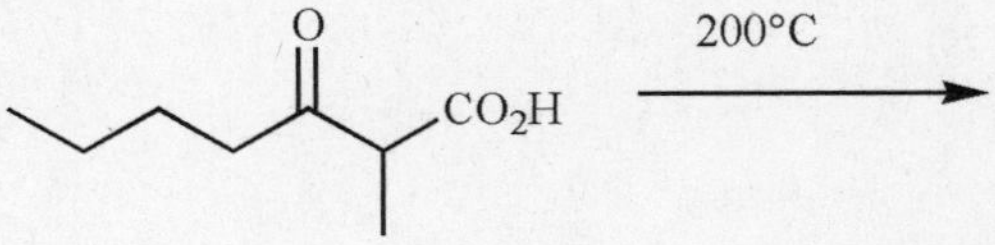 O, CO_2H — 200°C →

(w) 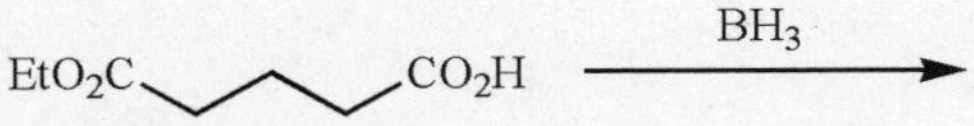 EtO_2C CO_2H — BH_3 →

(x) 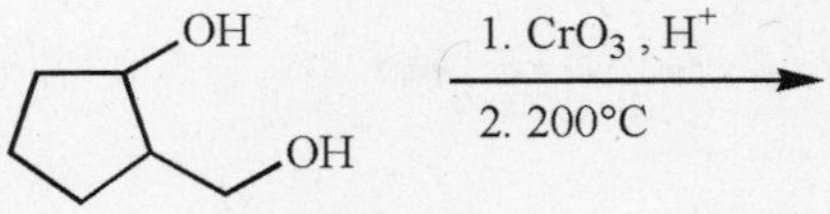 OH, OH — 1. CrO_3, H^+ / 2. 200°C →

(y) 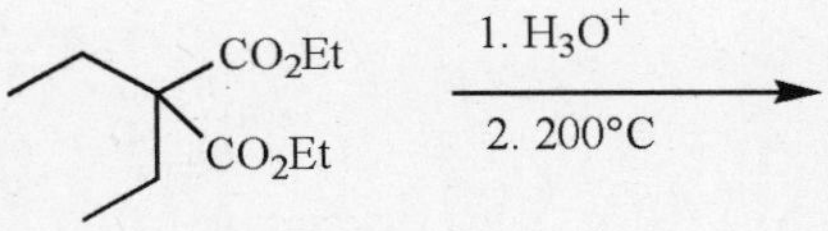 CO_2Et, CO_2Et — 1. H_3O^+ / 2. 200°C →

14. Provide a suitable synthesis for each of the following. Provide all reagents and show all intermediate products.

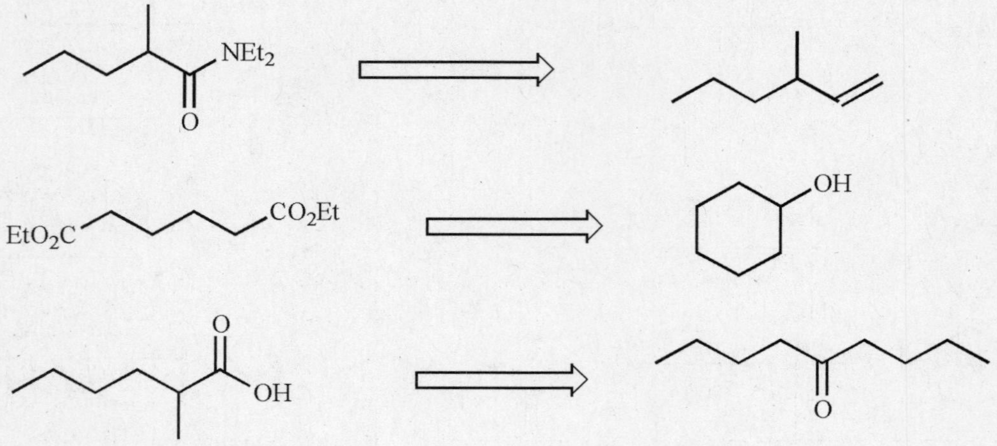

NEt$_2$ O ⟹

EtO$_2$C CO$_2$Et ⟹ OH

O OH ⟹ O

CHAPTER 25

Carboxylic Acid Derivatives and Acyl Substitution Reactions

What Are Carboxylic Acid Derivatives?

Carboxylic acids can be easily converted into several important derivatives: acid chlorides, anhydrides, esters, amides and nitriles. Each of these derivatives can be converted back to its parent carboxylic acid or transformed in a variety of other ways. Of particular interest are the chemical transformations of acid derivatives, not only into other acid derivatives and back into carboxylic acids, but also into functional groups such as alcohols, alkenes, ketones and aldehydes and amines.

25.1. STRUCTURE AND NOMENCLATURE OF ACID DERIVATIVES

There are several types of acid derivatives. These include acid halides, acid anhydrides, esters, and amides. Nitriles are considered to be acid derivatives as well. There are two important cyclic derivatives. Cyclic esters are called lactones, and cyclic amides are called lactams. Derivatives that have a O=C–N–C=O unit are known as imides.

Acid Chlorides

Acid chlorides were introduced in Section 24.4.

What Is the Basic Structure of an Acid Chloride?

An acid chloride has a chlorine attached directly to a carbonyl [Cl–C=O].

What Is the IUPAC Nomenclature System for an Acid Chloride?

The *-oic acid* ending of the parent carboxylic acid is dropped and replaced with *-oyl chloride*. The acid chloride of hexanoic acid is hexanoyl chloride. Using the common name acetic acid, the acid chloride is acetyl chloride.

Give the Structure of Each of the Following Compounds.
(A) 5-Phenyl-3,3-Dimethyloctanoyl Chloride; (B) Malonyl Chloride;
(C) Oxalyl Chloride; (D) 4-Methoxy-Benzoyl Chloride; (E) Phenylacetyl Chloride.

The structure of (a) is ***25.1***, (b) is ***25.2***, (c) is ***25.3***, (d) is ***25.4***, and (e) is ***25.5***.

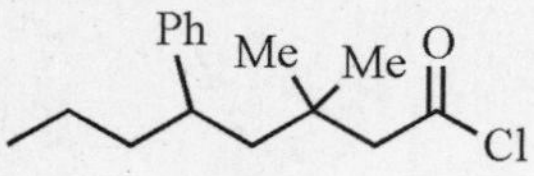

25.1 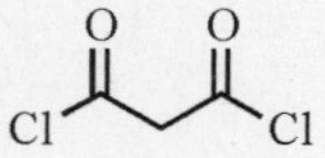 *25.2* 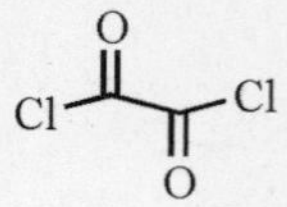 *25.3* 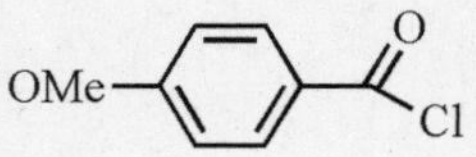 *25.4* 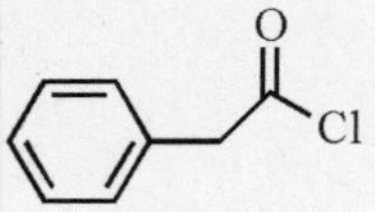 *25.5*

Acid Anhydrides

What Is the Basic Structure of an Acid Anhydride?

An acid anhydride is characterized by an O=C–O–C=O structure. A generic example is ***25.6***, where R is alkyl or aryl.

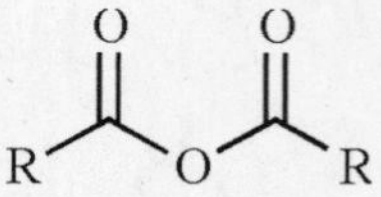

25.6

What Is the IUPAC Nomenclature System for a Symmetrical Acid Anhydride?

Acid anhydrides are essentially composed of two acid units that are joined together. There are, therefore, two acyl groups (RCO). If both of the acyl groups are the same, the *acid* ending of the acid is dropped and replaced with the word *anhydride*. Anhydride ***25.7*** is ethanoic anhydride (the common name is acetic anhydride), and ***25.8*** is butanoic anhydride.

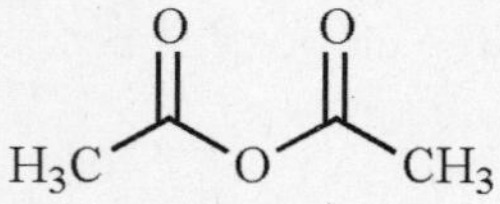

25.7 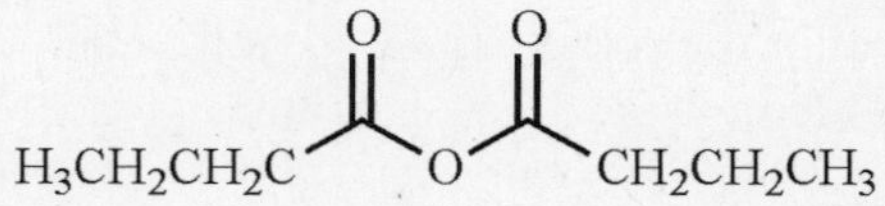 *25.8*

What Is a 'Mixed' Anhydride?

A mixed anhydride has two different acyl groups as part of the anhydride, as in ***25.9***.

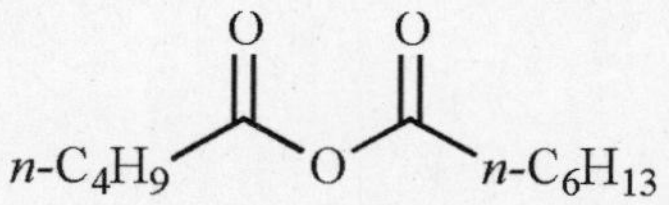

25.9

How Are Mixed Anhydrides Named?

The two acyl groups are arranged in alphabetical order followed by the word *anhydride*. Anhydride ***25.9*** is named butanoic hexanoic anhydride.

What Are the Common Cyclic Anhydrides Derived from Succinic Acid, Maleic Acid and Glutaric Acid? Give Their Names.

Dehydration of succinic acid gives succinic anhydride (***25.10***), maleic acid gives maleic anhydride (***25.11***) and glutaric acid gives glutaric anhydride (***25.12***). In general, dicarboxylic acids of four carbons up to six carbons give cyclic anhydrides. Diacids of less than four carbons would lead to highly strained three- or four-membered rings, which are difficult to form. Larger chain diacids must form eight and larger membered rings, and this is very difficult for C8–C13 compounds.

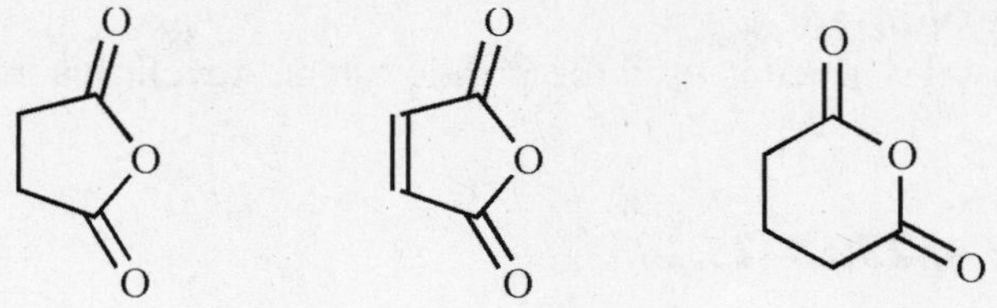

25.10 *25.11* *25.12*

What Is Phthalic Acid? What Is Phthalic Anhydride?

Phthalic acid is benzene 1,2-dicarboxylic acid (see ***25.13***), and the corresponding anhydride is ***25.14***.

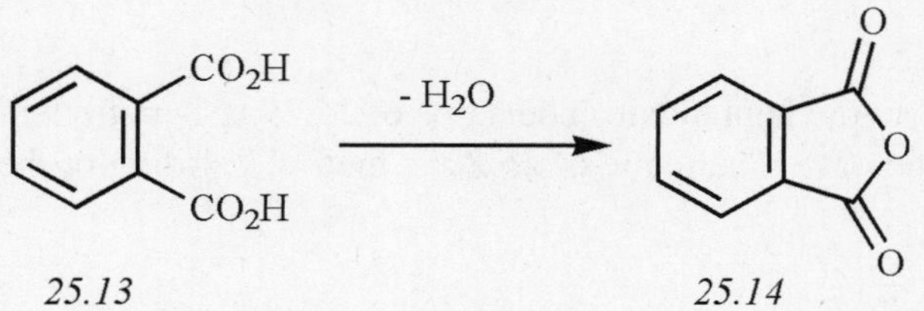

CO_2H CO_2H $- H_2O$

25.13 *25.14*

Esters

Esters were briefly introduced in Section 24.4.

What Is the Basic Structure of an Ester?

An ester has an OR group attached to a carbonyl [RO–C=O].

What Is the IUPAC Nomenclature System for an Ester?

The *-oic acid* ending of the parent acid is dropped and replaced with *-oate*. The 'alcohol' portion of the ester is put in front of the 'acid portion'. $CH_3CO_2CH_3$ is an example where the 'acid part' is ethanoic acid (acetic acid), and the 'alcohol part' is methanol. The IUPAC name is methyl ethanoate. Methyl acetate is the name based on the common name for the acid.

What Is the Structure of 2-Phenylbutyl 4-Chloro-3,3-Dimethylhexanoate?

The structure of 2-phenylbutyl 4-chloro-3,3-dimethylhexanoate is ***25.15***.

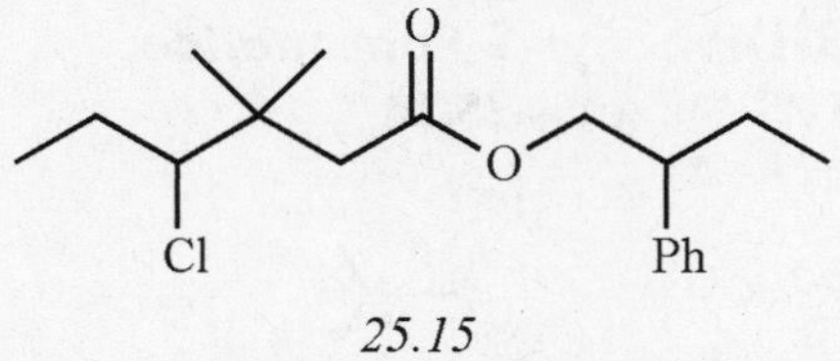

25.15

What Is the Structure of Ethyl Acetate? Of Methyl Benzoate?

Ethyl acetate is ***25.16***, and it is often abbreviated EtOAc. Methyl benzoate is ***25.17***, and benzoate is often abbreviated Bz, so ***25.17*** would be MeOBz.

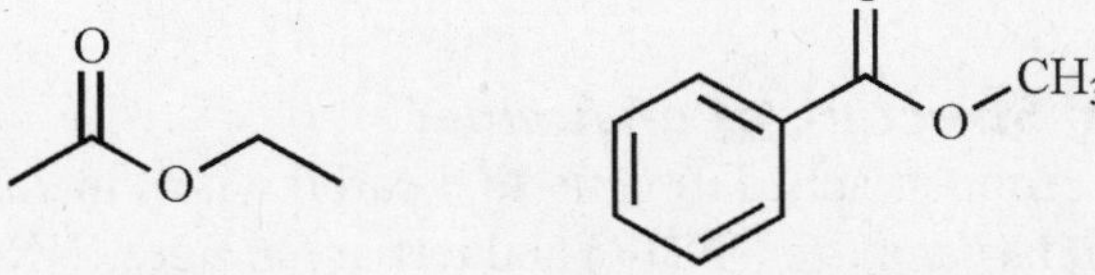

25.16 *25.17*

How Are Acyclic Esters Named?

The IUPAC system just described is used for acyclic esters. Acyclic esters do not contain a ring that includes the CO_2 group.

Give the IUPAC Name of 25.18–25.20.

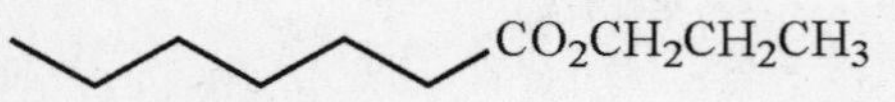

$CO_2CH_2CH_2CH_3$

25.18

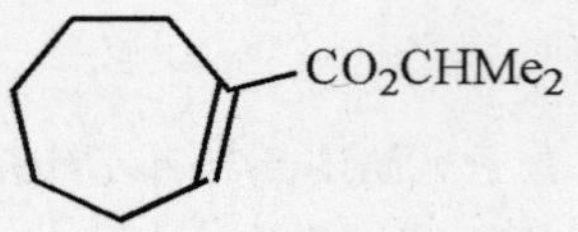

CO_2CHMe_2

25.19

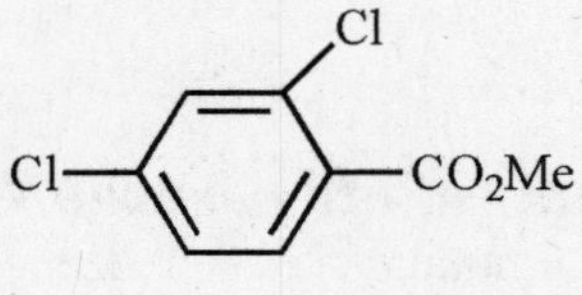

Cl Cl CO_2Me

25.20

The name of ***25.18*** is propyl heptanoate. The name of ***25.19*** is 1-methylethyl 1-cycloheptenoate, also called isopropyl cycloheptenoate. The name of ***25.20*** is methyl 2,4-dichlorobenzoate.

Lactones

What Is the Definition of a Lactone?

A lactone is a cyclic ester, as in structure ***25.21***.

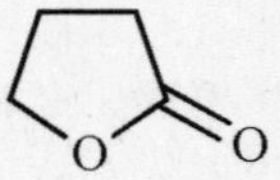

O O

25.21

How Are Lactones Named?

Lactones are named by dropping the *-oic acid* ending and replacing it with *-olide*. It is noted that ***25.21*** is named 4-butanolide by the IUPAC system, but is called γ-butyrolactone by the common nomenclature.

Give the Structure of Each of the Following. (A) 6-Hexanolide; (B) 2,3-Diphenyl-5-Pentanolide; (C) 4-But-2-Enolide

Lactone (a) is ***25.22***, (b) is ***25.23*** and (c) is **25.24**.

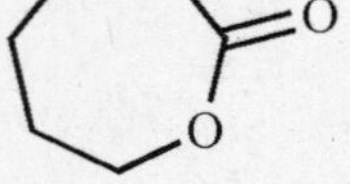

O O

25.22

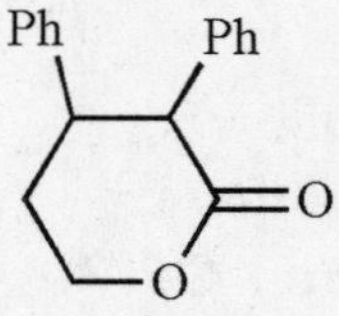

Ph Ph O O

25.23

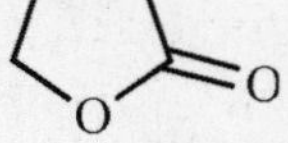

O O

25.24

Amides

What Is the Fundamental Structure of an Amide?

An amide has an amino group attached directly to a carbonyl, as in ***25.25***[$O{=}C{-}NR_2$]. There are primary amides ($-NH_2$), secondary amides (-NHR) and tertiary amides (NR_2).

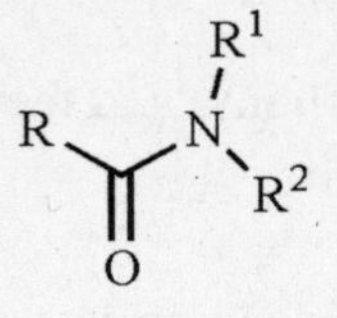

25.25

What Is the IUPAC Nomenclature System Used to Name an Amide?

The *-oic acid* of the parent carboxylic acid is dropped and replaced with *amide*. The amide of propanoic acid is, therefore, propanamide [$CH_3CH_2CONH_2$]. Amide ***25.26*** is named pentanamide.

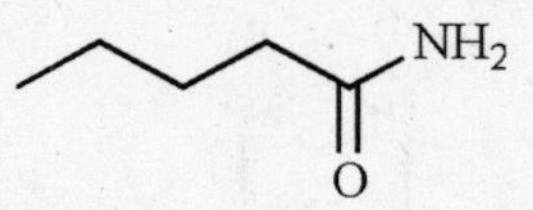

25.26

What Is the Nomenclature Protocol When Substituents Appear on the Nitrogen of an Amide?

When substituents appear on the nitrogen (as in ***25.27***), the position of the substituent is given by using *N*- in front of the substituent. Amide ***25.27*** is *N*-ethyl-*N*-methylpentanamide, and ***25.28*** is *N,N*-dimethyl-4-methylbenzamide. Note that the N is used to position substituents on the nitrogen just as the traditional number (4) is used to position the methyl group on the benzene ring.

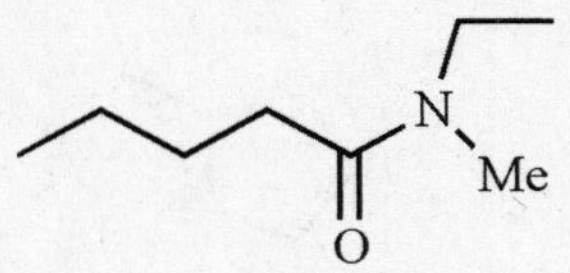

25.27

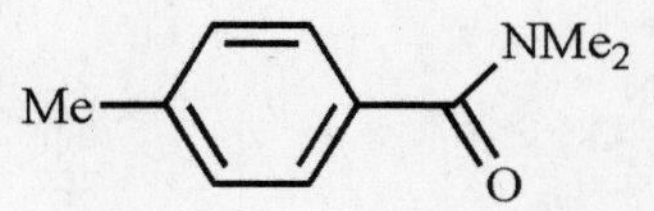

25.28

How Are Acyclic Amides Named?

The nomenclature system described above is for acyclic amides (the –CON unit is not part of a ring). If pentanoic acid is the parent, replacing the –OH with NR_2 leads to dropping the *-oic acid* and replacing it with *amide*. If the common name of the acid is used (such as acetic acid), the *-ic acid* is dropped and replaced with *amide*. The NH_2 amide of acetic acid is, therefore, ethanamide or acetamide [CH_3CONH_2].

Give the Structure for Each of the Following. (A) (N-Phenyl)-3,3,N-Diiphenylheptanamide; (B) N-Ethyl-3-Chlorobenzamide; (C) N,N-Diethylacetamide

The structure of (a) is ***25.29***, (b) is ***25.30*** and (c) is ***25.31***.

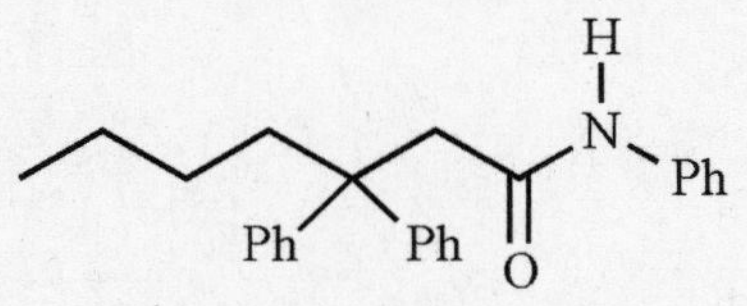

25.29

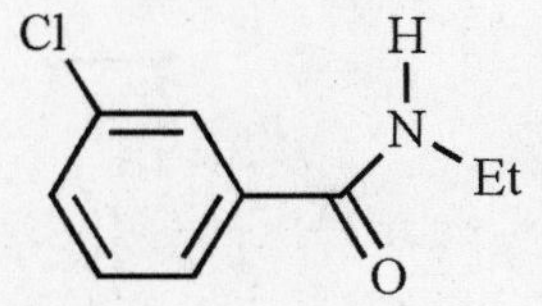

25.30

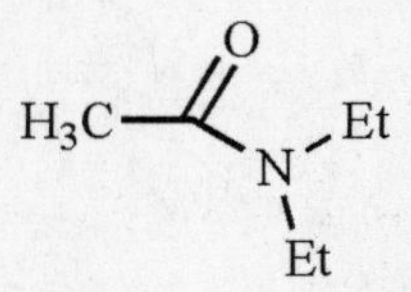

25.31

Lactams

A lactam is a cyclic amide where the carbonyl and the nitrogen are part of a ring.

How Are Lactams Named?

The names are based on a few specialized cyclic structures derived from the reduced form of a heterocyclic base. The base names focus on 2-pyrrolidinone (***25.32***), 2-piperidone (***25.33***, also called tetrahydropyridin-2-one) and hexahydroazepin-2-one (***25.34***). There is also a common nomenclature system, based on the common name of the acid, which includes the word *lactam*. Lactam ***25.32*** is γ-butyrolactam (2-pyrrolidinone), ***25.33*** is δ-valerolactam (2-piperidone), and ***25.34*** is caprolactam (hexahydroazepin-2-one).

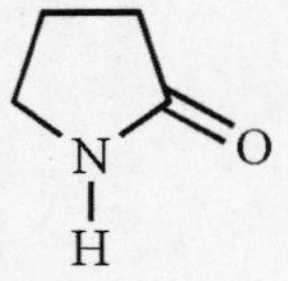

N–H, O

25.32 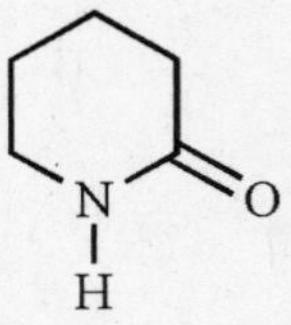 *25.33* 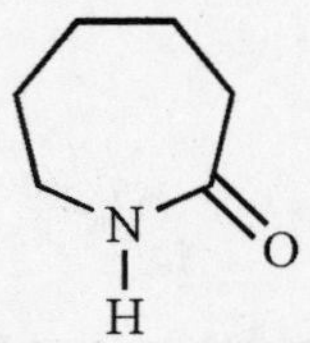 *25.34*

Give the Names of Lactams 25.35–25.37.

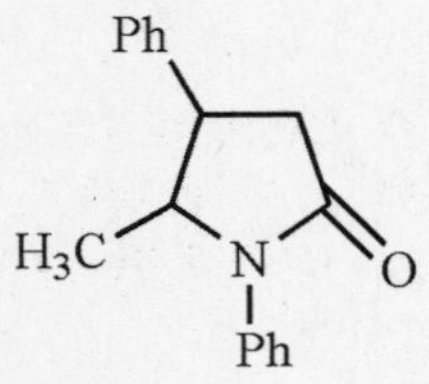

Ph, H_3C, N, O, Ph; CH_3, CH_3, N, O; CH_3, N–H, O

25.35 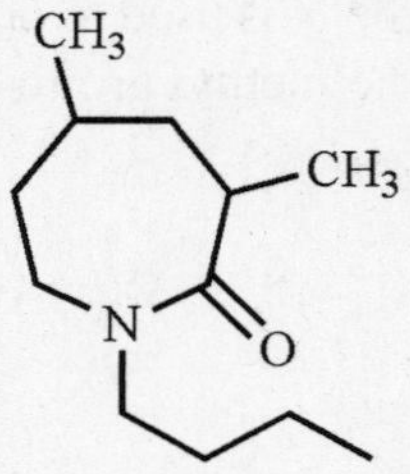 *25.36* 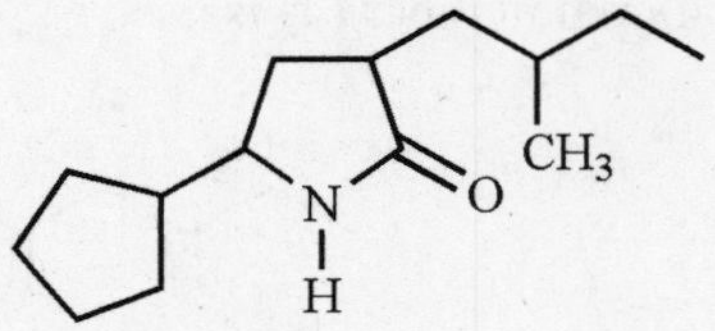 *25.37*

Lactam ***25.35*** is 5-methyl-1,4-diphenyl-2-pyrrolidinone. Lactam ***25.36*** is *N*-butyl-3,5-dimethylhexahydroazepin-2-one. Lactam ***25.37*** is 3-(2-methyl-butyl)-5-cyclopentyl-2-pyrrolidinone.

Imides

An imide is the nitrogen equivalent of an anhydride (see below) and is characterized by the O=C–*N*–C=O group.

How Are Imides Named?

The fundamental name of an imide uses the word *imide*. EtCONHCOEt is known as diethylimide. Cyclic imides such as ***25.38*** are also known, derived from αω-dicarboxylic acids.

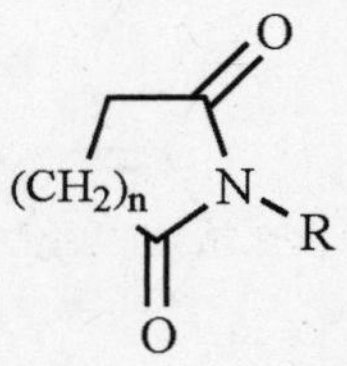

O, $(CH_2)_n$, N, R, O

25.38

Give the Structures and Names of the Cyclic Imides Derived from the Diacids Succinic Acid, Glutaric Acid and Phthalic Acid.

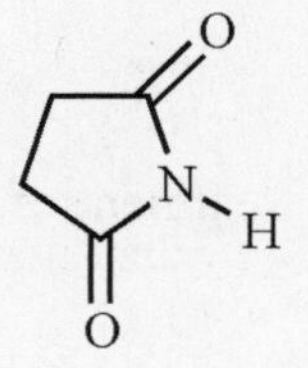

25.39 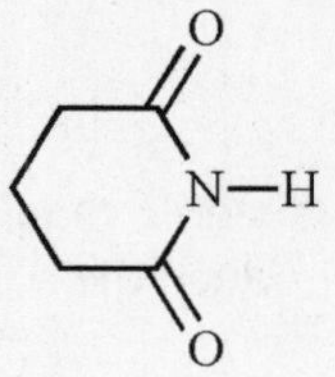 25.40 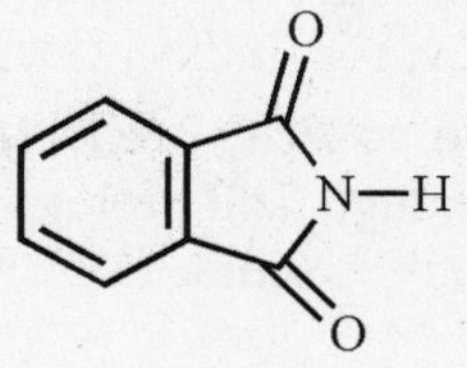 25.41

The imide derived from succinic acid is succinimide (***25.39***), glutaric acid gives glutarimide (***25.40***) and phthalic acid leads to phthalimide (***25.41***).

Nitriles

Nitriles were introduced in Sections 7.5 and 9.2.

What Is the Basic Structure of a Nitrile?

A nitrile is characterized by the presence of a cyanide group -C≡N. An example of a nitrile is $CH_3CH_2CH_2CH_2C{\equiv}N$.

How Are Nitriles Named by the IUPAC System?

Nitriles are considered to be carboxylic acid derivatives. They are named by dropping the *-oic acid* ending of the parent acid and replacing it with the word *nitrile*. The C5 nitrile shown above is derived from pentanoic acid and is named pentanonitrile. It can also be named by using the alkane, alkene or alkyne base and adding the word *nitrile*. In this system, the C5 nitrile is pentanenitrile.

Give the Structure of Each of the Following Nitriles.

(A) Nonanonitrile; (B) Benzonitrile; (C) 3,5,Diphenylhexanonitrile; (D) 3,3-Dimethylpentanenitrile

The structure of (a) is ***25.42***; (b) is ***25.43***; (c) is ***25.44*** and (d) is ***25.45***.

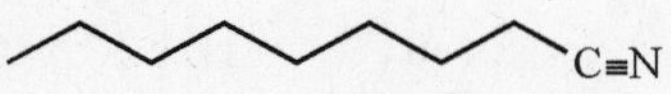

25.42 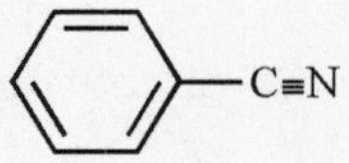 25.43 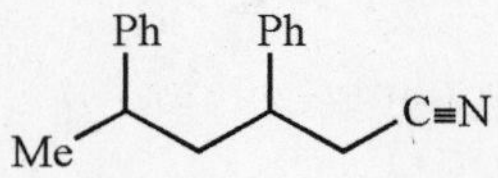 25.44 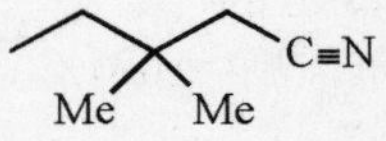 25.45

25.2. PREPARATION OF ACID DERIVATIVES

The various acid derivatives presented above can be prepared from the corresponding carboxylic acid or from a more reactive acid derivative.

Acid Chlorides

Give At Least Three Different Reagents That Transform a Carboxylic Acid into an Acid Chloride..

The same reagents used for conversion of alcohols to chlorides are effective with acids: thionyl chloride, ($SOCl_2$), PCl_3, PCl_5 and $POCl_3$ are all effective.

What Is the Product When 3-Methylhexanoic Acid Reacts with Phosphorus Trichloride?

The product is 3-methyl hexanoyl chloride.

Is It Possible to Prepare Acid Bromides?

Yes! Although acid bromides are usually less stable than acid chlorides, treatment of a carboxylic acid with thionyl bromide ($SOBr_2$) or PBr_3 will produce the acid bromide.

Why Are Acid Fluorides and Acid Iodides Almost Never Observed?

Both are very unstable and difficult to isolate. It is also true that thionyl fluoride and iodide are not usable reagents, and phosphorus iodides are relatively unstable. The latter reagents are usually formed *in situ* by reaction of phosphorous with iodide. Acid fluorides are generally too unstable to isolate, and their formation can be quite difficult.

What Are the UV Properties of Acid Chlorides?

There are no diagnostic absorption peaks in the UV spectrum that are used to identify this class of compounds.

Acid Anhydrides

What Is the Basic Structure of an Acid Anhydride?

The basic functional group is O=C–O–C=O, where alkyl or aryl groups are attached to each carbonyl, RCO_2OR' (see *25.46*). If R and R' are the same, this is called a symmetrical anhydride. If R and R' are different, it is called an unsymmetrical anhydride.

Give a Generic Reaction for the Preparation of an Anhydride.

In the simplest reaction type, two carboxylic acids combine, with loss of a molecule of water, to form the anhydride. The anhydride therefore consists of two acid fragments.

$$RC(=O)OH + R^1C(=O)OH \xrightarrow{-H_2O} RC(=O)OC(=O)R^1$$

25.46

What Is the Product When Butanoic Acid Is Heated to 200°C? The Product When Succinic Acid Is Similarly Heated?

The first product is butanoic anhydride, and the second product is succinic anhydride.

How Are Symmetrical Anhydrides Prepared?

A carboxylic acid can be heated with a dehydrating agent such as phosphorus pentoxide (P_2O_5) to produce the anhydride. The carboxylate salt of an acid can be reacted with the acid chloride of that acid.

What Is the Product When Butanoyl Chloride Is Heated with Butanoic Acid?

The product butanoyl anhydride.

Why Is an Acid Chloride More Useful for Preparation of an Anhydride Than a Carboxylic Acid?

An acid chloride is more reactive than the acid to acyl addition reactions. It is also true that mixed anhydrides are easier to prepare via the acid chloride/carboxylate salt route.

Why Is the Preparation of Mixed Anhydrides Often Accompanied by Low Yields and Significant Amounts of Symmetrical Anhydrides?

If two different acids such as ethanoic acid and butanoic acid are heated with a dehydrating agent, ethanoic acid can condense with itself or with butanoic acid. Likewise, butanoic acid can condense with itself or with ethanoic acid. This reaction, therefore, produces three anhydrides in a 1:2:1 ratio: ***25.47***, ***25.48*** and ***25.49***.

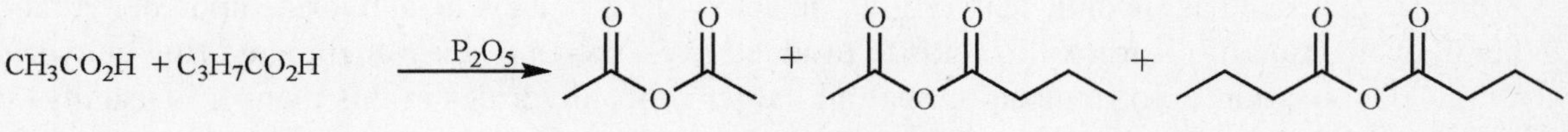

$CH_3CO_2H + C_3H_7CO_2H \xrightarrow{P_2O_5}$ *25.47* + *25.48* + *25.49*

How Can These Problems Be Overcome?

The best method for producing mixed anhydrides is to condense the acid chloride of one acid (such as butanoyl chloride) with the carboxylate salt of the other (such as sodium ethanoate). In this example, reaction of these two species will produce ***25.48***, exclusively.

Give the Major Product of the Reactions of 25.50–25.53.

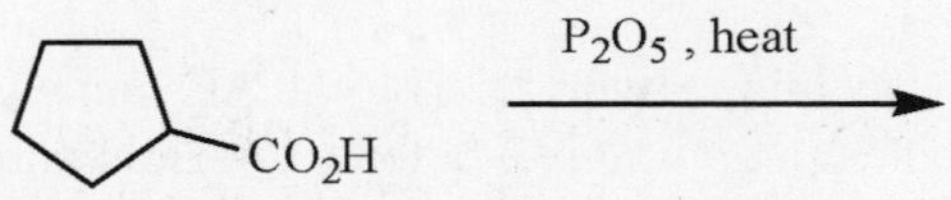

25.50 (cyclopentane–CO_2H) $\xrightarrow{P_2O_5\text{, heat}}$

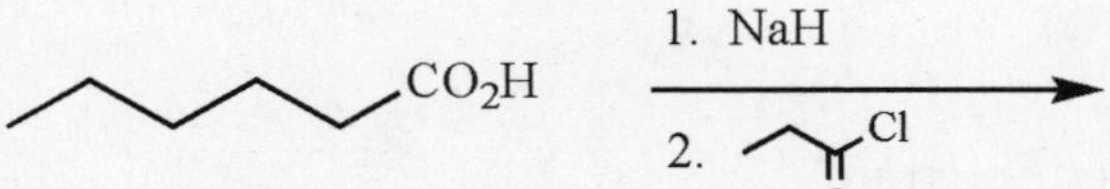

25.51 (CO_2H) $\xrightarrow[\text{2. propanoyl chloride (Cl, O)}]{\text{1. NaH}}$

25.52 (benzoyl chloride: O, Cl) $\xrightarrow[\text{heat}]{CO_2^-}$

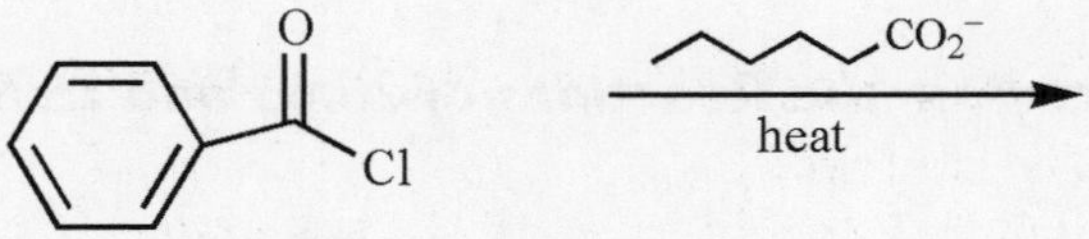

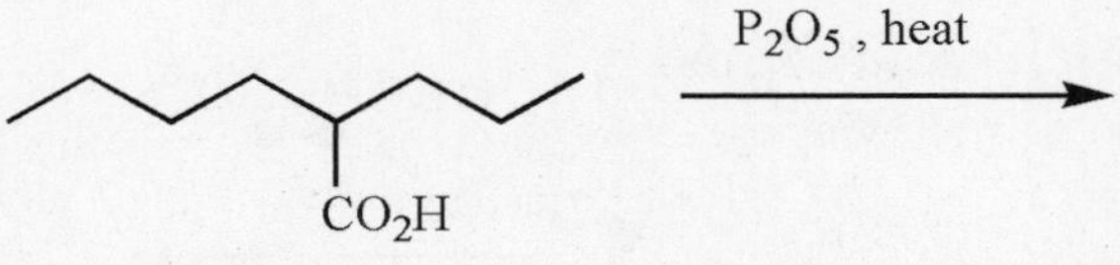

25.53 (CO_2H) $\xrightarrow{P_2O_5\text{, heat}}$

In the first case, cyclopentanecarboxylic acid (***25.50***) condenses with itself to form the symmetrical anhydride, ***25.54***. The second example generates mixed anhydride ***25.55*** by reaction of the carboxylic salt of hexanoic acid (***25.51***), formed by reaction with sodium hydride [NaH]) with propanoyl chloride. The third example also produces a mixed anhydride (***25.56***) by reaction of benzoyl chloride (***25.52***) with the sodium salt of hexanoic acid. The last example condenses ***25.53*** with itself to produce the symmetrical anhydride, ***25.57***.

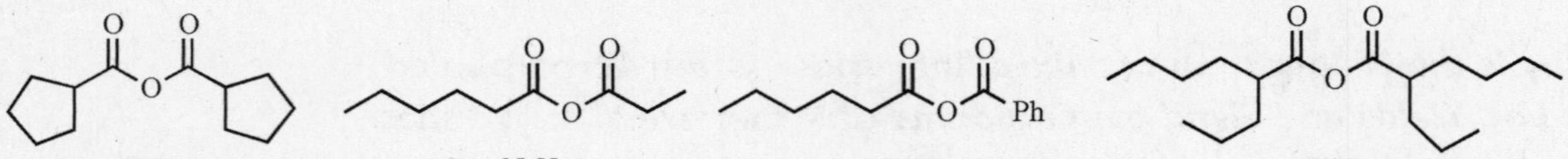

25.54 *25.55* *25.56* *25.57*

Cyclic Anhydrides

What Type of Acid Derivative Leads to a Cyclic Anhydride?

A dibasic acid such as succinic acid or glutaric acid will form a cyclic anhydride upon dehydration. In general, cyclization of dicarboxylic acids to produce five-, six- and seven-membered ring anhydrides is favorable. Cyclization to form most, but not all, larger ring anhydrides in this manner is usually very difficult.

Give the Name of 25.58–25.62.

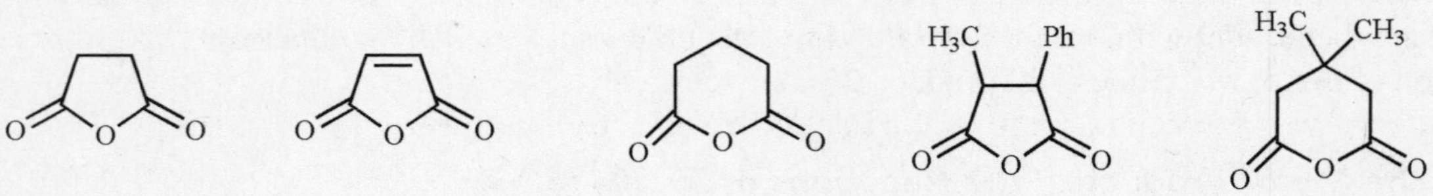

25.58 *25.59* *25.60* *25.61* *25.62*

The first anhydride (***25.58***) is succinic anhydride. The IUPAC name is 1,4-butanedioic anhydride. The second anhydride (***25.59***) is maleic anhydride (2-butene-1,4-dioic anhydride). The third (***25.60***) is glutaric anhydride (1,5-pentanedioic anhydride). Anhydride ***25.61*** is 2-methyl-3-phenyl-1,4-butanedioic anhydride (2-methyl-3-phenylsuccinic anhydride), and ***25.62*** is 3,3-dimethyl-1,5-pentanedioic anhydride (3,3-dimethylglutaric anhydride).

How Are Cyclic Anhydrides Formed?

Generally, heating dicarboxylic acids will lead to the anhydride.

Give the Major Product from the Reactions of 25.63 and 25.64.

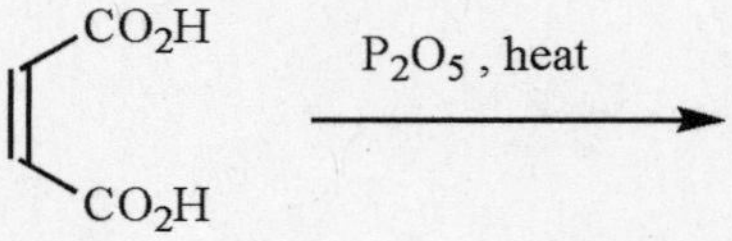

P_2O_5, heat

25.63

$HO_2C-(CH_2)_3-CO_2H$ →(P_2O_5, heat)

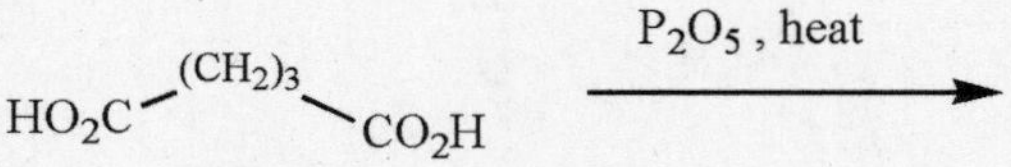

25.64

Maleic acid (***25.63***) is dehydrated to give maleic anhydride (***25.59***), and glutaric acid (***25.64***) is dehydrated to give glutaric anhydride, ***25.60***.

Explain Why Fumaric Acid (Trans-Butenedioic Acid) Does Not Give a Cyclic Anhydride upon Treatment with P_2O_5.

Unlike maleic acid (***25.63***), which has the carboxyl groups on the same side and close enough for those groups to interact to lose water, fumaric acid has the carbonyl groups on opposite sides of the double bond. They cannot get close enough to interact, and if water is lost to form an anhydride, it will be an acyclic anhydride.

Esters

Esters were first described in Section 24.4 and can be prepared by several important methods.

What Is the Product When Pentanoyl Chloride Reacts with Propanol in Benzene Solution in the Presence of Triethylamine?

The product is propyl pentanoate [$CH_3CH_2CH_2O_2C(CH_2)_4CH_3$]. The triethylamine reacts with the HCl by-product to produce triethylammonium chloride.

What Is the Product When Butanoyl Anhydride Reacts with Ethanol in the Presence of Triethylamine?

The anhydride reacts with ethanol, in the presence of an amine, to produce ethyl butanoate [$CH_3CH_2O_2CCH_2CH_2CH_3$] and butanoic acid. The triethylamine will react with butanoic acid to produce triethylammonium butanoate (the acid salt), driving the reaction towards the ester.

Which Is More Reactive, an Acid Chloride or an Ester? Explain.

An acid chloride is somewhat more reactive than an acid anhydride, but they are close in reactivity. The reason for the enhanced reactivity of the acid chloride is that chlorine is a better leaving group than an acyl group in acyl addition reactions.

Give the Major Product of the Reactions of 26.65 and 25.66.

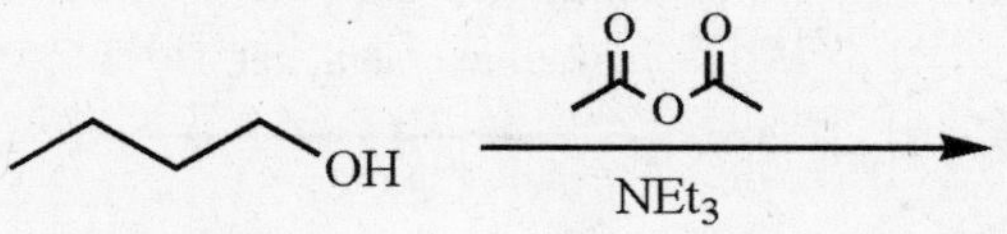

OH → (acetic anhydride, NEt_3)

25.65

OH → 1. $SOCl_2$; 2. cycloheptanol (OH), NEt_3

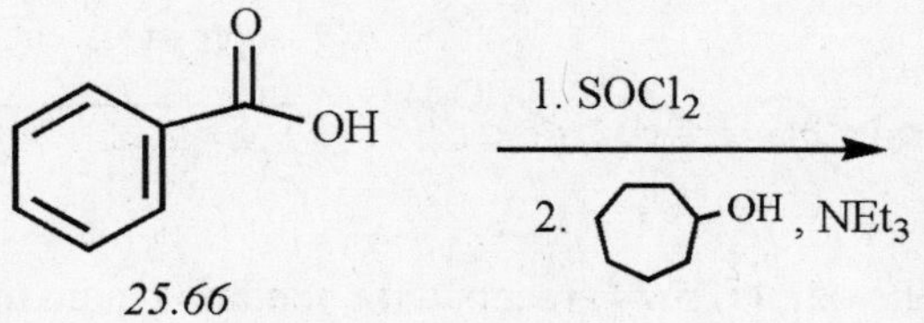

25.66

1-Butanol (***25.65***) reacts with acetic anhydride to form the ester, butyl butanoate (***25.67***). Initial reaction of benzoic acid (***25.66***) with $SOCl_2$ generates the acid chloride, which reacts with cycloheptanol to form cycloheptyl benzoate, ***25.68***.

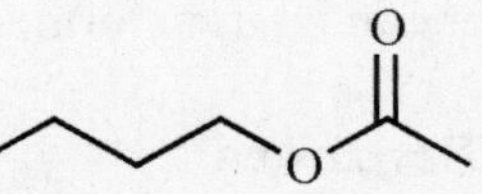

O O Ph O

25.67 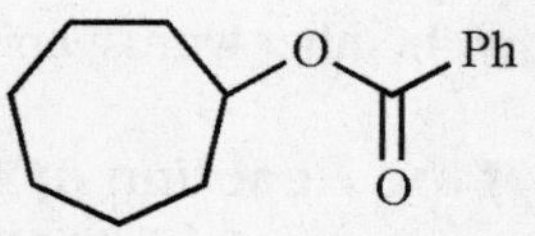 *25.68*

When Butanoic Acid Is Dissolved in Ethanol and Heated, in the Presence of an Acid Catalyst, What Is the Major Product?

The product is ethyl butanoate.

What Is the Mechanism of this Reaction?

The reaction begins by protonation of the carbonyl oxygen, which requires an acid with a pK_a much lower than that of a carboxylic acid, to give ***25.69***. This can add a molecule of ethanol to give the oxonium ion, ***25.70***. When this ion loses a proton, usually to the ethanol solvent, the highly reactive *ortho* acid derivative ***25.71*** is formed. This is quickly protonated to give ***25.72*** and loses water to give the resonance stabilized ion ***25.73***. Loss of a proton gives the final product, ethyl butanoate. As shown, all steps in this reaction are reversible. An excess of ethanol, or any alcohol, and removal of water will drive the reaction to the ester. Conversely, the use of aqueous acid will drive the equilibrium back towards the acid.

O OH +H⁺ H ⊕O OH OH ⊕OH +EtOH OH OH O Et ⊕ H -H⁺

25.69 *25.70*

OH OH O Et +H⁺ OH ⊕ OH₂ O Et - H₂O⁺ OH ⊕ O–Et ⊕ O-H O Et -H⁺ O OEt

25.71 *25.72* *25.73*

What Is the Major Product of the Reactions of 25.74 And 25.75?

OH pentanoic acid, cat. H^+

25.74

CO_2H isopropanol, cat. H^+

25.75

In the first reaction, cyclohexanol (***25.74***) reacts with pentanoic acid to form cyclohexyl pentanoate. In the second reaction, octanoic acid (***25.75***) reacts with isopropanol to form isopropyl octanoate.

What Is Transesterification?

Transesterification is the conversion of one ester to a different ester, but reacting the ester with an alcohol and an acid catalyst. In other words, an ethyl ester can be transformed into a methyl ester.

What Is the Product of the Reaction of Methyl Benzoate with Ethanol, in the Presence of an Acid Catalyst?

The product is ethyl benzoate.

What Is the Product When Phenol Reacts with Acetic Anhydride?

Phenol reacts with acetic anhydride to form phenyl acetate (CH_3CO_2Ph).

What Is the Product When Decanol Reacts with Ethanoyl Chloride in the Presence of Triethylamine?

The product is decyl ethanoate [$CH_3(CH_2)_8CH_2O_2CCH_3$].

What Is the Purpose of the Triethylamine?

The by-product of this reaction is HCl. The triethylamine functions as a base to react with HCl, forming $Et_3NH^+ Cl^-$, (triethylammonium chloride).

What Is DCC?

The structure of DCC (dicyclohexylcarbodiimide) is ***25.76***.

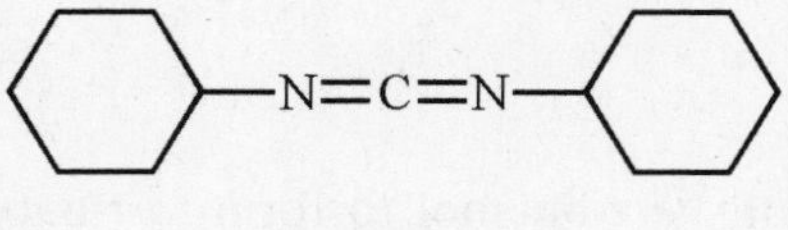

25.76

What Is the Product of the Reaction Between Propanoic Acid and Ethanol in the Presence of DCC?

The products are the ester, ethyl propanoate and dicyclohexyl urea (***25.77***).

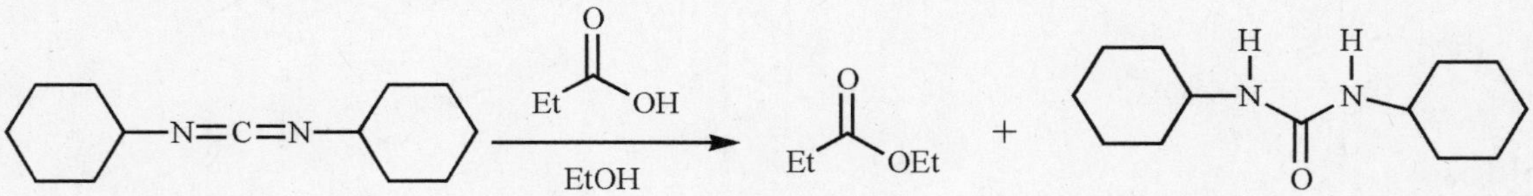

25.76 *25.77*

What Is the Mechanism of This Reaction?

Dicyclohexylcarbodiimide essentially acts as a dehydrating agent in this process. The initial reaction between DCC and the acid involves attack of the carboxyl oxygen on the carbon of DCC to give ***25.78*** after a proton transfer to nitrogen. The acyl carbon is attacked by the alcohol oxygen, displacing dicyclohexyl urea (***25.79***), again with a proton transfer to nitrogen, and generating the ester. For all practical purposes, the DCC converts the OH of the carboxylic acid into an excellent leaving group (the urea).

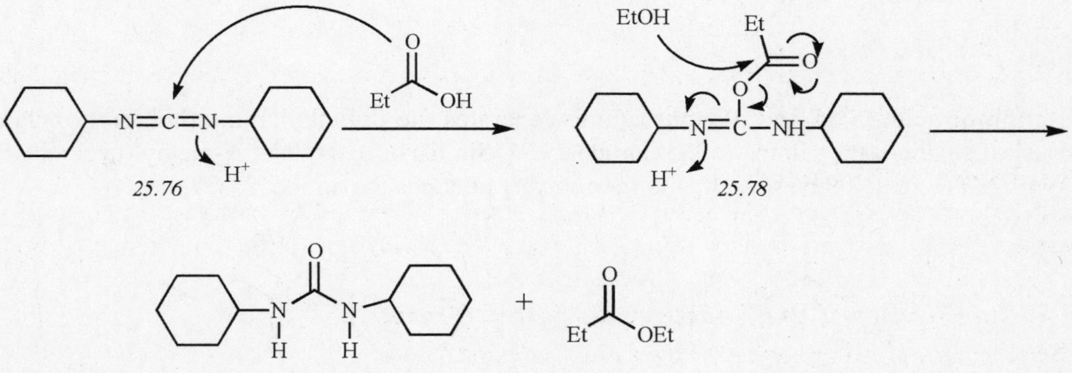

25.76 *25.78*

25.77

What Is the Product from the Reactions of 25.79–25.81?

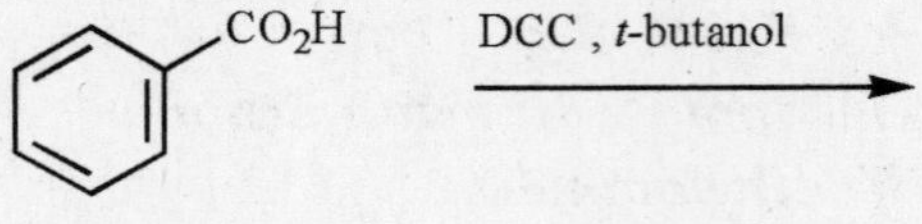

CO_2H DCC , *t*-butanol →

25.79

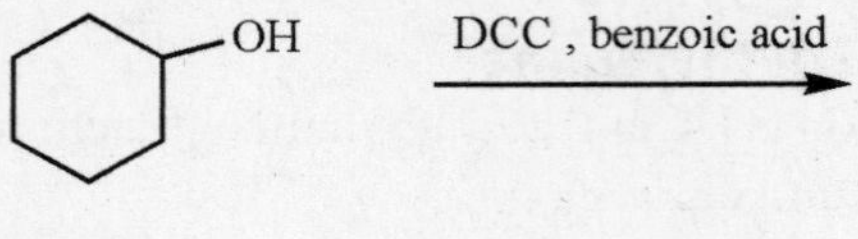

OH DCC , benzoic acid →

25.80

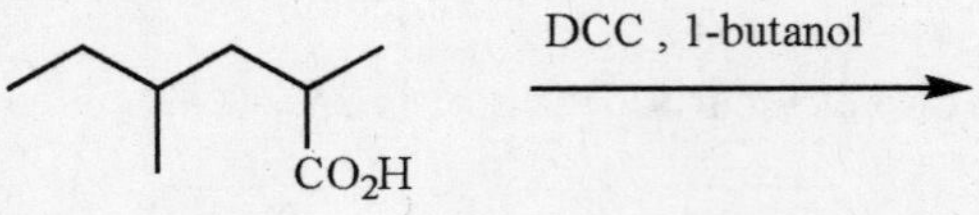

DCC , 1-butanol →

CO_2H

25.81

Benzoic acid (***25.79***) reacts with *tert*-butanol to form *tert*-butyl benzoate (***25.82***). Cyclohexanol (***25.80***) reacts with benzoic acid to form cyclohexyl benzoate (***25.83***), and 2,4-dimethyl-hexanoic acid (***25.81***) reacts with 1-butanol to give butyl 2,4-dimethylhexanoate (***25.84***).

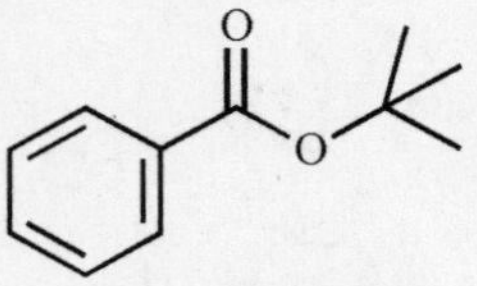

O O O Ph O O O

25.82 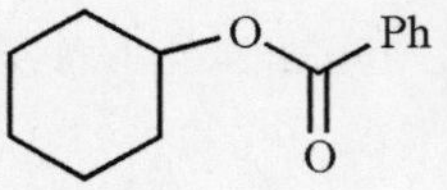 25.83 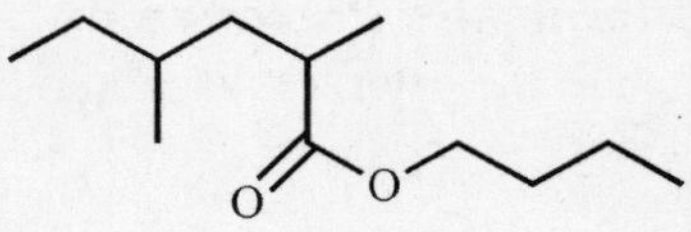 25.84

Amides

What Is the Product of the Reactions of 25.85 and 25.86?

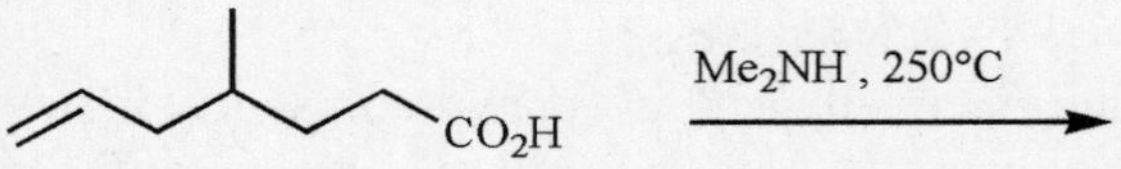

CO_2H Me_2NH , 250°C →

25.85

NHMe $CH_3CH_2CH_2CO_2H$ → 225°C

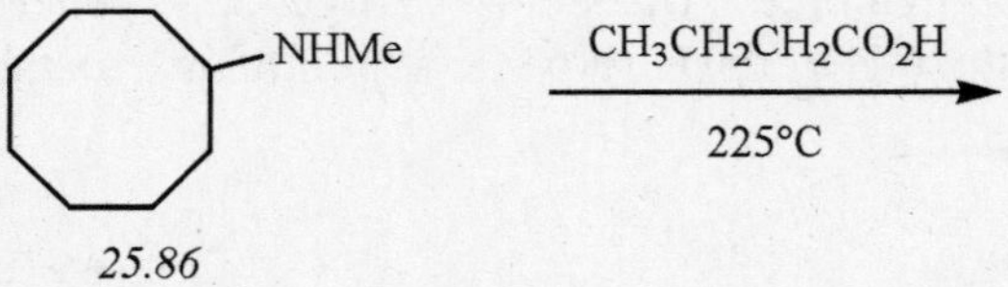

25.86

Reaction of acid ***25.85*** and dimethylamine generates the dimethylammonium carboxylate salt. Thermolysis dehydrates the salt to produce amide ***25.87***. Similarly, reaction of *N*-methyl-cyclooctylamine (***25.86***) with butanoic acid gives a salt, and thermolysis liberates the amide, ***25.88***.

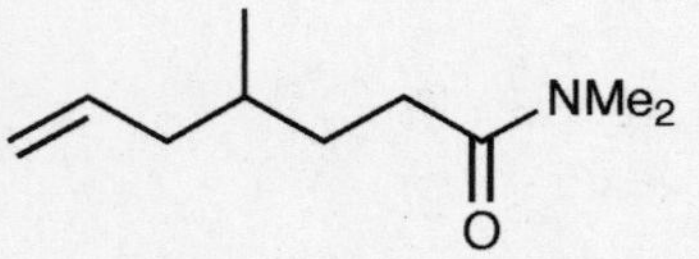

NMe_2 O Me N O

25.87 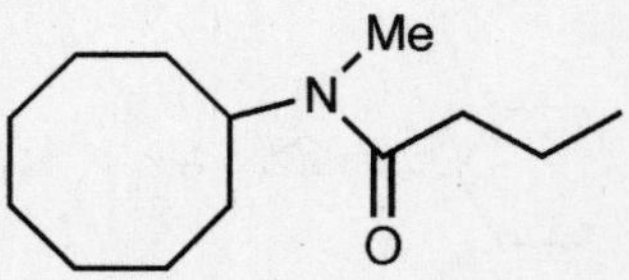 25.88

What Is the Major Product When Butanoyl Chloride Reacts with Butylamine in the Presence of Triethylamine?

The product is the amide, *N*-butylbutanamide [BuNH(C=O)Bu].

What Is the Product When Pentylamine Reacts with Acetic Anhydride?

Acetic anhydride behaves similarly to an acid chloride and provides the acyl portion of an amide. Reaction of acetic anhydride and pentylamine, therefore, gives *N*-pentylethanamide.

What Is the Product When Aniline Reacts with Acetic Anhydride? With Propanoyl Chloride?

When aniline reacts with acetic anhydride, the amide, 'N-acetylaniline' or N-phenylethanamide (***25.89***) is formed, and propanoyl chloride generates 'N-propanoyl aniline' (N-phenylpropanamide, ***25.90***).

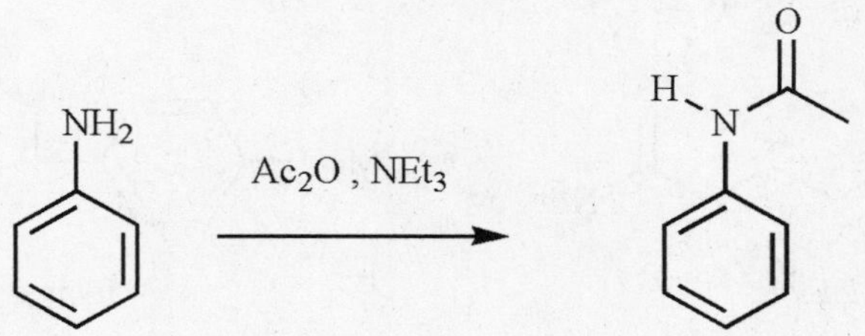

NH_2 $\xrightarrow{Ac_2O\text{ , }NEt_3}$ H–N–C(=O)CH₃

25.89

NH_2 $\xrightarrow{\text{propanoyl chloride , } NEt_3}$ H–N–C(=O)CH₂CH₃

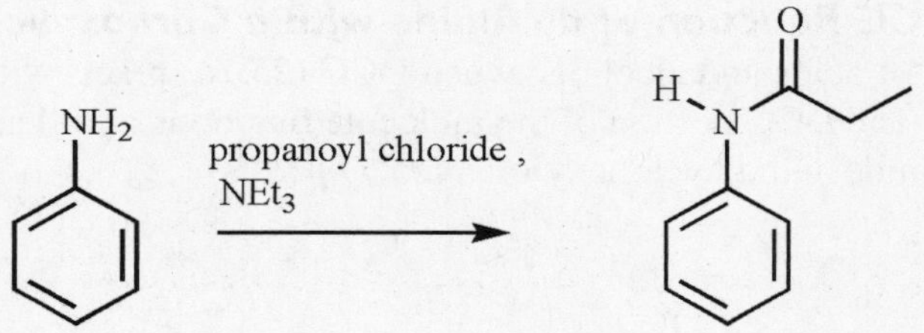

25.90

What Is the Product When Ethyl Pentanoate Reacts with Ammonia Under Conditions of High Heat and Pressure?

Under these conditions, an amide is formed (pentanamide) where the ammonia displaces ethanol via acyl addition and substitution.

What Is the Product When CH_3NH_2 Reacts with γ-Butyrolactone?

Once again, the nitrogen displaces the 'alcohol' via acyl addition. The OH is converted to $^{+}OH_2$ and displacement by the nitrogen (under high heat and pressure) forms the lactam, ***25.91*** (*N*-methyl-2-pyrrolidinone).

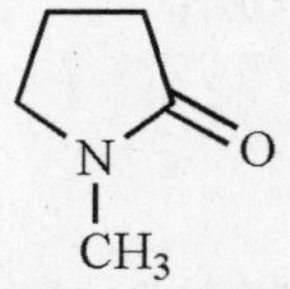

N O CH_3

25.91

Give the Major Product of the Reactions of 25.92 and 25.93.

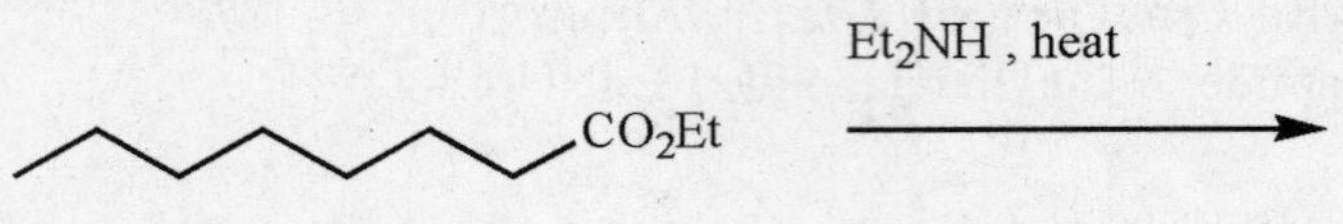

Et_2NH, heat

CO_2Et

25.92

MeO— —CO_2H

1. $SOCl_2$
2. EtOH, NEt_3
3. pyridine, NH_2

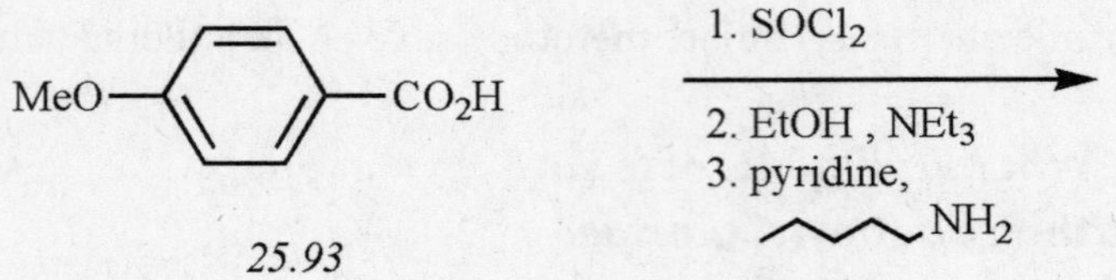

25.93

In the first reaction, ***25.92*** reacts with diethylamine to form the *N,N*-diethyl amide, ***25.94***. In the second reaction, conversion of ***25.93*** to the acid chloride allows esterification. Subsequent reaction with pentylamine leads to the amide, ***25.95***.

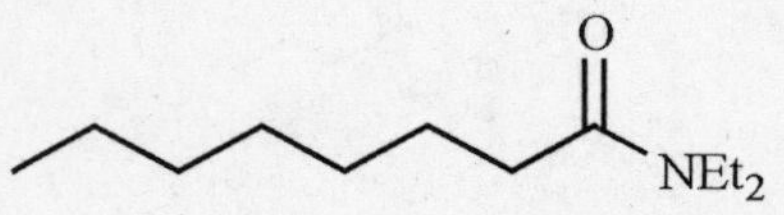

O NEt$_2$

25.94

MeO O NHC_5H_{11}

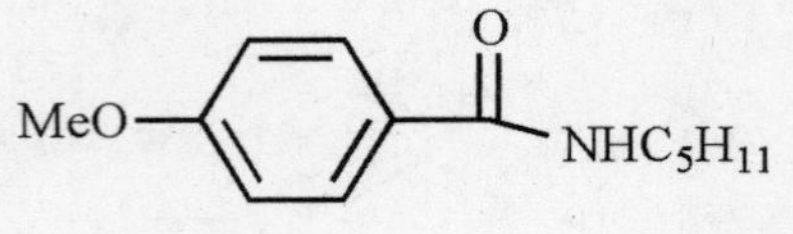

25.95

By Analogy with the DCC Reaction of Alcohols and Acids, What Is the Key Intermediate in the DCC Reaction of an Amine with a Carboxylic Acid?

As with the coupling of acids and alcohols, when DCC (***25.76***) reacts with the carboxylic acid, the key intermediate is ***25.96***. The DCC portion of the molecule functions as a 'leaving group' and attack by the amine generates the amide and dicyclohexylurea (***25.77***).

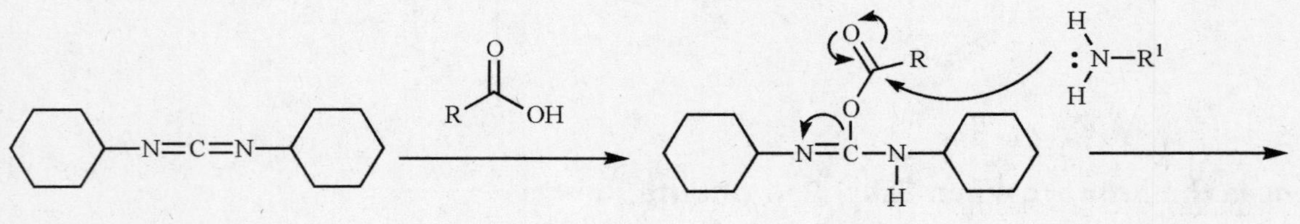

—N=C=N—

25.76

R OH O

N=C—N— H O R O H :N—R[1] H

25.96

R NHR[1] O + H H N N O

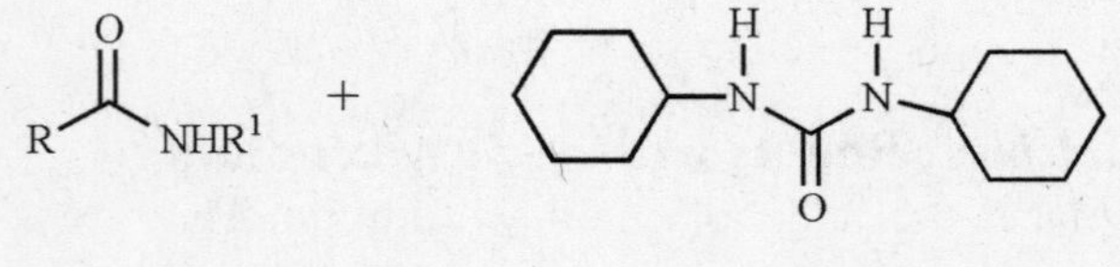

25.77

What Is the Final Product of This Reaction?

An amide, along with dicyclohexylurea, ***25.77***.

Nitriles

What Is the Major Product When 1-Iodopentane Reacts with KCN in Hot DMF?

The product is pentanenitrile, $CH_3CH_2CH_2CH_2C{\equiv}N$.

What Is the Product When 3S-Bromoheptane (25.97) Reacts with NaCN in DMF?

Since the reaction is S_N2, it proceeds with 100% inversion of configuration. The product is 2-ethylhexanenitrile, and the chiral center is R (R-2-ethylhexanenitrile, ***25.98***).

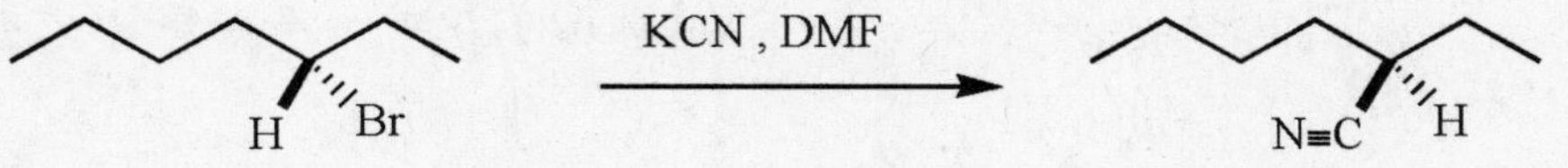

KCN , DMF

H Br N≡C H

25.97 *25.98*

What Is the Product When Cyclohexanone Reacts with HCN?

The product is a cyanohydrin, 1-cyanocyclohexanol, ***25.100***. Initial reaction of the carbonyl oxygen, and HCN generates the cation (***25.99***), which is then attacked by the nucleophilic cyanide ion.

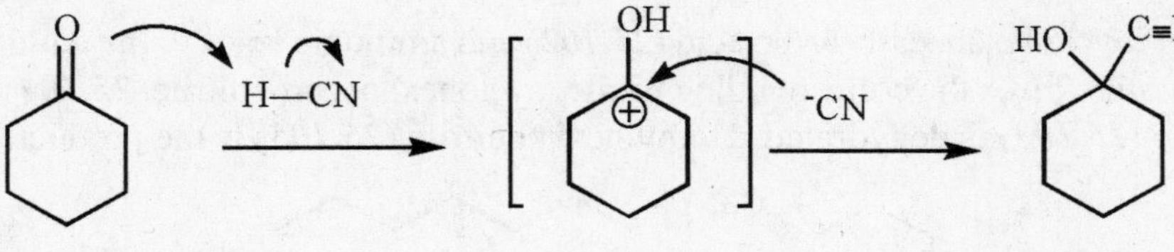

O H—CN OH ⊕ ⁻CN HO C≡N

25.99 *25.100*

What Is the Product of the Reaction Between Butanoyl Chloride and NaCN in DMF?

The cyanide behaves as a nucleophile and displaces the chloride (nucleophilic acyl substitution) to produce 2-oxopentanenitrile, ***25.101***.

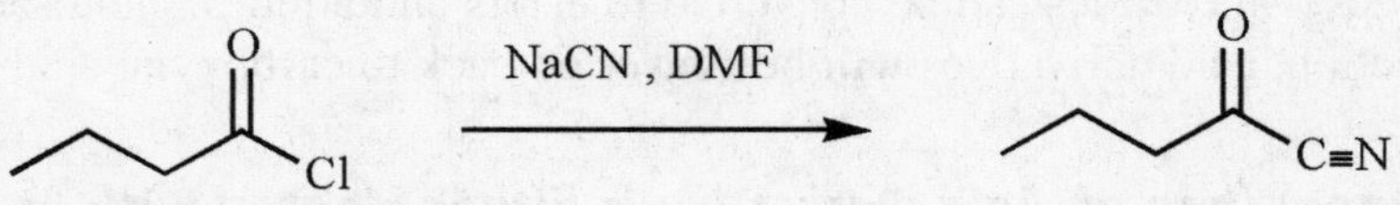

O Cl NaCN , DMF O C≡N

15.101

What Is a Dehydrating Agent?

A dehydrating agent is a reagent that reacts with water and removes it from the reaction medium or reacts with a molecule to remove the elements of water.

What Are Common Dehydrating Agents?

The most common dehydrating agent is probably phosphorus pentoxide (P_2O_5 - note the actual formula of this reagent is P_4O_{10}; the other formula was thought to be correct and accounts for the name.) Sulfuric acid is sometimes used as a dehydrating agent in relatively simple molecules.

What Is the Reaction Product When Butanamide Is Heated with Phosphorus Pentoxide?

Under these conditions, butanamide is converted to butanenitrile, $CH_3CH_2CH_2C{\equiv}N$.

What Is the Product When Benzoic Acid Is Treated with: i. $SOCl_2$ ii. NH_3, Heat iii. P_2O_5 , Heat?

The product is benzonitrile, Ph–C≡N.

What Is the Major Product of the Reactions 25.102 and 25.103?

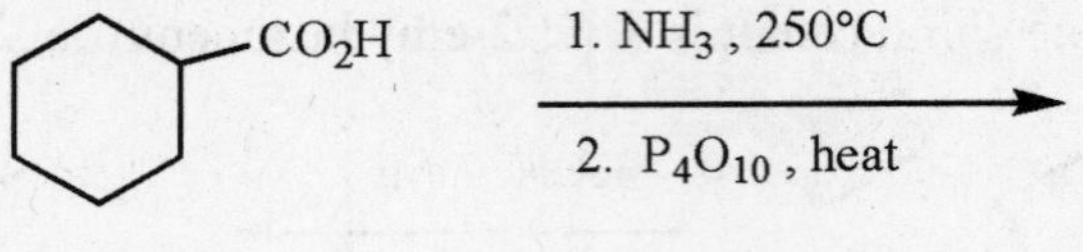

CO_2H 1. NH_3, 250°C 2. P_4O_{10}, heat

25.102

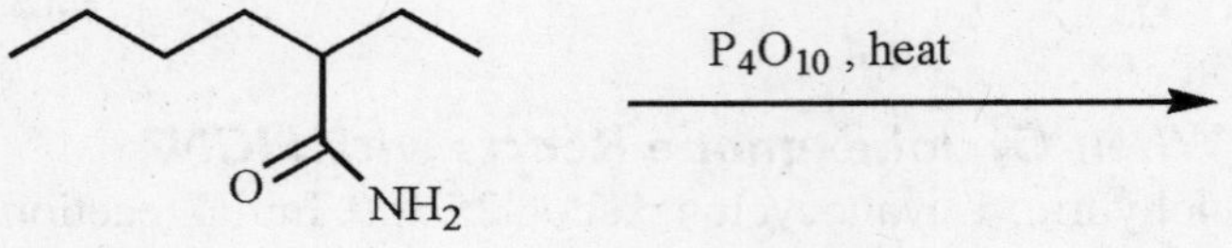

P_4O_{10}, heat

O NH_2

25.103

The reaction of cyclohexanecarboxylic acid (***25.102***) and ammonia leads to the amide (with heating). Subsequent dehydration gives the corresponding nitrile, cyclohexanecarbonitrile, ***25.104***. Similarly, when 2-ethylhexanamide (***25.103***) is dehydrated, 2-ethylhexanenitrile (***25.105***) is the product.

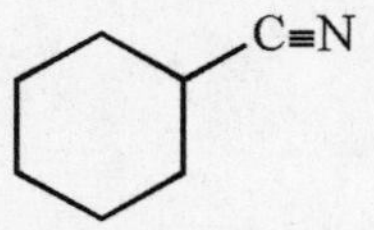

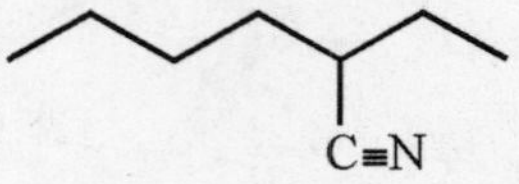

C≡N C≡N

25.104 25.105

25.3. CONVERSION OF ACID DERIVATIVES TO CARBOXYLIC ACIDS

Carboxylic acids can be converted to acid derivatives. Acid chlorides can be converted to esters, anhydrides and amides. Anhydrides can be converted to esters and amides, and esters can be converted to amides. In this section, acid derivatives will be converted back to carboxylic acids.

What Is the Product When an Acid Derivative Is Either Heated with Aqueous Acid or Treated with Aqueous Hydroxide and Then Neutralized with Acid?

Virtually all acid derivatives can be converted to their parent carboxylic acid by either acid or base hydrolysis.

What Is the Mechanism for the Base Hydrolysis of Butanoyl Chloride?

Hydroxide attacks the carbonyl carbon (acyl addition) to form the tetrahedral intermediate ***25.106***. The electrons on the alkoxide oxygen are transferred back to the acyl carbon, which expels chlorine, a good leaving group, and forms the carboxylic acid (***25.107***). Under the basic conditions of the reaction, this acid reacts with NaOH to form the carboxylate, ***25.108***.

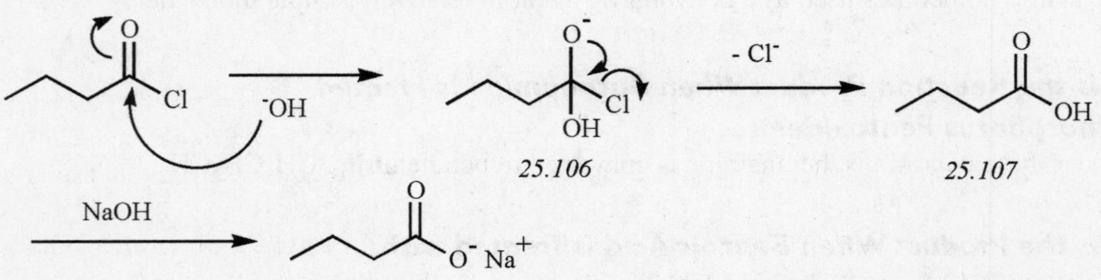

O Cl $^-$OH O^- Cl OH - Cl^- O OH

25.106 25.107

NaOH O O^- Na^+

25.108

What Is the Mechanism of Acid Hydrolysis of Butanoyl Chloride to the Acid?

The initial reaction is protonation of the carbonyl oxygen to give the resonance stabilized ion, ***25.109***, followed by reaction with water to give **25.110**. Loss of a proton gives ***25.111***, which allows expulsion of chloride ion to give ***25.112***. Final loss of a proton gives the acid, ***25.107***. The loss of H^+ and Cl^- 'is formally equivalent to losing HCl.

H+ · O-H · O-H · + H_2O · O-H · Cl · Cl · Cl · Cl · H · H · O

25.109 *25.110*

- H^+ · O-H · - Cl^- · O-H · - H^+ · O · Cl · OH · OH · OH

25.111 *25.112* *25.107*

What Is the Product When Ethanoic Anhydride Is Treated with Aqueous Acid?

Acid hydrolysis converts an anhydride to a mixture of the two component acids. In this case, ethanoic anhydride (acetic anhydride) gives two equivalents of ethanoic acid (acetic acid) upon aqueous acid hydrolysis.

Why Is It Not Important to the Final Product Distribution Which Carbonyl of Ethanoic Propionic Anhydride Is Attacked by Hydroxide in Base Hydrolysis?

Ethanoic anhydride is symmetrical and hydroxide can attack either carbonyl to give the same product.

What Are the Products When Butanoic Hexanoic Anhydride Is Treated with i. AQ. NaOH ii. Dilute Acid?

Base hydrolysis cleaves the anhydride into the two carboxylate components and dilute acid reprotonates the carboxylates to give a 1:1 mixture of butanoic acid and hexanoic acid.

What Is the Product When Glutaric Anhydride Is Treated with Aqueous Acid?

Acid hydrolysis cleaves the anhydride to give glutaric acid.

What Is the Mechanism for the Conversion of Ethyl Butanoate to Butanoic Acid and Ethanol Under Acidic Conditions?

The mechanism for ester hydrolysis is essentially identical to that given for acid hydrolysis of an acid chloride (butanoyl chloride; see above). In this case, the leaving group is OEt rather than Cl. Initial reaction with acid gives the resonance stabilized cation (***25.113***), which is attacked by water to give ***25.114***. Loss of a proton gives ***25.116***, and this step is followed by transfer of the proton to the OEt moiety to give ***25.117***. This allows expulsion of ethanol, the leaving group, to give ***25.112***, and final loss of a proton gives butanoic acid, ***25.107***, the by-product is ethanol, as shown. This hydrolysis mechanism is effectively the exact reverse of the esterification mechanism presented above.

O O-Et H+ ⊕O-H O-Et O-H ⊕ O-Et $+H_2O$ O-H O-Et H ⊕ H - H^+

25.113 *25.114*

O-H O-Et OH + H^+ O-H O-Et OH ⊕ H - EtOH ⊕O-H OH - H^+ O OH

25.116 *25.117* *25.112* *25.107*

What Is Saponification?

Saponification is a term used for the manufacture of soap. It involves treatment of fats (triglyceryl esters) with lye (hydroxide) to form the sodium salt of fatty acids (soap). Triglycerides are formed from glycerol (the alcohol) and long chain fatty acids. The term *saponification* has come to be used for the reaction: base hydrolysis of esters. The mechanism is identical to that shown for acid chlorides where Cl is replaced with OR.

What Is the Mechanism for the Acid Catalyzed Hydrolysis of N,N-Dimethylpentamide?

The mechanism is identical to that for acid chlorides and esters, where the 'leaving group' is now the amine ($HNEt_2$). The reaction begins the same way with protonation of the carbonyl oxygen to give ***25.118*** and addition of water to give ***25.119***. Loss of a proton will give ***25.120***, but the proton must be transferred to the nitrogen to convert it into a good leaving group, ***25.121***. It is noted that this hydrogen transfer may occur intramolecularly, avoiding the need to form ***25.120***. Loss of diethylamine gives ***25.112***, and loss of a proton gives butanoic acid, ***25.107***.

O NEt_2 H+ ⊕O-H NEt_2 O-H ⊕ NEt_2 + H_2O O-H NEt_2 H ⊕ H - H^+

25.118 *25.119*

O-H NEt_2 OH H+ O-H NEt_2 OH ⊕ H - $NHEt_2$ ⊕O-H OH - H^+ O OH

25.120 *25.121* *25.112* *25.107*

Why Is the Hydrolysis of an Amide More Difficult Than the Hydrolysis of an Ester?

The amine group is a poorer leaving group than OEt and the amide group is bulkier, making attack of water at the carbonyl more difficult.

What Is the Product of the 'Partial Hydrolysis' of Hexanenitrile?

The complete hydrolysis of a nitrile gives the carboxylic acid. 'Partial hydrolysis' refers to limiting the hydrolysis conditions to isolate the amide, which is an intermediate product on the path to the carboxylic acid. In this case, partial hydrolysis leads to hexanamide. In general, relatively dilute acid and relatively mild conditions give the amide.

What Conditions Are Required for the Conversion of a Nitrile to a Carboxylic Acid?

If concentrated aqueous acid and vigorous reaction conditions (refluxing conditions for a long period of time) are used, the nitrile is converted directly to the carboxylic acid.

25.4. CONVERSION OF ACID DERIVATIVES TO ESTERS

Carboxylic acids can be converted to acid derivatives. Acid chlorides can be converted to esters, anhydrides and amides. Anhydrides can be converted to esters and amides, and esters can be converted to amides. In this section, acid derivatives will be converted to esters.

What Is the Product of a Reaction Between an Acid Chloride and an Alcohol?

As noted in Section 25.2, the reaction of an acid chloride or an anhydride with an alcohol leads to an ester.

What Is the Product When Benzoyl Chloride Reacts with 2-Propanol?

The product is 2-propanoyl benzoate, or isopropyl benzoate.

What Does the Term Transesterification Mean?

Transesterification refers to exchanging the OR group of an ester with a new alcohol group (OR^1). Conversion of a methyl ester to an ethyl ester is an example of transesterification.

What Is the Product When Ethyl Butanoate Is Dissolved in Methanol with a Catalytic Amount of Acid?

The product is methyl butanoate.

What Is the Mechanism of This Reaction?

Initial protonation forms the usual ***25.122***, but the attacking species is now methanol rather than water, forming ***25.123***. Loss of a proton generates ***25.124***. If OMe is protonated again, the reaction is driven back towards the ethyl ester, but if OEt is protonated to give ***25.125***, ethanol is lost to give ***25.126***. Final loss of a proton gives the methyl ester (methyl butanoate). This is an equilibrium process, and an excess of ethanol favors the ethyl ester, whereas an excess of methanol favors the methyl ester.

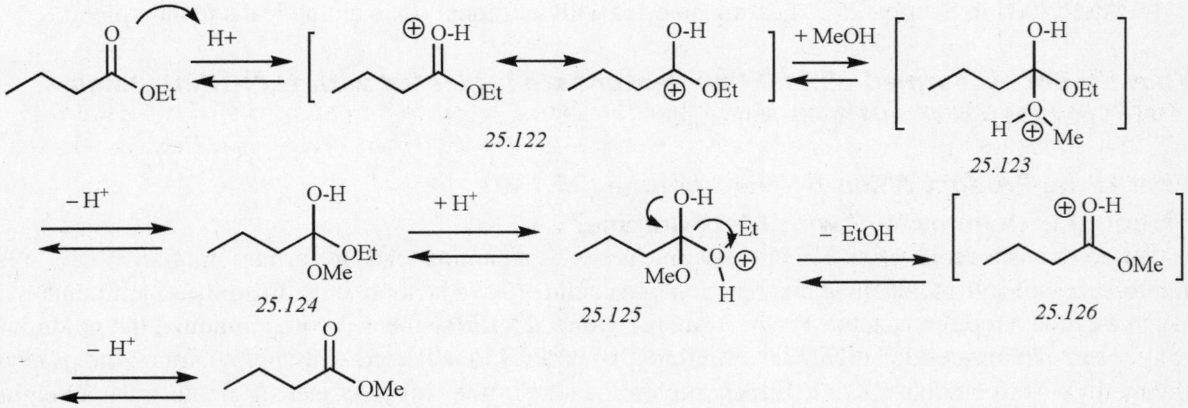

25.122

25.123

25.124

25.125

25.126

What Is the Product When Succinic Acid Is Dissolved in Ethanol with a Catalytic Amount of Acid?

The product is the diester, diethyl succinate, ***25.127***.

CO_2H / CO_2H —(EtOH , cat. H^+)→ CO_2Et / CO_2Et

25.127

Is It Possible to Form the 'Half-Ester' of a Dibasic Acid?

Yes. It is usually difficult to stop an equilibrium reaction (catalytic H^+ and an alcohol, for example), but if the half-acid chloride can be made, the half ester can be prepared by standard techniques.

25.5. CONVERSION OF ACID DERIVATIVES TO AMIDES

Carboxylic acids can be converted to acid derivatives. Acid chlorides can be converted to esters, anhydrides and amides. Anhydrides can be converted to esters and amides, and esters can be converted to amides. In this section, acid derivatives will be converted to amides.

What Is the Product When an Acid Chloride Reacts with a Primary or Secondary Amine?

As described in Section 25.2, reaction of amines with acid chlorides and anhydrides leads to amides.

What Is the Product Formed When Hexanoyl Chloride Reacts with Pyrrolidine (25.128)?

The product is ***25.129***.

O, Cl + HN (*25.128*) → O, N (*25.129*)

What Is the Product Formed When an Ester Is Heated with an Amine?

As described in Section 25.2, heating an ester with ammonia or an amine leads to an amide.

What Product Is Formed When Ethyl Pentanoate Is Heated with N,N-Diethylamine?

The product is N,N-dimethylpentanamide.

What Is the Product When δ-Valerolactone (25.130) Is Heated with Ammonia? with Methylamine?

A lactone is a cyclic ester. Reaction of an ester with an amine leads to formation of an amide. It is, therefore, reasonable to conclude that reaction of a cyclic ester (a lactone) with an amine or ammonia will lead to a cyclic amide (a lactam). When δ-valerolactone (***25.130***) is heated with ammonia, the product is δ-valerolactam (2-piperidone, ***25.131***). Similarly, when ***25.130*** is heated with methylamine, the product is *N*-methyl-2-piperidone, ***25.132***. Depending on the size of the ring, lactones of greater than an eight-member atom are opened by this reaction to give an open chain amino acid.

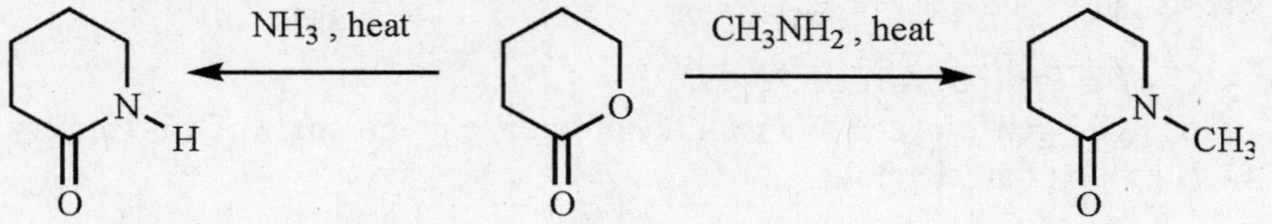

NH_3, heat ← 25.130 → CH_3NH_2, heat

25.131 25.130 25.132

Why Does Heating the Nine-Membered Ring Lactone with Ammonia Lead To an Open Chain Hydroxy Amide Rather Than a Lactam?

Nine-membered rings are high in energy due to **transannular** steric effects, the result of crowding hydrogen atoms inside the cavity of the ring. If a nine-membered ring is being formed, the molecule assumes a conformation analogous to a nine-membered ring in the transition state. The high transannular strain destabilizes that transition state and inhibits formation of the ring. Since cyclization to the nine-membered ring lactam is very difficult, the molecule simply opens to an amino acid (9-aminononanoic acid in this case).

When Succinic Anhydride Is Heated with Ammonia, What Is the Product?

Just as lactones are converted to lactams, anhydrides are converted to **imides** upon heating with ammonia or an amine. In this case, succinic anhydride is converted to succinimide, ***25.133***.

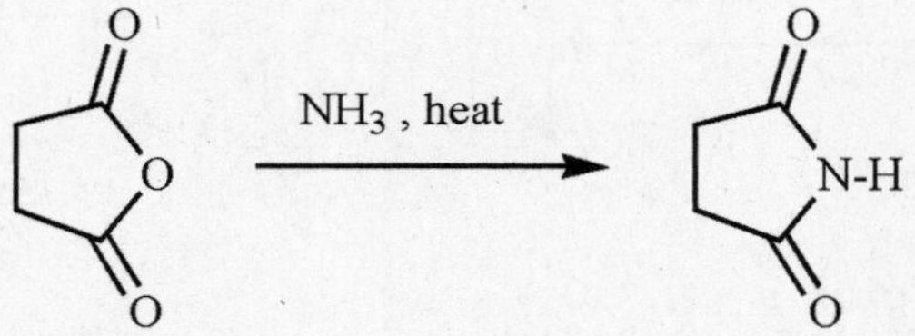

NH_3, heat

N-H

25.133

Why Is It Difficult to Convert Diethyl Malonate into a lactam?

If diethyl malonate were to form a lactam, it would be a four-membered ring lactam, a so-called β-lactam. In general, the strain inherent to four-membered rings makes their formation by this route difficult, but not impossible. There are many alternative synthetic routes to β-lactams, but they will not be discussed in this book.

If Diethyl Malonate Is Heated with Dimethylamine, What Is the Product if an Excess of the Amine Is used?

As with any other ester, heating with an amide will produce the amide. In this case, excess dimethylamine will convert both ester moieties to the amide and the product will be *bis*-(*N,N*-dimethylamino) propanediamide, ***25.134***.

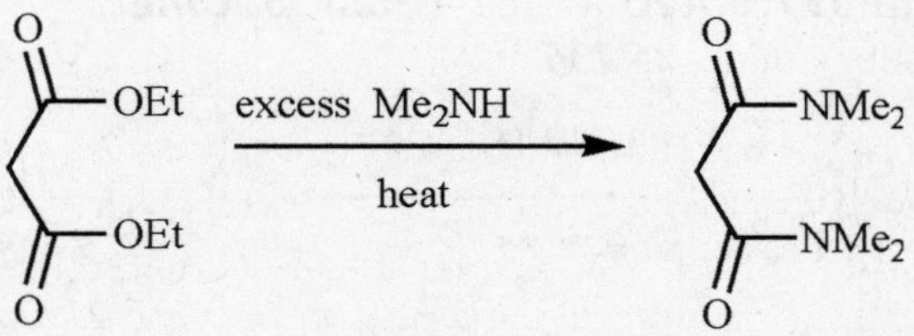

OEt excess Me_2NH heat → NMe_2

OEt NMe_2

25.134

Acidity of Amides

What Is the pK_a of a Typical Amide N–H?

The pK_a of the hydrogen attached to an amide nitrogen is in the range of 15–17, very similar to the acidity of the O–H group of an alcohol.

Is the Amide N–H More or Less Acidic Than an Amine?

An amide is significantly more acidic than the NH of an amine (a typical amine *N*–H shows a pK_a of about 25–30, and ammonia is 38). The carbonyl group on the nitrogen of the amide serves as an electron withdrawing group to stabilize the amide base, after removal of a hydrogen, and to make the *N*–H more polarized (more acidic).

What Type of Base Is Required to Make the N–H Bond of an Amide an Acid?

A relatively strong base is required such as $NaNH_2$, $NaNR_2$, an organolithium reagent such as *n*-butyllithium, or a base such as sodium hydride.

What Is the Structure of an Amide Anion ?

Removal of the acidic proton of an amide generates an resonance stabilized amide anion , ***25.135***.

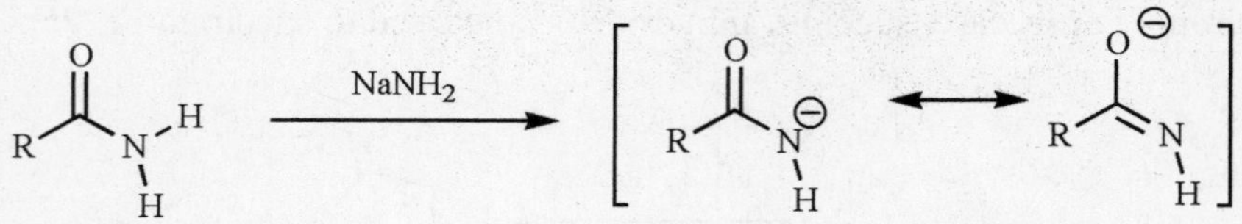

$NaNH_2$

25.135

How Stable Is an Amide Anion ?

It is quite stable due to the resonance contributors that are possible, dispersing the change on both nitrogen and the carbonyl oxygen.

Is an Amide Anion a Strong Base?

Yes! Since an acid with a pK_a of 15–17 is a relatively weak acid, the conjugate base must be relatively strong. In general, the anion of an amide is slightly less basic than the anion of an alcohol (an alkoxide).

If an Amide Anion Is Considered to Be a Reagent, How Would It Be Classified?

It is both a base and a nucleophile. It will, therefore, react with a suitable acid to regenerate the amide or with an electrophilic carbon, such as an alkyl halide, to form an *N*-substituted amide.

What Is the Reaction Product When the Amide Anion Derived from Butanamide Is Treated with 1-Bromobutane?

The product is *N*-butylbutanamide, ***25.136***.

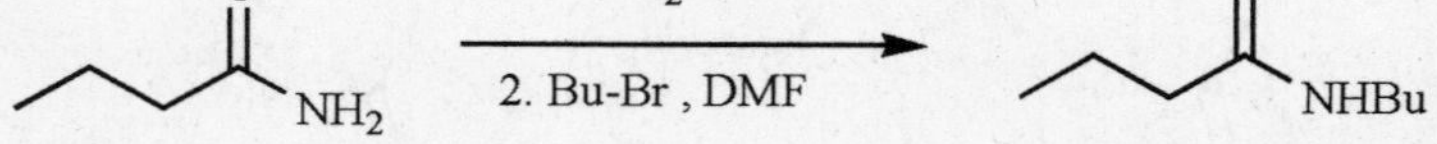

1. $NaNH_2$
2. Bu-Br , DMF

25.136

25.6. REDUCTION

In general, acid derivatives are reduced to alcohols. Amides and nitriles are exceptions, where reduction usually leads to an amine if the reducing agent is powerful enough.

Reduction to Alcohol

What Is the Major Product When Ethyl Benzoate Is Reduced with $LiAlH_4$ and Then Treated with Water?

The 'acid portion' is reduced to benzyl alcohol ($PhCH_2OH$) and the 'alcohol portion' is 'released,' giving, in this case, ethanol.

How Efficient Is the Reduction of an Ester to an Alcohol with $NaBH_4$?

Sodium borohydride is a much weaker reducing agent than $LiAlH_4$, and $NaBH_4$ reduces esters to alcohols with difficulty. It is usually, but not always, a slow reaction. The structure of the ester is important in determining the relative rate of reduction.

What Acid Derivatives Can Be Reduced to Alcohols with Hydrogen and a Catalyst?

Most acid derivatives resist reduction by catalytic hydrogenation. Exceptions are acid chlorides and acid anhydrides, which are readily reduced to alcohols. Acids, esters and amides are very difficult to reduce by this method and usually give no reaction.

What Is the Product When δ-Valerolactone Is Reduced with $LiAlH_4$?

The product is 1,5-pentanediol ($HOCH_2CH_2CH_2CH_2CH_2OH$).

What Is the Product When Diethyl Succinate Is Reduced with $LiAlH_4$?

Reduction of diethyl succinate gives 1,4-butanediol ($HOCH_2CH_2CH_2CH_2OH$).

Reduction to Amines

What Is the Product When N-Acetyl Aniline Is Reduced with $LiAlH_4$?

In most cases, the carbonyl portion of the amide is reduced to a hydrocarbon [$C{=}O \rightarrow -CH_2-$]. In this case, the product is *N*-ethylaniline.

What Is the Product When N,N-Diethylpentanamide Is Reduced with Hydrogen and Palladium Catalyst?

Amides are very resistant to reduction with hydrogen. The correct answer here is, therefore, no reaction.

What Is the Product When Benzonitrile Is Reduced with $LiAlH_4$?

Benzonitrile is reduced to benzylamine ($PhCH_2NH_2$). In general, nitriles are reduced to primary amines upon reaction with $LiAlH_4$.

Give the Major Product When 25.137–25.139 Are Reduced with $LiAlH_4$.

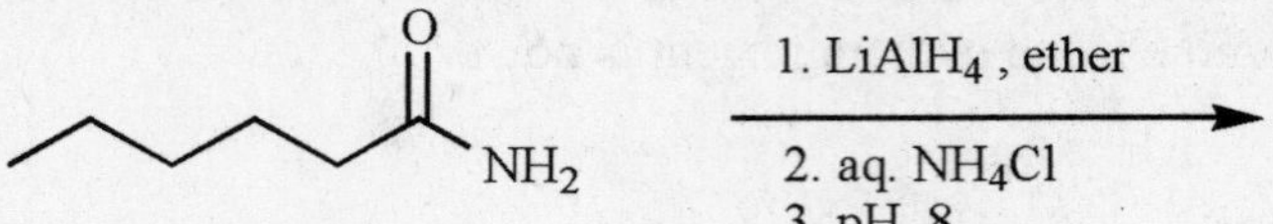

25.137 — 1. $LiAlH_4$, ether; 2. aq. NH_4Cl; 3. pH. 8

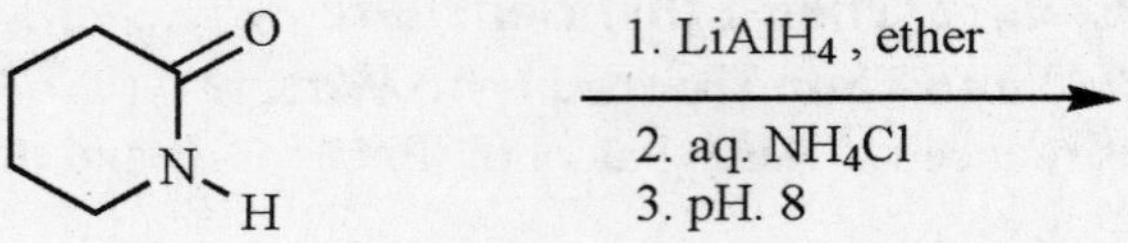

25.138 — 1. $LiAlH_4$, ether; 2. aq. NH_4Cl; 3. pH. 8

25.139 — 1. $LiAlH_4$, ether; 2. aq. NH_4Cl; 3. pH. 8

Reduction of hexanamide (***25.137***) leads to hexanamine (1-aminohexane, ***25.140***). Lactams are also reduced to amines with $LiAlH_4$. In this example, 2-piperidone (***25.138***) is reduced to piperidine (***25.141***). This result is typical for lactams, where the carbonyl is reduced to give the cyclic amine. In the last case, hexanenitrile (***25.139***) is reduced to 1-aminohexane (***25.137***). In all cases, after the acid workup of the hydride, the pH is adjusted to about 8 to facilitate isolation of the basic amine product.

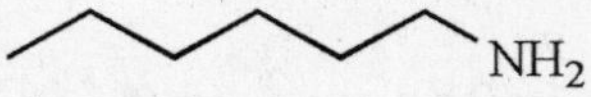

25.140 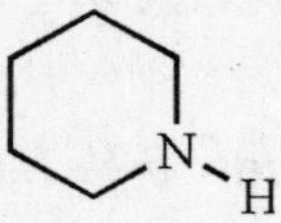 *25.141*

25.7. SULFONIC ACID DERIVATIVES

Sulfonic acids are essentially the sulfur analogs of carboxylic acids, and they form similar acid derivatives. Specifically, one can form sulfonyl halides, sulfonate esters, and sulfonamides.

What Is the Structure of a Sulfonic Acid?

The generic structure of a sulfonic acid is RSO_3H, where R - alkyl or aryl. See structure ***25.142***.

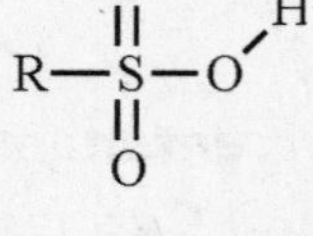

25.142

What Is a Sulfonyl Halide?

A sulfonyl halide is the acid halide of a sulfonic acid. The generic structure is RSO_2Cl (or Br). The name of CH_3SO_2Cl is methanesulfonyl chloride. The name of $PhSO_2Cl$ is benzenesulfonyl chloride. These two sulfonyl chlorides are derived from the corresponding sulfonic acids, methanesulfonic acid (CH_3SO_3H) and benzenesulfonic acid ($PhSO_3H$).

How Are Sulfonyl Halides Produced?

Sulfonyl halides are produced by the reaction of halogenating reagents such as thionyl chloride and a sulfonic acid. $RSO_3H + SOCl_2 \rightarrow RSO_2Cl$.

What Is the Product When Benzenesulfonyl Chloride Reacts with 1-Butanol in the Presence of Pyridine?

Just as an acid chloride derived from a carboxylic acid reacts with an alcohol to produce an ester, a sulfonyl chloride reacts with an alcohol to produce a sulfonic ester, RSO_3R.' In this case, the product is butyl benzenesulfonate, $PhSO_2OCH_2CH_2CH_2CH_3$.

What Is Tosyl Chloride? Mesyl Chloride?

Tosyl chloride is the sulfonyl chloride, *para*-toluenesulfonyl chloride (***25.143***), derived from *para*-toluenesulfonic acid. The abbreviation for this group is Ts and TsCl for tosyl chloride. Mesyl chloride is the sulfonyl chloride, methanesulfonyl chloride (***25.144***), derived from *methane*sulfonic acid. The abbreviation for this group is Ms and MsCl for mesyl chloride.

$CH_3C_6H_4SO_2Cl \equiv CH_3C_6H_4S(=O)_2Cl$

25.143

$CH_3SO_2Cl \equiv H_3C{-}S(=O)_2{-}Cl$

25.144

What Is a Tosylate? A Mesylate?

A tosylate is the sulfonyl ester derived from *para*-toluenesulfonic acid, such as the tosylate derivative of propanol, ***25.145***. [$CH_3CH_2CH_2OTs$]. A mesylate is the sulfonyl ester derived from methanesulfonic acid, such as the mesylate derivative of propanol, ***25.146*** [$CH_3CH_2CH_2OMs$].

$H_3C{-}C_6H_4{-}SO_2{-}O{-}CH_2CH_2CH_3$

25.145

$H_3C{-}SO_2{-}O{-}CH_2CH_2CH_3$

25.146

Give the Major Product for the Reactions of 25.147–25.149.

1-hexanol (OH) $\xrightarrow{PhSO_2Cl,\ Py}$

25.147

cyclohexanol (OH) $\xrightarrow{CH_3SO_2Cl,\ Py}$

25.148

2-butylcyclopentanol (OH) $\xrightarrow{TsCl,\ Py}$

25.149

In the first reaction, 1-hexanol (***25.147***) reacts with benzenesulfonyl chloride to give the benzenesulfonate ester, ***25.150***. Cyclohexanol (***25.148***) reacts with methanesulfonyl chloride, also called **mesyl chloride**, to give the methanesulfonate ester, ***25.151***, which is also called a **mesylate**. In the last case, alcohol ***25.149*** reacts with tosyl chloride to give the tosyl ester, ***25.152***. In each case, pyridine was added as a base to absorb the HCl by-product.

OSO_2Ph OSO_2CH_3 $OSO_2(p\text{-}MePh)$

25.150 *25.151* *25.152*

Test Yourself

1. Give the complete mechanism for the following reaction.

$$\text{CH}_3(\text{CH}_2)_4\text{CO}_2\text{Me} \xrightarrow{H_3O+} \text{CH}_3(\text{CH}_2)_4\text{CO}_2\text{H} + \text{MeOH}$$

2. For each of the following reactions, give the major product. Remember stereochemistry where appropriate, and if there is no reaction, indicate this by N.R.

(a)  $\text{C}_6\text{H}_5\text{–CO}_2\text{H}$ → 2-propanol / cat. H^+

(b) 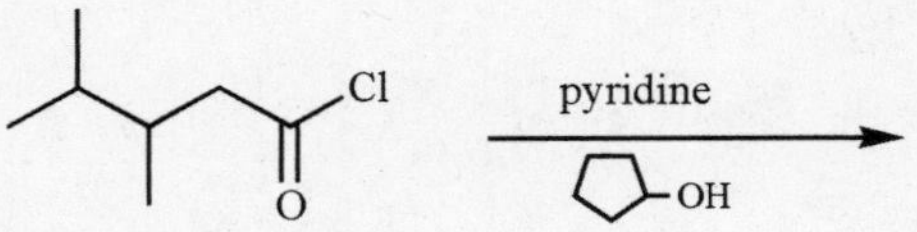 acid chloride (Cl, O) → pyridine / cyclopentyl–OH

(c) $\text{–SO}_3\text{H}$ → 1. $SOCl_2$; 2. cyclopentyl–OH, pyridine

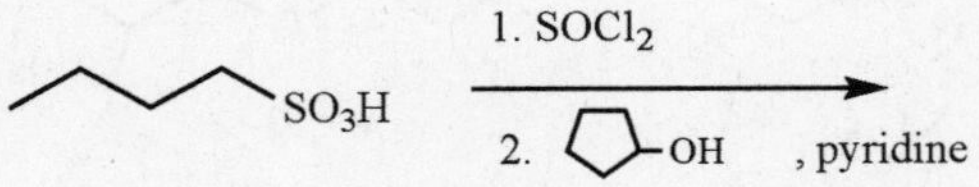

(d) 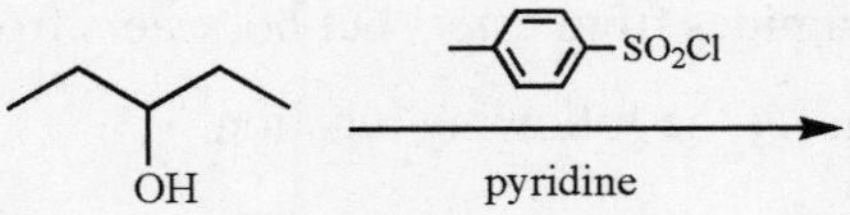 alcohol (OH) → $\text{CH}_3\text{–C}_6\text{H}_4\text{–SO}_2\text{Cl}$ / pyridine

3. When butanoyl chloride is treated with one equivalent of butylmagnesium bromide, why is it difficult to obtain 4-octanone as the final major product?

4. In each case, give the major product. Remember stereochemistry where appropriate, and if there is no reaction, indicate this by N.R.

(a) 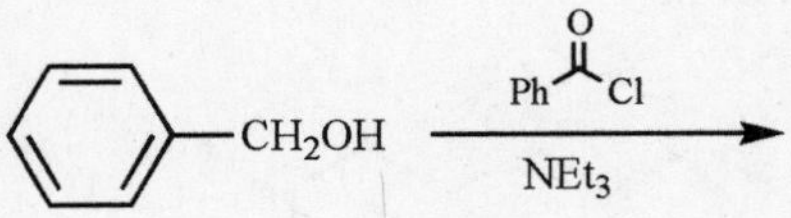 $\text{C}_6\text{H}_5\text{–CH}_2\text{OH}$ → PhC(O)Cl / NEt_3

(b) $\text{–CO}_2\text{H}$ → 1. $SOCl_2$; 2. $AlCl_3$, 1,4-diethylbenzene

(c) $\text{NO}_2\text{–C}_6\text{H}_4\text{–NO}_2$ → $AlCl_3$ / acid chloride (O, Cl)

(d) 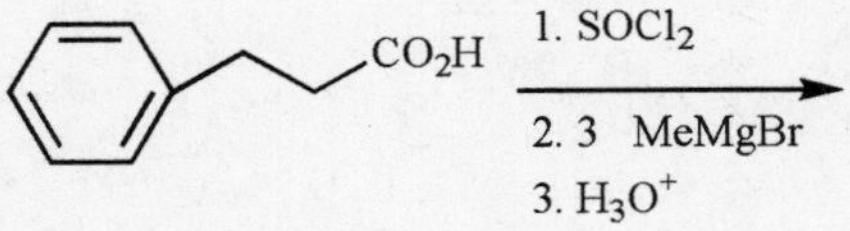 $\text{C}_6\text{H}_5\text{–CH}_2\text{CH}_2\text{–CO}_2\text{H}$ → 1. $SOCl_2$; 2. 3 MeMgBr; 3. H_3O^+

(e) 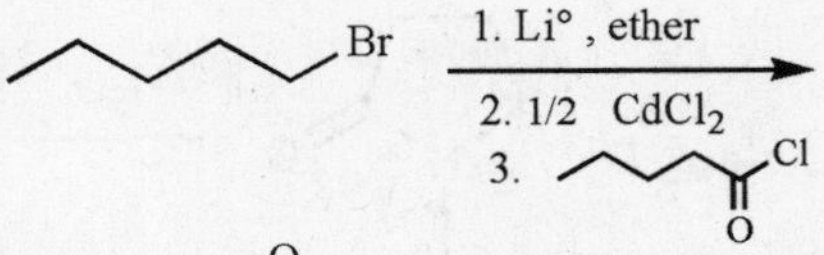 alkyl bromide (Br) → 1. Li°, ether; 2. 1/2 $CdCl_2$; 3. acid chloride (O, Cl)

(f) benzoyl chloride (O, Cl) → Bu_2CuLi, ether / -10°C

(g) 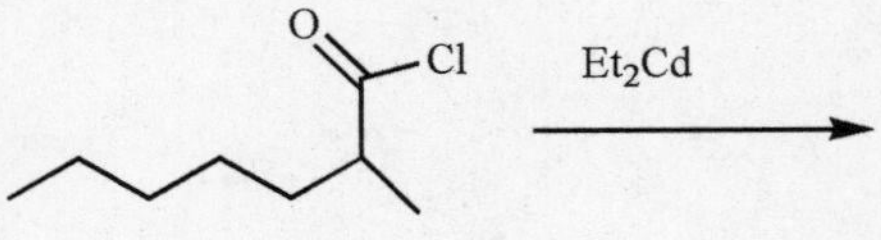 acid chloride (O, Cl) → Et_2Cd

(h) alkyl iodide (I) → 1. KCN ,THF; 2. $LiAlH_4$, ether; 3. H_2O

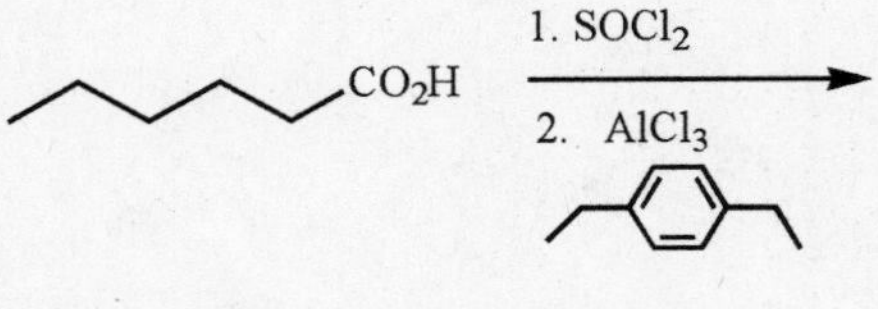

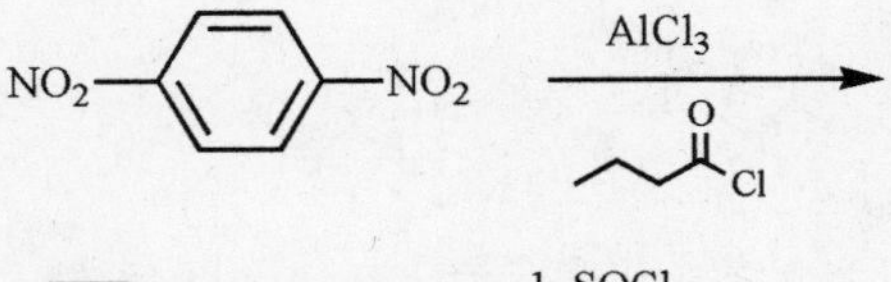

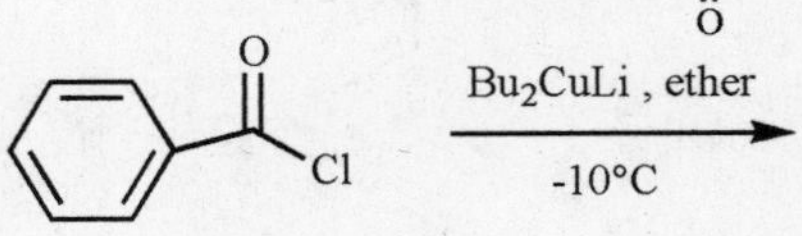

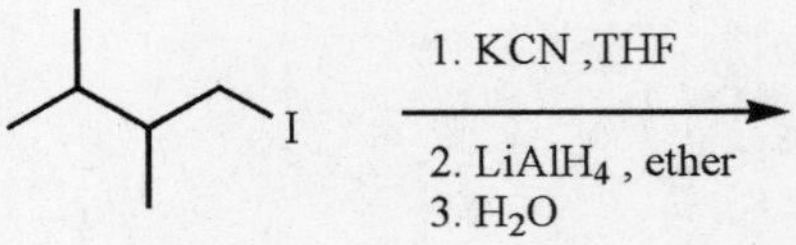

5. Give the IUPAC name for the following molecules.

(a) Cl O (b) O O Ph Ph (c) O N O

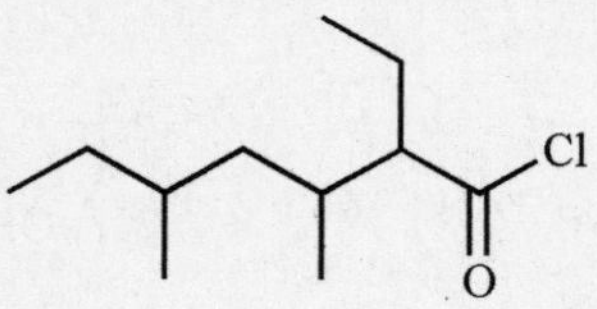

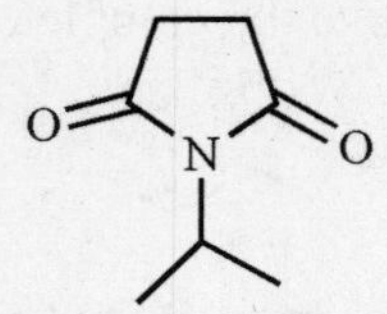

(d) O NH$_2$ (e) Cl O O Ph Ph (f) O O O (g) O Me Me O O Me Me

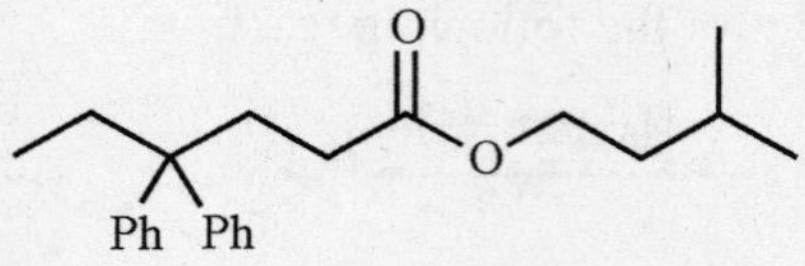

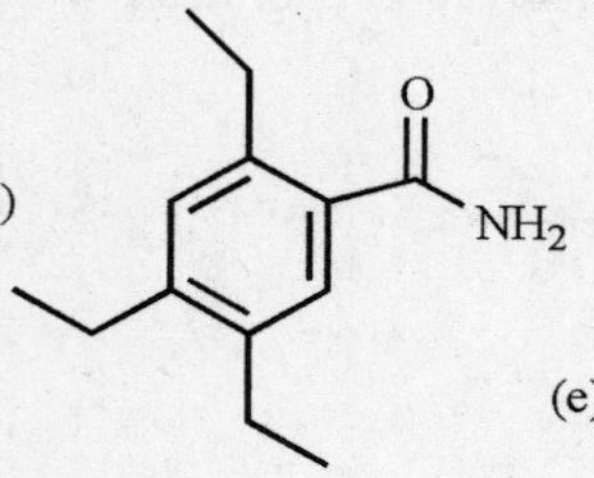

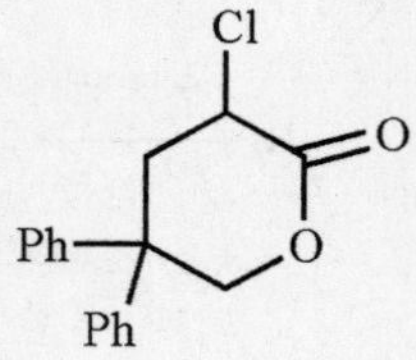

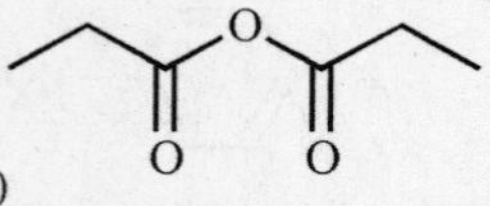

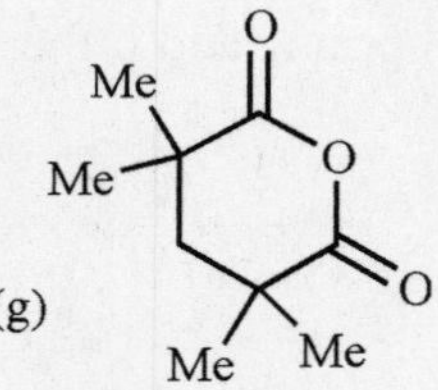

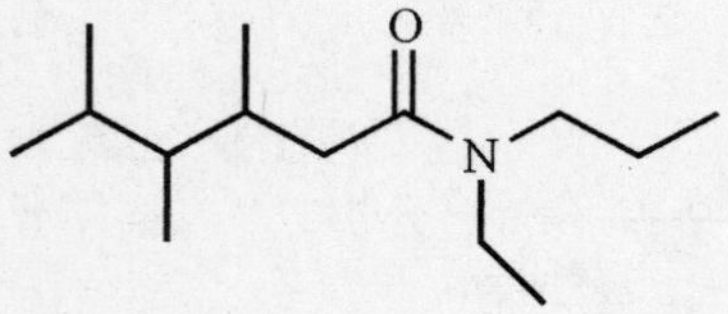

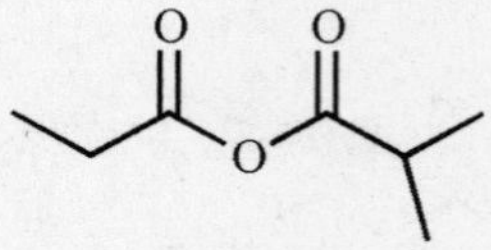

(h) O N (i) O O O (j) Me N O C_4H_9

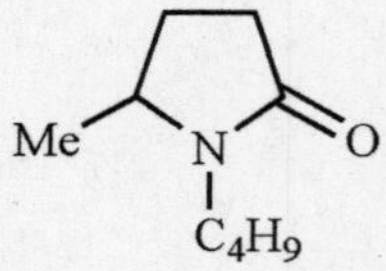

6. Why is it possible to prepare amides from esters but not esters from amides?

7. Give the complete mechanism for the following reaction.

H_3O+

O O O → CO_2H CO_2H

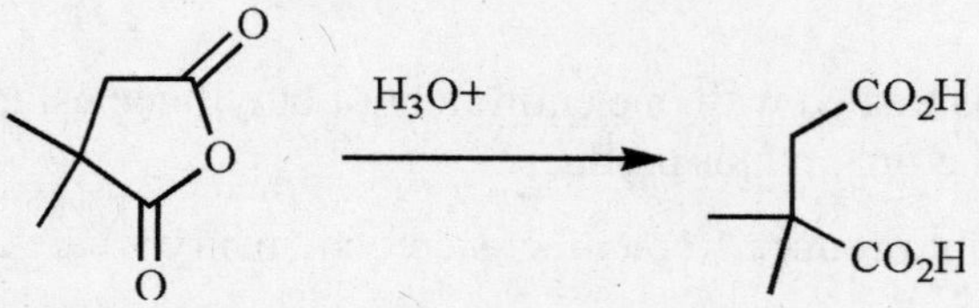

8. Give the complete mechanism for the following reaction.

H_3O^+, heat

N O → N H OH O

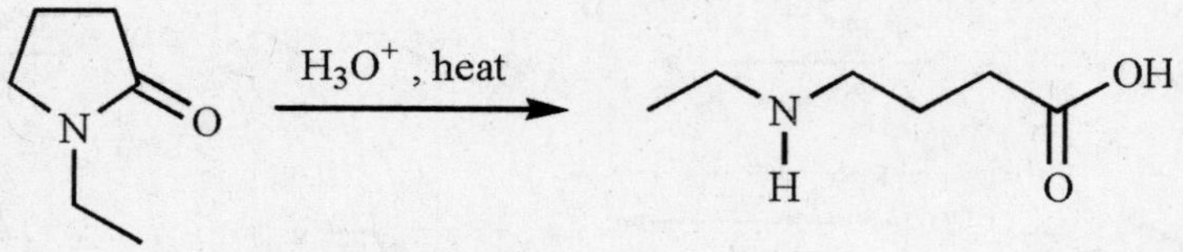

9. In each case, give the major product. If there is no reaction, indicate this by N.R.

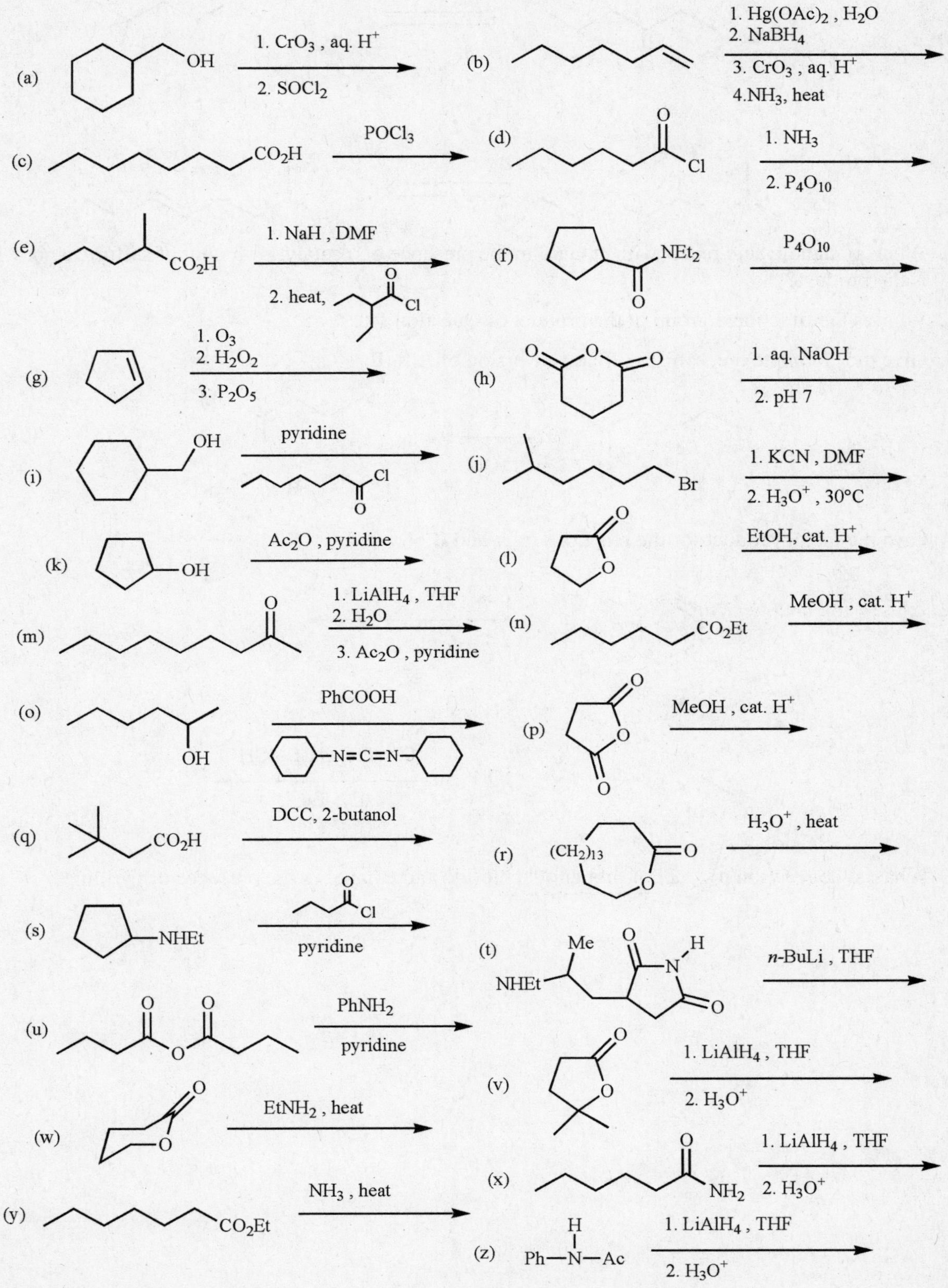

(a) [cyclohexylmethanol] OH —— 1. CrO_3, aq. H^+; 2. $SOCl_2$ ——→

(b) [1-heptene] —— 1. $Hg(OAc)_2$, H_2O; 2. $NaBH_4$; 3. CrO_3, aq. H^+; 4. NH_3, heat ——→

(c) [nonanoic acid] CO_2H —— $POCl_3$ ——→

(d) [hexanoyl chloride] O, Cl —— 1. NH_3; 2. P_4O_{10} ——→

(e) [2-methylpentanoic acid] CO_2H —— 1. NaH, DMF; 2. heat, [2-ethylbutanoyl chloride] O, Cl ——→

(f) [N,N-diethylcyclopentanecarboxamide] NEt_2, O —— P_4O_{10} ——→

(g) [cyclopentene] —— 1. O_3; 2. H_2O_2; 3. P_2O_5 ——→

(h) [glutaric anhydride] O, O, O —— 1. aq. NaOH; 2. pH 7 ——→

(i) [cyclohexylmethanol] OH —— pyridine, [octanoyl chloride] Cl, O ——→

(j) [1-bromo-5-methylheptane] Br —— 1. KCN, DMF; 2. H_3O^+, 30°C ——→

(k) [cyclopentanol] OH —— Ac_2O, pyridine ——→

(l) [γ-butyrolactone] O, O —— EtOH, cat. H^+ ——→

(m) [2-decanone] O —— 1. $LiAlH_4$, THF; 2. H_2O; 3. Ac_2O, pyridine ——→

(n) [ethyl hexanoate] CO_2Et —— MeOH, cat. H^+ ——→

(o) [2-hexanol] OH —— PhCOOH, [cyclohexyl]–N=C=N–[cyclohexyl] ——→

(p) [succinic anhydride] O, O, O —— MeOH, cat. H^+ ——→

(q) [3,3-dimethylbutanoic acid] CO_2H —— DCC, 2-butanol ——→

(r) [macrocyclic lactone] $(CH_2)_{13}$, O, O —— H_3O^+, heat ——→

(s) [N-ethylcyclopentylamine] NHEt —— [butanoyl chloride] O, Cl; pyridine ——→

(t) NHEt, Me, O, H, N, O [succinimide derivative] —— n-BuLi, THF ——→

(u) [butanoic anhydride] O, O, O —— $PhNH_2$; pyridine ——→

(v) [5,5-dimethyl-γ-butyrolactone] O, O —— 1. $LiAlH_4$, THF; 2. H_3O^+ ——→

(w) [δ-valerolactone] O, O —— $EtNH_2$, heat ——→

(x) [hexanamide] O, NH_2 —— 1. $LiAlH_4$, THF; 2. H_3O^+ ——→

(y) [ethyl heptanoate] CO_2Et —— NH_3, heat ——→

(z) Ph—N(H)—Ac —— 1. $LiAlH_4$, THF; 2. H_3O^+ ——→

10. In each case, provide a suitable synthesis. Show all reagents and intermediate products.

(a) O ⟹ Br

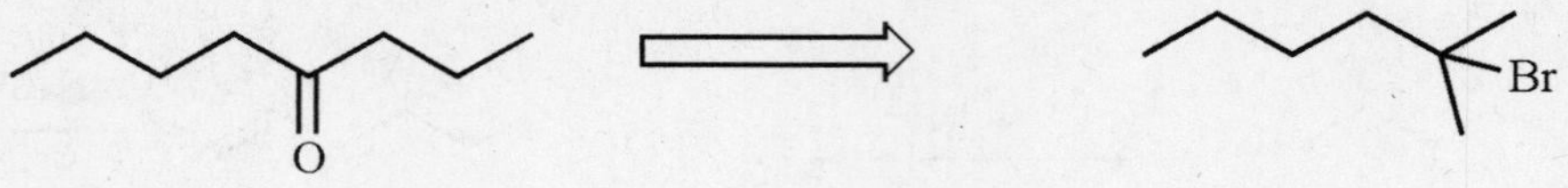

(b) Ph O ⟹

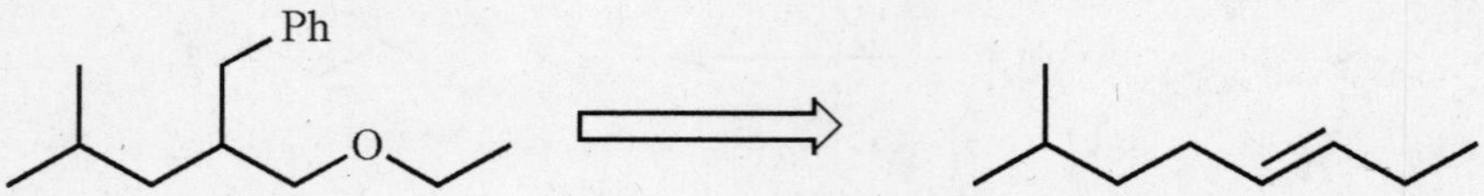

11. When pentanoic acid reacts with ethanol in the presence of a catalytic amount of sulfuric acid, what is the product?

12. What is the functional group in the product of Question 11?

13. Give the complete mechanism of the conversion of **A** to **B**.

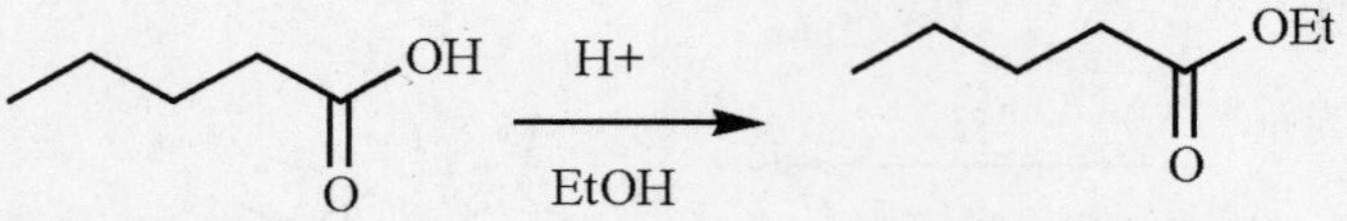

OH O **A** —H+, EtOH→ OEt O **B**

14. Give the major product for the reactions of **A** and **B**.

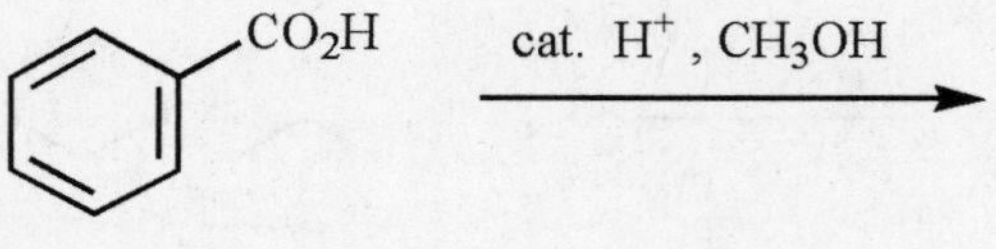

CO_2H **A** —cat. H^+, CH_3OH→

OH **B** —$CH_3CH_2CH_2COOH$, cat. *p*-TsOH→

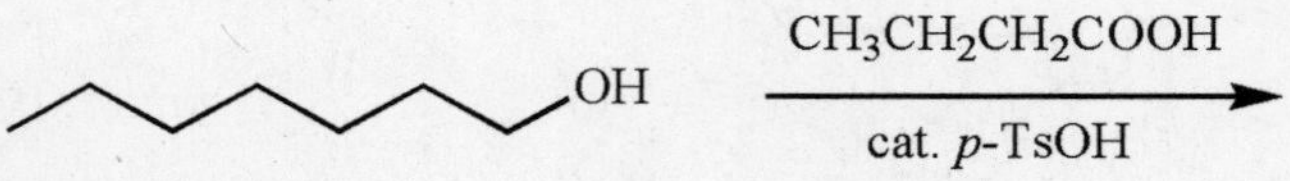

15. What is the reaction product of propanoyl chloride and ethanol in the presence of pyridine?

CHAPTER 26

Enolate Anions of Acid Derivatives and the Claisen Condensation

Several new carbon-carbon bond forming reactions will be presented that are clearly related to those first seen with aldehydes and ketones in Chapter 13. The Claisen Condensation, the Dieckmann condensation, the Knoevenagel Condensation and the Reformatsky reaction are notable carbon-carbon bond forming reactions that involve acid derivatives.

26.1. ENOLATE ANIONS OF ACID DERIVATIVES

The carbonyl of an acid derivative makes the α-hydrogen acidic, just as in ketones and aldehydes. This hydrogen can be removed by a suitable base to form an enolate anion, which can react with alkyl halides in an alkylation reaction or with carbonyl derivatives in a condensation reaction.

Is the α-Hydrogen in an Ester Acidic?

Yes! The α-proton of a typical ester has a pK_a of about 22–24.

What Bases Can Be used to Remove the α-Proton of an Ester?

A somewhat stronger base than is used for ketones must be employed. However, amide bases such as LDA are quite good. In addition, alkoxide bases (RO^-) are commonly employed.

Draw the Structure of the Enolate Anion of Ethyl Acetate.

The enolate anion derived from ethyl acetate is the resonance stabilized ion ***26.1***.

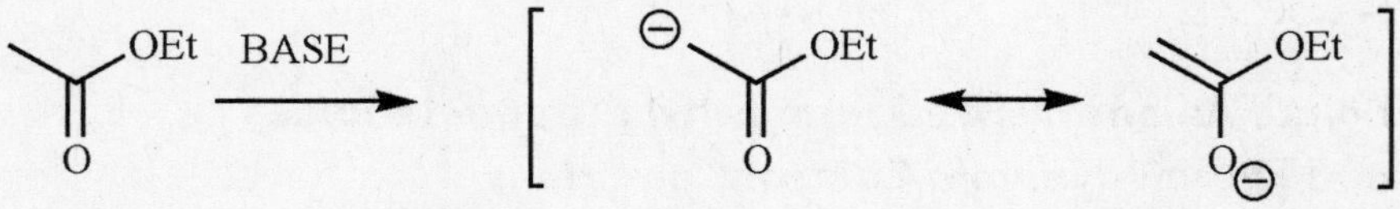

OEt BASE ⊖ OEt OEt O O O⊖

26.1

What Is the Product When Butanoic Acid Is Reacted with Sodium Ethoxide?

Since sodium ethoxide is a strong base, the acidic hydrogen (O–H) of butanoic acid is removed to form the sodium salt of the acid, sodium butanoate (***26.2***) and ethanol, the conjugate acid of sodium ethoxide.

NaOEt → + EtOH

26.2

What Is the Most Favorable Reaction When Ethyl Butanoate Is Reacted with Sodium Methoxide in Methanol?

The most electropositive atom is the carbonyl carbon. It is, therefore, reasonable to assume that sodium methoxide will attack the carbonyl via a reversible nucleophilic acyl addition reaction to give the usual tetrahedral intermediate, ***26.3***. Ethoxide can be lost from ***26.3*** to form the methyl ester. An excess of methoxide will lead to the methyl ester. Similarly, if the methyl ester were treated with sodium ethoxide in ethanol, the product would be the ethyl ester via the same intermediate, ***26.3***.

+ MeO^- / - MeO^- ; - EtO^- / + EtO^-

26.3

What Is the Product When Ethyl Butanoate Is Reacted with Lithium Diethylamide in THF, at −78°C?

Lithium diethylamide is a powerful but non-nucleophilic base. The usual nucleophilic attack at the carbonyl is, therefore, slow. The hydrogen on the carbon adjacent to the carbonyl (the α-carbon) is acidic and can be removed by this strong base. The product is the enolate anion, ***26.4***.

$LiNEt_2$, ether, -78°C

26.4

What Is the Relative pK_a of the α-Hydrogen of Ethyl Butanoate When Compared to the Two Acidic Protons of 2-Butanone?

The α-hydrogens of 2-butanone have a pK_a of 20 and 21 (for C_1 and C_3, respectively). The α-hydrogen of ethyl butanoate (the carbon next to the carbonyl, **not** next to the oxygen of the ester) has a pK_a of about 24–25.

Discuss the Enolate Anions Derived from Ethyl Butanoate under Both Kinetic and Thermodynamic Control Conditions.

In both cases, the enolate anion that is formed is ***26.4***. Under kinetic control (aprotic solvents, low temperature, strong non-nucleophilic base), ***26.4*** is formed in an effectively irreversible manner. Under thermodynamic control (protic solvent, alkoxide bases, higher reaction temperatures), the reaction is reversible, and ***26.4*** is in equilibrium with the starting ester. If methoxide is used as the base, the addition product (*26.3*) can be present. Similarly, if ethoxide is used, the OEt derivative can be present.

If Thermodynamic Conditions Are Used to Deprotonate an Ester, What Are the Limitations on the Base That Can Be Used?

Examination of *26.3* clearly shows that a transesterification reaction can occur if the alkoxide base is different from the alcohol portion of the ester. For this reason, NaOEt is used with ethyl ester, NaOMe with methyl esters, etc.

26.2. ENOLATE ANION ALKYLATION

Just as with the enolate anions derived from aldehydes and ketones, ester enolate anions are nucleophilic and react with alkyl halides to give the substitution product.

What Is the Product When Methyl Propionate Is Reacted First with Sodium Methoxide in Methanol and Then with Iodomethane?

The initial reaction gives the enolate anion (*26.5*), which reacts with iodomethane to give methyl 2-methylpropionate, *26.6*.

NaOMe , MeOH; reflux; MeI

26.5 *26.6*

What Is the Product When the Isopropyl Ester of Phenylacetic Acid (2-Phenyl Ethanoic Acid, 26.7) Is Reacted First with LDA (−78°C, THF) and Then with Benzyl Bromide?

The enolate anion is formed and alkylation with benzyl bromide gives *26.8*.

1. $LiN(iPr)_2$, THF, -78°C; 2. $PhCH_2Br$

26.7 *26.8*

Discuss the Acidity of the α-Hydrogen of a Dibasic Acid Such As Diethyl Malonate.

The hydrogens on the carbon between the two carbonyl groups have a pK_a of 12.9. The inductive effects of two carbonyl groups greatly enhance the acidity of those hydrogens.

Discuss the Relative Stability of the Enolate Anion Derived from Diethyl Malonate.

The presence of the second carbonyl in the enolate anion of diethyl malonate (*26.9*) leads to an additional resonance structure and greater stability when compared to enolate anion derived from a monocarboxylic acid.

NaOMe , MeOH; reflux

26.9

Is 26.9 More Stable or Less Stable Than 26.5?

Since *26.9* has greater resonance stability, it is more stable.

Is 26.9 Easier to Form or More Difficult, Relative to 26.5? How Does This Correlate with the pK_a of the α-Proton of Malonic Ester vs. Ethyl Propionate?

Since it is more stable, it is easier to form. This observation is consistent with the use of a weaker base to deprotonate malonic esters. Since the α-proton of malonic ester is more acidic, a weaker base is required for its deprotonation. The greater resonance stability of ***26.9*** shifts the acid-base equilibrium towards the enolate anion, contributing to the greater acidity of the α-proton of malonic ester.

Are Ester Enolate Anions Derived from Dibasic Acids More or Less Reactive Than Ester Enolate Anions Derived from Monobasic Acids? Explain.

They are more stable, due to the resonance stability of the anion (see ***26.9***, the enolate anion of diethyl malonate), which makes them *less reactive.*

What Is the Product When Diethyl Malonate Is Reacted with: I. NaOEt, EtOH II. $PhCH_2Br$? If This Product Is Then Reacted with I. NaOEt II. CH_3I, What Is the Final Product?

The first product is the 2-alkylated product, diethyl 2-benzylmalonate (***26.10***). If ***26.10*** is treated with additional base, a new enolate anion is formed (***26.11***), which then reacts with iodomethane to form diethyl 2-benzyl-2-methyl malonate, ***26.12***.

CO_2Et / CO_2Et —(1. NaOEt; 2. $PhCH_2Br$)→ Ph, CO_2Et, CO_2Et (*26.10*) —NaOEt→ Ph, ^-O, OEt, CO_2Et (*26.11*) —MeI→ Ph, CH_3, CO_2Et, CO_2Et (*26.12*)

What Is the Malonic Ester Synthesis?

The sequence that converted diethyl malonate to ***26.12*** is usually followed by hydrolysis of the diester to the dicarboxylic acid. To be formally correct, conversion of the diester to the dicarboxylic acid, followed by thermal decarboxylation to give the mono-carboxylic acid completes the **malonic ester synthesis** as this entire sequence is known as the **malonic ester synthesis**.

Show How Diethyl Malonate Can Be Converted into 2-Ethylpentanoic Acid.

The first step is to convert diethyl malonate (or another ester) into the 2-propyl derivative (***26.13***) by treatment with NaOEt followed by iodopropane. Repetition of this base-halide sequence with iodopropane leads to ***26.15***, via ***26.14***. When this ester is saponified (1. aq. NaOH 2. pH 5), the corresponding malonic acid derivative (***26.16***) is produced. The IUPAC name of ***26.16*** is 2-ethyl-2-propyl-1, 3-propanedioic acid. Heating to >200°C induces decarboxylation to give the final product, 2-ethylpentanoic acid, ***26.17***.

CO_2Et / CO_2Et —(1. NaOEt; 2. $CH_3CH_2CH_2Br$)→ CO_2Et, CO_2Et (*26.13*) —NaOEt→ ^-O, OEt, CO_2Et (*26.14*) —EtI→ CO_2Et, CO_2Et (*26.15*)

—(1. aq NaOH; 2. pH 5)→ CO_2H, CO_2H (*26.16*) —250°C→ CO_2H (*26.17*)

What Is the Name of This Overall Transformation?

As noted above, the reaction sequence that converts malonic acid into a mono carboxylic acid via alkylation-decarboxylation is called the **malonic ester synthesis**. This sequence is very powerful for the synthesis of highly substituted carboxylic acids.

Provide a Suitable Synthesis for 26.18 and 26.19 from Malonic Acid.

CO_2H ⟹ CO_2H / CO_2H

26.18

O, H ⟹ CO_2H / CO_2H

Ph

26.19

In the first step, malonic acid is converted to the diester and then alkylated with bromobutane to give ***26.20***. A second alkylation inserts the pentyl group via enolate anion alkylation. Saponification leads to the malonic acid derivative (***26.21***), and thermal decarboxylation gives the final target, ***26.18***. In the second synthesis, malonic acid is again esterified and then alkylated with benzyl bromide to give ***26.22***. Once again, a second alkylation sequence is required, giving ***26.23***. Saponification followed by decarboxylation gives the mono acid, ***26.24***. In this case, the carboxylic acid must be transformed into an aldehyde. One way to do this is to first reduce the acid to an alcohol (***26.25***) with $LiAlH_4$. Subsequent treatment with PCC (pyridinium chlorochromate; see Section 21.2) gives the final aldehyde target, ***26.19***.

CO_2H, CO_2H —(1. H_3O^+, MeOH; 2. NaOMe; 3. *n*-BuBr)→ Bu–CH(CO_2Me)$_2$ (*26.20*) —(1. NaOMe; 2. $C_5H_{11}Br$)→ Bu(C_5H_{11})C(CO_2Me)$_2$ (*26.21*) —(1. H_3O^+; 2. heat)→ Bu(C_5H_{11})CHCO_2H (*26.18*)

CO_2H, CO_2H —(1. H_3O^+, MeOH; 2. NaOMe; 3. $PhCH_2Br$)→ $PhCH_2$–CH(CO_2Me)$_2$ (*26.22*) —(1. NaOMe; 2. Br)→ $PhCH_2$, CO_2Me, CO_2Me (*26.23*) —(1. H_3O^+; 2. heat)→ $PhCH_2$, CO_2H (*26.24*)

—(1. $LiAlH_4$; 2. H_2O)→ $PhCH_2$, CH_2OH (*26.25*) —(PCC)→ $PhCH_2$, CHO (*26.19*)

If the Diethyl Ester of Succinic Acid Is Treated with i. NaOEt ii. Allyl Bromide, What Is the Major Product?

Succinic esters behave as any other ester, since the carbonyls are not conjugated. This sequence uses diethyl succinate ($EtO_2CCH_2CH_2CO_2$ Et), and enolate alkylation will produce the 2-allyl derivative, ***26.26***.

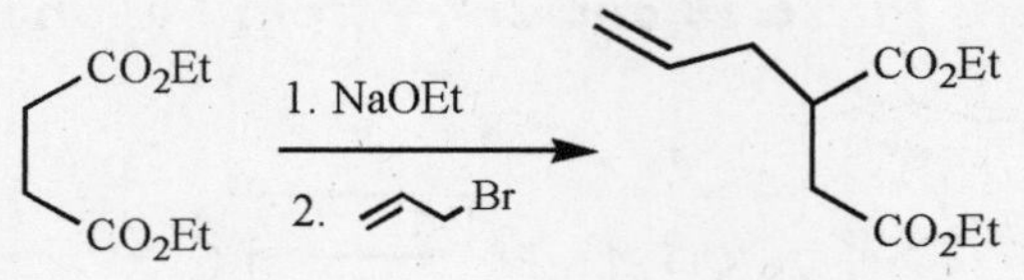

CO_2Et 1. NaOEt 2. Br CO_2Et → CO_2Et CO_2Et

26.26

What Is the pK_a of the α-Hydrogen of Diethyl Succinate?

It is very close to that of propionic acid esters. Ethyl propionate has a pK_a of 22–23 and diethyl succinate also has a pK_a of about 22.

If 2-Butyl-1,4-Butanedioic Acid Is first Esterified (Ethyl Ester) and Then Reacted with i. LDA, THF, −78°C ii. CH_3I, What Is the Major Product? Explain.

The product is the ethyl ester of diethyl 2-butyl-3-methyl-1,4-butanedioate (***26.27***). These are kinetic control conditions, and the most acidic hydrogen will be removed (attached to the less substituted carbon). Alkylation will then lead to the 2,3-dimethyl derivative rather than the 2,2-dimethyl derivative.

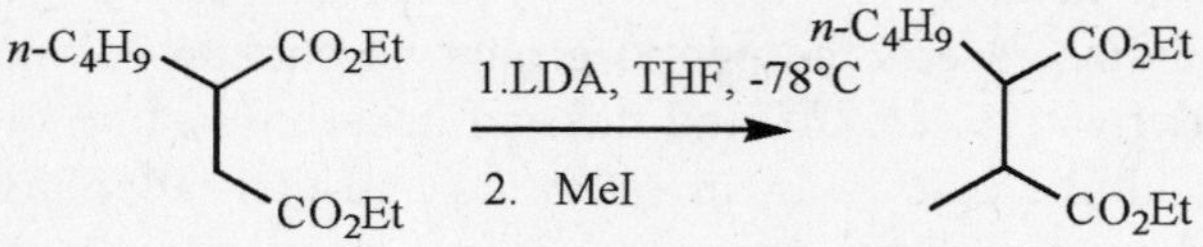

n-C_4H_9 CO_2Et CO_2Et 1.LDA, THF, -78°C 2. MeI → *n*-C_4H_9 CO_2Et CO_2Et

26.27

What Is Acetoacetic Acid? What Is the IUPAC Name?

Acetoacetic acid has the structure ***26.28***. The IUPAC name is 3-keto-butanoic acid.

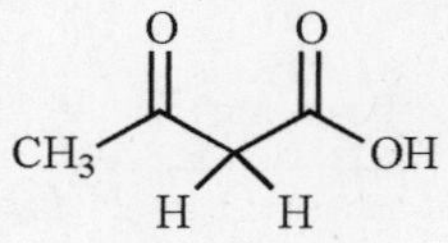

O O CH_3 OH H H

26.28

What Is the pK_a of the most Acidic Hydrogen in the Ethyl Ester of This Molecule? Identify That Hydrogen.

The most acidic hydrogens are those between the two carbonyl groups, as in malonic acid esters. The pK_a of those hydrogen atoms in ***26.28*** is about 11.

What Is the Product if Ethyl Acetoacetate Is Treated with i. LDA, THF, −78°C ii. Bromoethane iii. AQ. KOH iv. Adjust to pH 6 v. Heat to 200°C?

The final product is a ketone, 2-pentanone. The initial reaction generates the enolate anion (***26.29***), and alkylation with bromoethane gives ***26.30***. Saponification of the ester leads to the acid (***26.31***). This acid can be decarboxylated (it is a 1,3-dicarbonyl compound) to give the ketone as the final product (2-pentanone).

LDA , THF / -78°C → 26.29 → EtBr →

26.30 → 1. OH^- 2. H^+ → 26.31 → 250°C →

What Is the Name of This Overall Process?

This synthetic sequence is known as the **acetoacetic ester synthesis.**

26.3. THE CLAISEN CONDENSATION

Just as the enolate anions of aldehydes and ketones react with other aldehydes and ketones in the Aldol condensation, ester enolate anions react with a variety of carbonyl derivatives.

What Is the Product When Ethyl Propanoate Is i. Refluxed with Sodium Ethoxide in Ethanol ii. Treated with Dilute Aqueous Acid?

The product is a β-keto ester, ***26.32***.

1. NaOEt , EtOH , reflux 2. H_3O^+ → 26.32

Discuss How and Why This Reaction Works?

Initial reaction with NaOEt gives the ester enolate anion (***26.33***). The carbanionic carbon attacks the carbonyl of the second molecule of ethyl propionate to give the addition product, ***26.34***. Since OEt is a good leaving group, the alkoxide moiety displaces OEt to generate the ketone group in the final keto-ester product, ***26.35***.

NaOEt → 26.33 → 26.34 → 26.35

What Is the Name of This Reaction if Ethyl Propionate Is Refluxed with Sodium Ethoxide in Ethanol?

This is called the **Claisen condensation**.

What Is the Product When Ethyl Propanoate Is Reacted with i. LDA, THF, −78°C i. Ethyl propanoate?

The product is a β-keto ester, ***26.32***.

1. $LiN(iPr)_2$, THF , -78°C
2. CO_2Et

CO_2Et

O

26.32

Discuss the Possible Products if Ethyl Propionate and Ethyl Butanoate are Refluxed in Ethanol Containing Sodium Ethoxide.

Under these conditions, *both* esters give an enolate anion in the equilibrium conditions. Ethyl propionate gives ***26.33***, and ethyl butanoate will give ***26.37***. Since these enolate anions are in equilibrium with the free ester, ***26.33*** can react with ethyl propionate to give ***26.35***, but it can also react with ethyl butanoate to give ***26.36***. Similarly, ***26.37*** can react with ethyl propionate to give ***26.38***, or with ethyl butanoate to give ***26.39***. The equilibrium conditions with two different esters, therefore, lead to four possible ester products.

CO_2Et NaOEt O⁻ OEt 26.33 CO_2Et CO_2Et O CO_2Et 26.35 O CO_2Et 26.36

CO_2Et NaOEt O⁻ OEt 26.37 CO_2Et CO_2Et O CO_2Et 26.38 O CO_2Et 26.39

What Is a 'Mixed' Claisen Condensation?

The reaction of two different esters to produce the Claisen product is a mixed Claisen. The enolate anion of one ester condenses with the carbonyl of the other ester.

Which Are Better Conditions for a Mixed Claisen, Kinetic Control or Thermodynamic Control Conditions? Explain.

Although both conditions will generate the same enolate, kinetic control conditions are better for mixed Claisen condensations. Reaction of ethyl propionate with LDA, for example, will give ***26.33*** as the only enolate anion, and either ethyl propionate or ethyl butanoate can be added to give the appropriate mixed Claisen product (***26.36*** or ***26.38***, respectively). Under these conditions, both esters are not present at the time the enolate anion is formed, and the chemist has control of which ester is added as the carbonyl partner, and in what order.

What Is the Product or Products When Ethyl Butanoate Is Treated with Sodium Methoxide in Methanol?

Under these equilibrium conditions, enolate anion ***26.37*** is formed, but since ethyl butanoate is the only ester present, ***26.37*** is the only enolate anion, and it can only react with itself to give ***26.39***.

Give the Major Product of the Following Sequence with Ethyl Butanoate: i. LDA, THF, −78°C; ii. Methyl Propionate; iii. Saponification; iv. Heating to 200°C.

Under these kinetic control conditions, the initial Claisen product is ***26.38***. Saponification liberates the free carboxylic acid and heating leads to decarboxylation (this is a 1,3-dicarbonyl compound) to give the ketone, 3-hexanone.

Give the Major Product of the Reactions of 26.40 and 26.41.

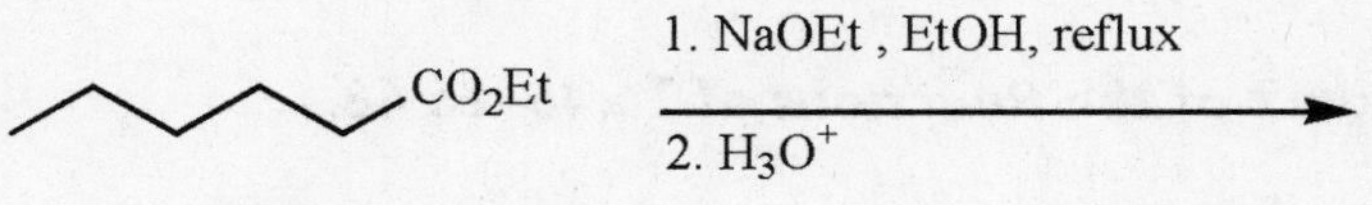

CO_2Et — 1. NaOEt , EtOH, reflux; 2. H_3O^+ →

26.40

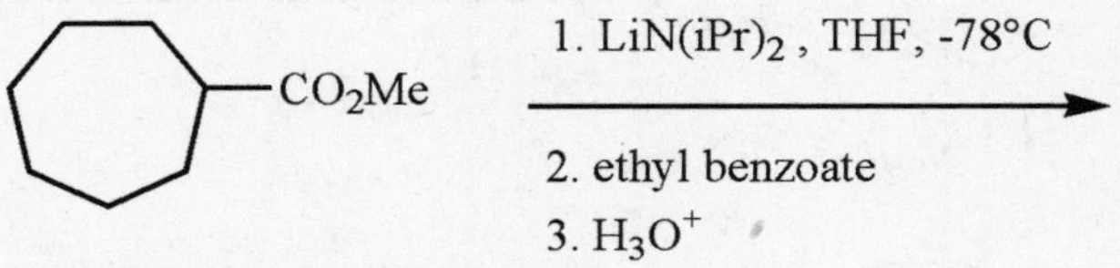

CO_2Me — 1. $LiN(iPr)_2$, THF, -78°C; 2. ethyl benzoate; 3. H_3O^+ →

26.41

In the first reaction, ethyl hexanoate (***26.40***) is condensed with itself under thermodynamic conditions to give the 'symmetrical' Claisen product, ***26.42***. The second example generates the enolate anion from ethyl cycloheptane carboxylate (***26.41***) and then reacts it with ethyl benzoate [$PhCO_2Et$] to give the 'mixed' Claisen product, ***26.43***.

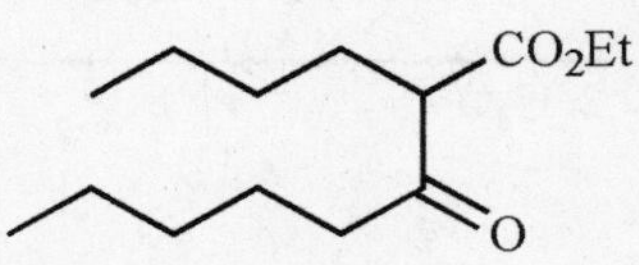

CO_2Et, O

26.42

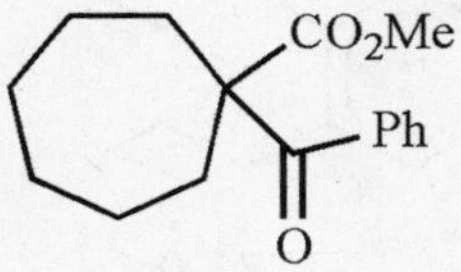

CO_2Me, Ph, O

26.43

26.4. THE DIECKMANN CONDENSATION

The intramolecular version of the Claisen condensation is known as the Dieckmann condensation.

What Is the Major Product When the Diethyl Ester of 1,6-Hexanedioic Acid (26.44) Is Reacted with Sodium Ethoxide in Ethanol?

This is an intramolecular Claisen condensation, and the product is the cyclic keto-ester, ***26.45***.

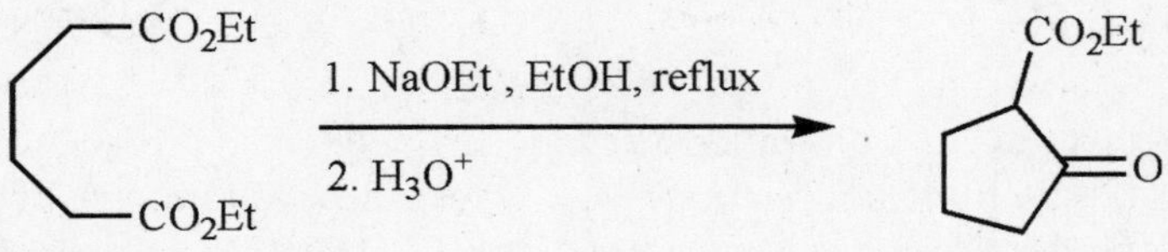

CO_2Et, CO_2Et — 1. NaOEt , EtOH, reflux; 2. H_3O^+ → CO_2Et, O

26.44 *26.45*

What Is the Name of This Reaction?

It is called the **Dieckmann Condensation.**

What Ring Sizes Are Formed by This Cyclization Reaction?

Rings of three-seven members can be formed by this technique. Formation of cyclic ketones of 8–13 members is very difficult by this method, although larger rings can be prepared by high dilution techniques.

Give the Major Product of the Following Sequence with the Diethyl Ester of 1,7-Heptanedioic Acid: i. LDA, THF, −78°C ii. Saponification iii. Heating to 200°C.

The Claisen product is the keto ester, but saponification and decarboxylation lead to the final product, cyclohexanone.

Give the Major Product of the Reactions of 26.46–26.48.

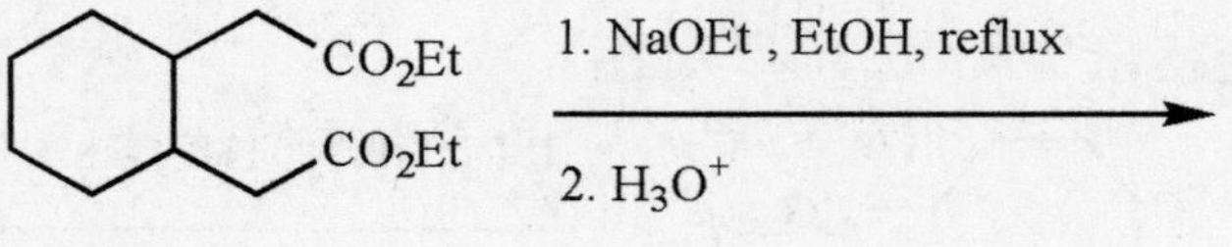

CO_2Et
CO_2Et
1. NaOEt , EtOH, reflux
2. H_3O^+

26.46

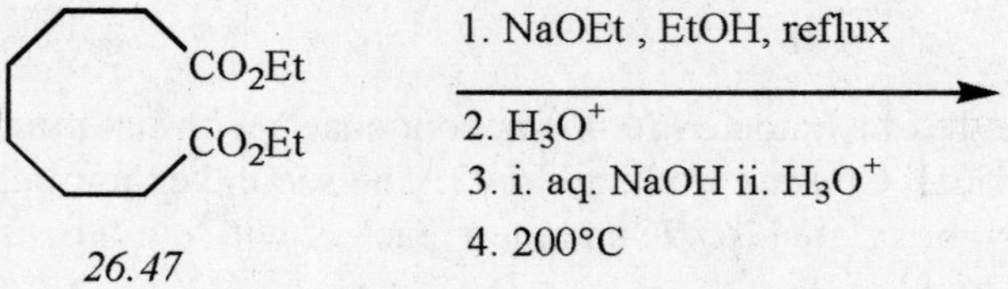

CO_2Et
CO_2Et
1. NaOEt , EtOH, reflux
2. H_3O^+
3. i. aq. NaOH ii. H_3O^+
4. 200°C

26.47

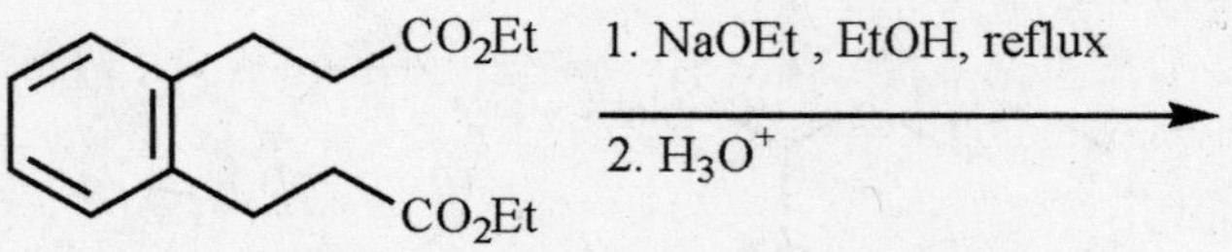

CO_2Et
CO_2Et
1. NaOEt , EtOH, reflux
2. H_3O^+

26.48

Dieckmann condensation with diester ***26.46*** leads to the bicyclic keto-ester, ***26.49***. In the second case, diethyl 1,8-octanedioate (***26.47***) is cyclized under Dieckmann conditions, but saponification and decarboxylation give cycloheptanoate (***26.50***) as the final product. In the last example, diester ***26.48*** is cyclized to keto-ester ***26.51***.

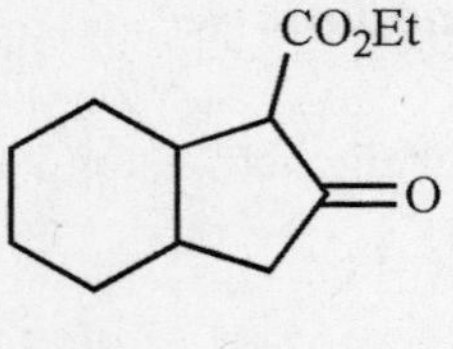

CO_2Et
=O

26.49

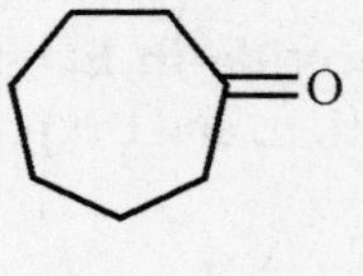

=O

26.50

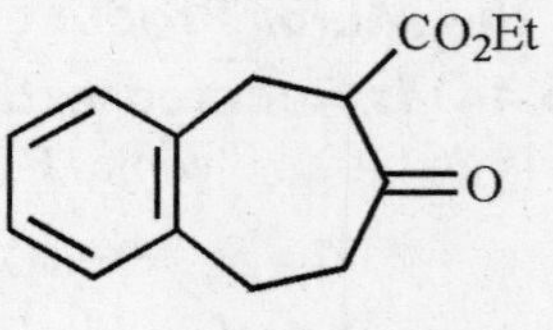

CO_2Et
=O

26.51

26.5. CONDENSATION WITH ALDEHYDES AND KETONES

The enolate anions derived from esters react with aldehydes and ketones via acyl addition to give β-hydroxy esters.

If the Enolate Anion of Ethyl 2-Phenylethanoate Is Treated with Cyclohexanone, What Is the Major Product after Hydrolysis?

The ester enolate anion (*26.52*), formed by reaction of ethyl 2-phenylethanoate and an appropriate base (such as LDA), reacts with cyclohexanone to give the β-hydroxy-ester ***26.53*** after hydrolysis.

Ph⁀CO_2Et —LDA→ Ph–CH=C(OEt)(O^-) (*26.52*) + cyclohexanone → —H_3O+→ Ph–CH(CO_2Et)–C(OH) cyclohexane (*26.53*)

26.52 *26.53*

Will the Enolate Anion of an Ester React with an Aldehyde?

Yes! Both ketones and aldehydes react to give the corresponding β-hydroxy ester.

Which Is Better for the Condensation of an Ester Enolate Anion and an Aldehyde, Kinetic Control Conditions or Thermodynamic Control Conditions?

In order to prevent a Claisen condensation of the ester, kinetic control conditions are better. Under thermodynamic control conditions, especially when both aldehyde or ketone and the ester are present in the same reaction, an Aldol condensation could also compete with the Claisen condensation, in addition to the 'mixed' condensation.

With What Type of Aldehyde Can Thermodynamic Control Conditions Best Be Utilized?

If an aldehyde does not have an α-carbon with acidic hydrogens (such as benzaldehyde), it can be added directly into the flask with the ester under thermodynamic control conditions.

Give the Major Product of the Following Sequence with Ethyl Butanoate: i. LDA, THF, −78°C ii. 5-Methyl-2-Hexanone iii. Saponification iv. Heating to 200°C.

The product is 4,7-dimethyl-4-octanol.

26.6. THE REFORMATSKY REACTION

When an α-halo ester is treated with zinc, the resulting zinc enolate anion can react with aldehydes and ketones to give β-hydroxy esters.

What Is the Reformatsky Reaction?

The **Reformatsky** reaction is the condensation of a *zinc enolate anion* of an ester formed from an α-halo ester, and an aldehyde, usually an aldehyde with no α-hydrogens.

Give an Example of the Reformatsky Reaction.

When ethyl 2-bromoacetate (***26.54***) reacts with zinc, the zinc enolate anion is formed (***26.55***). This then reacts with benzaldehyde via ***26.56*** to give the hydroxy-ester (***26.57***) after hydrolysis.

$$\underset{26.54}{Br\diagdown\!\diagup CO_2Et} \xrightarrow{Zn^\circ} \underset{26.55}{\left[BrZn^+ \ ^-CH_2CO_2Et \longleftrightarrow H_2C{=}C(O^-\ ^+ZnBr)(OEt)\right]} \xrightarrow{PhCHO}$$

$$\underset{26.56}{[\ldots]} \longrightarrow \xrightarrow{H_2O} \underset{26.57}{Ph\text{–}CH(OH)\text{–}CH_2CO_2Et}$$

In What Ways Is the Reformatsky Reaction Similar to the Enolate Anion Reactions Described Above?

The Reformatsky reaction is an enolate anion condensation reaction of an ester with an aldehyde. It is a metal enolate anion, but zinc is used rather than lithium or sodium, as in the usual Claisen type condensations. Since the zinc enolate anion is usually generated in the presence of the aldehyde partner, aldehydes with no enolizable hydrogens are usually required, and reaction with ketones is difficult.

How Are α-Halo Esters Prepared?

A common method for the preparation of α-halo acids is the reaction of an acid such as acetic acid with phosphorus and chlorine or bromine (P, Cl_2 or P, Br_2). Under these conditions, PCl_3 or PBr_3 is formed and reacts to give 2-chloroacetyl chloride (***26.58***) or 2-bromoacetyl chloride (***26.59***), which is then hydrolyzed to the corresponding acid. Such α-halo-acids are converted to the corresponding ester by the usual methods. The conversion of an acid to an α-halo acid chloride is called the **Hell-Volhard-Zelinsky** reaction.

$$\underset{26.58}{ClCH_2CO_2Cl} \xleftarrow{P\,,\,Cl_2} CH_3CO_2H \xrightarrow{P\,,\,Br_2} \underset{26.59}{BrCH_2COBr}$$

26.7. THE KNOEVENAGEL CONDENSATION

When the enolate anions derived from malonate esters (or related compounds such as CH_2X_2 where X = CO_2R, CN, SO_2R, etc.) react with ketones or aldehydes, the resulting products are β-hydroxy compounds that often eliminate water to give conjugated derivatives.

What Is the Major Product When Ethyl Malonate Is Treated with i. NaOEt ii. Acetone iii. H_3O^+?

The initially formed malonate anion (***26.60***) reacts with acetone to give the alkoxide (***26.61***) via acyl addition. Aqueous hydrolysis gives the alcohol (***26.62***), but alcohols of this type commonly dehydrate under these reactions conditions to the alkylidene derivative, ***26.63***. The α- protons of malonic ester are more acidic due to their proximity to two carbonyl groups, and the resulting enolate anion is more stable due to greater resonance delocalization. Since these protons are more acidic, a weaker base such as pyridine could be used in the initial deprotonation reaction. In this example, NaOEt was used, but the same reaction would occur (formation of ***26.61***) if pyridine were used rather than NaOEt.

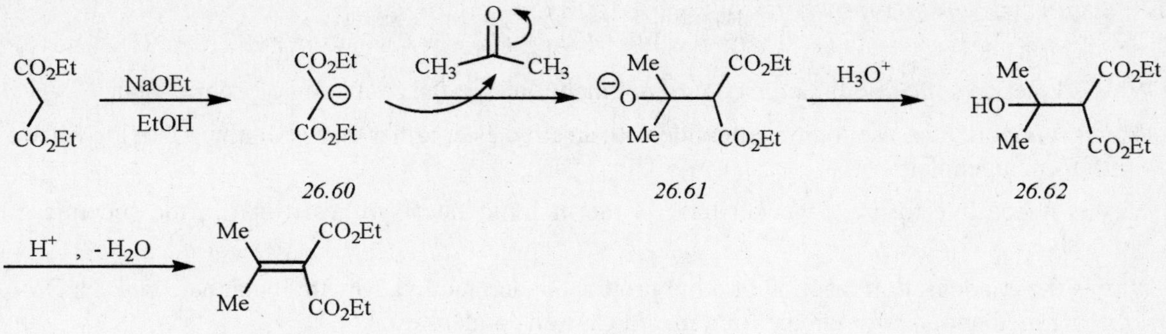

26.60 26.61 26.62

26.63

Why Is the Major Product Usually an Alkene?

The proximity of the alcohol moiety to two carbonyl leads to extensive hydrogen bonding and facile loss of water under the acidic conditions of the workup (conversion of the alkoxide product to the alcohol).

What Is the Name of This Type of Condensation?

This is known as the **Knoevenagel condensation**.

Can an Amine Be Used for the Base Rather Than NaOEt?

Yes! The pK_a of the $-CH_2-$ moiety of malonic ester is about 11. This is a sufficiently strong acid that pyridine be used as a base, rather than the stronger NaOEt.

What Is the Major Product of the Reactions of 26.64–26.65?

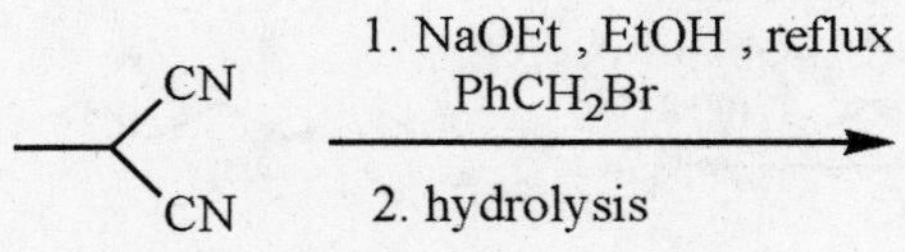

1. NaOEt , EtOH , reflux
$PhCH_2Br$
2. hydrolysis

26.64

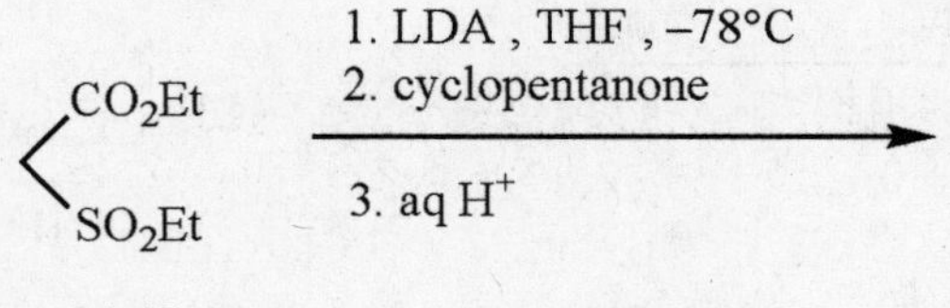

1. LDA , THF , –78°C
2. cyclopentanone
3. aq H^+

26.65

In the first reaction, the enolate anion derived from ***26.64*** reacts with benzyl bromide to give ***26.66***. This reaction is **not** an example of the Knoevenagel condensation, but illustrates that such compounds undergo alkylation just as any other enolate anion. In the second example, ***26.65*** undergoes a Knoevenagel condensation to give ***26.67***.

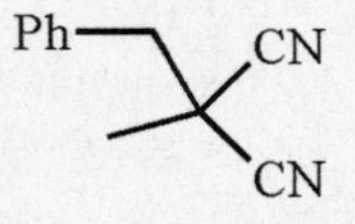

26.66 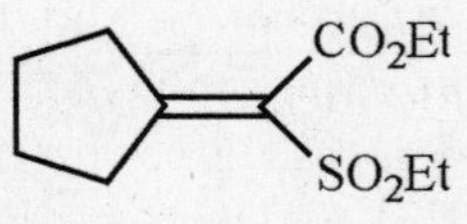 26.67

Test Yourself

1. Why is kinetic vs. thermodynamic control not mentioned in the formation of ester enolates?
2. Why is it important to use sodium ethoxide with an ethyl ester rather than sodium methoxide when trying to form an enolate?
3. Why is it possible to use a weaker base in the malonic ester synthesis than in the succinic ester synthesis?
4. Why is the condensation reaction of ethyl propionate and methyl 2-methylbutanoate with NaOMe in refluxing methanol a *poor* choice for a mixed Claisen condensation?
5. In each case, give the major product. If there is no reaction, indicate this by N.R.

(a) CO_2Et → 1. LDA, THF, -78°C 2. MeI

(b) CO_2Me → 1. LDA, -78°C, THF 2. S-2-iodopentane

(c) CO_2Me → 1. LDA, -78°C, THF 2. ethyl benozate

(d) CO_2Et → 1. NaOEt , EtOH, reflux 3. H_3O^+

(e) → 1. O_3 2. H_2O_2 3. $SOCl_2$; EtOH/pyridine 4. NaOEt, EtOH; H_3O^+

(f) CO_2Et → 1. LDA, -78°C, THF 2. 4-phenyl-3-methyl-2-hexanone

(g) → 1. O_3 2. H_2O_2 3. EtOH, H^+ 4. LDA, -78°C, THF 5. 4-phenyl-2-hexanone

(h) EtO_2C CO_2Et → 1. NaOEt , EtOH, reflux 3. H_3O^+

(i) CO_2Et → PhCHO NaOEt

(j) CO_2Et CO_2Et → 1. NaH , THF 2. benzyl bromide 3. NaH , THF 4. iodoethane

(k) CO_2Me → $PhCO_2Me$, NaOMe

(l) CO_2Et CO_2Et → 1. NaH, THF 2. iodohexane 3. aq. NaOH; H_3O^+ 4. 200°C

(m) Br CO_2Me → Zn° , EtOH; CHO

(n) HO_2C CO_2H → 1. H^+ , EtOH 2. LDA, THF, -78°C 3. Br

(o) CO_2H → P° , Br_2

(p) O CO_2Et → 1. NaH, THF 2. CHO 3. H_3O^+

(q) CO_2Et CO_2Et → 1. NaOEt , PhCHO 2. H_3O^+

(r) O CO_2Et → 1. NaH , THF 2. I 3. aq. NaOH; H_3O^+ 4. 200°C

(s) CO_2Et CH_3 O → 1. NaOEt , PhCHO 2. H_3O^+ 3. saponification 4. 200°C

(t) CO_2Et → 1. $SOCl_2$ 2. MeOH, pyridine 3. NaOEt, EtOH, reflux

6. In each case, provide a suitable synthesis. Show all reagents and intermediate products.

(a) Ph, O, O ⇨ CO_2H, CO_2H

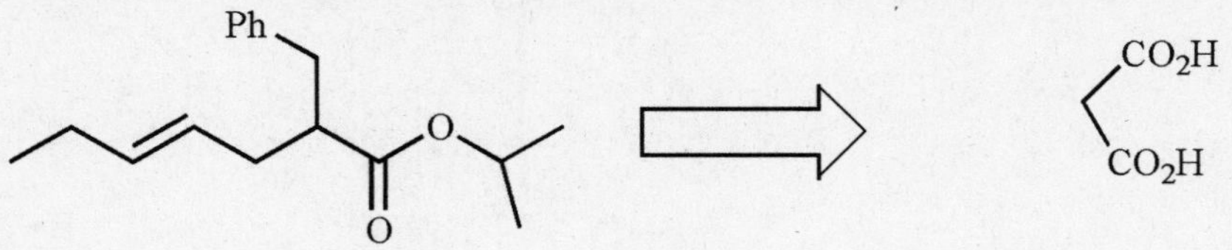

(b) CH_2 ⇨ OH, CO_2H

(c) O ⇨ Br

(d) O ⇨ Br

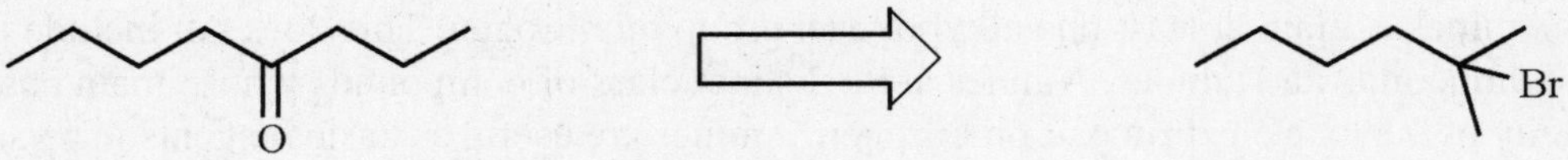

(e) Ph, O ⇨

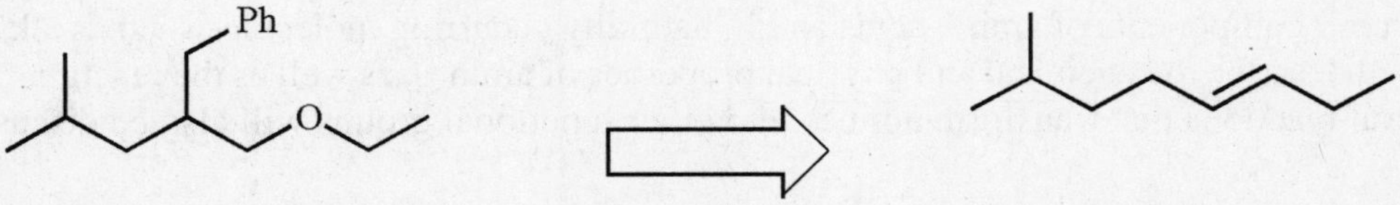

7. Provide a suitable synthesis for the preparation of each target from the indicated starting material..

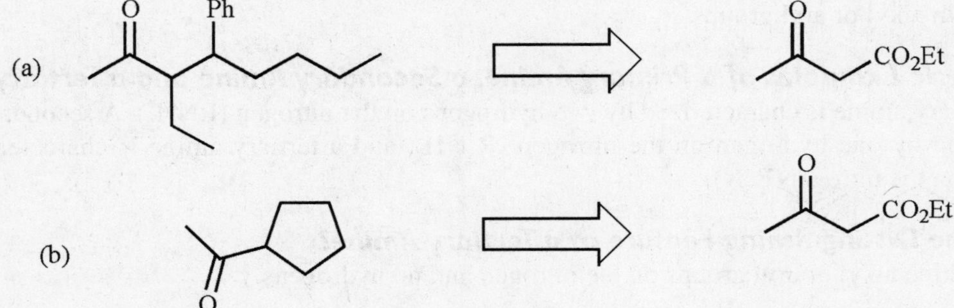

CHAPTER 27

Amines

Amines are organic molecules with alkyl groups, aryl groups or hydrogens attached to nitrogen. Amines contain at least one alkyl or aryl group on nitrogen. This does not include nitriles, nitro compounds and amides. Amines are a distinct class of compounds whose main characteristic is the basicity of the lone electron pair on nitrogen. Amines are useful as basic reagents in a variety of reactions. They also serve as intermediates in many reactions. The most notable occurrences of amines are as structural components of amino acids and in naturally occurring molecules such as alkaloids. This chapter will describe the chemical and physical properties of amines, as well as the reactions that form amines. Several reactions that transform amines into other functional groups will also be discussed.

27.1. STRUCTURE AND PROPERTIES

In a simple analogy, amines can be considered to be derivatives of ammonia where the hydrogens are replaced with alkyl or aryl groups.

Give Generic Examples of a Primary Amine, a Secondary Amine and a Tertiary Amine.

A primary amine is characterized by two hydrogens on the nitrogen (RNH_2). A secondary amine is characterized by one hydrogen on the nitrogen (R_2NH), and a tertiary amine is characterized by no hydrogens on the nitrogen(R_3N).

What Is the Distinguishing Feature of a Tertiary Amine?

It has three alkyl or aryl groups on the nitrogen and no hydrogens.

Classify an Amine As a Reagent.

An amine is both a nucleophile (if it reacts with carbon) and a base (if it reacts with a proton or a Lewis acid).

What Is the Product When an Amine Reacts with HCl?

The product is an ammonium chloride. For a secondary amine, the reaction is:

$R_2NH + HCl \rightarrow R_2N^+H_2\ Cl^-$

Which Is More Basic, a Primary Amine, a Secondary Amine or a Tertiary Amine? Explain.

A secondary amine is the most basic. The electron releasing alkyl groups suggest that the more groups on nitrogen the more basic it will be (3° > 2° > 1°), but there is a steric effect of the alkyl groups around the nitrogen. With a tertiary amine, the alkyl groups inhibit approach of the nitrogen to the acid, decreasing the basicity. Another important factor is the presence of N–H groups in the ammonium salt product (the product after the amine reacts with the acid). These N–H groups are capable of hydrogen bonding with the solvent [N-----H-----OH_2], further stabilizing the product. This suggests 1° > 2° > 3°. With secondary amines the electronic effects, solvent effects and steric effects are balanced to make it the most basic. The usual order of basicity is 2° > 1° ≈ 3°).

What Is the Product When an Amine Reacts with $AlCl_3$?

The product is the usual Lewis acid-Lewis base complex. For a secondary amine, the complex is: R_2HN^+:$^-AlCl_3$.

Which Is More Basic, R_2NH or R_2N-? Explain.

In R_2N-, the charge is concentrated on the nitrogen with a formal charge of -1. This is clearly a stronger base than the neutral amine, which has only a δ– charge on nitrogen due to the unshared electron pair.

Which Is the Strongest Nucleophile, a Primary Amine, Secondary Amine or a Tertiary Amine? Explain.

For essentially the same reasons described for basicity, secondary amines are usually the most nucleophilic.

What Is the Product of the Reaction Between Trimethylamine (a Tertiary Amine) and Iodomethane?

The product is tetramethylammonium iodide: Me_4N^+ I^-.

What Is a Distinguishing Feature of Amines That Can Be Correlated with their Solubility Characteristics?

Amines have a polarized C–N bond and also a polarized *N*–H bond for primary and secondary amines. The strong dipole of the C–N bond in small molecular weight amines promotes solubility in polar solvents. For primary and secondary amines, the ability to hydrogen bond to the *N*–H moiety makes solubility in water and other protic solvents very high, if the molecular weight of the amine is not too great.

If an Amine Contains Three Different Groups (RR^1R^2N), Is the Nitrogen a Stereogenic Center? Why or Why Not?

Such amines are considered to be chiral, racemic molecules. Although there are three different alkyl groups and the lone electron pair can be considered a fourth 'group,' there is rapid inversion of configuration around nitrogen (see ***27.1***) that generates a racemic mixture of the possible enantiomers. For this reason, the nitrogen of such amines is not considered when determining the absolute configuration of stereogenic centers. They are chiral, but exist as a racemic mixture.

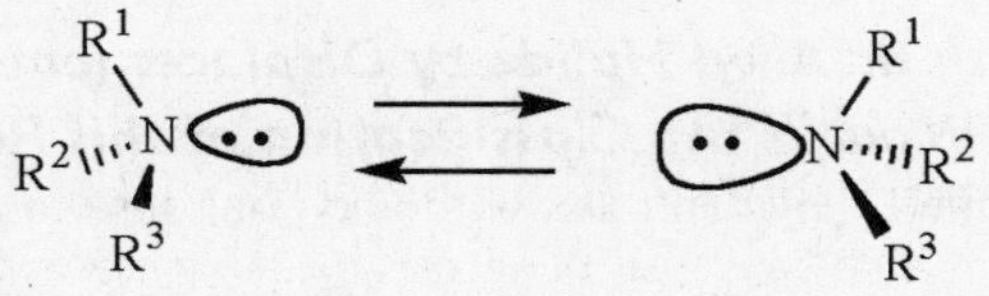

R^1 R^2 R^3 N ⇌ N R^1 R^2 R^3

27.1

How Can the Rapid Inversion Characteristic of an Amine Be Prevented by Structural Modification of the Amine?

If the alkyl groups are connected ('tied back') in such a way that there is no fluxional inversion, or limited fluxional inversion, as in 1-azabicyclo[2.2.2]octane (*27.2*), i.e., inversion around nitrogen is not possible, the nitrogen is a stereogenic center, and *27.2* is a chiral, non-racemic molecule.

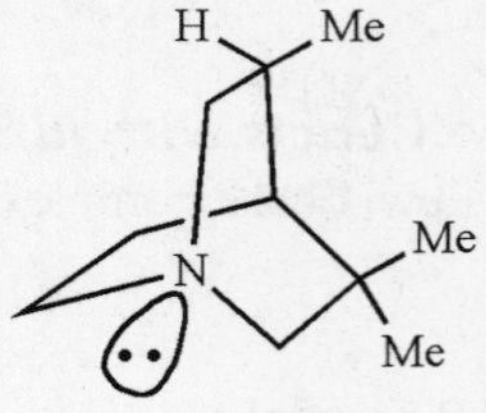

H Me

Me

N

Me

27.2

Which Is Expected to Have the Higher Boiling Point, a Primary Amine or a Tertiary Amine?

Since primary amines have two N–H units that can hydrogen bond, it is expected to have a higher boiling point than the tertiary amine, which cannot hydrogen bond if the molecular weights are approximately equal.

Do Amines Have a Distinguishing UV Absorption?

Non-aromatic amines do not have a distinguishing absorption in the UV.

Nomenclature

What Is the IUPAC Nomenclature System for Alkyl Amines?

Amines can be named as with an alkyl amine or as an alkanamine. The primary amine $CH_3CH_2CH_2CH_2NH_2$ is, therefore, butylamine or butanamine. Note the *-e* of *-ane* is dropped.

What Is the IUPAC Nomenclature System for Aryl Amines?

They are generally named as the parent aromatic amine. Aniline, 2-bromoaniline and 3-nitro-aniline are typical examples. If a methyl group is also attached to the benzene ring, the common name is *ortho*-toluidine, *meta*-toluidine and *para*-toluidine.

How Are Substituents on Nitrogen Treated in the IUPAC System?

When a group is attached to the nitrogen, it is placed in the name with an *N*- preceding it. An example is $CH_3CH_2NHCH_2CH_2CH_3$, which is named *N*-ethylpropanamine. If there are two groups on nitrogen, each uses the *N*- designation. The amine $CH_3CH_2CH_2N(CH_3)_2$ is named *N,N*-dimethyl-propanamine, and $PhN(CH_3)Et$ is *N*-ethyl-*N*-methylaniline.

27.2. PREPARATION OF ALKYL AMINES

Alkyl amines are prepared by several different routes involving several different functional groups. In general, substitution and reduction reactions are the most important.

When an Amine Reacts with an Alkyl Halide by Displacement of the Halide by Nitrogen, What Is the Classification of this Reaction?

Since amines are nucleophiles, they can react with alkyl halides via an S_N2 process to produce new amines.

When an Amine Reacts with an Alkyl Halide by Elimination of the Halide to Form an Alkene, What Is the Classification of This Reaction?

Since amines are also basic, however, E^2 reactions may be competitive with the substitution with secondary amines and will dominate with tertiary amines.

What Is the Expected Product When Butanamine Reacts with Iodomethane?

Since the nitrogen of butanamine is a nucleophile, it will displace iodide (in an S_N2 reaction) to give the ammonium salt, $BuNH_3^+ \, I^-$.

Which Is the Most Nucleophilic, Butanamine or N-Methylbutanamine?

The secondary amine, *N*-methylbutanamine, is more nucleophilic since there are two alkyl groups releasing electrons to nitrogen.

Discuss Why Monoalkylation of a Primary Amine Is Very Difficult.

If ethanamine reacts with iodomethane, the initial product is the ammonium iodide of N-methylethanamine ($EtNHMe^+ \, I^-$). Since ethanamine is also a base, it will deprotonate the ammonium salt to give the secondary amine, EtNHMe. Secondary amines are more nucleophilic than primary amines, and EtNHMe will likely react with MeI faster than will $EtNH_2$, leading to a tertiary amine. Since the products of the initial reaction with $EtNH_2$ are more reactive than the starting material, it is difficult to stop the reaction at the secondary amine stage.

How Can Polyalkylation Be Minimized in Reactions of Amines?

If a large excess of the primary amine is used, polyalkylation is minimized.

Competitive Elimination

Why Is Elimination a Problem When Amines React with Alkyl Halides?

With secondary and especially tertiary halides, elimination of the halide leaving group will lead to an alkene via an E2 reaction. Since all amines are relatively good bases, this can initiate the E2 reaction if the rate of the S_N2 reaction is relatively slow.

What Is the Product When 2-Bromo-2-Methylpentane Reacts with Triethylamine?

This is an E2 reaction, and the product is 2-methyl-2-pentene.

Is Elimination a Major Problem When Primary Amines React with Alkyl Halides? Why or Why Not?

In general, elimination is slower than substitution for primary halides. The ammonium intermediate ($RCH_2NH_2R^+$) is susceptible to both S_N2 and E2 reactions since $-NH_2R^+$ is a good leaving group. For primary amines, substitution is faster than elimination, and elimination is usually not a significant problem in alkylation reactions of primary amines.

Explain Why Reaction of a Tertiary Alcohol with Thionyl Chloride and Triethylamine Leads Directly to an Alkene.

Thionyl chloride first converts the alcohol to the tertiary chloride and triethylamine induces an E2 elimination, *in situ*, to give the alkene directly.

The Gabriel Synthesis

This reaction uses phthalimide as an amine surrogate. Reaction of an alkyl halide with the amide derived from phthalimide gives the *N*-alkylation product. Subsequent conversion of the phthalimide unit to an amine completes the **Gabriel synthesis.**

What Is the Structure of Phthalimide?

Phthalimide has the structure ***27.3***. It is the imide of phthalic acid.

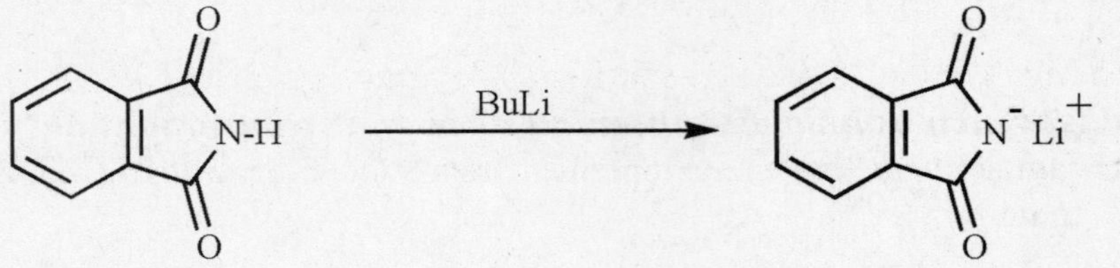

27.3 —BuLi→ 27.4

What Is the Product When Phthalimide Reacts with N- Butyllithium?

The imide N–H is somewhat acidic (pK_a of about 17), and treatment with a strong base such as butyllithium will give the imide anion, ***27.4***.

What Is the Product When the Sodium Salt of Phthalimide Reacts with 1-Bromopentane?

The phthalimide anion ***27.4*** is a good nucleophile and reacts with 1-bromopentane to give *N*-pentylphthalimide, ***27.5***. Hydrolysis liberates the amine (***27.6***) and the salt of phthalic acid, ***27.7***.

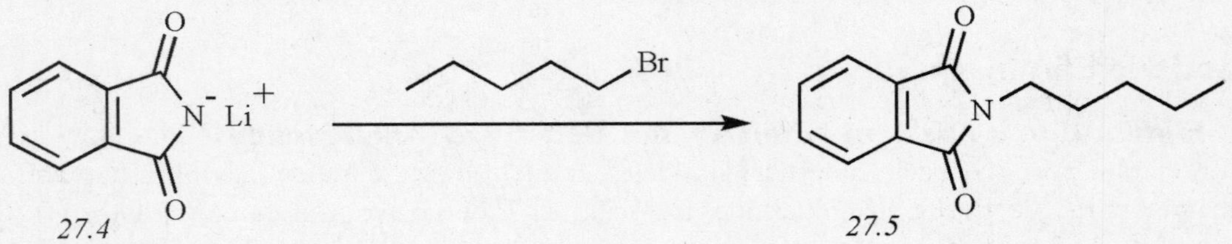

27.4 —Br→ 27.5

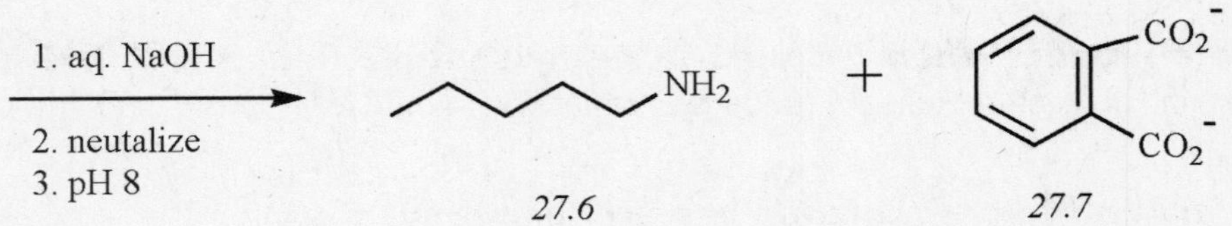

1. aq. NaOH
2. neutalize
3. pH 8

NH_2 27.6 + CO_2^- CO_2^- 27.7

Explain Why Polyalkylation Is Not a Problem with Phthalimide.

The product of the alkylation is an imide, which is significantly less basic than the imide anion. There is, therefore, virtually no chance of the product reacting competitively with ***27.5***.

Explain Why Polyalkylation Is Not a Problem when Phthalimide reacts with 1-bromopentane.

There are two products. The primary amine (pentanamine) is 'released' along with phthalic acid. The initial product is phthalic acid and the ammonium salt. Mild basification (to pH 8, for example) deprotonates the ammonium salt to give the amine (*27.6*) and converts phthalic acid into the dianion, *27.7*.

If N-Butyl Phthalimide Is Hydrolyzed with Acid, Why Must the pH of That Solution Be Adjusted to About 8 Prior to Attempts to Isolate the Amine?

Under slightly acidic conditions, amines are protonated to give the ammonium salt. Adjusting the pH to about 8 removes that proton and 'liberates' the free amine.

What Is the Product When 27.5 Is Treated with Hydrazine (NH_2NH_2)?

The product is the amine (*27.6*) and *27.8*, which is known as phthalhydrazide. Since *27.8* is easily separated from the amine, treatment with hydrazine offers a convenient and facile modification of the Gabriel synthesis.

NH_2NH_2 | NH_2 | + | NH | NH | O | O | N | O | O

27.5 *27.6* *27.8*

Show a Reaction Sequence That Will Convert 2-Pentanol to 2-Aminopentane?

Initial reaction of 2-pentanol with PBr_3 (see Section 11.3) leads to 2-bromopentane. Subsequent reaction with the phthalimide anion gives *27.9*, and treatment with hydrazine gives *27.8* and the targeted 2-aminopentane (*27.10*).

OH | PBr_3 | Br | $N^- Li^+$ | O | O | N | O | O

27.9

NH_2NH_2 | NH_2 | + | NH | NH | O | O

27.10 *27.8*

Reductive Amination of Carbonyls

What Is the Product When an Aldehyde or Ketone Reacts with a Primary Amine?

Amines react with aldehydes and ketones to form imines (see Section 21.3). These imines can be reduced to form new amines.

What Is the Product of the Reaction Between Pentanamine and Formaldehyde?

In general, aldehydes react with primary amines to form an imine. In this example, the initial product is an iminium salt (*27.11*), which can be deprotonated to give the imine, ***27.12***. In the case of imines derived from formaldehyde, such as ***27.12***, isolation is difficult due to their great reactivity with a variety of reagents..

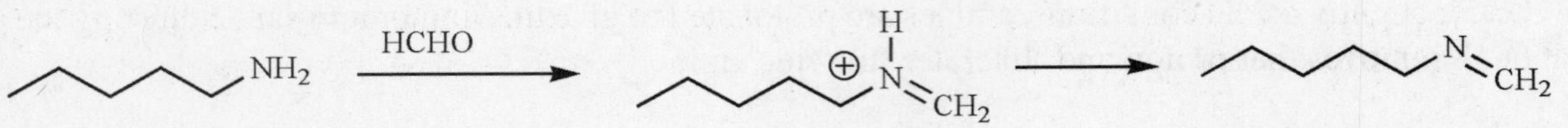

HCHO

27.11 *27.12*

What Is the Mechanism for Formation of 27.12?

Often, this reaction is done in the presence of an acid catalyst. This protonates the amine, after a proton transfer from nitrogen to oxygen, so the actual 'catalyst' is the ammonium salt. Initial reaction of the amine with formaldehyde gives the amino alcohol (***27.13***). If the oxygen is then protonated to give ***27.14***, water is lost to give the iminium salt, ***27.11*** (with a proton transfer to nitrogen). Loss of the proton from nitrogen to the amine or during a slightly basic workup will 'liberate' the imine product, ***27.12***.

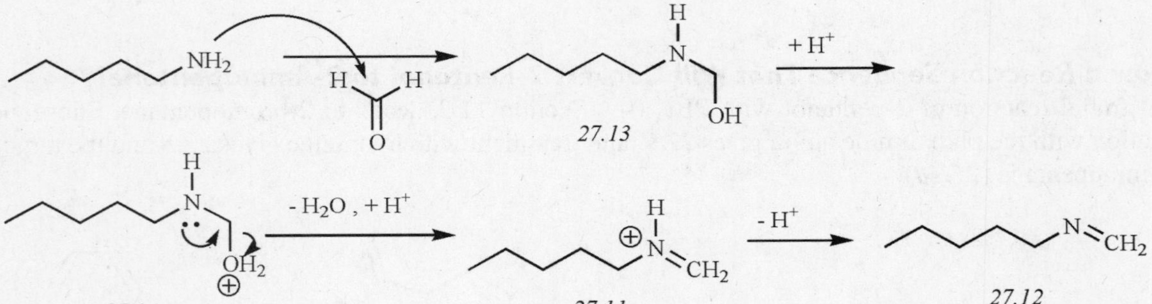

$+H^+$

27.13

$-H_2O, +H^+$ $-H^+$

27.14 *27.11* *27.12*

If Formaldehyde Were Reacted with Pentanamine in the Presence of Hydrogen and a Palladium Catalyst, What Is the Product?

With a palladium catalyst, the iminium salt (***27.11***) will be reduced to the corresponding amine. In this case, the product is *N*-methylpentanamine.

What Is the Generic Name for this Process?

Reductive amination.

What Is a Schiff Base?

A **Schiff Base** is the product derived from the reaction of an amine with an aromatic or aliphatic aldehyde (an *N*-substituted imine).

What Is the Product Formed When Ethanamine Reacts with Ethanal?

Reaction of ethanamine and ethanal, for example, leads to $CH_3CH_2N{=}CHCH_3$.

What Is the Product of the Reaction Between an Imine and $LiAlH_4$?

The powerful reducing agent $LiAlH_4$ is capable of reducing the C=N bond to the amine, via delivery of hydride to the δ+ carbon of the C=N bond. If R–N=CR'$_2$ is reacted with $LiAlH_4$ and hydrolyzed under slightly basic conditions, the amine product will be RNHCHR'$_2$.

What Is the Product Between an Imine and Hydrogen, in the Presence of a Catalyst?

The product will be the amine, via reductive amination (see reaction with formaldehyde).

What Is the Product When the Imine Derived from Aniline and Butanal Is Treated with $NaBH_4$?

Aniline and butanal will form the imine, $PhN{=}CHCH_2CH_2CH_3$. Sodium borohydride reduces imines to the amine, $PhNHCH_2CH_2CH_2CH_3$, although the reaction can be slow in the absence of an acid catalyst, which generates the iminium salt.

What Is the Product When the Aniline and Butanal React in the Presence of Hydrogen and a Palladium Catalyst?

The product will be the amine, $PhNHCH_2CH_2CH_2CH_3$.

Reduction of Nitriles

What Is the Product When Hexanenitrile Is Reduced with $LiAlH_4$?

When hexanenitrile is reduced, the product is hexanamine.

What Is the Product When Benzonitrile (PhC≡N) Is Reduced with Hydrogen and a Palladium Catalyst?

The product is benzylamine ($PhCH_2NH_2$).

What Is the Product of the $LiAlH_4$ Reduction of 1,6-Hexanedinitrile?

The product is the diamine, 1,6-hexanediamine [$H_2N\text{-}(CH_2)_6\text{-}NH_2$].

Reduction of Amides

What Is the Product When N-Ethylpentanamide Is Treated with $LiAlH_4$?

As first mentioned in Section 25.6, amides are reduced to amines. In this case, the product is *N*-ethylpentanamine.

What Is the Product When 2-Pyrrolidinone Is Reacted with $LiAlH_4$?

Reduction of lactams gives cyclic amines. Reduction of 2-pyrrolidinone gives pyrrolidine.

From Azides

What Is the Structure of Sodium Azide?

Sodium azide has the structure NaN_3. The azide anion is N_3^-.

Discuss the Relative Stability of the Azide Anion.

Azide anion is a resonance stabilized structure, *27.15*.

$$\left[{}^{\ominus}N{=}\overset{\oplus}{N}{=}N^{\ominus} \longleftrightarrow {}^{\ominus}_{\ominus}N{-}\overset{\oplus}{N}{\equiv}N \right] \equiv N_3^{\ominus}$$

27.15

Classify Azide As a Reagent.

Azide anion is a strong nucleophile, but a rather weak base. It will react with primary and secondary halides via a S_N2 reaction to form alkyl azides (RN_3).

What Is the Major Product of the Reaction Between Sodium Azide and 1-Iodohexane?

The product is 1-azidohexane, $CH_3CH_2CH_2CH_2CH_2CH_2N_3$.

What Is the Major Product When 1-Azidopentane Reacts with $LiAlH_4$? with H_2 and a Catalyst?

In both cases, the azide is reduced to the primary amine. In this case, the product is pentanamine.

Can the Reaction of Azide and Alkyl Halides Be Used to Produce Secondary or Tertiary Amines?

Not directly. Reduction of $-N_3$ always leads to $-NH_2$, the primary amine.

27.3. PREPARATION OF ARYL AMINES

The methods used to prepare aryl amines are slightly different from those used to prepare alkyl amines, due to the different chemical properties of aromatic compounds.

Aromatic Amines are Generally Considered to Be Derivatives of What Simple Compound?

Aniline, $PhNH_2$.

How Is Nitrobenzene Synthesized from Benzene?

Reaction of benzene with nitric and sulfuric acid gives nitrobenzene. This chemistry was discussed in detail in Section 17.2.

What Is the Major Product When Nitrobenzene is Reacted with Hydrogen and a Palladium Catalyst?

Reduction with hydrogen leads to aniline as the product.

Why Do These Conditions Not Reduce the Benzene Ring?

The benzene ring is aromatic (resonance stabilized), and an excess of hydrogen and very vigorous conditions (heat and pressure) are required to reduce it. The nitro group is relatively easy to reduce, and the mild conditions used will not affect the benzene ring.

What Is the Major Product of the Reaction Between $LiAlH_4$ and Nitrobenzene?

The product is **not** the amine (aniline), but rather a diazo compound, ***27.16.***

1. $LiAlH_4$, ether
2. H_2O

$-NO_2$ → $-N{=}N-$

27.16

What Is the Major Product of the Reaction Between 1-Nitrobutane and $LiAlH_4$?

Unlike aromatic nitro compounds, alkyl nitro derivatives are cleanly reduced to the amine. In this case, the product is 1-butanamine ($CH_3CH_2CH_2CH_2NH_2$).

Is Reduction of Aromatic Nitro Compounds with $LiAlH_4$ a Viable Synthetic Route to Aromatic Amines?

No. In general, diazo compounds rather than amines are produced..

What Reaction Conditions Are Required for Ammonia to React with Bromobenzene?

High temperatures and pressures (200–300°C, 2000–3000 psi) and high concentrations of ammonia. Under these nucleophilic aromatic substitution conditions, bromobenzene is converted into aniline.

Why Does the Presence of a Nitro Group at the Ortho and Para Position of a Benzene Ring Increase the Reactivity of Ammonia with the Aryl Halide?

As discussed in Section 20.1, the intermediate for this reaction is a carbanion, and the nitro groups will delocalize that negative charge onto the nitro groups. This leads to additional resonance structures and increased stability. The increased stability of the intermediate makes the overall reaction faster.

Give the Major Product of the Following Reactions.

-Br → NH_3, heat, pressure

1. HNO_3, H_2SO_4
2. excess H_2, Pd

-CHO →
1. $NaBH_4$ then aq NH_4Cl
2. PBr_3
3. NaCN, DMF
4. $LiAlH_4$ then H_3O^+

27.17

In the first reaction, the reaction of bromobenzene with ammonia under heat and pressure gives aniline (***27.18***) via nucleophilic aromatic substitution. The reaction of 1,4-dimethylbenzene (*para*-xylene) with nitric acid and sulfuric acid gives the nitro compound, *ortho* to a methyl. Subsequent catalytic hydrogenation reduces the nitro group to an amine in ***27.19***. When benzaldehyde (***27.17***) is reduced with sodium borohydride to benzyl alcohol, treatment with phosphorus tribromide gives benzyl bromide. Subsequent S_N2 reaction with cyanide gives benzyl cyanide and reduction with $LiAlH_4$ gives benzylamine, ***27.20***.

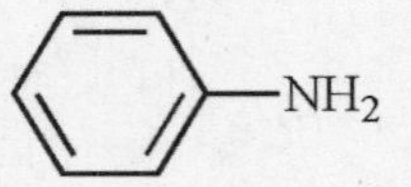

NH2 NH2 NH2

27.18 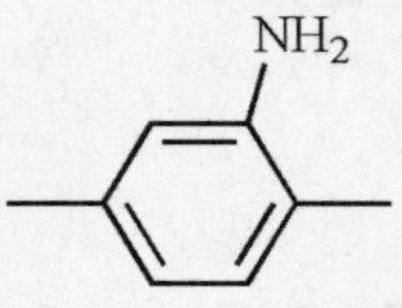 *27.19* 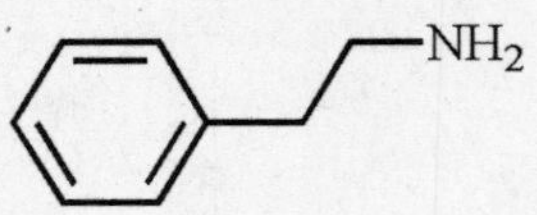 *27.20*

27.4. REACTIONS OF ALKYL AMINES

Alkyl amines react with a variety of reagents, sometimes to give a new amine and sometimes to give other compounds.

Substitution

See Chapter 7 for an introduction of aliphatic substitution reactions.

What Is the Product When a Primary or Secondary Amine Reacts with an Alkyl Halide?

Amines react with aliphatic alkyl halides to form new alkyl amines via a S_N2 reaction. Polyalkylation is a serious problem.

What Is the Product When a Tertiary Amine (R_3N) Reacts with an Alkyl Halide (R^1X)?

The product is an ammonium salt, $R_3N^+–NR^1\ X^-$.

What Is the Major Product of Each of the Following Reactions? (A) 1-Bromobutane and Dimethylamine; (B) 2-Iodopentane and Butylamine.

In reaction (a), the initial product is the tertiary amine, *N,N*-dimethylbutanamine, and this is likely the major product. In the second reaction, elimination to 2-pentene is likely to be a significant reaction, but the substitution process will give *N*-butyl-2-methylbutanamine and the major product. The steric hindrance of this second amine will make the subsequent reaction rather slow, especially if an excess of butylamine is used.

Elimination

What Is the Product of a Reaction Between an Amine and a Tertiary Halide?

Since a tertiary halide cannot undergo a S_N2 reaction, and amines are rather basic, the product is usually an alkene via an E2 reaction.

What Is the Product When 2-Methyl-2-Butanol Reacts with Thionyl Chloride?

The product is 2-chloro-2-methylbutane, as discussed in Section 11.3.

What Is the Product When 2-Methyl-2-Butanol Reacts with Thionyl Chloride in the Presence of Pyridine?

When a tertiary amine such as triethylamine or pyridine is used in the reaction of a tertiary (and occasionally a secondary) alcohol with thionyl chloride, the alkene is the product via an E2 pathway. See Section 9.1. In this example, the product is 2-methyl-2-butene.

What Is the Major Product from the Reactions of 27.21 and 27.22?

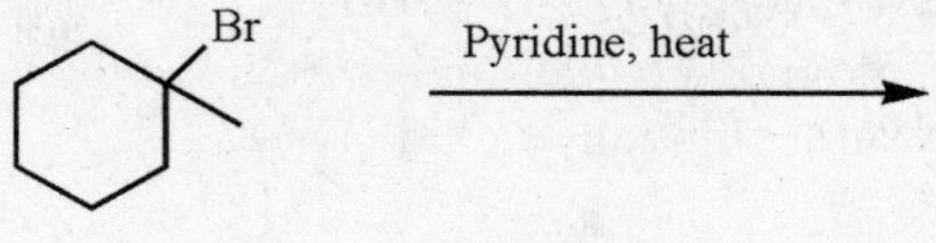

Br

Pyridine, heat

27.21

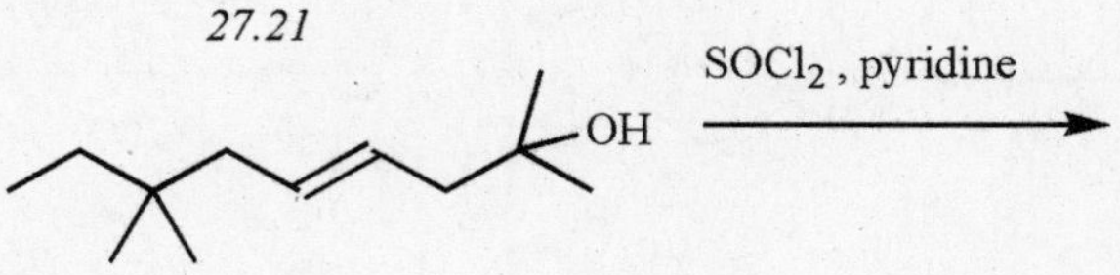

$SOCl_2$, pyridine

OH

27.22

In the first reaction, 1-bromo-1-methylcyclohexane (***27.21***) is treated with the pyridine, which acts as the base, to produce 1-methylcyclohexene (***27.23***) via an E2 reaction. In the second case, diene ***27.24*** is formed from alcohol ***27.22*** by the thionyl chloride/pyridine combination.

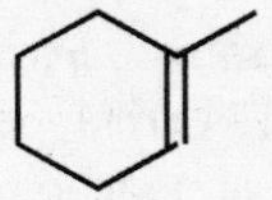

27.23 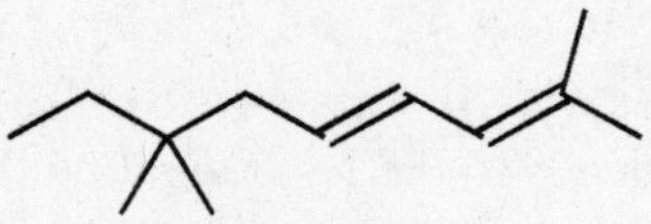 *27.24*

Hofmann Elimination

The bimolecular elimination reaction (E2) was discussed in Section 9.1.

What Is the Product When 2-Bromopentane Is Reacted with KOH in Ethanol?

The product is 2-pentene via an E2 reaction.

If the Base Required for an E2 Reaction Were Somehow 'Tethered' to the Molecule, So an Intermolecular Reaction Was Impossible, What Relationship Must the β-Hydrogen and the Base Have in Order to React?

They must have a 'syn' relationship (eclipsed).

What Is the Relationship of the β-Hydrogen and the Leaving Group in This Rotamer?

The answer depends on whether we use the normal anti elimination model with hydroxide attacking the β-hydrogen of the less substituted carbon atom via the lower energy conformation, or whether we invoke formation of a nitrogen ylid species ($Me_2RN^+–CH_2^-$), which demands removal of the β-hydrogen via a syn (eclipsed) conformation state. There is another model in which the base (hydroxide) is associated with the ammonium ion as a tight ion pair and can only remove a β-hydrogen via a syn conformation. We will examine both, beginning with the normal anti elimination model.

Give the Major Product When Trimethylamine Is Treated with 2-Bromopentane?

Although there could be some elimination, the major substitution product will be trimethylpentylammonium bromide, ***27.25***.

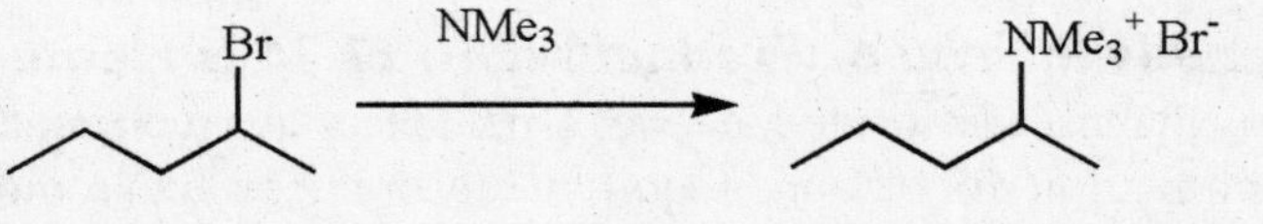

Br

NMe_3

$NMe_3^+ Br^-$

27.25

If This Product Were Reacted with Silver Oxide and Only One Equivalent of Water, What Is the Product?

Under these conditions, the bromide counterion in *27.25* is replaced with a hydroxide counterion, and the product is trimethylpentylammonium hydroxide, *27.26*.

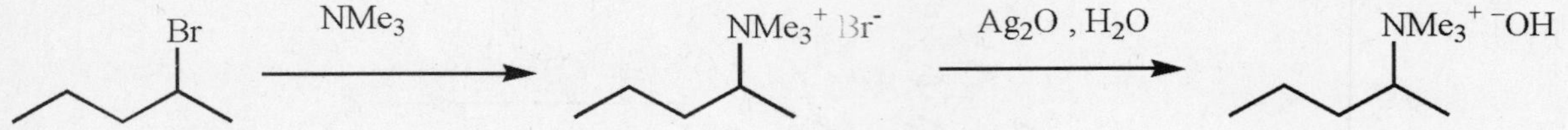

Br NMe3 NMe3+ Br- Ag2O , H2O NMe3+ -OH

27.25 *27.26*

What Is the Name of This Reaction Sequence?

This sequence is referred to as the **Hofmann Elimination**.

If This Ammonium Hydroxide Were Heated to about 150–200°C, Explain Why the Less Substituted Alkene Is Formed Rather Than the More Substituted Alkene, Assuming an Anti Transition State.

The anti conformation that leads to the less substituted alkene (1-pentene) is *27.27* and that for the more substituted alkene (2-pentene) is *27.28*. These conformations lead to the proper transition states. In *27.28*, there is a gauche interaction (see circled ethyl groups) that raises the energy of the resulting transition state relative to that from *27.27*, where this interaction is significantly less. Conformation *27.27* is lower in energy and leads to the transition state to give the less substituted alkene (1-pentene) as the major product.

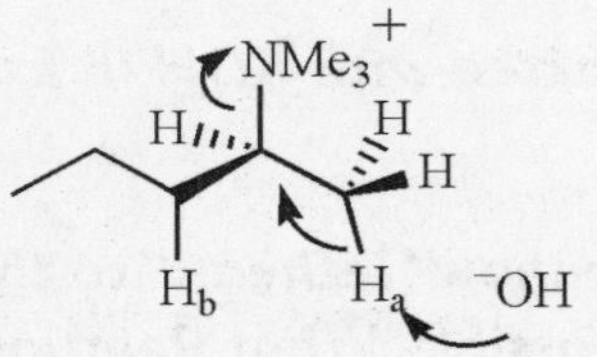

27.27 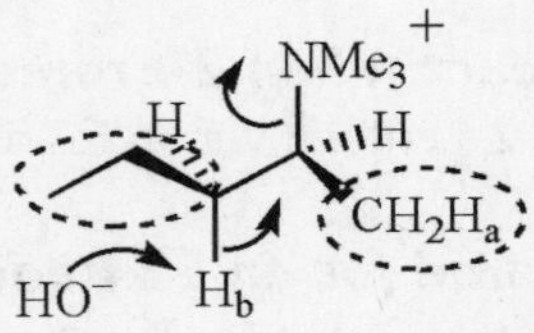 *27.28*

Draw a Diagram to Indicate the Conformation Required If a Syn Transition State Was Utilized for Removal of H_a.

If the leaving group (NMe_3) and H_a are syn, *27.29* is the proper conformation.

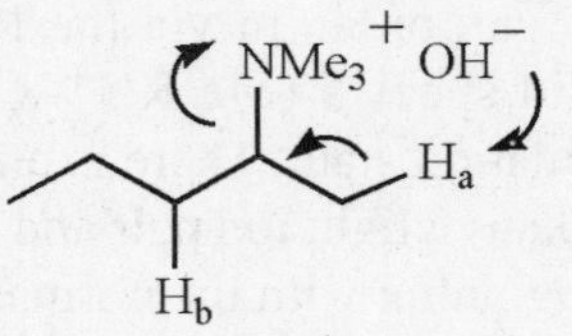

27.29

If 27.29 Represents a Tight Ion Pair (No Solvent Separation), and the Hydroxide Cannot Migrate Away from the Nitrogen, Discuss Why 1-Pentene Is the Major Product When 27.26 Is Heated.

The major product is the less substituted alkene, 1-pentene via syn elimination of H_a in *27.30* rather than H_b. The term syn elimination refers to the requirement for the base to remove the β-proton via a syn rotamer. The tethered base can only react the β-hydrogen via an eclipsed conformation. Examination of the Newman projections for removal of the two β-hydrogens (H_a and H_b) reveals that both of the high energy eclipsed rotamers have significant steric hindrance. In *27.31*, there is a steric interaction due to the methyl-alkyl interaction, which is missing in *27.30*. The lower energy rotamer will predominate, and loss of H_a from this rotamer will lead to the less substituted alkene, 1-pentene.

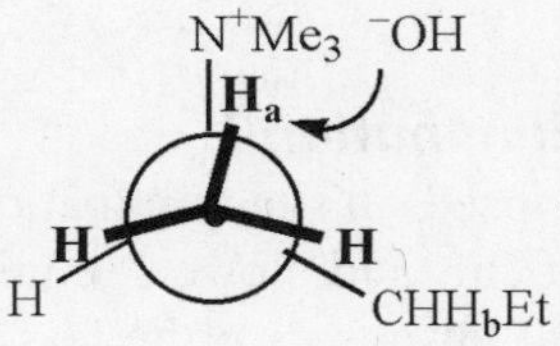

27.30 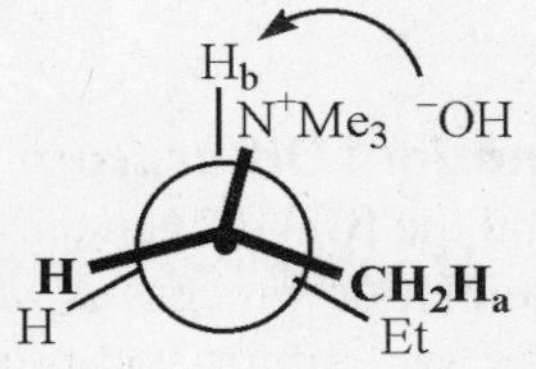 27.31

Draw the Nitrogen Ylid That Would Be Formed If the Basic Hydroxide Ion in 27.26 Removed a Proton from One of the Methyl Groups Attached to Nitrogen.

Removal of a proton from the *N*-methyl leads to a *N*–CH_2 unit, with a positively charged nitrogen and a negatively charged carbon - an ylid (see *27.32*).

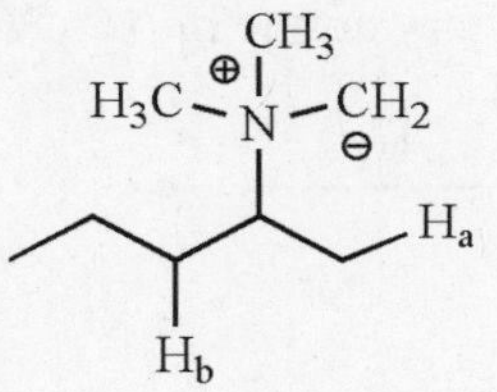

27.32

Using 27.32 As a Model, Draw the Syn Conformation Required for Removal of H_A and Formation of 1-Pentene, Showing the Electron Flow That Leads to the Products.

In **27.33** the CH_2^- unit of the ylid removes H_a from the less substituted carbon via a syn conformation. A syn conformation is the only way the base (-CH_2^-) can attack the proton in an acid-base reaction. Such a reaction may be concerted rather than stepwise. The products of the reaction are 1-pentene and trimethylamine.

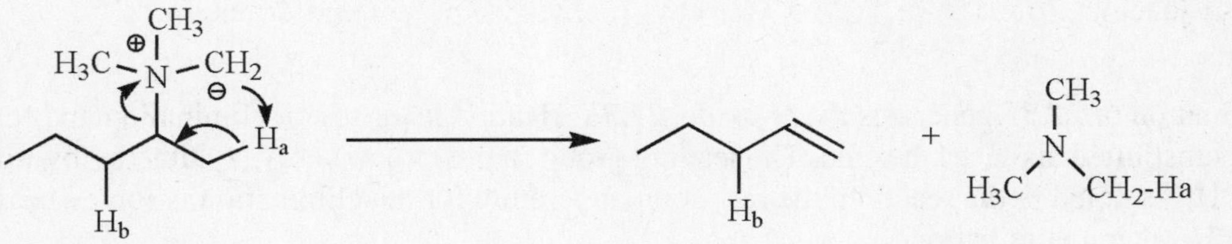

27.33

Give the Major Product If 2-Bromopentane Is Treated with i. Triethylamine ii. Ag_2O, H_2O iii. 200°C. Explain Your Answer.

The major product is ethene and 2-(*N,N*-diethylamino)pentane, *27.35*. The initial ammonium hydroxide (*27.34*) formed by reaction of 2-bromopentane with triethylamine will give syn elimination upon heating at the less sterically hindered and less substituted site, *which is on one of the ethyl groups*.

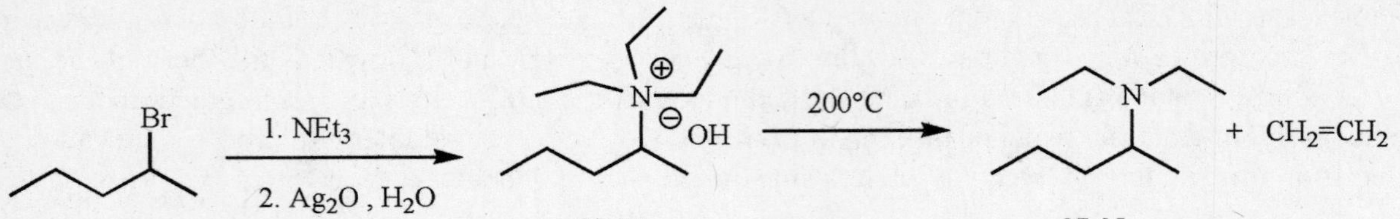

27.34 27.35

How Does 'Syn Elimination' Differ from 'Anti Elimination'?

The leaving group and the β-hydrogen are *syn* (eclipsed) in syn elimination, but they are *anti* in the E2 reaction. The *syn* elimination process is more difficult and slower, requiring significantly higher reaction temperatures. The *syn* elimination process will always give the less substituted alkene as the major product, whereas *anti* elimination (E2) will always give the more substituted alkene as the major product.

Cope Elimination

What Is the Product When a Tertiary Amine Is Oxidized with a Reagent Such As Hydrogen Peroxide, a Peroxyacid or $NaIO_4$?

These oxidizing agents convert a tertiary amine into an *N*-oxide such as ***27.36***.

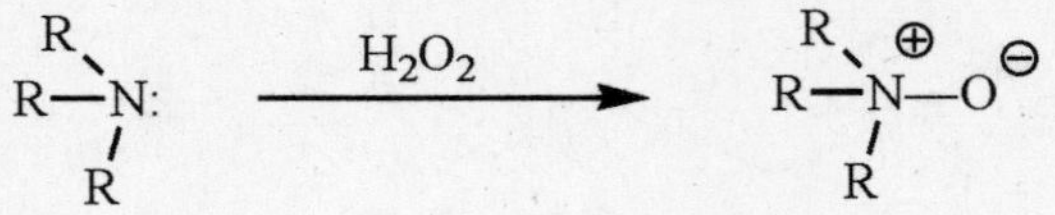

27.36

If the N-Oxide of 2-(N,N-Dimethylamino)Hexane (27.38) Is Heated, What Is the Expected Product?

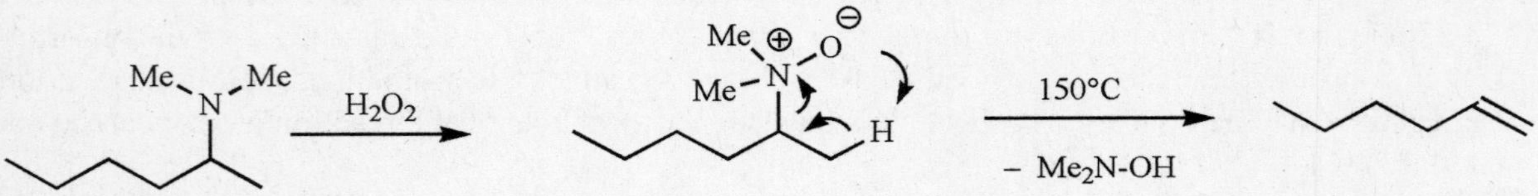

27.37 27.38

Oxidation of ***27.37*** generates the *N*-oxide, ***27.38***. Heating leads to *syn* elimination and formation of the less substituted alkene, 1-hexene. The leaving group in this process is *N,N*-Dimethylhydroxylamine, Me_2N-OH. As noted in the reaction, the reaction temperature for the elimination is somewhat lower than with the Hofmann elimination.

What Is the Name of This Process?

This specific type of elimination is called the **Cope elimination.**

Formation of Imines or Enamines

This reaction was presented in Section 21.3.

What Is the Product When Cyclopentanone Reacts with 1-Butanamine?

The product is *N*-butylcyclopentylimine, ***27.39***.

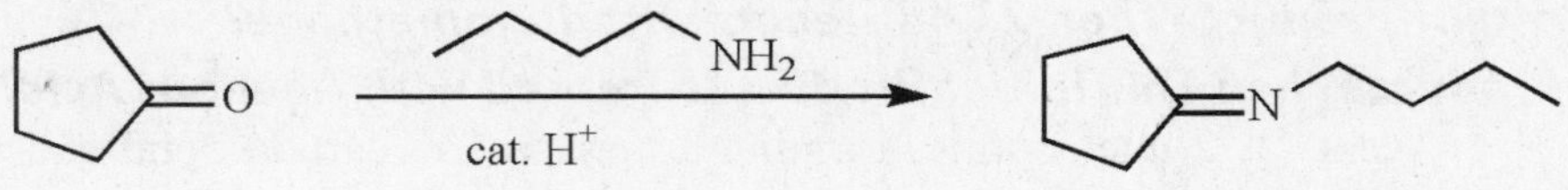

27.39

What Is an Enamine?

An enamine is a molecule that has an amino group (NR_2) attached directly to a carbon-carbon double bond [$C=C-NR_2$], as discussed in Section 21.3.

When Cyclopentanone Reacts with Diethylamine, What Is the Product?

This reaction produces the enamine, ***27.40***.

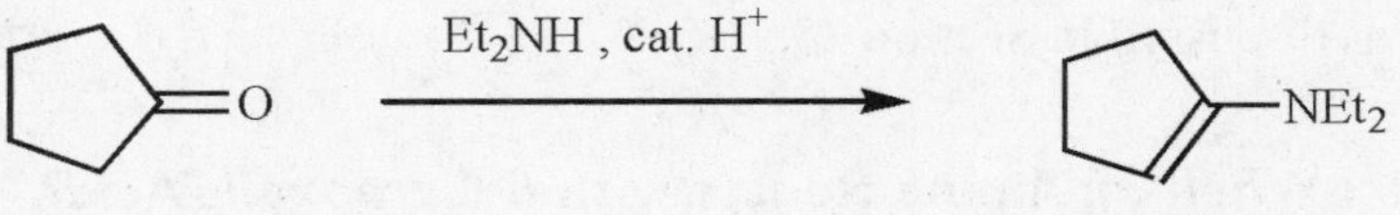

27.40

What Is the Product When an Aldehyde Reacts with a Secondary Amine?

When an aldehyde such as butanal reacts with a secondary amine such as diethylamine, the initially formed iminium salt (***27.41***) reacts with diethylamine to form the geminal diamine, ***27.42***. In the presence of a base (usually the amine), ***27.42*** eliminates a molecule of diethylamine to form the enamine, ***27.43***.

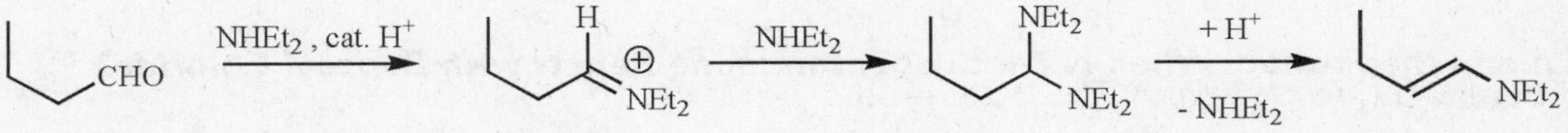

27.41 27.42 27.43

When Diethylamine Reacts with 2-Pentanone, Will the Enamine Have a More Highly Substituted or the Less Substituted Double Bond? Explain.

The major product of this reaction is enamine ***27.44***, where the double bond has the fewest substituents. This is explained by a comparison of enamine ***27.44*** (re-drawn as ***27.454***) with the other possible enamine regioisomer, ***27.46***. The planar nature of this molecule around nitrogen makes the steric interaction of the ethyl groups on nitrogen with the groups on the alkenyl carbons severe in ***27.46***. This interaction is minimized in ***27.45***, making it the major product.

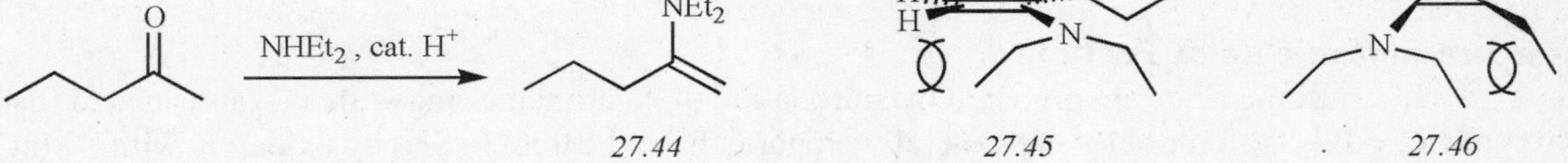

27.44 27.45 27.46

Predict How Enamines Should React with Alkyl Halides or with Ketones and Aldehydes.

The electron pair on nitrogen can be donated to the C=C group, making the terminal carbon of the $R_2N-C=C$ system nucleophilic. The best way to think about an enamine is to think of it behaving as a nitrogen 'enolate,' forming a new carbon-carbon bond with the terminal carbon of C=C.

What Is the Major Product When 27.44 Reacts with Iodomethane? What Is the Product When This Initial Product Is Treated with Aqueous Acid?

Enamine *27.44* reacts as a nucleophile, forming a new carbon-carbon bond and generating an iminium salt, *27.47*. Hydrolysis of the iminium salt (as observed in previous sections) generates a ketone with loss of diethylamine to give 3-hexanone, *27.48*.

NEt$_2$ CH$_3$I → ⊕NEt$_2$ CH$_3$ →(H$_3$O+) O CH$_3$

27.44 *27.47* *27.48*

With Acid Derivatives

The chemistry was discussed in Section 25.5.

What Is the Product When an Amine Reacts with a Carboxylic Acid?

An amine is a base and reacts with the acidic hydrogen of the carboxyl group to give an ammonium carboxylate [R_3NH^+ ^-O_2CR].

Why Does Heating the Reaction Product of an Amine and a Carboxylic Acid Produce an Amide?

Thermally, the ammonium salt protonates another molecule of the ammonium salt, allowing thermal dehydration (loss of water) to form the amide.

What Is the Product When N-Methylpentanamine Reacts with Benzoyl Chloride?

In this case, the product is an amide, *27.49*.

NHMe →(PhCOCl) N(Me)C(=O)Ph

27.49

What Is the Structure of a Sulfonic Acid?

A sulfonic acid has the structure RSO_2OH. The R group can be alkyl or aryl.

How Are Sulfonic Acids Prepared?

Aromatic sulfonic acids are prepared by sulfonation of an aromatic ring with SO_3 and an acid (usually sulfuric). Alkyl sulfonic acids are generally prepared from the thiol (RSH) by oxidation with $KMnO_4$ (HNO_3 can also be used).

What Is the Formula of Benzenesulfonic Acid? Of Tosic Acid? Of Methanesulfonic Acid?

Benzenesulfonic acid has the formula $PhSO_3H$, tosic acid is *para*-toluenesulfonic acid, *p*-MeC_6H_4–SO_3H. Methanesulfonic acid has the formula CH_3SO_3H.

How Is a Sulfonyl Chloride Made?

Just as treatment of a carboxylic acid with $SOCl_2$ or PCl_3 gives the acid chloride, so reaction of a sulfonic acid with $SOCl_2$ (usually in the presence of a base such as pyridine to react with the HCl by-product) gives the sulfonyl chloride: $RSO_3H + SOCl_2 \rightarrow RSO_2Cl$, where RSO_2Cl is the generic structure of the sulfonyl chloride.

What Is Tosyl Chloride? Mesyl Chloride?

Tosyl chloride is the common name for *para*-toluenesulfonyl chloride (***27.50***), and mesyl chloride is the common name for methanesulfonyl chloride (***27.51***).

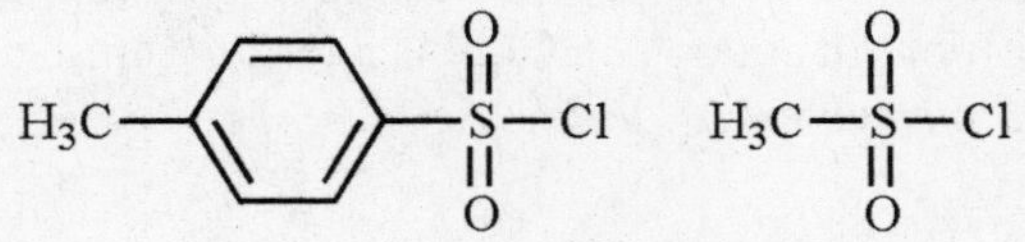

$H_3C-C_6H_4-S(=O)_2-Cl$ $H_3C-S(=O)_2-Cl$

27.50   *27.51*

What Is the Structure of a Sulfonamide?

A sulfonamide is the amide form of the sulfonic acid and has the generic structure $R'SO_2NR_2$. The R group on the nitrogen can be H,H (primary amide), alkyl or aryl,H = secondary amide or alkyl,aryl/alkyl,aryl = tertiary amide.

What Is the Product When Diethylamine Reacts with Benzenesulfonic Acid?

Just as with carboxylic acids, amines react with the acidic hydrogen of a sulfonic acid to produce the ammonium salt: $R_3NH^+\ ^-O_3SR'$.

What Is the Product When Tosyl Chloride Reacts with Diethylamine?

If tosyl chloride is represented as TsCl, the product is the sulfonamide, $TsNEt_2$ [p-$MeC_6H_4SO_2NEt_2$].

What Is the Product When Benzenesulfonyl Chloride Reacts with Ammonia?

The product is benzene sulfonamide, $PhSO_2NH_2$.

Hinsberg Test

What Is the Product When a Primary Amine Reacts with Benzenesulfonyl Chloride in the Presence of Hydroxide?

The product is the sulfonamide: RCH_2NHSO_2Ph.

What Is the Product When a Secondary Amine Reacts with Benzenesulfonyl Chloride in the Presence of Hydroxide?

The product is the sulfonamide: R_2NSO_2Ph.

What Is the Product When a Tertiary Amine Reacts with Benzenesulfonyl Chloride in the Presence of Hydroxide?

The product is the ammonium salt: $R_3NH^+\ ^-O_3SPh$.

Describe the Hinsberg Test.

The Hinsberg test is used to determine if an amine is primary, secondary, or tertiary. It is not always reliable, but in many cases can be used successfully. The amine is, presumably, insoluble in water. The amine is converted to the benzenesulfonamide and placed in aqueous sodium hydroxide. A sulfonamide derived from a primary amine will have an acidic *N*–H that is deprotonated by NaOH to give a soluble product. A secondary amine will give a sulfonamide with no acidic hydrogen and will not be soluble in aqueous NaOH. If the sulfonamide is insoluble in aqueous NaOH, it is then treated for solubility in aq. HCl. If the amine is a tertiary amine, no sulfonamide was formed and $R_3NH^+Cl^-$ will be soluble in aqueous HCl. If the amine is secondary, the sulfonamide product will not be soluble in aqueous HCl. The test is, therefore: If the product of the amine and $PhSO_2Cl$ is soluble in aqueous NaOH it is a primary amine. If it is insoluble in NaOH but soluble in aqueous HCl, it is a tertiary amine. If it is insoluble in both aqueous NaOH and aqueous HCl, it is a secondary amine.

If an Amine Is Treated with $PhSO_2Cl$ and the Product Is Soluble in AQ. NaOH and Insoluble in AQ. HCl, Is It Primary, Secondary or Tertiary?

It is a primary amine.

If an Amine Is Reacted with $PhSO_2Cl$ and the Product Is Insoluble in AQ. NaOH, but Soluble in AQ. HCl, Is It Primary, Secondary or Tertiary?

It is a tertiary amine.

27.5. REACTIONS OF ARYL AMINES

Aryl amines can undergo most of the reactions involving nitrogen described in previous sections, as well as aromatic substitution reactions. Electrophilic Aromatic Substitution was described in Chapter 17.

What Is the Reaction Product When Aniline Is Reacted with $AlCl_3$?

Since $AlCl_3$ is a powerful Lewis acid, aniline will function as a Lewis base, generating the usual complex: $PhNH_2{:}AlCl_3$.

If Aniline Is Reacted with Bromine and No Lewis Acid Catalyst, Explain Why a Tribromoaniline Is Formed.

The NH_2 group on the benzene ring strongly activates that ring to electrophilic aromatic substitution. So strong is the activation that bromine reacts, without a catalyst, to give not only the monosubstitution product, but the tribromo derivatives. (The specific product is 2,4,6-tribromoaniline.)

What Is the Product When N-Acetylaniline Is Treated with a Mixture of Nitric Acid and Sulfuric Acid?

The amide group is a strong activating group and is an *ortho/para* director. The products of this nitration are, therefore, a mixture of the 2-nitro- and 4-nitroacetanilide derivatives. Nucleophilic Aromatic Substitution was described in Chapter 20.

What Is the Formula of Sodium Nitrite?

$NaNO_2$.

What Is the Reactive Species in an Aqueous Solution of Sodium Nitrite and HCl?

The reactive species formed by this reaction is nitrous acid, HONO.

What Is the Product When Aniline Is Treated with $NaNO_2$ in AQ. HCl?

The product is benzenediazonium chloride: $PhN_2^+ \ Cl^-$.

How Stable are Benzenediazonium Salts?

Aromatic diazonium salts are usually stable when kept in aqueous solution. If one attempts to isolate them in pure, anhydrous form, they are very dangerous and often, if not usually, decompose violently (explode). Diazonium salts should **always** be handled only in aqueous solution.

How Reactive Are Benzenediazonium Salts?

They are very reactive to nucleophilic aromatic substitution reactions where water, CuX (X = CN, Br, Cl, but not I - use KI for iodide), or molecules possessing activated aromatic rings are the nucleophiles.

Give an Example of a Reaction That Produces Phenols from Benzenediazonium Chloride.

If benzenediazonium chloride is heated with water, phenol is the product.

Show a Reaction That Produces Chlorobenzene from Benzenediazonium Chloride.

If benzenediazonium chloride is heated with cuprous chloride, chlorobenzene is the product: $PhN_2^+ \ Cl^- + CuCl \rightarrow Ph{-}Cl$.

What Is the Product of the Reaction Between Nitrobenzene and i. HNO_3, H_2SO_4 ii. $NaNO_2$, HCl iii. CuBr?

The product is bromobenzene: Ph–Br.

What Is the Product of the Reaction Between Aniline and i. HONO ii. CuCN?

The product is benzonitrile, Ph–C≡N.

What Is the Product When Benzenediazonium Chloride Is Reacted with H_3PO_2? With $NaBH_4$?

In both cases, these reducing agents convert PhN_2^+ to Ph–H.

What Is the Product When Benzenediazonium Chloride Is Treated with Anisole?

The product is a diazo compound, *27.53*.

Explain How 27.53 Is Formed.

It is formed by the anisole attacking the electrophilic nitrogen of the diazonium salt. This reaction forms the coupling product (cation *27.52*), which loses a proton, usually to the water solvent, to reform the aromatic ring in the final product, *27.53*. Since the OMe activates the *ortho* and *para* positions by an electron releasing inductive effect, these are the carbons that attack the nitrogen. Attack via the *ortho* position is usually sterically hindered, and the major product is attack via the *para* carbon to give *27.53*.

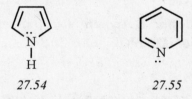

27.52

27.53

What Is the Family Name for Products with the Structure Ar–N=N–Ar?

This type of reaction is called **azo coupling.** Since many of these compounds have strong absorption in the visible region of the electromagnetic spectrum, they can be used as dyes. A common name for this type of compound is **azo dye.**

27.6. HETEROCYCLIC AMINES

A heterocyclic amine is defined as an amine in which nitrogen is part of a ring. Many heterocyclic amines are aromatic, including pyrrole and pyridine.

Representative Amines

What Is the Structure of Pyrrole?

Pyrrole is a five-membered aromatic ring compound that contains one nitrogen. Its structure is ***27.54***.

27.54 27.55

What Is the Structure of Pyridine?

Pyridine is a six-membered aromatic ring that contains one nitrogen. Its structure is ***27.55***.

Discuss the Nomenclature System for Substituted Pyrroles and Pyridines.

The names *pyrrole* and *pyridine* constitute the IUPAC base name of all derivatives of these compounds. The nitrogen always receives the lowest number (1). The ring is numbered to give the smallest combination of substituent numbers.

Give the Name of 27.56–27.59.

(a) 27.56 (b) 27.57 (c) 27.58 (d) 27.59

Amine ***27.56*** is named 3-cyanopyrrole. Amine ***27.57*** is 2-bromo-4-ethylpyrrole. Amine ***27.58*** is 3-isopropylpyridine [3-(1-methylethyl)pyridine is the proper IUPAC name], and amine ***27.59*** is 2,3-dinitropyridine.

Aromatic Character

Why is Pyridine Aromatic?

There are two formal sigma bonds to nitrogen. One electron pair is involved in a π-bond and with the other two π-bonds in the ring, there are the requisite six π-electrons that can be delocalized and pyridine is aromatic (see ***27.60***). The remaining lone electron pair on nitrogen is perpendicular to the aromatic π-cloud. See Section 16.2 for a discussion of aromaticity.

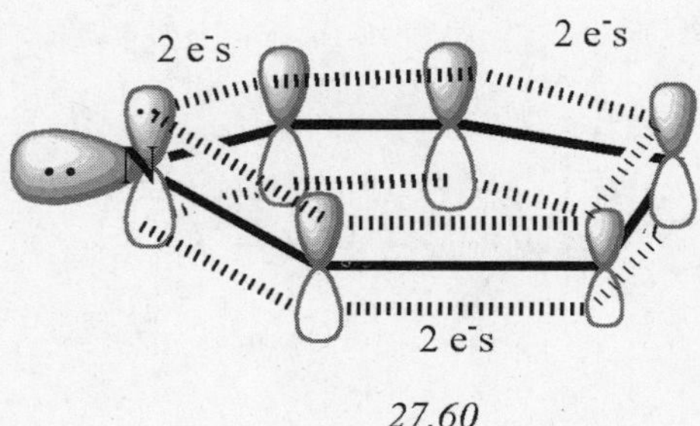

27.60

Is the Hydrogen Attached to Nitrogen in Pyrrole Perpendicular or Parallel to the π-Bonds in the Ring? Explain.

The hydrogen of the *N*–H unit is perpendicular to the π-bonds. The lone electron pair of the nitrogen is part of the six-electron aromatic cloud, forcing the hydrogen to be perpendicular.

Which Is More Basic, Pyridine or Pyrrole?

Pyridine is more basic. The electron pair on nitrogen in pyrrole is 'tied up' as part of the aromatic cloud, but the electron pair on nitrogen in pyridine is not part of the aromatic system. In pyridine, the electron pair on nitrogen is perpendicular to the aromatic system and is relatively unhindered (easily accessible).

Which Is More Basic, Pyridine or Piperidine?

Pyridine is about as basic an aniline, which is generally less basic than secondary alkyl amines. On this basis, piperidine would be expected to be the more basic of the two.

Reactions

What Is the Major Product When Pyrrole Reacts with Bromine and Acetic acid?

The product is that of electrophilic aromatic substitution (see Section 9.5) and the major product is 2-bromopyrrole, ***27.61***.

N–H (pyrrole) $\xrightarrow{Br_2\text{, AcOH}}$ N–H (2-bromopyrrole, Br)

27.61

Give a Mechanistic Rationale for the Formation of 27.61.

When a bromonium ion is attacked by pyrrole, both the C2 and the C3 positions are susceptible to formation of the new C–Br bond. Attack at C2 generates resonance intermediate ***27.62*** where there are three canonical forms, including one that includes delocalization of the charge onto nitrogen. Attack at C3, however, can only give two canonical forms (***27.63***) and is less stable. This usually leads to electrophilic aromatic substitution at the C2 position in five-membered heterocyclic rings, including furan and thiophene.

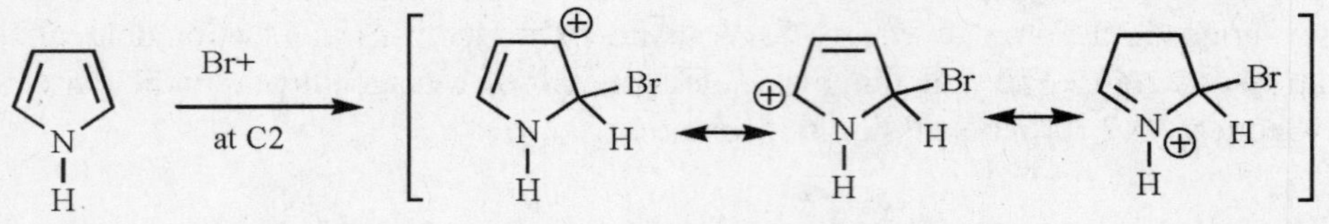

Br+ at C2

27.62

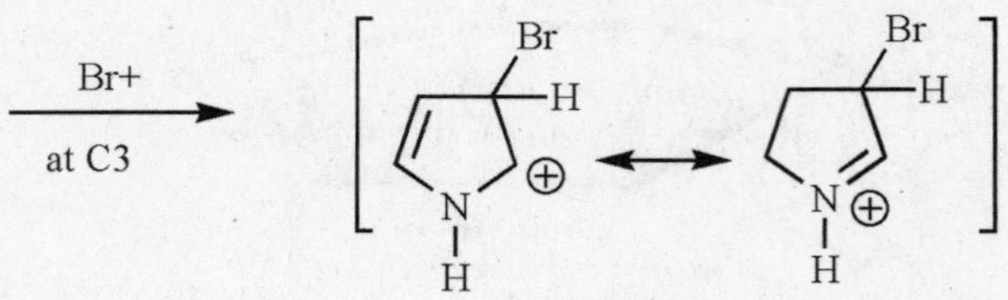

Br+ at C3

27.63

What Is the Major Product When Pyridine Reacts with Nitric Acid and Sulfuric Acid?

Electrophilic aromatic substitution of pyridine usually gives the 3-substituted product. In this case, the major product is 3-nitropyridine, ***27.64***. It is important to note that the reaction requires high temperatures. In general, pyridine reacts poorly in electrophilic aromatic substitution reactions and often requires very vigorous conditions.

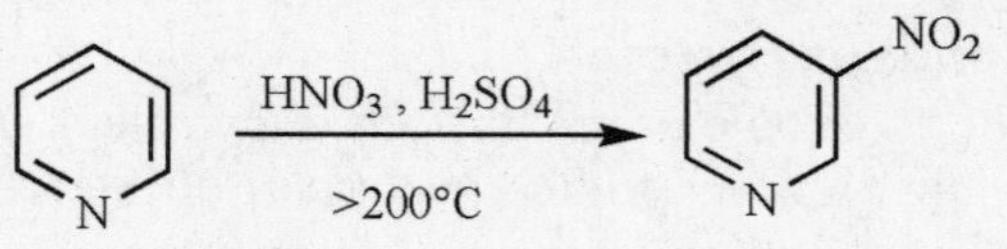

HNO_3, H_2SO_4 >200°C

27.64

What Is the Mechanistic Rationale for This Reaction?

When NO_2^+ reacts with pyridine at C2 (or at C4), one of the resonance forms places a positive charge on an electron deficient nitrogen (see ***27.65***) and is particularly unstable. Attack at C3, however, does not place the positive charge directly on nitrogen. Attack at C3 (to give ***27.66***) is preferred, since it gives the most stable cationic intermediate.

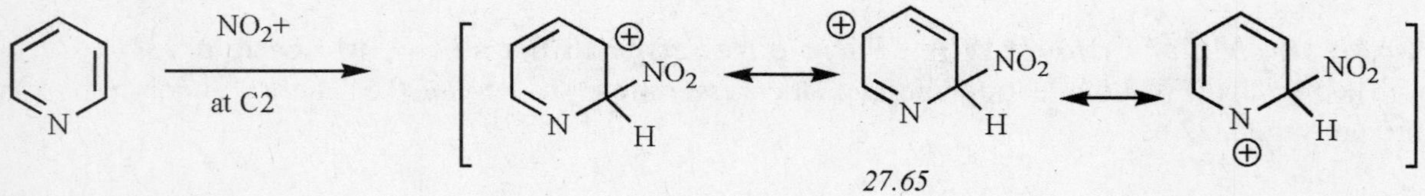

NO_2+ at C2

27.65

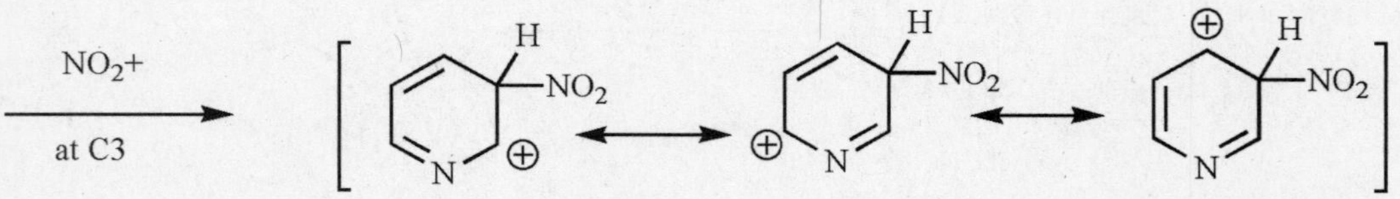

NO_2+ at C3

27.66

Alkylation and Elimination

Explain Why Pyridine and Pyrrole Do Not Give Good Yields of Alkylation Products.

The proximity of the nitrogen lone pair in pyridine to the aromatic system makes pyridine a poor nucleophile. Pyrrole is even weaker since the electron pair on nitrogen is tied up in the aromatic system.

Why Is Pyridine Often Used in E2 Elimination Reactions?

Pyridine is sufficiently basic to remove a β-hydrogen of an alkyl halide and induce an E^2 reaction. It is generally unreactive to other reactions and is not very nucleophilic. Its poor nucleophilic strength makes it particularly attractive for an E^2 reaction, where S_N2 is usually competitive if a nucleophilic base is used with a secondary halide.

Why Is Pyridine Added to the Reaction of a Carboxylic Acid and Thionyl Chloride?

Pyridine is a base and reacts with the HCl by-product, producing pyridinium hydrochloride.

Test Yourself

1. Which of the following is the more basic?

PH_2 $(\ )_2PH$ $(\ )_3P$

2. Choose the more basic atom in each pair. Explain your choice.

(a) N–H   N   (b) N   N

3. Give the IUPAC name for each of the following.

(a) N–H   (b) $(\ )_3N$ (c) Ph, NH_2

(d) Me, N, Br   (e) Ph, N, Ph, Ph   (f) Me, N, Me, Me

4. 1-Bromopentane is reacted with a large excess of methylamine. How can one prove the major product is a secondary amine without using spectroscopy?

5. Give the correct name for each of the following.

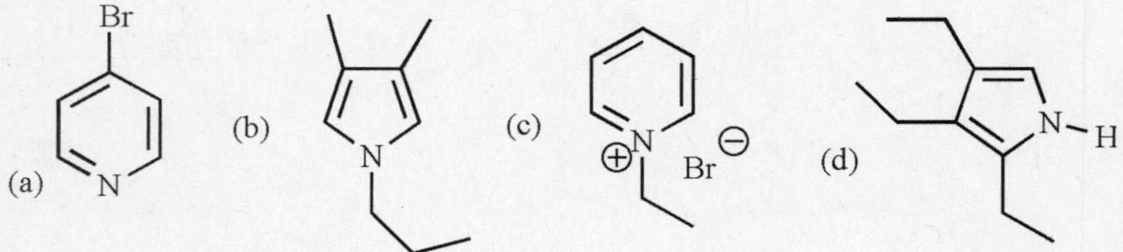

6. Give the major product for each of the following reactions. If there is no reaction, indicate this by N.R.

(a) aniline ($C_6H_5NH_2$) — $AlCl_3$ →

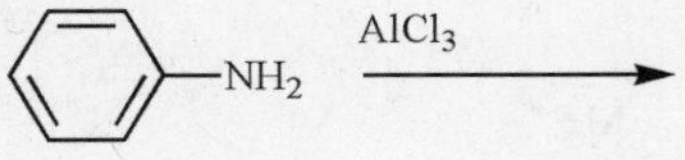

(b) (1-bromoethyl)cyclohexane (Br) — 1. NMe_3; 2. Ag_2O, H_2O; 3. 150°C →

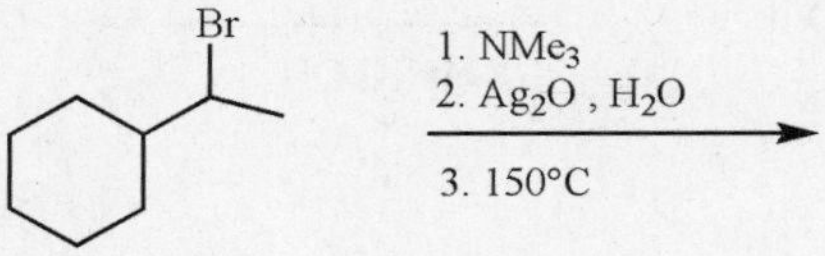

(c) $(C_4H_9)_3N$ — allyl bromide (Br) →

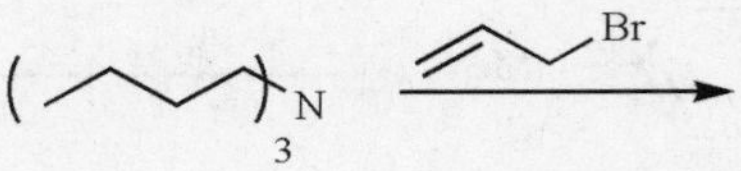

(d) NMe_2-substituted alkane — 1. MeI; 2. Ag_2O, H_2O; 3. 150°C →

(e) PPh_3 — iodopentane →

(f) 1-methylcyclohexanol (OH) — $POCl_3$, pyridine →

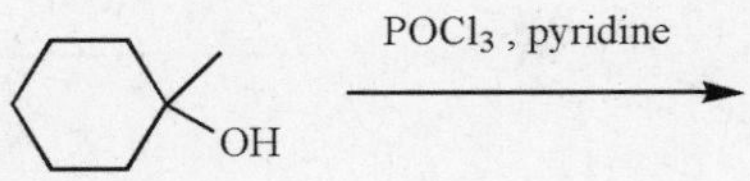

(g) $C_6H_5NMe_2$ — H_2O_2, 25°C →

(h) tertiary alkyl bromide (Br) — pyridine, heat →

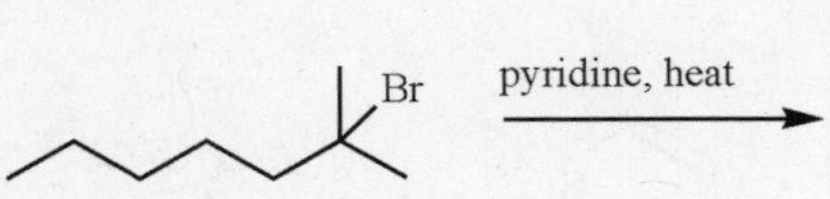

(i) NMe_2-substituted branched alkane — H_2O_2, 180°C →

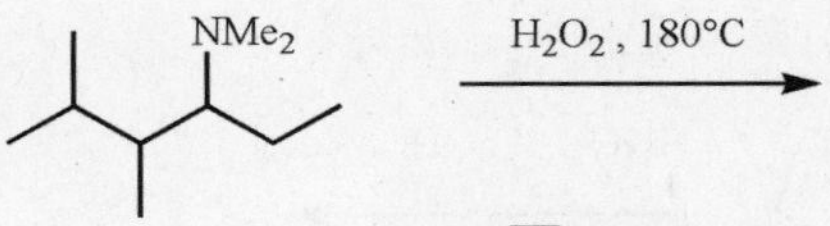

(j) diethyl phthalate (CO_2Et, CO_2Et) — 1. NH_3, heat; 2. BuLi; 3. benzyl bromide; 4. saponification →

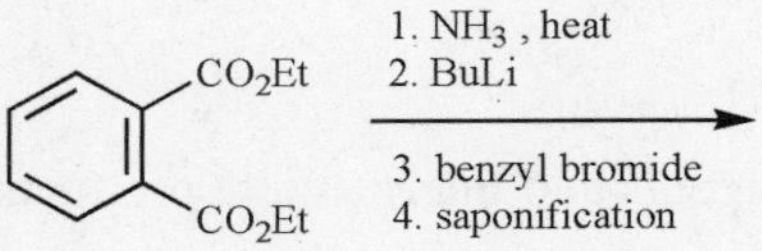

(k) 3-pentanone (O) — morpholine (O, N-H), cat. H^+ →

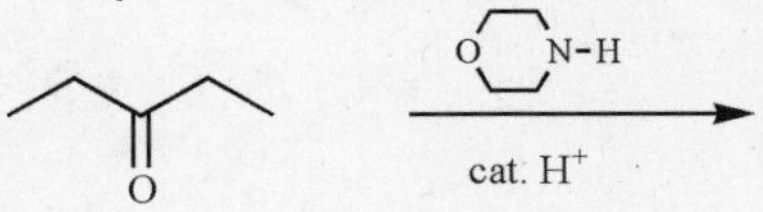

(l) phenylacetaldehyde (CHO) — $C_3H_7NH_2$; H_2, Pd-C →

(m) pyrrolidine enamine (N) — 1. iodopropane; 2. H_3O^+ →

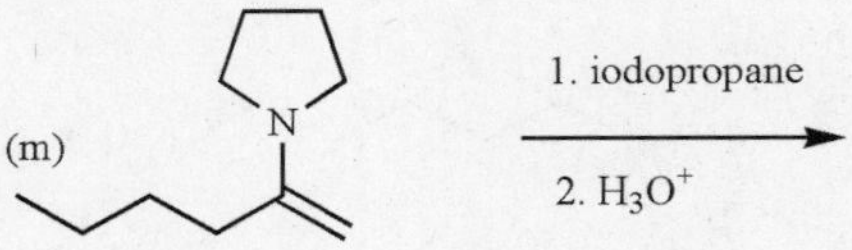

(n) NH_2-substituted alkane — PhCHO, cat. H^+ →

(o) pyrrolidine (N-H) — CO_2H carboxylic acid, 250°C →

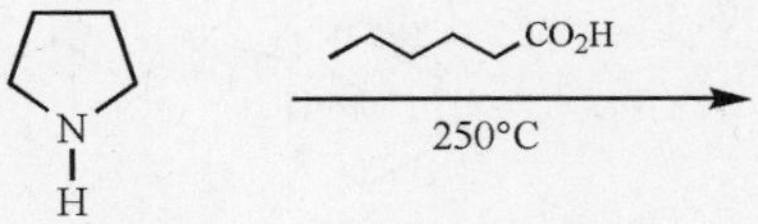

(p) N-butylcyclohexanimine (C_4H_9, N) — 1. HCl; 2. $NaBH_4$ →

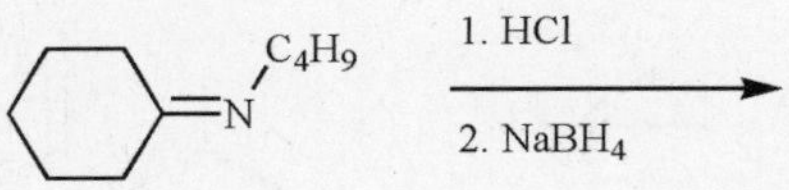

(q) NHEt-substituted alkane — CH_3SO_2Cl →

(r) alkyl bromide (Br) — KCN, DMF →

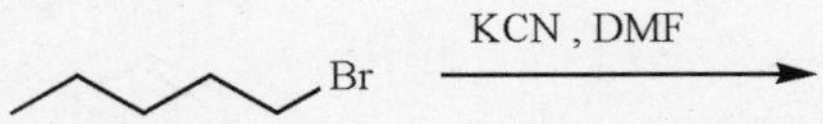

(s) aniline (NH_2) — TsCl, aq. NaOH →

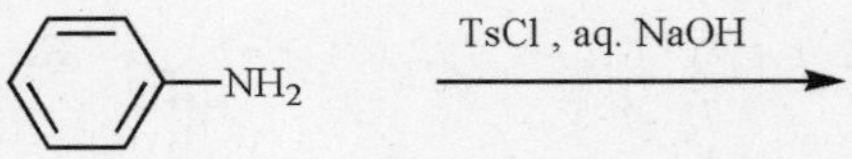

(t) N-methyl-2-piperidinone (O, N, Me) — 1. $LiAlH_4$, THF; 2. aq. NaOH →

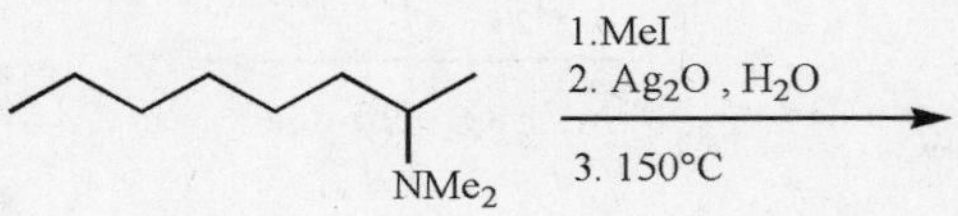

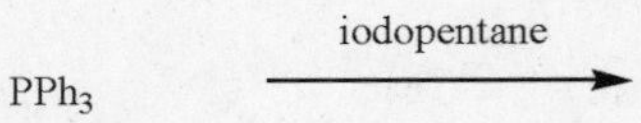

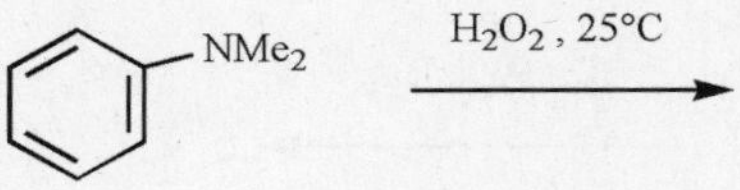

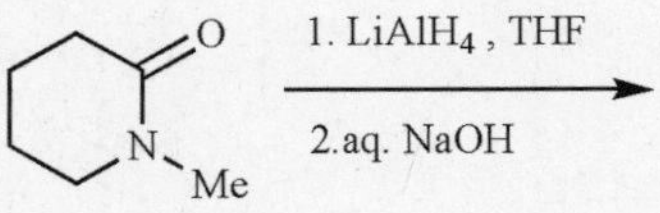

(u) aniline (NH_2) — 1. $NaNO_2$, HCl; 2. H_2O, reflux →

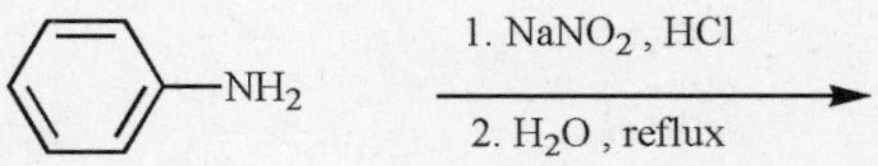

(v) alkyl bromide (Br) — 1. NaN_3, THF; 2. $LiAlH_4$, THF; 3. aq. NaOH →

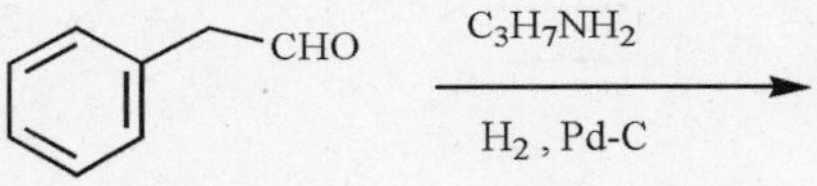

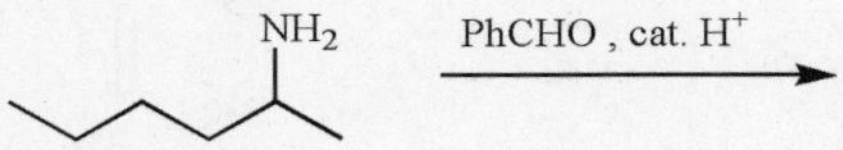

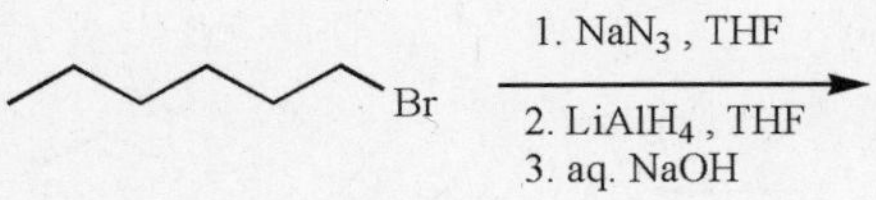

(w) aniline (NH_2) — 1. Ac_2O; 2. $AlCl_3$, butanoyl chloride; 3. saponification →

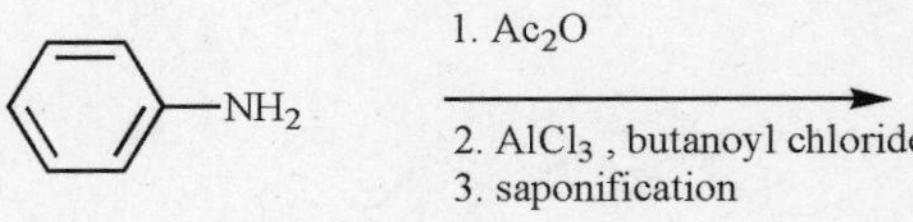

(x) —NO_2 → 1. $LiAlH_4$, ether; 2. aq. NaOH

(y) —$N_2^+\ Cl^-$ → CuBr

(z) C_4H_9, NO_2 → 1. H_2, Pd-C; 2. Ac_2O, pyridine

(aa) Me, OMe, Me → PhN_2+ Cl-

(ab) N → Br_2, AcOH

(ac) Me, Me, Me, N, Me, Me → Cl_2, AcOH

(ad) N, Me → HNO_3, AcOH

(ae) N → $AlCl_3$, Cl_2

7. In each case, provide a suitable synthesis. Show all reagents and intermediate products.

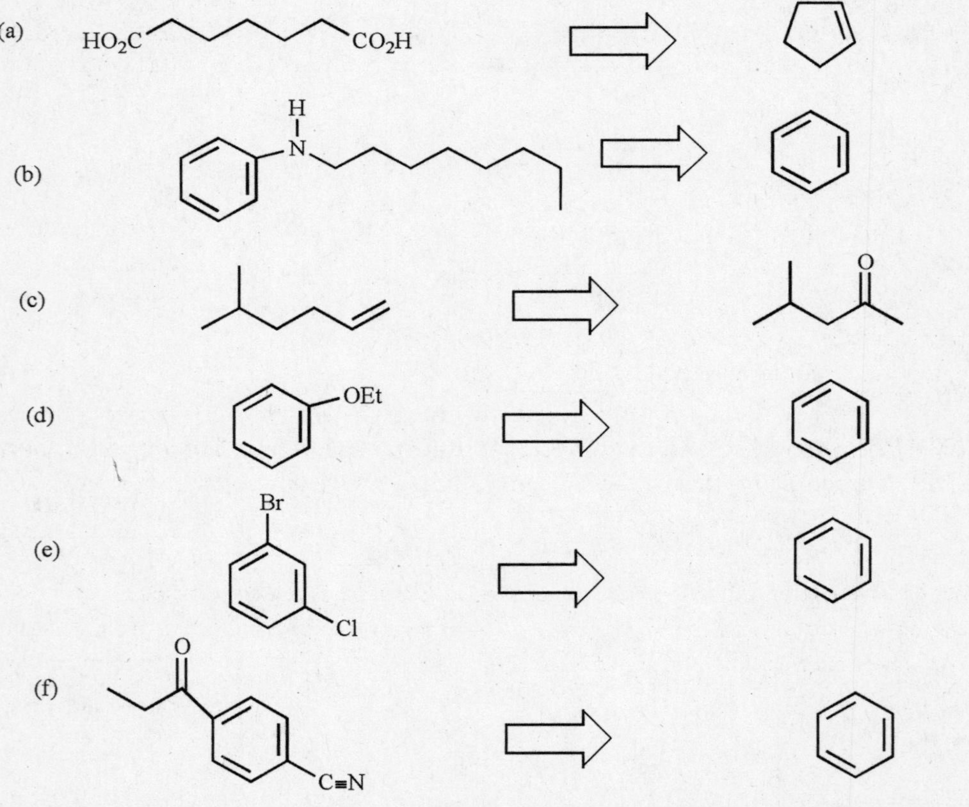

8. Give a mechanistic rationale for the formation of the *N,N*-diethyl enamine of cyclopentanone.

CHAPTER 28

Amino Acids, Peptides and Proteins

Organic molecules that contained two or more functional groups have been seen only sparingly in this book. This chapter discusses one of the most important classes of organic molecules, amino acids. Amino acids are important biologically as constituents of mammalian peptides and proteins. They are important chemically, since their properties and chemical reactions graphically illustrate the problems that arise when two different functional groups are in a single molecule and what happens when those groups interact with each other. This chapter will present the chemical and physical properties of amino acids, their reactions and methods of their preparation. Since amino acids are the important building blocks of proteins and peptides, a brief overview of that chemistry will be given as well.

28.1. AMINO ACIDS

An amino acid is a difunctional molecule that contains an amino group (NH_2) and a carboxylic acid group (CO_2H).

Give the Generic Structure of an α-Amino Acid.

The general structure will be $HO_2CCH(R)NH_2$.

If R Is Not Equal to H, Does the Amino Acid Shown Have a Stereogenic Center?

Yes! The carbon connected to the NH_2 group, the COOH group, and the R group is a stereogenic carbon.

What Is the Absolute Configuration of Most Essential Amino Acids?

These natural amino acids will have the S-configuration, as in ***28.1***. This amino acid is also shown in its Fisher projection (***28.2***).

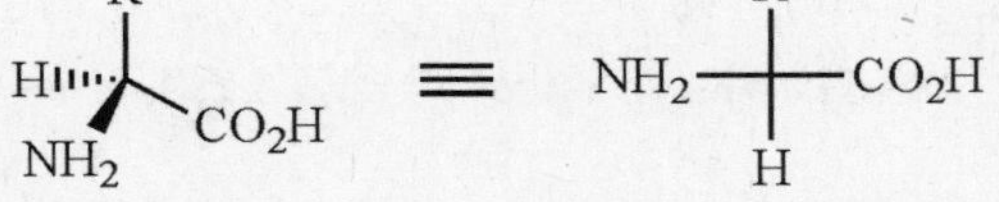

R; H; CO_2H; NH_2 ≡ R; NH_2; CO_2H; H

28.1 *28.2*

Why Are These Amino Acids Called α-Amino Acids?

The NH_2 group is attached to the carbon that is α- to the carboxyl group. They are 2-amino-alkanoic acids. It is important to note that there are amino acids other than α-amino acids. In a long chain carboxylic acid, the NH_2 group can appear on any carbon of the chain to give a 'non-α-amino acid.'

Give the Structure of Each of the Following: A) 4-Aminobutanoic Acid; B) 5-Aminopentanoic Acid; C) 6-Aminohexanoic Acid.

Structure (a) is H_2N-$(CH_2)_3$-CO_2H. Structure (b) is H_2N-$(CH_2)_4$-CO_2H. Structure (c) is H_2N-$(CH_2)_4$-CO_2H.

What Is a Zwitterion?

A **zwitterion** is a dipolar ion that has a positive and a negative charge in the same molecule.

Define the Term Amphoteric.

Amphoteric refers to a molecule that can function as both an acid and a base. At the proper pH, an amino acid can be an acid (via COOH), but changing the pH allows the amino group to function as a base.

Why Are Amino Acids Neutral at Neutral pH?

They are **zwitterions**. An internal acid-base reaction makes the structure of an amino acid an internal salt: H_3N^+-CHR-CO_2-. The formal charge of this molecule is zero (electrically neutral), since the + and − charges cancel.

Give the Zwitterionic Structure of a Generic α-Amino Acid.

H_3N^+-CHR-CO_2^-.

Why Does an α-Amino Acid Exist As a Zwitterion?

At neutral pH, the NH_2 group is a base and reacts with the acidic COOH group to form the zwitterion.

What Is a Neutral Amino Acid?

A neutral amino acid is an α-amino acid that, at neutral pH, has a side chain (R in ***28.1***) that is neither acidic nor basic (no amine or carboxylic acid groups). In general, these are simple alkyl or aryl groups (methyl, ethyl, isopropyl, phenyl, *p*-methoxyphenyl, etc.).

List the Common Neutral Amino Acids by Name, 3-Letter Code, and Give Their Structure.

The neural amino acids are: glycine (**Gly**, ***28.3***), alanine (**Ala**, ***28.4***), isoleucine (**Ile**, ***28.5***), leucine (**Leu**, ***28.6***), valine (**Val**, ***28.7***), phenylalanine (**Phe**, ***28.8***), serine (**Ser**, ***28.9***), cysteine (**Cys**, ***28.10***), methionine (**Met**, ***28.11***), threonine (**Thr**, ***28.12***), asparagine (**Asn**, ***28.13***), glutamine (**Glu**, ***28.14***), proline (**Pro**, ***28.15***), tyrosine (**Tyr**, ***28.16***) and tryptophan (**Trp**, ***28.17***). Note that the indole nitrogen in tryptophan is not very basic.

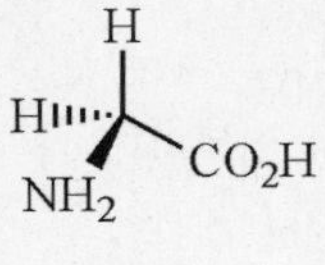

28.3 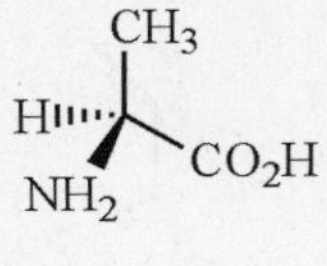 *28.4* 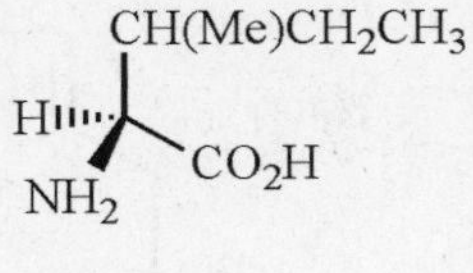 *28.5* 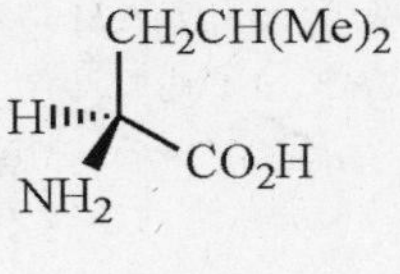 *28.6* 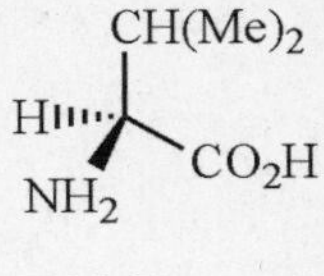 *28.7*

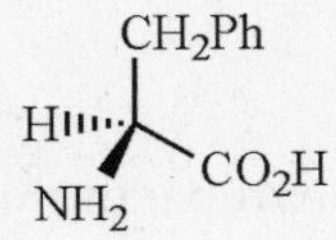

28.8 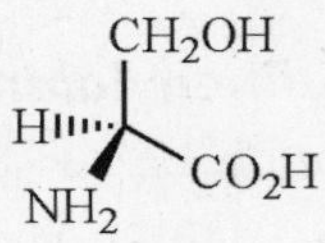 *28.9* 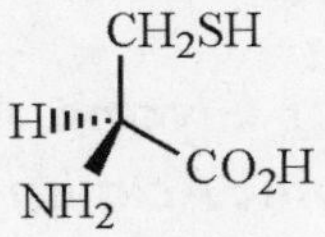 *28.10* 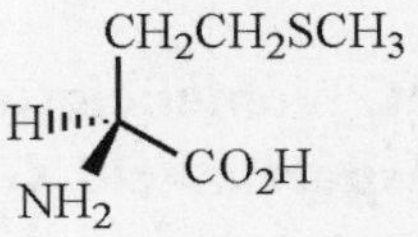 *28.11* 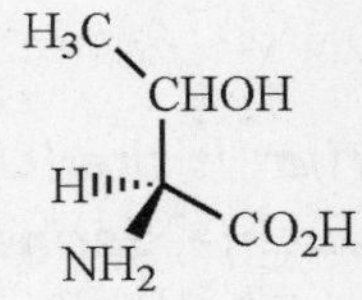 *28.12*

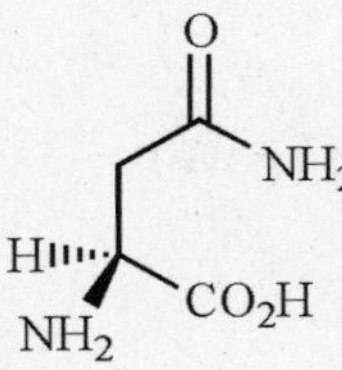

28.13 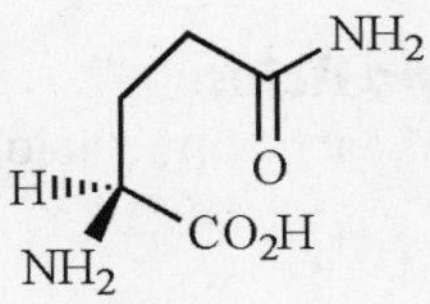 *28.14* 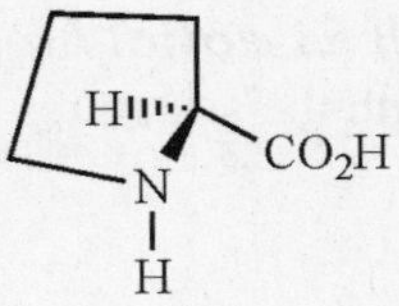 *28.15* 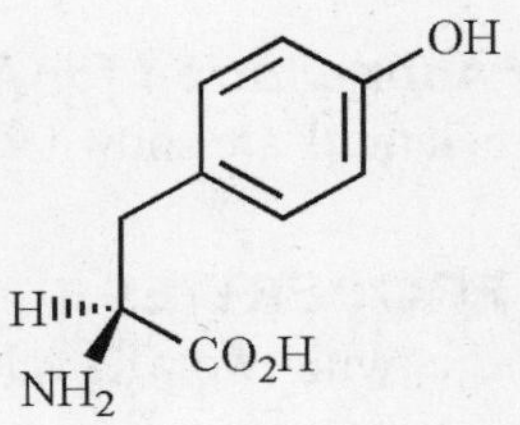 *28.16* 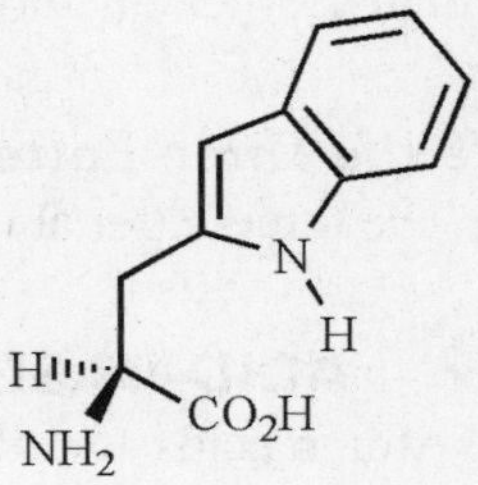 *28.17*

What Is an Acidic Amino Acid?

An acidic amino acid is an α-amino acid that has a side chain (R in ***28.1***), which contains a carboxylic acid unit, COOH.

List the Common Acidic Amino Acids by Name, 3-Letter Code, and Give Their Structure.

The two common acidic amino acids are aspartic acid (**Asp**, ***28.18***) and glutamic acid (**Glu**, ***28.19***).

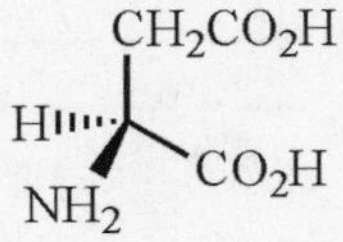

28.18 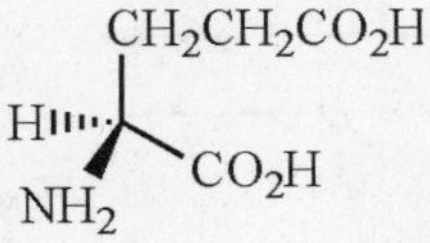 *28.19*

What Is a Basic Amino Acid?

A basic amino acid is an α-amino acid that has a side chain (R in ***28.1***), which contains a free amino group (-NH_2 or NHR).

List the Common Basic Amino Acids by Name, 3-Letter Code, and Give Their Structure.

The most common basic amino acids are arginine (**Arg**, **28.20**), histidine (**His**, ***28.21***) and lysine (**Lys**, ***28.22***).

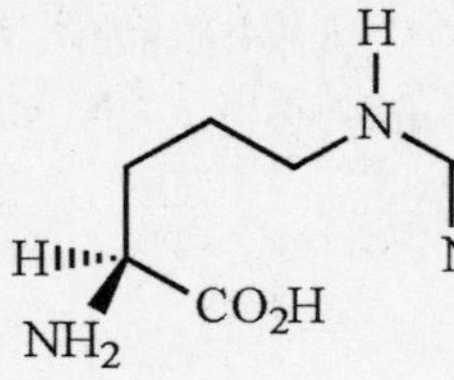

H NH₂ NH CO₂H

28.20 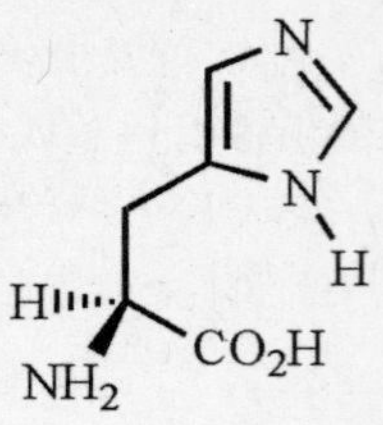 28.21 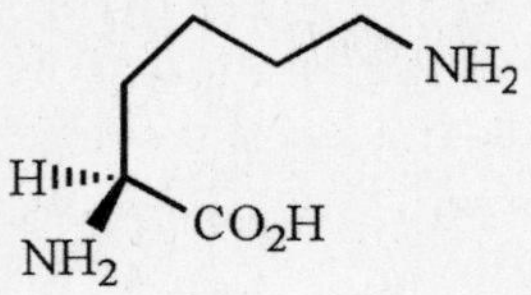 28.22

What Is the IUPAC Nomenclature for Glycine, Alanine, Phenylalanine, Leucine, Serine, Aspartic Acid, Glutamic Acid and Lysine.

The IUPAC name for glycine is 2-aminoethanoic acid. Alanine is 2-aminopropanoic acid. Phenylalanine is 2-amino-3-phenylpropanoic acid. Leucine is 2-amino-4-methylpentanoic acid. Serine is 2-amino-3-hydroxypropanoic acid. Aspartic acid is 2-amino-1,4-butanedioic acid. Glutamic acid is 2-amino-1,5-pentanedioic acid and lysine is 2,5-diaminohexanoic acid.

Give the Three-Letter Abbreviation for All Essential Amino Acids.

The three-letter abbreviations are shown with the structures of each amino acid (see above).

28.2. ACID-BASE PROPERTIES

Amino acids are dibasic, with two pK_a values.

There Are Two pK_a Values for Neutral Amino Acids. Draw the Reactions That Illustrate Each Acid-Base Process.

Glycine is used to illustrate this process. The neutral form of glycine is ***28.23***. In acidic solutions, the NH_2 group is converted to the ammonium salt, and the carboxylate is protonated as the acid (in ***28.25***). Neutralization with base removes the most acidic hydrogen, from the carboxyl, to generate the zwitterion, ***28.23***. Further basification removes the proton from the ammonium ion, which is also acidic. This reaction liberates the free amine, along with the carboxylate salt, ***28.24***.

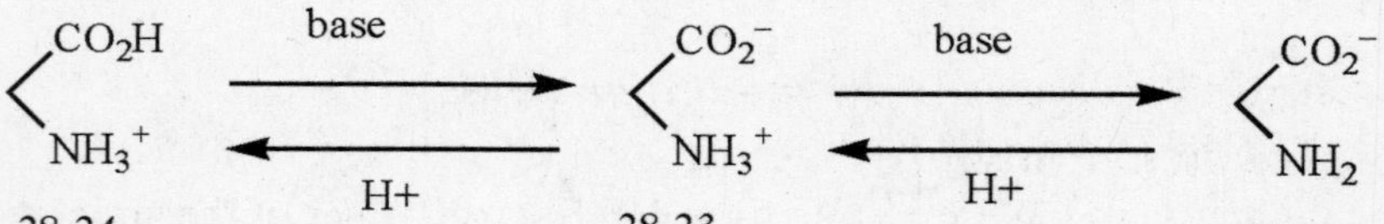

$$\underset{28.24}{\overset{CO_2H}{\underset{NH_3^+}{\langle}}} \underset{H+}{\overset{base}{\rightleftharpoons}} \underset{28.23}{\overset{CO_2^-}{\underset{NH_3^+}{\langle}}} \underset{H+}{\overset{base}{\rightleftharpoons}} \underset{28.25}{\overset{CO_2^-}{\underset{NH_2}{\langle}}}$$

Why is the First pK_a of Glycine 2.34 and the Second pK_a 9.6?

The presence of an amino group of the α-carbon allows internal hydrogen bonding, which makes the carboxyl proton more acidic than acetic acid, hence the relatively low pK_a of 2.34. Removal of the proton from ***28.24*** is analogous to most other ammonium salts, which have pK_a values around 9–10. The 9.6 pK_a observed is, therefore, typical.

Why Is the pK_a of the Carboxyl Group of Glycine Lower Than the pK_a of Acetic Acid?

The nitrogen on the α-carbon can hydrogen bond with the acidic proton (through-space interaction), weakening the O–H bond and making it a stronger acid.

In Aspartic Acid, the pK_a of the Carboxyl of the α-Amino Acid is 2.09, and the pK_a of the 'Side Chain' Carboxyl is 3.86. Why Is the Side Chain Carboxyl Less Acidic?

The electron withdrawing nitrogen group is close to the COOH of the 'α-amino acid unit,' and the effect is rather large. This nitrogen is much further away from the COOH on the side chain, and the effects are minimal.

What Is the Isoelectric Point for an Amino Acid?

The **isoelectric point** is the pH at which the amino acid is completely neutral (no longer exhibits a charge). The structure at the isoelectric point is the zwitterion form, such as *28.23*.

Relate Isoelectric Point to the Acid/Base Equilibrium Species Present in Glycine.

At the isoelectric point, the zwitterion *28.23* (or the analogous structure for the other amino acids) represents the major species in solution.

How Does the Presence of an Acidic Side Chain Influence the Isoelectric Point of an Amino Acid?

The presence of the second carboxyl group will lower the pH of the isoelectric point (to pH 3.2–3.5, typically).

How Does the Presence of a Basic Side Chain Influence the Isoelectric Point of an Amino Acid?

The presence of the basic amino groups will raise the pH of the isoelectric point (to pH 7.6–10.8, in most cases).

28.3. CHIRALITY OF NATURALLY OCCURRING AMINO ACIDS

With the exception of glycine, the naturally occurring amino acids have a stereogenic center, and are chiral molecules. Due to their importance, a system of nomenclature was developed prior to the IUPAC rules that must be mentioned.

What Is the Absolute Configuration of the Amino Acids Most Commonly Found in Proteins?

The major enantiomer found in most proteins is the (S) enantiomer of the amino acid.

What Is a (d) Amino Acid?

The symbol (d) is used for amino acids that have a positive (+) specific rotation.

What Is an (l) Amino Acid?

The symbol (l) is used for amino acids that have a negative (−) specific rotation.

Is There Any Correlation Between the d,l Designator and the R/S Configuration?

No! The R/S configuration is a name based on arbitrary, but universally accepted, rules (see Section 3.3). The d,l label represents specific rotation, which is an unchangeable physical property of the molecule.

What is the Structure of Glyceraldehyde?

The structure of glyceraldehyde is $HOCH_2CH(OH)CHO$.

What Is the D Nomenclature?

The structure of R-(+)-glyceraldehyde is ***28.26*** (in Fisher projection). A German chemist, Emil Fisher invented a system of nomenclature, which compared amino acids to glyceraldehyde. He assigned the letter D to ***28.26*** (D-glyceraldehyde, where D is dextrorotatory and is a nomenclature designator). The D represents the absolute configuration when placed in the Fisher projection (OH is on the right and CHO is on the 'top'). Fisher guessed at the absolute configuration; he was later proved to be correct. Using glyceraldehyde as a model, α-amino acids can be converted to the analogous Fischer projection and assigned a D or L configuration. The D-nomenclature system places the amino acid in a Fisher projection (see ***28.27*** for alanine) where the COOH of the amino acid correspond to the CHO of glyceraldehyde. The amino acid side chain is compared with the CH_2OH of glyceraldehydes, and the amino group is compared with OH. If the COOH is at the 'top' and the side chain is at the 'bottom', the NH_2 group will be on the left (L) or the right (D). As drawn, ***28.27*** is D-alanine (R-alanine).

28.26	*28.27*	*28.28*
CHO	CO_2H	CO_2H
H—+—OH	H—+—NH_2	NH_2—+—H
CH_2OH	CH_3	CH_3

What Determines If an Amino Acid Is in the L Nomenclature?

As described above, the NH_2 group in the Fisher projection will appear on the left side, and the COOH at the top as in ***28.28***, which is L-alanine (S-alanine).

Draw D-(+)-Glyceraldehyde in Fisher Projection.

D-(+)-glyceraldehyde is ***28.26***.

How Does D-(+)-Glyceraldehyde Structurally Correlate with D-Alanine?

The CHO correlates with CO_2H and the NH_2 correlates with the OH.

Draw the Fisher Projection of L-Phenylalanine? of D-Leucine? of L-Serine? of D-Cysteine?

The Fisher projection of L-phenylalanine is ***28.29***. D-leucine is ***28.30***. L-serine is ***28.31***, and D-cysteine is ***28.32***.

28.29	*28.30*	*28.31*	*28.32*
CO_2H	CO_2H	CO_2H	CO_2H
NH_2—+—H	H—+—NH_2	NH_2—+—H	H—+—NH_2
CH_2Ph	CH_2CHMe_2	CH_2OH	CH_2SH

What Is Allothreonine? Alloisoleucine?

Two of the amino acids have a second stereogenic center in the side chain, threonine (***28.33*** is *L-Thr* in Fisher projection) and isoleucine (***28.34*** is *L-Ile* in Fisher projection). The diastereomer of threonine is called **allothreonine** (*28.35*), and the diastereomer of isoleucine is **alloisoleucine** (*28.36*).

CO_2H; NH_2—H; H—OH; CH_3

28.33

CO_2H; NH_2—H; H—CH_3; CH_2CH_3

28.34

CO_2H; NH_2—H; HO—H; CH_3

28.35

CO_2H; NH_2—H; H_3C—H; CH_2CH_3

28.36

Is There Any Correlation Between the d,l Designator and the D,L Designators?

No. The (d,l) nomenclature refers to specific rotation (+,−) whereas the (D,L) refers to the name of the absolute configuration (analogous to R,S).

Is There Any Correlation Between the D,L Designator and the R,S Designators?

Both refer to absolute configuration, but (R,S) is based on the Cahn-Prelog-Ingold selection rules, and (D,L) is based on a comparison with glyceraldehyde.

28.4. SYNTHESIS OF AMINO ACIDS

Amino acids are synthesized by relatively conventional methods. However, the presence of two functional groups in one molecule forces certain constraints on the reagents and reactions that can be used, and the order in which each functional group is incorporated.

What Is the Product When 2-Bromoethanoic Acid Is Heated with an Excess of Ammonia?

The product is glycine.

What Is the Product When 2-Bromo-3-Methylpentanoic Acid Is Heated with Excess Ammonia?

The product is isoleucine.

Does This Reaction Produce a Single Enantiomer or a Racemic Mixture? A Single Diastereomer or a Mixture? Explain.

If the starting bromide (*28.37*) is racemic, the final product (*28.38*) will also be racemic. There may be some enantioselectivity if the bromide is chiral, since displacement will be via a S_N2 reaction. Since there is no stereocontrol in this reaction, a mixture of diastereomers in ***28.37*** will result in a diastereomeric mixture of isoleucine and alloisoleucine (***28.38***). If the methyl-bearing carbon is chiral and the bromine-bearing carbon is not, a mixture of diastereomers will result.

Br; CO_2H; CH_3 —(2 NH_3)→ NH_3^+; CO_2^-; CH_3

28.37 *28.38*

The Strecker Synthesis

What Is the Product When Phenylacetaldehyde Is Reacted with Sodium Cyanide and Ammonium Chloride?

The initial product of the reaction with phenylacetaldehyde (***28.39***) is the aminonitrile, ***28.40***.

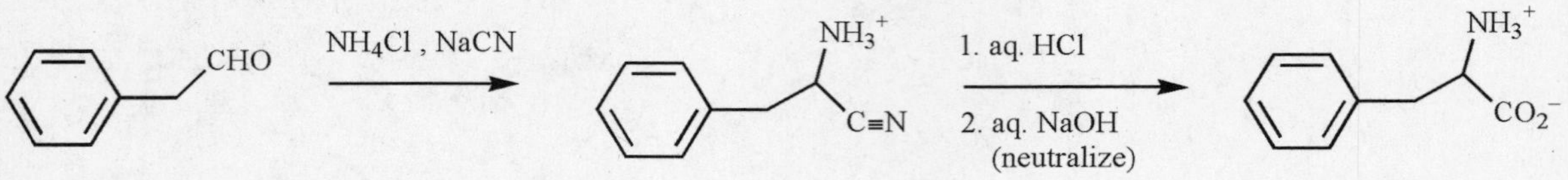

CHO $\xrightarrow{NH_4Cl,\ NaCN}$ NH_3^+ C≡N $\xrightarrow[\text{2. aq. NaOH (neutralize)}]{\text{1. aq. HCl}}$ NH_3^+ CO_2^-

28.39 *28.40* *28.41*

If 28.40 Is Treated with a). AQ. HCl b) NaOH, What Is the Product?

When ***28.40*** is treated with acid, the cyano group in hydrolyzed to a carboxylic acid (see Section 25.3), and neutralization with base will give the zwitterionic amino acid, phenylalanine (***28.41***).

What Is the Name of This Process?

This reaction sequence is called the **Strecker Synthesis,** and it adds a carbon in the course of the reaction sequence.

Give the Major Product for the Reactions of 28.42–28.44.

(a) CHO *28.42* — 1. NaCN , NH_4Cl 2. aq. HCl 3. aq. NaOH

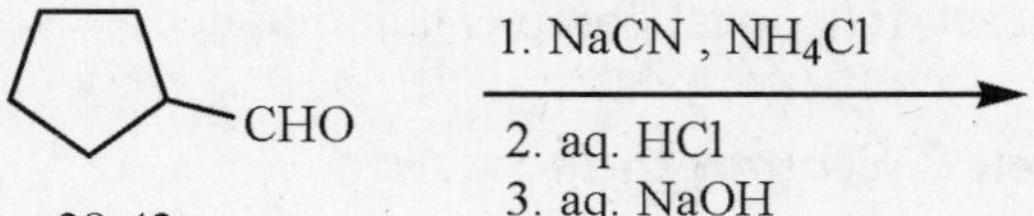

(b) Ph CHO *28.43* — 1. NaCN , NH_4Cl 2. aq. HCl 3. aq. NaOH

(c) CHO *28.44* — 1. NaCN , NH_4Cl 2. aq. HCl 3. aq. NaOH

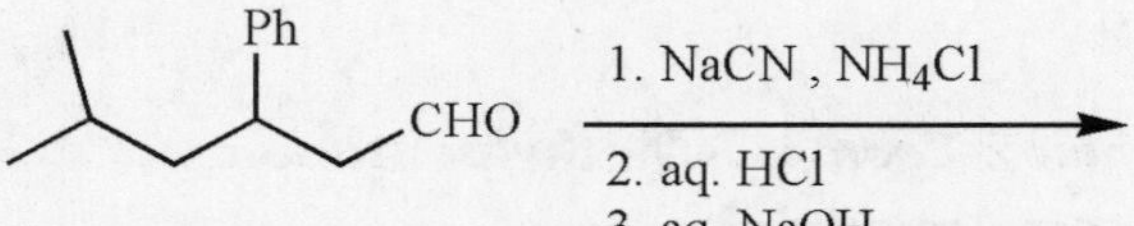

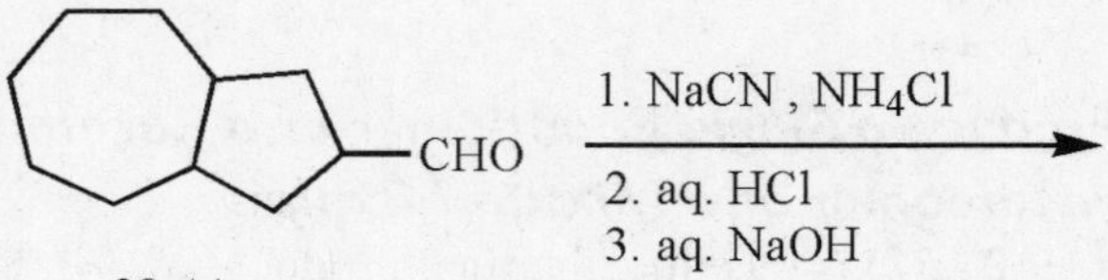

In all three cases, the product is a racemic amino acid. In (a) aldehyde ***28.42*** is converted to ***28.45***. In (b) aldehyde ***28.43*** is converted to ***28.46***, and in (c) aldehyde ***28.44*** is converted to ***28.47***.

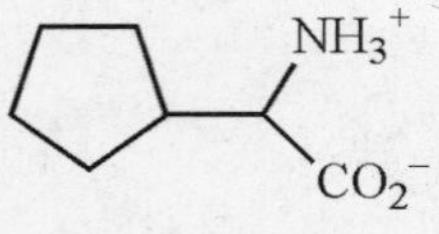

NH_3^+ CO_2^- *28.45*

Ph NH_3^+ CO_2^- *28.46*

NH_3^+ CO_2^- *28.47*

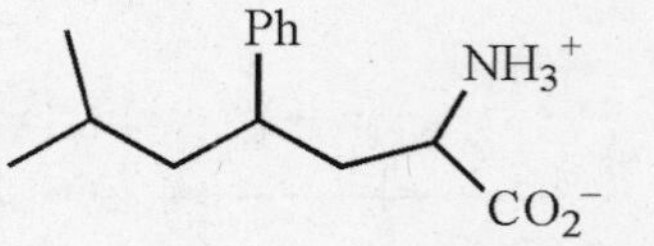

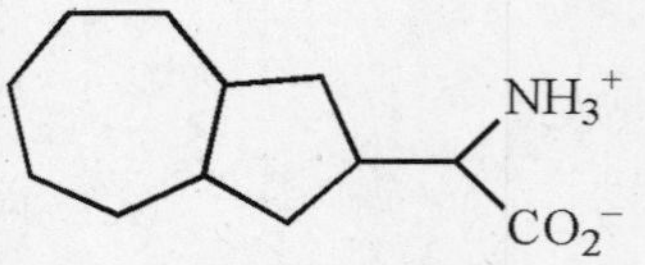

The Gabriel Synthesis

What Is the Product When the Sodium Salt of Phthalimide Is Reacted with Diethyl 2-Bromomalonate (28.48)?

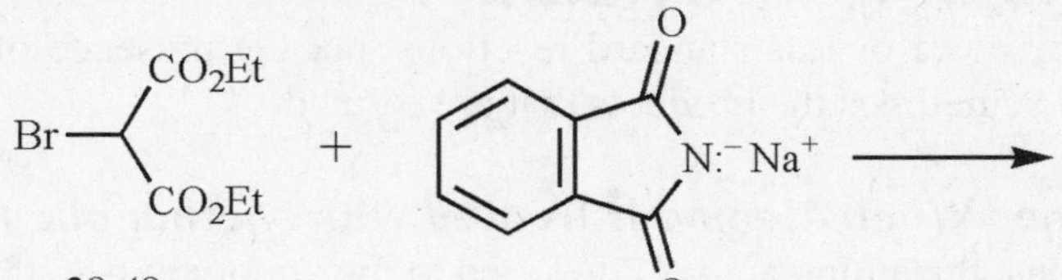

28.48

The nucleophilic phthalimide anion displaces the bromide to give ***28.49***.

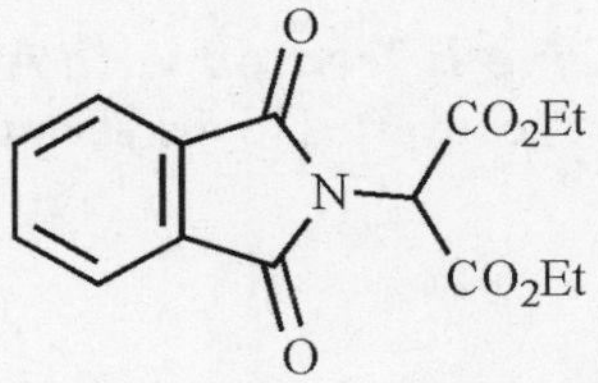

28.49

If 28.49 Is Reacted with Na/EtOH, What Is the Resulting Product? If Iodomethane Is Added, What Is the Product?

Treatment of ***28.49*** with sodium in ethanol, which produces sodium ethoxide *in situ*, leads to deprotonation of the acidic α-proton to form the enolate anion (***28.50***). Subsequent reaction with iodomethane gives the alkylated product, ***28.51***.

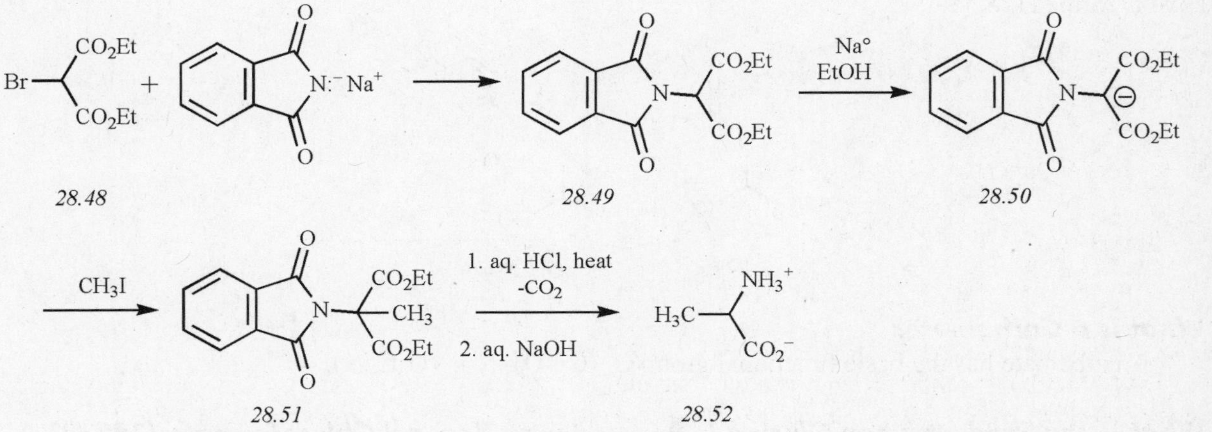

28.48 *28.49* *28.50*

28.51 *28.52*

Hydrolysis of 28.51 and Neutralization with Base Gives What Product?

The hydrolysis sequence converts the phthalimide unit into an amino group. The ester groups are hydrolyzed to the corresponding acid, and heating the 1,3-diacid leads to decarboxylation. The overall sequence generates an amino acid, alanine (***28.52***).

What Is the Name of This Overall Process?

The name of this synthetic sequence is the **Gabriel Synthesis**.

28.5. REACTIONS OF AMINO ACIDS

Amino acids react by more or less standard reactions, but the presence of two functional groups modify the reactivity and sometimes the products that are formed.

What Product is Formed When Alanine Is Treated with Methanolic HCl?

Under these conditions, the amino acid is converted to the ammonium acid NH_3^+–CH(R)–COOH, and in the presence of methanol, the methyl ester is formed. In this case, the product is the methyl ester of alanine [H_3N^+–CH(Me)CO_2Me].

What Is the Product When Isoleucine Is Treated with Acetic Anhydride?

Under these conditions, isoleucine is converted to the acetamide derivative (***28.53***).

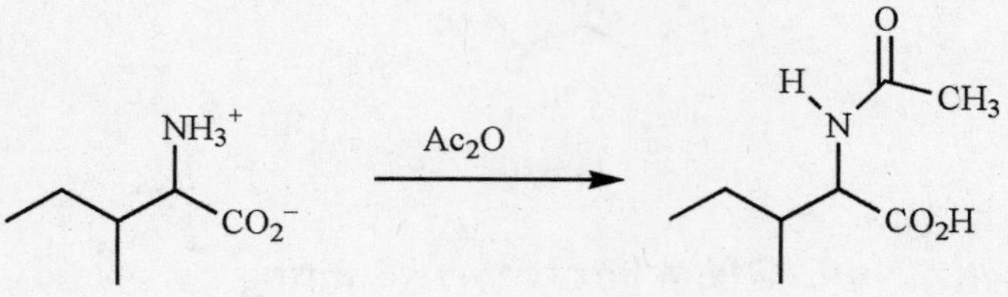

NH_3^+ CO_2^- —Ac_2O→ H–N–C(=O)CH_3 CO_2H

28.53

What Is the Product When Alanine Is Treated with Benzyl Chloride in the Presence of an Amine? With Tosyl Chloride in the Presence of an Amine?

In the first reaction, the benzamide derivative (***28.54***) is formed and in the second, the *N*-tosyl derivative is formed (***28.55***).

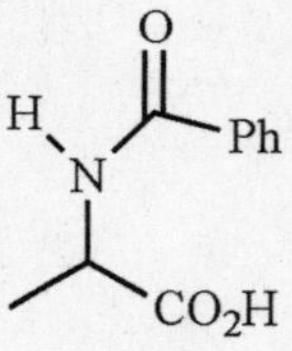

H–N–C(=O)Ph CO_2H H–N–S(=O)(=O)–C_6H_4–Me CO_2H

28.54 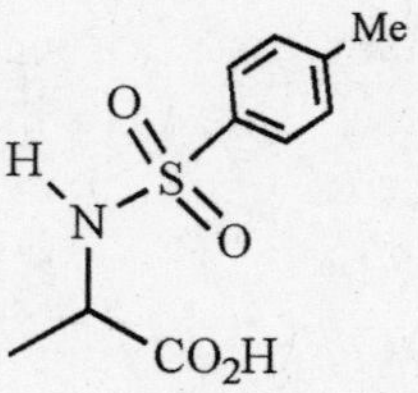   *28.55*

What Is a Carbamate?

A carbamate has the basic functional group O–(C=O)–N = (O_2C–N).

What Is the Product when Glycine Is Reacted with Benzoyl Chloroformate (28.56)?

The product is the benzyl carbamate (the so-called CBz derivative), ***28.57***.

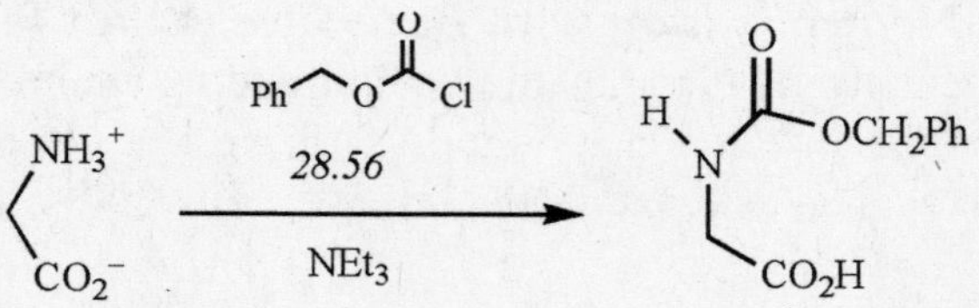

NH_3^+ CO_2^- —(Ph–CH_2–O–C(=O)–Cl *28.56*; NEt_3)→ H–N–C(=O)–OCH_2Ph CO_2H

28.57

What Is the Product When Glycine Is Reacted with tert-Butyl Carbonate (28.58)?

The product is the *tert*-butyl carbamate, the so-called t-BOC or BOC derivative, ***28.59***.

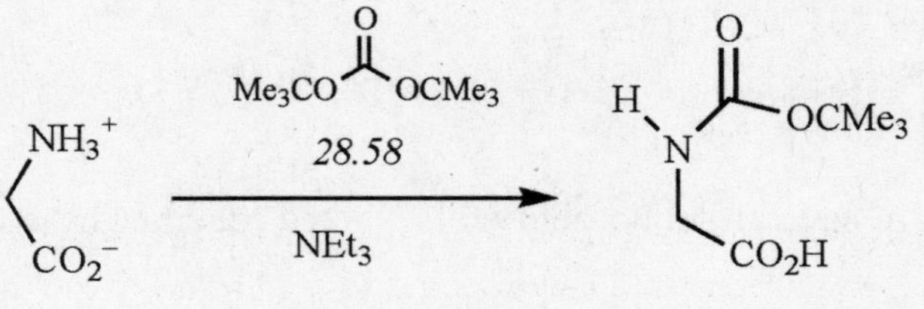

NH_3^+ CO_2^- → (Me_3CO–C(=O)–$OCMe_3$, *28.58*; NEt_3) → H–N(–C(=O)–$OCMe_3$)–CH_2–CO_2H

28.59

What Is the Structure of Ninhydrin?

The structure of ninhydrin is ***28.60***.

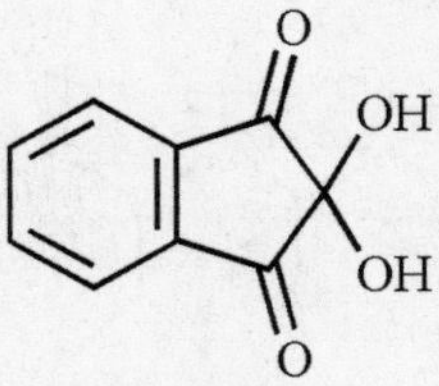

28.60

What Is the Initial Product When Ninhydrin Reacts with Leucine?

In the initial reaction, ninhydrin (***28.60***) reacts with the amine portion of the amino acid to produce an imine. When ninhydrin reacts with leucine, the product is ***28.61***.

The initially formed imine (***28.61***) decarboxylates under the reaction conditions to form a new imine, ***28.62***. This imine can react with more ninhydrin to form a new imine (***28.63***) which is known as **Ruhemann's Purple** (absorbs strongly at 570 nm in the visible spectrum; it has a bluish-purple color). The alkyl side chain of the amino acid is lost as an aldehyde (***28.64***). This reaction is diagnostic for amino acids that contain a primary amino function ($-NH_2$). It does not work with secondary amines. Proline, therefore, does not react in this way with ninhydrin. Proline reacts, but does not give Ruhemann's Purple.

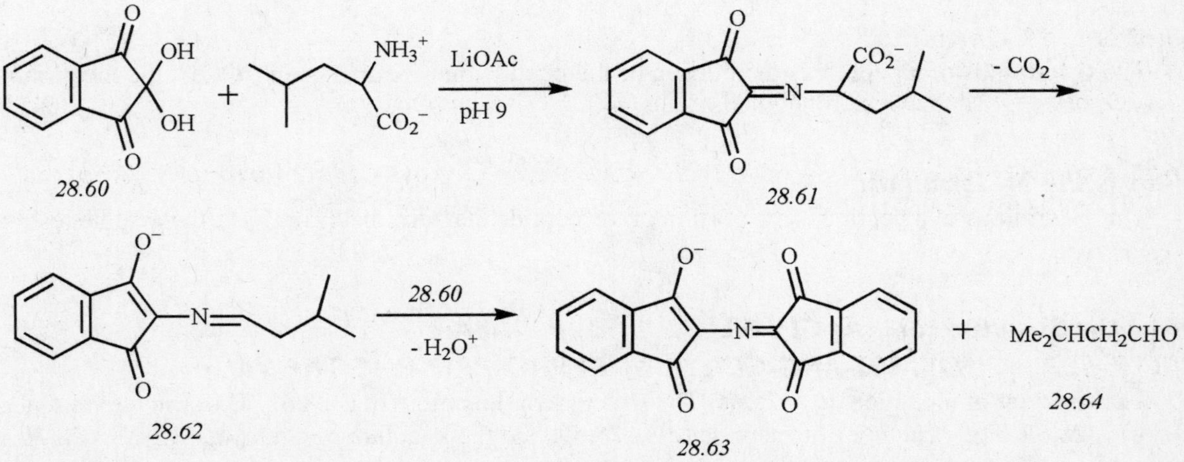

28.60 + leucine (NH_3^+, CO_2^-) → (LiOAc, pH 9) → *28.61* → ($-CO_2$)

28.62 → (*28.60*, $-H_2O^+$) → *28.63* + Me_2CHCH_2CHO (*28.64*)

What Is the Final Product of the Reaction Between Ninhydrin and Leucine? Of Any Amino Acid?

The final product is Ruhemann's Purple (*28.63*) and an aldehyde, if the amino acid contained a primary amino functionality.

28.6. PEPTIDES

Molecules that contain several amino acid units, linked together by amide bonds, are known as peptides.

What Is a Peptide?

A peptide is a biologically important polymer composed of several amino acids, linked together by amide bonds called peptide bonds. An example is ***28.65***, a tetrapeptide (alanine-serine-leucine-methionine). A polypeptide can be composed of hundreds of amino acid residues.

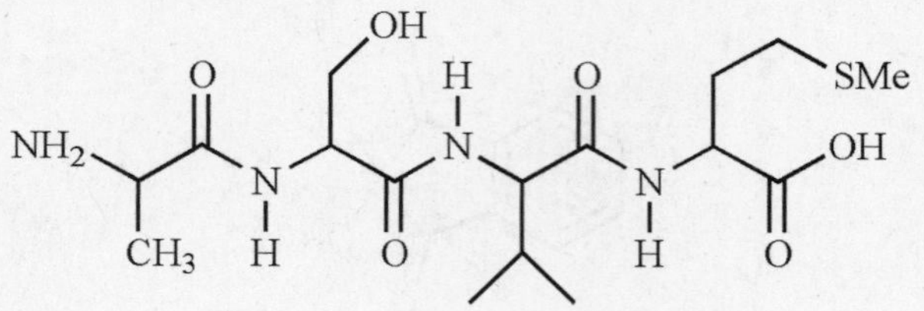

28.65

What Is a Residue?

Residue is the term used for each amino acid unit in a peptide.

Is There a Shorthand Method for Showing the Structure of 28.65?

Yes! The three-letter codes for the amino acids can be used. Therefore, ***28.65*** is *ala-ser-val-met*.

What Is a Dipeptide? A Pentapeptide?

A dipeptide is a molecule composed of two amino acid residues. A pentapeptide is a molecule composed of five amino acid residues.

What Is the C-Terminus?

The C-terminus of a peptide is the portion of the peptide that terminates in COOH, the methionine residue in ***28.65***.

What Is the N-Terminus?

The *N*-terminus of a peptide is the portion of the peptide that terminates in -NH_2, the alanine residue in ***28.65***.

Draw the Structure of A) GLY-GLU; B) ILE-ALA; C) TYR-SER; D) MET-ARG-GLY; E) PRO-PRO-PHE-TRP-VAL.

The structure of dipeptide (a) is ***28.66***. The structure of dipeptide (b) is ***28.67***. The structure of dipeptide (c) is ***28.68***. The structure of tripeptide (d) is ***28.69***, and the structure of pentapeptide (e) is ***28.70***.

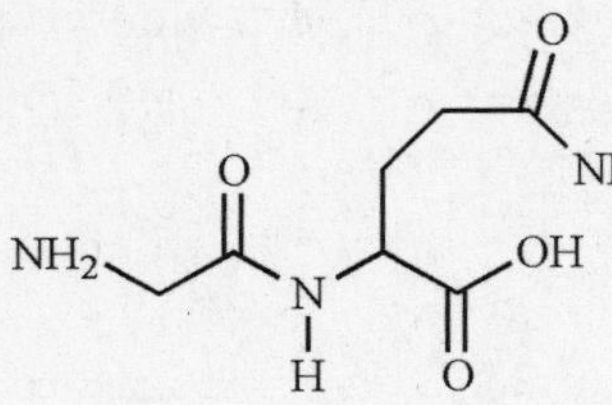

28.66 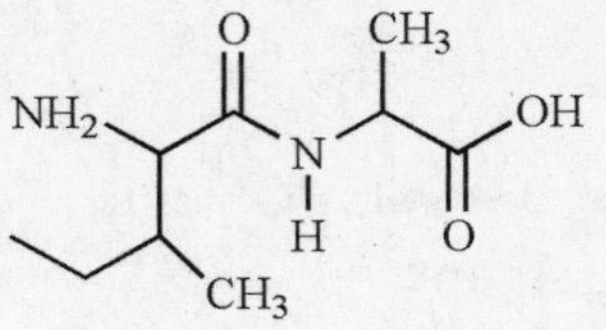 *28.67* 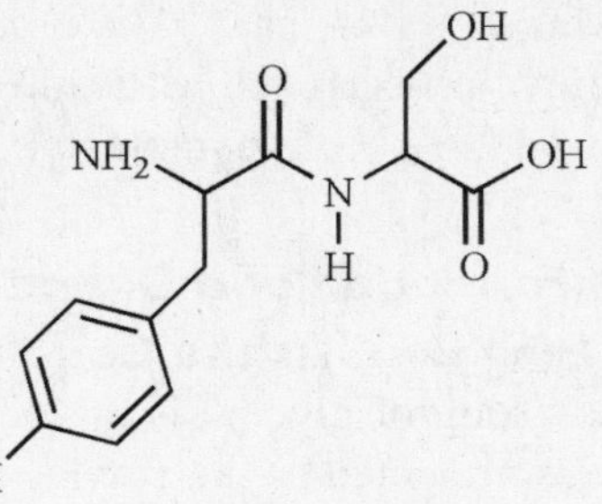 *28.68*

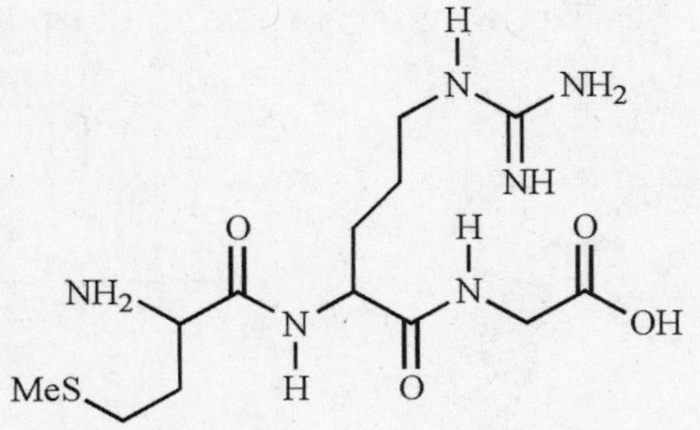

28.69 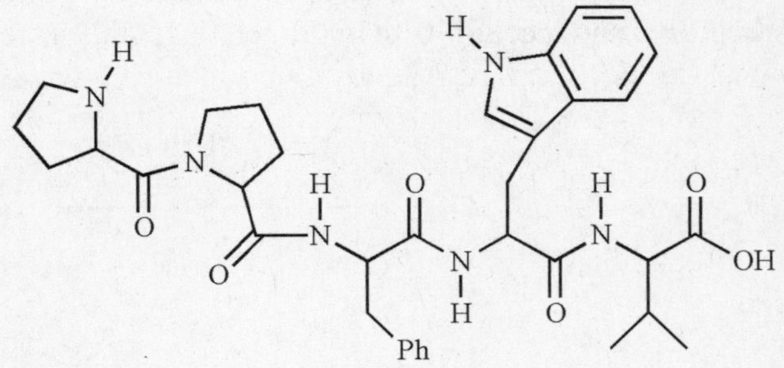 *28.70*

Amino Acid Coupling Reactions

Why Is It Necessary to Protect the Amino Group of an Amino Acid or Peptide If Coupling Is to Occur at the C-Terminus?

When a peptide is formed, the amino group on one amino acid is reacted with the carboxyl of a second to form the amide (peptide) bond. The amino acid that is to couple via the COOH group **must** have its amino group protected (blocked) so the only amide bond that will be formed is with the second amino acid. In the same vein, the amino acid that is to be coupled via the -NH_2 group must have its carboxyl group protected (blocked) in order to prevent unwanted coupling.

If Two Amino Acids Are to Be Coupled, Describe the Necessary Protection, in General Terms.

The amine group of one amino acid is protected, usually as an amide or a carbamate, and the carboxyl group of the second amino acid is blocked, usually as an ester. The free carboxyl group on one amino acid is then coupled to the free amino group of the second amino acid to give the amide bond. There are several methods for doing this coupling. In one, the COOH is converted to an acid chloride and then coupled with the amine. Alternatively, the carbonyl can be 'activated' (by DCC, for example), allowing reaction with the amino group.

What Are Suitable N-Protecting Groups?

The most common amine protecting groups for this purpose are amide (NHAc, NHCOPh, etc.) and carbamates ($NHCO_2CH_2Ph$ [called Cbz] and $NHCO_2CMe_3$ [called t-BOC or just BOC].

What Are Suitable C-Protecting Groups?

The acid group is usually protected as a methyl or ethyl ester.

What Is DCC?

DCC is **dicyclohexylcarbodiimide** (see Section 25.2): *c*-C_6H_{11}–N=C=*N*-*c*-C_6H_{11}, where -*c*- indicates cyclo, so -*c*-C_6H_{11} is cyclohexyl.

Describe the Formation of a Dipeptide Between Glycine and Serine Using DCC As the Coupling Agent.

The amino group of glycine is protected as the benzyl carbamate to give ***28.71***, and the carboxyl group of serine is protected as the ethyl ester. The acid moiety of ***28.71*** reacts with DCC to form ***28.72***, and the carbonyl of this intermediate is attacked by the amino group of the serine ethyl ester. This reaction displaces dicyclohexylurea and produces the protected dipeptide, ***28.73***. Saponification removes the ester group, and reaction with hydrogen (Pd catalyst) removes the CBz group to give the dipeptide, **28.74**.

28.71 → (DCC) → 28.72 → (H_2N–CH(CH_2OH)–CO–OEt) →

28.73 → 1. aq. NaOH; 2. H_3O^+; 3. H_2, Pd → 28.74

How Are C-Terminus Protecting Groups Removed?

If the C-terminus (the COOH) is protected as an ester, saponification (i. aq. OH^- ii. aq. H^+) will convert the ester to the carboxylic acid.

How Are N-Terminus Protecting Groups Removed?

If an amide group is used (acetamide, benzamide), basic hydrolysis followed by neutralization with acid usually removes the group. If the benzylic carbamate is used (CBz), catalytic hydrogenation with a palladium catalyst removes the protecting group.

What Is the Merrifield Synthesis?

This is a solid-phase peptide synthesis where the growing peptide is bound to a polymer; the polymer is usually attached at the C-terminus. A C-terminus amino acid, bound to a chloro-methylated polystyrene polymer, is coupled via DCC to an amino protected amino acid to form a peptide. The peptide ester is cleaved to liberate a 'free' carboxyl, which is coupled with a new polymer protected (C-terminus protected) amino acid to give a tripeptide. This process is repeated over and over again until the requisite polypeptide has been prepared. When the peptide synthesis is complete, the *N*-terminus is deprotected.

Give a Generalized Merrifield Synthesis of GLY-PHE-ALA.

The BOC protected glycine (***28.75***) is coupled with the polymer-bound phenylalanine to give ***28.76***. Treatment with trifluoroacetic acid removes the polymer, allowing DCC coupling with a polymer-bound alanine molecule, giving *28.77*. Treatment with trifluoroacetic acid removes the polymer, and aqueous HF cleaves the BOC group to give the tripeptide (*28.78*). Very often, aqueous HF cleaves both the polymer and the BOC group, which is why BOC is used as the protecting group in this sequence. The **Merrifield synthesis** can be automated and can be used to produce rather large polypeptides.

CO_2H (28.75) —DCC, Ph/NH_2 O-Polymer→ Boc-Gly-Phe-O-Polymer (28.76) —CF_3COOH→

—DCC, NH_2 O-Polymer→ Boc-Gly-Phe-Ala-O-Polymer (28.77) —1. CF_3COOH 2. aq. HF→ Gly-Phe-Ala (28.78)

End Group Analysis

These are chemical reactions that cleave the amino acid residues from the *N*-terminus or the C-terminus of the peptide, allowing them to be identified.

If a Polypeptide Is Heated with 6N HCl, What Is the Result?

If a peptide is heated in 6N aqueous HCl (usually for one day at greater than 100°C), all amide (peptide) bonds in the molecule are cleaved to give all constituent amino acid residues as discrete units. Complete degradation to the amino acids occurs with most peptides, but occasionally partial degradation occurs to give a mixture of amino acids and small peptides.

What Is a Carboxypeptidase?

A carboxypeptidase is an enzyme that selectively cleaves the peptide bond to the C-terminal amino acid in a peptide. It can be used to cleave and identify what that amino acid is. An enzyme that cleaves only terminal amino acid residues and not internal amino acid residues is called an *exopeptidase.*

When a Peptide Is Digested with the Enzyme Trypsin, What Is the Result?

Trypsin is an enzyme that cleaves peptide bonds at the carbonyl group of arginine or lysine, if these amino acids are **not** the *N*-terminus of the peptide, and if they are **not** followed by a proline. Under these conditions, however, treatment of a peptide with trypsin will lead to cuts in the peptide chain specifically at arginine or lysine. An enzyme that cleaves a peptide bond within the chain rather than at a terminal amino acid residue is called an *endopeptidase*.

When a Peptide Is Digested with the Enzyme Chymotrypsin, What Is the Result?

Chymotrypsin is an enzyme that cleaves peptides at amino acids containing aromatic side chains, such as phenylalanine, tyrosine and tryptophan. This enzyme will also cleave amino acids with large aliphatic side chains, such as leucine or isoleucine in some cases. Incubation of a peptide with this enzyme will, therefore, cleave the peptide at specific and predictable locations.

What Is Sanger's Reagent?

Sanger's reagent is 2,4-dinitrofluorobenzene (*28.79*). It reacts selectively with the *N*-terminal amino acid residue of a peptide via nucleophilic aromatic substitution. When *28.79* reacts with the glycine residue of a peptide, ***28.80*** is formed. Aqueous acid hydrolysis with 6N HCl will 'release' all amino acids, but only the *N*-terminal amino acid will be attached to the Sanger's reagent (forming ***28.81***). These *N*-aryl amino acids usually form a derivative with a yellow color that is easily identified.

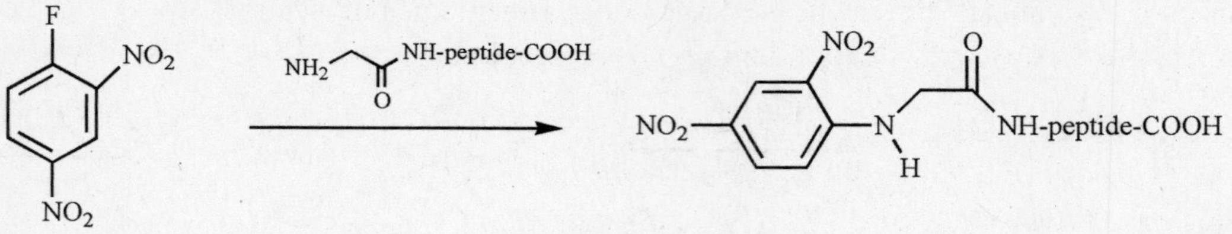

28.79 → 28.80 (reagent: NH$_2$CH$_2$C(O)NH-peptide-COOH)

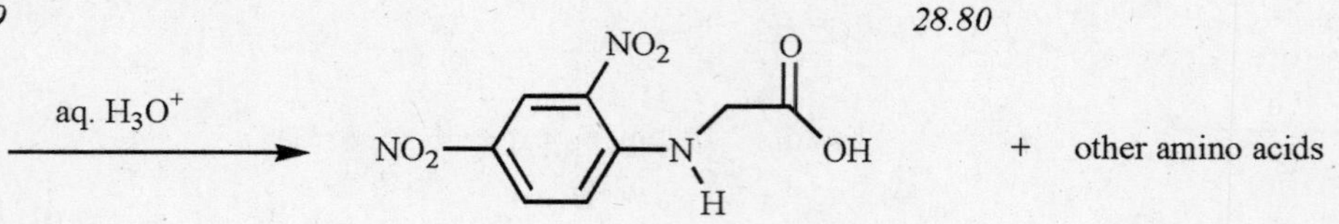

aq. H_3O^+ → 28.81 + other amino acids

What Is the Structure of Dansyl Chloride?

Dansyl chloride is 5-dimethylamino-1-naphthalenesulfonyl chloride (***28.82***).

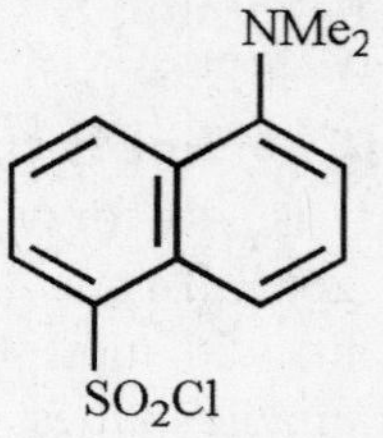

28.82

What Is the Result of Reacting a Peptide with Dansyl Chloride?

The *N*-terminal amino acid will displace the fluorine to form an *N*-aryl derivative, such as ***28.83***.

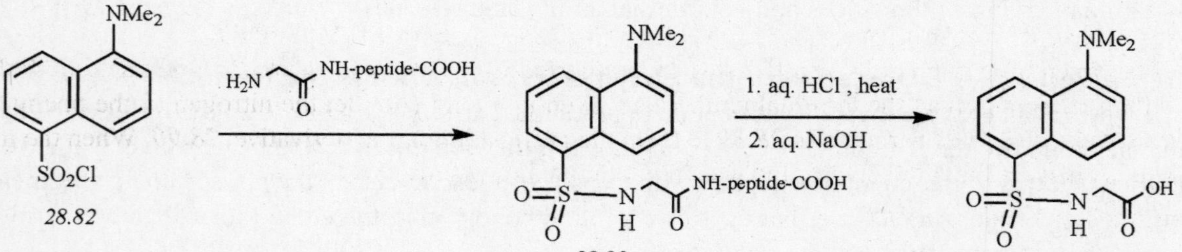

28.82 → 28.83 (reagent: H_2NCH$_2$C(O)NH-peptide-COOH); 1. aq. HCl, heat 2. aq. NaOH → 28.84

What Is the Result of Treating a Peptide with Dansyl Chloride and Then Heating This Product with Aqueous Acid?

The result is initial formation of sulfonamide ***28.83***. Hydrolysis leads to cleavage of the *N*-terminal amino acid residue from the peptide as a dansylamino acid (***28.84***), which is fluorescent and easily detected in the presence of the other amino acids liberated in the hydrolysis step.

What Is the Edmund Degradation?

The **Edmund Degradation** is the process for derivatizing and cleaving the *N*-terminal amino acid of a peptide by first treating the peptide with phenyl isothiocyanate [Ph–N=C=S] and then subjecting the peptide to acid hydrolysis. The *N*-terminal amino acid is converted to an *N*-phenyl-thiohydantoin by this procedure, cleaving only the terminal amino acid from the peptide and allowing easy identification of that amino acid.

What Is the Key Reagent in the Edmund Degradation?

Phenylisothiocyanate, Ph–N=C=S.

Draw the Phenylthiocarbamoyl Derivative Resulting from the Reaction of ALA-PHE and Phenyl Isothiocyanate.

The initial product of the reaction between dipeptide ***28.85*** and *N*-phenylisothiocyanate is an *N*-phenylthiourea derivative, ***28.86***.

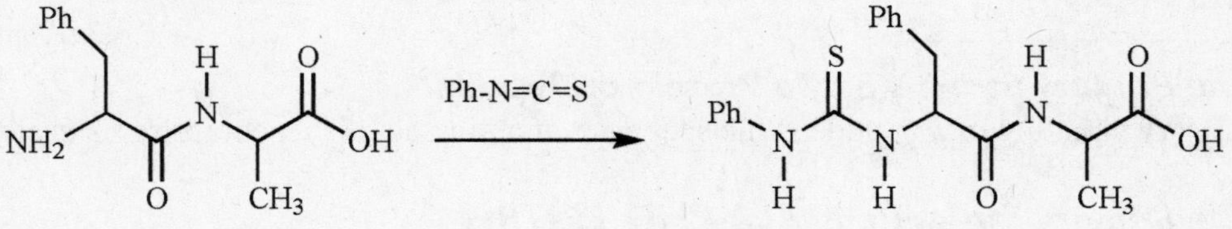

Ph-N=C=S

28.85 *28.86*

What Is the Thiazolone Derivative of Isoleucine?

A thiazolone is the IUPAC name for a thiohydantoin. If isoleucine were the *N*-terminal amino acid, Ph–N=C=S would convert it into ***28.87***.

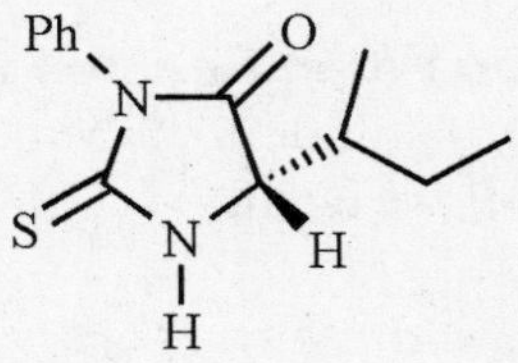

28.87

Draw a Complete Sequence for the Edmund Degradation of ALA-MET.

The dipeptide ala-met (***28.88***) is first treated with Ph–N=C=S to form ***28.89***. Upon acid hydrolysis, which cleaves only at the terminal amino acid, even in a long peptide, the nitrogen of the phenylthiourea attacks the amide carbonyl in ***28.89*** to form the cyclic ammonium derivative, ***28.90***. When the met residue is cleaved, the *N*-phenylthiohydantoin of alanine is formed (***28.91***).

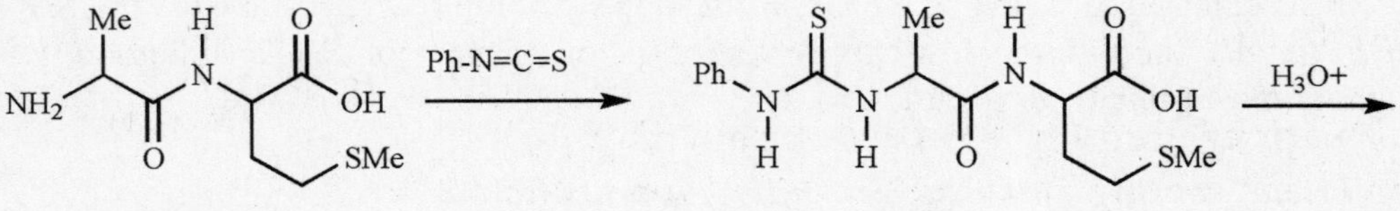

Ph-N=C=S H_3O+

28.88 *28.89* ***28.90*** *28.91*

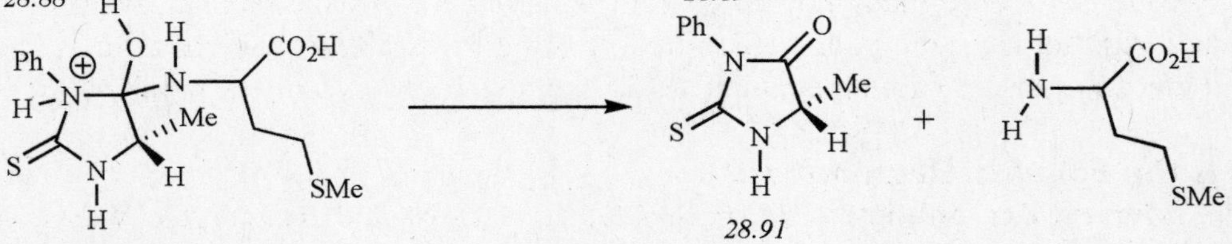

28.7. PROTEINS

Proteins are one of the most important classes of biomolecules, and they are polypeptides.

What Is a Protein?

A protein is a long chain biopolymer composed of amino acids joined together by amide bonds, usually called peptide bonds. There are two basic types of proteins: *simple proteins*, which yield only amino acids and no other organic compounds upon hydrolysis, and *conjugated proteins*, which give other compounds along with amino acids upon hydrolysis.

What Is an Enzyme?

Enzymes are proteins that are the biological catalysts required to initiate chemical reactions in biological systems.

What Is the Primary Structure of a Protein or Peptide?

The primary structure of a peptide is the sequence of amino acids that comprise its basic structure.

What Is the Primary Structure of ALA-PHE-ILE-TRP?

The sequence ala-phe-ile-trp *is* the primary structure.

What Are Disulfide Bonds?

When cysteine is incorporated in a peptide or protein, two SH units can come together and react to form a disulfide bond (R–S–S–R).

What Does Dithiothreitol (28.92) Do When Reacted with a Disulfide Bond?

Dithiothreitol (***28.92***) reacts with a disulfide to 'liberate' two thiol moieties and form ***28.93***. Dithiothreitol is sometimes called **Cleland's reagent.**

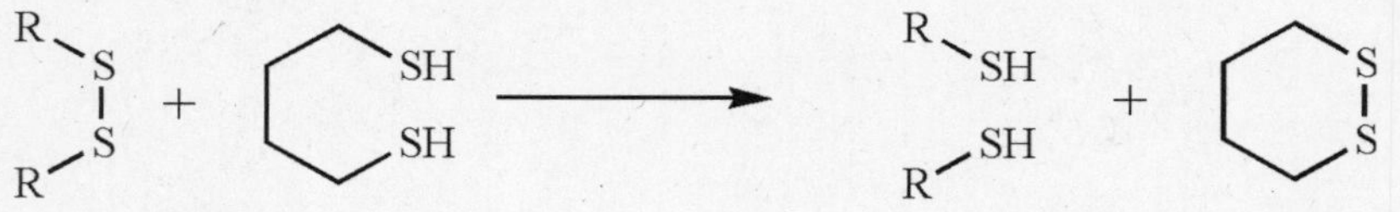

R–S–S–R + 28.92 (SH, SH) ⟶ R–SH + R–SH + 28.93 (S–S)

28.92 *28.93*

Draw the Structure of ALA-PHE, Showing the Relative Positions of the Carbonyl Groups, Amino Groups and Side Chains.

The basic structure of a protein involves amide bonds where the alkyl side chain of one amino acid residue is effectively 'anti' to the alkyl side chain of the adjacent amino acid residue (as in ***28.94***). The amide carbonyls are also 'anti' in this low energy conformation of the peptide. This alternating pattern appears throughout most proteins or peptides.

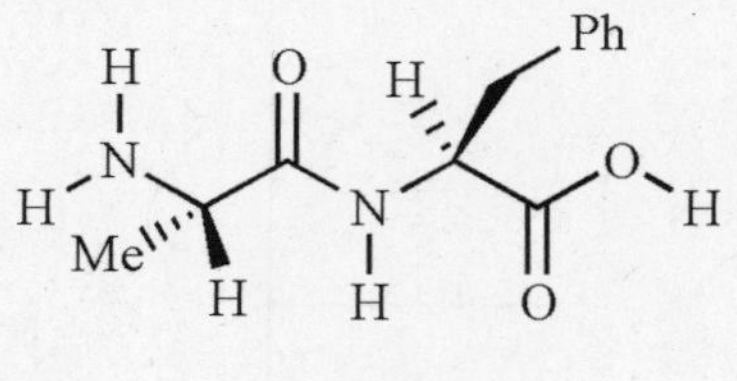

H, O, H, Ph, N, H, N, O, H, Me, H, H, O

28.94

What Is the Secondary Structure of a Protein or Peptide?

The secondary structure of a protein or peptide is the amount of structural regularity, which results from intermolecular hydrogen bonding. The most common secondary structure features are formation of an α-helix or a β-pleated sheet structure.

What Is an α-Helix?

An α-helix is the shape a peptide assumes due to the chirality of the individual amino acids, forming a spiral type structure (see ***28.95***, taken from Garrett, R.H.; Grisham, C.M. *Biochemistry*, Saunders, Fort Worth, *1995*; see Figure 5.18, p. 91), held in that shape by hydrogen bonding between NH, OH, C=O and other heteroatom functional groups.

What Is a β-Strand?

A β-strand is an open strand of the peptide where intermolecular hydrogen bonding is more important, as in ***28.96*** (see Garrett, R.H.; Grisham, C.M. *Biochemistry*, Saunders, Fort Worth, *1995*; see Figure 5.18, p. 91).

α-Helix
only the N-Cα-C backbone is represented. The vertical line is the helix axis

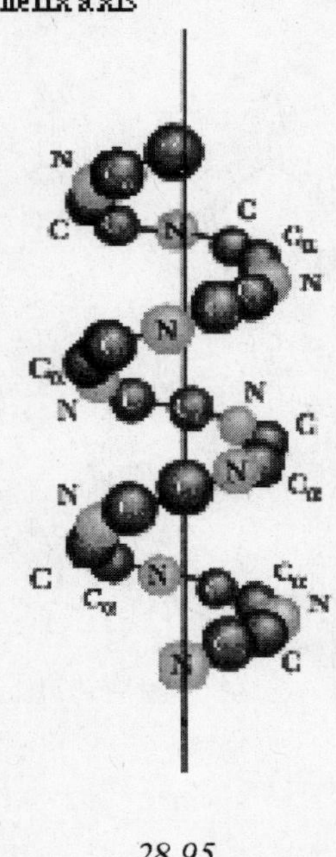

28.95

β-Strand
Note that the amide planes are perpendicular to the page

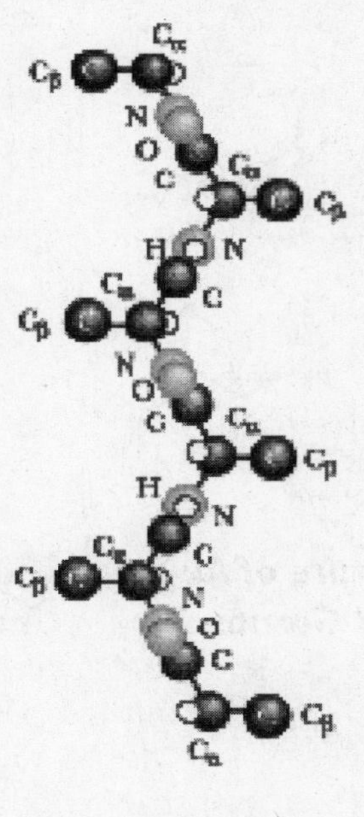

28.96

What Is the Importance of Hydrogen Bonding in the Secondary Structure of a Protein or Peptide?

Intramolecular hydrogen bonding between various NH, OH, SH, C=O and C=N groups of the peptide stabilize the helical structure. Each 'turn' of the α-helix is held in that position by intramolecular hydrogen bonding.

What Is the β-Structure of a Protein or Peptide?

The β-pleated sheet structure is shown in ***28.97***. The antiparallel pleated sheet is shown in ***28.98***. In both cases, intermolecular hydrogen bonding allows 'stacking' of the peptide chains. In the pleated sheet, the peptide chains align in a parallel manner, with all chains oriented N→C. In the antiparallel structure, the chains alternate N→C, C→N, N→C, etc., as shown.

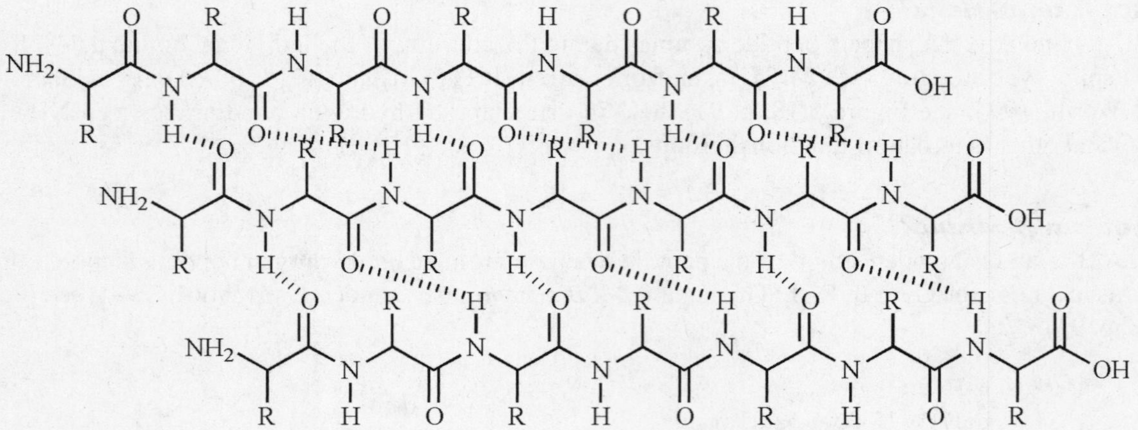

28.97

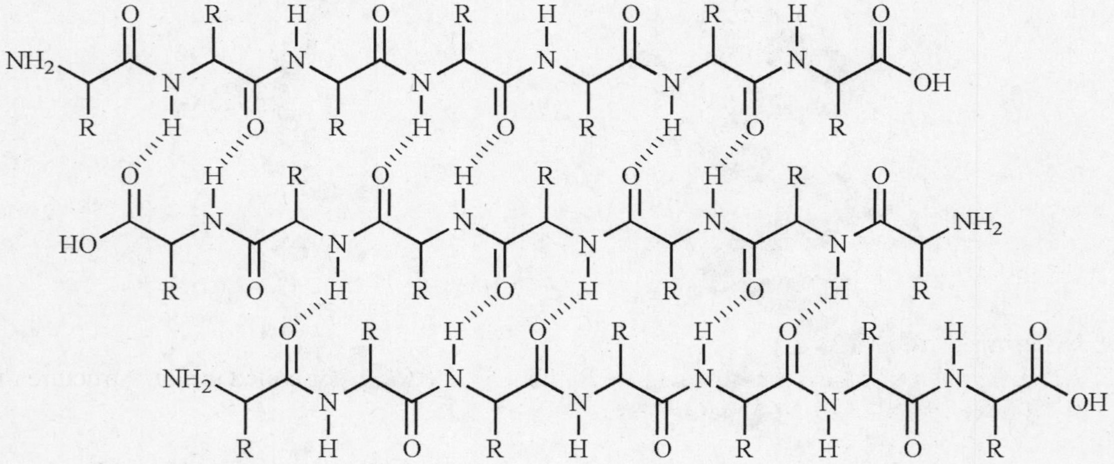

28.98

What Is a Random Coil?

A **random coil** is a type of secondary structure that is 'random' and does not conform to a distinct structure. The peptide chains arrange in a random manner, held together by hydrogen bonding.

Are Most Proteins Composed of One of the Above Mentioned Secondary Structures or Mixtures of Several?

Most proteins assume several different secondary structures, and a typical protein is composed of various percentages of each of the structural types.

What Is the Tertiary Structure of a Protein or Peptide?

The tertiary structure of a protein is its complete three-dimensional structure due to folding and coiling of the peptide chain. The tertiary structure will be the result of both the primary and secondary structure, as well as 'folding' of the peptide chains, loosely illustrated by the structure ***28.99***. Another view is ribbon structure ***28.100*** (see Garrett, R.H.; Grisham, C.M. *Biochemistry*, Saunders, Fort Worth, *1995*; see Figure 5.18, p. 144).

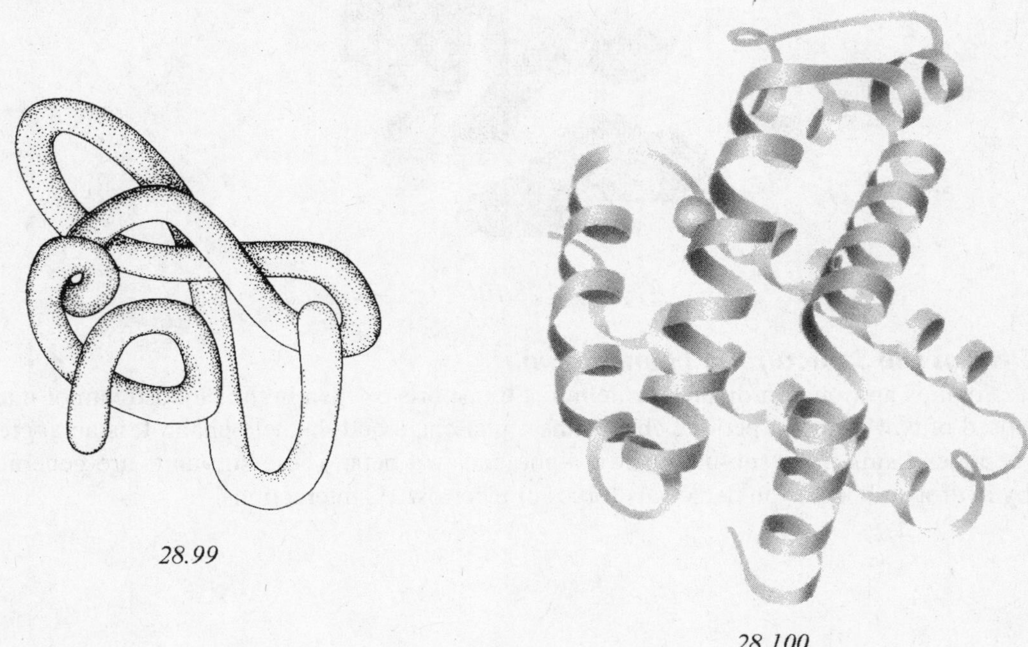

28.99

28.100

What Is Primarily Responsible for the Tertiary Structure?

A combination of hydrogen bonding, disulfide linkages (R–S–S–R) between cysteine residues in different parts of the peptide chain, electrostatic interactions, dipole-dipole interactions, and 'π-stacking' of aromatic rings in those amino acid residues containing aromatic rings.

What Is Denaturation?

Denaturation is the process that disrupts the bonding of the folded or coiled tertiary structure of the protein, leading to a random coil (denatured protein).

What Are Some Common Denaturants?

Organic solvents, detergents, and concentrated urea solutions all act as denaturants. Heating a protein can also lead to denaturation.

Is a Denatured Protein Biologically Active?

The biological activity of a protein is often a function of its tertiary structure, and denaturation usually deactivates the protein.

What Are Hydrophobic Residues?

When two hydrocarbon fragments of different amino acids, such as the interaction of two isopropyl groups of two different valine residues, come in close proximity, the 'like-dissolves-like' rule suggests these residues will interact with each other.

What Is the Quaternary Structure of a Protein or Peptide?

The quaternary structure of a protein or peptide is the interaction of two or more peptide chains that join together to form 'clusters' of peptides. This cluster is usually necessary for the biological activity. A schematic example is ***28.101***, which is composed of three separate 'units.'

28.101

Briefly Discuss the Structure of Hemoglobin.

Hemoglobin is an iron-coordinated protein that transports oxygen in the bloodstream of mammals. It is composed of two different peptide chains, the alpha-chain and the beta-chain. It is an aggregate of four polypeptide chains (four sub-units), two alpha and two beta. These sub-units are generally held together by hydrogen bonds, van der Waals forces or electrostatic interactions.

Test Yourself

1. Give the IUPAC name for each of the following.

(a) 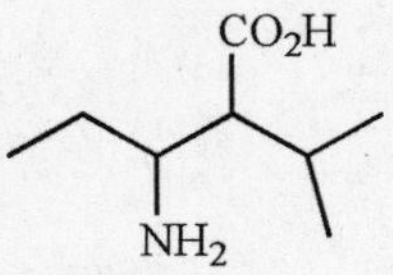 CO_2H NH_2 (b) 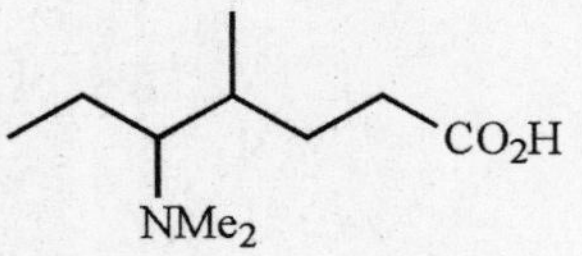 CO_2H NMe_2 (c) 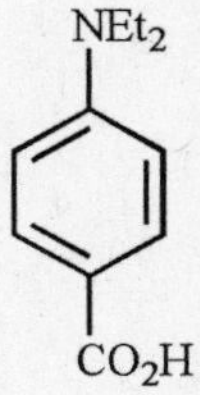 NEt_2 CO_2H

2. Give the structure for each of the following in Fisher projection.
(a) glu; (b) val; (c) ile; (d) ser; (e) pro; (f) ala; (g) asp; (h) his; (i) arg; (j) met.

3. In each case, give the major product. If there is no reaction, indicate this by N.R.

(a) 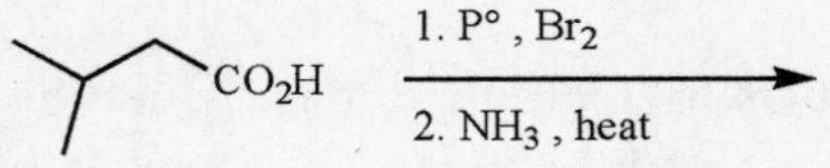 CO_2H — 1. P°, Br_2; 2. NH_3, heat →

(b) 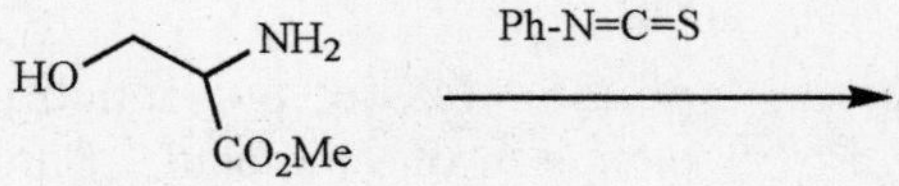 HO, NH_2, CO_2Me — Ph-N=C=S →

(c) OH — 1. PCC, CH_2Cl_2; 2. NaCN, NH_3; 3. aq. HCl; 4. neutralize →

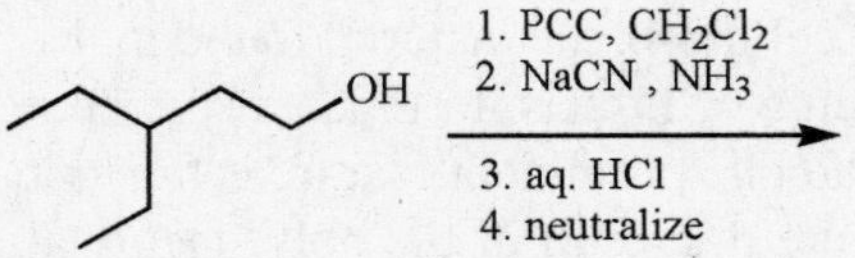

(d) 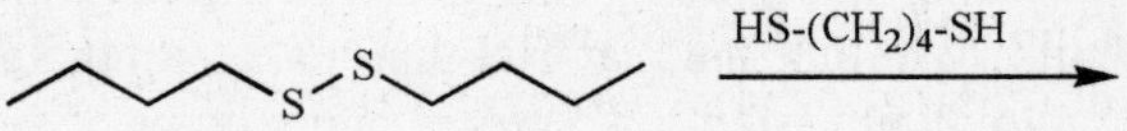 S, S — HS-$(CH_2)_4$-SH →

(e) 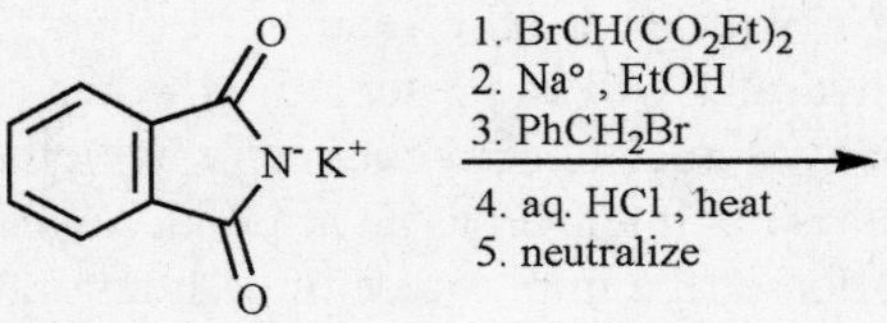 O, N^-K^+, O — 1. $BrCH(CO_2Et)_2$; 2. Na°, EtOH; 3. $PhCH_2Br$; 4. aq. HCl, heat; 5. neutralize →

(f) 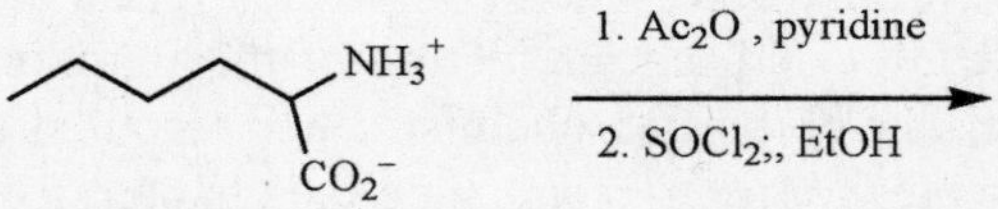 NH_3^+, CO_2^- — 1. Ac_2O, pyridine; 2. $SOCl_2$;, EtOH →

(g) 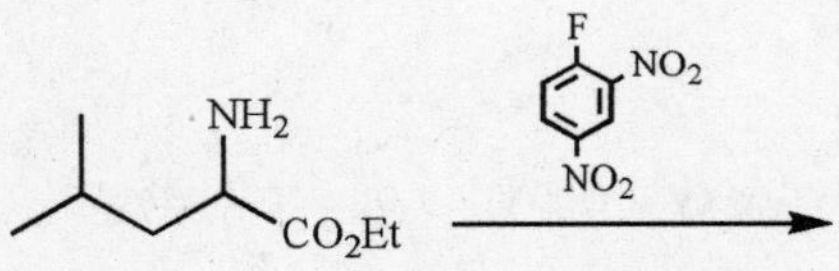 NH_2, CO_2Et — F, NO_2, NO_2 →

(h) 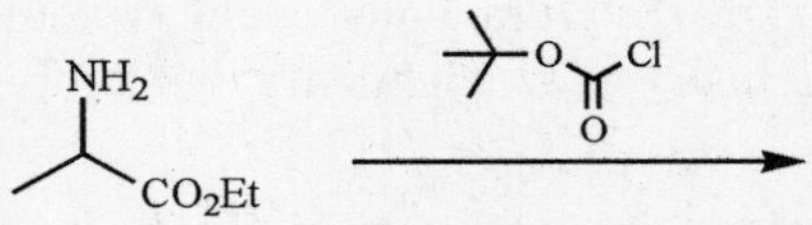 NH_2, CO_2Et — O, Cl, O →

(i) 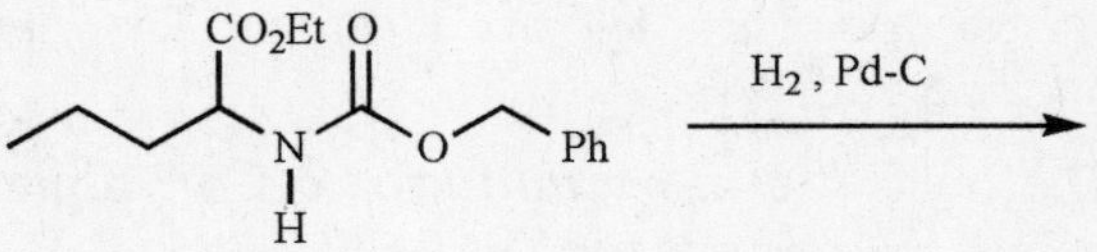 CO_2Et O, N, H, O, Ph — H_2, Pd-C →

(j) 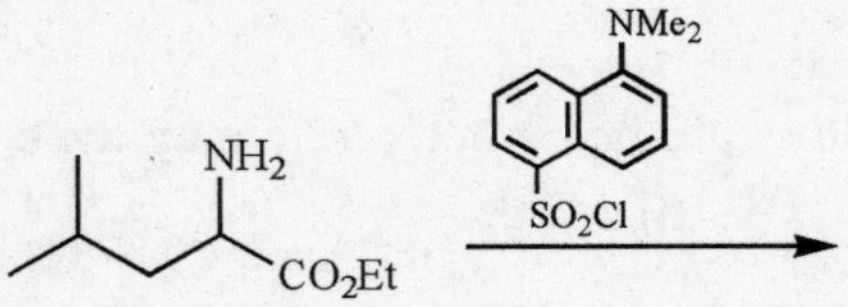 NH_2, CO_2Et — NMe_2, SO_2Cl →

(k) H_3C, S, NH_3^+, CO_2^- — LiOAc; O, OH, OH, O →

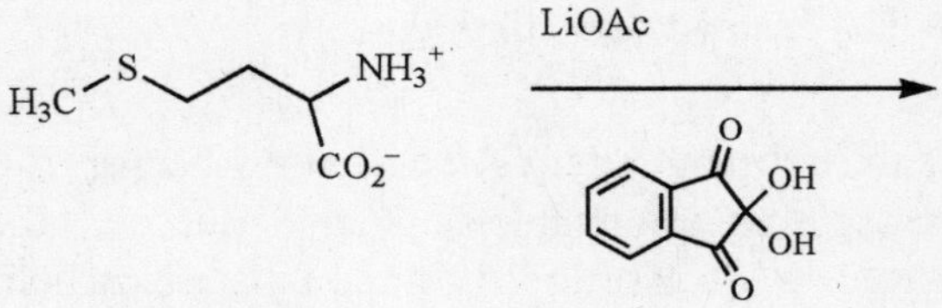

4. In each case, provide a suitable synthesis. Show all reagents and intermediate products.

(a)  NH_2, CO_2H ⇨

(b) gly-glu-ala-phe ⇨ CO_2H, NH_2

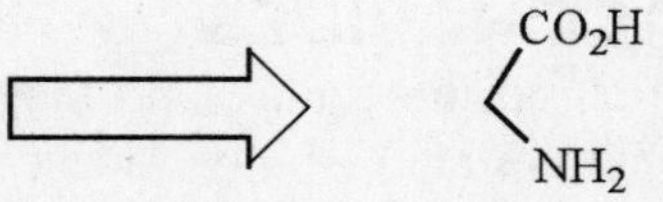

CHAPTER 29

Carbohydrates and Nucleic Acids

Another important class of polyfunctional molecules are carbohydrates, which have several hydroxyl groups in one molecule. Carbohydrates are very important components of naturally occurring molecules. Their occurrence ranges from cellulose, the material found in the cell walls of plants, to chitin, the material that makes up the exoskeleton of insects, to glucose, the energy source for mammalian systems. There are many types of sugars, and one prominent feature is the unique chemistry and properties of these compounds. These properties are the result of the polyfunctional nature of the molecules and the interaction of the functional groups, with each other and with other chemical reagents. The properties, chemistry, and reactions of carbohydrates will be presented, as well as methods for their preparation. To conclude the chapter, a brief review of nucleic acid chemistry, which contains sugars as a key structural component, will be presented. This is not intended as an in-depth review of biological chemistry or biochemistry, but simply as an illustration of the importance of carbohydrates.

What Is the Definition of a Carbohydrate?

A carbohydrate is a 'hydrate of carbon.' They are polyhydroxy aldehydes or ketones.

What Is the General Formula for a Carbohydrate?

The general formula for a carbohydrate is $C_nH_{2n}O_n$, although the carbohydrate may have more or fewer hydrogens and more oxygens.

Why Are Carbohydrates Also Called Sugars?

Sucrose is common table sugar, and glucose is the common sugar used as an energy source in mammalian systems. Both of these compounds are carbohydrates and typical examples of this class of compounds. For this reason, carbohydrates are often referred to as 'sugars.'

What Is a Saccharide?

Saccharide is another term for carbohydrates or sugars.

29.1. MONOSACCHARIDES

Carbohydrates are chiral molecules, and they can be subdivided into categories that reflect their cyclic form. Monosaccharides are carbohydrates consisting of one sugar unit.

Chirality

Given the Structure of Glucose (29.1), How Many Stereogenic Centers Are Present? Identify Each Using the R/S Nomenclature.

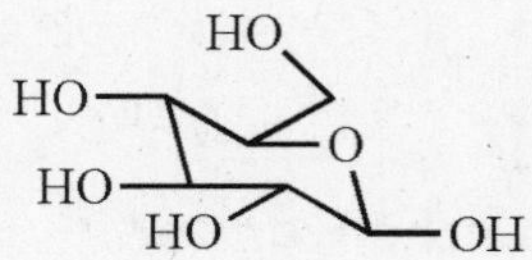

29.1

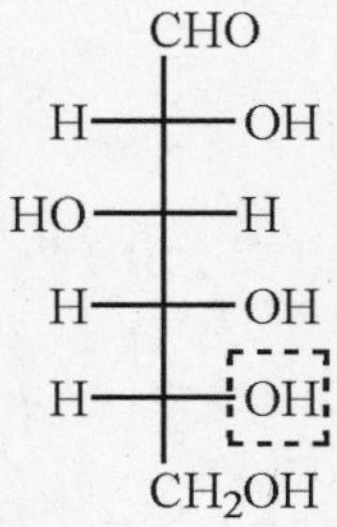

29.2

Glucose is drawn in two forms, as the cyclic hemiacetal, ***29.1*** and as the open-chain aldehyde, **29.2**. The reason for this will be discussed below. Using ***29.2***, drawn in Fisher projection, C2 (attached to CHO) is R, C3 is S, C4 is R, and C5 is R. In the cyclic form (***29.1***), the CHOH group attached to the oxygen in the ring is also a stereogenic center, but can exist as both R or S (see below).

Describe the D/L System for Sugars.

As with amino acids, the D,L system is based on comparison to (+)- or (−)-glyceraldehyde. When the OH of (+)-glyceraldehyde is 'on the right' (in the box in ***29.3***) of the Fisher projection, it is given the label D-glyceraldehyde. If the OH on the next to last carbon in the open chain form of the carbohydrate (C5 in ***29.2***; the OH group is shown in a box) is on the right, it is a D-sugar. If the OH is on the left, it is an L sugar. In ***29.2***, that OH (in the box) is on the right, and this molecule is called D-glucose.

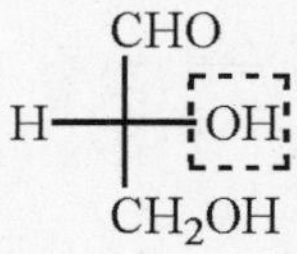

29.3

How Is D-Glucose Related to D-Glyceraldehyde?

The C5 carbon in ***29.2*** has the H–C–OH group oriented exactly as in D-glyceraldehyde (see the boxed OH on H–C–OH) and is, therefore, assigned the label D.

What Is a Monosaccharide?

Monosaccharides are sugars that can **not** be hydrolyzed into more simple sugars.

What Is the Molecular Formula for Glucose? For Fructose?

In both cases, the formula is $C_6H_{12}O_6$.

What Is a Haworth Projection?

A **Haworth projection** takes the cyclic hemiacetal form of the sugar (***29.1***) and 'flattens' it, making the H and OH groups appear either on the 'top' or the 'bottom,' as in ***29.4***. This is a general way in which to present the structure of sugars.

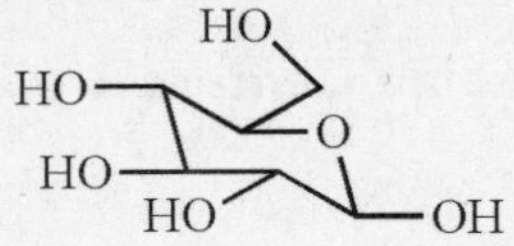

HO, HO, HO, HO, O, OH

29.1

CH_2OH, H, O, OH, OH, H, H, OH, H, OH, H

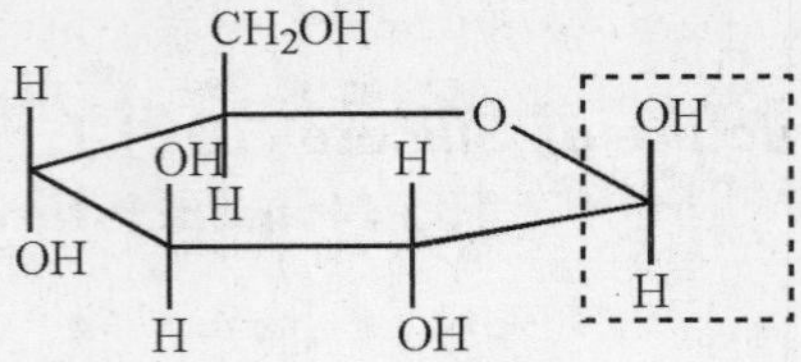

29.4

What Is the Acyclic (Open Chain) Form of a Carbohydrate?

The carbohydrates drawn are hemi-acetals and are in equilibrium with their aldehyde precursors. The aldehyde form of the carbohydrate is acyclic (no ring) and is referred to as the open-chain form.

What Is the Usual Reaction of an Aldehyde and an Alcohol?

In general, aldehydes react with alcohols to form acetals (RCH[OR']$_2$), but the reaction proceeds by formation of an unstable hemiacetal [RCH(OH)OR']. See Section 21.3.

What Is the Structure of a Hemiacetal?

A hemiacetal has an OH and an OR on the same carbon [RCH(OH)OR'].

Draw D-Fructose in Fisher Projection As the Open Chain Aldehyde.

The Fisher projection of D-fructose is ***29.5***.

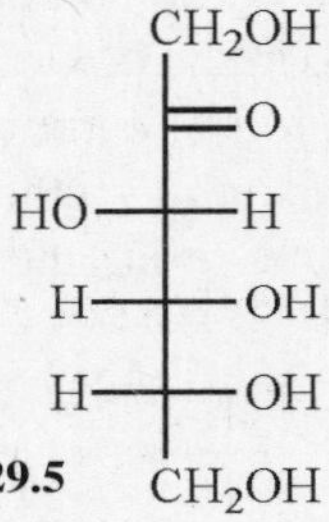

CH_2OH

=O

HO—H

H—OH

H—OH

29.5 CH_2OH

How Can an Aldohexose Form a Hemiacetal?

An aldohexose such as glucose forms a hemiacetal by cyclization to a pyranose (***29.4*** above) for D-glucose. The appropriate alcohol moiety attacks the acyl carbon of the aldehyde carbon.

Draw D-Glucose in Fisher Projection As the Open Chain Aldehyde.

The Fisher projection of D-glucose is ***29.2***.

Draw L-Glucose in Fisher Projection As the Open Chain Aldehyde.

The Fisher projection of L-glucose is ***29.6***.

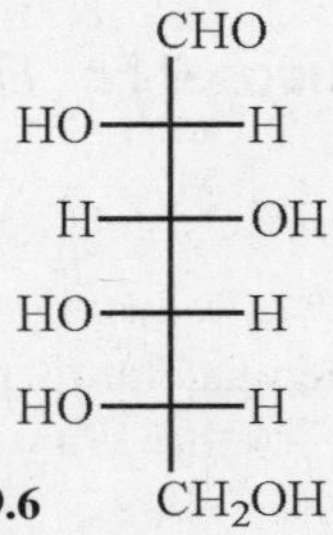

CHO

HO—H

H—OH

HO—H

HO—H

29.6 CH_2OH

If the Absolute Configuration of All Alcohol Groups Is Fixed in Glucose, Why Does the Cyclic Hemiacetal Form Two Different Cyclic Structures? Draw Both in Haworth Projection.

When the hemiacetal forms, cyclization can occur from two faces to give the OH 'up' or the OH 'down' (***29.4***). Note the boxed area indicating the 'up' OH group. The molecule with OH 'up' is a different molecule than the one with the OH 'down.' This concept will be discussed later in this chapter.

What Is an Aldohexose? A Ketohexose? An Aldopyranose? A Ketofuranose?

An aldose is a polyhydroxy aldehyde, and a ketose is a polyhydroxy ketone. A hexose is a six-carbon sugar, and a pentose is a five-carbon sugar. A furanose is the cyclic hemiacetal or hemiketal form of the sugar that exists as a five-membered tetrahydrofuran ring. A pyranose is the cyclic hemiacetal or hemiketal form of the sugar that exists as a six-membered pyran ring. An example of an aldohexose is glucose (***29.2***), and an example of a ketohexose is fructose (***29.5***). An aldopyranose is the cyclic form of glucose (***29.1***), and a ketofuranose is the cyclic form of fructose (***29.7***).

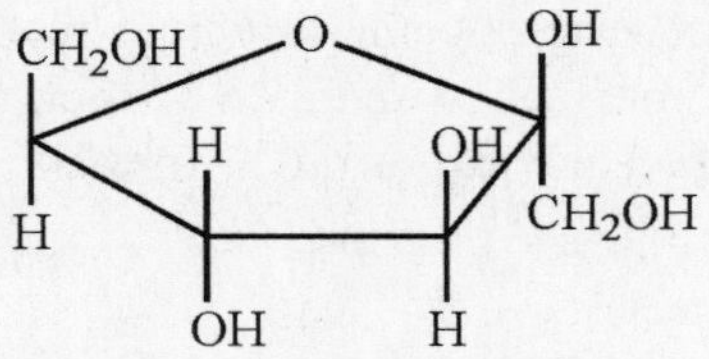

CH_2OH O OH H OH H CH_2OH OH H

29.7

What Is the Structure of α-Glucose and β-Glucose in Haworth Projection?

The Haworth formula for β-D-glucose is ***29.4***, and the Haworth structure of α-D-glucose is ***29.8*** (also shown in the chair form for comparison).

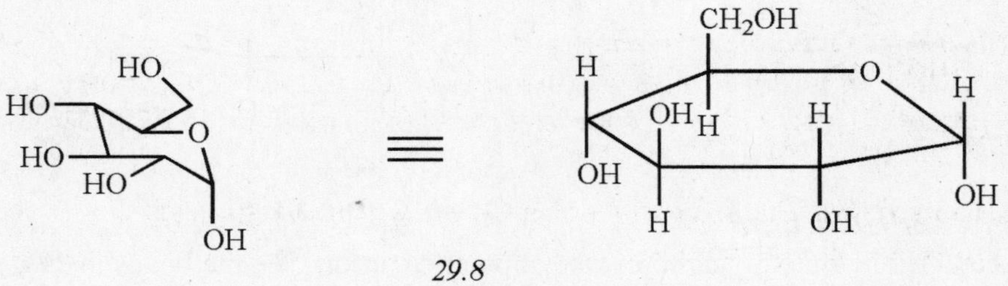

HO HO O HO HO OH ≡ CH_2OH H O H OH H H OH OH H OH

29.8

Pyranoses

A pyranose is the six-membered hemiacetal structure formed by aldohexoses such as glucose. The structure ***29.8*** is an example of a pyranose.

Draw the Eight Different D-Isomers of Glucose (Including Glucose) in Their D-Pyranose Form. Give the Name of Each Isomer.

The eight isomers are D-(+)-glucose (***29.9***), D-(+)-allose (***29.10***), D-(+)-mannose (***29.11***), D-(+)-altrose (***29.12***), D-(−)-gulose (***29.13***), D-(−)-idose (***29.14***), D-(+)-galactose (***29.15***), and D-(+)-talose (***29.16***).

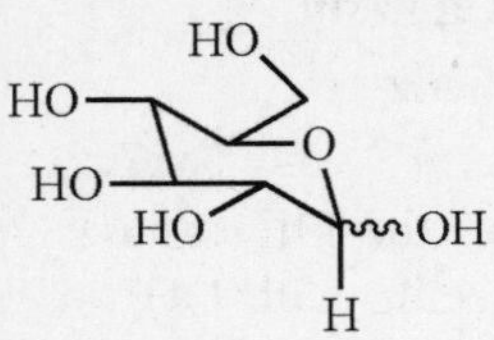

29.9 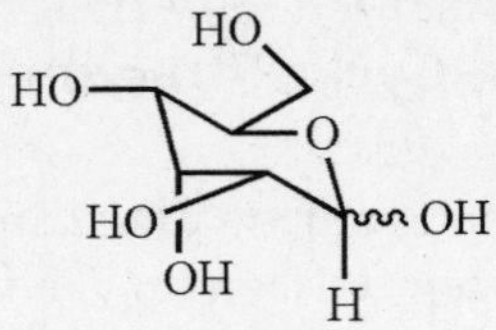 29.10 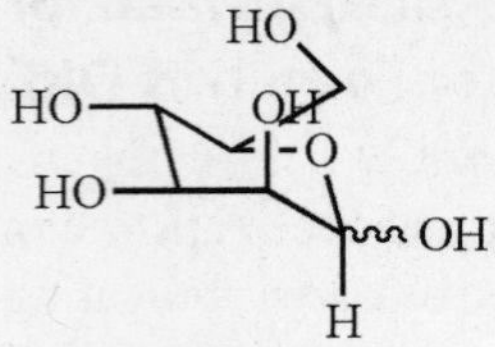 29.11 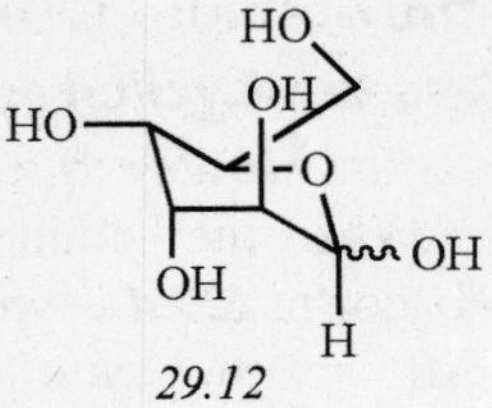 29.12

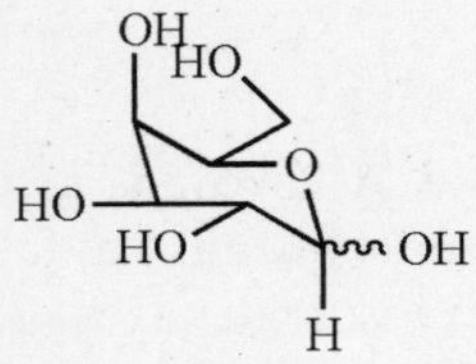

29.13 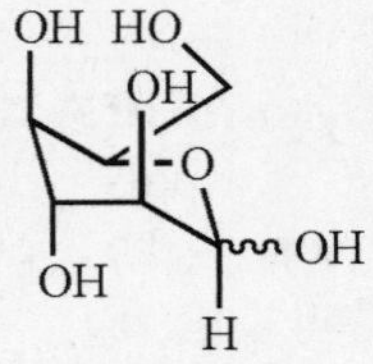 29.14 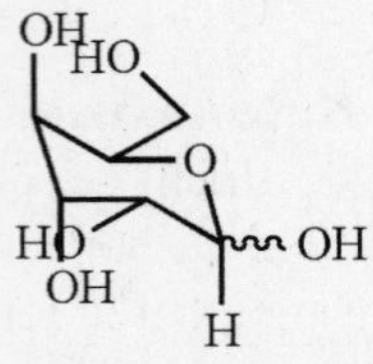 29.15 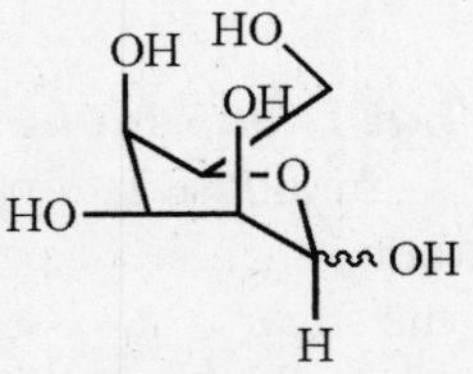 29.16

What Is an Anomer?

When the hemiacetal forms, the new stereogenic carbon formed can be either R or S. These diastereomeric products are referred to as anomers (**29.4** and **29.8** are anomers). The anomeric carbon is shown in a box for both the chair form and the Haworth formulas of **29.4** and **29.8**.

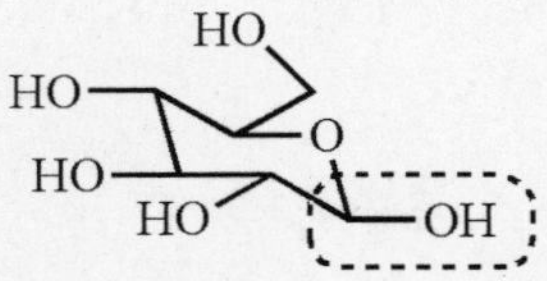

29.1 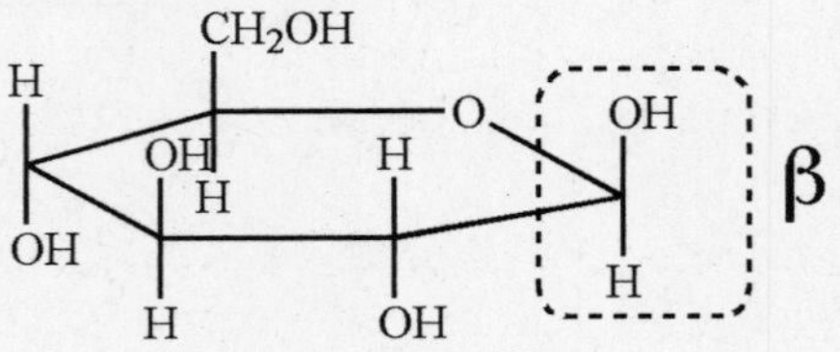 29.4 β

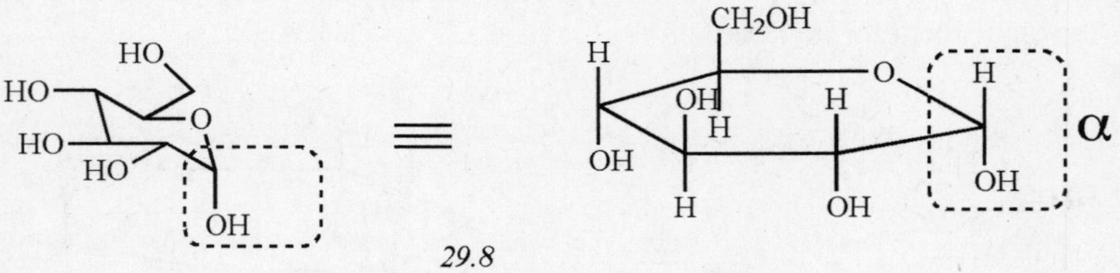

29.8 α

What Is an Anomeric Carbon?

The hemiacetal (or hemiketal) carbon is the anomeric carbon. See the boxes in **29.4** and **29.8**.

Does D-Glucose Exist 100% As a Pyranose? Explain.

No! There is a small percentage of the open-chain aldehyde in equilibrium with the α- and β-D-glucose anomers (usually less than 1%).

Identify the Structure of α-D-Glucose and β-D-Glucose.

The α-form of D-glucose is **29.8**, and the β-form is **29.4**.

If Pure-D-Glucose Is Dissolved in Water, Why Does the Specific Rotation of the Solution Change to a Constant but Different Value over Time?

Once dissolved in water, the β-anomer opens to the aldehyde and then can close again to either the β-form or the α-form. Similarly, if the pure α-anomer is dissolved in water, it will equilibrate to the same mixture of the two anomers.

Which Is More Stable, α-D-Glucose or β-D-Glucose? Explain.

The most stable anomer is the β-anomer (*29.4*), where the OH group is in the axial position. The interaction of the lone electron pairs on the oxygen when it is in the equatorial position makes the axial orientation more stable. This minimization of electronic repulsion to make the axial substituent more stable is called the **anomeric effect**.

Does D-Glucose Exist Primarily in the α-Form or the β-Form? Explain.

D-glucose exists primarily as the β-anomer (*29.4*). There is actually an equilibrium mixture of 64% of ***29.4*** and 36% of the α-anomer, ***29.8***.

What Is Mutarotation?

Mutarotation is the change in specific rotation of a sugar when dissolved in water. This occurs when the hemiacetal form of a carbohydrate (such as ***29.8***) equilibrates with the other forms in such as way that the configuration of the O–CHOH group from C1–R to C1–S (or from C1–S to C1–R). In water, a sample of pure (C1–R) will equilibrate to a mixture of ***29.8*** and ***29.4***,i.e., C1–R + C1–S, and this equilibrium mixture will be reflected in a change in the specific rotation.

What Properties of a Sugar Such as Glucose Lead to Mutarotation?

Taking glucose as an example, the open chain form (***29.2***) will close to the hemiacetal (***29.8***) by attack of the C5 OH on the carbonyl of the aldehyde. In this hemiacetal, the carbon bearing the C1–OH can assume either the R or S configuration by equilibration (***29.1***).

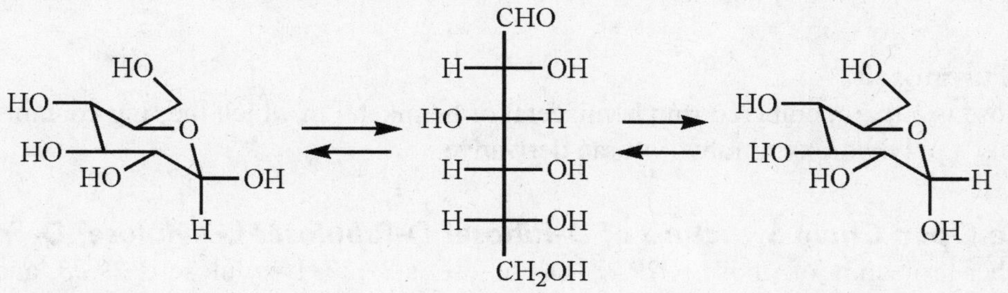

29.1 *29.2* *29.8*

Why Does D-Glucose Have More Than One Value for Specific Rotation?

D-Glucose exists in two anomeric forms, β-D-glucose (***29.1***) and α-D-glucose (***29.8***). In pure form, both ***29.1*** and ***29.8*** exhibit a characteristic specific rotation that does not change unless put into water.

What Is a Deoxy Sugar?

A deoxy sugar is a carbohydrate in which at least one of the OH groups is missing.

Give the Open Chain Fisher Projections of 2-Deoxy-D- Glucose. 6-Deoxy-L-Mannose (Rhamnose).

The Fisher projection of 2-deoxy-D-glucose is *29.17*, and the Fisher projection of rhamnose is *29.18*.

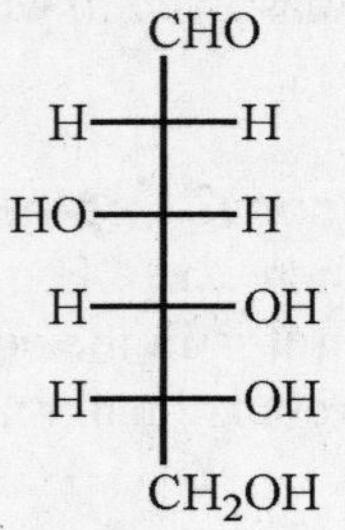

29.17 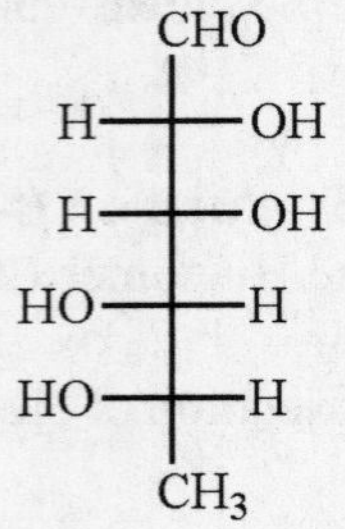 *29.18*

Give the Pyranose Form of 2-Deoxy-D-Glucose. 6-Deoxy-L-Mannose (Rhamnose).

The pyranose form of 2-deoxy-D-glucose is *29.19*, and rhamnose is *29.20*.

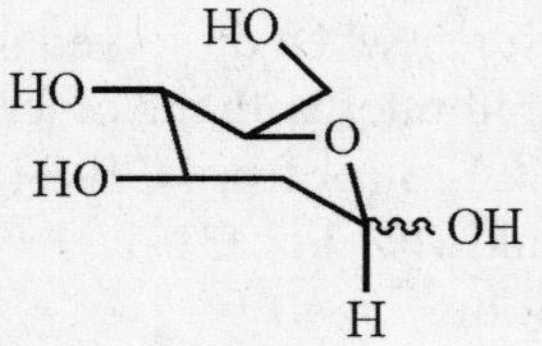

29.19 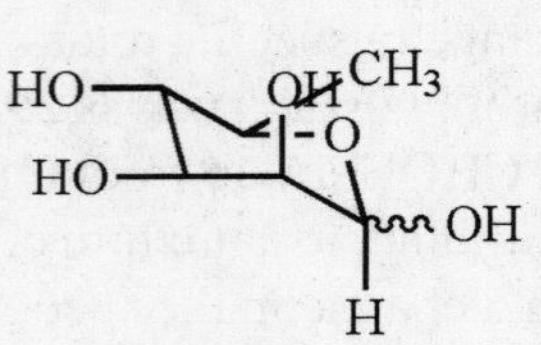 *29.20*

Furanoses

What Is a Ketose?

A ketose is a polyhydroxy ketone.

What Is a Furanose?

A furanose is a five-membered ring hemiacetal or hemiketal in which the ring contains an oxygen. It is essentially a polyhydroxy tetrahydrofuran derivative.

What Is the Open Chain Structure of D-Ribose? D-Ribulose? L-Xylulose? D-Fructose?

The Fisher projection of ribose is *29.21*, D-ribulose is *29.22*, L-xylulose is *29.23*, and D-fructose is *29.24*.

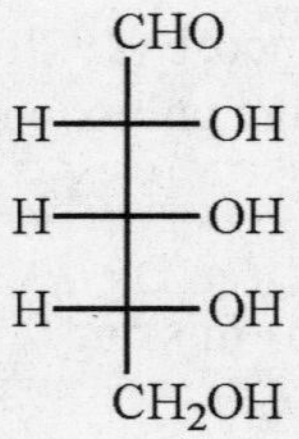

29.21 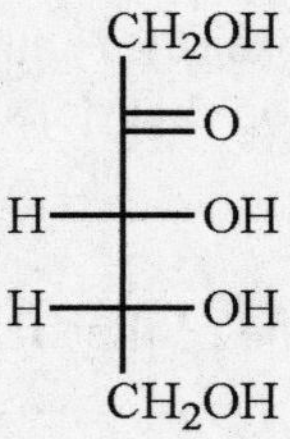 *29.22* 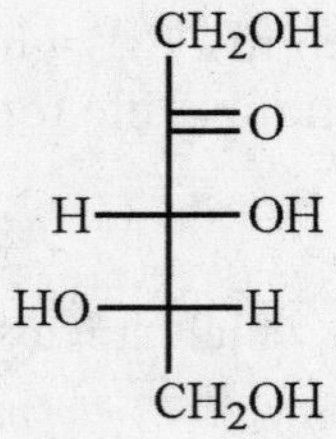 *29.23* 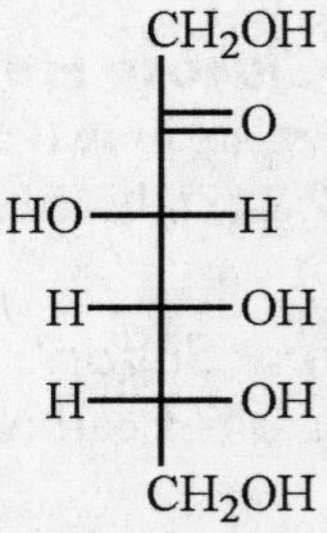 *29.24*

Draw the β-Furanose Form of D-Ribose, D-Ribulose, L-Xylulose and D-Fructose Using Haworth Formulas.

In each case, the β-form of these sugars is shown. The Haworth formula ***29.25*** is β-(D)-ribose, ***29.26*** is β-(D)-ribulose, ***29.27*** is β-(L)-xylulose, and ***29.28*** is β-(D)-fructose.

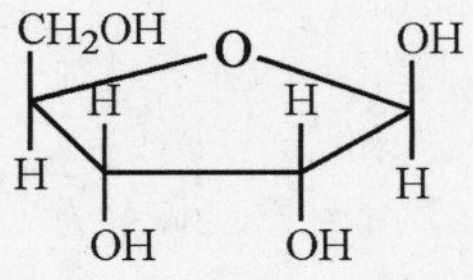

29.25 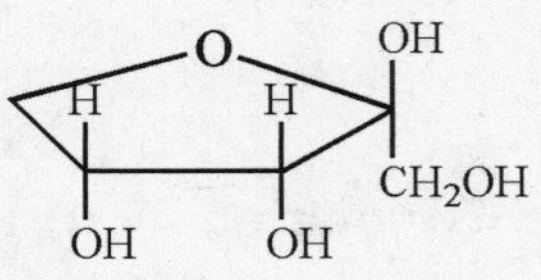 *29.26* 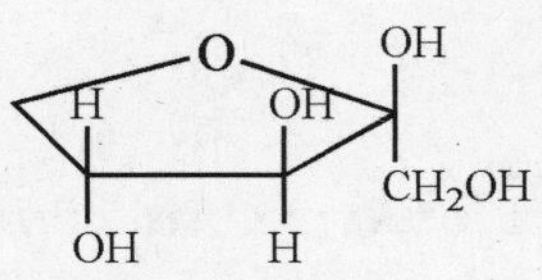 *29.27* 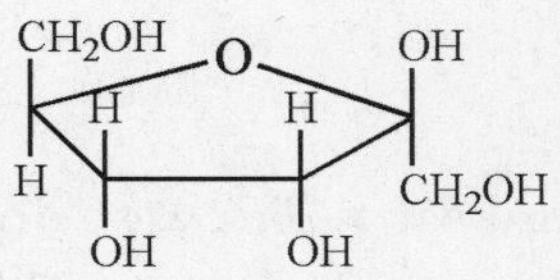 *29.28*

What Is the Furanose Form of 2-Deoxy-D-Ribose?

The structure of β-(D)-2-deoxyribose is ***29.29***.

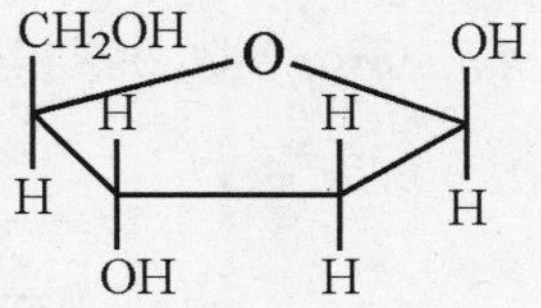

29.29

Does D-Fructose Undergo Mutarotation?

Yes! At equilibrium there is about 4-9% of α-D-fructofuranose and 21-31% of β-D-fructofuranose. Complicating this picture is the fact that fructose also exists in the pyranose form. At equilibrium, there is 0-3% of α-D-fructopyranose, and the dominant isomer is 57-75% of β-D-fructopyranose (***19.2.29***).

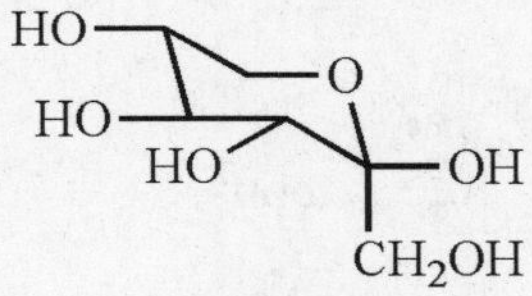

29.30

29.2. DISACCHARIDES

Disaccharides are molecules in which two monosaccharide units are joined together. There are several ways in which these units can be connected, leading to different molecules.

What Is a Disaccharide?

A disaccharide is a carbohydrate that gives two monosaccharides upon hydrolysis.

What Is an α or β Linkage?

If we use ***29.31*** as an example of a disaccharide, the C2—O—C2' linkage (C2 is part of one sugar unit, and C2' is part of the other sugar unit) is axial, or α (down as drawn). The β-linkage would have the C2—O—C2' linkage with the oxygen equatorial (up as drawn).

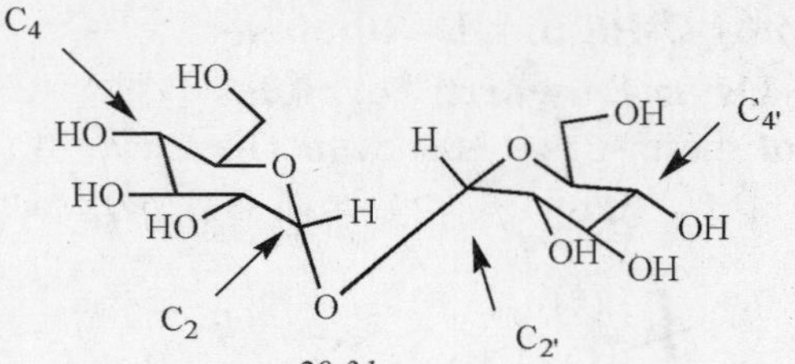

29.31

Draw Two Glucose Molecules As a Disaccharide Connected by a β-Linkage Between C2 and C2'. Draw a Disaccharide with a β-Linkage Between C2 and C4'.

In ***29.31***, the C2 and C4 carbons of one glucose molecule are marked, as well as the C2/C4 carbons of the second glucose (C2' and C4'). These carbons are also marked in ***29.32***. In ***29.32***, the term 'β-linkage' refers to the C2–C2' connection (the anomeric carbons) being axial (β). This 'linkage' appears to be an ether type linkage, but is, in fact, a linked acetal. The second type of disaccharide connects the anomeric OH with the C4' OH to give ***29.31***, which is a β-linkage with the anomeric C2 oxygen being equatorial. The name of ***29.31*** is maltose, and ***29.32*** is cellobiose.

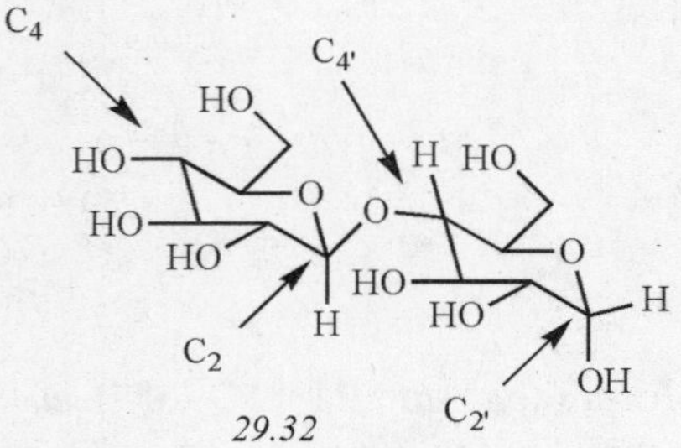

29.32

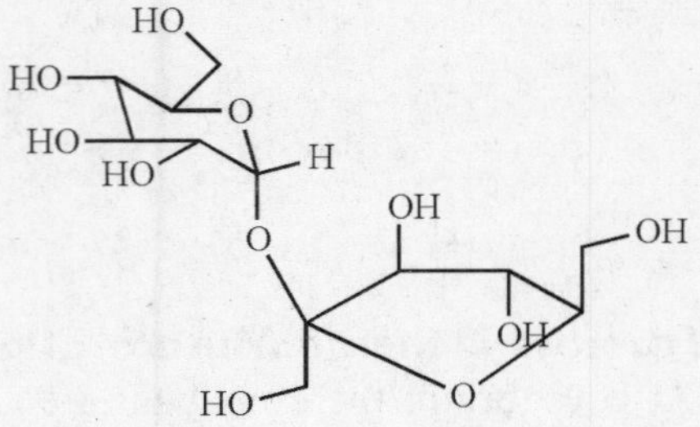

29.33

Draw a D-Glucose and a D-Fructose Connected by an α-Linkage between C2 and C2'.

This disaccharide has the structure ***29.33***. This molecule is called sucrose.

What Is a Head-to-Head Disaccharide?

A head-to-head disaccharide is a molecule composed of two monosaccharides linked by the C2–C2' atoms, as in ***29.31***.

What Is a Head-to-Tail disaccharide?

A head-to-tail disaccharide is a molecule composed of two monosaccharides linked by the C2–C4' atoms, as in ***29.32***.

Draw the Structure of A) Maltose; B) Cellobiose; C) Lactose; D) Sucrose.

The structure of maltose is ***29.31***, cellobiose is ***29.32***, lactose is ***29.34***, and sucrose is ***29.33.*** Lactose is composed of a D-galactose and a D-glucose, with a C2–C2' β-linkage.

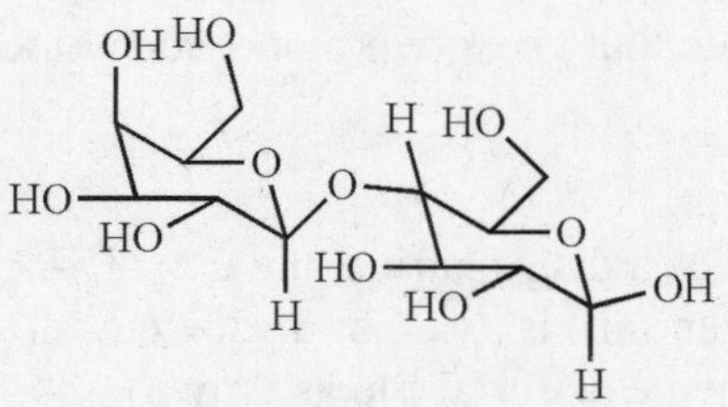

29.34

29.3. POLYSACCHARIDES

Polysaccharides are molecules composed of many sugar units, which gives them special chemical, biological, and conformational properties.

What Is a Polysaccharide?

A polysaccharide is a carbohydrate composed of many monosaccharide units. It is essentially a polymer composed of monosaccharide monomers.

What Is an Oligosaccharide?

An oligosaccharide is a polysaccharide composed of about 3-10 monosaccharide units.

What Is the Structure of Starch?

Starch is actually a mixture of two polysaccharides. One is a water soluble polysaccharide called amylose, which is a linear polymer (100-several thousand D-glucose units) attached by 1,4-β-linkages. The second polysaccharide is called amylopectin and is a branched polymer (100-several thousand D-glucose units) attached by 1,4-β-linkages.

What Is the Structure of Amylose? Of Amylopectin?

The linear polysaccharide amylose has the structure ***29.35***, with repeating D-glucose molecules with a 1,4-β-linkage. The branched polymer amylopectin can be represented as ***29.36***, again with repeating D-glucose units.

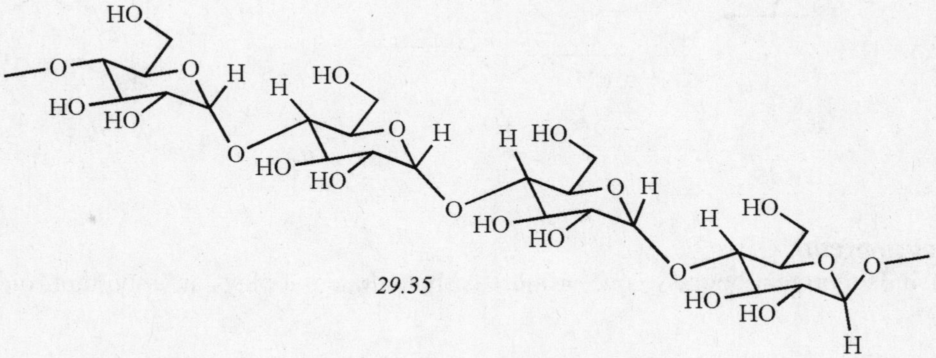

29.35

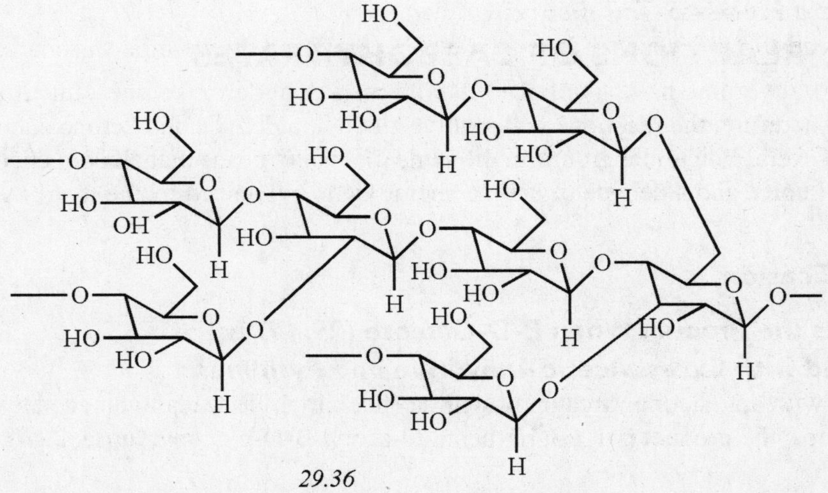

29.36

What Is the Structure of Cellulose?

Cellulose has a structure similar to amylose (***29.35***) except that the linear D-glucose units are attached by a β-linkage (see ***29.37***). Cellulose may be the most abundant organic material on earth and is the structural material that composes most plants.

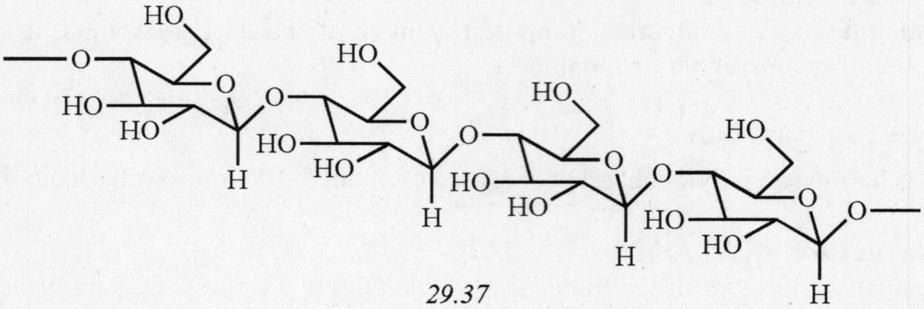

29.37

What Is the Structure of Chitin?

Chitin (***29.38***) is a polysaccharide that comprises the exoskeleton of insects and is also found in crustaceans. It is a linear polymer of *N*-acetylglucosamine (***29.39***).

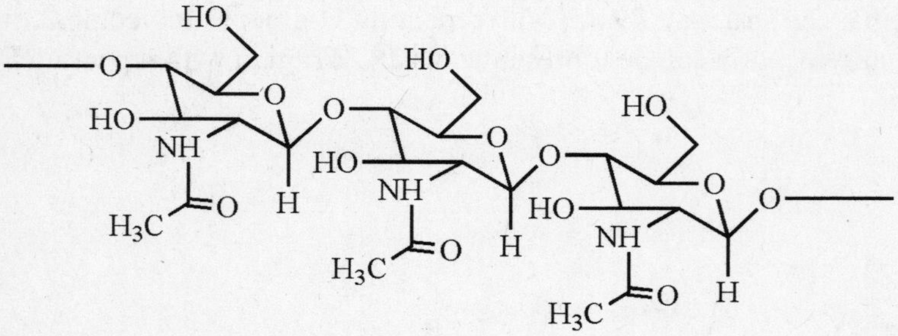

29.38

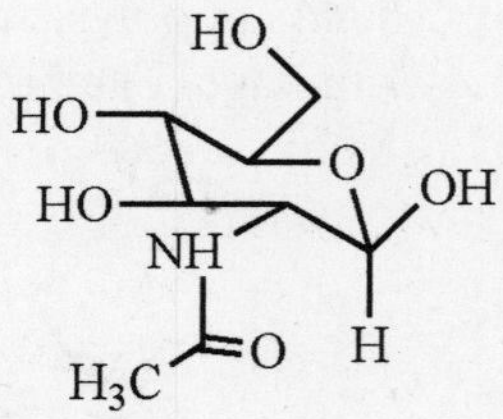

29.39

What Is a Glycoprotein?

A glycoprotein is a protein bound to one or more carbohydrates. It plays an important role in biological interactions.

29.4. REACTIONS OF CARBOHYDRATES

Carbohydrates have alcohol units and an aldehyde or ketone unit in the various equilibrating forms. Therefore, the reactions will involve alcohol, aldehyde and ketone chemistry. However, the presence of several functional groups in the same molecule means that alcohol chemistry is influenced by the carbonyl units, and aldehyde or ketone chemistry is influenced by the hydroxyl groups.

Esterification

What Is the Product When B-D-Glucose (29.40) Is Reacted with Excess Acetic Anhydride and Pyridine?

As with any alcohol, treatment with acetic anhydride leads to an acetate ester. Since there are five OH groups, the product is 1,2,3,4,6-penta-O-acetyl-β-D-glucopyranose, ***29.41***.

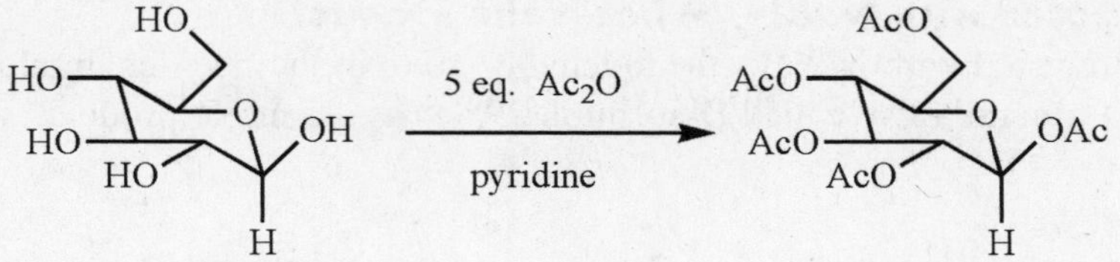

HO, HO, HO, HO, OH, H — 5 eq. Ac_2O / pyridine → AcO, AcO, AcO, AcO, OAc, H

29.40 *29.41*

Is It Possible to Convert Only One of the Hydroxyl Groups to the Corresponding Ester?

Yes! However, there are differences in reactivity. The anomeric OH is probably the most reactive and would give the corresponding mono-ester. Selective conversion of one of the other hydroxyl units usually requires protection-deprotection sequences and will not be discussed here. The same comments apply, more or less, to the other reactions presented below.

Etherification

What Is the Product When A-D-Glucose Is Treated with Dimethyl Sulfate?

Dimethyl sulfate (Me_2SO_4) converts alcohols into methyl ethers (R-O-Me), and reaction of α-D-glucose with dimethyl sulfate forms 1,2,3,4,6-pentamethoxy-α-D-glucopyranose, ***29.42***.

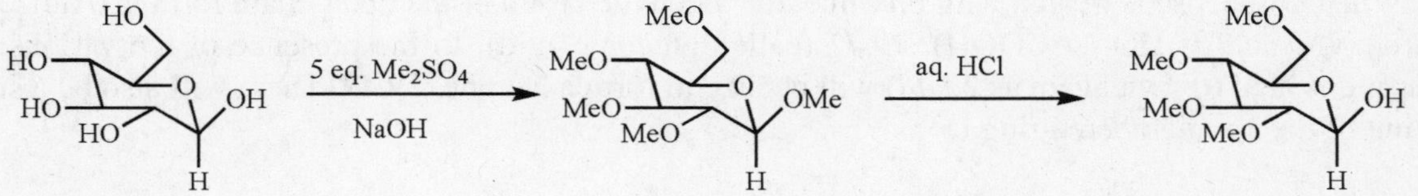

HO, HO, HO, HO, OH, H — 5 eq. Me_2SO_4 / NaOH → MeO, MeO, MeO, MeO, OMe, H — aq. HCl → MeO, MeO, MeO, MeO, OH, H

29.42 *29.43*

If Pentamethoxy-D-Glucose (29.42) Is Reacted with Aqueous HCl, What Is the Expected Reaction Product?

In general, methyl ethers are resistant to aqueous acid hydrolysis. The OMe at the anomeric carbon, however, is part of an acetal structure and, as such, subject to acid hydrolysis. Treatment of ***29.42*** with aqueous HCl will give the hemiacetal, ***29.43*** (2,3,4,6-tetra-O-methyl-D-glucopyranose).

Reduction

What Is the Product of the Reaction Between D-Galactose and Hydrogen in the Presence of a Nickel Catalyst?

Catalytic hydrogenation will reduce the open-chain aldehyde form of the carbohydrate to the alcohol. In this case, D-galactose (***29.44***) is reduced to ***29.45***.

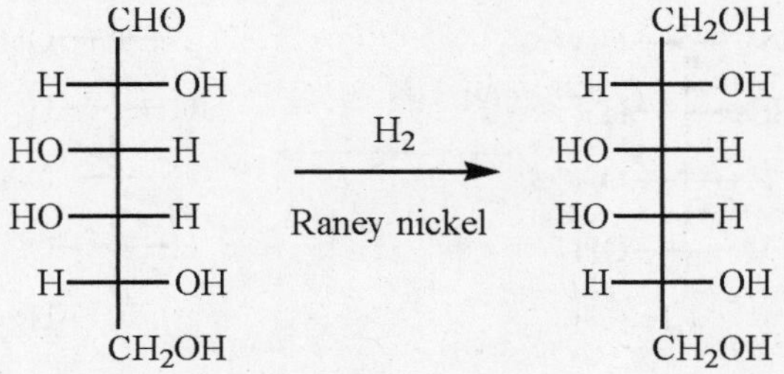

29.44		*29.45*
CHO		CH_2OH
H—OH	H_2	H—OH
HO—H	→	HO—H
HO—H	Raney nickel	HO—H
H—OH		H—OH
CH_2OH		CH_2OH

If D-Glucose Is Treated with $NaBH_4$, What Is the Product?

If D-glucose is reduced with $NaBH_4$, the aldehyde group in the open-chain aldehyde is reduced to the alcohol, giving D-glucitol (also called D-sorbitol, ***29.46***) as the major product.

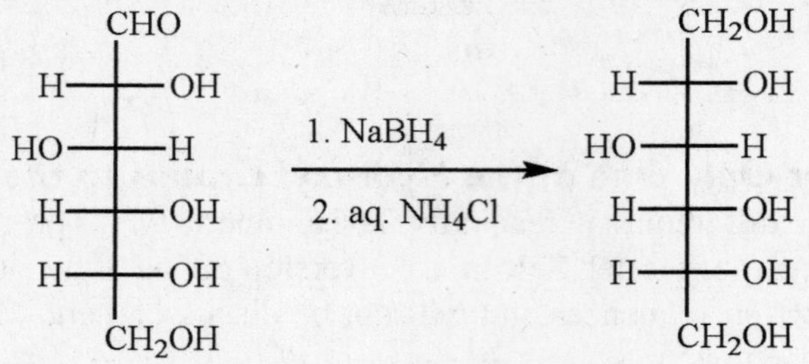

1. $NaBH_4$
2. aq. NH_4Cl

29.46

Oxidation

What Is the Product of D-Glucose and Aqueous Bromine Buffered with Calcium Carbonate?

When D-glucose is treated with bromine, the aldehyde group of the open-chain form is oxidized to a carboxylic acid (CHO → COOH), ***29.47*** (called gluconic acid). In the presence of the various OH groups, this acid (drawn again as ***29.48***) will cyclize to form a lactone, ***29.49***. There will also be a small amount of the six-membered ring lactone.

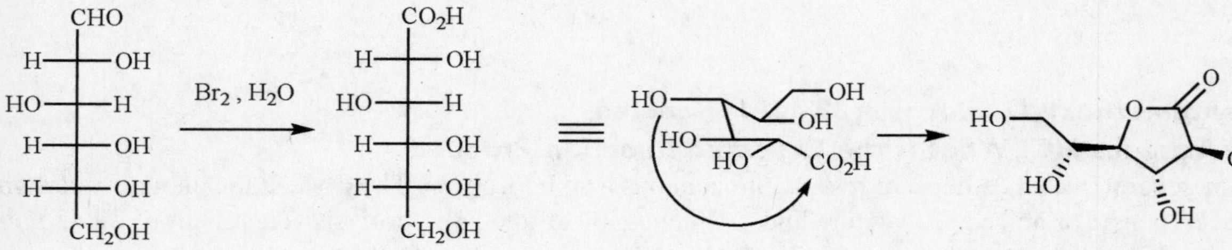

Br_2, H_2O

29.47 *29.48* *29.49*

What Is the Product of D-Altrose and Dilute Nitric Acid When Heated?

Under these conditions, nitric acid is a sufficiently strong oxidizing agent to convert not only the CHO group to a carboxylic acid, but also to convert the terminal CH_2OH group to CO_2H. The final product is, therefore, the diacid, ***29.50*** (glucaric acid).

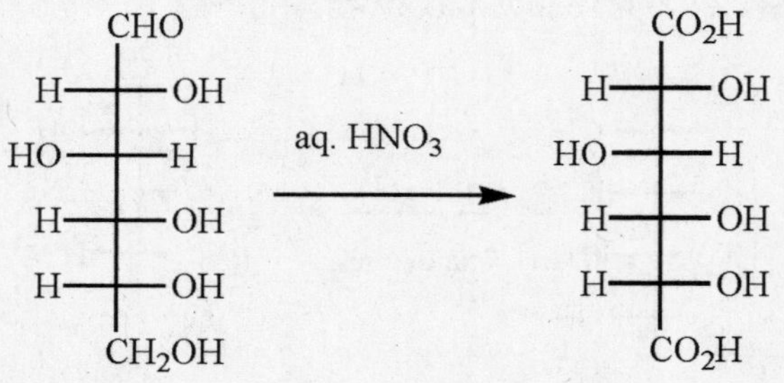

aq. HNO_3

29.50

What Is Fehling's Solution?

Fehling's solution is an aqueous solution of copper (II) sulfate complexed with tartaric acid.

Give the Product of the Reaction Between D-(−)-Arabinose and Fehling's Solution.

D-(−)-Arabinose (***29.51***) is oxidized by this reagent (***29.52*** is tartaric acid) to give ***29.53***. Mono acids of this type are generically known as aldonic acids. Fehling's solution is, therefore, a mild and selective oxidizing agent.

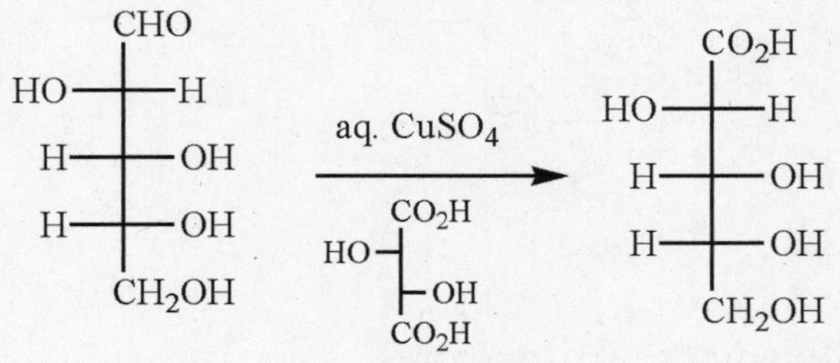

CHO; HO—H; H—OH; H—OH; CH_2OH — aq. $CuSO_4$ / CO_2H; HO—; —OH; CO_2H → CO_2H; HO—H; H—OH; H—OH; CH_2OH

29.51 *29.52* *29.53*

What Is the Fehling's Test?

When an aldose or ketose [both α-hydroxy aldehydes (CHOH-(C=O)H) and ketones as well as 'normal' aldehydes react] is treated with Fehling's solution, the oxidation to the acid is accompanied by disappearance of the bluish color of the cupric solution and precipitation of a reddish-copper precipitate of cuprous oxide. This is taken as diagnostic of the presence of an aldehyde moiety (or an α-hydroxy aldehyde or ketone) in the carbohydrate.

What Is the Name Given to Carbohydrates That Give a Positive Test with Fehling's Solution?

These types of sugars are called **reducing sugars**.

What Is Benedict's Reagent?

Benedict's reagent is a solution of cupric sulfate using citric acid as the complexing agent rather than tartaric acid.

What Is the Product When Benedict's Reagent Reacts with Maltose? With D-Fructose?

Benedict's reagent also oxidizes an aldose to the aldonic acid (monocarboxylic acid) and is used to detect reducing sugars. In this example, maltose is oxidized to ***29.54***.

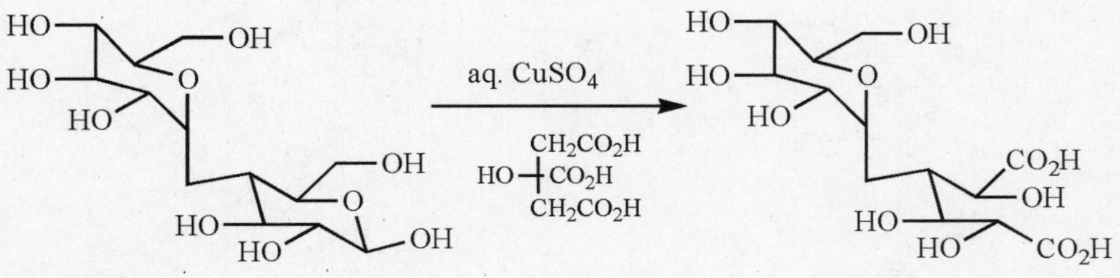

HO, HO, HO, OH, O, OH, O, HO, HO, OH — aq. $CuSO_4$ / CH_2CO_2H; HO—CO_2H; CH_2CO_2H → HO, HO, HO, OH, O, CO_2H, OH, HO, HO, CO_2H

29.54

What Is the Product When Benedict's Reagent Reacts with D-Fructose?

Fructose also reacts by interconversion of the hydroxyketone to an enol (*29.55*), which equilibrates to the aldehyde (*29.56*), and is then oxidized to the aldonic acid (*29.57*).

Note that Fehling's solution oxidizes α-hydroxy ketones in a similar manner.

$$\text{D-fructose} \xrightarrow[\text{citrate}]{\text{aq. } CuSO_4} \textbf{29.55} \text{ (enediol)} \longrightarrow \textbf{29.56} \text{ (aldehyde)} \longrightarrow \textbf{29.57} \text{ (aldonic acid)} + Cu_2O$$

29.55 *29.56* *29.57*

What Does a Positive Benedict's Test Indicate?

A positive Benedict's test (loss of the blue color and precipitation of the red-copper cuprous oxide) indicates the presence of an aldehyde group or an α-hydroxy ketone or aldehyde moiety in the carbohydrate.

What Is the Tollen's Test?

The **Tollen's test** oxidizes aldehydes to carboxylic acids using silver oxide (Ag_2O) in aqueous ammonium hydroxide. The Tollen's test oxidizes reducing sugars to the aldonic acid. A positive Tollen's test is accompanied by precipitation of silver on the sides of the reaction vessel, usually a test tube; a silver mirror.

What Does the Tollen's Test Indicate?

As with other oxidizing tests, the Tollen's test indicates the presence of an aldehyde or an α-hydroxy ketone.

Osazone Formation

An osazone is a *bis*-hydrazone formed by reaction of carbohydrates with a hydrazine such as phenylhydrazine ($PhNHNH_2$). An example is the reaction of D-(+)-glucose with phenylhydrazine to give the osazone, ***29.58***.

$$\text{D-glucose} \xrightarrow{3\ PhNHNH_2} \textbf{29.58}$$

29.58

What Reagents React with Carbohydrates to Give an Osazone?

An excess of a hydrazine ($RNHNH_2$) is required.

D-Ribose (29.59) and D-Arabinose (29.60) Have Opposite Absolute Configurations for the 2-Hydroxyl Group. What Is the Product When Each Is Treated with Three Equivalents of Phenylhydrazine?

The C2 hydroxyl is oxidized to a ketone group and in both cases the product is ***29.61***.

CHO
H—OH
H—OH
H—OH
CH_2OH

29.59

3 $PhNHNH_2$ →

=NNHPh
=NNHPh
H—OH
H—OH
CH_2OH

29.61

← 3 $PhNHNH_2$

CHO
HO—H
H—OH
H—OH
CH_2OH

29.60

29.5. SYNTHESIS OF CARBOHYDRATES

Carbohydrates are prepared by modification of standard procedures that take the multiple functionality into account.

What Is the Kiliani-Fischer Synthesis?

The Kiliani-Fisher synthesis extends the chain length of a carbohydrate by reacting an aldose with HCN to form the cyanohydrin. Reduction of the nitrile then leads to a new aldose, with one additional carbon relative to the starting carbohydrate.

Does the Kiliani-Fisher Synthesis Provide Pure D- or Pure L-Carbohydrates?

No! The initial step of the synthesis converts the aldehyde unit in ***29.62***, for example, into the cyano unit in ***29.63***. The cyanohydrin is formed as a mixture of diastereomers. These diastereomers must be separated prior to conversion of the nitrile to the aldehyde in order to obtain pure D or pure L carbohydrates.

CHO
H—OH
H—OH
H—OH
CH_2OH

29.62

HCN →

C≡N
H—OH
H—OH
H—OH
H—OH
CH_2OH

29.63

What Is the Product When D-Ribose Is Treated with i. HCN ii. Aqueous Acid iii. Na(Hg)?

When D-ribose (***29.62***) reacts with HCN, a cyanohydrin is formed (***29.63***). Nitriles are hydrolyzed to carboxylic acids with aqueous acid, and treatment of ***29.63*** with aqueous acid gives the aldonic acid, ***29.64***. When the acid unit in ***29.64*** reacts with sodium amalgam (Na[Hg]), a powerful reducing agent, the acid group is reduced to an aldehyde, ***29.65***, a mixture of allose and altrose.

$C{\equiv}N$ / H–OH / H–OH / H–OH / H–OH / CH_2OH (29.63) $\xrightarrow{H_3O^+}$ CO_2H / H–OH / H–OH / H–OH / H–OH / CH_2OH (29.64) $\xrightarrow{Na(Hg)}$ CHO / H–OH / H–OH / H–OH / H–OH / CH_2OH (29.65)

29.63 *29.64* *29.65*

What Is the Ruff Degradation?

The **Ruff degradation** involves oxidation of an aldose to an aldonic acid with bromine in water, followed by oxidative cleavage to a new aldose with hydrogen peroxide (H_2O_2) and ferric sulfate [$Fe_2(SO_4)_3$]. Ferric acetate is also used. This procedure gives a carbohydrate with one less carbon than the starting carbohydrate. An example is the oxidation of D-glucose to ***29.66*** with bromine water, followed by cleavage to D-arabinose (***29.67***) with hydrogen peroxide and ferric sulfate.

CHO / H–OH / HO–H / H–OH / H–OH / CH_2OH $\xrightarrow{Br_2,\ H_2O}$ CO_2H / H–OH / HO–H / H–OH / H–OH / CH_2OH (29.66) $\xrightarrow[H_2O_2]{Fe_2(SO_4)_3}$ CHO / HO–H / H–OH / H–OH / CH_2OH (29.67)

29.66 *29.67*

What Is the Wohl Degradation?

The **Wohl degradation** is virtually the opposite process to the Kiliani-Fisher synthesis. The aldehyde group in an aldose is converted to a nitrile via conversion to the oxime and dehydration with acetic anhydride. The nitrile is then treated with base to give an aldehyde with loss of HCN and one carbon from the carbohydrate chain.

What Is the Major Product When D-Xylose Is Treated with I. Hydroxylamine II. Acetic Anhydride and III. Sodium Methoxide?

Hydroxylamine (NH_2OH) reacts with the aldehyde of the aldose (D-xylose, ***29.68***) to give the oxime (***29.69***). Dehydration with acetic anhydride gives the nitrile, ***29.70***. When the α-hydroxy nitrile group is treated with base, HCN is lost to give the new aldose (***29.71***, D-threose). This overall process is called the Wohl degradation.

CHO / HO–H / HO–H / H–OH / CH_2OH (29.68) $\xrightarrow{H_2N\text{-}OH}$ =N-OH / HO–H / HO–H / H–OH / CH_2OH (29.69) $\xrightarrow[NaOAc]{Ac_2O}$ $C{\equiv}N$ / HO–H / HO–H / H–OH / CH_2OH (29.70) $\xrightarrow{NaOMe}$ CHO / HO–H / H–OH / CH_2OH (29.71)

29.68 *29.69* *29.70* *29.71*

29.6. NUCLEIC ACIDS, NUCLEOTIDES AND NUCLEOSIDES

The critically important biological molecules DNA are RNA are polymers composed of sugar units with a heteronuclear amine base attached.

What Is a Nucleoside?

A nucleoside is a carbohydrate, usually a cyclic furanose, that is attached to a heterocyclic amine base. The amine base is attached at the anomeric carbon of the sugar.

What Sugars Are Usually Involved in the Structure of a Nucleic Acid?

The most common sugars are ribose (***29.72***) and 2-deoxyribose (***29.73***), although other sugars are often seen. Both ribose and 2-deoxyribose are shown in their cyclic forms.

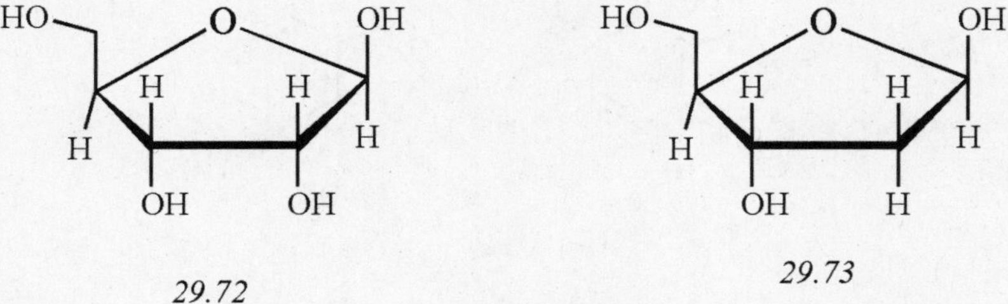

29.72 *29.73*

Nucleobases

What Is the Amine Part of a Nucleoside?

The amine base is usually a pyrimidine or a purine attached at the anomeric carbon of the ribose or 2-deoxyribose. These amines are commonly called nucleobases. In the case of nucleosides, the nucleobase is attached to a sugar.

Give the Structure of Pyrimidine.

Pyrimidine is a six-membered aromatic ring containing two nitrogens at the 1- and 3-position (***29.74***).

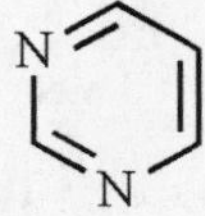

29.74

Give the Structure of A) Cytosine; B) Uracil; C) Thymine.

Cytosine is ***29.75***, uracil is ***29.76***, and thymine is ***29.77***. All are considered to be pyrimidine bases.

29.75 *29.76* *29.77*

Give the Structure of Purine.

Purine is a bicyclic aromatic amine, ***29.78***.

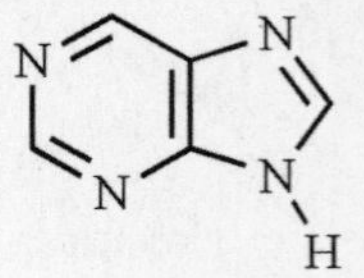

29.78

What Is the Structure of (A) Guanine; (B) Adenine?

Guanine is ***29.79***, and adenine is ***29.80***. Both are considered to be purine bases.

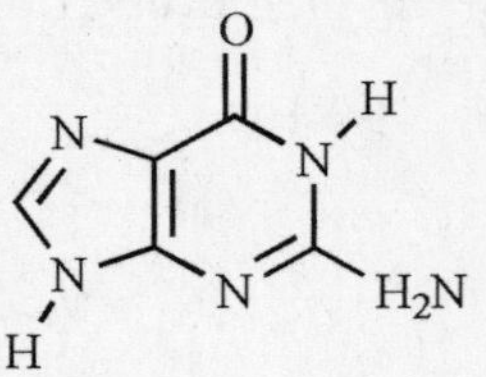

29.79 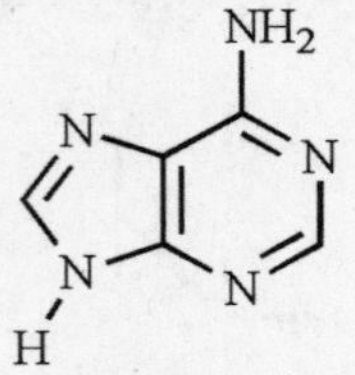 *29.80*

Nucleosides

What Are the Two Most Common Types of Nucleoside?

The most common nucleosides are purine and pyrimidine nucleosides. When cytosine is attached to a ribose, a nucleoside, it is called cytidine. Uracil gives uridine, thymine gives thymidine, guanine gives guanosine, and adenine gives adenosine.

What Is the Structure of (A) Adenosine; (B) Guanosine; (C) Uridine; (D) Thymidine?

Adenosine is a nucleoside with structure ***29.81***, guanosine has the structure ***29.82***, uridine is ***29.83***, and thymidine is ***29.84***.

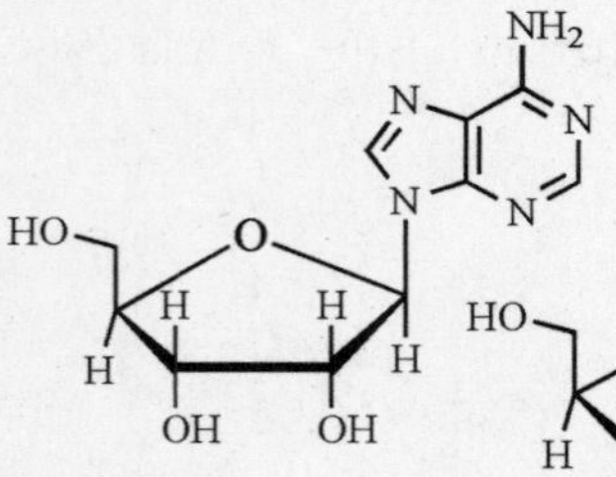

29.81 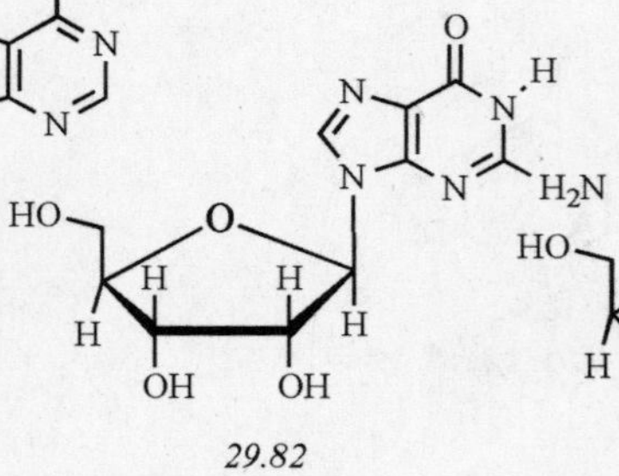 *29.82* 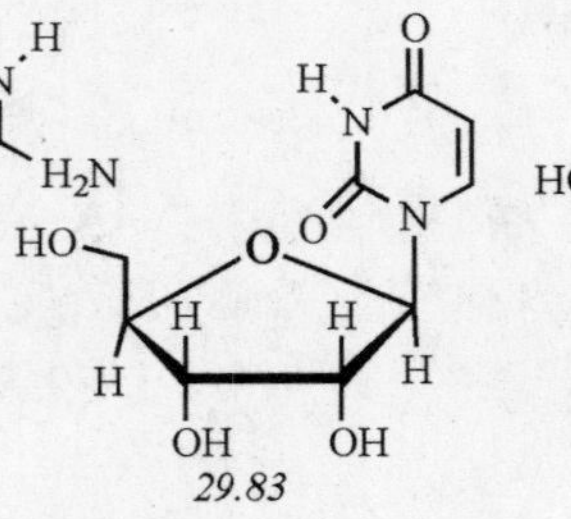 *29.83* 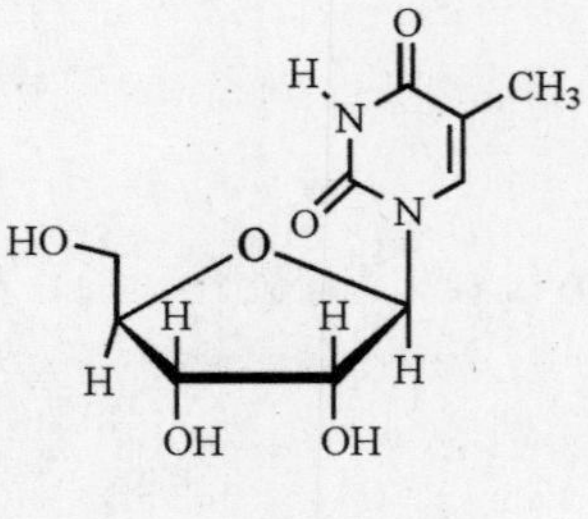 *29.84*

What Are the Single-Letter Codes for the Important Nucleosides?

Each nucleoside has a single-letter code: **A** for adenosine, **G** for guanosine, **U** for uridine, **T** for thymidine, and **C** for cytidine.

Nucleotides

What Is a Nucleotide?

A nucleotide is the phosphoric acid ester of a nucleoside. The $(HO)_2P(=O)$-O unit is attached at the C_5 CH_2OH moiety of the sugar. If two phosphoric acids are attached, the molecule is called a diphosphate, and if three phosphoric acids are attached, it is a triphosphate.

What Is the Structure of Adenosine Monophosphate? What Is the Abbreviation for This Molecule?

The structure of the monophosphate nucleotide is ***29.85***. It is given a three-letter abbreviation, AMP —for adenosine monophosphate).

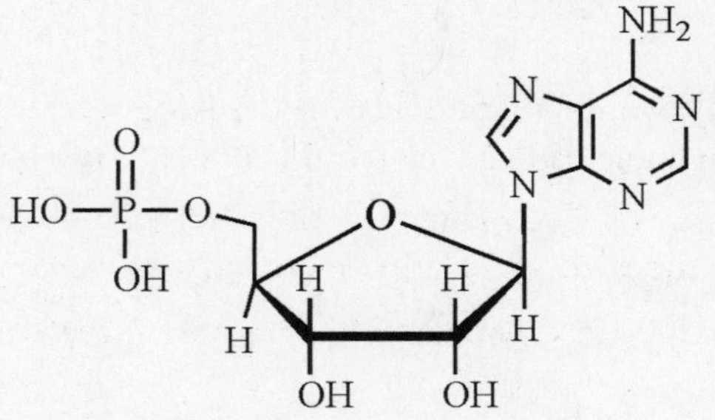

29.85

What Is the Structure of Uridine Diphosphate? What Is the Abbreviation for This Molecule?

Uridine diphosphate is a nucleotide diphosphate and has the structure ***29.86***. The three-letter code for this molecule is UDP.

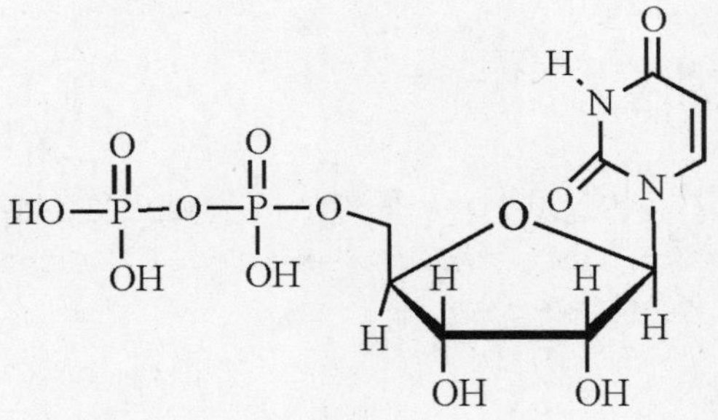

29.86

What Is the Structure of Thymidine Triphosphate? What Is the Abbreviation for This Molecule?

The triphosphate nucleotide thymidine triphosphate has the structure ***29.87*** and is given the three-letter code TTP.

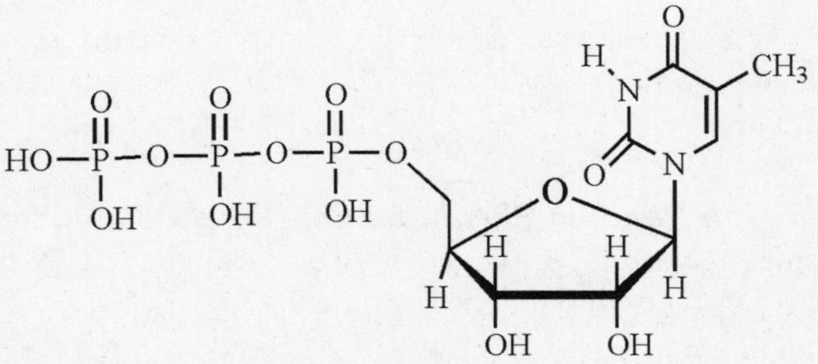

29.87

What Are the Three-Letter Codes for All Five Important Nucleotide Triphosphates Derived from the Important Purine and Pyrimidine Bases?

Adenosine triphosphate is ATP, uridine triphosphate is UTP, thymidine triphosphate is TTP, guanosine triphosphate is GTP, and cytidine triphosphate is CTP.

What Is a Polynucleotide?

A polynucleotide is a polymer of nucleosides linked together by phosphate linkages, usually monophosphate linkages. A polynucleotide of this type is called a **nucleic acid**.

What Is a Deoxyribonucleotide?

This is a nuclei acid that uses 2-deoxyribose as the carbohydrate portion of the nucleotide.

Base Pairing

Purine bases in a nucleic acid will form strong hydrogen bonds when in close proximity to certain pyrimidine bases in another nucleic acid or within the same nucleic acid. The two bases that form these hydrogen bonds are said to be **base paired** and are referred to as **complementary bases**. The usual complementary bases are: C–G, T–A. These hydrogen bonds are shown in ***29.88*** for C–G and ***29.89*** for T–A.

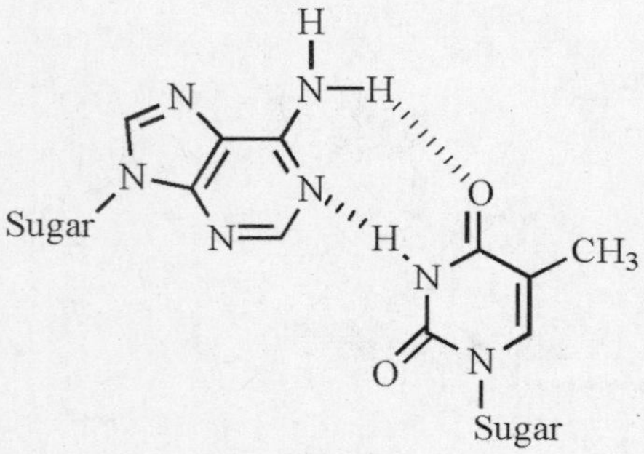

29.88

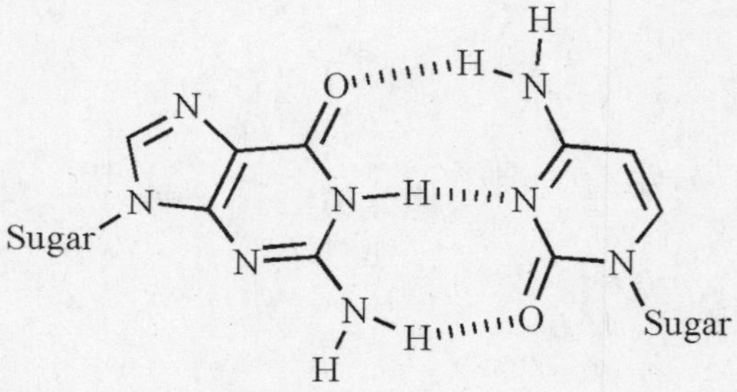

29.89

What Does the Term 'Double Stranded' Mean?

Two nucleic acids that join together in an antiparallel manner, as in ***29.90***. Complementary bases are usually important, i.e., a T nucleotide in one nucleic acid will be matched with an A nucleotide in the second nucleic acid, as shown. The base containing U is usually not base-paired and can have virtually any other base as its complement, as shown in ***29.90***.

A-T-A-A-G-U-T-U--C-T-G

T-A-T-T-C-C-A-A-G-A-C

29.90

What Role Does Hydrogen Bonding Play in Double Stranded Nucleotides?

The hydrogen bonding between the C–G and T–A base pairs is largely responsible for 'binding together' the two nucleic acid strands into the 'double strand.'

What Base Pairs Can Hydrogen Bond in a Nucleotide?

The most common hydrogen bonding pairs (complementary bases) are C–G (cytidine and guanosine) and T–A (thymidine and adenosine).

29.7. RNA

Ribonucleic acids are key biological nucleotides where ribose is the carbohydrate unit.

What Is RNA?

RNA is a ribonucleic acid where the nucleotide backbone of the polymer is composed of ribose units.

What Is the General Structure of RNA?

A general structure of RNA is shown in ***29.91***. Each nucleotide, linked by a monophosphate unit, is usually attached at the 3'-OH and the 5'-OH. The 5' OH is the CH_2OH unit. Each sugar is a ribose unit, and the 'BASE' is one of the five bases described above. This is a polymeric structure and may be relatively short with only a few nucleotides, or it can be composed of hundreds of nucleotides. Most RNA has thymidine rather than uridine in the structure.

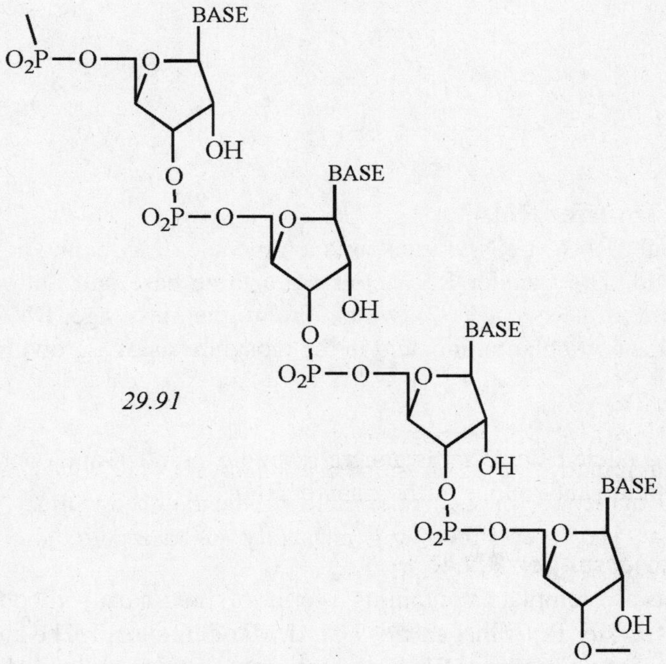

BASE
O_2P—O—O
OH
O
BASE
O_2P—O—O
OH
O
BASE
O_2P—O—O
OH
O
BASE
O_2P—O—O
OH
O—

29.91

Is RNA Usually Single Stranded or Double Stranded?

RNA is usually single stranded.

In Single Stranded RNA, What Is the Role of Hydrogen Bonding and Base Pairing?

The base pairing occurs, but the C–G and T–A pairing occurs intramolecularly, causing the RNA molecule to 'bend' and 'fold' into a relatively complex structure.

Transfer RNA

The usual representation of RNA is ***29.92*** [taken from Theil, E.C., Ed., *Principles of Chemistry in Biology,* American Chemical Society, Washington DC, *1998*, p. 229]. The single stranded nucleic acid is folded in a relatively specific pattern, a 'cloverleaf,' allowing it to interact with messenger RNA.

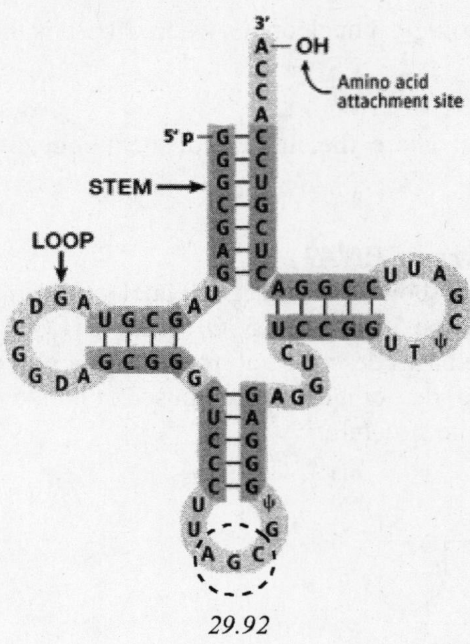

29.92

What Is the Role of Transfer RNA?

A molecule of transfer RNA (t-RNA) transports a molecule of an amino acid, which is attached as an ester at the 3' OH end. The transfer RNA contains a three base-pair anticodon, the three circled nucleotides at the 'bottom' of t-RNA (***29.92***), which bind to the messenger RNA at a specific site. This allows transfer of only this particular amino acid in the biosynthesis of a growing peptide chain.

Messenger RNA

Messenger RNA is a nucleic acid that is generally single stranded and contains the three base-pair codons that correlate with the anticodons of the transfer RNA.

What Is the Role of Messenger RNA?

Messenger RNA acts as a template, containing a series of three-base-pair codons, each of which correlates with a particular transfer RNA that carries a particular amino acid. The anti-codon of the transfer RNA binds to the codon of the messenger RNA, and the amino acid is 'unloaded' from the transfer RNA by formation of a peptide bond to the growing peptide chain. The transfer RNA is then 'released,' and the next transfer RNA, loaded with an amino acid, will attach itself to the appropriate codon. In this way, peptides, enzymes, etc., are synthesized so the chemical integrity and overall primary, secondary, tertiary and quaternary structures are maintained.

What Is the Genetic Code?

The genetic code is the message carried by messenger-RNA in the three-letter (three base pair) codon that defines which amino acid is to be used in the biosynthesis of a peptide. Each amino acid usually has two or more codons (three-letter code) that will allow the proper transfer RNA to interact with the messenger RNA.

29.8. DNA

Deoxyribonucleic acids are key biological nucleotides where 2-deoxyribose is the carbohydrate unit.

What Is DNA?

DNA is deoxyribonucleic acid and is a double stranded pair of nucleic acids composed of nucleotides using 2-deoxyribose as the sugar portion.

What Is the General Structure of DNA?

A fragmentary structure of DNA (single stranded) is represented by structure ***29.93***. This single strand will be phase-paired (complementary bases) in a double stranded array.

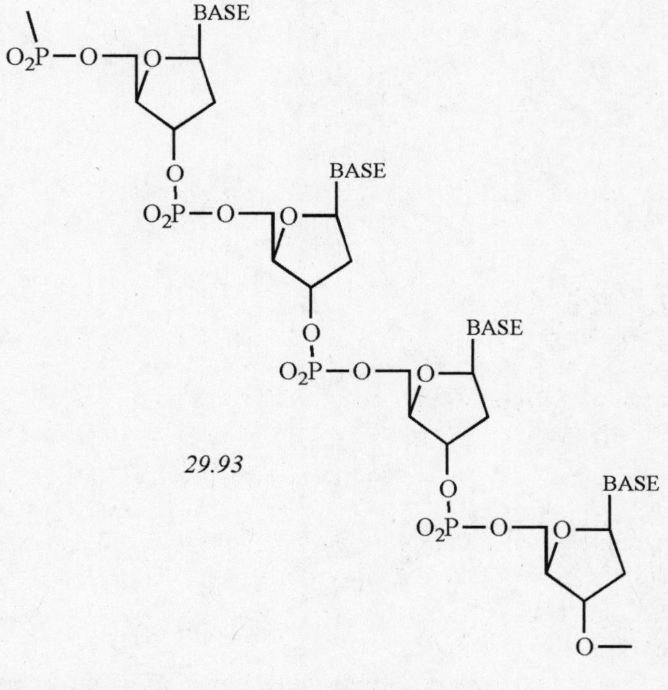

BASE
O_2P—O—O
O
BASE
O_2P—O—O
O
BASE
O_2P—O—O
29.93
O
BASE
O_2P—O—O
O—

Is DNA Usually Single Stranded or Double Stranded?

In most cases, biologically active DNA is double stranded. In order to replicate, however, DNA must become dissociated and single stranded (partially or totally). After replication, DNA will again resume its double stranded corm.

What Is the α-Helix?

Double stranded DNA has a helical structure that resembles a spiral in its natural conformation. This spiral structure tends to 'rotate' to the left and is referred to as the α-helix [see ***29.94***; taken from Theil, E.C., Ed., *Principles of Chemistry in Biology,* American Chemical Society, Washington DC, 1998, p. 228].

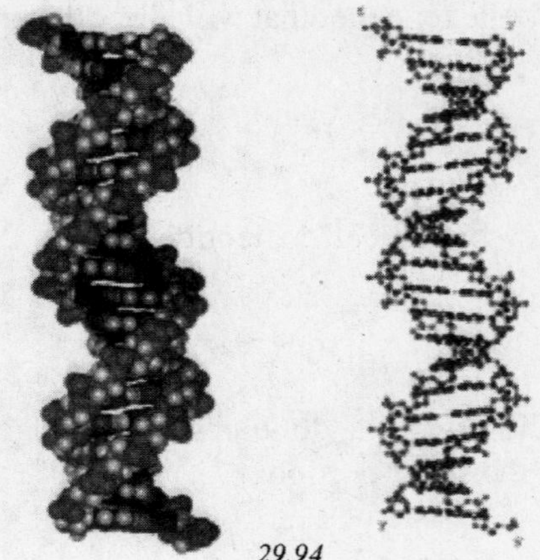

29.94

What Is the Watson-Crick Model?

The **Watson-Crick model** is the double stranded, helical model of DNA.

Test Yourself

1. Assign the absolute configuration to all stereogenic centers.

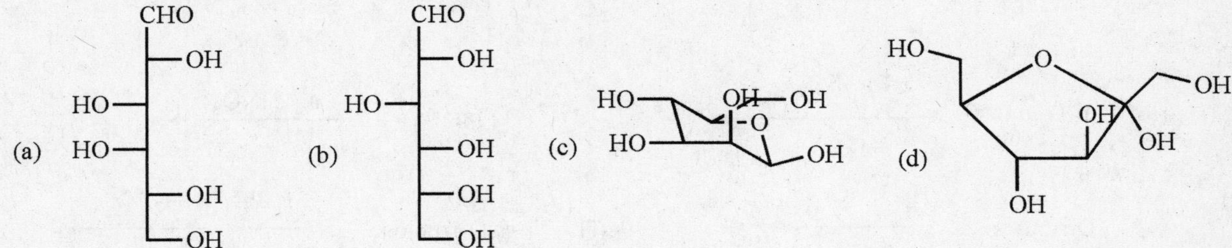

2. Identify each of the following as a D or an L sugar.

3. Draw each of the following in its Fischer Projection.
   (a) D-glucose; (b) L-mannose; (c) D-gulose; (d) L-altrose.
4. Draw cyclic form of
   (a) 3-deoxy-D-altrose; (b) 2-deoxy-L-idose;
   (c) 6-deoxy-D-galactose; (d) 5-deoxy-D-talose.
5. Draw
   (a) α-D-fructofuranose;
   (b) β-D-fructofuranose;
   (c) α-D-fructopyranose;
   (d) β-D-fructopyranose.
6. Draw the following disaccharides:
   (a) head-to-head α-D-gulose-L-altrose;
   (b) head-to-tail-α-D-talose-D-galactose;
   (c) head-to-tail-β-D-glucose-D-mannose.
7. Give the structures of:
   (a) dimethyl sulfate; (b) diethyl sulfate; (c) Ac_2O.
8. Draw the structures of
   (a) CDP; (b) ATP; (c) GMP.
9. What is the complementary strand for G–A–A–T–C–C–A–C–T–T–G–C?

10. In each case, give the major product. Remember stereochemistry.

(a) α-L-mannose $\xrightarrow{5\ Ac_2O\ ,\ pyridine}$

(b) L-altrose $\xrightarrow[2.aq.\ NH_4Cl]{1.\ NaBH_4}$

(c) β-D-gulose $\xrightarrow{5\ Me_2SO_4}$

(d) D-talose $\xrightarrow{aq.\ HNO_3}$

(e) D-ribulose $\xrightarrow{H_2,\ Pd\text{-}C}$

(f) β-D-fructose $\xrightarrow{3\ PhNHNH_2}$

11. In each case, provide a suitable synthesis. Show all reagents and intermediate products.

(a)

CHO		CHO
~OH	⇒	—OH
~OH		—OH
—OH		—OH
—OH		
—OH		

(b)

CHO		CHO
HO—	⇒	HO—
HO—		—OH
—OH		HO—
		HO—
		—OH

Test Yourself Answers

Chapter 1

Atomic Orbitals and Bonding

1. A 3s orbital is spherically symmetrical, as are all s orbitals. It is further from the nucleus than the 1s or 2s orbital. A 3p orbital is 'dumbbell' shaped, as are all p orbitals. It is further from the nucleus than the 2p orbitals.
2. The only difference is the three-dimensional direction of the orbital. Both are identical in energy. A P_x orbital is directed along the x-axis of a three coordinate system (x-y-z), and the p_y orbital is directed along the y-axis.
3. Oxygen is $1s^22s^22p^4$. Fluorine is $1s^22s^22p^5$. Chlorine is $1s^22s^22p^63s^23p^5$.

   Sulfur is $1s^22s^22p^63s^23p^4$. Silicon is $1s^22s^22p^63s^23p^2$.
4. Bonds (a), (b), (e) and (f) are ionic. Bonds (c), (d) and (g) are covalent.
5. (a) Carbon forms 4 bonds. (b) Nitrogen forms 3 bonds with one electron pair remaining.

   (c) Fluorine forms 1 bond, with three electron pairs remaining. (d) Boron forms 3 bonds.

   (e) Oxygen forms 2 bonds with two electron pairs remaining.
6. Boron is in Group III and requires 5 electrons to complete the octet. It only has three valence electrons to form covalent bonds, however. When those three bonds are formed, boron remains electron deficient. It can react with a molecule that can donate an electron pair, a Lewis base, making BF_3 a Lewis acid.
7. Since fluorine is the most electronegative, the C-F bond is the most polarized.
8. In (a), the 2s and 2p electrons are valence electrons, and this is N. Atom (b) is Na and the 3s orbitals are the valence electrons. In (c), the 1s electrons are valence, and this is He. In (d), the 3s and 3p electrons are valence, and this is P.
9. (a) C–O–H (H bonding) (b) C–F (dipole-dipole)

   (c) C–C (van der Waals)(d) N—H (H-bonding)

10.

(a) $C=C-C^{+}$ ⟷ $^{+}C-C=C$ **resonance**

(b) $^{+}Cl=C$ ⟷ $:Cl-C^{+}$ **resonance**

(c) $C-C=C-C^{-}$ ⟷ $C-C^{-}-C=C$ **resonance**

(d) H–O–C(–C)=O **no resonance**

Chapter 2

Structure and Molecules

1. In (a), the third molecule (1-butanol) can hydrogen bond whereas the others cannot and has the highest boiling point. In (b), the first compound (butanoic acid) hydrogen bonds to a much greater extent than the other compound and has the higher boiling point. In (c), there is no hydrogen bonding or dipole-dipole interactions, and the higher mass 7-carbon molecule (heptane) has the higher boiling point.

2. Molecule (a) is angular about the oxygen (bent), as is (c). Molecule (b) is tetrahedral about the central carbon. Molecule (d) is pyramidal with N at the apex of the pyramid and the H and two carbon groups at the other corners.

3. In (a), the dipole bisects the C–O–C bond, as it does in (c). In (b), the dipole is along the C–H bond (towards H). In (d), the dipole is along the *N*-lone electron pair line, and in (e), the dipole bisects the Br–C–Cl bond.

4. The alcohol functional group is C–O–H. The ketone functional group is C=O where the carbonyl is connected to two carbon groups. The alkyne functional group is C≡C, and the aldehyde functional group is C=O, where at least one of the groups attached to the carbonyl carbon is a hydrogen. The thiol functional group is S–H, and the nitrile functional group is C≡N.

5. For molecule (a), C^1, $C^2 = 0$; $N = +1$; $H^1-H^5 = 0$; $O^2 = 0$ and $O^1 = -1$. The formal charge for the molecule is 0 (+1–1). For (b), C^1, C^2 and $C^3 = 0$ but $C^4 = -1$. The $N = +1$ and $H^1-H^6 = 0$. The $O = -1$. For the molecule, the formal charge is -1 ($-1+1-1 = -1$).

6. In (a), the alcohol has the higher boiling point due to hydrogen bonding. In (b), the diol has the higher boiling point, since two OH groups can hydrogen bond more extensively than one OH group. In (c), the carboxylic acid hydrogen bonds more than the alcohol and has the higher boiling point.

7. Molecule (a), –118.6°C; molecule (b) –90.6°C; molecule (c) –33.1°C. Molecule (c) has a slightly higher mass, and is more symmetrical and will 'pack' into a crystal lattice more efficiently. For these reasons, it will have the higher melting point.

Chapter 3

Alkanes, Isomers, and Nomenclature

1. Formulae (b) and (f)
2. The longest continuous chain is 13, not 11, so this molecule is a tridecane. The correct name should be 5-ethyl-7-methyltridecane.
3. Molecule (a) is 7-(1-methylethyl)-3,3,5-trimethylundecane. Molecule (b) is 2,4-dibromo-2,3,3,4-tetramethyl-pentane. Molecule (c) is 3,4-diethyl-4,5,9-trimethyltetradecane. Molecule (d) is 4-bromo-6-chloro-5-propyl-5-(1,1,2-trimethylpropyl)tridecane.
4. (a) (b) 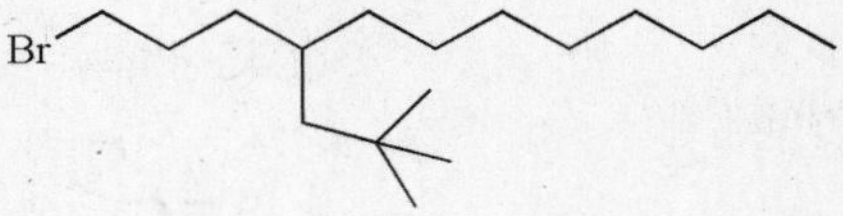 Br (c) 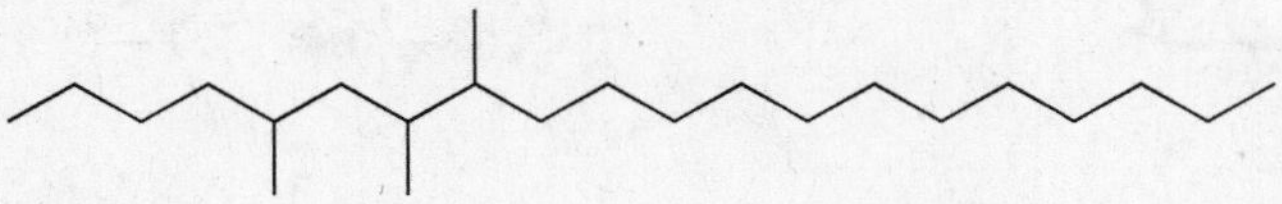 (d) 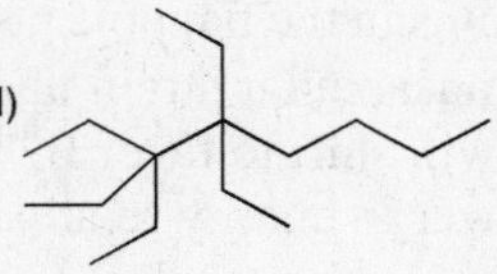 (e) 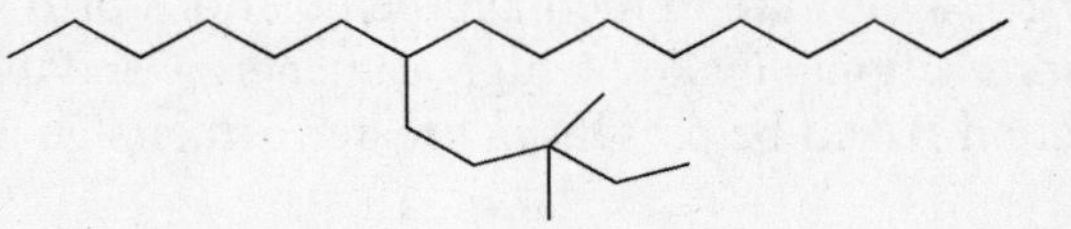
5. (a) Br (b) 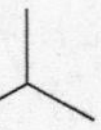 (c) 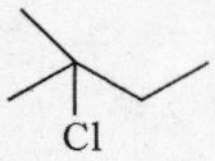 Cl (d) 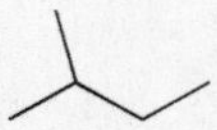 (e)  I
6. Eight isomers of (a) and (b) are:

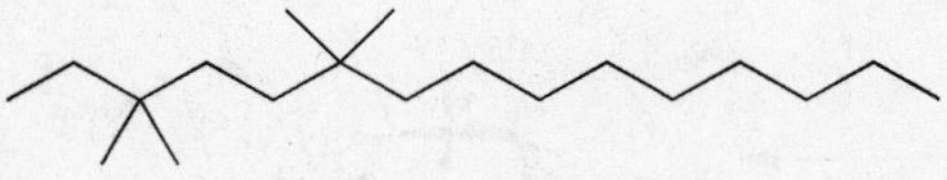

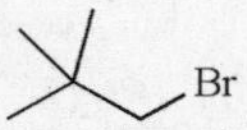

(a)

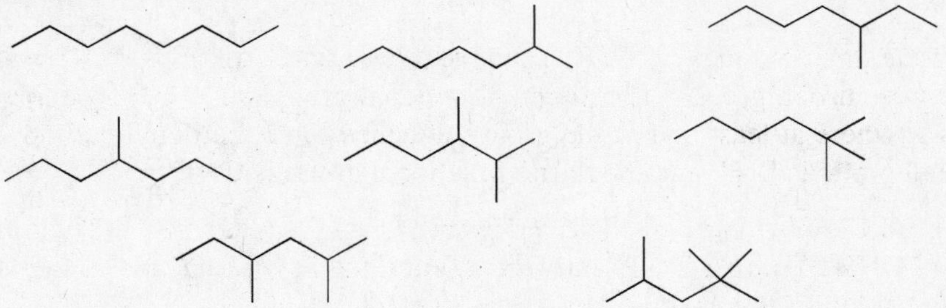

(b)

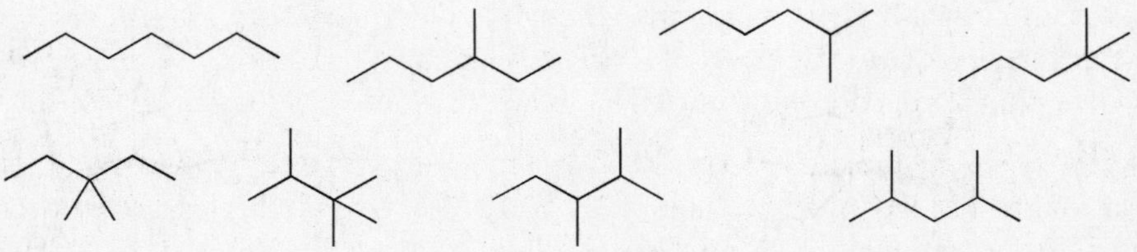

7. (a) 1,1-diethyl-3,3-dimethylcyclopentane, (b) 1,2,2,4-tetramethyl-3-propylcylcohexane, (c) 1-methyl-4-(1-methylbutyl)cycloundecane.

Chapter 4

Conformations

1.

lowest highest highest lowest

2.

A B

3. In the answer for Question 2, conformation A has both chlorine atoms in the flagpole position, whereas hydrogen atoms are in the flagpole positions of B. The greater transannular strain in A (greater flagpole interactions) makes A higher in energy, and the equilibrium will shift to favor B. Therefore, conformation B will be present in greater percentage because it is lower in energy relative to A.

4.

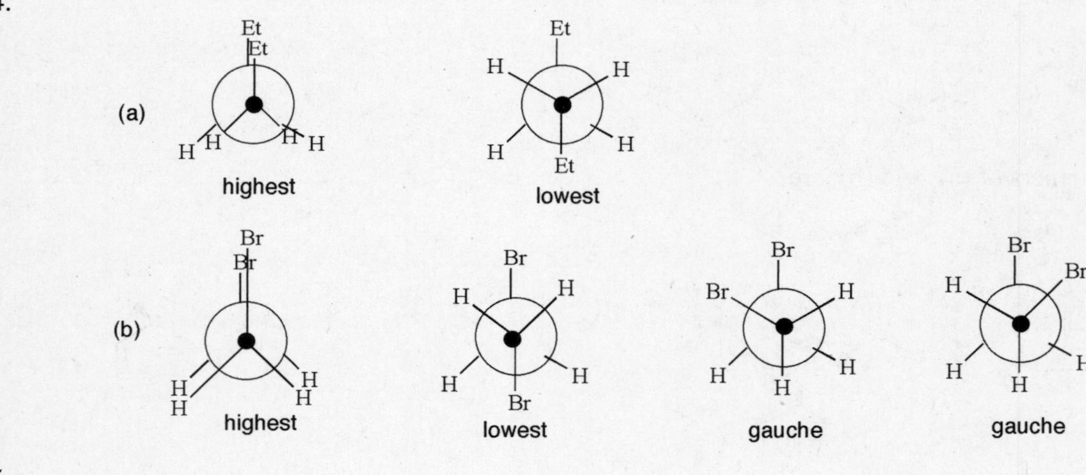

5.

all methyl groups equatorial

all methyl groups axial

6. Cyclopropane has the greatest angle strain and also the greatest torsion strain. Cyclopentane, in the envelope conformation, has more angle strain than torsion strain. The fact that the envelope conformation is more preferred shows that the molecule relieves torsional strain despite the fact that angle strain is slightly increased.

7.

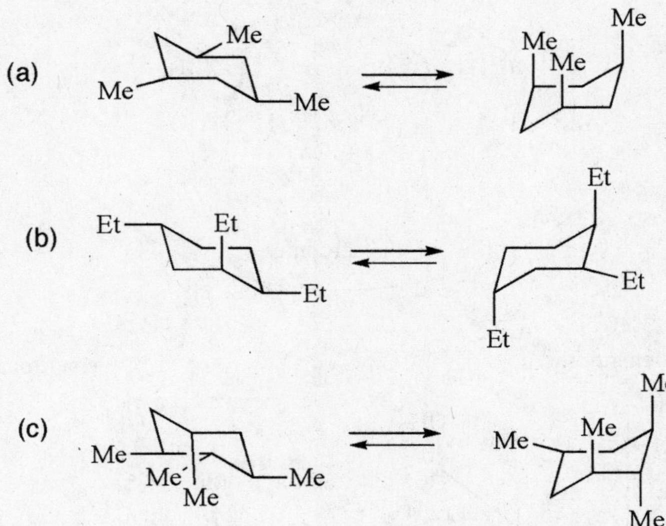

Chapter 5

Stereochemistry

1. (a) C(Br,Me,Et) = R, (b) C(Br,Et,H) = S; C(Br,Me,H) = S, (c) achiral, (d) C(OH,H) = S, (e) S(f) R(g) R(h) R

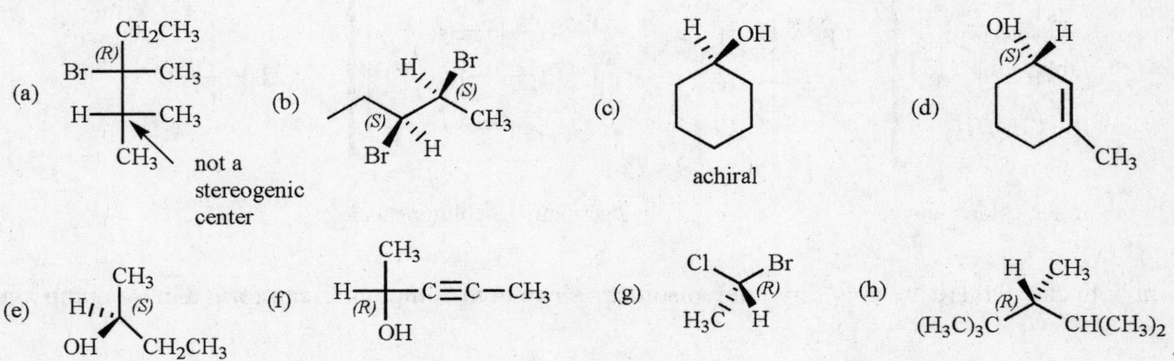

2. The two alcohols that have a stereogenic center are unsuitable. They would rotate plane polarized light in the polarimeter and possibly obscure or interfere with the rotation of the molecule of interest.

3. (a) –5.23°; (b) +10.95°; (c) +0.12°; (d) –1.67°

4. (a) 44%R, 56%S; (b) 95.5%R, 4.5%S; (c) 52%R, 48%S; (d) 42%R, 58%S.

5.

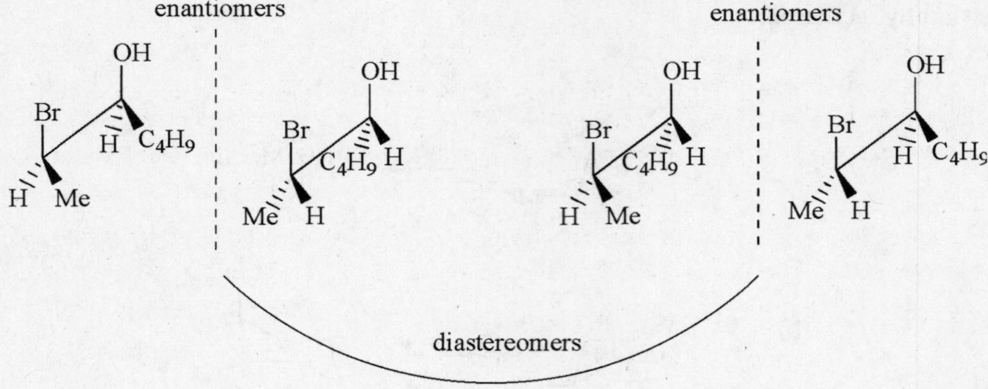

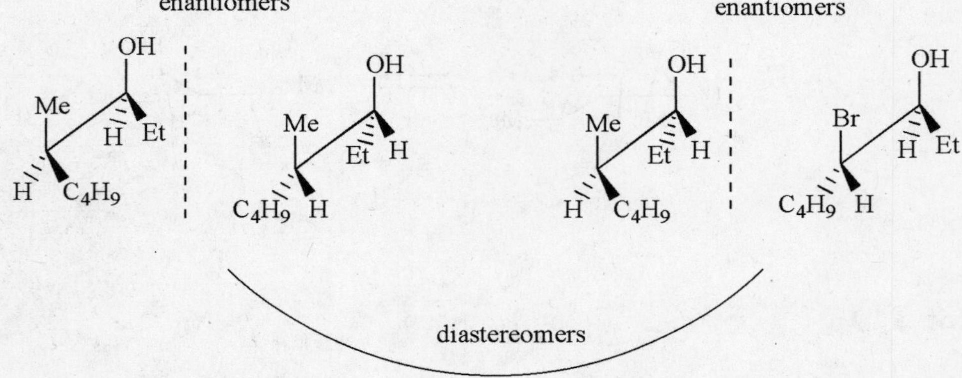

6.

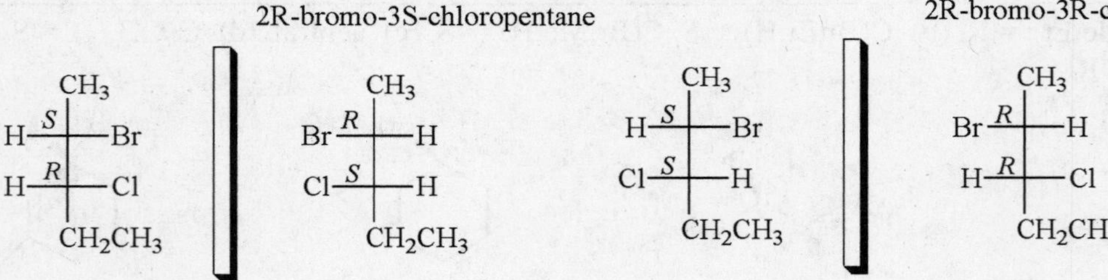

7. In both cases, there are only three stereoisomers, since both compounds can form a meso compound.

Chapter 6

Acids and Bases

1. HCOOH (formic acid) is more acidic. Acetic acid has an electron releasing carbon group (the methyl) directly attached to the carbonyl (C=O) carbon, whereas formic acid has a H attached to the carbonyl carbon. The electron releasing nature of the methyl group makes the O–H bond stronger via a through-bond inductive effect, and acetic acid is weaker because of it.

2. Since oxygen is more electronegative than nitrogen, nitrogen in CH_3NHCH_3 should be better able to donate electrons to a hydrogen atom than the oxygen in CH_3OCH_3. Therefore, the nitrogen compound is a better base.

3. (a) –4.53; (b) 8.63; (c) 1.24; (d) –7.9

4. (a) 4.68×10^{-3}; (b) 2.82×10^{-24}; (c) 8.91×10^{-18}; (d) 1.66×10^{-4}; (e) 7.08×10^{-11}

5. Although all three structures look identical, they represent delocalization of the charge on all three oxygen atoms. A model of the carbonate dianion is shown to give a better idea of the dispersion of electron density due to resonance.

CO_3^{-2} =

6. In this question, acid A is more acidic. In A, the OH unit of the carboxyl group is held close in space to the electron withdrawing chorine, so there is a significant hydrogen bond (through-space effect), that makes the OH unit more polarized and more acidic. The COOH unit is held on the opposite side of the molecule from the chorine atoms in B, so the internal hydrogen bonding is ineffective or completely missing. Therefore, B does not have the enhancement in acidity and A is more acidic than B.

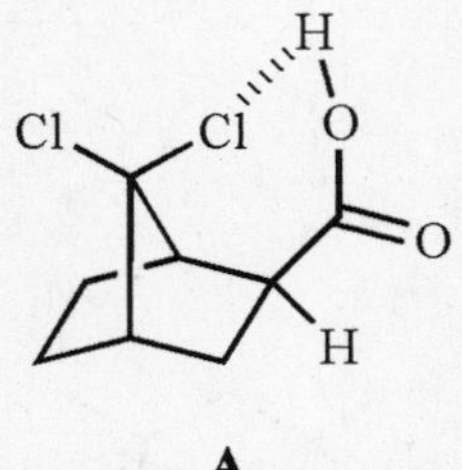

A

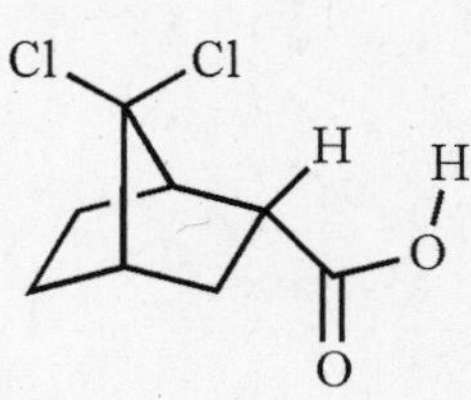

B

7. In this question, acid A is more acidic. In A, there are two bromine atoms, and the OH unit of the carboxyl group is close to the electron withdrawing bromine, so there is a significant hydrogen bond (through-space effect) that makes the OH unit more polarized and more acidic. In B, the bromine atoms are too far away from the COOH unit, so the internal hydrogen bonding is completely missing. Therefore, B does not have the enhancement in acidity and A is more acidic than B.

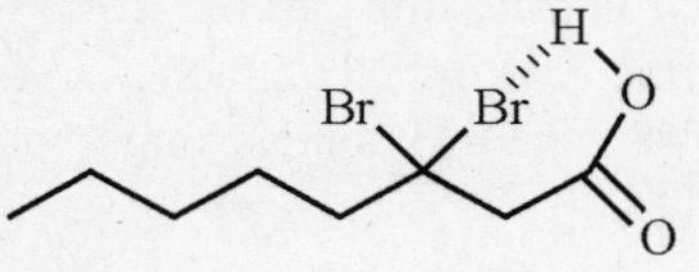

Br Br H O O
A

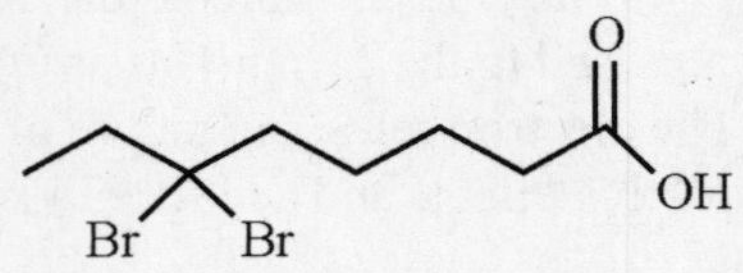

O OH Br Br
B

8.

Cl H O O

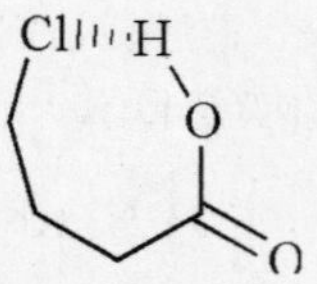

9. Water can solvate the two ions that are formed (H_3O^+ and acetate) much better than ethanol.

Chapter 7

Alkyl Halides and Substitution Reactions

1. (a) 1,1,-dibromocyclopentane;
   (b) 3-bromo-5-chloro-5,6-dimethyloctane;
   (c) 9-cyclopentyl-4,4-dimethyl-8(1-iodo-1-methylpentyl)hexadecane.

2.

(a) F

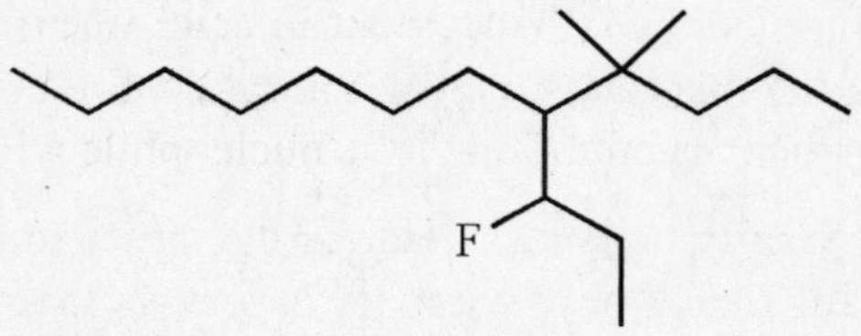

(b) I

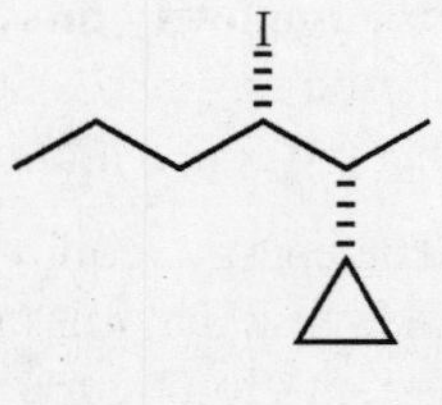

(c) Br Cl

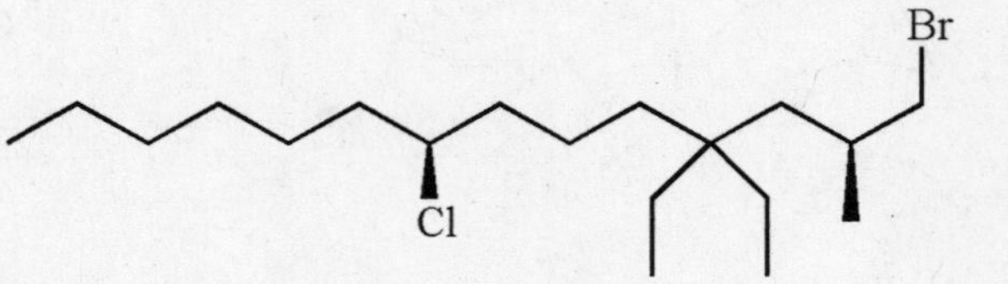

3. Cation B has two allylic C=C groups, and the positive charge can be delocalized over five carbons. Cation A is tertiary, but cannot delocalize the charge by resonance. For this reason, B is more stable.

4.

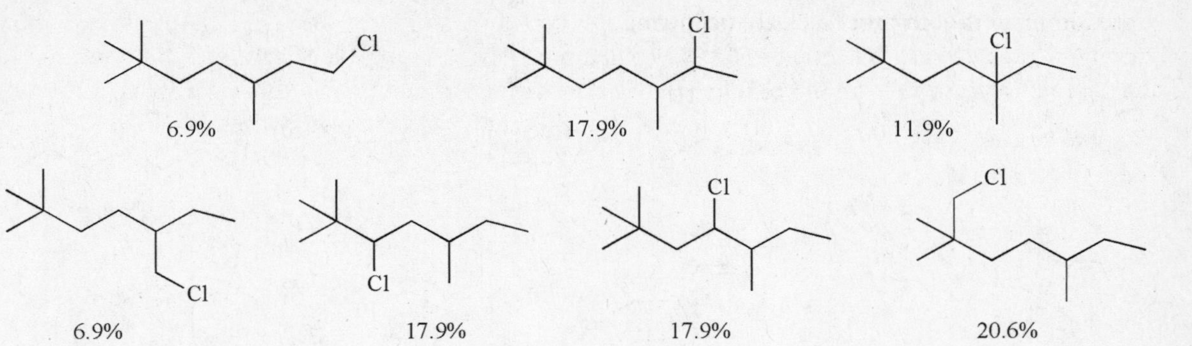

6.9% 17.9% 11.9%

6.9% 17.9% 17.9% 20.6%

5. This is an S_N2 reaction, and the rate depends on the concentration of **both** nucleophile and halide. If the concentration of KI (the nucleophile) is increased 10 fold, the rate of the reaction is expected to increase by 10.

6. Although 1-bromo-2,2-dimethylpropane is technically a primary halide, it is a neopentyl halide, and the large *tert*-butyl group provides enormous steric hindrance to formation of the S_N2 transition state (therefore a very slow reaction). 1-Bromomethane is, of course, a normal methyl halide and reacts very fast.

7. In the S_N2 transition state, the π-electrons of the carbon-carbon double bond help 'push out' the departing bromide, slightly increasing the rate of the reaction. This assist is not possible unless the π-bond is in close proximity to the carbon being attacked by the nucleophile (the allylic carbon).

8.

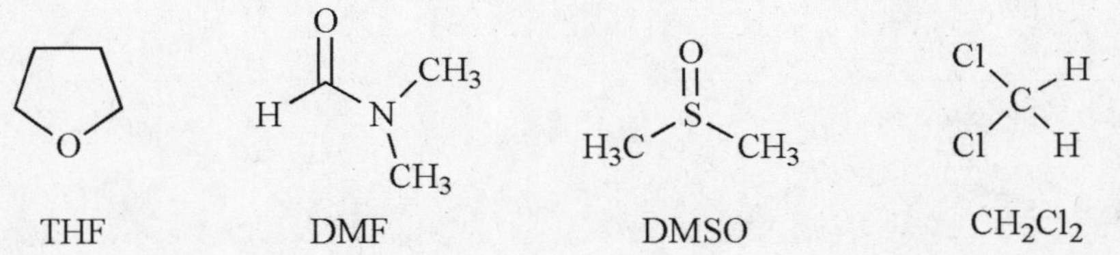

THF DMF DMSO CH_2Cl_2

9. rate = k [RX]. Since the relative ratio of nucleophile to halide has no place in the rate expression (a first order reaction), increasing the concentration has little effect. A more concentrated solution of the halide will probably react faster, however, and in this case, the reaction rate will increase by increasing the concentration of the halide. Increasing the concentration of the nucleophile will have no effect.

10. The energy required for the small hydrogen to migrate is less than for the larger methyl group. It is also true that migration of a methyl group leads to another secondary cation, so there is no energetic driving force for a rearrangement.

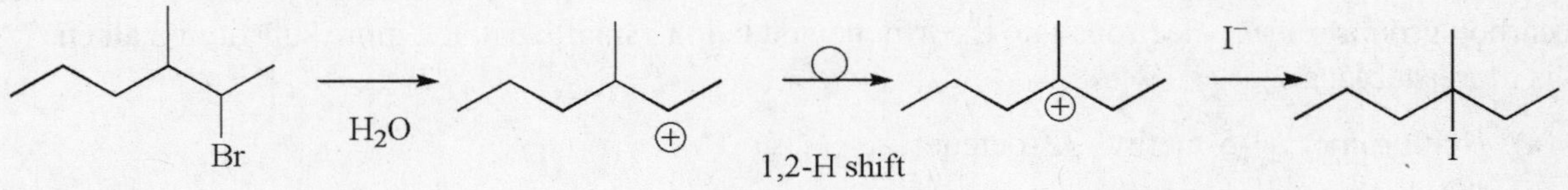

H_2O 1,2-H shift I^-

11. The initially formed oxonium ion (by protonation of the OH) can either lose water to give the secondary cation (leading to a racemic mixture of chlorides) or have water displaced by chloride in an S_N2 reaction, giving the inverted chloride. The two are in competition, but the cation reaction gives

an equal mixture of inversion and retention and S_N2 gives only inversion. The inverted product must predominate if both processes are occurring.

12.

(a) No Reaction (b) 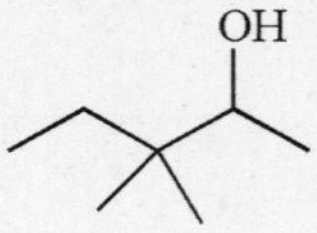 OH (c) 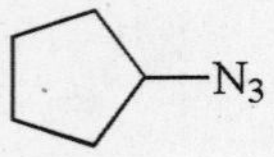 N_3 (d) 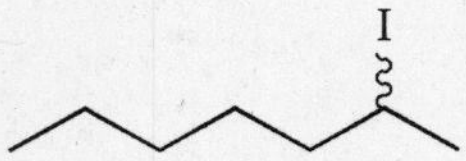 I racemic

(e) 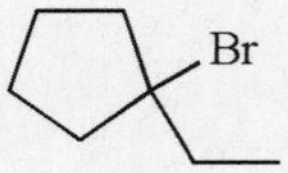 Br (f) see below (g) 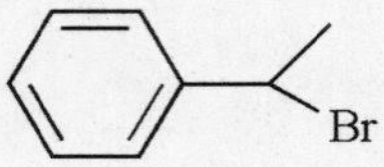 Br (h) 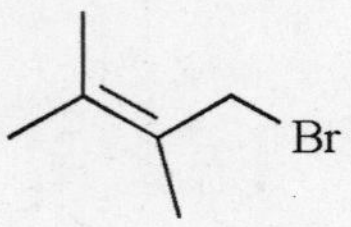 Br (i) 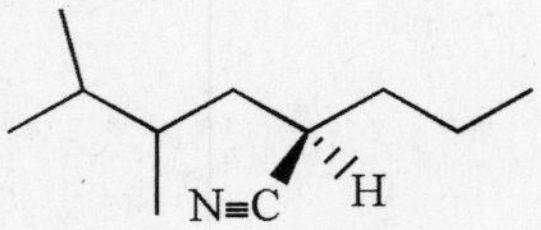 N≡C H

(j) No reaction (k) 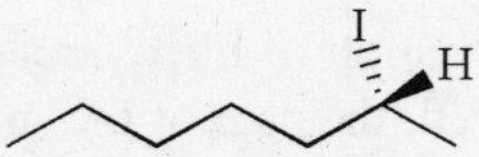 I H (l) $C_5H_{11}—C{\equiv}C—C_4H_9$ (m) 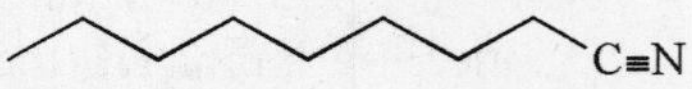 C≡N

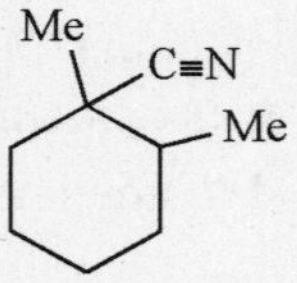

(n) Me C≡N Me (o) 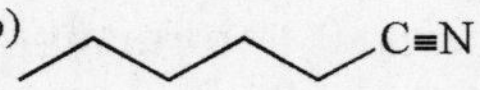 C≡N (p) 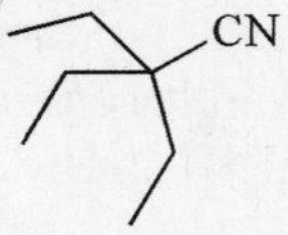 CN

Reaction (f) shows four different chloride products (1-chloro-3-methylpentane, 2-chloro-3-methylpentane, 3-chloro-3-methylpentane and 1-chloro-2-ethylbutane), resulting from replacing each of the four different hydrogens. Reactions (a) and (j) give no reaction because they are tertiary halides under S_N2 conditions.

Chapter 8

Alkenes and Alkynes: Structure and Nomenclature

1. All of the carbon atoms are coplanar since each methyl-C=C unit is trigonal planar.
2. The name is 1-bromo-1-chloro-1-pentene. No cis/trans or E/Z isomers are designated, so those terms are not included in the name.
3. The endocyclic (the C=C unit is within the ring) alkene (a) has three carbon groups attached to the C=C group, whereas the exocyclic (the C=C unit is outside the ring) alkene (B) has only two. Since carbon groups release electrons into the π-bond and help to stabilize it, the more substituted alkene (A) is more stable.
4. (a) 4-chloromethyl-5-methyl-3Z-octene;
   (b) 1,2-diethylcyclohexene;
   (c) 3-(2-methylbutyl)-2,4,4-trimethyldec-1-ene;
   (d) trans-3,4-dichloro-3-hexene;
   (e) 5-chloro-1,5-diphenylundec-1-ene;
   (f) 1,5,5-tribromo-3-butyl-3-ethylcycloheptene.

5. This is a Z alkene, since the priority groups are Cl and ethyl. There are two identical groups (Et), however, on opposite sides of C=C, making it a trans alkene.

6. (a) 5,9,9-trimethyl-2-decyne; (b) 6-phenyl-1-hexyne; (c) tridec-12-en-2-yne; (d) 7-phenyl-3-propyl-4-(1,1,-dichloropropyl)tridec-1-yne; (e) 1-cyclohexyl-1-pentyne.

7. The most acidic hydrogen is the O-H hydrogen ($pK_a \approx 4.5$) and that is removed much faster than the alkyne C-H ($pK_a \approx 25$).

8.

(a) (b) Bu

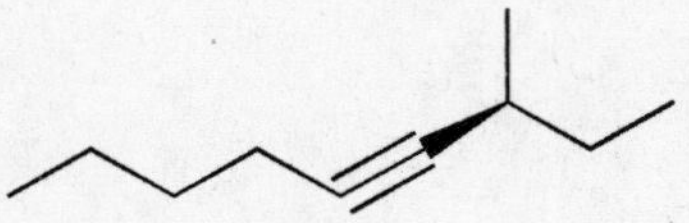

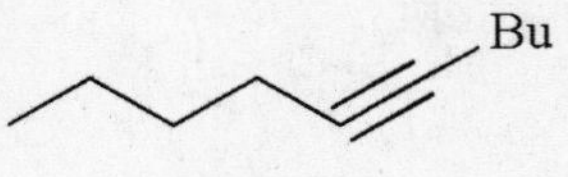

(c) no reaction (d) H C≡C-Ph

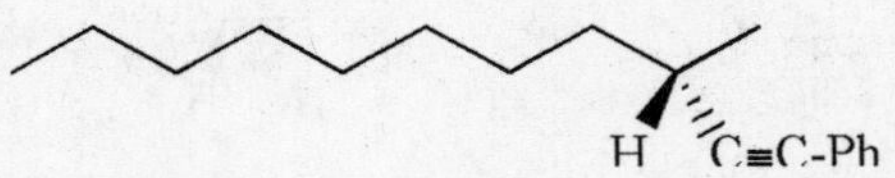

Note that reaction (a) proceeds with inversion of configuration of the stereogenic center in 2-bromobutane and that the stereogenic center in (d) also undergoes inversion of configuration.

Chapter 9

Elimination Reactions

1.

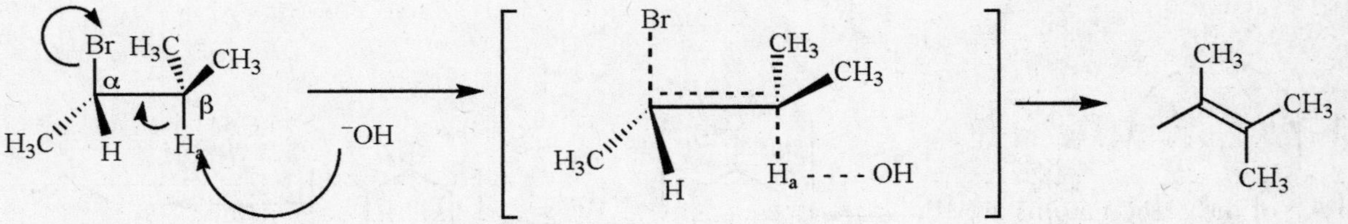

Br H_3C CH_3 α β H_3C H H_a ^-OH → [Br CH_3 CH_3 H_3C H H_a - - - - OH] → CH_3 CH_3 CH_3

2. 3-Bromo-2,2,4,4-tetramethylpentane does not have a β-hydrogen atom relative to the bromine so there is no possibility of an E2. In addition, the bromine-bearing carbon is very sterically hindered, precluding any S_N2-type products.

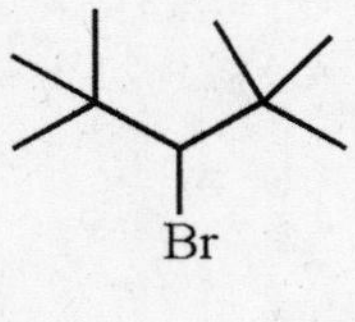

Br

3.

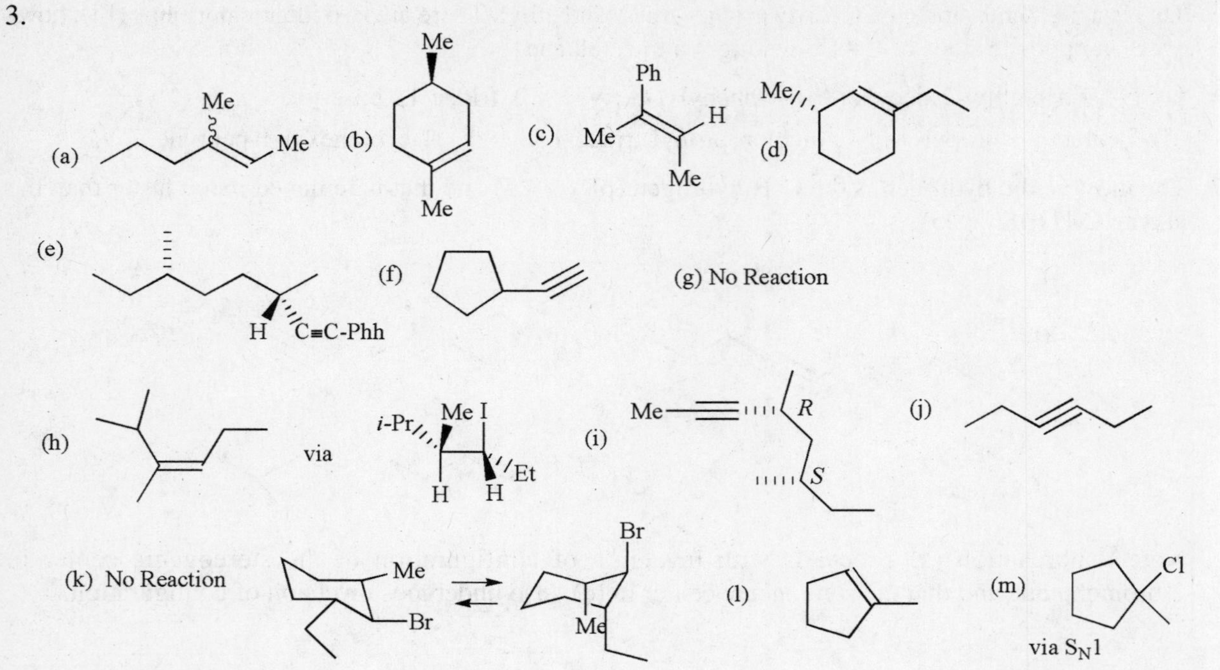

(a) Me, Me

(b) Me, Me

(c) Ph, H, Me, Me

(d) Me

(e) H C≡C-Phh

(f)

(g) No Reaction

(h) via i-Pr, Me I, Et, H H

(i) Me, R, S

(j)

(k) No Reaction — Me, Br ⇌ Br, Me

(l)

(m) Cl via S_N1

(l) No Reaction

Reaction (g) gives no reaction because there is no acidic hydrogen to be removed. Reaction (k) gives no reaction because in the two equilibrating chair conformations, a β-hydrogen never has a trans-diaxial relationship to the bromine leaving group. Reaction (l) gives no reaction because it is a tertiary halide, which cannot undergo a S_N2 reaction with the nucleophilic chloride ion (which would give the same compound), and chloride ion is too weak a base to induce an E2 reaction.

4.

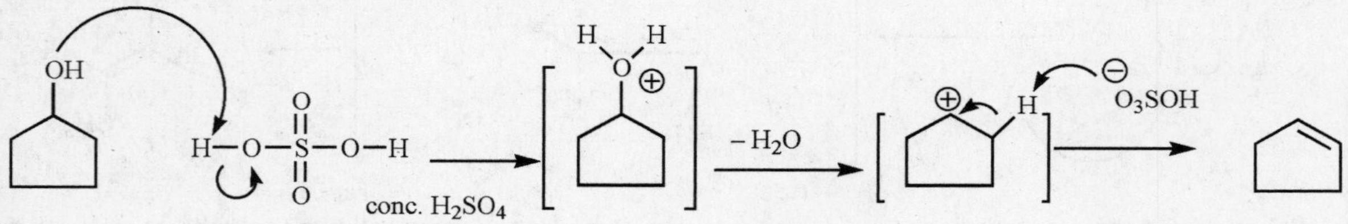

OH; H—O—S(=O)(=O)—O—H; conc. H_2SO_4 → [H, H, O⊕] → $-H_2O$ → [⊕, H, ⊖ O_3SOH] →

Chapter 10

Organometallic Compounds

1. Ethanol is a strong acid in the presence of the strongly basic Grignard reagent. Ethylmagnesium bromide reacts with ethanol to form ethane and the magnesium ethoxide.

2. Butyllithium is a powerful base, sufficiently strong to remove the weakly acidic hydrogen of an alkyne (pK_a about 25). In addition, the conjugate acid of the reaction is butane, which escapes from the medium, drives the reaction to the desired product and doses not interfere with isolation or further reaction of the alkyne anion product.

3.

(a) 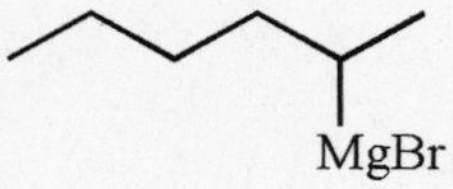 MgBr (b)  MgI (c) 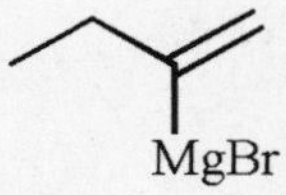 MgBr

(d) 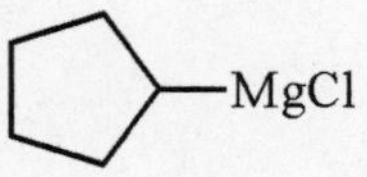 MgCl (e) 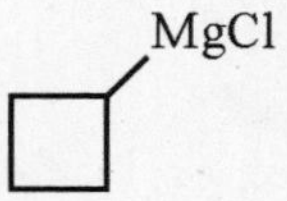 MgCl (f) 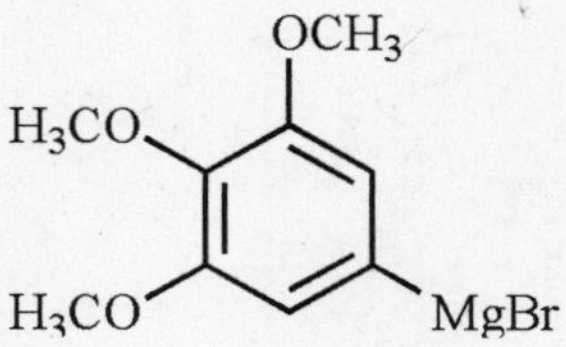 OCH_3, H_3CO, H_3CO, MgBr (g) 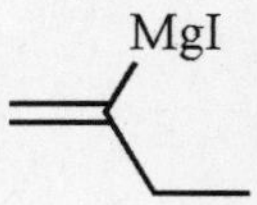 MgI

4.

(a) 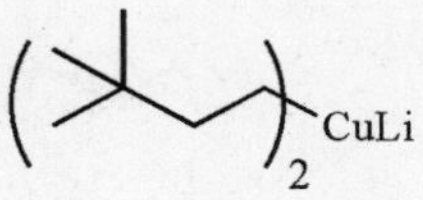 CuLi 2 (b) 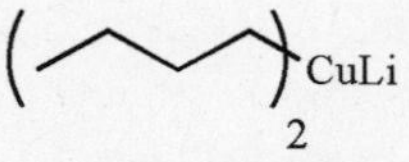 CuLi 2 also - Bu_2CuLi (c) 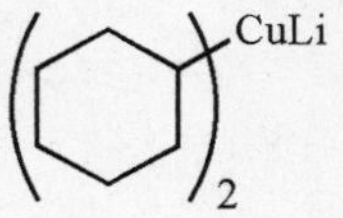 CuLi 2 (d) 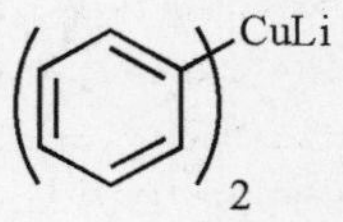 CuLi 2 also - Ph_2CuLi

5. Butyllithium is an exceptionally strong base (pK_a of the conjugate acid, butane, is greater than 40), and an amine *N*-H (pK_a = 25) will be a strong acid in this system.

6.

(a)  + $CH_3NH^- \ ^+MgI$ (d) 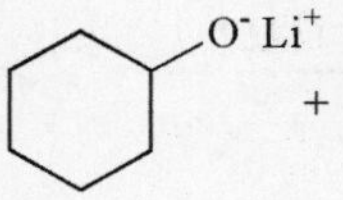 $O^- Li^+$ + butane

(b) 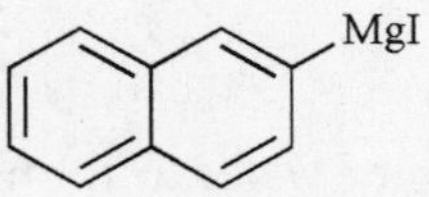 MgI (e) 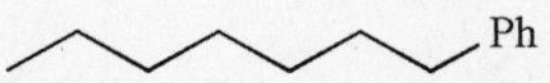 Ph

(c) 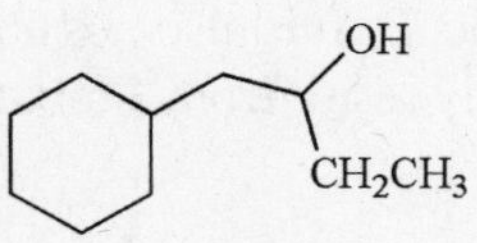 OH CH_2CH_3 (f)  (g)

Chapter 11

Alcohols. Preparation, Substitution and Elimination Reactions

1.

(a) O, Cl : C : Cl   (b) O, Br : S : Br

2.

(a) OH, H+, OH_2^+, - H_2O, ⊕, Br-, Br

(b) OH, H+, OH_2^+, - H_2O, ⊕, Cl-, Cl

3. The basic oxygen of the alcohol (***A***) reaction with the acidic hydrogen of HCl to form an *oxonium* ion (***B***). The energy required for this cation to ionize (by losing water) to form a primary cation is too high. The nucleophilic chloride displaces H_2O, which is a good leaving group, in a S_N2 reaction, giving 1-chloropentane (***C***) as the product.

OH, HCl, H, O⊕, H, Cl-, Cl, S_N2, ***A***, ***B***, ***C***

4. Tertiaryalcohol ***A*** reacts with HCl to form an oxonium ion, ***B***. Ionization of water to form a secondary cation (***C***) is possible, and chloride can attack ***C*** to form the chloride, ***D***. It is also possible for chloride to attack ***B*** in an S_N2 reaction, also giving ***D***. The products probably result from a mixture of these two mechanistic processes.

OH, HCl, H O⊕ H, $-H_2O$, ⊕, Cl-, Cl, ***A***, ***B***, ***C***, ***D***

5.

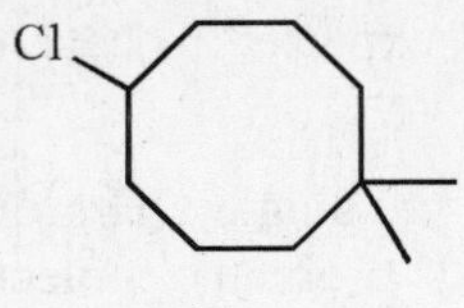

In the first reaction, alcohol (***A***) is converted to the chloride (***E***) with phosphorus pentachloride. The allylic alcohol (***B***) is converted to the allylic chloride (***F***) with phosphorus trichloride. In this case, migration of the double bond leads to 3-chlorohept-1-ene as a minor product, presumably via an allylic cation. Reaction of alcohol (***C***) with phosphorus oxychloride leads to chloride (***G***). Without an amine base, one might expect the stereochemistry of the chlorination to proceed with retention of configuration. With most phosphorus halides, however, the reaction proceeds with low stereoselectivity, and (***G***) is a mixture of *cis*- and *trans*-isomers. In the final reaction, alcohol (***D***) is converted to bromide (***H***).

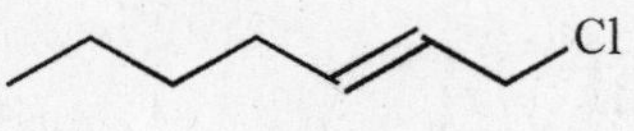

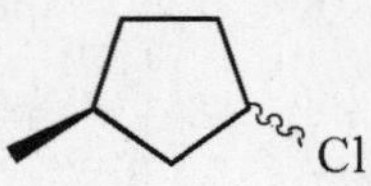

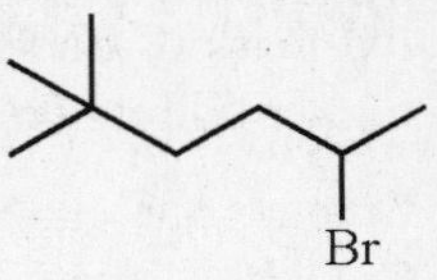

6. The initially formed oxonium ion (by protonation of the OH) can either lose water to give the secondary cation (leading to a racemic mixture of chlorides) or have water displaced by chloride in an S_N2 reaction, giving the inverted chloride. The two are in competition, but the cation reaction gives an equal mixture of inversion and retention, and S_N2 gives only inversion. The inverted product must predominate if both processes are occurring.

7. Reaction of these alcohols with thionyl chloride reveals that treatment of 5-methyl-1-hexanol (***A***) with thionyl chloride generates 5-methyl-1-chlorohexane, *D*. Reaction of 3,3-dimethyl-1-cyclohexanol (***B***) with thionyl chloride generates 3,3-dimethyl-1-chloro-cyclohexane, *E*. Treatment of ***C*** with thionyl chloride, in the presence of the basic pyridine, gives chloride ***F*** with inversion of configuration at the C–OH carbon.

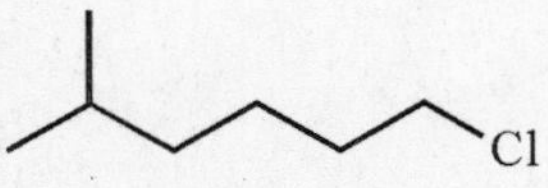

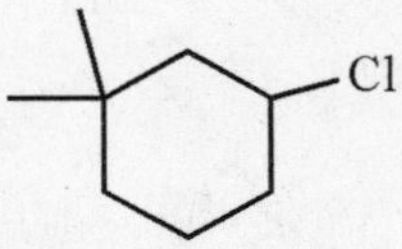

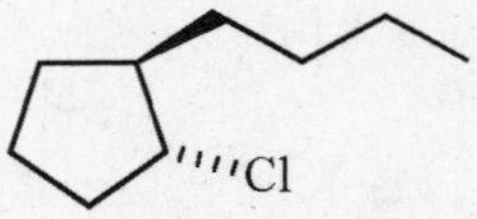

8. Ethanol is a strong acid in the presence of the strongly basic Grignard reagent. Ethylmagnesium bromide reacts with ethanol to form ethane and the magnesium ethoxide.

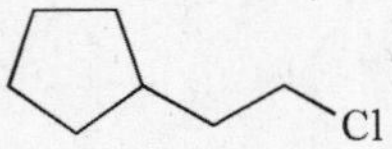

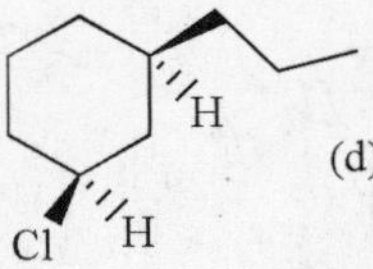

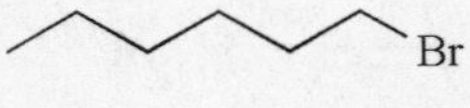

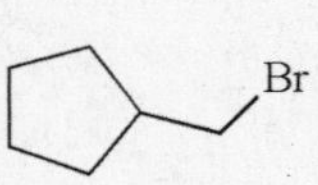

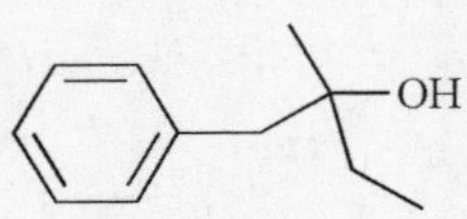

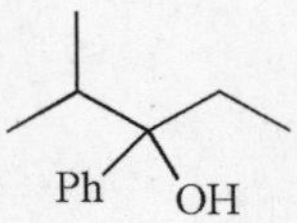

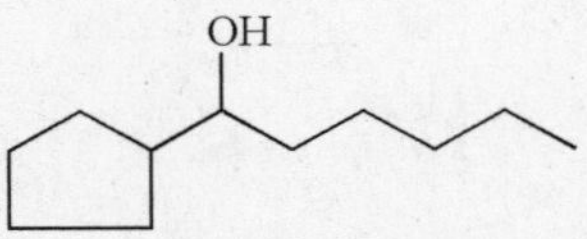

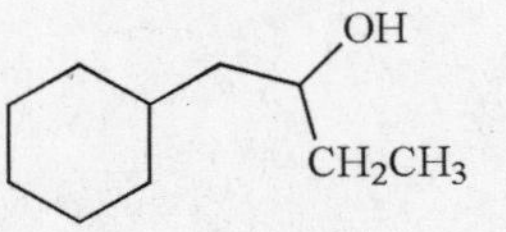

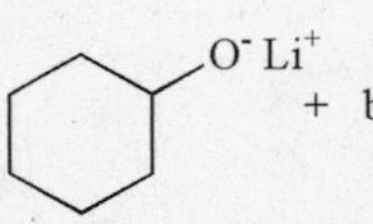

9.

10. Both (b) and (d) are alcohols and will have higher boiling points due to hydrogen bonding. Since the mass of dodecanol (d) is higher than that of pentanol (b), dodecanol is expected to have the higher boiling point.

11. Octanol has enough carbons to outweigh the 'water solubility' provided by the single OH group. Octanol is, therefore, essentially insoluble in water (0.06 g/100 g water), and an organic molecule that is insoluble in water would reasonably be expected to be soluble in octanol.

12. (a) 3-chloro-1-ethyl-3-methylcyclohexanol; (b) 3,7-dimethyl-5,9-diphenyl-4-nonanol; (c) 5,5-dimethyl-3,6-nonanediol. (d) 3-bromo-8-chlorocyclooct-3-en-1-ol; (e) 2,4,4,5,5,7,7-heptamethyl-1-octanol. (f) dec-6-yn-2-ol.

13. Methanol (pK_a 15.5) is more acidic than *tert*-butanol (pK_a 19.0). The bulky alkyl groups in *tert*-butanol inhibits solvation, diminishing the acidity of the O-H.

14. No. The charge on oxygen is too far away from the C=C group ($CH_2=CHCH_2O^-$).

15. Since 2-methyl-2-iodopropane is a tertiary halide, an S_N2 reaction (required by the Williamson ether synthesis) is not possible.

16. Hydroboration of 1-hexene (***A***) followed by oxidation leads to 1-hexanol (***D***). 1-Methylcyclopentene (***B***) is converted to *trans*-2-methylcyclopentanol (***E***). *cis*-Addition of borane leaves the boron and methyl group on opposite faces, leading to the *trans*-geometry (remember the *cis*-addition refers to the B and the H of the borane). Alkene ***C*** is converted to secondary alcohol ***F***.

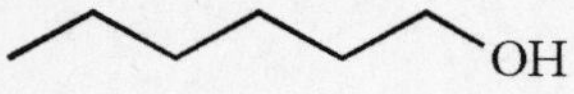

OH Me OH Me Me OH

D 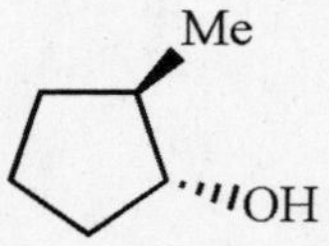 ***E*** 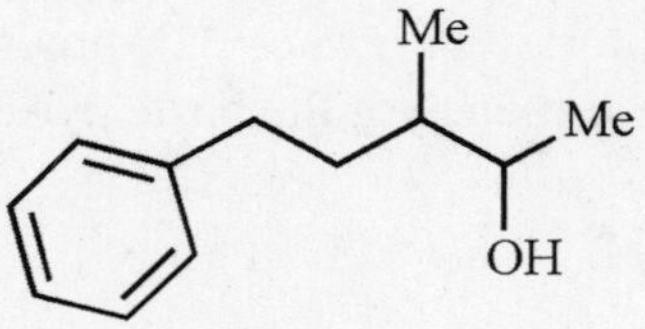 ***F***

17.

(a) 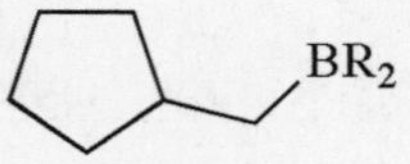 BR_2 (b) 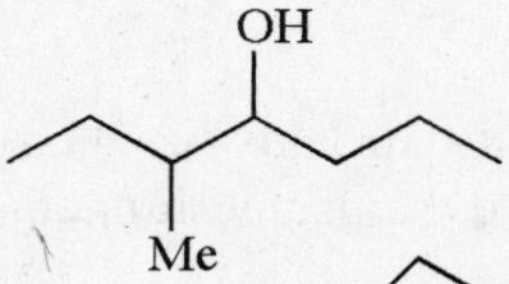 OH Me (c) 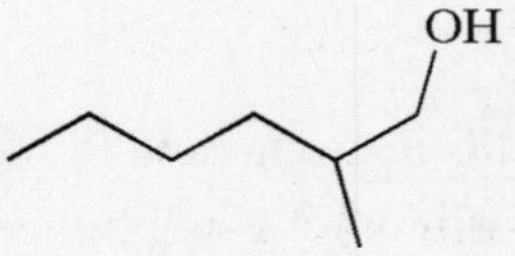 OH

(d)  Cl (e) 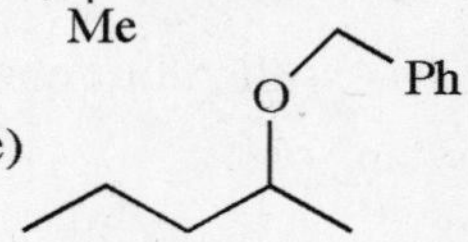 O Ph (f) 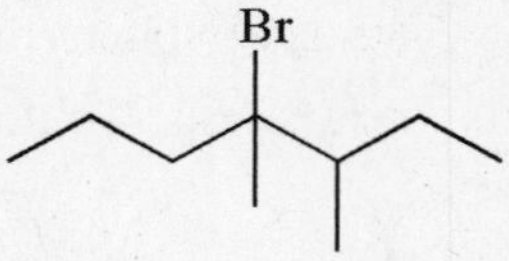 Br

(g) 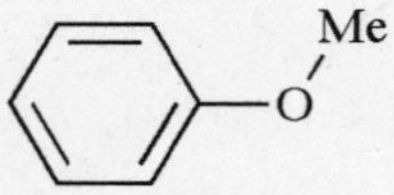 Me O (h) 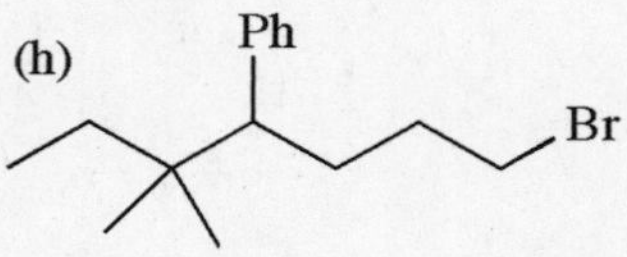 Ph Br (i) 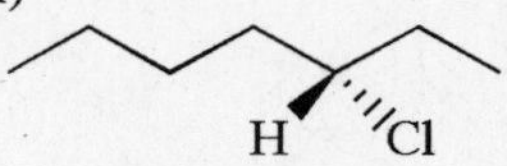 H Cl (j) 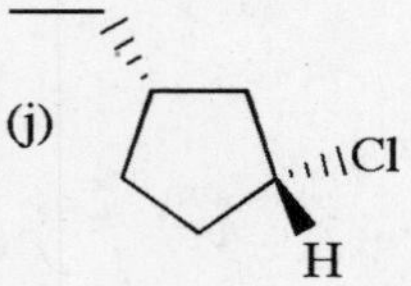 Cl H

(k) 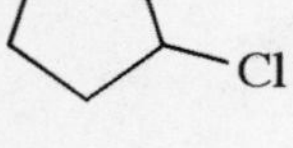 Cl (l) 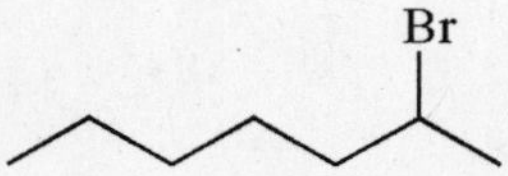 Br (m) 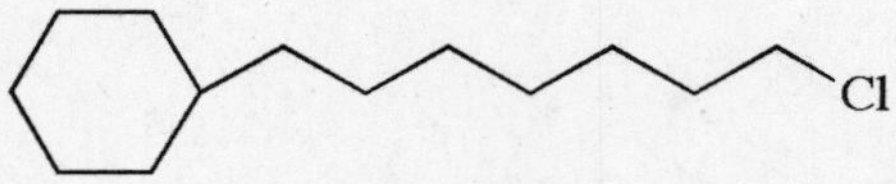 Cl

18.

(a) 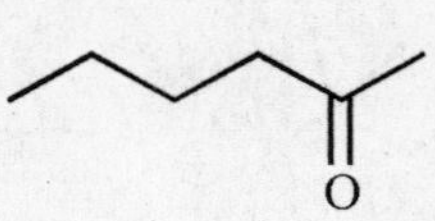 (b)

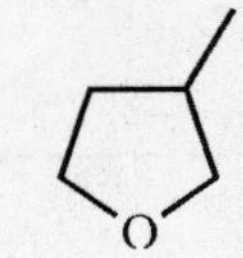

19. In the presence of hydroxide, hydrogen peroxide is deprotonated to give the hydroperoxide anion (oOOH). This anion attacks the boron of ***A*** to form an 'ate' complex, ***B***. A boron → oxygen alkyl shift (a 1,2 shift) leads to formation of a B-O bond in ***C***. Sequential reactions of OOH with boron leads to the alcohol product, ***D*** (1-pentanol) and boric acid [$B(OH)_3$].

HO-OH + NaOH ⟶ Na^+ $^-$OOH

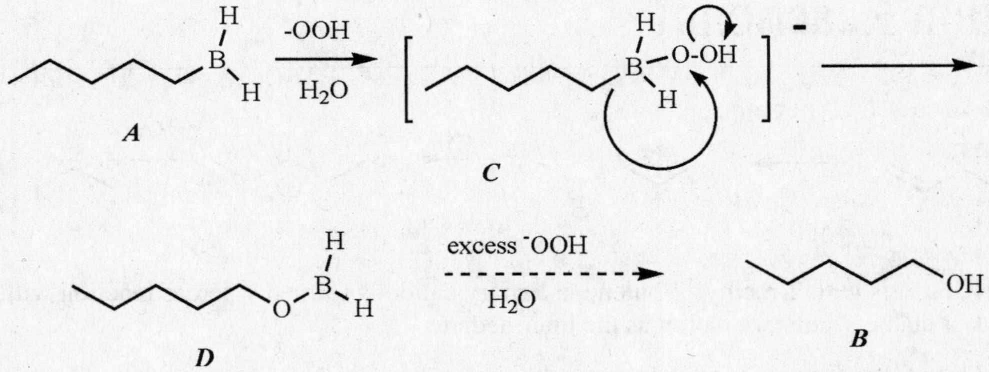

A —(-OOH, H_2O)→ [ ]$^-$ ***C*** ⟶

D —(excess $^-$OOH, H_2O)⇢ OH ***B***

Chapter 12

Ethers

1. (a) 4,4-dimethyl-2-phenyltetrahydrofuran; (b) 2-ethoxy-3-methylpentane; (c) 6-chloro-3-ethoxy-8-phenylnonane; (d) 4,4-dichlorotetrahydropyran; (e) 1-phenoxyhexane; (f) 4-chloro-3-ethylanisole.

2.

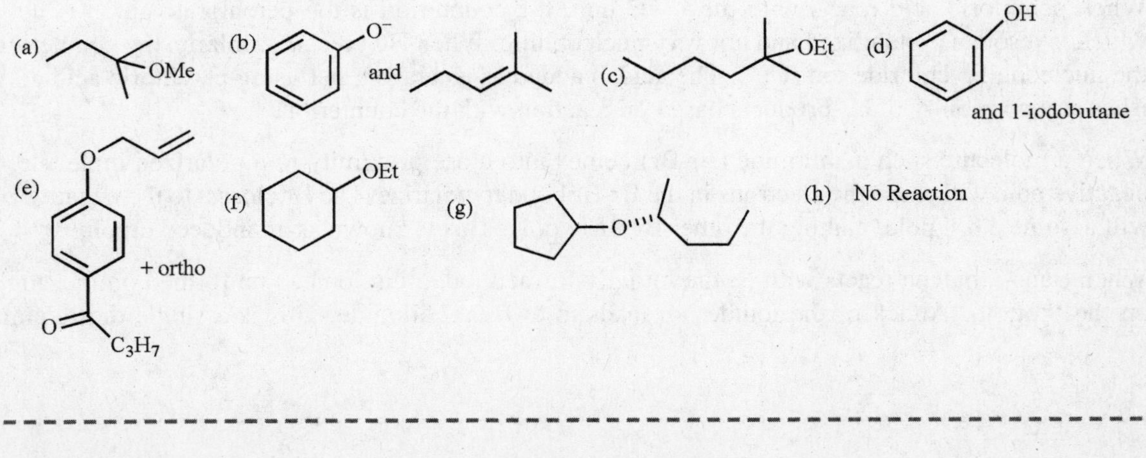

(a) OMe (b) O^- and (c) OEt (d) OH and 1-iodobutane

(e) + ortho C_3H_7 (f) OEt (g) (h) No Reaction

3.

H^+ $+t$-BuOH   H   O⊕ $-H^+$ O   OH

Chapter 13

Addition Reactions

1.

H+   ⊕   1,2-H shift   ⊕   Cl-   Cl

2. When HCl reacts with 2-methyl-2-butene, a tertiary cation is formed whereas reaction with 1-butene gives a less stable secondary cation as the intermediate.

3. If H^+ adds to C1, a benzyl cation is formed (the charge can be delocalized into the π-bonds of the benzene ring), whereas addition of H^+ to C2 generates a secondary cation that is not resonance stabilized.

4. Ni, Pt, Pd, Rh, Ir, Ru.

5.

H+   ⊕   1,2-Et shift   ⊕ H_2O OH_2 ⊕

-H+   OH

When perchloric acid reacts with the C=C unit, the counterion is the perchlorate anion (ClO_4^-), which is resonance stabilized and not very nucleophilic. When HCl reacts similarly, the counterion is the nucleophilic chloride ion (Cl^-). The acid is added as a catalyst, but using perchloric acid diminishes any possibility of by-products based on reaction with the counterion.

6. When a molecule such as bromine (Br-Br) comes into close proximity to a polarized molecule, the negative pole will cause the electrons in the Br-Br bond to polarize. The Br closest to the negative pole will assume a $\delta+$ pole, making the other Br a $\delta-$ pole. This is known as an induced dipole.

7. When trans-2-butene reacts with I_2, the initially formed iodonium ion can be formed on the 'top' or on the 'bottom.' Attack of the iodide ion leads to a '*trans*' diiodide. This is a single diastereomer,

although it is racemic. This diastereomer is the meso compound, not the d,l pair. If *cis*-2-butene reacted with I_2, only the d,l diastereomer would be formed.

8. Once the bromonium ion is formed, the nucleophile bromide ion will attack the less sterically hindered carbon in what is essentially an S_N2 process.

9.

(a) No Reaction (b) 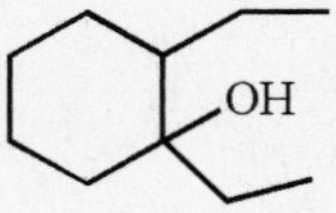 OH (c) 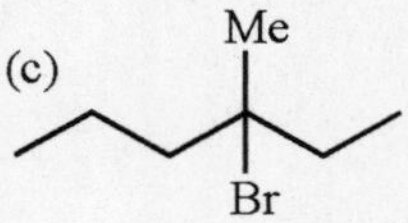 Me Br (d) 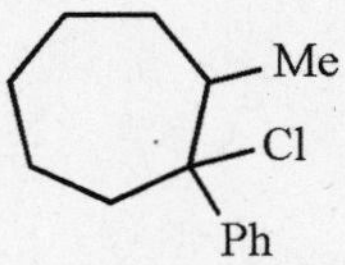 Me Cl Ph

(e) 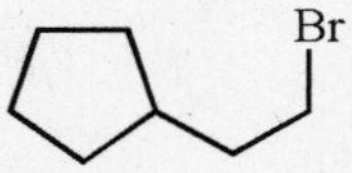 Br (f)  (g) 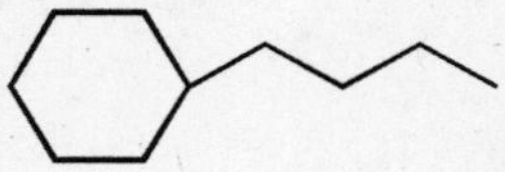 (h) 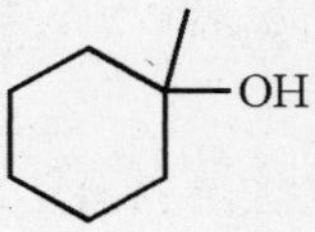 OH

(i) 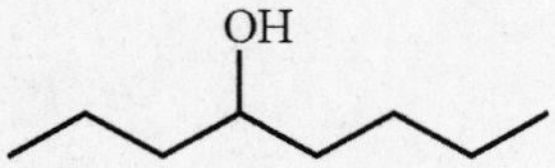 OH (j) 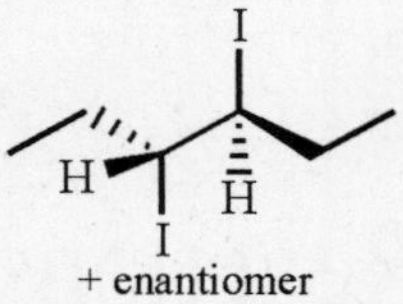 I H H I + enantiomer (k) 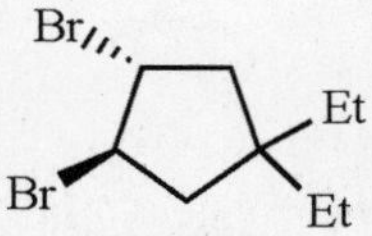 Br Br Et Et + enantiomer (l) 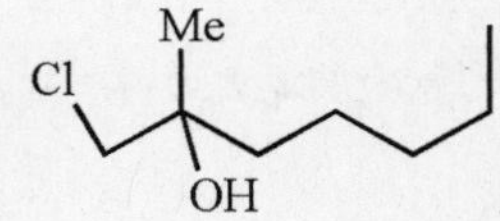 Me Cl OH

Reaction (a) gives no reaction, since it is a tertiary sulfonate ester under S_N2 conditions.

10.

(a) 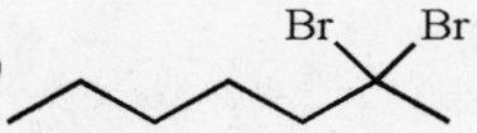 Br Br (b) 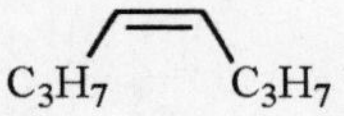 C_3H_7 C_3H_7 (e) 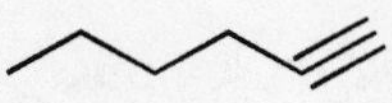 (f) 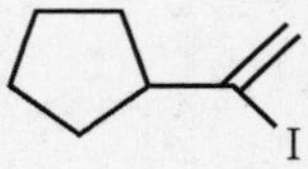 I

(g) 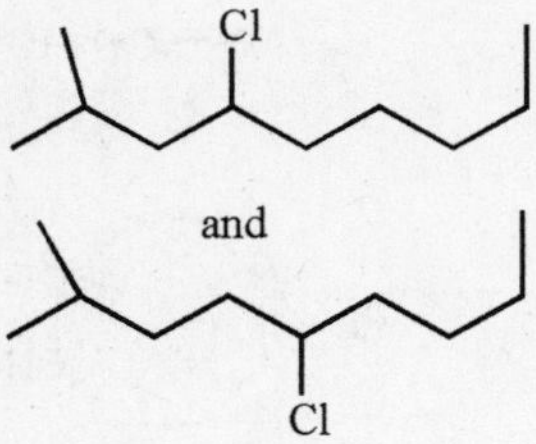 Cl and Cl (h) 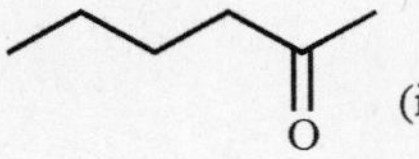 O (i) 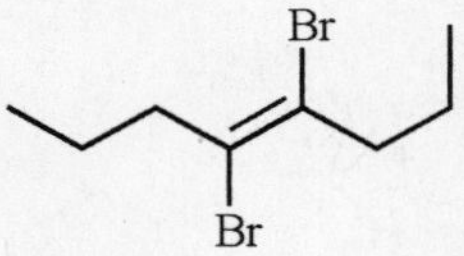 Br Br (j) 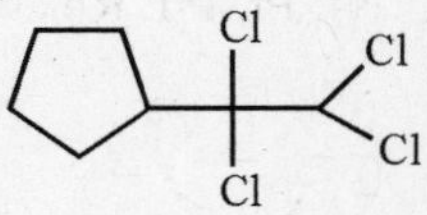 Cl Cl Cl Cl

(k) 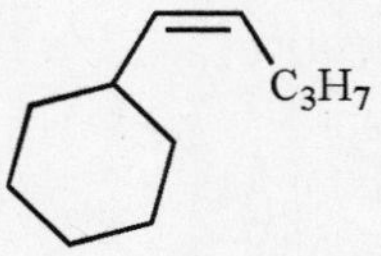 C_3H_7

Both (c) and (d) give no reaction. In (c), there is no acidic hydrogen to be removed, and in (d) there is no catalyst for the hydrogen.

11.

(a) 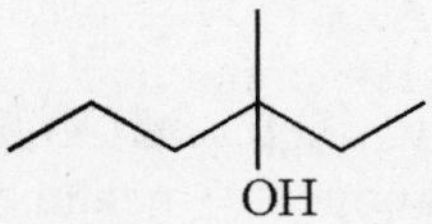 OH (b) 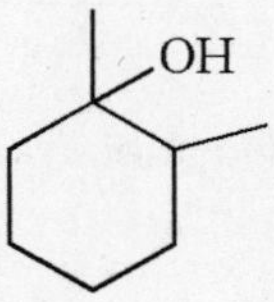 OH (c) 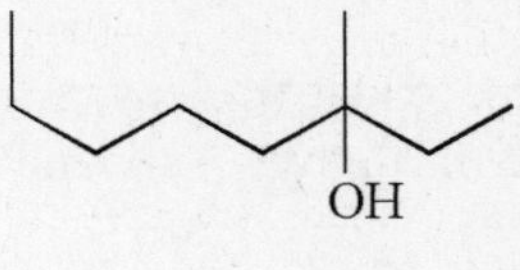 OH

(d) 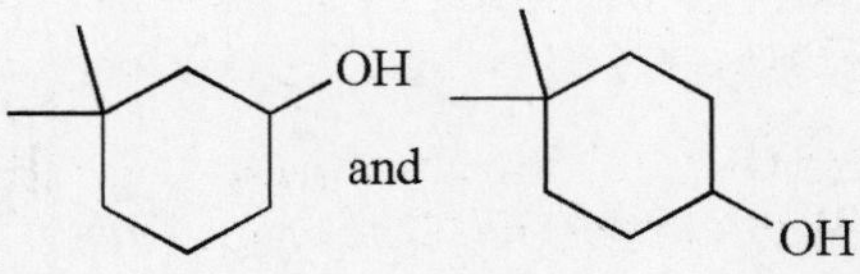 OH and OH

12.

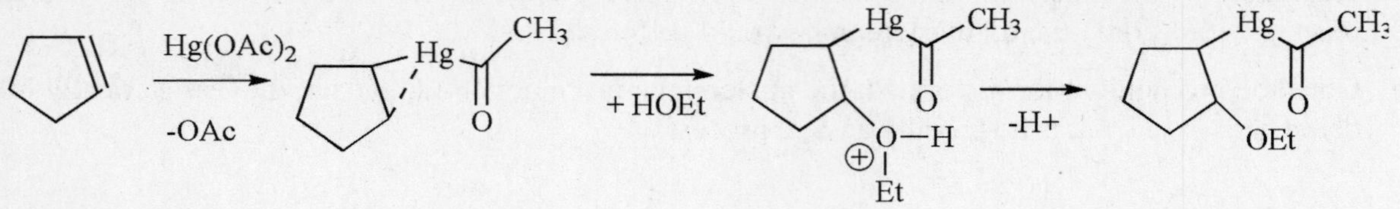

Hg(OAc)$_2$ / -OAc; + HOEt; -H+

13.

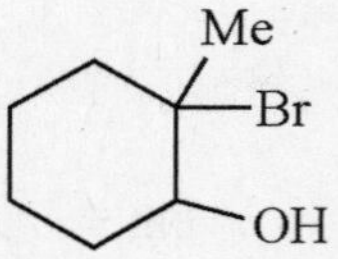

Chapter 14

Oxidation and Reduction Reactions of Alkenes and Alcohols

1.

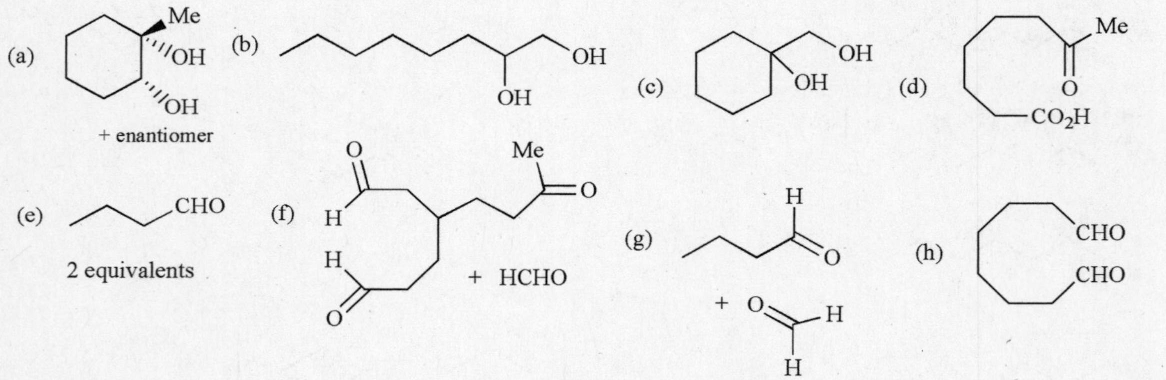

(a) + enantiomer (b) (c) (d)

(e) CHO 2 equivalents (f) + HCHO (g) (h) CHO CHO

2. Under these conditions, this alkene is cleaved to a mixture of acetone and 2-pentanone.
3. The ozonolysis with reductive workup gives acetone and propanal as the products.
4. Cleavage of 2-octene with ozone with an oxidative workup leads to a mixture of ethanoic acid and hexanoic acid.
5. In reaction (a), oxidative cleavage of methylcyclohexene leads to a ketoacid (**A**). In reaction (b), oxidative cleavage of 3-ethyl-3-nonene leads to a mixture of two products, 3-pentanone (**B**) and hexanoic acid (**C**).

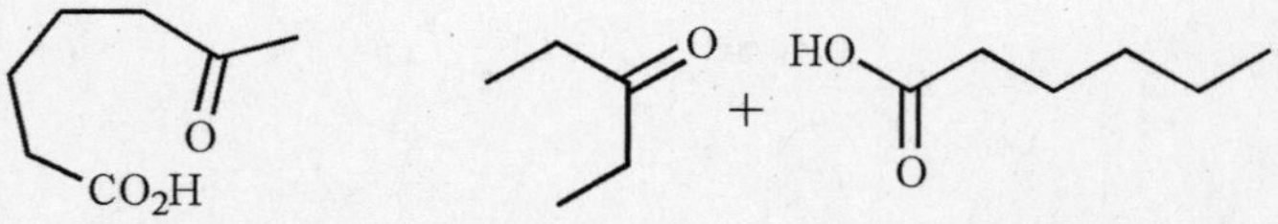

A **B** + **C**

6.

(a) 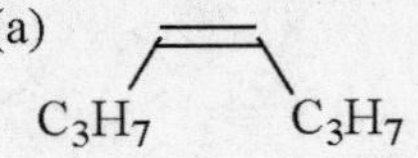 C_3H_7 C_3H_7 (b) No Reaction (c) 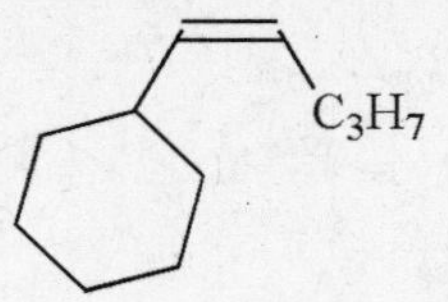 C_3H_7

(d) 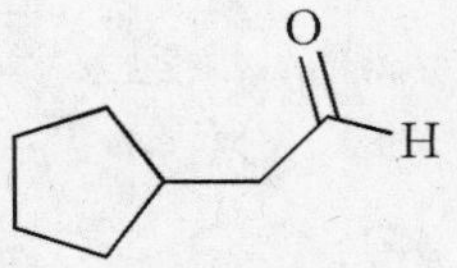 H (e) 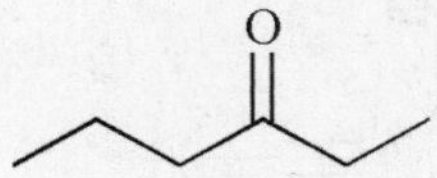 O

Reaction (d) gives no reaction, since there is no catalyst.

7. Since 3-hexene is symmetrical, oxidative cleavage leads to propanoic acid as the major product.

8. As described in Section 5.5.E, these conditions lead to a 1,2-diol rather than oxidative cleavage. The product of this reaction is 3,4-hexanediol.

Chapter 15

Epoxides

1. (a) 1-ethyl-2-methyloxirane;
   (b) 1,2-dimethyl-1,2-epoxycyclooctene;
   (c) 5-chloro-3,3-dimethyl-6-phenyl-1,2-epoxynonene;
   (d) 2-methyl-1-phenyl-1,2-epoxyoctene;
   (e) 5,5-dichloro-7-(4-chloro-2-methoxyphenyl)-1,2-epoxyheptene.

2.

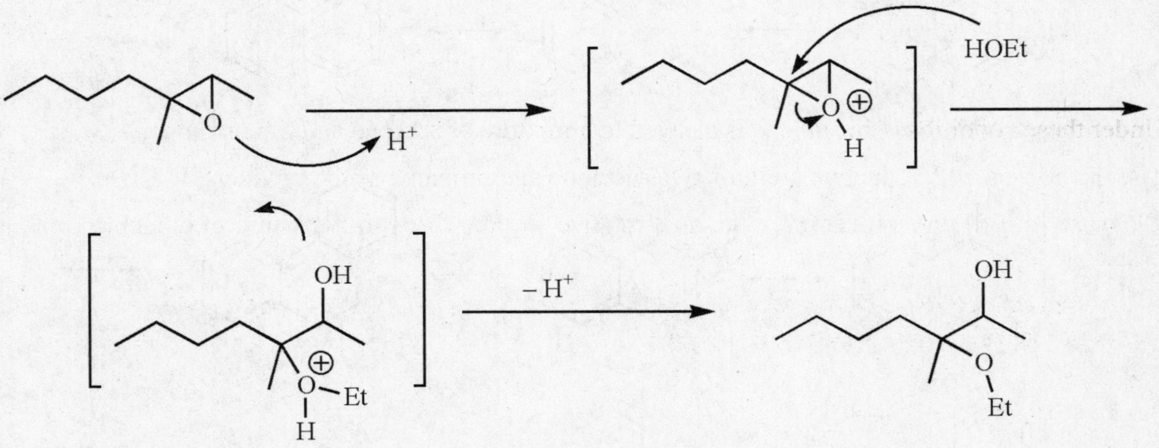

O H^+ O⊕ H HOEt

OH O⊕ Et H $-H^+$ OH O Et

3.

(a) 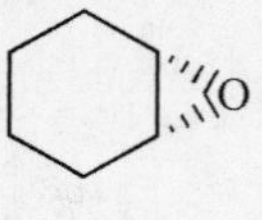 O (b) 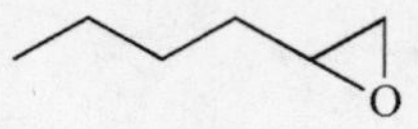 O (c) 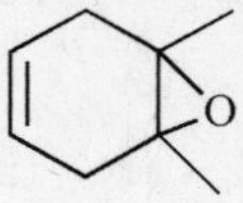 O (d) 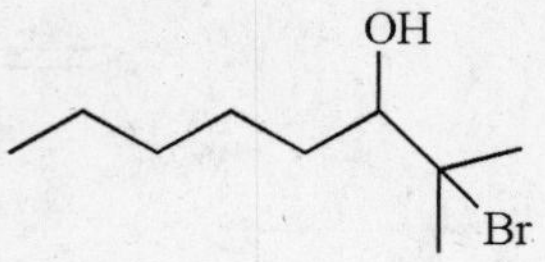 OH Br

(e) 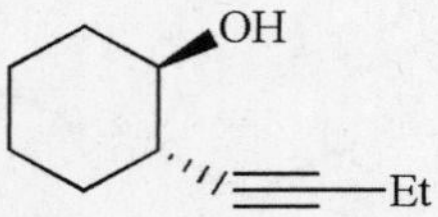 OH Et (f) 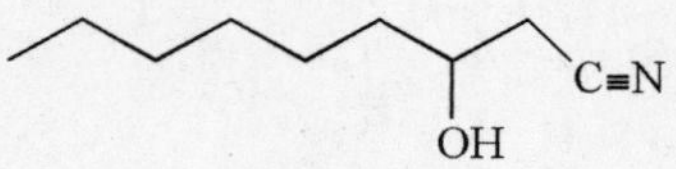 C≡N OH (g) 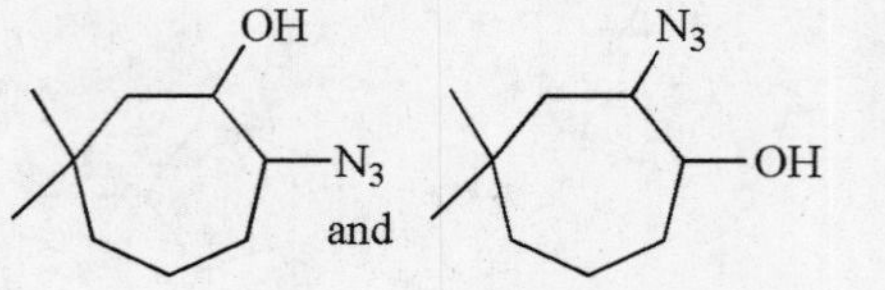 OH N_3 and N_3 OH

(h) 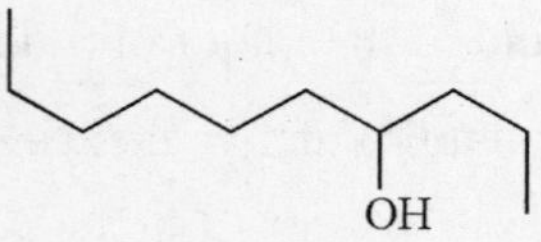 OH (i) 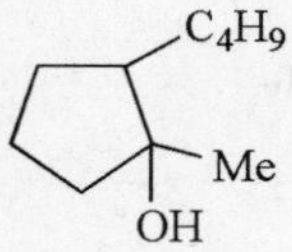 C_4H_9 Me OH (j) 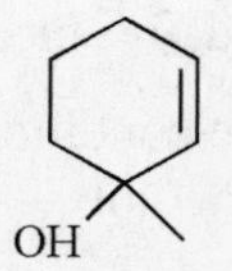 OH (k) 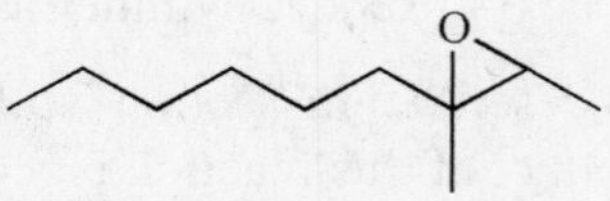 O

Chapter 16

Benzene, Aromaticity, and Benzene Derivatives

1. The aromatic molecules are (a), (b), (d), (h), (i) and (j).

2.

(a) 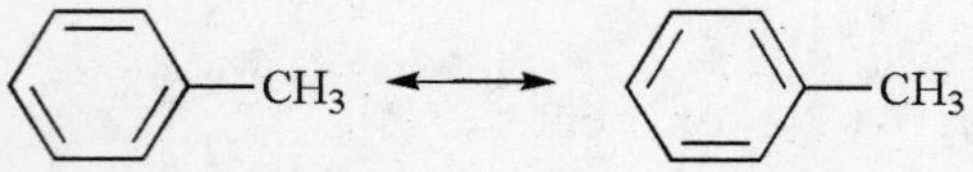 CH_3 ⟷ CH_3

(b) 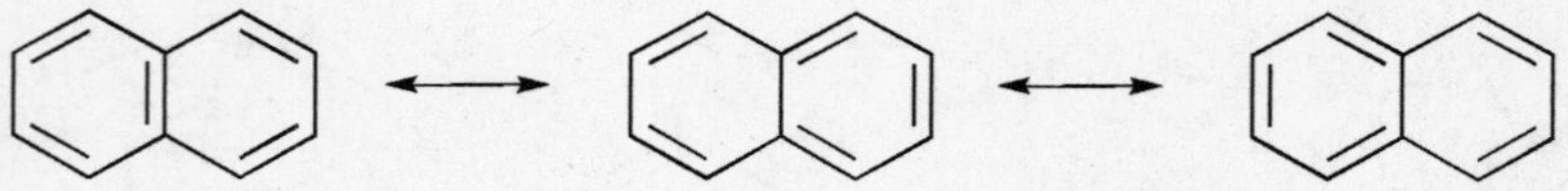 ⟷ ⟷

(c) 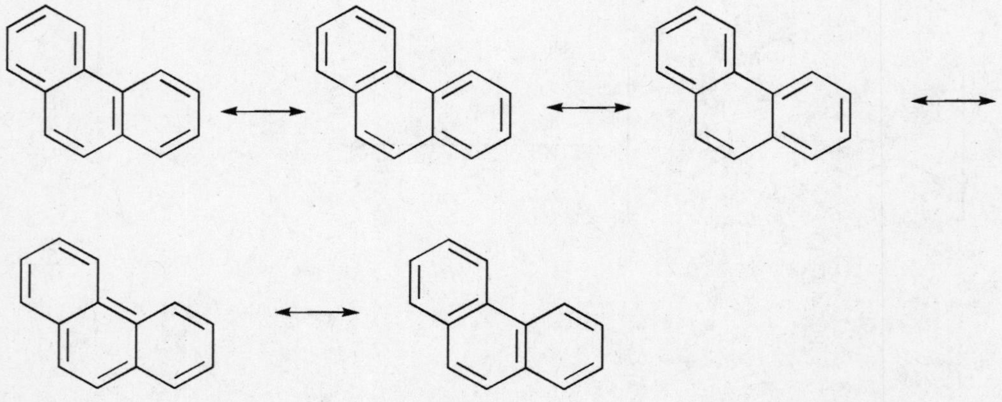 ⟷ ⟷ ⟷

⟷

3. (a) 3,5-dichlorophenol;
   (b) 1,3-dinitrobenzene;
   (c) 4-methylanisole;
   (d) 3-ethyl-5-methylbenzoic acid;
   (e) hexachlorobenzene;
   (f) 3-(2-methylbutyl)phenol;
   (g) N,3,5-trimethylaniline;
   (h) 4-bromo-3-butylbenzenesulfonic acid.

Chapter 17

4. This molecule is named 2-phenyloctane. The eight-carbon alkane chain takes priority over the -carbon benzene ring.

Electrophilic Aromatic Substitution

1. (a) *para*-dichlorobenzene; (b) *meta*-diethyl benzene; (c) *ortho*-chloroanisole;
   (d) *meta*-ethylnitrobenzene; (e) *para*-nitrobenzoic acid.
2. (a) 1,4-dichlorobenzene; (b) 1,3-diethylbenzene; (c) 2-chloroanisole;
   (d) 3-ethylnitrobenzene; (e) 4-nitrobenzoic acid.
3. Since carbons can release electrons to an electron deficient center, the presence of a carbon group adjacent to the positive charge in the Wheland intermediate will stabilize the charge and make carbon groups activating.
4.

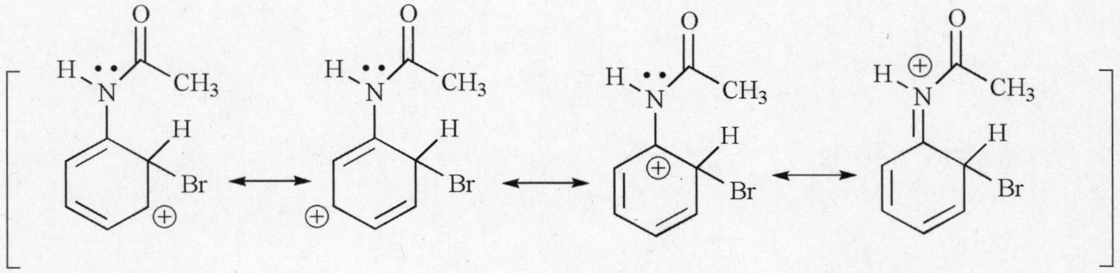

H N CH3 O H Br ⊕ ⟷ H N CH3 O H Br ⊕ ⟷ H N CH3 O H Br ⊕ ⟷ H ⊕N CH3 O H Br

5. The reaction will produce a mixture of both the *ortho* and the *para* products. The only way to obtain a pure sample of the *ortho* product is to separate the two products, usually via chromatography or, in some cases, fractional distillation. These products are liquids, although the *para* product is a low melting solid, mp 7.2°C. Since the *para* melts at 7.2°C, and the *ortho* at -72°C, it is possible that fractional crystallization might separate them.
6. (a) 3,5-dibutylbenzoic acid; (b) 4-methyl-4-phenylbutanoyl chloride;
   (c) 4,5,5-trimethyl-2-propylheptanoyl chloride; (d) 3-bromo-4-methylbenzoyl chloride.
7. The presence of the polar methoxy group will stabilize the *ortho* attack intermediate more than the *para* attack intermediate due to dipole-dipole interactions (the *ortho* effect).

8.

(a)

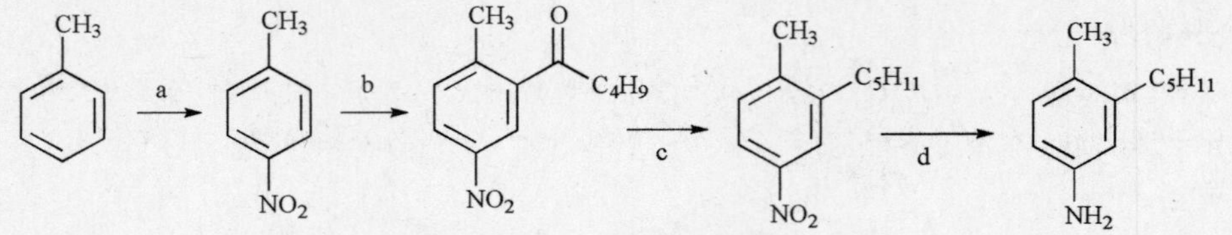

(a) HNO_3, H_2SO_4 (b) butanoyl chloride, $AlCl_3$ (c) N_2H_4, aq. KOH (d) H_2, Ni

(b)

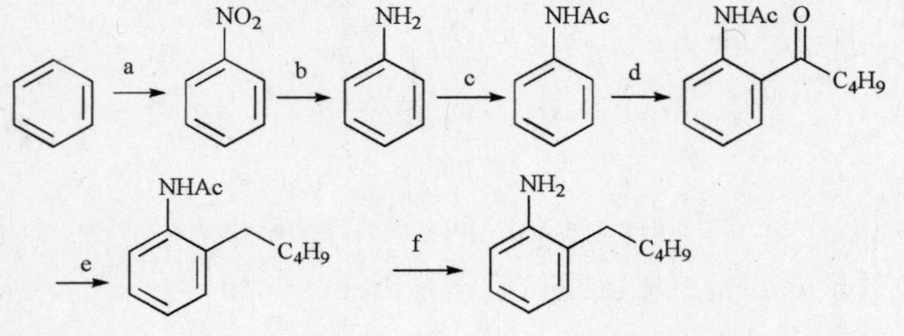

(a) HNO_3, H_2SO_4 (b) H_2, Ni (c) Ac_2O, (d) pentanoyl chloride/$AlCl_3$ - then separate the ortho isomer from the para isomers (e) NH_2NH_2, KOH (f) H_3O^+

(c)

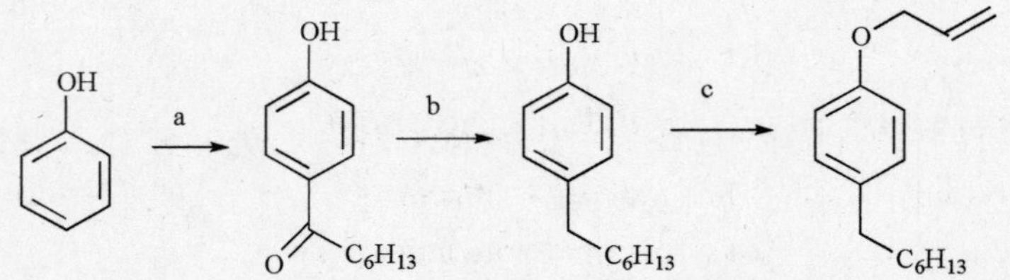

(a) heptanoyl chloride, $AlCl_3$ (b) Zn(Hg), HCl (c) allyl bromide, NaH

9.

(a) 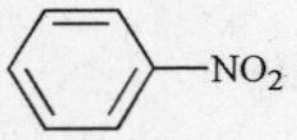 (b) 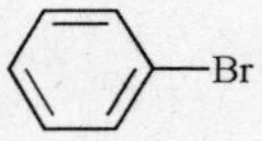 (c) 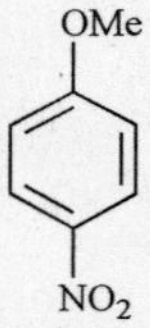 + ortho (d) 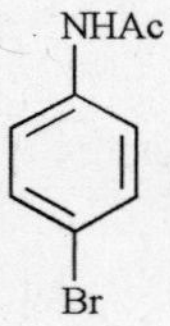 + ortho

(e) 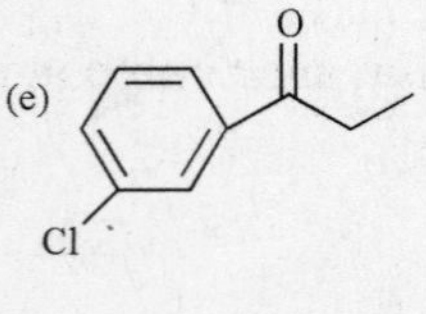 (f) 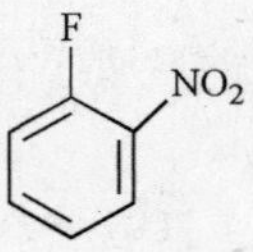 + para (g) 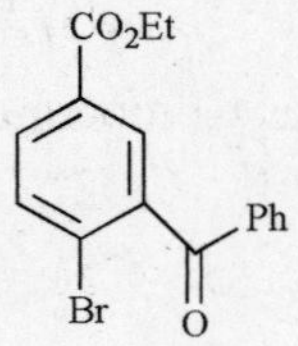 (h)

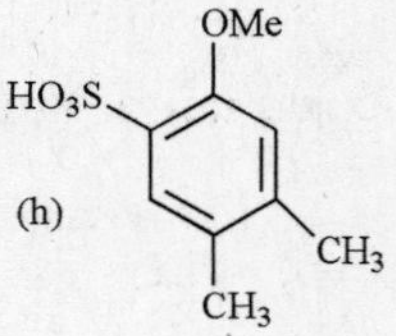

(j) 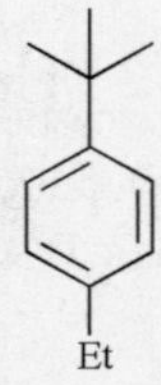 + ortho (k) 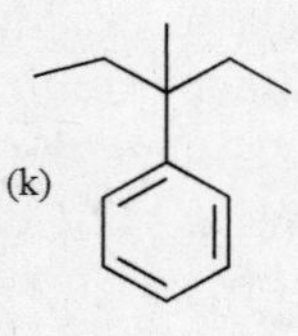 (l) 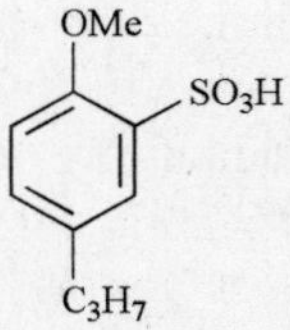 (m)

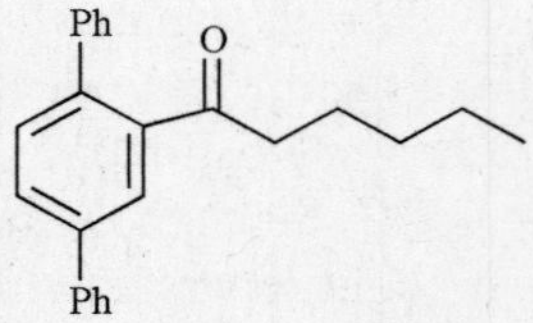

(n) 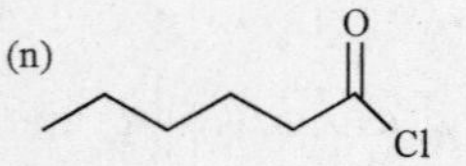 (o)  (p) 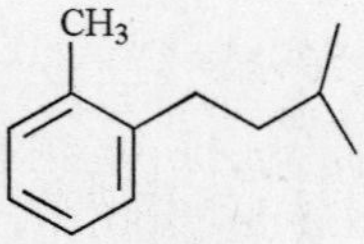 + para (q)

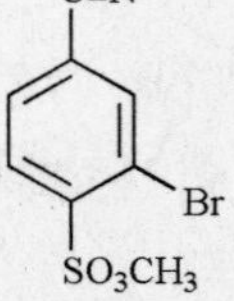

Reaction (i) gives no reaction because the aromatic ring is too deactivated for a Friedel-Crafts acylation reaction.

10. The product is a mixture of the *ortho* and *para* bromides in the ring bearing the OR group. The OR group is much more activating than the alkyl group and leads almost exclusively to the products shown.

O, Br and Br, O

Chapter 18

Spectroscopy

1. (a) 6.16×10^{15} Hz; (b) 2.35×10^{-5} cm; (c) 4.34×10^{2} nm; (d) 3.8×10^{-4} μ; (e) 4.67×10^{7} Hz; (f) 6.09×10^{-2} m; (g) 1.14×10^{8}
2. (a) 1.61×10^{3} kcal, 6.74×10^{3} kJ; (b) 6.9×10^{2} kcal, 2.89×10^{3} kJ; (c) 52.2 eV; (d) 9.34 eV.
3. (a) $[\ \ddot{O}\ ]^{\bullet +}$ (b) $[\ CH_3\text{—}\dot{N}\text{—}CH_3\ ]^{\bullet +}$ (c) $[\ \ ]^{\bullet +}$ (d) $[CH_3\text{—}C{\equiv}N^{\bullet}]^{\bullet +}$
4. In all cases, P = 100%. (a) M+1 = 5.55% of P, M+2 = 0.35% of P; (b) M+1 = 9.25% of P, M+2 = 0.39% of P; (c) M+1 = 5.55% of P, M+2 = 0.55% of P; (d) M+1 = 11.47% of P, M+2 = .0.82% of P.
5. (a) $C_8H_8O_2$; (b) $C_7H_{17}N$; (c) $C_5H_{12}N_2$; (d) $C_6H_{12}O$.
6. Compound (a) contains one bromine. Compound (b) contains one sulfur, and compound (c) contains one chlorine.
7. (a) $-CH_3$; (b) $-H_2O$; (c) $-CH_2{=}CH_2$; (d) $-CH_2CH_3$; (e) $-C_3H_7$; (f) $-CO_2$.
8. Both (b) and (c) can undergo a McLafferty rearrangement.

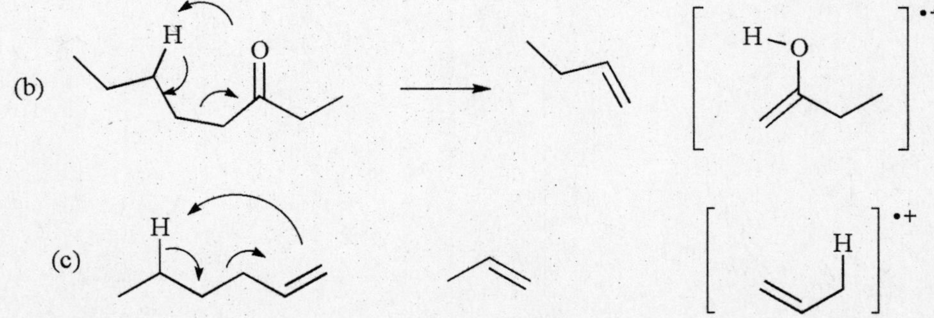

9. Water can eventually dissolve pressed KBr. Brief exposure will 'etch' or otherwise damage the surface of the plates, interfering with the transmission of light and the quality of the infrared absorption peaks. It also drastically reduces the lifetime of the plates.

10. A bending vibration describes a bond vibration in which the two atoms connected to the bond move 'up and down' more or less in unison. A stretching vibration describes a bond vibration in which the two atoms connected to the bond move alternately away and towards each other, along the line between the two atoms (along the bond).

11. The strong C≡C bond gives a much less intense signal. The C–O absorption band is rather strong. The C-O signal appears at lower energy since it takes less energy to make that bond vibrate.

12. (a) 1.488×10^{-23}; (b) 1.073×10^{-23}; (c) 0.996×10^{-23}.

13. In (a), the most prominent band is the carbonyl at 5.80 μ (1724 cm^{-1}). In (b), the most prominent band is the O–H band at about 2.8 μ (3571 cm^{-1}). The bromine in (c) appears at about 16 μ (625 cm^{-1}) but is not diagnostic. The acid band (COOH) between 3-4 μ (3333 - 2500 cm^{-1}) is the most prominent, and the carbonyl band at 5.80 μ (1724 cm^{-1}) is also diagnostic.

14. (a) 3; (b) 1; (c) 1; (d) 1; (e)1;

(f) none; this nucleus does not lead to the NMR phenomenon.

15. (a) 122.5 MHz; (b) 538.6 MHz; (c) 1273.3 MHz.

16. (a) 5.21 ppm. (b) 16.39 ppm. (c) 7.2 ppm. (d) 4.24 ppm.

17.

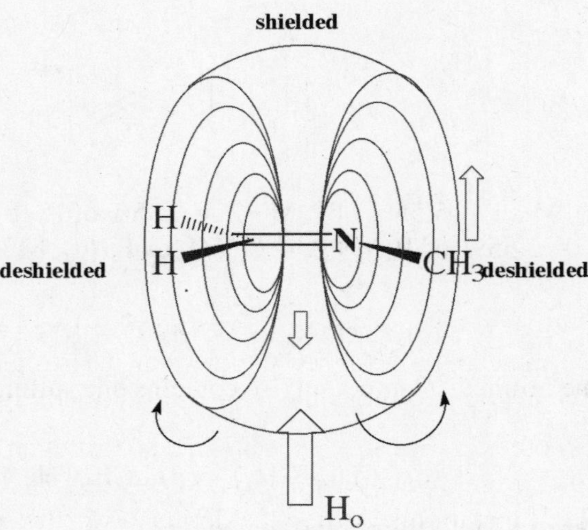

18. (a) HC-NR_2; (b) H-C-C=O; (c) C=C-C-H; (d) C≡C-H.

19.

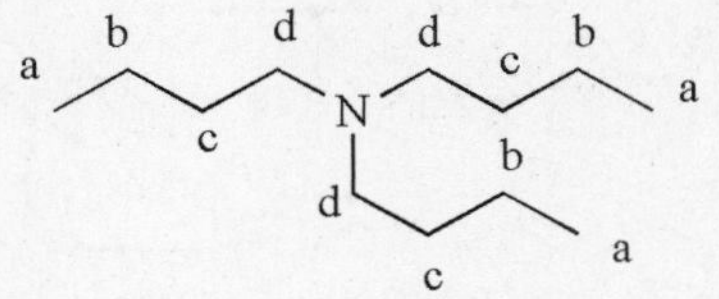

20. (a) 4; (b) 2; (c) 6; (d) 0; (e) 3.

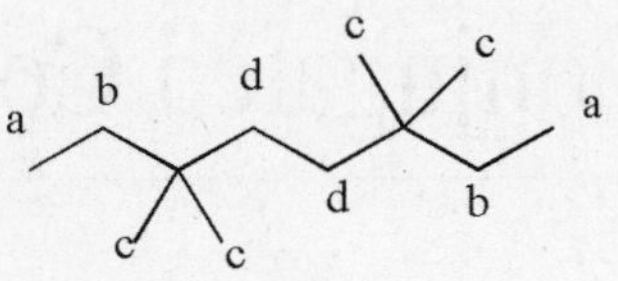

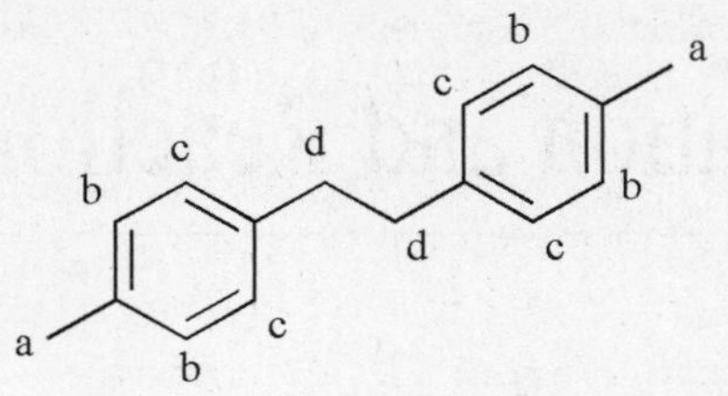

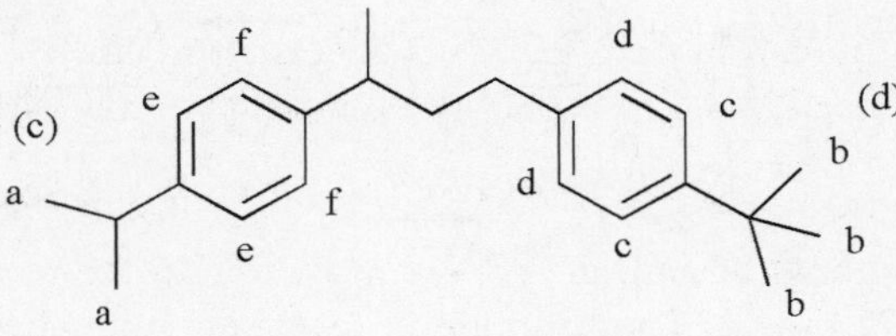

21.

22.

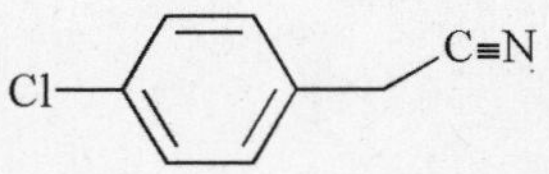

23. The formula is C_4H_9Cl. The M+2 peak indicates the presence of one Cl. There are no signals in the IR that are helpful. The one signal in the NMR at 0.9 ppm indicates a signal type of proton in the molecule, probably a *tert*-butyl group. This molecule is 2-chloro-2-methylpropane.

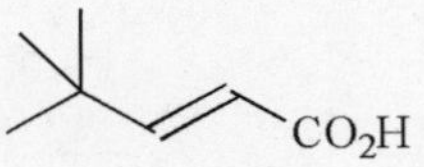

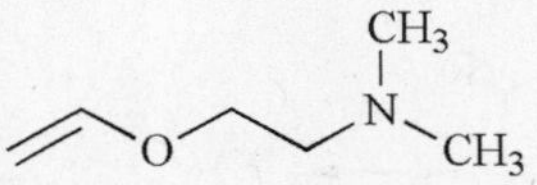

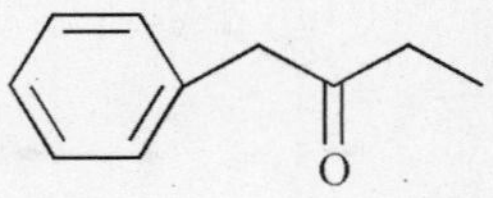

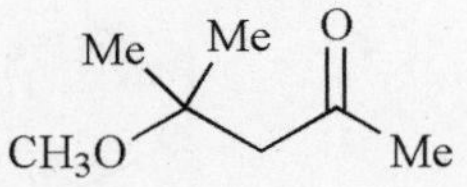

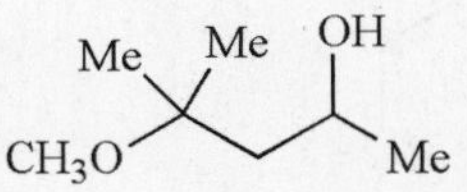

24. The formula is $C_5H_{12}O$, and the IR indicates a strong OH, so this is an alcohol. There is a triplet at about 1 ppm worth 6 H, and a multiplet at about 3.3 ppm worth 1 H. The singlet at 3.2 ppm is the proton on oxygen (the OH). The multiplet at about 1.3 ppm is worth 4 H. The key is the single proton downfield that indicates an HC–OH unit. That proton is downfield so it has to be attached to the carbon bearing the oxygen. The triplet at about 1 ppm worth 6H has to be two identical methyl groups that have two neighbors. This indicates two ethyl groups. Putting this information together, the molecule is 3-pentanol.

25. The formula is $C_{11}H_{14}O$, and the IR indicates a carbonyl but no aldehyde peaks. In the NMR, there is a phenyl group (monosubstituted), a doublet at 1 ppm worth 6H, and a multiplet at 2.5 ppm worth 1H. These two signals are indicative of an isopropyl group, and the proton of the CH signal is consistent with it being attached to the carbon of a carbonyl group. There is also a singlet at 3.4 ppm worth 2H, and this CH_2 group must be between the phenyl and the carbonyl (no neighboring protons). Therefore, the molecule is 1-phenyl-3-methyl-2-butanone.

Chapter 19

Conjugation and Reactions of Conjugated Compounds

1.

(a) 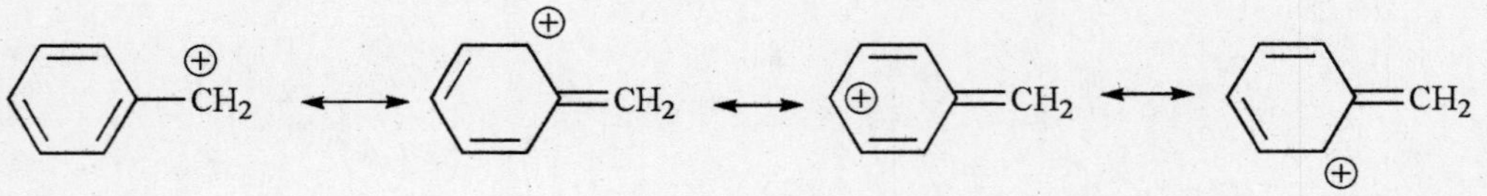 CH_2 ⟷ CH_2 ⟷ CH_2 ⟷ CH_2

(b) 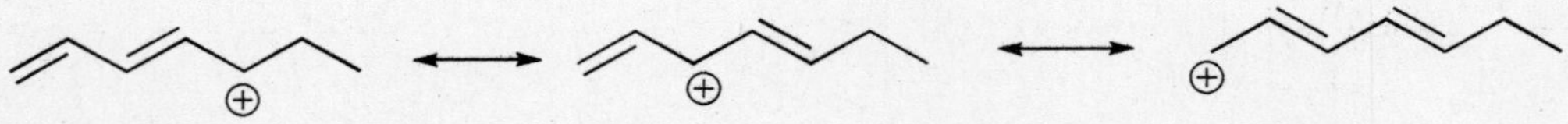 ⟷ ⟷

(c) 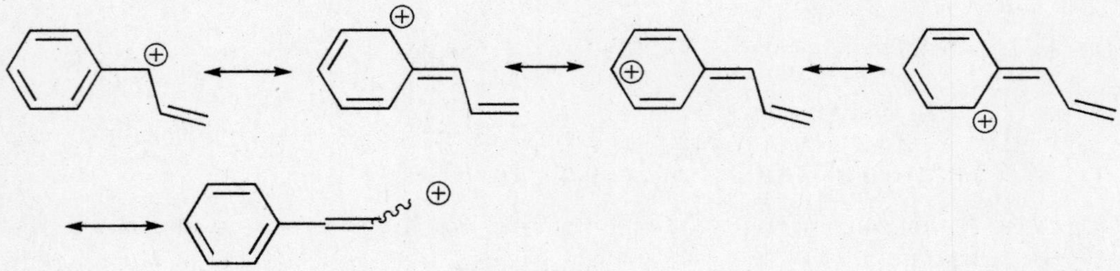 ⟷ ⟷ ⟷

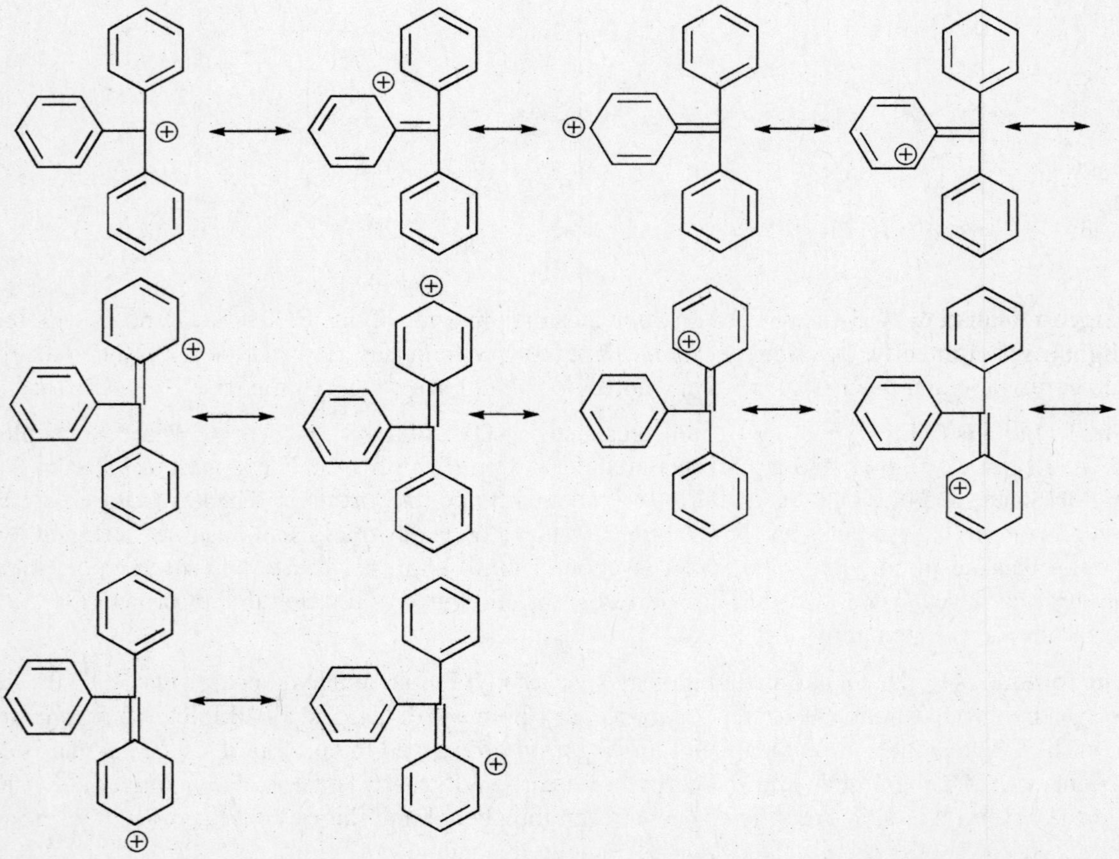

⟷

⟷ ⟷ ⟷ ⟷

⟷ ⟷ ⟷ ⟷

⟷

2. Molecules (b), (c), (e), (f) and (h) contain conjugated double bonds.
3. (a) 1,5-hexadiene
   (b) cyclopentadiene
   (c) hex-1-en-3-one
   (d) 1,2,4,5-tetramethyl-1,4-cyclohexadiene
   (e) penta-1,4-dien-3-one
   (f) 1,4E,6-heptatriene
   (g) cyclohexane
   (h) ethenylbenzene (styrene). The requested structures are:

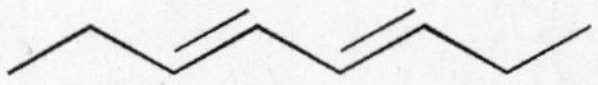

octadiene

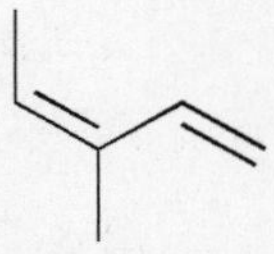

pentadiene

4. The cisoid dienes are (a), (b) and (e). The only transoid diene is (d). Only the acyclic (a) has both cisoid and transoid rotamers in equilibrium.
5. Both the 1,2-products (E+Z 1,2-dichloro-3-pentene), and the 1,4- products (E+Z 1,4-dichloro-2-pentene) are formed. At -78°C, the kinetic product (1,2-addition) is the major product.
6. 83 kcal = 347.4 kJ = 2929.9 cm^{-1} = 2,373,800 nm.

   132 kJ = 901863 nm = 9,018,633.5 Å = 31.5 kcal.

   0.45 nm = 1.57×10^{-5} kcal = 5.55×10^{-3} cm^{-1}.

   2.4 kcal = 68640 nm = 686400 Å = 847.2 cm^{-1}.
7. Only the conjugated double bonds will absorb strongly in the UV: (a), (b), (d), (e) and (f).
8. If A = (ε×)(l)×(c), then A = 6457×0.3×5 = 9685.5 for 3-buten-2-one. Similarly,

   A = 17×0.3×5 = 25.5 for 2-butanone.
9. Since 1,4-diphenyl-1,3-butadiene is more extensively conjugated than 1,3-butadiene, it will absorb light more efficiently, leading to a larger extinction coefficient.

10.

(a) Cl and Cl

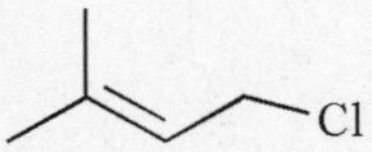

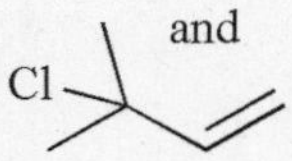

(b) HO Ph Ph

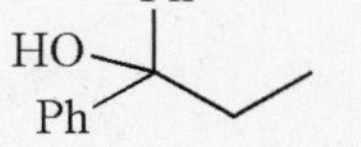

(c) Ph Ph Br

(d) H Br

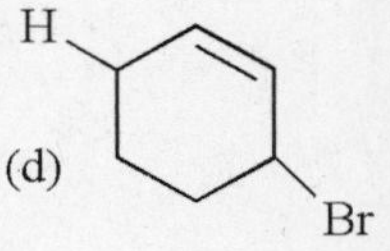

(e) H Cl

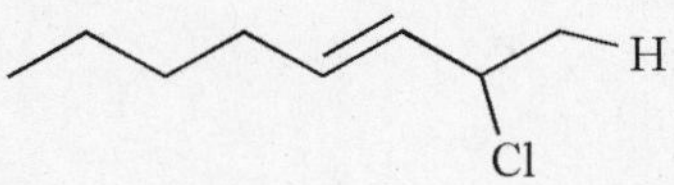

(f) Br Br

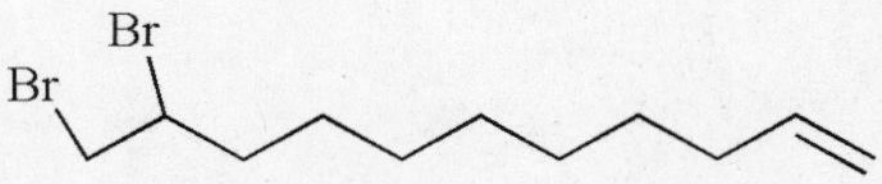

(g) Br Br

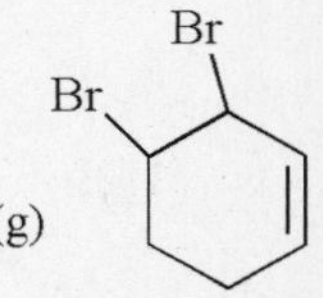

11.

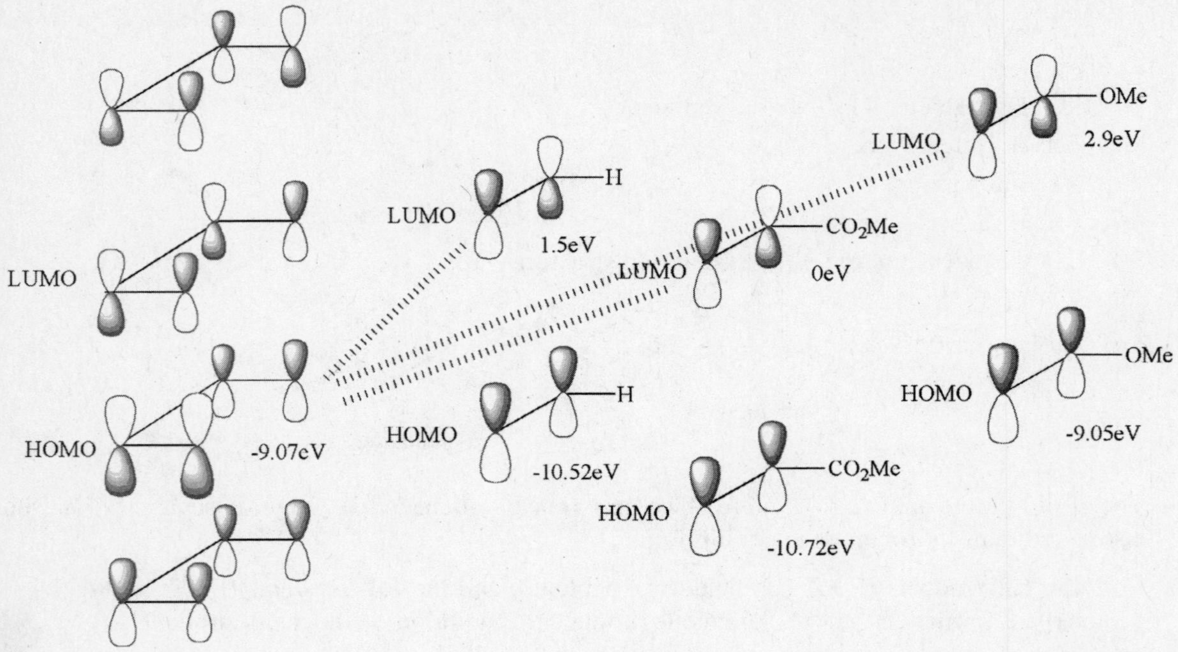

LUMO

HOMO

-9.07eV

LUMO

1.5eV

HOMO

-10.52eV

LUMO

0eV

HOMO

-10.72eV

LUMO

2.9eV

HOMO

-9.05eV

The ΔE for butadiene-ethene is 10.57 eV and is 9.07 eV for butadiene-methyl acrylate and 11.07 eV for butadiene-methyl vinyl ether. Methyl acrylate reacts faster since it has the smallest ΔE.

12. (a), (b) and (c). Triene (d) might react with a diene in a Diels-Alder reaction but it cannot react as a diene (with another alkene).

13.

(a)

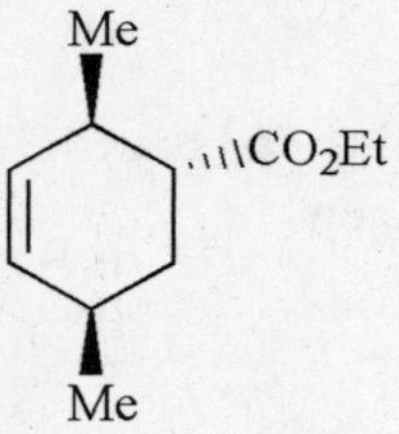

(b)

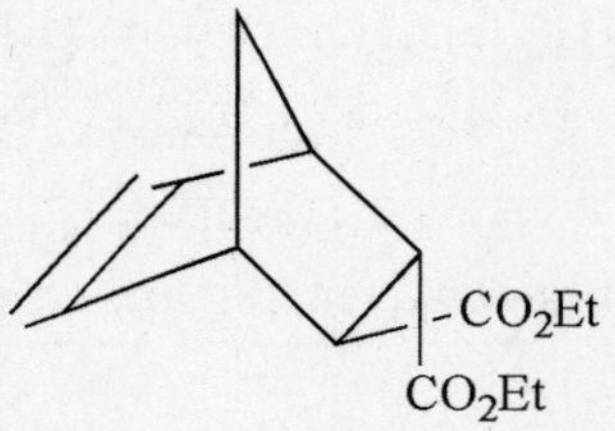

(c)

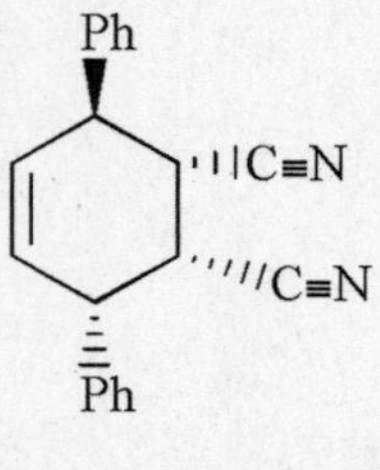

(d)

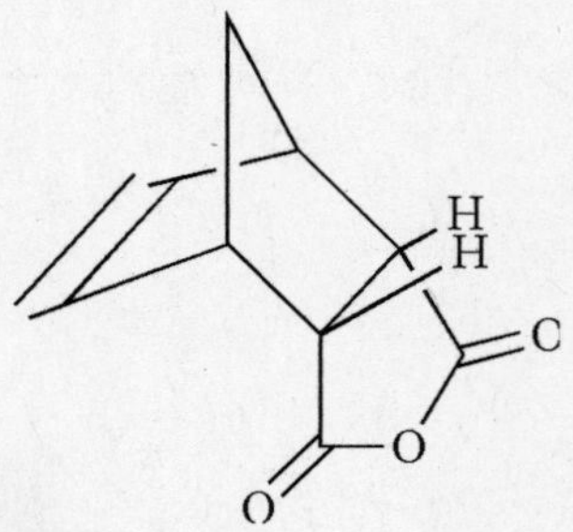

14.

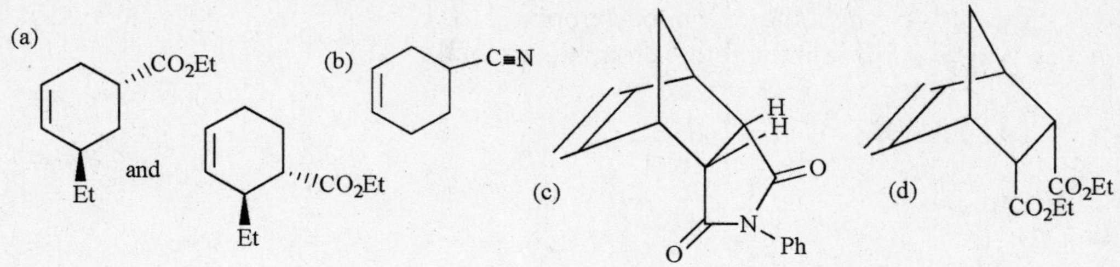

(a) CO_2Et Et and CO_2Et Et (b) C≡N (c) H H O N Ph O (d) CO_2Et CO_2Et

15.

(a) CO_2Et —a→ CO_2Et —b→ OH —c→ CHO —d→

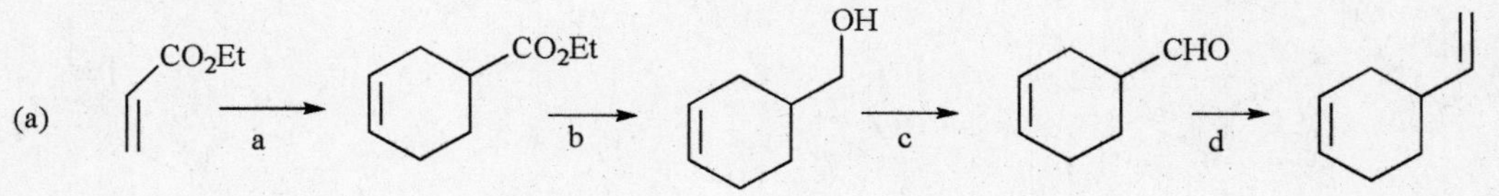

(a) 1,3-butadiene, heat (b) i. $LiAlH_4$ ii. H_3O^+ (c) PCC, CH_2Cl_2 (d) $Ph_3P{=}CH_2$

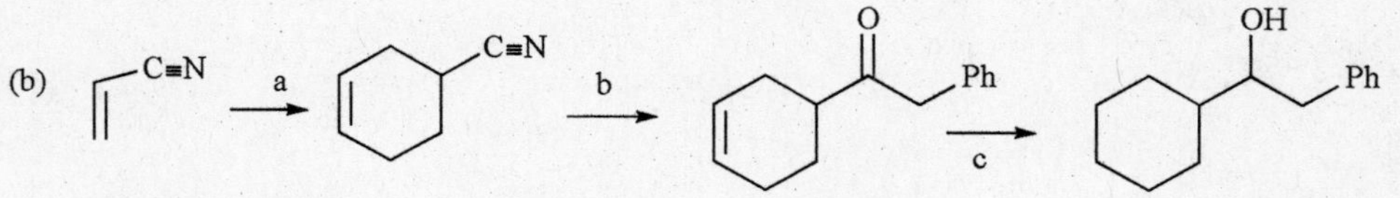

(b) C≡N —a→ C≡N —b→ O Ph —c→ OH Ph

(a) 1,3-butadiene, heat (b) i. $PhCH_2MgBr$ ii. H_3O^+, heat (c) excess H_2, PtO_2

Chapter 20

Nucleophilic Aromatic Substitution

1.

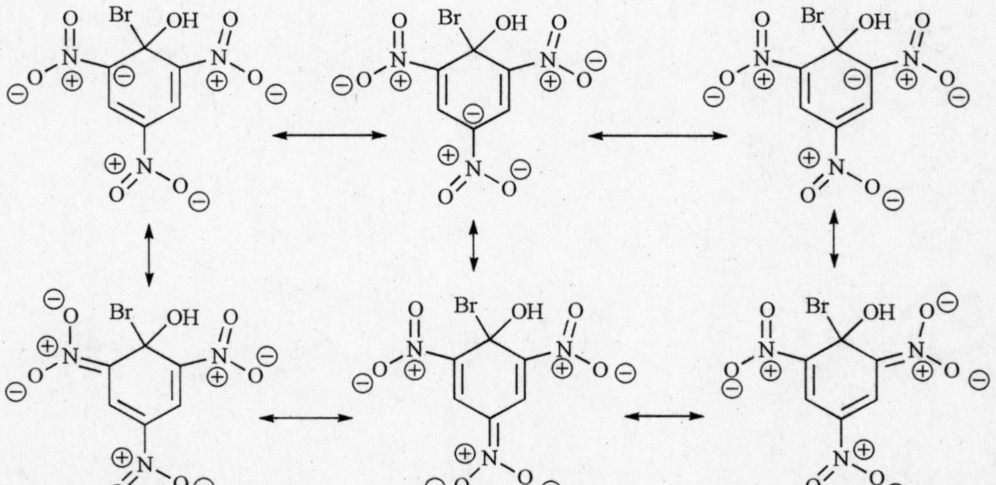

2. When the C-Li bond is formed, it is perpendicular to the aromatic π-cloud, but is parallel to the adjacent C-Cl. Since it is parallel, the carbanion carbon (C-Li) can displace the chlorine to form a new π-bond, which will be perpendicular to the aromatic π-cloud.

3.

(a)

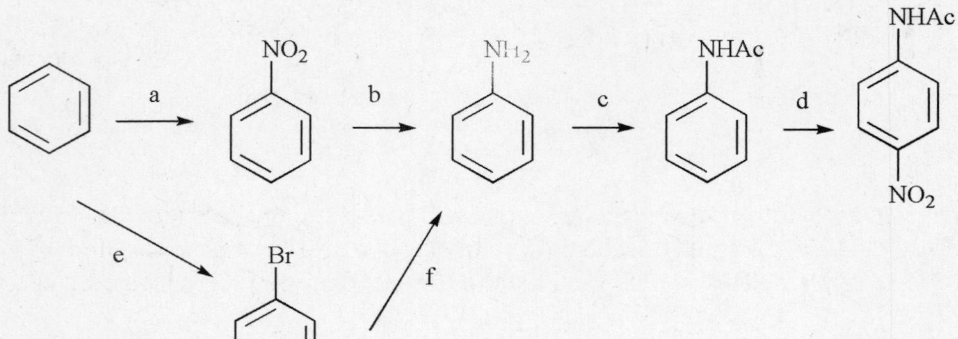

NO$_2$ NH$_2$ NHAc NHAc NO$_2$ Br

a b c d e f

(a) HNO_3, H_2SO_4 (b) H_2, Pd (c) Ac_2O, pyridine (d) HNO_3, H_2SO_4 (e) Br_2, $AlCl_3$ (f) NH_3, heat, pressure

(b)

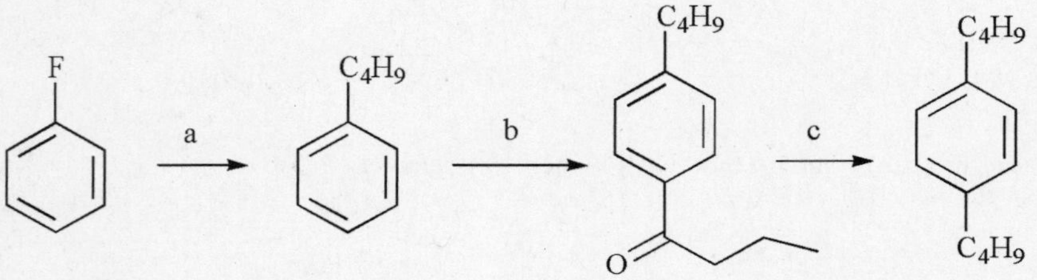

F C_4H_9 C_4H_9 O C_4H_9 C_4H_9

a b c

(a) BuLi (b) butanoyl chlaride, $AlCl_3$ (c) N_2H_4, aq. KOH

4.

(a) OH CH$_3$

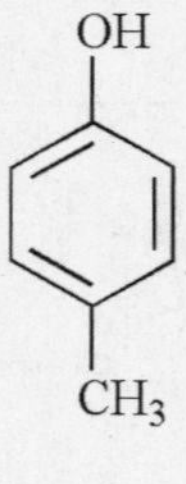

(b) OH NO$_2$ NO$_2$ NO$_2$

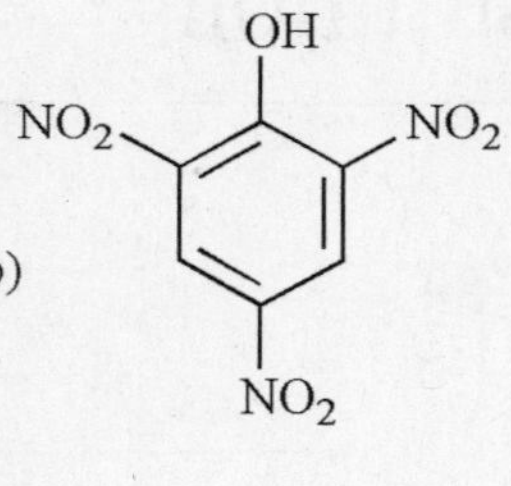

(c) OMe NO$_2$ NO$_2$ NO$_2$

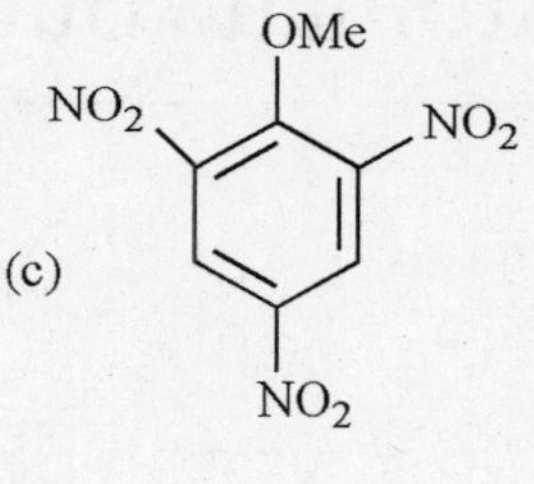

(d) OH OMe OMe

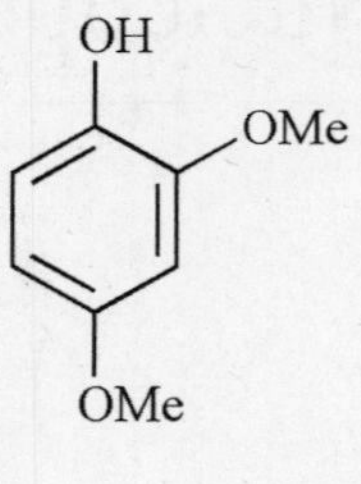

(e) NH$_2$

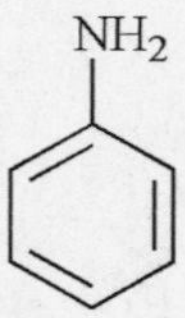

(f) OH NO$_2$

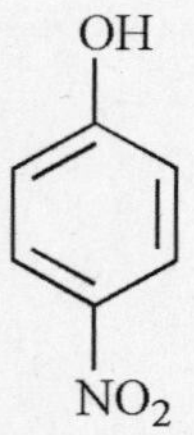

(g) C_4H_9

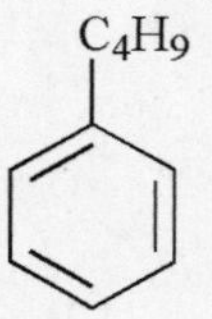

(h) NH$_2$ + NH$_2$

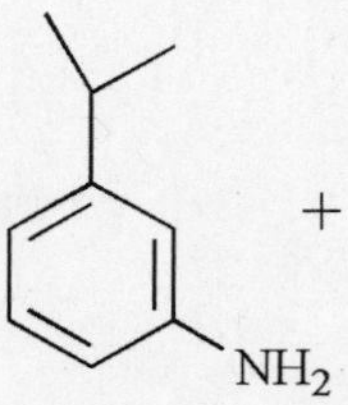

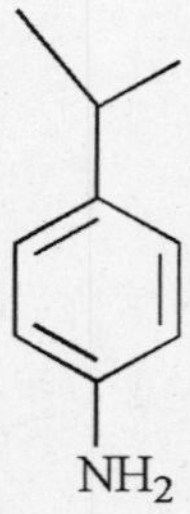

Chapter 21

Aldehyde and Ketones. Acyl Addition Reactions

1.

(a) Ph OH (b) OH (c) OH

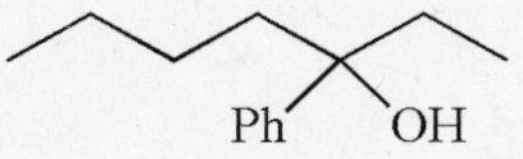

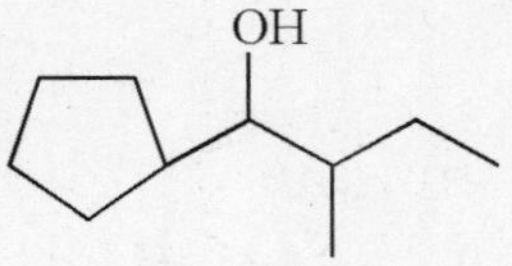

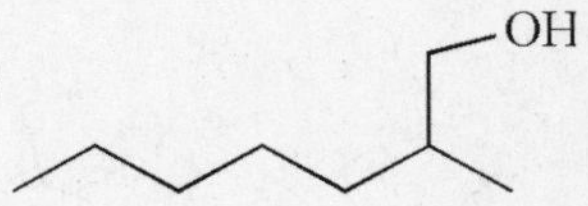

2. The conjugated carbonyl group in cyclohexenone will give a strong absorption band (extinction coefficient - 8230) at 225 nm in the UV, whereas cyclohexanone gives a weak absorption (extinction coefficient - 27) at 280 nm. This is a characteristic difference between a conjugated and unconjugated ketone.

3. (a) 3-butyl-2-heptanone; (b) 4-methyl-3-(1-methylethyl)-2-(1,2-dimethylpropyl)pentanal; (c) 2,5-dichloro-4-(1-methylpropyl)benzaldehyde; (d) 3,4-diphenylcyclopentanone; (e) 5,5-dimethyl-2,6-nonanedione; (f) cyclohexene carboxaldehyde; (g) 2-benzyl-2-ethylcyclohexanone; (h) 3-methylhex-2-enal.

4. The ketone product is also very reactive with the Grignard reagent. As it is formed, the ketone can compete with the acid chloride for reaction with butylmagnesium bromide.

5.

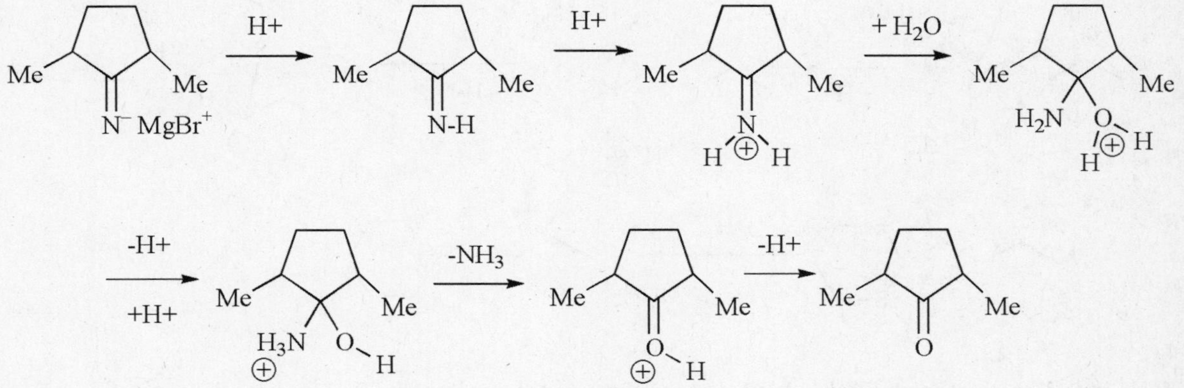

Me Me $N^- MgBr^+$ —H+→ Me Me N-H —H+→ Me Me $N^{\oplus}$ H H —$+ H_2O$→ Me Me H_2N $O^{\oplus}$ H H

—-H+ / +H+→ Me Me $H_3N^{\oplus}$ O H —$-NH_3$→ Me Me $O^{\oplus}$ H —-H+→ Me Me O

6.

(a) + H^+ → → + H_2O → - H^+ → + H^+ → - $HOCH_2CH_2OH$ → $-H^+$ →

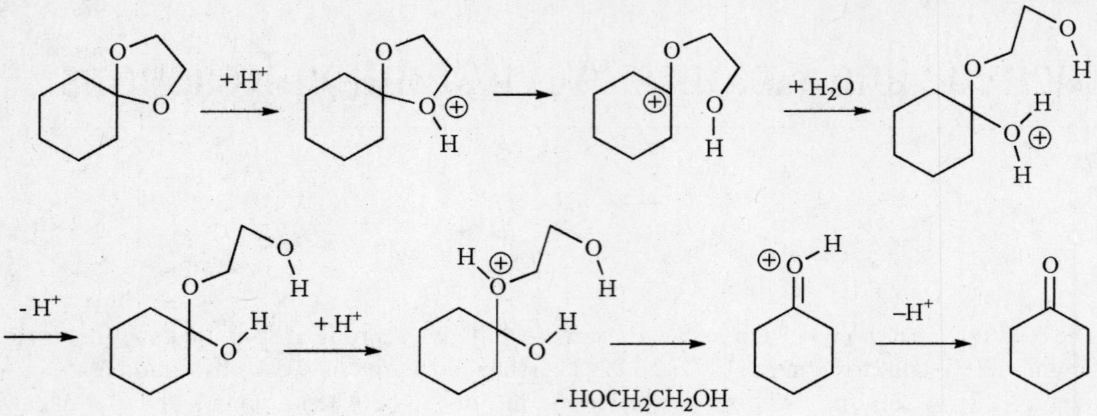

(b) + H^+ → SH SH → $-H^+$ → + H^+ → $- H_2O$ → → $- H^+$ →

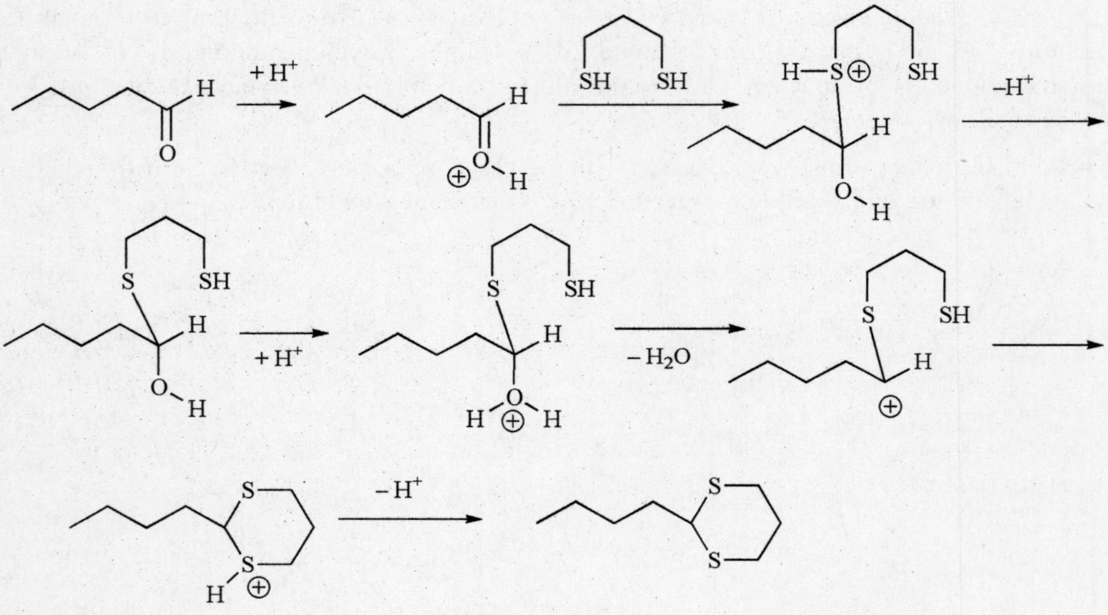

7. The initial product is an enol, which tautomerizes to the ketone, the isolated product.

8.

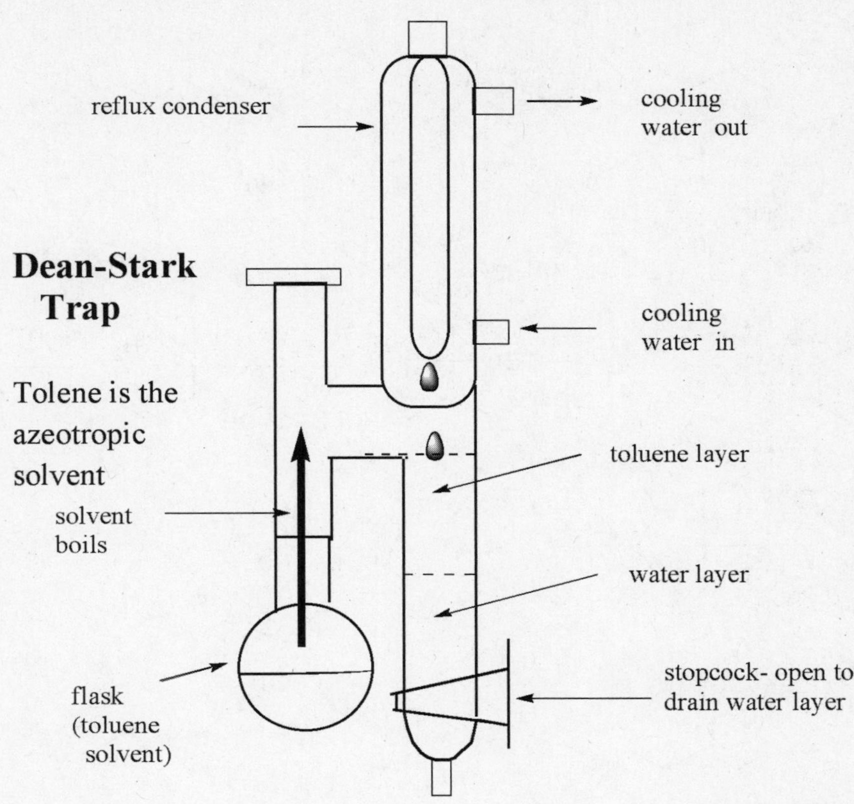

9. Compounds (a) and (d) have no hydrogen atoms on the carbon adjacent to the carbonyl group and also have electron withdrawing halogen atoms on those carbons. Therefore, compounds (a) and (d) might form stable hydrates. Since (c) and (d) have hydrogen atoms on at least one carbon adjacent to the carbonyl and do not have electron withdrawing groups, the hydrates formed from these molecules would not be expected to be stable.

10.

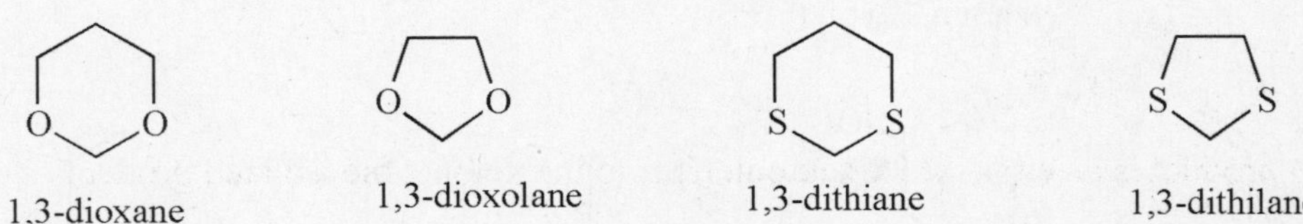

11. (a) 3-methyl-1-pentanethiol;

(b) cyclopentanethiol;

(c) 4,4-dimethyl-5-phenylhexane-2-thiol;

(d) 1,3-propanedithiol.

12.

(a) 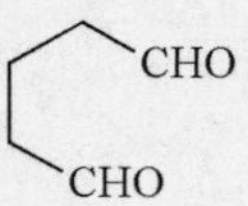 CHO CHO (b) 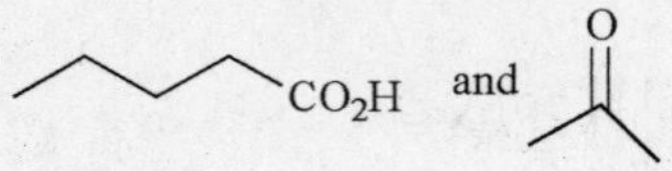 CO_2H and O (c) 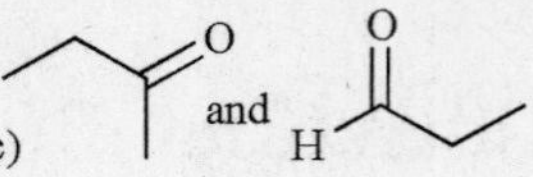 O and O H (d) 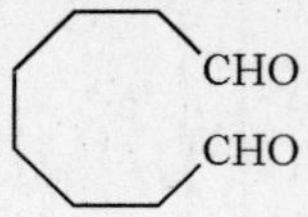 CHO CHO

(e) 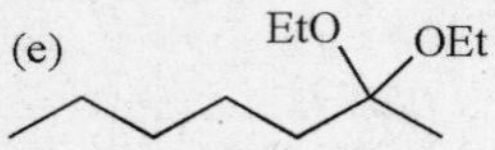 EtO OEt (f) 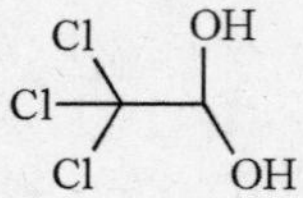 Cl Cl Cl OH OH (g) 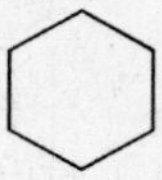 (h) 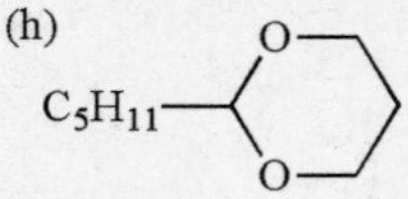 C_5H_{11} O O (i) 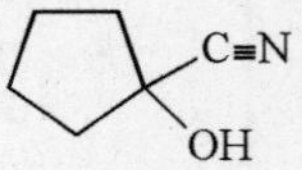 C≡N OH

(j) 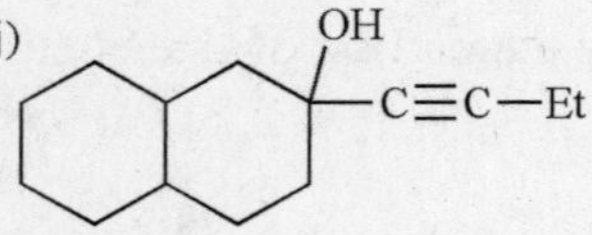 OH C≡C—Et (k) 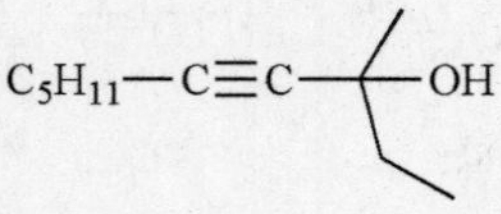 C_5H_{11}—C≡C— OH (l) 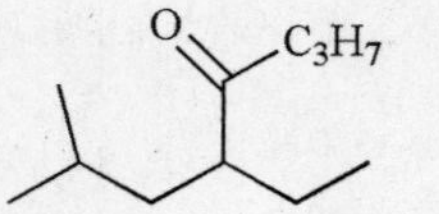 O C_3H_7 (m) 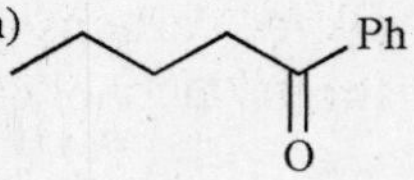 Ph O

(n) 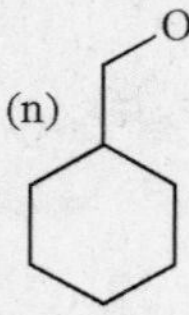 OH (o) 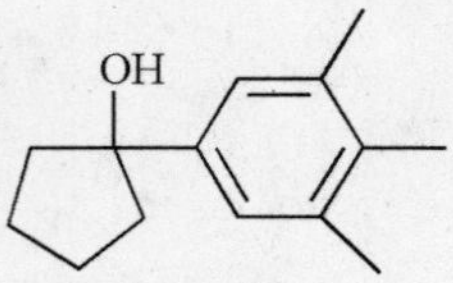 OH (p) 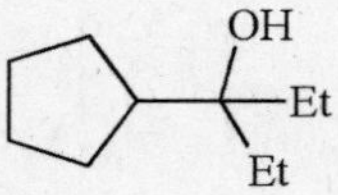 OH Et Et (q) 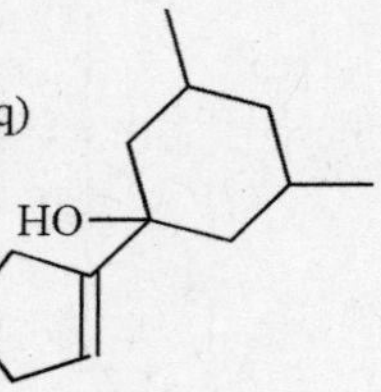 HO (r) 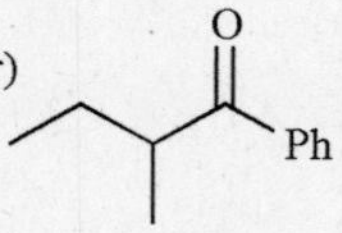 O Ph

(s) 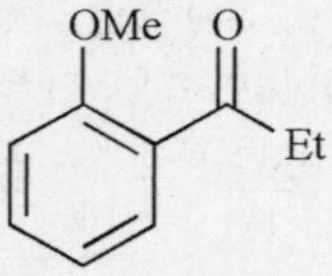 OMe O Et (t) 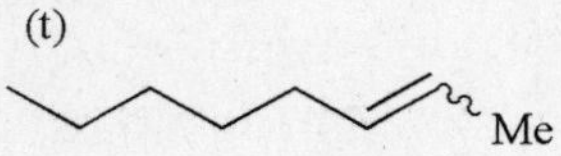 Me (u) 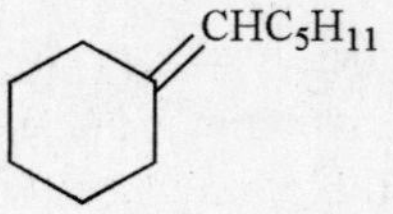 CHC_5H_{11} (v) 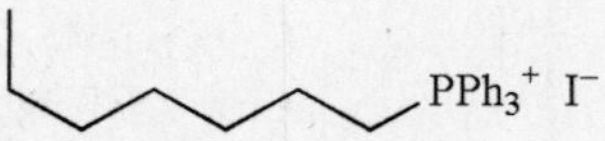 PPh_3^+ I^-

13.

(a) 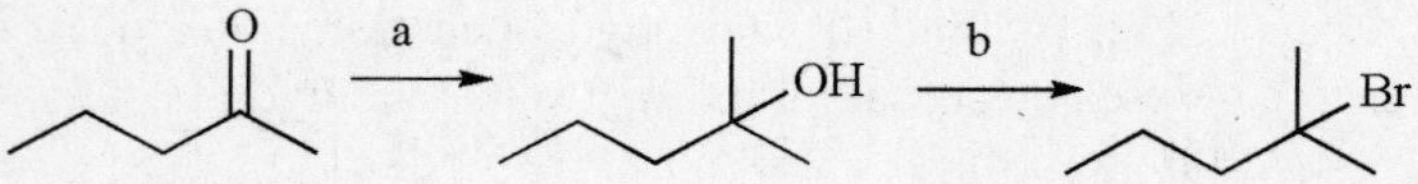 O —a→ OH —b→ Br

(a) i. MeMgBr ii, H_3O^+ (b) PBr_3

(b) 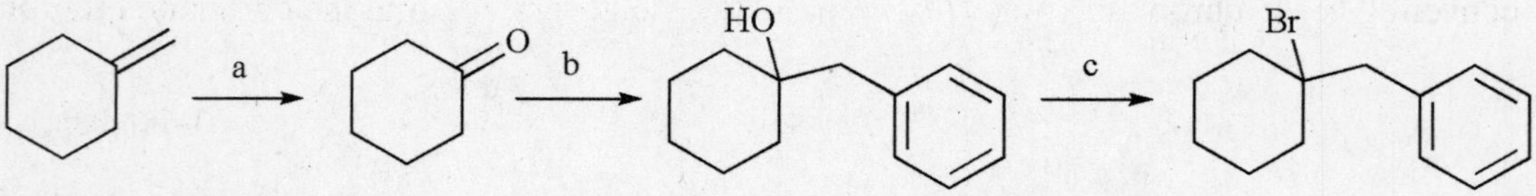 —a→ O —b→ HO —c→ Br

(a) i. O_3, −78°C ii. H_2O_2 (b) i. $PhCH_2MgBr$ ii. H_3O^+ (c) PBr_3

(c) 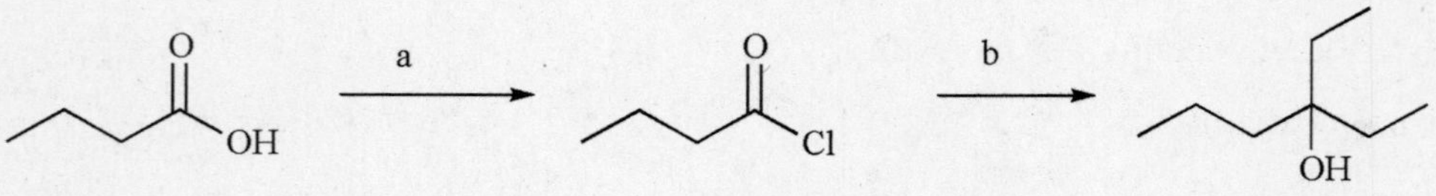 O OH —a→ O Cl —b→ OH

(a) $SOCl_2$ (b) i. excess EtMgBr ii. H_3O^+

Chapter 22

Alcohols, Aldehydes and Ketones. Oxidation and Reduction

1. Reactions (d), (e) and (f) are reductions. (b) is -1 for both carbons of the alkene and -2 for both carbons of the alkane (Δ is -2). (c) is 0 for the carbon of the alcohol and +2 for the carbonyl carbon (Δ is +2). (d) is +2 for the carbonyl carbon and 0 for the alcohol carbon (Δ is +2). (e) is 0 for both alkyne carbons and -2 for both alkane carbons (Δ is -4). (f) is +3 for the carbon of the ester and -1 for the alcohol carbon (Δ is -4).

2.

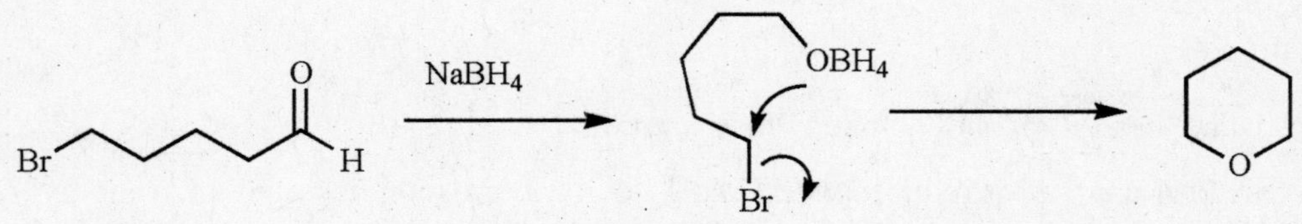

Br H O NaBH$_4$ OBH$_4$ Br O

A *B* *C*

Initial reduction of aldehyde **A** will generate an alkoxide **B**, which can react with the terminal chloride to give tetrahydropyran as a by-product by the Williamson ether reaction. Since the product **B** is not a completely ionic alkoxide, but is bound to the boron, the oxygen is not as nucleophilic, accounting for only a small amount of the ether tetrahydropyran **C** as a by-product.

3.

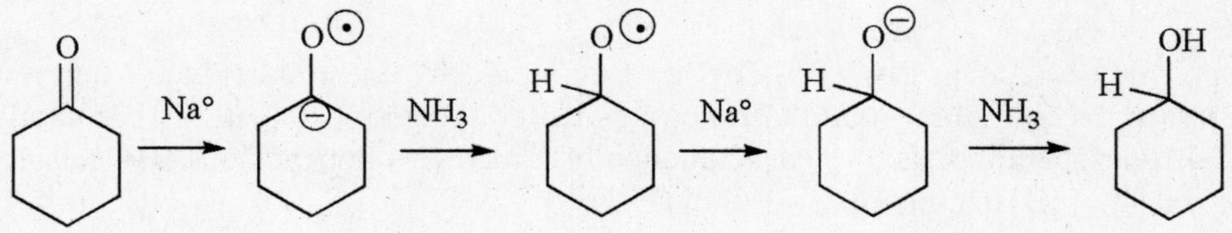

O Na° NH$_3$ H Na° H NH$_3$ H OH

4. The bulky *tert*-butyl groups in A inhibit approach of a base that can remove the α-hydrogen after A is converted to the chromate ester. This steric hindrance is very small in the chromate ester of B.

5.

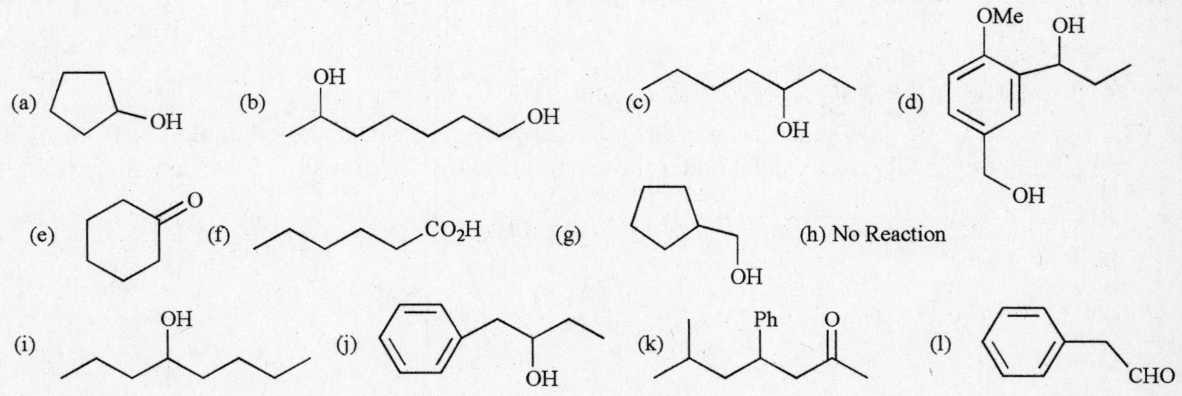

(a) OH (b) OH OH (c) OH (d) OMe OH OH

(e) O (f) CO$_2$H (g) OH (h) No Reaction

(i) OH (j) OH (k) Ph O (l) CHO

Reaction (h) gives no reaction.

6.

(a)  CHO (b) 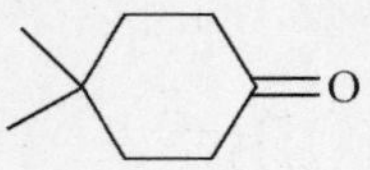 =O (c) 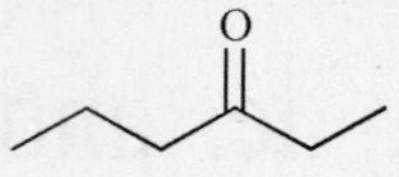 O (d) 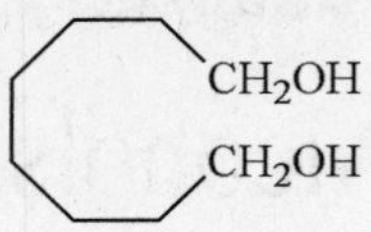 CH_2OH CH_2OH

(e) 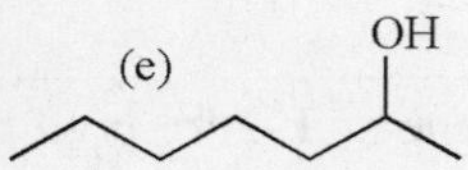 OH (f) 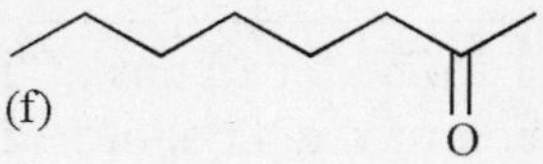 O (g) 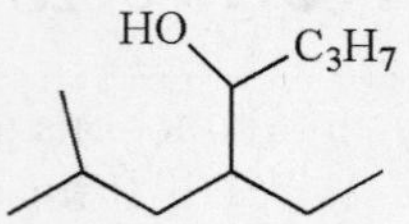 HO C_3H_7 (h) 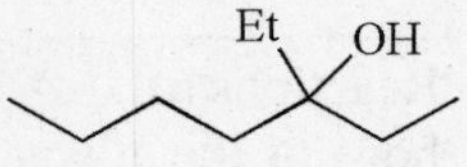 Et OH

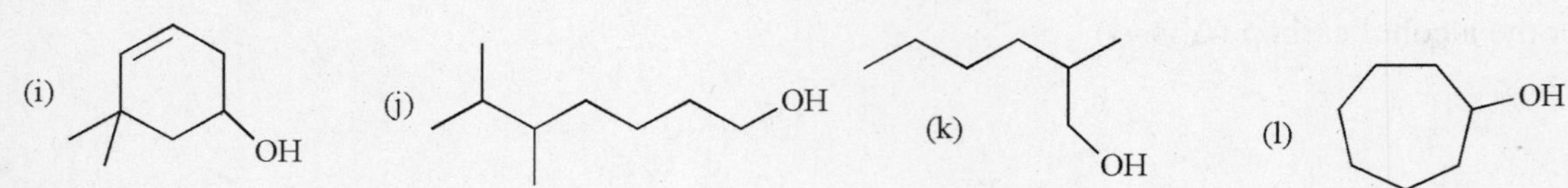

7. The product is benzyl alcohol (1-phenylmethanol).

8. 5-Methylhept-4-enal (**A**) is reduced with $LiAlH_4$ to give corresponding alcohol, 5-methyl-hept-4-en-1-ol (**C**). In the second example, conjugated ketone (**B**) is reduced to the alcohol (**D**).

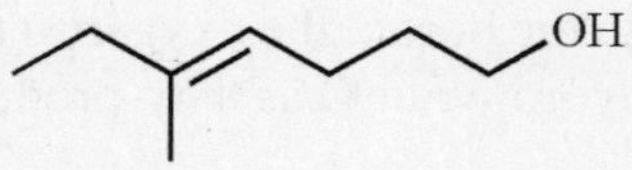

OH OH

C 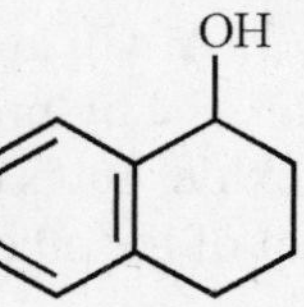 **D**

9. Keto-aldehyde (**A**) contains two carbonyl groups, an aldehyde and a ketone. When reduced with excess hydrogen (platinum oxide - sometimes called Adam's catalyst - is the catalyst) a diol (2-ethyl-1,5-hexanediol, **C** is formed. Reduction of 2-methylcyclopentanone (**B**) leads to a mixture of the *cis*-alcohol (**D**) and the *trans*-alcohol (**E**).

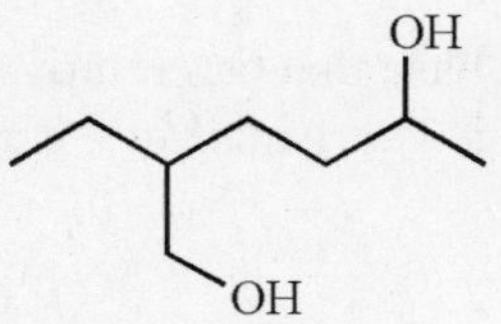

OH OH OH OH

C 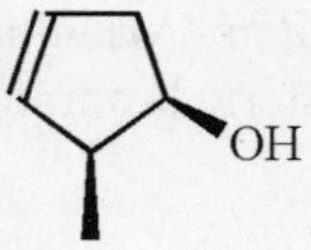 **D** 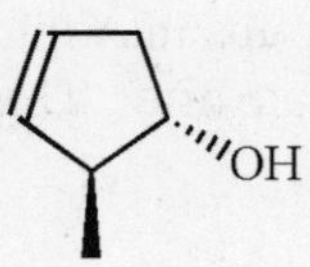 **E**

10. The reduction of cyclohexane carboxaldehyde (**A**) gives cyclohexanemethanol (**C**). Both of the ketone functions in (**B**) are reduced under these conditions to 4,5-dimethyl-2,6-octanediol (**D**).

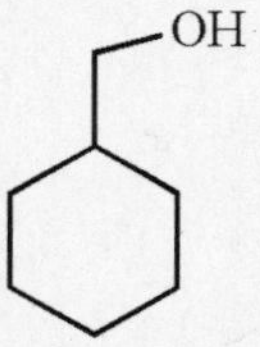

OH CH_3 OH CH_3 OH

C 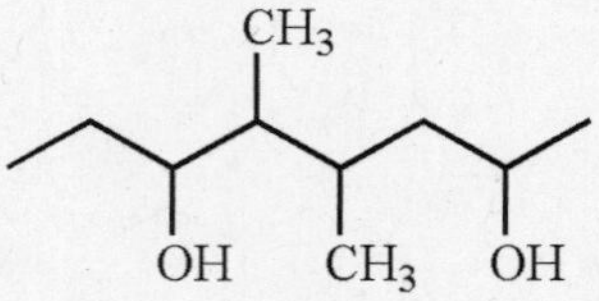 **D**

Chapter 23

Enolate Anions and the Aldol Condensation

1. The most acidic hydrogens are H_b in (a), H_a in (b), H_b in (c) and H_a in (d).
2. The 1,3-carbonyl compound [compound (c), 2,4-heptanedione] will have the highest enol content.
3. An aprotic solvent such as ether or THF, low reaction temperature such as -78°C, a strong non-nucleophilic base that generates a weak conjugate acid (such as lithium diisopropyl amide), the use of a relatively covalent counterion such as lithium and relatively short reaction times.
4.

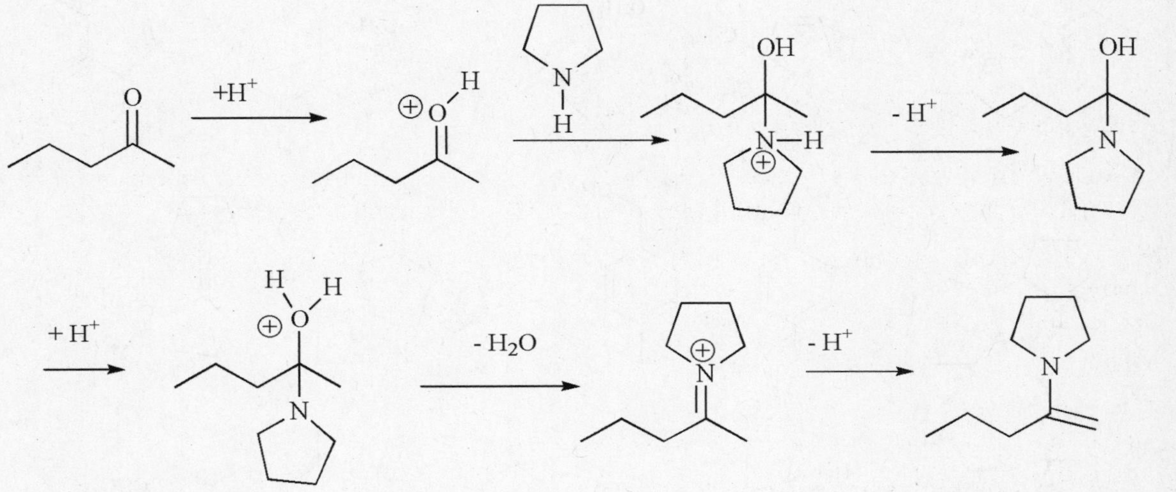

+H$^+$ ... - H$^+$ ... + H$^+$ ... - H_2O ... - H$^+$

5.

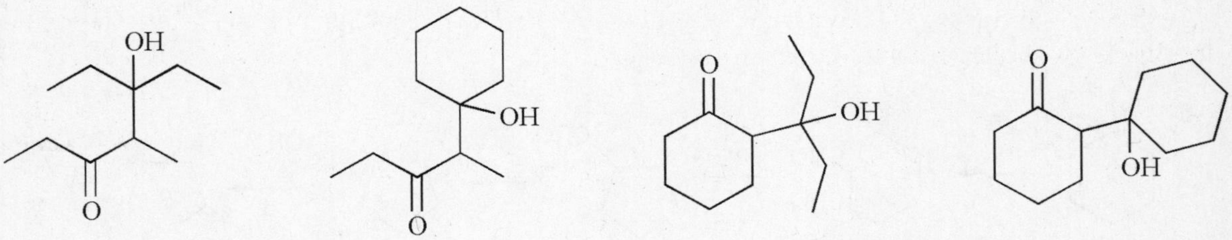

6. The product is 2-octanone, resulting from 1,4-addition of the butyl cuprate to the conjugated carbonyl derivative.
7. In the presence of base, pentanal, which has an α-hydrogen, will undergo an Aldol condensation faster than it can undergo the Cannizzaro reaction.

8.

(a) OH, C_4H_9, CHO

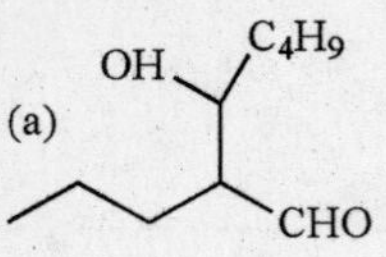

(b) OH, C_4H_9, C_4H_9, CHO

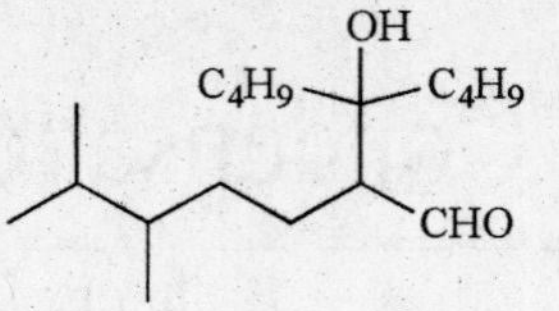

(c) O, N

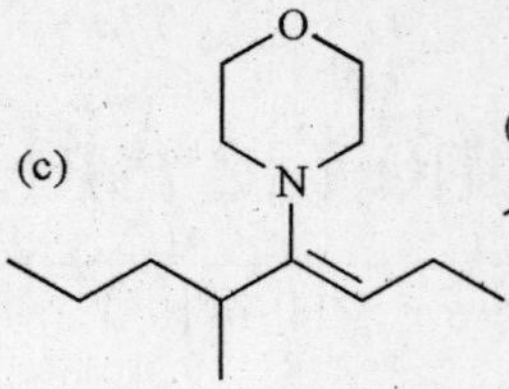

(d) N

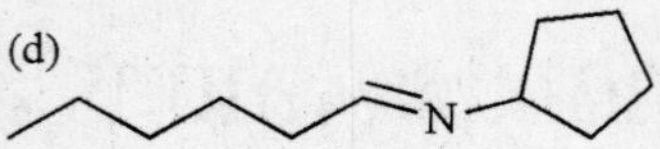

(e) Li, N

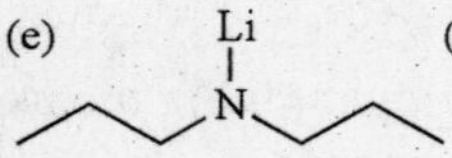

(f) NEt_2

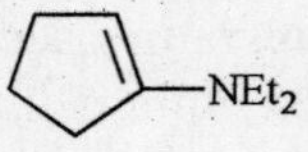

(g) O, OH

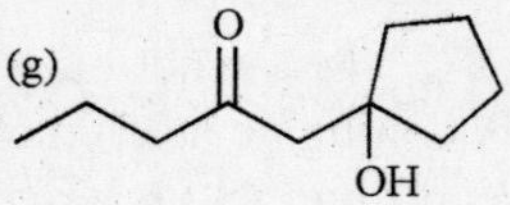

(h) O, OH

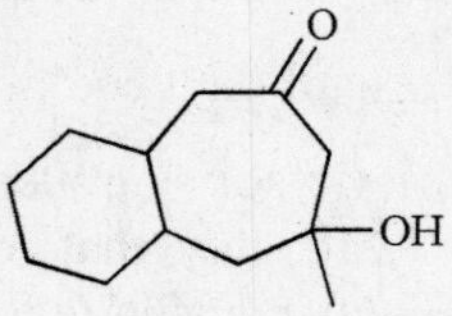

(i) H, O

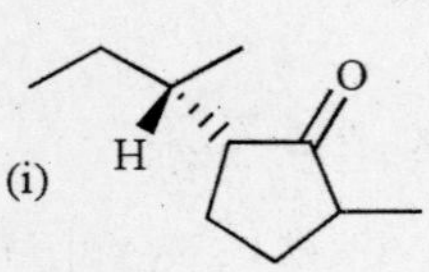

(j) C_6H_{13}, OH or C_6H_{13}, O

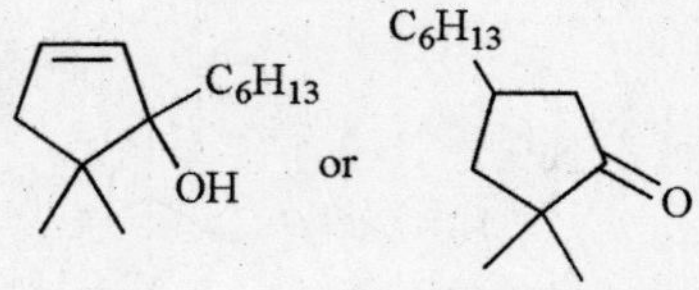

(k) C_4H_9, OH, Me, O

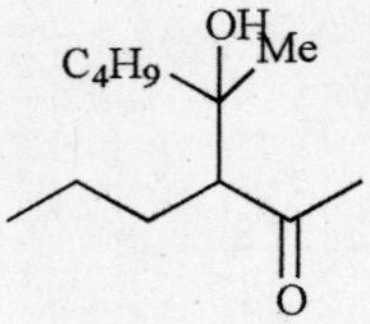

(l) O, O

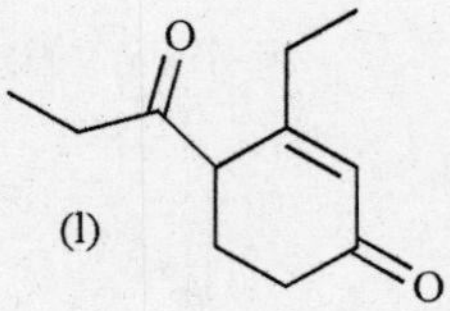

(m) OH, O

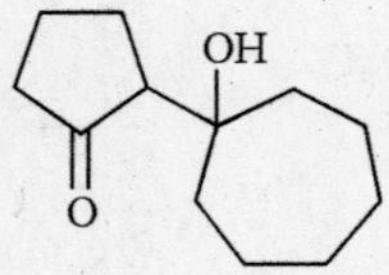

(n) CO_2H and OH

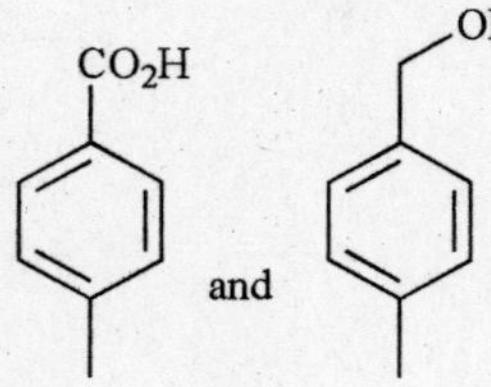

(o) O, OH

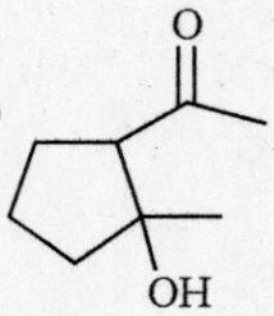

(p) N, I-, C_5H_{11}

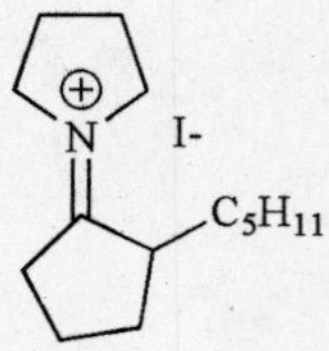

(q) OH, CHO

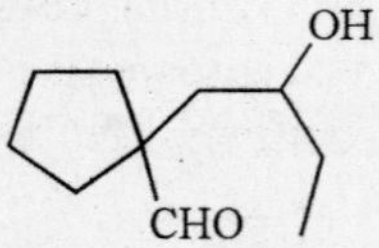

(r) HO, O

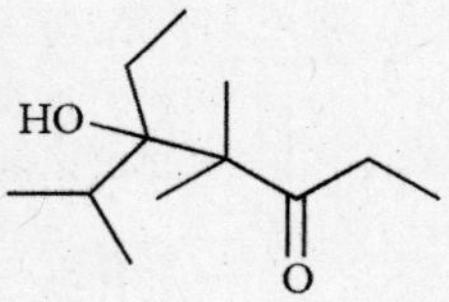

(s) O, Ph

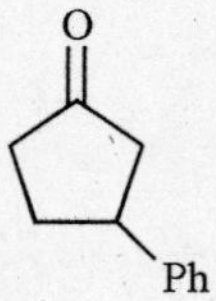

(t) CHO, Ph

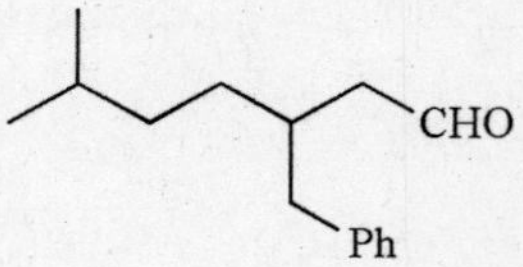

9.

(a) a → CHO, CHO; b → CHO; c → OH; d → OMe

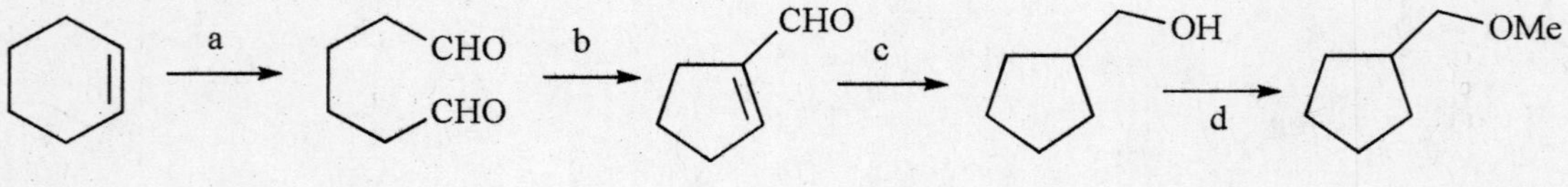

(a) i. O_3 ii. Me_2S (b) i. NaOEt, EtOH ii. H_3O^+, heat (c) 2 H_2, PtO_2 (d) i. NaH, THF ii. MeI

(b) a → CHO; b → CHO; c → Ph, OH

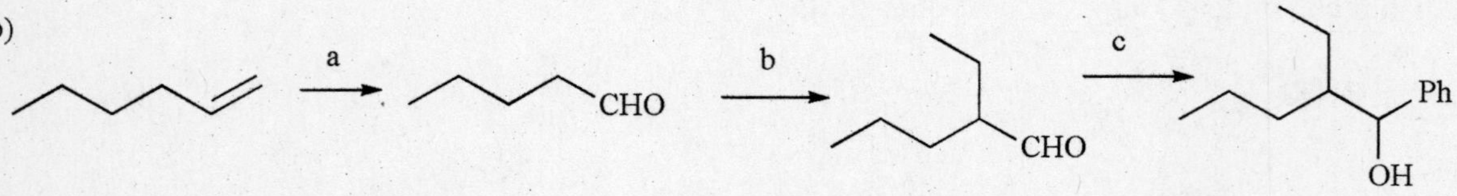

(a) i. O_3 i. Me_2S (b) i. LDA, THF, -78°C i. EtBr (c) i. PhMgBr, THF ii. H_3O^+

Chapter 24

Carboxylic Acids

1. (a) 4-ethyl-2-(1-methylbutyl)heptanoic acid
   (b) 3,4,5-trimethylbenzoic acid
   (c) 5,5-dichloro-2-methyl-7-phenylheptanoic acid
   (d) hex-2-ynoic acid
   (e) 4-methylhex-3Z-enoic acid
   (f) 2-ethylhep-6-ynoic acid
   (g) 3-butyl-6-ethyl-2-pentyldecanoic acid
2. The strongest acid is 2-chlorohexanoic acid, where the Cl is closest to the carboxyl group.
3. When the Cl is in the *ortho* position, there is a strong through-space inductive effect that strengthens the acid. This through-space effect is not possible when the Cl is in the *para* position.
4.

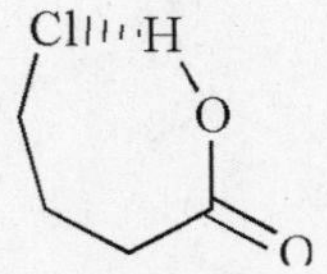

Cl H O O

5. Water can solvate the two ions that are formed (H_3O^+ and acetate) much better than ethanol.
6. The iodoform test is specific for methyl ketones and methyl carbinols (CH[OH]Me), since a methyl group is required to eventually generate iodoform (CHI_3). A positive iodoform test (precipitation of the yellow solid, iodoform) indicates the presence of a methyl ketone unit in an unknown ketone.
7. When the OH of an acid is in the C_4 position, it can attack the carbonyl and cyclize to form a lactone. The product of this reaction is γ-butyrolactone, which does not have an acidic hydrogen.
8.

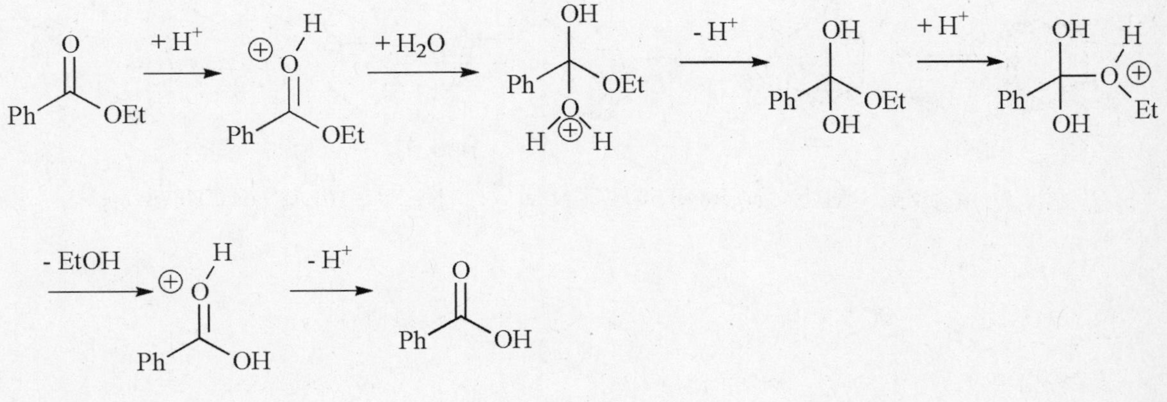

O Ph OEt $+ H^+$ → ⊕ H O Ph OEt $+ H_2O$ → OH Ph OEt O⊕ H H $- H^+$ → OH Ph OEt OH $+ H^+$ → OH H O⊕ Ph Et OH

$- EtOH$ → ⊕ H O Ph OH $- H^+$ → O Ph OH

9.

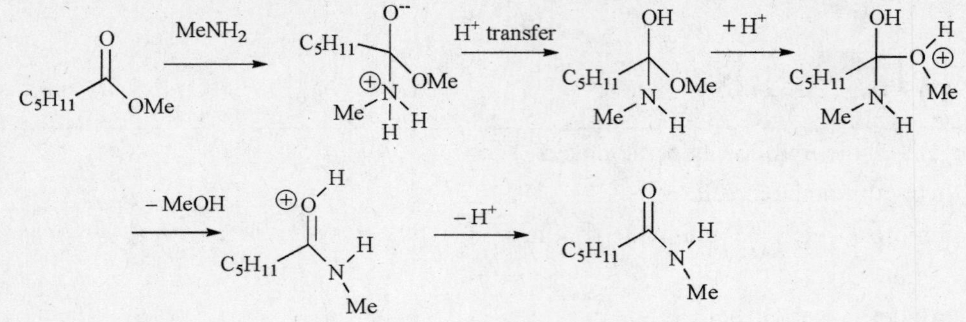

$C_5H_{11}C(=O)OMe$ —$MeNH_2$→ $C_5H_{11}C(O^-)(OMe)(N^+H_2Me)$ —H⁺ transfer→ $C_5H_{11}C(OH)(OMe)(NHMe)$ —$+H^+$→ $C_5H_{11}C(OH)(O^+HMe)(NHMe)$

—$-MeOH$→ $C_5H_{11}C(=O^+H)NHMe$ —$-H^+$→ $C_5H_{11}C(=O)NHMe$

10. (a) 2-propyl-1,3-propanedioic acid (2-propylmalonic acid); (b) 2-ethyl-1,4-butanedioic acid (2-ethylsuccinic acid); (c) 3,4-diphenyl-1,6-hexanedioic acid (3,4-diphenylglutaric acid).

11. The first pK_a of malonic acid is lower (stronger acid) because the electron withdrawing carboxyl group is closer to the first carboxyl group.

12. (a) Phthalic acid; (b) Terephthalic acid

13.

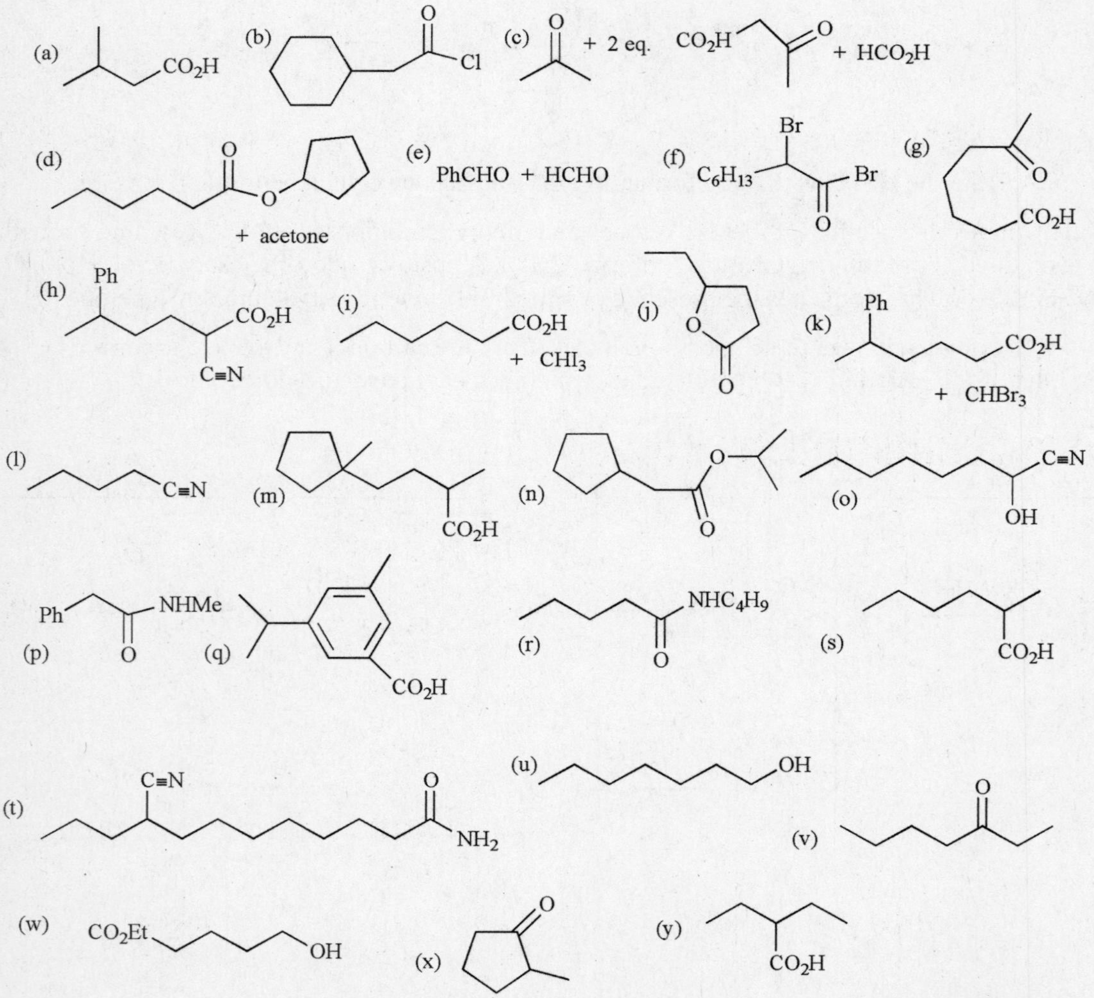

(a) CO_2H

(b) Cl

(c) + 2 eq. CO_2H + HCO_2H

(d) + acetone

(e) PhCHO + HCHO

(f) C_6H_{13} Br Br

(g) CO_2H

(h) Ph CO_2H C≡N

(i) CO_2H + CHI_3

(j)

(k) Ph CO_2H + $CHBr_3$

(l) C≡N

(m) CO_2H

(n)

(o) C≡N OH

(p) Ph NHMe

(q) CO_2H

(r) NHC_4H_9

(s) CO_2H

(t) C≡N NH_2

(u) OH

(v)

(w) CO_2Et OH

(x)

(y) CO_2H

14.

(a)

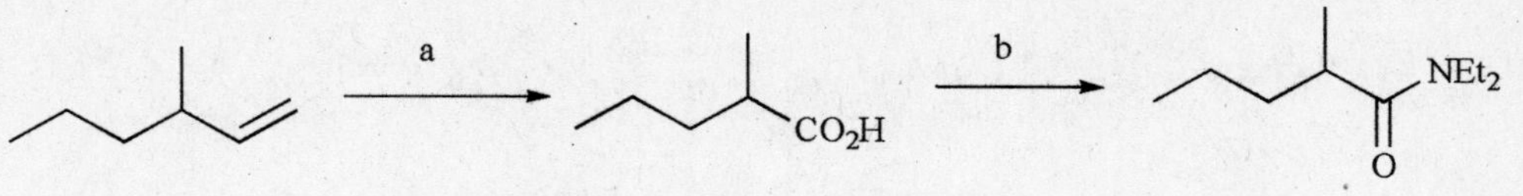

CO_2H NEt$_2$ O

(a) i. O_3 ii. H_2O_2 (b) i. $SOCl_2$ ii. $NHEt_2$/pyridine

(b)

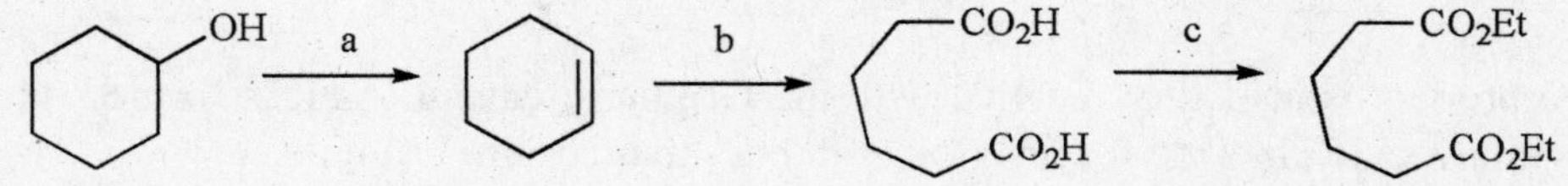

OH CO_2H CO_2H CO_2Et CO_2Et

(a) conc. H_2SO_4, heat (b) i. O_3 ii. H_2O_2 (c) EtOH, H^+

(c)

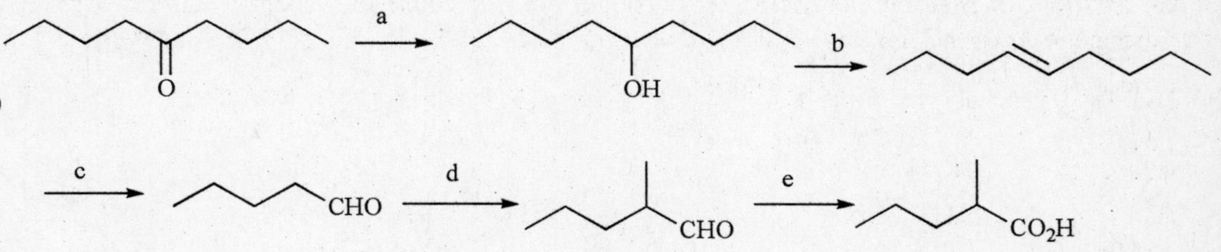

O OH CHO CHO CO_2H

(a) i. $LiAlH_4$ ii. H_2O (b) conc. H_2SO_4 (c) i. O_3 ii. Me_2S (d) i. LDA, THF, -78°C ii. MeI (e) CrO_3

Chapter 25

Carboxylic Acid Derivatives and Acyl Substitution Reactions

1.

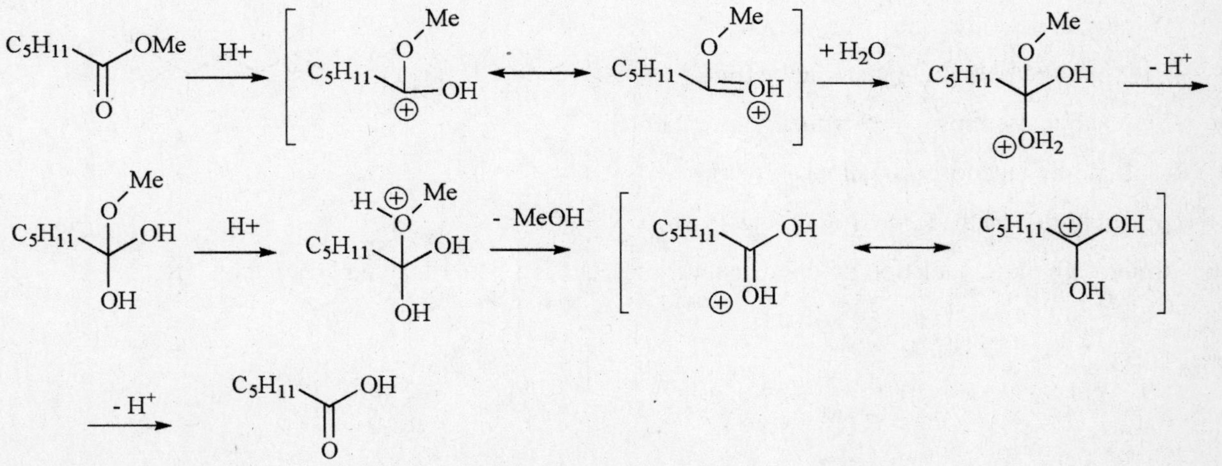

C_5H_{11} OMe O, H+, Me O C_5H_{11} OH ⊕, Me O C_5H_{11} =OH ⊕, + H_2O, Me O C_5H_{11} OH ⊕OH_2, - H^+

Me O C_5H_{11} OH OH, H+, H ⊕ Me O C_5H_{11} OH OH, - MeOH, C_5H_{11} OH ⊕ OH, C_5H_{11} ⊕ OH OH

- H^+, C_5H_{11} OH O

2.

(a) 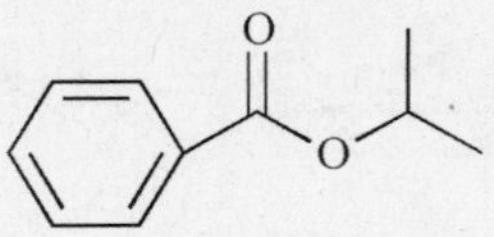 (b)

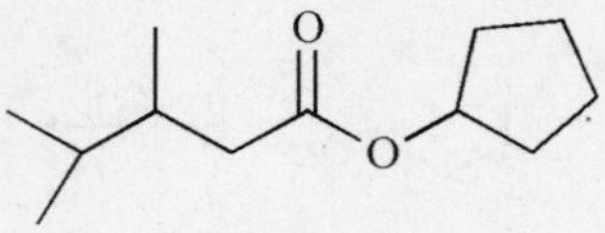

(c) 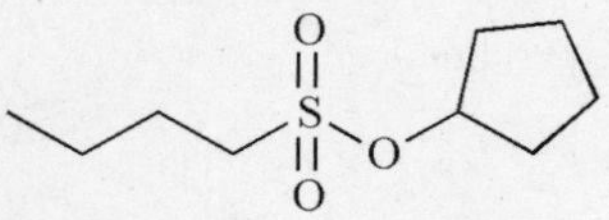 (d) 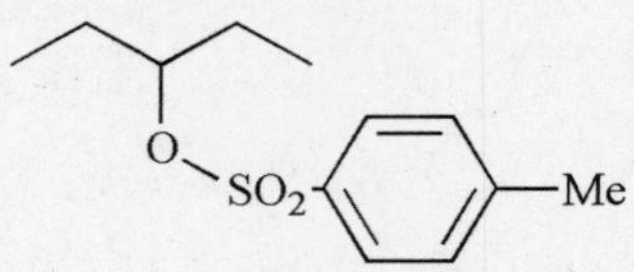 O SO$_2$ Me

3. The ketone product is also very reactive with the Grignard reagent. As it is formed, the ketone can compete with the acid chloride for reaction with butylmagnesium bromide.

4.

(a) 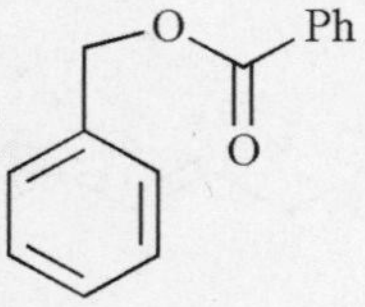 O Ph O (b) 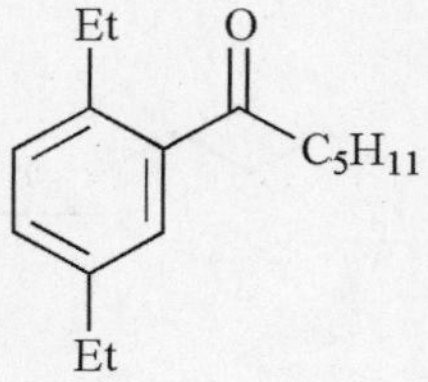 Et O C_5H_{11} Et (c) No Reaction (d) 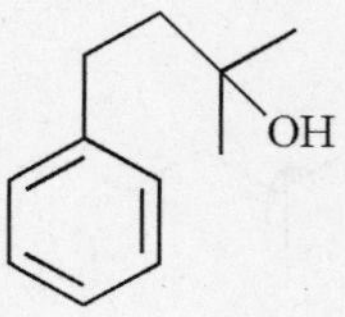 OH

(e) 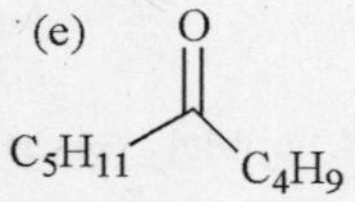 O C_5H_{11} C_4H_9 (f) 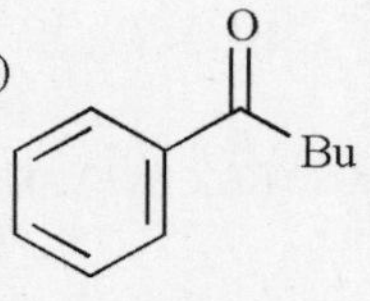 O Bu (g) 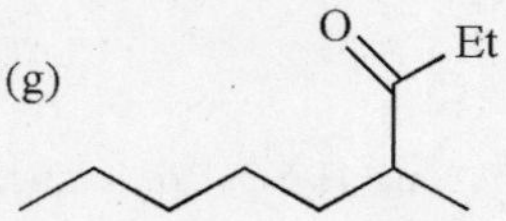 O Et (h) 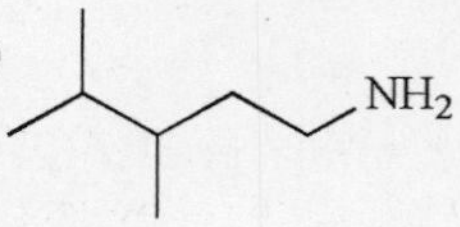 NH_2

Reaction (c) gives no reaction, since the two nitro groups deactivate the benzene ring too much for a Friedel-Crafts acylation reaction to occur.

5. (a) 2-ethyl-3,5-dimethylheptanoyl chloride
   (b) 3-methylbutyl-4,4-diphenylhexanoate
   (c) *N*-(1-methylethyl)-succinimide
   (d) 2,4,5-triethylbenzamide
   (e) 2-chloro-4,4-diphenylpentanolactone
   (f) propanoic anhydride
   (g) 2,2,4,4-tetramethylglutaric anhydride
   (h) *N*-ethyl, *N*-propyl-3,4,5-trimethylhexanamide
   (i) 2-methylethanoic propanoic anhydride
   (j) *N*-butyl-5-methyl-2-pyrrolidinone
6. Amines are more nucleophilic than alcohols, and OR is a better leaving group than NR_2.

7.

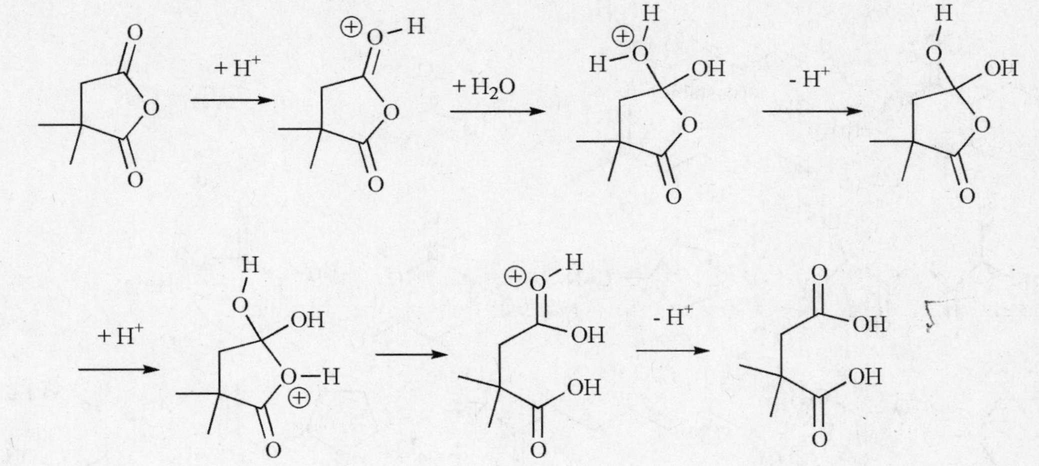

$\xrightarrow{+H^+}$ $\xrightarrow{+H_2O}$ $\xrightarrow{-H^+}$

$\xrightarrow{+H^+}$ $\longrightarrow$ $\xrightarrow{-H^+}$

8.

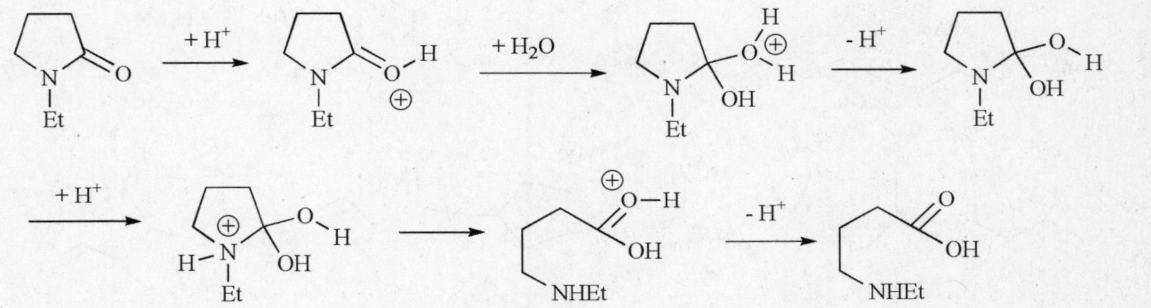

$\xrightarrow{+H^+}$ $\xrightarrow{+H_2O}$ $\xrightarrow{-H^+}$

$\xrightarrow{+H^+}$ $\longrightarrow$ $\xrightarrow{-H^+}$

NHEt NHEt

Although the mechanism is drawn with the NHEt group being neutral, in acid solution this basic amine will certainly be protonated to give the ammonium salt. It is likely that proton transfers can occur to the amine rather than intermolecularly as shown. It is also likely that the ammonium salt is the actual acid catalyst in this reaction.

9.

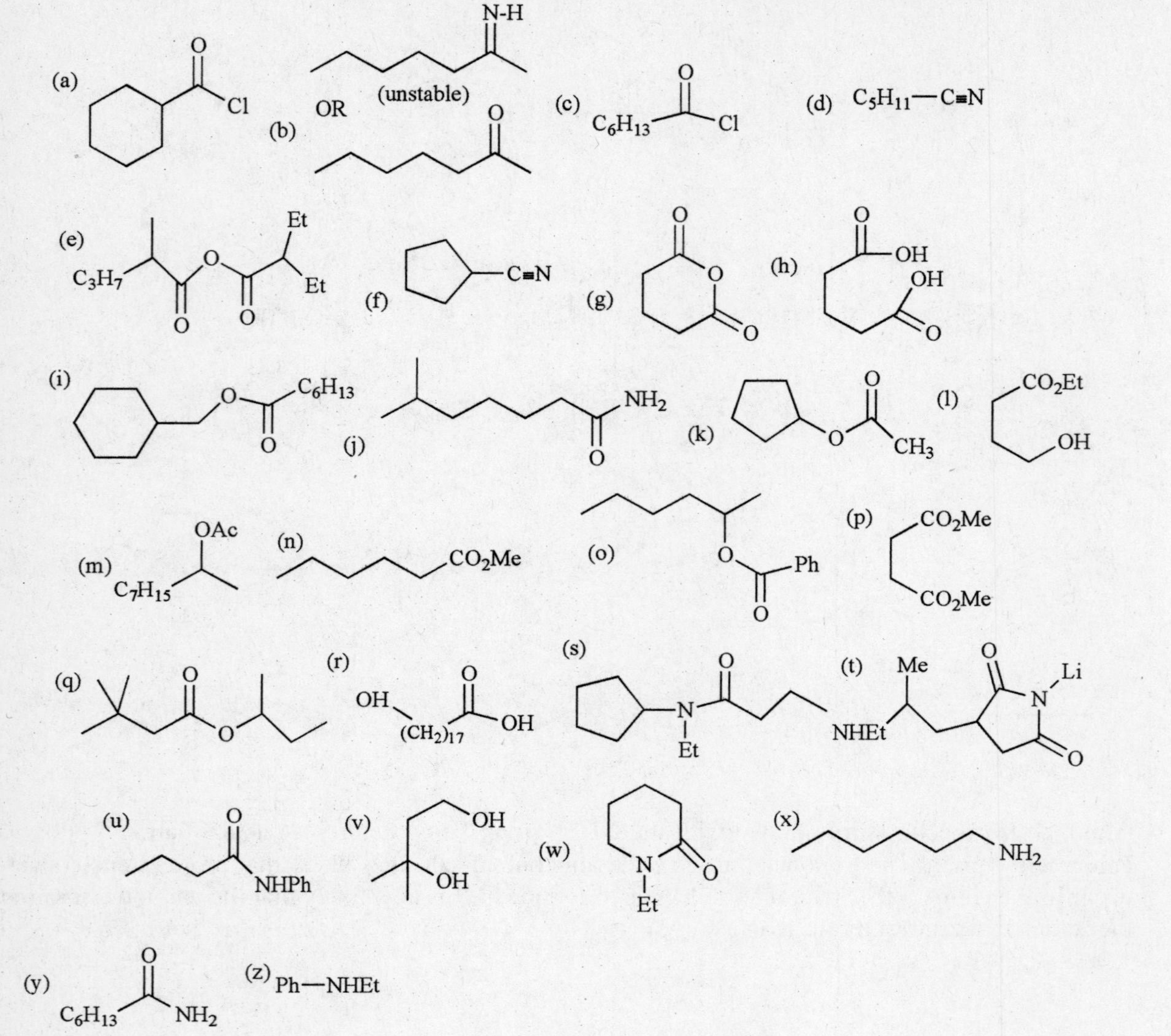

(a) O, Cl

(b) N-H (unstable) OR O

(c) O, C_6H_{13}, Cl

(d) C_5H_{11}—C≡N

(e) Et, C_3H_7, O, Et, O, O

(f) C≡N

(g) O, O, O

(h) O, OH, OH, O

(i) O, C_6H_{13}, O

(j) NH_2, O

(k) O, O, CH_3

(l) CO_2Et, OH

(m) OAc, C_7H_{15}

(n) CO_2Me

(o) O, Ph, O

(p) CO_2Me, CO_2Me

(q) O, O

(r) OH, $(CH_2)_{17}$, O, OH

(s) O, N, Et

(t) Me, O, Li, N, NHEt, O

(u) O, NHPh

(v) OH, OH

(w) N, O, Et

(x) NH_2

(y) O, C_6H_{13}, NH_2

(z) Ph—NHEt

10.

(a)

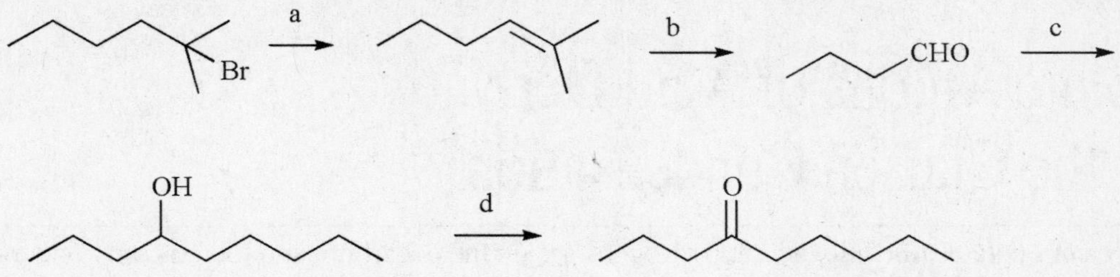

(a) KOH, EtOH (b) i. O_3 ii. Me_2S (c) i. $C_5H_{11}MgBr$ ii. H_3O^+ (d) CrO_3, H^+

(b)

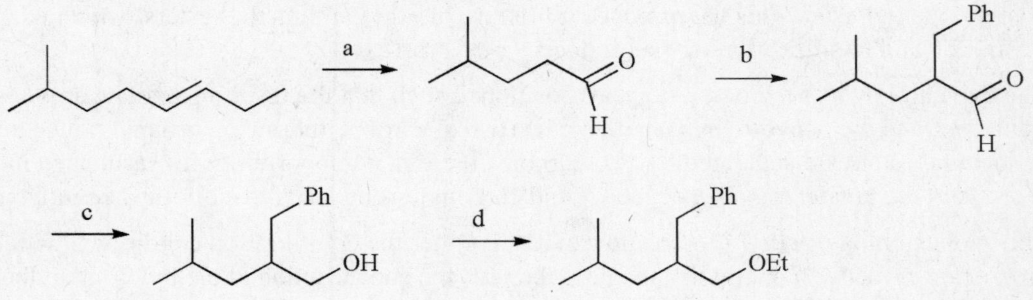

(a) i. O_3 ii. Me_2S (b) i. LDA, THF, -78°C ii. $PhCH_2Br$
(c) i. $NaBH_4$ ii. aq. NH_4Cl (d) i. NaH, THF ii. EtBr

11. The product is the ethyl ester of pentanoic acid, ethyl pentanoate, **A**

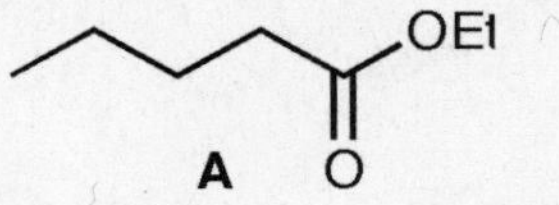

12. The functional group is the ester group, $R'CO_2R$, where R is the alcohol part of the ester and R' is the carboxylic acid part of the ester.

13.

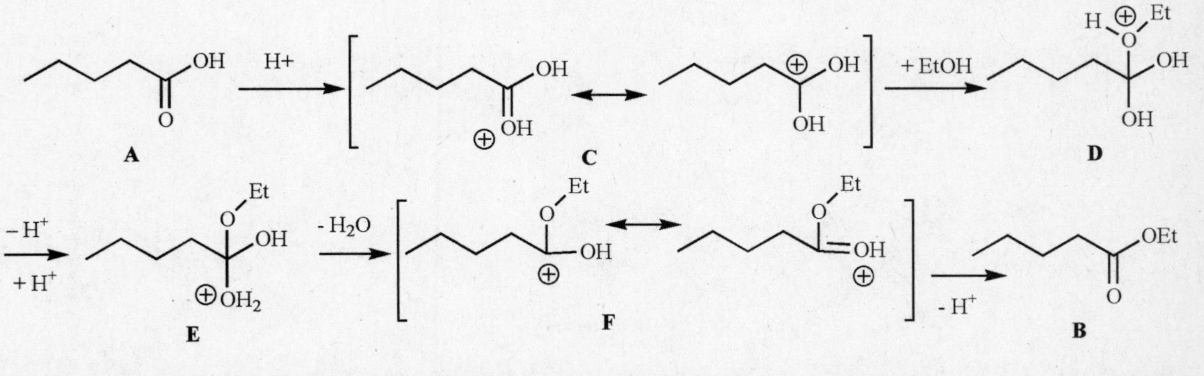

14. In the first reaction, benzoic acid (**A**) reacts with methanol to form the ester, methyl benzoate [$PhCO_2CH_3$]. In the second reaction, 1-heptanol (**B**) reacts with butanoic acid to give heptyl butanoate, [$CH_3CH_2CH_2CO_2(CH_2)_6CH_3$].

15. The product of this reaction is ethyl propanoate [$CH_3CH_2CO_2Et$].

Chapter 26

Enolate Anions of Acid Derivatives and the Claisen Condensation

1. For ester enolates, the only acidic hydrogens are on the α-carbon, and there is only one α-carbon. The kinetic and thermodynamic enolates are the same. The different conditions may be important for mixed Claisen condensations, however, in order to prevent or minimize unwanted cross coupling reactions.
2. If methanolic sodium methoxide is used as a base with an ethyl ester, transesterification can occur to give the methyl ester. This can produce a mixture of ethyl and methyl esters, which can complicate separation and identification of the products.
3. The two carbonyls in malonic esters are positioned such that the resulting enolate anion is resonance stabilized, and the α-hydrogen is more acidic (two electron withdrawing groups). Since the carbonyls in succinic esters are separated by two carbons, there is no opportunity for resonance in the enolate anion, and the greater distance of the second carbonyl group leads to a diminished inductive effect.
4. The enolate anion of ethyl propionate can react, under the equilibration conditions, with either ethyl propionate or methyl 2-methylbutanoate. Likewise, the enolate anion of methyl 2-methylbutanoate can react with ether ethyl propionate or methyl 2-methylbutanoate. This leads to four different Claisen condensation products.
5.

(a) No reaction (no α-proton)

(b) 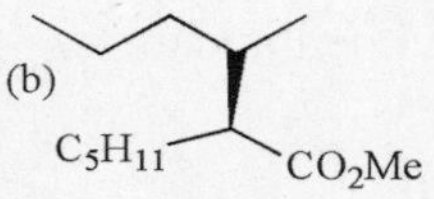 C_5H_{11} CO_2Me

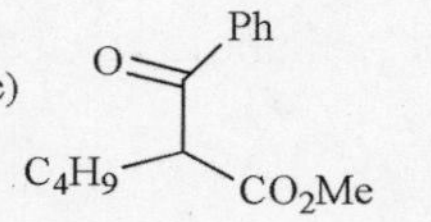

(c) Ph O C_4H_9 CO_2Me

(d) 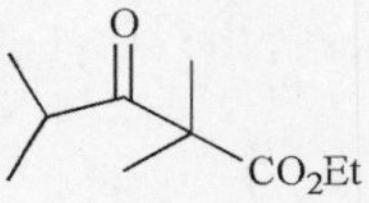 O CO_2Et

(e) 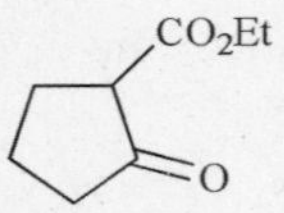 CO_2Et O

(f) 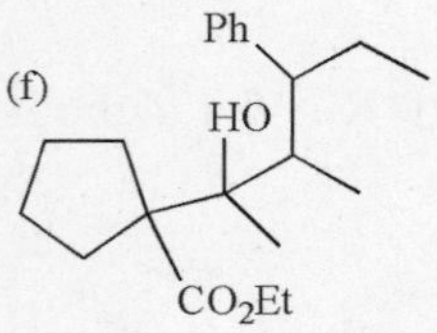 Ph HO CO_2Et

(g) 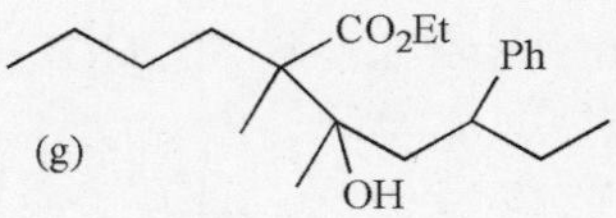 CO_2Et Ph OH

(h) 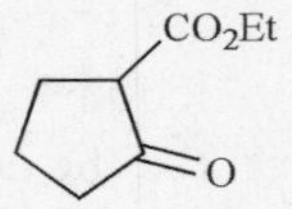 CO_2Et O

(i) 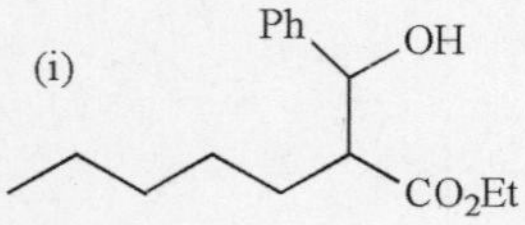 Ph OH CO_2Et

(j) 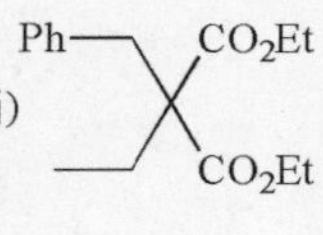 Ph CO_2Et CO_2Et

(k) 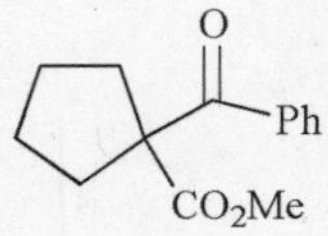 O Ph CO_2Me

(l) 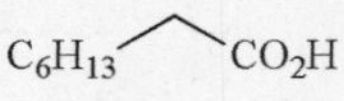 C_6H_{13} CO_2H

(m) 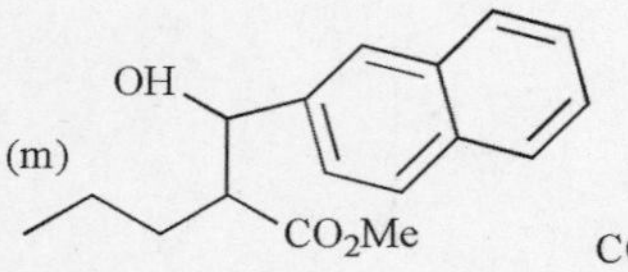 OH CO_2Me

(n) 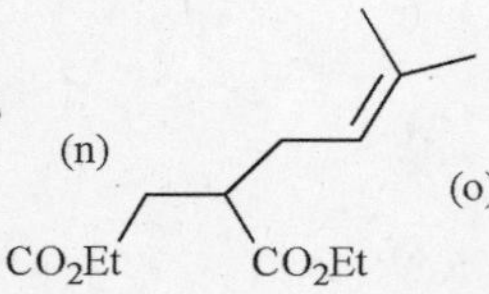 CO_2Et CO_2Et

(o) Br C_5H_{11} Br O

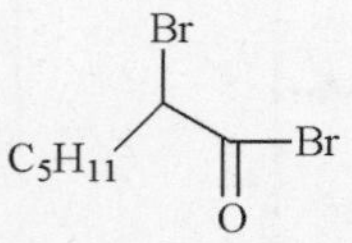

(p) 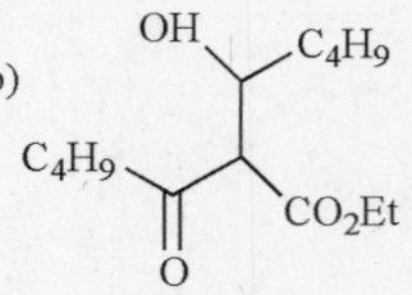 OH C_4H_9 C_4H_9 O CO_2Et

(q) 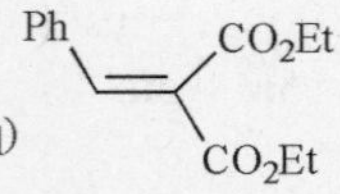 Ph CO_2Et CO_2Et

(r) 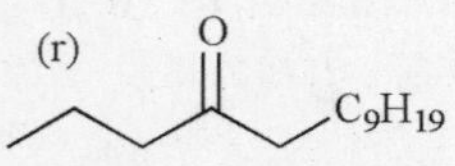 O C_9H_{19}

(s) 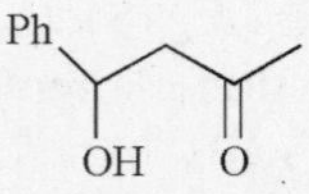 Ph OH O

(t) 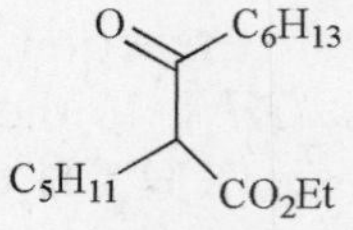 O C_6H_{13} C_5H_{11} CO_2Et

Reaction (a) gives N.R.; there is no α-hydrogen.

6.

(a)

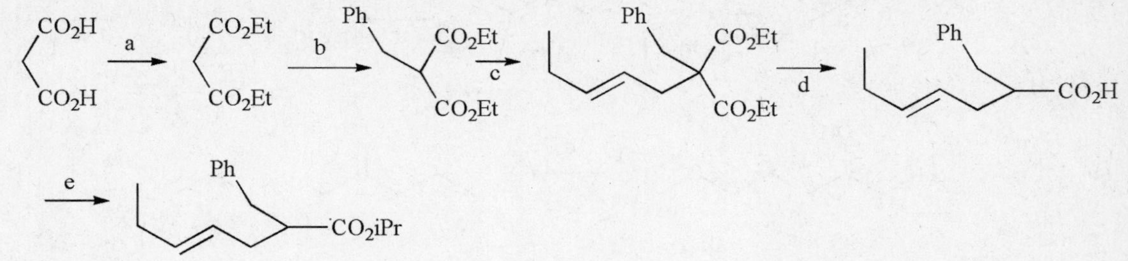

a, b, c, d, e

Ph CO_2H CO_2Et CO_2iPr

(a) EtOH, H^+ (b) i. NaOEt ii. $PhCH_2Br$ (c) i. NaOEt ii. 1-bromo-2E-pentene
(d) i. saponification (ii) 200°C (e) iPrOH, cat. H^+

(b)

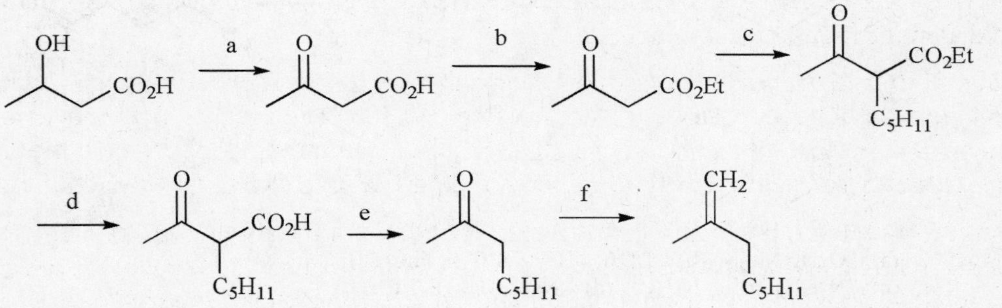

OH O CO_2H CO_2Et C_5H_{11} CH_2

a, b, c, d, e, f

(a) CrO_3, H^+ (b) EtOH, H^+ (c) i. NaOEt ii. $C_5H_{11}Br$ (d) saponification (e)200°C (f) $Ph_3P{=}CH_2$

(c)

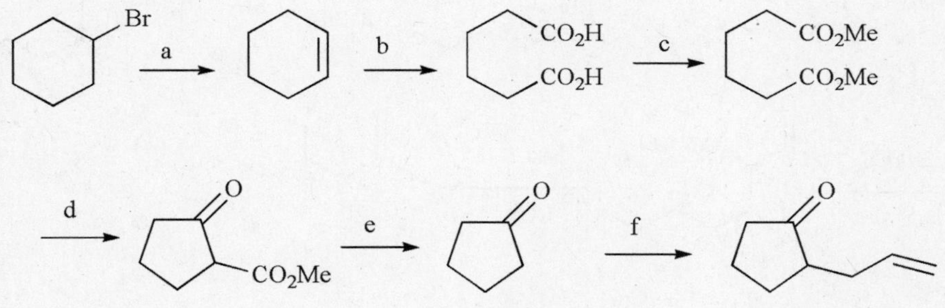

Br CO_2H CO_2Me O

a, b, c, d, e, f

(a) KOH, EtOH (b) i. O_3 ii. H_2O_2 (c) MeOH, H^+ (d) i. NaOMe ii. dil H_3O^+
(e) i. saponification ii. 200°C (f) i. LDA, THF, -78°C ii. allyl bromide

(d) Br —a→ —b→ CHO —c→

OH —d→ O

(a) KOH, EtOH (b) i. O_3 ii. Me_2S (c) i. $C_5H_{11}MgBr$ ii. H_3O^+ (d) CrO_3, H^+

(e) —a→ O H —b→ Ph O H

—c→ Ph OH —d→ Ph OEt

(a) i. O_3 ii. Me_2S (b) i. LDA, THF, -78°C ii. $PhCH_2Br$
(c) i. $NaBH_4$ ii. aq. NH_4Cl (d) i. NaH, THF ii. EtBr

7.

(a) O CO_2Et —1. NaOEt; 2. Ph Br→ O CO_2Et Ph —1. NaOEt; 2. EtBr→ O CO_2Et Et Ph

—1. aq. NaOH; 2. H_3O^+→ O COOH Et Ph —heat; $-CO_2$→ O Ph

(b) O CO_2Et —1. NaOEt, EtOH; 2. Br Br→ [O CO_2Et Br] —NaOEt→ [O ⊖ CO_2Et Br]

—$-Br^-$→ O CO_2Et —1. aq. NaOH; 2. H_3O^+; 3. heat, $-CO_2$→ O

Amines

1. Using the analogy of amines, the secondary phosphine (Pr_2PH) will be the most basic.
2. In (a), pyridine is more basic since the electron pair on nitrogen in pyrrole is part of the aromatic system. In (b), the electron pair on nitrogen is 'tied back' in the bicyclic amine, which is more basic than triethylamine, which exists as the fluxional isomers discussed in this chapter.
3. (a) *N*-ethyl-2,3-dimethylbutanamine (b) triheptylamine
   (c) 1-benzyl-2-methylpropanamine (d) *N*-methyl, *N*-pentyl-4-bromoaniline
   (e) *N*-benzyl-N,2-diphenylhexanamine (f) N-2,2-trimethylpyrrolidine
4. The Hinsberg test will distinguish this product as a secondary amine.
5. (a) 4-bromopyridine (b) 3,4-dimethyl-*N*-propylpyrrole
   (c) *N*-ethylpyridinium bromide (d) 2,3,4-triethylpyrrole
6.

(a) 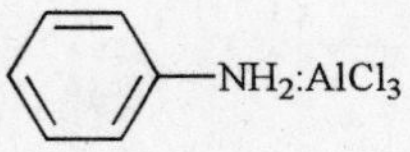 $NH_2{:}AlCl_3$ (b) 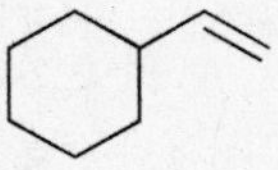 (c) 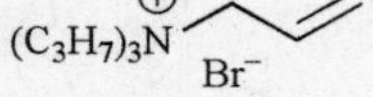 $(C_3H_7)_3N^{\oplus}$ Br^- (d)

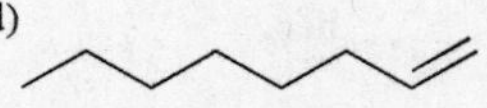

(e) 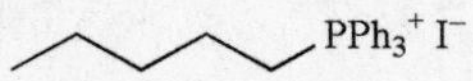 $PPh_3^+\ I^-$ (f) 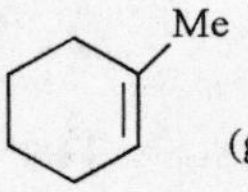 Me (g) 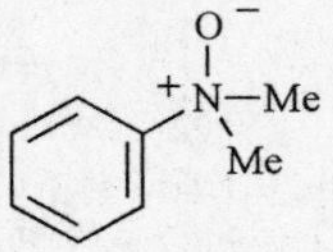 O^-, $^+N{-}Me$, Me (h)

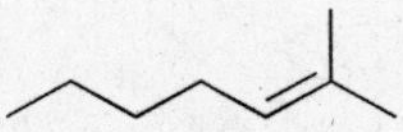

(i) 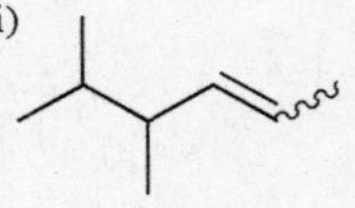 (j) 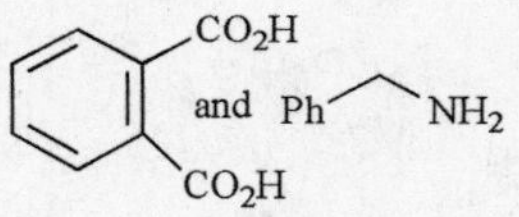 CO_2H, CO_2H and Ph NH_2 (k) 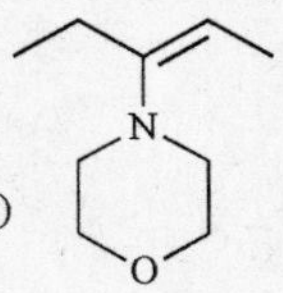 N, O (l) 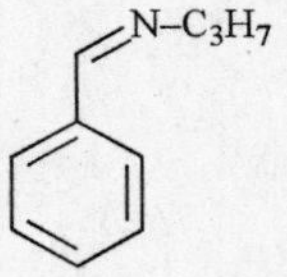 $N{-}C_3H_7$

(m) 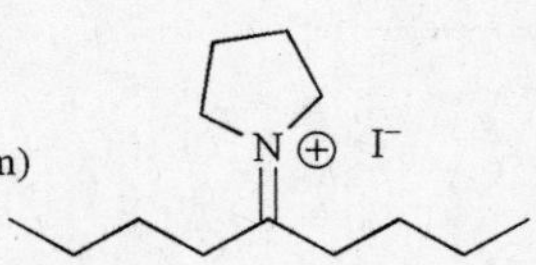 $N\oplus\ I^-$ (n) 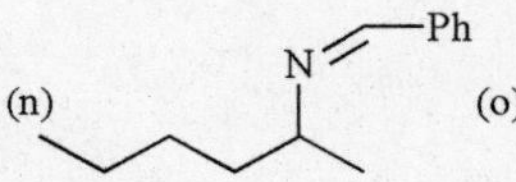 N, Ph (o) O, N

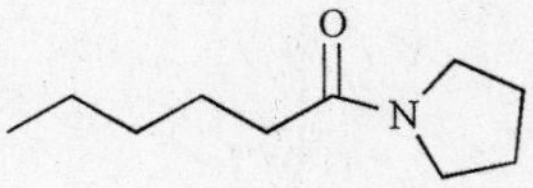

(p) 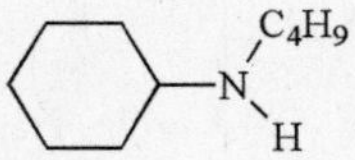 C_4H_9, N, H (q) 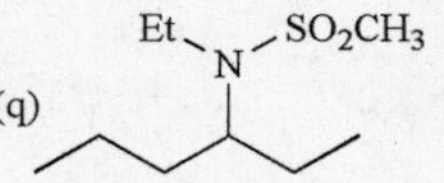 Et, SO_2CH_3, N (r)  C≡N (s) 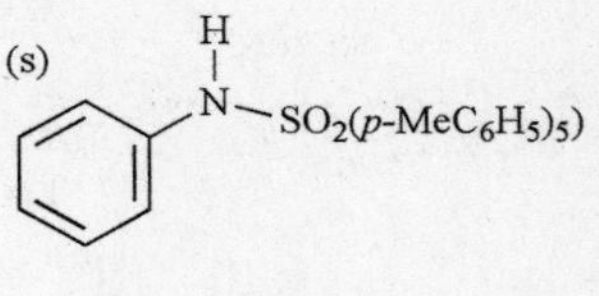 H, N, $SO_2(p\text{-}MeC_6H_5)_5)$

(t) 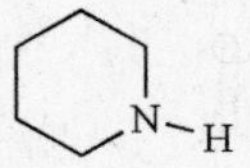 N–H (u) 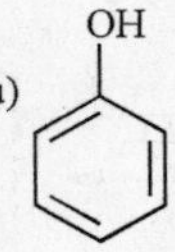 OH (v) 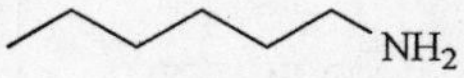 NH_2 (w) 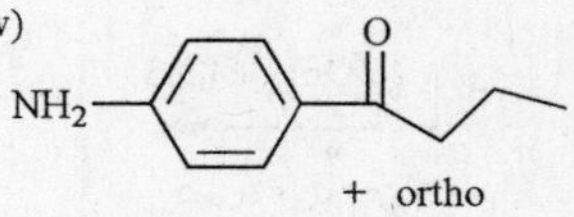 NH_2, O, + ortho

(x) 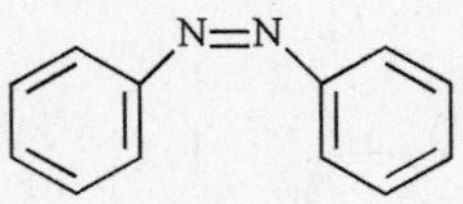 N=N (y)  Br (z) 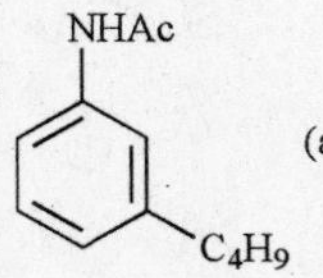 NHAc, C_4H_9 (aa) 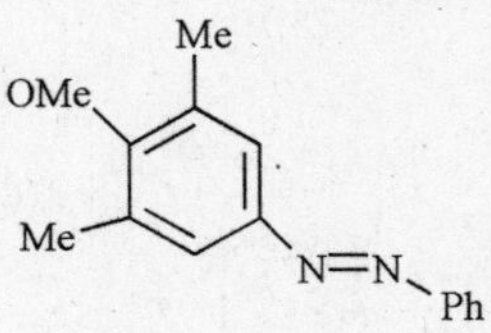 Me, OMe, Me, N=N, Ph

(ab) 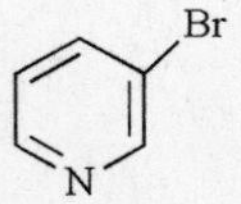 Br, N (ad) 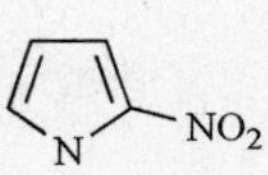 N, NO_2 (ae) 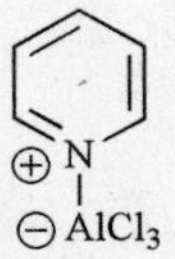 $\oplus N$, $\ominus AlCl_3$

In (ac), there is no reaction, since there are no aromatic hydrogens to be replaced.

7.

(a) [scheme]

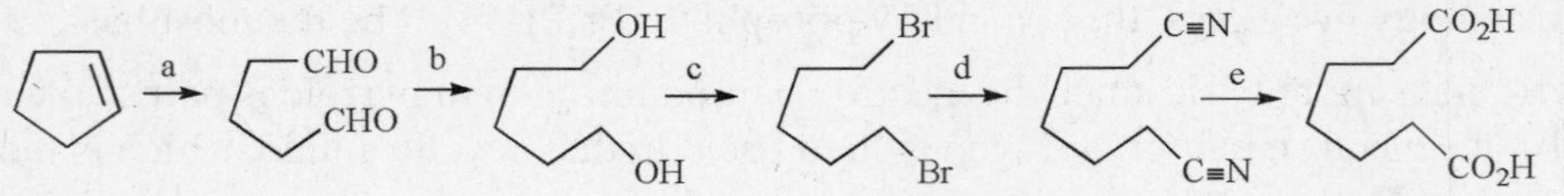

(a) i. O_3 ii. Me_2S (b) i. $NaBH_4$ ii. aq. NH_4Cl (c) excess PBr_3 (d) excess KCN, DMF (e) H_3O^+, heat

(b) [scheme]

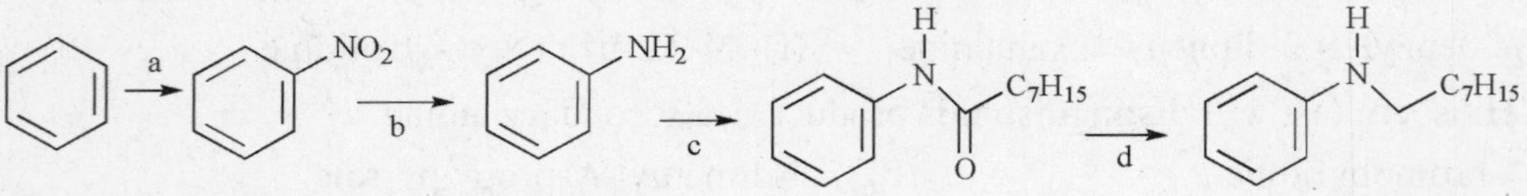

(a) HNO_3, H_2SO_4 (b) H_2, Ni (c) octanoyl chloride, pyridine (d) i. $LiAlH_4$ ii. aq. NaOH

(c) [scheme]

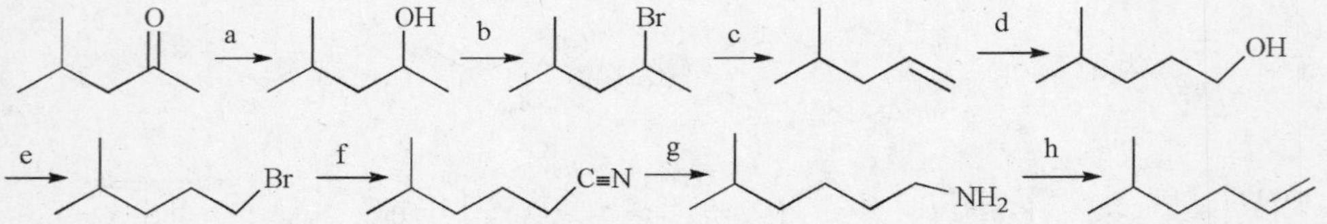

(a) i. $NaBH_4$ ii. aq. NH_4Cl (b) PBr_3 (c) i. NMe_3 ii. Ag_2O, H_2O iii. 150°C (d) i. B_2H_6 ii. NaOH, H_2O_2
(e) PBr_3 (f) KCN, DMF (g) i. $LiAlH_4$ ii. aq. NaOH (h) i. excess MeI ii. H_2O_2 iii. 150°C

(d) [scheme]

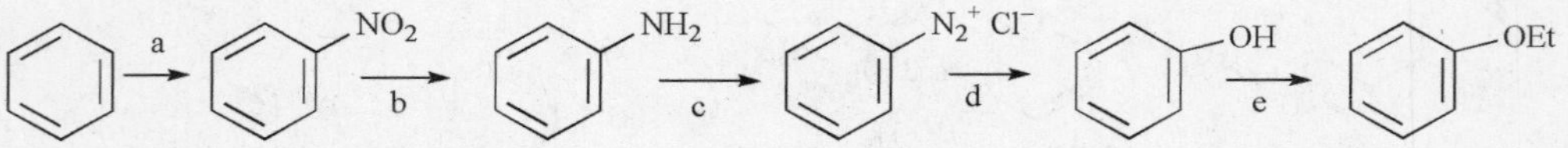

(a) HNO_3, H_2SO_4 (b) H_2, Ni (c) $NaNO_2$, HCl (d) H_2O, reflux (e) i. NaH, THF ii. Et-I

(e) [scheme]

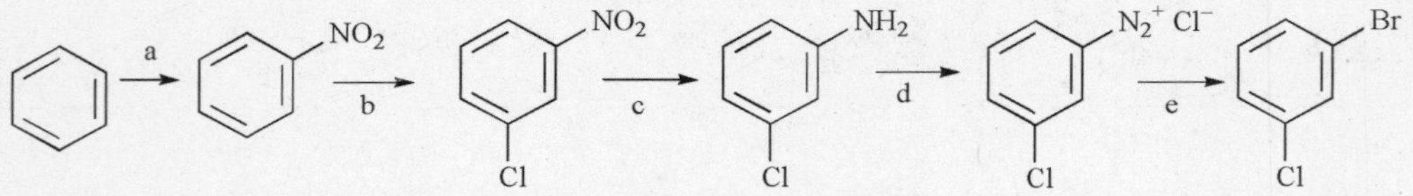

(a) HNO_3, H_2SO_4 (b) Cl_2, $AlCl_3$ (c) H_2, Pd (d) $NaNO_2$, HCl (e) CuBr

(f) [scheme]

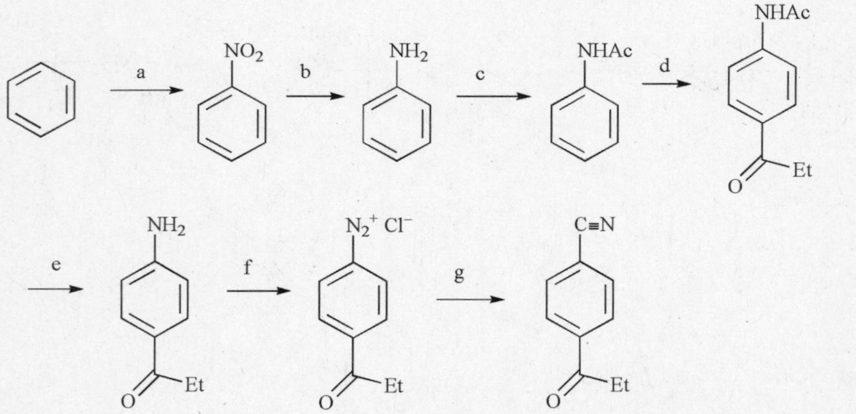

(a) HNO_3, H_2SO_4 (b) H_2, Ni (c) Ac_2O, pyridine (d) propanoyl chloride, $AlCl_3$ (e) saponification
(f) $NaNO_2$, HCl (g) CuCN

8. Initial protonation of cyclopentanone by the acid catalyst promotes the reaction by forming ***A***. This cation is attacked by diethylamine to form ammonium salt **B**. Proton transfer to the OH group produces **C**, which loses water to give 'cation' **D**. This 'cation' does not exist in this form, since the nitrogen donates the electron pair to form iminium salt **E**. When secondary amines are used, this iminium salt does not have a hydrogen attached to the nitrogen that can be removed to form an imine. Only primary amines will produce an iminium salt that can be converted to an imine. In the case of iminium salt **E**, the α-hydrogen is acidic (much as in a ketone that produces an enolate anion), and removal of that hydrogen with a base (usually the amine) generates the enamine, **F**. In fact, depending on the solvent, enamine **F** and iminium salt **E** are in equilibrium, with the equilibrium generally favoring the enamine.

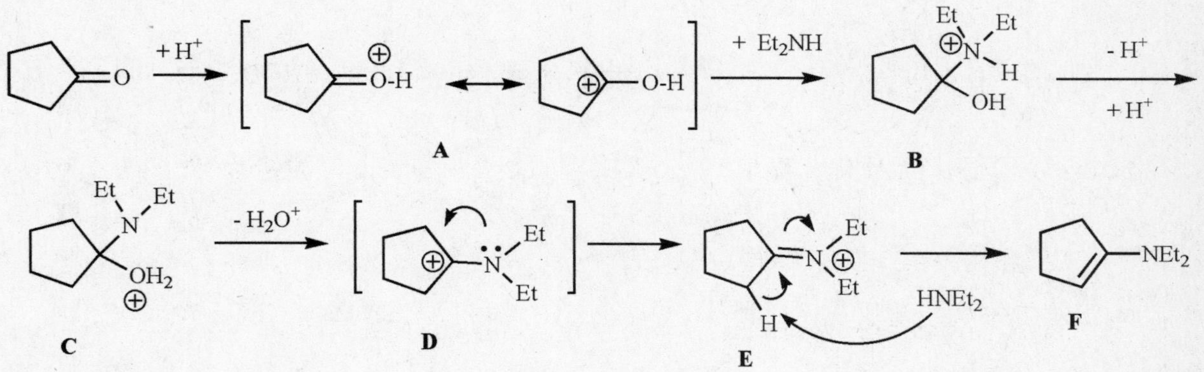

$+H^+$; $\oplus$ O-H ; $\oplus$ O-H ; **A** ; $+ Et_2NH$; Et, Et, $\oplus$N, H, OH ; **B** ; $- H^+$; $+H^+$

Et, Et, N, OH_2 $\oplus$; **C** ; $- H_2O^+$; $\oplus$ N Et Et ; **D** ; N$\oplus$ Et Et, H ; **E** ; $HNEt_2$; NEt_2 ; **F**

Chapter 28

Amino Acids, Peptides and Proteins

1. (a) 3-amino-2-(1-methylethyl)pentanoic acid;

   (b) 5-(*N,N*-dimethylamino)-4-methylheptanoic acid; (c) 4-(*N,N*-diethylamino)benzoic acid.

2.

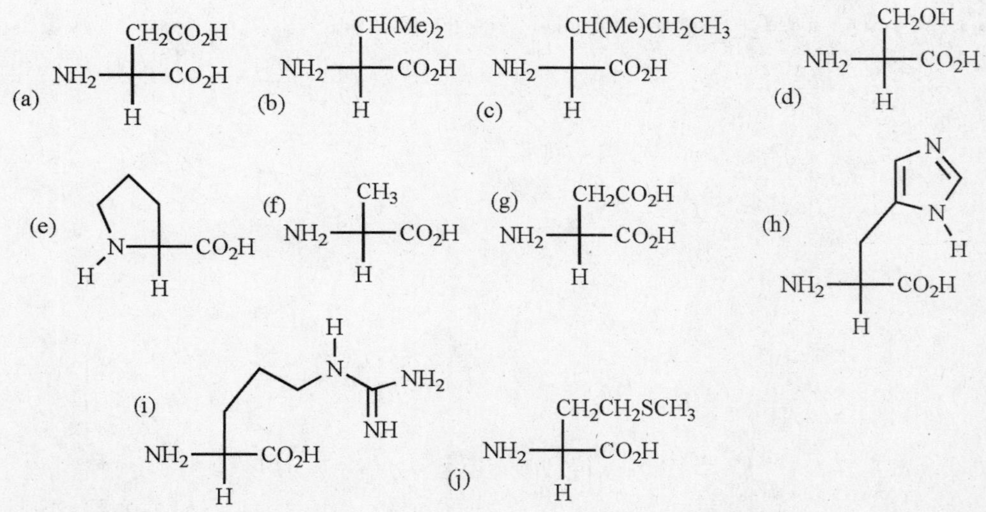

(a) CH_2CO_2H, NH_2, CO_2H, H

(b) $CH(Me)_2$, NH_2, CO_2H, H

(c) $CH(Me)CH_2CH_3$, NH_2, CO_2H, H

(d) CH_2OH, NH_2, CO_2H, H

(e) N, H, CO_2H, H

(f) CH_3, NH_2, CO_2H, H

(g) CH_2CO_2H, NH_2, CO_2H, H

(h) N, N, H, NH_2, CO_2H, H

(i) H, N, NH_2, NH, NH_2, CO_2H, H

(j) $CH_2CH_2SCH_3$, NH_2, CO_2H, H

3.

(a) 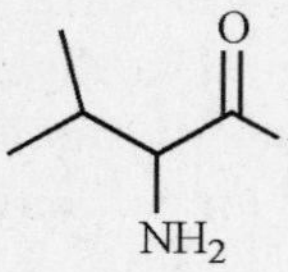 (b) 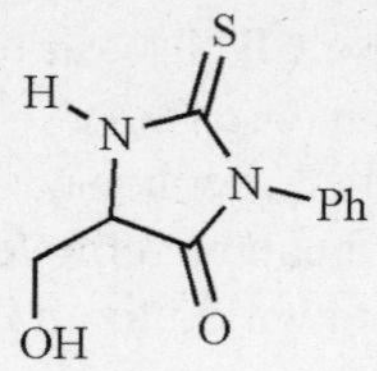 (c) 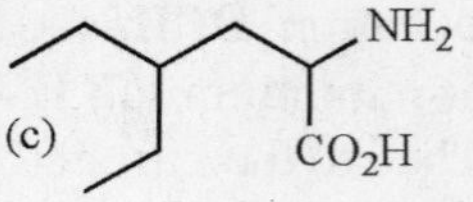 (d)

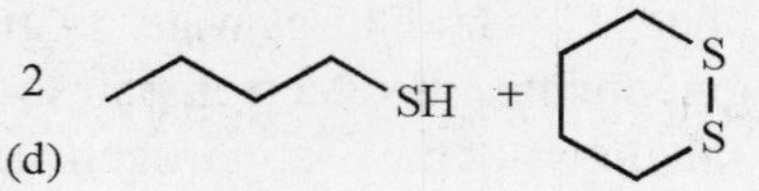

NH_2, NH_2, O; H, N, S, N–Ph, OH, O; NH_2, CO_2H; 2 SH + S S

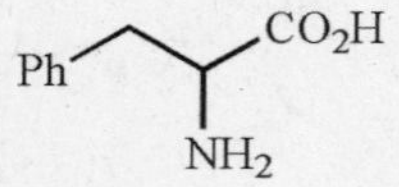

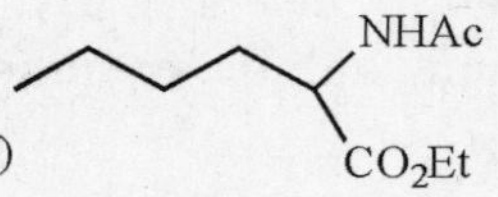

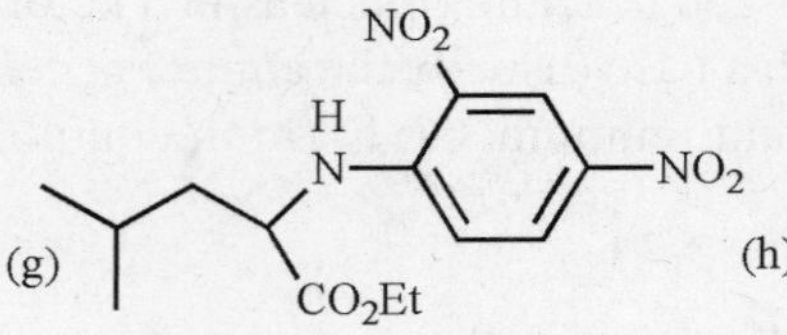

(e) Ph, CO_2H, NH_2 (f) NHAc, CO_2Et (g) NO_2, H, N, NO_2, CO_2Et (h) H, N, O, O, CO_2Et

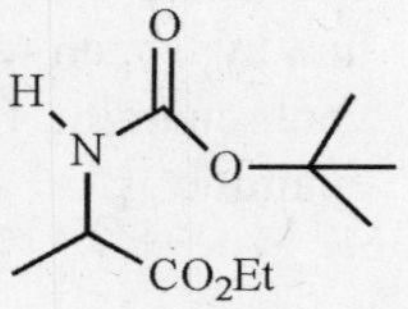

(i) 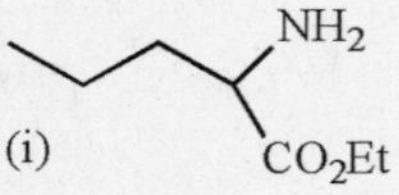 NH_2, CO_2Et (j) 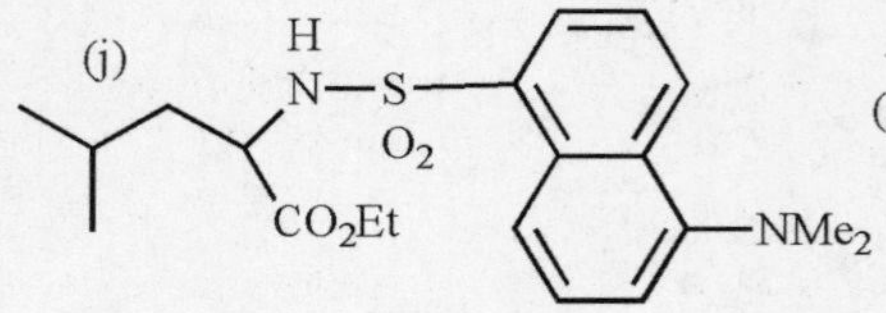 H, N–S, O_2, CO_2Et, NMe_2 (k) 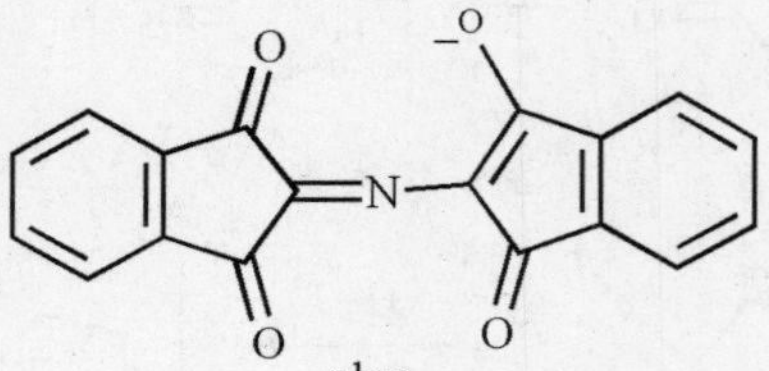 O, O, =N, ⁻O, O

plus

SMe, CHO

4.

(a)

C_4H_9 →(a) C_4H_9 OH →(b) C_4H_9 CHO →(c) C_4H_9 OH C≡N →(d) C_4H_9 NH_2 CO_2H

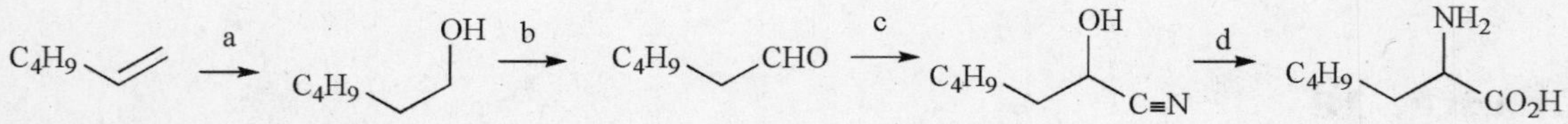

(a) i. B_2H_6 ii. NaOH, H_2O_2 (b) PCC, CH_2Cl_2 (c) NaCN, H^+ (d) i. aq. NH_4Cl, NH_3 ii.H_3O^+, heat ii. pH 8

(b) gly →(a) CBZ-gly-glu-OH →(b) CBZ-gly-glu-ala-OH →(c) NH_2-gly-glu-ala-phe-OH

(a) i. $PhCH_2COCl$ ii. glu-OEt iii. H_3O^+ (b) i. DCC, ala-OEt ii. H_3O^+ (c) i. DCC,phe-OEt ii. H_3O^+ iii. H_2, Pd

Chapter 29

Carbohydrates and Nucleic Acids

1.

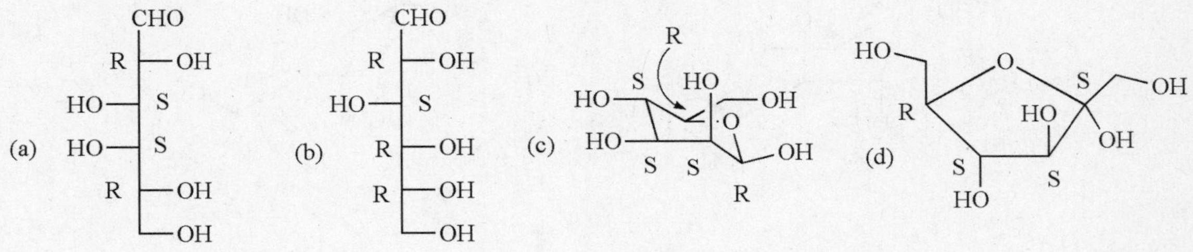

2. (a) D; (b) D; (c) L; (d) D; (e) L.

3.

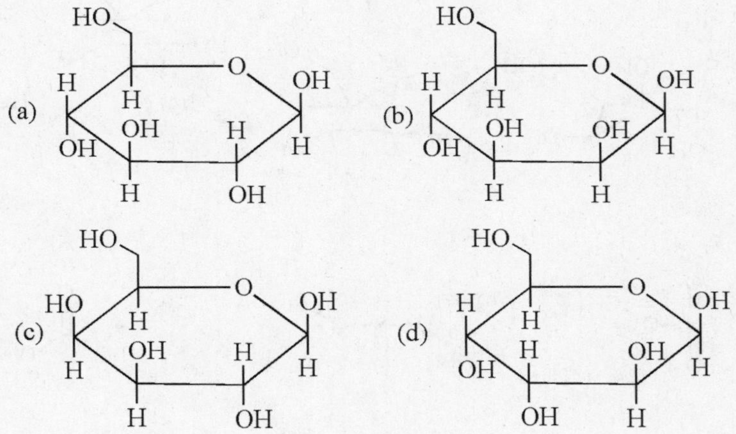

4.

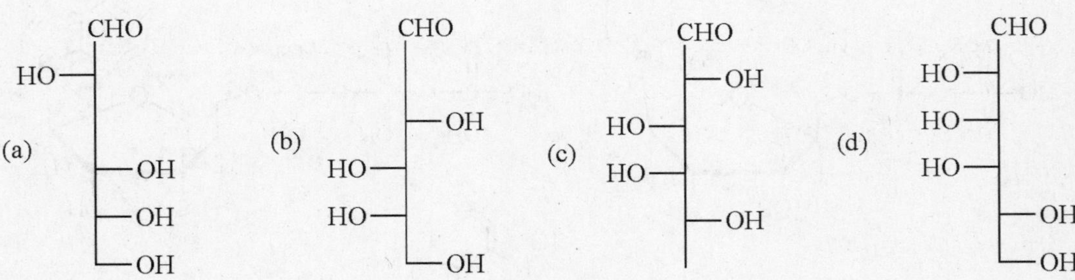

5.

(a) HO, OH, O, OH, OH, OH

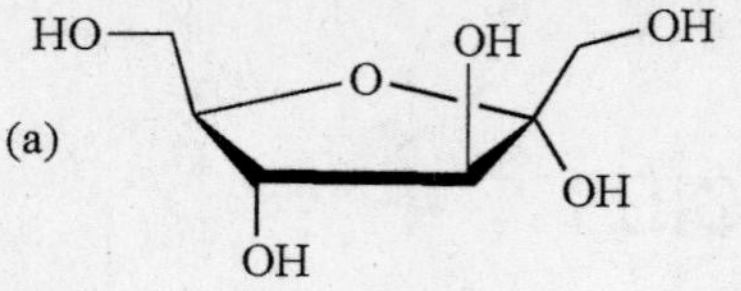

(b) HO, O, OH, OH, OH, OH

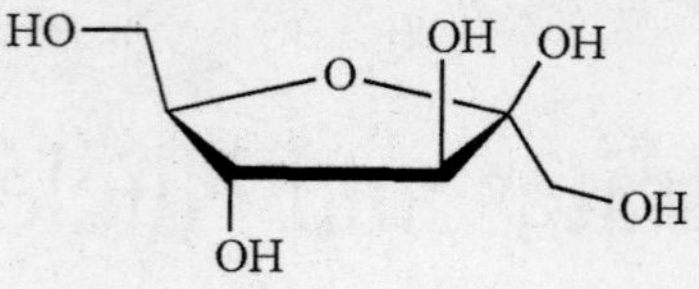

(c) HO, OH, O, OH, OH, OH

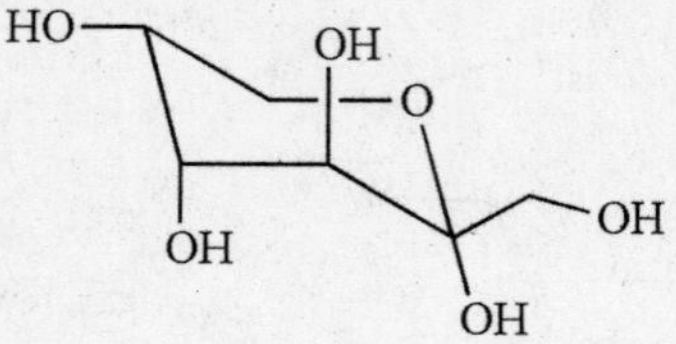

(d) HO, OH, O, OH, OH, OH

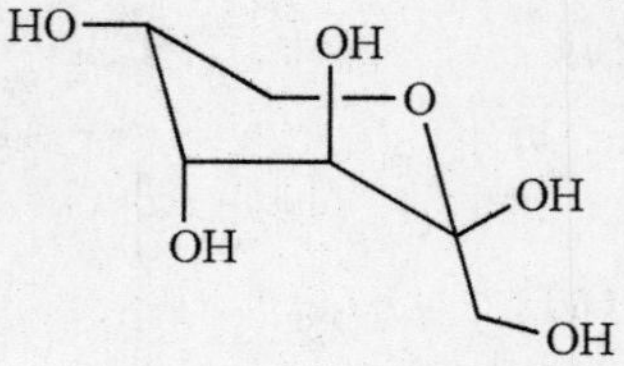

6.

(a) OH, HO, O, HO, HO, H, H, O, OH, O, OH, OH, OH

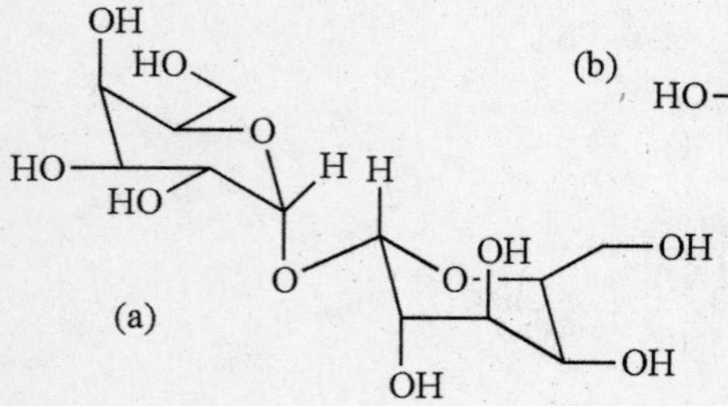

(b) OH, HO, OH, O, HO, H, O, H, OH, O, HO, HO, OH

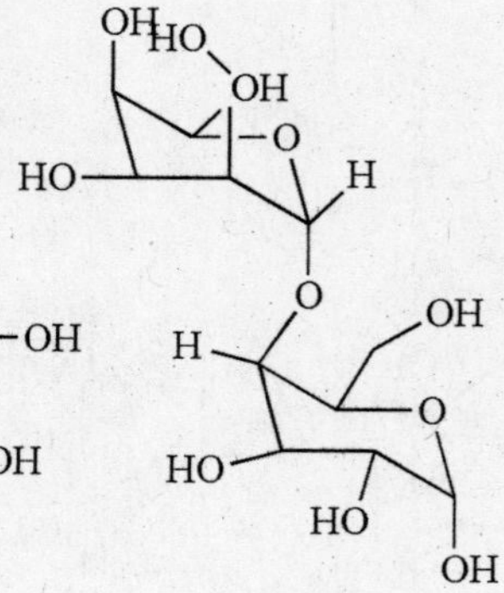

(c) HO, HO, O, HO, HO, O, HO, OH, O, HO, OH

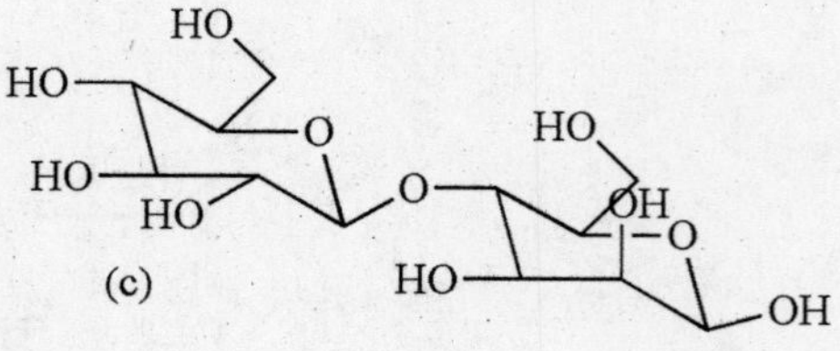

7.

(a) $H_3C-O-SO_2-O-CH_3$

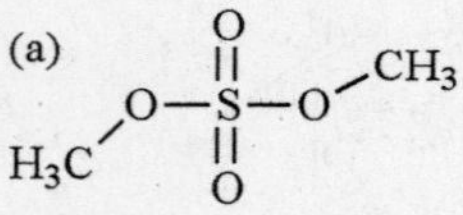

(b) $Et-O-SO_2-O-Et$

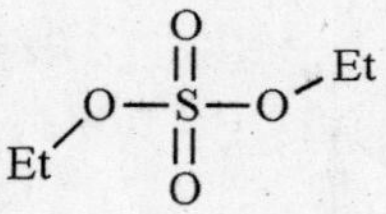

(c) $H_3C-CO-O-CO-CH_3$

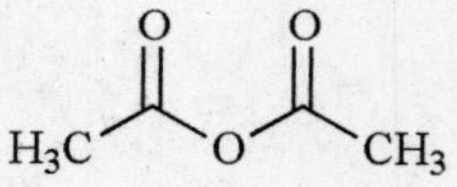

8.

(a) NH_2, N, O, N, HO–P(O)(OH)–O–P(O)(OH)–O, O, H, H, H, H, OH, OH

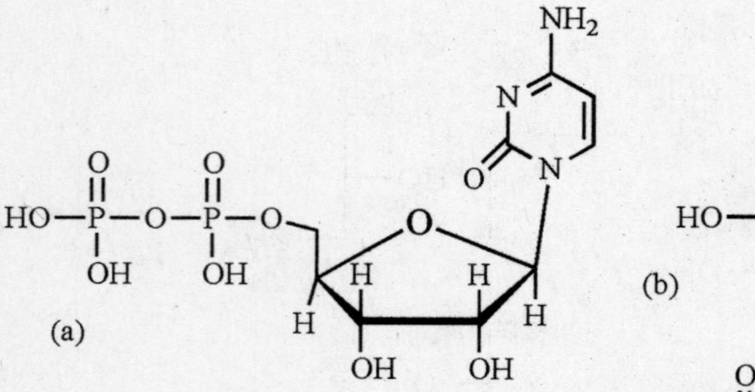

(b) NH_2, N, N, N, N, HO–P(O)(OH)–O–P(O)(OH)–O–P(O)(OH)–O, O, H, H, H, H, OH, OH

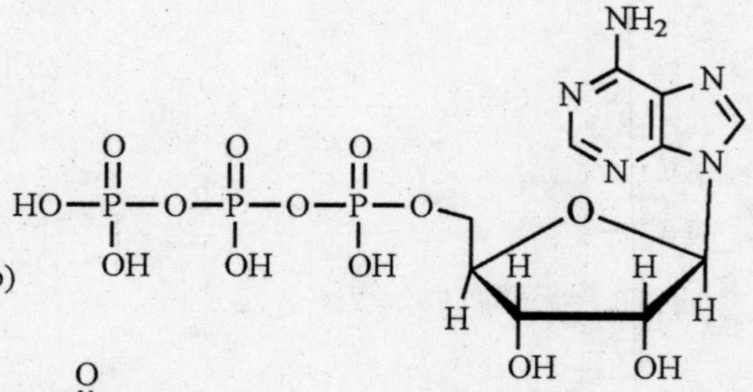

(c) O, H, N, N, H_2N, N, N, HO–P(O)(OH)–O, O, H, H, H, H, OH, OH

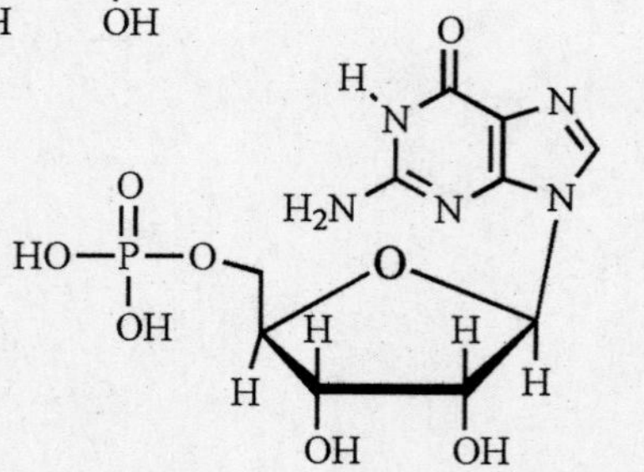

9. C-T-T-A-G-G-T-G-A-A-C-G

10.

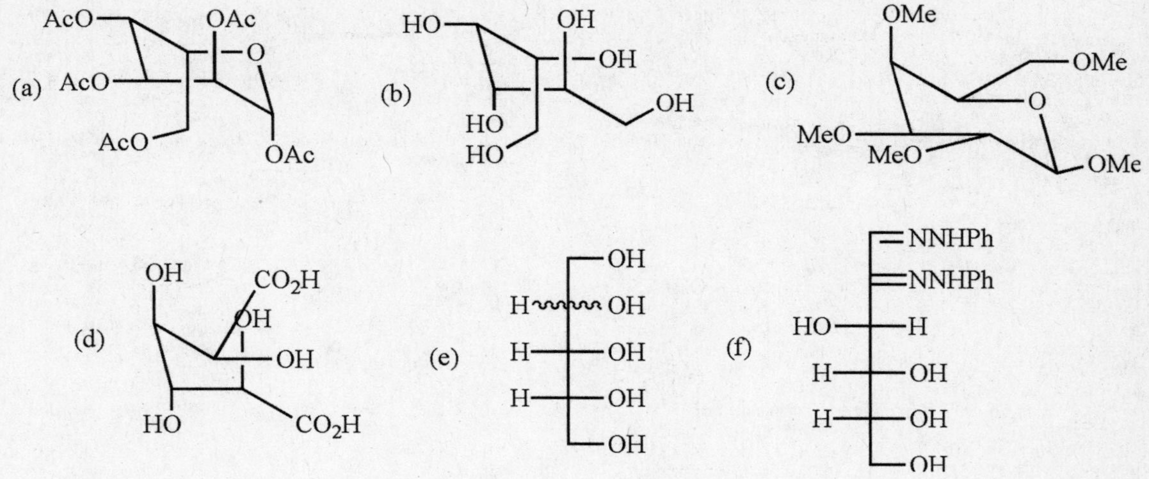

(a) AcO, OAc, O, AcO, AcO, OAc

(b) HO, OH, OH, OH, HO, HO

(c) OMe, OMe, O, MeO, MeO, OMe

(d) OH, CO_2H, OH, OH, HO, CO_2H

(e) OH, H~OH, H–OH, H–OH, OH

(f) =NNHPh, =NNHPh, HO–H, H–OH, H–OH, OH

11.

(a) CHO, OH, OH, OH → (a) → CHO, ~OH, OH, OH, OH → (b) → CHO, ~OH, ~OH, OH, OH, OH

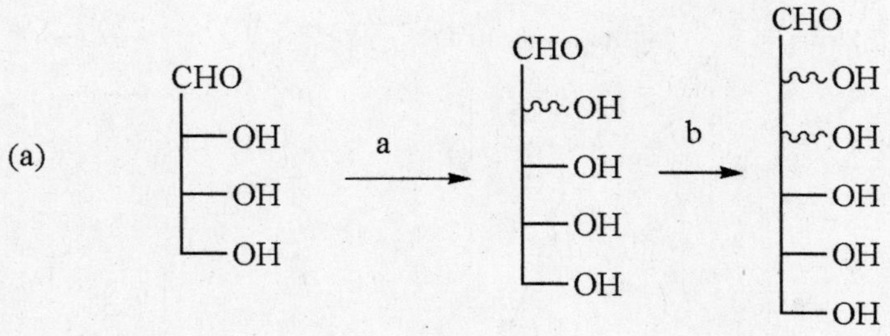

(a) i. HCN ii. H_3O^+ iii. Na(Hg) (b) i. HCN ii. H_3O^+ iii. Na(Hg)

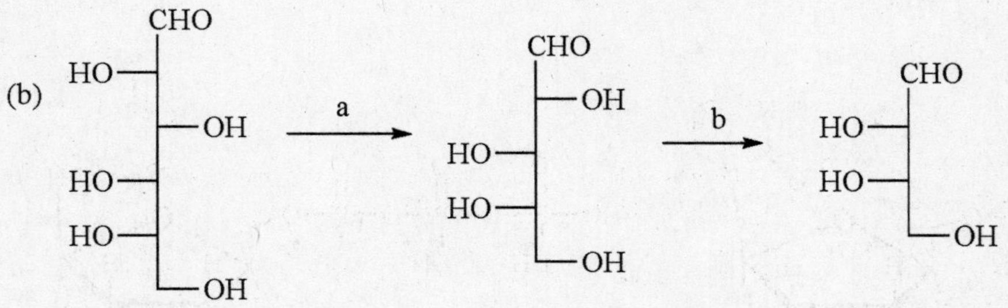

(b) CHO, HO, OH, HO, HO, OH → (a) → CHO, OH, HO, HO, OH → (b) → CHO, HO, HO, OH

(a) i. NH_2OH ii.NaOAc, Ac_2O iii. NaOMe (b) i. NH_2OH ii.NaOAc, Ac_2O iii. NaOMe

Index